30°E
60°E
90°E
120°E
150°E
180°
ARCTIC OCEAN
Barents Sea
RUSSIA AND THE POST-SOVIET STATES
RUSSIA
Aldan
Sea of Okhotsk
60°N
Bering Sea
SWEDEN
FINLAND
Helsinki
St. Petersburg
Moscow
Novosibirsk
Warsaw
Minsk
Kiev
Astana
KAZAKHSTAN
Budapest
Bucharest
GEORGIA
Istanbul
Rome
Ankara
ARMENIA
AZERBAIJAN
UZBEKISTAN
Tashkent
Bishkek
KYRGYZSTAN
Ashgabat
TURKMENISTAN
TAJIKISTAN
TURKEY
Athens
Ulan Bator
MONGOLIA
Urumqi
EAST ASIA
CHINA
Harbin
Shenyang
Beijing
Tianjin
N. KOREA
Seoul
S. KOREA
Qingdao
Tai'an
Zaozhuang
Xian
Sea of Japan
JAPAN
Tokyo
Yokohama
Osaka
NORTH PACIFIC OCEAN
Mediterranean Sea
CYPRUS
LEBANON
ISRAEL
SYRIA
Damascus
Amman
JORDAN
Baghdad
IRAQ
Tehran
Mashhad
IRAN
AFGHANISTAN
Kabul
Islamabad
Lahore
PAKISTAN
Karachi
Tripoli
LIBYA
Alexandria
Cairo
CA AND SOUTHWEST ASIA
EGYPT
KUWAIT
BAHRAIN
Riyadh
SAUDI ARABIA
QATAR
Abu Dhabi
UAE
Muscat
OMAN
Red Sea
Lhasa
NEPAL
Kathmandu
BHUTAN
New Delhi
INDIA
BANGLADESH
Dhaka
Kolkata (Calcutta)
Chengdu
Chongqing
Wuhan
Ningbo
Shanghai
East China Sea
Guangzhou (Canton)
Taipei
TAIWAN
Hong Kong (Xianggang)
30°N
Ahmadabad
Mumbai (Bombay)
SOUTH ASIA
Hyderabad
MYANMAR (BURMA)
Nay Pyi Taw
LAO
Hanoi
Vientiane
THAILAND
VIETNAM
Bay of Bengal
Bangkok
CAMBODIA
Phnom Penh
Ho Chi Minh City
South China Sea
PHILIPPINES
Manila
Philippine Sea
OCEANIA: AUSTRALIA, NEW ZEALAND, AND THE PACIFIC
Khartoum
SUDAN
ERITREA
CHAD
N'Djamena
Sana'a
YEMEN
Arabian Sea
Bangalore
Chennai (Madras)
SRI LANKA
Colombo
MALDIVES
DJIBOUTI
Addis Ababa
ETHIOPIA
SOUTH SUDAN
Juba
CENTRAL AFRICAN REPUBLIC
Bangui
CAMEROON
Yaounde
Libreville
SOMALIA
Mogadishu
KENYA
UGANDA
Kampala
Nairobi
RWANDA
BURUNDI
DEM. REP. OF CONGO
Kinshasa
CONGO
TANZANIA
Dodoma
Dar es Salaam
SEYCHELLES
BRUNEI
SOUTHEAST ASIA
Kuala Lumpur
MALAYSIA
SINGAPORE
PALAU
FEDERATED STATES OF MICRONESIA
MARSHALL ISLANDS
NAURU
KIRIBATI
0°
Jakarta
Bandung
Surabaya
INDONESIA
PAPUA NEW GUINEA
Port Moresby
SOLOMON ISLANDS
TUVALU
TIMOR-LESTE
Timor Sea
Darwin
Luanda
ANGOLA
SUB-SAHARAN AFRICA
ZAMBIA
MALAWI
MOZAMBIQUE
COMOROS
INDIAN OCEAN
Coral Sea
SAMOA
VANUATU
FIJI
TONGA
Harare
ZIMBABWE
MADAGASCAR
Antananarivo
Mozambique Channel
NAMIBIA
Windhoek
BOTSWANA
Gaborone
Pretoria
Maputo
Mbabane
ESWATINI
Bloemfontein
SOUTH AFRICA
LESOTHO
Maseru
MAURITIUS
Cape Town
AUSTRALIA
Perth
30°S
Great Australian Bight
Canberra
Sydney
Melbourne
Tasman Sea
Auckland
Wellington
NEW ZEALAND
60°S
TARCTICA
EUROPE
20°W
10°W
0°
50°N
40°N
NORWAY
FINLAND
Helsinki
Oslo
Stockholm
Tallinn
ESTONIA
St. Petersburg
RUSSIA
SWEDEN
LATVIA
Riga
UNITED KINGDOM
DENMARK
Copenhagen
Baltic Sea
LITHUANIA
Vilnius
Dublin
IRELAND
Minsk
BELARUS
London
Amsterdam
NETHERLANDS
Berlin
GERMANY
Warsaw
POLAND
Channel Islands (U.K.)
Brussels
BELGIUM
LUXEMBOURG
Prague
CZECHIA
Kiev
UKRAINE
ATLANTIC OCEAN
Paris
FRANCE
LIECHTENSTEIN
SLOVAKIA
Vienna
AUSTRIA
Bern
SWITZERLAND
Budapest
HUNGARY
MOLDOVA
Chisinau
SLOVENIA
CROATIA
ROMANIA
Bucharest
MONACO
SAN MARINO
BOSNIA AND HERZEGOVINA
SERBIA
SPAIN
ANDORRA
PORTUGAL
Madrid
Lisbon
Corsica (FR.)
ITALY
MONTENEGRO
KOSOVO
BULGARIA
Sofia
Balearic Is. (SP.)
Sardinia (IT.)
Rome
VATICAN CITY
N. MACEDONIA
ALBANIA
Istanbul
GREECE
TURKEY
Mediterranean
Sicily (IT.)
Athens
MOROCCO
ALGERIA
TUNISIA

World Regional Geography

Global Patterns, Local Lives

EIGHTH EDITION

Lydia Mihelič Pulsipher
Geography Professor Emeritus, University of Tennessee

Alex Pulsipher
Geographer and Independent Scholar

Ola Johansson
Geography Professor, University of Pittsburgh at Johnstown

with the assistance of

Conrad "Mac" Goodwin
Anthropologist/Archaeologist and Independent Scholar

macmillan international
HIGHER EDUCATION

Dedicated to Sam Pulsipher

Vice President, STEM: Daryl Fox
Program Director: Andrew Dunaway
Senior Program Manager: Jennifer Edwards
Developmental Editors: Marion Castellucci, Debbie Hardin
Director of Development: Lisa Samols
Editorial Assistant: Nathan Livingston
Executive Media Editor: Amy Thorne
Senior Media Editors: Emily Marino, Alexandra Gordon
Director of Content: Jennifer Driscoll-Hollis
Marketing Manager: Leah Christians
Marketing Assistant: Madeleine Inskeep
Director, Content Management Enhancement: Tracey Kuehn
Senior Managing Editor: Lisa Kinne
Senior Content Project Manager: Kerry O'Shaughnessy
Senior Workflow Project Manager: Paul Rohloff
Media Project Manager: Brian Nobile
Senior Photo Editor: Sheena Goldstein
Photo Researcher: Brittani Morgan, Lumina Datamatics, Inc.
Director of Design, Content Management: Diana Blume
Design Services Manager: Natasha A. S. Wolfe
Interior Designer: Patrice Sheridan
Art Manager: Matthew McAdams
Production Supervisor: Robert Cherry
Maps: International Mapping
Visual Development Director: Emiko Paul
Photo Essay Design: Kelly Murphy, Science Arts
Composition: Lumina Datamatics, Inc.
Printing and Binding: LSC Communications

ISBN-13: 978-1-319-32485-8
ISBN-10: 1-319-32485-1

Printed in the United States of America
1 2 3 4 5 6 24 23 22 21 20 19

Macmillan Learning
One New York Plaza
Suite 4600
New York, NY 10004-1562
www.macmillanlearning.com

Brief Contents

Dear Reader xx

Preface xxi

CHAPTER 1 Geography: An Exploration of Connections 1

CHAPTER 2 North America 63

CHAPTER 3 Middle and South America 129

CHAPTER 4 Europe 191

CHAPTER 5 Russia and the Post-Soviet States 257

CHAPTER 6 North Africa and Southwest Asia 305

CHAPTER 7 Sub-Saharan Africa 365

CHAPTER 8 South Asia 425

CHAPTER 9 East Asia 485

CHAPTER 10 Southeast Asia 551

CHAPTER 11 Oceania: Australia, New Zealand, and the Pacific 613

EPILOGUE 1 Polar Regions 659

EPILOGUE 2 Space 669

Glossary G-1

Index I-1

Contents

DEAR READER xx

PREFACE xxi

Sonali Pal Chaudhury/NurPhoto via Getty Images

CHAPTER 1
Geography: An Exploration of Connections 1

THE STUDY OF GEOGRAPHY 3

Geographers' Visual Tools 4
Understanding Maps 4 • Geographic Information Science (GISc) 8 • The Detective Work of Photo Interpretation 8

The Region as a Concept 9

Geographic Themes in This Book 11

ENVIRONMENT: PHYSICAL AND HUMAN 12

Physical Geography 12
Landforms: The Sculpting of Earth 12 • Plate Tectonics 12 • External Landscape Processes 12 • Climate 13

Human Impact on the Environment 15
Global Climate Change 18

PHOTO ESSAY Human Impact on the Biosphere 20
Vulnerability to Climate Change 22

PHOTO ESSAY Vulnerability to Climate Change 24
Responding to Climate Change 26

Water 26
Calculating Water Use per Capita 27 • Water: Who Gets Access to It? Who Owns It? 27 • Water Quality 28 • Water and Urbanization 28

Food 28
Agriculture: Early Human Impacts on the Physical Environment 28 • Modern Food Production and Food Security 29

GLOBALIZATION AND DEVELOPMENT 30

The Transitions of Development 31

Classifications of Development 31

Measuring Development 31
Measures of Economic Development 31 • Measures of Human Development 32

Sustainable Development and Political Ecology 32
Globalization 33
What Is the Global Economy? 33
PHOTO ESSAY Globalization and Development 34
The Debate over Globalization and Free Trade 37 • Reforming Free Trade and Development Institutions 38 • Globalization, Development, and Government 38
POWER AND POLITICS 38
How Deep Do Political Freedoms Go? 39
What Factors Encourage the Expansion of Political Freedoms? 39
PHOTO ESSAY Power and Politics 40
Geopolitics 42
International Cooperation 42
URBANIZATION 43
Why Are Cities Growing? 43
Urban Growth in Less Developed Countries 43 • Urban Migration, Opportunity, and Gender 43
PHOTO ESSAY Urbanization 44
Urban Planning and the Environment 46
POPULATION, GENDER, AND CULTURE 46
Global Spatial Variations in Population Density and Growth 46
Global Patterns of Population Growth 46
The Demographic Transition: Population Growth Rates and Human Development 47
Age and Sex Structures 50
Gender 50
PHOTO ESSAY Population, Gender, and Culture 52
Gender Equality and Development 54 • Gender-Based Violence 54 • Human Trafficking and Modern Slavery 54
Culture 54
Ethnicity and Culture: Slippery Concepts 55 • Values 55 • Religion and Belief Systems 56 • Language 56 • Race 59

Justin Sullivan/Getty Images

CHAPTER 2 North America 63
What Makes North America a Region? 65 • Terms in This Chapter 65
ENVIRONMENT: PHYSICAL AND HUMAN 66
Landforms 66
Climate 67
Environment 69
Climate Change and Air Pollution 69
PHOTO ESSAY Vulnerability to Climate Change in North America 70
Water Resources 72 • Loss of Natural Habitat and Urbanization 73 • Oil Extraction 73 • Logging 74 • Coal Mining 75
Human Patterns over Time 75
The Peopling of North America 75 • The European Transformation 77 • Expansion West of the Mississippi and Great Lakes 78 • European Settlement and American Indians and Aboriginal Peoples 79 • The Changing Regional Composition of North America 80
GLOBALIZATION AND DEVELOPMENT 80
From Manufacturing to Services 80
The Knowledge Economy 81
Globalization and Free Trade 81
The Asian Link to Globalization 83 • Backlash Against Globalization and Free Trade 83 • U.S.–Canadian Economic Interdependence 83
Women in the Economy 83
PHOTO ESSAY Globalization and Development in North America 84

North America's Changing Food-Production Systems 86

Family Farms Give Way to Agribusiness 86 • Food Production and Sustainability 86

Changing Transportation Networks and the North American Economy 88

POWER AND POLITICS 88

Influence of the United States and Canada Abroad 88

Challenges to the United States' Global Power 89

The Expansion of Political Freedoms in the United States and Canada 89

PHOTO ESSAY Power and Politics in North America 90

Disillusionment and the Role of Money in Politics 92

Debt and Politics in the United States 92

Relationships Between Canada and the United States 93

Asymmetries 93 • Similarities 93 • Democratic Systems of Government: Shared Ideals, Different Trajectories 93 • The Social Safety Net: Canadian and U.S. Approaches 94 • Gender in National Politics 94

Drugs and Politics in North America 95

The War on Drugs 96 • New Approaches to Managing Drugs 96

URBANIZATION 97

PHOTO ESSAY Urbanization in North America 98

Farmland and Urban Sprawl 99

Smart Growth and Livability 99

Housing Segregation, Redlining, and Gentrification 100

POPULATION, GENDER, AND CULTURE 100

Gender and Fertility 101

Aging in North America 101

Dilemmas of Aging 101

Population Distribution 101

The Geography of Population Change in North America 101

PHOTO ESSAY Population, Gender, and Culture in North America 102

Cultural Issues 104

Immigration and Diversity 104 • Why Do People Decide to Immigrate to North America? 105 • Do New Immigrants Cost U.S. Taxpayers Too Much Money? 106 • Do Immigrants Take Jobs Away from U.S. Citizens? 106 • Are Too Many Immigrants Being Admitted to the United States? 107 • Illegal Immigrant Children 107

Race and Ethnicity in North America 107

A Backlash Against Growing Diversity 107 • Income Discrepancies and the Culture of Poverty 109 • Race and Law Enforcement in the Spotlight 109

Religion 109

The Geography of Christian Subgroups 110 • The Relationship Between Religion and Politics 110

The American Family 111

The Nuclear Family Becomes a Shaky Norm 111 • The Current Diversity of North American Family Types 111

SUBREGIONS OF NORTH AMERICA 111

New England and the Atlantic Provinces 112

The Past Lives in the Present 112 • Strong and Diverse Human Resources 113

Québec 114

Origins of French Settlement 114 • The Quiet Revolution 115

The Old Economic Core 115

Loss of Economic Dominance 115 • The Impact of the Economic Decline on People 115

The American South (The Southeast) 116

Multiple Images Characterize the South 117 • Poverty Persists for Some, but Return Black Migration Is a Boon 117

The Great Plains Breadbasket 118

Adapting to the Challenges of the Great Plains 118 • Globalization Impacts in the Great Plains 119 • Changing Population Patterns and Land Use in the Great Plains 120 • Connecting Bees, Global Corn, the Northern Plains, and Almond Groves in California 120

The Continental Interior 121

Four Physical Zones 121 • Native Americans and Aboriginal Peoples 122 • Nonindigenous Inhabitants 122 • Federal Lands 122

The Pacific Northwest 122

Older Extractive Industries 123 • Newer IT Industries 123

Southern California and the Southwest 124
Agriculture, Energy, Transportation, and Information Service Industries 124 • Water: Scarcities and Disparities of Access 125 • Cultural and Economic Interests Shared with Mexico 125 • Contentious Border Issues 126

RODRIGO BUENDIA/AFP/Getty Images

CHAPTER 3 Middle and South America 129

What Makes Middle and South America a Region? 132 • Terms in This Chapter 132

ENVIRONMENT: PHYSICAL AND HUMAN 133

Landforms 133
Highlands 133 • Lowlands 134

Climate 135
Temperature-Altitude Zones 135 • El Niño 135 • Hurricanes 135

Environment 137
Tropical Forests, Climate Change, and Globalization 138 • The Water Management Crisis 138

PHOTO ESSAY Vulnerability to Climate Change in Middle and South America 140
Integrating Environmental Protection with Economic Development: Ecotourism 141

Human Patterns over Time 142
The Peopling of Middle and South America 142 • European Conquest 143 • A Global Exchange of Crops and Animals 144 • The Independence Era 144 • The Legacy of Resource Extraction: Inequality and Instability 144

GLOBALIZATION AND DEVELOPMENT 148

Neoliberalism and Socialism 148
Era of Socialistic Import Substitution Industrialization 149 • Era of Neoliberalism and Structural Adjustment 149

PHOTO ESSAY Globalization and Development in Middle and South America 151
Persistent Inequality 152

China's Rising Influence 152

Regional Trade Agreements 153
Export Processing Zones 153

Migration and Remittances 153

The Informal Economy 154

Rapid Regional Expansion of Internet Use 154

Food Production and Development 154

POWER AND POLITICS 157

Foreign Involvement in the Region's Politics 157

Revolutionary Movements 157

PHOTO ESSAY Power and Politics in Middle and South America 158
Cuba 159 • The Zapatista Rebellion 159 • Brazil's Landless Movement 159

The Drug Trade, Conflict, and Legalization 159

URBANIZATION 161

Rural-to-Urban Migration 161
Gender and Rural–Urban Migration 161

Inadequate Urban Planning 161

PHOTO ESSAY Urbanization in Middle and South America 162

Elite Landscapes 164

Urban Transportation 164

POPULATION, GENDER, AND CULTURE 164

Population Trends 164

Population Distribution 165

Gender 167
LGBTQ Issues 167

Culture 167

PHOTO ESSAY Population, Gender, and Culture in Middle and South America 168
Race and Skin Color 169 • Religion in Contemporary Life 170

SUBREGIONS OF MIDDLE AND SOUTH AMERICA 171

The Caribbean 171
Political, Social, and Economic Change 171 • Cuba and Puerto Rico Compared 172 • Haiti and Barbados Compared 173

Mexico 174
Formal and Informal Economies 174 • Geographic Distribution of Economic Sectors 174 • Migration from the Mexican Perspective 175
Central America 176
Social and Economic Conditions 176 • The Exceptions: Costa Rica and Panama 176 • Decades of Civil Conflict 178 • CAFTA and Migration to the United States 178 • Nicaragua's Conflicts 178
The Northern Andes and Caribbean Coast of South America 179
The Guianas 180 • Venezuela 180 • Colombia 180
The Central Andes 181
Settlement Patterns 181 • Climate Change, Water, and Development Issues 182 • Agricultural Restoration in the Peruvian Highlands 182 • Neoliberalism and Socialism in Recent Decades 182
The Southern Cone 183
The Economies of the Southern Cone 183 • Buenos Aires: A Primate City 184
Brazil 185
Brazil's Size and Varied Topography 185 • Managing Brazil's Large and Varied Economy 185 • Agriculture, Activism, and the Amazon 185 • Urbanization 186

Matej Leskovsek/SIPA/newscom

CHAPTER 4
Europe 191
What Makes Europe a Region? 193 • Terms in This Chapter 193
ENVIRONMENT: PHYSICAL AND HUMAN 194
Physical Geography 194
Landforms 194 • Vegetation and Climate 195
Environment 196
Europe's Vulnerability to Climate Change 198 • Europe Leads the Response to Global Climate Change 198 • Europe's Everyday Green Behavior 198 • Transportation Impacts on Europe's Environment 198
PHOTO ESSAY Vulnerability to Climate Change in Europe 199
Europe's Energy Issues 201 • Air Pollution 202 • Freshwater and Seawater Pollution 202
Human Patterns over Time 202
Sources of European Culture 203 • The Inequalities of Feudalism 204 • Early Urbanization Transformed Europe 204 • The Age of Exploration: Europe Goes Global 206 • Revolutions in Industry and Politics 207 • Political Awakening 207 • The Impact of Communism 207 • Two World Wars Affect Europe's Stance Globally 208 • The Cold War 208 • Decolonization 209 • The Birth of the European Union 209
GLOBALIZATION AND DEVELOPMENT 210
Economic Comparisons of the EU and the United States 211
Economic Integration and a Common Currency 211
PHOTO ESSAY Globalization and Development in Europe 213
The Euro and Debt Crises 214
Food Production for the European Union and the World 214
The Common Agricultural Program (CAP) 214
The Growth of Corporate Agriculture and Food Marketing 215
The Wide Reach of Europe's Environmental Impact 216
POWER AND POLITICS 217
Popular Democracy and Nationalism 217
The Politics of EU Expansion 217
PHOTO ESSAY Power and Politics in Europe 218
Postcolonial Tribulations in Europe 220
EU Governing Institutions 220
NATO and the European Union as Global Peacemakers 221
The Future of the European Union 221
URBANIZATION 222
Europe's Growing Service Economies 222
PHOTO ESSAY Urbanization in Europe 223

POPULATION, GENDER, AND CULTURE 224

Europe's Rapidly Aging Population 224

Active Aging 226

Migrants, Refugees, and Borders: Needs and Fears 226

The Rules for Assimilation in Europe 229

Changing Gender Roles in Europe 230

The Gender Pay Gap and Double Day 231 • Influence of Women on Policy Formation 231

Social Protection Systems: Goals and Outcomes 232

PHOTO ESSAY Population, Gender, and Culture in Europe 233

Social Protection, Social Cohesion, and Migrants in Europe 235

Human Well-Being in Europe 235

THE SUBREGIONS OF EUROPE 236

West Europe 236

Benelux 236 • France 236 • Germany 239 • The British Isles 240

South Europe 242

Spain and Portugal 243 • Italy 244 • Greece 245

North Europe 246

Scandinavia 247 • The Baltic States 248

Central Europe 249

A Time of Troubles in Southeastern Europe 249 • Transitions in Agriculture, Industry, and Society 251 • Progress in Market Reform 252

CHAPTER 5

Russia and the Post-Soviet States 257

What Makes Russia and the Post-Soviet States a Region? 260 • Terms in This Chapter 260

ENVIRONMENT: PHYSICAL AND HUMAN 261

Landforms 261

Climate and Vegetation 261

Environment 263

Nuclear Power and Pollution 264 • Resource Use and the Environment 264

Climate Change 266

Human Patterns over Time 266

The Rise of the Russian Empire 266

PHOTO ESSAY Vulnerability to Climate Change in Russia and the Post-Soviet States 267

The Communist Revolution and Its Aftermath 269 • The Stalin Era 270 • World War II and the Cold War 271

GLOBALIZATION AND DEVELOPMENT 272

Oil and Gas Development: Fueling Globalization 272

Russia, BRICS, and the Global Economy 274

Institutionalized Corruption 274

Patterns of Work 274

The Informal Economy 275

Geographic Patterns of Human Well-Being 275

PHOTO ESSAY Globalization and Development in Russia and the Post-Soviet States 276

Food Production in the Post-Soviet Era 277

Changing Agricultural Production 277

POWER AND POLITICS 278

Crisis in Ukraine 278

PHOTO ESSAY Power and Politics in Russia and the Post-Soviet States 279

Cultural Diversity and Russian Domination 280

The Conflict in Chechnya 281

The Conflict in Georgia 282

The Media Landscape 282

URBANIZATION 283

Urban Legacies of Soviet Regional Economic Development 283

Urban Central Asia and Caucasia 283

PHOTO ESSAY Urbanization in Russia and the Post-Soviet States 284

POPULATION, GENDER, AND CULTURE 286

Life Expectancy and Population Structure 286

Population Distribution 287

Sociocultural Issues 288

Gender: Challenges and Opportunities 288 • Religious and Nationalist Resurgence 290

PHOTO ESSAY Population, Gender, and Culture in Russia and the Post-Soviet States 291

SUBREGIONS OF RUSSIA AND THE POST-SOVIET STATES 292

Russia 293

European Russia 293 • Central Siberia 295 • The Russian Far East 296

Belarus, Ukraine, and Moldova 297

Belarus 298 • Moldova and Ukraine 298

Caucasia: Georgia, Armenia, and Azerbaijan 299

The Central Asian States 300

Physical Setting 301 • Past and Present in Central Asia 301

Daniel Berehulak/Getty Images

CHAPTER 6

North Africa and Southwest Asia 305

What Makes North Africa and Southwest Asia a Region? 307 • Terms in This Chapter 308

ENVIRONMENT: PHYSICAL AND HUMAN 309

Climate 309

Landforms and Vegetation 309

Environment 309

An Ancient Heritage of Water Conservation 311 • Water and Food Production 311 • Vulnerability to Climate Change 313

PHOTO ESSAY Vulnerability to Climate Change in North Africa and Southwest Asia 314

Human Patterns over Time 316

Agriculture and the Development of Civilization 316 • The Coming of Monotheism: Judaism, Christianity, and Islam 317 • Western Domination and State Formation 321

GLOBALIZATION AND DEVELOPMENT 324

Fossil Fuel Exports and Development 324

PHOTO ESSAY Globalization and Development in North Africa and Southwest Asia 327

Economic Diversification and Growth 328

POWER AND POLITICS 328

The Arab Spring and Beyond 328

Different Paths Toward Reform 329 • Islamism in a Globalizing World 330 • The Iranian Theocracy 331

The Role of Media and the Internet in Political Change 331

PHOTO ESSAY Power and Politics in North Africa and Southwest Asia 332

Democratization and Women 334

Four Geopolitical Situations in the Region 334

Situation 1: Israel and Palestine 334 • Situation 2: Iraq 338 • Situation 3: Syria 338 • Situation 4: The Sunni–Shiʿite Rivalry 339

URBANIZATION 339

Urban Migration 340

PHOTO ESSAY Urbanization in North Africa and Southwest Asia 341

POPULATION, GENDER, AND CULTURE 342

Population Distribution 343

Sociocultural Issues 343

Families and Gender 343 • Gender Roles and Gendered Spaces 344

PHOTO ESSAY Population, Gender, and Culture in North Africa and Southwest Asia 345

The Rights of Women 346 • The Lives of Children 348

SUBREGIONS OF NORTH AFRICA AND SOUTHWEST ASIA 349

The Maghreb 349

The Berber–Arab Cultural Mix 350 • Violence in Algeria 350 • Trouble in Western Sahara 351

The Nile: Sudan and Egypt 352

Sudan 352 • Egypt 353

The Arabian Peninsula 354

Wealth in Saudi Arabia and the Gulf States 354 • Changes in Attitudes Toward Work in the Gulf States 356 • Yemen: History, Poverty, and Conflict 356

The Eastern Mediterranean 357

The Northeast 358

Turkey 358 • Iran 360 • Iraq 361

YASUYOSHI CHIBA/AFP/Getty Images

CHAPTER 7 Sub-Saharan Africa 365

What Makes Sub-Saharan Africa a Region? 367 • Terms in This Chapter 368

ENVIRONMENT: PHYSICAL AND HUMAN 368

Landforms and Rivers 369

Climate and Vegetation 370

Environment 372

Deforestation and Climate Change 372 • Alternatives to Deforestation 372 • Agriculture and Climate Change 373 • Water and Climate Change 375

PHOTO ESSAY Vulnerability to Climate Change in Sub-Saharan Africa 376

Wildlife and Climate Change 378

Human Patterns over Time 379

The Peopling of Africa and Beyond 379 • Early Agriculture, Industry, and Trade in Africa 379 • From European Companies to African Colonies 382 • Power and Wealth in the Aftermath of Independence 385

GLOBALIZATION AND DEVELOPMENT 386

Successive Eras of Globalization 386

The Colonial Exploitation Era 386 • The Era of Debt Accumulation and Structural Adjustment 387

PHOTO ESSAY Globalization and Development in Sub-Saharan Africa 388

The Current Era of Diverse Globalization 390

Regional Economic Development 391

Local Economic Development 392

Transportation Needs 392 • Energy Needs 393

POWER AND POLITICS 394

Corruption, Authoritarianism, and Weak States 394

The Case of Congo (Kinshasa) 395

The Role of Geopolitics 395

PHOTO ESSAY Power and Politics in Sub-Saharan Africa 396

Radical Islamism in the Sahel 398

Conflict and Refugees 398

Successes and Failures of Democratization 398

Gender, Power, and Politics 399

URBANIZATION 399

Recent Rapid Urban Growth 399

Causes and Consequences of Unplanned Urban Growth 399

PHOTO ESSAY Urbanization in Sub-Saharan Africa 400

The Scarcity of Clean Water in Sub-Saharan Cities 402 • Sachet Water: Effective Urban Adaptation or Plastic Plague? 402 • Food Security in Urban Areas 402

POPULATION, GENDER, AND CULTURE 402

Population Growth 403

Population Growth in Nigeria and South Africa Compared 404 • Migration 404

Population and Public Health 405

HIV/AIDS in Sub-Saharan Africa 405 • HIV and Gender 406 • HIV/AIDS Education 406

Gender Issues 406

Female Genital Mutilation 406 • Lesbian, Gay, Bisexual, and Transgender Rights 407

PHOTO ESSAY Population, Gender, and Culture in Sub-Saharan Africa 408

Culture: Religion 409
Indigenous Belief Systems 409 • Islam and Christianity 409

Culture: Ethnicity and Language 410

SUBREGIONS OF SUB-SAHARAN AFRICA 411

West Africa 412
Horizontal Zones of Physical and Cultural Difference 412 • Regional Conflict in Coastal West Africa 413

Central Africa 413
Wet Forests and Dry Savannas 413 • A Colonial Heritage of Violence 414 • Postcolonial Exploitation 414

East Africa 415
Conflict in the Horn of Africa 415 • Mixed Farming in the Interior Valleys and Uplands 417 • Coastal Lowlands 417 • Contested Space in Kenya and Tanzania 417 • The Islands 418

Southern Africa 419
Large-Scale European Settlement 419 • Commodity Dependence 420 • Food Insecurity in Southern Africa 421 • Reasons for Optimism in Southern Africa 421

Sebastian D'Souza/AFP/Getty Images

CHAPTER 8
South Asia 425

What Makes South Asia a Region? 427 • Terms in This Chapter 427

ENVIRONMENT: PHYSICAL AND HUMAN 428

Landforms 429

Climate and Vegetation 430

Environment 431
Vulnerability and Resilience to Climate Change 431

PHOTO ESSAY Vulnerability to Climate Change in South Asia 434
Contributions to Climate Change 437 • Air Pollution 437 • Solar and Wind Power Generation 438 • Electric Vehicles 438 • Resistance to Deforestation 439 • Protected Areas 439

Human Patterns over Time 440
The Indus Valley Civilization 440 • Wealth, Invasions, and Trade 441 • Language and Ethnicity 442 • Religious Traditions 442 • British Colonial Rule and Its Legacies 445 • Independence and Partition 446 • The Postindependence Period 447

GLOBALIZATION AND DEVELOPMENT 447

From Self-Sufficiency to Globalization 448
Central Planning and Self-Sufficiency 448 • Economic Reforms 448

Economic Growth in India 448
Incomes and Human Development Across Indian Society 449 • Outsourcing and the Middle Class 450

Rural Economic Development 450
The Green Revolution 450 • Hunger and Malnutrition 450 • Farm Size and Income 450

PHOTO ESSAY Globalization and Development in South Asia 451
Growth in the Rural Nonfarm Economy 452

Microcredit: A South Asian Innovation for the Poor 452

POWER AND POLITICS 454

Religious Nationalism 454

PHOTO ESSAY Power and Politics in South Asia 455

Caste and Politics 456

Movements Against Government Inefficiency and Corruption 456

Regional Conflicts 457
Conflict in Kashmir 457 • War and Reconstruction in Afghanistan 457 • Nepal's Rebels 458

URBANIZATION 458

PHOTO ESSAY Urbanization in South Asia 459
Urban Disparities in Mumbai 460
Solutions for Some Problems Faced by Slum Dwellers 460
Addresses 460 • Services 460 • Information 460
Risks to Children in Cities 460
Risks to Adults in Cities 461
The Urban Village 461

POPULATION, GENDER, AND CULTURE 462
Population 462
Slowing Population Growth, Health Care, and Urbanization 463 • Slowing Population Growth and Gender 463 • Population Pyramids 463 • Sex Ratio Imbalance 464
Cultural Issues 465
The Texture of Village Life 465 • Social Patterns in the Status of Women 465
Gender, Politics, and Power 468
Women and the Taliban in Afghanistan 468
PHOTO ESSAY Population, Gender, and Culture in South Asia 469
LGBTQ Issues in South Asia 470

SUBREGIONS OF SOUTH ASIA 471
Afghanistan and Pakistan 471
Afghanistan 472 • Pakistan 472
Himalayan Country 474
Northwest India 475
Rajasthan 475 • Punjab and Haryana 475 • Delhi 476 • Uttar Pradesh 477
Northeastern South Asia 477
The Ganga–Brahmaputra Delta 477 • West Bengal 477 • Kolkata 477 • Far Eastern India 478 • Bangladesh 479
Central India 479
South India and Sri Lanka 481
Economic Growth in Bangalore and Chennai 481 • Kerala 482 • Sri Lanka 482 • Sri Lanka's Civil War 482

Lou Linwei/Alamy Stock Photo

CHAPTER 9
East Asia 485
What Makes East Asia a Region? 487 • Terms in This Chapter 487

ENVIRONMENT: PHYSICAL AND HUMAN 488
Landforms 488
Climate 488
The Dry Interior 488 • The Monsoon East 490 • Natural Hazards 490
Environment 490
Water Shortages 490 • Flooding in China 491
PHOTO ESSAY Vulnerability to Climate Change in East Asia 492
Natural Hazards and Energy Vulnerability 494
Food and Sustainability 494
Food Production 494 • Rice Cultivation 494 • Fisheries and Globalization 496 • People and Animals 496
Three Gorges Dam: The Power of Water 496
Air Pollution: Choking on Success 497
Air Quality Elsewhere in East Asia 498
Human Patterns over Time 499
Imperial China 499 • Confucianism and East Asian Culture 500 • Why Did China Not Colonize an Overseas Empire? 502 • European and Japanese Imperialism 502 • China's Turbulent Twentieth Century 502 • Japan Becomes a World Leader 503 • Chinese and Japanese Influences on Korea, Taiwan, and Mongolia 504

GLOBALIZATION AND DEVELOPMENT 505

PHOTO ESSAY Globalization and Development 506

The Japanese Miracle 508

Japan Rises from the Ashes 508 • Productivity Innovations in Japan 508

Mainland Economies: Communists in Command 508

The Commune System 508 • Focus on Heavy Industry 509 • Struggles with Regional Disparity 509

Economic Reforms in China 509

East Asia's Role as Mega-Financier 510

POWER AND POLITICS 511

Japan's Political Shifts 511

PHOTO ESSAY Power and Politics in East Asia 512

Political Change in China? 514

Information Technology and Political Freedom 514 • Protests and Political Freedom 514

Geopolitics of the Sea 516

URBANIZATION 516

Spatial Disparities 517

International Trade and Investment 517

Transportation Improvements in China 517

Economic Transitions 517

PHOTO ESSAY Urbanization in East Asia 518

Hong Kong's Special Role 521

Shanghai's Latest Transformation 521

POPULATION, GENDER, AND CULTURE 522

Responding to an Aging Population 522

Japan's Options 522 • China's Options 523

The Legacies of China's One-Child Policy 524

Gender Imbalance and the Cultural Preference for Sons 524 • A Shortage of Brides 524

Population Distribution 524

PHOTO ESSAY Population, Gender, and Culture in East Asia 525

Sociocultural Issues 526

Western China's Muslims 527 • The Tibetans 528 • Indigenous Diversity in Southern China 529 • Taiwan's Many Minorities 529 • The Ainu in Japan 529 • East Asia's Most Influential Cultural Export: The Overseas Chinese 529

SUBREGIONS OF EAST ASIA 530

China's Northeast 530

The Loess Plateau 531 • The North China Plain 532 • China's Far Northeast 532 • Beijing and Tianjin 533

Central China 533

Sichuan Province 533 • The Central and Coastal Plains 534

Southern China 535

The Yunnan–Guizhou Plateau 536 • The Southeastern Coast 536 • Hong Kong 537 • Macao 537

Far North and West China 537

Xinjiang 537 • The Plateau of Tibet 539

Mongolia 539

The Communist Era in Mongolia 540 • Postcommunist Boom and Bust 540 • Gender Roles in Mongolia 541

Korea, North and South 541

Contrasting Political Systems 542 • Contrasting Economies 543

Japan 544

A Wide Geographic Spectrum 544 • The Challenge of Living Well on Limited Resources 544 • Japan's Culture of Contrasts 546 • Working in Japan 546

Taiwan 547

Yvan Cohen/LightRocket via Getty Images

CHAPTER 10 Southeast Asia 551

What Makes Southeast Asia a Region? 553 • Terms in This Chapter 553

ENVIRONMENT: PHYSICAL AND HUMAN 555

Landforms 555

Climate and Vegetation 557

Deforestation and Climate Change 559
Food Production and Climate Change 561
Shifting Cultivation 562 • Wet Rice Cultivation 563 • Commercial Agriculture 563
Water and Climate Change 563
PHOTO ESSAY Vulnerability to Climate Change in Southeast Asia 564
Melting Glaciers and Increased Evaporation 566 • Coral Reef Bleaching 566 • Storm Surges and Flooding 566
Responses to Climate Change in the Mekong River System 566
The Effects of Hydroelectric Dams 567
Human Patterns over Time 567
The Peopling of Southeast Asia 568
Diverse Cultural Influences 568
Religious Legacies 568 • European Colonization 568
Struggles for Independence Aided by World War II 570
The Vietnam War 570 • The "Killing Fields" in Cambodia 571

GLOBALIZATION AND DEVELOPMENT 572
Borrowed Ideas for Economic Growth 572
Working Conditions 572
Export Processing Zones 573
The Feminization of Labor 573
Growth of the Service Sector 573
Economic Crises and Recovery: The Perils of Globalization 573
PHOTO ESSAY Globalization and Development in Southeast Asia 574
The IMF Bailout and Its Aftermath 576 • The Impact of China's Growth 576 • Global Recessions from 2007 to the Present 576
Regional Trade, ASEAN, and the Trans-Pacific Partnership 576
Regional and Global Integration 576 • Regional Inequalities 577 • Piracy on the High Seas 577 • Tourism and Development 578

POWER AND POLITICS 579
Southeast Asia's Authoritarian Tendencies 579
PHOTO ESSAY Power and Politics in Southeast Asia 580
China's Soft Power Brings Worries and Militarism 582
Will Expansion of Political Freedoms Bring Peace to Indonesia? 583
Terrorism, Politics, and Economic Issues 584

URBANIZATION 584
The Forces of Change 584
Emigration Related to Urbanization and Globalization 585
The Merchant Marine 585 • The Maid Trade 585
PHOTO ESSAY Urbanization in Southeast Asia 586

POPULATION, GENDER, AND CULTURE 588
Population Dynamics 588
Population Pyramids 591
Gender Roles in Southeast Asia 591
Family Organization, Traditional and Modern 591 • Political and Economic Empowerment of Women 592 • LGBTQ Rights in Southeast Asia: Thailand and Vietnam 592 • Globalization and Gender: The Sex Industry 592 • Southeast Asia's Encounter with HIV/AIDS 593
Sociocultural Issues 593
Cultural and Religious Pluralism 593
PHOTO ESSAY Population, Gender, and Culture in Southeast Asia 594
Globalization Brings Cultural Diversity but Also Homogeneity 598

SUBREGIONS OF SOUTHEAST ASIA 599
Mainland Southeast Asia: Myanmar (Burma) and Thailand 599
Myanmar (Burma) 601 • Thailand 602
Mainland Southeast Asia: Lao, Cambodia, and Vietnam 602
Island and Peninsular Southeast Asia: Malaysia, Brunei, and Singapore 604
Malaysia 604 • Brunei 606 • Singapore 606
Indonesia and Timor-Leste 607
The Philippines 608
Population, Urbanization, and Disaffection 608 • Cultural and Economic Disparities 609 • Filipinos Envision a Better Future 610

Josh Haner/The New York Times/Redux

CHAPTER 11 Oceania: Australia, New Zealand, and the Pacific 613

What Makes Oceania a Region? 615 • Terms in This Chapter 615

ENVIRONMENT: PHYSICAL AND HUMAN 616

Landforms: Continent Formation 616

Landforms: Island Formation 617

Climate 618

Moisture, Air Movement, and Rainfall 618 • El Niño 620

Fauna and Flora 621

Animal and Plant Life in Australia 621 • Animal and Plant Life in New Zealand and the Pacific Islands 621

Climate Change in Oceania 622

Sea Level Rise and Storm Surges 622 • Wildfires and Other Water-Related Vulnerabilities 622 • Coral Bleaching 623 • Responses to Potential Climate Change Crises 623

Invasive Species and Food Production 623

Australia 623

PHOTO ESSAY Vulnerability to Climate Change in Oceania 624

New Zealand 626

Environmental Awareness in New Zealand and the Pacific Islands 627

Globalization and Environmental Damage 628

Mining in Papua New Guinea and Nauru 628 • Nuclear Pollution 628 • The Great Pacific Garbage Island 629 • The UN Convention on the Law of the Sea 630

Human Patterns over Time 630

The Peopling of Oceania 630 • The Saga of Rapa Nui (Easter Island) 632 • Early Contact Between Pacific Peoples and the Americas 632 • Arrival of the Europeans in the Pacific 632 • The Colonization of Australia and New Zealand 633 • Oceania's Shifting Global Relationships 634

GLOBALIZATION AND DEVELOPMENT 634

Oceania's New Asian Orientation 634

Asia's Economic Development "Miracle" Stresses Australia and New Zealand 635

Maintaining Raw Materials Exports as Service Economies Develop 636

Economic Change in the Pacific Islands 636

PHOTO ESSAY Globalization and Development in Oceania 637

Subsistence Affluence Practices Could Go Global 638 • Patterns of Human Well-Being 638

The Advantages and Stresses of Tourism 638

Conflict over Tourism in Hawaii 638 • Sustainable Tourism 639

The Future: Diverse Global Orientations 639

POWER AND POLITICS 640

Parliamentary System in Australia and New Zealand 640

The Pacific Way in the Islands 640

The Pacific Way in Fiji 640 • The Value of the Pacific Way to Grassroots Sustainability 641

The Challenge of Migrants and Refugees 641

PHOTO ESSAY Power and Politics in Oceania 642

URBANIZATION 643

PHOTO ESSAY Urbanization in Oceania 644

POPULATION, GENDER, AND CULTURE 646

Population Numbers and Densities 646

Contrasting Population and Gender Patterns 646

PHOTO ESSAY Gender, Population, and Culture in Oceania 648

Gender Pay Gaps 649

Sociocultural Issues 650

Ethnic Roots Reexamined 650 • Gender Myths and Realities 652 • Forging Unity in Oceania 653

EPILOGUE 1
Polar Regions 659

What Makes the Arctic and Antarctic Regions? 661

ENVIRONMENT: PHYSICAL AND HUMAN 661
Physical Geography 661
Climate and Vegetation 661
Environmental Issues 661
Polar Melting and Sea Level Rise 662 • Methane and Permafrost 662 • New Protected Areas 662

GLOBALIZATION AND DEVELOPMENT 663

POWER AND POLITICS 664

URBANIZATION 665

POPULATION, GENDER, AND CULTURE 665
Gender Imbalance 666
Indigenous Peoples and Cultural Survival 666

EPILOGUE 2
Space 669

What Makes Outer Space a Region? 670

ENVIRONMENT: PHYSICAL AND HUMAN 670
Space Weather 670
Environmental Issues 671
Environmental Hazards from Space 671 • Environmental Benefits of Satellites 672

GLOBALIZATION AND DEVELOPMENT 672
The Satellite Economy 672
The Impacts of Satellites in Rural Areas 673
Space Launches 673
Manufacturing in Space 674
Mining in Space 674

POWER AND POLITICS 674
Space and the Law 674
The Militarization of Space 675

URBANIZATION 675
The International Space Station 675
Human Settlements Beyond Earth's Orbit 675
Remote Sensing of Urban Areas 676

POPULATION, GENDER, AND CULTURE 676
Fertility in Space 677
Using Satellites to Study Population 677
Gender in Space 677
Cultural Issues 677
What Will Space Exploration Do to Life on Earth? 677 • Is Exploration Beyond Geospace Wise? 678

Glossary G-1
Index I-1

Dear Reader

The world is out there waiting, and you should see it as soon as possible! In so doing you will help us share and preserve this small planet. The study of geography will be a preview to this monumental but enjoyable task; it will challenge you to actively watch the world around you, broadening your ideas about why places are as they are, and letting you rub elbows with people who often think and act very differently from you. We, the authors of this book, try to help you connect with unfamiliar people and places by putting them in a broad geographic context. But who, you might ask, are we to think we can undertake such a huge task?

Lydia Pulsipher (in collaboration with archaeologist Mac Goodwin) spent years studying the enslaved African experience in the Caribbean, learning how Africans and their descendants created a life for themselves despite enslavement. She has traveled widely in Asia, Oceania, Europe, and the Americas. She has made a lifelong study of the global diffusion of food plants, and more recently has followed the effect on women of the transition from communism to capitalism in Central Europe. In addition, she serves as Honorary Consul for the Republic of Slovenia, a position that keeps her engaged with events in the European Union.

Alex Pulsipher is an independent scholar in Knoxville, Tennessee, interested in cultural and technological changes that support sustainable development in the era of climate change. He has traveled throughout South and Southeast Asia, Europe, Mexico, and Ecuador, and shares knowledge gathered from travel and the researching and writing of this book through local community involvement and his YouTube channel.

Ola Johansson is a professor of geography at the University of Pittsburgh at Johnstown. Born in Sweden but a long-time resident of the United States, his perspective on the world is multifaceted and intercultural. His outlook is not limited to the global North, but is informed by travels in South America, the Caribbean, North Africa, Southwest Asia, and East Asia. His current research explores the geographies of music, ranging from local scenes to the global exchange of musical styles and knowledge.

Conrad "Mac" Goodwin is an anthropologist and archaeologist who lent his experience and insights to the development of this book and continues to advise the authors and contribute to the research process.

We have enjoyed working on this eighth edition. The world is continually changing, and this book will help you grasp the character and impact of those changes. Bon voyage!

Lydia Mihelič Pulsipher
Alex Pulsipher
Ola Johansson
Mac Goodwin

Preface

FIVE THEMES PROVIDE A FRAMEWORK FOR EXPLORING EACH REGION

Many instructors have found that focusing their courses on a few key ideas makes their teaching more effective and helps students retain information. With input from instructors like you, we've woven throughout each chapter five common themes that provide a framework for students to use as they expand their knowledge of the world and each of its regions.

Power and Politics

What are the different ways that power is wielded in societies? Where are authoritarian modes of governance dominant? Where have political freedoms expanded? What kind of changes is the expansion of political freedoms bringing to different world regions? How are changes in the geopolitical order affecting current events?

Environment: Physical and Human

What are the physical characteristics of a region? What are the impacts of human use? How do water scarcity, water pollution, and water management affect people and environments? How might global climate change and changes in food production systems affect soil and water resources? What are the highlights of human history in the region?

Urbanization

What forces are driving urbanization in a particular region? How have cities responded to growth? How are regions affected by the changes that accompany urbanization—for example, the growth of slums; changes in access to jobs, education, and health care; and depopulation and decline in rural areas?

Globalization and Development

How are regions changing as flows of people, ideas, products, and resources become more global? How do shifts in economic, social, and other dimensions of development affect human well-being? What paths have been charted by the so-called developed world? What solutions are emerging from the so-called less developed countries?

Population, Gender, and Culture

What are the major forces driving population growth or decline in a region? How are changes in life expectancy, family size, gender balance, and the age of the population influencing population change? How have changes in gender roles influenced population trends? What are the important sociocultural issues in the region?

ENGAGING FEATURES ALLOW STUDENTS TO UNDERSTAND GEOGRAPHY'S IMPORTANCE IN THEIR LIVES

PHOTO ESSAYS

In each chapter, **five photo essays** provide engaging illustrations of the issues in the five theme sections. Two essays, Globalization and Development and Population, Gender, and Culture, are completely new for each chapter. Each photo essay builds off of a thematic map, and the corresponding photos have been selected to illustrate key points in the section. **Thinking Geographically** questions embedded in the photo essays challenge students to link what they see in the maps and images to the section content.

- Vulnerability to Climate Change, based on a map that combines physical vulnerabilities with resilience factors
- Globalization and Development, based on a map of the Human Development Index (NEW)
- Power and Politics, based on a map showing regional conflicts and the extent of democracy
- Urbanization, based on a map showing city populations
- Population, Gender, and Culture, based on a map of the Gender Development Index (NEW)

PHOTO ESSAY

4.36 Population, Gender, and Culture in Europe

The Gender Development Index (GDI) reveals that while European countries generally treat males and females fairly equally, some countries (such as the UK, the Netherlands, Germany, Austria, Italy, Greece, Malta, Cyprus, two of the Baltic states, and much of southeastern Europe) have a GDI ranking that is markedly lower than that of the rest of Europe. This raises concerns about gender equity and the well-being of women in these countries. Notice also that Bosnia-Herzegovina, North Macedonia, Turkey, the Middle East, and North Africa (see world inset map) have notably lower rankings. These are predominantly Muslim areas. It is data like this, revealing deep cultural differences, that worries Europeans about the prospects of assimilation of immigrants.

THINKING GEOGRAPHICALLY

Thinking about Europe as a whole, what about the gender development map is most surprising to you?

A If you were a woman with only a high school education, where in Europe would you aspire to settle for a well-paying job and why?

B C D If a country ranks high on the gender development scale (GDI), is it necessarily an affluent place? Why or why not?

D Sweden is a popular country for immigrants to settle in. What factors are immigrants likely to be weighing when they decide on Sweden?

A A businessman in Venice, Italy. Italy, like a number of other affluent EU countries, ranks lower on gender development than poorer, newer EU members because of tenacious patriarchal attitudes toward women. Women are paid less for equal work and are rarely found in executive positions. There is not yet a national consensus that the genders should aim for equality. [Sam Edwards/Getty Images]

B A scientist in Poland. A biologist collects plant specimens from the forest floor in Bialowieza National Park. Poland ranks high in gender development primarily because during the communist era, women were encouraged to get educated and move into male-dominated occupations, especially science. They were still underrepresented in supervisory positions. [Raymond Gehman/Getty Images]

C An apprentice in Romania. Romania ranks high in gender development. Under former communist policies, women were urged to take even hard manual labor jobs previously filled by males. Here, an ethnic minority woman is learning a trade in a program specifically aimed at the previously underemployed. [Stephen Bisgrove/Alamy Stock Photo]

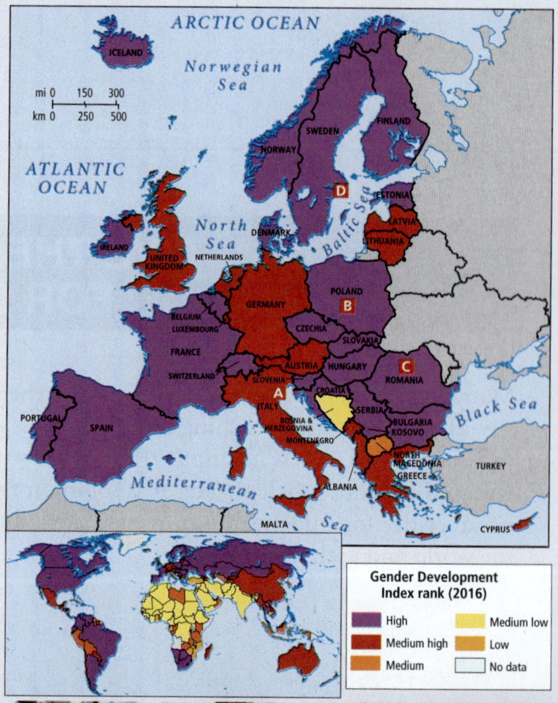

D A father and his children in Stockholm, Sweden. Sweden, which ranks high in gender development, has emphasized equalizing gender roles, including encouraging men to take major roles in child rearing and other domestic duties. The Swedish government plays an active role in influencing how children are raised. [Maskot/DigitalVision/Getty Images]

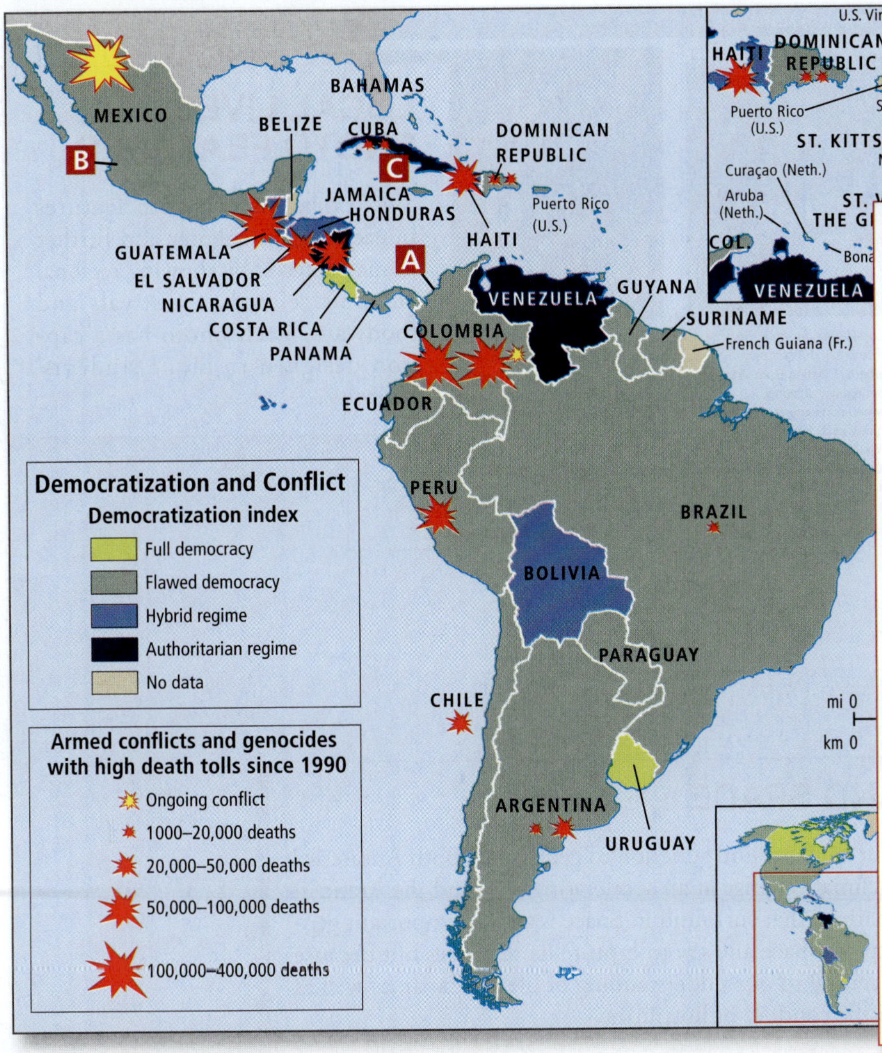

CONSISTENT BASE MAPS

In this edition, we continue to improve what has often been cited as a principal strength of this textbook: high-quality, relevant, and **consistent maps**. To help students make conceptual connections and to compare regions, every chapter contains the following:

- Regional physical features map at the beginning of each chapter
- Political map of the region
- Climate map with photos of different climate zones
- Map of the region's vulnerability to climate change, with photo essay
- Map of the region's performance on the Human Development Index, with photo essay
- Map of regional trends in power and politics, with photo essay
- Urbanization map, with photo essay
- Map of population density
- Map of the region's performance on the Gender Development Index, with photo essay
- Maps of subregions

VISUAL HISTORIES

Visual timelines for each region use images to illustrate key points in the region's history, and **Thinking Geographically** questions test students' understanding of the region's development.

Figure 2.16 VISUAL HISTORY OF NORTH AMERICA

THINKING GEOGRAPHICALLY

A How did food crops such as corn, beans, and squash influence the development of settlements like Cahokia?

B Why did the plantation system inhibit the establishment of roads, communication networks, and small enterprises that could have boosted other forms of economic activity in the Southern colonies?

C What economic activities now dominate the coastal Pacific Northwest?

D How did the steel industry stimulate development in the "economic core" of North America?

E What activities have become economically important in California other than agriculture and the oil industry?

F What were the causes of the Dust Bowl?

Figure 6.14 LOCAL LIVES: Festivals of North Africa and Southwest Asia

(A) Passover is a holiday that is celebrated by Jews in Israel and around the world. It commemorates the biblical event of the Jewish population escaping enslavement in Egypt. Passover is celebrated every spring and involves a ritual meal called a *Seder*. In the picture, members of a tiny ethno-religious group in the Jewish tradition, the Samaritans, engage in a Passover pilgrimage at Mount Gerizim, which is located on the West Bank. [GALI TIBBON/AFP/Getty Images]

(B) A banquet is set in Gaza, Palestine, for *Iftar*, the sunset meal that Muslims eat to break the fast each night in the holy month of Ramadan. During Ramadan, observant Muslims refrain from eating, drinking, quarreling, or having sex between sunrise and sunset. They offer extra prayers each day as a demonstration of their submission to Allah. Wealthy Muslims often sponsor public banquets, such as the one shown here. [MOHAMMED ABED/AFP/Getty Images]

(C) An Iranian woman celebrates *Na*… an ancient holiday of Zoroastrian ori… celebrated throughout Iran, parts of Turkey, and much of Central Asia (se… Figure 5.25). It occurs at the March equinox and celebrates the coming o… spring. People jump over bonfires while singing a verse of purification that is meant to remove sickness and problems, replacing them with warmth and energy. [BEHROUZ MEHRI/AFP/Getty Images]

LOCAL LIVES PHOTO FEATURES

Two **Local Lives** photo features in each region chapter add further human interest by showing regional customs related to festivals and foodways. Each photo has a caption designed to pique students' curiosity.

EPILOGUES: POLAR REGIONS AND SPACE

Unique sections on the polar regions and space round out students' introduction to geography. Both Antarctica and the Arctic are experiencing modifications linked to climate change at an accelerating rate, and the Arctic in particular is the focus of increasing attention from the nations that surround it. Space is also an important geographical topic not only because nations continue to explore space and try to capture its resources but because the parts of space closest to Earth have already become central to our understanding of life on Earth, as well as to the management of global communications and even the conduct of hostilities.

Cosmonauts aboard the *Salyut 7* spacecraft in 1982. [SVF2/Sovfoto/Universal Images Group via Getty Images]

2

Epilogue: Space

On June 8, 1985, Russian cosmonauts Vladimir Dzhanibekov and Viktor Savinikh floated into the dark and frozen *Salyut 7* space station and found the crackers and salt tablets left by the previous crew. A traditional Russian greeting, guests are welcomed with bread, symbolizing wealth, and salt, symbolizing protection. The salt was particularly relevant as *Salyut 7* had lost all power and communications systems 4 months previously and was drifting out of orbit. Dzhanibekov and Savinikh had to find out what was wrong, fix it, and prepare the station for its next crew. While working, the cosmonauts had to take frequent breaks due to the absence of ventilation in the frozen station, which created a risk of carbon dioxide poisoning from their own exhaled breath. Eventually, the cause of the power loss was found—a faulty battery charger for the solar power system—and easily mended.

Salyut 7 was the most advanced Soviet space station, though at roughly half the size of a small mobile home, living quarters were cramped relative to the current International Space Station (ISS), which is the size of a six-bedroom house. With no laundry, cosmonauts wore the same underwear for several days in a row, after which it was jettisoned out the waste hatch and descended toward Earth, burning up on reentry.

Salyut 7 was celebrated for its many photographic observations of the Earth, which contributed to oil and gas exploration, agriculture, Earth science, and mapping, and were estimated to have had a substantial economic impact. The crew also kept a log of their observations, which later became the basis for a major Russian film about *Salyut 7* in 2017.

After its last manned mission in 1986, *Salyut 7* was placed into a high "storage" orbit, from which it would be retrieved for later use. However, disruptions after the fall of the USSR delayed its retrieval, and higher-than-expected solar wind drag started *Salyut 7* on a descent toward Earth. In 1991 it burned up reentering the Earth's atmosphere.

The story of *Salyut 7* captures the difficulties of life in space and how events on Earth can wreak havoc with even the best-laid plans for space exploration.

Learning Objectives

Environment: Physical and Human

E2.1 Investigate the effects on Earth of solar weather, asteroids, and comets and the usefulness of human-made satellites in better understanding processes on Earth.

Globalization and Development

E2.2 Evaluate how commercial activity in space is expanding and propelling future space exploration.

Power and Politics

E2.3 Analyze current disagreements about the militarization of space and the rights of individuals and corporations to own property in space.

Urbanization

E2.4 Assess the usefulness of space-based remote sensing technologies to better understand urban processes on Earth and to help cities adapt to change.

Population, Gender, and Culture

E2.5 Describe how gender issues have shaped space exploration, and how space exploration might change cultures across Earth.

669

EMPHASIS ON EVIDENCE-BASED CRITICAL THINKING PEDAGOGY

Learning Objectives

Environment: Physical and Human

2.1 Describe how landforms influence the movement of air masses in North America, shaping the region's climate.

2.2 Explain how North America's massive consumption of resources impacts environments within the region and globally.

2.3 Identify historical patterns of subregional interaction within North America, and explain how they relate to current trends in economic development.

Globalization and Development

2.4 Describe North America's position in the global economy and how globalization has transformed the region.

Power and Politics

2.5 Identify the major similarities and differences between Canada and the United States in the roles each country's government plays domestically and internationally.

Urbanization

2.6 Explain how No changed since Wor driven these chang

Population, Gende

2.7 Identify how th society has contribu populations.

2.8 Describe how N distribution is chan

2.9 Explain how the increase in migration from Middle and South America and parts of Asia is changing the culture and politics of North America.

Subregions of North America

2.10 Identify key characteristics of the subregions of North America.

LEARNING OBJECTIVES (NEW)

Each chapter now features learning objectives to help in student learning and assessment. Learning objectives are presented at the beginning of each chapter and then repeated at the start of each major theme and subregions heading. Practice assessments designed to help students master each learning objective are available in SaplingPlus.

CHECK YOUR UNDERSTANDING (NEW)

At the close of every main section, students can review the content by answering a few questions covering important points in the section. The questions emphasize key themes and help students achieve the learning objectives that begin each chapter.

THINKING GEOGRAPHICALLY

Each photo essay and visual history timeline is accompanied by **Thinking Geographically** questions that challenge students to connect what they see in the maps and images to the issues discussed in the theme sections.

CRITICAL THINKING QUESTIONS

At the end of each chapter, a **new theme diagram** illustrates the potential linkages among the five themes, and a series of questions encourages students to more broadly analyze the chapter content and connect the themes. These questions can be used for assignments, group projects, or class discussion.

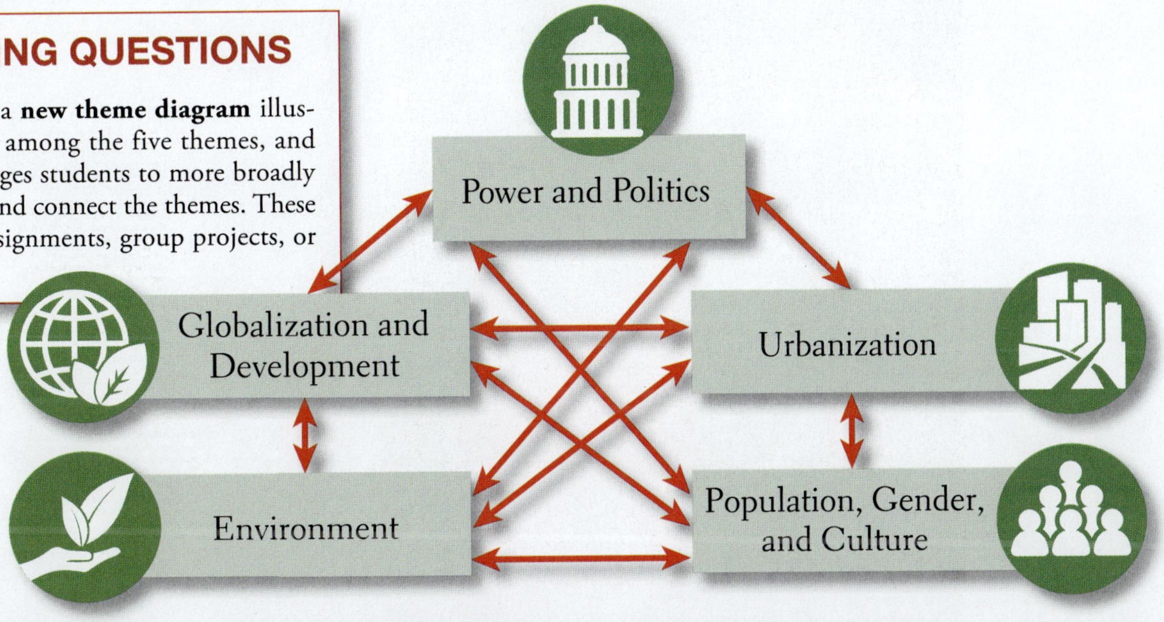

NEW TO THE EIGHTH EDITION

In this eighth edition, we continue to portray the rich diversity of human life across the world and humanize geographic issues by representing the lives of women, men, and children in the various regions of the globe. We also continue to update the data we use in the text and maps and the analysis we offer, in particular reorienting discussions from focusing on problems to presenting possible solutions. In striving to reach these goals, we have made this eighth edition of *World Regional Geography* as up-to-date, instructive, and visually appealing as possible. Highlights of the eighth edition revision include:

- **Two new photo essays in each chapter.** Because instructors have reported that students find the photo essays so engaging, we have added two more to each chapter. Each chapter thus has five photo essays, one illustrating the issues in each theme.
  - A new **Globalization and Development photo essay** is based on a map of the region's performance on the Human Development Index
  - A new **Population, Gender and Culture photo essay** is based on a map of the region's performance on the Gender Development Index.
- **New thematic structure.** We have folded Human Patterns over Time into the Environment theme, and have combined Population and Gender with Sociocultural Issues, so that all the content now falls into a theme section or the subregions section. Each chapter now has five themes—Environment: Physical and Human; Globalization and Development; Power and Politics; Urbanization; and Population, Gender, and Culture.
- **New chapter openers.** Each chapter now opens with a dramatic photo and vignette to engage students' interest in the region.
- **New learning objectives.** Learning objectives appear on the chapter opener and provide students with a framework for what they are expected to learn by studying the chapter.
- **New Check Your Understanding quizzes.** Each major section now ends with a brief quiz that tests students' knowledge of what they have just read.
- **New emphasis on thematic connections.** In the Critical Thinking Questions we have added a diagram showing the linkages among themes, and many of the questions now help students consider issues from multiple thematic angles.
- **Revised Power and Politics maps.** The maps showing conflict zones and the extent of democratization have been thoroughly updated to reflect the developments of the last few years.
- **New streamlined design.** We have redesigned the text to improve its appearance and readability.

SaplingPlus RESOURCES TARGET THE MOST CHALLENGING CONCEPTS AND SKILLS IN THE COURSE

World Regional Geography: Global Patterns, Local Lives is accompanied by a media and supplements package that facilitates student learning and enhances the teaching experience. Fully loaded with our interactive e-book and all student and instructor resources, SaplingPlus is organized around a set of pre-built units available for each chapter of *World Regional Geography: Global Patterns, Local Lives.*

Created and supported by educators, SaplingPlus's instructional online homework drives students' success and saves educators time. Every homework problem contains hints, answer-specific feedback, and solutions to ensure students find the help they need.

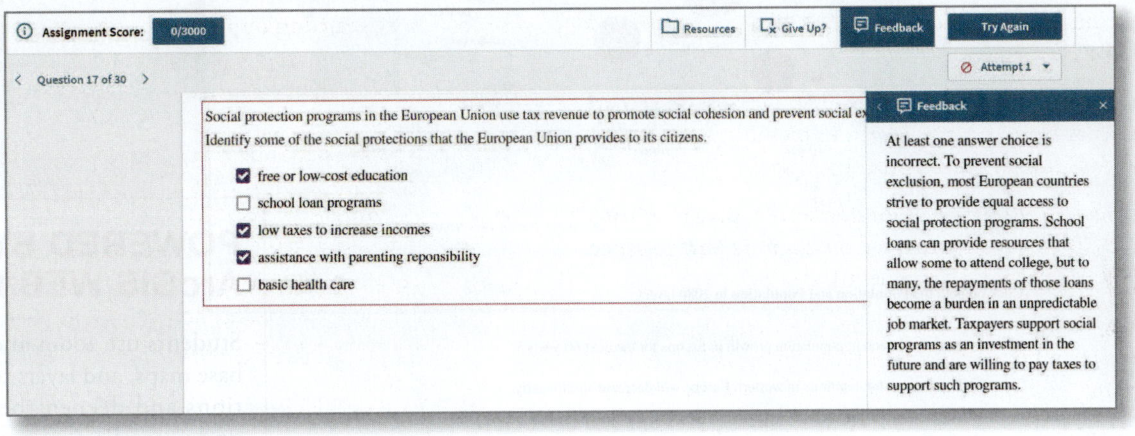

LearningCurve

Put "testing to learn" into action. Based on educational research, LearningCurve really works: Game-like quizzing motivates students and adapts to their needs based on their performance. It is the perfect tool to get them to engage before class and review after class. Additional reporting tools and metrics help teachers get a handle on what their class knows and doesn't know. Available in SaplingPlus and Achieve Read & Practice.

Achieve READ & PRACTICE

For instructors who want an affordable digital solution, Achieve Read & Practice is available with *World Regional Geography.* Achieve Read & Practice is the marriage of our LearningCurve adaptive quizzing and our mobile, accessible e-book, in one easy-to-use and affordable product.

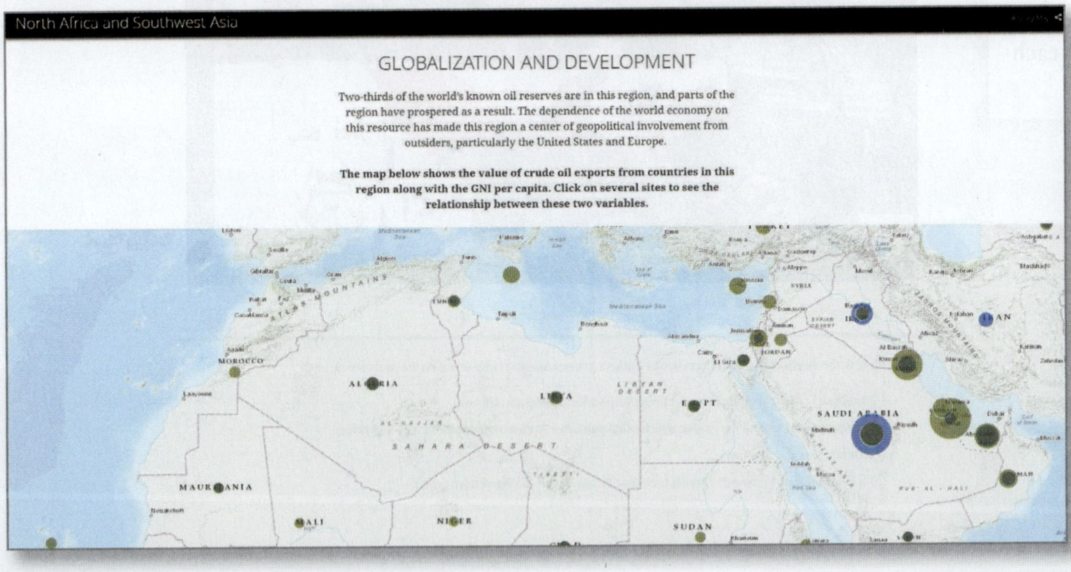

POWERED BY ArcGIS STORY MAPS

One story map per chapter helps students further their comprehension of key topics. Assessment questions for each story map are available in SaplingPlus.

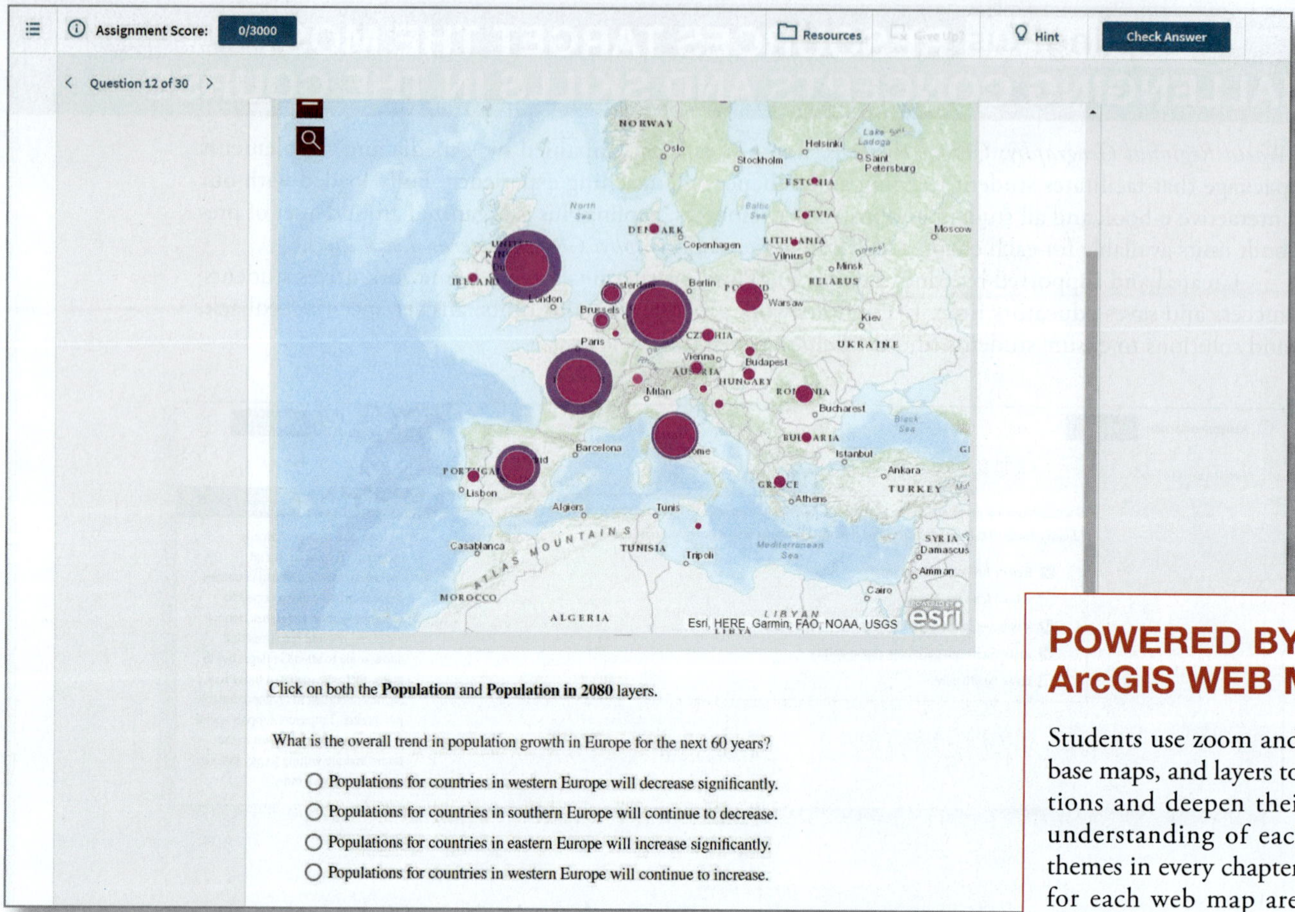

POWERED BY ArcGIS WEB MAPS

Students use zoom and search tools, base maps, and layers to answer questions and deepen their geographic understanding of each of the five themes in every chapter. Assessments for each web map are available in SaplingPlus.

VIDEO ACTIVITIES

50+ video activities have been carefully curated to highlight key topics from each region.

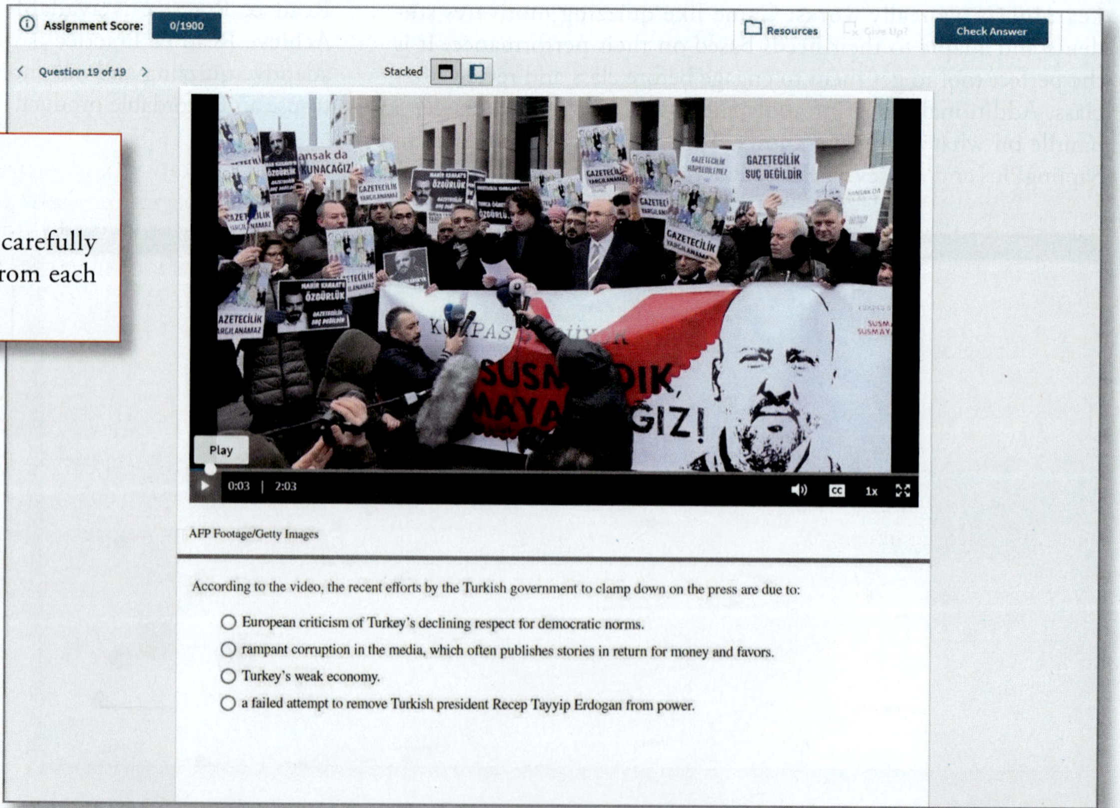

INSTRUCTOR RESOURCES

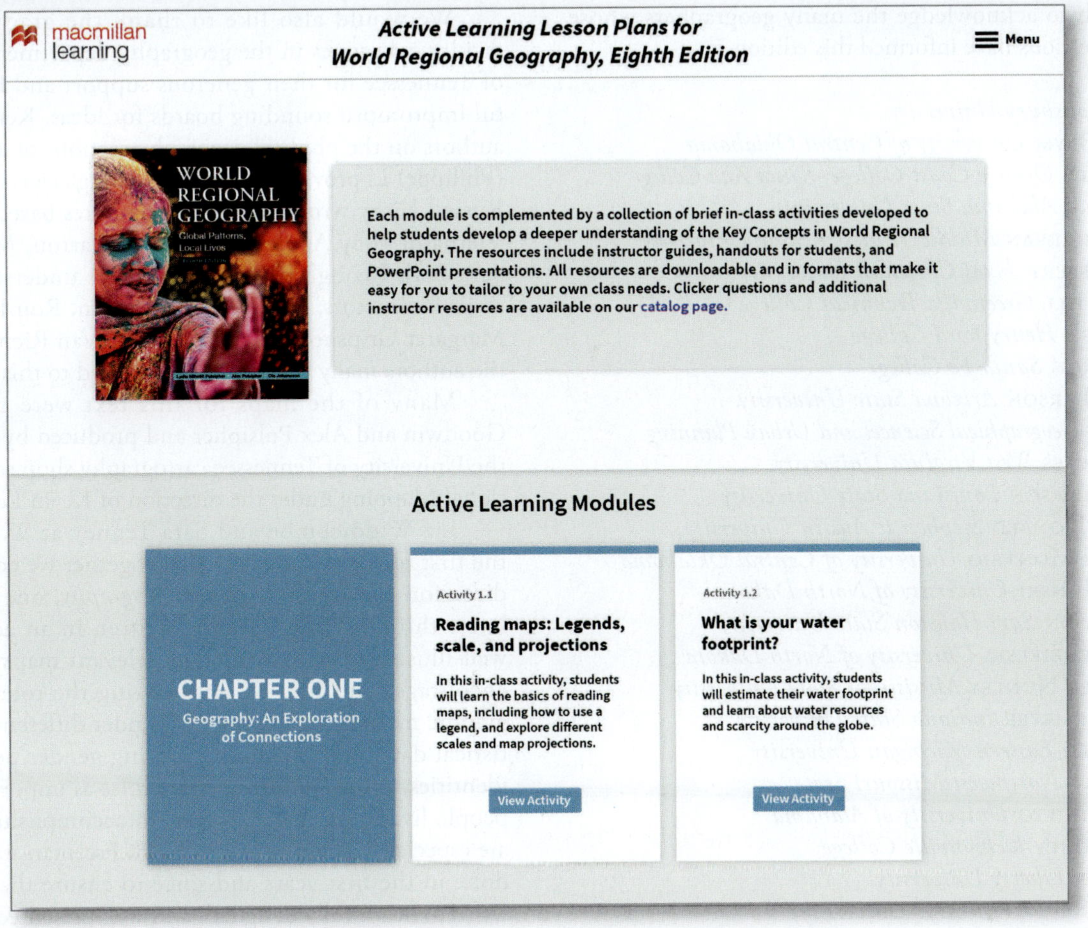

THE HUB FOR ACTIVE LEARNING

For this eighth edition, we've developed a suite of active learning resources to help instructors make their classes more engaging. Each chapter has a variety of activities based on the content in the text. The resources include handouts for students, instructor guides for each activity, and PowerPoint slide shows for each chapter.

TEST BANK

The Test Bank, written by the author, provides a wide range of questions appropriate for assessing student comprehension, interpretation, analysis, and synthesis skills. Each question is tagged for difficulty, Bloom's Taxonomy level, book sections and page numbers, and learning objectives.

LECTURE OUTLINES FOR POWERPOINT

Seasoned veterans and instructors who are new to the discipline will appreciate these detailed companion lectures, perfect for walking students through the key ideas in each chapter. These rich, prebuilt lectures include all figure images, making it easy for instructors to transition to using the book in their classrooms.

CLICKER QUESTIONS

Clicker questions allow instructors to jump-start discussions, illuminate important points, and promote better conceptual understanding during lectures.

ACKNOWLEDGMENTS

The authors wish to acknowledge the many geographers whose insights and suggestions have informed this edition.

Amal Ali *Salisbury University*
Michelle Brym *University of Central Oklahoma*
John Conley *Orange Coast College, Santa Ana College*
Elisha Dung *Alabama State University*
Hari Garbharran *Middle Tennessee State University*
Lauren Geffert *Texas Christian University*
Nicholas Hill *Greenville Technical College*
Tarek Joseph *Henry Ford College*
Heidi Lannon *Santa Fe College*
Elizabeth Larson *Arizona State University, School of Geographical Sciences and Urban Planning*
Joshua Lohnes *West Virginia University*
Kent Mathewson *Louisiana State University*
Darrel McDonald *Stephen F. Austin University*
Maria Diaz Montejo *University of Central Oklahoma*
Douglas Munski *University of North Dakota*
Velvet Nelson *Sam Houston State University*
Michal Niedzielski *University of North Dakota*
Christopher Nunley *Mississippi State University*
Nancy Obermeyer *Indiana State University*
John Oswald *Eastern Michigan University*
Kevin Romig *Northwest Missouri State*
Jeffrey Richetto *University of Alabama*
Patricia Richey *Jacksonville College*
Rob Ritchie *Liberty University*
Amy Siciliano *Saint Mary's University*
Dudley Shurlds *East Mississippi Community College*
Andrew Sluyter *Louisiana State University*
Trevor Smith *Mississippi Gulf Coast Community College*
Jose Torres *Central Connecticut State University*
Amy Trauger *University of Georgia*
Jean Vincent *Santa Fe College*
Timothy Vowles *University of Northern Colorado*
Larry Wade *Trinity Valley Community College*
Kyle Walker *Texas Christian University*
April Watson *Florida Atlantic University*
Mark Welford *Georgia Southern University*
Qihao Weng *Indiana State University*
Donald Williams *Western New England University*
Keith Yearman *College of DuPage*

We would also like to thank the graduate students and faculty colleagues in the geography department at the University of Tennessee for their generous support and for serving as helpful impromptu sounding boards for ideas. Ken Orvis advised the authors on the physical geography sections of all editions. Yingkui (Philippe) Li provided information on glaciers and climate change; Russell Kirby wrote one of the vignettes based on his research in Vietnam; Toby Applegate, Melanie Barron, Michelle Brym, and Sara Beth Keough helped the authors understand how to better assist instructors; and Derek Alderman, Ron Kalafsky, Tom Bell, Margaret Gripshover, and Micheline Van Riemsdijk chatted with the authors many times on issues related to this textbook.

Many of the maps for this text were conceived by Mac Goodwin and Alex Pulsipher and produced by Will Fontanez and the University of Tennessee cartography shop staff and by International Mapping under the direction of Kevin Lear.

Liz Widdicombe and Sara Tenney at W. H. Freeman were the first to suggest the idea that together we could develop a new direction for *World Regional Geography*, one that included the latest thinking in geography written in an accessible style and well-illustrated with attractive, relevant maps and photos. They encouraged our idea of emphasizing the role of gender in geographic matters, taking note of gender differences revealed by statistical data, and explaining varying gender perspectives, gender identities, and traditional gender roles as important factors in how people live in particular places. In accomplishing these goals, we are especially indebted to the W. H. Freeman staff for all they have done in the first years and since to ensure that this book is well written, beautifully designed, and well presented to the public.

We would also like to gratefully acknowledge the efforts of the following people at Macmillan Learning for this eighth edition: Andrew Dunaway, program director; Jennifer Edwards, senior program manager; Marion Castellucci and Debbie Hardin, developmental editors; Kerry O'Shaughnessy, senior content project manager; Leah Christians, marketing manager; Deborah Heimann, copyeditor; Natasha Wolfe, design services manager; Matthew McAdams, art manager; Paul Rohloff, senior workflow project manager; Amy Thorne and Emily Marino, senior media editors; Sheena Goldstein, senior photo editor; Brittani Morgan, photo researcher; and Karen Misler, project manager. We are also grateful to the supplements and media authors, who have created unusually useful, up-to-date, and labor-saving materials for instructors and students who use our book.

World Regional Geography

Men celebrate Holi by throwing colorful pigments and carrying Radha-Krishna, who represent the feminine and masculine aspects of God for many Hindus.
[Sonali Pal Chaudhury/NurPhoto via Getty Images]

1

Geography: An Exploration of Connections

Clouds of colorful pigments fill the air while raucous music and singing reverberate through the streets of Kolkata, India, during Holi, the ancient spring festival of colors. Celebrated primarily in South Asia, Holi has many origin stories, such as the god Krishna's longing for the milk maid Radha, whom he feared would reject him because of his blue skin. On his mother's advice, Krishna presented Radha with bright colored pigments and asked her to paint his face whatever color she liked. Her gleeful response is celebrated by the throwing of colors during Holi, which is a festival of love and a day to meet others, forget and forgive, and repair broken relationships.

What might explain the absence of women from this photograph of Holi celebrants? Despite the joyousness of Holi, there is a dark side to this holiday, which has become notorious for sexual harassment. Women are harassed, groped, and injured to such an extent that some universities in India now forbid female students to leave their dorms during the festival.

Sexual harassment is on the rise in India, and some research suggests the problem relates to the country's growing sex-ratio imbalance, with 110 men for every 100 women. The imbalance is the result of age-old beliefs that female children are liabilities (see the Sex Ratio Imbalance section in Chapter 8). With so many families having boys, this has resulted in a larger population of young men who are now facing difficulty in finding marriage partners. Recent research suggests that these populations are more prone toward violence against women in general, not just during festivals like Holi. Much of this research is geographic, involving analysis of male-on-female violence that compares places with a higher or lower sex-ratio imbalance.

Learning Objectives

The Study of Geography

1.1 Explain what geographers study, and describe the maps and other tools they use.

1.2 Discuss the various definitions of a region.

Environment

1.3 Identify the main concerns of physical geography.

1.4 Evaluate how multiple environmental factors interact to influence the vulnerability of a location to climate change.

Globalization and Development

1.5 Understand the transitions of development.

1.6 Analyze how global flows of information, goods, and people are transforming patterns of economic development.

Power and Politics

1.7 Distinguish the different ways that power is wielded in societies, from more authoritarian modes of governing to those based on notions of political freedom.

Urbanization

1.8 Analyze the various factors that are contributing to the growth of urban areas, especially in the developing world.

Population, Gender, and Culture

1.9 Evaluate the multiple reasons that population growth is slowing throughout the world.

1.10 Describe some of the global consistencies in gender disparity and how they are changing.

1.11 Evaluate the biological significance of race and contrast it with the cultural significance of race.

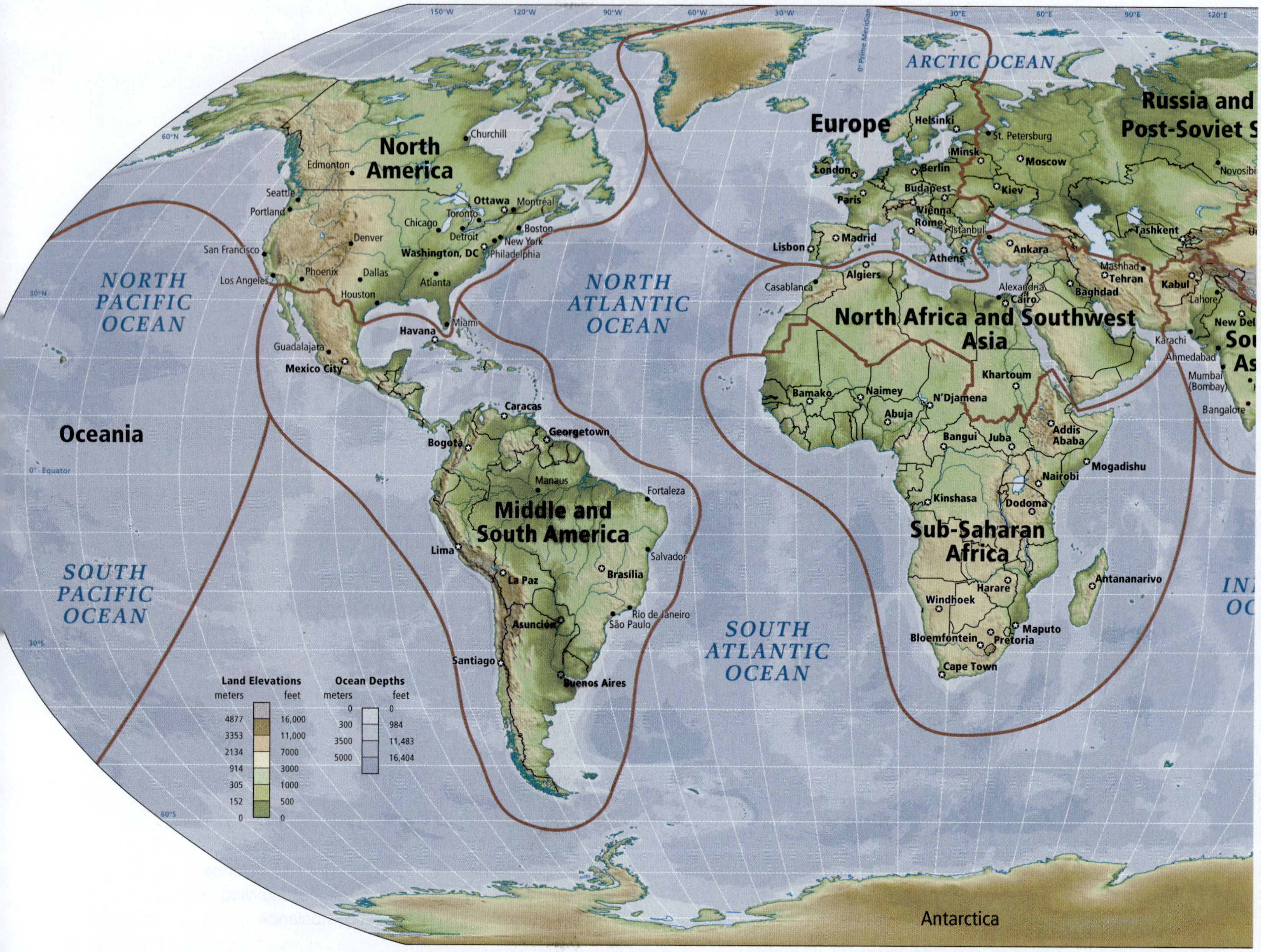

Where are you? You may be in a house or a library or sitting outside on a fine fall afternoon. You are probably in a community (perhaps a college or university), and you are in a country (perhaps the United States) and a region of the world (perhaps North America, Southeast Asia, or the Pacific; see **Figure 1.1**). Why are you where you are? Some answers are immediate, such as "I have an assignment to read." Other explanations are more complex, such as your belief in the value of an education, your career plans, and your or someone's willingness to sacrifice to pay your tuition. Even past social movements that opened up higher education to more than a fortunate few may help explain why you are where you are.

The questions *where* and *why* are central to geography. Geographers seek to understand why different places have different sights, sounds, smells, and arrangements of features. They study what has contributed to the look and feel of a place, to the standard of living and customs of the people, and to the way people in one place relate to people in other places. Furthermore, geographers often think on several scales, from the local to the global. For example, when choosing the best location for a new grocery store, a geographer might consider the physical characteristics of potential sites, the socioeconomic circumstances of the neighborhood, traffic patterns, the store's location relative to the city's main population concentrations, and the location of competitors. She would probably also consider national or even international transportation routes, possibly to determine cost-efficient connections to suppliers.

To make it easier to understand a geographer's many interests, try this exercise. Draw a map of your most familiar childhood landscape. Relax, and recall the objects and experiences that were most

Figure 1.1 Regions of the world.

important to you there. If the place was your neighborhood, you might start by drawing and labeling your home. Then fill in other places you encountered regularly, such as your backyard, your best friend's home, or your school. **Figure 1.2** shows the childhood landscape remembered by Julia Stump in Franklin, Tennessee.

Maps almost always have a depth of meaning that is not immediately apparent. Consider how your map reveals the ways in which your life was structured by space. What is the scale of your map? That is, how much space did you decide to illustrate on the map? The amount of space your map covers may represent the degree of freedom you had as a child, or how aware you were of the world around you. Were there places you were not supposed to go? Does your map reveal, perhaps subtly, such emotions as fear, pleasure, or longing? Does it indicate your sex, your ethnicity, or the makeup of your family? Did you use symbols to show certain features? How did you represent vegetation, water, or pavement? How does your childhood landscape differ from the one drawn by Julia Stump?

In making your map and analyzing it, you have engaged in several aspects of geography:

- Analysis of the landscapes, including human and natural features
- Spatial analysis (the study of how people, objects, and ideas are related to one another across space)
- The use of different scales of analysis (your map probably shows the spatial features of your childhood at a detailed *local scale*)
- Cartography (the making of maps)

As you progress through this book and this course, you will acquire geographic information and skills that will help you achieve your goals, whatever they are. If you want to travel or work outside your hometown or simply understand local events within the context of world events, knowing how to practice geography will make your task easier and more engaging.

THE STUDY OF GEOGRAPHY

1.1 Explain what geographers study and describe the maps and other tools they use.

1.2 Discuss the various definitions of a region.

The primary concern of both physical geography and human geography is the study of Earth's surface and the interactive physical and human processes that shape the surface.

Geography is the study of our planet's surface and the processes that shape it. Yet this definition does not begin to convey the fascinating interactions of human and environmental forces that have given Earth its diverse landscapes and ways of life.

Geography is unique in that it links the physical sciences—such as geology, physics, and biology—with the social sciences—such as anthropology, economics, and political science, yielding a "big picture" perspective that can be hard to find in today's world of hyper-specialization. **Physical geography** generally focuses on how Earth's physical processes work independently of humans, but physical geographers have become increasingly interested in how humans affect these processes and are affected in return. **Human geography** is the study of the various aspects of human life that create the distinctive landscapes and regions of the world. Physical geography and human geography are often tightly linked. For example, geographers might try to understand:

- How and why people came to occupy a particular place
- How people use the physical aspects of that place (climate, landforms, and resources) and then modify them to suit their particular needs

physical geography the study of Earth's physical processes: how they work and interact, how they affect humans, and how they are affected by humans

human geography the study of patterns and processes that have shaped human understanding, use, and alteration of Earth's surface

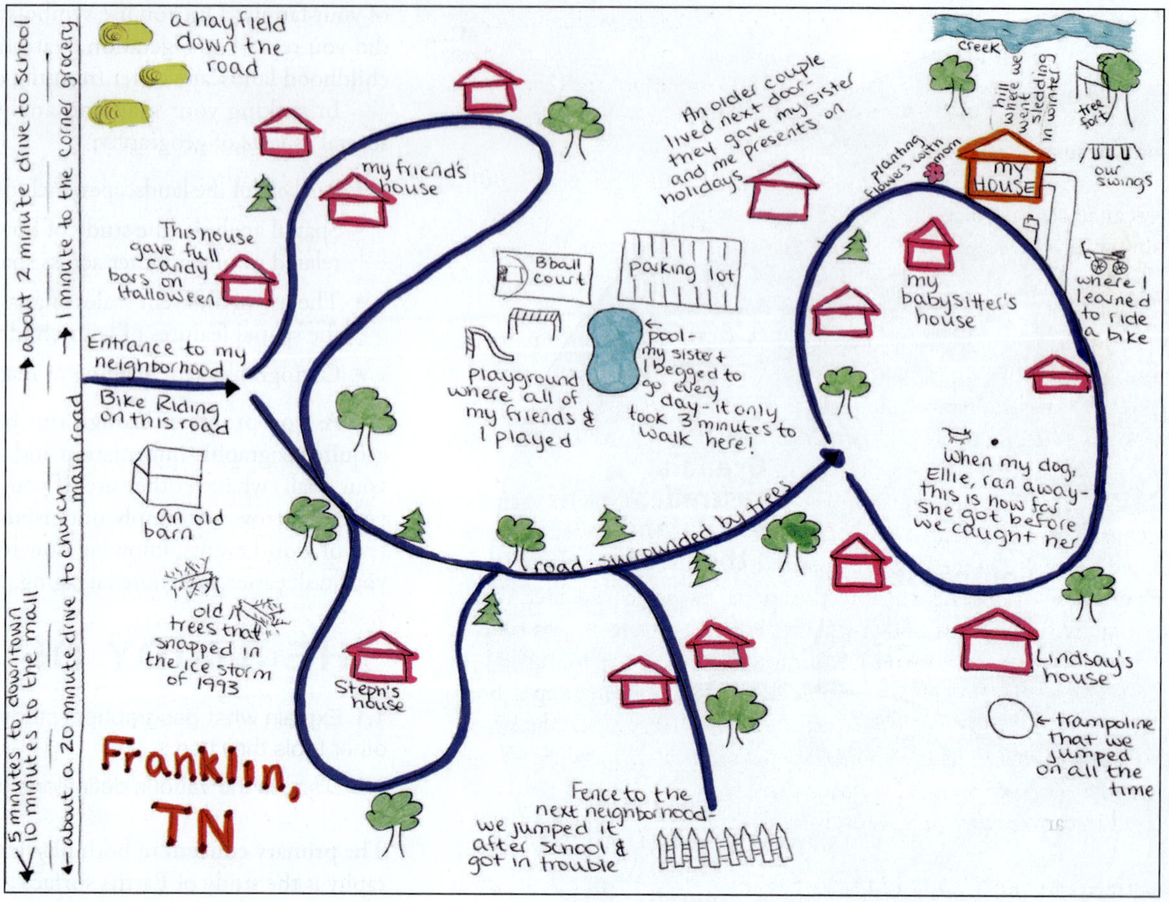

Figure 1.2 A childhood landscape map. Julia Stump drew this map of her childhood landscape in Franklin, Tennessee, as an exercise in Dr. Pulsipher's world geography class. [Courtesy Julia Stump]

- How people create environmental problems
- How human and natural features in a landscape come together to create a **cultural landscape** that has aspects designed intentionally by people and others that evolved unintentionally out of human interactions with nature
- How people interact across space

Geographers usually specialize in one or more fields of study, or subdisciplines. Some of these particular types of geography are mentioned over the course of the book. Despite their individual specialties, geographers often cooperate in studying *spatial interactions* between people and places and the *spatial distribution* of relevant phenomena. For example, in the face of increasing global warming, climatologists, cultural geographers, and economic geographers work together to understand the spatial distribution of carbon dioxide emissions, as well as how to limit such emissions. This could take the form of redesigning urban areas so that people can live closer to where they work and farmers can produce food in or near cities where the food will be consumed.

cultural landscape the combination of human and natural features in a landscape, including aspects designed intentionally by people and parts that evolved unintentionally out of human interactions with nature

scale the proportion that relates the dimensions of the map to the dimensions of the area it represents; also, variable-sized units of geographical analysis from the local to the regional to the global

GEOGRAPHERS' VISUAL TOOLS

Among geographers' most important tools are maps, which they use to record, analyze, and explain spatial relationships, as you did on your childhood landscape map. Geographers who specialize in depicting geographic information on maps are called *cartographers*.

Understanding Maps

A map is a visual representation of space used to record, display, analyze, and explain spatial relationships. **Figure 1.3** explains the various features of maps.

Legend and Scale The first thing to check on a map is the *legend*, which is usually a small box on the map that provides basic information about how to read the map, such as the meaning of the symbols and colors used (see parts A–C and the Legend box in Figure 1.3). Sometimes the scale of the map is also given in the legend.

In cartography, *scale* has a slightly different meaning than it does in general geographic analysis. **Scale** on a map refers to the relationship between the size of objects on the map and the actual size they have on the surface of Earth. It is usually represented by a scale bar (see Figure 1.3D–G) but is also sometimes represented by a ratio (for example 1:8000) or a fraction (1/8000), which indicates that one unit of measure on the map equals a particular number of units on the ground. For example, 1:8000 inches means that 1 inch on the map represents 8000 inches (or about an eighth of a mile) on the surface of Earth.

A scale of 1/800 is considered larger than a scale of 1/8000 because the features on a 1/800 scale map are larger and can be shown in more detail. The larger the scale of the map, the smaller the area it covers. Objects look larger on a larger-scale map and smaller on a smaller-scale map.

In Figure 1.3, different *scales of imagery* are demonstrated using maps, photographs, and satellite images. Read the captions carefully to understand the scale being depicted in each image. In the Scale box in Figure 1.3, the largest-scale map is that on the left (D); the smallest is on the right (G).

It is important to keep the two types of scales used in geography—*map scale* and *scale of analysis*—distinct, because they have opposite meanings! In spatial analysis of a region, such as Southwest Asia, scale refers to the spatial extent of the area that is being discussed. Thus a large-scale analysis means a large area is being explored. But in cartography, a large-scale map shows a smaller area enlarged so that fine detail is visible, while a small-scale map shows a larger area in much less detail. In this book, when we talk about scale we are referring to its meaning in spatial analysis (larger scale = larger area), unless we specifically indicate that we are talking about scale as used in cartography (larger scale = smaller area).

Longitude and Latitude Most maps contain lines of latitude and longitude, which enable a person to establish a position on the map relative to other points on the globe. Lines of **longitude** (also called *meridians*) run from pole to pole; lines of **latitude** (also called *parallels*) run around Earth parallel to the equator (see Figure 1.3H).

Both latitude and longitude lines describe circles. There are 360° (the symbol ° refers to degrees) in each circle of latitude and 180° in each pole-to-pole semicircle of longitude. Each degree spans 60 minutes (minutes are designated with the symbol ′), and each minute has 60 seconds (which are designated with the symbol ″). These are measures of relative linear space on a circle, not measures of time. They do not even represent real distance because the circles of latitude get successively smaller to the north and south of the equator until they become virtual dots at the poles.

The globe is also divided into hemispheres. The Northern and Southern hemispheres are on either side of the equator. The Western and Eastern hemispheres are defined by the prime meridian and the International Date Line. The prime meridian, 0° longitude, runs from the North Pole through Greenwich, England, to the South Pole. That location was chosen by a committee representing 26 nations in 1884, a time when global shipping was growing to the point where all navigation charts needed a common reference point to avoid confusion. Greenwich, England, was chosen mainly because it was the location of the British Royal observatory and was already being used as the 0° longitude reference point by the British Royal Navy, which was the largest at the time. The half of the globe's surface west of the prime meridian is called the Western Hemisphere; the half to the east is called the Eastern Hemisphere. The longitude lines both east and west of the prime meridian are labeled from 1° to 180° by their direction and distance in degrees from the prime meridian. For example, 20 degrees east longitude would be written as 20° E. The longitude line at 180° runs through the Pacific Ocean and is used roughly as the International Date Line; the calendar day officially begins when midnight falls at this line.

The equator divides the globe into the Northern and Southern hemispheres. Latitude is measured from 0° at the equator to 90° at the North or South poles.

Lines of longitude and latitude form a grid that can be used to designate the location of a place. In Figure 1.3H, notice the dot that marks the location of Khartoum below the 20th parallel in eastern Africa. The position of Khartoum is 15° 35′ 1″ N latitude by 32° 32′ 3″ E longitude.

Map Projections Printed maps must solve the problem of showing the spherical Earth on a flat piece of paper. Imagine drawing a map of Earth on an orange, peeling the orange, and then trying to flatten out the orange-peel map and transfer it exactly to a flat piece of paper. The various ways of showing the spherical surface of Earth on flat paper are called **map projections**. All projections create some distortion. For maps of small parts of Earth's surface, the distortion is minimal. Developing a projection for the whole surface of Earth that minimizes distortion is much more challenging.

For large midlatitude regions of Earth that are mainly east/west in extent (North America, Europe, China, Russia), an *Albers projection* is often used. As you can see in Figure 1.3I, this is a conic, or cone-shaped, projection. The cartographer chooses two standard parallels (lines of latitude) on which to orient the map, and these parallels have no distortion. Areas along and between these parallels display minimal distortion. Areas farther to the north or south of the chosen parallels have more distortion.

The *Mercator projection* (see Figure 1.3J) has long been used by the general public and the media, but geographers rarely use this projection because of its gross distortion near the poles. Remember the poles are actually just points (see Figure 1.3H); but to make his flat map, the Flemish cartographer Gerhardus Mercator (1512–1594) stretched out the poles, depicting them as lines equal in length to the equator—a very large distortion! As a result, for example, Greenland appears about as large as Africa, even though it is only about one-fourteenth Africa's size. Nevertheless, the Mercator projection is still useful for navigation because a straight line between two points on this map gives the compass direction between them.

The *Robinson projection* (see Figure 1.3K) shows the longitude lines bending toward the poles to give an impression of Earth's curvature, and it has the advantage of showing an uninterrupted view of land and ocean; however, as a result, the shapes and sizes of landmasses can be quite distorted (but much less than in Mercator maps) and the poles are still stretched out, not the points they are in reality.

Maps are not politically unbiased. Most currently popular world

longitude the distance in degrees east and west of Greenwich, England; lines of longitude, also called meridians, run from pole to pole (the line of longitude at Greenwich is 0° and is known as the prime meridian)

latitude the distance in degrees north or south of the equator; lines of latitude run parallel to the equator, and are also called parallels

map projections the various ways of showing the spherical Earth on a flat surface

1.3 Understanding maps.

The Legend

The colors in the legend convey information about what the map is representing. In the population density map below, the lowest density (0–3 persons per square mile), is colored light tan. A part of North America with this density is shown in the map inset to the right of the legend. On the far right is a picture of this area. Two other densities (27–260 per square mile and more than 2600 people per square mile) are also shown in this manner. [A: Witold Skrypczak/Lonely Planet Images/Getty Images; B: David DeHetre/Moment Open/Getty Images; C: Mike Powell/DigitalVision/Getty Images]

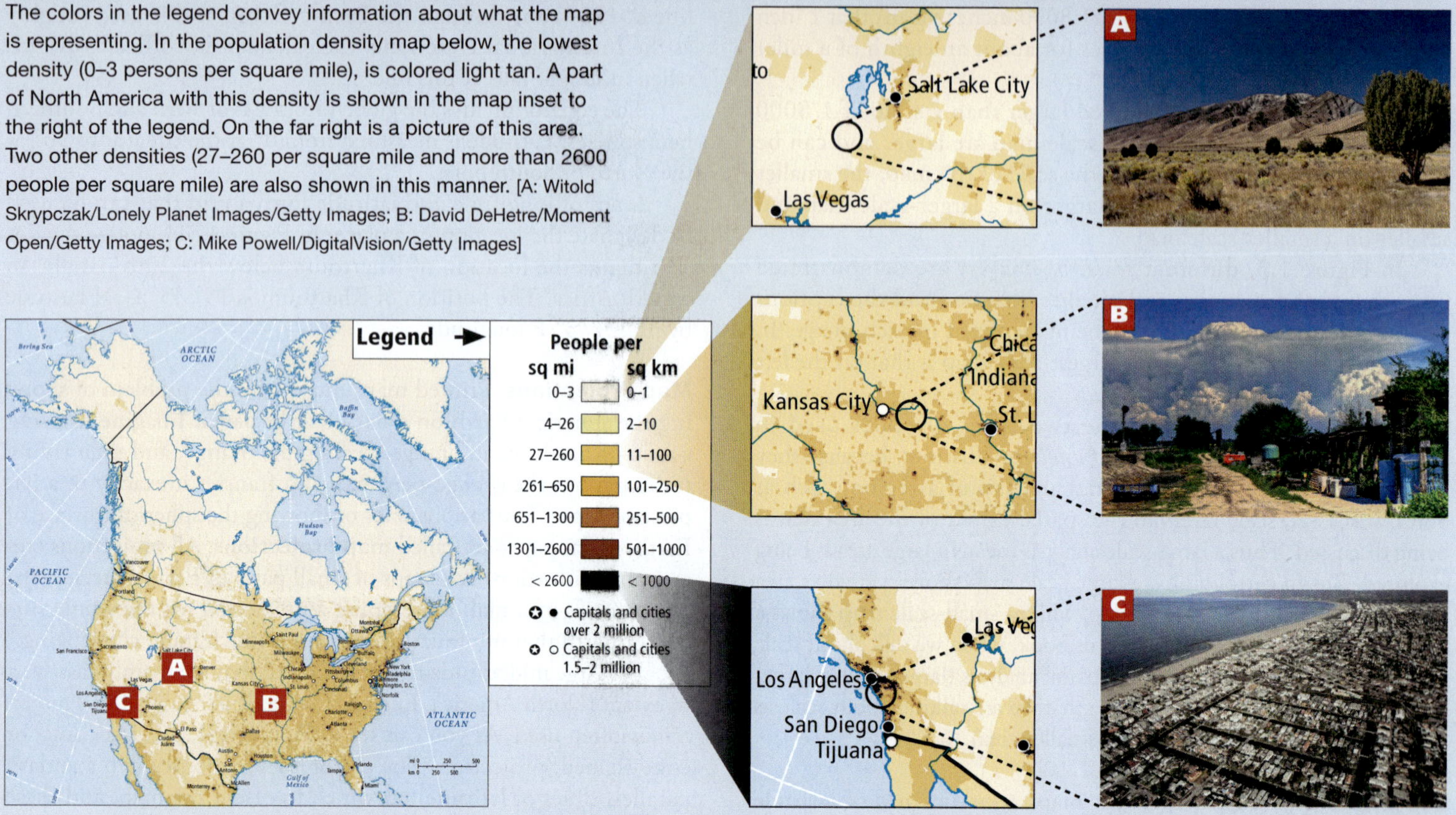

Scale

Maps often display information at different spatial scales, which means that lengths, areas, distances, and sizes can appear dramatically different. This book often combines maps at several different scales with photographs taken by people at Earth's surface and photographs taken by satellites or astronauts in space. All of these visual tools convey information at a spatial scale. Here are some of the map scales you might encounter in this book. The scale is visible below each image. [D–G: USGS]

Representation of Scale

Here are some representations of map scale that you may encounter on maps in this book and elsewhere.

(i) This scale bar means that the length of the entire box represents 8000 feet on the ground.

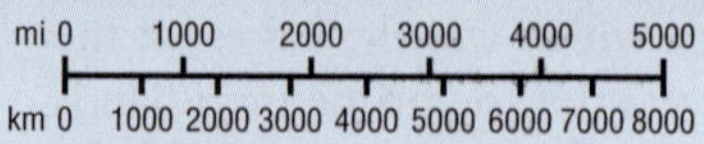

(ii) This scale bar works like the one on the left, but also gives lengths in miles and kilometers.

1:8000

(iii) This means that 1 unit of measure on the map (an inch, for example) equals 8000 similar units of measure on the ground.

Latitude and Longitude

H Lines of longitude and latitude form a global scale grid that can be used to designate the location of any place on the planet.

The distance between lines of longitude decreases toward the poles.

Lines of latitude decrease in length as they approach the poles.

Lines of latitude and longitude intersect at right angles.

Lines of latitude are parallel to each other.

The equator is at zero degrees latitude.

The prime meridian is at zero degrees longitude and passes through Greenwich, England. The counterpart to the prime meridian on the opposite side of the globe is the international date line.

The equator divides the globe into Northern and Southern hemispheres.

All lines of longitude or meridians are of equal length.

The half of the globe's surface west of the prime meridian is called the Western Hemisphere; the half to the east is called the Eastern Hemisphere.

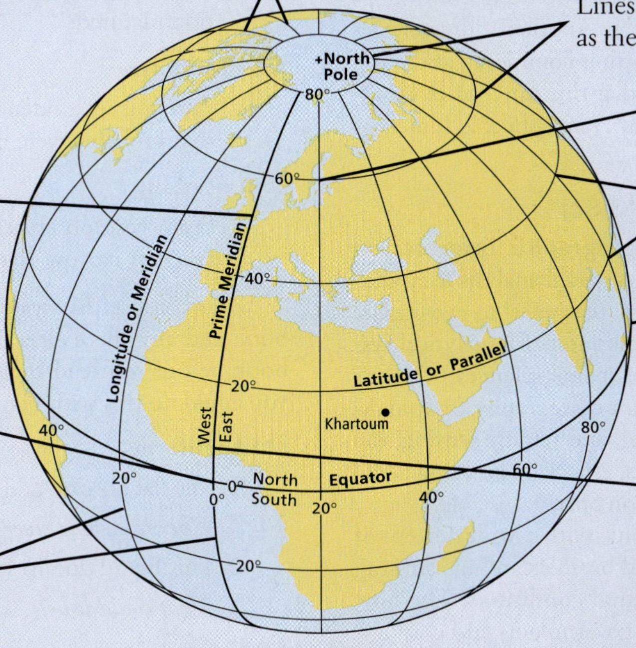

Projections

I **Albers Projection**

Two standard parallels (selected by mapmaker)

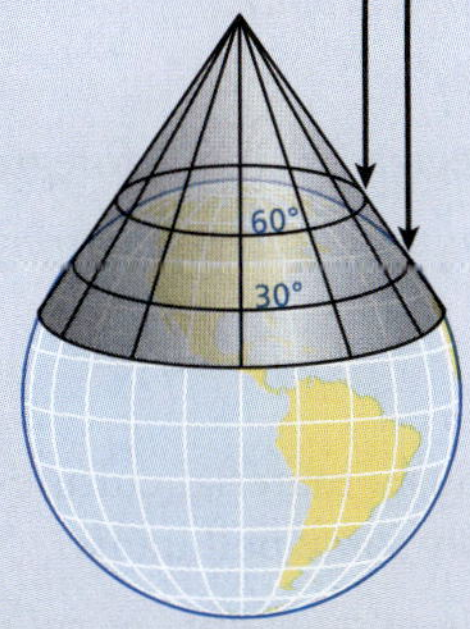

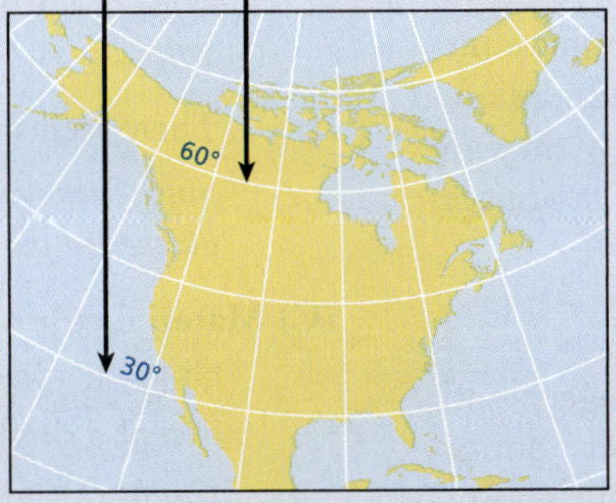

Pros: Minimal distortion near two parallels (lines of latitude)

Cons: Areas farther away from these lines are distorted.

J **Mercator Projection**

North Pole

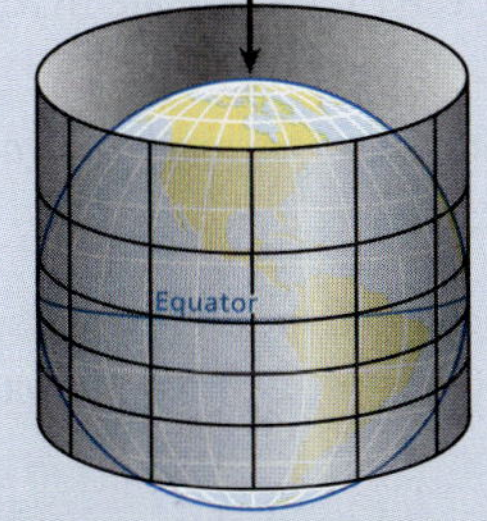

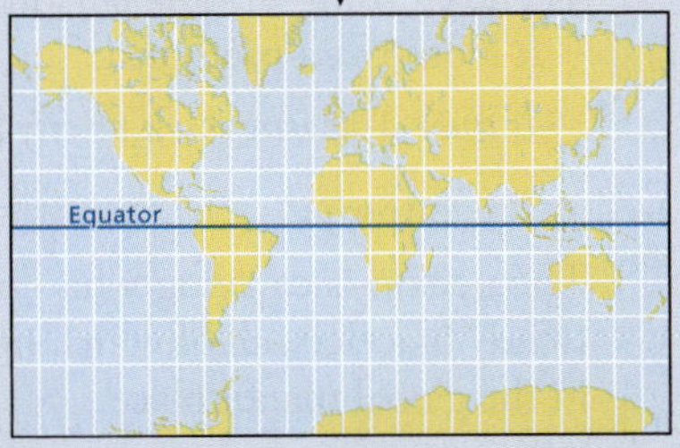

Pros: A straight line between two points on this map gives an accurate compass direction between them. Minimal distortion within 15 degrees of the equator.

Cons: Extreme distortion near the poles, especially above 60° latitude.

K **Robinson Projection**

North Pole

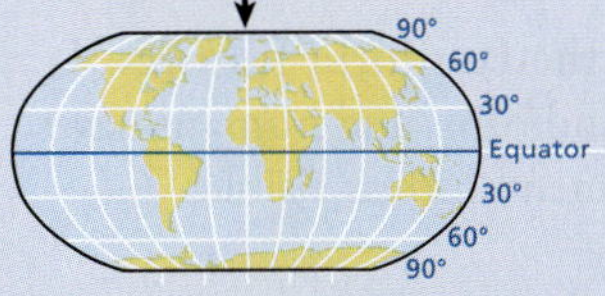

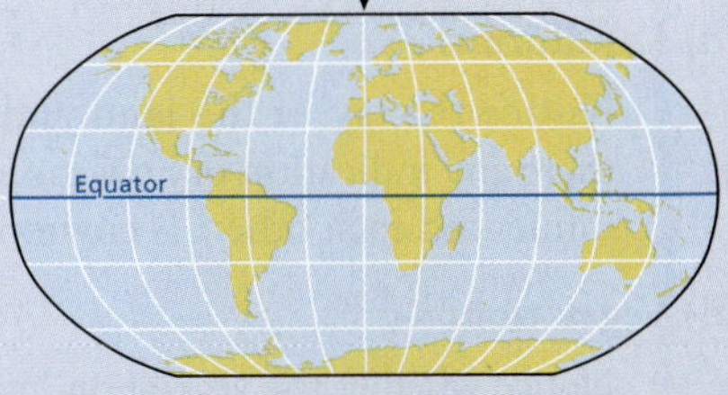

Pros: Uninterrupted view of land and ocean. Less distortion in high latitudes than in the Mercator projection.

Cons: The shapes of landmasses are slightly distorted due to the curvature of the longitude lines.

map projections reflect the European origins of modern cartography. For example, Europe is often placed near the center of the map, where distortion is minimal; other population centers, such as East Asia, are placed at the highly distorted periphery. Therefore, for a less biased study of the modern world, we need world maps that center on different parts of the globe. Another source of bias in world maps is the convention that north is commonly at the top of the map. Some cartographers think that this can lead to a subconscious assumption that the Northern Hemisphere is somehow superior to the Southern Hemisphere.

Geographic Information Science (GISc)

The acronym **GISc** usually refers to **geographic information science**, the body of science that supports spatial analysis technologies. (*GIS* [without the c] is an older term that refers to geographic information *systems* and describes the computerized analytical systems that are the tools of this newest of spatial sciences.) GISc is multidisciplinary, using techniques from cartography (mapmaking), geodesy (measuring Earth's surface), and remote sensing, the science of making reliable measurements, especially by using aerial or space-based sensors (see the epilogue on Space).

GISc is a growing field in geography, with a wide variety of practical applications in government and business and in assessing and improving human and environmental conditions. The now widespread use of GISc, particularly by governments and corporations, has dramatically increased the amount of information that is collected, analyzed, and distributed. There are new possibilities for solving problems but also serious ethical questions. What rights do people have over information about their location and movements gathered from their cell phones? Should this information reside in the public domain? Should a government or corporation have the right to sell information to anyone, without special permission, about where people spend their time and how frequently they go to particular places? Progress on these questions has not kept pace with the technological advances in GISc.

The Detective Work of Photo Interpretation

In addition to creating and analyzing maps, satellite imagery, and other visual data, geographers often make and interpret photos. Understanding the geographic information contained in photos can sometimes be like detective work. Below are some points to keep in mind as you look at the pictures throughout this book. Try them out first with the photo in **Figure 1.4**.

(A) Landforms:

Notice the lay of the land and the landform features. Is there any indication of how the landforms and humans have influenced each other? Is environmental stress visible?

(B) Vegetation:

Notice whether the vegetation indicates a wet or dry, or warm or cold environment. Can you recognize specific species? Does the vegetation appear to be natural or influenced by human use?

(C) Material culture:

Are there buildings, tools, clothing, agricultural products, plantings, or vehicles that give clues about the cultural background, wealth, values, or aesthetics of the people who live where the picture was taken?

(D) People:

What does the person in the photo suggest about the situation pictured?

(E) Economy:

Can you see evidence of the global economy, such as goods that probably were not produced locally?

(F) Location:

From your observations, can you tell where the picture was taken or narrow down the possible locations?

You can use this system to analyze any of the photos in this book and anywhere else. Practice by analyzing the photos in this book before you read their captions. Here is an example of how you could do this with Figure 1.4:

(A) Landforms:

1. The flat horizon and abundance of water suggest a river delta.
   Where are some river deltas?
2. This liquid doesn't look natural.
   What could it be?

(B) Vegetation:

3. This looks like fairly thick and varied vegetation. These could be palm trees and other types of plant life found in tropical climates.
   Must be fairly wet and warm, possibly tropical.
   Where are some tropical river deltas?
   There seems to be a lot of dead vegetation in the foreground.
   Since the vegetation in the background is still green, perhaps the blackened plants were poisoned?

(C) Material culture:

There is not much that is obviously material culture here, except a single person and possibly some pollutants.

Could this area be abandoned due to some kind of pollution? Maybe it's a remote area?

(D) People:

4. The clothing on this person doesn't look like he made it. It looks mass-produced.
   Is there a nearby town or city that sells mass-produced goods?

(E) Global economy:

The person is carrying what look like Crocs, and has some trendy bracelets, all of which indicate that he is participating in the global marketplace.

What other connections can be seen to the global economy?
Could the black substance be a resource that is traded globally?

(F) Location:

This could be somewhere tropical where there could have been an oil spill. *Hint:* Use this book! Look at Figure 6.20 to see the member countries of OPEC (the Organization

geographic information science (GISc) the body of science that supports multiple spatial analysis technologies and keeps them at the cutting edge

Figure 1.4 Resources and the environment. A multinational corporation began extracting resources from this area 50 years ago. A recent United Nations (UN) report stated that the area now needs a massive cleanup, which could take up to 30 years and cost more than a billion dollars, making it one of the biggest such efforts in the world. The area has had approximately 300 incidences of such pollution each year since the 1970s, causing an unknown number of deaths. Can you guess where this is? [PIUS UTOMI EKPEI/AFP/Getty Images]

of the Petroleum Exporting Countries). The combination of the possible oil spill and the vegetation suggests that the photo could be of Venezuela, Ecuador, Nigeria, Angola, the American South, or Indonesia. Suggestion: To further narrow your guesses, quickly look through Chapters 3, 7, and 10! Check your answer on the last page of the chapter.

THE REGION AS A CONCEPT

The concept of *region* is useful to geographers because it allows them to break up the world into manageable units in order to analyze and compare spatial relationships (Figure 1.1). Nonetheless, regions do not have rigid definitions and their boundaries are fluid.

A **region** is a unit of Earth's surface that contains distinct patterns of physical features and/or distinct patterns of human development. It could be a **formal region**, defined by a single specific trait that has been mapped and described, such as the area in which Spanish is the primary language of government. It could also be a **functional region**, which is an area defined by an activity organized around a single place, such as the area served by a public utility. Many regions we encounter in daily life are **vernacular regions** based on perceptions of shared characteristics. Perceptions often differ, which makes it hard for people to agree precisely on the boundaries of vernacular regions. For example, some people might define the region of "the South" in the United States by its distinctive vegetation, architecture, foods, and music, while others would define the region by its religious and political leanings. These two different perceptions of "the South" would give different boundaries to the region if it were mapped.

Another issue in defining regions is that they may shift over time. The people and the land they occupy may change so drastically in character that they can no longer be thought of as belonging to a certain region, and become more closely aligned with another, perhaps adjacent, region. Examples of this are countries in Central Europe, such as Poland and Hungary, which, for more than 40 years, were closely aligned with Russia and the Soviet Union, a vast region that stretched across northern Eurasia to the Pacific (**Figure 1.5A**).

region a unit of Earth's surface that contains distinct patterns of physical features and/or distinct patterns of human development

formal region an area defined by a single specific trait that has been mapped and described, such as the area in which Spanish is the primary language of government

functional region an area defined by an activity organized around a single place, such as the area served by a single post office

vernacular region an area defined by perceptions of shared characteristics

Figure 1.5 Changing country alliances and relationships in Europe (A) before 1989 and (B) in 2016.

(A) Pre-1989 alignment of countries in Europe and the Soviet Union.

Poland and Hungary's borders with western Europe were highly militarized and shut to travelers. The Soviet Union collapsed in the early 1990s, bringing drastic political and economic changes to the region, and in 2004, Poland and Hungary became members of the European Union (EU; see **Figure 1.5B**). Their western borders are now open, while their eastern borders are now more heavily guarded and harder to cross. Nevertheless, people share cultural features (language, religion, historical connections) even if their governments have very different regional allegiances. For this reason, regional borders are often "fuzzy," meaning that they are hard to determine precisely.

If regions are so difficult to define and describe, why do geographers use them? To discuss the whole world at once would be impossible, so we divide the world into manageable parts. There is nothing sacred or unchanging about the criteria or the boundaries for the world regions we use. They are just practical aids to learning. In defining each of the world regions for this book, we have considered such factors as physical features, political boundaries, cultural characteristics, history, how the places now define themselves, and what the future may hold. We are constantly reevaluating regional boundaries and, in this edition, have made some changes. For example, the troubled new country of South Sudan, once part of North Africa and Southwest Asia, is now considered part of sub-Saharan Africa, to which it is more culturally aligned (see Chapter 7).

This book organizes the material into three regional scales of analysis: the *global scale*, the *world regional scale*, and the *local scale*. The term *scale of analysis* refers to the relative size of the area under discussion. At the global scale, explored in this chapter, the entire world is treated as a single area—a unity that is more and more relevant as our planet operates as a global system. We use the term *world region* for the largest divisions of the globe, such as East Asia and North America (see Figure 1.1). We have defined ten world regions, each of which is covered in a separate chapter. In each regional chapter, we consider the interactions of human geography and physical geography in relation to the five thematic

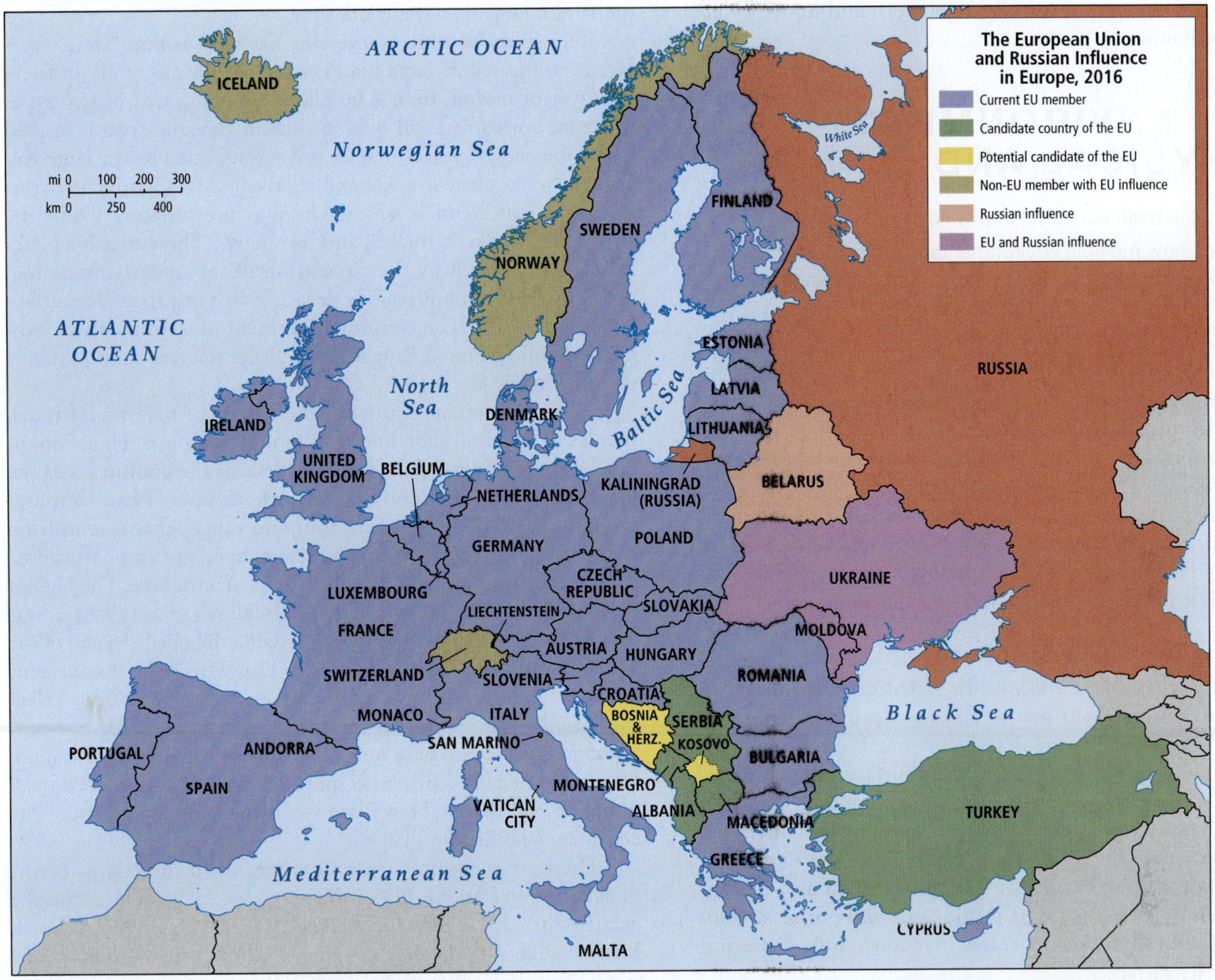

(B) Post-2016 alignments of the European Union and of Russia and the post-Soviet states.

concepts: environment; globalization and development; power and politics; urbanization; and population, gender, and culture. These concepts then allow the reader to make comparisons across regions. Because people live their daily lives at the local scale—in villages, towns, and city neighborhoods—this book shows through vignettes how global or regional patterns affect individuals where they live.

In summary, regions have the following traits:

- A region is a unit of Earth's surface that contains distinct environmental or cultural patterns.
- No two regions are necessarily defined by the same set of attributes.
- Regional definitions and the territory included can change.
- On the ground, the boundaries of regions are usually indistinct and hard to agree upon.
- Regions can vary greatly in size (scale).

CHECK YOUR UNDERSTANDING

1. What two simple questions are central to geography?
2. What is the primary concern of physical geography?
3. Why are maps important tools for geographers?
4. What makes photo interpretation a useful tool for a geographer?
5. What is a region?

GEOGRAPHIC THEMES IN THIS BOOK

Within the world regional framework, this book is also organized around five thematic concepts of special significance in the modern world. These concepts are the focus of the geographic themes offered in every chapter in this book: the ***environment: physical and human***, ***globalization and development***, ***power and politics***, ***urbanization***, and ***population, gender, and culture***. The sections that follow explain each of these five thematic concepts, how they tie in with the main concerns of geographers, and how they are related to each other.

ENVIRONMENT: PHYSICAL AND HUMAN

1.3 Identify the main concerns of physical geography.

1.4 Evaluate how multiple environmental factors interact to influence the vulnerability of a location to climate change.

PHYSICAL GEOGRAPHY

Physical geography is concerned with the processes that shape Earth's landforms, climate, and vegetation. Physical geographers often look at problems spatially, depicting the results of their analysis on maps, and synthesize insights from a wide variety of disciplines. While physical geography provides a backdrop for the many aspects of human geography discussed in this book, it is also a large and growing field of study worth exploring in greater detail. What follows are just the basics of landforms and climate.

Landforms: The Sculpting of Earth

The processes that create the world's varied **landforms**—such as mountain ranges, continents, and the deep ocean floor—are some of the most powerful and slow-moving forces on the planet. Originating deep beneath Earth's surface, these *internal processes* can move entire continents, often taking hundreds of millions of years to do their work. However, it is the more rapid *external processes* taking place on the surface of the Earth that create the landscape features we encounter, such as a beautiful canyon, often on a scale of millions of years or less. Geomorphologists study these processes that constantly shape and reshape Earth's surface.

Plate Tectonics

Two key ideas related to internal processes in physical geography are the *Pangaea hypothesis* and *plate tectonics*. The geophysicist Alfred Wegener first suggested the Pangaea hypothesis in 1912, proposing that all the continents were once joined in a single vast continent called Pangaea (meaning "all lands"), which then fragmented into the continents we know today (**Figure 1.6**). As one piece of evidence for his theory, Wegener pointed to the neat fit between the west coast of Africa and the east coast of South America.

For decades, most scientists rejected Wegener's hypothesis. We now know, however, that Earth's continents have been assembled into supercontinents a number of times, only to break apart again. All of this activity is made possible by plate tectonics, a process of continental motion discovered in the 1960s, long after Wegener's time.

The study of **plate tectonics** has shown that Earth's surface is composed of large plates that float on top of an underlying layer of molten rock. The plates are composed of two types of "crust," or cooled and hardened rock. Oceanic crust is located under the oceans, where intense pressures created by the large volumes of water make it dense and relatively thin. Continental crust forms the continents where much lower pressures exerted by the atmosphere make it thicker and less dense. These massive plates drift slowly, driven by the circulation of the underlying molten rock flowing from hot regions deep inside Earth to cooler surface regions and back. The creeping movement of tectonic plates fragmented and separated Pangaea into pieces that are the continents we know today (see Figure 1.6E).

Plate movements influence the shapes of major landforms, such as continental shorelines and mountain ranges. Huge mountains have piled up on the leading edges of the continents as the plates carrying them collide with other plates. Plate tectonics accounts for the long, linear mountain ranges that extend from Alaska to Chile in the Western Hemisphere and from Southeast Asia to the European Alps in the Eastern Hemisphere. The highest mountain range in the world, the Himalayas of South Asia, was created when what is now India, situated at the northern end of the Indian-Australian Plate, ground into Eurasia. The only continent that lacks these long, linear mountain ranges is Africa. Often called the "plateau continent," Africa is believed to have been at the center of Pangaea and to have moved relatively little since the breakup. However, as Figure 1.6E shows, parts of eastern Africa—the Somali Subplate and the Arabian Plate—continue to separate from the continent (the African Plate).

Humans encounter tectonic forces most directly as earthquakes and volcanoes. Plates slipping past each other create the catastrophic shaking of the landscape we know as an earthquake. Plates collide and one may slip under the other in a process called *subduction*. Volcanoes arise at zones of subduction or sometimes in the middle of a plate, where gases and molten rock (called *magma*) can rise to Earth's surface through fissures and holes in the plate. Volcanoes and earthquakes are particularly common around the edges of the Pacific Ocean, an area known as the *Ring of Fire* (**Figure 1.7**).

External Landscape Processes

The landforms created by plate tectonics have been further shaped by the external processes that we can observe daily. One such process is *weathering*. Rock fractures and decomposes into tiny pieces as it is exposed to sunlight, wind, rain, snow, ice, freezing and thawing, and the effects of life-forms (such as plant roots). *Erosion*, during which weathered rock particles and dirt are carried away by wind and water and deposited elsewhere, is a major force shaping landscapes. Eroded material can raise and flatten the land around a river, where periodic flooding spreads huge quantities of silt over thousands of years. As valleys between hills are filled in by silt, a **floodplain** is created. Where rivers meet the sea, floodplains often fan out roughly in the shape of a triangle, creating a *delta*. External processes tend to smooth out the dramatic mountains and valleys created by internal processes.

landforms physical features of Earth's surface, such as mountain ranges, river valleys, basins, and cliffs

plate tectonics the theory that Earth's surface is composed of large plates that float on top of an underlying layer of molten rock; the movement and interaction of the plates create many of the large features of Earth's surface, particularly mountains

floodplain the flat land along a river where sediment is deposited during flooding

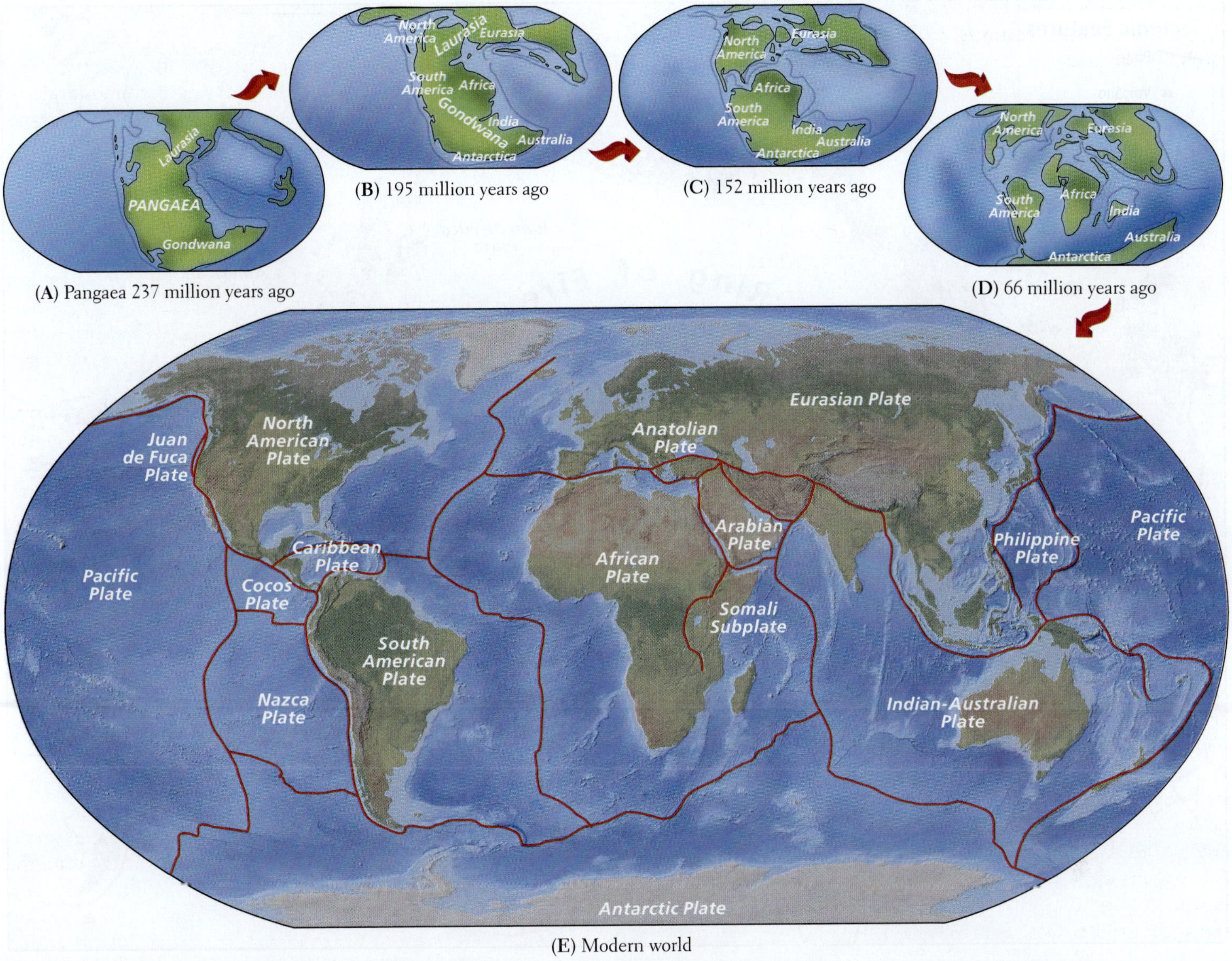

Figure 1.6 The breakup of Pangaea. **(A)** through **(D)** show the changing arrangement of Earth's land surfaces, Pangaea, starting over 237 million years ago. The modern world map **(E)** depicts the current boundaries of the major tectonic plates. Pangaea is only one of many tectonic plate configurations that have coalesced and then fragmented over the last billion years. [Research from: Frank Press, Raymond Siever, John Grotzinger, and Thomas H. Jordan, *Understanding Earth*, 4th ed. (New York: W. H. Freeman, 2004), pp. 42–43]

Human activity often contributes to external landscape processes. The clearing of forests for farming exposes more soil to sunlight, wind, and rain. These agents in turn increase weathering and erosion. Flooding becomes more common because the removal of trees and other vegetation limits the land's ability to absorb rainwater. As erosion increases, rivers may fill with silt, and deltas may extend far into the oceans. When humans dam rivers or build levees to control flooding, as has happened in the Mississippi River Delta (see Figure 2.7), they interrupt the natural processes of deposition and removal of silt.

Climate

The processes associated with climate are generally more rapid than those that shape landforms. **Climate** is the long-term balance of temperature and moisture that keeps weather patterns fairly consistent from year to year across the planet. The last major change in climate at the global scale took place about 15,000 years ago, when the glaciers of the last ice age began to melt. **Weather**, the short-term and spatially limited expression of climate, can change in a matter of minutes.

Solar energy is the engine of climate. Earth's atmosphere, oceans, and land surfaces absorb large amounts of solar energy, and the differences in the amounts they absorb account for part of the variations in climate we observe.

climate the long-term balance of temperature and precipitation that characteristically prevails in a particular region

weather the short-term and spatially limited expression of climate that can change in a matter of minutes

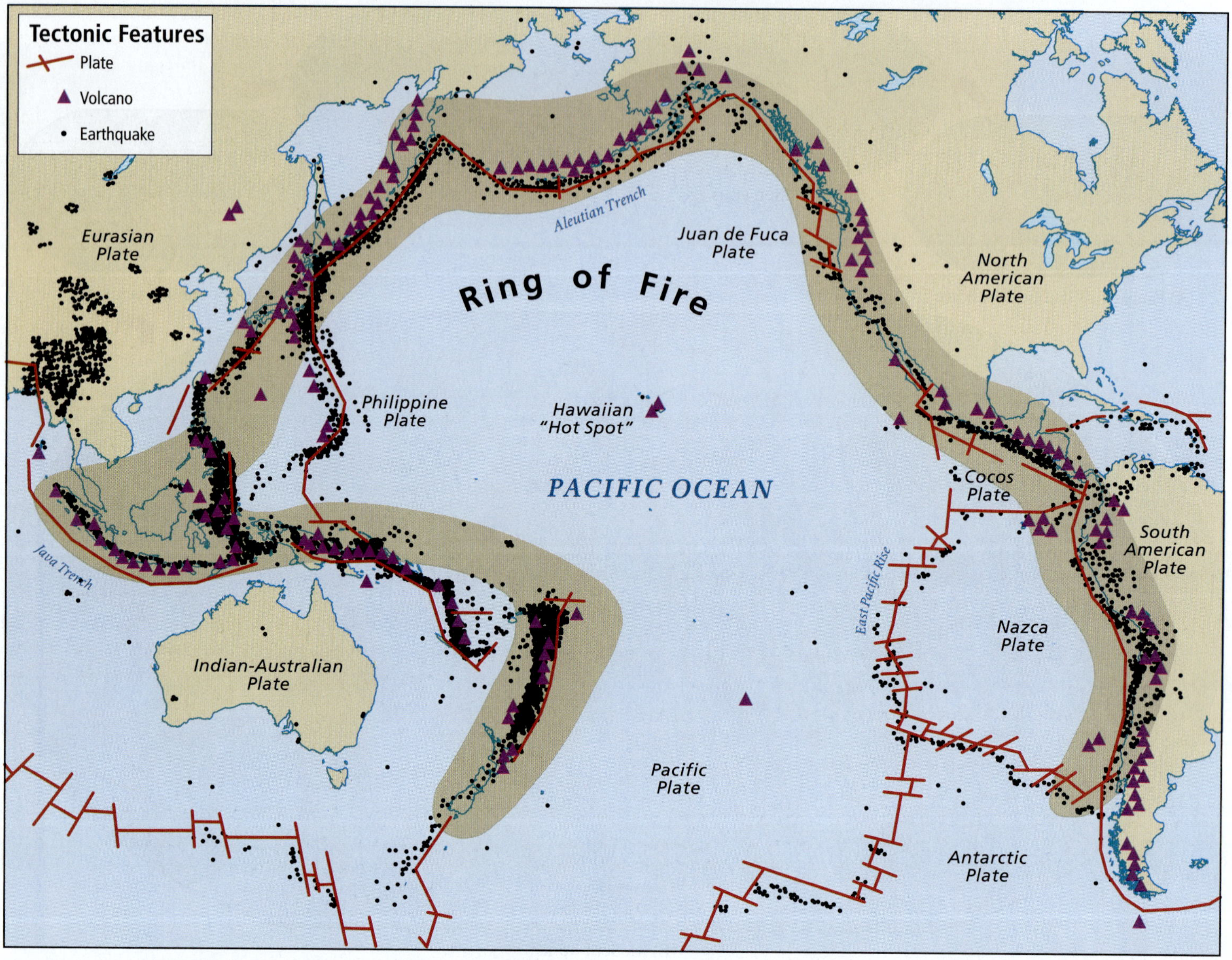

Figure 1.7 Ring of Fire. Volcanic formations encircling the Pacific Basin form the Ring of Fire, a zone of frequent earthquakes and volcanic eruptions. [Research from: United States Geological Survey, Active Volcanoes and Plate Tectonics, "Hot Spots" and the "Ring of Fire," at http:/vulcan.wr.usgs.gov/Glossary/PlateTectonics/Maps/map_plate_tectonics_world.html; and Frank Press, Raymond Siever, John Grotzinger, and Thomas H. Jordan, *Understanding Earth*, 4th ed. (New York: W. H. Freeman, 2004), p. 27]

The most intense direct solar energy strikes Earth more or less head-on in a broad band stretching about 30° north and south of the equator. The highest average temperatures on Earth's surface are found near this band, with temperatures falling as one moves north or south due to solar energy striking the Earth's surface less directly—at more of an obtuse (wide) angle. This wide angle causes a deflection of some of the Sun's rays, resulting in less absorption of solar energy and lower average annual temperatures.

The fact that Earth's axis sits at a 23° angle as it orbits the Sun means that the band of greatest solar intensity that strikes Earth varies in its location over the course of a year, creating seasons. Just how this yearly seasonal pattern works is illustrated by **Figure 1.8**.

Climate Region Geographers have several systems for classifying the world's climates, based on patterns of temperature and precipitation. This book uses a modification of the widely known *Köppen classification system*, which divides the world into several types of climate regions, labeled (A) through (E) on the climate map in **Figure 1.9**. As is the case with many maps, the sharp boundaries shown on climate maps are in reality much more gradual transitions. As you look at the regions on this map, examine the photos, and read the accompanying climate descriptions, notice the importance of climate to vegetation. Each regional chapter includes a climate map; when reading these maps, refer to the written descriptions in Figure 1.9 as necessary.

Temperature and Air Pressure Daily weather patterns are largely a result of variations in solar energy absorption that create complex patterns of air temperature and *air pressure*. To understand air pressure, think of air as existing in a particular unit of space—for example, a column of air above a square foot of ground. Air pressure is the weight of the column of air, pushing down on the square foot of ground. Air pressure and temperature are closely related. The gas molecules in warm air are relatively far apart, so

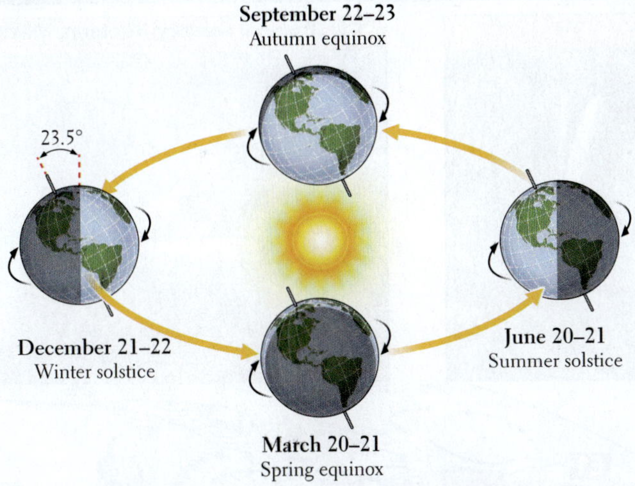

Figure 1.8 Seasonal changes in Earth's orientation to the Sun. Diagram of the angle of Earth's orientation to the Sun, showing seasonal changes in the Northern Hemisphere over the course of one year.

when the air in the column is warm, there are fewer molecules in the column and it weighs less. This results in less pressure on the ground underneath the column. When the air in the column is cool, the opposite happens. The gas molecules are closer together, so more of them can fit in the column and it weighs more. This results in more pressure on the ground underneath the column.

Air tends to move from areas of higher pressure to areas of lower pressure, creating wind. If you go to the beach on a hot day, you will notice a cool breeze blowing in off the water. This happens because land heats up (and cools down) faster than water, so on a hot day, the air over the land warms and becomes less dense than the air over the water. This causes the air over the water to flow inland. At night the breeze often reverses direction, blowing from the now cooling land onto the now relatively warmer water.

Air movements have a continuous and important influence on global weather patterns and are closely associated with land and water masses. Because continents heat up and cool off much more rapidly than the oceans that surround them, the wind tends to blow from the ocean to the land during summer and from the land to the ocean during winter. It is almost as if the continents were breathing once a year, inhaling in summer and exhaling in winter.

Precipitation Perhaps the most tangible way we experience changes in air temperature and pressure is through rain or snow. *Precipitation* (dew, rain, sleet, hail, and snow) occurs primarily because warm air holds more moisture than cool air. When this warmer moist air rises to a higher altitude, its temperature drops, reducing its ability to hold moisture. The moisture condenses into drops that form clouds and may eventually fall as rain or some other form of precipitation. This happens when moisture-bearing air is forced to rise as it passes over mountain ranges: the air cools, and the moisture condenses to produce rainfall (**Figure 1.10**). This process, known as **orographic precipitation**, is most common in coastal areas where wind blows moist air from above the ocean onto the land and up the side of a coastal mountain range. Most of the moisture falls as rain as the cooling air rises along the coastal side of the range. On the inland side, the descending air warms and ceases to drop its moisture. The drier side of a mountain range is said to be in the **rain shadow**. Rain shadows may extend for hundreds of miles across the interiors of continents, as they do on the east side of California's Pacific coastal ranges, or north of the Himalayas of Eurasia.

A central aspect of Earth's climate is the *rain belt* (also known as the intertropical convergence zone, or ITCZ; see Chapter 7 and Chapter 8). Near the equator, moisture-laden tropical air is heated by the strong direct sunlight and rises to the point where it releases its moisture as rain. Neighboring nonequatorial areas also receive some of this moisture when seasonally shifting winds move the rain belt north and south of the equator. The huge downpours of the Asian summer monsoon are an example of this equatorial rain belt.

In the summer *monsoon* season, the Eurasian continental landmass heats up, causing the overlying air to expand, become less dense, and rise. The somewhat cooler, yet moist, air of the Indian Ocean is drawn inland. The effect is so powerful that the equatorial rain belt is sucked onto the land (see Figure 8.10). This results in tremendous, sometimes catastrophic, summer rains throughout virtually all of South Asia and Southeast Asia and much of coastal and interior East Asia. The reverse happens in the winter as similar forces pull the equatorial rain belt south during the Southern Hemisphere's summer.

Much of the moisture that falls on North America and Eurasia is *frontal precipitation* caused by the interaction of large air masses of different temperatures and densities. These masses develop when air stays over a particular area long enough to take on the temperature of the land or sea beneath it. Often when we listen to a weather forecast, we hear about warm fronts or cold fronts. A *front* is the zone where warm and cold air masses come into contact, and it is always named after the air mass whose leading edge is moving into an area. At a front, the warm air tends to rise over the cold air, carrying warm clouds to a higher, cooler altitude. Rain or snow may follow. Much of the rain that falls along the outer edges of a hurricane is the result of frontal precipitation.

CHECK YOUR UNDERSTANDING

1. What are the main concerns of physical geography?
2. What is the difference between internal and external processes?
3. How is climate different from weather?
4. Why do seasons occur?
5. What causes variations in air pressure, and why is this important?
6. How does precipitation occur?

HUMAN IMPACT ON THE ENVIRONMENT

Throughout our history humans have overused resources and degraded environments, sometimes with disastrous consequences. In the past this has played out on a local scale. For example, the Maya civilization of the Yucatan Peninsula declined in the ninth century in part due to the depletion of local soil nutrients, which made agriculture much less productive. Today humans are affecting environments on a global scale and with such intensity that scientists now refer to the present era as the

orographic precipitation precipitation produced when a moving moist air mass encounters a mountain range, rises, cools, and releases condensed moisture that falls as rain

rain shadow an area of low rainfall on the drier side of a mountain range affected by orographic rainfall

1.9 Climate Regions of the World

A Tropical humid climates. In *tropical wet climates*, rain falls predictably every afternoon and usually just before dawn. The *tropical wet/dry climate*, also called a *tropical savanna*, has a wider range of temperatures and less precipitation than the tropical wet climate. [A1: yenwen/E+/Getty Images; A2: Education Images/Universal Images Group/Getty Images]

B Arid and semiarid climates. *Deserts* generally receive very little rainfall (10 inches or less per year), mostly in rare downpours. *Steppes* have similar but more moderate climates, usually receiving about 10 inches more rain per year than deserts. While deserts often feature rocks, bare earth, or sand, steppes are covered with grass or scrubby vegetation. [B1: Hoberman Collection/Universal Images Group/Getty Images; B2: Palani Mohan/Getty Images News/Getty Images]

C Temperate climates. *Midlatitude temperate climates* receive rainfall throughout the year and have short, mild winters and long, hot summers. *Subtropical temperate climates* are similar, but with dry winters. *Mediterranean climates* have moderate temperatures but are dry in summer and wet in winter. [C1: RDImages/Epics/Hulton Archive/Getty Images; C2: ALEXANDER JOE/AFP/Getty Images; C3: Tim Graham/Getty Images News/Getty Images]

D Cool humid climates. Stretching across the broad interiors of Eurasia and North America are *continental climates*, which either have dry winters (northeastern Eurasia) or receive rain all year (North America and north-central Eurasia). Summers in cool humid climates are short but can have very warm days. [D1: Ovchinnikova Irina/Shutterstock; D2: Nikki Kahn/The Washington Post/Getty Images]

E Coldest climates. *Arctic and high-altitude climates* are the coldest and driest. Although moisture is present, there is little evaporation because of the low temperatures. A low-lying vegetation called *tundra* covers the ground. The high-altitude climate is more widespread and subject to larger daily fluctuations in temperature. As one ascends in altitude, the changes in climate loosely mimic those found as one moves from lower to higher latitudes. These changes are known as temperature-altitude zones (see Figure 3.12). [E1: PAUL J. RICHARDS/AFP/Getty Images; E2: P. Morris/Hulton Archive/Getty Images]

A1 Tropical wet, Hawaii

A2 Tropical wet/dry, Yucatán, Mexico

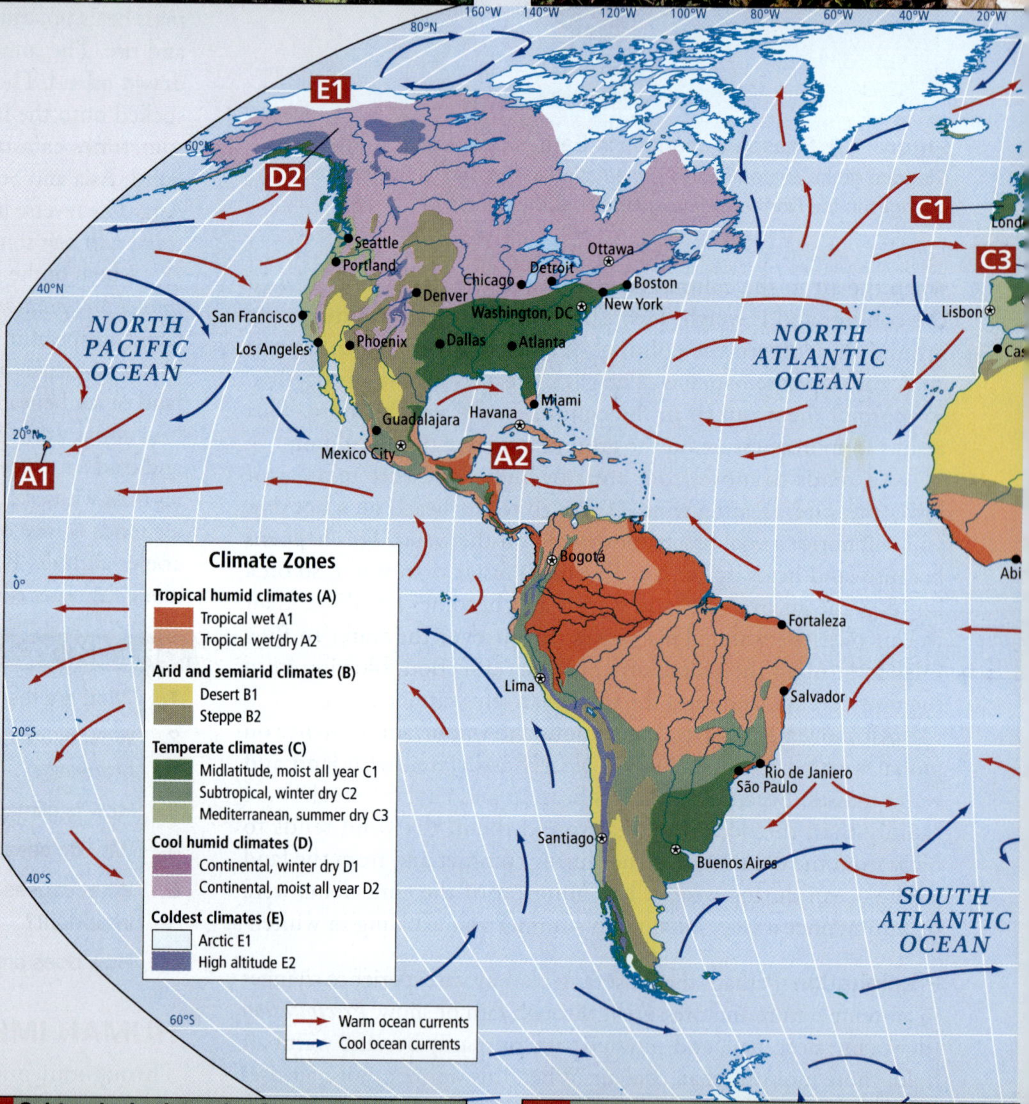

C2 Subtropical, winter dry, South Africa

C3 Mediterranean, summer dry, Italy

B1 Desert, Namibia

B2 Steppe, Mongolia

C1 Midlatitude, moist all year, United Kingdom

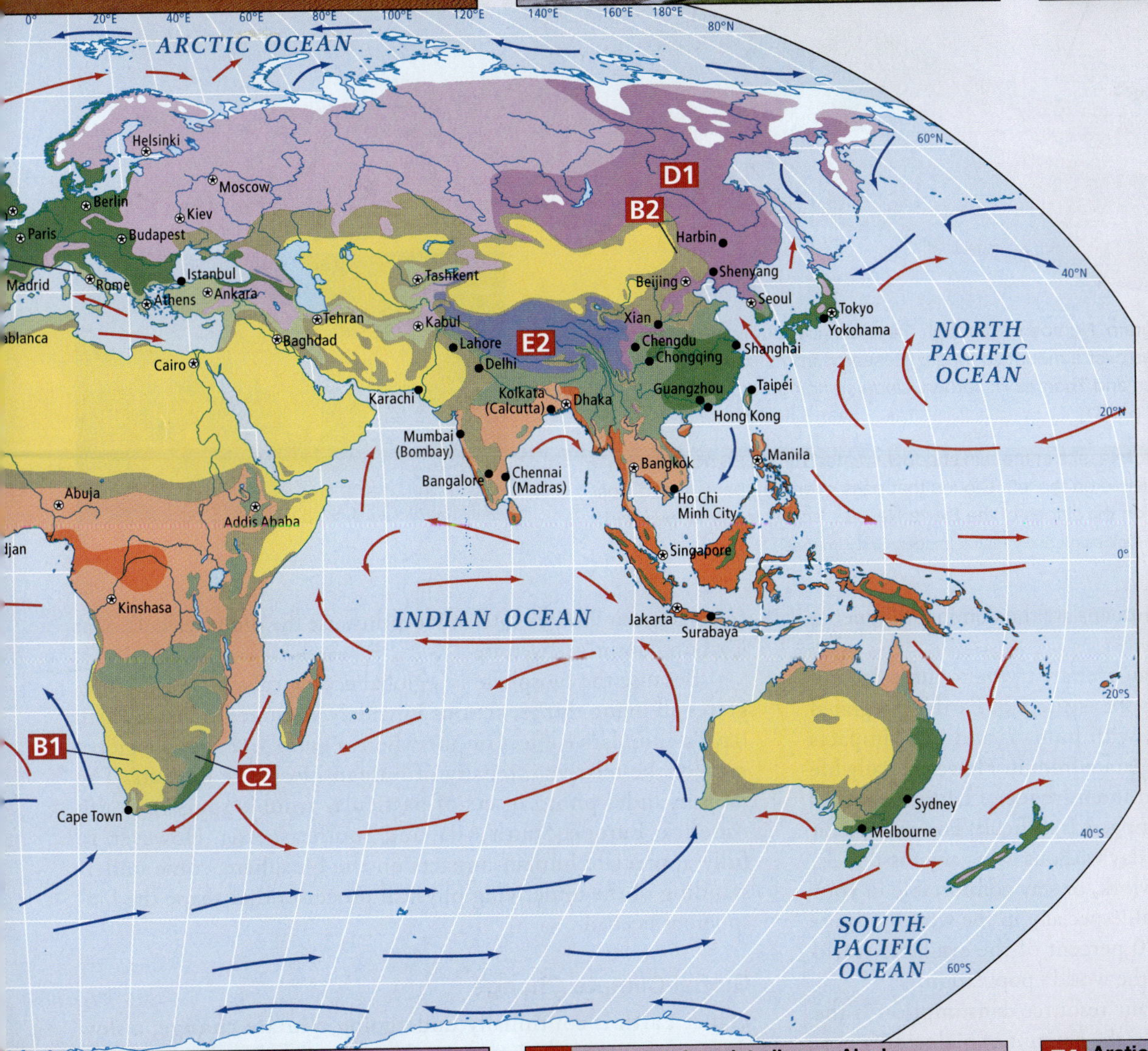

E2 High altitude, Tibet

D1 Continental, winter dry, Russia

D2 Continental, moist all year, Alaska

E1 Arctic tundra, Canada

Figure 1.10 Orographic rainfall (and rain shadow diagram).

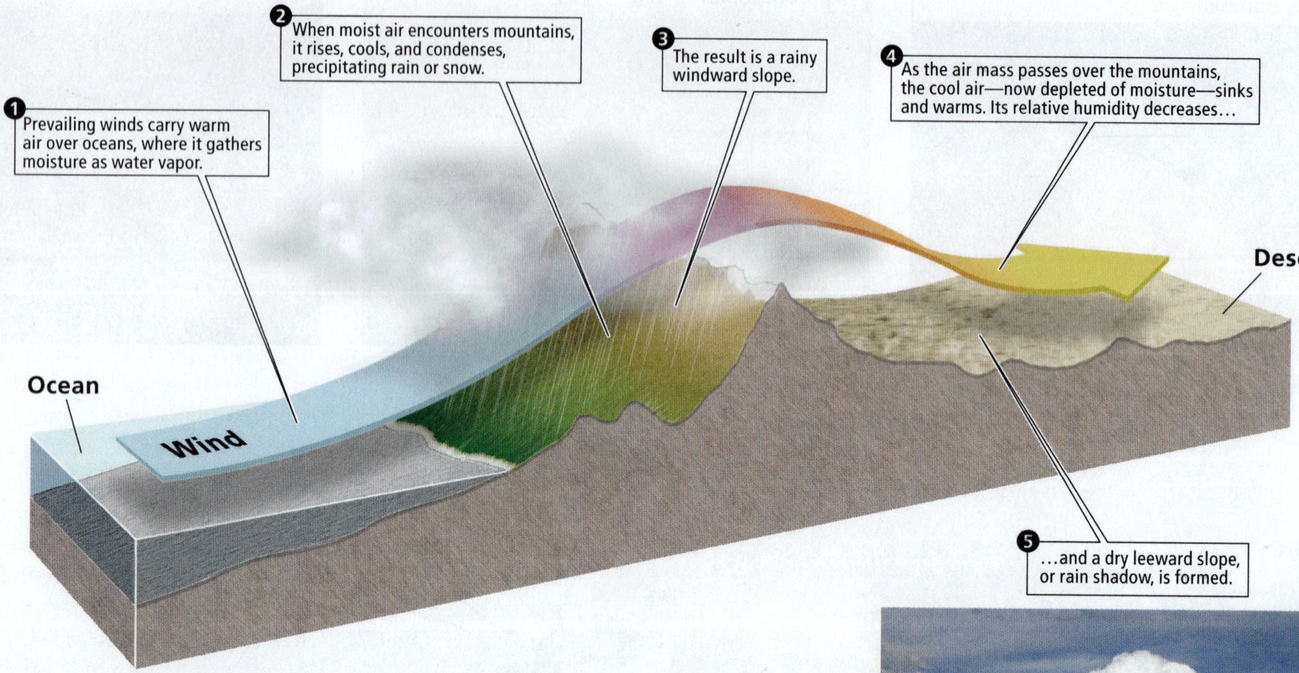

(A) A rain shadow diagram. Numbers 1 through 5 explain how humid wind is depleted of moisture as it passes over a mountain range. [Research from: Frank Press, Raymond Siever, John Grotzinger, and Thomas H. Jordan, *Understanding Earth*, 4th ed. (New York: W. H. Freeman, 2004), p. 281]

(B) Owens Valley, California, sits in the rain shadow created by the Sierra Nevada mountain range, just visible on the left. Clouds that pass over here have already lost most of their moisture on the slopes of the Sierra Nevada, so the lower slopes and valleys are dry. [Sabrina Dalbesio/Lonely Planet Images/Getty Images]

Anthropocene—the period when humans are the dominant influence on climate and the environment.

In response, more people than ever are trying to limit damage to the **biosphere**, defined here as the entirety of Earth's integrated physical and biological systems, with humans and their impacts included as part of nature, not separate from it. However, daily life has become so transformed by our intensive use of Earth's resources that substantially reducing our impacts is difficult. For example, if you tried to live for even just one day without using any fossil fuels, how would you get to school or work, or stay comfortable in your home? And yet change is essential, especially in the wealthy countries that consume more than 80 percent of the available world resources with just 20 percent of the world's population.

A tool for understanding your resource consumption is the *ecological footprint*, which estimates the amount of biologically productive land and sea area needed to sustain a person at their current lifestyle in their country, and then calculates how many Earths would be required if everyone on the planet lived this lifestyle. This is particularly useful for drawing comparisons between countries (see **Figure 1.11**). You can calculate your own footprint using the online Global Footprint Network's calculator. A similar concept more closely related to global warming is the *carbon footprint*, which measures the greenhouse gas emissions a person's activities produce. To calculate your family's carbon footprint, use the online calculator at Carbon Footprint's website.

biosphere the entirety of Earth's integrated physical spheres, with humans and other impacts included as part of nature

Because the biosphere is a global ecological system that integrates all living things, actions in widely separated parts of Earth have a cumulative effect on the whole. **Figure 1.12** shows a global map of the relative intensity of human biosphere impacts. The map includes photo insets of particular trouble spots in South America, Europe, South Asia, and Southeast Asia. However, to fully appreciate human impacts on the biosphere, some understanding of the underlying physical processes that shape the biosphere is needed.

Global Climate Change

Planet Earth is continually undergoing **climate change**, a slow shifting of climate patterns caused by the general cooling or warming of the atmosphere. The current trend of **global warming**—which refers to the observed warming of Earth's climate as atmospheric levels of greenhouse gases increase—is extraordinary because it is happening more quickly than climate changes in the past. Scientific evidence indicates that it is extremely likely that the changes are linked to human activities. **Greenhouse gases (GHG)**, which include carbon dioxide (CO_2), methane, water vapor, and other gases, are essential to keeping Earth's incoming and outgoing thermal radiation balanced so that Earth's surface temperature remains hospitable to life. These gases absorb solar radiation, keeping Earth's surface (like the interior of a

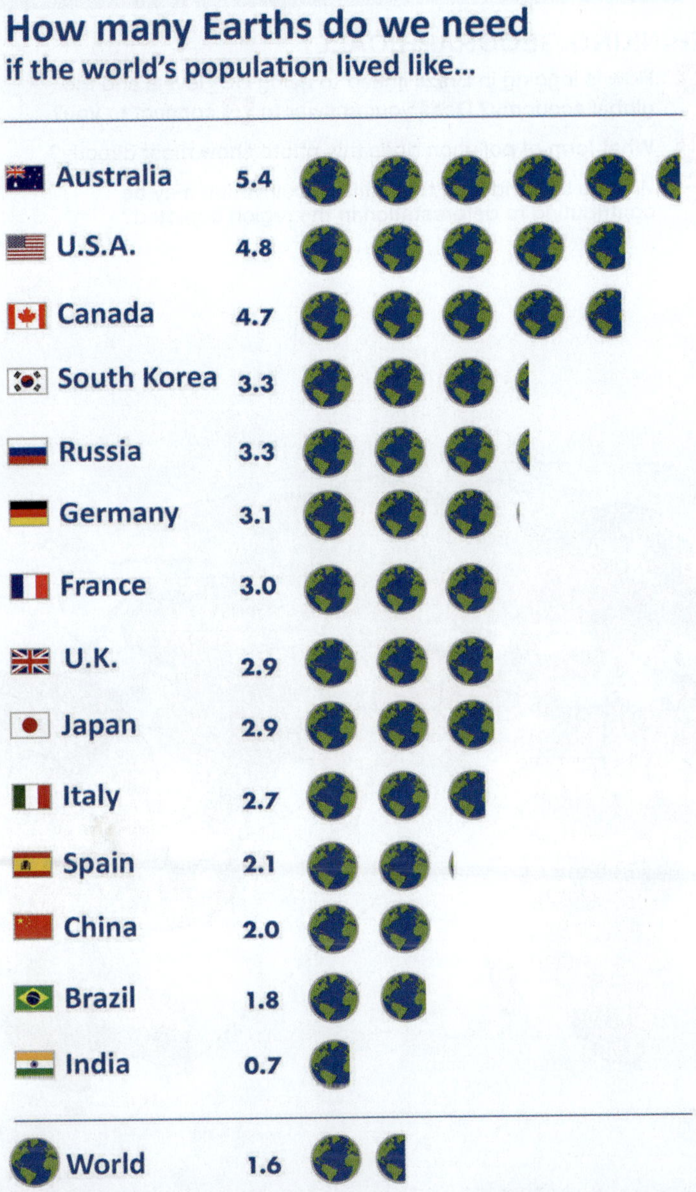

Figure 1.11 How many Earths would we need if the world's population lived like . . . ? [Global Footprint Network National Footprint Accounts 2016]

greenhouse) warmer than it would otherwise be. **Figure 1.13** shows this system of incoming and outgoing radiation and the role of GHGs in trapping some of the outgoing radiation and sending it back to warm Earth.

The vast majority of scientists agree that the warming trend in Earth's climate is caused by GHGs being released by humans at accelerating rates through the burning of fossil fuels (for example, in coal-fired power plants and car engines) and by the effects of deforestation, which reduces the amount of carbon dioxide that is absorbed by trees (see Figure 1.12A, C). The most recent evidence indicates that Earth is warming very rapidly.

Scientists also agree that GHG emissions must be reduced to avoid catastrophic climate change in coming years. Climatologists, biogeographers, and other scientists are documenting long-term global warming and cooling trends by examining evidence in tree rings, fossilized pollen and marine creatures, and glacial ice. These data indicate that the twentieth century was the warmest century in 600 years, and that the current decade is the hottest on record. Evidence is mounting that these are not normal fluctuations. By 2100 average global temperatures could rise between 1.5 and 10 degrees. The result will be multiple impacts on the biosphere, such as flooding, desertification, rising sea levels, loss of agricultural land, and loss of species. These impacts could uproot hundreds of millions of people, mostly in the developing world, creating multiple global political and economic crises that will make GHG reduction much more difficult.

Drivers of Global Climate Change A wide variety of human activities emit GHGs. Electricity generation for the heating, cooling, and lighting of businesses, industries, and homes and gas- and diesel-powered vehicles burn large amounts of fossil fuels such as coal, natural gas, and oil, which emit GHGs. Widespread deforestation worsens the situation. Living forests take in CO_2 from the atmosphere via photosynthesis, releasing oxygen and storing the carbon in the biomass of their woody trunks and branches. As more trees are cut down and their wood is used for fuel, more carbon enters the atmosphere, less is taken out, and less is stored. The loss of trees and other forest organisms has produced as much as 30 percent of the buildup of CO_2 in the atmosphere. The use of fossil fuels accounts for the other 70 percent. The large quantities of GHGs from these sources have already led to significant warming of the Earth's climate.

The industrialized countries (the United States, Canada, the European Union, Russia, and Australia) and the large, rapidly developing countries (notably China and India) have the highest percentages of total GHG emissions. **Figure 1.14** shows the global patterns in 2016 of total CO_2 emissions in total tons (A) and per capita (B). Note that the United States is among the leaders in both categories.

Climate-Change Impacts The current and future impacts of rising global temperatures are not uniform across the globe. In fact, a key problem in reducing GHGs is that those most responsible for global warming (the world's wealthiest and most industrialized countries) have the least incentive to reduce emissions because they are the least vulnerable to the changes global warming causes in the physical environment because their wealth, power, and high human development make them more able to adapt. Those most vulnerable to these changes (poor countries with low levels of human development) are the least responsible for the growth in GHG emissions but have little power in the global political arena; hence, they do not have the power (politically and economically) to affect the level of emissions (**Figure 1.15**).

Sea Level Rise. One of the most severe climate change impacts is sea level rise, which is driven by the melting of the polar ice caps, as well as the expansion of the oceans as they warm. Satellite imagery from the National Aeronautics and Space Administration (NASA) shows that the ice caps are shrinking by 13 percent per decade and that melting is happening faster than predicted. The most extreme predictions for sea level rise are a

climate change a slow shifting of climate patterns caused by the general cooling or warming of the atmosphere

global warming the warming of Earth's climate as atmospheric levels of greenhouse gases increase

greenhouse gases (GHGs) gases, such as carbon dioxide and methane, released into the atmosphere by human activities; these gases are harmful when released in excessive amounts

PHOTO ESSAY

1.12 Human Impact on the Biosphere

Humans have such enormous impacts on the biosphere that scientists refer to the present era as the *Anthropocene*—the period when humans are the dominant influence on the environment. More people than ever are trying to limit these impacts, but daily life has become so transformed by our intensive use of Earth's resources that substantially reducing our impacts is difficult. The map shows varying levels of human impact, using data from hundreds of studies. High and medium-high impact areas often have roads, agriculture, or other intensive uses, while low-medium impact areas have biodiversity loss and other disturbances related to human activity. [Research from: UN Environment Programme, "Human Impact, Year 1700 (Approximately)," and "Human Impact, Year 2002" (New York: United Nations Development Program, 2002–2006), at http://www.grida.no /graphicslib/detail/human-impact-year-1700-approximately_6963 and http://www.grida .no/graphicslib/detail/human-impact-year-2002_157a]

THINKING GEOGRAPHICALLY

A How is logging in Brazil linked to rising CO_2 levels and the global economy? Does your answer to **A** connect to you?

B What form of pollution does this photo show most directly?

C What is the evidence that shifting cultivation may be contributing to deforestation in the region depicted?

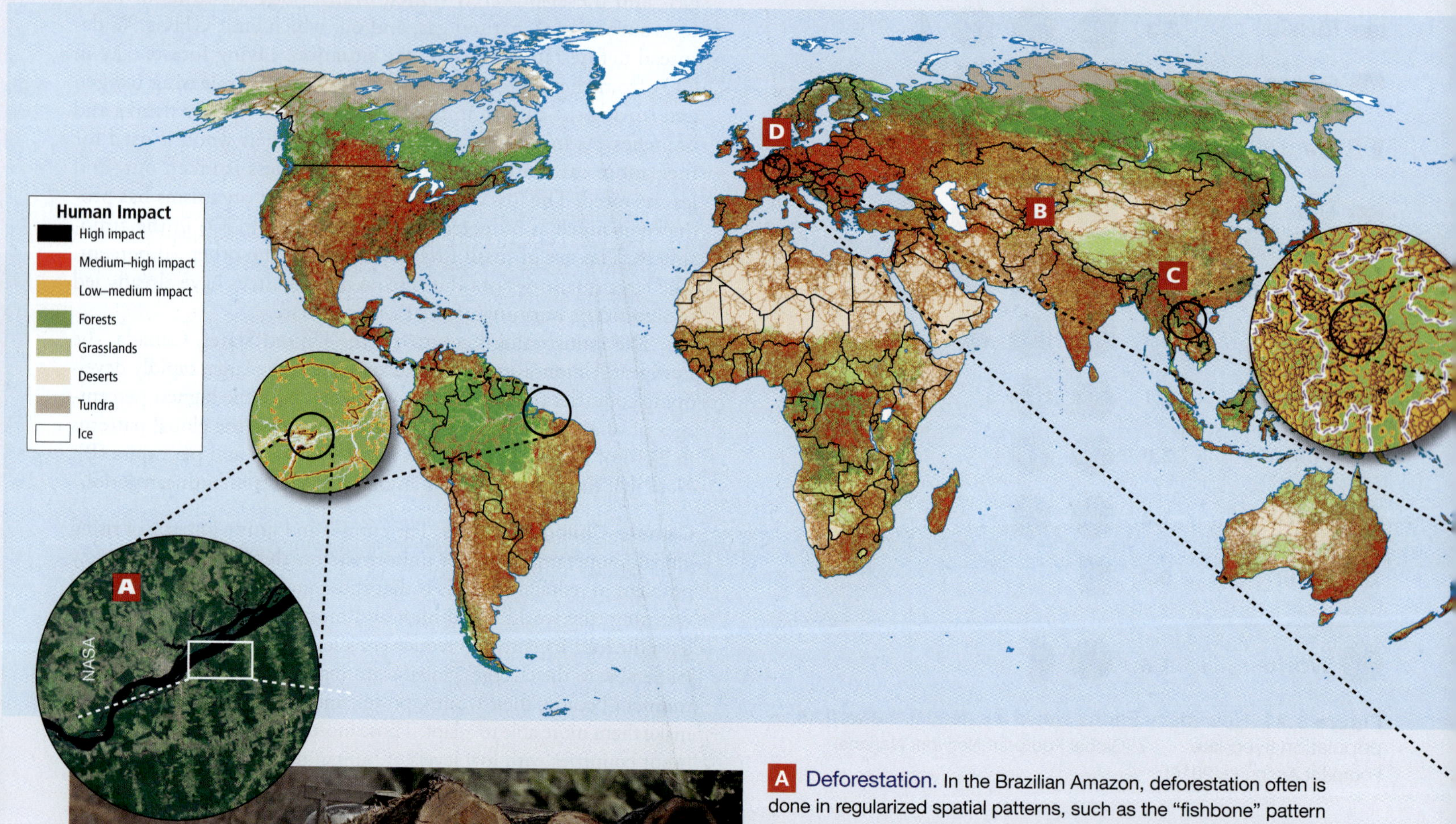

A Deforestation. In the Brazilian Amazon, deforestation often is done in regularized spatial patterns, such as the "fishbone" pattern (see satellite image inset A). This pattern results from regulations that determine the location of roads used for settlement and logging. Whole logs are brought by road to rivers, where they are put on barges and taken to a port for export. [Mario Tama/Getty Images News/Getty Images]

B War and political conflict. War can have a devastating effect on the environment. Women outside Kabul, Afghanistan, must now carry water by hand from distant sources that have not been polluted or damaged by war. Long trips for water place these women in danger of assault. Between the women and their small mud houses can be seen a shallow muddy stream. [MASSOUD HOSSAINI/AFP/Getty Images]

C Food and deforestation. Farmers practicing shifting cultivation plant "hill rice" in Myanmar (Burma). Shifting cultivation is an ancient technique that can be sustained indefinitely (see Chapter 3), given sufficient land and fallow periods long enough for forest to regrow (20 years or more). Today, more and more forest is being turned over to short-fallow cultivation—3 to 6 years of cultivation—resulting in a loss of habitat and biodiversity (see inset). [China Photos/Getty Images News/Getty Images; inset: NASA]

D Development and mining. Perhaps no other human activity has as striking an impact on the landscape as mining. This open-pit coal mine is located in one of the most industrialized and densely inhabited parts of Germany. [PATRIK STOLLARZ/AFP/Getty Images]

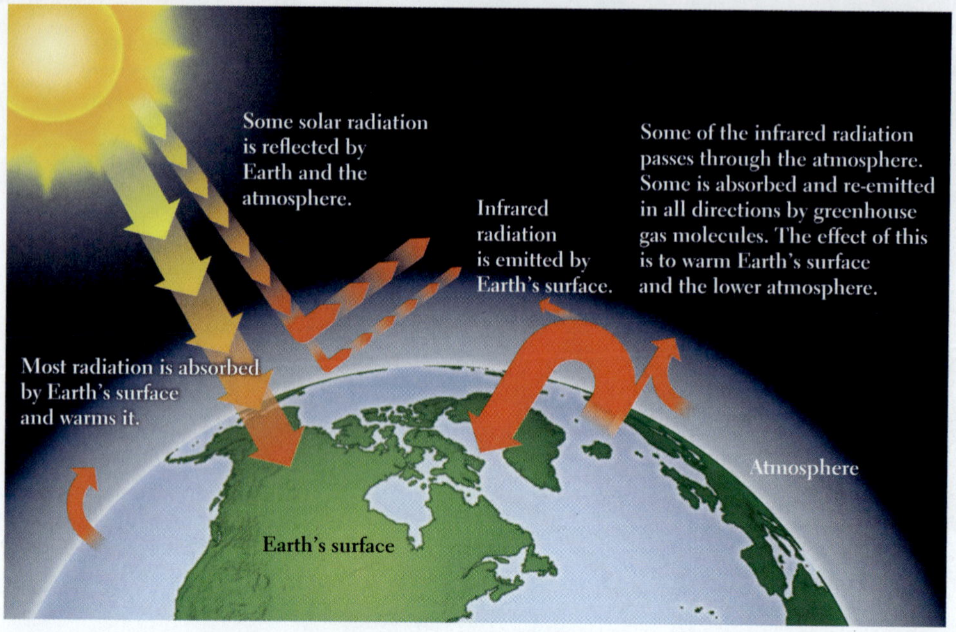

Figure 1.13 The balance of incoming and outgoing radiation and the greenhouse effect. To maintain an even temperature, Earth has to balance energy coming in with energy going out. Energy coming in is mostly sunshine, and energy going out is mostly radiant heat.

10–12 foot (3–4 meter) rise by the end of this century, but most studies suggest a rise of 1–6 feet (0.3–2 meters). While impacts from a 1-foot rise in sea level would be felt in every coastal city in the world, 6 feet would render major parts of wealthy low-lying coastal cities, such as Boston, Massachusetts, or Miami, Florida, uninhabitable without the installation of huge and expensive antiflooding infrastructure. In less well-off cities, such as Lagos, Nigeria, or Alexandria, Egypt, a 1-foot rise in sea level would force millions to relocate, and a 6-foot rise could force abandonment of the city.

Melting Glaciers. The melting of high mountain glaciers in the Himalayas, which are a major source of water for many of the world's large rivers, threatens millions living in river valleys such as the Ganga, the Indus, the Brahmaputra, the Huang He (Yellow), and the Chang Jiang (Yangtze). Scientists have monitored mountain glaciers across the globe for more than 30 years and while some are growing, the majority are melting rapidly. Over the short term, melting mountain glaciers will increase the flow of many rivers, but eventually river flows will decrease as mountain glaciers shrink or disappear entirely.

Drought and Water Scarcity. Over time, higher average annual temperatures will bring warmer climate zones to most parts of the planet. Some areas will experience drought and water scarcity as higher temperatures increase evaporation rates from soils, vegetation, and bodies of water. Animal and plant species that cannot adapt will disappear, and large numbers of people will be displaced as the zones where crops grow shift.

Stronger Storms. Higher temperatures also will lead to stronger tornados and hurricanes because these storms are powered by warm ocean water and warm, rising air (see Figure 1.15). One example is Hurricane Maria, the unusually large and powerful hurricane that struck Puerto Rico in the autumn of 2017, killing over 3000 people. In 2015 a record 22 hurricanes (typhoons) reached category 4 or 5 strength in the Northern Hemisphere.

Changing Ocean Currents. Another effect of climate change is likely to be a shift in ocean currents, caused by warming of the ocean itself. For example, the Gulf Stream that warms northwestern Europe and brings rain there is currently at its weakest in 1400 years. This could result in a cooler North Atlantic, resulting in colder temperatures in Europe, and more severe storms. Because ocean currents are all connected, a weakening Gulf Stream would wreak havoc on climates across the globe, but especially along the west coast of the Americas (see Chapter 2), the west coast of Africa (see Chapter 7), and northwestern Europe (see Chapter 4).

Vulnerability to Climate Change

The vulnerability a place has to climate change can be thought of as the amount of risk its human or natural systems have of being damaged by climate-related hazards such as sea level rise, drought, flooding, or increased storm intensity. Many of these vulnerabilities are water related; others are not. Scientists who study climate change agree that while we can take measures to minimize temperature increases, we can't stop them entirely, much less reverse those that have already occurred. We are going to have to live with and adapt to the impacts of climate change for quite some time. The first step in doing this is to understand how and where humans and ecosystems are especially vulnerable to climate change.

Three concepts are important in understanding a place's vulnerability to climate change: *exposure*, *sensitivity*, and *resilience*. Here we explore them in the context of the vulnerability to water-related climate-change impacts in Mumbai, India.

Exposure refers to the extent to which a place is exposed to climate-change impacts. A low-lying coastal city like Mumbai, India (see the photo on the Chapter 8 opening page), is highly exposed to sea level rise. *Sensitivity* refers to how sensitive a place is to those impacts. Many of Mumbai's inhabitants, for example, are very sensitive to the impacts of sea level rise because they are extremely poor and can only afford to live in low-lying slums that have no sanitation and therefore have polluted waterways nearby. If these waterways were to flood, they would spread epidemics of waterborne illnesses that could kill millions of people in the city and neighboring areas. *Resilience* refers to a place's ability to bounce back from the disturbances that climate-change impacts create. Mumbai's resilience to sea level rise is bolstered because despite its widespread poverty, it is the wealthiest city in South Asia. This wealth enables it to afford relief and recovery systems that could help it deal with sea level rise over the short and long term.

Over the short term, Mumbai benefits from the fact that it has more and better hospitals and emergency response teams than any other city in South Asia. Over the long term, Mumbai's well-trained municipal planning staff can create and execute plans to help sensitive populations, like people living in slums, adapt to sea level rise. This could be done, for example, through planned relocation to higher ground or by building sea walls and dikes that could keep sea waters out of low-lying slum areas. Of course, Mumbai's overall vulnerability is more complicated than these examples suggest because the city faces many more climate-change impacts than just sea level

Figure 1.14 Carbon dioxide (CO_2) emissions around the world in 2016. [Research from: http://edgar.jrc.ec.europa.eu/overview.php?v=CO2andGHG1970-2016&sort=des8 and http://edgar.jrc.ec.europa.eu/overview.php?v=CO2andGHG1970-2016&dst=CO2pc]

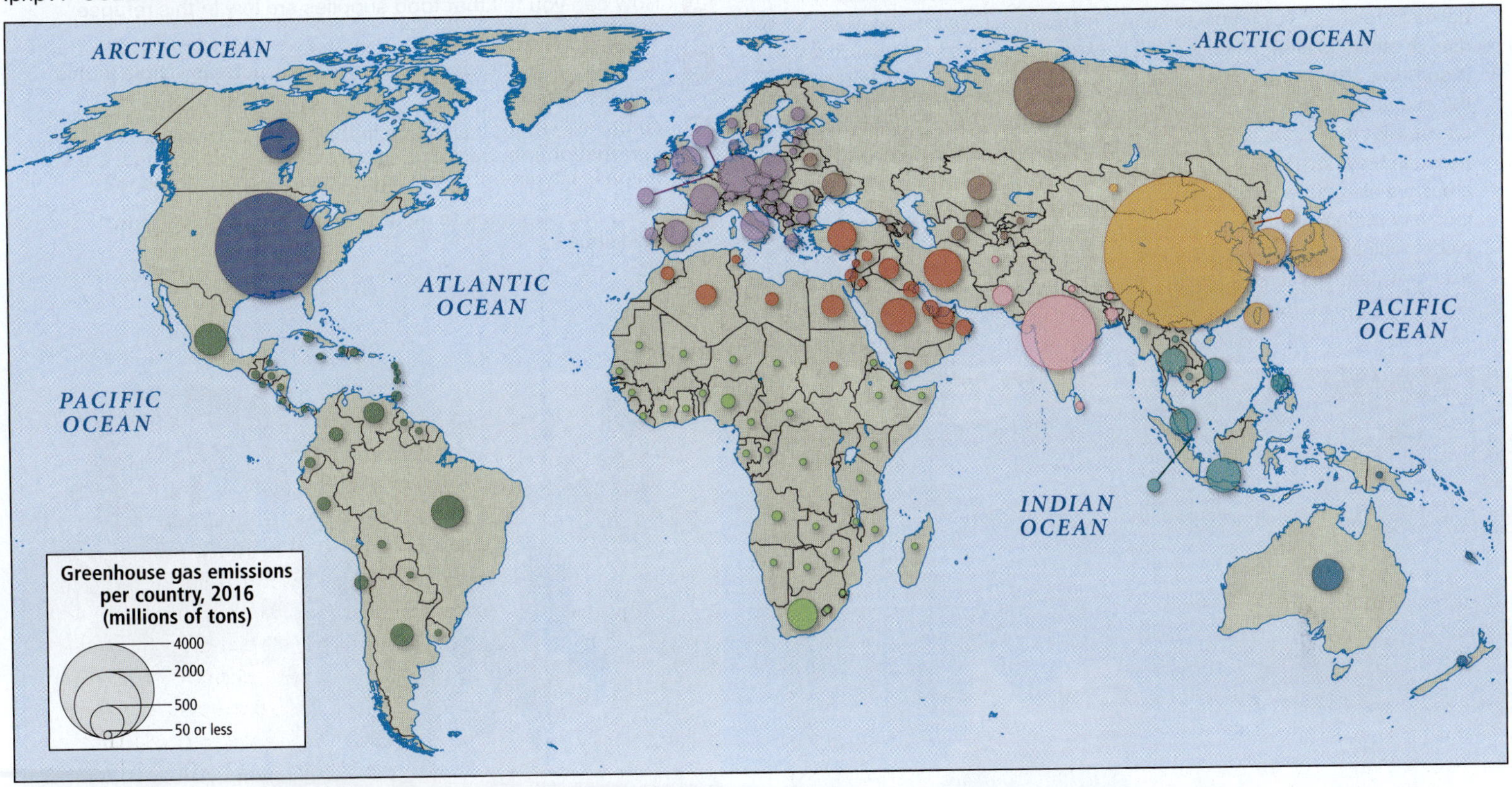

(A) Total emissions per country by millions of tons, 2016. China emits the most tons of CO_2, though the increase in its rate of use has been slowing in recent years. The United States emits the second most, having had a small emissions reduction since 2007. India's CO_2 emissions are increasing, and Russia's have slightly decreased, as have Japan's. These top five countries contribute more than 50 percent of the world's CO_2 emissions.

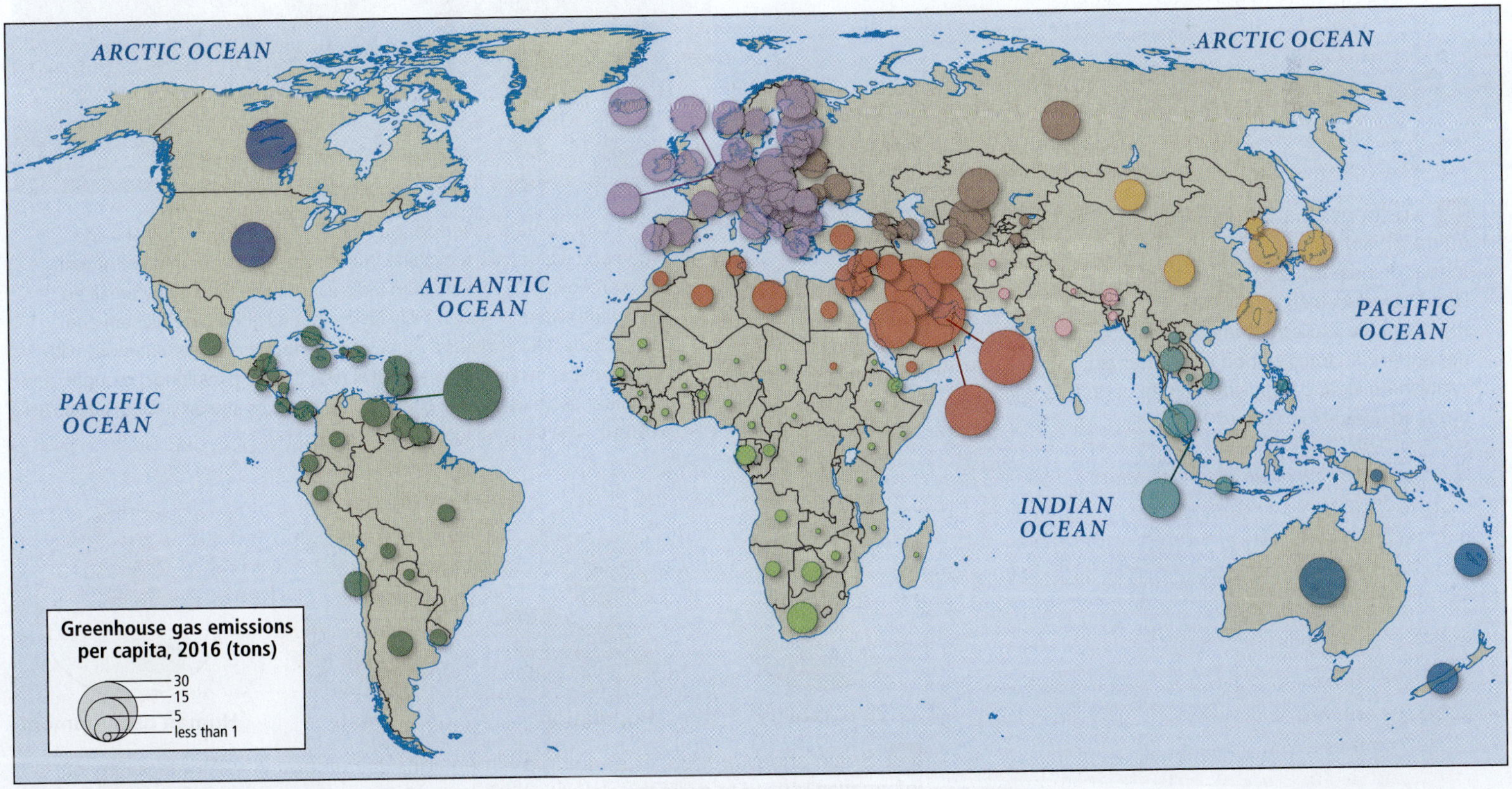

(B) Tons of emissions per capita, 2016. The United States, Canada, Australia, Trinidad and Tobago, and the Gulf states (Bahrain, Kuwait, Qatar, Saudi Arabia, and the United Arab Emirates) have the highest rates of CO_2 emissions per capita. China and India have much lower emissions per capita.

PHOTO ESSAY
1.15 Vulnerability to Climate Change

The map shows overall global patterns of vulnerability to climate change, based on a combination of human and environmental factors. Areas in darkest brown are vulnerable to floods, hurricanes, droughts, sea level rise, or other hazards. When a place is exposed to a hazard to which it is sensitive and for which it has little resilience, it becomes vulnerable. For example, many places are exposed to drought, but generally, those places with the poorest populations are the most sensitive to drought because they have more farmers without access to irrigation water. Poor areas are also the least resilient to drought due to inadequate relief and recovery systems, such as emergency water and food distribution. A place's vulnerability can be thought of as a combination of its exposure, sensitivity, and resilience in the face of multiple climate hazards.

THINKING GEOGRAPHICALLY

A How can you tell that food supplies are low in this refugee camp?

B What sign of an orderly response to disaster is visible in this picture?

C Of the four thumbnail maps in this graphic, which depicts the information that best explains why Spain and Portugal are so much less vulnerable to climate change than Morocco?

E Does this photo relate most to short-term or long-term resilience?

A Northern Uganda and South Sudan: High to extreme vulnerability. The overall situation in these areas is somewhat similar to that of India (B) but for different reasons. Better access to water in Uganda reduces sensitivity to drought, but armed conflict has reduced the country's resilience, with many people living in refugee camps and dependent on foreign food and water aid. Shown here are refugees in northern Uganda, picking up bits of donated grain that has been dropped. [Jean-Marc Giboux/Getty Images News/Getty Images]

B India: Moderate resilience, high vulnerability. Rural Indians line up for food and water after a cyclone (hurricane). Advances in government-led disaster recovery have increased India's resilience, and many areas no longer have the extreme vulnerability levels that neighboring Pakistan and Afghanistan have. Nevertheless, much of India remains very vulnerable, with high exposure and sensitivity to sea level rise, flooding, cyclones, drought, and other disturbances that climate change can create or intensify. [DESHAKALYAN CHOWDHURY/AFP/Getty Images]

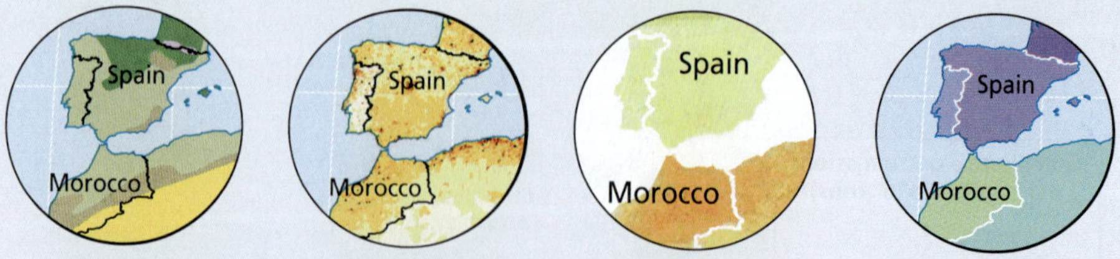

C Spain and Morocco: The multiple dimensions of vulnerability. A wide variety of information is used to make the global map of vulnerability on the next page. For example, the contrast in vulnerability between Spain and Morocco relates to (among other things) differences in population density and distribution, climate, and human development (which is itself based on many factors).

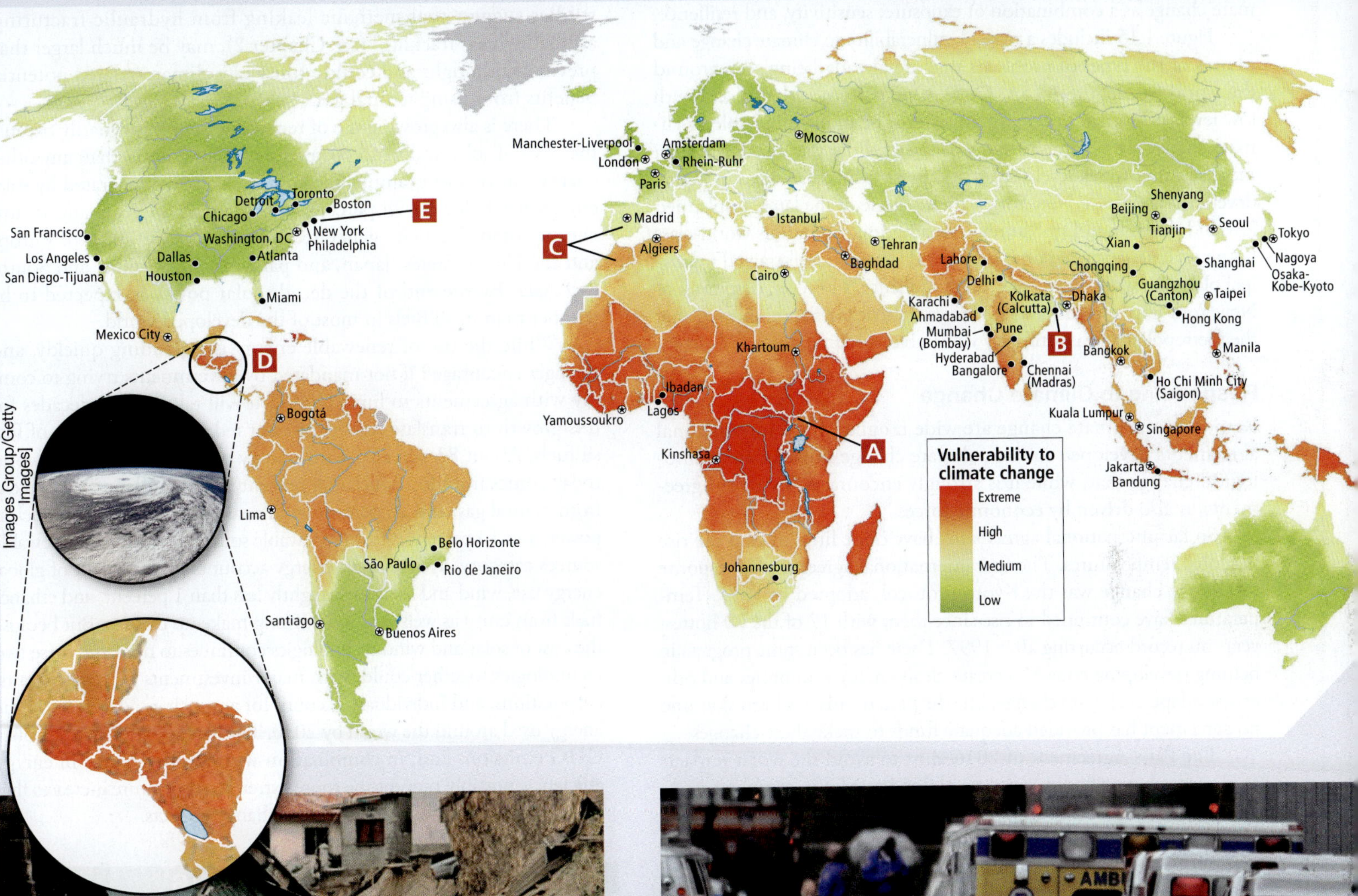

[Universal History Archive/Universal Images Group/Getty Images]

D Honduras: High vulnerability, low resilience. Climatologists predict that hurricanes will increase in intensity as the planet warms, and there has been an increase in powerful storms in recent decades. Poverty and inadequate recovery systems make much of Central America particularly vulnerable to hurricane-related hazards. Shown here is flood damage along the Aguán River in Honduras that was caused by Hurricane Mitch in 1999. More than 9000 people died, making Mitch the second-deadliest hurricane in history. [ORLANDO SIERRA/AFP/Getty Images]

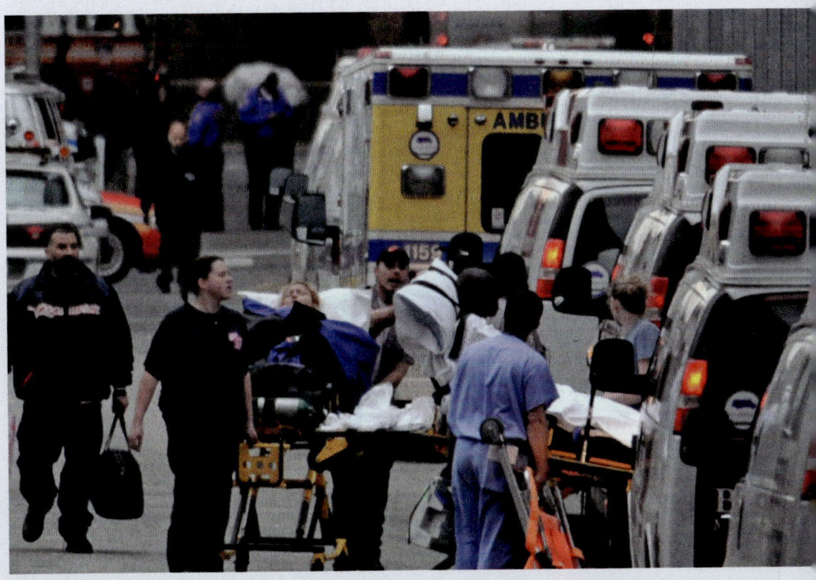

E United States: High resilience, low vulnerability. Effective and well-funded recovery and relief systems give the United States high resilience to climate hazards. This contributes to its generally low vulnerability. Shown here are ambulances in New York City responding to a hurricane. [STAN HONDA/AFP/Getty Images]

rise. However, these examples help us understand vulnerability to climate change as a combination of exposure, sensitivity, and resilience.

Figure 1.15 includes a map of vulnerability to climate change and photos of the types of problems that are already being seen around the world (see Figure 1.15A, C–E). One pattern is that places with low levels of human development tend to be more vulnerable to climate change. For example, many of the qualities that make Mumbai more sensitive to sea level rise are less present in urban areas in highly developed countries. New York City has none of the large, unplanned lowland slums found in Mumbai. In addition, numerous world-class hospitals, emergency response teams, and large and well-trained municipal planning staffs boost New York's resilience. However, despite all of New York City's wealth and resources, Hurricane Sandy showed that the metropolitan area is still very vulnerable to a large storm.

Responding to Climate Change

Responses to climate change are wide ranging, from international agreements developed to limit climate change impacts, to technological change that, while it is strongly encouraged by these agreements, is also driven by economic forces.

So far international agreements have done little to stop the rise in global temperatures. The first international agreement in response to climate change was the Kyoto Protocol, adopted in 1997. Temperatures have continued to rise since then, with 17 of the 20 hottest years on record occurring after 1997. There has been some progress in helping developing countries create clean-energy economies and otherwise adapt to climate change, but the pace of this has been slow and no agreement has provided adequate funds to make these changes.

The Paris Agreement of 2016 aims to avoid the worst impacts of climate change by limiting the global increase in temperature to 2 degrees Celsius (3.6 degrees Fahrenheit) above preindustrial average temperatures by the year 2030. The agreement requires all signatories to develop and implement plans to reduce greenhouse gas emissions, and requires wealthy countries to fund greenhouse gas emissions in poorer countries. One hundred eighty countries representing 90 percent of global emissions signed onto the agreement in 2016, although the Trump administration later moved to withdraw the United States, objecting to perceived negative economic impacts.

Recent years have seen some good news regarding efforts to curtail GHG emissions. To avoid a temperature rise of more than 2°C, global GHG emissions need to be in decline now. In most wealthy countries this has been occurring since 2014, including the 28 countries that make up the European Union, Australia, Canada, and the United States. These countries account for about a quarter of global emissions. In China emissions appear to be leveling off, and in India the growth of emissions is slowing down. Across the globe a "decoupling" of economic growth and GHG emissions is occurring, as countries are increasingly able to grow their economies and raise living standards while GHG emissions have leveled off or grown much more slowly. While global temperatures are still rising, there is reason for optimism that the worst impacts of climate change might be avoided.

These signs are related to changes in energy sources, especially declining use of coal, the fuel currently responsible for most GHG emissions, and increasing use of natural gas, a fossil fuel that is now cheaper than coal and produces fewer GHG emissions. However, cheap natural gas is a result of new extraction methods that can leak methane, a potent GHG, into the atmosphere; result in earthquakes; and cause other environmental problems. Recent studies suggest that methane leaking from hydraulic fracturing, also known as "fracking" (see Chapter 2), may be much larger than previously thought and could nullify any climate-related potential benefits from using natural gas.

There is also growing use of renewable energy, primarily because the costs of solar and wind power are declining faster than any other energy source. For example, the cost of electricity generated by solar energy has fallen by 99 percent over the past 25 years and is now cheaper than electricity generated by fossil fuels in most of Europe and the United States, Japan, and parts of China, India, and Southeast Asia. By the end of the decade solar power is expected to be cheaper than fossil fuels in most of the developed world.

While the use of renewable energy is expanding quickly, and strongly encouraged if not mandated by governments trying to comply with agreements to limit GHGs, it will take several decades for this growth to translate into significant reductions in the use of fossil fuels. About 83 percent of the energy used throughout the world today comes from fossil fuels (34 percent from petroleum, 23 percent from natural gas, and 26 percent from coal); 4 percent is from nuclear power; and 13 percent is from renewable sources. Within the renewable sources category, hydroelectric energy accounts for 7 percent of global energy use, wind and solar each slightly less than 1 percent, and ethanol fuels from crops as well as wood burning make up the rest. But because the cost of solar and wind technologies continues to decrease, these two technologies together could, with major investments by governments, corporations, and individuals, account for as much as 34 percent of the energy used around the world by 2030. This would significantly reduce GHG emissions and, in combination with improvements in energy efficiency, possibly prevent the more extreme temperature increases that would lead to the most severe climate-change impacts.

CHECK YOUR UNDERSTANDING

1. What human activities create large amounts of GHG emissions?
2. What factors influence the vulnerability of a location to climate change?
3. How effective have global agreements to control GHG emissions been?
4. Why is a rapid increase in the use of solar and wind resources likely in coming decades?

WATER

Water is emerging as the major resource issue of the twenty-first century, in part because it is so intimately linked with energy, food, and other agricultural products. Demand for clean water skyrockets as populations grow and more people move out of poverty. This is related in large part to increasing meat consumption. Irrigated agriculture now accounts for 70 percent of all global freshwater use, and 36 percent of the calories produced go to feed animals raised for meat. Meanwhile growing industrial demands for water, larger cities, and more intensive agricultural systems are worsening *water pollution*. With fresh, clean water becoming scarce in so many parts of the world, water is becoming a commodity rather than a free good. In some places water prices have increased so much that poor people's access becomes severely reduced, negatively affecting their health. Water disputes between people, cities, and even nations are proliferating.

Calculating Water Use per Capita

Humans require an average of 5 to 13 gallons (20 to 50 liters) of clean water per day for basic domestic needs: drinking, cooking, and bathing/cleaning. Per capita domestic water consumption tends to increase as incomes rise; the average person in a wealthy country consumes as much as 20 times the amount of water, per capita, as the average person in a very poor country. Much of this extra water use in developed countries goes for nonessential activities such as landscape watering. However, domestic water consumption is only a fraction of a person's actual water consumption. *Virtual water* is the volume of water required to produce, process, and deliver a good or service that a person consumes. To grow an apple and ship it from the orchard to the consumer, for instance, requires many liters of water. When we add an individual's domestic water consumption to her virtual water consumption, we have that person's total *water footprint*. The more one consumes, the larger one's virtual water footprint. **Table 1.1** shows the amounts of water used to produce some commonly consumed products. (As you look at Table 1.1 and read further, note that there are 1000 liters, or 263 gallons, in a cubic meter (m^3).)

Personal water footprints vary widely according to physical geography, standards of living, and agricultural and industrial technology and efficiency. For example, to produce 1 ton of corn in the United States requires 489 m^3 of virtual water, on average, whereas in India the same amount of corn requires 1935 m^3 of virtual water; in Mexico, 1744 m^3; and in the Netherlands, just 408 m^3. To estimate your individual water footprint, you can use the calculators on the Water Footprint Network or *National Geographic* websites.

Water: Who Gets Access to It? Who Owns It?

The availability of water varies greatly across the world, as the map in **Figure 1.16** indicates. This variability is due to natural conditions, human use of water, and population density.

Table 1.1 The global average virtual water content of everyday products*

Product†	Virtual water content (in liters)
1 potato	25
1 cup tea	35
1 kilogram of bread	1608
1 apple	125
1 glass of beer	75
1 glass of wine	120
1 egg	135
1 cup of coffee	140
1 glass of orange juice	170
1 pound of chicken meat	2000
1 hamburger	2400
1 pound of cheese	2500
1 pair of bovine leather shoes	8000

*Virtual water is the volume of water used to produce a product.

†To see the virtual water content of additional products, go to www.waterfootprint.org/?page=files/productgallery [Research from: Arjen Y. Hoekstra and Ashok K. Chapagain, *Globalization of Water—Sharing the Planet's Freshwater Resources* (Malden, MA: Blackwell, 2008), p. 15, Table 2.2]

Though many people consider water a human right that should not cost anything to access, water has become the third most valuable commodity, after oil and electricity. Water in wells, streams, and rivers is increasingly being *privatized*, with its ownership transferred from governments—which can be held accountable for protecting the rights of all citizens to water—to individuals, corporations, and other private entities that manage the water for profit.

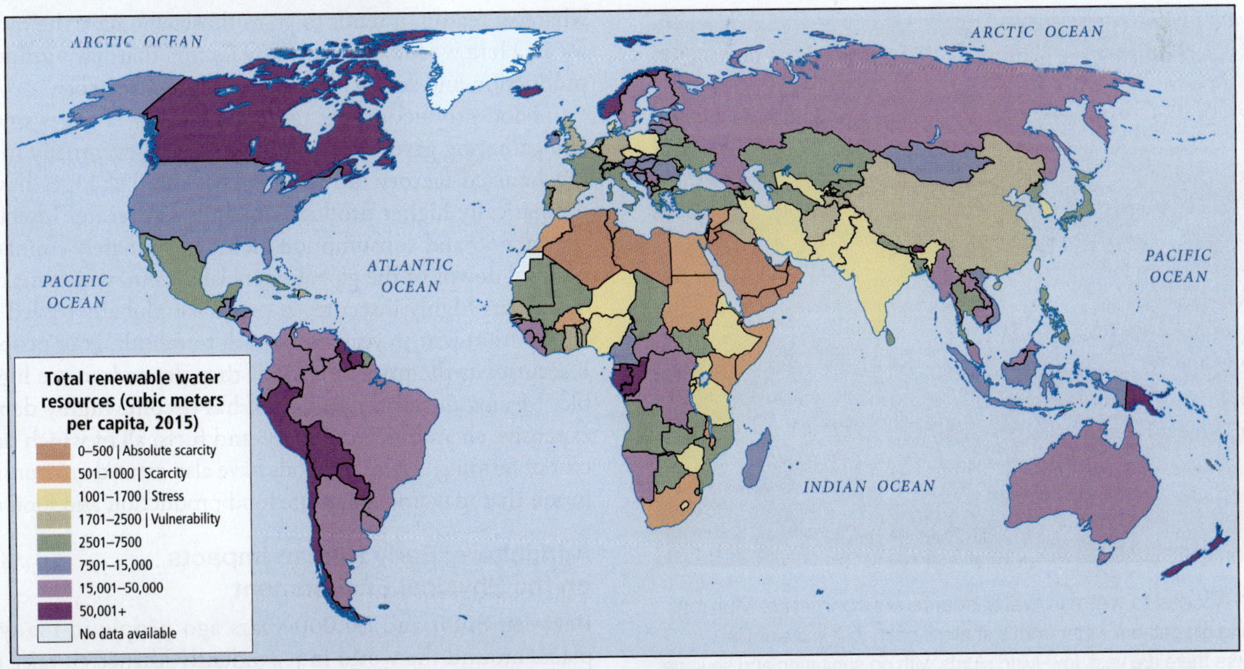

Figure 1.16 Map of national renewable water resources. Access to water varies greatly from place to place. This map shows the total renewable water resources per capita in cubic meters. Purple and blue countries are relatively water secure, while those shown in orange are subject to scarcity. [Research from: WWAP, with data from the FAO AQUASTAT database: http://www.fao.org/nr/water/aquastat/maps/World-Map.TRWR.cap_eng.htm, aggregate data for all countries except Andorra and Serbia]

Governments often privatize water under the rationale that private enterprise will make needed investments that will boost the efficiency of water distribution systems and the overall quality of the water supply. Regardless of whether or not these potential gains are realized, privatization usually raises the cost of water for consumers. This can become quite controversial. For example, in 1999 the city of Cochabamba, Bolivia, sold control of its water to a group of multinational corporations led by the California-based Bechtel, which then raised the price of water beyond what the urban poor could afford. The result was a nationwide series of riots that led finally to the abandonment of privatization. This event inspired the 2008 James Bond movie, *Quantum of Solace*.

Water Quality

About one-sixth of the world's population does not have access to clean drinking water, and dirty water kills more than 6 million people each year. In an average year, more people die this way than in all of the world's armed conflicts. In the poorer parts of cities in the developing world, many people draw water with a pail from a communal spigot, sometimes from shallow wells, or simply from holes dug in the ground (**Figure 1.17**). Usually this water should be boiled before use, even for bathing; but boiling requires fuel, a scarce resource for the poor. These water-quality and access problems help explain why so many people are chronically ill and why 24,000 children under the age of 5 die every day from waterborne diseases.

Water and Urbanization

Urban development patterns dramatically affect the management of water. In most urban slum areas in poor countries, and even in some places in the United States and Canada, crucial water-management technologies such as sewage treatment systems are sometimes entirely absent. Germ-laden human waste from toilets and kitchens may be deposited in urban gutters that drain into creeks, rivers, and bays. In poor countries, installing wastewater collection and treatment systems in cities already housing several million inhabitants is often considered prohibitively costly.

Figure 1.17 Access to water in water-scarce environments: Mumbai, India. A young girl gathers water from a shallow open well in a slum in Mumbai, where there are vast, low-lying slums with no sanitation and sewage that pollutes nearby waterways. If these waterways were to flood, they would spread deadly epidemics of waterborne illnesses via open wells such as the one shown here. What factors may have led this child to endanger her health by drinking this water? [SAJJAD HUSSAIN/AFP/Getty Images]

However, cost-effective and innovative solutions are possible. For example, Windhoek, the capital of Namibia, one of the most arid countries in sub-Saharan Africa, uses semi-purified sewage effluent to water parks, gardens, and sports fields, lowering the need for potable water. Since 1969, the city has been increasing its use of reclaimed water for potable uses, and today recycled sewage water supplies more than 14 percent of its drinking water. It is thus possible for Windhoek, with a population of 322,000, to recharge its aquifer, which helped it to withstand 2 years with no rainfall.

Independent of sewage, water inevitably becomes polluted in cities as parking lots and rooftops replace areas that were once covered with natural vegetation. Rainwater quickly runs off these hard surfaces, collecting in low places and becoming stagnant instead of being absorbed into the ground. In urban slums, flooding can spread polluted water over wide areas, carrying it into homes and into local freshwater sources, as well as to places where children play (see Figure 1.17). Diseases such as malaria and cholera, carried in this water, can spread rapidly as a result. Fortunately, new technologies and urban planning approaches are being developed that can help cities avoid these problems, but they are not yet in widespread use.

FOOD

According to the UN Food and Agriculture Organization (FAO), one-fifth of humanity subsists on a diet too low in total calories and vital nutrients to sustain adequate health and normal physical and mental development (see **Figure 1.18**). As we will see in later chapters, this alarming hunger problem is partly due to political instability, corruption, and inadequate distribution systems rather than simple food production. However, when food is scarce for whatever reason, it tends to go to those who have the money to pay for it. Here we look at the many factors that have influenced food production and scarcity throughout human history.

Food production has undergone many changes since hunting and gathering gave way to agriculture, and eventually to the highly mechanized factory farms of today. These changes have brought dramatically higher productivity, but also greater insecurity. Food production and consumption are now intimately connected to the ups and downs of the global financial system, which makes the lives of farmers highly insecure, as prices for globally traded foods vary widely from year to year and month to month. Poor people also face insecurity as the prices for foods they depend on are highly unstable. Meanwhile food production has become highly dependent on expensive chemicals, machinery, and fuels, all of which drive up the cost of farming. These methods have also created environmental pollution that may impede future food production and food security.

Agriculture: Early Human Impacts on the Physical Environment

Between 8000 and 20,000 years ago people in many different places around the world independently learned to raise plants and animals for food and other uses through selective breeding. This eventually became the labor-intensive, small-scale "subsistence" agriculture still practiced in some areas today, which is aimed at producing enough for a single family unit. Over time farmers were

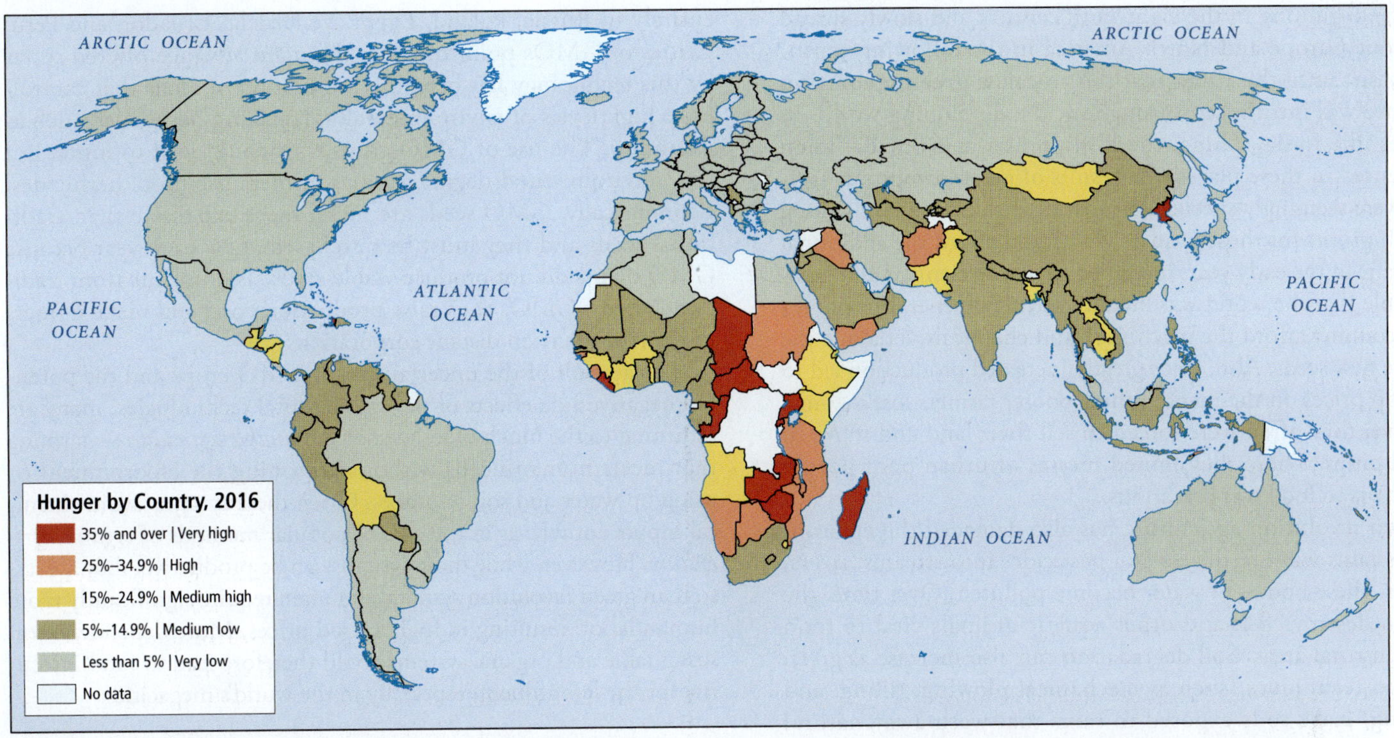

Figure 1.18 Global map of undernourishment. This Food and Agriculture Organization of the United Nations (FAO) map is based on 2017 data. The proportion of people suffering from undernourishment—the lack of adequate nutrition to meet their daily needs—has been declining in the developing world but remains a problem in much of sub-Saharan Africa; parts of South, Southeast, and East Asia; and in Central America, Bolivia in South America, and Haiti in the Caribbean. [Research from: Food and Agriculture Organization of the United Nations, "FAO state of food security," at http://www.fao.org/state-of-food-security-nutrition/en]

able to produce enough to feed more than just themselves, making possible larger populations, some of which developed more specialized occupations and began to live in towns and cities.

While agriculture solved some problems, it created others. For example, the transition from hunting and gathering to agriculture often brought a decline in the nutritional quality of human diets. As people stopped eating diverse wild food species in favor of one or two species of domesticated plants and animals, they became, on average, shorter and more malnourished. Land clearing for agriculture also increased vulnerability to drought and other natural disasters that could wipe out entire harvests. Thus, as ever-larger populations depended solely on agriculture, episodic famine became more common and affected more people.

Agriculture made the growth of cities possible, which led to cultural and technical advances that created many amazing civilizations. Dense urban settlements also facilitated the spread of disease, the development of social hierarchies where the powerful oppressed the weak, more gender-specific divisions of labor, the monopolization of wealth by a small number of people, and the raising of armies for large-scale warfare.

Modern Food Production and Food Security

The **Industrial Revolution**, which took place from about 1750 to 1850, led directly to the development of mechanized, chemically intensive commercial agriculture. Modern processes of food production, distribution, and consumption have greatly increased the supply and, to some extent, the security of food systems. However, this has come at the cost of soil degradation and environmental pollution that may impede future food production and food security.

Over the past five centuries of global interaction and trade, people have become ever more removed from their sources of food. Today, occupational specialization means that most food is mass-produced. Far fewer people work in agriculture than did in the past, and most people know little about cultivation and animal tending. They work for cash to buy food and other necessities.

One aspect of this dependence on money is that the *food security*—the ability of a society to consistently supply a sufficient amount of basic food to the entire population—can be threatened by economic disruptions, even in distant places. As countries become more involved with the global economy, they import more food, and their own food production becomes vulnerable to price changes in global markets. For example, a crisis in food security occurred in 2007–2008 and again in 2011–2013 when the world price of corn increased dramatically. Financial speculators responded to U.S. policies encouraging the making of ethanol (a substitute for gasoline) out of corn, investing heavily in it. As a result, global corn prices rose beyond the reach of many poor people, leading to widespread protests and political instability throughout the developing world. Some household economies were so ruined that parents sold important assets to feed their children or went without food themselves. Figure 1.18 identifies countries in which undernourishment remains an ongoing problem and periodic food insecurity is especially intense.

Another way that modernized agriculture influences food security is through its reliance on machines, chemical fertilizers, pesticides, and specially bred seeds. The shift to this kind of agriculture happened first

Industrial Revolution a series of innovations and ideas that occurred broadly between 1750 and 1850, which changed the way goods were manufactured in Europe beginning in the eighteenth century and slowly spread throughout Europe and North America in the following centuries. It wasn't until the 1960s that this so-called **green revolution** agriculture was introduced throughout the developing world, in countries like India, China, the Philippines, and Brazil. When implemented in these places, the results of green revolution agriculture were seemingly spectacular: soaring production levels along with high profits for those farmers who could afford the additional investment. In the early years it seemed to many scientists and business people that the world was literally getting greener. But poorer farmers couldn't afford the machinery and chemicals, let alone the expensive new seeds. Also, since greatly increased production led to lower crop prices on the market, these poorer farmers lost income. To survive, they often were forced to sell their land and move to crowded cities. There they joined masses of urban poor people whose access to food was precarious.

Green revolution agriculture has also damaged the environment. As rains wash fertilizers and pesticides into streams, rivers, and lakes, these bodies of water become polluted. Over time, the pollution destroys fish and other aquatic animals vital to food security in rural areas. Soil degradation can also increase as green revolution techniques (such as mechanical plowing, tilling, and harvesting) leave soils exposed to rains that wash away natural nutrients and the soil itself. Indeed, many of the most agriculturally productive parts of North America, Europe, and Asia have already suffered moderate to serious loss of soil through erosion. Globally, soil erosion and other problems related to food production affect about 7 million square miles (2000 million hectares), putting at risk the livelihoods of a billion people.

At least in the short term, green revolution agriculture raised *carrying capacity*, the maximum number of people that could be supported on a given piece of land. However, it is unclear how sustainable these green revolution gains in food production are. Scientists from many disciplines estimate that within the next 50 years, environmental problems such as water scarcity and global climate change will limit increases in food production or even reverse them.

The increase in food also contributed to a population boom by enabling more people to survive to reproductive age. In the 25 years between 1965, when the green revolution was implemented, and 1990, total global food production rose between 70 and 135 percent (varying from region to region). The world's population rose about 70 percent during the same time period. It is now estimated that to feed the population projected for 2050, global food output must increase by another 70 percent.

The technological advances that could make current agricultural systems more productive are increasingly controversial. *Genetic modification (GMO)*, the practice of splicing together the genes from widely divergent species to create particular characteristics, is being used to boost productivity. GMOs were first developed in the United States, where they are now in widespread use; however, elsewhere many worry about the side effects of GMOs, fearing ecological damage and catastrophic crop failures. Foods containing GMOs must be labeled in much of Europe, South America, Asia, and Oceania, and are banned entirely in Russia, Poland, Egypt, Venezuela, Ecuador, and Peru. Critics of GMOs point out that the main advance offered so far by this technology has been the production of seeds that can tolerate high levels of environmentally damaging herbicides, such as Roundup. The use of GMO crops thus could lead to more, not less, environmental degradation as farmers use more herbicides. Economically, GMO seeds are much more expensive than traditional seeds and they must be purchased anew each year because GMO plants do not produce viable seeds, as do plants from traditional seeds. GMOs thus raise production costs and make farmers more dependent on distant corporations.

green revolution increases in food production brought about through the use of new seeds, mechanized equipment, fertilizers, pesticides, and herbicides

As a result of the uncertainties of GMO crops and the potential negative side effects of new agricultural technologies, many are returning to the much older idea of *sustainable agriculture*—farming that meets human needs without poisoning the environment or using up water and soil resources. Often these systems avoid chemical inputs entirely, as in the case of popular methods of *organic agriculture*. However, while these systems can be productive, they are less so than green revolution systems and often require significantly more human labor, resulting in higher food prices. More dependence on sustainable and organic systems could therefore lead to food insecurity for the less affluent, especially in the world's megacities.

CHECK YOUR UNDERSTANDING

1. When did people learn to use selective breeding to develop plants and animals for food and other agricultural products?
2. What changes in agriculture led to increases in the supply of food and other agricultural products, beginning in the eighteenth century?
3. What aspects of modern agriculture reduce food security and contribute to environmental damage?
4. What has happened to the many farmers unable to afford the chemicals and machinery required for modern commercial agriculture?
5. What is sustainable agriculture?

GLOBALIZATION AND DEVELOPMENT

1.5 Understand the transitions of development.

1.6 Analyze how global flows of information, goods, and people are transforming patterns of economic development.

Currently over 70 percent of humanity lives on less than $27 a day, or $10,000 a year, an income well below the U.S. poverty line. Over 10 percent of the world lives in extreme poverty, living on less than $1.90 a day, and over 22,000 children each day die from malnutrition and poverty-related diseases. That's the bad news. The good news is that poverty is decreasing. In the past 30 years the number of people living in extreme poverty has been cut in half, from 1.85 billion to 760 million people, and the number of children who die from poverty-related illnesses has also been cut in half. These changes are due to a wide range of transitions that are together called *development*.

THE TRANSITIONS OF DEVELOPMENT

Development is a complex transition in how people make a living, characterized in part by a shift from economies based on extractive resources to those based on human resources.

Extractive resources are those that must be mined from Earth's surface (mineral ores) or grown from its soil (timber and plants). *Human resources*, such as skills and brainpower, are used to transform extractive resources into useful products (such as refrigerators or bread) or bodies of knowledge (such as books or computer software). Development can also be described as a transition through dependence on three different economic sectors: the **primary sector** is based on *extractive* economic activities such as mining, fishing, forestry, and agriculture; the **secondary sector** is based on *industrial* economic activities such as processing, manufacturing, and construction that require more skilled human resources; and the **tertiary sector** is based on *service-oriented* economic activities based almost entirely on human skills, such as education, health care, transport, tourism, and finance. Recently a fourth, the **quaternary sector**, has been broken off from the tertiary sector and includes knowledge-based pursuits such as information technology (IT) and research and development. Generally speaking, as people in a society shift from extractive activities, such as farming and mining, to industrial and service activities, both human resources and material standards of living rise.

CLASSIFICATIONS OF DEVELOPMENT

Countries and regions are often classified according to their level of development. Poorer countries are often referred to as **less developed countries**, or **LDCs**, where the primary sector (extraction) tends to dominate and a large share of people work in agriculture. Most of sub-Saharan Africa (see Chapter 7) is made up of LDCs. Some LDCs have shifted to more employment in the secondary sector (industry), which is higher-wage but still labor-intensive, as well as higher-wage tertiary sector (services) jobs, which include many jobs in IT. These former LDCs, which include countries like China, Brazil, and Mexico, are often called **middle income countries**, or **MICs**, where living standards are higher and social services, such as health care and education, are much improved. The richest countries, often referred to as *developed* or **more developed countries (MDCs)**, are now lessening their dependence on labor-intensive manufacturing and shifting toward more highly skilled mechanized production or knowledge-based service and technology (quaternary) industries. Generally, MDCs provide adequate education, health care, and other social services to help their people contribute to economic development. These classifications describe the dominant activities in countries, and as such they hide much complexity. For example, there are often a few extremely wealthy people in LDCs whose living standards and access to health care and education are better than what most people have in MICs or MDCs. Similarly there are many poor people in MDCs whose living standards and access to health care and education are more characteristic of MICs.

MEASURING DEVELOPMENT

The most long-standing measures of development are economic, but in recent decades more comprehensive measures of development have come into use.

Measures of Economic Development

Gross domestic product (GDP) per capita is an economic measure of development that refers to the total market value of all goods and services produced in a country in a given year. A closely related index that is now used more often by international agencies is **gross national income (GNI) per capita**, the sum of a nation's gross domestic product plus net income received from overseas. When GDP or GNI is divided by the number of people in the country, the result is per capita GDP or GNI. This book now uses primarily the GNI per capita statistics.

GNI data are usually corrected to account for variations in the purchasing power of currency around the globe. A GNI of U.S.$18,000 per capita in Barbados might represent a middle-class standard of living, whereas that same amount in New York City could not buy even basic food and shelter. Because of these purchasing power variations, in this book GNI (and occasionally GDP) per capita figures have been adjusted to **purchasing power parity (PPP)**. PPP, usually indicated in parentheses after GDP or GNI, is the amount that the local currency equivalent of U.S. dollars purchases in a given country. The *Economist* magazine does an annual survey of the purchasing power parity of different currencies, using a McDonald's Big Mac because it is a standardized product for sale all over the world. In July 2018, a Big Mac at McDonald's in the United States cost U.S.$5.51, while in India it costs U.S.$2.51. Of course, for the consumer in India, where annual per capita GNI (PPP) is U.S.$7060, this would be a rather expensive meal. On the other hand, at $5.51, the Big Mac would be an economy meal in the United States, where the GNI per capita (PPP) is U.S.$60,200 (see "Big Mac Index" on the *Economist's* website). **Figure 1.19** shows a global map of GNI per capita PPP.

development a complex transition in how people make a living, characterized in part by a shift from economies based on extractive resources to those based on human resources

primary sector extractive economic activity such as mining, forestry, and agriculture

secondary sector industrial economic activity such as processing, manufacturing, and construction

tertiary sector service-based economic activity such as transportation, education, health care, tourism, and financial services

quaternary sector knowledge-based economic activities such as information technology and research and development

less developed countries (LDCs) countries that have labor-intensive and low-wage, often agricultural, economies

middle income countries (MICs) countries that have shifted to more employment in the secondary sector (industry) and higher-wage tertiary sector (services)

more developed countries (MDCs) countries that have economies often generated by tertiary and quaternary sectors and that generally provide adequate education, health care, and other social services to help their people contribute to economic development

gross domestic product (GDP) per capita the total market value of all goods and services produced within a particular country's borders and within a given year, divided by the number of people in the country

gross national income (GNI) per capita the sum of a country's gross domestic product plus all net income received from overseas, divided by the midyear population

purchasing power parity (PPP) the amount that the local currency equivalent of U.S.$1 purchases in a given country

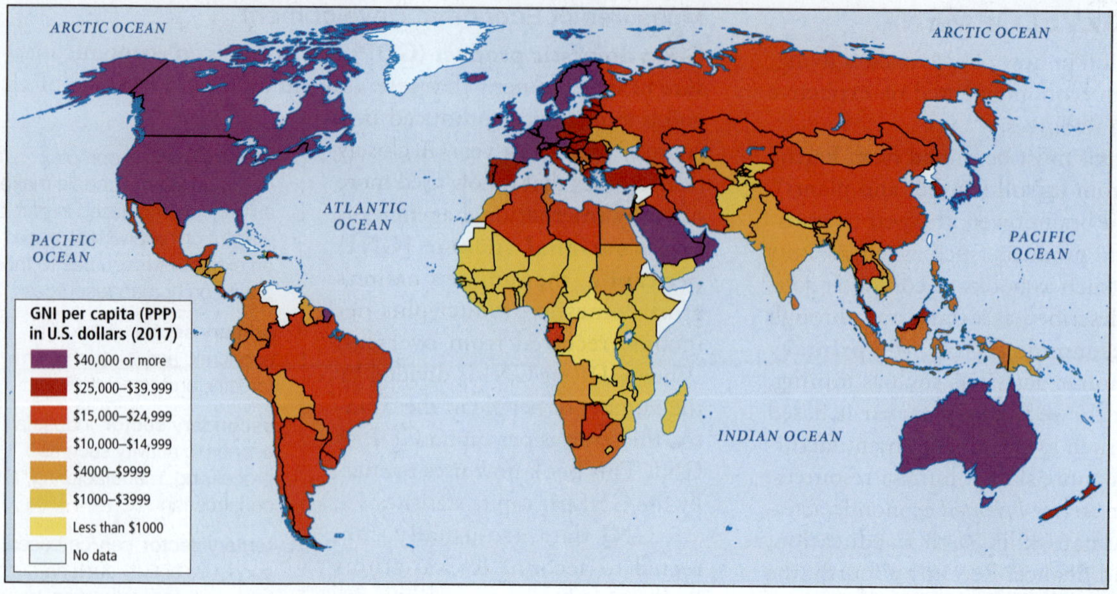

FIGURE 1.19 Gross national income (GNI) per capita, adjusted for purchasing power parity (PPP).

GNI per capita (PPP) is often used as a measure of how well people are living, but it has many disadvantages. First is its inability to describe wealth distribution. Because GNI per capita is an average, it can hide the fact that a country can have a few fabulously rich people and a great mass of abjectly poor people. Another disadvantage of GDP (PPP) and GNI (PPP) per capita is that neither takes into consideration whether these levels of income are achieved at the expense of environmental sustainability, human well-being, or human rights.

Measures of Human Development

The shortcomings of purely economic measures of development led to broader concepts of *human development.* This term generally means a healthy and socially rewarding standard of living in an environment that is safe and sustainable. The **United Nations Human Development Index (HDI)** calculates a country's level of development with a combination of statistical indicators for per capita income (PPP), life expectancy, and education (see **Figure 1.20**). These indicators capture a much broader picture of development than GNI per capita PPP because they include information about health care and education, not just income.

A disadvantage of statistical indicators, including HDI, is that they measure only what goes on in the *formal economy*—all the activities that are officially recorded as part of a country's production. Many goods and services are produced outside formal markets, in the **informal economy**. Here, work is often bartered for food, housing, or services, or for cash payments made "off the books"—payments that are not reported to the government as taxable income. It is estimated that one-third or more of the world's work takes place in the informal economy. Examples of workers in this category are those who earn a living from bartered or "off-the-books" services such as housework, gardening, herding, animal care, or elder and child care.

There is a gender aspect to informal economies. Researchers studying all types of societies and cultures have shown that, on average, women perform about 60 percent of all the work done, and that much of this work is unpaid and in the informal economy. Yet only the work women are paid for in the formal economy appears in the statistics, so economic figures per capita ignore much of the work women do. Statistics also neglect the contributions of millions of men and children who work in the informal economy as street vendors, craftspeople, traders, service workers, or seasonal laborers.

SUSTAINABLE DEVELOPMENT AND POLITICAL ECOLOGY

The UN defines **sustainable development** as improving current living standards in ways that will not jeopardize those of future generations. Without sustainable development strategies, efforts to improve living standards for those who need it most will increasingly be foiled by degraded or scarce resources. Many geographers who focus on these issues use concepts from **political ecology**, the study of the interactions among development, politics, human well-being, and the environment. Political ecologists examine how the power relationships in a society affect the ways in which development proceeds. "Who is actually benefiting from so-called development projects?" is a common question asked. For instance, in Malaysia, a country in Southeast Asia, the clearing of forests to grow oil palm trees might at first seem to benefit many people. It would create some jobs, earn profits for the growers, and raise tax revenues for the government through the sale of palm oil, an important and widely used edible oil and industrial lubricant.

United Nations Human Development Index (HDI) calculates a country's level of development with a combination of statistical indicators for per capita income (PPP), life expectancy, and education

informal economy all aspects of the economy that take place outside official channels or "off the books"

sustainable development the improvement of current standards of living in ways that will not jeopardize those of future generations

political ecology the study of the interactions among development, politics, human well-being, and the environment

However, these gains must be balanced against the loss of highly biodiverse tropical forest ecosystems and the human cultures that depend on them. Not only are forest dwellers losing their lands and means of livelihood to oil palm plantations; valuable knowledge that could be used to develop more sustainable uses of forest ecosystems is being lost as forest dwellers are forced to migrate to crowded cities, where their woodland skills are useless and therefore soon forgotten.

Political ecologists are raising awareness that development should be measured by the improvements brought to overall human well-being and long-term environmental quality, not just by the income created. By these standards, converting forests to oil palm plantations might appear less attractive, since only a few benefit from it and it carries a cost of widespread and often irreversible ecological and social disruption.

CHECK YOUR UNDERSTANDING

1. What does the term *development* refer to?
2. What are some ways development is measured? What are their advantages and disadvantages?
3. What are some social services that support development?
4. Why is the informal economy important?
5. What kinds of questions do political ecologists ask about development projects?

GLOBALIZATION

The term **globalization** refers to the worldwide changes brought about by flows of many things: money, resources, products, people, information, and ideas, just to name a few. For example, a person's attitudes and values may be modified because of being in contact with people and ideas from foreign cultures. Political activities are becoming more global in scope as international agreements are necessary to manage an interconnected world. And the causes and impacts of climate change illustrate how local behavior can have implications on a global scale. Globalization is among the most complex and far-reaching of the concepts described in this book.

Across the world, local self-sufficiency is giving way to global interdependence and international trade. While for a time there was anticipation that globalization would lead to more prosperity for all, the economic recession that began in 2007 cast a spotlight on the unpredictable and destabilizing effects of global interdependence. In that year, poorly regulated banks made bad mortgage loans to new homeowners, resulting in foreclosures and the failure of some U.S. and European banks that had invested in U.S. mortgages. Many people in the United States and Europe saw their savings wiped out, lost their homes, faced pay cuts, and lost jobs; many bought less and cancelled vacations, and as a result, people across the world also lost their jobs or saw their incomes decline.

What Is the Global Economy?

The *global economy* is the worldwide system in which goods, services, and labor are exchanged. Most of us participate in the global economy every day. For example, books like this can be made from trees cut down in Southeast Asia or Siberia and shipped to a paper mill in Oregon. Many books are now printed in Asia because labor costs are lower there. Globalization is not new. At least 2500 years ago, silk and other goods were traded along the Central Asian Silk Road that connected Greece and then Rome in the Mediterranean with distant China, an expanse of more than 4000 miles (6437 km).

European **colonialism** was an early undertaking related to globalization. Starting in about 1500 C.E., European countries began conquering distant parts of the world and extracting their resources. The colonizers organized systems to process those resources into higher-value goods to be traded wherever there was a market. Sugarcane, for example, was grown on Caribbean and Brazilian plantations with slave labor from Africa (**Figure 1.21**) and made locally into crude sugar, molasses, and rum. It was then shipped to Europe and North America, where it was further refined and sold at considerable profit. The global economy grew as each region produced goods for export rather than just for local consumption. At the same time, regions also became more dependent on imported food, clothing, machinery, energy, and knowledge.

The new wealth derived from the resources of the colonies helped finance Europe's Industrial Revolution. European colonizers also integrated the economies of their colonial possessions in the Americas, Africa, and Asia with their own. For example, in the British Caribbean colonies, hundreds of thousands of enslaved Africans wore rough garments made of cheap cloth woven in England from cotton grown in British India. The sugar that slaves produced on British-owned plantations with iron equipment from British foundries was transported to European markets in ships made in the British Isles of trees and resources from the North American colonies and other parts of the world.

Until the early twentieth century, much of the activity of the global economy took place within the colonial empires of a few European nations (Britain, France, the Netherlands, Spain, and Portugal). By the 1960s, global economic and political changes brought an end to these empires, and now almost all colonial territories are independent countries.

Nevertheless, the global economy persists. Banks and *multinational corporations*—such as Shell, Chevron, IBM, Walmart, Coca-Cola, Bechtel, Apple, British Petroleum, Toyota, Google, and Cisco—operate across international borders. These corporations extract resources (including intellectual properties, such as knowledge of medicinal plants) from many places, make products in factories located to take advantage of cheap labor and transportation facilities, and market their products wherever they can make the most profit. Their global influence, wealth, and importance to local economies enable the multinational companies to powerfully influence politics in the countries in which they operate.

globalization the growth of worldwide linkages and the changes these linkages are bringing about in daily life, especially in economies, but also in culture and biology

colonialism a period starting in about 1500 C.E., when European countries began conquering distant parts of the world and extracting their resources

PHOTO ESSAY

1.20 Globalization and Development

Transitions away from employment in labor-intensive agriculture to better-paying manufacturing and services sector jobs are increasing funds available for human development. Some countries have emerged as leaders in overall human development, while others lead in particular areas, and some countries lag far behind. The HDI map shows that in the most populous parts of the world, China and India, human development is now at high and medium HDI, respectively. This is a dramatic shift reflecting decades of strong economic growth, driven in part by globalization, and governments willing to invest in schools, health care, and other infrastructure that supports human development. The map also shows that sub-Saharan Africa has the lowest HDI, and Europe, Japan, Australia, and New Zealand the highest. In the Americas, Canada's HDI is higher than that of the United States, and Argentina, Chile, and Uruguay have fairly high HDI, while the rest of the countries have middle to low HDI.

THINKING GEOGRAPHICALLY

A What about this photo suggests rising incomes in China?

B What details in this photo suggest high human development in Norway?

C What suggests that the migrant workers in this photo have low incomes?

E What about this photo suggests that population growth is high in Niger?

F What suggests that Finland has a different educational philosophy than does the United States?

G What about this photo suggests low levels of spending on health care in India?

A China: High human development. Migrants to Guangdong, China, head home for the holidays on motorcycles bearing gifts. China's GNI PPP per capita has grown dramatically since the country opened up its economy to foreign markets and investors in the 1980s, with a boom in manufacturing quadrupling average incomes between 2006 and 2016. Life expectancy is now near U.S. levels of 20 years ago, but education levels remain relatively moderate. [VCG/VCG via Getty Images]

B Norway: Investing in human development. A control center on an oil platform in Norway, which has ranked at or near the top of the HDI for decades. While its wealthy economy is based largely on oil extraction, this is a state-run industry and much of the profits have been spread throughout the population with heavy investments in health care, education, and other social services. [Ulrich Baumgarten/Getty Images]

C Qatar: High inequality. Flush with revenues from huge natural gas and oil reserves, Qatar ranks at or near the top of all countries for GNI per capita PPP. However, it ranks significantly lower for HDI due in part to the meager incomes, low educational attainments, and poor health care provided to its largely migrant workforce. [Sam Tarling/Corbis via Getty Images]

D Japan: Habits, health care, and life expectancy. Elderly people work out in Japan in celebration of "Respect for the Aged Day." Japan has the world's highest life expectancy due to a diet largely free of processed foods and sugar and a mainly state-run health-care system that spends less per capita than most industrialized countries and less than half what the United States spends. [YOSHIKAZU TSUNO/AFP/Getty Images]

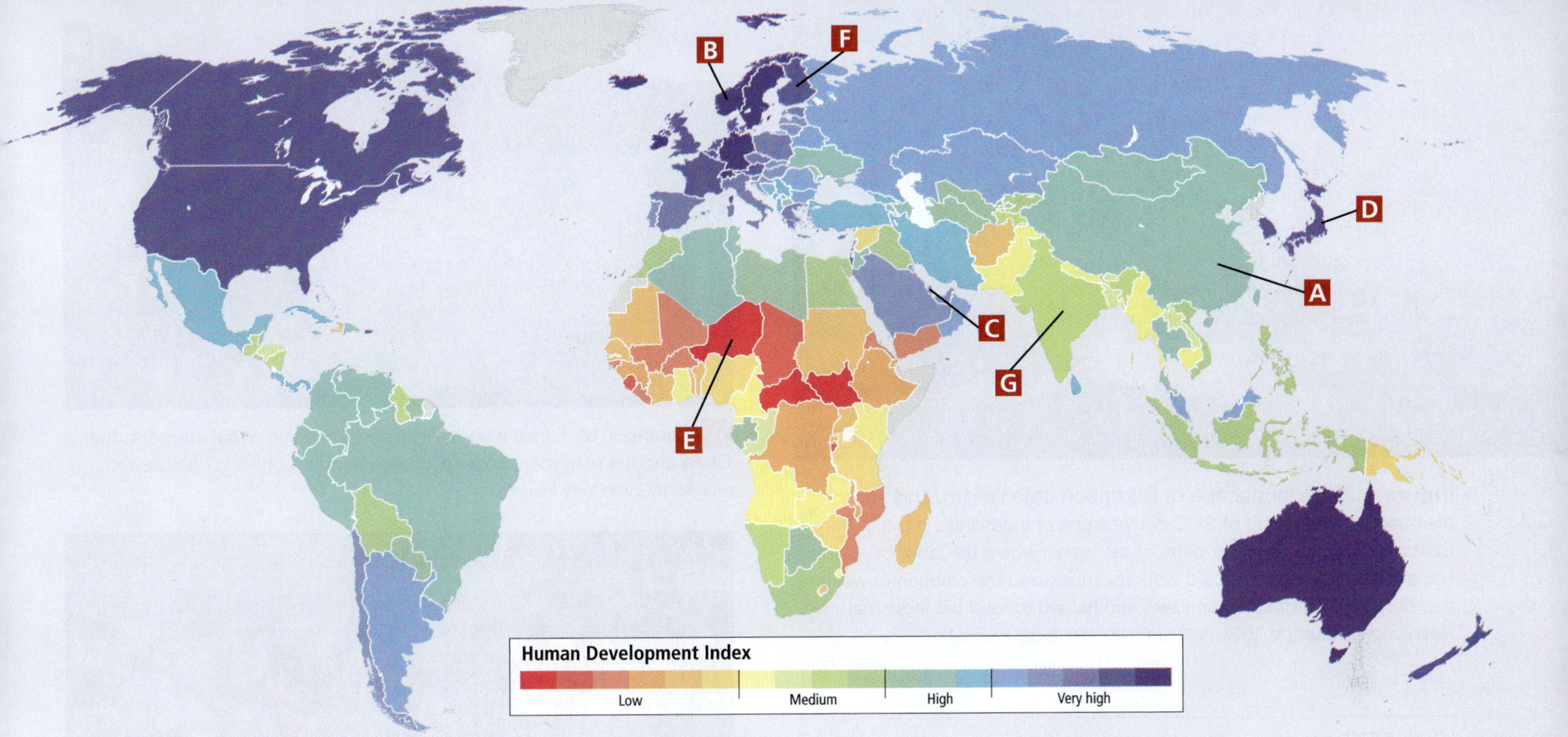

E Niger: Low human development. A woman processes grain in Niger, which consistently ranks near the bottom of the HDI. Over 80 percent of Niger's population are rural farmers, with the economy providing a GNI per capita PPP that is one-sixtieth of Norway's. An underdeveloped health-care system provides a life expectancy of 61 years, equivalent to what India achieved more than 2 decades ago. The adult literacy rate is around 20 percent. [Jim Richardson/National Geographic/Getty Images]

F Finland: Education. Finland achieves among the highest educational attainments of any country, despite rarely assigning homework, standardized tests, or even grades before eighth grade; requiring less time in school; and spending less money per pupil than similarly ranked countries. Finland's highly educated workforce supports globally competitive manufacturing and high-tech sectors. [OLIVIER MORIN/AFP/Getty Images]

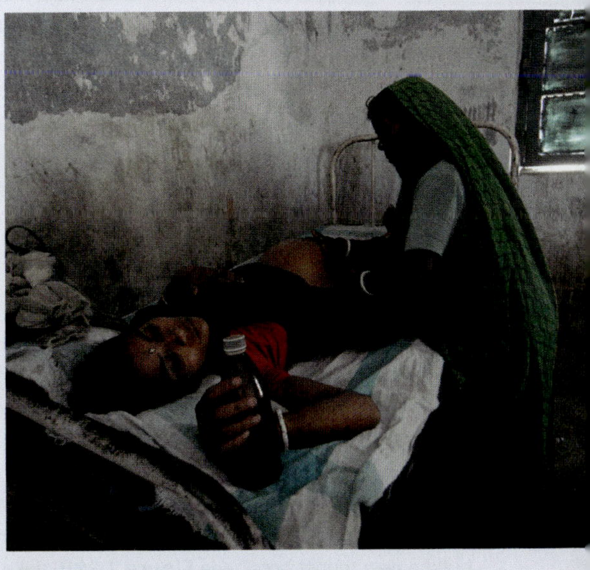

G India: Medium human development. A pregnant woman receives a massage at a rural clinic in Bihar, India. 60 percent of Indians live on less than $3.20 a day, and most work in agriculture in rural areas. India has some of the lowest per capita spending on health care in the world, with which it achieves a life expectancy of 69 years, just slightly lower than what the United States had about 60 years ago. [Priyanka ParAshar/Mint via Getty Images]

Figure 1.21 A landscape of European colonialism. This sugar plantation on the island of St. Croix was one of thousands in the Caribbean (and many other parts of the world) where the labor of enslaved people and low-paid workers subsidized the creation of wealth that benefited European colonizers and helped to fund the Industrial Revolution in Europe. [DEA PICTURE LIBRARY/Getty Images]

Figure 1.22 Workers in the global economy.

A Soufrière, St. Lucia, where Olivia and Anna live. What suggests that Olivia and her neighbors share a similar standard of living? [Education Images/UIG via Getty Images]

B Setiya, along with other migrant workers, boards a bus that will take him to a construction site in Malacca, Malaysia. What about this photo suggests that Setiya is a low-paid worker? [SAEED KHAN/AFP/Getty Images]

C The trailer at the back of Tanya's lot, where her daughter Reyna lives. How does this photo suggest that Tanya and Reyna have low incomes? [Yalonda M. James/Charlotte Observer/MCT via Getty Images]

VIGNETTE Sixty-year-old Olivia lives near Soufrière on St. Lucia, an island in the eastern Caribbean (**Figure 1.22A**). She, her daughter Anna, and her three grandchildren live in a wooden house surrounded by a leafy green garden dotted with fruit trees. Anna has a tiny shop at the side of the house, from which she sells various small everyday items and preserves that she and her mother make from the garden fruits.

On days when the cruise ships dock, Olivia heads to the market shed on the beach with a basket of homegrown goods. She calls out to the passengers as they near the shore, offering her spices and snacks for sale. In a good week she makes U.S.$70. Her daughter makes about U.S.$100 per week in the shop and is constantly looking for other ways to earn a few dollars.

Olivia and Anna support their family of five on about U.S.$170 a week (U.S.$8840 per year). From this income they take care of their expenses, including school fees for the granddaughter who will go to high school in the capital next year and perhaps college if she succeeds. Their livelihood puts them at or above the standard of living of most of their neighbors.

On the other side of the world, in Malacca, Malaysia, 30-year-old Setiya, an undocumented immigrant from Tegal, Indonesia, is helping build a new tourist hotel (see **Figure 1.22B**). Like a million other Indonesians attracted by the booming economy, he snuck into Malaysia, risking arrest, because in Malaysia average wages are four times higher than at home.

This is Setiya's second trip to Malaysia. His first trip was to work legally on a Malaysian oil palm plantation. Upon arrival, however, Setiya found that he would have to work for 3 months just to pay off his boat fare from Indonesia. Not one to give in easily, he quietly went to another city and found a construction job earning U.S.$11 a day (about U.S.$2600 a year), which allowed him to send money to his family in Indonesia.

Periodically, Malaysia deports foreign workers as local authorities respond to political pressure from locals who complain that wages are being driven down by so many people willing to work for so little money. In January 2018, Setiya was deported back to Indonesia, but after a few months he was once again on a boat to enter Malaysia illegally in order to support his family.

In the United States, 50-year-old Tanya works at a fast-food restaurant outside Charleston, South Carolina, making less than U.S.$7.50 an hour. She had been earning U.S.$8.00 an hour sewing shirts at a textile plant until it closed and moved to Indonesia. Her husband is a delivery truck driver for a snack-food company.

Between them, Tanya and her husband make U.S.$30,000 a year, but from this income they must cover all their expenses, including their mortgage, car payments, and gasoline. In addition, they help their daughter, Rayna, who quit school after eleventh grade and married a man who is now out of work. They and their baby live at the back of Tanya's lot in an old mobile home (see Figure 1.22C).

With Tanya's lower wage (almost U.S.$1000 less a year), there will not be enough money to pay college tuition for her son, who is in high school. He had hoped to become an engineer, and would have been the first in the family to go to college. For now, he is working at the local gas station.

These people, living worlds apart, are all part of the global economy. Workers around the world are paid startlingly different rates for jobs that require about the same skill level. Varying costs of living and varying local standards of wealth make a difference in how people live and how they perceive their own situation. Though Tanya's family has the highest income by far, they live in poverty compared to their neighbors, and their hopes for the future are dim. Olivia's family, on the other hand, are not well off, but they do not think of themselves as poor because they have what they need, others around them live in similar circumstances, and their children seem to have a future. They can subsist on local resources, and the tourist trade promises continued cash income. But their subsistence depends on circumstances beyond their control; in an instant, the cruise-line companies can choose another port of call. Setiya, by far the poorest, seems trapped by his status as an undocumented worker, which robs him of many of his rights. Still, the higher pay that he can earn in Malaysia offers him a possible way out of poverty. *[Source: From Lydia Pulsipher's and Alex Pulsipher's field notes]* ■

The Debate over Globalization and Free Trade

The term **free trade** refers to the unrestricted international exchange of goods, services, and capital. Free trade is an ideal that has not been achieved and probably never will be. Currently, all governments impose some restrictions on trade to protect their own national economies from foreign competition, although there are far fewer such restrictions than there were in the 1980s. Restrictions take several forms:

- *Tariffs* are taxes imposed on imported goods that increase the cost of those goods to the consumer, thus giving price advantages to competing, locally made goods.
- *Export subsidies* reduce costs for home-country producers so they can compete better with foreign producers in the global market.
- *Import quotas* set limits on the amount of a given good that may be imported over a set period of time, curtailing supply and keeping prices high, again to protect local producers.
- *Investment restrictions* keep outsiders from building competitive businesses in a particular country.

These and other forms of trade protection are subjects of contention. Proponents of free trade argue that the removal of all such protective measures encourages efficiency, lowers prices, and gives consumers more choices. Companies can sell to larger markets and take advantage of mass-production systems that lower costs further. As a result, businesses can grow faster, thereby providing people with jobs and opportunities to raise their standard of living. These pro–free trade arguments have been quite successful, and in recent decades, restrictions on trade imposed by individual countries have been greatly reduced. Several *regional trade blocs* have been formed; these are associations of neighboring countries that agree to lower trade barriers for one another. The main ones are the United States-Mexico-Canada Free Trade Agreement (USMCA) (formerly known as the North American Free Trade Agreement, or NAFTA), the European Union (EU), the Southern Common Market in South America (Mercosur), and the Association of Southeast Asian Nations (ASEAN).

One of the main global institutions that supports the ideal of free trade is the *World Trade Organization (WTO)*, whose stated mission is to lower trade barriers and to establish ground rules for international trade. Related institutions, the *World Bank* and the *International Monetary Fund (IMF)*, both make loans to countries that need money to pay for economic development or to avoid financial crises. Before approving a loan, the World Bank or the IMF may require a borrowing country to reduce and eventually remove tariffs, import quotas, and other trade-restricting regulations. These requirements are part of larger "belt-tightening" policies, often called *structural adjustment programs (SAPs)*, that the IMF imposes on countries seeking loans, such as the requirement to close or privatize government enterprises and to reduce government services, including education and health-care programs that benefit the poor.

Those opposed to free trade argue that it leads to a less regulated global economy that can be chaotic, resulting in rapid cycles of growth and decline that increase the disparity between rich and poor worldwide. They note that the rules of trade have been made by the more powerful countries, often in secretive negotiations that sideline the interests of poor people and poor countries, and favor corporate interests. That is, free trade is assured on goods and services that benefit more developed countries (MDCs) but is less assured when lesser developed countries (LDCs) or middle income countries (MICs) have a competitive edge. Labor unions point out that as corporations relocate factories and services to poorer countries where wages are lower, jobs are lost in richer countries, creating poverty there. In the poorer countries, multinational corporations often work with governments to prevent workers from organizing labor unions that could bargain for *living wages*, wages high enough to provide food, shelter, clothing, health care, and child care for a family.

In the many LDCs and MICs that often lack effective enforcement of environmental protection laws, multinational corporations tend to use highly polluting and unsafe production methods to lower costs. Some fear that a "race to the bottom" in wages, working conditions, government services, and environmental quality is underway as countries compete for profits and potential investors. Multinational corporations that have recently agreed to address worker abuses include Apple, Nike, and Walmart.

free trade the unrestricted international exchange of goods, services, and capital

Reforming Free Trade and Development Institutions

In response to the many criticisms of free trade, and the widely recognized failures of policies that rely on the power of markets to guide development, the IMF and the World Bank have made some changes. One of the worst problems of SAPs is that instead of making it easier to pay off debt, sometimes they send a country into an economic decline that intensifies poverty and makes loan repayment impossible. SAPs have been replaced with *Poverty Reduction Strategy Papers*, or PRSPs, that plan for both economic growth *and* poverty reduction. Critics claim that PRSPs are simply a new name for SAPs with few actual changes, and indeed PRSPs still involve market-based solutions and are aimed at reducing the role of government in the economy. It remains to be seen if the stated goals of PRSPs, of maintaining education, health care, and social services, lead to a noticeably different impact on societies by the IMF and the World Bank.

Fair trade, proposed as an alternative to free trade, is intended to provide a fair price to producers and to uphold environmental and safety standards in the workplace. Economic relationships surrounding trade are rearranged in order to provide better prices for producers from LDCs. For example, "fair trade" coffee and chocolate are now sold widely in North America and Europe. Prices are somewhat higher for consumers, but the extreme profits of middlemen are eliminated. As a result, growers of coffee and cocoa beans who produce for fair trade companies can receive living wages and work under better conditions.

In evaluating free trade, globalization, and fair trade, consider how many of the things you own or consume were produced in the global economy—your computer, clothes, furniture, appliances, car, and foods. These products are cheaper for you to buy and your standard of living is higher as a result of lower production costs as well as competition among many global producers. However, you or someone you know may have lost a job because a company moved to another location where labor and resources were cheaper. Underpaid workers (even children) working under harsh conditions that possibly generate high levels of pollution may have made those cheap products. If workers can't earn living wages, often some family members end up migrating, perhaps without the proper papers, to earn a better wage. For example, NAFTA, the precursor to the USMCA, helped U.S. corporate farmers export crops grown in the United States to Mexico. These large producers, who often benefited from subsidies from the U.S. government, were able to sell corn at a lower price than small Mexican farmers, millions of whom were forced out of business. Many of these impoverished ex-farmers migrated illegally to the United States to support their families, where they found jobs but also discrimination and hardship.

state an organized political community in a defined territory and under one government that has *sovereignty*, the ability to conduct its internal affairs as it sees fit without interference from outside

nation a community of people usually having a common culture, origin, and language, occupying a specific territory; states often are composed of multiple nations

nation-state a state that coincides with a single nation, occupying roughly the same territory

Globalization, Development, and Government

Globalization has helped some countries make the shift from extractive resources to human resources, but only with the help of effective governments able to develop and execute well-designed policies to help navigate this complex transition. For example, China has benefited from a strong government that has been able to adapt policies formulated elsewhere in East Asia to develop a highly profitable globalized manufacturing sector (see Chapter 9). This and other industries have funded major investments in human development that are propelling China toward greater dependence on human resources. Countries with weaker governments, such as Niger, have been unable to navigate the complex transition toward a human resource economy and remain extractive resource–based economies, with agriculture providing most employment. The much lower earnings from agriculture mean less money is available to invest in human development.

CHECK YOUR UNDERSTANDING

1. How are global flows of information, goods, and people transforming patterns of economic development?
2. Why are workers around the world paid different rates for jobs that require about the same skill level?
3. What are some arguments for and against free trade?
4. What is an alternative to free trade?
5. How might a trade agreement result in migration?

POWER AND POLITICS

1.7 Distinguish the different ways that power is wielded in societies, from more authoritarian modes of governing to those based on notions of political freedom.

Geographers have long studied the ways in which people use their power over themselves and each other to change their circumstances. The political institutions that shape how this process unfolds have become globalized to an extent, though they are highly adapted to local conditions. The one most clearly present on the global political map is the **state**, which is an organized political community in a defined territory and under one government that has *sovereignty*, or the ability to conduct its internal affairs as it sees fit without interference from outside. The term **nation** is often used interchangeably with *state*, as it refers to a community of people usually having a common culture, origin, and language, occupying a specific territory: for example, the French nation. However, this can be misleading, because states often are composed of multiple nations, as is France with its large populations of Algerians, Moroccans, and Congolese. When a state and a nation occupy roughly the same territory, it is called a **nation-state**: for example Japan. When a nation has no state, it is called a *stateless nation*, such as the Kurds of northern Iraq, western Iran, and southeastern Turkey.

The concept of the state in the global political order that exists today is largely of European origin, and the global map of states contains many legacies of European colonialism. However, the first states originated in the fertile crescent in modern-day Iraq and were based around cities, hence the term *city-state*. And the European ideal of sovereignty, in which states are not supposed to interfere in the internal politics of neighboring or rival states, is rarely practiced in the world today, with every state having intelligence agencies that often deliberately interfere in the internal affairs of other states.

The ways in which power is actually wielded within states are similarly complex, with major differences across the world in how states operate, and many modes of governing that are rapidly changing (**Figure 1.23**). Over the past two centuries there has been a trend toward political systems guided by competitive elections, a process often called **democratization**. This trend is of particular interest because it runs counter to **authoritarianism**, a form of government that subordinates individual freedom to the power of elite regional and local leaders. In democratic systems of government, average individuals have the right to participate in free elections as well as many other **political freedoms**. These include:

- *freedom of speech*, the right to express oneself in public and through the media;
- *freedom of assembly*, the right to gather together in groups to pursue common interests;
- *freedom of movement*, the right to travel, live, and work in any part of a state that a person is a citizen of;
- *freedom from unreasonable searches and seizures*, the right to privacy and protection from searches and seizures of individuals and their property by anyone not possessing a warrant granted by a court of law;
- *freedom of the press*, the right to communicate through any media without interference from the government or other entities; and
- *freedom of religion*, the right to practice or not practice any faith or spiritual path.

Geographers do not necessarily conclude that democracy and respect for political freedoms are the best political arrangement. Indeed, many geographers are critical of the imposition of democracy, often by foreign governments or organizations, in places where local people have not chosen democratic systems and where long-standing cultural traditions support other political arrangements. Few would deny that the shift toward more democratic systems of government and greater political freedom over the past century and into recent times is extremely significant.

Nevertheless, there have been recent changes of course. After the Cold War ended with the fall of the Soviet Union in the early 1990s, it seemed that there was consistent movement toward democratization around the world. More recently, though, there has been a significant retreat from democratic practice. To better understand this trend, geographers and other scholars have become particularly interested in the role of political freedoms at the local level and the roles that social movements, international organizations, and the media play in the exercise of power and politics in particular places.

HOW DEEP DO POLITICAL FREEDOMS GO?

There was a steady expansion in the twentieth century of some political freedoms in many states, with more and more leaders coming to power through elections. However, the status of other political freedoms is more complicated. For example, in the United States, recent revelations by former employees of federal government intelligence agencies have led many to question how well protected some political freedoms are in a country generally considered to be one of the more democratic and "politically free" places on Earth. The revelations disclosed secret mass surveillance programs in which the phone and Internet communications of more than a billion people, including all U.S. citizens, are collected, stored, and analyzed by the federal government and various corporations that it hires. These activities have been criticized as undermining many political freedoms, including the freedom from unreasonable search and seizure and, through intimidation, freedom of the press.

The Arab Spring movements, which commenced in early 2011 in a number of countries in North Africa and Southwest Asia, further highlight the complexity of the expansion of political freedoms. These movements showed that massive public demonstrations can bring about changes in government, as was the case in Tunisia, Egypt, and Libya, but the outcomes are more complex and ambiguous than a straightforward expansion of political freedoms. While the demonstrations were an expression of a desire for certain political freedoms (freedom of assembly, freedom of speech) and have resulted in political reforms and elections in some cases, they have also provoked extremely repressive responses from authoritarian states (Egypt and Syria) and similarly forceful tactics from groups involved in Libya's ongoing civil war. Although the passage of time may tell a different story, the result of the Arab Spring has so far been widespread violence and constraints on political freedoms that many now call the *Arab Winter*.

What Factors Encourage the Expansion of Political Freedoms?

Here are some of the most widely agreed-upon factors that support the expansion of political freedoms:

- **Peace.** Peace is essential to creating an environment in which people can, among other things, vote in free and fair elections, speak and gather freely, and use print and electronic media to voice their concerns.
- **Broad prosperity.** As a broader segment of the population gains access to more than the bare essentials of life, there is often a shift toward greater political freedom. Whether general prosperity must be in place before this occurs is still widely debated, as is the question of whether prosperity necessarily leads to any expansion of political freedoms.
- **Education.** Better-educated people tend to want a stronger voice in how they are governed. Although democracy has spread to countries with relatively undereducated populations, leaders in such places sometimes become more authoritarian once elected.

democratization the transition toward political systems that are guided by competitive elections

authoritarianism a political system that subordinates individual freedom to the power of the state or of elite regional and local leaders

political freedoms the rights and capacities that support individual and collective liberty and public participation in political decision making

1.23 Power and Politics

The map and accompanying photos show two related trends in political power. Generally, countries with fewer political freedoms and lower levels of democratization also have the most violent conflict. Countries are colored on the map according to their score on a "democracy index" created by the *Economist* magazine, which uses a combination of statistical indicators to capture political freedoms crucial to the process of democratization, as well as the ability of governments to enact the will of their citizens and be free of corruption. Also displayed are major conflicts initiated or in progress since 1990 that have resulted in at least 10,000 casualties. Connections between democratization and armed conflict are explored in the photo captions.

THINKING GEOGRAPHICALLY

A How might the use of cell phone cameras among the demonstrators in the refugee camp contribute to or detract from political freedoms?

B, **D** Of the factors mentioned in the text as necessary for democracy to flourish, which are obviously present in **B** and missing in **D**?

C What factors frustrated attempts to end Colombia's civil war?

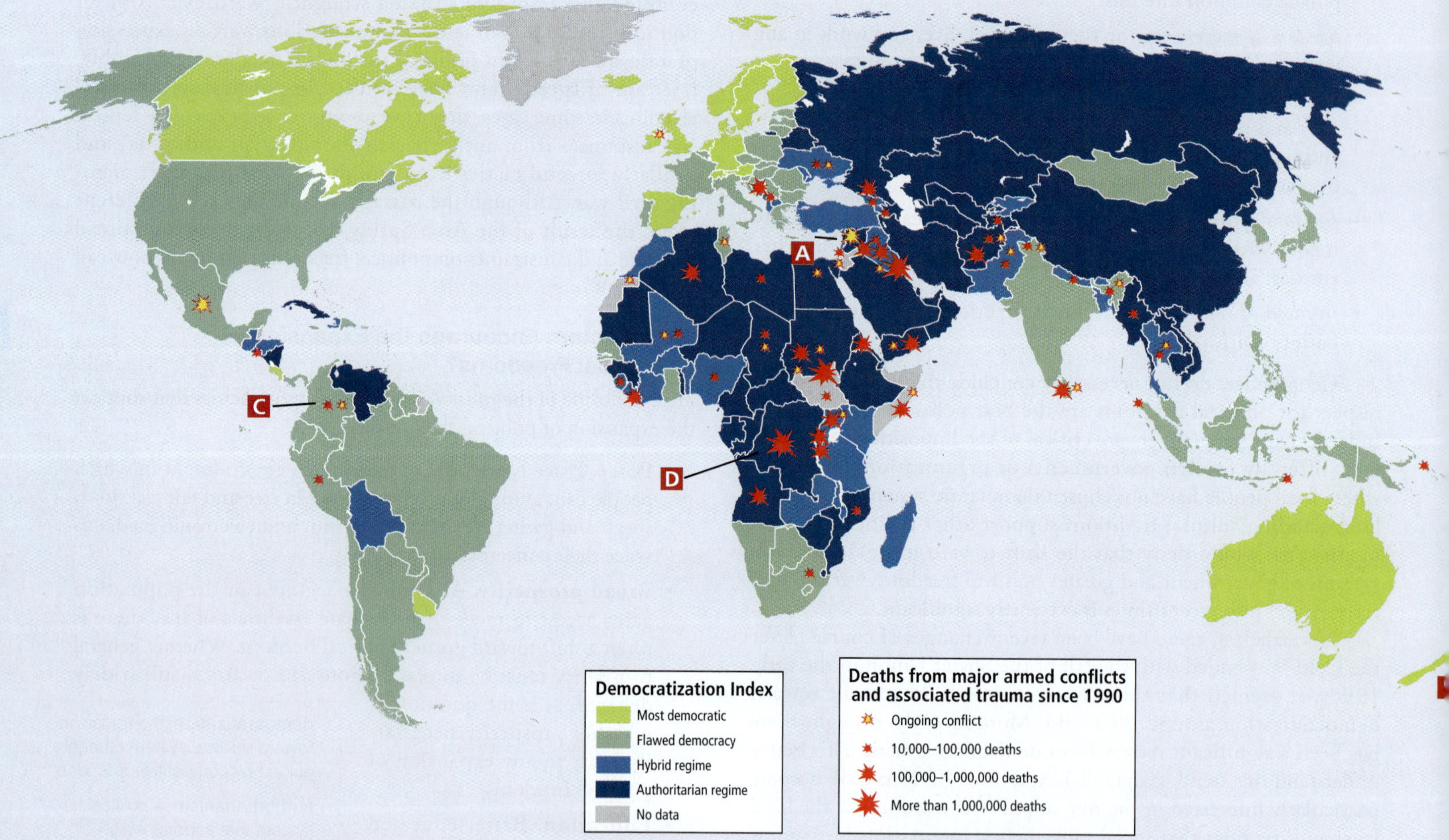

A Refugees. Syrian refugees take part in a demonstration at the Zaatari refugee camp in Jordan, near the border with Syria. Civil war broke out when Syria's government responded to citizen protests with a harsh military crackdown on the civilian population. By 2018, the war had led to over 500,000 deaths, more than a million war refugees crossing into Europe, and 12 million displaced people within Syria (see Chapter 6). [KHALIL MAZRAAWI/AFP/Getty Images]

B Civil society. A member of the Service and Food Workers Union in New Zealand solicits support from passing motorists for better wages and working conditions for the union's largely immigrant and female members (see Chapter 11). New Zealand has well-protected political freedoms and is among the world's most democratized nations. [Michael Bradley/Getty Images]

C Political freedom. Government repression and lack of political freedoms frustrated attempts to end Colombia's civil war for 5 decades until 2016, when a historic peace agreement between rebel groups and the administration of Colombian president Juan Manuel Santos (shown above) formally ended the country's more than 50-year civil war. Santos was awarded the Nobel Peace Prize and major rebel groups have disbanded and become political parties, but violence has continued as gangs and militants battle for control of cocaine-producing areas and illegal mining operations. [RAUL ARBOLEDA/AFP/Getty Images]

D Political violence. Political campaigns often become violent during times of war or conflict. Supporters of the political opposition to Joseph Kabila in September 2015 in Kinshasa, Congo, beat this man, who was suspected of being a pro-government supporter. Kabila was trying to change the constitution so that he could run for a third term in 2016, but then delayed the election until 2018. The country's ongoing civil war has resulted in more than 5.4 million deaths, mostly caused by malnutrition among people displaced by the violence (see Chapter 7). [JUNIOR KANNAH/AFP/Getty Images]

- **Civil society.** **Civil society** is made up of the social groups and traditions that function independently of the state and its institutions to foster a sense of unity and an informed common purpose among the general population. Civil society institutions can include the media, nongovernmental organizations (NGOs; discussed below and in Figure 1.24), political parties, universities, unions, parent/teacher associations, and in some cases religious organizations (see Figure 1.23C).

GEOPOLITICS

Geopolitics is the study of how geography and politics influence each other. Its focus is often at a large spatial scale, for example looking at how leaders ensure that their own country's interests are advanced in politics, trade, and confrontations with other countries. As the map in Figure 1.23 shows, there is a lack of political freedom and an abundance of violence in many parts of the world. A possible explanation for this is that the expansion of political freedoms is sometimes at odds with how leaders understand geopolitics, especially when large-scale violence is present.

In the modern era, geopolitics was perhaps most obviously at work during the *Cold War*, the period from 1946 to the early 1990s when the United States and its allies in western Europe faced off against the Union of Soviet Socialist Republics (USSR) and its allies in eastern Europe and Central Asia. Ideologically, the United States promoted a version of free market **capitalism**—an economic system based on the private ownership of the means of production and distribution of goods, driven by the profit motive, and characterized by a competitive but state-regulated marketplace. By contrast, the USSR and its allies favored what was called **communism** but what was actually a state-controlled economy, a socialized system of public services, and a centralized government in which citizens participated only indirectly through the Communist Party. Communism is also an ideology, based in part on the writings of the German revolutionary Karl Marx, that calls for the overthrow of capitalism and establishment of an egalitarian society in which workers share what they produce.

civil society the social groups and traditions that function independently of the state to foster a sense of unity and common purpose among the general population

geopolitics the study of how geography and politics influence each other

capitalism an economic system based on the private ownership of the means of production and distribution of goods, driven by the profit motive, and characterized by a competitive marketplace

communism an economic system based on a state-controlled economy, a socialized system of public services, and a centralized government in which citizens participate only indirectly through the Communist Party

United Nations (UN) an assembly of 193 member states that focuses on economic development, general health and well-being, democratization, peacekeeping assistance, humanitarian aid, and scientific research

nongovernmental organizations (NGOs) associations outside the formal institutions of government that promote activism on political, social, economic, or environmental issues

The Cold War became a race to attract the loyalties of unallied countries and to arm them, eventually influencing the internal and external policies of virtually every country in the world. Complex local issues were often oversimplified into a contest of democracy and capitalism versus communism.

In the post–Cold War period of the 1990s, geopolitics shifted. The Soviet Union dissolved, creating many independent states, nearly all of which began to implement some democratic and free market reforms. For a while it looked like a new era of trade and amicable global prosperity was beginning. But throughout the 1990s, while the developed countries were in a period of unprecedented prosperity, many unresolved political conflicts flared up in southeastern Europe (often referred to as the Balkans), Central and South America, Africa, Southwest Asia, and South Asia.

The new geopolitical era ushered in by the terrorist attacks on the United States on September 11, 2001, is still evolving. Because of the size and the global power of the United States, the attacks and the U.S. reactions to them affected virtually every international relationship, public and private. The ensuing adjustments, which will continue for years, are directly or indirectly affecting the daily lives of billions of people around the world.

Among the geopolitical adjustments taking shape is the challenge posed to the persistent international superpower status of the United States and the European Union by a still-evolving association between five developing economies: Brazil, Russia, India, China, and South Africa, known collectively as BRICS. All five are leaders in their regions, have advanced industrialized economies, and show signs of being able to spread general well-being to their populations. Politically Brazil, South Africa, and India are what are known as flawed democracies, and Russia and China are considered authoritarian regimes (see Figure 1.23). Together they represent over 41% of the earth's population and 32% of its GNI PPP. Each is dealing with serious internal adjustments to the global economy and with political upheavals, which are covered in the respective regional chapters of this book.

INTERNATIONAL COOPERATION

The trend toward more international cooperation on such things as climate change, peacekeeping, and emergency relief in disasters has the potential to expand political freedoms. The prime example of this today is the **United Nations (UN)**, an assembly of 193 member states. The member states sponsor programs and agencies that focus on, among other things, economic development, general health and well-being, democratization, peacekeeping, and humanitarian aid. However, the UN rarely challenges a country's *sovereignty*, its right to conduct its internal affairs as it sees fit without interference from outside. Consequently, the UN often can enforce its rulings only through economic sanctions. While the UN does play a role in peacekeeping in troubled areas, there are no true UN military forces. Rather, there are peacekeeping-only troops from member states that wear UN designations on their uniforms and take orders from temporary UN commanders. In some instances, the UN can endorse military action by non-UN troops, via the *Security Council*—a division of the UN with five permanent members (the United States, United Kingdom, China, France, and Russia) and ten rotating members. The Security Council is intended to respond quickly to international crises.

In addition to the UN, **nongovernmental organizations (NGOs)** are an important embodiment of international cooperation. NGOs generally act on a specific set of goals, for example environmental protection. Some, such as Doctors Without Borders and Gift of the Givers (**Figure 1.24**), provide medical care to those who need it most, especially in conflict or disaster zones, where they sometimes help provide a level of stability that governments cannot.

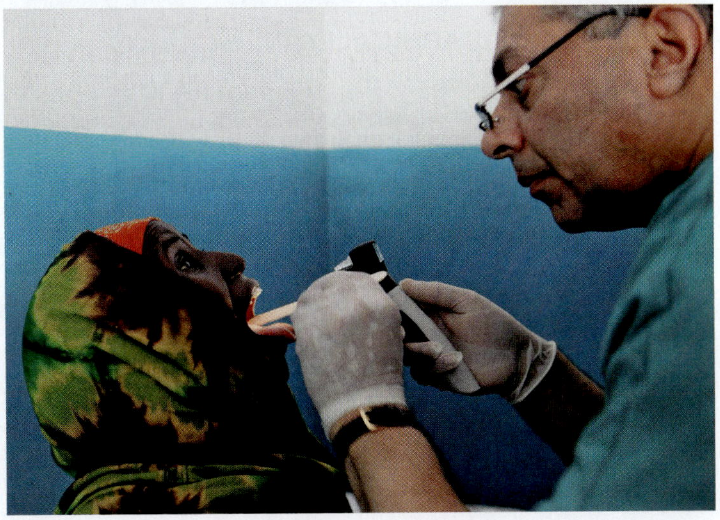

Figure 1.24 Nongovernmental organizations in action. A doctor, a volunteer with the South Africa–based NGO Gift of the Givers, examines a patient in Somalia. Gift of the Givers, which is the largest NGO started and staffed by Africans, provides medical help and food to famine-stricken parts of Somalia. [The Times/Gallo Images/Getty Images]

CHECK YOUR UNDERSTANDING

1. What are some contrasting ways that power is wielded in societies?
2. What has been the trend regarding political freedoms over the past two centuries?
3. What are some factors that support the expansion of political freedoms?
4. What is geopolitics?

URBANIZATION

1.8 Analyze the various factors that are contributing to the growth of urban areas, especially in the developing world.

Human life on Earth is increasingly urban. Over 4 billion people, more than half of the world's population, now live in urban places. There are more than 400 cities of over 1 million people and 28 cities of over 10 million people (**Figure 1.25**). This is quite a change from even the recent past. In 1700, fewer than 7 million people, or just 10% of the world's total population, lived in cities, and only five cities had populations of several hundred thousand people or more. The world we live in today has been transformed by **urbanization**, the process whereby cities, towns, and suburbs grow as populations shift from rural to urban livelihoods. Because the process of urbanization tends to go hand in hand with economic development, the average urbanization rate among MDCs is close to 80%. Meanwhile, in many LDCs only 10% to 50% of the population is urban.

WHY ARE CITIES GROWING?

Cities are centers of innovation and culture and these features attract new migrants, who are looking not just for jobs but also for education, the freedom to express themselves, and excitement (see Figure 1.25B); thus cities are said to *pull* in migrants. Meanwhile, the mechanization of agriculture has drastically reduced the need for rural labor, resulting in people being *pushed* off the land. Increasing the food supply to the point that large nonfarming populations can be fed has made it possible for people to move out of rural areas and into cities, where the development of manufacturing and service economies has created many jobs (see Figure 1.25A). This *push/pull* phenomenon of urbanization from farm to city often is lopsided because the growth of urban jobs rarely keeps up with in-migration: push forces are stronger than pull forces. One of the push forces is climate change, which can bring extended droughts or floods that force people off the land or otherwise limit rural sustenance.

Urban Growth in Less Developed Countries

The most rapidly growing cities are in developing countries (LDCs) in Asia, Africa, and Middle and South America (see the map in Figure 1.25). For example, in sub-Saharan Africa countries are on average 40% urban, but are expected to become 56% urban by 2050. The settlement patterns of these cities bear witness to their rapid and often unplanned growth, fueled in part by the steady arrival of masses of poor rural people looking for work. Cities like Mumbai (in India), Cairo (in Egypt), Nairobi (in Kenya), and Rio de Janeiro (in Brazil) sprawl out from a small, affluent core, often the oldest part, where there are upscale businesses, fine old buildings, banks, shopping centers, and residences for wealthy people. Surrounding these elite landscapes are sprawling mixed commercial, industrial, and middle-class residential areas, interspersed with pockets of extremely poor neighborhoods.

Numerous cities have been unprepared for the massive inflow of rural migrants, about a billion of whom now live in polluted **slum** areas. One in eight humans, and about one-third of the world's urban population, lives in these run-down neighborhoods, plagued by natural and human-made hazards, poor housing, out-of-date infrastructure, and inadequate access to food, clean water, education, and social services (see Figure 3.30 and the Chapter 7 opening photo). Also known in various locales as *barrios*, *favelas*, *shantytowns*, *ghettos*, and *tent villages*, the slum settlements provide housing for the poorest of the poor, who provide low-wage labor for the city. Housing is often self-built out of any materials the residents can find: cardboard, corrugated metal, masonry, scraps of wood and plastic. There are usually no toilets with sewer connections and little access to clean water. Electricity is often obtained from illegal and dangerous connections to nearby power lines. Schools are few and overcrowded, and transportation is provided only by informal, nonscheduled van-based services. Life in these areas can be insecure and chaotic, with police unwilling to enter slums and criminal organizations and gangs often maintaining control through violence. The UN estimates that more than 2 billion people will live in urban slums by 2030.

Urban Migration, Opportunity, and Gender

While many migrants to urban areas end up in slums, many are also able to take advantage of the opportunities cities offer. Those who are financially able to come to urban

urbanization the process whereby cities, towns, and suburbs grow as populations shift from rural to urban livelihoods

slum densely populated area characterized by crowding, run-down housing, and inadequate access to food, clean water, education, and social services

1.25 Urbanization

Most people now live in cities, with over a quarter in cities of over 500,000 people, most of which are in Asia. Because urbanization tends to increase with economic development, the urbanization rate among MDCs is close to 80%, but only 10% to 50% in many LDCs. Many cities in LDCs have been unprepared for massive migration from rural areas, leading to the expansion of unplanned shantytowns that now house about a billion people. On the map, the color of the country indicates the percentage of the population living in urban areas. The blue circles represent the populations of the world's largest urban areas in 2015. [Research from: Population Reference Bureau, 2018 World Population Data Sheet, at https://www.prb.org/2018-world-population-data-sheet-with-focus-on-changing-age-structures/; and http://www.demographia.com/db-worldua.pdf, p. 22]

THINKING GEOGRAPHICALLY

A In what kind of neighborhood of Dhaka would you guess this man lives?

B To what group of urban migrants does this skydiving young man probably belong?

C How does Copenhagen contribute to the resources it is relying on to achieve carbon neutrality?

D This photo exemplifies what problem commonly faced by rapidly growing cities?

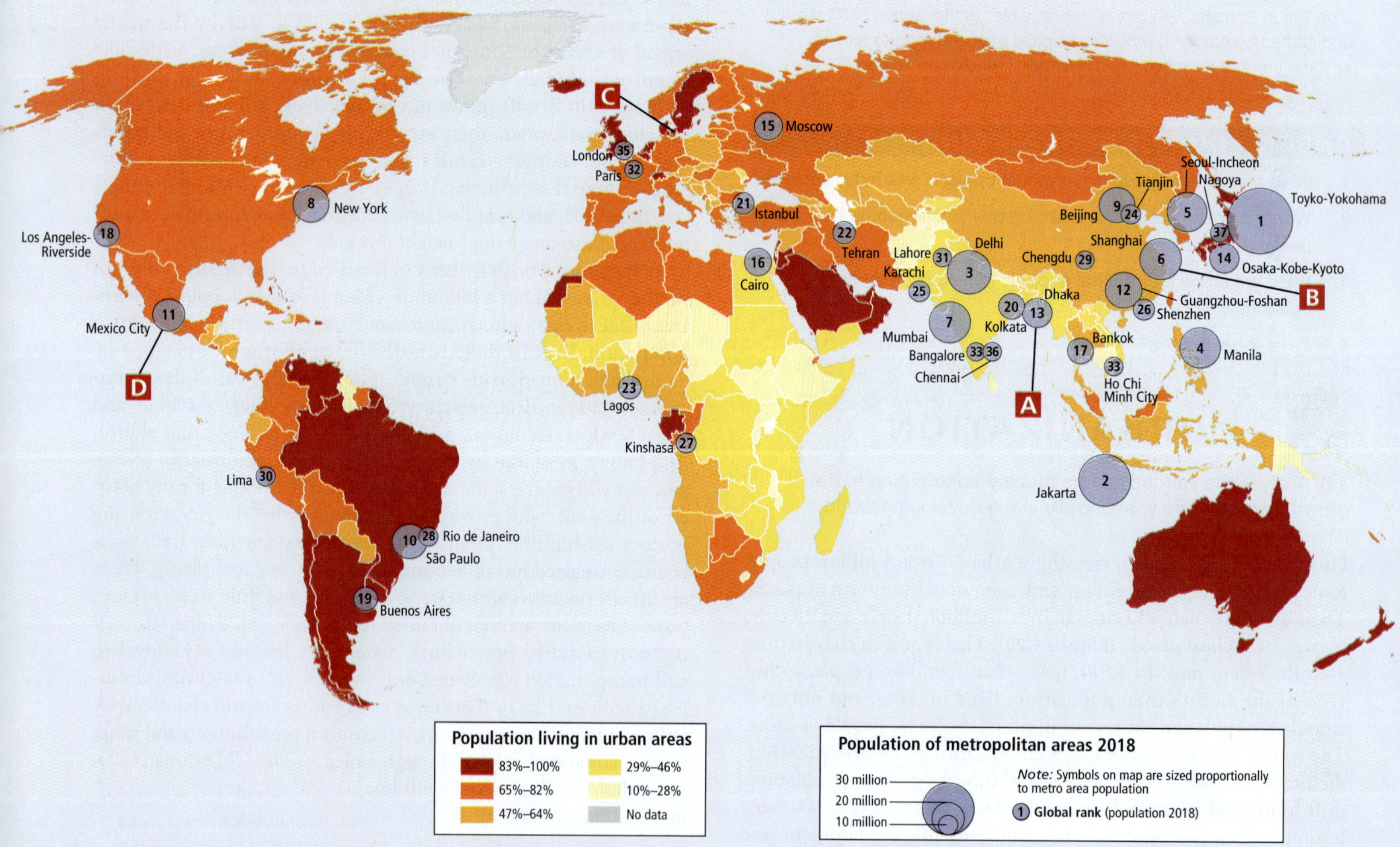

A Migrants. A recent migrant to Dhaka pulls a cart loaded with goods. Many of the world's fastest-growing cities are attracting more people than they can support with decent jobs, housing, and infrastructure. [FARJANA K. GODHULY/AFP/Getty Images]

B Youth and talent. Cities have always been centers of innovation, entertainment, and culture, in large part because they draw both money and talented people. This resident of Shanghai is engaging in a current fad: parachuting off one of the new skyscrapers that now dominate the city's skyline. [LIU JIN/AFP/Getty Images]

C Urban environmentalism. Cyclists represent over a third of the commuting population in Copenhagen, Denmark, which aims to be the world's first "carbon-neutral" capital by 2025. Efficient district heating systems provide heat from a centralized waste incineration plant to over 275,000 homes, and much of the city's electricity is provided by coastal wind farms. [Francis Dean/Corbis via Getty Images]

D Infrastructure. Some cities struggle with major environmental problems. Mexico City, currently the world's twelfth-largest city, occasionally suffers from severe flooding because of its location on an old, now-sinking lake bed and its antiquated drainage and sewage infrastructure. [ALFREDO ESTRELLA/AFP/Getty Images]

areas for education and complete their studies tend to find employment in modern industries and business services. They constitute the new middle class and leave their imprint on urban landscapes via the high-rise apartments they occupy and the shops and entertainment facilities they frequent. Cities such as Mumbai in India, São Paulo in Brazil, Cape Town in South Africa, and Shanghai in China are now home to this more educated group of new urban residents, many of whom may have started life on farms and in villages.

In the past, most migrants in cities were young men, but these days, the number of young women migrating is growing, and this is having an impact on population growth. Cities offer women more than better-paying jobs: they provide access to education, better health care, and more personal freedom. Once in cities, women often choose to have fewer children as they delay childbearing in favor of advancing their education or career. Within urban households, people are choosing to have fewer children for a variety of reasons, such as the fact that without farms to run, the labor provided by a large family is no longer needed. Raising a child in an urban area is usually more expensive, involving more years of formal education before they can make a contribution to their family's income. Hence, urbanization is linked to slower population growth.

URBAN PLANNING AND THE ENVIRONMENT

Careful attention to planning often gives cities significant environmental advantages. For example, in most European and East Asian cities and many older U.S. cities, such as Boston or New York, people live in compact multi-unit housing that uses much less energy for heating and cooling compared to the detached single-family units common in much of the United States. The compactness of these cities also saves energy spent on transport since people don't have to commute as far and can often use buses, trains, or other more energy-efficient public transportation. Dense urban settlement has resulted in reduced GHG emissions in Europe and in some North American cities. Los Angeles, San Francisco, Boston, Washington, DC, and New York City all reduced their per capita emissions between 1980 and 2010 primarily by improving their public transportation systems.

In most of North America and much of the developing world, unplanned or poorly managed growth, known as *urban sprawl*, is increasing the spatial extent of cities, resulting in greater environmental impacts from urban living. Lower-density residential and commercial developments in suburbs place a greater burden on the environment than more compact urban areas, both because of increased heating and cooling costs, but also because public transportation is often not available or convenient. In wealthier cities, poorly planned growth means that people drive long distances for work and play, adding pollution to the areas where they live. For example, since 2000, per capita emissions in many North American cities, such as Atlanta, Phoenix, and Detroit, have risen as more people live in spread-out suburbs. In poorer cities, such as Kinshasa, Congo, or Karachi, Pakistan, unplanned growth means that urban infrastructure of all kinds gets badly overloaded. Major roads are clogged with so much traffic that the commute to work can take many hours. Sewers are so overloaded that they often discharge human waste directly into urban streams, and millions of people have no access to sewers at all, so they are forced to defecate in streets and public places. Major health hazards are a direct result of unplanned growth in cities in the developing world.

CHECK YOUR UNDERSTANDING

1. What factors contribute to the growth of urban areas, especially in the developing world?
2. How much of the world's population lives in cities?
3. How does urbanization contribute to slower population growth?
4. How does urban form influence greenhouse gas emissions?

POPULATION, GENDER, AND CULTURE

1.9 Evaluate the multiple reasons that population growth is slowing throughout the world.

1.10 Describe some of the global consistencies in gender disparity and how they are changing.

1.11 Evaluate the biological significance of race and contrast it with the cultural significance of race.

Many geographers are interested in the growth and decline of human populations. Gender is a cultural feature so central to the ways populations change that it is often taken for granted. Its influences are as powerful and deep as economic development, human development, politics, urbanization, religion, language, and race. All of these are central aspects of human culture shaping humanity's responses to the enormous challenges it now faces.

GLOBAL SPATIAL VARIATIONS IN POPULATION DENSITY AND GROWTH

If the more than 7 billion people on Earth today were evenly distributed across the land surface, they would produce an *average population density* of about 121 people per square mile (47 per square kilometer). However, people are not evenly distributed but rather tend to live where more resources are available (**Figure 1.26**). Most people live where climates are warm and wet enough to support agriculture, along rivers that provide irrigation and a means of transportation, in lowland regions that have rich soils, or close to the sea that provides both food and transportation opportunities. This means that people are concentrated on roughly 20% of the available land. Because of the arrangement of land masses, nearly 90% of all people live north of the equator, and most of them live between 20° N and 60° N latitude.

GLOBAL PATTERNS OF POPULATION GROWTH

It took between 1 million and 2 million years (at least 40,000 generations) for humans to reach a population of 2 billion, which happened around 1945. Then, remarkably, in less than a century—by 2018—the world's population more than tripled to 7.6 billion (**Figure 1.27**).

Growth rates are now slowing in most societies and in a few have even begun to decline (a process called *negative growth*). But what happened to make the population grow so quickly in such a short time?

The explanation lies in the changing relationships between humans and the environment. For most of human history, fluctuating food availability, natural hazards, and disease kept human death rates high, especially for infants and children. Out of many pregnancies, a couple might raise only one or two children.

A rapid upsurge in human population began about 1500 as food became more abundant due to the exchange of productive food crops between many civilizations during the age of European colonization. In the centuries that followed, advances in health care, science, and urban sanitation reduced the incidence of disease, allowing humans to live longer as well. With more people living long enough to reproduce successfully, populations grew exponentially, resulting in a pattern of population growth often called a *J curve* when depicted on a graph.

Today, the human population is growing in most regions of the world, more rapidly in some places as more young people reach the age of reproduction. Nevertheless, the *rate* of global population growth is slowing. Per capita, people are having fewer children. In 1968, the global growth rate was 2.1% per year, and it fell by 2018 to roughly 1.2% per year. If slower growth trends continue, the world population may level off at about 9–10 billion by 2070, and then possibly decline to 8.5 billion by 2100. However, this projection depends on whether levels of human development can advance in LDCs enough to where people have the economic security, education, and access to health care and birth control to choose to have smaller families. Without these factors in place, human population could hit 11 billion by the end of this century and keep growing. Regardless of which projection actually occurs, over half of the growth will occur in sub-Saharan Africa, with most of the rest in South Asia.

THE DEMOGRAPHIC TRANSITION: POPULATION GROWTH RATES AND HUMAN DEVELOPMENT

A **demographic transition** is a shift in population growth from low, to high, to again low or negative growth. It occurs when a period of high birth and death rates transitions to a period of high birth and low death rates, and finally to a period of low birth and death rates (see **Figure 1.28**). Driving these changes are shifts in economic and human development that can be thought of as having three stages, the last two of which have two parts.

The first or "subsistence stage" is characterized by a subsistence economy, usually agricultural, in which people produce most of what they consume and human development is generally low, with short life expectancy due to low access to health care and education. Birth rates are high because families need lots of children to help with agricultural work. Death rates are also high due to irregular food supplies related to drought, flooding, or other natural disasters, and low access to health care. There are currently no countries at this stage, mostly due to the spread of at least rudimentary modern health care to almost all parts of the planet.

Stage 2 is a transition from agriculture to industry. The first part of this stage (2A on Figure 1.28) involves the introduction of more productive farming techniques that can feed more people, and improved access to health care resulting in longer life expectancy. The result is rapidly growing rural populations mostly employed in agriculture, with low to medium human development that is rising. Afghanistan and many countries in sub-Saharan Africa are at this stage of the demographic transition.

The second part of stage 2 (2B on Figure 1.28) begins when more people start working in industries, such as manufacturing, that provide higher incomes but that require at least a few years of formal education. Because children have to spend more time being educated, they are more of a financial drain on families, which tend to remain smaller, resulting in declining birth rates. Access to health care improves further, continuing the decline in death rates; however, population growth tends to level off toward the end of this stage. Human development can be at medium to high levels during this stage depending on political decisions to invest in social services. Countries at this stage of development include India, Bangladesh, and much of Middle and South America.

Stage 3 is a transitional phase from an industrial to a post-industrial economy. The first part (3A on Figure 1.28) occurs as birth and death rates come into balance again, resulting in very slow or no growth at all. Economies are transitioning from industrial to higher-skill service sector and knowledge-based jobs that pay better but require more education, so families have fewer children and birth rates continue to decline. Higher incomes support higher investments in human development, bringing even better access to health care, so life expectancies continue to increase but only slightly, resulting in low to no growth. Countries at this stage include the United States, Canada, much of western Europe, Australia, and New Zealand.

The second part of stage 3 (3B on Figure 1.28) is a period of population decline that only a few countries have gone through and many might not go through at all. Sustained high levels of human development result in longer life expectancies and overall older populations. Birth rates may fall below replacement levels due to the high cost of raising and educating children, and many people feeling that they don't need children to support them in their old age due to high-functioning social services for the elderly. Countries at this stage include Japan, much of eastern Europe, and parts of Russia and Ukraine.

A useful concept for understanding the demographic transition is the **rate of natural increase (RNI)**, (often called the *growth rate*), which is the difference between the number of people being born (the **birth rate**) and the number dying (the **death rate**). The rate of natural increase is expressed as a percentage per year. For example, in 2018, the annual birth rate in the United States (population 328 million) was 12 per 1000 people, and the death rate was 9 per 1000 people. Therefore, the annual rate of natural increase was 3 per 1000 (12 – 9 = 3), or 0.3%. For comparison, consider Afghanistan (in South Asia, population 36 million). In 2018, Afghanistan's birth rate was 35 per 1000, and the death rate was 7 per 1000. Thus the annual rate of natural increase was 28 per 1000 (35 – 7 = 28), or 2.8% per year.

Gender plays a large role in the demographic transition. Opportunities for work and education outside the home grow for women throughout the demographic transition. This tends to result in delaying of childbearing and fewer children born overall. A useful statistic for understanding

demographic transition the change from high birth and death rates to low birth and death rates usually accompanied by economic changes and rising human development, such as increased access to health care, sanitation, urbanization, and education

rate of natural increase (RNI) the rate of population growth measured as the excess of births over deaths per 1000 people per year, without regard to the effects of migration

birth rate the number of births per 1000 people in a given population, per unit of time (usually per year)

death rate the ratio of total deaths to total population in a specified community, usually expressed in numbers per 1000 or in percentages the contribution of gender to population is the **total fertility rate (TFR)**, the average number of children a woman is likely to have during her reproductive years (15–49 years). As education and formal employment rates for women increase, and as they postpone childbearing into their late twenties, total fertility rates tend to decline. For example, U.S. women often seek higher education and careers, and couples choose to marry late, if at all. The TFR in 2018 for a woman in the United States was 1.8. By way of comparison, women in Afghanistan have much fewer opportunities to study or work outside the home, and tend to marry early, contributing to a high fertility rate of 4.8.

Better access to health care can also move societies toward slower population growth, because parents can have fewer children and still be assured that enough will survive into adulthood to care for them in their old age. A useful statistic for understanding this aspect of the demographic transition is the *infant mortality rate*, which is the number of deaths of children under the age of one for every 1000 live births. The United States has a relatively low infant mortality rate of 5.6 per 1000 live births, while Afghanistan's is very high at 57 per 1000. Improved health care also means better access to birth control, which gives women more control over if and when they become pregnant. In the United States, 73% of women use some form of birth control, while in Afghanistan only 23% do.

The transition away from *subsistence economies*, such as those found in most of rural Afghanistan, involves huge cultural changes, with a shift from skills being learned around the home, farm, and surrounding places, to attendance at expensive schools and universities. When measured in monetary terms, subsistence economies seem quite poor. For example, Afghanistan's GNI per capita PPP in 2018 was $2000, compared to $60,200 for the United States. However, subsistence economies are often especially rich in nonmonetary terms, such as the relationships they nourish, and the many different skills that people learn from an early age.

total fertility rate (TFR) the average number of children that women in a particular population are likely to have at the present rate of natural increase

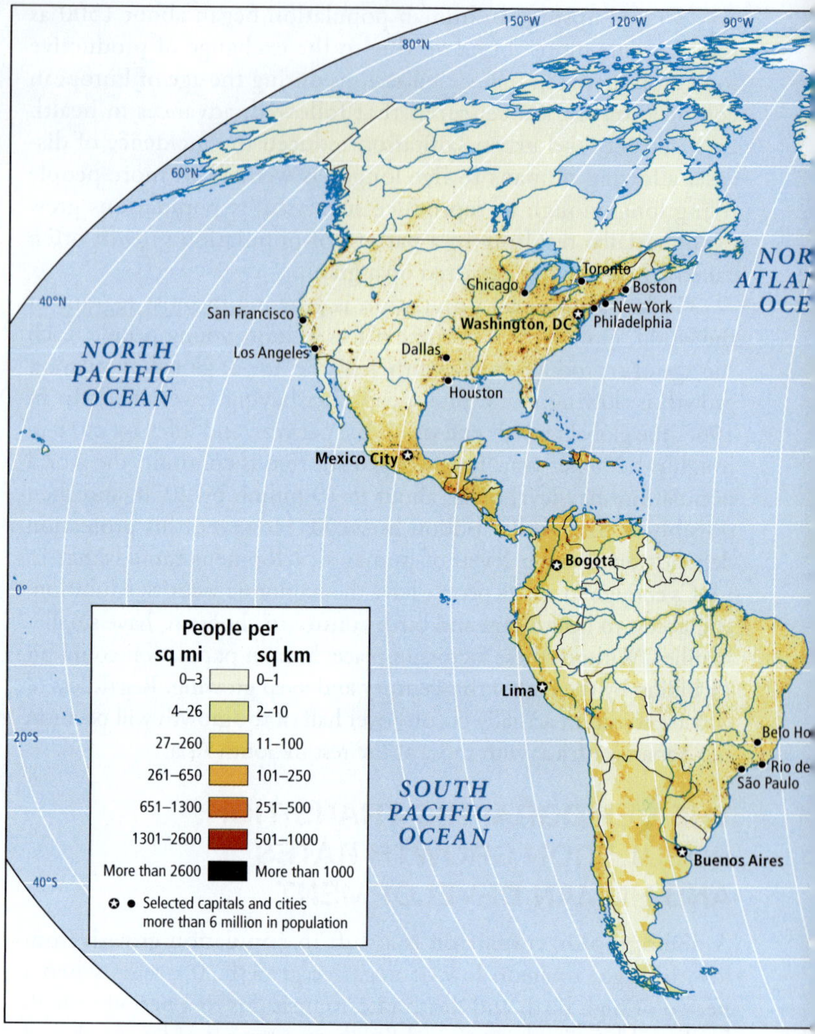

Figure 1.26 World population density. Two major population concentrations are immediately visible on this map: in China, especially in the North China Plain (south of Beijing), and in South Asia, especially in the Ganga River Valley stretching from Delhi, India, to Dhaka, Bangladesh, and in Pakistan's Indus River Valley (stretching from Lahore to Karachi). Another area of high population density is in Indonesia on the Island of Java (near Jakarta). These are areas of ancient human occupation and cultural development. Other major population concentrations can be found in western Europe (around Amsterdam) and west Africa (near Lagos, Nigeria).

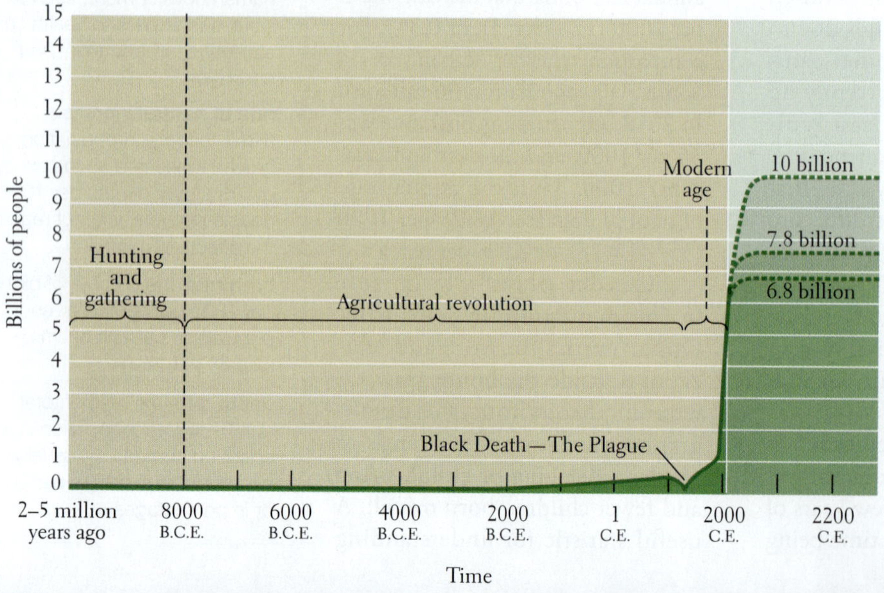

Figure 1.27 Exponential population growth: The J curve. The curve's J shape is a result of successive doublings of the population within ever-shorter time periods (exponential growth). It starts out nearly flat, but as doubling time shortens, the curve bends ever more sharply upward. Note that B.C.E. (before the common era) is equivalent to B.C. (before Christ); C.E. (common era) is equivalent to A.D. (anno Domini). [Research from: G. Tyler Miller, Jr., *Living in the Environment*, 8th ed. (Belmont, CA: Wadsworth, 1994), p. 4]

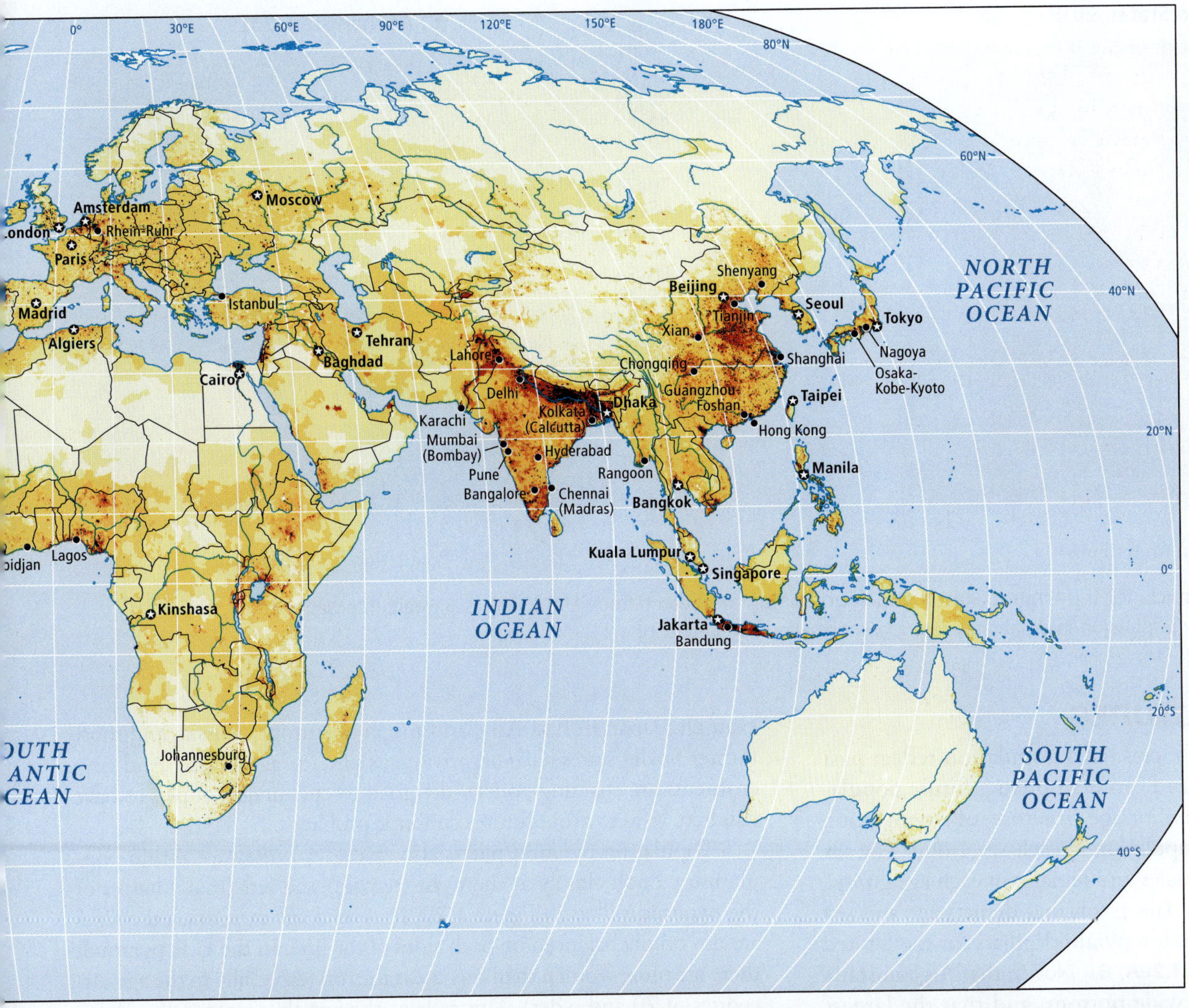

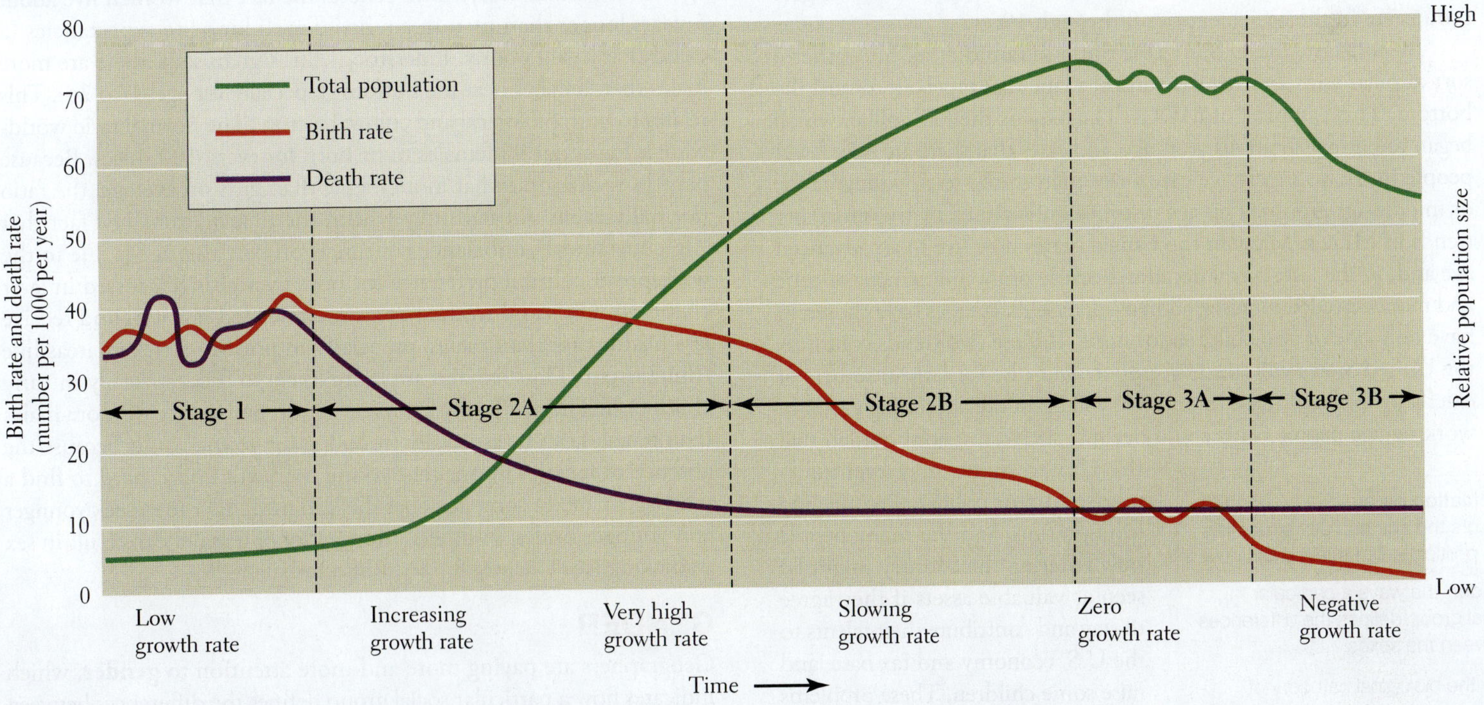

Figure 1.28 Demographic transition diagram. See the text for a detailed explanation. [Research from: G. Tyler Miller, Jr., *Living in the Environment*, 8th ed. (Belmont, CA: Wadsworth, 1994), p. 218]

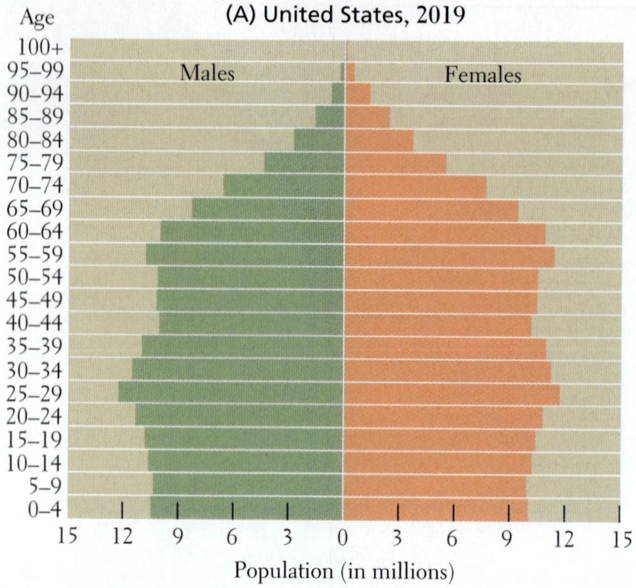

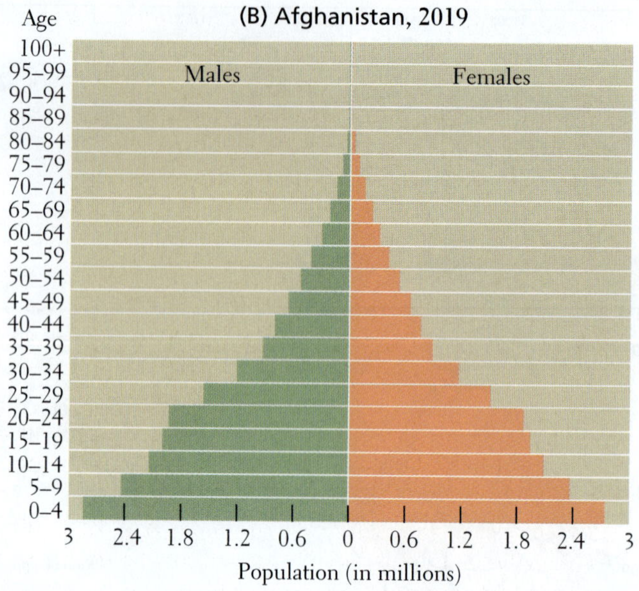

Figure 1.29 Population pyramids for the United States and Afghanistan. [Research from: U.S. Census Bureau, International Programs, http://www.census.gov/population/international/data/idb/informationGateway.php]

AGE AND SEX STRUCTURES

The age and sex structures of a country's population reflect past and present social conditions and can help predict future population trends. The age distribution, or age structure, of a population is the proportion of the total population in each age group. The sex structure is the proportion of males and females in each age group.

The **population pyramid** is a graph that depicts age and sex structures. Consider the population pyramid/pillars for the United States and Afghanistan (**Figure 1.29A, B**). Notice that Afghanistan's (B) is a true pyramid, with a wide bottom, and that the largest groups are in the age categories 0 through 19.

By contrast, the U.S.'s (A) is not a pyramid at all but rather a sort of pillar with an irregular vertical shape that tapers in toward the bottom. This is typical of MDCs. The base of the U.S. pillar, which began to narrow about 40 years ago, indicates that there are now fewer people in the youngest age categories than in the teen, young adult, or middle-age groups. This age distribution is caused by two emerging trends in MDCs. Many in the United States now live to an advanced age and, in the last several decades, because of the investment of time and money needed to raise children, many U.S. couples have chosen to have only one or two children, or none. If these two trends continue, the United States will need to support and care for large numbers of elderly people, and those responsible will be an ever-declining group of working-age people. This could become such a financial burden that the United States may eventually reverse current policies discouraging immigration, and the many people now entering the country might be seen as valuable assets if they agree to stay and contribute their talents to the U.S. economy and tax base, and raise some children. These problems are already present in some countries where populations are declining due to low birth and death rates. For example, Japan in East Asia, and much of Europe, Russia, and most former Soviet states have negative rates of natural increase. And governments are enacting policies designed to encourage couples to have children, so as to avoid future economic problems.

population pyramid a graph that depicts the age and sex structures of a political unit, usually a country

gender the ways a particular social group defines the differences between the sexes

sex the biological category of male, female, or intersex; does not indicate how people may behave or identify themselves

Population pyramid/pillars also reveal *sex imbalance* within populations. Look closely at the right (female) and left (male) halves of the pyramid/pillars in Figure 1.29. In several age categories, the sexes are not evenly balanced on both sides of the line. In the U.S. pyramid, there are more women than men near the top (especially in the age categories of 70 and older). This reflects the fact that women live about 5 years longer than men in countries with long life expectancies (a trend that is still poorly understood). In Afghanistan, there are more males than females near the bottom (especially for ages 0 to 24). This relates to both biological and cultural factors. The normal ratio worldwide is for about 95 females to be born for every 100 males. Because baby boys are somewhat more fragile than girls on average, the ratio normally evens out naturally within the first 5 years. The fact that Afghanistan's sex imbalance continues on past this age is due to the widespread cultural preference for boys over girls (discussed in later chapters). Girls and women are sometimes fed less well and receive less health care than males, especially in poverty-stricken areas like Afghanistan. Thus females are more likely to die, especially in early childhood. The artificially achieved gender imbalance of more males than females can be especially troubling for young adults because the absence of females means that young men will find it hard to find a marriage mate. In some societies this can mean that men seek younger and younger brides; in others, the scarcity of females can result in sex trafficking (see Chapter 8, "Sex Ratio Imbalance").

GENDER

Geographers are paying more and more attention to **gender**, which indicates how a particular social group defines the differences between the sexes. It is different from **sex**, which refers to the biological category of male, female, or *intersex*, the state of being in variation from typical male or female biology. Sex does not indicate how people may behave or identify themselves, while gender does. There is increasing acceptance of *transgender* people, whose gender identity may not necessarily align with their sex, and a variety of accommodations for transgender people are helpful to their social functioning. Careful attention to the roots of gender-based biases, phobias, and violence may help to create a more just future for all people.

In virtually all parts of the world, and for at least tens of thousands of years, the biological fact of maleness and femaleness has been translated into specific roles for each sex. The activities assigned to men and to women can vary greatly from culture to culture and from era to era, but they remain central to the ways societies function. The existence of transgender people and social roles for people of genders other than male or female are also ancient, and acceptance varies across the globe. Increasing attention to gender roles by geographers has been driven largely by interest in the shifts that occur when traditional ways are transformed by modernization.

For women, the historical and modern global gender picture is puzzlingly negative. In nearly every culture, in every region of the world, and for a great deal of recorded history, women have had (and still have) an inferior status. Exceptions are rare, and the intensity of this second-class designation varies considerably. On average, women have less access to education, medical care, and basic resources including food and water. They start work at a younger age and work longer hours than men. Around the world, many people of both sexes still routinely accept the idea that men are more productive and intelligent than women. The puzzling question of how and why women became subordinate to men has not yet been well explored because, oddly enough, few thought the question significant until the last 50 years.

However, focusing exclusively on women's issues misses the strict gender-based expectations of males. For most of human history, young men have borne a disproportionate share of burdensome physical tasks and dangerous undertakings. Until recently, it was mostly young men who left home to migrate to distant, low-paying jobs and send money back home to their families. Overwhelmingly, it has been young men (the majority of soldiers) who die in wars or suffer physical and psychological injuries from combat.

In nearly all cultures, families prefer boys over girls because, as adults, due to social norms, boys will have greater earning capacity (**Table 1.2**) and more power in society. This preference for boys has some unexpected side effects. As mentioned above, the preference may lead to fewer girls being born, which then leads to an eventual shortage of marriageable women, leaving many men without the hope of forming a family. Currently, there is concern in several Asian societies that the scarcity of young women, which in some places is extreme (the shortage in China equals the entire population of Canada), is already leading to antisocial behavior on the part of discouraged young men.

There are some striking consistencies regarding traditional gender roles over time and space. Men are expected to fulfill public roles, while women fulfill private roles and have limited freedom to access public spaces. Certainly there are exceptions in every culture, and customs are changing, especially in wealthier countries. But generally, men work outside the home in positions such as executive jobs, animal herders, hunters, farmers, warriors, or government leaders. Women keep house, bear and rear children, care for the elderly, grow and preserve food, and prepare the meals, among many other tasks. In nearly all cultures, women are defined as dependent on men—their fathers, husbands, brothers, or adult sons—even when the women may produce most of the family sustenance.

Table 1.2 Comparison of male and female incomes for selected countries

Country	HDI rank	Female GNI per capita (PPP U.S.$)	Male GNI per capita (PPP U.S.$)	Female income as percent of male income
Austria	23	29,598	58,826	50.31
Barbados	57	10,245	14,739	69.51
Botswana	106	15,179	18,096	83.88
Canada	9	33,587	50,843	66.06
Japan	20	24,975	49,541	50.41
Jordan	80	3587	18,831	19.04
Kuwait	48	42,292	111,968	37.77
Poland	36	18,423	28,271	65.16
Russia	50	17,269	28,287	61.05
Saudi Arabia	39	20,094	77,044	26.08
Sweden	14	40,222	51,084	78.74
United Kingdom	14	27,259	51,628	52.80
United States	8	43,054	63,156	68.17

Data from the UN Human Development Report 2018, Table 4; see http://hdr.undp.org/en/composite/GDI

Because their activities are focused on the home, women tend to marry early. One-quarter of the girls in developing countries are mothers before they are 18. This is crucial in that pregnancy is the leading cause of death among girls ages 15 to 19 worldwide, primarily because immature female bodies are not ready for the stress of pregnancy and birth. Globally, babies born to women under 18 have a 60% higher chance of dying in infancy than do those born to women over 18.

Typically, women also have less access to education than men (around the world, 70% of youth who leave school early are girls). They are less likely to have access to information and paid employment, and so have less access to wealth and political power. When they do work outside the home (as is increasingly the case in every world region), women tend to fill lower-paid positions, often as laborers, service workers, or lower-level professionals. And even when they work outside the home, most women retain their household duties, so they work a *double day*.

The extent to which physical differences between males and females affect their social roles is still being debated. On average, men are stronger and taller, but women have more endurance and pain tolerance relative to men. Women's physical capabilities are somewhat limited during pregnancy and nursing, but from the age of about 45, women are no longer subject to the limitations related to pregnancy, and most contribute in some significant way to the well-being of their adult children and grandchildren, an important social and economic role. A growing number of biologists suggest that the evolutionary advantage of menopause in midlife is that it gives women the time, energy, and freedom to help succeeding generations thrive. This notion is called the *grandmother hypothesis* (**Figure 1.30C**). Certainly, grandfathers can play nurturing roles, but women tend to live 5 years longer than men.

PHOTO ESSAY

1.30 Population, Gender, and Culture

Gender roles are shifting as societies develop economically and politically, and as access to education, careers, and health care increases for women. Despite the clear benefits of women leaving domestic spaces and entering public life, gender inequality persists almost everywhere, manifesting in many ways, some of which can be measured more easily than others. The map displays the UN Gender Development Index (GDI), and the photos illustrate some of the gender issues that the map captures and others that it doesn't.

THINKING GEOGRAPHICALLY

A Why is it difficult to compare countries in terms of the incidence of gender-based violence?

B What percentage of human-trafficking victims are women?

D If these girls receive 7 or more years of education, what are some likely "ripple effects" according to research on gender equality?

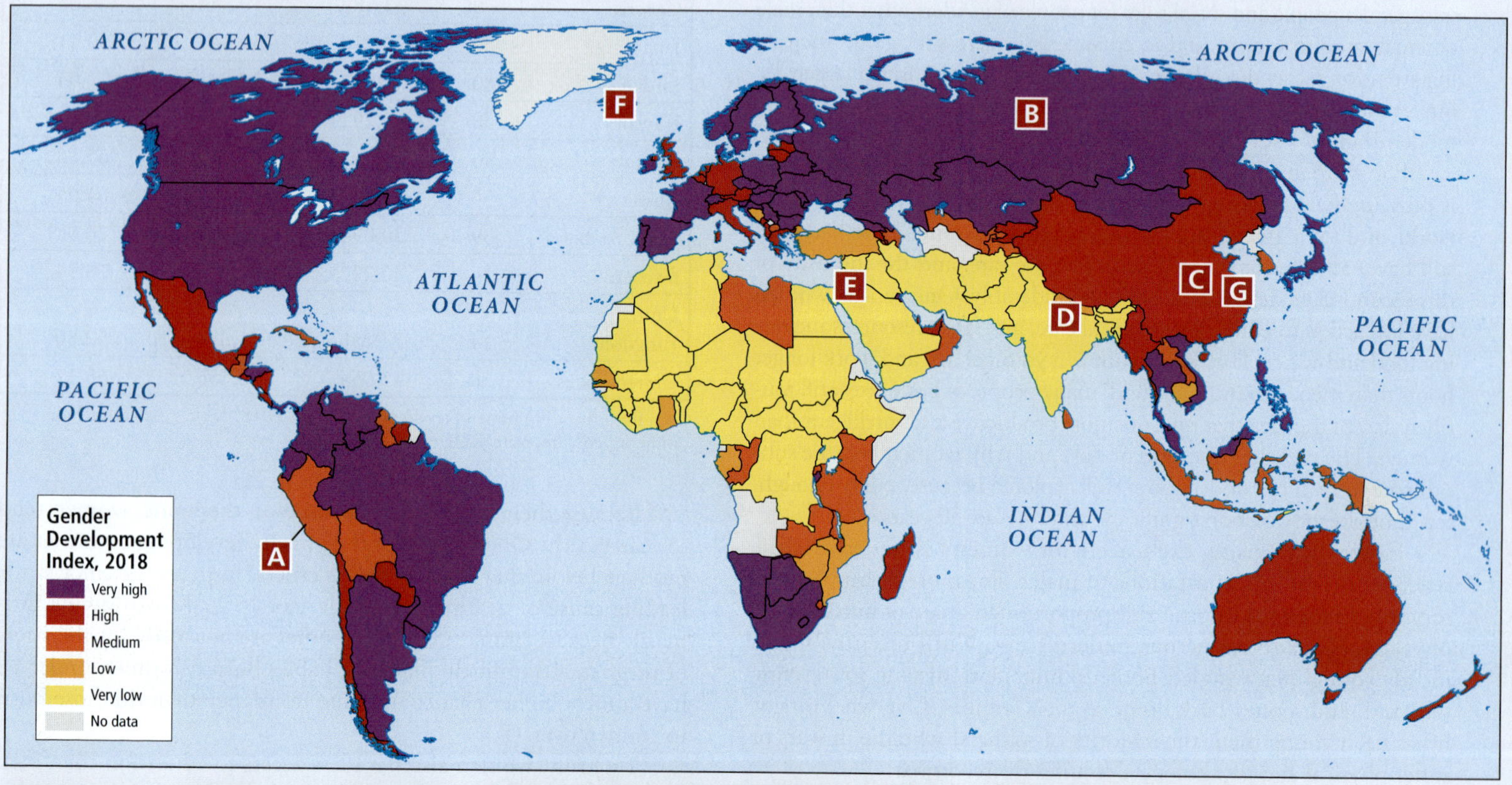

A Gender-based violence. Women protest against gender-based violence in Lima, Peru, where some surveys suggest that half of women have experienced such violence, including rape and assault. [Guillermo Gutierrez/SOPA Images/LightRocket via Getty Images]

B Human trafficking. Nikolai Rantchev holds up a portrait of his daughter, who was kidnapped by sex traffickers posing as a translation agency and taken to Cyprus, where she was eventually murdered. Corruption in government and the power of criminal organizations have limited efforts to combat human trafficking in Russia. [STEFANOS KOURATZIS/AFP/Getty Images]

C The grandmother hypothesis. A grandmother in Xian, China, helps her toddler granddaughter learn to walk. In every culture and community worldwide, grandmothers contribute to the care and education of their grandchildren. The idea that this has played an essential role in human evolution is called the grandmother hypothesis. [Tim Graham/Getty Images]

D **Education.** Girls at a school in Gorakhpur, India. Surveys suggest that while support for girls' education is strong in India, 63% think university education is more important for boys than girls. Lower educational opportunities for girls are a major source of gender inequality in India. [Andrew Aitchison/In Pictures Ltd./Corbis via Getty Images]

E **Child marriage and pregnancy.** A 13-year-old girl at her engagement party in a Syrian refugee camp in Jordan. Sexual harassment of single females in the camps has motivated parents to marry their girl children. Childbirth at such a young age is risky, with complications from pregnancy the leading cause of death globally for girls between the ages of 15 and 19. [Sam Tarling/Corbis via Getty Images]

F **Economic inequality.** Under Prime Minister Katrin Jakobsdottir, Iceland made it illegal to pay men and women differently for the same work. Iceland has been a consistent world leader in gender equality. [Marlene Awaad/Bloomberg via Getty Images]

G **Transgender issues.** Xiaomi, a self-described transgender woman, crosses a street in Shanghai, China. After generations of repression, transgender people are becoming more visible in East Asia, especially in larger cities. [JOHANNES EISELE/AFP/Getty Images]

Gender Equality and Development

Traditional notions of gender roles are now being challenged by demands for greater equality seemingly everywhere. In many countries, including most MDCs, women are acquiring education at higher rates than men. It will take women a while to catch up in the job market, where throughout the world women are paid less than men for doing the same work. This *gender pay gap* has fallen significantly in the past few decades in most countries, due to declining discrimination against women.

Research data suggest that there is a "ripple effect" that benefits the whole group when developing countries pay attention to the need for gender equality:

- Girls in developing countries who get 7 or more years of education marry 4 years later than average and have 2.2 fewer children.
- An extra year of secondary schooling over the average for her locale boosts a girl's lifetime income by 15% to 25%.
- The children of educated mothers are healthier and more likely to finish secondary school.
- When women and girls earn income, 90% of their earnings are invested in the family, compared to just 40% of males' earnings.

A new measure of gender equality is the *UN Gender Development Index (GDI)*, a composite measure reflecting the degree to which women and men are equal within a particular country in three dimensions: longevity, education, and income (see Figure 1.30). Ranks are from most equal (1) to least equal (5). Not all countries are ranked because many did not submit data.

Gender-Based Violence

An emerging body of research suggests that gender-based violence is widespread but rarely reported, with, for example, around 80% of rapes going unreported in the United States. Comparisons between countries are complicated by differing laws and social norms about reporting violence. For example, rape statistics would suggest that more rape occurs in Sweden than anywhere else in the world. Actually the opposite is closer to the truth, in that Sweden has such strong laws discouraging rape that women and men are more comfortable reporting it.

Gender-based violence is often assumed to be perpetrated exclusively by men against women, and most data indicate that women are much more likely to be victims. A recent survey of 10,000 men in six Asian countries indicated that one in four participants had raped a woman, usually a romantic partner. Seventy percent of the men who had committed rape said they felt entitled to do so. However, recent surveys in the United States suggest that males are victims in between 10% to 40% of gender-based violence, most of which occurs during childhood, and that women are the perpetrators in up to a third of these cases. The most likely group to experience gender-based violence is lesbian, gay, bisexual, transgender, and questioning (LGBTQ) people, with recent surveys in the United States suggesting that up to 44% of lesbians and 40% of gay men are victims, versus 35% of heterosexual women and 21% of heterosexual men. Surveys of transgender people suggest roughly half are sexually assaulted, with racial minorities significantly more at risk. These and similar studies done in other countries suggest that gender-based violence is a widespread global phenomenon deeply rooted in societal power relations and that preventative measures are most effective during childhood and adolescence.

culture all the ideas, materials, and institutions that people have invented that are not directly part of our biological inheritance

Human Trafficking and Modern Slavery

The trade of humans for the purpose of forced labor, sexual slavery, or prostitution is known as human trafficking. The International Labor Organization of the UN estimates that over 40 million people are in "modern slavery" with 16 million working as domestic servants, construction laborers, or agricultural laborers, 5 million in forced sexual exploitation, 5 million in forced labor imposed by state authorities, and 15 million in forced marriage. Women make up 71% of trafficked humans, and children 25%. Of those forced into the sex trade, 99% are women. Human trafficking is a global problem that is most common in sub-Saharan Africa, Southwest Asia, South Asia, and North Korea. Over $150 billion is generated by human trafficking annually.

Government-led efforts to combat trafficking are often ineffective due to the illegal and clandestine nature of this activity. Human trafficking is more common in the informal economy, where activities are often hidden and unregulated. Hence, increasing human development, enforcing labor laws, and improving the security of women and children are general strategies to combat trafficking. Because migration, especially in high-conflict zones and areas where governments are unstable, is a common context for trafficking, humanitarian efforts by NGOs and governments can be effectively coordinated in these areas with antitrafficking efforts. International pressure on states that practice forced labor, such as North Korea and China, has been only somewhat effective.

CHECK YOUR UNDERSTANDING

1. Why is population growth slowing throughout the world?
2. What are some implications of aging populations?
3. What is the difference between gender and sex?
4. What are some of the global patterns related to gender? How are they changing?
5. What group is most likely to be a target of gender-based violence?

CULTURE

Many geographers are interested in the cultural practices of humans and the spatial and environmental patterns these factors create. Cultural geography focuses on culture as a complex of important distinguishing characteristics of human societies. **Culture** comprises everything people use to live that is not directly part of biological inheritance. Culture is represented by the ideas, materials, methods, and social arrangements that people have invented and passed on to subsequent generations, such as methods of producing food and shelter. Culture includes language, music, tools and technology, clothing, gender roles, religion, and belief systems such as those prescribed in Confucianism, Hinduism, Islam, Judaism, Christianity, and a variety of indigenous belief systems.

Ethnicity and Culture: Slippery Concepts

A group of people who share a location of origin, a set of beliefs, a way of life, a technology, and usually a common ancestry and sense of common history form an **ethnic group**. The term *culture group* is often used interchangeably with *ethnic group*. Both of the concepts of culture and ethnicity are imprecise, especially as they are popularly used. For instance, as part of the modern globalization process, migrating people often move well beyond their customary cultural or ethnic boundaries to cities or even distant countries. In these new places they take on many new ways of life and beliefs—their culture actually changes, yet they still may identify with their cultural or ethnic origins.

The Kurds in Southwest Asia are an example of the tenacity of this ethnic or cultural group identity. Long before the U.S. war in Iraq, the Kurds were asserting their right to create their own country in the territory where they have lived as nomadic herders since before the founding of Islam (600 C.E.). The Kurds are a stateless nation and Syria, Iraq, Iran, and Turkey all claim parts of the traditional Kurdish area, while Kurds have de facto control over part of this homeland. Many Kurds are now educated urban dwellers, living and working in modern settings in Turkey, Iraq, Iran, or even London and New York, yet they actively support the cause of establishing a permanent Kurdish homeland. Although urban Kurds think of themselves as ethnic Kurds and are so regarded in the larger society, they do not follow the traditional Kurdish way of life. We could argue that these urban Kurds have a new identity within the Kurdish culture or ethnic group. Or they may be in a *transcultural* position, moving from one culture to another. Some scholars, such as Benedict Anderson, argue that a sense of belonging to any community that is not based on regular interaction between individuals is largely grounded in the individual's imagination. This may help explain how a person's beliefs and ways of living can change so profoundly, even while they may still identify with a particular culture or ethnicity.

Another problem with the imprecision of the concept of culture is that it is often applied to a very large group that shares only the most general of characteristics. For example, one often hears the terms American culture, African American culture, or Asian culture. In each case, the group referred to is far too large to share more than a few broad characteristics. It might fairly be said, for example, that U.S. culture is characterized by beliefs that promote individual rights, autonomy, and individual responsibility. But when we look beyond these broad abstractions and get to specifics, just what constitutes the rights and responsibilities of the individual are quite debatable. In fact, U.S. culture encompasses many subcultures that share some of the core set of beliefs but disagree over parts of the core and over a host of other matters. The same is true, in varying degrees, for all other regions of the world. When many culture groups live in close association more or less amicably, the society may be called *multicultural.*

Values

Occasionally you will hear someone say, "After all is said and done, people are all alike," or "People ultimately all want the same thing." It is a heartwarming sentiment, but an oversimplification. True, we all want food, shelter, health, love, and acceptance, but global research into values shows that not only are people quite different, the act of studying and understanding values on a global scale is a hugely difficult task that necessarily involves the simplification of complex information about our differences. The products of such research are often criticized as revealing more about the people doing the research than about the actual values being studied.

For example, the World Values Survey is a global network of social scientists who have been collecting data about values from countries that contain 90% of the world's population. Its main findings are that the variation in what people value can be understood in terms of two main continuums: "traditional values versus secular rational values" and "survival values versus self-expression values." **Figure 1.31** displays World Values Survey data on a graph of these two continuums, with country-level data grouped into regions. From a regional geography perspective, it is interesting to note how the data suggest that world regions have some coherence when measured in this way (even if the regions identified in the survey data are slightly different from those used in this book).

The World Values Survey defines "traditional values" as centering around religion, deference to authority, marriage, and traditional family values, while "secular rational values" emphasize science, materialism, and personal autonomy. "Survival values" prioritize security, political nonaction, patriarchy, distrust of outsiders, and a weak sense of happiness, while "self-expression values" support political empowerment, gender and LGBTQ nondiscrimination, acceptance of outsiders, and a strong sense of happiness. The authors' interpretation of the data suggests that values become more oriented toward self-expression as the sense of security and autonomy increases. On a societal scale this involves, and possibly even causes, a transition from subsistence agriculture and resource extraction livelihoods to industrial and manufacturing employment, and on to service and knowledge sector dominance of the economy.

Critics of the World Values Survey claim the study itself is *ethnocentric*, meaning that its authors judge a wide variety of cultures according to the standards of their own culture. Many of the studies' authors are indeed from western and northern Europe ("Protestant Europe" in Figure 1.31), and critics argue that the wording of questions asked in the surveys reveals a subtle bias in favor of "self-expression" and "secular rational values" that are, by the study's own findings, much more prevalent in the authors' native cultures. Critics claim that this ethnocentrism extends into the interpretation of the huge volume of information collected in the surveys, with the simplification of complex data into two continuums of "traditional values versus secular rational values" and "survival values versus self-expression values" reflecting the values of "Protestant Europe" more than any deeper insights into human values.

Many efforts to study values and other aspects of human culture, especially on a broad global scale, run into similar problems as the World Values Survey, and this book is no exception. Especially when trying to digest complex information into a few comprehensible statements, cultural bias and a certain level of ethnocentrism may be inescapable, even for highly trained social scientists. Perhaps the best we can do when trying to understand other cultures and how their values differ from our own is to be aware of our own biases and values and not pretend they are absent.

ethnic group a group of people who share a common ancestry and sense of common history, a set of beliefs, a way of life, a technology, and usually a common geographic location of origin

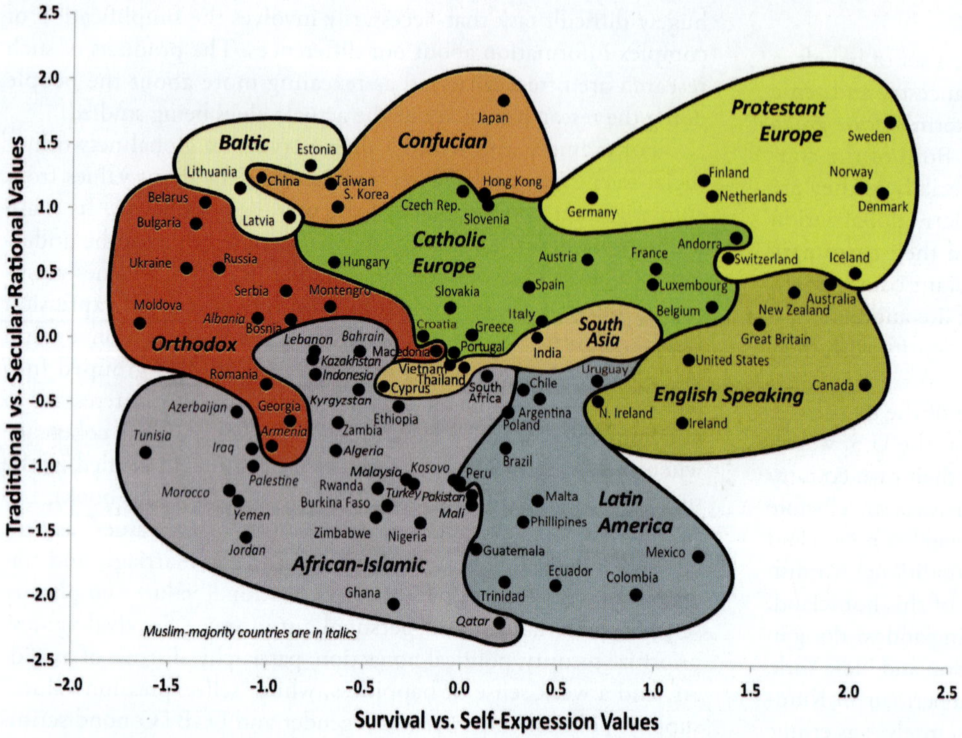

Figure 1.31 World Values Survey cultural map of the world, 2017.

The regions depicted in this graphic correspond to the regions in this book as follows:
Protestant Europe: Europe (western and northern: Chapter 4)
Catholic Europe: Europe (southern and eastern: Chapter 4)
English Speaking: North America (Chapter 2), Oceania (Chapter 11)
Confucian: East Asia (Chapter 9)
Orthodox: Russia and the post-Soviet states (Chapter 5), Europe (eastern: Chapter 4)
South Asia: South Asia (Chapter 8), Southeast Asia (Chapter 10)
African Islamic: North Africa and Southwest Asia (Chapter 6), sub-Saharan Africa (Chapter 7), Southeast Asia (Chapter 10)
Latin America: Middle and South America (Chapter 3)
[World Values Survey]

Religion and Belief Systems

The religions of the world are formal and informal institutions that embody value systems. Most have roots deep in history, and many include a spiritual belief in a higher power (such as God, Yahweh, or Allah) as the underpinning for their value systems. Today, religions often focus on reinterpreting age-old values for the modern world. Some formal religious institutions—such as Islam, Buddhism, and Christianity—proselytize; that is, they try to extend their influence by converting others. Others, such as Judaism and Hinduism, accept converts only reluctantly. Informal religions, often called *belief systems,* have no formal central doctrine and no firm policy on who may or may not be a practitioner. The fact that many people across the world combine informal religious beliefs with their more formal religious practices, of whatever persuasion, accounts for the very rich array of personal beliefs found in the world today.

Religious beliefs are often reflected in the landscape. For example, settlement patterns can demonstrate the central role of religion in community life: village buildings may be grouped around a mosque, a temple, a synagogue, or a church, and the same can be said for urban neighborhoods. In some places, religious rivalry is a major feature of the landscape. Certain spaces may be clearly delineated for the use of one group or another, as in Northern Ireland's Protestant and Catholic neighborhoods.

Religion is often used to support powerful interests in society. For example, during the era of European colonization, religion (Christianity) was used, sometimes forcibly, as a way to weaken resistance to rule by outsiders. Today religion is used by many leaders, political parties, and governments throughout the world to appeal to voters on the basis of their supposed piety or other values shared by a particular religious group. In recent decades major religious political movements have transformed politics in the United States, South Asia, Russia, North Africa, Southwest Asia, and much of sub-Saharan Africa.

Figure 1.32 shows the distribution of the major religious traditions on Earth today; it demonstrates some of the religious consequences of colonization. Note, for instance, the distribution of Roman Catholicism in parts of the Americas, Africa, and Southeast Asia, all places colonized by European Catholic countries.

Religion also plays a major role in economies. Throughout human history religions have spread through trade contacts. In the seventh and eighth centuries, Islamic people used a combination of trade and political power (and, less often, actual conquest) to extend their influence across North Africa, throughout Central Asia, and eventually into South and Southeast Asia (see Figure 6.15). Recent studies may help explain how religion and trade are linked, by demonstrating how people who engage in commerce often are more trusting of members of their own faith. Other research suggests a more complicated relationship, showing that religious belief tends to decline with economic development and urbanization, and that higher attendance at religious services diminishes economic activity. Meanwhile belief in an afterlife correlates with slightly higher economic growth in developing countries. These findings may reflect the growth of fundamentalist evangelical Christian churches in developing regions (see Chapters 3 and 7), which often teach that God rewards faith with wealth.

Language

Language is central to almost every aspect of human culture and every type of geography discussed in this book. Today there are roughly 7000 languages in the world, half of which will be lost in the next 100 years, mostly because of globalization and the

1.32 Major Religions of the World

The small symbols on the map indicate a localized concentration of a particular religion within an area where another religion is predominant.

A Indigenous religion, Mexico [LUIS ACOSTA/AFP/Getty Images]

B Islam, Egypt [YEHUDA RAIZNER/AFP/Getty Images]

C Christianity, Ethiopia [PETER DELARUE/AFP/Getty Images]

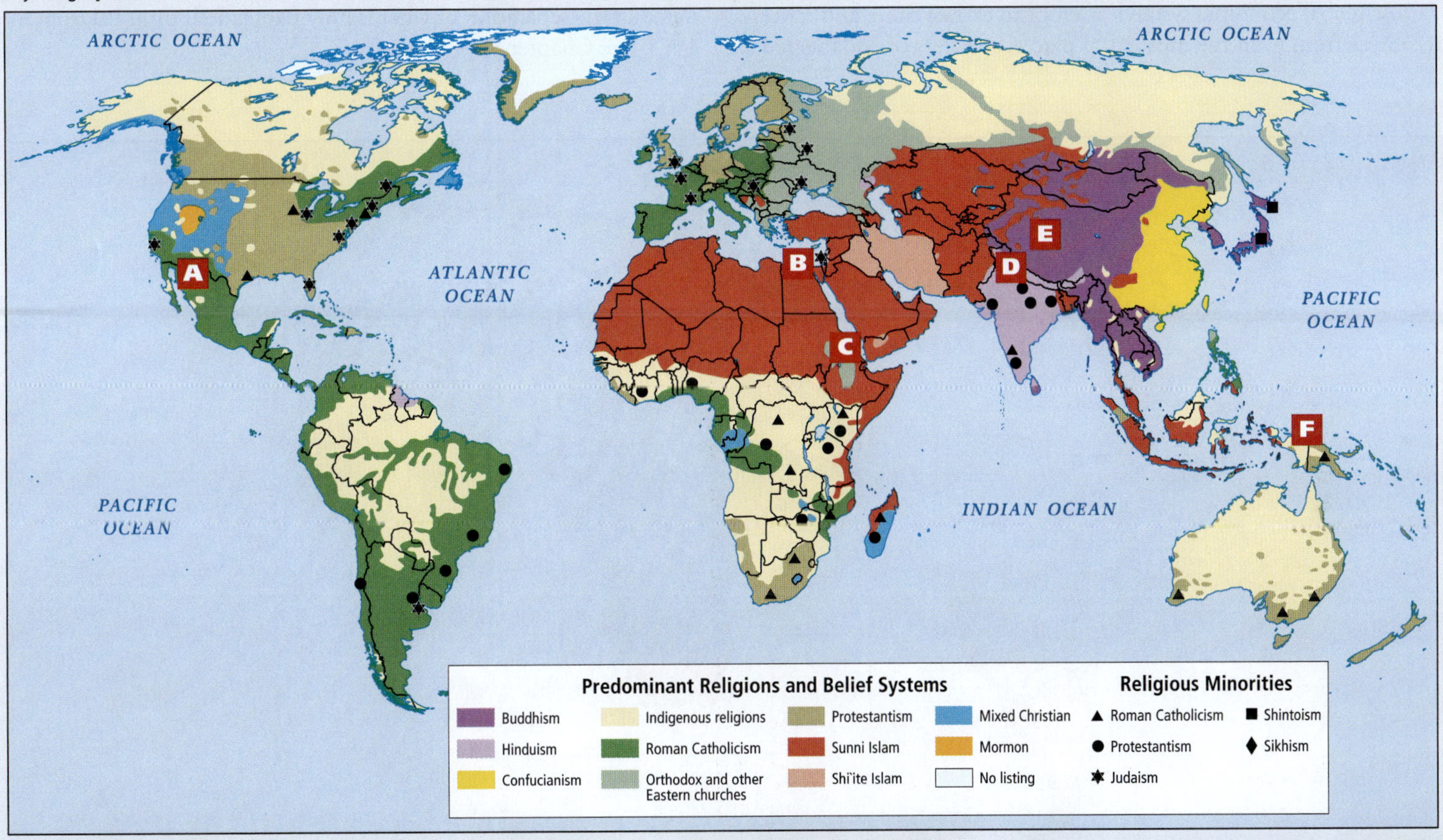

D Hinduism, India [Sonu Mehta/Hindustan Times via Getty Images]

E Buddhism, Tibet [TPG/Getty Images]

F Indigenous religion, West Papua, Indonesia [ROMEO GACAD/AFP/Getty Images]

tendency it has of favoring larger culture groups with more people to sell products to, living under larger, more powerful governments that can shape global trade and politics to their advantage.

Why should we care that we are losing languages at the rate of one every 2 weeks? Environmental knowledge is a major reason. Many languages now going extinct contain vast knowledge of the environments in which they developed, with names for plants and animals, and descriptions of the functioning of natural systems that simply don't exist in other languages. For example, the Maya people of the Lacondon forest in southern Mexico practice a highly productive system of farming within intact forests, the success of which depends on concepts contained in their native language. Losing this and other languages means losing important ecological knowledge that may help humans survive the many challenges we face. The global distribution of the major groups of dominant languages, and the larger language families, can be seen in **Figure 1.33**.

As global economic systems attempt to extract more and more resources from even the most rural places, speakers of endangered languages are usually on the losing end, culturally speaking. As they lose access to land and resources and are forced to urbanize, they have to learn new languages, which their children often learn instead of the language their parents spoke. However, sometimes younger generations try to revitalize dying languages, as in the case of Darrick Baxter, who created a cell phone app to teach his daughter their ancestral Ojibwe language. Baxter made the code for the app available for free download, and it has since been adapted to teach hundreds of endangered languages worldwide.

Language is often central to politics, with every country having had to make the difficult decision of which languages to favor in education, government, and commerce. This decision inevitably benefits certain groups over others, creating tensions that sometimes lead to armed conflict. For example, when in 1947 Urdu and English were declared the only official languages of Pakistan, a movement of linguistic and cultural pride grew in Bengali-speaking East Pakistan that played a major role in the civil war that culminated in the separation of what is now Bangladesh from Pakistan in 1971 (see Chapter 8).

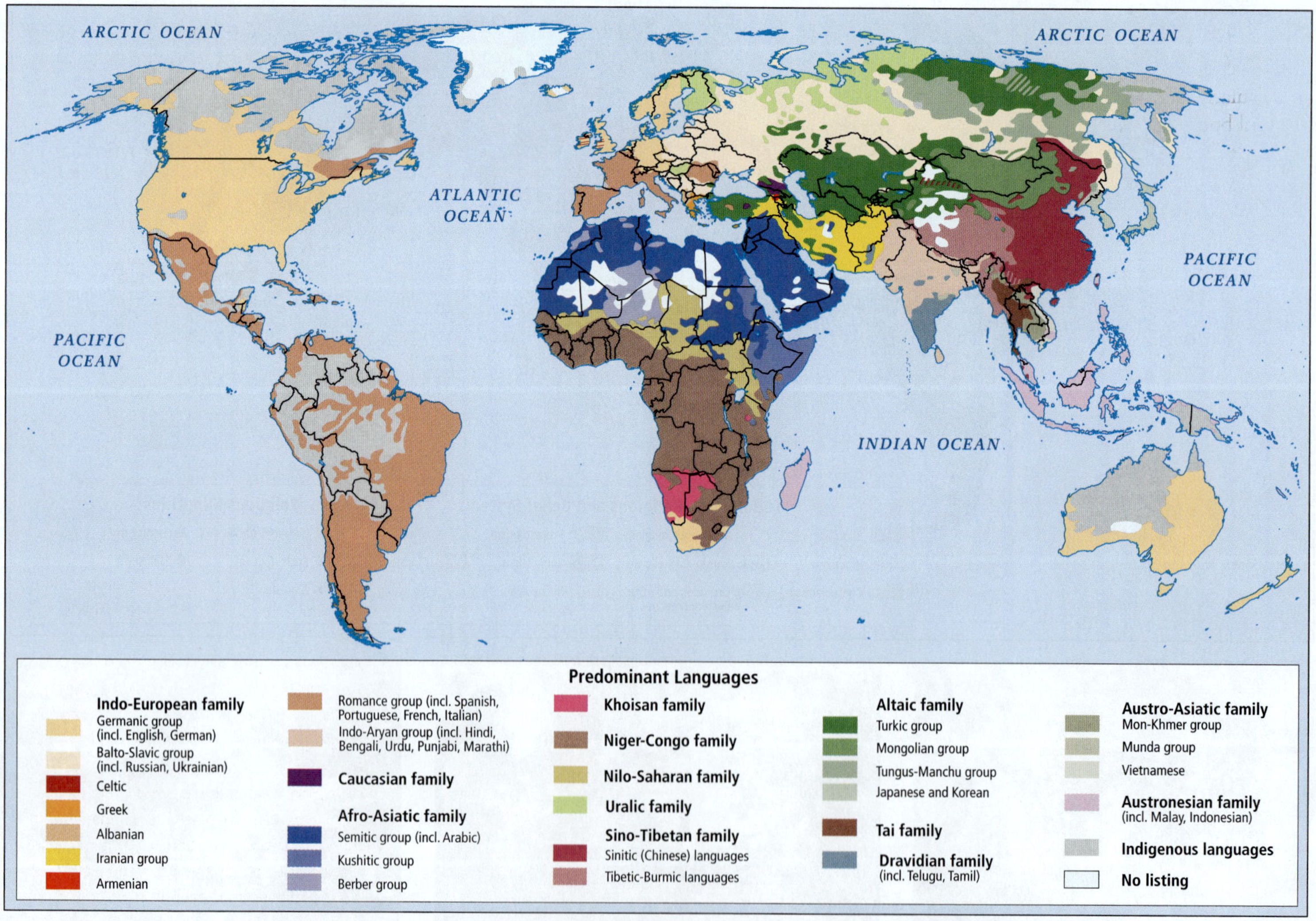

Figure 1.33 World's major language families. Distinct languages (Spanish and Portuguese, for example) are part of a larger group (Romance), which in turn is part of a language family (Indo-European).

English is the most spoken language in the world, with 1.13 billion speakers, having in the last two years overtaken Mandarin Chinese (1.11 billion speakers), which had been the world's most spoken language for more than 1000 years. Trade, migration, and imperialism fueled the spread of English and other European languages after 1500, when the languages of Spanish, Portuguese, French, Dutch, and English colonists began to replace the languages of the people they traded with and conquered. English became dominant due to the vast British Empire, and the dominance of the United States after World War II. Today it is the primary language of international trade, finance, diplomacy, foreign aid, and the Internet, where more than half the world's websites are created in English. Outside of the countries where English is the primary language, such as the United States, the United Kingdom, Canada, Australia, and New Zealand, it is a language of political and economic elites, whose privilege is, intentionally or not, upheld by powerful global corporations and financial institutions like the IMF, the World Bank, and the WTO whose activities are usually conducted in English.

Race

Like ideas about gender roles, ideas about race affect human relationships everywhere on Earth. However, while race is of enormous social significance across the world, biologists tell us that from a scientific standpoint, race is a meaningless concept. The characteristics we popularly identify as **race** markers—skin color, hair texture, and face and body shape—have no significance as biological categories. All people now alive in the world are members of one species, *Homo sapiens sapiens*. Many invisible biological characteristics, such as blood type and DNA patterns, cut across cultural race-based groupings made on the basis of skin color or hair texture or facial features. In fact, over the last several thousand years there has been such massive gene flow among moving human populations that no modern group presents a discrete set of biological characteristics. Although any two of us may look quite different, from the biological point of view we are all simply *Homo sapiens sapiens* and are closely related.

Some of the easily visible features of particular human groups evolved to help them adapt to environmental conditions. For example, biologists have shown that people with darker skin (containing a high proportion of protective melanin pigment) evolved in regions close to the equator, where sunlight is most intense (**Figure 1.34**). All humans need the nutrient vitamin D, and sunlight striking the skin helps the body absorb vitamin D. Too much of the vitamin, however, can result in improper kidney functioning. Dark skin absorbs less vitamin D than light skin and thus would be a protective adaptation in equatorial zones. In higher latitudes, where the Sun's rays are more dispersed, light skin facilitates the sufficient absorption of vitamin D; darker-skinned people at these higher latitudes may need to supplement vitamin D to be sure they get enough, since D deficiencies can result in several health risks. Light-skinned people with little protective melanin in their skin—if they live in equatorial or high-intensity sunlit zones (parts of Australia, for example)—need to protect against too much vitamin D. All skin types need to protect against serious sunburn, and skin cancer. Similar correlations have been observed between skin color, sunlight, and another essential vitamin, folate, which if deficient can result in birth defects.

Despite its biological insignificance, over time, race has acquired enormous cultural significance as humans from different parts of the world have encountered each other in situations of unequal power. *Racism*—the idea that skin color is a primary determinant of human abilities and even cultural traits—has often been invoked to justify the oppression, enslavement, or genocide of particular groups, as well as confiscation of their land and resources. Race and its implications in North America will be covered in Chapter 2, and the topic will be discussed in several other world regions as well.

race a social or political construct that is based on apparent characteristics such as skin color, hair texture, and face and body shape, but that is of no biological significance

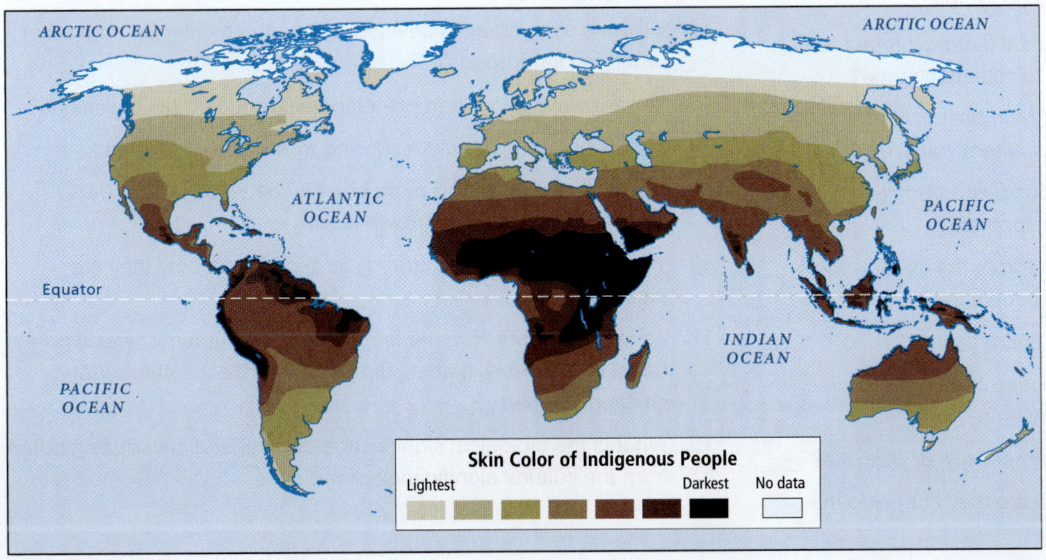

Figure 1.34 Skin color map for indigenous people as predicted from multiple environmental factors. Skin color has a biological role completely separate from any other human characteristics. That is, skin plays a twin role with respect to the Sun: protection from excessive UV radiation and absorption of enough sunlight to trigger the production of vitamin D. [Research from: UNEP/GRID-Arendal, at http://www.grida.no/graphicslib/detail/skin-colour-map-indigenous-people_8b88, Emmanuelle Bournay, cartographer; and G. Chaplin (2004), "Geographic Distribution of Environmental Factors Influencing Human Skin Coloration," *American Journal of Physical Anthropology*, 125, 292–302; map updated in 2007]

Still, recognizing all the ills that have emerged from racism and similar prejudices, we need not infer that human history has been marked primarily by conflict and exploitation or that these conditions are inevitable. The science of psychology has shown, through numerous experiments, that humans have strong inclinations toward *altruism*, the willingness to sacrifice one's own well-being for the sake of others. Studies across many scientific disciplines suggest that altruism, especially within groups, has been a strong force supporting the development of human societies since prehistoric times. It could be argued that altruism is the norm, and that one reason inhumane behavior is so distressing is that it is an anomaly.

CHECK YOUR UNDERSTANDING

1. What are some of the lenses through which cultural geographers seek to understand humanity?
2. What is an ethnic group?
3. How was religion used during the era of European colonization?
4. What are some languages that have become dominant due to global trade?
5. Why is race biologically meaningless?

CRITICAL THINKING QUESTIONS

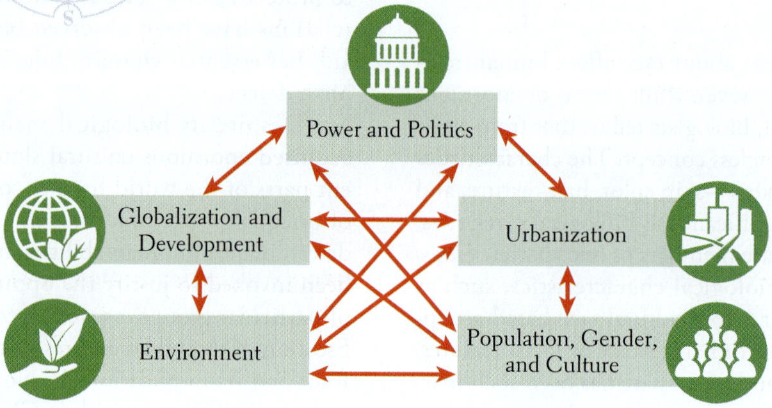

The diagram represents connections among the five geographic themes around which this book is structured. Listed below are some important questions that have been addressed in this chapter. Answer each question, and indicate which themes are involved in your response.

1. What are the main concerns of physical geography?
2. How do multiple human and environmental factors interact to influence the vulnerability of a location to climate change?
3. How do the vulnerabilities of the city of Mumbai to climate change compare to the vulnerabilities of where you live?
4. How are global flows of information, goods, and people transforming patterns of economic development?
5. How is globalization evident in your life—from the clothes you wear, to your favorite foods, to your career plans?
6. What is the difference between a state and a nation?
7. What are some examples of shifts in modes of government? Take your examples from recent world events.
8. How do the political freedoms you enjoy shape your daily life?
9. What are the BRICS countries, and how are they changing the global geopolitical order?
10. How has urbanization been influenced by the development of manufacturing, growth in service economies, and the mechanization of agriculture?
11. How far back in your family history would you need to go to find ancestors who lived in a rural area and grew almost all their own food?
12. How are slum areas in developing countries different from other urban neighborhoods?
13. In what ways can urban planning help reduce GHG emissions?
14. Why is population growth slowing throughout the world?
15. How is the shift toward more equality between the genders influencing population growth rates?
16. What are some global patterns in gender and how they are changing?
17. Ask your parents who the first woman in your family was who had a career. How, if at all, did this influence the number of children she had?
18. What is the biological significance of race, and how does it differ from the cultural significance of race?

Key Terms

authoritarianism 39
biosphere 18
birth rate 47
capitalism 42
civil society 42
climate 13
climate change 18
colonialism 33
communism 42
cultural landscape 4
culture 54
death rate 47
democratization 39
demographic transition 47
development 31
ethnic group 55
floodplain 12
formal region 9
free trade 37
functional region 9
gender 50
geographic information science (GISc) 8
geopolitics 42
global warming 18
globalization 33
green revolution 30
greenhouse gases (GHG) 18
gross domestic product (GDP) per capita 31
gross national income (GNI) per capita 31
human geography 3
Industrial Revolution 29
informal economy 32
landforms 12
latitude 5
less developed country (LDC) 31
longitude 5
map projections 5
middle income country (MIC) 31
more developed country (MDC) 31
nation 38
nation-state 38
nongovernmental organizations (NGOs) 42
orographic precipitation 15
physical geography 3
plate tectonics 12
political ecology 32
political freedoms 39
population pyramid 50
primary sector 31
purchasing power parity (PPP) 31
quaternary sector 31
race 59
rain shadow 15
rate of natural increase (RNI) 47
region 9
scale 4
secondary sector 31
sex 50
slum 43
state 38
sustainable development 32
tertiary sector 31
total fertility rate (TFR) 48
United Nations (UN) 42
United Nations Human Development Index 32
urbanization 43
vernacular region 9
weather 13

Answer to question about Figure 1.4: Nigeria

More Practice at SaplingPlus

Read the interactive e-text, review key concepts, and check your understanding.

Migrant workers.
[Justin Sullivan/Getty Images]

2 North America

Like the migrant in the photo, Carlos Hernandez works in the fields surrounding Mendota, the "Canteloupe Capital of the World," in California's Central Valley. A 67-year-old father of three, Carlos has been here since fleeing El Salvador during a civil war. While tending the crops that feed North America, Carlos has endured drought, recession, unemployment, and hunger. Now he has a new worry as the U.S. government, alarmed by Latino gang activity, has been deporting people like Carlos who don't have legal immigration status. "I've worked here for 40 years. My wife, kids, everybody is here. I don't have anyone back in El Salvador."

Downtown at Pupuseria Morenita, Mendota's city manager is munching on Salvadoran pupusas and writing to his congressional representatives, describing what Mendota would face if everyone without citizenship were deported: "Economically it would be devastating. Culturally our diversity would be lost." While his senator may be sympathetic, his representative, himself a child of immigrants, recently voted to deport more illegal immigrants.

Now a new storm looms for Carlos: mechanization. Farmers throughout California are turning to machines to harvest crops as labor becomes more scarce and expensive. Mechanization places over half of Mendota's agricultural jobs at risk. The remaining jobs will pay better and be easier since instead of stooping to pick cantaloupes in the hot sun, workers will sort them while riding on a harvester under a shade cloth. Having been in Mendota so long, Carlos thinks he can get one of these jobs, but there's a catch. Better paying, easier jobs will be more attractive to U.S. citizens.

Learning Objectives

Environment: Physical and Human

2.1 Describe how landforms influence the movement of air masses in North America, shaping the region's climate.

2.2 Explain how North America's massive consumption of resources impacts environments within the region and globally.

2.3 Identify historical patterns of subregional interaction within North America, and explain how they relate to current trends in economic development.

Globalization and Development

2.4 Describe North America's position in the global economy and how globalization has transformed the region.

Power and Politics

2.5 Identify the major similarities and differences between Canada and the United States in the roles each country's government plays domestically and internationally.

Urbanization

2.6 Explain how North America's urban areas have changed since World War II, and the factors that have driven these changes.

Population, Gender, and Culture

2.7 Identify how the changing role of women in society has contributed to the aging of North American populations.

2.8 Describe how North America's population distribution is changing.

2.9 Explain how the increase in migration from Middle and South America and parts of Asia is changing the culture and politics of North America.

Subregions of North America

2.10 Identify key characteristics of the subregions of North America.

RUSSIA
Bering Strait
Bering Sea
ARCTIC OCEAN
Lincoln Sea
Ellesmere Island
Queen Elizabeth Islands
Thule
Aleutian Islands
Yukon
Brooks Range
ALASKA
Mt. McKinley elev. 20,320
Fairbanks
Alaska Range
Anchorage
Valdez
Gulf of Alaska
Beaufort Sea
Tuktoyaktuk
Banks Island
Viscount Melville Sound
Resolute
Mackenzie
Klondike Region
Great Bear Lake
Victoria Island
Baffin
YUKON
Whitehorse
Dogrib Territory
NORTHWEST TERRITORIES
NUNAVUT
Foxe Basin
Yellowknife
Great Slave Lake
Cape Dorset
BRITISH COLUMBIA
Rocky Mountains
Peace
Athabasca Lake
Athabasca
Reindeer Lake
Hudson Bay
Churchill
ALBERTA
CANADA
Coast Mountains
Edmonton
SASKATCHEWAN
MANITOBA
Vancouver Island
Vancouver
Victoria
Calgary
Saskatoon
Saskatchewan
James Bay
Canadian
Seattle
WASHINGTON
Regina
Lake Winnipeg
Mt. St. Helens elev. 9678
Mt. Rainier elev. 14,410
Winnipeg
Lake of the Woods
ONTARIO
Lake Nipigon
Columbia
Cascade Range
Missouri
OREGON
Columbia Plateau
MONTANA
NORTH DAKOTA
Lake Superior
IDAHO
Great
MINNESOTA
Mt. Shasta elev. 14,162
Yellowstone
Black Hills
SOUTH DAKOTA
Huron
PACIFIC OCEAN
WYOMING
Minneapolis
WISCONSIN
San Francisco
Snake
NEVADA
Great Salt Lake
Front Range
Platte
Sand Hills
Lake Michigan
MICHIGAN
Central Valley
Sierra Nevada
Great Basin
UTAH
Plains
NEBRASKA
IOWA
Chicago
Detroit
Denver
COLORADO
Pikes Peak elev. 14,110
ILLINOIS
INDIANA
OHIO
CALIFORNIA
Colorado
KANSAS
MISSOURI
Ohio
UNITED STATES
St. Louis
Los Angeles
Death Valley elev. -282
Grand Canyon
Colorado Plateau
ARIZONA
KENTUCKY
San Diego
Tijuana
Phoenix
NEW MEXICO
OKLAHOMA
Arkansas
Knoxville
TENNESSEE
Tennessee
ARKANSAS
Ouachita Mtns.
Red
Mississippi
Atlanta
Nogales
Ciudad Juárez
Dallas
MISSISSIPPI
ALABAMA
Baja California
Rio Grande
Pecos
TEXAS
Hermosillo
Brazos
Houston
LOUISIANA
New Orleans
Mulege
Gulf of California
MEXICO
Gulf of Mexico
Monterrey
Matamoros
Durango
Land Elevations
meters
feet
4877
16,000
3353
11,000
2134
7000
914
3000
305
1000
152
500
0
0
mi 0 100 200 300 400 500
km 0 200 400 600 800
1:24,000,000
Azimuthal Equidistant Projection

Stories like that of Carlos Hernandez and places like Mendota are not what people usually expect when they think about North America, a vast region renowned for its wealth, innovation, and economic opportunities (**Figure 2.1**). These stories reflect the experiences of many in this region where people and communities are struggling to adapt to economic and political changes driven by globalization. The United States has been the world's largest economy for 130 years and its strongest military power for more than half a century. But after decades of high-paying jobs being moved overseas or replaced by technology, of wealth becoming more concentrated in the hands of the upper middle class and the rich, and of multiple financial crises, the U.S. economy has lost much of its momentum. By some measures, the United States is now no longer the world's largest economy; in 2014, China's economy outgrew that of the United States in terms of its purchasing power. China will surpass the United States in all measures of economic size within the next decade. In the midst of these shifts, there is renewed interest in helping poor people navigate difficult economic times. For example, health care for low-income people received major government investment in the United States with the passage of the Affordable Care Act in 2010. These trends exist to a lesser extent in Canada, due largely to more stable and substantial support for government programs that help poor people.

What Makes North America a Region?

In this book we define North America as Canada and the United States, two countries that are linked because of their geographic proximity, similar history, and many common cultural, economic, and political features (**Figure 2.2**). This is a cultural definition of the region, not a physical or political one. Mexico and much of Central America are physically part of the continent of North America, but culturally they are more connected to Middle and South America. For similar reasons, Hawaii is part of Oceania, and Puerto Rico part of Middle and South America, even though both are politically part of the United States.

Terms in This Chapter

The term *North America* is used to refer to both countries. Even though it is common on both sides of the border to call the people of Canada "Canadians" and people in the United States "Americans," this text uses the terms *United States* or *U.S.*, rather than *America*, for the United States. Other terms relate to the cultural diversity of the region. The text uses the term **Latino** to refer to all Spanish-speaking people from Middle and South America, although their ancestors may have been European, African, Asian, or Native American. Native American people have different terms they prefer: some in the United States still favor *American Indian* while in Canada *Aboriginal peoples* is an umbrella term covering people who identify as *First Nations*, *Nunavut*, and *Métis* (people of mixed Native American and European ancestry). In this text we will most often use *Native American* for those in the United States and Aboriginal peoples for those in Canada.

Latino a term used to refer to all Spanish-speaking people from Middle and South America, although their ancestors may have been European, African, Asian, or Native American

◀ **Figure 2.1** Physical features of North America.

Figure 2.2 Political map of North America.

Figure 2.3 The Rockies in Banff Springs, Alberta, Canada. [George Rose/Getty Images]

Figure 2.4 The Appalachian Mountains in Virginia. [KAREN BLEIER/ Getty Images]

ENVIRONMENT: PHYSICAL AND HUMAN

2.1 Describe how landforms influence the movement of air masses in North America, shaping the region's climate.

2.2 Explain how North America's massive consumption of resources impacts environments within the region and globally.

2.3 Identify historical patterns of subregional interaction within North America, and explain how they relate to current trends in economic development.

The continent of North America has almost every type of climate and a wide variety of landforms (Figure 2.1). Huge expanses of mountain peaks, ridges, and valleys meet expansive plains, long, winding rivers, myriad lakes, and extraordinarily lengthy coastlines. Here we focus on a few of the most significant landforms, the major climates, the many environmental consequences of the North American lifestyle, and the history of human settlement in the region.

LANDFORMS

A wide mass of mountains and basins, known as the Rocky Mountain zone, dominates western North America (**Figure 2.3**). It stretches down from the Bering Strait in the far north, through Alaska, and into Mexico. This zone formed about 200 million years ago when, as part of the breakup of the supercontinent Pangaea (see Figure 1.6), the Pacific Plate pushed against the North American Plate, thrusting up mountains. These plates still rub against each other, causing earthquakes along the Pacific coast of North America.

The much older and hence more eroded Appalachian Mountains stretch along the eastern edge of North America from New Brunswick and Maine to Georgia (**Figure 2.4**). This range resulted from very ancient collisions between the North American Plate and the African Plate.

Between these two mountain ranges lies the huge central lowland of undulating plains that stretches from the Arctic to the Gulf of Mexico. This landform was created by the deposition of deep layers of material eroded from the mountains and carried to this central North American region by wind and rain and by the rivers flowing east and west into what is now the Mississippi drainage basin.

During periodic ice ages over the last 2 million years, glaciers have covered the northern portion of North America. In the most recent ice age (between 10,000 and 25,000 years ago), the glaciers, sometimes as much as 2 miles (about 3 kilometers) thick, moved south from the Arctic, picking up rocks and soil and scouring depressions in the land surface. When the glaciers later melted, these depressions filled with water, forming the Great Lakes. Thousands of smaller lakes, ponds, and wetlands that stretch from

Figure 2.5 Central lowlands near Grand Marais, Minnesota. [Jeffrey Phelps/Getty Images]

Figure 2.6 Coastal lowlands near Larose, Louisiana. [Alex Ogle/AFP/Getty Images]

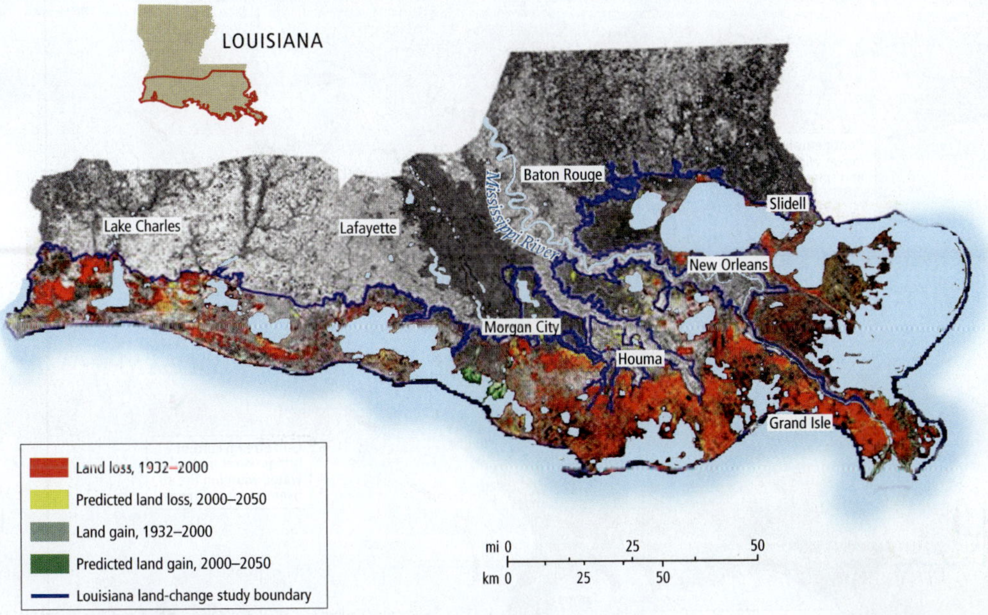

Figure 2.7 Wetland loss in Louisiana's Mississippi River Delta. The Louisiana coastline and the lower Mississippi River basin are the end point for the vast Mississippi drainage basin and provide coastal wildlife habitats, recreational opportunities, and transportation lanes that connect the vast interior of the country to the ocean and to offshore oil and gas. Most importantly, the wetlands provide a buffer against damage from hurricanes. Unfortunately, Louisiana has lost one-quarter of its total wetlands over the last century, largely because of human activity. The remaining 3.67 million acres constitute 14 percent of the total wetland area in the lower 48 states. [Research from: USGS/National Wetlands Research Center]

Minnesota and Manitoba to the Atlantic were formed in the same way (**Figure 2.5**). Melting glaciers also dumped huge quantities of soil throughout the central United States. This soil, often many meters deep, provides the basis for large-scale agriculture but remains susceptible to wind and water erosion.

East of the Appalachians, the Atlantic coastal lowland stretches from New Brunswick to Florida. It then sweeps west to the southern reaches of the central lowland along the Gulf of Mexico (**Figure 2.6**). In Louisiana and Mississippi, much of this lowland is filled in by the Mississippi River delta—a low, flat, swampy transition zone between land and sea (**Figure 2.7**). The delta was formed by massive loads of silt deposited during floods over the past 150-plus million years by the Mississippi, North America's largest river system. The delta deposit originally began at what is now the junction of the Mississippi and Ohio rivers at Cairo, Illinois; slowly, as ever more sediment was deposited, the delta advanced 1000 miles (1600 kilometers) into the Gulf of Mexico.

Over the centuries, human activities such as deforestation, deep plowing, and heavy grazing have led to erosion and added to the silt load of the rivers. The construction of levees during the last 300 years along riverbanks has drastically reduced flooding. Because of this flood control, much of the silt that used to be spread widely across the lowlands during floods is being carried to the southern part of the Mississippi delta—a low, flat zone characterized by swamps, lagoons, and sandbars. Some of these wetland areas are getting overloaded with silt, the weight of which is causing parts of the delta to sink into the Gulf of Mexico, a process called *subsidence*.

CLIMATE

The landforms across this continental expanse influence the movement and interaction of air masses and contribute to its enormous climate variety (**Figure 2.8**). Along the southern west coast of

2.8 Climates of North America

A **Arctic, Alaska** [Lowell Georgia/National Geographic/Getty Images]

B **Temperate midlatitude, Maryland** [Linda Davidson/The Washington Post via Getty Images]

C **Desert, Utah** [DEA/F. Barbagallo/De Agostini/Getty Images]

North America, the climate is generally mild (Mediterranean)—dry and warm in summer, cool and moist in winter. North of San Francisco, the coast receives moderate to heavy rainfall. East of the Pacific coastal mountains, climates are much drier because as the moist air sinks into the warmer interior lowlands, it tends to hold its moisture. This interior region becomes increasingly arid moving eastward across the Great Basin (see Figure 2.8C) and Rocky Mountains. Many dams and reservoirs for irrigation projects have been built to make agriculture possible. Because of the low level of rainfall, however, the amount of water that is being extracted exceeds the capacity of ancient underground water basins **(aquifers)** to replenish themselves.

On the eastern side of the Rocky Mountains, the main source of moisture is the Gulf of Mexico. When the continent is warming in the spring and summer, the air masses above it rise, sucking in warm, moist, buoyant air from the Gulf. This air interacts with cooler, drier, heavier air masses moving into the central lowland from the north and west, often creating violent thunderstorms and tornadoes (see Figure 2.8A). Generally, central North America is wettest in the eastern and southern parts and driest in the west and north (see Figure 2.8B). Along the Atlantic coast, moisture is supplied by warm, wet air above the Gulf Stream—a warm ocean current that flows north from the eastern Caribbean and Florida and follows the coastline of the eastern United States and Canada before crossing the North Atlantic Ocean.

The large size of the North American continent creates wide temperature variations. Because land heats up and cools off more rapidly than water, temperatures in the interior of the continent are hotter in the summer and colder in the winter than in coastal areas, where temperatures are moderated by the oceans.

aquifers ancient natural underground reservoirs of water

CHECK YOUR UNDERSTANDING

1. What are North America's two main mountain ranges? Where are they located?
2. How do the major landforms of this region influence the movement of air masses to contribute to its climatic variety?

ENVIRONMENT

North America's wide range of resources and its seemingly limitless stretches of forest and grasslands sometimes divert attention from the environmental impacts of its settlement and development. By global standards, most North Americans are relatively wealthy and can access vast resources both from within the region and globally. The United States in particular uses its status as the world's second-largest national economy, and the largest global military power, to protect and extend the access of its major corporations to resources throughout the world. The North American tendency toward intensive resource consumption has many environmental consequences. This section focuses on a few of those consequences: climate change and air pollution, depletion and pollution of water resources and fisheries, and habitat loss.

Climate Change and Air Pollution

On a per capita basis, North Americans contribute among the largest amounts of greenhouse gases (GHGs) to Earth's atmosphere. Home to just under 5 percent of the world's population, the region produces 16 percent of humanity's GHG emissions. No other region has higher per capita emissions. This is largely a result of North America's high consumption of fossil fuels, which is related to several factors. One of these is North America's dominant pattern of urbanization, characterized by vast and still-growing suburbs where automobiles are the main mode of transportation. The mostly single-story freestanding buildings across the region's urban landscapes require more energy to heat and cool than do the densely packed, high-rise buildings typical of cities in other world regions. North American industrial and agricultural production also depends very heavily on fossil fuels.

Canada's government was one of the first to commit to international agreements aimed at reducing the consumption of fossil fuels. The U.S. government has wavered, at times supporting such agreements and enacting policies to reduce its high GHG emissions, only to reverse course later, fearing damage to its economy. Because the United States wields so much power on the global stage both economically and politically, its unsteady support for international agreements to reduce GHG emissions has made it harder for other countries to commit wholeheartedly to GHG reductions.

GHG emissions have been falling in both the United States and Canada since 2008 due to the economic recession, though increased energy conservation has played a role in recent years. Many utility companies are also switching from coal to natural gas, which is cheaper and produces less CO_2 when burned.

Both Canada and the United States have long explored alternative sources of energy, such as geothermal, nuclear power, wind, and solar. Solar has grown at exponential rates in recent years, and the United States now has the fourth-largest installed solar power–generating capacity in the world, after China, Japan, and Germany. The United States has the world's second-largest wind power–generating capacity, and wind has several times more installed generating capacity than solar. Moreover, the U.S. solar and wind power industries now employ more people than the coal industry, even though the latter still supplies around a quarter of all electric power. This could give solar and wind increased political leverage, especially as these industries continue to grow rapidly in coming years. While Canada has much less solar- or wind-generating capacity, almost 64 percent of its energy comes from renewable resources, mostly hydroelectric power. Renewable energy accounts for around 11 percent of U.S. energy consumption.

Vulnerability to Climate Change Both Canada and the United States are vulnerable to the effects of climate change. Dense population centers on the Gulf of Mexico and on the Atlantic coast are very exposed to hurricanes, which may become more violent as oceans warm (**Figure 2.9B, C**). Sea level rise and coastal erosion, caused by the thermal expansion of the oceans, are already affecting Arctic coastal areas (see Figure 2.9A). Here, at least 26 coastal communities are being forced to relocate inland, at an estimated cost of U.S.$130 million per village.

2.9 Vulnerability to Climate Change in North America

North America's wealth and its well-developed emergency response systems make it very resilient and reduce its overall vulnerability to climate change. Nevertheless, climate-change impacts are already forcing some people to move, especially those exposed to hurricanes and sea level rise. Currently, more than 90 U.S. coastal communities are experiencing chronic flooding, and in 20 years that number will double.

THINKING GEOGRAPHICALLY

A Describe the coastal erosion illustrated by this photo.

B Why are stronger hurricanes more likely as the climate warms?

C What elements of this community's emergency response system may be compromised by this flooding?

A The location of Shishmaref, Alaska, on the Arctic Sea leaves it exposed to coastal erosion, which may be increasing in the area because of warmer air and sea temperatures. [GABRIEL BOUYS/AFP/Getty Images]

B Hurricanes gain strength with warmer temperatures. Many low-lying coastal cities are very exposed and sensitive to hurricanes, though their resilience varies. Shown here is a fire that destroyed several homes in New Orleans when the flooding caused by Hurricane Katrina made it impossible for fire trucks to reach the fire. [Craig Warga/NY Daily News Archive via Getty Images]

C Florida's low elevations expose it to sea level rise and flooding during and after hurricanes. [Josh Ritchie/Getty Images]

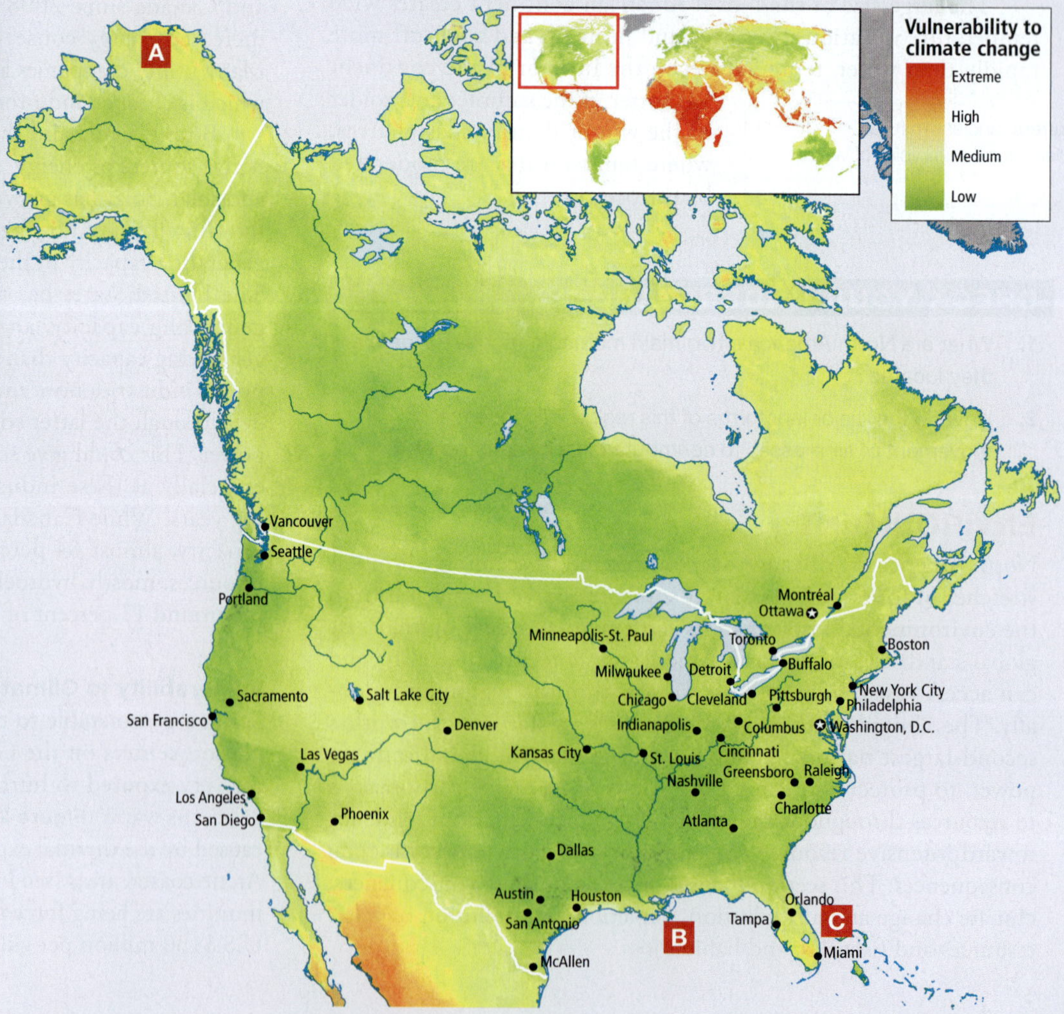

▶ This map shows patterns of vulnerability to climate change, based on a combination of human and environmental factors. Areas in darker red are the most vulnerable to hazards related to climate change.

The stark contrast in vulnerability along the U.S.–Mexico border reflects different sensitivity and resilience to water scarcity. The U.S. side has a much better water infrastructure, reducing its sensitivity to drought, and its emergency response systems make it more resilient to water shortages than Mexico.

Regardless of how much humans decrease their GHG output, climate-change impacts will force some people to move. Currently more than 90 U.S. coastal communities are experiencing chronic flooding that is forcing some people to move, and in 20 years that number will double. By the end of the century 670 coastal towns and cities in North America will be battling chronic flooding related to sea level rise. Another major area of vulnerability is arid farming zones, where higher temperatures are reducing soil moisture, making irrigation crucial.

In all of these areas, resilience to climate change is bolstered over the short term by the region's relatively strong emergency response and recovery systems. Long-term resilience is boosted by careful planning as well as by North America's large and diverse economy, which can provide alternative livelihoods to people who face significant exposure to climate impacts.

Air Pollution In addition to climate change, most GHGs contribute to various forms of air pollution, such as smog and acid rain. **Smog** is a combination of industrial emissions, car exhaust, and water vapor that frequently hovers as a yellow-brown haze over cities, including many in North America, and causes a variety of health problems. These same emissions also result in **acid rain**, which is created when pollutants dissolve in precipitation and make the rain acidic. Acid rain can kill trees and, when concentrated in lakes and streams, poison fish and wildlife.

The United States, with its large population and extensive range of industries, is responsible for the vast majority of acid rain in North America. Because of continental weather and wind patterns, however, the area most affected by acid rain encompasses a wide swath on both sides of the eastern U.S.–Canada border (**Figure 2.10**). The eastern half of the continent, which includes the entire Eastern Seaboard from the Gulf Coast to Newfoundland, also is significantly impacted by acid rain.

Solutions to North America's air pollution problems are based

smog a combination of industrial emissions, car exhaust, and water vapor that frequently hovers as a yellow-brown haze over many cities, causing a variety of health problems

acid rain precipitation that has formed through the interaction of rainwater or moisture in the air with sulfur dioxide and nitrogen oxides emitted during the burning of fossil fuels, making it acidic

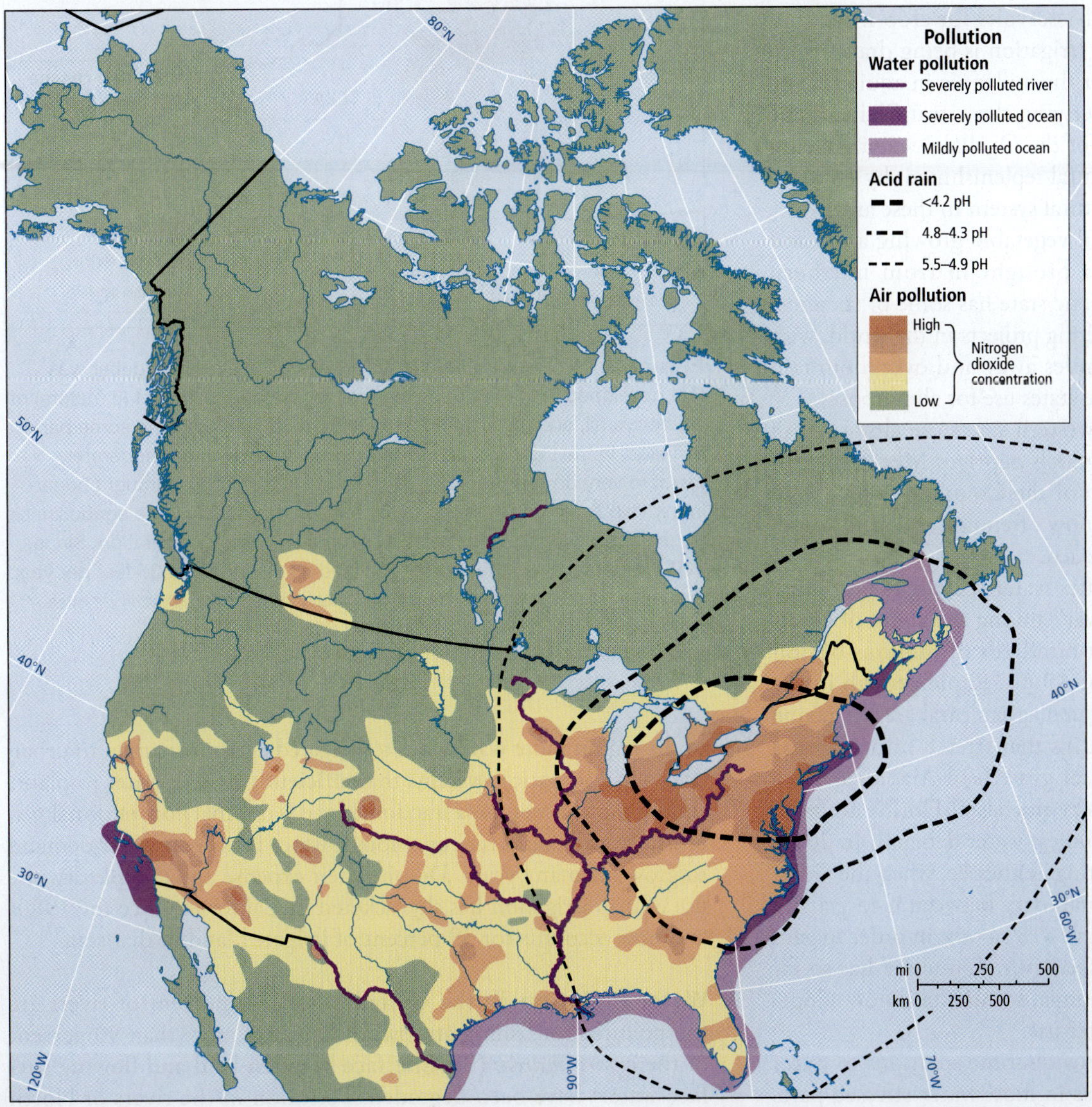

Figure 2.10 Air and water pollution in North America. This map shows two aspects of air pollution, as well as polluted rivers and coastal areas. Red and yellow indicate concentrations of nitrogen dioxide (NO_2), a toxic gas that comes primarily from the combustion of fossil fuels by motor vehicles and power plants. This gas interacts with rain to produce nitric acid, a major component of acid rain, as well as toxic organic nitrates that contribute to urban smog. The map also shows polluted coastlines (including all of the coastline from Texas to New Brunswick) as well as severely polluted rivers, which include much of the Mississippi River and its tributaries.

on greater use of clean renewable resources and moving away from fossil fuels. Electric vehicles are an important way to reduce emissions. Studies show that even when the electricity used by electric vehicles is generated with coal, the most polluting fossil fuel, these vehicles still pollute significantly less than those powered by gas or diesel. Currently, electric vehicles are used in parts of North America that rely more on renewables and cleaner fossil fuels, such as natural gas, and hence in these regions the average electric vehicle emits roughly two to three times less than the average conventional vehicle over its entire life cycle.

Water Resources

Water generally becomes more precious the farther west one goes in North America. As populations and per capita water usage grow, conflicts over water are becoming more and more common, especially in the west. People who live in the humid eastern areas often find it difficult to believe that water is becoming scarce, but even there water conflicts are increasing.

Water Depletion On the North American Great Plains, rainfall varies considerably from year to year. To make farming more secure and predictable, taxpayers across the continent have subsidized the building of pumps, aqueducts, and reservoirs for crop irrigation. However, more and more water for irrigation is being drawn from fossil water that has been stored over the millennia in aquifers. The *Ogallala aquifer* (**Figure 2.11**) underlying the Great Plains is the largest in North America. In parts of the Ogallala, water is being pumped out at rates that exceed natural replenishment by 10 to 40 times, threatening the entire agricultural system in these areas.

Southern California's fruit- and vegetable-growing areas and its major cities depend on water brought in from northern California and surrounding states. The state has some of the most expensive and massive water-engineering projects in the world, with water pumped from hundreds of miles away and over mountain ranges, using more energy than some states use for all purposes.

California and six other southwestern states use almost all of the water in the Colorado River, which deprives Mexico of this much-needed resource. The mouth of the Colorado (which is in Mexico) was once navigable but is now often dry and sandy, with only a trickle of water getting to Mexico.

Even in the eastern United States water scarcity is becoming an issue in some places. A "water war" among the states of Georgia, Florida, and Alabama has cost hundreds of millions of dollars in legal and court fees since the 1990s. Rapid growth in the city of Atlanta and in Georgia's irrigated agricultural areas is taxing water resources within two river basins that stretch into Alabama and Florida. Urban and agricultural growth in Alabama and a multi-million-dollar shellfish industry in coastal Florida are both threatened by Georgia's seemingly endless water demand. In 2013, the water dispute spread to neighboring Tennessee, when the Georgia legislature voted to redraw the boundary between Georgia and Tennessee, correcting an "error" in an 1818 survey, in order to gain access to the waters of the Tennessee River. Tennessee has so far refused to hand over any land to Georgia, so Atlanta is now adopting stricter regulations on urban water use.

Advances in irrigation could provide some solutions to water depletion in North America, because in most water-stressed parts of the region more water goes to irrigated agriculture than to urban uses. Drip irrigation systems that efficiently deliver water to plants through drip hoses use a fraction of the water that conventional systems use and could free up enough water for urban water demand to grow for many years. Despite their expense and complexity, the use of drip irrigation has skyrocketed in California since the 1980s and now accounts for 39 percent of irrigated land in that state.

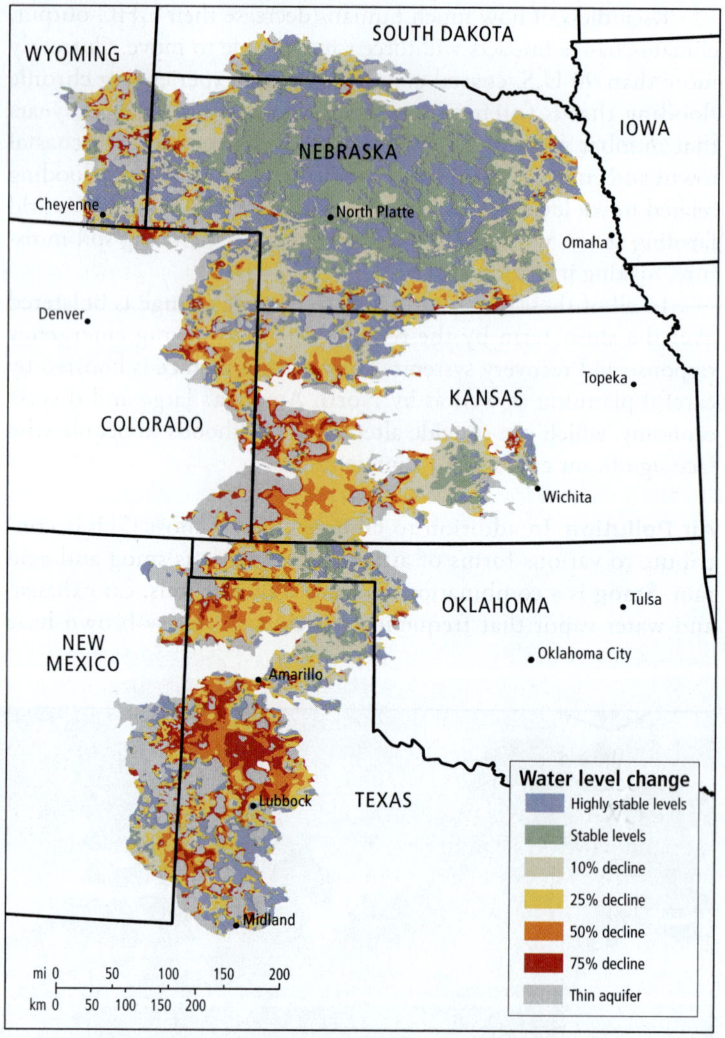

Figure 2.11 The Ogallala aquifer. Since the Ogallala aquifer was first developed in the 1950s, it has lost an average of 15 feet (5 meters) of water overall, and more than 234 feet (71 meters) of water in some parts of Texas. In the Ogallala area, the climate fluctuates from moderately moist to very dry, and the dry periods are lengthening. A drought began in mid-1992 and has returned every few years, causing large agribusiness firms to pump Ogallala water to supplement scarce precipitation. Since 1992, water levels in the aquifer declined an average of 1.35 feet per year and now exceed replenishment rates many times over. [Data from: https://pubs.usgs.gov/sir/2017/5040/sir20175040.pdf]

Water Pollution In the United States, 40 percent of rivers are too polluted for fishing and swimming, and more than 90 percent of the *riparian areas* (the interface between land and flowing surface water) have been degraded. Pollution in the rivers of North

America comes mainly as storm-water runoff from agricultural areas, urban and suburban developments, and industrial sites.

In the 1970s, scientists studying coastal areas began noticing *dead zones* where water is so polluted that it supports almost no life. Dead zones occur near the mouths of major river systems that have been polluted by fertilizers and pesticides washed from farms and lawns when it rains. A large dead zone is in the Gulf of Mexico near the mouth of the Mississippi, and similar zones have been found in all U.S. coastal areas. Even Canada, where much lower population density means that rivers are generally cleaner, has dead zones on its western coast.

A recently discovered type of water pollution involves pharmaceuticals, such as antibiotics, that are excreted by humans and are not removed during water purification processes. These chemicals then make their way into rivers and lakes, where they enter the food system in drinking water or through fish.

Solutions to water pollution in North America are many, but the highest priorities currently involve changes to agricultural systems and better management of storm-water runoff. The United States was one of the first countries to attempt to comprehensively address water pollution with the passage of the *Clean Water Act* (CWA) in the 1970s. The law aims to continually improve and maintain water quality throughout U.S. water bodies, and it has had dramatic impacts on urban and industrial areas, where pollution is occurring at much lower levels than before implementation of the CWA. While the CWA is a major success, and has been imitated in many other countries, it has always struggled to deal with agricultural sources of pollution, in part because they are much more widely dispersed "non-point" sources that are harder to monitor than urban and industrial "point sources" such as factories and wastewater treatment plants. Recent CWA water pollution control efforts focus on getting farmers to better manage the use of agricultural fertilizers, pesticides, and herbicides so that they are less likely to be washed into nearby water bodies during rain events.

Figure 2.12 Contestants in the annual Redneck Fishing Tournament on the Illinois River near Bath, Illinois. Asian carp are an invasive species imported by fish farmers in the southern United States in the 1970s to help clean their ponds. They escaped and now make up 90 percent of the biomass in the Illinois River near the town of Bath. The tournament takes advantage of the fact that when startled the fish tend to jump out of the water and can be caught in nets. Started as an attempt to eradicate the invasive fish, the festival has become a major celebration for the surrounding area. [Benjamin Lowy/Getty Images]

Loss of Natural Habitat and Urbanization

Before the European colonization of North America, which began soon after 1500, the environmental impact of humans in the region was relatively low. Major Native American impacts stemmed from the use of fire to clear forests for agriculture and to remove forest underbrush to improve hunting. These significant landscape changes pale in comparison to the environmental impact of Europeans who cleared millions of acres of forests and grasslands to make way for farms, cities, and industries. This was particularly true in the area that became the United States.

While agriculture is the primary driver of habitat loss in North America, invasive species are also a major concern. As North American native plants and animals have been forced into ever-smaller territories, many have died out entirely and been replaced by nonnative species (European and African grasses and the domestic cat, for example) brought in by humans either purposely or inadvertently. Estimates vary, but at least 4000 nonnative species have invaded North America. An example is the Asian carp, which is rapidly invading rivers throughout the Mississippi drainage basin, and now threatens the Great Lakes (**Figure 2.12**).

Urbanization is another major cause of habitat loss, with **urban sprawl** (see the "Farmland and Urban Sprawl" section) a concern in many communities. For several decades, middle- and upper-income urbanites throughout this region have wanted more spacious homes and yards, which has resulted in sprawling lower-density suburban neighborhoods that are car-dependent. Farms, forests, and other "undeveloped" land have given way to pavement, lawns, homes, shopping centers, office complexes, and golf courses. In the process, natural habitats are being degraded even more intensely than they were by farming.

Oil Extraction

In many coastal and interior areas of North America, oil extraction is a large and potentially environmentally devastating industry. This often-overlooked problem was made clear in the spring and summer of 2010, when U.S. waters in the Gulf of Mexico became the site of the largest accidental marine oil spill in world history when the Deepwater Horizon, an offshore oil-drilling rig leased by British Petroleum (BP), leaked 200 million gallons of oil over a period of 4 months. Studies almost a decade later show oil from the spill still contaminating wildlife throughout the Gulf of Mexico, impacting fish, shellfish, and shrimp industries and marine ecosystems for decades to come.

In other places, too, oil extraction has a dramatic effect on the environment. Along the northern coast of Alaska, the Trans-Alaska Pipeline runs southward for 800 miles to the Port of Valdez (**Figure 2.13**). Often running above ground to avoid shifting as Earth freezes and thaws, the pipeline poses a constant risk of rupture, which could potentially result in devastating oil spills. The pipeline also interferes with migrations of caribou and other animals that Alaska's indigenous people have depended on for food in the

urban sprawl the encroachment of suburbs on agricultural land

Figure 2.13 The Trans-Alaska Pipeline. The pipeline runs for 800 miles through remote areas where mining and oil industries can have a significant effect on the landscape. [Daniel Acker/Bloomberg via Getty Images]

past. Protests about the environmental impact of oil extraction in Alaska tend to be quieted by the yearly rebate of several thousand dollars from oil revenues received by each Alaskan.

With one-tenth the population of the United States, Canada has the largest proven oil reserves in the world after Saudi Arabia, sufficient to meet its needs plus provide export capacity. It is the largest foreign supplier of oil to the United States, but more than 90 percent of this oil is in hard-to-access oil sands in western Canada. The costs of extracting this oil are high in terms of environmental impacts, energy expended, and water resources used. Transport of the extracted oil across the North American continent through the Keystone Pipeline system (part of which is already in use) also poses environmental risks.

While Canada may have more proven oil reserves, the United States is now the largest producer of oil and gas, thanks largely to new drilling technologies known as **fracking** that use high-pressure injections of water to fracture rock formations, releasing natural gas and oil that the formations contain. First developed in the 1940s, fracking became widely used only in the past 10 years, after some important drilling innovations were made. Since then, large portions of the central United States have been opened up to fracking, especially in North Dakota, Texas, and Oklahoma.

While fracking has produced large amounts of fuel at a relatively low cost, it is controversial because of its environmental impacts, which include much higher GHG emissions than conventional drilling, significant local air pollution, water pollution, and earthquakes. In 2015, scientific studies concluded that a dramatic increase in earthquakes in Oklahoma, which had become a near-daily occurrence in some areas, was triggered by the practice of injecting wastewater produced during fracking into the ground. In 2016, Oklahoma started regulating the amount of wastewater injected by fracking operations after a lawsuit was filed by environmental groups.

fracking a drilling technology that uses high-pressure injection of water to fracture rock formations, releasing natural gas and oil that they contain

clear-cutting a method of logging that involves cutting down all trees on a given plot of land regardless of age, health, or species

Logging

Though widespread forest clearing for agriculture is now rare, logging is common throughout North America. It is especially important along the northern Pacific coast and in the southeastern United States. Logging in these areas provides most of the construction lumber and much of the paper used in Canada, the United States, and parts of Asia.

Although the logging industry provides jobs and an exportable commodity, it has been depleting the continent's forests. Environmentalists have focused on the damage created by the logging industry, especially via **clear-cutting,** the main logging method used throughout North America. In this method, all trees on a given plot of land are cut down, regardless of age, health, or species (**Figure 2.14**). Clear-cutting destroys wild animal and plant hab-

Figure 2.14 Clear-cutting.

(A) A logger just outside the Olympic National Park in Washington State makes the first cut in the process of felling an 800-year-old 120-foot cedar. Part of an irreplaceable old-growth forest, the tree is worth about U.S.$10,000 at the sawmill. [James P. Blair/National Geographic/Getty Images]

(B) A clear-cut in Washington State's Olympic Peninsula, where more than 30 percent of the forests have been clear-cut since 1970. Logging in old-growth forests has been declining in recent years but still continues. [Wild Horizons/UIG via Getty Images]

itats, thereby reducing species diversity. It also leaves forest soils uncovered and very susceptible to erosion.

Concerns about the environmental impacts of logging are growing in the major logging states and provinces in part because of a shift away from employment in forestry and toward service sector and manufacturing jobs. For example, even in many remote areas of the Pacific Northwest where logging was once the backbone of the economy, residents now depend on tourism and other occupations that rely on the beauty of intact forest ecosystems.

Coal Mining

Coal mining is a major industry, especially in Wyoming and parts of the Appalachian Mountains (from Pennsylvania through West Virginia and Kentucky), that is damaging to the environment. Strip mining, in which vast quantities of soil and rock are removed in order to extract underlying coal, can result in visual wastelands and in huge piles of mining waste called *tailings* that pollute waterways and threaten communities that depend on well water. Particularly damaging is a form of strip mining known as *mountaintop removal*, in which the whole top of a mountain is leveled and the tailings are pushed into surrounding valleys, resulting in the pollution of entire watersheds (**Figure 2.15**).

Figure 2.15 Mountaintop removal coal mining.

(A) Kayford Mountain, West Virginia, has been the site of years of mountaintop removal coal mining. The top of the mountain has been extensively mined and the tailings pushed into the valley, shown in the foreground. [Rick Eglinton/Toronto Star via Getty Images]

(B) Polluted water seeps out of the Kayford Mountain mine site. [MANDEL NGAN/AFP/Getty Images]

Recent years have seen lower coal consumption in North America and in major coal export markets in Europe and China. This resulted from the cancellation of plans to build new coal-fired electrical power plants, and the closure of old plants and the mines that feed them as power generators shift over to natural gas and renewable energy.

CHECK YOUR UNDERSTANDING

1. How does North America's massive consumption of resources impact environments within the region and globally?
2. How does North America compare to other regions in terms of its vulnerability to climate change?
3. What are some solutions to major environmental problems in North America, such as air pollution, water depletion, and water pollution?
4. What percentage of rivers in the United States are too polluted for fishing or swimming?
5. In what parts of North America is logging most common, and where is it in decline as a major employer?
6. How is coal mining influenced by changes in North America's overall energy economy?

HUMAN PATTERNS OVER TIME

Perhaps more than any other, North America is a region of immigrants. In prehistoric times, humans came from Eurasia via Alaska, dispersing to the south and east. Beginning in the 1600s, waves of European immigrants and enslaved Africans spread over the continent, primarily from east to west. Today, immigrants are coming mostly from Asia and Middle and South America, arriving mainly in the Southwest and West, where immigrant populations are most concentrated. Migration within the region is still a defining characteristic of life for most North Americans, who are among the world's most mobile people, moving an average of 12 times in a lifetime.

The Peopling of North America

Recent evidence suggests that humans first came to North America from northeastern Asia at least 25,000 years ago and perhaps earlier, most arriving during an ice age. At that time, the global climate was cooler, polar ice caps were thicker, and sea levels were lower. The Bering land bridge, a huge, low landmass more than 1000 miles (1600 kilometers) wide, connected Siberia to Alaska. Bands of hunters crossed by foot or small boats into Alaska and traveled down the west coast of North America.

By 15,000 years ago, humans had reached nearly to the tip of South America and had moved deep into that continent. By 10,000 years ago, global temperatures began to rise. As the ice caps melted, sea levels rose and the Bering land bridge was submerged.

Over thousands of years, the people settling in the Americas domesticated plants, created paths and roads, cleared forests, built permanent shelters, and sometimes created elaborate social systems. About 3000 years ago, corn was introduced from Mexico (into what is now the southwestern U.S. desert), as were other Mexican domesticated crops, particularly squash and beans. Such

A Cahokia, Illinois, a community of 30,000 that lasted from 700 to 1400 C.E. [Cahokia Mounds State Historic Site, painting by Michael Hampshire]

B Enslaved people and workers in North America pick leaves and operate machines at a tobacco factory in 1750. [MPI/Getty Images]

C A log raft being floated down Oregon's Columbia River in 1902. [Library of Congress]

10,000 B.C.E. — 0 C.E. — 700 C.E. — 1400 C.E. — 1700 C.E.

25,000–10,000 B.C.E. Bering land bridge

700–1400 C.E. Cahokia flourishes

1492 Arrival of Europeans

1600–1900 Plantations in Southern colonies; industries in New England

NOTE: Timeline range is not to scale

Figure 2.16 VISUAL HISTORY OF NORTH AMERICA

THINKING GEOGRAPHICALLY

A How did food crops such as corn, beans, and squash influence the development of settlements like Cahokia?

B Why did the plantation system inhibit the establishment of roads, communication networks, and small enterprises that could have boosted other forms of economic activity in the Southern colonies?

food crops are thought to have been closely linked to settled life and to North America's prehistoric population growth.

These foods provided surpluses that allowed some community members to engage in trade and specialized activities other than agriculture, hunting, and gathering, making possible large, city-like regional settlements. For example, by 1000 years ago, the urban settlement of Cahokia, including suburban settlements (in what is now central Illinois, across the Mississippi from St. Louis), covered 5 square miles (12 square kilometers) and was home to an estimated 30,000 people (**Figure 2.16A**). Here people specialized in crafts, trade, and other activities beyond the production of basic necessities.

The Arrival of the Europeans North America was completely transformed by the sweeping occupation of the continent by Europeans. As early as 1497, English, Italian, and Portuguese sailors and fishermen were surveying the Eastern Seaboard of North America and establishing seasonal camps in places like St. John's, Newfoundland, in Canada. Over the next century English, French, and Spanish explorers attempted to establish colonies, of which the Spanish were the first to succeed. Within current U.S. territory, San Juan, Puerto Rico, was the first successful permanent settlement in 1521. A failed Spanish plantation settlement in coastal Georgia in 1526 introduced the first Africans to North America, some of whom escaped and joined indigenous groups in the sea islands of Georgia and South Carolina. In the 1540s, Spanish explorer Hernando de Soto made his way from Florida deep into the heartland of the continent, becoming the first European to cross the Mississippi River. The Spanish founded St. Augustine in Florida in 1565, and in 1598, Espanola, New Mexico, the first European town established west of the Mississippi. In the early seventeenth century, the British established colonies along the Atlantic coast in what is now Virginia (1607) and Massachusetts (1620), and the Dutch founded New Amsterdam (now New York) in 1624 at the mouth of the Hudson River. The French explored the northern interior of the continent, founding Québec City (1608) at the mouth of the St. Lawrence River and Montréal further upstream. Assisted by enslaved Africans, colonists and settlers from northern Europe built villages, towns, port cities, and plantations along the eastern coast over the next two centuries. By the mid-1800s, they had occupied most lands of Native American and Aboriginal peoples into the central part of the continent.

Disease, Technology, and Native Americans The rapid expansion of European settlement was facilitated by the vulnerability of Native American and Aboriginal populations to European diseases. Having been isolated from the rest of the world for many years, Native American and Aboriginal populations had no immunity to diseases such as measles and smallpox. Transmitted by Europeans and Africans who had built up immunity to them, these diseases killed up to 90 percent of Native American and Aboriginal populations within the first 100 years of contact. It is now thought that diseases spread by early expeditions, such as de Soto's into Florida, Georgia, Tennessee, and Arkansas, so decimated populations in the North American interior that fields and villages were abandoned, the forest grew back, and later explorers erroneously assumed the land had never been occupied.

Technologically advanced European weapons, trained dogs, and horses also took a large toll on Native American and Aboriginal peoples, who often had only bows and arrows, clubs, and spears. Some Native Americans in the Southwest acquired horses and guns from the Spanish and learned to use them in warfare against the Europeans, but their other technologies could not compete. Numbers reveal the devastating effect of European settlement on Native

D Steel workers in Pittsburgh, Pennsylvania, in 1905. [Library of Congress]

E Oil wells encroach on agricultural land in Long Beach, California, in 1923. [Library of Congress]

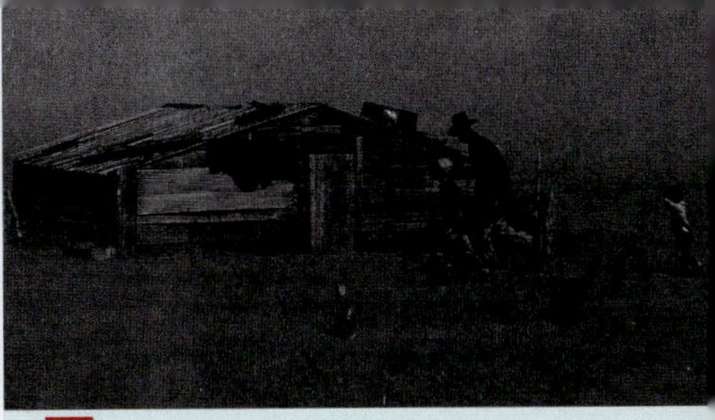

F A farmer and his children walk through a dust storm in Oklahoma in 1936. [Library of Congress]

1800 C.E. — 1900 C.E. — 2000 C.E.

- **1830s** Trail of Tears
- **1849** California Gold Rush
- **1861–1865** U.S. Civil War
- **1902**
- **1905**
- **1923**
- **1930s** Dust Bowl

C What economic activities now dominate the coastal Pacific Northwest?

D How did the steel industry stimulate development in the "economic core" of North America?

E What activities have become economically important in California other than agriculture and the oil industry?

F What were the causes of the Dust Bowl?

American and Aboriginal populations: There were roughly 18 million Native Americans and Aboriginal peoples in North America in 1492. By 1542, after just a few Spanish expeditions, there were only half that number. By 1907, slightly more than 400,000, or a mere 2 percent of the original population, remained.

The European Transformation

European settlement erased many of the landscapes familiar to Native American and Aboriginal peoples and imposed new ones that fit the varied physical and cultural desires of the new occupants.

The Southern Settlements By the late 1600s, large plantations in the colonies of Virginia, the Carolinas, and Georgia were cultivating crops such as tobacco, rice, and cotton, which became valuable exports. During this time large numbers of enslaved Africans were brought into the region, becoming the dominant labor force on many plantations (see Figure 2.16B). By the start of the U.S. Civil War in 1861, enslaved people made up about one-third of the population in the Southern states and were often a majority in the plantation regions. Working and living conditions on these plantations were often brutal and hazardous, with execution, beatings, and rape being regular occurrences. Enslaved people were usually denied formal education, and gatherings of any kind were often banned in order to make slave rebellions more difficult to organize. Even so, there were more than 250 documented slave rebellions in North America. The largest concentrations of African Americans in North America are still in the southeastern states (**Figure 2.17**).

The plantation system consolidated wealth into the hands of a small class of landowners who made up just 12 percent of Southerners in 1860. These elite members of the planter class kept taxes low and invested money from their exported crops in Europe or the more prosperous Northern colonies instead of in **infrastructure** at home. As a result, the road, rail, and communication networks and other facilities necessary for further economic growth in the South were rarely built.

More than half of Southerners were poor white farmers.

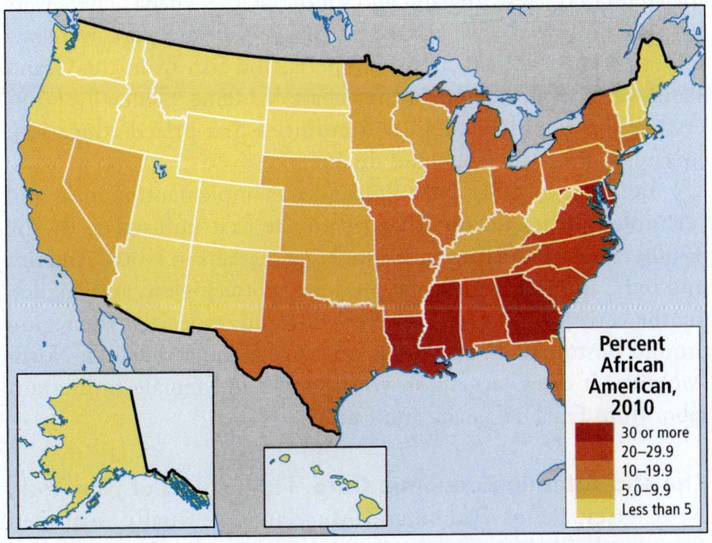

Figure 2.17 Percent of African American population in each state. The percent refers only to those persons who selected "Black, African Am., Negro" as their only race in the 2010 census. It does not include those who selected more than one race that included black. [Data from: http://www.census.gov/quickfacts/map/RHI225214/00]

infrastructure road, rail, and communication networks and other facilities necessary for economic activity

Both they and the enslaved African population lived simply, and their meager consumption did not provide much demand for goods. Because of this, there were few market towns and almost no industries. Plantations tended to import from Europe and the northern United States whatever they couldn't produce for themselves. The result was much slower economic growth in the South relative to areas in the North, where a more diverse range of agriculture and industries supported a larger service economy of small shops, garment making, restaurants and bars, and transportation services.

Along with the abolition movement to free enslaved Africans and their descendants, the inability of the federal government to reconcile the different needs of the agricultural export–oriented Southern economy and the industrializing Northern economy was one of the causes of the Civil War (1861–1865) in the United States. After the war, while the victorious North returned to promoting its own industrial development, the plantation economy declined, and the South sank deeply into poverty. The South remained economically and socially underdeveloped well into the 1970s, and even today it remains the poorest subregion of North America.

The Northern Settlements Throughout the seventeenth century, relatively poor subsistence farming communities dominated the colonies of New England and southeastern Canada. There were no plantations and few enslaved people, and most exports were raw materials such as timber, animal pelts, and fish from the Grand Banks off Newfoundland and the coast of Maine. Generally, farmers lived in interdependent communities that prized education, ingenuity, self-sufficiency, and thrift.

By the late 1600s, New England was implementing ideas and technology from Europe that led to the first industries. By the 1700s, diverse industries were supplying markets in North America and the Caribbean with metal products, pottery, glass, and textiles. By the early 1800s, southern New England, especially the region around Boston, had become the center of manufacturing in North America. It drew largely on young male and female immigrant labor from French Canada and Europe.

The Mid-Atlantic Economic Core The colonies of New York, New Jersey, Pennsylvania, and Maryland eventually surpassed New England and southeastern Canada in population and in wealth. This mid-Atlantic region benefited from more fertile soils, a slightly warmer climate, multiple deepwater harbors, and better access to the resources of the interior. By the end of the Revolutionary War in 1783, the mid-Atlantic region was on its way to becoming the **economic core**, or the dominant economic region, of North America. Port cities such as New York, Philadelphia, and Baltimore prospered as the intermediaries for trade between Europe and the vast American continental interior.

In the early nineteenth century, both agriculture and manufacturing grew and diversified, drawing immigrants from much of northwestern Europe. As farmers became more successful, they bought more and more mechanized equipment, appliances, and consumer goods made in nearby cities. By the mid-nineteenth century, the economy of the core was increasingly based on the steel industry, which spread westward to Pittsburgh and the Great Lakes' industrial cities of Cleveland, Detroit, and Chicago. The steel industry relied on the mining of coal and iron ore deposits throughout the region and beyond. Steel became the basis for mechanization, and the region was soon producing heavy farm and railroad equipment (see Figure 2.16D).

economic core the dominant economic region within a larger region

By the early twentieth century, the economic core stretched from the Atlantic to St. Louis on the Mississippi (including many small industrial cities along the river, to Milwaukee), and from Ottawa to Washington, DC. It dominated North America economically and politically well into the middle of the twentieth century. Most other areas produced food and raw materials for the core's markets and depended on the core's factories for manufactured goods.

Expansion West of the Mississippi and Great Lakes

The east-to-west trend of settlement continued as land in the densely settled eastern parts of the continent became too expensive for new immigrants. By the 1840s, immigrant farmers from central and northern Europe, as well as European descendants born in eastern North America, were pushing their way beyond the Great Lakes and across the Mississippi River, north and west into the Great Plains of Canada and the United States (**Figure 2.18**).

The Great Plains Much of the land west of the Great Lakes and the Mississippi River was dry grassland or prairie. The soil usually proved very productive in wet years, and the area became known as North America's *breadbasket.* But the naturally arid character of this land eventually created an ecological disaster for Great Plains farmers. In the 1930s, after 10 especially dry years, a series of devastating dust storms blew away topsoil by the ton. This hardship was made worse by the widespread economic depression of the 1930s. Many Great Plains farm families packed up what they could and left what became known as the Dust Bowl (see Figure 2.16F), heading west to California and other states on the Pacific Coast.

The Mountain West and Pacific Coast Some immigrants and settlers from east of the Mississippi, alerted to the possibilities farther west, skipped over the Great Plains entirely. By the 1840s, they were coming to the valleys of the Rocky Mountains, to the Great Basin, and to the well-watered and fertile coastal zones of what was then known as the Oregon Territory and California. In 1848, only 1 year after California was forcibly annexed from Mexico to the United States, news of the discovery of gold in California drew thousands of people with the prospect of getting rich quickly. The vast majority of gold seekers were unsuccessful, however, and by 1852 they had to look for employment elsewhere. Farther north, logging eventually became a major industry (see Figure 2.16C).

The extension of railroads across the continent in the nineteenth century facilitated the transportation of manufactured goods to the West as well as raw materials and eventually fresh

Figure 2.18 Nineteenth-century transportation. [Research from: James A. Henretta, W. Elliot Brownlee, David Brody, and Susan Ware, *America's History*, 2nd ed. (New York: Worth, 1993), pp. 400–401; James L. Roark, Michael P. Johnson, Patricia Cline Cohen, Sarah Stage, Alan Lawson, and Susan M. Hartmann, *The American Promise: A History of the United States*, 3rd ed. (Boston: Bedford/St. Martin's, 2005), p. 601]

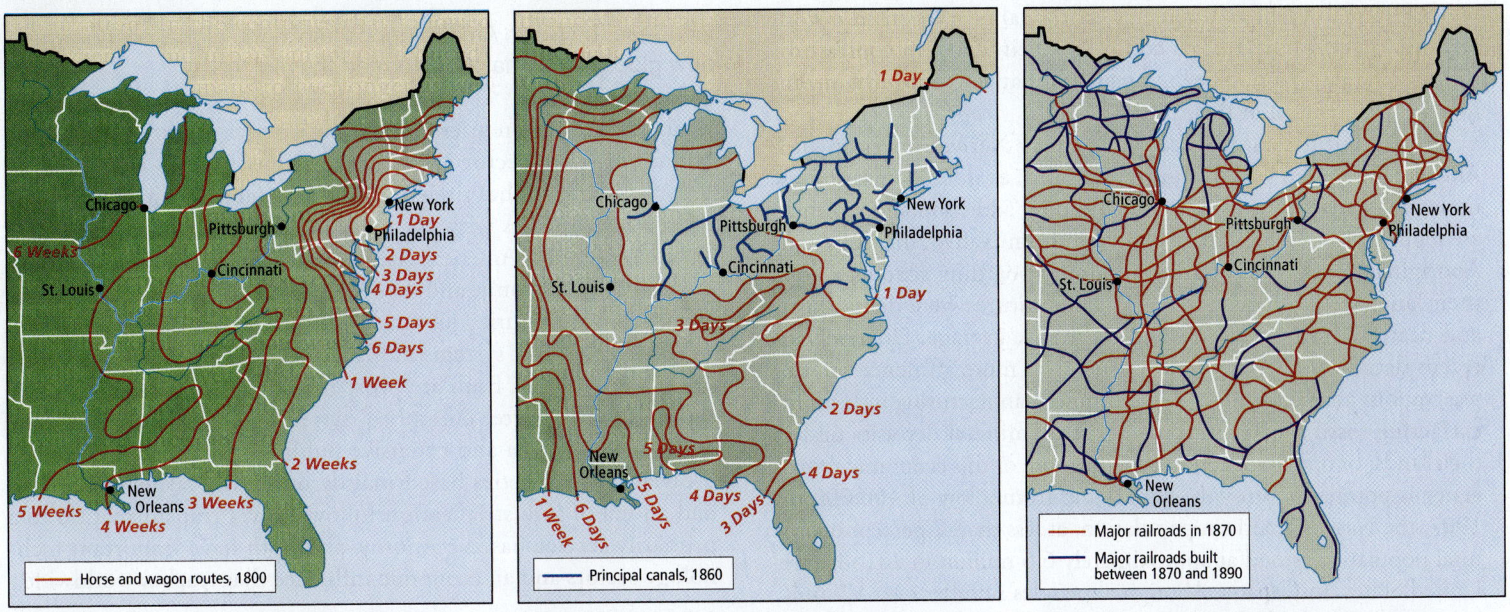

(A) Travel times from New York City, 1800. It took a day to travel by wagon from New York City to Philadelphia and a week to go to Pittsburgh.

(B) Travel times from New York City, 1860. The travel time from New York to Philadelphia was now only 2 or 3 hours and to Pittsburgh less than a day because people could go part of the way via canals (dark blue). Using the canals, the Great Lakes, and rivers, they could easily reach principal cities along the Mississippi River, and travel was less expensive and onerous than it had been.

(C) Railroad expansion by 1890. With the building of railroads, which began in the decade before the Civil War, the mobility of people and goods increased dramatically. By 1890, railroads crossed the continent, though the network was densest in the eastern half.

produce to the East. Today, the coastal areas of this region, often called the Pacific Northwest, have thriving, diverse, high-tech economies and flourishing populations. Partially in response to the long history of resource extraction in the Pacific Northwest, residents are on the forefront of so many efforts to reduce human impacts on the environment that the region has been nicknamed "Ecotopia."

The Southwest People from the Spanish colony of Mexico first colonized the Southwest in the late 1500s. Their settlements were sparse. As immigrants from the United States moved into the region, drawn by the cattle-raising industry, Mexico found it progressively more difficult to maintain control, and by 1850, nearly the entire Southwest was under U.S. control.

By the twentieth century, a vibrant agricultural economy had developed in central and southern California, supported by massive government-sponsored water-movement and irrigation projects. The mild Mediterranean climate made it possible to grow vegetables almost year-round. With the advent of refrigerated railroad cars, fresh California vegetables could be sent to the major population centers of the East. Southern California's economy rapidly diversified to include oil (see Figure 2.16E), entertainment, and a variety of engineering- and technology-based industries.

European Settlement and American Indians and Aboriginal Peoples

As settlement relentlessly expanded west, Native American and Aboriginal peoples who had survived early encounters with Europeans and were living in the eastern part of the continent occupied land that European newcomers wished to use. During the 1800s, almost all the surviving Native American and Aboriginal peoples were killed in innumerable skirmishes with European newcomers, were absorbed into the societies of the Europeans (through intermarriage and acculturation), or were forcibly relocated west to relatively small reservations with few resources. The largest relocation, in the 1830s, involved the Choctaw, Seminole, Creek, Chickasaw, and Cherokee of the southeastern states. These people had already adopted many European methods of farming, building, education, government, and religion. Nevertheless, they were rounded up by the U.S. Army and marched to Oklahoma along a route that became known as the Trail of Tears because of the more than 4000 Native Americans who died along the way.

As Europeans occupied the Great Plains and prairies, many of the reservations were further shrunk or relocated onto even less desirable land. Today, reservations cover just over 2 percent of the land area of the United States.

In Canada the picture is somewhat different. Reservations now cover 20 percent of Canada, mostly because of the creation of the

Nunavut Territory (now known simply as Nunavut) in 1999 in the far north and the ceding of Northwest Territory land to the Tłįchǫ Aboriginal people (formerly known as the Dogrib) in 2003 (see the Figure 2.1 map). These Canadian Aboriginal peoples stand out as having won the right to legal control of their lands. In contrast to the United States, it had been unusual for native groups in Canada to have legal control of their territories.

After centuries of mistreatment, many Native American and Aboriginal peoples still live in poverty and, as in all communities under severe stress, rates of alcohol and drug addiction and violence are high. Until the 1980s many children from Native American and Aboriginal families were forced to attend boarding schools where their languages were made illegal, their cultures were denigrated, and death rates from disease were far above average. However, in recent decades some tribes have developed more affluence on the reservations and territories by establishing manufacturing industries; extracting fossil fuel, uranium, and other mineral deposits under their lands; or opening casinos. One measure of this economic resurgence is population growth. Expanding from a low of 400,000 in 1907, the Native American population, at less than 2 percent of the total population, stood at approximately 6.6 million in 2015 in the United States. In Canada, Aboriginal peoples number over 1.7 million, or 3.8 percent of the population in 2016.

The Changing Regional Composition of North America

The regions of European-led settlement still remain in North America, but they are now less distinctive. The economic core region is less dominant in industry, which has spread to other parts of the continent. Some regions, such as the Pacific Northwest, that were once dependent on agriculture, logging, or mineral extraction now have high-tech industries as well. The West Coast, in particular, has blossomed with a high-tech economy and a rapidly expanding population that includes many immigrants from Asia and Middle and South America. The West Coast also benefits from North America's trade with Asia, which now surpasses trade with Europe in volume and value.

CHECK YOUR UNDERSTANDING

1. When did humans first come to North America and by what route?
2. What was the first European power to explore west of the Mississippi?
3. What biological factor facilitated European expansion into North America?
4. Where in North America did Europeans first bring Africans, and for what purpose?
5. What region in North America was the first to develop industries?
6. What was North America's early-twentieth-century economic core, and where was it located?
7. What parts of North America were settled relatively later?
8. How many Native American and Aboriginal peoples lived in North America in 1492? How many remained by 1907? How many were there in 2015–2016?

GLOBALIZATION AND DEVELOPMENT

2.4 Describe North America's position in the global economy and how globalization has transformed the region.

Globalization has reoriented North America toward knowledge-intensive service sector jobs that require education and training. Income inequality has risen as most manufacturing jobs have been moved to countries with cheaper labor or have been replaced by technology. North America's size and wealth have made it the center of the global economy, and its demand for imported goods and its export of manufacturing jobs make it a major engine of globalization.

The economic systems of Canada and the United States have much in common. Both are prosperous mixed economies, where capitalist market "free enterprise" systems coexist with strong government regulation and extensive public services. Both countries evolved from societies based mainly on family farms. Both then had an era of industrialization followed by a transformation to a primarily service-based economy, and both have important technology sectors and an economic influence that reaches worldwide.

FROM MANUFACTURING TO SERVICES

After industrializing throughout the nineteenth and early twentieth centuries, manufacturing employment peaked in North America during World War II, when 38 percent of the U.S. workforce was employed in manufacturing, much of it in the Old Economic Core. Even then services employed almost twice as many people, and by the 1960s, manufacturing jobs were in decline. The geography of manufacturing was also changing as corporations tried to boost profits by relying more on technology and cheaper labor outside the Old Economic Core. In the Old Economic Core, higher pay and benefits and better working conditions won by labor unions led to increased production costs. This lowered the high profits demanded by the owners and shareholders of manufacturing corporations. A number of companies began moving their factories to the southeastern United States, where wages were lower and corporate profits higher because of the absence of labor unions. Today, more than half a century into this process of deindustrialization, many cities of the Old Economic Core have large abandoned industrial districts.

Since the late 1980s, the United States and Canada have entered into free trade agreements (see Chapter 1) that have enabled many manufacturing industries (such as clothing, electronic assembly, and auto parts manufacturing), to move south to Mexico or overseas to China and other places where labor is significantly cheaper than it is in North America. Companies have also saved on production costs because laws that in the United States and Canada regulate environmental protection and safe and healthy workplaces are absent or less strictly enforced in those other countries.

Technology and automation have also contributed at least as much and possibly more to the decline of manufacturing employment than has the movement of jobs to other countries. The steel industry provides an illustration. In 1980, huge steel plants, most of them in the economic core, employed more than 500,000 workers. At that time, it took about 10 person-hours and cost about U.S.$1000 to produce 1 ton of steel. Spurred by more efficient foreign competitors, the North American steel industry applied new technology to lower production costs, improve efficiency, and increase production. By 2017, steel was being produced at the rate of 1.5 person-hours per ton and at a cost of about U.S.$200 per ton. The reorganized steel industry in the United States now produces much of the steel in small, highly efficient minimills distributed throughout the United States. In total, the steel industry now employs fewer than half the workers it did in 1980. Throughout North America, this trend toward higher efficiency has resulted in fewer people producing more of a given product at a far lower cost than was the case 40 years ago. Remarkably, even as employment in manufacturing has declined over the last three decades, now standing at around 8 percent of total employment, the actual amount of manufacturing output has steadily increased.

The economic base of North America is now a broad *tertiary sector* in which people are engaged in various services such as transportation, utilities, wholesale and retail trade, health, leisure, maintenance, finance, government, information, and education. As of 2018, in both Canada and the United States about 80 percent of jobs and a similar percentage of the gross national income (GNI) were in the tertiary, or service, sector. There are high-paying jobs in all the service categories, but low-paying jobs are far more common. The largest private employer in the United States is the discount retail chain Walmart (1.4 million employees), where the average wage is U.S.$14 an hour, or U.S.$28,000 a year, full time. This is just slightly above the poverty level for a family of four in the United States. Moreover, roughly half of Walmart employees are part time, earning roughly U.S.$11 an hour on average and receiving no company benefits or health care. Because of low wages, many Walmart workers are forced to turn to government programs to meet their basic needs, costing taxpayers as much as U.S.$6 billion a year. Walmart creates primarily retail jobs because, for the most part, its wares are manufactured abroad.

THE KNOWLEDGE ECONOMY

An important and rapidly growing subcategory of the service sector involves the creation, processing, and communication of information—what is often called the *knowledge economy.* The knowledge economy includes workers who manage information, such as those employed in finance, journalism, higher education, research and development, and many aspects of health care. It also includes the *information technology* or *IT* sector, which deals with computer software and hardware and the management of digital data.

Industries that rely on the use of computers and the internet to process and transport information are freer to locate where they wish than were the manufacturing industries of the Old Economic Core, which needed locally available material resources such as steel and coal. These newer industries are more dependent on skilled managers, communicators, thinkers, and technicians, and are often located near major universities and research institutions.

Crucial to the knowledge economy is the internet, which was first widely available in North America and has emerged as an economic force more rapidly there than in any other region in the world. With only 5 percent of the world's population, North America accounted for 8.2 percent of the world's internet users in 2017. Roughly 88 percent of the population of both the United States and Canada uses the internet, compared to 86 percent of the European Union (EU) and 52 percent of the world as a whole. The total economic impact of the internet in North America is hard to assess, but retail internet sales increase every year, accounting for 13 percent of total retail sales in 2017 and roughly half of the growth in retail sales since 2016.

The growth of internet-based activity makes access to the internet crucial in North America. Unfortunately, a **digital divide**—a discrepancy in access to information technology between small, rural, and poor areas and large, wealthy cities—has developed, because about a quarter of the North American population is not yet able to afford computers and home internet connections.

GLOBALIZATION AND FREE TRADE

The United States and Canada are major engines of globalization through the size and technological sophistication of their economies. Their combined economy is almost as large as that of the entire European Union. North America's advantageous position in the global economy is also a reflection of its geopolitical influence—its ability to mold the pro-globalization free trade policies that suit the major corporations and the governments of North America. However, free trade has not always been emphasized, and public support for it has lagged in recent years as so many poor people have seen their incomes decline while wealthy people get even richer.

Before North America's rise to prosperity and global dominance, trade barriers were important aids to the region's development. For example, when it became independent of Britain in 1776, the new U.S. government imposed tariffs and quotas on imports and gave subsidies to domestic producers. This protected fledgling domestic industries and commercial agriculture, allowing its economic core region to flourish.

Because Canada and the United States are both wealthy and globally competitive exporters, leaders in business and government generally see tariffs and quotas as obstacles to North America's economic expansion. Thus, they usually advocate for trade barriers to be reduced worldwide. Critics of these free trade policies are many. Some argue that shrinking or defunct manufacturing industries could grow again with the assistance of tariffs like the ones that rapidly growing countries, such as China, currently use. Others argue that the benefits of free trade go mostly to well-paid financiers, deal brokers, and managers of large businesses, while many workers end up losing their jobs to cheaper labor overseas, or see their incomes stagnate. Furthermore, there is some evidence of protectionist policies working in favor of businesses in North America. For example, both the United States and Canada give significant subsidies to their farmers, which have enabled them to compete in the global food trade.

Despite considerable opposition, free trade has been a guiding principle in economic relations between the United States and Canada for many years. The process of reducing trade barriers began formally with the Canada–U.S. Free Trade Agreement of 1989. Mexico was included when the North American Free Trade Agreement (NAFTA) was created in 1994, and the agreement was renegotiated and renamed the **United States–Mexico–Canada Agreement (USMCA)** in 2018.

digital divide the discrepancy in access to information technology between small, rural, and poor areas and large, wealthy cities that contain major governmental research laboratories and universities

Figure 2.19 Walmart on the global scale. Today, Walmart has more than 11,000 retail stores in 28 countries and employs 2.2 million people around the world (about 1.4 million in the United States). [Data from: http://corporate.walmart.com/our-story/our-locations#]

The major long-term goal of USMCA is to increase the amount of trade between Canada, the United States, and Mexico. Today, it is the world's largest trading bloc in terms of the gross domestic product (GDP) of its member states. USMCA is only 1 of 14 free trade agreements that the United States and Canada have with countries around the world. Both Canada and the United States are members of the World Trade Organization (see Chapter 1).

The impacts of free trade agreements are hard to assess because it is difficult to tell whether the many observable changes in the North American economy that have occurred over the past three decades, when most free trade agreements were entered into, have been caused by the agreements or by other changes in regional and global economies. However, a few things are clear. Trade has increased, and many companies are making higher profits because they now have larger markets. Since 1990, exports among the USMCA countries have increased in value by more than 300 percent. By value, USMCA's exports to the world economy have increased by about 300 percent for the United States and Canada and by 600 percent for Mexico. Some U.S. companies, such as Walmart, expanded aggressively into Mexico after 1994 when NAFTA was passed. Mexico now has more Walmarts (2306 retail stores) than any country except the United States, which has 5249 (**Figure 2.19**). Canada has 411 Walmarts.

United States-Mexico-Canada Agreement (USMCA) a revised version of the North American Free Trade Agreement made in 1994 that added Mexico to the 1989 economic arrangement between the United States and Canada

trade deficit the extent to which the money earned by exports is exceeded by the money spent on imports

Free trade has increased the tendency of the United States to spend more money on imports than it earns from exports. This imbalance is called a **trade deficit**. Before NAFTA, the United States usually had much smaller trade deficits with Mexico and Canada. After the agreement was signed, these deficits rose dramatically, especially with Mexico. For example, between 1994 and 2017, the value of U.S. exports to Mexico increased from roughly U.S.$50 billion to $276 billion, while the value of imports increased from U.S.$49 billion to $339 billion. This amounts to a shift from a slight trade surplus in 1994 of U.S.$1 billion to a deficit in 2017 of U.S.$63 billion.

Free trade may also have resulted in a net loss of jobs in the United States. Since 1994, increased imports from Mexico and Canada have displaced about 2 million U.S. jobs, while increased exports to these countries have created only about 1 million jobs. The export-related jobs that free trade agreements facilitate pay up to 18 percent more than the average North American wage. However, those jobs are usually in different locations than the ones that were lost and the people who take them tend to be younger and with different skills than those who lost jobs. Former manufacturing workers often end up with short-term contract jobs or low-skill, often service sector, jobs that pay the minimum wage and include no benefits. In Canada, these impacts of free trade have been moderated by Canada's more extensive assistance to low-income people.

The Asian Link to Globalization

Low trade barriers with Asia have been a powerful force for globalization in North America. The seemingly endless variety of goods imported from China—everything from underwear to the chemicals used to make prescription drugs—is made possible in large part by China's lower wages and wide variety of policies that subsidize exports. Many factories that first relocated to Mexico from the southern United States later moved to China to take advantage of its enormous supply of cheap labor. By some estimates as many as 2.4 million U.S. jobs were lost between 1999 and 2011 as a result of low trade barriers with China.

While political pressure to raise trade barriers with Asia has grown in recent years, the effects of trade barriers are more complex than many voters think. For example, some barriers may actually encourage Asian investment in North America. For example, Japanese and Korean automotive companies that want to import whole cars into North America have for decades had to pay tariffs and face other restrictions. To avoid these, they have located plants in North America, often establishing them in the rural mid-South of the United States or in southern Canada.

Backlash Against Globalization and Free Trade

Concerns about globalization and free trade were heightened by the worldwide economic downturn that began in 2007. Now referred to as the *Great Recession*, this event came on the heels of an economic expansion fueled by a booming U.S. housing industry. This growth was based on banks allowing millions of buyers to purchase homes with mortgages that were well beyond their means. By 2007 it became clear that much of the growth was unstable and unsustainable. Growth in the U.S. housing industry came to a sudden halt, and when too many homebuyers could no longer afford their mortgage payments, the banks that had lent them money started to fail. This produced worldwide ripple effects because many foreign banks were involved in the U.S. housing market. Between September 2008 and March 2009, the U.S. stock market fell by nearly half, wiping out the savings and pensions of millions of Americans. Similar plunges followed in foreign stock markets, ultimately resulting in a worldwide economic downturn because businesses could no longer find money to fund expansion. As the recession intensified, job losses in the United States caused a sharp drop in consumption, which further affected world markets.

Thanks to its strong regulatory controls, Canada did not have bank failures. However, because so much of the Canadian economy is linked to exports and imports from the United States, Canada underwent a slowdown and many Canadians lost their jobs. Canada's recovery was quicker than that of the United States, though, possibly because household consumption was buttressed by Canada's stronger social safety net.

Efforts to deal with the causes of the recession, and the disparity in wealth that it worsened, haven't yet been very successful. The financial industry has funded extensive lobbying to counter attempts to better regulate the banks that, through their lending practices, sparked the recession. Meanwhile millions of U.S. voters who saw their incomes fall blamed globalization and free trade, not a poorly regulated financial industry, focusing on the movement of jobs from the United States to China, Mexico, and elsewhere. Donald Trump made use of this and other sources of discontent in the United States to win the presidency in 2016.

After years of criticizing the movement of U.S. jobs abroad due to cheaper labor, the Trump administration began a "trade war" with China in 2018 by placing tariffs on a wide range of Chinese exports to the United States. China retaliated by raising tariffs on many U.S. goods. The tariffs were subsequently extended to other countries, including Canada, Mexico, the European Union, Japan, and Iran, with the Trump administration holding out the possibility of making the tariffs apply to all countries. Most U.S. industries have been highly critical of the tariffs, claiming they will raise costs for U.S. products that depend on imported materials subject to the tariffs, such as steel and aluminum. While the overall economic impact of the trade war is not yet clear, it has resulted in the renegotiation of trade relationships, with the United States pushing for greater protections for its economy or, alternatively, fewer tariffs placed on its goods by other countries, but above all a reduction of its trade imbalances with most countries.

Supporters of tariffs can point to many examples where tariffs would seem to be justified out of a sense of fairness. For example, the United States has historically maintained low barriers for the wide array of Chinese goods entering the country, but China has long had high tariffs on U.S. manufactures, such as cars. China has also forced U.S. companies to transfer patents and technology to Chinese companies in return for access to cheap labor or to the country's huge domestic markets. At times Chinese companies have also stolen technology from U.S. companies.

The current backlash against free trade is partially a response to these concerns, but it is also a result of growing disparities in wealth in North America that have been underway since the 1970s (**Figure 2.20**). This trend is global, but it is extreme in the United States, where in 2017 the wealthiest 1 percent of households owned 39 percent of the country's total wealth, and the bottom 90 percent owned only 23 percent of the wealth. While there is scholarly evidence for and against the notion that free trade and globalization have resulted in greater income disparity, key members of the Trump administration believe that it does and are willing to support a trade war to try to get back jobs lost to overseas competitors.

U.S.-Canadian Economic Interdependence

The trade war that began in 2018 highlighted the economic interdependence of Canada and the United States. The two countries engage in mutual tourism, direct investment, migration, and, most of all, trade in a wide variety of goods. Canada is the second-largest market for U.S. goods, after the European Union. By 2017, that trade relationship had evolved into a two-way flow of U.S.$673 billion annually (**Figure 2.21**). In 2017, 14 percent of U.S. imports came from Canada and 17 percent of U.S. exports went to Canada. In the same year Canada sold 75 percent of its exports to the United States and bought 58 percent of its imports from the United States. With such a greater proportion of its economy tied to the United States, the trade war is a much greater threat to Canada than it is to the United States.

WOMEN IN THE ECONOMY

While North American women have made steady gains in terms of pay equity and overall participation in the labor force, there are

PHOTO ESSAY

2.20 Globalization and Development in North America

Globalization has reoriented North America toward knowledge-intensive service sector jobs that require education and training. Income inequality has risen as manufacturing jobs have been moved abroad or replaced by technology. While North America's size and wealth mean it is still at the center of the global economy, there are also areas of relatively low human development.

THINKING GEOGRAPHICALLY

A What percentage of U.S. workers earns below $15 an hour?

B What are two causes of the decline in manufacturing employment in North America?

C The United States spends more on its social safety net than any country in the world except for__________?

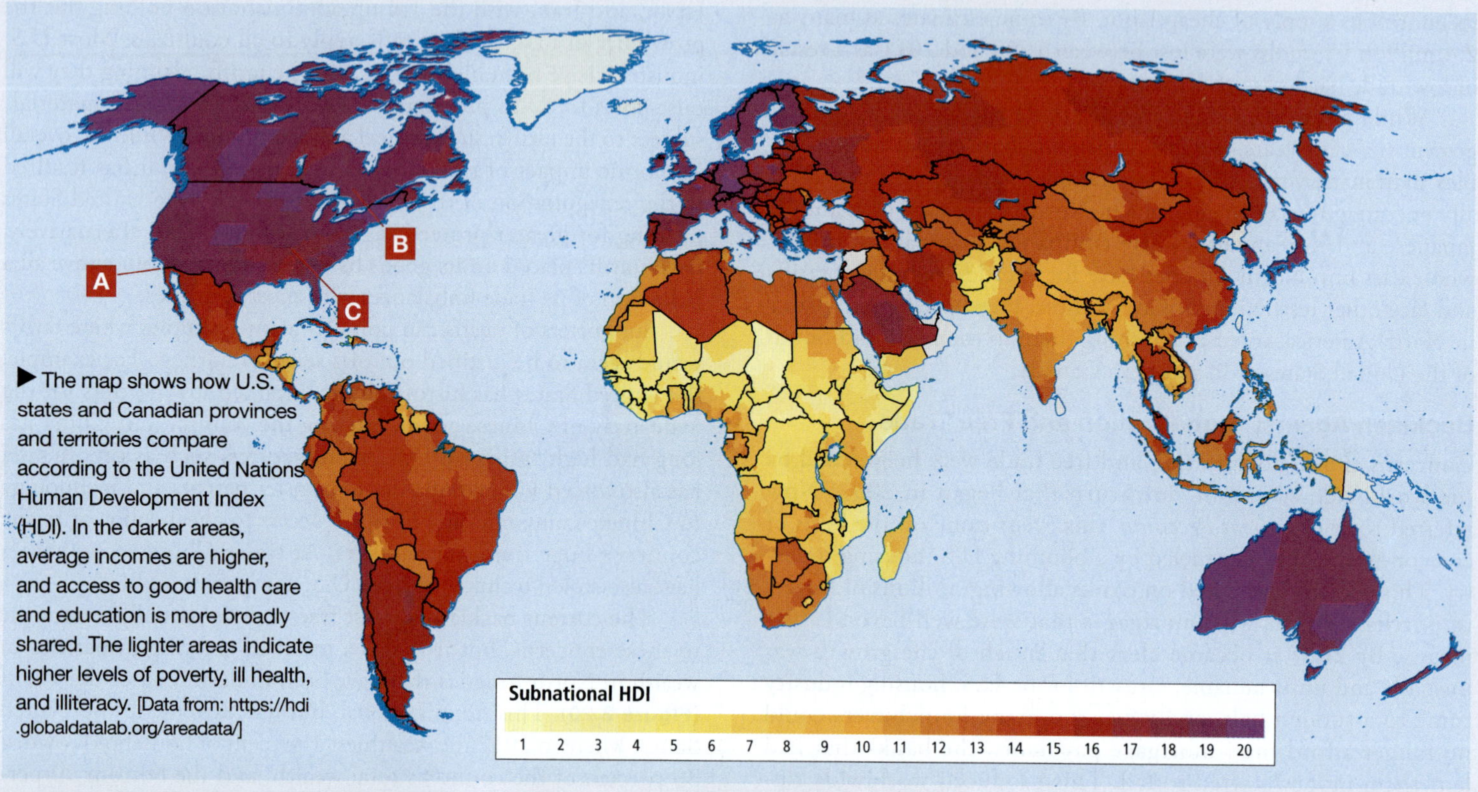

▶ The map shows how U.S. states and Canadian provinces and territories compare according to the United Nations Human Development Index (HDI). In the darker areas, average incomes are higher, and access to good health care and education is more broadly shared. The lighter areas indicate higher levels of poverty, ill health, and illiteracy. [Data from: https://hdi.globaldatalab.org/areadata/]

A Service workers protest. A striking McDonald's employee is arrested during a protest for a higher minimum wage in Los Angeles. Many service industry jobs pay wages well below what is necessary to support a family. [David McNew/Getty Images]

B Automation. Robots weld car bodies on a General Motors assembly line in Oshawa, Ontario, in Canada. Automation has been occurring for decades in North America and may account for more job losses in manufacturing than globalization. This photo was taken in 1985. [Keith Beaty/Toronto Star via Getty Images]

C Income disparity. An uninsured family waits to receive treatment at a temporary free health-care clinic in Bristol, Tennessee. Access to health care is a major component of human development, which is lower in the southeastern United States. [Mario Tama/Getty Images]

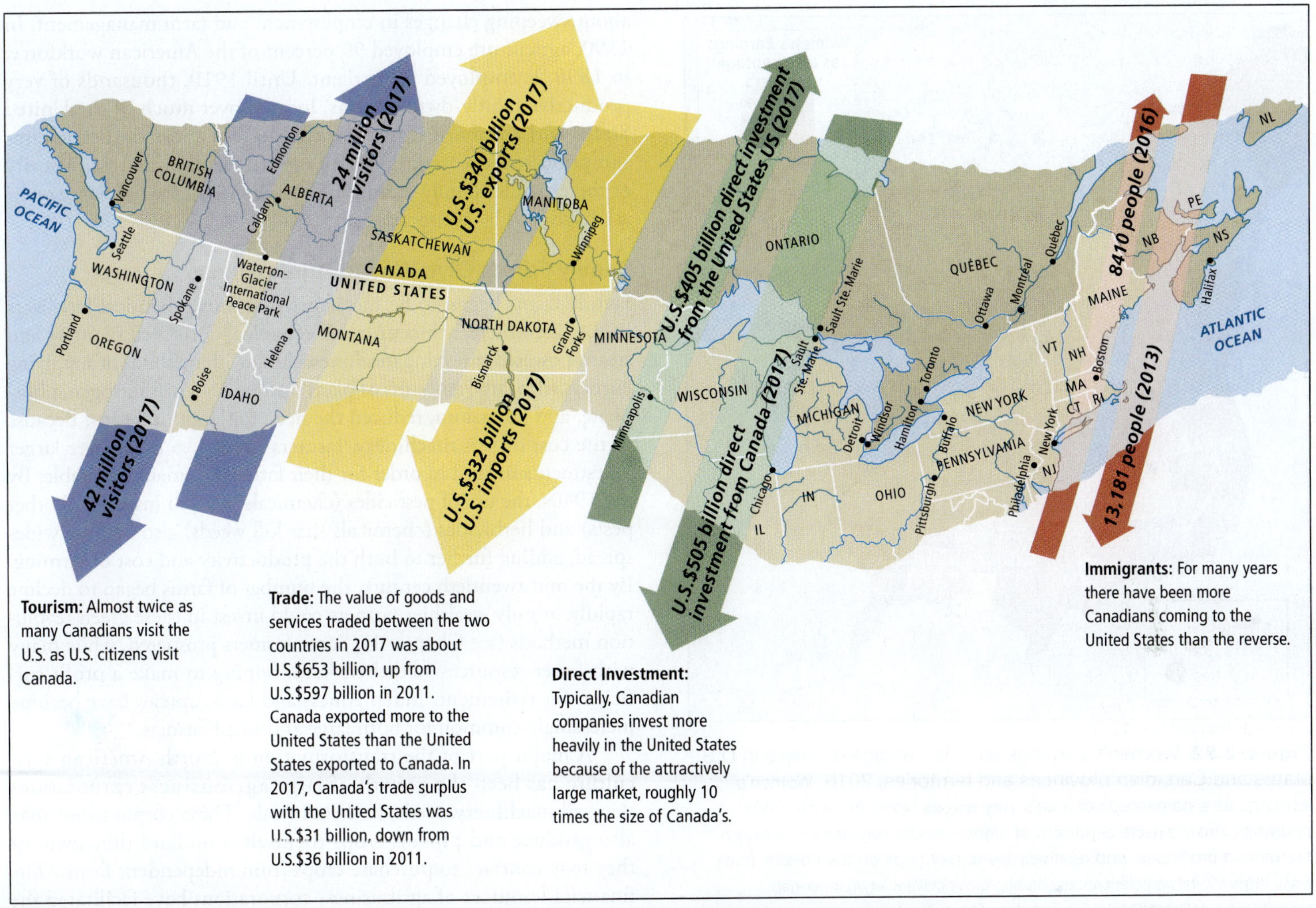

Figure 2.21 Transfers of tourists, goods, investment, and immigrants between the United States and Canada. Canada and the United States have one of the world's largest trading relationships. The flows of goods, money, and people across the long Canada–U.S. border are essential to both countries. However, because of its relatively small population and economy, Canada is more reliant on the United States than the United States is on Canada. All amounts shown are in U.S. dollars. [Research from: *National Geographic*, February 1990: 106–107; and augmented with data from https://ustr.gov/countries-regions/americas/canada, http://canadaimmigrants.com/canada-immigrants-by-source-country-2016/, https://www150.statcan.gc.ca/n1/daily-quotidien/180425/dq180425a-eng.htm, and https://www150.statcan.gc.ca/n1/daily-quotidien/180220/dq180220c-eng.htm]

still important ways in which they lag behind their male counterparts. On average, North American female workers earn about 82 cents for every dollar that male workers earn, with both the United States and Canada having about the same gender pay gap, though there is considerable variation among states and provinces (**Figure 2.22**). This is actually an improvement over previous decades. During World War II, when large numbers of women first started working in male-dominated jobs, North American female workers earned, on average, only 57 percent of what male workers earned. The advances made by this older generation of women and the ones that followed have transformed North American workplaces. For the first time in history, women now represent more than half of the North American labor force, though most still work for male managers.

Throughout North America, the number of women entrepreneurs is on the rise, and women start nearly half of all new businesses. While women-owned businesses tend to be small and less financially secure than those in which men have most of the control, credit opportunities for businesswomen have been improving as more upper-level jobs in banking are held by women. Women are also facing less discrimination in hiring; a 2012 study shows that many male executives now prefer to hire qualified women because they are particularly ambitious and willing to gain advanced qualifications.

In secondary and higher education, North American women have equaled or exceeded the level of men in most categories. In 2016 in the United States, 40 percent of women between the ages of 25 and 34 held an undergraduate degree, compared to just 33 percent of men.

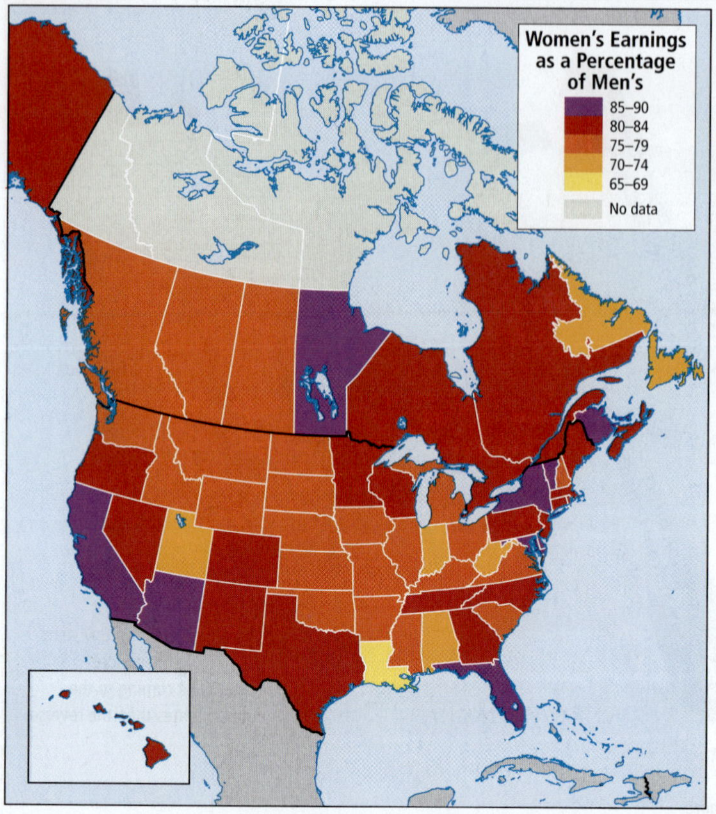

Figure 2.22 Women's earnings as a percentage of men's in U.S. states and Canadian provinces and territories, 2016. Women's earnings as a percentage of men's vary across North America. Both countries show a general pattern of higher gender pay gaps in western states and provinces, and relatively lower pay gaps on the coasts. [Data from: https://www.conferenceboard.ca/hcp/provincial/society/gender-gap.aspx?AspxAutoDetectCookieSupport=1 and https://nwlc.org/resources/wage-gap-state-state/]

NORTH AMERICA'S CHANGING FOOD-PRODUCTION SYSTEMS

Agriculture is somewhat paradoxical in this region, occupying huge areas of land, producing enormous amounts of food, resulting in gigantic environmental impacts, but contributing relatively little to each country's GDP, mostly because other sectors of the economy are so large and prosperous. Agriculture now accounts for less than 1.2 percent of the United States' GDP and less than 2 percent of Canada's. Due to mechanization, fewer than 2 percent of North Americans are employed directly in agriculture. This is a dramatic shift for a region where agricultural exports were once the backbone of the economy. Nevertheless, North America remains a key part of the global food trade, with the United States the world's largest food exporter, and Canada the eighth largest. Agriculture is more predominant in the United States, where 44 percent of the total land area is under cultivation. Only 7 percent of land is cultivated in Canada, due to much harsher winters (**Figure 2.23**).

agribusiness the business of farming conducted by large-scale operations that purchase, produce, finance, package, and distribute agricultural products

The shift to mechanized agriculture in North America brought about sweeping changes in employment and farm management. In 1790, agriculture employed 90 percent of the American workforce; in 1890, it employed 50 percent. Until 1910, thousands of very productive family-owned farms, located over much of the United States and southern Canada, provided for most domestic consumption and the majority of all exports. Today, the vast majority of these smaller family-run farms have been replaced by large operations owned by corporations.

Family Farms Give Way to Agribusiness

Family farms began to be mechanized and use chemical fertilizers in the late nineteenth century. Mechanical corn-seed planters and steam-powered threshing machines, along with methods of supplying farmers with large amounts of plant nutrients such as nitrogen, phosphate, and potassium, reduced the need for labor on farms. Because of the cost of this machinery, farmers needed to make ever-larger investments in land in order for their farms to remain profitable. By the 1940s, the use of pesticides (chemicals that kill insects and other pests) and herbicides (chemicals that kill weeds) also became widespread, adding further to both the productivity and cost of farming. By the mid-twentieth century, the number of farms began to decline rapidly, as only wealthier farmers could invest in these green revolution methods (see Chapter 1). Some farmers prospered, while many with fewer resources sold their land, hoping to make a profit sufficient for retirement. Indebtedness and bankruptcies have become increasingly common for both large and small farmers.

A major part of the transformation of North American agriculture has been the growth of large **agribusiness** corporations that sell machinery, seeds, and chemicals. These corporations may also produce and process crops themselves on land they own, or they may contract to purchase crops from independent farms. The financial resources of agribusiness corporations have facilitated the transition to green revolution methods, enabling large investments in the research and development of new products. The corporations can also provide loans and cash to individual farmers, often as a part of contracts that leave farmers with little actual control over which crops are grown and what methods are used.

While green revolution agriculture provides a wide variety of food at low prices for North Americans, the shift to these production methods has depressed local economies and created social problems in many rural areas. Communities in places such as the Great Plains of both Canada and the United States were once made up of farming families with similar middle-class incomes, social standing, and commitment to the region. Today, farm communities are often composed of a few wealthy farmer-managers amid a majority of poor, often migrant Latino or Asian laborers who work on large farms and in food-processing plants for wages that are too low to provide a decent standard of living. These workers also struggle to be accepted into the communities where they live.

Food Production and Sustainability

Can green revolution agriculture, like that practiced by successful North American corn farmers, persist over time? Many modern strategies to increase yields, including the use of chemical fertilizers, pesticides, and herbicides, and the large-scale production of meat on "factory farms," can have negative effects. These methods can threaten the health of farmworkers and nearby residents,

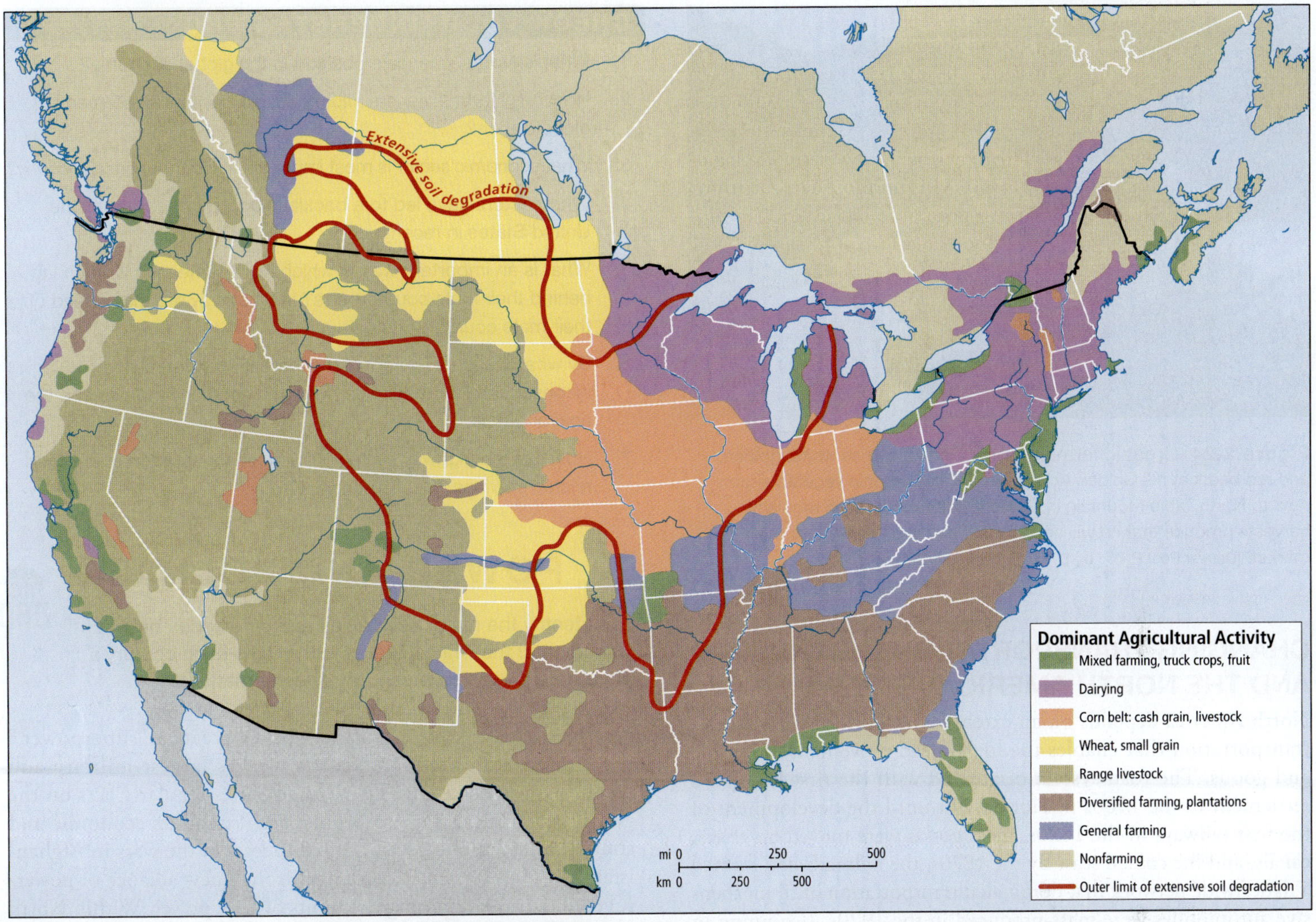

Figure 2.23 Agriculture in North America. Throughout much of North America, some type of agriculture is possible. The major exceptions are the northern parts of Canada and Alaska and the dry mountain and basin region (the continental interior) that lies between the Great Plains and the Pacific coastal zone. However, in some marginal areas, such as Southern California, southern Arizona, and the Utah Valley, irrigation is needed for cultivation. [Research from: Arthur Getis and Judith Getis, eds., *The United States and Canada: The Land and the People* (Dubuque, IA: William C. Brown, 1995), p. 165]

pollute nearby streams and lakes, degrade soils, and even affect distant coastal areas (see the "Environment" section).

Recently, researchers have been studying the long-term impacts of genetically modified (GMO) crops on human health and the environment. While most scientific studies suggest that GMO plants have little impact on human health, many of these studies have been funded or even conducted by the agribusiness corporations that create GMO crops, making it possible that the studies might be biased. The biggest documented impact that GMO plants have had on the environment is in the way they have influenced the use of herbicides and pesticides. While some varieties of GMO crops have been developed to be more resistant to insects and other pests, and so require fewer pesticides, others are designed to tolerate and even require more intensive use of herbicides, which in turn pollutes waterbodies nearby and downstream.

Critics of these food production systems argue that the government subsidies currently being given to farmers should be used as a tool for reform. They advocate directing subsidies away from factory farms and large-scale, chemically intensive crop production and toward small farmers who are willing to use methods that have fewer negative impacts.

Throughout North America, there is a revival of small family farms that supply **organically grown** (produced without chemical fertilizers, herbicides, or pesticides) vegetables and fruits and grass-fed meat directly to consumers. Farmers' markets have popped up across the country (**Figure 2.24**), and more and more people are buying locally grown organic foods, paying higher prices than those in traditional grocery stores. This movement is gaining such favor that large corporate farms are seeing the potential for high profits in sustainable (that is, organic), if not locally grown, food production. Organic food production can reduce harmful impacts on the environment because it uses less-polluting methods. Consumers also gain from higher-quality, toxin-free vegetables and fruits.

organically grown products produced without chemical fertilizers, herbicides, and pesticides

Figure 2.24 Organic farming. Jason Plotkin carries white, golden, and red beets at his Golden Acre Farm, a small organic vegetable farm next to North Table Mountain in Golden, Colorado. Plotkin regularly harvests produce that he will sell the next day at a nearby farmer's market. [Denver Post Photo by Cyrus McCrimmon/Getty Images]

CHANGING TRANSPORTATION NETWORKS AND THE NORTH AMERICAN ECONOMY

North America depends on an extensive network of road and air transportation that enables the high-speed movement of people and goods. The road system originated with the first European settlement in the 1500s in Florida, but until the development of the first railways in the 1820s, most goods were moved on rivers, canals, and the coastal seas. By the 1870s, the railroads had formed a nationwide network that dominated transportation until inexpensive automobiles were mass-produced in the 1920s. Beginning in the 1950s, the growth of automobile- and truck-based transportation was helped by the U.S. Interstate Highway System and the Trans-Canada Highway System—a huge network of high-speed, multilane roads that continues to dominate transportation in North America. Because this network is connected to the vast system of local roads, it can be used to deliver manufactured products faster and with more flexibility than can be done using the rail system. The highways have thus made it possible to disperse thousands of jobs in industry and related services into suburban and semi-rural locales across the country, where land is cheaper and living costs are lower. An often-noted aspect of the North American system is the absence of effective intercity passenger rail for most of the region.

After World War II, air transportation became a central part of transportation in North America. The primary niche of air transportation is business travel, because face-to-face contact remains essential to American business culture despite the growth of telecommunications and the internet. With many industries widely distributed across numerous medium-sized cities, air service is organized as a *hub-and-spoke network*. Hubs are strategically located airports, such as those in Atlanta, Chicago, Dallas, and Los Angeles. These airports serve as collection and transfer points for passengers and cargo continuing on to smaller cities and towns. Most airports are also located near major highways, which provide an essential link for high-speed travel and cargo shipping.

CHECK YOUR UNDERSTANDING

1. What is North America's position in the global economy?
2. How has North America influenced, and been transformed by, globalization?
3. What economic sector is most important in North America and why?
4. What factors have led to a backlash against free trade in the United States in recent years?
5. What is an important way in which North American women lag behind their male counterparts? In what way are they ahead of their male counterparts?
6. By what measure is North American agriculture relatively inconsequential, despite its huge influence on land use and the environment?
7. How has the highway system influenced the location of jobs and related services in North America?

POWER AND POLITICS

2.5 Identify the major similarities and differences between Canada and the United States in the roles each country's government plays domestically and internationally.

North America is home to the world's greatest "superpower," the United States, which has the world's largest military and second-largest economy. The United States regularly acts on the global geopolitical stage to protect and extend its economic and strategic interests; however, limited success in the wars in Afghanistan and Iraq, as well as challenges from a new alliance of powerful countries, are exposing the limits of U.S. power. Within North America, political freedoms are continuing to expand, but so is disillusionment with the power of money in the political process. Major differences remain between Canada and the United States in how they assist lower-income members of society, and in their approach to illegal drugs, especially marijuana.

INFLUENCE OF THE UNITED STATES AND CANADA ABROAD

The United States and Canada have dramatically different roles in the global geopolitical order. The United States is recognized as the most powerful country in the world, with a military budget equivalent to the next seven highest-spending countries combined. While the promotion of democracy and political freedoms is an official goal of U.S. foreign policy, in practice, the United States tends to focus its activities abroad on the protection and extension of its own economic and strategic military interests, which are often linked. This can be seen, for example, in the recent U.S. war in Iraq (2003), which numerous planners of the war within the Bush administration (2000–2008) now acknowledge was motivated more by a desire to control access to Iraq's lucrative oil reserves than to bring democracy to Iraq (**Figure 2.25**).

U.S. economic and strategic interests are reflected in the global distribution of its military bases and its spending on aid to foreign governments. Four main concentrations of bases and spending, where the United States has for years had strong strategic and economic interests, can be seen on the map in Figure 2.25 in Europe (mainly Germany, Italy, and the UK), East Asia, North Africa, and Southwest Asia (especially Egypt, Israel, Iraq, Jordan), and Afghanistan and Pakistan in South Asia. The concentrations of bases and spending in Europe and East Asia relate to strategic and economic interests dating from World War II that remained relevant due to the Cold War (see Chapter 4), the subsequent collapse of the Soviet Union, the large volume of trade between the United States and Europe, and the rapid growth of U.S. trade with East Asia. The concentrations in Iraq, Iran, Afghanistan, Pakistan, Israel, and Egypt relate to the "War on Terror" (see the "Challenges to the United States' Global Power" section) and the oil and mineral resources of South and Southwest Asia in general.

The map in Figure 2.25 also shows that there are many places where the United States does not have many bases and where spending on foreign aid is at low or moderate levels. These are generally places where the United States has fewer strategic and economic interests. If U.S. policies to support democracy abroad were a strong factor in its allocation of assistance, one might expect that the focus of U.S. spending on military assistance and foreign aid would be in the parts of sub-Saharan Africa and elsewhere that have low levels of democratization. However, due largely to the poverty and political instability of sub-Saharan Africa, the United States has few economic or strategic interests in the region (see Figure 2.25D). (The same could be said of Haiti in the Caribbean; see Figure 2.25C.) The United States has few bases and spends only a modest amount on military assistance and foreign aid in these parts of the world. Recent moves to increase spending on foreign aid have been made with an eye toward countering the influence of China in the developing world.

Canada takes a more "live and let live" approach on the world stage. While trade is a similarly strong motivator for Canada's foreign policies and foreign aid projects, Canada has far fewer strategic military interests abroad. Arguably, Canada's greatest geopolitical impact comes from its ability to influence the United States. As the second-largest trading partner of the United States, and one with considerable energy resources, Canada has helped to reduce U.S. dependence on overseas oil. This relationship was highlighted by the 9/11 attacks in 2001 and the subsequent Iraq War. At that time, about 25 percent of the oil imported into the United States came from the Organization of the Petroleum Exporting Countries (OPEC), which was mostly made up of the countries along the Persian Gulf, where the terrorists had come from. Recent data show that overall U.S. imports of oil have declined sharply thanks to new oil extraction techniques being used in the United States, from 60 percent of the total amount of crude oil used in 2012 to 25 percent in 2016. Canada now supplies about 40 percent of this oil (**Figure 2.26**).

The extraction of fuel from the oil sands of western Canada is motivated in part by U.S. uneasiness about being dependent on oil from OPEC countries. The oil sands could potentially double the amount of oil that Canada exports to the United States, substantially reducing the need for imports from OPEC countries. The United States has had antagonistic relationships with some Persian Gulf states for a number of years, which to some justifies the many environmental costs of developing Canada's oil sands.

CHALLENGES TO THE UNITED STATES' GLOBAL POWER

A number of challenges to U.S. global power have emerged in recent decades. The wars in Iraq and Afghanistan exposed the limits of the ability of the United States to bring about change in other countries, and a new alliance is challenging the United States and its allies in Europe.

Immediately after the attacks on New York City and Washington, DC, on September 11, 2001 (9/11), the international community extended warm sympathy to the United States. Many governments, despite the protests of large numbers of their citizens, supported then-president George W. Bush in his launching of the *War on Terror*, defined as a defense of the American way of life and of democratic principles. The first target was Afghanistan, which was then thought to be host to Osama bin Laden and the elusive Al Qaeda network that claimed credit for masterminding the 9/11 attacks. The aim was to capture bin Laden—accomplished a decade later in 2011 when bin Laden was killed by U.S. Army Special Forces in Abbottabad, Pakistan—and remake Afghanistan into a stable ally, which still had not been done by 2018.

Although NATO (North Atlantic Treaty Organization) forces (including Canadian troops) joined U.S. forces in Afghanistan, the war proved difficult to resolve because of heavy resistance from tribal leaders within the country and from insurgents in adjacent Pakistan. Also, after the spring of 2003, attention and troop support were diverted as President Bush brought the War on Terror to Iraq, which was later shown to have had no role in the 9/11 attacks.

The wars in Iraq and Afghanistan resulted in few concrete gains, despite heavy losses of civilian life. U.S. combat troops withdrew from Iraq in 2011 and from Afghanistan in 2014, only to return shortly thereafter as both countries have remained plagued by violence. In the case of Iraq, international public opinion polls showed that the U.S. image abroad was badly damaged by what was seen as an unjust war aimed mainly at controlling Iraq's oil resources (see Figure 2.25A, B, and the figure map).

Since 2001, an alliance of five countries has emerged as a symbol of resistance to the global power of the United States and its allies. Together, these countries—Brazil, Russia, India, China, and South Africa (known collectively as BRICS)—account for over a quarter of the world's land area and more than 40 percent of its population, and by 2027 they will together form a larger group of economies than the United States and its major allies (Canada, France, Germany, Italy, Japan, and the United Kingdom), which are collectively known as the G7. Meeting at yearly summits, the BRICS countries are committed to establishing a "more multipolar world order" in which the United States and its allies are less dominant. The BRICS countries are the most prominent of several groups of rapidly developing countries that are publicly challenging the current global geopolitical order.

THE EXPANSION OF POLITICAL FREEDOMS IN THE UNITED STATES AND CANADA

Compared to many world regions, North America has relatively high levels of political freedom. This is part of a long-term trend toward more openness in political decision-making processes. However, there is also widespread dissatisfaction with the political

PHOTO ESSAY

2.25 Power and Politics in North America

Political freedoms are generally well protected and democratization is at a relatively high level in North America. The United States is often thought of as using its power to promote political freedoms and democratization on a global scale. There is some truth to this. However, strategic and economic interests often play a greater role in shaping U.S. actions abroad. For example, some U.S. officials who planned the Iraq War point out that the desire to control Iraq's oil resources and those of its neighbors influenced U.S. actions there more than did the promotion of democracy. If the main U.S. interest abroad were the promotion of democracy, then sub-Saharan Africa would be the major focus of global U.S. foreign aid and military installations, given its large size, population, and the many violent conflicts that plague this region.

THINKING GEOGRAPHICALLY

A What economic interests does the United States have in Europe?

B What strategic and economic interests make it unlikely that the U.S. will withdraw completely from Afghanistan and Iraq in the near future?

C What has kept U.S. economic interests in Haiti at a low level relative to those in other countries?

D What has kept U.S. economic interests in Africa at a low level relative to those in other regions?

A A Canadian soldier wounded in Afghanistan is lifted off a plane at Ramstein Air Base, one of 260 U.S. military bases in Germany, which are a legacy of World War II and Cold War era efforts to discourage potential aggression from the (now-defunct) Soviet Union against U.S. allies in Western Europe. They have since been used to project U.S. military power throughout Asia and Africa. [Tara Walton/Toronto Star via Getty Images]

B A boy and his donkey transport election supplies, paid for in part by U.S. foreign aid, to a rural polling station in Afghanistan. Afghanistan became a major recipient of U.S. aid and military intervention only after the attacks of September 11, 2001 (masterminded by the Al Qaeda terrorist network, based in part in Afghanistan). Before 9/11, U.S. foreign aid to Afghanistan was much lower and U.S. military bases there were nonexistent. [SHAH MARAI/AFP/Getty Images]

C Forty percent of Haiti's population lives in urban slums like this one outside Port-au-Prince. There is no public sanitation or running water, and residents steal electricity from nearby power lines via improvised wiring. Haiti is far from being a functional democracy and is plagued by economic and political instability, but compared to Israel, Egypt, or Afghanistan, it receives relatively little U.S. foreign or military aid. [HECTOR RETAMAL/AFP/Getty Images]

▼ On the map, countries are colored according to their score on a democracy index that uses a combination of statistical indicators to capture elements crucial to the process of democratization (see Figure 1.23 for more).

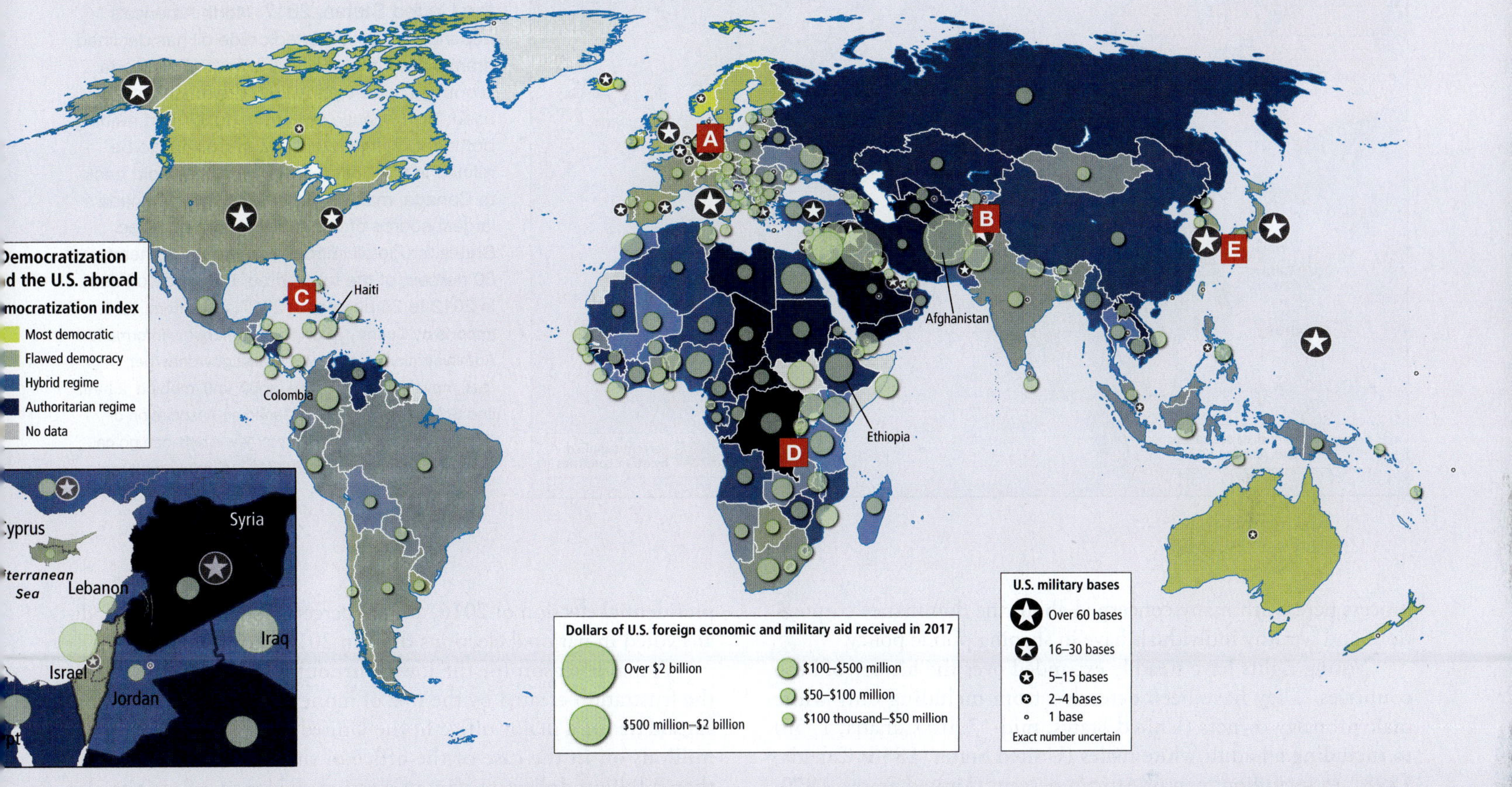

D A U.S. army instructor and a Malian soldier greet each other during a training exercise in 2018. Democracy is fragile throughout much of sub-Saharan Africa, and elections are often plagued by violence. While U.S. financial support for democracy in sub-Saharan Africa is growing, U.S. foreign aid to the area is relatively small, and historically the United States has done little to promote democracy in the region. The United States has few strategic or economic interests, and few military bases, in the region. [ISSOUF SANOGO/AFP/Getty Images]

E Japanese environmentalists protest against relocation and expansion of a U.S. military base in Okinawa, Japan. The more than 130 U.S. military bases in Japan are first and foremost a projection of U.S. power designed to counter any future aggression by China, North Korea, Russia, South Korea, or Japan itself. [Kyodo News/Getty Images]

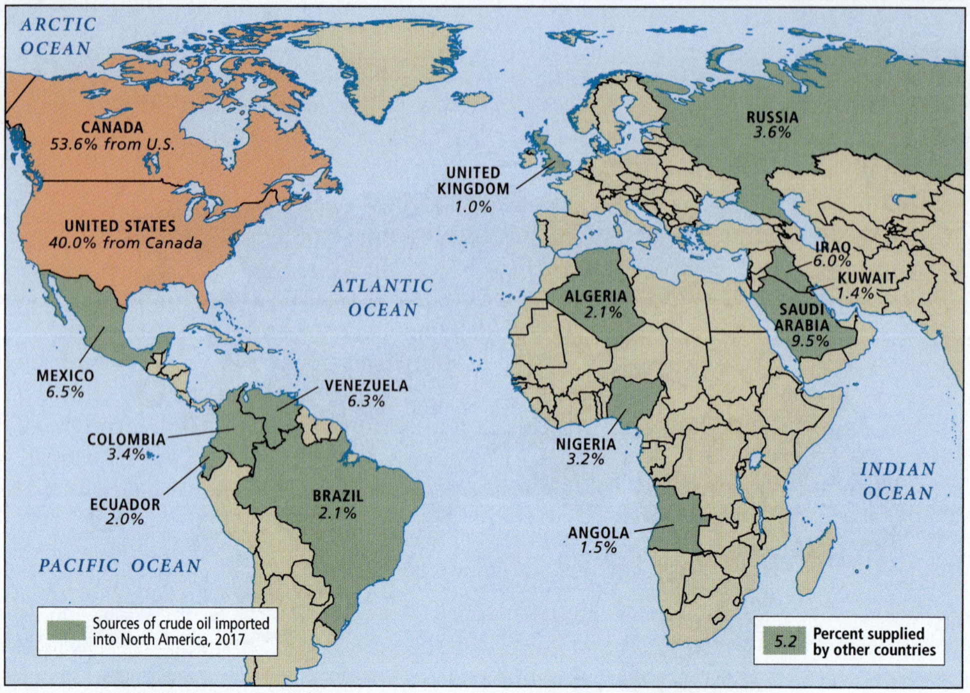

Figure 2.26 Top 15 sources of average daily crude oil imports into Canada and the United States, 2017. North America's dependence on imported crude oil has declined dramatically in recent years. In 2017, Canada produced more than twice as much oil as it used; but because of transportation and refining bottlenecks, much of Canada's oil has to be refined in the United States and then sold back to Canada, making the United States Canada's largest source of imported oil. In the United States, crude oil imports dropped from nearly 60 percent of the total amount of crude oil used in 2012 to 25 percent in 2017. [Data from: "U.S. Imports by Country of Origin," U.S. Energy Information Administration, at https://www.eia.gov/dnav/pet/pet_move_impcus_a2_nus_ep00_im0_mbblpd_a.htm; and *Statistical Handbook*, Canadian Association of Petroleum Producers, at https://www.neb-one.gc.ca/nrg/ntgrtd/mrkt/snpsht/2017/02-04cndncrlmprtsdcln-eng.html]

process here, with many concerned about the dominance corporations and wealthy individuals have in shaping public policy.

Voting rights have steadily expanded over the history of both countries. They have been extended from including only white male property owners (United States, mid-1700s; Canada, 1758) to including all adult white males (United States, 1856; Canada, 1898), to including men of African descent (United States, 1870; Canada, 1837), women (United States, 1920; Canada, 1918), and indigenous peoples (United States, 1924; Canada, 1960).

The ways in which candidates for elected office are selected has also opened up in both countries. In the past, a few political insiders selected the candidates that they wanted to run for a particular office. Starting in the 1920s in the United States, candidates for office began to be selected via primaries, which are elections that determine who a political party will nominate to run as their candidate for a particular office. For example, in 2016, Hillary Clinton was chosen to be the Democratic Party's candidate for president of the United States only after she won primaries in enough states to defeat democratic socialist Bernie Sanders and move on to the general election. Primaries have only recently been adopted in Canada, in part because they are a much more expensive way to choose candidates.

Disillusionment and the Role of Money in Politics

While there has been a trend toward more openness in the political process, there is widespread disillusionment about politics in both the United States and Canada. Evidence of this can be seen in *voter turnout*—the percentage of people who decide to cast a vote in an election—which is low relative to many other counties with similarly high levels of economic development, especially those of Europe. The United States has a generally lower voter turnout than Canada: 55 percent of voting-age U.S. citizens voted in the presidential election of 2016, and 68 percent of voting-age Canadians voted in national elections there in 2015.

A major reason for the lower turnout in the United States is the frustration created by the role of money in politics. A potential candidate for a major office in the United States must now spend millions or, in the case of the office of the U.S. president, more than a billion dollars to win an election. Expensive elections also mean that successful candidates have to spend more of their time raising money in order to be reelected and less time doing their job representing the people who elected them. As a result, politicians tend to focus on wealthy people and corporations who can make the donations that will help them get reelected. Recent political science research comparing how public policy in the United States reflects the preferences of the voting public suggests that the policies that the federal government adopts are generally those supported by high-income "economic elites," corporations, and their associated interest groups. The research also suggests that the preferences of lower-income people have a much lower impact on federal policy.

DEBT AND POLITICS IN THE UNITED STATES

There is much focus in the United States on the national debt. The debt consists of the money the United States borrows by issuing Treasury securities to cover the expenses it has that exceed its income from taxes and other revenues.

To whom does the United States owe its debt? As of 2018 about 43 percent of the total debt is owed to U.S. citizens who own Treasury securities, and 27 percent is owed to various federal government entities, such as the Social Security Trust Fund. About 30 percent is owed to foreign investors, mostly governments, the largest of which are China (5 percent), Japan (5 percent), and Brazil (1.4 percent).

Responses to the national debt highlight the differences between the two major political parties in the United States. Most Republicans believe that the debt is dangerous and needs to be reduced immediately by drastically reducing government spending, especially on social services such as the Affordable Care Act, education, and various kinds of aid to the poor. Many Democrats are less troubled by the debt and believe that some government spending, especially military spending and subsidies to large corporations, should be cut in order to fund social services, and that further revenues should be raised by taxing the very rich.

These differences between the two major political parties show up repeatedly in the United States, as the government's ability to raise revenue via taxation and the ways it spends this revenue are central to almost every major political issue.

RELATIONSHIPS BETWEEN CANADA AND THE UNITED STATES

Citizens of Canada and the United States share many characteristics and concerns. Indeed, in the minds of many people—especially those in the United States—the two countries are like one. Yet that is hardly the case. Three key factors characterize the interaction between Canada and the United States: *asymmetries*, *similarities*, and *interdependencies*.

Asymmetries

Asymmetry means "lack of balance." Although the United States and Canada occupy about the same amount of space, much of Canada's territory is cold and sparsely inhabited. The U.S. population is about ten times the Canadian population. While Canada's economy is one of the largest and most productive in the world, producing U.S.$1.6 trillion (PPP) in goods and services in 2016, it is dwarfed by the U.S. economy, which is more than ten times larger, at U.S.$20 trillion (PPP) in 2018.

In international affairs, Canada quietly supports civil society efforts abroad, while the United States is an economic, military, and political superpower preoccupied with maintaining its role as a world leader. In framing foreign affairs policy, the United States regards Canada's position only as an afterthought, in part because the country is such a secure ally. But managing its relationship with the United States is a top foreign policy priority for Canada. As former Canadian Prime Minister Pierre Trudeau once told the U.S. Congress, "Living next to you is in some ways like sleeping with an elephant: No matter how friendly and even-tempered the beast, one is affected by every twitch and grunt."

Similarities

Notwithstanding the asymmetries, the United States and Canada have much in common. Both are former British colonies and have retained English as the dominant language. Both experienced settlement and exploration by the French. From their common British colonial heritage, they developed comparable democratic political traditions. Both are federations (of states or provinces), and both are representative democracies, with similar legal systems.

Not the least of the features they share is a 4200-mile (6720-kilometer) border, which until 2009 contained the longest sections of unfortified political boundary in the world. For years, the Canadian border had just 1000 U.S. guards, while the Mexican border, which is half as long, had nearly 10,000 agents. In 2009, the Obama administration decided to equalize surveillance of the two borders for national security reasons. Where some rural residents once passed in and out of Canada and the United States unobserved many times in the course of a routine day, now there are drone aircraft with night-vision cameras and cloud-piercing radar scanning the landscape for smugglers, undocumented immigrants, and terrorists.

Canada and the United States share many other landscape similarities. Their cities and suburbs look much the same. The billboards that line their highways and freeways advertise the same brand names. Shopping malls and satellite business districts have followed suburbia into the countryside, encouraging comparable types of mass consumption and urban sprawl. The two countries also share similar patterns of ethnic diversity that developed in nearly identical stages of immigration from abroad.

Democratic Systems of Government: Shared Ideals, Different Trajectories

Canada and the United States have similar democratic systems of government, but there are differences in the way power is divided between the federal government and provincial or state governments. There are also differences in the way the division of power has changed since each country became independent.

Both countries have a federal government in which a union of states (United States) or provinces (Canada) recognizes the sovereignty of a central authority, while states/provinces and local governments retain many governing powers. In both Canada and the United States, the federal government has an elected executive branch, elected legislatures, and an appointed judiciary. In Canada, the executive branch is more closely bound to follow the will of the legislature. The Canadian federal government has more and stronger powers (at least constitutionally) than does the U.S. federal government.

Over the years, both the Canadian and U.S. federal governments have moved away from the original intentions of their constitutions. Canada's initially strong federal government has become somewhat weaker, largely in response to demands by provinces, such as the French-speaking province of Québec (see Figure 2.47), for more autonomy over local affairs.

The more limited federal government that the United States had at its outset has expanded its powers. The U.S. federal government's original source of power was its mandate to regulate trade between states. Over time, this mandate has been interpreted ever more broadly. Now the U.S. federal government has a powerful effect on life even at the local level, primarily through its ability to dispense federal tax monies via such means as grants for school systems, federally assisted housing, military bases, drug and food regulation and enforcement, urban renewal, and the rebuilding of interstate highways. Money for these programs is withheld if state and local governments do not conform to federal standards. This practice has made some poorer states dependent on the federal government. However, it has also encouraged some state and local governments to enact more enlightened laws than they might have done otherwise. For example, in the 1960s the federal government promoted civil rights for African American citizens by requiring states to end racial segregation in schools in order to receive federal support for their school systems.

The Social Safety Net: Canadian and U.S. Approaches

The Canadian and U.S. governments differ significantly in the role they play in society. In Canada there is broad political support for a robust **social safety net**, the services provided by the government—such as welfare, unemployment benefits, and health care—that prevent people from falling into extreme poverty. In the United States there is much less support for these programs and a great deal of contention over nearly all efforts to strengthen the U.S. social safety net. Many in the United States are concerned that the cost of these programs would raise their taxes, which they prefer to keep low. Ironically, while U.S. citizens do pay slightly lower taxes than Canadians, they actually spend more money for their social safety net and generally get less in return than their neighbors to the north.

For many decades, Canada's government spent a higher percentage of its GDP than the United States on social programs. These programs have generally made the financial lives of Canadians more secure. Compared to Canada, the U.S. government provides much less of a social safety net: less generous income support to poor people (welfare), lower unemployment benefits, and health-care coverage for fewer people. Within the United States the prevailing political culture favors a less robust social safety net that makes possible lower taxes for businesses, corporations, and the wealthy. These groups are assumed to invest their profits in job-creating activities that contribute to the tax base, with further benefits *trickling down* to those most in need.

However, looking at government expenditures of tax dollars on the social safety net only provides part of the picture. If you include private social spending such as tax breaks given by the U.S. government to companies that provide health insurance benefits to their employees, which is how just under half of the United States gets its health insurance, the United States spends more of its GDP on its social safety net than any other country in the world except for France (**Figure 2.27**). Even when considering only government expenditures of tax dollars on the social safety net, the United States has still been spending more on social programs than Canada has since 2008, in part due to efforts to provide health insurance to more U.S. citizens.

Two Health-Care Systems The contrasts between health-care systems in the two countries are striking. Reformed in the 1970s, Canada's system uses tax dollars to cover 100 percent of the population. In the United States, health care has generally been private, tied to employment, and relatively expensive. Significant reform of this system began with the Affordable Care Act (ACA), which has been dubbed "Obamacare" after President Obama, who pushed for the legislation, and which reduced the percentage of people who have no health-care coverage from 18 percent to 11.4 percent.

social safety net the services provided by the government—such as welfare, unemployment benefits, and health care—that prevent people from falling into extreme poverty

Despite the ACA, the U.S. system is still much less efficient than Canada's. In 2017, the United States spent more per capita on health care than any other industrialized country—U.S.$10,209 per capita, for a total of 17.1 percent of its GDP per capita. Canada spent U.S.$4826 per capita, or just 10.4 percent of its GDP per capita (GNI figures are not available)—yet Canada had better health outcomes, outranking the United States on most indicators of overall health, such as infant mortality, maternal mortality, and life expectancy (**Table 2.1**). Explanations for Canada's better health-care outcomes revolve around the simplicity of its system (a "single payer" national system versus many private insurers in the United States), lower salaries for its doctors, lower expenditures on pharmaceuticals, and its greater ability to deliver quality health care to low-income people.

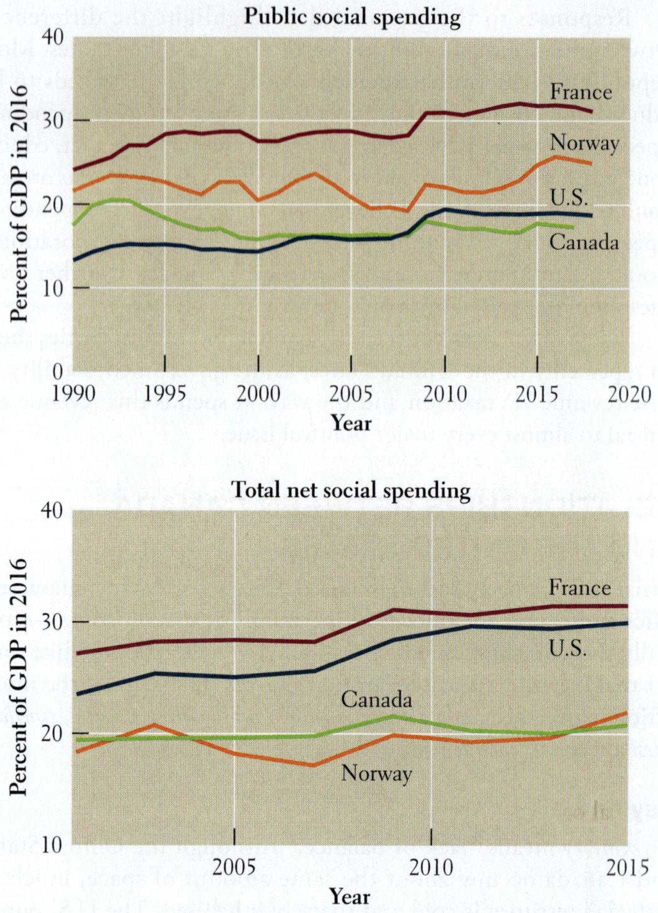

Figure 2.27 The United States spends more on its social safety net and gets less. The United States pays more for its social safety net than Canada, when considering only government expenditures of tax dollars. Including private social spending, the United States spends more of its GDP on its social safety net than any other country in the world except for France, despite a prevailing political culture that favors a less-robust social safety net. [Data from: http://www.oecd.org/social/expenditure.htm]

Gender in National Politics

Gender issues are increasingly in the spotlight in North American politics, where contradictions regarding the role of women in politics are becoming more widely acknowledged, and openness to LGBTQ candidates is growing. While women voters are a potent political force, outnumbering male voters in both Canada and the

Table 2.1 Health-related indexes for Canada and the United States

Country	Public health expenditures as a percent of GDP in 2017[2]	Percentage of population with no insurance, 2017[3]	Deaths per 1000 people in 2017[1]	Infant mortality per 1000 live births, 2017[1]	Maternal mortality per 100,000 births, 2017[1]	Life expectancy at birth (years), 2017[1]	Public health expenditures per capita (PPP U.S.$), 2017[2]
Canada	10.4	0	8	4.3	11	82	4862
United States	17.1	8.8	8	5.8	28	79	10,209

[1] Population Reference Bureau, *2017 World Populatiåon Data Sheet,* at http://www.prb.org/Publications/Datasheets
[2] https://data.oecd.org/healthres/health-spending.htm#indicator-chart
[3] https://www.census.gov/library/publications/2017/demo/p60-260.html

United States, successful female politicians are much less numerous than male politicians. Of the 535 people in the U.S. Congress as of 2019, only 121 members, or 23 percent, were women. Gender equity was somewhat closer at the state level, where 28 percent of legislators were women. In Canada, women have had slightly more success. As of 2018, Canadian women constituted 28 percent of the House of Commons in Parliament and held 33 percent of provincial legislative seats. Neither country compares well to the world at large in that, as **Figure 2.28** shows, over the last decade other countries have added substantially more women to legislatures.

Things may be opening up slightly for women in executive political offices. In 2018, out of Canada's ten provinces and three territories, only one had a female executive. In 2018, the United States had 9 out of a possible 50 female governors. At the national level, Canada briefly had a female prime minister in 1993, and in 2016, Hillary Clinton won the popular vote against Donald Trump but ultimately lost the presidential election.

As more women are elected, research is showing that they govern differently. In the U.S. Congress, women are twice as likely to get new laws passed and more likely to focus on civil rights and civil liberties, domestic violence, health care, and equal pay between men and women. Women are also better at making sure federal tax dollars are spent in the districts they represent, securing on average about 10 percent more federal spending than men. However, women are no less likely than men to work with members of opposing political parties.

LGBTQ issues have come more strongly into focus in recent years. Canada legalized same-sex marriage in 2005, and its prime minister delivered a historic apology for discrimination against LGBTQ people in 2017. The United States legalized same-sex marriage in 2017. In both the United States and Canada, LGBTQ people account for less than 2 percent of Congress or Parliament, which is far below the 4.5 percent of the population that identifies as LGBTQ in surveys, or the probably much larger population that remains "closeted."

DRUGS AND POLITICS IN NORTH AMERICA

Drugs have had a significant influence on society in North America, as drug use is relatively high in this region on a per capita basis. The United States is the world's largest importer of illegal drugs. Its trade in illegal drugs is worth U.S.$400 billion to $500 billion per year, equal to roughly 2.6 percent of the U.S. economy or roughly twice the amount of the entire agriculture sector.

Governmental responses to the illegal drug trade highlight inequalities in wealth and power, with hundreds of thousands of poor people sentenced to years in jail, often for possessing small amounts of illegal drugs, while wealthier drug users are usually placed in drug treatment programs and sentenced to community service. Meanwhile the four largest banks in North America—JPMorgan Chase, Bank of America, Wells Fargo, and Citigroup—have all recently been caught helping Mexican and Colombian drug cartels to conceal hundreds of billions of dollars from U.S. law enforcement. These are highly illegal activities at the heart of

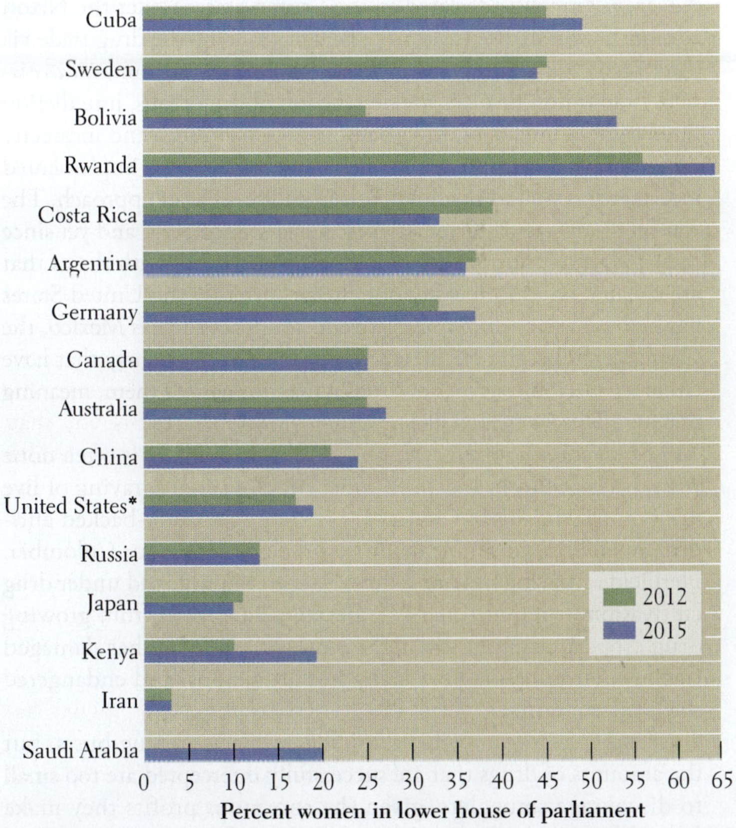

* Percentage for the entire U.S. Congress is used, which has 100 senators (20 are women) and 435 congresspeople (77 are women)

Figure 2.28 Country comparison of women in national legislatures, 2012 and 2015. [Data from: http://unstats.un.org/unsd/gender/chapter5/chapter5.html]

the drug economy, for which the banks have merely been fined. In the 1980s the U.S. government was itself involved in the illegal drug trade, with U.S. Senate investigations revealing that the CIA cooperated with cocaine traffickers as part of its effort to fund a rebellion against the Nicaraguan government.

Illegal drugs are available almost everywhere and drug overdose is the leading cause of accidental death in the United States, topping gun violence and even motor vehicle accidents, at more than 72,000 deaths per year. Mexico has been plagued by decades of violence associated with the vast trade in illegal drugs destined for North America.

In recent years, there has been an escalation in prescription drug abuse, primarily of opioid-based painkillers, such as fentanyl and oxycodone, which are responsible for more than two-thirds of all deaths from overdose in the United States. Heroin has also surged in popularity as policies that reduce the availability of prescription painkillers have taken effect. Methamphetamine (meth) is also a growing concern, as it is highly addictive and extremely toxic, and can be made from household chemicals. Drug cartels in Mexico have recently shifted over to heroin and meth after decades of declining use of cocaine, and heroin and meth addiction is often severe along the major transportation routes that connect northern Mexico with North American cities.

The War on Drugs

A *war on drugs* was declared in the United States under the Nixon administration in 1971, with the goal of reducing the drug trade via three tactics discussed below: *eradication*, *interdiction*, and *incarceration*. Eradication involves the destruction of drug crops, interdiction intercepts shipments of drugs before they reach users, and incarceration imprisons drug offenders. These policies have been implemented in Canada as well, though with a generally less strict approach. The war on drugs has cost over a trillion dollars since 1971 and yet since then the drug trade has become more widespread, with drugs that are cheaper and more easily obtained. Currently the United States has higher rates of drug use relative to countries such as Mexico, the Czech Republic, the Netherlands, Portugal, and Uruguay that have either legalized most types of drugs or decriminalized them, meaning that an offense is treated like a parking violation.

Eradication is often accomplished by burning, as is often done with drugs found in North America, or else aerial spraying of live drug crops with herbicides, as is often done in U.S.-backed antidrug operations in source countries, such as Mexico or Colombia. Herbicide spraying has not reduced the amount of land under drug cultivation or reduced harvests because the profits from growing drug crops are simply too high. However, spraying has damaged fragile environments, hurt many nondrug crops, and endangered the health of farmers.

Interdiction can yield seemingly spectacular drug busts, but the amounts of drugs that are successfully intercepted are too small to discourage smugglers, given the enormous profits they make from the many shipments that do get through.

Incarceration has done relatively little to reduce drug use or decrease the flow of drugs into the United States, but it has played a major role in giving the United States the largest prison population in the world. With 4 percent of the world's population, the United States has almost a quarter of the world's prisoners, more than China and India combined (which together account for over a third of the world's population). Half of federal prisoners and around 16 percent of state prisoners are drug offenders.

Racism influences incarceration rates in the United States, where minorities are jailed more often and for longer terms than whites. Roughly 68 percent of U.S. drug users are white, 14 percent are African American, and 16 percent are Latino. However, African Americans make up 39 percent of those incarcerated for drugs, and Latinos, 29 percent. Studies suggest several important sources of racial bias in drug-related incarceration. Federal drug laws impose far longer prison terms for crack cocaine, which is more common in low-income minority communities, than for powder cocaine, which is more common in higher-income white communities.

Local police departments also target minority neighborhoods for drug law enforcement disproportionately. In part this is because their antidrug funding from the federal government is based on the numbers of people they arrest on drug offenses, and the quickest and cheapest way for them to raise those numbers is to focus on poor and generally minority neighborhoods, where drug dealing and use are often more visible on the street relative to higher-income majority-white neighborhoods, where drug use usually happens behind closed doors.

Incarceration has had a particularly profound effect on African American communities, where 1 in 9 children has a parent in prison. In 2014 California passed a law aimed at reducing racial discrimination in drug laws by equalizing prison terms for crack cocaine and powder cocaine.

New Approaches to Managing Drugs

Because of the ineffectiveness and high cost of the war on drugs, new approaches are being tried. Drug courts, which supervise treatment for nonviolent drug addicts instead of sending them to jail, are a big component of this. Incarcerating a drug user costs around U.S.$25,000 per year, and most users return to drugs upon their release. In contrast, drug courts get users off drugs about 60 percent of the time for a cost of U.S.$900–$3500 per person. Begun in the late 1980s in Miami in the midst of a crack cocaine epidemic, drug courts have since spread to every state, saving billions as a result. For example, Texas, a strong proponent of incarceration and home to the largest prison population in the United States (larger than the prison population of the United Kingdom, France, and Germany combined), has been able to close three of its bigger prisons since implementing drug courts. While significantly more funding has been given to drug treatment in recent years, the United States still spends more money on eradication, interdiction, and incarceration than it does on treatment.

The legalization of drugs is another tactic. In North America most of the focus has been on legalizing marijuana, though other countries, such as neighboring Mexico, have decriminalized cocaine, heroin, and marijuana. Roughly half of North America's population currently has access to legal marijuana for medicinal use. The use of marijuana for medical reasons in North America was documented as early as 1764, and recreational and medicinal use spread throughout much of urban North America during the nineteenth century. Efforts to regulate marijuana were initiated only in the early twentieth century, when campaigns to outlaw its use in the United States associated marijuana with "dangerous" Mexican immigrants. Arrests and incarceration for possessing and selling marijuana did not become common until the 1970s. Use surged during the Vietnam War, when many soldiers first encountered it in Southeast Asia, not far from South Asia, where marijuana was first domesticated.

Marijuana use is on the rise in North America, where in 2018 Canada became the second country in the world, after Uruguay, to legalize marijuana for recreational use. The highest rates of use are in northern Canada, the west coast, and New England. These are all areas where marijuana is legal for either recreational or medical use.

In the United States, over 60 percent of the population supports the legalization of marijuana for recreational use. Arguments in favor tend to focus on the financial benefits of legalizing marijuana, noting the billions of dollars that could be saved in law enforcement costs and earned by taxing marijuana at rates similar to those on alcohol and tobacco. Some studies suggest that marijuana legalization is reducing rates of violent crime, possibly by freeing up police to pursue these crimes, or as a side effect of reduced alcohol consumption in areas that have legalized.

Opponents of marijuana legalization point to research showing that marijuana has some adverse effects, such as an increased risk of cardiovascular disease, and that it can impede concentration and reduce motivation to reach or set goals. Marijuana remains illegal under federal U.S. law, and around one-third of incarcerated federal drug offenders are serving prison sentences for marijuana-related offenses.

CHECK YOUR UNDERSTANDING

1. What are the major similarities and differences between Canada and the United States in the roles each country's government plays domestically and internationally?
2. What ideal does the United States officially promote about its foreign policy, and how does this influence the global distribution of U.S. spending on foreign aid and military assistance?
3. Does Canada or the United States have a more robust social safety net? Which country spends more on their social safety net?
4. How does the percentage of elected female politicians reflect the percentage of women in the general population of the United States and Canada?
5. What are the three main strategies of the war on drugs and two more recent alternatives?

URBANIZATION

2.6 Explain how North America's urban areas have changed since World War II, and the factors that have driven these changes.

Close to 80 percent of North Americans live in **metropolitan areas**—cities of 50,000 or more, plus their surrounding suburbs and towns. This is a reflection of the dominance of the North American economy by activities located in cities, especially larger ones. The largest North American city, greater New York, has a GDP roughly equal to all of Canada. Within the United States, the six largest metro areas (New York, Los Angeles, Chicago, Washington, Dallas, and Houston) account for one-quarter of GDP, and the 23 largest cities have a combined GDP equal to the rest of the country. This urban economic dominance is typical of most world regions, as cities are generally where most service sector, knowledge economy, and manufacturing jobs are located. Rural areas generally depend on "extractive" primary sector industries, such as agriculture and mining, which are much less profitable and employ fewer people.

North America is somewhat unique in that around half of North Americans live in car-dependent suburbs that were built after World War II. Only in Australia do suburbs dominate like they do in North America, where urban life bears little resemblance to that in the central cities of the past. North America's urban areas are now much less densely populated and more energy intensive, having expanded much faster than their populations have increased numerically.

Urban life has changed dramatically since World War II. In the nineteenth and early twentieth centuries, cities in Canada and the United States consisted of dense inner cores and less-dense urban peripheries that graded quickly into farmland. Starting in the early 1900s, central cities began losing population and investment, while urban peripheries—the **suburbs**—began growing (**Figure 2.29**). Workers were drawn by the opportunity to raise their families in single-family homes in secure and pleasant surroundings on lots large enough for recreation. Most continued to work in the city, traveling to and from on streetcars.

After World War II, suburban growth dramatically accelerated as cars became affordable to more people (**Figure 2.30**). Especially in the United States, this shift brought declining interest in publicly funded, non-car-based forms of transportation for expanding urban areas. Change was hastened by a consortium of automobile manufacturers, tire companies, and oil companies that purchased, dismantled, and replaced many urban streetcars and light-rail trains with bus systems.

In the United States, suburban growth was also encouraged by the elaborate, publicly funded Interstate Highway System, a network of free, high-speed roads that passed through most cities, providing easy access to surrounding land along the interstate. Canada, with a less extensive interstate highway system and slightly fewer people who could afford cars, had less suburban growth.

The U.S. 2010 census data indicates that 51 percent of people live in suburbs, 32 percent are urbanites, and 17 percent live in rural areas. The 2011 census data for Canada indicates that 39 percent of Canadians live in suburbs, 41 percent are urbanites, and 20 percent are rural. Both urban and suburban populations are considered urban in Figure 2.29 and in similar maps in other chapters.

As North American suburbs grew and spread out, nearby cities often coalesced into a single urban mass, or **conurbation**. The largest conurbation in North America, sometimes called a *megalopolis*, is the 500-mile (800-kilometer) band of urbanized area stretching from Boston through New York City, Philadelphia, and Baltimore, to south

metropolitan areas cities of 50,000 or more and their surrounding suburbs and towns

suburbs populated areas along the peripheries of cities

conurbation an area formed when several cities expand so that their edges meet and coalesce

PHOTO ESSAY

2.29 Urbanization in North America

North America is very urbanized, with 81 percent of the population living in cities. Cities dominate the North American economy, and some of the wealthiest cities in the world are located here. However, only a few have high levels of "livability" compared to cities in Europe or Oceania. About half of all North Americans live in suburbs with sprawling development patterns that make people dependent on their automobiles. These residents are much greater consumers of energy than those in regions with more densely populated cities that allow for more walking and public transportation. [Data from: http://www.demographia.com/db-worldua.pdf; and 2018 World Population Data Sheet, Population Reference Bureau, http://www.prb.org/pdf15/2015-world-population-data-sheet_eng.pdf]

THINKING GEOGRAPHICALLY

A **What aspects of growth in Dallas are typical of U.S. cities? What sets it apart?**

B **What factors reduce livability in New York?**

C **Is Vancouver's growth pattern typical for North America?**

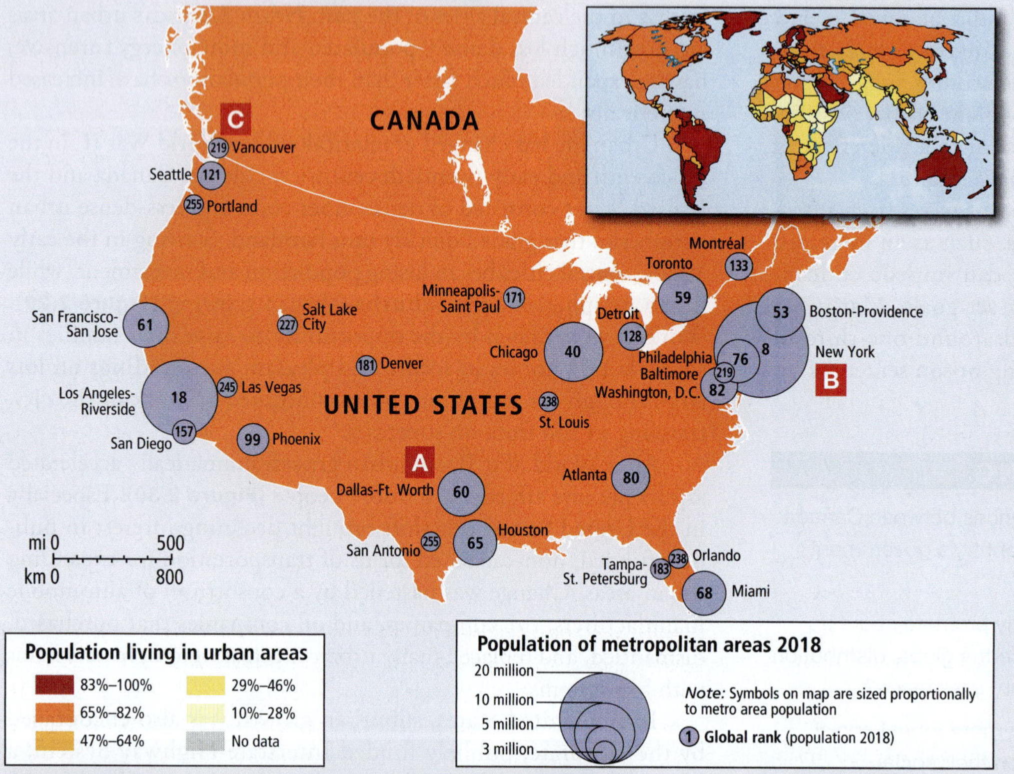

A The Alan Ross Texas Freedom parade takes place in Dallas, home to the sixth-largest LGBTQ community in the United States. Dallas has added more population than any city in the region in recent years, and is now the fourth-largest metro area in the United States. It is among the least-dense and most-sprawling cities in North America and depends almost entirely on cars for transportation. Dallas is generally in the lower-middle ranking of livability for North America. [MICHAEL MATHES/AFP/Getty Images]

B New York City is the second-wealthiest city in the world (after Tokyo) and is North America's financial capital and largest city. Known for its extensive mass transit system (shown here), high population density, and world-class music and arts scene, the high cost of living has driven away almost 900,000 residents since 2010. In terms of livability, New York ranks toward the middle for the region. [Mario Tama/Getty Images]

C Vancouver, Canada, is considered one of the best places to live in North America. It is a leader in controlling sprawl and in developing and maintaining its urban core. Over the past 10 years, the use of public transportation has risen by 50 percent, while the use of cars has fallen by 30 percent. [Christopher Morris—Corbis/Getty Images]

Figure 2.30 Urban sprawl in Phoenix, Arizona.

(A) Phoenix grew rapidly between 1950 and 2000, resulting in a demand for housing. The photo is of Sun City, a new, car-oriented suburb designed for senior citizens, located about 30 miles from Phoenix's center. [Emory Kristof/National Geographic/Getty Images]

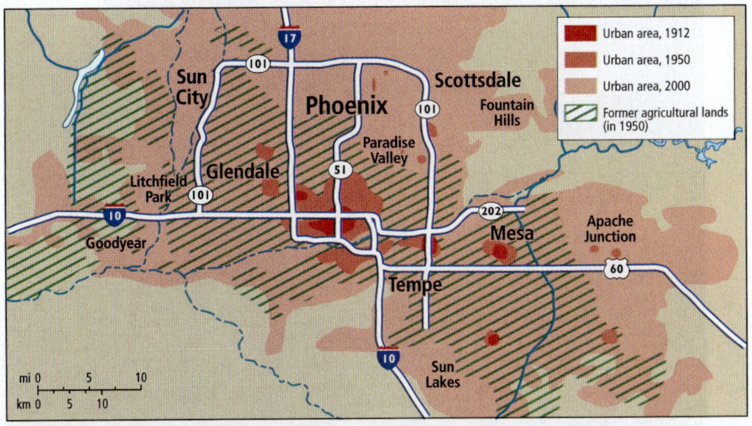

(B) This map shows the original urban settlement and how it has grown from 1912 to the present. A preference for single-family homes means that the city is sprawling into the surrounding desert rather than expanding vertically into high-rise apartments.

of Washington, DC. Other large conurbations in North America include the San Francisco Bay Area, Los Angeles and its environs, the region around Chicago, and the stretch of urban development from Seattle–Tacoma to Vancouver, British Columbia.

This pattern of *urban sprawl* requires residents to drive automobiles to complete most daily activities such as grocery shopping and commuting to work. Two major side effects of the dependence on vehicles that comes with urban sprawl are air pollution and emission of the greenhouse gases that contribute to climate change. Another important environmental consequence is the habitat loss that results when suburban development expands into farmland, forests, grasslands, and deserts.

FARMLAND AND URBAN SPRAWL

Urban sprawl drives farmers from land that is located close to urban areas, because farmland on the urban fringe is very attractive to real estate developers. The land is cheap compared to urban land, and it is easy to build roads and houses on since farms are generally flat and already cleared. As farmland is turned into suburban housing, property taxes go up and surrounding farmers who can no longer afford to keep their land sell it to housing developers. In North America each year, 2 million acres of agricultural and forest lands make way for urban sprawl.

VIGNETTE For Canada's Aboriginal peoples, the chance to indulge in the extravagance of urban sprawl is viewed positively. The Tsawwassen are one of 630 native groups that are negotiating with the Canadian government to gain full control over their ancestral territories. The land they recently received by a treaty lies 20 miles south of Vancouver, near the Strait of St. George. Eager to finally enjoy some of the fruits of development, they are planning a 2-million-square-foot "big box" and indoor shopping mall in the midst of what has been rural farmland. The surrounding communities, which had previously been able to shield their village-like settlements against urban sprawl, feel that there are enough shopping opportunities already. The Tsawwassen see the mall as an overdue chance for the prosperity that their neighbors already enjoy. ■

SMART GROWTH AND LIVABILITY

In recent years the pace of urban sprawl has slowed as center cities have become more livable and drawn in more people. The term *smart growth* has been coined for a range of policies that shift urban growth away from car-dependent suburbs toward compact, walkable urban centers with mixed commercial and residential development and access to transit. *Livability* relates to factors that lead to a higher quality of life in urban settings, such as safety, good schools, affordable housing, quality health care, numerous and well-maintained parks that offer recreational opportunities, and well-developed public transportation systems. While smart growth and the shift toward livability are just starting to take hold throughout North America, a few cities, such as Vancouver, British Columbia, and Portland, Oregon, are much further along in implementing these principles (see Figure 2.29C).

Some older cities are enjoying a renaissance as people move back to them in search of more livability. Places like New York City, which developed long before the term *smart growth* became popular, are arguably the furthest along in crucial aspects of smart growth, given their high density and widespread use of mass transit (see Figure 2.29B). However, the high cost of living in these places can reduce their livability.

It may be a while before smart growth and livability make much of an impact in some of North America's fastest-growing large cities, such as Dallas (see Figure 2.29A), Atlanta, and Phoenix, where in recent decades car-dependent suburbs have spread out over enormous areas. Not surprisingly, none of these cities ranks high on livability indexes, with car dependence and rapid growth resulting in massive traffic jams, long commutes, and declining air quality.

Inner-city decay is another impact of sprawl on livability. This is especially true in the United States, where many inner cities are dotted with large tracts of abandoned former industrial land and neighborhoods debilitated by persistent poverty and loss of jobs. Old industrial sites that once held factories or rail yards

Figure 2.31 The Greening of Detroit Reynaldo Medina weeds squash plants near downtown Detroit. Once the center of North America's auto industry, Detroit began to decline in the 1950s as factories were moved to the city's suburbs and to other parts of the United States. Competition from car makers in Japan forced companies such as Ford and General Motors to cut thousands of jobs, causing the city's population to shrink by two-thirds over the next 70 years. [Fabrizio Constantini/Bloomberg via Getty Images]

are called **brownfields**. Because they are often contaminated with chemicals and covered with obsolete structures, they can be very expensive to redevelop for other uses. Nevertheless, many of these sites are being redeveloped, at times on a large scale, but also on a small scale and more informally (**Figure 2.31**).

Today the depressing decay of a now much smaller Detroit is beginning to sprout with greenery and flowers as inner-city dwellers grow gardens on urban plots once occupied by houses that have since burned down. These green oases draw neighbors together and reconnect people with natural cycles. Children learn that food is something one can grow, not just buy.

HOUSING SEGREGATION, REDLINING, AND GENTRIFICATION

Also left behind in the inner cities are people, many of whom were drawn in generations ago by the promise of jobs that have since moved out to the suburbs or overseas. Often the majority of the population in these inner cities is a mixture of African Americans, Asian Americans, Latinos, and new immigrants. Many are struggling to revitalize their communities, where a lot of the people are poor and in need of jobs and social services, such as health care and education, that have moved to the suburbs.

A major feature of many U.S. cities is **housing segregation**, in which impoverished minority groups are denied fair access to housing by a variety of practices instituted by banks, landlords, and local governments. Banks in many U.S. cities have a history of denying mortgage loans on homes in the suburbs to minority borrowers even when they can afford them, and also of **redlining**—a practice that involves denying mortgages in minority neighborhoods in inner cities. The effect has been that many minorities have been shut out of homeownership and can only rent homes from landlords in poor, inner-city neighborhoods. There, rental units are often in a state of disrepair because landlords can't get loans to finance renovation. Landlords in suburbs will often refuse to rent to minorities, which allows landlords in inner cities to increase their rental prices. Local governments often concentrate low-income rental housing, subsidized by the federal government, in poor, inner-city neighborhoods where minorities are in the majority.

brownfields old industrial sites whose degraded conditions pose obstacles to redevelopment

housing segregation a common phenomenon in cities, in which ethnic or racial or economic classes live in separate neighborhoods and the segregation is maintained by discriminatory customs, not law

redlining The practice used by banks of drawing red lines on urban maps to show where they will not approve loans for home buyers regardless of qualifications

gentrification the renovation of old urban districts by affluent investors, a process that often displaces poorer residents

Legislation prohibiting redlining and other practices that create housing segregation has helped some inner-city minority communities get more access to mortgages and enabled more minority households to move to more affluent suburbs. However, similar but less formal practices still exist and many U.S. cities remain highly segregated, with whites dominating affluent suburban areas and minorities dominating poorer inner cities and their immediate suburbs. Canadian cities have had much less housing segregation, as redlining there focused more on unregulated suburban development.

The new emphasis on smart growth and livability has often perpetuated housing segregation in the United States through the **gentrification** of old inner-city neighborhoods. When affluent, usually white, people invest substantial sums of money in renovating old houses and buildings, poor, often minority, inner-city residents can be displaced in the process. The effect of gentrification on the displaced poor appears to be somewhat less harsh in Canada than in the United States, primarily because Canada's stronger social safety net better ensures social services, housing, and help with housing maintenance.

CHECK YOUR UNDERSTANDING

1. How have North America's urban areas changed since World War II?
2. Why has Canada had less suburban growth than the United States?
3. How has urban sprawl impacted farmland on the urban fringe?
4. What is *smart growth*?
5. How has redlining contributed to housing segregation in the United States?

POPULATION, GENDER, AND CULTURE

2.7 Identify how the changing role of women in society has contributed to the aging of North American populations.

2.8 Describe how North America's population distribution is changing.

2.9 Explain how the increase in migration from Middle and South America and parts of Asia is changing the culture and politics of North America.

As the population of this region ages, due in part to changing gender roles, it is also becoming more diverse, largely from immigration. Many are struggling to adapt to the reality that in the next 20 years, people of European descent will no longer be a majority of the population in North America. This will bring a host of cultural and political changes likely to challenge how racial and ethnic minorities have been treated in this region.

GENDER AND FERTILITY

North America is well along the demographic transition, with the number of children the average woman will have in her lifetime (also known as the *fertility rate*) declining dramatically since the early 1800s. This has happened in response to higher human development: increasing economic development, improved health care, urbanization, and more women participating in the workforce. Greater gender equality has been a major factor in the decline of North American birth rates as more women delay childbearing to pursue education and careers (see **Figure 2.32**).

All of these factors are connected in how they influence fertility. For example, social scientists have documented that increasing economic development is associated with decreased fertility. It is also usually associated with more women participating in the workforce, higher rates of urbanization, and improved health care. Similarly, urbanization is also associated with more women in the workforce, as families that move to the city are more likely to need the cash income that an adult woman who works outside the home can provide. This in turn is associated with lower fertility because women who have careers tend to delay childbearing. With regard to health care, improvements in the quality of health care and better access to medical professionals and medicines tend to lower the fertility rate as families choose to have fewer children because more will survive into adulthood. This choice is made easier with better access to birth control. All of these factors have come together to influence both gender and fertility in North America.

By the early 1800s, small numbers of North American women were starting to work in urban factories, where they had better access to health care. As more women chose this path, the fertility rate fell steadily from a high of seven births per woman in 1800 to three births per woman in the 1920s. After the rapid rise in fertility following World War II, there was a sharp fertility decline as North American women worked more and gained access to birth control via "the pill," which became widespread in the 1960s. Since a low in the 1970s, North American fertility rates have remained more or less flat, at a level just below replacement level, contributing to an overall aging of the population.

AGING IN NORTH AMERICA

The aging of the region's population relates both to longer life expectancy, which increases the number of older people, and to a declining fertility rate, which decreases the number of younger people (**Figure 2.33**). During the twentieth century, the number of older North Americans grew rapidly. In 1900, 1 in 25 individuals was over the age of 65; by 2017, the number was 1 in 7. By 2050, when most of the current readers of this book will be over 50, it is likely that 1 in 5 North Americans will be elderly.

Dilemmas of Aging

Aging populations in developed regions like North America present some as-yet unresolved dilemmas. On the one hand, it is widely agreed that globally, population growth should be reduced to lessen the environmental impact of human life on Earth, especially in countries like the United States and Canada that consume a lot of resources. On the other hand, slower population growth means that populations will age and there will be fewer working-age people to keep economies growing.

Concerns about the financial burdens imposed by aging populations have inspired the United States to raise its legal retirement age, which determines the age at which people can start receiving "social security" retirement income from the government, from 65 to 67. Supporters of these policies argue that if those over 65 remain self-supporting longer, there will be more working people spending their incomes and creating jobs by so doing. Critics of this strategy point out that it disproportionately hurts lower-income people, who have lower life expectancies and tend to retire earlier than wealthier people. Canada's government made this argument in 2016 when it reversed its plans to raise its legal retirement age from 65 to 67, also arguing that there are better ways of encouraging economic growth than forcing people to work longer.

Eventually most elderly people's living arrangements have to change as physical frailty increases and incomes shrink (**Figure 2.34**). North Americans of all ages are choosing to live alone more than they did in the past, and this can be problematic for the elderly as their need for physical assistance increases. One solution to the problem of providing affordable living arrangements for the elderly is *cohousing*, where residents live in housing designed to promote social interactions, and where elderly people look after each other with the aid of a small paid staff.

POPULATION DISTRIBUTION

The population map of North America (**Figure 2.35**) shows the uneven distribution of the more than 362 million people who live here. Canadians make up just over one-tenth (37 million) of North America's population. They live primarily in southeastern Canada, close to the border with the United States. The population of the United States is more than 325 million, with many people moving from the Old Economic Core into other regions of the country that are now growing much faster. The U.S. Census Bureau predicts that the Northeast and Middle West will grow more slowly than the South and West. Canada's national statistics agency predicts a similar pattern of western and southward movement of population.

The Geography of Population Change in North America

North Americans are among the most mobile people in the world. Every year, almost one-fifth of the U.S. population and two-fifths of Canada's population relocate. Some people are changing jobs and moving to cities; others are attending school or retiring to a warmer climate or a smaller city or town; and others are merely moving across town or to the suburbs or the countryside. Still other people are arriving from outside the region as immigrants.

In many farm towns and rural areas in the *Middle West*, or *Midwest* (the large central farming region of North America), populations are shrinking. As family farms are consolidated under

PHOTO ESSAY

2.32 Population, Gender, and Culture in North America

The Gender Development Index (GDI) measures disparities in HDI by gender. Greater gender equality has been a factor in the decline of North American birth rates as more women delay childbearing to pursue education and careers. Declining birth rates play a role in the aging of North American populations, which may slow economic growth.

THINKING GEOGRAPHICALLY

A What is driving lower fertility rates in North America?

B Which problems does cohousing potentially address?

C How much of the North American workforce is female?

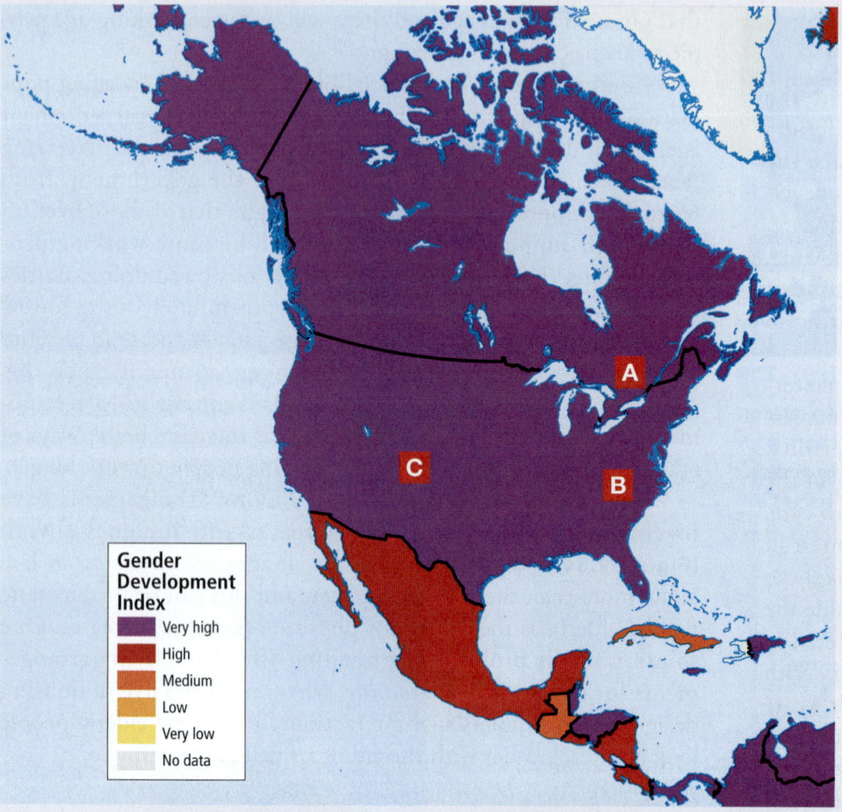

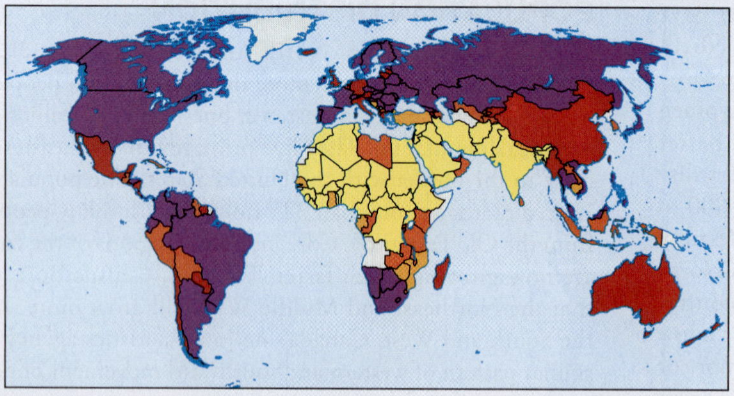

▲ To create the GDI, HDI values are estimated separately for women and men, and the ratio of these two values is the GDI. The closer the ratio is to 1, the smaller the gap between women and men. The map shows that the United States and Canada have fairly equitable human development across genders, relative to most countries; however, Australia and North and West Europe are doing better at equalizing pay and opportunities for men and women.

A A family with 12 children in York, Canada. Average family size, including parents, in Canada has fallen from over 5 in 1900 to around 2.5 today. [Toronto Star Archives/ Getty Images]

B Elderly residents do chores at a cohousing community in northern Virginia. Cohousing, which is designed to promote social interactions, is an affordable way for elderly and retired people to live in a social setting. [Jahi Chikwendiu/The Washington Post via Getty Images]

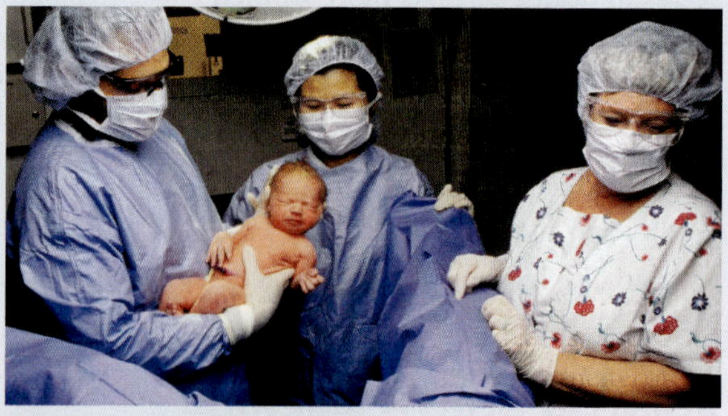

C Health care is one of the most popular employment areas for women in North America. Women frequently are in lower-paid health-care jobs such as nursing, while higher-paid jobs, such as surgery, tend to be male-dominated. [John Greim/LightRocket via Getty Images]

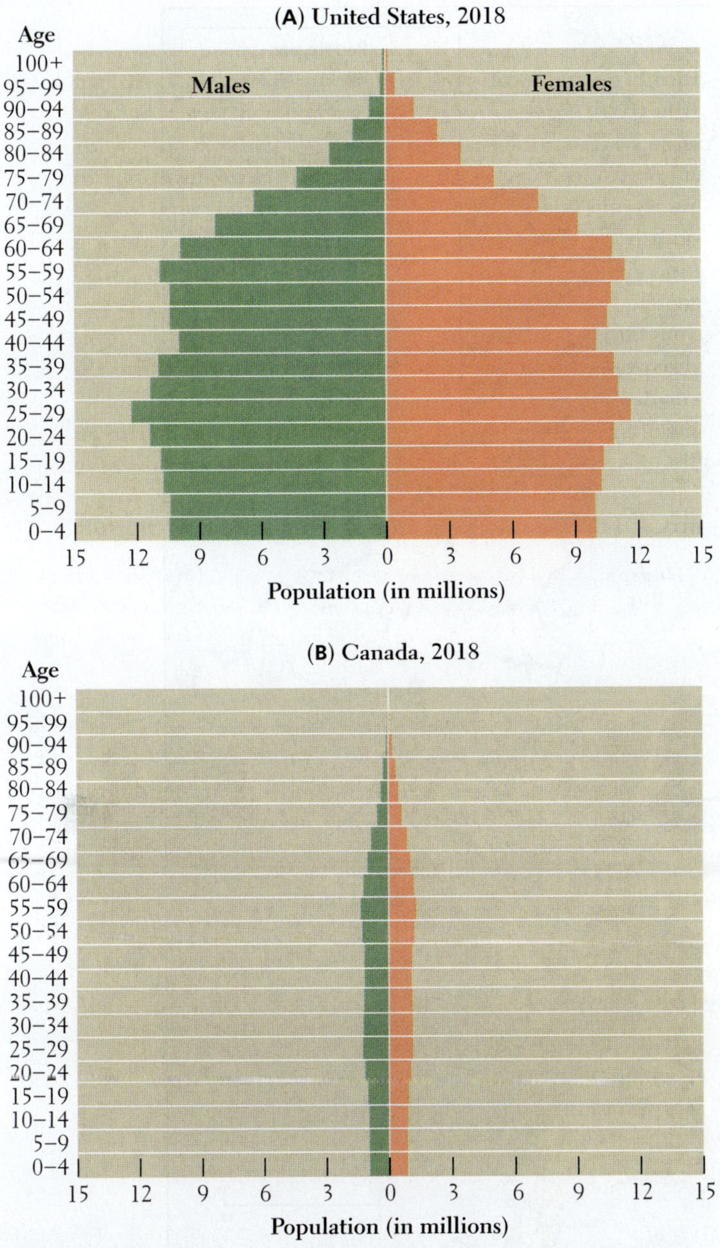

Figure 2.33 Population pyramids for the United States and Canada, 2015. The "baby boomers" (those born between 1947 and 1964) constitute the largest age group in North America, as indicated by the wider middle portion of these population pyramids. [Data from: International Data Base, U.S. Census Bureau, at https://www.census.gov/data-tools/demo/idb/region.php?N=%20Results%20&T=12&A=separate&RT=0&Y=2018&R=-1&C=CA and https://www.census.gov/data-tools/demo/idb/region.php?N=%20Results%20&T=12&A=separate&RT=0&Y=2018&R=-1&C=US]

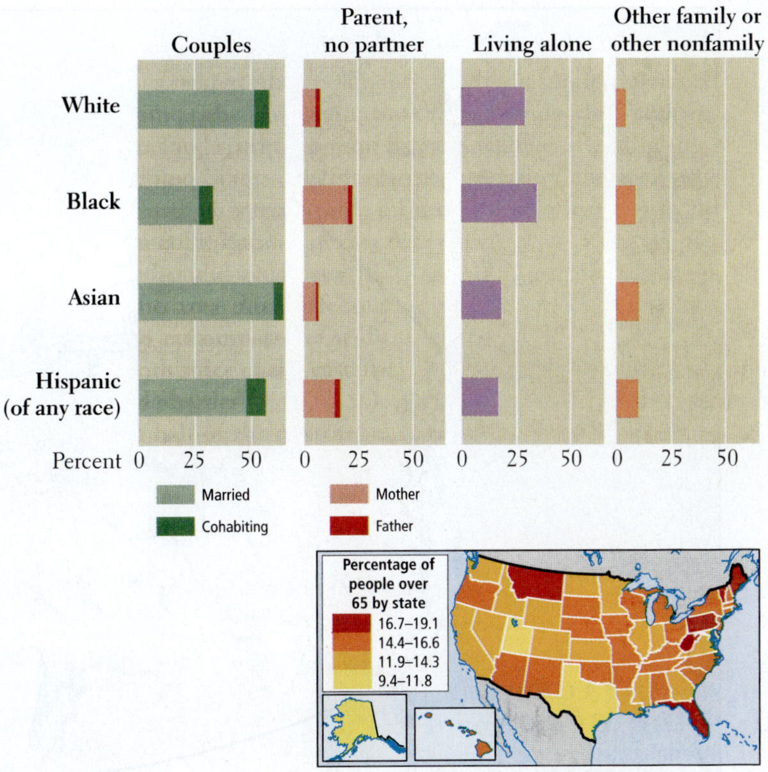

Figure 2.34 Living arrangements of U.S. people age 65 and over by ethnicity and type of companion, 2015. [Data from: https://www.census.gov/hhes/families/files/graphics/HH-7a.pdf, and http://www.census.gov/quickfacts/map/AGE775214/00]

corporate ownership, labor needs are decreasing and young people are choosing better-paying careers in cities. Midwestern cities are growing only modestly but are becoming more ethnically diverse, with rising populations of Latinos and Asians in places such as Indianapolis, St. Louis, and Chicago.

In the western mountainous interior, historically settlement has been light (see Figure 2.35) due to the lack of rain, rugged topography, and, in northern or high-altitude zones, a growing season too short to sustain agriculture. There are some population clusters in irrigated agricultural areas, such as in the Utah Valley, near rich mineral deposits or resort areas. The gambling economy and frenetic construction activity generated by real estate speculation account for several knots of dense population at the southern end of the region.

Along the Pacific coast, a band of population centers stretches north from San Diego to Vancouver (see Figure 2.29A) and includes Los Angeles, San Francisco, Portland, and Seattle. These are all port cities engaged in trade around the **Pacific Rim** (all the countries that border the Pacific Ocean). Over the past several decades, these North American cities have become centers of technological innovation with many new jobs that have attracted larger populations.

The rate of natural increase in North America (0.4 percent per year) is low, less than half the rate of the rest of the Americas (1.2 percent). Still, North Americans are adding to their numbers fast enough through births and immigration that the population could reach 401 million by 2030 and 445 million by 2050.

Pacific Rim a term that refers to all the countries that border the Pacific Ocean

CHECK YOUR UNDERSTANDING

1. How has the changing role of women in society contributed to the aging of North American populations?
2. What are some economic implications of the aging of North America?
3. In what ways is North America's population distribution changing?

Figure 2.35 Population density in North America.

CULTURAL ISSUES

North America is increasingly diverse, with new immigrants from across the globe adding to a dominant culture that has European roots.

Immigration and Diversity

Immigration has played a central role in populating both the United States and Canada. Since the 1700s, a majority of people in North America have descended primarily from European immigrants, but also from enslaved Africans. However, in this highly globalized region most new immigration is coming from East and South Asia and Middle and South America. Migration from these regions promises to make North America a region where people of non-European or mixed descent will become the majority within a few decades (**Figure 2.36**).

Most major North American cities are already characterized by ethnic diversity, with the once numerically dominant Caucasian population now a minority in many big cities. However, most large cities are still highly segregated at the neighborhood level, with different ethnic groups living mostly among themselves. There are many methods of measuring urban segregation, but generally speaking the least-segregated large cities are on the West Coast of North America and the most segregated are in the Old Economic Core and the South.

In the United States, the spatial pattern of immigration is also changing. For decades, immigrants settled mainly in coastal or border states such as New York, Florida, Texas, or California. However, since

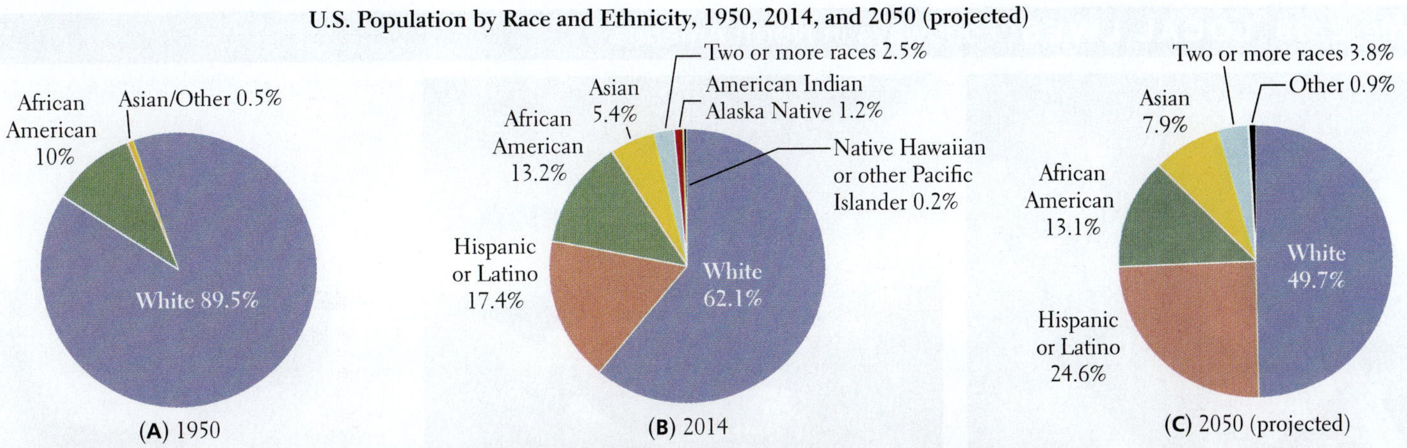

Figure 2.36 The changing ethnic composition in the United States, 1950, 2014, and 2050 (projected). As the percentage of ethnic minorities increases in the United States, the percentage of whites decreases. By 2050, if present reproductive trends continue, whites will constitute only slightly less than half of the population. [Data from: Jorge del Pinal and Audrey Singer, "Generations of Diversity: Latinos in the United States," *Population Bulletin 52* (October 1997): 14; U.S. Census Bureau, "Race by Sex, for the United States, Urban and Rural, 1950, and for the United States, 1850 to 1940," *Census of Population, 1950*, Vol. 2, Part 1, United States Summary, Table 36, 1953, at http://www2.census.gov/prod2/decennial/documents/21983999v2p1ch3.pdf; http://2010.census.gov/news/releases/operations/cb11-cn125.html; http://quickfacts.census.gov/qfd/states/00000.htm; and http://www.baltimoresun.com/sdut-us-whites-falling-to-minority-in-under-5-age-group-2013jun13-story.html/]

Figure 2.37 Percent of total foreign-born people within each state in (A) 2000 and (B) 2016. [Data from: "Percent of People Who Are Foreign Born—United States—Places by State; for Puerto Rico Universe: Total Population, 2010 American Community Survey 1-Year Estimates," American FactFinder, U.S. Census Bureau, at http://factfinder2.census.gov/faces/tableservices/jsf/pages/productview.xhtml?pid=ACS_10_1YR_GCT0501.US13PR&prodType=table; and 2016 data: https://www.brookings.edu/blog/the-avenue/2017/10/02/recent-foreign-born-growth-counters-trumps-immigration-stereotypes/]

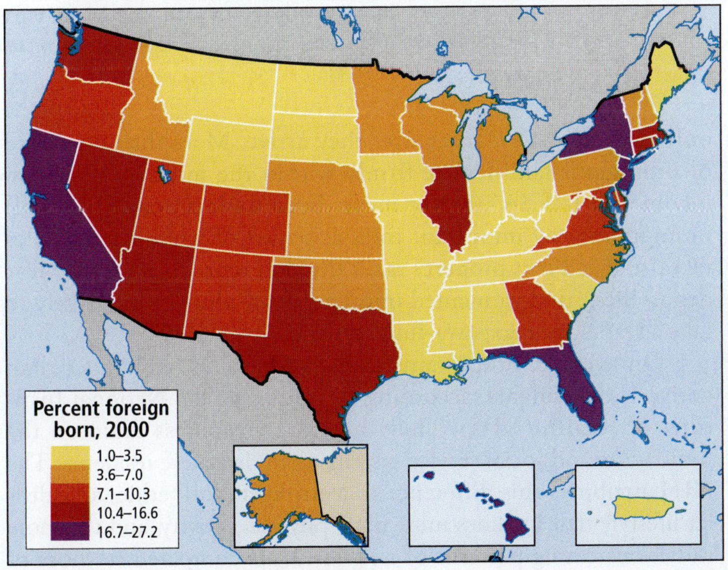

(A) Percentage of foreign-born of total population within each state in 2000.

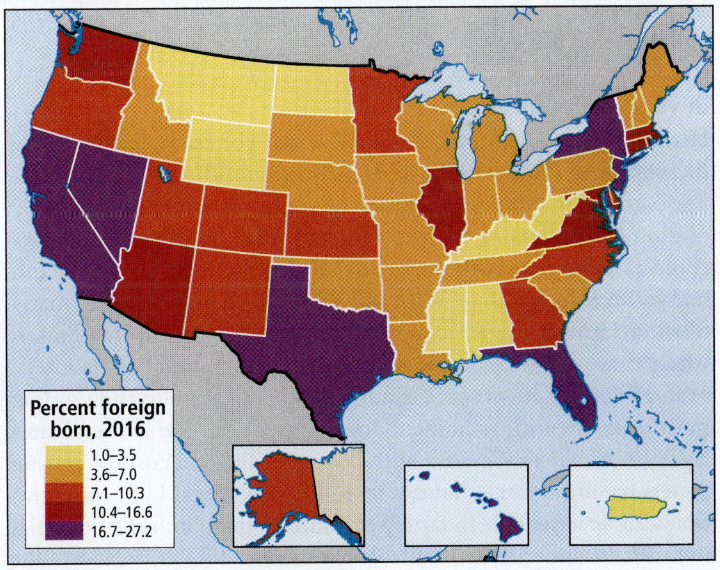

(B) Percentage of foreign-born of total population within each state in 2016.

about 1990, more immigrants have been settling in interior states such as Illinois, Colorado, Nevada, and Utah (**Figure 2.37**). The influence of immigrants can also be seen in many aspects of life in North America, including the popularity of ethnic cuisines (**Figure 2.38**).

Why Do People Decide to Immigrate to North America?

Given the diverse economic opportunities that North America offers, immigration is often considered in terms of positive **pull factors** that draw in migrants. However, the decision to leave one's homeland is usually not an easy one. There are almost always negative **push factors** that cause people to leave home, family, and friends, and strike out into the unknown with what are usually limited resources. These push factors can be civil or political unrest or some kind of discrimination, but most often they involve a lack of economic opportunity. Many push

pull factors positive factors that draw in migrants

push factors negative factors that cause people to leave home, family, and friends

Figure 2.38 LOCAL LIVES: Foodways in North America

(A) Pork ribs are smoked and slow cooked in the "pit" at Spoon's Barbecue in Charlotte, North Carolina. Originally, Native Americans would first dig a pit or trench and fill it with burning coals or hot rocks. They would place meat on top and either slow cook it in the open at a low temperature for several hours or cover the entire pit and let it cook for a day or more. Most barbeque is now cooked in a metal smoker, still called a pit, which is often attached to a trailer so that it can be brought to special events. [Scott Olson/Getty Images]

(B) Poutine, a French Canadian dish, now hugely popular throughout Canada, consists of French fries served with cheese curds (a tasty by-product of cheese making) and covered in gravy. One story holds that in the 1950s, a take-out customer of Fernand LaChance's restaurant in Warwick, Québec, asked for the combination of fries and cheese curds, to which the owner responded "Ça va faire une maudite poutine!" ("That's going to make a cursed mess!"), thus giving the dish its name. The gravy was later added to keep the fries warm. [Christinne Muschi/Toronto Star via Getty Images]

(C) New England–style clam chowder served in a sourdough bread bowl is a classic San Francisco dish that springs from two American traditions. New England clam chowder was developed by fishers and consists of clams and other seafood, potatoes, and seasonings, cooked in a broth to which hard "sea biscuits" were mixed to thicken the chowder. In San Francisco, sourdough bread came via pioneer families and was later popularized by French bakers. It has a particularly sour, tangy flavor that is produced by local bacteria of the San Francisco Bay Area. [Andrew McKinney/Getty Images]

factors relate to the forces of globalization. For example, beginning in 1994 NAFTA (now USMCA) enabled subsidized U.S. agribusinesses to sell their corn in Mexico on a massive scale. As many as 2 million Mexican corn farmers who had small holdings found they could not compete with the artificially low price of U.S. corn and had to give up farming. With families to feed and educate, many of these farmers migrated across the border to the United States, where they now work in construction, as agricultural field laborers, or in a wide variety of service industries, living frugally and sending most of their earnings home. Some have entered the United States legally, but many—because of the complexities and costs of getting papers to immigrate—have entered illegally. Legal U.S. immigration visas are available in far fewer numbers than both the potential immigrants and the jobs available for them to fill.

Do New Immigrants Cost U.S. Taxpayers Too Much Money?

Many North Americans are concerned that immigrants burden schools, hospitals, and government services. And yet numerous studies have shown that, over the long run, immigrants contribute more to the U.S. economy than they cost. Legal immigrants have passed an exhaustive screening process that assures they will be self-supporting. As a result, most start to work and pay taxes within a week or two of their arrival in the country. Most immigrants who draw on taxpayer-funded services tend to be legal refugees fleeing a major crisis in their homeland and are dependent only in the first few years after they arrive. More than one-third of immigrant families are firmly within the middle class, with incomes of U.S.$45,000 or more. Even undocumented (illegal) immigrants play important roles as payers of payroll taxes, sales taxes, and indirect property taxes through rent. Because they fear deportation, undocumented immigrants are also the least likely to take advantage of taxpayer-funded social services.

On average, immigrants are healthier and live longer than native U.S. residents, according to studies by the National Institutes of Health (NIH). They therefore create less drain on the health-care and social service systems than do native residents. The NIH attributes this difference to a stronger work ethic, a healthier lifestyle that includes more daily physical activity, and the more nutritious eating patterns of new residents compared to those of U.S. society at large. Unfortunately, these healthy practices tend to diminish the longer immigrants are in the country, and the healthier status does not carry over to immigrants' children, who are nearly as likely to suffer from obesity as native-born children.

Do Immigrants Take Jobs Away from U.S. Citizens?

The least educated, least skilled American workers are the most likely to end up competing with immigrants for jobs. In a local area, a large pool of immigrant labor can drive down wages in fields like roofing, landscaping, and general construction. Immigrants with little education now fill many of the very lowest-paid service, construction, and agricultural jobs.

It is often argued that U.S. citizens have rejected these jobs because of their low pay, which results in immigrants being needed to fill the jobs. Others say that these jobs might pay more and thus be more attractive to U.S. citizens if there were not a large pool of immigrants ready to do the work for less pay. Research has failed to clarify the issue. Some studies show that immigrants have driven down wages by 7.4 percent for those U.S. natives who do not have a high school diploma. However, other studies show no drop in wages at all.

Are Too Many Immigrants Being Admitted to the United States?

Many people are concerned that immigrants are coming in such large numbers that they will strain resources here. There is some validity to this. Immigrants and their children accounted for 78 percent of the U.S. population growth in the 1990s. At the current growth rates, by the year 2050, the U.S. population could reach 445 million, with immigration accounting for the majority of the increase. But one of the problems with making these projections is that no one really knows how many undocumented immigrants there are. Most research indicates that illegal immigration possibly exceeded legal immigration rates during the mid-1990s. While estimates of the undocumented immigrant population currently in the United States range from 7 to 20 million, reports since 2012 indicate that migration—legal and illegal—is down to less than half that of previous years, and that for the past decade more undocumented immigrants have been leaving the United States than entering.

Undocumented immigrants tend to lack skills, and they are not screened for criminal background, as are all legal immigrants. However, research also shows that undocumented immigrants are not only less likely to partake of social services, but are also less likely than the general population to participate in criminal behavior, with only tiny percentages of them having committed offenses. Nevertheless, recent years have seen a surge in anti-immigrant sentiment in U.S. politics, with Donald Trump winning the 2016 presidential election in part on promises to build a wall along the entire U.S.–Mexico border in order to combat illegal immigration.

Illegal Immigrant Children

Children have recently become a focus of U.S. immigration policy and enforcement. In 2018 the Trump administration began a policy of criminally prosecuting all immigrants caught crossing the border illegally. One aspect of this policy was the separation of children from their families during the legal process. A media scandal resulted as people throughout the world objected to the thousands of children being separated from their families, and the policy was quickly reversed. A longer-standing problem that is proving harder to deal with is the 13 percent of illegal immigrants caught crossing the U.S. border with Mexico who are children unaccompanied by an adult. Most are coming from Central America, where economic instability and violence have torn many families apart. Several types of U.S. policies also play a role in this migration. The *Central America Free Trade Agreement* (*CAFTA*) is an expansion of USMCA that has created many of the same problems for poor farmers in Central America as USMCA did in Mexico. U.S. policies that grant special rights to children also created the belief among some Central American families that their children would be granted asylum in the United States if they arrived alone. Thousands of children have been deported back to their home countries, but many more are being held in prison-like detention facilities run by U.S. corporations. Relatively little action has been taken by the U.S. government or governments in Central America to address the root causes of the migration.

VIGNETTE Every youth who arrives at the border detention shelter in Arizona writes a life story in a creative writing class. A 14-year-old indigenous girl from Honduras started hers: "At 10 years old, my papa started to tell me the good things and the bad things." The next day the teacher asked what this meant. After sitting in silence for 30 minutes, the girl wrote: "While my father is a good man, he did not always do good things. Some of these things created a bad situation for me." Afraid to tell her mom, she called her brothers to help her leave Honduras.

Three years before, at the peak of the global recession, her then-teenage brothers, unable to find work in their rural area, decided to take the month-long journey to the United States by jumping trains through Mexico and then walking for days across the arid terrain of Texas. Once out of Texas, they found migrant farm work in Southern states. The brothers were working in Florida when they spoke to their sister on the phone.

Within days, the brothers sent their sister all of their savings—U.S.$8000—and arranged for a *coyote* to accompany the girl to the United States. As an unaccompanied young woman, she was vulnerable to exploitation, abuse, and sex trafficking. Like most detained in the border shelter, she is unwilling to discuss what happened on her journey, describing it as "the necessary suffering to become someone." Luckily, she made it without falling victim to the drug trafficking and fatal gang violence along the border.

In the 2 months it took her to arrive and be captured at the border, she lost contact with her brothers. Lacking family contacts and money, her deportation is likely. Her biggest desire is "to work hard and give the money back to my brothers." *[Material for this vignette was pieced together by Lydia Pulsipher from personal correspondence with Elizabeth Kennedy and Stuart Aitken, geographers who participate in the ISYS Unaccompanied Minors project.]* ■

RACE AND ETHNICITY IN NORTH AMERICA

Numerous surveys show that a large majority of North Americans of all backgrounds favor equal opportunities for racial and ethnic minority groups. Moreover, social norms, festivals, and other aspects of daily life are increasingly inclusive of all people, regardless of race or ethnicity (**Figure 2.39**). Nonetheless, in the United States, and to a lesser extent Canada, people of color regularly encounter discrimination, and statistics show that, with the notable exception of Asians, they have higher poverty rates, lower life expectancies, higher infant mortality rates, lower levels of academic achievement, and higher unemployment than whites. This is despite the removal of many legal barriers to equality over the past 60 years.

A Backlash Against Growing Diversity

The United States has a long history of racially polarizing politics, but a backlash against growing diversity has occurred in recent years as many whites have become fearful of becoming a minority. Look

Figure 2.39 LOCAL LIVES: Festivals in North America

(A) Mardi Gras Indian Chief Golden Comanche leads a procession through a neighborhood in New Orleans, Louisiana. Mardi Gras, which means "Fat Tuesday," marks the last day before the start of Lent, a 40-day period of fasting and abstinence in Christianity that precedes Easter. Mardi Gras Indians originated as a tribute to Native American communities that had harbored runaway enslaved African Americans. [Mario Tama/Getty Images]

(B) A temporary sculpture at Burning Man, a weeklong "art event and transitory community based on radical self-expression and self-reliance." Begun in 1986 as a bonfire on the beach in San Francisco, Burning Man is guided by principles of inclusive participation, anticommercialism, and the goal of not leaving any physical trace of the festival after it happens each year. [David McNew/Newsmakers/Getty Images]

(C) Bull riding is an event at the Calgary Stampede, a 10-day rodeo, exhibition, and festival that was started in 1886 to help draw settlers from the east. Billed as the "Greatest Outdoor Show on Earth," the Stampede draws more than a million people each year. The highlight of today's Stampede is its rodeo, which offers more prize money than any other. [Photo by Todd Korol /Sports Illustrated/Getty Images]

again at Figure 2.36, which shows the changes in the ethnic composition of the U.S. population from 1950 to 2014 and the projected changes for 2050, by which time whites will be a minority. While Asians are the fastest-growing ethnic group, Latinos are the largest minority group, having overtaken African Americans in 2001, because of a higher birth rate and a high immigration rate. Current demographic studies suggest that the United States will be majority nonwhite by 2050 or earlier. Donald Trump's campaign for the presidency responded to the anxiety that many white U.S. voters feel about being a minority. He appealed to people and organizations that support "white supremacy," the notion that white people are superior to people of other races and should dominate them.

While whites will eventually be a minority in North America, they have dominated for so long that the economic and political systems they have created will continue to favor them long after they are a minority. For example, the practice of manipulating the boundaries of political districts to benefit candidates of a particular group or political party, known as **gerrymandering**, has long been used by political parties in the United States to ensure the success of candidates of a particular race (**Figure 2.40**). Because people of color are much more likely to vote for Democrats (84 percent more likely in the case of African Americans), and whites are more likely to vote for Republicans (51 percent vote Republican and 43 percent vote Democratic), both parties have altered the boundaries of political districts to give more representation to their favored group.

gerrymandering the practice of manipulating the boundaries of political districts to benefit candidates of a particular group or political party

Gerrymandering became more widespread and controversial in 2010, when Republicans made a major effort to win control of state legislatures in order to influence the adjustment of the boundaries

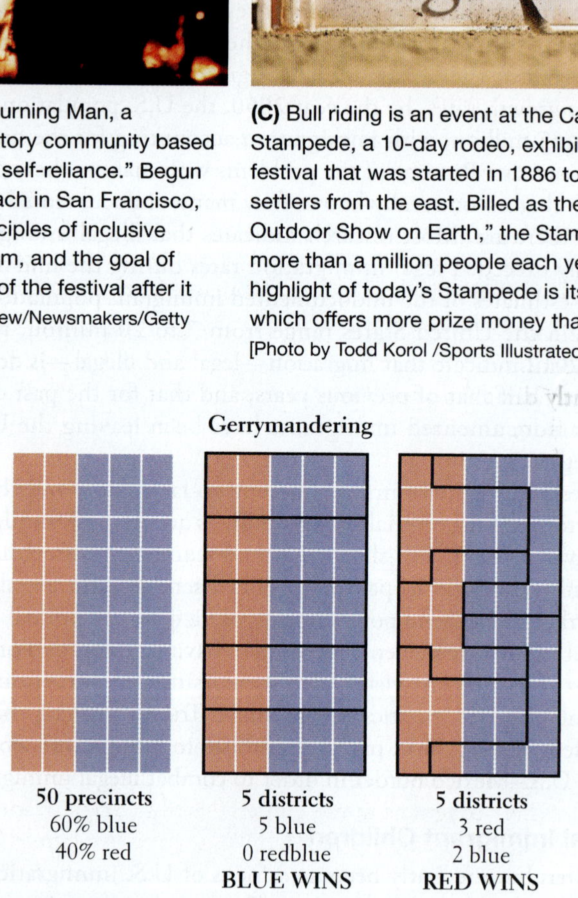

Figure 2.40 Gerrymandering [Data from: https://commons.wikimedia.org/wiki/File%3AHow_to_Steal_an_Election_-_Gerrymandering.svg]

of U.S. congressional districts, a process that occurs every 10 years as new census data become available. The result was rampant gerrymandering that led to overrepresentation of Republicans in the U.S. House of Representatives and in state legislatures. For example, in 2016 Republicans won control of 55 percent of seats in the U.S. House with only 49 percent of the vote. With the Republican party over 80 percent white, and the Democratic party over 40 percent people of color, the use of gerrymandering by Republicans is one way in which whites' political power will continue to prevail in the United States even when people of color are a majority of the population.

Issues of race and politics are somewhat more muted in Canada than in the United States, even though both countries are on a similar trajectory toward whites being a minority within the next 25 years. People of color (often referred to in Canada as visible minorities) are predominantly Asian, and like Asians in the United States, they tend to be more educated and earn higher incomes than other people of color. Immigrants make up 22 percent of Canada's population versus 14 percent in the United States, but anti-immigrant sentiment is slightly lower than in the United States and all major political parties are pro-immigration. No Canadian political parties are attempting to extend the political power of the declining white majority with gerrymandering, because the practice has been illegal in Canada since the 1960s, as it has been in most of Europe.

Income Discrepancies and the Culture of Poverty

Over the past few decades, many non–Euro-Americans and those of mixed heritage have joined the middle class, finding success in the highest ranks of government and business. In particular, African Americans have completed advanced degrees in large numbers, and more than one-third now live in the suburbs. Yet, as overall groups, African Americans, Latinos, and Native Americans remain the country's poorest and least formally educated people (**Figure 2.41**). Similar patterns exist in Canada, though income differences are less.

Is anything other than prejudice holding back some in these ethnic minority groups? Some social scientists suggest that persistently disadvantaged Americans of all ethnic backgrounds may suffer from a *culture of poverty*, meaning that poverty itself forces coping strategies that are counterproductive to economic and social advancement. For example, most people living in poverty do not have the opportunity to complete a college education. They work at jobs with low wages, and because of housing segregation few can find decent affordable housing. Meanwhile, low social status can create the perception that there is no point in trying to succeed.

One component in the culture of poverty is the growing numbers of low-income, single-parent families among people of color. In 2015, only 25 percent of white children lived in single-parent families, while 67 percent of African Americans, 52 percent of Native Americans, and 42 percent of Latinos did. One reason for these differences is that social welfare programs in the United States give significantly more income support to single-parent families than to two-parent families. This has the effect of penalizing married couples with children and rewarding those that live separately. The assistance packages usually fall far short of meeting actual needs, and the enormous responsibilities of both child rearing and breadwinning are often left in the hands of young mothers who, being undereducated and overworked, are less able to help their children rise out of poverty.

One result of the proliferation of single-parent households in the United States with lower education levels and incomes is high rates of child poverty. In 2014 in the United States, 20 percent of children aged 0 to 17 (14.5 million) lived in poverty, whereas only 14.5 percent of adults did. In Canada, these figures are about the same, though that country's more extensive social programs reduce the worst impacts of poverty such as hunger and illness. By comparison, in Denmark, just 3.8 percent of children lived in poverty; in the UK, 9.5 percent, and in France, 10.8 percent.

Race and Law Enforcement in the Spotlight

In 2008 and again in 2012, the United States made history by electing its first African American president. Some heralded this as the end of racism and of the need even to talk about race. As if to prove them wrong, an avalanche of news stories about police shootings of unarmed people of color occurred. The statistics are disturbing, with African Americans roughly three times as likely to be shot by police as whites, and with fewer than one-third of the African Americans killed by police suspected of a crime and armed.

Lost in most discussion is the fact that unwarranted shootings by police that disproportionately affect people of color, even though they are quite common, have historically rarely made even the local news, much less the national news. Another sign of progress is the fact that the current rate of police shootings involving people of color is low relative to the 1970s, as is the rate at which police are being killed in the line of duty. Most importantly, there are proven policies that police departments throughout North America have implemented and that have resulted in fewer shootings of both citizens and police, though to date relatively few departments have done so. It may be a long time before any person's or organization's claims to be "beyond race" ring true in the United States, but at least the subject is receiving more attention, being more openly discussed, and in some cases effective action is being taken.

RELIGION

Because so many early immigrants to North America were Christian in their home countries, Christianity is currently the predominant religious affiliation in North America. In surveys taken in

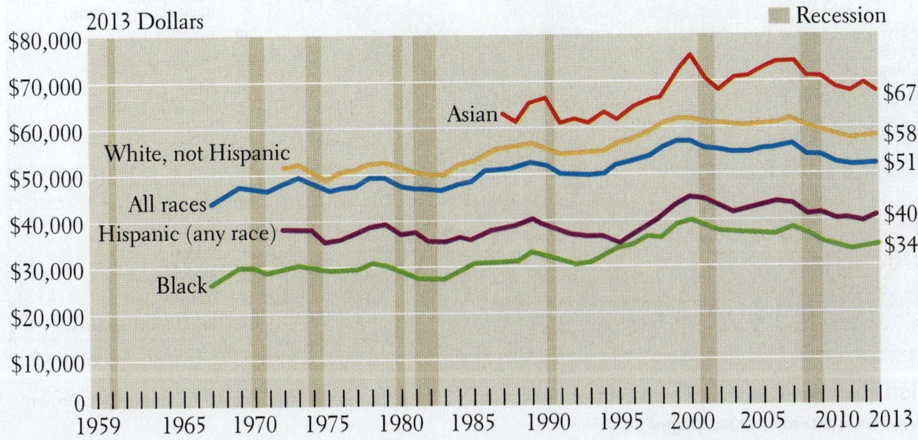

Figure 2.41 Median household income by race and ethnicity, in U.S. dollars, 1967–2013. [Data from: http://www.census.gov/content/dam/Census/library/publications/2015/demo/p60-252.pdf]

2017, 73 percent of those living in the United States identified themselves as Christian, as did 67 percent of those in Canada. In the United States, Protestants are the most numerous religious group, making up 48 percent of the population; in Canada Catholics are more numerous at 39 percent. Despite the numerical dominance of Christianity, about 23 percent of people in both countries reported having no religious affiliation, and the number who identify as Christian is declining in both countries. There has been an increase in the number of people who identify with no tradition at all and of those who identify as Muslim, Hindu, and Jewish.

The Geography of Christian Subgroups

There are many versions of Christianity in North America, and their geographic distributions are closely linked to the settlement patterns of the immigrants who brought them here (**Figure 2.42**). Roman Catholicism dominates in regions where Latino, French, Irish, and Italian people have settled—in Québec, the Southwestern United States, southern Louisiana, and New England. Lutheranism is dominant where Scandinavian people settled, primarily in Minnesota and the eastern Dakotas. Mormons dominate in Utah.

Baptists, particularly Southern Baptists, and other evangelical Christians are prominent in the *Bible Belt*, which stretches across the Southeast from Texas to the Atlantic coast. Evangelical Christianity is such an important part of community life in the South that frequently the first question newcomers to the region are asked is what church they attend. Among all the subregions in North America, rates of weekly attendance at a religious service are highest in the southern United States.

The Relationship Between Religion and Politics

Just how interactive religion and politics should be in North American life has long been a controversial issue in the United States, more so than in Canada. While framers of the U.S. Constitution declared that church and state should remain separate to ensure religious freedom, religion has had a powerful influence on politics in the form of Christian political movements surrounding the abolition of slavery (1770s to 1865); the prohibition of the sale, production, and distribution of alcohol (1840s to 1933); the movement to ban abortion (1970s to the present) and the promotion of prayer in the public schools (1960s to the present); and the failed movement to ban same-sex marriage (1970s to 2015), just to name a few.

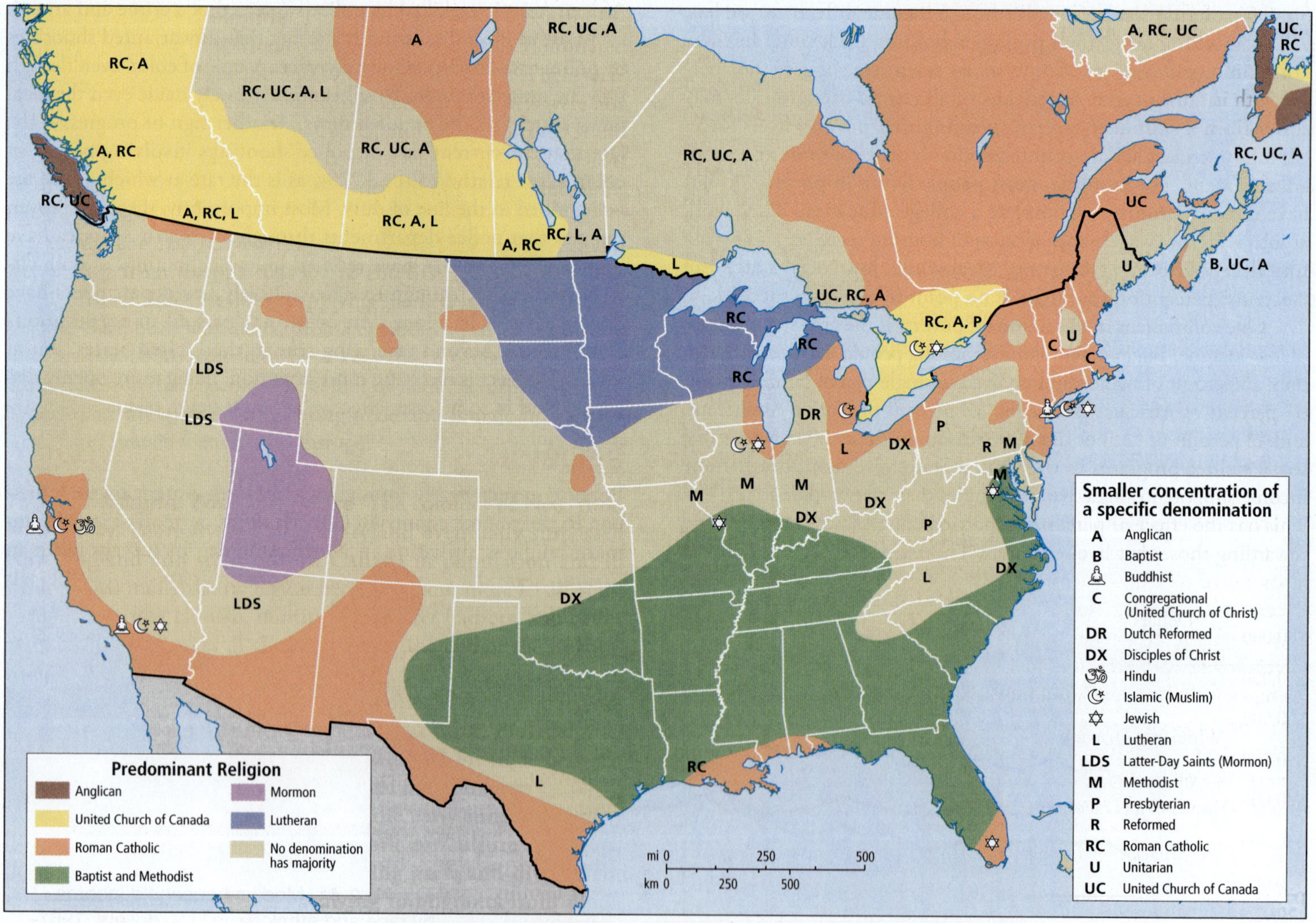

Figure 2.42 Religious affiliations across North America. [Data from: Jerome Fellmann, Arthur Getis, and Judith Getis, *Human Geography* (Dubuque, IA: Brown & Benchmark, 1997), p. 164; and http://www.pewresearch.org/topics/religious-affiliation/2015/pages/3/]

National surveys have consistently indicated that a substantial majority of North Americans favor the separation of church and state and support personal choice in belief and behavior. And yet since the 1970s, U.S. Evangelical Christians have become highly politically mobilized, to the point where this group now accounts for over one-quarter of the voting public even though it represents only about 15 percent of the population. The racial divide among U.S. evangelicals is stark, with whites overwhelmingly supporting the Republican Party and nonwhites supporting Democrats.

THE AMERICAN FAMILY

The family is often seen as being endangered by North America's fast-changing culture. A century ago, most North Americans lived in large, extended families of several generations. Families pooled their incomes and shared chores. Aunts, uncles, cousins, siblings, and grandparents were almost as likely to provide daily care for a child as were the mother and father. Though the **nuclear family**, consisting of a married father and mother and their children, has been a part of society going back through history, it became especially widespread in the post-1900 industrial age.

The Nuclear Family Becomes a Shaky Norm

After World War I, and especially after World War II, many young people began to leave their large kin groups on the farm and migrate to distant cities, where they established new nuclear families. Suburbia, with its many similar single-family homes, soon seemed to provide the perfect domestic space for the emerging nuclear family.

This compact family type suited industry and business because it had no firm ties to other relatives and so was portable. Many North Americans born since 1950 moved as many as ten times before reaching adulthood. The grandparents, aunts, and uncles who were left behind missed helping raise the younger generation and had no one to look after them in old age. This separation of older from younger generations is one reason nursing homes for the elderly have proliferated.

In the 1970s, the nuclear family encountered challenges from many areas, including changing gender roles. Suburban sprawl meant onerous commutes to jobs for men and long, lonely days at home for women. More women began to want their own careers, and rising consumption patterns made their incomes useful to family economies. By the 1980s, 70 percent of working-age women were in the paid workforce, compared with 30 percent of their mothers' generation.

Once both parents are employed, however, a family cannot as easily move to a new location for one spouse's upwardly mobile career. Also, working women could not manage all of the family's housework and child care, as well as a job, so the demand for commercial child care grew sharply. With family no longer around to strengthen the marital bond, and with more women able to support themselves and not financially bound to unhappy marriages, divorce rates rose. This continued into the mid-1970s, after which the rate of divorce declined, possibly as a result of higher levels of education, which decreases the likelihood of divorce.

The Current Diversity of North American Family Types

There is no longer a typical American household, only a diversity of household forms and ways of family life (**Figure 2.43**). In 1960, the nuclear family—households consisting of married couples with children—comprised 74.3 percent of U.S. households.

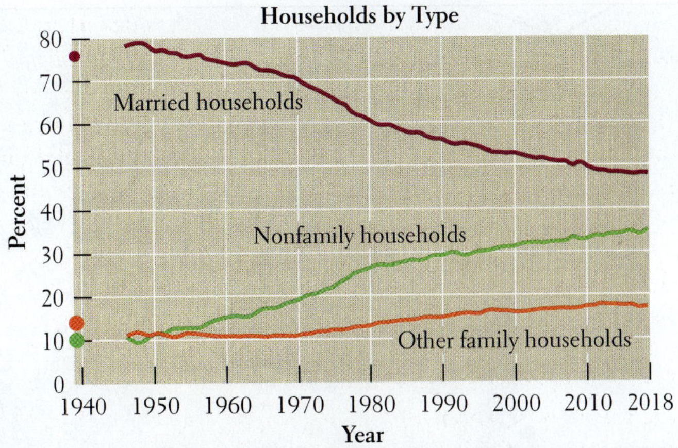

Figure 2.43 U.S. households by type, 2014. [Data from: "Selected Population Profile in the United States: 2007–2009 American Community Survey 3-Year Estimates," American FactFinder, U.S. Census Bureau, at http://factfinder2.census.gov/faces/tableservices/jsf/pages/productview.xhtml?pid=ACS_09_3YR_S0201&prodType=table; and http://www.census.gov/content/dam/Census/library/publications/2014/demo/p60-249.pdf?cssp=SERP]

This number declined to 56 percent by 1990 and to 49.3 percent by 2010. Family households headed by a single person (female or male) rose from 10.7 percent in 1960, to 15 percent in 1990, to 18 percent in 2010. Nonfamily households (unrelated by blood or marriage) rose from 15 percent of the total in 1960, to 29 percent in 1990, to nearly 34 percent in 2010. Canada has similar patterns of change: nuclear family households shrank from 91.6 percent of all Canadian households in 1961 to 67 percent in 2011; family households headed by a single parent rose from 8.4 percent in 1961 to 16.3 percent in 2011; and nonfamily households rose from 8.6 percent in 1961 to 17.1 percent in 2011.

nuclear family a family consisting of a married father and mother and their children

CHECK YOUR UNDERSTANDING

1. How is migration changing the ethnic and racial makeup of North America?
2. Generally speaking, how do race and ethnicity impact access to health care, education, and financial services in North America?
3. How does the percentage of children living in poverty in the United States and Canada compare to other developed countries?
4. What forces have transformed North American families in the past century?

SUBREGIONS OF NORTH AMERICA

2.10 Identify key characteristics of the subregions of North America.

When geographers try to understand patterns of human geography in a region as large and varied as North America, they usually impose some sort of subregional order on the whole (**Figure 2.44**).

Figure 2.44 Subregions of North America. The division of North America into subregions, as shown here, is partly based on *The Nine Nations of North America* (1981) by Joel Garreau. He proposed a set of regions that cut across state and national boundaries.

The order used in this book divides North America into eight subregions, partly inspired by Joel Garreau's *The Nine Nations of North America* (1981). Each section on a subregion sketches in the features that give the subregion its distinct "character of place." As we noted in Chapter 1, however, geographers rarely reach consensus on just where regional boundaries should be drawn or specifically how to define a given region.

NEW ENGLAND AND THE ATLANTIC PROVINCES

New England and Canada's Atlantic Provinces (**Figure 2.45**) were among the earliest parts of North America to be settled by Europeans. Of all the regions of North America, this one may maintain the strongest connection with the past, and it holds a reputation as a North American *cultural hearth*, meaning the place from which much culture has emanated. Philosophically, New Englanders laid the foundation for religious freedom in North America through their strong conviction, which eventually became part of the U.S. Constitution, that there should be no established church and no requirement that citizens or public officials hold any particular religious beliefs. New England in particular is the source of a classic American village style that is arranged around a town square, or *common*. Once used for grazing animals, many commons are now public parks, such as Boston Common (**Figure 2.46**). Many interior furnishing styles also originated in New England; in all classes of American homes today, there are copies of New England–designed furniture and accessories.

The Past Lives in the Present

Many of the continuities between the present and past economies of New England and the Atlantic Provinces derive from the region's geography. During the last ice age, glaciers scraped away much of the topsoil, leaving behind some of the finest building stone in North America but only marginally productive land. Although farmers settled here, they struggled to survive. Although many areas in the region now try to capitalize on their rural and village ambiance and historic heritage by enticing tourists and retirees, outside of a few large cities on the coast or near New York City,

Figure 2.45 New England and the Atlantic Provinces subregion.

many areas, especially in northern New England and the Atlantic Provinces of Canada, remain relatively poor.

After being cleared in the days of early settlement, New England's evergreen and hardwood deciduous forests have slowly returned to fill in the fields abandoned by farmers. Some of these second-growth forests are now being clear-cut by logging companies, and in rural areas, paper milling from wood is still supplying jobs as other blue-collar occupations die out.

Abundant fish was the major attraction that drew the first wave of Europeans to New England. Around the time of Columbus, fishers from Europe's Atlantic Coast were coming to the Grand Banks, offshore of Newfoundland and Maine, to take huge catches of cod and other fish. The fishing lasted for more than 500 years, but eventually the Grand Banks fish stocks were depleted by modern industrial fishing practices (some of these large, factory-type vessels came from Japan and Russia) that have a severe impact on marine ecosystems. Canada and the United States extended their legal boundaries to 200 miles offshore, but unsustainable fishing

Figure 2.46 A New England common. Most towns in New England are centered around an open space called a common. Once used for grazing animals, most commons are now parks used for recreation and community events. This common is in Boston, Massachusetts. [Joe Raedle/Getty Images]

practices were retained until, in the 1980s, there was a crash in cod and other fish species. Tens of thousands of New England and Atlantic Province fishers had to find alternative employment. By 2015 science-based catch limits, set by the New England Fishery Management Council, had failed to curtail the crash of cod fisheries. New research suggests that higher ocean temperatures related to climate change are now preventing cod populations from rebounding.

Strong and Diverse Human Resources

Many parts of southern New England now thrive on service- and knowledge-based industries that require skilled and educated workers: insurance, banking, high-tech engineering, and genetic and medical research, to name but a few. New England's considerable human resources derive in large part from the strong emphasis placed on education, hard work, and philanthropy by the earliest Puritan settlers, who established many high-quality schools, colleges, and universities. The cities of Boston and Cambridge have some of the nation's foremost institutions of higher learning (Harvard, the Massachusetts Institute of Technology, and Boston University, for example). The Boston area has capitalized on its supply of university graduates to become among North America's most important high-technology centers, comparable to California's Silicon Valley.

In its cities New England is no longer the Anglo-American stronghold that people sometimes imagine. The dominant Anglo-American population has for years shared New England with Native Americans, African Americans, and immigrants from Portugal and southern Europe, but today, in cities such as Boston and Providence, people of color are more numerous than whites. Corner groceries may be owned by Koreans or Mexicans;

Jamaicans may be law students or street vendors selling meat patties and hot wings; restaurants serve food from Thailand; and schoolteachers may be Filipino, Brazilian, or West Indian. The blossoming cultural diversity of New England and the entrepreneurial skills of new migrants are helping the subregion keep pace with change across the continent. These trends are less obvious in the Atlantic Provinces of Canada, but there too, immigrants from Asia and elsewhere are influencing ways of life.

CHECK YOUR UNDERSTANDING

1. How has the ideal of religious freedom influenced New England and the region as a whole?
2. What type of village/town pattern of spatial organization originated in New England?
3. What New England city is a major North American high-tech center?
4. How has the racial/ethnic makeup of some New England cities changed?

QUÉBEC

Québec is the most culturally distinct subregion in all of North America (**Figure 2.47**). For more than 300 years, most of the population has been French-speaking. French Canadians are now struggling to resolve Québec's relationship with the rest of Canada.

Origins of French Settlement

In the seventeenth century France founded key cities in Québec and began encouraging its citizens to settle in Canada. By 1763, when France ceded its Canadian colonies to Britain at the conclusion of the Seven Years War, there were 65,000 French settlers in Canada. Most lived along the St. Lawrence River, which linked the Great Lakes (and the interior of the continent) to the *world ocean* and global trade. The settlers lived on long, narrow strips of land that stretched back from the river's edge. This *long-lot system* gave the settlers access to resources from both the river and the land: they fished and traded on the river, they farmed the fertile soil of the floodplain, and they hunted on higher forested ground beyond. Because of the orientation of the long lots to the riverside, early French colonists joked that one could travel along the St. Lawrence and see every house in Canada.

Through the first half of the twentieth century, Québec remained a land of farmers with little formal education eking out a living on poor soils similar to those of New England and the Atlantic Provinces, growing only enough food to support their families. After World War II, Québec's economy grew steadily, propelled by increasing demand for the natural resources of northern

Figure 2.47 The Québec subregion.

A Québec City was founded in 1608, making it one of the oldest cities in North America. The cityscape is dominated by the Château Frontenac hotel, which opened in 1893. Québec City is the provincial capital and the seat of Québec's parliament, and has become a biomedical high-tech center featuring leading-edge software development. [David Boily/AFP/Getty Images]

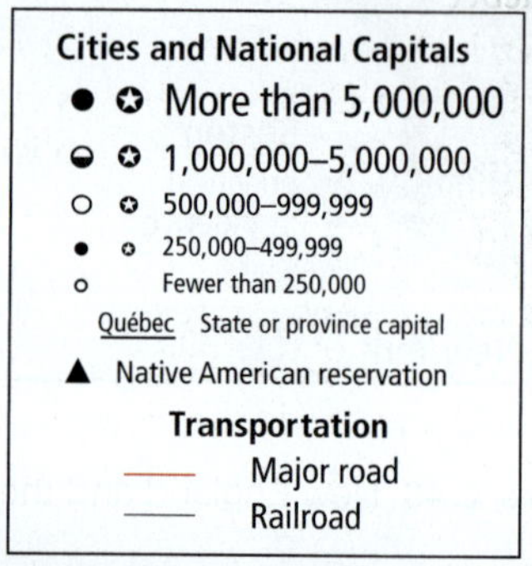

B Aboriginal peoples from various ethnic groups live throughout Québec, but one-fourth of them live in far northern and eastern rural Québec.

Québec, such as timber, iron ore, and hydroelectric power. In the St. Lawrence River valley, cities prospered from the river transport and the processing of resources (see Figure 2.47). Most of these enterprises were in the hands of Anglo-Canadians (those with ancestry in the British Isles) who tended to be better educated, wealthier, and politically dominant in the key city of Montréal, Québec's most prosperous city.

The Quiet Revolution

In the mid-twentieth century, as rural French Canadians moved into Québec's cities, the so-called Quiet Revolution began. Because their access to education and training was growing, the Québécois were for the first time able to compete with English-speaking residents of Québec for higher-paying jobs and political power. Gradually, their conservative Catholic, rural-based culture gave way to a more cosmopolitan one that began to challenge discrimination at the hands of English speakers. The Québécois resisted Anglo-Canadian disparagement of Québécois culture and efforts to block access to education and economic participation. In the 1970s, the Québécois pushed for more autonomy from Canada and then for outright national sovereignty (separation from Canada). The province passed laws that heavily favored the French language in education, government, and business. In response, many English-speaking natives of Québec left the province. A referendum on Québec sovereignty failed narrowly in 1996. By 2009, a new, more articulate and less radical movement had emerged among young adults, seeking Québécois control over all manner of social policies and even over foreign policy. Specifically, they sought foreign policy control because of strong opposition to Canada's participation with the United States in the war in Afghanistan. In recent years the issue of sovereignty for Québec has remained on the back burner.

Québec extends north into areas around Hudson Bay that are rich in timber and mineral deposits (iron ore, copper, and oil) and are also the homelands of many Aboriginal peoples (see Figure 2.47B). The resources are hard to reach in the remote, difficult, glaciated terrain of the Canadian Shield. Although these are the native lands of the Cree and Inuit Aboriginal peoples, the Québec provincial government has legal control of resources and since the 1970s has pushed through the development of a series of hydroelectric power projects in the vicinity of James Bay (part of Hudson Bay) in order to run mineral-processing plants, sawmills, and paper mills. There is significant public Québécois support for the project, which is seen as serving the popular goals of eliminating oil imports and shifting to hydroelectric power and wind power. However, since the 1970s, the Cree and Inuit have protested the clear-cutting of their forests for paper production, the diversion of pristine rivers for hydropower dams, and the loss of natural ecosystems. By 2015, the James Bay project had been largely completed, with protests quieted by large cash payments to the Cree and Inuit. Some electric power generated in northern Québec is sold to the northeastern United States, accounting for about 10 percent of its power consumption.

CHECK YOUR UNDERSTANDING

1. Why is Québec the most culturally distinct subregion in North America?
2. How have the Québécois responded to discrimination against them in Canada?
3. Which groups have disagreed with the Québec provincial government over development of the subregion's natural resources via the James Bay project?

THE OLD ECONOMIC CORE

The Old Economic Core—an area that includes southern Ontario and the north-central part of the United States, from Illinois to New York (**Figure 2.48**)—represents less than 5 percent of the total land area of the United States and Canada, but it was once the economic core of North America. As recently as 1975, its industries produced more than 70 percent of the continent's steel and a similar percentage of its motor vehicles and parts, an output made possible by the availability of energy and mineral resources in the region or just beyond its boundaries. The industrial economy of this region has gone into severe decline, and large cities dependent on manufacturing, such as Detroit and Cleveland, and hundreds of smaller cities and towns, have suffered near economic collapse. Detroit's decline has been the most spectacular of any large city in North America, with its population (just over 600,000) now around a third of what it was in 1950 (1.8 million). Dotted with gigantic abandoned factories and public buildings (**Figure 2.49**), Detroit is a symbol of the *Rust Belt*, a popular term for this subregion.

Loss of Economic Dominance

Plants started closing in the 1970s, and by the 1990s, most manufacturing jobs had moved outside the old industrial heartland to the South, Middle West, Pacific Northwest, and abroad. Industrial resources for factories now come from elsewhere: coal is mined in Wyoming, British Columbia, Alberta, and Appalachia; steel, the mainstay of the automotive and construction industries, can be more cheaply obtained from scrap metal or purchased from China, India, or Brazil.

The Impact of the Economic Decline on People

After World War II, industrial jobs drew millions of rural men and women, both black and white, from the South. Many of their sons and daughters, now without work and without funds to retrain or relocate, constitute some of the more than 46 million North Americans living in poverty as of 2016. Consider the case of one inner-city neighborhood on the west side of Chicago, called North Lawndale, which has had repeated massive job losses.

In 1960, there were 125,000 people living in North Lawndale. It had two large factories employing 57,000 workers, and a large retail chain's corporate headquarters that had thousands of secretaries and office workers. One by one, the large employers closed their doors, and by the 1980s North Lawndale was disintegrating. The housing stock deteriorated as families and businesses left to look for opportunities elsewhere. By 2010, North Lawndale had shrunk to just 36,000 people, and the median household income was roughly half of that for Chicago as a whole.

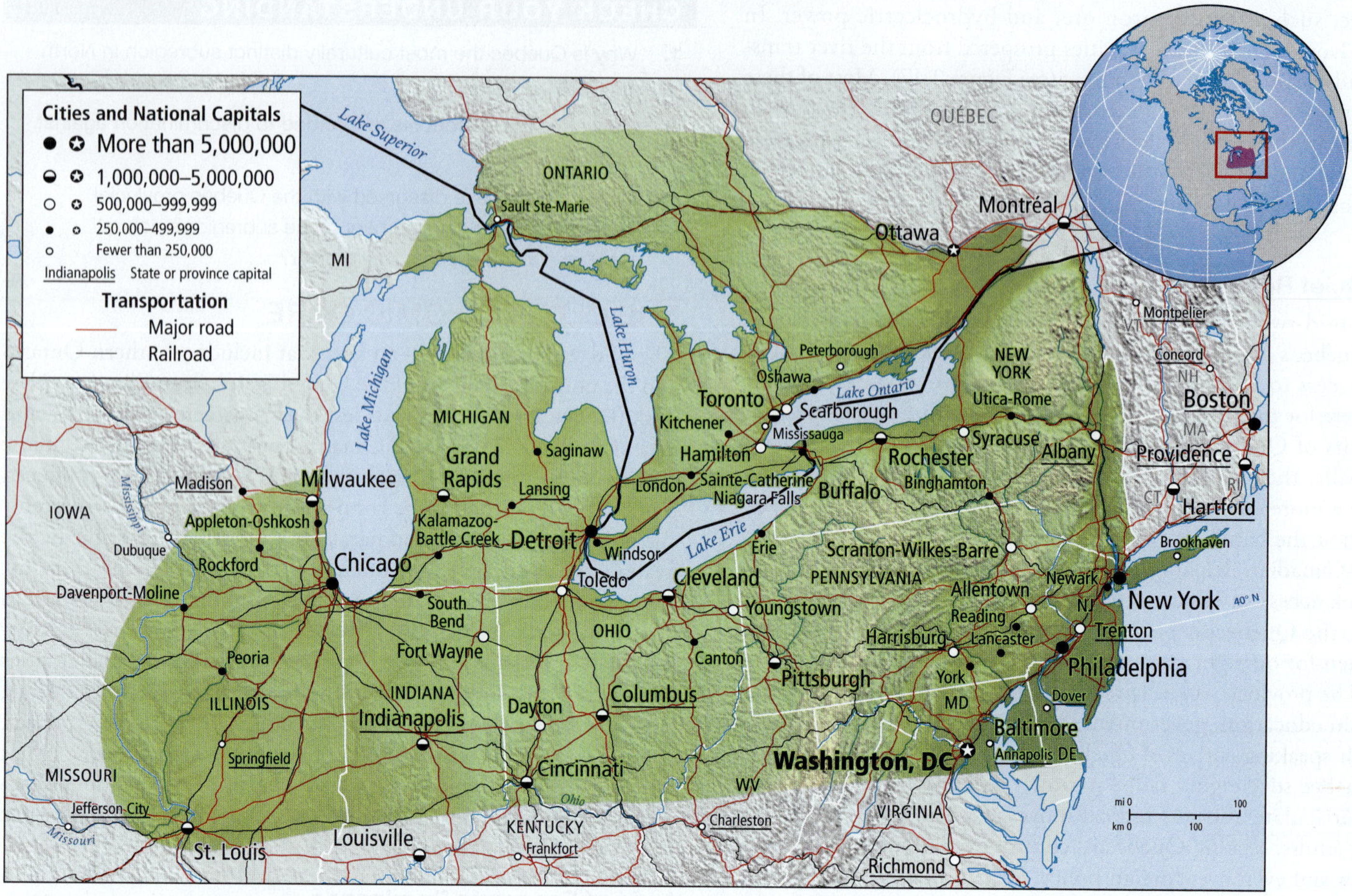

Figure 2.48 The Old Economic Core subregion.

Figure 2.49 Michigan Central Railroad Station in Detroit. When it was built in 1913, Michigan Central Station was the tallest train station in the world. With the rise of automobile-based transportation, though, by 1956 the building was virtually unused. It finally closed in 1988 and has been deteriorating ever since. Now the question is: Will concerns about fuel costs and CO_2 emissions decrease automobile use and reintroduce an era of rail transportation? [J.D. Pooley/Getty Images]

Despite the importance of industry, it would be wrong to think of the Old Economic Core as solely based on factories and mines. The region's largest cities, such as New York, Toronto, and Chicago, continue to prosper, largely because of their strong service industries, which connect them to global economies and make them highly influential throughout North America. Between the great industrial cities of this region are thousands of acres of some of the best temperate-climate farmland in North America. And thousands of miles of shoreline around the Great Lakes and oceanfront along the Eastern Seaboard provide both winter and summer recreational landscapes, as do the mountains of New York State and Pennsylvania.

CHECK YOUR UNDERSTANDING

1. What was this subregion's relationship to the rest of North America?
2. What factors brought on economic decline in the Old Economic Core?
3. How are this subregion's largest cities faring economically today?

THE AMERICAN SOUTH (THE SOUTHEAST)

The regional boundaries of the American South are perhaps less distinct and based more on a perceived state of mind and way of life than are those of most other U.S. regions. This region, in fact,

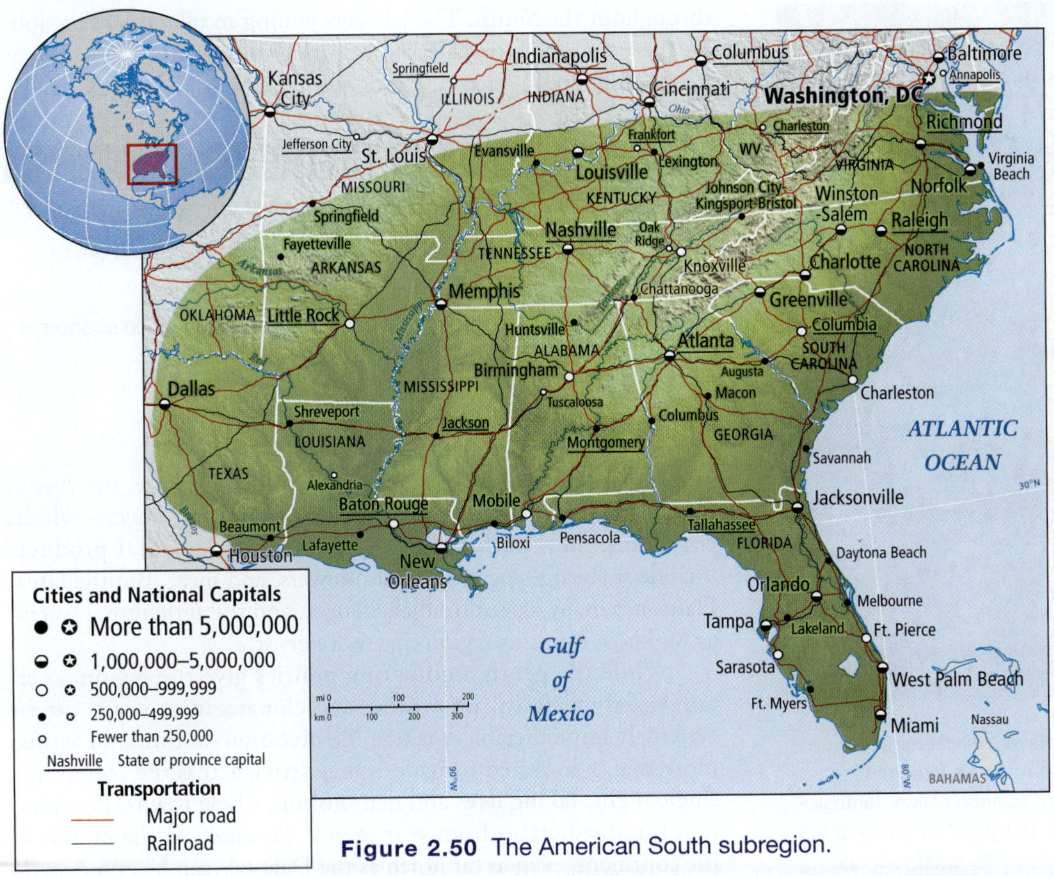

Figure 2.50 The American South subregion.

covers only the southeastern part of the country, not the whole of the southern United States (**Figure 2.50**). Somewhere west of Houston, Texas, the American South grades into the Southwest, a region with noticeably different environmental and cultural features.

Multiple Images Characterize the South

The American South is dominated by a complex of features that many people would identify as *Southern*: a wide range of unique dialects; barbecue and other comfort foods from both white and African American traditions (see Figure 2.38A); stock car racing; horse farms; bluegrass, jazz, country, and blues music; the open friendliness of the people; conservative stances on politics, gun ownership, and abortion; and the prominent role of religion in public life.

Some places in the South have few recognizable Southern qualities (**Figure 2.51**). Parts of Missouri, Oklahoma, and Texas have strong ties to the Great Plains; Miami, on the far southern tip of Florida, with its cosmopolitan Latino culture and its trade and immigration ties to the Caribbean and South America, hardly seems Southern at all. New Orleans, also in the heart of the South, has a unique set of cultural roots—French, African, Cajun, Creole, Caribbean—that set it apart from other Southern cities.

North Americans and foreign visitors to the South tend to hold many outdated images of this subregion from the civil rights era of the 1960s and even from the Civil War era. It is true that significant racial segregation still persists in that black people and white people tend to live and worship separately, but in fact, many more whites and blacks share neighborhoods in the South than in the Old Economic Core. In rural and urban settings, schools are becoming more integrated as a result of residential patterns as well as busing plans and magnet schools. The workplace is now integrated, and it is not uncommon to find African Americans and other minorities in supervisory and administrative positions, in business, government, and education. Biracial families are common.

Since the 1990s, Latino immigrants have come by the thousands into Southern cities. Of the 11 U.S. cities with the largest Hispanic populations, 3 are in the Southeast. Much of the restaurant service, road maintenance, landscaping, roofing, and construction work is now done by Spanish-speaking workers from Mexico and Central America, many of whom are sinking roots in the region and saving to establish businesses and buy homes.

Poverty Persists for Some, but Return Black Migration Is a Boon

The South has the nation's highest concentration of families living below the poverty line, and most of them are white. On the other hand, Southerners of any race or ethnicity are able to maintain a substantially better standard of living than their parents did.

Many African Americans left the South in the 1950s and 1960s for factory jobs in the Old Economic Core in what was called the *Great Migration*. Now increasing numbers are coming back to the region because of jobs, business opportunities, the lower cost of living, a milder climate, as well as safer, more spacious, and friendlier neighborhoods than the ones they left in places such as New York, Illinois, or California. This return is being called the *New Great*

Figure 2.51 Atlanta's Dragon Con. Cosplayers (short for costume players) dressed as *Star Wars* stormtroopers at Atlanta's Dragon Con, a 4-day multi-genre convention that focuses on science fiction, fantasy, comics, cartoons, and associated fan cultures. Dragon Con is one of the city's largest conventions, drawing 50,000 participants each year and contributing more than U.S.$25 million to the metropolitan economy. Atlanta is widely considered the capital of the American South subregion, given its large population (just under 6 million people), its central location within the region, and its historical importance. [John Amis/AFP/Getty Images]

Migration. In the 1990s, 3.5 million people who self-identified as black moved to the South from other parts of the United States. Seven of the ten metropolitan areas that gained the most African American migrants during the 1990s were in the South, principally in Florida, Georgia, North Carolina, Virginia, Tennessee, and Texas. Other Southern states had no increase, and Louisiana experienced black emigration even before Hurricane Katrina. Since 2000, half a million new black residents have moved to Atlanta, many of them young, educated professionals. Another 500,000 of Atlanta's newest citizens are Hispanic, and Asians constitute another substantial minority.

The South is steadily improving its position as a region of growth. The federally funded Interstate Highway System opened the South to auto and truck transportation. Inexpensive industrial locations, close to arterial highways, have drawn many factories to the South, including those involved in automobile and modular-home manufacturing. For several decades, tourists and retirees by the hundreds of thousands have been driving south on the interstate highways, many attracted by the bucolic rural landscapes, historic sites, and warm temperatures.

Southern agriculture is now highly mechanized and more specialized than ever before. Strawberries, blueberries, peaches, tomatoes, mushrooms, and wines from local vineyards are replacing the traditional cash crops of tobacco, cotton, and rice. Most of the country's broiler chickens are now raised by large operations throughout the South. The laborers willing to take low-wage jobs on these factory farms are often immigrants from places such as Russia, Ukraine, Vietnam, Haiti, and Honduras.

CHECK YOUR UNDERSTANDING

1. What are some defining characteristics of the American South?
2. Most families living below the poverty line in the South are of what racial/ethnic background?
3. What role has the Interstate Highway System played in economic development in the South?

THE GREAT PLAINS BREADBASKET

The Great Plains (**Figure 2.52**) received its nickname, the *Breadbasket,* from the immense quantities of grain it produces—wheat, corn, sorghum, barley, and oats. Other agricultural products include soybeans, sugar beets, sunflowers, and meat. But the Great Plains has many agricultural challenges, and the agricultural system as we know it in this region may not persist.

While the gently undulating prairies give the region a certain visual regularity, its weather and climate, in contrast, can be extremely unpredictable, making life precarious at times. In spring, more than a hundred tornadoes may strike across the region in a single night, taking lives and demolishing whole towns. Precipitation is unpredictable from year to year. Summers in the middle of the continent, even as far north as the Dakotas, can be oppressively hot and humid yet lacking in sufficient rainfall, while winters can be terribly cold and dry, or extremely snowy.

Adapting to the Challenges of the Great Plains

The people of the plains learned to adapt to these challenges long before European settlers began to stake claims there in the 1860s. As people do in even drier regions in the Continental Interior, Great Plains farmers often irrigate their crops with water pumped from deep aquifers and delivered with central-pivot sprinklers that irrigate circular-shaped areas (**Figure 2.53**). However, to produce enough grain for one slice of bread in irrigated fields takes 10.4 gallons of water, a rate of water usage that is not sustainable (see Table 1.2 in Chapter 1). Most wheat is harvested by an energy-intensive system that involves traveling teams of combines that are used to start harvesting in the south in June and move north over the course of the summer. To select the most profitable crops and find the best time to sell them, plains farmers keep close tabs on the global commodities markets, usually with computers installed in barns, homes, or even in the cabs of the huge (and very costly) farm machines they use to work their land.

Animal raising, the other important activity in the Great Plains, has become problematic from environmental and animal-rights perspectives (see Figure 2.52A). Rather than being herded on the open range, as in the past, cattle are now raised in fenced pastures and then shipped to feedlots, where they are fattened for market on a diet rich in sorghum and corn, but often in muddy, inhospitable pens. And like wheat, meat production uses a great deal of water in an area prone to drought. For example, to grow and process the amount of beef in one hamburger requires 640 gallons of water.

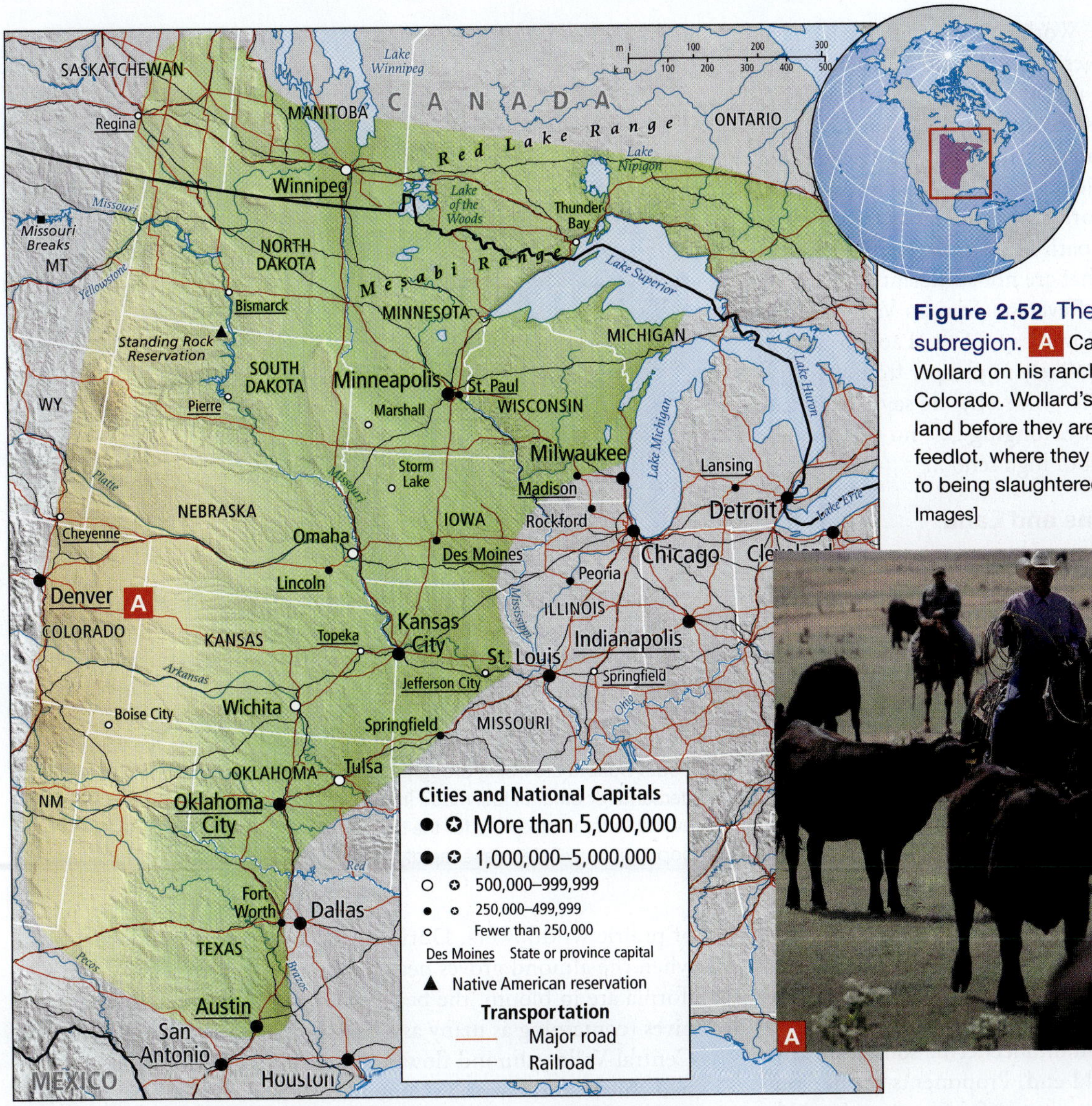

Figure 2.52 The Great Plains subregion. **A** Cattleman Gary Wollard on his ranch in eastern Colorado. Wollard's cattle graze on his land before they are shipped off to a feedlot, where they are fattened prior to being slaughtered. [John Moore/Getty Images]

Figure 2.53 Irrigation in the Great Plains and Continental Interior. A center-pivot irrigation system as seen from the air in eastern Oregon. Often a well at the center of each irrigator taps an underground aquifer at an unsustainable rate. [David Boyer/National Geographic/Getty Images]

Erosion is a serious problem on the plains: soil is disappearing more than 16 times faster than it can form. Grain fields and pasture grasses do not hold the soil as well as the original dense prairie grasses once did. Plowing and the sharp hooves of the cattle loosen the soil so that it is more easily carried away by wind and water erosion. Experts estimate that each pound of steak produced in a feedlot results in 35 pounds of eroded soil. Given the costs in soil and water, raising cattle in this region is by no means a sustainable economic activity.

Globalization Impacts in the Great Plains

Cattle and other livestock, such as hogs and turkeys, are slaughtered and processed for market in small plants across the plains by low-wage, often immigrant, labor. In the 1970s, a meatpacking job in one of the Great Plains major cities provided a stable annual income of U.S.$30,000 or more, relatively high for the time. The workforce was unionized, and virtually all workers were descendants of German, Slavic, or Scandinavian immigrants who arrived in North America in the nineteenth century. But in the 1980s, a number of unionized meatpacking companies closed their doors. Other nonunion plants opened, often in isolated small towns in

Iowa, Nebraska, and Minnesota. Workers are often recent immigrants from Mexico, Central America, Laos, and Vietnam. Pay is often minimum wage, and with union work rules gone, safety is often haphazard. Working hours are long or short, at the convenience of the packing-house manager, overtime pay is rare, and those who protest may lose their jobs.

Many of the immigrant workers are refugees from war in their home countries, and their path to assimilation into American society is blocked by wages that are not sufficient to provide a decent life for their children. The Reverend Tom Lo Van, a Laotian Lutheran pastor in Storm Lake, Iowa, says of the Laotian families in his church that the youth are unlikely to prosper from their parents' toil. "This new generation is worse off," he says. "Our kids have no self-identity, no sense of belonging . . . no role models. Eighty percent of [them] drop out of high school."

Changing Population Patterns and Land Use in the Great Plains

Mechanization has reduced the number of jobs in agriculture and encouraged the consolidation of ownership. Increasingly, corporations own the farms of the Great Plains. Individuals who still farm may own several large farms in different locales. They often choose to live in cities, traveling to their farms seasonally. As a result of these trends, rural depopulation is severe. Thousands of small prairie towns are dying out. Between 1900 and 2010, two-thirds of the counties in the Great Plains lost population, with rural counties as a whole losing more than 40 percent of their population. The area now has fewer than 6 people per square mile (2 per square kilometer). Most larger cities in the region (Denver, Minneapolis, Dallas, Kansas City, Oklahoma City, Tulsa, Omaha, Des Moines) are growing because young people are leaving the small towns on the prairies.

In fact, some suggest that modern agriculture, with its high demand for water and its depletion of once fertile soils, is an environmental experiment that should end. Proponents of the "Buffalo Commons" argue that the drier parts of this region should be made into a vast nature preserve, where crops would be replaced by native grasses on which herds of American bison would graze.

These days some of the only rural Great Plains counties with increasing populations are those with significant numbers of Native Americans. In the western Great Plains, some Native Americans are returning to work in newly established gambling casinos, new ranching establishments, and other enterprises. Although still only a fraction of the total plains population, the Native American population in this region grew by 40 percent between 1990 and 2010.

Connecting Bees, Global Corn, the Northern Plains, and Almond Groves in California

Cattle ranches and grain farms still prevail in the middle and southern plains, but in the northwestern plains, many farmers have been putting their land into the Federal Conservation Reserve program, which is aimed at stopping soil erosion and saving water by paying farmers to take land out of production.

As a consequence, the return of native grasses and wildflowers brought beekeepers and their hives to the northern plains. For some years, beekeepers "parked" their beehives in this area so that their bees could produce honey by feeding on the nectar and pollen of prairie wildflowers. During a few lucrative weeks in February, when the almond groves near Fresno in the Central Valley of California are in bloom, the beekeepers truck their thousands of beehives (containing as many as a billion bees) to pollinate California's Central Valley almond flowers (**Figure 2.54**). California produces two-thirds of the world's almonds; without the bees to pollinate, there would be no almonds. When the almond bloom is over, the bees need to return to their Great Plains wildflower sanctuary.

Figure 2.54 Bees, almonds, and ethanol. Flowering almond trees in California are pollinated by high plains honeybees, which live in hives (the white boxes shown in the picture) that are usually trucked in from South Dakota, where they flourish among the fields of wildflowers that have sprung up on abandoned farmland. However, California's almond industry has recently been struggling to find enough bees because the fields in South Dakota are being brought back into corn production to satisfy the demand for ethanol fuel made from corn, reducing the amount of land available for wildflowers for the bees. With the wildflowers gone, bee populations will decrease rapidly. [Phil Hawkins/Bloomberg via Getty Images]

Here is the problem: Since 2008, when corn became a commodity in high demand for making ethanol, producing corn-based sweeteners, and feeding the world's hungry, Great Plains farmers who had their land in the Federal Conservation Reserve suddenly could make more money putting their land back into corn. They plowed up the prairie, and the bees lost at least half of the wildflower cover on which they had previously thrived. This then threatened almond production in California.

CHECK YOUR UNDERSTANDING

1. Why do plains farmers keep close tabs on the global commodities markets?
2. What population patterns are changing the Great Plains?
3. How are soil erosion and water depletion affecting this subregion?
4. Give an example of how landscape change in the Great Plains is connected to global demands for certain crops. How are Great Plains landscape changes impacting agricultural production elsewhere?

THE CONTINENTAL INTERIOR

Among the most striking features of the Continental Interior are its huge size, its physical diversity, and its very low population density (**Figure 2.55**). This is a land of extreme physical environments, characterized by rugged terrain, extreme temperatures, and lack of water (for more photos of this region, see Figures 2.3 and 2.8C). These extreme physical features restrict many economic enterprises and account for the low population density of fewer than 2 people per square mile (1 person per square kilometer) in most parts of the region (see Figure 2.35).

Four Physical Zones

Physically, there are four distinct zones within the Continental Interior: the Canadian Shield, the frigid and rugged lands of Alaska, the Rocky Mountains, and the Great Basin. The *Canadian Shield* is a vast, glaciated territory north of the Great Plains that

Figure 2.55 The Continental Interior subregion. **A** The Bingham Canyon open-pit copper mine is the largest human-made excavation in the world. It is 0.75 miles deep, 2.5 miles wide, and pollutes groundwater over a 75 square mile area. It has been mined for 150 years. [Ray Ng/The LIFE Images Collection/Getty Images]

today is characterized by thin or nonexistent soils, innumerable lakes, and large meandering rivers. The rugged lands of Alaska lie to the northwest of the shield. The shield and Alaska have northern coniferous (*boreal*) and subarctic (*taiga*) forests along their southern portions. Farther north, the forests give way to the **tundra**, a region of winters so long and cold that the ground is permanently frozen several feet below the surface. Shallow-rooted, ground-hugging plants such as mosses, lichens, dwarf trees, and some grasses are the only vegetation.

The Rocky Mountains stretch in a wide belt from southeastern Alaska to New Mexico. The highest areas are generally treeless, with glaciers or tundra vegetation, while forests line the rock-strewn slopes on the lower elevations. Between the Rockies and the Pacific coastal zone is the *Great Basin*, which was formed by an ancient volcanic mega-crater. It is a dry region of widely spaced mountains, covered mainly by desert scrub and occasional woodlands.

Native Americans and Aboriginal Peoples

The Continental Interior has some of the largest populations of native people in North America, most of them living on reservations in the United States, or "reserves" as native lands are called in Canada. Because of the removal of Native Americans and Aboriginal peoples from their ancestral lands in the nineteenth century, only those who live in the tundra and northern forests of the Canadian Shield still occupy most of their original territories. The Nunavut, for example, recently won rights to their territory (part of the former Northwest Territory) after 30 years of negotiation. They are now able to hunt and fish and generally maintain the ways of their ancestors, although most of them now use snowmobiles and modern rifles. In September 2003, the Tłı̨chǫ Aboriginal people, formerly known as the Dogrib, a group of 3500, reclaimed 15,000 square miles of their land just south of the Arctic Circle. The Nunavut and the Tłı̨chǫ, who have a strong sense of cultural heritage and an entrepreneurial streak as well, have websites and manage their lives and the resources of their ancestral lands, which include oil, gas, gold, and diamonds.

Nonindigenous Inhabitants

The Continental Interior is a fragile environment and one of the most intense battlegrounds in North America between environmentalists and resource developers.

Although much of the Continental Interior remains sparsely inhabited, nonindigenous people have settled in considerable density in a few places and now make up the vast majority of the population. Where irrigation is possible, agriculture has expanded—in Utah and Nevada; in the lowlands along the Snake and Columbia Rivers of Idaho, Oregon, and Washington; and as far north as the Peace River district in the Canadian province of Alberta (see Figure 2.47). There are many cattle and sheep ranches throughout the Great Basin. Overgrazing and erosion are problems, but more serious concerns include unsustainable groundwater extraction and pollution from chemical fertilizers. An additional pollutant is the malodorous effluent from feedlots where large numbers of beef cattle are fattened for market. Efforts by the U.S. government to curb abuses on federal land, which is often leased to ranchers in extensive holdings, have not been very successful.

tundra a region of winters so long and cold that the ground is permanently frozen several feet below the surface

Increasing numbers of people are migrating to the Continental Interior. The many national parks in the region attract both seasonal and permanent workers and millions of tourists. Other people come to exploit the area's natural resources. Towns such as Laramie, Wyoming, and both Calgary and Edmonton, Alberta, have swelled with workers in the mining and fossil fuel industries. These industries often find themselves in conflict with each other and with the Native American and Aboriginal peoples of the Continental Interior.

In the United States, environmental groups have pressured the federal government to set aside more land for parks and wilderness preserves and to stop building pipelines for oil and gas transport from Alberta, Canada. Environmentalists also want to limit or eliminate activities such as mining and logging, which they see as damaging to the environment and the scenery. A switch to more recreational and preservation-oriented uses throughout the region would change, and possibly lessen, employment opportunities, while the growing numbers of visitors would place stress on natural areas, particularly on water resources.

Federal Lands

More than half the land in the U.S. Continental Interior is federally owned. Mining and oil drilling, often on leased federal land, are by far the largest industries (see Figure 2.55A). Since the mid-nineteenth century, the region's wide range of mineral resources has supported the major cities and towns. Mineral resources link these places to the global economy, but fluctuating world market prices for minerals have resulted in alternating periods of booms and busts. In recent times, the most stable mineral enterprises have been oil-drilling operations along the northern coast of Alaska. From here, the Trans-Alaska Pipeline runs southward for 800 miles (1300 kilometers) to the Port of Valdez.

CHECK YOUR UNDERSTANDING

1. What are the four distinct zones of the Continental Interior?
2. What are some groups that often come into conflict in this subregion?
3. What are the largest industries in the Continental Interior?

THE PACIFIC NORTHWEST

Once a fairly isolated region, the Pacific Northwest (**Figure 2.56**) is now highly integrated into the global economy, and the focus of debates about how North America should deal with environmental and development issues. While the economy in much of this region is shifting from logging, fishing, and farming to information technology industries, attitudes about the environment are also changing. Forests that were once valued primarily for their timber are now also prized for their recreational value, especially their natural beauty and their wildlife. Energy, needed to run cities, homes, and industries, is now more than ever a crucial resource for the region.

The physical geography of this long coastal strip consists of fjords and islands, mountains and valleys. Most of the agriculture, as well as the largest cities, are located in the southern part, in a series of lowlands lying east of the long, rugged coastal mountain ranges that extend north and south. Throughout the region, the

Figure 2.56 The Pacific Northwest subregion.

climate is wet. Winds blowing in from the Pacific Ocean bring moist and relatively warm air inland, where it is pushed up over the mountains, resulting in usually copious orographic rainfall throughout the area and seasonal snowfall in the north. The close proximity of the ocean gives this region a milder climate than is found at similar latitudes farther inland. In this rainy, temperate zone, enormous trees and huge forests have flourished for eons (see Figure 2.14A). Especially in the southern parts of the subregion, the balmy climate and the spectacular scenery attract many vacationing or relocating Canadians and U.S. citizens.

Older Extractive Industries

Pacific Northwest logging provides much of the construction lumber and an increasing amount of the paper used in North America. Lumber and wood products are also important exports to Asia, and the lumber industry is responsible directly and indirectly for hundreds of thousands of jobs in the region. As the forests shrink, environmentalists are harshly critical of the logging industry for such practices as clear-cutting (see Figure 2.14B), which destroys wild animal and plant habitats, thereby reducing species diversity and leaving the land susceptible to erosion and the remaining adjacent forests susceptible to diseases and pests. The more expensive alternative logging method of selective cutting is still highly disruptive to habitats and does not preserve forest diversity, but it is healthier for soils and maintains some diversity of fauna and flora.

Disputes also rage over how to acquire sufficient energy for the region. Large hydroelectric dams in the Pacific Northwest have attracted industries that need cheap electricity, particularly aluminum smelting and associated manufacturing industries in aerospace and defense. These industries are major employers, but their demand for labor is erratic and involves frequent periodic layoffs. In addition, another major employer in the region, the fishing industry, finds fault with hydroelectric dams because they block the seasonal migrations of salmon to and from their spawning grounds, so some of the region's most valuable fish cannot produce young. Around 40 dams have been removed to improve ecological conditions in rivers in the Pacific Northwest over the past two decades.

Newer IT Industries

Given the environmental impacts of the logging and hydroelectric industries, it is not surprising that many people in the region are enthusiastic about the role of information technology within the major urban areas of the Pacific Northwest: San Francisco, Portland, Seattle, and Vancouver. With Silicon Valley just outside San Francisco and with a growing number of IT companies around Seattle, the Pacific Northwest is a world leader in information technology. However, IT firms produce hazardous waste such as lead and other heavy metals used in circuit boards, and they are particularly dependent on secure and steady sources of electricity—whether from water power, gas, oil, coal, or nuclear fuels—that have significant negative environmental impacts.

The Pacific Northwest has growing connections to Asia. Large-scale immigration from Asia has been underway along the Pacific Coast for many decades. Meanwhile the countries of the Pacific Rim have become major trading partners for North America. Vancouver, Seattle, Portland, and San Francisco are orienting toward Asia, not just in the trade of forest products, but in banking and tourism and in the design and manufacture of IT equipment, much of which is outsourced to Asia.

CHECK YOUR UNDERSTANDING

1. What industry is challenging the traditional logging, fishing, farming, and hydroelectric industries for primacy in the Pacific Northwest?
2. What part of the world is this subregion increasingly connected to via immigration, trade, and outsourcing?
3. What are some environmental impacts of IT industries?

SOUTHERN CALIFORNIA AND THE SOUTHWEST

Southern California and the Southwest (**Figure 2.57**) are united primarily by long and deep cultural ties to Mexico that have intensified over the past three decades; their attractive warm, dry climate; and their increasing dependence on water from outside the region.

The varied landscapes of the Southwest include the Pacific coastal zone, the hills and interior valleys of Southern California, the mountains and dramatic mesas and canyons of southern Arizona and New Mexico, and the gentle coastal plain of south Texas on the Gulf of Mexico. This region has warm average year-round temperatures and an arid climate, usually receiving less than 20 inches of rainfall annually. The natural vegetation is scrub, bunchgrass, and widely spaced trees. Ranching is the most widespread form of land use, but other forms are more important economically and in numbers of employees.

Agriculture, Energy, Transportation, and Information Service Industries

Where irrigation is possible, farmers take advantage of the nearly year-round growing season to cultivate high-quality fruits, nuts, and vegetables. As discussed in the opening vignette of this chapter (see also Figure 2.57), California's Central Valley is a leading agricultural producer in the United States and is one of the most valuable agricultural districts in the world. Many of the crops are produced on massive, plantation-like farms. Migrant Mexican workers, often undocumented immigrants and sometimes mere children, supply most of the labor and lower the cost of North American food by working for very low wages.

Although irrigated agriculture is important, the bulk of the region's economy is nonagricultural. Information and service industries—high-tech, software, financial, and the entertainment industries—are the most profitable and employ the most people, but the coastal zones of Southern California and Texas are also home to

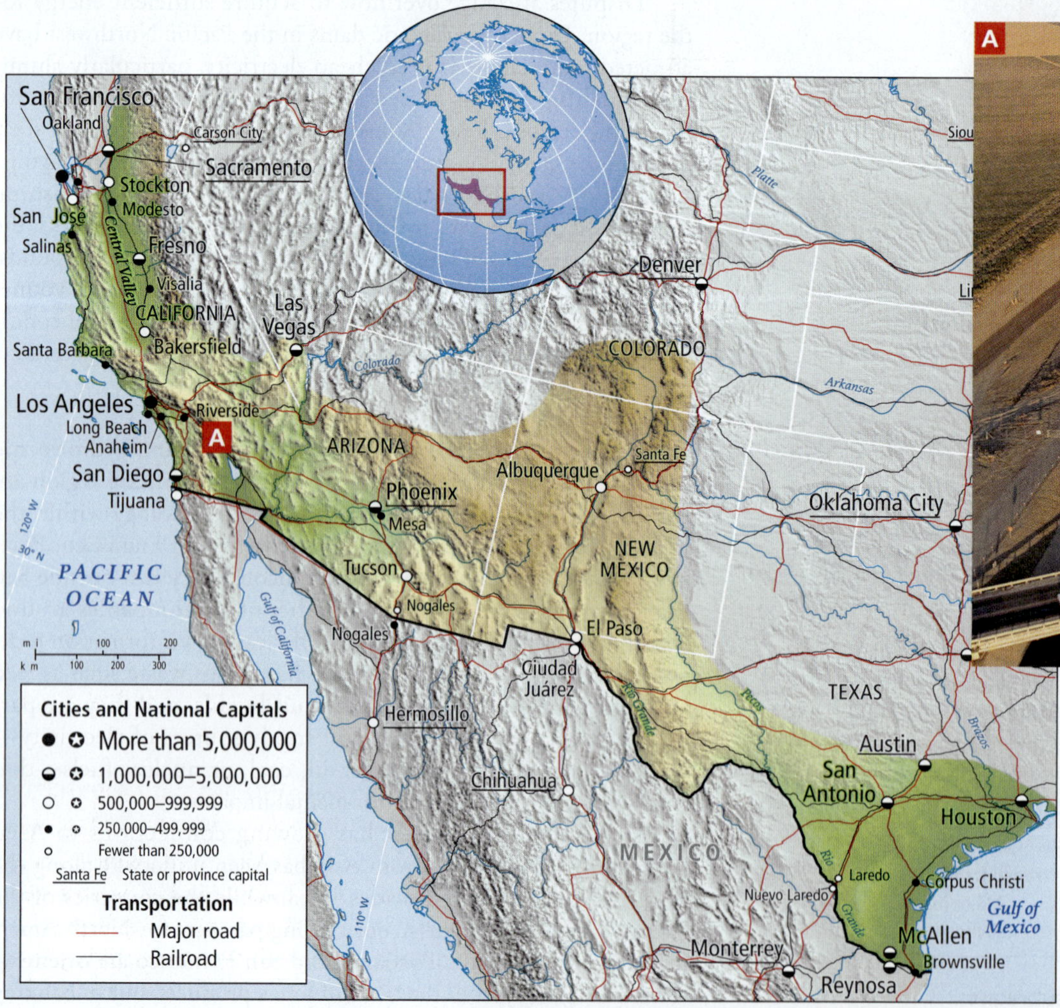

Figure 2.57 Southern California and the Southwest. **A** The Colorado River Aqueduct, which supplies much of southwestern California with drinking water diverted from the Colorado River, is a key resource in this dry subregion. Agriculture can be highly productive, given the nearly year-round growing season, but it relies heavily on irrigation. [Sarah Leen/National Geographic/Getty Images]

oil drilling, refining, and associated chemical industries, as well as other industries that need the cheap transport provided by the ocean. Los Angeles—with its major trade, transportation, media, entertainment, finance industries, and research facilities—is now the second most populous city in the United States. Like San Francisco, Seattle, and other Pacific coastal cities, Los Angeles is strategically located for trade with Pacific Rim countries and replaced New York as the largest port in the United States more than 20 years ago.

On the eastern flank of this region, Austin, Texas, has attracted IT industries, lured by research activities connected to the University of Texas, the warm climate, and the laid-back lifestyle of central Texas. Austin has over 2 million people and doubled in population between 1990 and 2010. The region as a whole draws large numbers of retirees, with entire planned communities of 10,000 and 20,000 people established in the deserts of New Mexico and Arizona yearly.

From the start of the first decade of the 2000s, residents of Southern California and the Southwest have had to worry about energy costs, congestion, smog, and water scarcity. In the case of energy, a number of factors have provoked questions about the region's environmental sustainability. Energy usage has been high, in part because of historically low energy prices. Deregulation of energy providers leads to escalating prices and corruption among some private generators and distributors, resulting in decreased energy supplies. Especially in Southern California and Arizona, people are turning to solar energy, making use of the subregion's abundant sunshine. However here, as elsewhere, much remains to be done to establish a more sustainable energy economy.

Figure 2.58 The Colorado River basins. The Colorado River, which flows through some of the driest land in the Continental Interior, is modified by many dams, reservoirs, and diversion canals that enable river water to supply a host of cities in the Southwest.

Water: Scarcities and Disparities of Access

Increasingly scarce water is the most worrisome environmental problem in this region. Because rainfall is light and irregular, water must be imported to serve the dense urban settlements and industries, of which agriculture is the most demanding water user. Most water comes from the Colorado River, which flows naturally southwest from near Denver, through portions of Utah, Arizona, Nevada, and along the California–Arizona border, through a small part of Mexico, and on to the Gulf of California (**Figure 2.58**). Since the middle of the twentieth century, the water has been captured by such dams as Glen Canyon, Hoover, Davis, and Parker, and held in reservoirs, where it is used first to generate electricity.

The water is also sent through aqueducts to irrigate agriculture and supply domestic and industrial users in Phoenix and Tucson, Arizona; Las Vegas, Nevada; and the Imperial Valley and cities in Southern California (see Figure 2.57A). Recently there has been more awareness that at current rates of development, the environment of the Colorado River basin will not be able to sustain even present rates of water diversion. Control of *wastage*—water lost through evaporation, percolation, and leaks in canals, as well as through nonessential uses such as lawn watering—would help significantly.

Raising the cost of water to users would control its use; now it is so cheap that few take the trouble to conserve because it seems almost a free commodity. But higher user costs could seriously impact those who use the least and are the poorest. A popular saying in this region is that although water flows downhill, it flows uphill to money—to those users who are the richest and most powerful politically.

Cultural and Economic Interests Shared with Mexico

Much of this region was originally a colony of Spain, and the area has maintained the Spanish language, a distinctive Latino culture, and other connections with Mexico. Today, the Latino culture is gaining prominence in the Southwest and spreading beyond it. As we have seen, large numbers of immigrants are arriving from Mexico, and the economies of the United States and Mexico have become more interdependent over the past three decades of reduced trade barriers.

Among these interdependencies are the factories, known as *maquiladoras*, set up by U.S., Canadian, European, and Asian companies in Mexican towns just across the border from U.S. towns in California, Arizona, and Texas. Maquiladoras in such places as Ciudad Juárez (across from El Paso), Nuevo Laredo (across from Laredo), Nogales (across from Nogales, Arizona), and Tijuana (across from San Diego) produce manufactured goods for sale primarily in the United States and Canada. They reduce costs by taking advantage of lower wages in Mexico, cheaper land and resources, lower taxes, and weaker environmental regulations. (The maquiladora phenomenon is discussed further in Chapter 3.) These factories are a key part of a larger transborder economic network that stretches across North America.

Until fairly recently, the U.S.–Mexico border was one of the world's most *permeable national borders*, meaning that people and goods flowed across it easily (though not without controversy). For years there were as many as 230 million legal border crossings each year through 35 points of entry. Since the terrorist attacks of 2001, the border has been made much less permeable—a high wall now runs along much of the border—and for undocumented travelers the trip has become much more dangerous. Smugglers who move workers north and illegal drugs and weapons both north and south across the more heavily guarded border charge high fees. They often cheat their passengers, trade the women and girls to sex traffickers, and in a number of cases have left whole truckloads of people to die in the desert heat without food or water.

Contentious Border Issues

As discussed earlier, several contentious issues surround the estimated 11 million undocumented immigrants currently residing in the United States, of whom perhaps 6 million are Mexican. One conflict is about language. Although English and bilingual speakers in the Southwest far outnumber those who speak only Spanish, some fear that English could be challenged in much the same way it has been challenged by French Canadians in Québec. Accordingly, Arizona in 2006 and California in 1986 made English the official language. Some bilingual programs, initiated in the 1970s to help migrant children make a smooth transition from Spanish to English, have been abolished by those who feel it is better that children learn English as quickly as possible. Scientists have found, however, that migrant children do better in math and science and have higher self-esteem if they can study for a period in their native language.

CHECK YOUR UNDERSTANDING

1. What agricultural products does this region supply to North America?
2. What economic activities are the most profitable in this subregion?
3. Why is language an issue in this subregion?

CRITICAL THINKING QUESTIONS

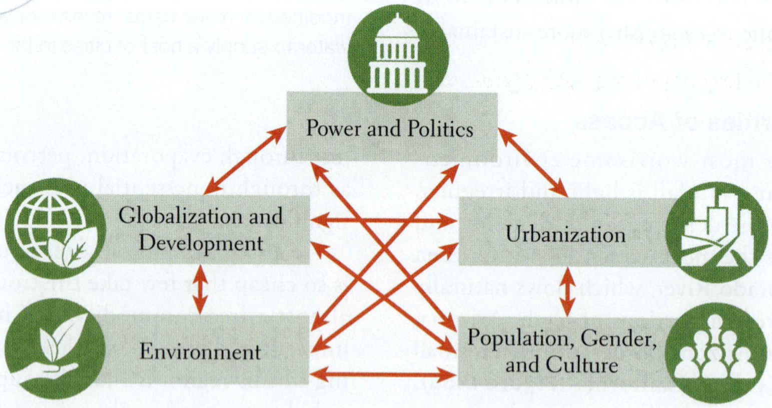

The diagram represents connections among the five geographic themes that structure this book. Listed below are some important questions that have been addressed in this chapter. Answer each question, and indicate where in the diagram you think the topics in each question belong.

1. How do the major landforms of this region influence the movement of air masses to contribute to its climatic variety?
2. How is North America's position in the global economy linked to its environmental impact?
3. What are three ways to reduce the environmental impacts of North America's resource consumption? Using an online ecological footprint calculator (see Chapter 1), assess your own impact. Of the three strategies for reducing environmental impacts, how many are you using?

4. How has the United States influenced international agreements to reduce GHG emissions?
5. How do historical patterns of subregional interaction within North America relate to current trends in economic development?
6. With ___________ of the world's population, North America produces ___________of the world's GHG emissions, giving it the highest per capita GHG emissions of any world region.
7. In what way does North America's dominant pattern of urbanization influence its energy use?
8. What is North America's position in the global economy, and how has globalization transformed this region?
9. What are four ways in which you are connected to the global economy?
10. How is the role of the United States in the global geopolitical order linked to its position in the global economy?
11. Where are challenges to U.S. economic and political dominance coming from, and how is the United States responding to them?
12. What are the major similarities and differences between Canada and the United States in the roles each country's government plays domestically and internationally?
13. How have urban areas changed differently in the United States and Canada since World War II?
14. What role do urban areas play in the North American economy?
15. How has the changing role of women in society contributed to the aging of North American populations?
16. In what ways is North America's population distribution changing?
17. What world regions account for the majority of new immigrants to North America, and how is this influencing the racial and ethnic makeup of this region? How is this process unfolding differently in the United States and Canada?
18. Using information from this chapter, argue for or against the following statement: "Immigrants are costing taxpayers too much money."

Key Terms

acid rain 71
agribusiness 86
aquifers 69
brownfields 100
clear-cutting 74
conurbation 97
digital divide 81
economic core 78
fracking 74
gentrification 100
gerrymandering 108
housing segregation 100
infrastructure 77
Latino 65
metropolitan areas 97
nuclear family 111
organically grown 87
Pacific Rim 103
pull factors 105
push factors 105
redlining 100
smog 71
social safety net 94
suburbs 97
trade deficit 82
tundra 122
United States–Mexico–Canada Agreement (USMCA) 81
urban sprawl 73

More Practice at SaplingPlus

Read the interactive e-text, review key concepts, and check your understanding.

Ecuadorians try to contain an oil spill in an Amazon Basin nature reserve.
[RODRIGO BUENDIA/AFP/Getty Images]

3

Middle and South America

The boat trip down the Aguarico River in Ecuador took me into a world of magnificent trees, river canoes, and houses built high up on stilts to avoid floods. I was there to visit the Secoya, a group of 350 indigenous people locked in negotiations with the U.S. oil company Occidental Petroleum over their plan to drill for oil on Secoya lands. Oil revenues supply 40 percent of the Ecuadorian government's budget and are essential to paying off its national debt. The government had threatened to use military force to compel the Secoya to allow drilling.

The Secoya wanted to protect themselves from cultural disruption and pollution like the oil spill in an Amazon basin nature reserve, shown here. As Colon Piaguaje, chief of the Secoya, put it to me, "A slow death will occur. Water will be poorer. Trees will be cut. We will lose our culture and our language, alcoholism will increase, as will marriages to outsiders, and eventually we will disperse to other areas." Given all the impending changes, Chief Piaguaje asked Occidental to use the highest environmental standards in the industry. He also asked the company to establish a fund to pay for the educational and health needs of the Secoya people. Chief Piaguaje made these requests because of what has happened in nearby places that have had oil development. Wildlife and people exposed to spills from wells, pipelines, and open pits are often poisoned or drowned. After heavy rains, the pits overflow, polluting nearby streams and wells.

Learning Objectives

Environment: Physical and Human

3.1 Distinguish among the four main temperature-altitude zones found in Middle and South America.

3.2 Explain how deforestation, climate change, and water scarcity are linked in this region.

3.3 Evaluate how this region's integration into the global economy left it with instability and the widest gap between rich and poor in the world.

Globalization and Development

3.4 Contrast the influences of neoliberalism and socialism on economic development in this region.

Power and Politics

3.5 Analyze the impacts of foreign interventions and the international drug trade on politics in this region.

Urbanization

3.6 Explain why this region has urban landscapes that often lack support services and infrastructure.

Population, Gender, and Culture

3.7 Examine the factors that created a population explosion in this region during the early twentieth century and reduced population growth in the late twentieth century.

3.8 Describe the major institutions that shape gender roles in this region.

3.9 Evaluate the forces challenging the dominance of the Roman Catholic Church.

Subregions of Middle and South America

3.10 Compare how the subregions of Middle and South America have been influenced differently by foreign powers.

U.S.
Tijuana
El Paso
Ciudad Juárez
San Antonio
Houston
New Orleans
Miss.
Jacksonville
Orlando
Tampa
Miami
Bermuda (U.K.)
ATLANTIC OCEAN
Rio Grande
Baja California
Hermosillo
Gulf of California
Sierra Madre Occidental
Sierra Madre Oriental
Mulegé
Nuevo Laredo
Matamoros
MEXICO
Monterrey
Durango
Iquitos
Mazatlán
San Luis Potosí
Tampico
Gulf of Mexico
BAHAMAS
Havana
GREATER ANTILLES
Puerto Vallarta
Guadalajara
Mérida
Cancún
Yucatán Channel
Bay of Campeche
Yucatán Peninsula
CUBA
DOMINICAN REPUBLIC
Veracruz
Mexico City
Puebla
Oaxaca
Belize City
BELIZE
Belmopan
Port-au-Prince
HAITI
Santo Domingo
San Juan
Puerto Rico (U.S.)
LESSER ANTILLES
JAMAICA
Kingston
PACIFIC OCEAN
Acapulco
Gulf of Tehuantepec
Guatemala City
GUATEMALA
HONDURAS
Tegucigalpa
Caribbean Sea
Montserrat (U.K.)
San Salvador
EL SALVADOR
NICARAGUA
Managua
Gulf of Venezuela
ST. LUCIA
ST. VINCENT
BARBADOS
GRENADA
Lake Nicaragua
Barranquilla
Maracaibo
Caracas
Valencia
TRINIDAD & TOBAGO
Port of Spain
San José
COSTA RICA
Panama Canal
Colón
Panamá
Lake Maracaibo
VENEZUELA
PANAMA
Gulf of Panama
Medellín
Llanos
Orinoco
Georgetown
GUYANA
Paramaribo
Bogotá
Guiana Highlands
SURINAME
Cali
COLOMBIA
FRENCH GUIANA (France)
Quito
Galápagos Islands (Ecuador)
ECUADOR
Guayaquil
Gulf of Guayaquil
Iquitos
Negro
Manaus
Amazon
Marañón
Solimões
Amazon Basin
Piura
Madeira
BRA
PERU
Ucayali
Xingu
Callao
Lima
Cusco
Andes
Mato Grosso
BOLIVIA
La Paz
Lake Titicaca
Cochabamba
Altiplano
Sucre
Santa Cruz
Potosí
Tropic of Cancer
Tropic of Capricorn
Equator
Antofagasta
Atacama Desert
PARAGUAY
Paraná
Asunción
CHILE
Mtns.
Easter Island (Chile)
Córdoba
Pôrto Alegre
Valparaíso
Santiago
Rosario
Paraná
URUGUAY
Buenos Aires
Montevideo
La Plata
Río de la Plata
ARGENTINA
Pampas
Monte Verde
Puerto Montt
Patagonia
Comodoro Rivadavia
Falkland Islands (U.K.)
Stanley
Strait of Magellan
Punta Arenas
Tierra del Fuego
30°N
20°N
10°N
0° Equator
10°S
20°S
30°S
40°S
50°S

Drilling for oil in South America is not a recent development. The U.S. oil company Texaco went to Ecuador in 1964, and before it left in 1992 its pipelines and waste ponds leaked almost 17 million gallons of oil and 16 billion gallons of toxic runoff and oil waste into local soil and waterways, resulting in as many as one-quarter of the people in surrounding communities developing cancer. Texaco sold its operations to the government in 1992, but its oil wastes continue to leak from over 1000 open pits and other oil infrastructure that is illegal in the United States but was used in Ecuador to save the company money.

In 1993, some 30,000 Ecuadorians sued Texaco in New York State, where the company (now owned by and called Chevron) is headquartered. In 2002, the suit against Chevron was dismissed, based on the argument that it had no jurisdiction in Ecuador. The case was re-filed in Ecuador in 2003, and in 2011 Chevron was assessed damages of U.S.$9.5 billion. Because Chevron no longer has any assets in Ecuador, the plaintiffs tried to collect the money in countries where Chevron does have assets, such as the United States, Canada, Argentina, and Brazil, all without success. In 2014, a U.S. court ruled that the 2011 Ecuadorian judgment against Chevron was too tainted by bribery and corruption to be honored, claims that were rejected unanimously by Ecuador's Constitutional Court in 2018.

Now one of the longest-running legal battles in history involving a major multinational corporation, the case has raised awareness about how industries often fail to take environmental precautions in developing countries, especially in remote areas where **indigenous** people often live. Meanwhile the Secoya and other Ecuadorians have to live with the pollution.

In fact, the rich resources of Middle and South America have attracted outsiders since the first voyage of Christopher Columbus in 1492 (**Figure 3.1**). For the first few centuries after Columbus, this region was considered by Europeans to be much more valuable than North America, and its "discovery" marked a major expansion of the global economy. However, this region occupied a disadvantaged position in global trade, supplying cheap raw materials that aided the Industrial Revolution in Europe and later in North America but reaping few of the profits. These extractive industries did little to advance economic development within the region, as most profits went to a few wealthy elites and foreign investors and corporations, and the negative environmental effects were not addressed by those responsible. Since World War II most countries have worked to control their own resources, develop local manufacturing and service-based industries, and share profits more broadly with investments in human development. These efforts have been opposed by local elites, foreign investors and corporations, and some foreign governments, resulting in many conflicts. Meanwhile successful efforts to limit environmental pollution have been few and far between.

indigenous native to a particular place or region

◀ **Figure 3.1** Regional landforms map of Middle and South America.

What Makes Middle and South America a Region?

This region is defined in large part by the legacies of its colonization by Spain and Portugal, which left the Spanish and Portuguese languages and Catholicism as major sources of unity. Culturally, this region has large indigenous populations whose cultures have blended with and changed the European, African, and Asian cultures introduced by the colonists. These cultural factors override the fact that Mexico and most of Central America are physically part of the North American continent, with only the **isthmus** (land bridge) of Panama providing connection to the continent of South America (**Figure 3.2**). Middle and South America, like North America, is a region of great physical and cultural diversity, but here there are wider disparities of wealth than in any other world region.

isthmus a narrow strip of land that joins two larger land areas

Middle America Mexico, Central America, and the islands of the Caribbean

South America the continent south of Central America

Terms in This Chapter

In this book, **Middle America** refers to Mexico, Central America, and the islands of the Caribbean. **South America** refers to the continent south of Central America. The term *Latin America* is not used in this book because it describes the region only in terms of the Roman (Latin-speaking) origins of the former colonial powers of Spain and Portugal. It ignores the region's large indigenous groups, its African, Asian, and Northern European populations, as well as the many mixed cultures, often called *mestizo* cultures, that have emerged. In this chapter, we use the term *indigenous groups* or *peoples* rather than *Native Americans* to refer to the native inhabitants of the region.

Figure 3.2 Political map of Middle and South America.

ENVIRONMENT: PHYSICAL AND HUMAN

3.1 Distinguish among the four main temperature-altitude zones found in Middle and South America.

3.2 Explain how deforestation, climate change, and water scarcity are linked in this region.

3.3 Evaluate how this region's integration into the global economy left it with instability and the widest gap between rich and poor in the world.

Middle and South America extend south from the midlatitudes of the Northern Hemisphere across the equator through the Southern Hemisphere, nearly to Antarctica. This long north–south expanse combines with variations in altitude to create the wide range of climates in the region. Tectonic forces have shaped the primary landforms of this huge territory to form an overall pattern of highlands to the west and lowlands to the east (Figure 3.1).

LANDFORMS

There is a wide variety of landforms in Middle and South America, and this variety accounts for the many different climatic zones in the region. But for ease in learning, landforms are here divided into just two categories: highlands and lowlands.

Highlands

A nearly continuous chain of mountains stretches along the western edge of the American continents for more than 10,000 miles (16,000 kilometers) from Alaska in the north to Tierra del Fuego at the southern tip of South America (**Figure 3.3**). The middle part of this long mountain chain is known as the Sierra Madre in Mexico (**Figure 3.4**), by various names in Central America, and as the Andes in South America, which is the longest mountain chain in the world (**Figure 3.5**). It was formed by a lengthy **subduction zone**, which runs thousands of miles along the western coast of the continents (see the illustration of the Ring of Fire in Figure 1.7).

Figure 3.3 Tierra del Fuego, Argentina. [Travel Images/UIG/Getty Images]

Figure 3.4 A woman rests while walking in the Sierra Madre mountains of Mexico. [Michel SETBOUN/Gamma-Rapho via Getty Images]

Figure 3.5 The Andes Mountains in Chile. [Hoberman Collection/UIG via Getty Images]

Here, two oceanic plates—the Cocos Plate and the Nazca Plate—plunge beneath three continental plates—the North American Plate, the Caribbean Plate, and the South American Plate.

In a process that continues today, the leading edges of the overriding plates crumple to create mountain chains. In addition, molten rock from beneath Earth's crust ascends to the surface through fissures in the overriding plate to form volcanoes. Such volcanoes are the backbone of the highlands that run through Middle America and the Andes of South America. Here millions of people live close to active volcanoes and in earthquake-prone zones, which can pose deadly hazards (for example, the 2010 and 2015 earthquakes in Chile).

The chain of high and low mountainous islands in the eastern Caribbean is also volcanic in origin, created as the Atlantic Plate thrusts under the eastern edge of the Caribbean Plate. On the island of Montserrat, people have been living with an active, and sometimes deadly, volcano for more than a decade (**Figure 3.6**). Eruptions have taken the form of violent blasts of superheated rock, ash, and gas

subduction zone a zone where one tectonic plate slides under another (known as *pyroclastic flows*) that move down the volcano's slopes at speeds upward of 450 miles (700 kilometers) per hour. The unusually strong earthquake in Haiti in January 2010 was also the result of plate tectonics.

Lowlands

Vast lowlands extend over most of the land to the east of the western mountains. In Mexico, east of the Sierra Madre, a coastal plain borders the Gulf of Mexico (**Figure 3.7**). Farther south, in Central America, wide aprons of sloping land descend to the Caribbean coast. In South America, a huge wedge of lowlands, widest in the north, stretches from the Andes east to the Atlantic Ocean. These South American lowlands are interrupted in the northeast and the southeast by two modest highland zones: the Guiana Highlands and the Brazilian Highlands (**Figure 3.8**). Elsewhere in the lowlands, grasslands cover extensive flat expanses, including the *llanos* of Venezuela, Colombia, and Brazil, and the *pampas* of Argentina (**Figure 3.9**).

The largest feature of the South American lowlands is the Amazon Basin, drained by the Amazon River and its tributaries (**Figure 3.10**). Earth's largest remaining expanse of tropical rain forest gives the Amazon Basin global significance as a reservoir of **biodiversity**, home to millions of plant and animal species.

biodiversity the variety of life forms to be found in a given area

Figure 3.6 Soufrière Hills Volcano, Montserrat. [Alain BUU/Gamma-Rapho via Getty Images]

Figure 3.7 The Yucatan Lowlands. A Mayan temple rises out of the Yucatan lowlands in Mexico. [DEA/G. DAGLI ORTI/De Agostini/Getty Images]

Figure 3.8 The Brazilian Highlands. Rio De Janeiro is built where the Brazilian Highlands meet the Atlantic Ocean. [Shaun Botterill–FIFA/Getty Images]

Figure 3.9 The pampas of Argentina. [James P. Blair/National Geographic/Getty Images]

Figure 3.10 The Amazon Basin as seen from the air. [ANTONIO SCORZA/AFP/Getty Images]

The basin's water resources are also astounding: 20 percent of Earth's flowing surface waters. The Amazon River itself runs so deep that ocean liners can travel 2300 miles (3700 kilometers) upriver from the Atlantic Ocean all the way to Iquitos, jokingly referred to as Peru's "Atlantic seaport." The Amazon River system starts as streams high in the Andes flowing eastward toward the Atlantic. Once the basin's rivers reach the flat land of the Amazon Plain, their velocity slows abruptly and fine soil particles, or *silt*, then sink to the riverbed. Seasonal floods transport the silt to surrounding areas, nourishing millions of acres of tropical forest. Not all of the Amazon Basin is rain forest, however. Variations in weather and soil types, as well as human activity, have created grasslands and seasonally dry deciduous tropical forests in some areas.

CLIMATE

From the jungles of the Caribbean and the Amazon to the high, glacier-capped peaks of the Andes to the parched moonscape of the Atacama Desert and the frigid fjords of Tierra del Fuego, the range of climates in Middle and South America is enormous (**Figure 3.11**). In this region, the wide range of temperatures reflects both the great distance the landmass spans on either side of the equator and the tremendous variations in altitude across the region's landmass (the highest point in the Americas is Aconcagua in Argentina, at 22,841 feet [6962 meters]). Patterns of precipitation are affected both by the local shape of the land and by global patterns of wind and ocean currents that bring moisture in varying amounts.

Temperature-Altitude Zones

Four main **temperature-altitude zones**, shown in **Figure 3.12**, are commonly recognized in the region. As altitude increases, the temperature of the air decreases by about 1°F per 300 feet (1°C per 165 meters) of elevation. Thus temperatures are highest in the lowlands, which are known in Spanish as the *tierra caliente*, or "hot land." The *tierra caliente* extends up to about 3000 feet (1000 meters), and in some parts of the region these lowlands cover large areas. Where moisture is adequate, tropical rain forests thrive, as does a wide range of tropical crops, notably bananas, sugarcane, cacao, and pineapples. Many coastal areas of the *tierra caliente*, such as northeastern Brazil, have become zones of plantation agriculture that support populations of considerable size.

Between 3000 and 6500 feet (1000 to 2000 meters) is the cooler *tierra templada* ("temperate land"). The year-round, spring-like climate of this zone drew large numbers of indigenous people in the distant past. Here, crops such as roses, corn, beans, squash, various green vegetables, wheat, and coffee are grown.

Between 6500 and 12,000 feet (2000 to 3600 meters) is the *tierra fria* ("cool land"). A variety of crops, such as wheat, fruit trees, potatoes, and cool-weather vegetables—cabbage and broccoli, for example—do very well at this altitude. Several modern population centers are in this zone, including Mexico City, Mexico, and Quito, Ecuador.

Above 12,000 feet (3600 meters) is the *tierra helada* ("frozen land"). In the highest reaches of this zone, vegetation is almost absent, and mountaintops emerge from under snow and glaciers. A remarkable feature of such tropical mountain zones is that in a single day of strenuous hiking, one can encounter many of the climate types found on Earth.

Precipitation The pattern of precipitation throughout the region is influenced by the interaction of global wind patterns with mountains and ocean currents (see Figure 1.9). The **trade winds** sweep off the Atlantic, bringing heavy seasonal rains to places roughly 23° north and south of the equator (see the Figure 3.11 map). Winds from the Pacific bring seasonal rains to the west coast of Central America, but mountains block those rains from reaching the Caribbean side, which receives heavy rainfall from the northeast trade winds.

The Andes are a major influence on precipitation in South America. They block the rains borne by the trade winds off the Atlantic into the Amazon Basin and farther south, creating a rain shadow on the western side of the Andes in northern Chile and southwestern Peru (see Figure 3.11B). Southern Chile is in the path of eastward-trending winds that sweep off of the Southern Ocean, bringing steady, cold rains that support forests similar to those of the Pacific Northwest in North America. The Andes block this flow of wet, cool air and divert it to the north. They thereby create another extensive rain shadow on the eastern side of the mountains along the southeastern coast of Argentina (Patagonia).

Adjacent oceans and their currents also influence the pattern of precipitation. Along the west coasts of Peru and Chile, the cold surface waters of the Peru Current bring cold air that cannot carry much moisture. The combined effects of the Peru Current and the central Andes rain shadow create the world's driest nonpolar desert, the Atacama of northern Chile.

El Niño

One aspect of the Peru Current, which is only partly understood, is its tendency to change direction every few years (on an irregular cycle, possibly linked to sunspot activity). When this happens, warm water flows eastward from the western Pacific, bringing torrential rains to parts of the west coast of South America. The phenomenon was named **El Niño,** or "the Christ Child," by Peruvian fishermen, who noticed that when it does occur, it reaches its peak around Christmas. Peru's major banks plan for slower economic growth during El Niño years because of the increased flooding and damage to roads and bridges, increased waterborne disease, and reduced fish catches.

El Niño also has global effects, bringing cold air and drought to normally warm and humid western Oceania and unpredictable weather patterns to Mexico and the southwestern United States. The El Niño phenomenon in the western Pacific is discussed further in Chapter 11, where Figure 11.8 illustrates its trans-Pacific effects.

Hurricanes

Many of the region's coastal areas are threatened by powerful and damaging storms that form annually, primarily in the Atlantic Ocean north of the equator and close to Africa but also in the southeastern Pacific. When enough warm wet air comes together, groups of thunderstorms join to form a swirling spiral of wind that moves across Earth's surface. The highest wind speeds are found at the edge of the eye, or center, of the storm. Once wind speeds reach 74 miles (121 kilometers) per hour, the storm is officially called a *hurricane.* Because they draw their energy from warm surface waters, hurricanes slow down and eventually dissipate as they move over cooler water or land, usually

temperature-altitude zones regions of the same latitude that vary in climate according to altitude

trade winds winds that blow from the northeast and the southeast toward the equator

El Niño periodic climate-altering changes in the circulation of the Pacific Ocean and associated airmasses, now understood to operate on a global scale

3.11 Climates of Middle and South America

A Tropical wet, Belém, Brazil [Holger Leue/Lonely Planet Images/Getty Images]

B Desert, Atacama, Chile [Veronique DURRUTY/Gamma-Rapho via Getty Images]

C Continental, moist all year, Tierra del Fuego [DEA/A. GAROZZO/Getty Images]

Figure 3.12 Temperature-altitude zones of Middle and South America. Temperatures tend to decrease as altitude increases, resulting in changes in the natural vegetation on mountainsides, as shown here. The same is true for crops, some of which are suited to lower, warmer elevations and some to higher, cooler ones.

lasting about one week. As the populations of coastal areas grow, more people are being exposed to hurricanes. Many scientists also think that climate change is increasing the number and intensity of hurricanes (see Figure 3.14C later in the chapter).

CHECK YOUR UNDERSTANDING

1. How far north and south do Middle and South America extend?
2. What is the major physical feature dominating the western edge of the American continents?
3. What are two aspects of the rain forests of the Amazon Basin that make them worth preserving?
4. What are the four temperature-altitude zones found in this region, and how do they influence where and how people live and the crops they can grow?

ENVIRONMENT

Environments in Middle and South America have long inspired concern about the use and misuse of Earth's resources. Millennia before Europeans arrived in this region, human settlements in the Americas had major environmental impacts (see the "Human Patterns over Time" section). However, modern impacts have intensified due to increased population density and higher per capita consumption, and because the region's environments now supply global demands.

Tropical Forests, Climate Change, and Globalization

Climate change dominates current efforts to control deforestation in the Amazon and the rest of this region, although the loss of biodiversity is also a huge concern, as the rain forests of the Amazon Basin are some of the most biodiverse on the planet.

As explained in Chapter 1, forests release oxygen and absorb carbon dioxide (CO_2), the greenhouse gas (GHG) most responsible for global warming. The loss of forests contributes to global warming through the release of CO_2 that happens as trees are burned to make way for crops and roads. Also, when there are fewer trees, less CO_2 can be absorbed from the atmosphere. Together, all of Earth's tropical rain forests absorb about 18 percent of the CO_2 added to the atmosphere yearly by human activity. Because 50 percent of Earth's remaining tropical rain forests are in South America, keeping these forests intact is crucial to minimizing climate change.

Middle and South American rain forests are being diminished by multiple human activities. One of the biggest of these is caused by the clearing of land for agriculture, mostly for soybeans, cattle, African oil palm (for cooking oil), and sugarcane (see **Figure 3.13C–E**). Brazil has in recent years ranked between the world's fourth and seventh largest emitter of GHGs, after the United States, China, and Indonesia. If deforestation continues in Middle America at the current rate, the natural forest cover could be entirely gone in 20 years.

Hardwood logging and the extraction of underlying minerals, including oil, gas, metals, and precious stones, also contribute to deforestation. Investment capital in logging industries is coming from Asian multinational companies that have turned to the Amazon forests after having logged as much as 50 percent of the tropical forests in Southeast Asia. The roads that have been built to support these activities, such as the Trans-Amazon Highway and the many roads that feed into it, have helped to accelerate deforestation by opening new forest areas to migrants. The governments of Peru, Ecuador, and Brazil encourage impoverished urban people to occupy cheap land along the newly built roads (see Figure 3.13A and B), and provide chainsaws to help remove the trees. However, the settlers often find it difficult to cultivate the soils of the Amazon, and after a few years their lands are eroded and depleted of nutrients. Frequently they move on to new plots, selling their land to ranchers as cattle pastures (see Figure 3.13D).

There is now an international effort to combat climate change by preserving the world's remaining forests—especially the crucial tropical rain forests found in this region. In response to international pressure, Brazil introduced new regulations that discouraged the cultivation of soybeans on once-forested lands. Enacted from 2005 to 2010, these policies reduced the rate of deforestation in the Amazon by 75 percent and Brazil's GHG emissions by 39 percent, making the country a leader in GHG reduction. However, Brazil's economy slowed in the following years, resulting in greater pressure to spur economic development with new road and hydroelectric projects in the Amazon. As a result, recent years have seen a return to high levels of deforestation in Brazil's portion of the Amazon.

maquiladoras foreign-owned, tax-exempt factories, often located in Mexican towns near the U.S. border, that hire workers at low wages to assemble manufactured goods, which are then exported mainly to the United States and Canada

Much of this region is particularly vulnerable to the impacts of climate change. **Figure 3.14** illustrates several such cases of how human systems are responding to environmental crises made worse by climate change.

The Water Management Crisis

Although Middle and South America receive more rainfall than any other world region and have three of the world's six largest rivers (in volume), parts of the region are experiencing water crises. The crisis is complicated by climate change, but is primarily human created and most visible in urban areas, where local water resources become too polluted by untreated sewage and unregulated industries to be drinkable, forcing cities to bring in potable water from distant areas. But water-related stresses are also severe in rural areas, especially in Haiti, the high Andean mountain zones of Peru and Bolivia, Central America, and northern Mexico. Of these areas, only northern Mexico and the high Andes are dry year-round.

Haiti and Central America's problems with freshwater storage and distribution relate mostly to poverty and the mismanagement of resources. These are the poorest areas in the region and are plagued by huge disparities in wealth and power that skew resources toward the rich and away from poor people. Corruption and poorly trained civil servants often complicate efforts to help poor people access water and other resources. For example, only 64 percent of the population in Haiti has access to clean water, and even this access is prone to lengthy interruptions, especially during times of drought. Haiti receives enough precipitation that simple rainfall collection and storage systems would provide low-cost access to safe water for many, but corruption and poor planning make these systems rare. Natural disasters, such as hurricanes (shown in Figure 3.14C), often vividly expose the weaknesses of infrastructure, including transportation and water systems, which may take decades to recover in these countries.

In northern Mexico, south of the U.S. states of Arizona and New Mexico, water scarcity is caused not only by a dry climate, but also by inadequate planning and lax environmental policies. Cities have grown rapidly due to the expansion of numerous factories, or **maquiladoras**, which have been set up to take advantage of NAFTA/USMCA-related trade with the United States and Canada (see the discussion of maquiladoras in "Export Processing Zones"). The factories have been allowed to pollute waterways along the U.S.–Mexico border with few restraints. Meanwhile many migrants who work in the maquiladoras live in *colonias*, or communities on the urban fringes of large cities such as Nogales, Tijuana, and Ciudad Juárez, where water has to be brought in via trucks at great expense and stored in rooftop cisterns (see Figure 3.14A) because all nearby water resources have been polluted.

Some sources of water in this region are directly threatened by climate change. In the Andean mountain zones of Peru and Bolivia, glaciers feed rivers that are the main source of water for millions of people (see Figure 3.14B). Should the glaciers actually disappear, many rivers will run much lower for parts of the year, straining communities, farms, and industries that depend on them.

Behind these immediate problems lie systemic policy failures and inadequate planning, as in the case of Cochabamba, Bolivia (discussed in Chapter 1), where efforts to improve the city's inadequate water supply system focused on "marketizing" the water

A A newly built road in Suriname. What environmental impact is clearly visible in this photo? [Wesley Bocxe/Getty Images]

B A family on a river raft in the Peruvian Amazon. [Michael Fairchild/Getty Images]

C African oil palm is planted on land recently cleared for agriculture. [Glowimages/Getty Images]

D Cattle on land that has just been burned in Rondônia, Brazil. What land use likely preceded cattle grazing here? [Michael Nichols/Getty Images]

E Soy fields recently cleared of forest in Mato Grosso, Brazil. Where are the soybeans grown on this land likely to end up? [JC Patricio/Getty Images]

Figure 3.13 Deforestation in the Amazon. Deforestation in the Amazon is driven by poor people's need for livelihoods, governments' desire to assert control over lightly populated areas, and the demands for wood, meat, and food in distant urban centers and the global market. Together, these forces have led to a rapid loss of forest, especially in Brazil, which loses more forest each year than does any other country on the planet.

3.14 Vulnerability to Climate Change in Middle and South America

Environmental stresses such as drought, hurricanes, flooding, and glacial melting combine with growing populations and persistent poverty to create a complex landscape of vulnerability to climate change. In this resource-rich region, huge disparities in wealth and power skew resources toward the rich. Meanwhile, efforts to improve poor people's access to resources, especially water, are complicated by corruption and poorly trained civil servants, often resulting in inadequate planning, underdeveloped infrastructure, and lax environmental policies. Only a few water resources in this region are directly threatened by climate change.

THINKING GEOGRAPHICALLY

A What clues can be seen in this photo that the neighborhood lacks a centralized water distribution system?

B Why is glacial melting of particular concern to cities in Bolivia?

C In addition to exposure to tropical storms and hurricanes, what else contributes to Haiti's vulnerability to climate change?

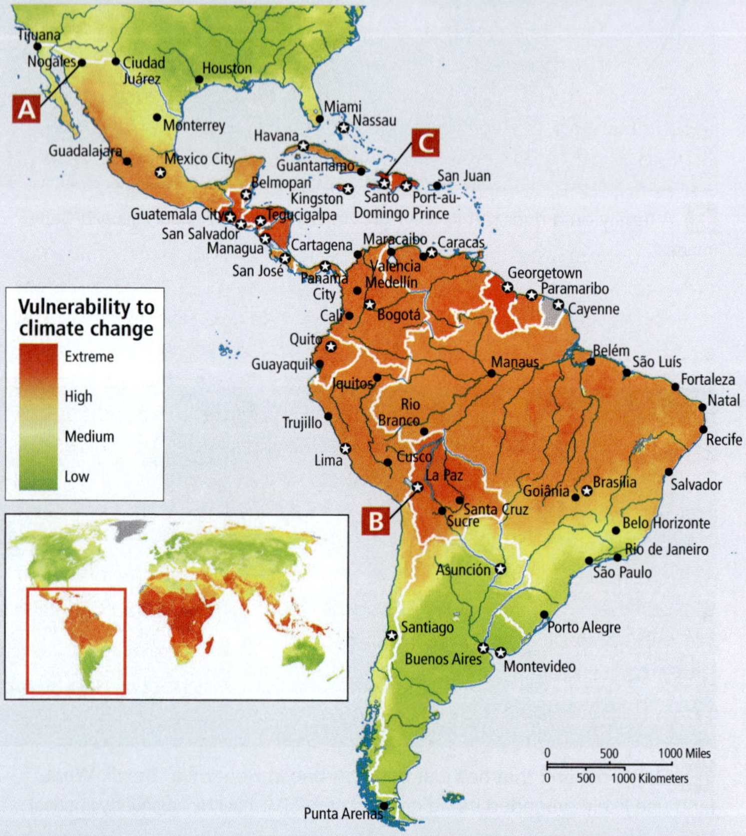

A Drought and pollution. A "colonia" in Nogales, Mexico, where there is no centralized water infrastructure. Water is gathered off of the roofs of homes or brought in by truck. The higher temperatures that climate change is bringing are making this area drier, while water pollution from factories worsens water scarcity. Millions of Mexicans who have moved to work in factories along the U.S.–Mexico border are thus highly vulnerable to climate change. [John Moore/Getty Images]

B Glacial melting. La Paz and many smaller cities and towns in Bolivia's drought-prone highlands receive much of their drinking water from glaciers in the Andes. Higher temperatures are causing these glaciers to melt rapidly. Many have already disappeared, and the rest could be gone in 15 years. [John Coletti/Getty Images]

C Flooding and poverty. Haitians examine damage to crops from flooding associated with Hurricane Sandy. Poverty and corruption have limited efforts to preserve Haiti's forest cover. So much has been removed that even mild tropical storms can cause catastrophic flooding. [THONY BELIZAIRE/AFP/Getty Images]

system. The water supply, long thought of as a public resource, was sold to a group of foreign corporations led by Bechtel of San Francisco, California. The hope was that Bechtel, in return for profits, would make investments in infrastructure that Cochabamba's notoriously corrupt water utility would not make. Unfortunately, Bechtel, unfamiliar with the actual living conditions of the majority of Cochabamba's citizens, immediately increased water prices to levels that few urban residents could afford, while doing little to improve water supply or delivery systems. At one point Bechtel even charged urban residents for water taken from their own wells and rainwater harvested off their roofs! Popular protests forced Bechtel to abandon Cochabamba's water utility, which remains plagued by corruption and an inadequate infrastructure.

Poor water management has led to crisis in Mexico City, the region's largest and wealthiest city, which faces both water shortages and seasonal flooding. Poor maintenance of the municipal water system means that 35 percent of water is lost through leaky pipes, and many areas on the urban fringe now depend on water trucks. Mexico City's main water supplies are aquifers deep under the ancient lakebed on which the city sits, and these are being rapidly depleted. In part this is because the rainwater that once replenished them is now being diverted away from the city by massive storm sewers built to avoid flooding. Some people are starting to use cisterns to capture the plentiful rain that falls on the city for roughly half the year, and if this became a widespread practice it could relieve pressure on the aquifer and also reduce the risk of flooding.

The lack of adequate sanitation is a major part of this region's water crisis, posing a major health hazard in many cities. Rio de Janeiro's failure to treat more than 40 percent of the city's wastewater was spotlighted during the 2016 Olympics, when swimmers, sailors, and rowers from around the world were forced to compete in water containing dangerous levels of raw sewage. In the slums of Rio and other cities, many urban residents have no access to toilets and are forced to relieve themselves on the street, allowing waterborne illnesses to become widespread. Significant improvements have been made in recent years, especially in Mexico City where close to 90 percent of wastewater is now being treated. However, most other cities and almost all towns treat only a tiny fraction of their wastewater. Access to toilets and other "improved sanitation" technologies has improved to above 80 percent of the population in all but the poorest countries in this region in the past two decades.

Integrating Environmental Protection with Economic Development: Ecotourism

Many parts of this region are trying to earn money from the beauty of still-intact natural environments through **ecotourism**—in which nature-oriented vacations are offered to travelers usually hailing from affluent cities or foreign countries (**Figure 3.15**). Middle and South America is the world leader in setting aside land as nature reserves, indigenous lands, or other areas "protected" from resource extraction and other intensive development. Brazil alone has more protected land area than any other country (more than twice the United States), though in reality many of these lands are regularly subjected to extractive resource development, both legal and illegal. High levels of poverty, especially in the remote areas where most protected lands are located, mean that they must provide concrete economic benefits if they are to avoid oil and gas, mining, forestry, or agricultural development. Ecotourism has the potential to provide these benefits while preserving this region's biodiversity, reducing emissions of greenhouse gases, and providing rural and indigenous people with a chance to use their skills to teach tourists about nature. Ecotourism is now the most rapidly growing segment of the global tourism and travel industry, which itself is a huge and fast growing industry. However, as the following vignette illustrates, ecotourism has its upsides and downsides.

ecotourism tourism built around natural environments

Figure 3.15 Ecotourism in the Amazon.

(A) A tourist poses at the bottom of a giant ceiba tree in the Ecuadorian Amazon. [Photograph by Michael Schwab/Getty Images]

(B) An Amazon river dolphin being fed by an ecotour guide. [Morales/AGE Fotostock]

(C) A tourist explores a walkway suspended in the canopy of tall rain forest trees in the Ecuadorian Amazon. [Inga Spence/Getty Images]

A Teotihuacan, Mexico, once home to an estimated 150,000 to 250,000 people. [Bjorn Holland/Getty Images]

B Machu Picchu, an estate built for the Incan emperor of the fifteenth century. [Lluís vinagre–world photography/Moment/Getty Images]

10,000 B.C.E — 5000 B.C.E — 0 C.E. — 1300 C.E. — 1400 C.E.

23,000 B.C.E.–12,000 B.C.E Bering land bridge

200 B.C.E.–800 C.E. Teotihuacan flourishes

1325 C.E. Aztec capital of Tenochtitlán founded

Figure 3.16 A VISUAL HISTORY OF MIDDLE AND SOUTH AMERICA

Thinking Geographically

A How does this image support the notion that some indigenous groups may have been better off than their contemporaries in Europe?

B What about Machu Picchu in the Andean highlands suggests that it was more than a mere summer residence?

VIGNETTE Puerto Misahualli, a small river boomtown in the Ecuadorian Amazon, is currently enjoying significant economic growth. Its prosperity is due to the many European, North American, and other foreign travelers who come for experiences that will bring them closer to the now-legendary rain forests of the Amazon.

The array of ecotourism offerings can be perplexing. One indigenous man offers to be a visitor's guide for as long as desired, traveling by boat and on foot, camping out in "untouched forest teeming with wildlife" with a guarantee that monkeys and birds will be eaten and offering ceremonies with a local shaman involving the consumption of powerful hallucinogenic Amazonian plants. At the other extreme, a well-known eco-lodge offers plush rooms with a river view, a chlorinated swimming pool, and a fancy restaurant serving "international cuisine." All of this is on a private, 740-acre nature reserve separated from the surrounding community by a wall topped with broken glass. It seems more like a fortified resort than an eco-lodge.

By contrast, the solar-powered Yachana Lodge has simple rooms and local cuisine. Its knowledgeable resident naturalist is a veteran of many campaigns to preserve Ecuador's wilderness. Profits from the lodge fund a local clinic, a high school, and various programs that teach sustainable agricultural methods that protect the fragile Amazon soils while increasing farmers' earnings from surplus produce. The nonprofit group running the Yachana Lodge—the Foundation for Integrated Education and Development—earns just barely enough to sustain the clinic, the school, and the agricultural programs. *[Source: Alex Pulsipher's field notes in Ecuador.]* ■

CHECK YOUR UNDERSTANDING

1. How does deforestation in this region contribute to global climate change?
2. Why does so much deforested land end up in cattle production?
3. What are some causes of the water management crisis this region faces?
4. Where is progress being made in the water crisis?
5. What can ecotourism offer this region's protected areas?

HUMAN PATTERNS OVER TIME

The conquest of Middle and South America by Europeans wiped out most of the indigenous civilizations, and the conquerors set up new societies in their place. Many cultural features from the time of European colonialism endure to this day, as do some vestiges of the precolonial era.

The Peopling of Middle and South America

Recent evidence suggests that between 14,000 and 25,000 years ago, groups of hunters and gatherers from northeastern Asia spread throughout North America after crossing the Bering land bridge on foot or moving along shorelines in small boats or both. Some of these groups ventured south across the Central American isthmus, reaching the tip of South America by about 13,000 years ago.

By 1492, there were 50 to 100 million indigenous people in Middle and South America. In some places, population densities were high

A mosaic in Lima, Peru, depicting Francisco Pizarro, nqueror of the Inca Empire. anita Delimont/Getty Images]

D Chile wins independence from Spain in 1818 with help from Argentina. [The Battle of Maipu on the 5th April, 1818, printed by Raffet (lithograph)/Spanish School, 19th century/ INDEX Fototeca/Private Collection/Bridgeman Images]

E Former slaves cultivate sugar cane in Puerto Rico in 1899. [Library of Congress Prints and Photographs Division]

1500 C.E. | 1600 C.E. | 1700 C.E. | 1800 C.E. | 1900 C.E. | 2000 C.E.

- 92 ival of ropeans
- **1519–1521** Aztec Empire conquered
- **1533** Inca Empire conquered
- **1750** Andean potato fuels population explosion in Europe
- **1791–1822** Wars of independence from Spain and Portugal
- **1821–1888** Slavery abolished throughout mainland Middle and South America

C Describe the mood of this depiction of the Spanish conquest of the Incas.

D In what way does this painting of Creole Argentinians and Chileans celebrating independence echo similar paintings of the American Revolution?

E Even after the end of slavery, who constituted the major labor force in sugar cultivation?

enough to threaten sustainability. People altered landscapes in many ways that are still visible. They modified drainage to irrigate crops, they terraced hillsides, and they built paved walkways across swamps and mountains. They constructed cities with sewer systems, freshwater aqueducts, and huge earthen and stone ceremonial structures that rivaled the pyramids of Egypt (**Figure 3.16A**).

The indigenous people also practiced the system of **shifting cultivation** that is still common in wet, hot regions in Central America and the Amazon Basin. In this system, also known as *slash and burn*, small plots are cleared in forestlands, the brush is dried and burned to release nutrients in the soil, and the clearings are planted with multiple crop species. Each plot is used for only 2 or 3 years and then abandoned for several decades—long enough to allow the forest to regrow. If there is sufficient land, this system is highly productive per unit of land and labor, and is sustainable for long periods of time. However, if population pressure increases to the point that a plot must be used before it has fully regrown and its fertility has been restored, its yields decrease drastically. Recent discoveries of extensive geometric earthworks in the Amazon suggest widespread human inhabitance there within the past several thousand years that may have been dependent on shifting cultivation or more intensive agriculture.

The **Aztecs** of the high central valley of Mexico had some technologies and social systems that rivaled or surpassed those of Asian and European civilizations of the time. Particularly well developed were urban water supplies, sewage systems, and elaborate marketing systems. Historians have concluded that by 1500 C.E., Aztecs probably lived more comfortably on the whole than their contemporaries in Europe.

In 1492, the largest state in the region was that of the **Incas**, stretching from what is now southern Colombia to northern Chile and Argentina. The main population clusters were in the Andean highlands, where the cooler temperatures at these high altitudes eliminated the diseases of the tropical lowlands, while proximity to the equator guaranteed mild winters and long growing seasons. For several hundred years, the Inca empire was one of the most efficiently managed empires in the history of the world. Most remarkably, the Inca did not use money, conducting trade instead through a barter system. Taxes were collected in the form of labor, which was used in highly organized public works projects, such as paved roads, elaborate terraces, irrigation systems, and great stone cities in the Andean highlands (see Figure 3.16B). Incan agriculture was very advanced, particularly in the development of crops, which included numerous varieties of potatoes and grains.

European Conquest

The European conquest of Middle and South America was one of the most significant events in human history (see Figure 3.16C). It rapidly altered

shifting cultivation an agricultural system in which small forested plots are cleared and the dried brush is burned to release nutrients. Plots are then planted with multiple species, cultivated for 2 or 3 years, and then left to reforest.

Aztecs indigenous people of high-central Mexico noted for their advanced civilization before the Spanish conquest

Incas indigenous people who ruled the largest pre-Columbian state in the Americas, with a domain stretching from what is now southern Colombia to northern Chile and Argentina landscapes and cultures and, through disease and slavery, ended the lives of millions of indigenous people. While the superior military technology of the Spanish and Portuguese sped the conquest of Middle and South America, a larger factor was the vulnerability of the indigenous people to diseases carried by the Europeans. In the 150 years following 1492, the total population of Middle and South America was reduced by more than 90 percent, to just 5.6 million. To obtain a new supply of labor to replace the dying indigenous people, the Spanish initiated the first shipments of enslaved Africans to the region in the early 1500s. Between 9 and 10 million enslaved people were brought to this region, compared to the less than 400,000 brought to North America.

Columbus established the first Spanish colony in 1492 on the Caribbean island of Hispaniola (which is now occupied by Haiti and the Dominican Republic). After learning of Columbus's exploits, other Europeans, mainly from Spain and Portugal on Europe's Iberian Peninsula, conquered the rest of Middle and South America.

The first part of the mainland to be invaded was Mexico, home to several advanced indigenous civilizations, most notably the Aztecs. The Spanish were unsuccessful in their first attempt to capture the Aztec capital of Tenochtitlán, but they succeeded a few months later after a smallpox epidemic decimated the native population. The Spanish demolished the grand Aztec capital in 1521 and built Mexico City on its ruins.

A tiny band of Spaniards, again aided by a smallpox epidemic, conquered the Incas in South America. Out of the ruins of the Inca empire, the Spanish created the Viceroyalty of Peru, which originally encompassed all of South America except Portuguese Brazil. The newly constructed capital of Lima flourished, in large part as a transshipment point for enormous quantities of silver extracted from mines in the highlands of what is now Bolivia.

Diplomacy by the Roman Catholic Church prevented conflict between Spain and Portugal over the lands of the region. The Treaty of Tordesillas of 1494 divided Middle and South America at approximately 46° W longitude (**Figure 3.17**). Portugal took all lands to the east and eventually acquired much of what is today Brazil (where Portuguese is still the primary language); Spain took all lands to the west.

By the 1530s, a mere 40 years after Columbus's arrival, all major population centers of Middle and South America had been conquered and were rapidly being transformed by Iberian colonial policies. The colonies soon became part of extensive regional and global trade networks, the latter with Europe, Africa, and Asia. What did not develop were strong governments that had motives beyond the facilitation of resource extraction. Leaders rarely pushed for policies that encouraged the development of a "social contract" that would inspire people to work together to solve the many complex problems that the conquest created. The Iberian colonies were set up to extract resources from the new world, not to create just or equitable societies.

Creoles people mostly of European descent born in the Americas

mestizos people of mixed European, African, and indigenous descent

The Iberian colonial system established patterns of government that contribute to present-day corruption. Colonial officials were allowed and even encouraged to use their power to enrich themselves and their friends and families. This had an enduring impact on the way both government and businesses are run in this region, where even today people in positions of power habitually use their influence to do favors for friends and relatives in ways that constantly erode any sense of "fairness" in how societies are run. Brazil's president from 1930 to 1951, Getulio Vargas, summed up the sentiment well: "For my friends anything, for my enemies, the law."

A Global Exchange of Crops and Animals

From the earliest days of the conquest, plants and animals were exchanged between Middle and South America, Europe, Africa, and Asia via the trade routes illustrated in Figure 3.17, in what is often called the *Columbian exchange.* Many plants essential to agriculture in Middle and South America today—rice, sugarcane, bananas, citrus, melons, onions, apples, wheat, barley, and oats, for example—were all originally imports from Europe, Africa, or Asia. When disease decimated the native populations of the region, the colonists turned much of the abandoned land into pasture for herd animals imported from Europe, including sheep, goats, oxen, cattle, donkeys, horses, and mules.

Just as consequential were plants first domesticated by indigenous people of Middle and South America. These plants have changed diets everywhere and have become essential components of agricultural economies around the globe. The potato improved the diet of the poor in western and northern Europe so much that by 1750 it had fueled a population explosion that would propel mass European migration to North America and elsewhere. Manioc (cassava) fueled a more modest population expansion in West Africa. Corn, peanuts, vanilla, and cacao (the source of chocolate) are globally important crops to this day, as are peppers, pineapples, and tomatoes (**Table 3.1**).

The Independence Era

In the early nineteenth century, wars of independence left Spain with only a few colonies in the Caribbean. **Figure 3.18** shows the European colonizing countries and the dates of independence for the various Middle and South American countries. The supporters of the nineteenth-century revolutions were primarily **Creoles** (people of mostly European descent born in the Americas) and some relatively wealthy **mestizos** (people of mixed European, African, and indigenous descent). The Creoles' access to the profits of the colonial system had been restricted by policies favoring the "mother" country, and the mestizos were excluded from many commercial and political opportunities by racist colonial policies. Once these groups gained power, however, they became a new elite who controlled the still very weak states of this region. Like their European colonial forebears they monopolized economic opportunities around resource extraction and allowed states to remain weak and unable to address the vast poverty created by this region's economic and political systems. Most people remained poor and relatively powerless.

The Legacy of Resource Extraction: Inequality and Instability

Middle and South America remained economically underdeveloped until well into the twentieth century, with most people impoverished

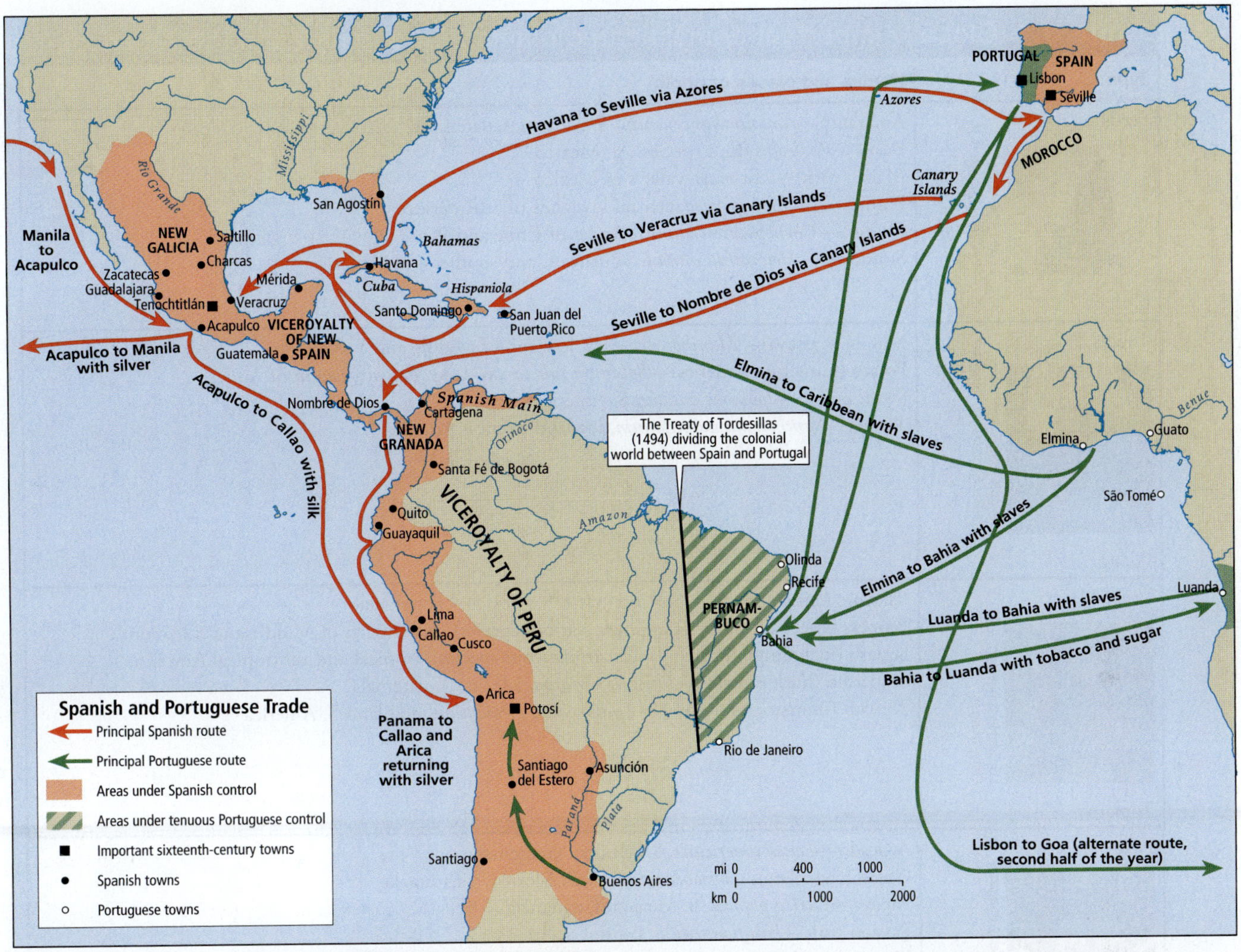

Figure 3.17 Spanish and Portuguese trade routes and territories in the Americas, circa 1600. The major trade routes from Spain led to the two main centers of its empire, Mexico and Peru. The Spanish colonies could trade only with Spain, not directly with one another. By contrast, there were direct trade routes from Portuguese colonies in Brazil to Portuguese outposts in Africa. Between 4 and 5 million Africans were enslaved and traded to Brazilian plantation and mine owners (as well as 1 million to the Spanish empire, 2 million to the British Caribbean, and 1.3 million to the French Caribbean and Dutch colonies in Middle and South America). [Research from: *Hammond Times Concise Atlas of World History* (Maplewood, NJ: Hammond, 1994), pp. 66–67]

and a tiny elite maintaining tight control of the region's resources and political power. Economic development was based on resource extraction that tended to increase the power and wealth of local elites, foreign investors, and foreign corporations. Beginning in the colonial era gold, silver, sugar, coffee, tobacco, cacao (chocolate), indigo (a blue dye), and tobacco were sent from this region in vast quantities to Europe, often with most of the profits going to European traders and other "middle men," as well as landowners within Middle and South America. This tended to produce high levels of economic inequality, as the people working to produce the raw materials were paid poorly, or not at all in the case of slaves.

Dependence on resource extraction also brought economic instability to the region due to wide swings in prices for raw materials that often occur in the global market. This happens because so many places across the globe can produce basic raw materials, such as coffee beans, that are interchangeable. If there is a huge crop of coffee in, for example, Indonesia, it can create an oversupply in global markets for coffee that can drive down the price worldwide. When this occurs the economies of other areas that produce coffee, such as Brazil, can be devastated.

In other parts of the Americas, such as the northern states of what became the United States, manufacturing industries provided a more stable base for economic development. Prices for manufactured goods are usually more steady because they tend to be more highly specialized, with only a few places producing a given item, for example the cast iron cooking stoves that became the basis of Detroit Michigan's early manufacturing economy. Historically the quantities of manufactured goods have been more limited and specialized, meaning that a global oversupply of stoves that drives down their price is much less likely than it is for a less specialized agricultural product, such as the coffee that Brazil became famous for. Meanwhile, innovations in design and manufacturing processes can create better products that create repeat customers and facilitate shifts into new products, such as automobiles in the case of Detroit, resulting in cycles of more stable long-term growth.

Table 3.1 Globally important domesticated plants that originated in the Americas

Type	Names and places of origin
Seeds  ALAIN JOCARD/ AFP/Getty Images Quinoa	Amaranth—*Amaranthus cruentus*, S. Mexico, Guatemala Beans—*Phaseolus* (four species), S. Mexico Maize (corn)—*Zea mays*, valleys of Mexico Peanut—*Arachis hypogaea*, central lowlands of S. America Quinoa—*Chenopodium quinoa*, Andes of Chile and Peru Sunflower—*Helianthus annuus*, southwest and southeast N. America
Tubers  François ANCELLET/Gamma-Rapho via Getty Images Potato	Manioc (cassava)—*Manihot esculenta*, lowland of Middle and S. America Potato (numerous varieties)—*Solanum tuberosum*, Lake Titicaca region of Andes Sweet potato—*Ipomoea batatas*, S. America Tannia—*Xanthosoma sagittifolium*, lowland tropical America
Vegetables  Louise Heusinkveld/ Getty Images Tomato	Chayote (christophene)—*Sechium edule*, S. Mexico, Guatemala Peppers (sweet and hot)—*Capsicum* (various species), many parts of Middle and S. America Squash (including pumpkin)—*Cucurbita* (four species), tropical and subtropical America Tomatillo (husk tomato)—*Physalis ixocarpa*, Mexico, Guatemala Tomato (numerous varieties)— *Lycopersicon esculentum*, highland S. America
Fruit  Travis Dove/Getty Images Pineapple	Avocado—*Persea americana*, S. Mexico, Guatemala Cacao (chocolate)—*Theobroma cacao*, S. Mexico, Guatemala Papaya—*Carica papaya*, S. Mexico, Guatemala Passion fruit—*Passiflora edulis*, central S. America Pineapple—*Ananas comosus*, central S. America Prickly pear cactus (tuna)—*Opuntia* (several species), tropical and subtropical America Strawberry (commercial berry)—*Fragaria* (various species), genetic cross of Chilean berry and wild berry from N. America Vanilla—*Vanilla planifolia*, S. Mexico, Guatemala, perhaps Caribbean
Ceremonial and drug plants  Neil Fletcher and Matthew Ward/ Getty Images Coca	Coca (cocaine)—*Erythroxylon coca*, eastern Andes of Ecuador, Peru, and Bolivia Tobacco—*Nicotiana tabacum*, tropical America

However, manufacturing areas require investments in the education and training of workers, in equipment, and in infrastructure such as power generation, that most wealthy people in this region were unaccustomed to making. Many local elites and foreign investors chose to take the profits earned from resource extraction in this region and fund business ventures in Europe that were seen as less risky. As a result dependence on raw materials continued for centuries, and while some countries have developed more diverse economies, most remain dependent on exports of raw materials to this day. Manufactured goods, such as machinery needed for farming or mining, must often still be purchased from abroad at great expense.

Infrastructure and Institutions of Resource Extraction To support resource extraction a modest flow of foreign investment and manufactured goods entered the region, first from Europe and later from North America. This included transportation systems, such as railroads and roads that connected interior resource production zones to coastal ports, but rarely to each other. Economic growth was driven by foreign demand for raw materials, not local consumption, as most people were too poor to buy much. The most profitable transportation systems, such as the still very lucrative Panama Canal, were bankrolled and owned by foreigners who invested most of the profits abroad, further depriving the region of

Figure 3.18 The colonial heritage of Middle and South America. Most of Middle and South America was colonized by Spain and Portugal, but important and influential small colonies were held by Britain, France, and the Netherlands. Nearly all the colonies had gained independence by the late twentieth century. Those for which no date appears on the map are still linked in some way to the colonizing country. [Research from: *Hammond Times Concise Atlas of World History* (Maplewood, NJ: Hammond, 1994), p. 69]

money that could have been invested in manufacturing and other industries.

Several economic institutions arose in the early colonial era of Middle and South America to extract raw materials to be sent to Europe and North America. Large rural estates called **haciendas** were granted to colonists as a reward for conquering territory and people for Spain. For generations these essentially feudal estates were passed down through families, often to heirs who lived in a distant city or in Europe and took little interest in the day-to-day operations. Productivity was often low, and most inhabitants of haciendas remained extremely poor, but they did produce a diverse array of products (cattle, cotton, rum, sugar).

As markets for meat, hides, and wool grew in Europe and North America, many haciendas became vast cattle and sheep ranches. Today, commercial ranches serving such global markets as the fast-food industry are found in the drier grasslands and savannas of South America, Central America, and northern Mexico, and even in the wet tropics on freshly cleared rain forest lands.

Plantations are large factory farms. In addition to growing crops such as sugar, coffee, cotton, or (more recently) bananas, crops are often processed on site for shipment. Plantations were established early on in the colonial era and many persist to this day, in part because they are more efficient and profitable than haciendas due to more intensive management and bigger investments in equipment. However, as was the case in the southern United States, plantations had relatively little local economic impact. Instead of employing local populations, plantation owners imported enslaved people from Africa. Plantation equipment was usually imported from Europe, which was also where the owners preferred to invest their profits. As a result, little money was available to support the development of local industries, such as manufacturing, that could have grown up around the plantations.

First developed by the European colonizers of the Caribbean and northeastern Brazil in the 1600s, plantations became common throughout Middle and South America. Unlike haciendas, which were established in the continental interior in a variety of climates, plantations were usually situated in coastal areas with year-round growing seasons and easier access to global markets via ocean transport.

hacienda a large agricultural estate in Middle or South America, more common in the past; usually not specialized by crop and not focused on market production

plantation a large factory farm that grows and partially processes a single cash crop

Figure 3.19 A copper mine in northern Chile, where some of the largest copper mines in the world are located. Copper accounts for 13 percent of Chile's GDP, and Chile produces one-third of the world's copper, more than any other country. Chile's copper mines are dependent on imported machinery, such as large dump trucks that are made in the United States. [WYSOCKI Pawel/Getty Images]

Mining was a major early extractive industry (**Figure 3.19**). Important mines (primarily gold and silver at first) were first built in north-central Mexico, the Andes, and the Brazilian Highlands, but eventually spread to many other places. Extremely inhumane labor practices were common in all of these mines. Today, oil and gas have been added to the mineral extraction industry, and rich mines throughout the region continue to produce gold, silver, copper, tin, precious gems, titanium, bauxite, and tungsten. Like plantations, equipment for mines tended to be imported, so few local industries developed around them, and profits went mostly to wealthy foreign investors.

While the societies of Middle and South America are today much more economically diverse, technologically sophisticated, and politically open than they once were, huge disparities in wealth and power remain. Around 30 percent of the population is very poor and lacks access to adequate food, shelter, water, sanitation, basic education, and land. Meanwhile, a small upper class has levels of affluence equivalent to those of the very wealthy in the United States. While corruption in government and business is now more limited, it is still a major obstacle to the rule of law. These conditions are in part the lingering effects of colonial economic policies that fostered privileges for wealthy elites and foreign investors who often spent their profits elsewhere rather than reinvesting them within the region, and favored the export of raw materials over the development of manufacturing industries. Current variations of these policies are further discussed in the "Globalization and Development" section below.

neoliberalism a capitalist economic and political system that favors the fulfillment of human needs by free and open markets strongly linked to global markets through free trade, with governments having a minimal role

socialism an economic system in which all people have access to basic goods and services and large-scale industries may be collectively owned, with any profits benefitting society as a whole

CHECK YOUR UNDERSTANDING

1. When did early humans arrive in this region, and how did they get there?
2. Under what conditions can shifting cultivation be considered sustainable?
3. What were the primary colonizing countries of Middle and South America, and which countries had smaller colonies?
4. Why does this region have such a wide gap between rich and poor?
5. Why can dependence on resource extraction lead to economic instability?
6. What are some important institutions of extractive economic development in this region?

GLOBALIZATION AND DEVELOPMENT

3.4 Contrast the influences of neoliberalism and socialism on economic development in this region.

Middle and South America is today one of the poorer world regions, though not the poorest (see Figures 1.19 and 1.20), and it is still economically dependent on resource extraction. Centuries of poverty, inequality, economic instability, corruption, and a massive flow of resources and money out of the region have inspired numerous reform efforts aimed at stabilizing economic growth with manufacturing industries and spreading wealth and opportunity more widely. However, these reform efforts have been strongly opposed by many wealthy and powerful people in the region, and by highly influential foreign entities (including many foreign investors, corporations, governments, and international institutions). As a result, change has been rare and contentious, and while several countries have had significant economic growth and some manufacturing industries have developed, ongoing dependence on the export of raw materials means that periodic economic downturns can be devastating for the region's poor and middle class.

NEOLIBERALISM AND SOCIALISM

Two economic and political systems have had a huge influence on development in this region since the early twentieth century. **Neoliberalism** is a capitalist system that favors the fulfillment of human needs by free and open markets that are strongly linked to global markets through free trade, with governments having a minimal role in society. **Socialism** favors governments taking a much more active role to ensure higher living standards for poor people, supplying services like health care and education free to all, and sometimes running part or all of the economy through state-owned enterprises and planning agencies. The interplay between these two systems is complex, with most governments having aspects of both. For example, in Ecuador there is both a state-run oil company and numerous private foreign-owned oil companies extracting oil from the Amazon region and selling it on global markets. Similarly Ecuador's health-care system has a state-run national network of hospitals that provides care to all Ecuadorians for free, but there are many private clinics that offer services for a fee.

The interplay between proponents of neoliberalism and socialism can be quite hostile, and tensions are increased by many foreign corporations and the U.S. government that habitually back neoliberalism in economic affairs and oppose movements toward socialism. Often governments switch between administrations strongly favoring one system or the other, with South America having gone through several phases in which aspects of neoliberalism or socialism have dominated economic development.

Era of Socialistic Import Substitution Industrialization

After World War II, there were numerous socialist political movements that challenged the economic domination of local elites and foreign investors. This was also a period of relative prosperity because prices for many of the region's raw material exports were rising as Europe rebuilt after years of war. Governments across the region pursued ambitious plans to modernize and industrialize their national economies with new manufacturing industries and other development projects. Many adopted a strategy of keeping money and resources within their borders by replacing imported manufactured goods with local products. Known as **import substitution industrialization (ISI)**, these policies financed local production of manufactured goods and placed high tariffs on imported manufactured items.

The money and resources kept within each country were supposed to fund further development in manufacturing and other industries that would replace the less stable extractive industries as the backbone of national economies. These new industries would help fund social programs giving poor people better access to housing, health care, and education. However, the new industries often were less profitable than expected due to corruption and weak governments that lacked the investment, managerial skills, and technological know-how needed to run efficient factories. Also, only the larger countries had enough consumers to fund these industries with their purchases. Brazil and Mexico both had success with some ISI programs: Mexico developed strong manufacturing industries for automobiles and a wide variety of consumer goods, and Brazil developed these industries as well as aircraft and armaments manufacturing. Even so, Mexico and Brazil still remained dependent on raw materials exports for much of their economy. Most countries did not have successful ISI programs and stayed dependent on the export of raw materials, which left them vulnerable to instability in the prices of these raw materials on global markets.

By the early 1970s, global prices of raw materials were falling, but instead of scaling back their ISI projects, governments borrowed billions of dollars from major international banks, most of which were in North America or Europe. When a worldwide economic recession hit in the 1980s, many governments were unable to repay their loans (**Figure 3.20**).

Era of Neoliberalism and Structural Adjustment

The large debts that ISI-era governments ran up led directly to the imposition of neoliberal policies. Alarmed over the large debts owed by governments throughout the developing world, the international banks took action in the early 1980s through the International Monetary Fund (IMF, see Chapter 1), which developed and enforced **structural adjustment programs (SAPs)** that ensured that sufficient money would be available to repay the loans despite widespread economic instability. Often described as part of the "Washington Consensus" because of strong support by the U.S. government for neoliberal policies, SAPs were based on several interdependent concepts: **privatization** (the selling of formerly government-owned industries and enterprises to private investors), **marketization** (the development of a free market economy in support of free trade), *globalization* (the opening of national economies to global investors; see Chapter 1), and **austerity** (reductions in government deficits through cuts in spending and/or higher taxes).

Neoliberalism via SAPs had the effect of rolling back many of the ISI-era reforms. Privatization resulted in many government enterprises being sold to foreign investors and corporations located in North America, Europe, and Asia, often at very low prices. Marketization required that governments remove tariffs on imported goods of all types, which caused many local industries to fail. Austerity reversed the ISI-era trend of expanding government social programs and building infrastructure, resulting in government services and often jobs being taken away from poor and middle-class workers. Governments were required to fire many civil servants and drastically reduce spending on public health, education, job training, day care, water systems, sanitation, public transportation, and infrastructure building and maintenance.

Alternation Between Neoliberalism and Socialism The neoliberal/SAP era did not produce the sustained economic growth or stability that was expected to relieve the debt crisis and create broader prosperity, in large part because countries remained dependent on the export of raw materials by industries that were often foreign-owned. The result was a socialist political backlash

import substitution industrialization (ISI) policies that encourage local production of machinery and other items that previously had been imported at great expense

structural adjustment programs (SAPs) economic reorganization toward less government involvement in industry, agriculture, and social services; sometimes imposed by the World Bank and the International Monetary Fund as conditions for receiving loans

privatization the selling of formerly government-owned industries and firms to private companies or individuals

marketization the development of a free market economy in support of free trade

austerity reductions in government deficits through cuts in spending and/or higher taxes

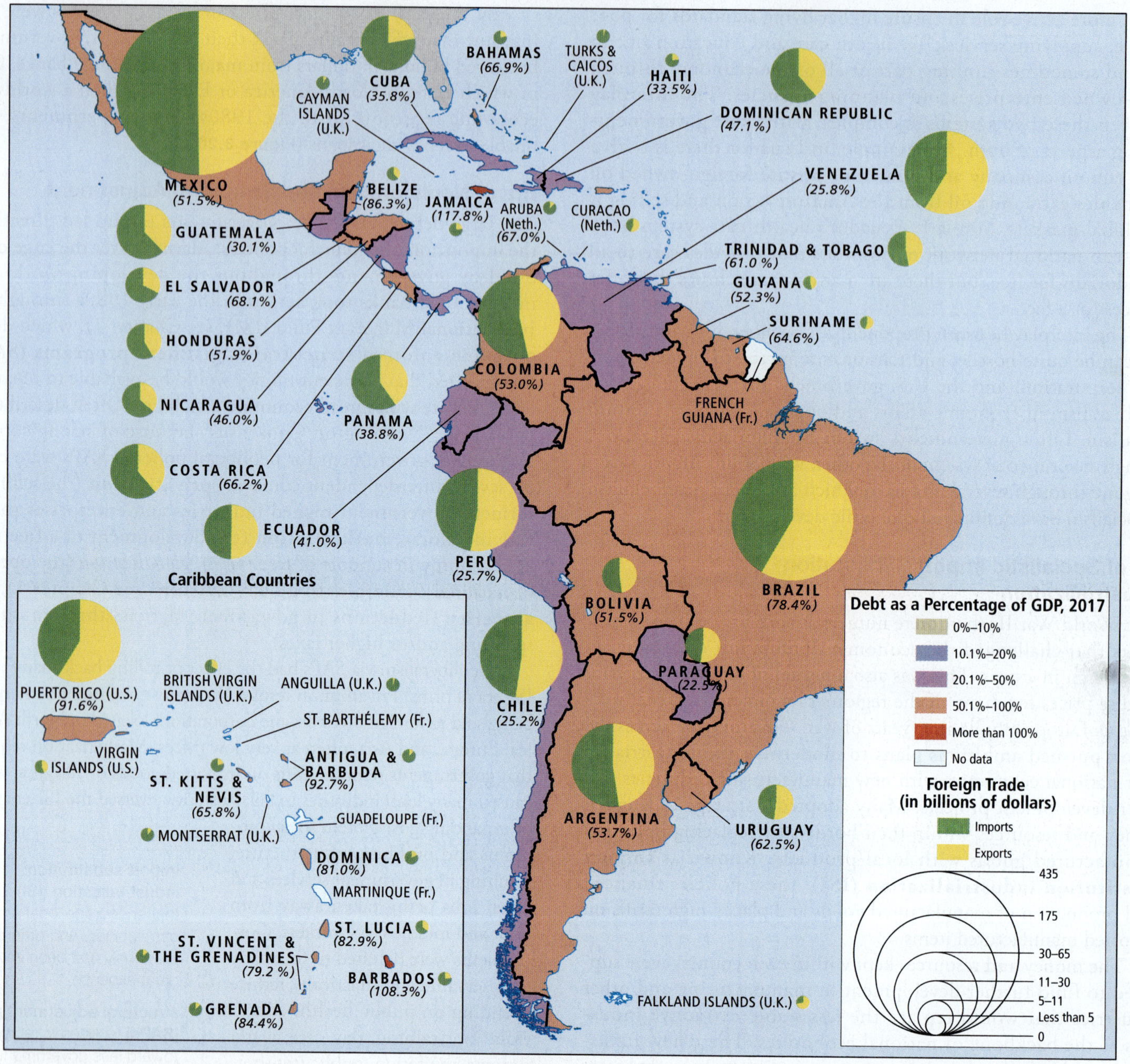

Figure 3.20 Debt, imports, and exports. The map shows government debt as a percentage of GDP, as well as the size of foreign trade, for each country. This type of information is used to determine how much money countries can borrow for development projects from foreign investors, international banks, the IMF, the World Bank, and other neoliberal institutions. Lower government debt and higher exports are favored because they indicate a country has more money to repay loans.

known as the "Pink Tide" that, starting in the late 1990s, elected socialist-leaning presidents who explicitly opposed neoliberalism in eight countries in the region: Venezuela, Argentina, Bolivia, Brazil, Chile, Ecuador, Nicaragua, and Uruguay. Venezuelan president Hugo Chávez emerged as a leader of the backlash, **nationalizing** foreign oil companies operating in Venezuela and using some of his country's oil wealth to help other countries, such as Argentina, Bolivia, and Nicaragua, pay off their debts to the IMF. Strong prices for this region's exports, which remain mostly raw materials, supported years of steady expansion of social policies, resulting in major improvements in human development (**Figure 3.21**). However, by 2015 falling prices for raw materials exports resulted in an economic downturn that brought neoliberal governments back into power in Brazil, Argentina, and Chile.

nationalize to seize private property and place it under government ownership, with some compensation

3.21 Globalization and Development in Middle and South America

The Human Development Index (HDI) map of Middle and South America shows that most countries rank in the medium and high ranges in providing the basics (education, health care, and income) for their citizens. Most of Africa, Central Asia, and South Asia rank lower, as can be seen in the inset map, while North America, Europe, Australia, Japan, and South Korea rank higher. Much of East Asia, Russia, and some of the states of the former Soviet Union are in the same range as Middle and South America. Of course, the UN HDI is a countrywide index and does not give any indication of how well-being is distributed within a given country. On this map and the Gini index map in Figure 3.22, the countries of the Central American isthmus stand out as poorer than Mexico, the Caribbean, and South America.

THINKING GEOGRAPHICALLY

A What about this photo suggests a once-prosperous country now in a state of civil unrest?

B Given the information on the map, what other country would you expect to have a national health-care system that covers all citizens?

C Based on what you have read in the text, what international agreement has facilitated the development of Mexico's manufacturing sector?

A Economic and political turmoil in Venezuela. A man holds his child while taking a break from a food line in Caracas. In the street is a barricade protesting the food scarcity and inflation brought on by lower production of the country's main export, oil. Venezuela's human development has fallen in recent years, but so far it maintains a medium level due to investments in health, education, and poverty reduction. [John Moore/Getty Images]

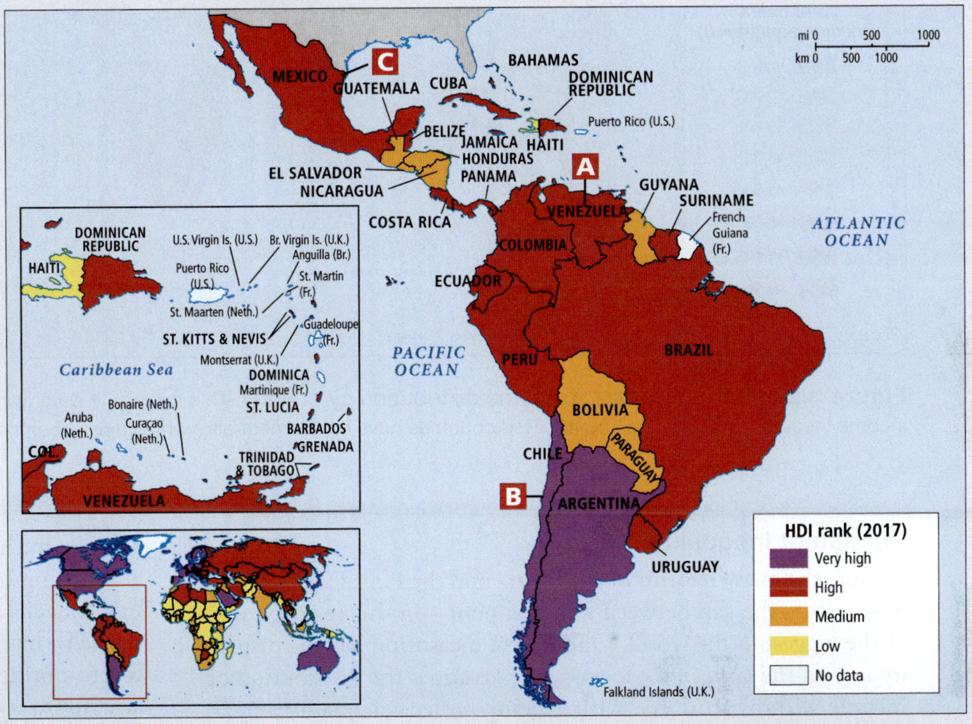

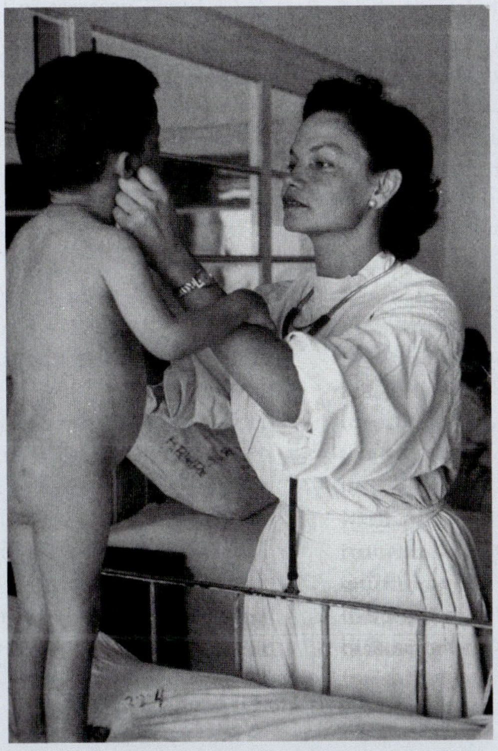

B Health care in Chile. A Chilean doctor examines a child with tuberculosis in 1950, when Chile became one of the first countries in the region to institute a national health-care system that covers all citizens. Despite its somewhat unstable raw materials–based economy, Chile has the region's highest human development, which reflects long life expectancy, high literacy, and high incomes. [Eliot Elisofon/The LIFE Picture Collection/Getty Images]

C Manufacturing in Mexico. Mexico has emerged as one of the more stable economies in the region largely because it has a large manufacturing sector that has reduced its dependence on raw materials exports. Its human development is at a medium level due to somewhat lower income, health, and education outcomes. [PEDRO PARDO/AFP/Getty Images]

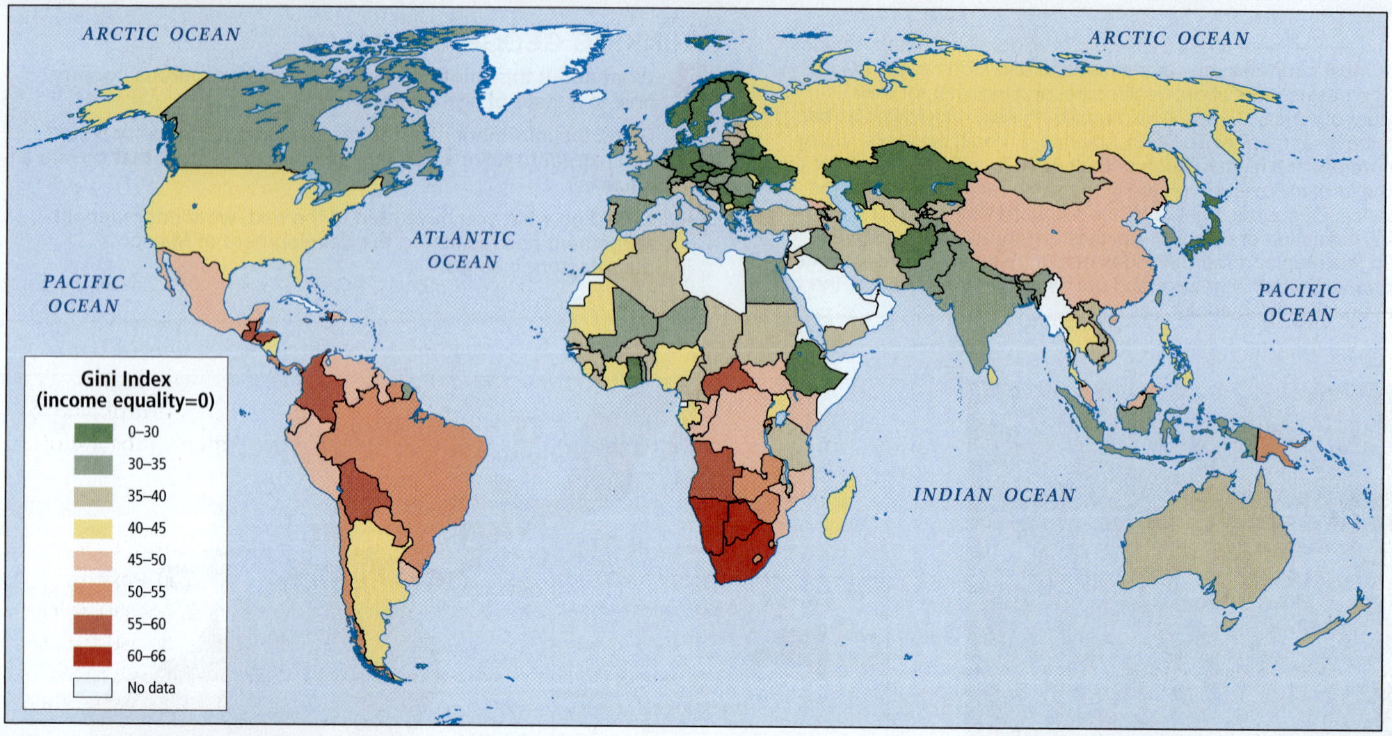

Figure 3.22 Gini index map of income distribution by country. This map, with data from the World Bank, shows the level of income inequality within each country. Red countries have higher inequality while green countries have lower inequality.

Persistent Inequality

Despite the many reforms of the ISI era and the Pink Tide, **income disparity**—the gap between rich and poor—in this region is one of the biggest in the world. One way of measuring income disparity is with the *Gini index*, a ratio that describes the distribution of income within a country, with 1 being perfect inequality and 0, perfect equality. The **Figure 3.22** map of the Gini index for countries around the world shows that Middle and South America have relatively high levels of income disparity. In recent years, disparities have been shrinking throughout the region, primarily because of new economic and social policies aimed at reducing disparities. The poverty rate for the region as a whole is around 33 percent, but in the poorer countries (Haiti, Guatemala, Honduras, Nicaragua, Peru, Bolivia, and Paraguay), more than half the population lives in poverty.

CHINA'S RISING INFLUENCE

China's demand for resources, and its willingness to lend directly to governments as a means of securing access to these resources, has transformed this region. International banks, the IMF and other neoliberal institutions, and the United States have seen their influence fall. Loans from the IMF dropped from U.S.$48 billion in 2003 to just U.S.$1 billion in 2008, recovering only since the election of neoliberal governments starting in 2015. In contrast to the years after the global recession of the early 1980s, when almost every country in this region turned to the IMF for loans, fewer countries are turning to the IMF for loans. Both in response to China's rising financial influence, and in recognition of the failures of SAPs, the IMF has shifted its policies worldwide from promoting SAPs to promoting *Poverty Reduction Strategy Programs* (PRSPs), which attempt to combine marketization with strong investment in social programs.

income disparity the gap in income between rich and poor

China's greatest influence in this region comes from trade. Strong economic growth in Brazil, Chile, and Peru during the Pink Tide years of the 2000s was based mainly on exports of raw materials to China, which is now their largest trading partner. China is the second- or third-largest trading partner for every other major country in the region, including Mexico, Colombia, Argentina, Venezuela, Ecuador, and the Dominican Republic. When China's growth slowed after the global recession that began in 2007, China's once-insatiable demand for raw materials also fell and growth slowed throughout this region. China will likely remain a major trade partner for this region, but it may reduce its imports in coming years as it shifts toward a model of more moderate economic growth that is based on serving the needs of China's huge and increasingly wealthy urban populations.

Regardless, a decade of trade with China has brought many changes to this region. Brazil is now part of the BRICS association between Russia, India, China, and South Africa (see Chapter 1). The BRICS countries are attempting to counter the influence of the United States and Europe on the global economy. Rich in resources and busily transforming themselves into large middle-income, global economic powerhouses, the BRICS countries recently formed their own bank, headquartered in China, as an alternative to the IMF and World Bank. Meanwhile, in 2014 the IMF asked for help from Brazil, Mexico, and Peru in designing and funding a bailout package for overly indebted countries in the European Union—the first time countries in Middle and South America have been asked to play such a role.

One change that trade with China has not brought is any significant shift away from dependence on raw materials. Of the major economies in this region only Mexico has growth that is driven by the production of manufactured goods, most of which are destined for the United States.

REGIONAL TRADE AGREEMENTS

There are a large number of regional trade agreements in this region, all of which reduce tariffs and other trade barriers among a group of neighboring countries. The two largest free trade agreements within the region are the USMCA and UNASUR.

The 1994 North American Free Trade Agreement (NAFTA), now known in the United States as the United States–Mexico–Canada Agreement (USMCA), established a free trade bloc consisting of the United States, Mexico, and Canada, and containing more than 450 million people. The main goal of the USMCA is to reduce barriers to trade, thereby creating expanded markets for the goods and services produced in the three countries. Since 1994, the value of the economies of these three countries has grown steadily; by 2018, it was worth more than U.S.$22 trillion, making it the world's largest trade bloc. An attempt by the United States to create a free trade bloc for all of the Americas (FTAA) has stalled in recent years, largely because of dissatisfaction with the effects of globalization and neoliberal policies such as SAPs that have been promoted by the United States.

A variety of other trade blocs have developed in this region over the years, the three most significant of which are UNASUR, a union of all South American nations, Mercosur (Chile, Argentina, Uruguay, Paraguay, and Brazil), and the Andean Community of Nations (Peru, Bolivia, Ecuador, and Colombia). Of the three, Mercosur is the most successful, having boosted trade between members more than tenfold during the 1990s, but none have attained lasting influence due in part to political disputes.

The overall record of regional free trade agreements so far is mixed. While they have increased the amount of trade, the benefits of that trade usually are not spread evenly among regions or among all sectors of society. In Mexico, for instance, the USMCA has brought somewhat greater economic stability by helping along the development of manufacturing industries, but it has introduced instability as well. As many as one-third of small-scale farmers throughout Mexico have lost their jobs because of the increased competition from subsidized corporate farms in the United States, which, through USMCA, now have unrestricted access to Mexican markets. In particular, corn imports from the United States have driven down the price of corn in local markets, pushing small farmers out of business.

Export Processing Zones

Regional free trade agreements, especially those involving the United States, have often included the creation of manufacturing industries in **export processing zones (EPZs)**. Also known as *free trade zones*, EPZs are specially created areas within a country where, in order to attract foreign-owned factories, taxes on imports and exports are not charged. The main benefit to the host country is the employment of local people, which eases unemployment and brings money into the economy. Products are often assembled strictly for export to foreign markets with few links to local economies.

There are EPZs in nearly all countries on the Middle and South American mainland and on some Caribbean islands (and in many other world regions). However, the largest of the EPZs is the conglomeration of assembly factories, called maquiladoras, that are located along the Mexican side of the U.S.–Mexico border. Although these factories do provide employment, jobs are often not secure, as the following vignette illustrates.

VIGNETTE In August 2003, Orbalin Hernandez returned to his self-built shelter in the town of Mexicali on the border between Mexico and the United States. Recently fired for taking off his safety goggles while loading TV screens onto trucks at Thomson Electronics, he had just gone to the personnel office to ask for his job back. Because there were no previous problems with him, he was rehired at his old salary of U.S.$300 per month ($1.88 per hour). Thomson is a French-owned electronics firm that took advantage of NAFTA (see "Regional Trade Agreements") when it moved to Mexicali from Scranton, Pennsylvania, in 2001. There, its 1100 workers had been paid an average of $20 per hour. Many of the Scranton workers were unable to find work after Thomson left and still feel bitter toward the Mexicali workers.

In 2006, Orbalin and his fellow Mexicali workers were told that their wages were too high to allow their employers to compete with companies located in China, where in 2005, workers with the same skills as Orbalin earned just U.S.$0.35 an hour. By 2007, 14 Mexicali plants had closed and moved to Asia, including the Thomson plant. Those firms remaining in Mexicali cut wages and reduced benefits.

After a few months Orbalin got hired at the Kellogg's plant, loading frozen waffles onto trucks bound for the United States for about U.S.$8 a day. He has kept this job ever since, even though his salary is barely enough for him to support his wife and four children. Like most in Mexicali he is skeptical of the new higher salaries of U.S.$16 an hour required for autoworkers by the renegotiation of NAFTA, now called the USMCA. None of the auto plants in Mexicali pay more than U.S.$3 or $4 an hour, and fewer than 1 percent of Mexicans earn U.S.$16 an hour. ■

MIGRATION AND REMITTANCES

Migration for the purpose of supporting family members with remittances (money sent home) is an extremely important economic activity throughout this region (see further discussion in "Rural-to-Urban Migration"). Each year, remittances amount to roughly double the U.S. foreign aid to the region. Most such "remittance migrations" are within

export processing zones (EPZs) specially created legal spaces or industrial parks within a country where, to attract foreign-owned factories, duties and taxes are not charged countries. While the individual amounts of cash sent home are small, these remittances are crucial for families that may not have much other income.

One report (by geographer Dennis Conway and anthropologist Jeffrey H. Cohen) suggests that couples who migrate to the United States from indigenous villages in Mexico typically work at menial jobs and live frugally in order to save a substantial nest egg. They may then return home for several years to build a house (usually a family self-help project) and buy furnishings. When the money runs out, the couple, or just one member of the couple, may migrate again to save up another nest egg.

THE INFORMAL ECONOMY

For centuries, many small business owners throughout Middle and South America have operated in the informal economy, supporting their families through inventive entrepreneurship but without their work being officially recognized in government statistics and without paying business, sales, or income taxes. The majority of working people are employed in the informal economy in most of Central America and the Andean countries (Colombia, Peru, Bolivia), and for the rest of the region about 30 to 40 percent of employment is informal. In the United States, around 16 percent of employment is informal. Most jobs are held by small-scale operators involved with street vending or recycling used items such as clothing, glass, or waste materials. Often people stay in the informal economy despite the job insecurity and lower pay because the work gives them more freedom to work when and where they want. Avoiding government oversight can also be a motivation, as some countries have elaborate regulations for even small businesses.

RAPID REGIONAL EXPANSION OF INTERNET USE

Overall this region is more advanced in terms of information technology than many other developing regions—a fact that could open up many new economic opportunities in the near future. Brazil ranks fifth in the world in terms of total number of internet users and has two world-class technology hubs in the environs of São Paulo.

Mexico, with 49 percent of its population connected, recently launched a program to give its citizens access to training and higher education via the internet (**Figure 3.23**). Similar efforts throughout the region are part of a long-term shift, especially in urban areas, toward more technologically sophisticated and better-paid service sector employment. In 2000, only 3 percent of the region's population used the internet, but that figure had risen to nearly 67 percent by 2018, an increase of over 2000 percent.

FOOD PRODUCTION AND DEVELOPMENT

Food production throughout Middle and South America is shifting away from small-scale, often subsistence-level production toward large-scale, green revolution agriculture that is aimed at earning cash. This shift has forced many small-scale farmers who cannot afford the investment in machinery or chemicals off their land and into cities or to wealthier countries where they often work illegally. In some cases, these people have effectively advocated for policies that help lift them out of landlessness and poverty (see "Revolutionary Movements").

For centuries, while people labored for very low wages on haciendas, the hacienda owners would at least allow them a bit of land to grow a small garden or some cash crops. SAPs encouraged a shift to green revolution agriculture, which was based on crops that could be exported for cash. This type of production, it was hoped, could help countries pay off debts faster. SAPs also made it easier for foreign multinational corporations, such as Del Monte, to use their financial resources to buy up many haciendas and other farms, converting them to plantations. Many rural people have been forced off lands they once cultivated and onto plantations where they now work as migrant laborers for low wages, as the following story illustrates.

VIGNETTE Aguilar Busto Rosalino used to work on a Costa Rican hacienda. He had a plot on which to grow his own food and in return worked 3 days per week for the hacienda. Since a banana plantation took over the hacienda, he rises well before dawn and works 5 days a week from 5:00 A.M. to 6:00 P.M., stopping only for a half-hour lunch break. Because he now lacks the time and land to farm, he must buy most of his food.

Aguilar places plastic bags containing pesticide around bunches of young bananas. He prefers this work to his last assignment of spraying a more powerful pesticide, which left him and 10,000 other plantation workers sterile. He works very hard because he is paid according to how many bananas he treats. Usually he earns between U.S.$5.00 and $14.50 a day.

It is common practice for these banana operations to fire their workers every 3 months so that they can avoid paying the employee benefits that Costa Rican law mandates. Although Aguilar makes barely enough to live on, he has no plans to press for higher wages because if he did he would be put on a "blacklist" of people that the plantations agree not to hire. ■

Because of the moneymaking potential and political power of large-scale agriculture, conflicts are only rarely resolved in favor of small farmers and agricultural workers. Moreover, in rapidly urbanizing countries, green revolution agriculture is seen as the only way to supply cheap food for the millions of city dwellers. Ironically, many of these new urbanites were once farmers capable of feeding themselves, who only moved to the cities because they were unable to compete with the new green revolution systems.

Many areas have seen haciendas converted to green revolution agriculture, similar to farms that can be found in the Midwestern United States. There is a wide belt of large-scale farming and ranching from the Argentine pampas into Brazil (**Figure 3.24**) and, as we saw in the earlier discussion of environmental issues (see "Tropical Forests, Climate Change, and Globalization"), soybeans and beef, which are often produced by large green revolution farms on once-forested lands in and around the Amazon Basin, are all major exports for countries like Brazil and Argentina.

At a medium scale are zones of modern mixed farming (those that produce meat, vegetables, and specialty foods for sale in urban centers), located on large and small plots around most major urban centers. **Figure 3.25** illustrates how the foodways of the area mix indigenous plants (corn, potatoes, tomatoes) and methods of

Figure 3.23 Internet use in Middle and South America, 2015. The figure shows the number of internet users in each country and the percentage of the country's population that uses the internet. By November 2018, more than 444 million people (67 percent of the region's population) were using the internet. Argentina has the highest percentage of internet users (93 percent) in South America; Costa Rica has the highest percentage (86 percent) in Middle America; and the Caribbean Netherlands has the highest percentage (94 percent) in the Caribbean.

cooking (baking, grilling), with European, Asian, and African foods (beef, pork, rice) and cooking methods (frying).

CHECK YOUR UNDERSTANDING

1. How have neoliberalism and socialism influenced economic development in this region?
2. Why did import substitution fail to work in some countries?
3. Why did SAPs fail to produce the sustained economic growth and stability that were expected to relieve the debt crisis and create broader prosperity?
4. What is the BRICS group of countries?
5. How have regional trade blocks influenced development in this region?
6. What shifts have occurred in food production?

Figure 3.24 Agricultural and mineral zones in Middle and South America. [Research from: *Goode's World Atlas*, 21st ed. (Chicago: Rand McNally, 2010), pp. 160–161, Minerals and Economic map]

Figure 3.25 LOCAL LIVES: Foodways in Middle and South America

(A) Corn tortillas are prepared in Mexico City. Corn has been a pillar of diets in Mexico and the countries of Central America for thousands of years. Now cultivated throughout the world, corn was first domesticated from a wild grass in southern Mexico. [Greg Elms/Getty Images]

(B) An Argentine *asado*, or barbeque. Here beef is roasted over an open bed of coals, using a special *parilla*, or grill. This method of cooking is a variant of methods first developed by the indigenous peoples of the grasslands of Argentina. Asado is the national dish of Argentina; it is prepared widely throughout Uruguay, Paraguay, and southern Brazil. [Virginia Sherwood/Bravo/NBCU Photo Bank via Getty Images]

(C) *Lomo saltado*, a Peruvian dish with Asian influences. First developed in restaurants started by Chinese immigrants, it is a mix of Chinese and Peruvian ingredients and culinary traditions. It consists of strips of beef or pork marinated in vinegar and stir-fried with onions, parsley, tomatoes, and other vegetables, and then served over rice and french fries. [Juanmonino/E+/Getty Images]

POWER AND POLITICS

3.5 Analyze the impacts of foreign interventions and the international drug trade on politics in this region.

For centuries this region was ruled by corrupt authoritarian economic elites and the military who generally resisted efforts to expand economic opportunities and political power, often with the support of foreign governments and corporations. The past 50 years have seen a significant expansion of political freedoms, and almost all countries now have multiparty political systems and elected governments, though foreign political interference remains significant. While a number of violent political conflicts are ongoing, the international illegal drug trade has become a leading source of violence and corruption in the region.

In the last 50 years, there have been repeated peaceful and democratic transfers of power in countries once dominated by rulers who seized power by force. These **dictators** often claimed absolute authority, governing with little respect for the law or the rights of their citizens. Their authority was based on alliances between the military, wealthy rural landowners, wealthy urban entrepreneurs, foreign corporations, and even foreign governments such as the United States.

Although a slow expansion of political freedoms has transformed the politics of the region, problems remain (**Figure 3.26**). Elections are sometimes poorly or unfairly run and their results are frequently contested. Elected governments are sometimes threatened with a **coup d'état**, in which the military takes control of the government by force. Such coups are usually a response to policies that are unpopular with large segments of the population, powerful elites, the military, or foreign powers such as the United States. In the last decade, coups have been attempted in Venezuela, Colombia, Ecuador, and Bolivia and successful in Honduras, but peaceful democratic elections are increasingly the norm. This is significant in a region where one country alone, Bolivia, has had over 190 coup attempts since its independence from Spain.

There is also a long tradition of *populist political movements*, which appeal to the preferences of the majority of people, who in this region of extreme inequalities tend to be poor. Such movements have emerged for centuries, but most were kept from power by alliances between the wealthy, the military, foreign investors, and foreign governments. However, the expansion of political freedoms over the past several decades opened the way for a wave of populist movements that gained control of national governments. The Pink Tide governments were among these, but so were populists who enacted neoliberal reforms in Peru, Argentina, and Brazil. Populists frequently come to power with promises of "cleaning out" corruption from government and business, though they are also often removed from power under accusations of corruption.

FOREIGN INVOLVEMENT IN THE REGION'S POLITICS

Interventions in the region's politics by outside powers have frequently compromised political freedoms and human rights. Although the former Soviet Union, Britain, France, Spain, and other European countries have wielded much influence over the past century, by far the most active foreign power has been the United States, which generally backs neoliberal governments that favor U.S. corporations and their investments in this region.

In 1823, the United States introduced the Monroe Doctrine to warn Europeans that no further colonization would be tolerated in the Americas. Subsequent U.S. administrations interpreted this policy more broadly to mean that the United States itself had the sole right to intervene in the affairs of the countries of Middle and South America, and it has done so many times. The official goal for such interventions was usually to make countries safe for democracy, but in most cases the driving motive was to protect U.S. political and economic interests.

At various times during the past 150 years, U.S.-backed unelected political leaders, many of them military dictators, have been installed in most countries in Central and South America. After World War II, worried that socialism would infiltrate Middle and South America, the United States intervened against socialist movements and governments. Perhaps the most infamous intervention took place in Chile in 1973 when, with U.S. aid, the elected socialist-oriented government of Salvador Allende was overthrown. Allende was killed, and a military dictator, General Augusto Pinochet, was installed in his place. This coup d'état was part of a broader U.S.-backed covert campaign known as *Operation Condor* that targeted socialists and other political opponents of neoliberal governments throughout South America from 1968 to 1989. With U.S. intelligence agencies providing funding, weapons, training, and technical support, neoliberal military governments in Argentina, Chile, Uruguay, Paraguay, and Brazil waged the "Dirty War" in which 60,000–80,000 civilians were killed (mostly in Chile and Argentina) and over 400,000 imprisoned. Some of those killed or imprisoned were socialist guerillas attempting to achieve political revolutions through violent means, but most were "suspected socialists" engaged in peaceful political opposition of unelected governments. Operation Condor and the Dirty War peaked in the 1970s and 1980s and gradually came to a close in the late 1980s as military governments were voted out of office.

REVOLUTIONARY MOVEMENTS

Efforts to address this region's poverty, inequality, and injustice have often taken the form of revolutionary movements. Almost every country in the region has had a popular revolutionary movement, often armed, at some point in its recent history. These movements were often partially shaped by the Cold War, with socialist revolutionary movements often receiving assistance from the Soviet Union, while more neoliberal and elite-based movements were supported by the United States and its allies. In recent decades two major broad-based rural political movements have arisen that have influenced change, mostly at the local level. The areas in which these movements take place are examples of what geographers call **contested space**, where various groups are in conflict over the right to use a specific territory as each sees fit.

dictator a ruler who claims absolute authority, governing with little respect for the law or the rights of citizens

coup d'état a military- or civilian-led forceful takeover of a government

contested space any area that two or more groups claim or want to use in different and often conflicting ways, such as the Amazon or Palestine

PHOTO ESSAY

3.26 Power and Politics in Middle and South America

Political freedoms are expanding in Middle and South America, although many barriers persist. Centuries of rule by authoritarian economic elites and militaries, combined with rampant corruption, political repression, and more recently drug trafficking, have at times inspired popular revolutionary movements, some of which have resulted in civil wars that have cost many lives. Nevertheless, in the region as a whole, there has been a decline in political violence and a shift toward democracy in recent decades. Elections are now the norm, although sometimes they are poorly or unfairly run, and their results are frequently contested. [Research from: Economist Intelligence Unit, Democracy Index 2017]

THINKING GEOGRAPHICALLY

A What are the indications that these men were once involved in violent combat?

B What evidence can you find in the photo of a violent situation in Mexico?

C Why would the Cuban government lionize these men in a mural?

A Violence in Colombia. Colombian police officers, crippled in combat with a rebel group, undertake a protest march destined for the capital in Bogotá. Popular revolutionary movements and a drug war drove this complex conflict, major aspects of which came to a close in 2016 with a ceasefire between the government and the largest rebel group. Political institutions and the rule of law have been badly damaged, with many high-ranking officials linked to corruption and extreme human rights abuses. [LUIS RAMIREZ/AFP/Getty Images]

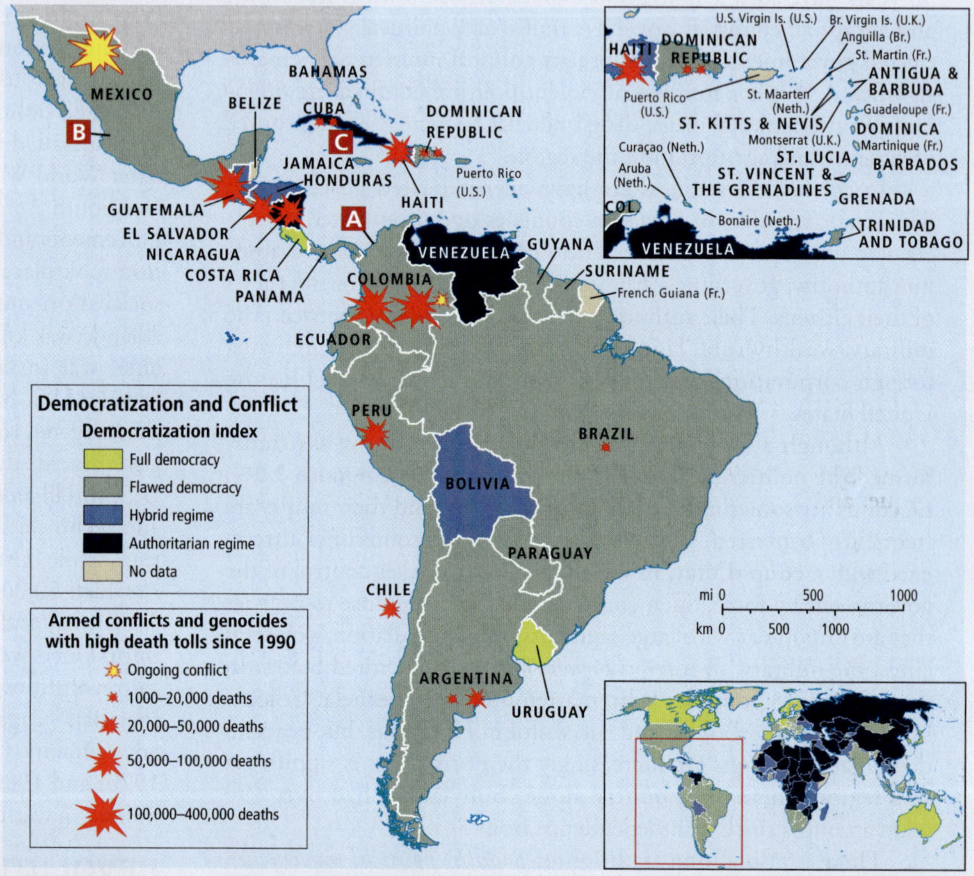

B Illegal drugs in Mexico. Federal agents search cars at a checkpoint in Juárez in Chihuahua, Mexico. Mexico's drug war has led to 260,000 deaths over the past decade and corruption that has eroded local political institutions, especially near the U.S. border. Most of Mexico's drug production and distribution supply the United States. [Shaul Schwarz/Edit by Getty Images]

C Revolutionaries honored in Cuba. Cuban revolutionaries Che Guevara (left), Camilo Cienfuegos (center), and Juan Antonio Mella are immortalized in an aging mural in Havana. Since coming to power in 1959, Cuba's communist government has been praised for achieving advances in human development and criticized for jailing and sometimes executing leaders of the political opposition. [Luis Gonzalo Vinagre Solans/Getty Images]

Cuba

Since 1959, the Caribbean island of Cuba has been governed by a revolutionary socialist government led first by Fidel Castro and then, starting in 2006, by Fidel's brother Raúl (see Figure 3.26C). Cuba's successes have inspired similar socialist governments throughout the region, most recently Venezuela, Bolivia, and other Pink Tide governments, while its failures combined with sustained hostility from the U.S. government have emboldened harsh repression of socialists throughout this region by neoliberal governments.

The revolution that brought Fidel Castro to power transformed a plantation and tourist economy once known for its extreme income disparities into one of the most egalitarian in the region. However, because Castro adopted socialism and, after a brief period of Cold War neutrality, allied Cuba with the Soviet Union, the United States became extremely hostile, funding many efforts to overthrow Cuba's government and instituting a trade embargo.

With help from the Soviet Union, Castro managed to dramatically improve the country's literacy and infant mortality rates and the life expectancy of its population. He also imprisoned or executed thousands of Cubans who disagreed with his policies, many of whom fled to southern Florida. From the 1960s to the late 1980s Castro, with Soviet aid, also actively supported socialist revolutions with troops and weapons in Bolivia, Grenada, El Salvador, Nicaragua, Congo (formerly Zaire), Ethiopia, Angola, Mozambique, and Namibia. Following the demise of the Soviet Union, Castro opened the country to foreign investment. Many countries responded, and Cuba is now a major European tourist destination. Formal diplomatic relations were restored between Cuba and the United States in 2014, and President Obama visited Cuba in 2016. However, political repression in Cuba persists and the economic embargo looks likely to remain in place for some time.

The Zapatista Rebellion

In the southern Mexican state of Chiapas, indigenous farmers have waged a decades-long rebellion against economic and political systems that have left them poor and powerless. Most farm tiny plots on infertile hillsides, many are malnourished, and roughly one-third of children do not attend school. In the early twentieth century, the Mexican government enacted land reforms that redistributed some hacienda lands to communities of poor farmers. These communal farms or "ejidos" still exist, and they are some of the best lands that poor farmers have access to, comprising roughly half of all farmland in Mexico in the 1990s. However, over the years many ejidos have been invaded by wealthier farmers, or sold off to loggers and mining companies by the Mexican government, which abandoned the process of recognizing new ejidos at the insistence of the U.S. government as part of the negotiations that created NAFTA (now the USMCA).

The Zapatista rebellion (named for the hero of the 1910 Mexican revolution, Emiliano Zapata) began on the day NAFTA took effect in 1994. The Zapatistas view NAFTA (and now USMCA) as a threat because it encourages the Mexican government to assist in the dismantling of ejidos and the spread of large-scale, export-oriented green revolution agriculture, logging, and mining. Starting as an armed rebellion in Chiapas, the movement was largely suppressed by the Mexican national army. However, the Zapatistas' support in local communities allowed them to continue in many parts of Chiapas. Their extensive use of the internet to promote their cause earned the Zapatistas a worldwide following of supporters.

After 9 years of armed resistance the Zapatista movement redirected its energies toward local nonviolent political campaigns, mainly in Chiapas, to set up people's governing bodies parallel to local official governments. The Zapatistas have made significant improvements in education, women's rights, and poverty reduction, thus deepening their support base in the areas they control. While the Zapatistas continue to focus more locally, the movement still has a strong influence on activists in other movements, a national voice via campaigns against the political establishment, and global reach through the internet and numerous training programs for supporters of the movement around the world.

Brazil's Landless Movement

Sixty-five percent of Brazil's arable and pasture land is owned by wealthy farmers who make up just 2 percent of the population. Since 1985, more than 2 million small-scale farmers have been forced to sell their land to larger farms that practice green revolution agriculture. Because the larger farms specialize in major export items, such as cattle and soybeans, they have been favored by governments wishing to increase exports. As a result, many poor farmers have been forced to migrate to urban areas.

To help these people, organizations such as the Movement of Landless Rural Workers (MST) began taking over unused portions of some large farms. Since the mid-1980s, the MST has coordinated the occupation of more than 51 million acres of Brazilian land (an area about the size of Kansas). Some 250,000 families have gained land titles, while the former owners have been paid off by the Brazilian government and have moved elsewhere. There are now 1.5 million members of the MST spread across 23 of Brazil's 26 states, and movements with similar goals have been launched in Ecuador, Venezuela, Colombia, Peru, Paraguay, Mexico, and Bolivia.

The Homeless Workers Movement In the late 1990s a group focusing on urban homelessness and affordable housing shortages grew out of the MST. Called the Homeless Workers Movement or MTST, it appeals to the roughly 6 million Brazilian families who are forced to live in shacks and other improvised dwellings due to chronic shortages of affordable housing. The MTST occupies abandoned buildings and vacant land, often near the centers of major Brazilian cities. It is part of broader resistance of poor communities to attempts by local governments to relocate them to the urban periphery and away from city centers, where most of the jobs are.

THE DRUG TRADE, CONFLICT, AND LEGALIZATION

The international illegal drug trade is a leading source of violence and corruption throughout the region. The primary drugs traded are cocaine, heroin, marijuana, and methamphetamine. Most drugs are produced in or pass through northwestern South America, Central America, and Mexico and are consumed in the United States or Western Europe. **Figure 3.27** illustrates some geographic aspects of the cocaine trade.

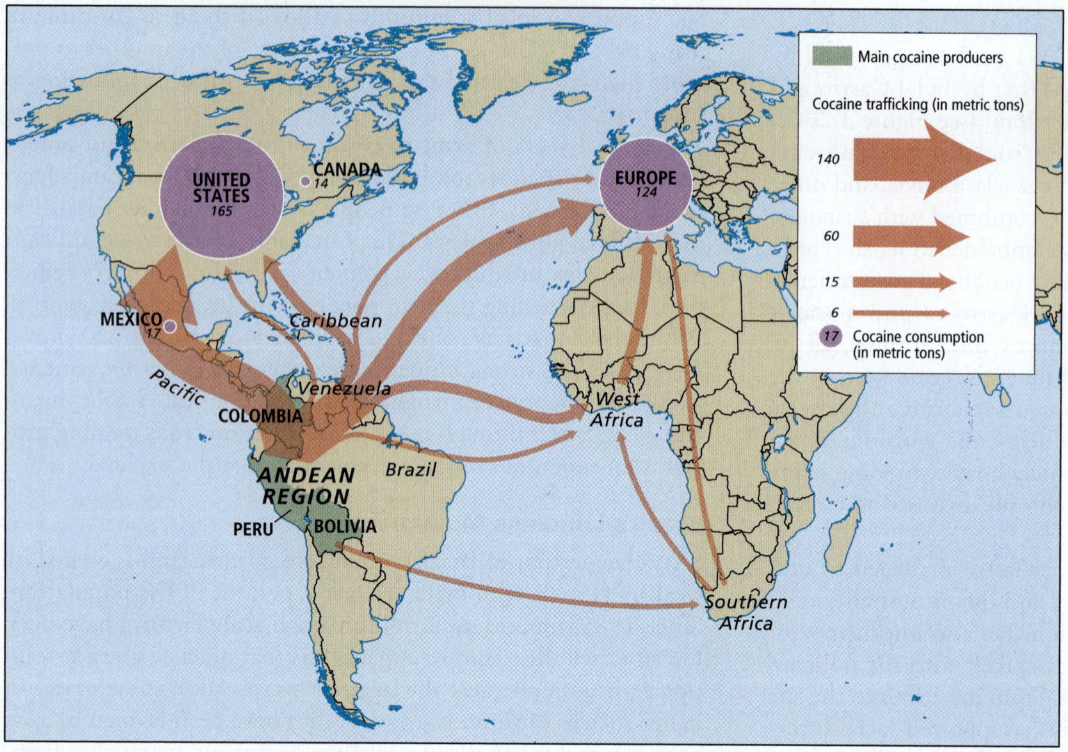

Figure 3.27 Linkages: Cocaine sources, trafficking routes, and seizures worldwide, 2016. Colombia, Peru, and Bolivia are the most important sources of cocaine cultivation and production in the world. The big cocaine markets are in North America and Western Europe. Use has declined by 36 percent in the United States since 1998 but is still high enough to support a large black market trade. Cocaine use in Western Europe and West Africa is on the rise, due in part to new trade routes through West Africa to Europe. [Research from: World Drug Report 2018, United Nations Office on Drugs and Crime, at https://www.unodc.org/wdr2018/prelaunch/WDR18_Booklet_1_EXSUM.pdf, p.18]

Production of cocaine, heroin, and methamphetamine is illegal in most of Middle and South America. However, public figures, from the local police on up to high officials, are paid to turn a blind eye to the industry. Most coca growers are small-scale farmers of indigenous or mestizo origin, working in remote locations where they can make a better income for their families from these plants than from other cash crops. Many rent land from the drug cartels that get the drugs to global markets. The cartels are monopolistic and violent, and those who oppose them—whether farmers, journalists, or law enforcement officials—often end up kidnapped, tortured, and brutally murdered. In Colombia, the illegal drug trade has financed all sides of a civil war that displaced more than 6.5 million people over the past several decades (see Figure 3.26B). Mexico has endured a decade of armed conflict between powerful drug cartels that control a hugely profitable U.S.-oriented drug trade. This "drug war" has resulted in over 260,000 deaths, 37,000 missing people, and at least 330,000 people forced from their homes and communities by violence.

The U.S.-led "war on drugs" has been so ineffective that despite generous U.S. funding and military assistance, most countries in the region are considering new approaches to limiting the drug trade. Since the war on drugs began in 1971, the U.S. military has collaborated with national militaries to eradicate drug crops through spraying of herbicides or burning and intercept of drug shipments before they reach the United States. Almost half a century later, illegal drug production levels in Middle and South America are higher than ever, exceeding demand in the United States, where street prices for many drugs, such as cocaine and heroin, have fallen dramatically since the 1970s. Militarized responses have been similarly ineffective in reducing the violence associated with the drug trade. For example, attempts by the Mexican police and military and U.S. authorities to control drug cartel territories (see Figure 3.26A) have resulted in much higher murder rates in these areas.

Many who have studied the drug trade between the United States and this region call for radical changes in international drug policies, including the legalization of drug use in the main (U.S.) market, which would make the drug trade unprofitable for criminals. Recent studies suggest that the legalization of marijuana in several U.S. states has reduced the flow of illegal marijuana across the U.S.–Mexico border by more than three-quarters. The cartels have tried to adjust by increasing their exports of methamphetamine and heroin, but the overall value of illegal drugs coming into the United States, especially across the border with Mexico, has been falling since 2012 when several large states legalized marijuana for recreational use.

Within this region many countries are legalizing drugs to combat the illegal drug trade and treating drug use as a public health issue instead of a crime. In 2009, Mexico decriminalized possession of small amounts of most drugs, including marijuana, cocaine, heroin, methamphetamine, and LSD. Most South American countries have made similar moves. In 2018 Mexico legalized marijuana use, cultivation, sale, and distribution, a move that Uruguay made in 2013, and Colombia in 2015. Medical use of marijuana in some form is now allowed in almost every country in the region.

Most studies of decriminalization and legalization of drugs, in this region and others, suggest that it does not lead to higher rates of drug use or violent crime, but does decrease the cost of law enforcement. In Mexico, decriminalization and legalization of drugs may have influenced some drug cartels to move into other illegal activities, such as fuel theft, illegal logging, illegal mining, and kidnapping.

CHECK YOUR UNDERSTANDING

1. How have foreign interventions influenced politics in this region?
2. Why are elected governments sometimes challenged with a coup d'état?
3. Why is this region home to so many revolutionary movements?
4. What strategies have been tried to control the international drug trade in this region?
5. How has the drug trade influenced democracy in this region?

URBANIZATION

3.6 Explain why this region has urban landscapes that often lack adequate support services and infrastructure.

Since the 1950s, cities have grown rapidly in this region as most rural people have migrated to cities and towns, which now hold more than 78 percent of the population. Often chronically short of funding, and at times subject to considerable corruption, cities have been unable to plan effectively for much of this growth, resulting in urban landscapes that often lack adequate support services and infrastructure.

These problems are intensified when one city becomes a **primate city** that is vastly larger than all the others in a country, accounting for a large percentage of the country's total population (**Figure 3.28**). Examples in Middle and South America are Mexico City (with 21 million people, a bit more than 17 percent of Mexico's total population); Santiago, Chile (7 million, 39 percent of that country's population); Lima, Peru (9.7 million, 31 percent); Buenos Aires, Argentina (13 million, 30 percent); and Managua, Nicaragua (2.5 million, more than 40 percent). The concentration of people into just one or two large cities can overwhelm urban infrastructure, especially in poor countries. On the other hand, with so much wealth and power concentrated in one place, rural areas and even other towns and cities have difficulty competing for talent, investment, industries, and government services.

RURAL-TO-URBAN MIGRATION

It always takes some resourcefulness to move from one place to another. Those people who already have some years of education and strong ambition are the ones who migrate to cities. The loss of these resourceful young adults in which the community has invested years of nurturing and education is referred to as **brain drain**. Brain drain in this region also happens at an international level when migrants move to North America and Europe. Migrants often want better access to jobs and education and to help their families with money they send back home.

Gender and Rural–Urban Migration

Interestingly, rural women are just as likely as rural men to migrate to the city. This is especially true when employment is available in foreign-owned factories that produce goods for export. Companies prefer women for such jobs because they will accept lower pay and are less likely to organize to demand better pay and working conditions. Factors that push people out of rural areas can also affect women disproportionately. For example, the shift toward green revolution agriculture has had a particularly hard impact on rural women because they are rarely considered for such jobs as farm-equipment operators or mechanics. In urban areas, unskilled migrant women who can't find work in factories often find work as street vendors or domestic servants.

Male urban migrants tend to depend on short-term, low-skill day work in construction, maintenance, small-scale manufacturing, and petty commerce. Many work in the informal economy as street vendors, errand runners, car washers, and trash recyclers, and some turn to crime.

Once in the city, migrants face huge obstacles but also many opportunities. Families may disintegrate because of extreme poverty and malnutrition, long commutes for both working parents, and poor quality of day care for children. Children may be left alone or sent into the streets to scavenge for food or to earn money for the family as street vendors (see Figure 3.28A). On the other hand, cities offer unmatched educational and employment opportunities, and personal freedoms that often don't exist in small rural communities.

INADEQUATE URBAN PLANNING

Sufficient resources have often not been directed to urban planning, especially in the poorer parts of cities. The need for affordable housing is regularly neglected in favor of more high-profile urban development projects, such as soccer stadiums and commercial real estate. Even the well-designed and planned urban infrastructure projects that wealthier areas get, such as new sewer systems, are often plagued by the corruption that is rampant within most of this region. As a result cities have been unable to cope with the massive rush to the cities, resulting in both affluent and working-class areas becoming unwilling neighbors to unplanned slums. Filled with very poor migrants often living in self-built housing, as depicted in the diagram in **Figure 3.29**, shantytowns often occupy parks, small patches of land, or abandoned buildings.

The best known of these shantytowns are Brazil's **favelas** (**Figure 3.30**). In other countries they are known as slums, colonias, barrios, or barriadas. The settlements often spring up overnight after a coordinated collective occupation by poor homeless families. They rarely occur with the landowner's permission, and are without city-supplied water, electricity, or sewer service. Housing is hastily constructed with whatever materials are available, and with no building codes in effect, is often unsafe, especially when electricity is delivered via illegal connections from nearby power lines. Fires and electrocution are common, as is waterborne disease (due to a lack of sewer service), flooding, and landslides. Crime is also a huge problem, with police often reluctant to enter the maze of narrow passageways that provide access to dwellings. Once the settlements are established, efforts to eject squatters usually fail, though

primate city a city, plus its suburbs, that is vastly larger than all others in a country and in which economic and political activity is centered

brain drain the migration of educated and ambitious young adults to cities or foreign countries, depriving the communities from which the young people come of talented youth

favelas Brazilian urban slums and shantytowns built by the poor; called colonias, barrios, or barriadas in other countries

PHOTO ESSAY

3.28 Urbanization in Middle and South America

In the last several decades, cities have grown extremely quickly, and 78 percent of the region's population is now urban. New opportunities have opened for migrants from rural areas, but the low-skilled jobs most migrants can get often don't pay well, and affordable housing is so scarce that many people are forced to live in unplanned shantytowns on the outskirts of vast cities.

THINKING GEOGRAPHICALLY

A Why might work in the informal economy be considered insecure?

B What about this picture suggests that there may be considerable wealth in Santiago?

C What are some defining characteristics of informal squatter settlements?

A Urban migrants in Brazil. A child works as a street vendor in Salvador, Brazil. Many urban migrants work in similar informal sector jobs that pay less and are much less secure than formal sector jobs. [Mathias T Oppersdorff/Getty Images]

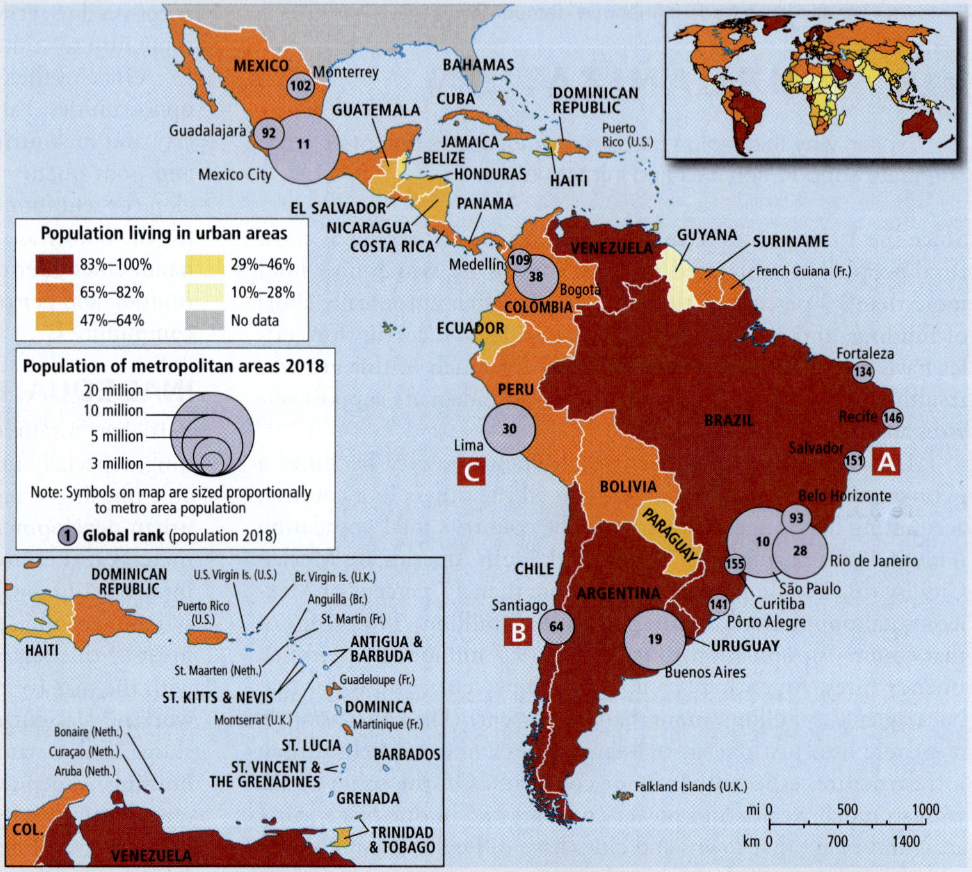

B Primate city in Chile. Santiago is one of the region's primate cities, home to 39 percent of Chile's population. It dominates Chile's economy, generating 40 percent of the country's GDP. [Jumper/Photodisc/Getty Images]

C Shantytown violence in Peru. Police attempt to evict 2500 people from a shantytown outside Lima, Peru. Structures made out of woven grass mats are the first to be built, followed by wooden shacks and eventually cement or brick houses. Water and electricity may eventually be extended to such settlements, but only after years or decades of occupation. [ERNESTO BENAVIDES/AFP/GettyImages]

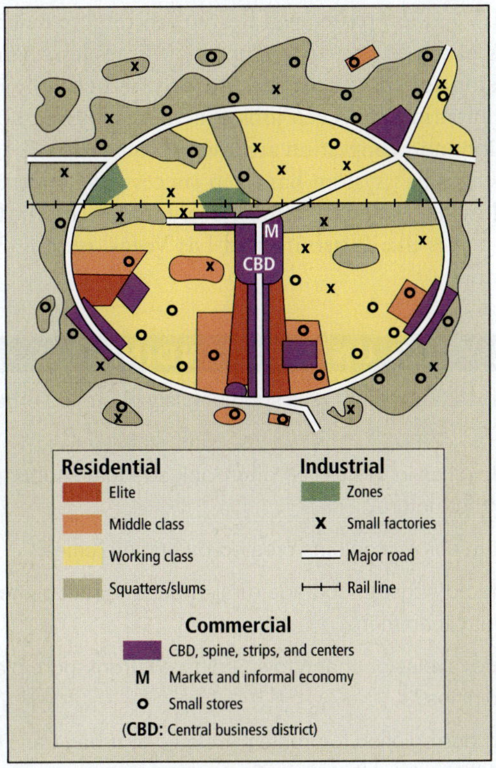

Figure 3.29 Crowley's model of urban land use in mainland Middle and South America. William Crowley, an urban geographer who specializes in Middle and South America, developed this model to depict how residential, industrial, and commercial uses are mixed together, with people of widely varying incomes living in close proximity to one another and to industries. Squatters and slum-dwellers ring the city in an irregular pattern. [Crowley, William K. "Order and Disorder—A Model of Latin American Urban Land Use." *Yearbook of the Association of Pacific Coast Geographers*, vol. 57, 1995, pp. 9–31. Project MUSE, doi:10.1353/pcg.1995.0010]

police are often called in to attempt to evict them (see Figure 3.28C). Poor people are such a huge portion of the population that even those in positions of power will not challenge them directly. Nearby wealthy neighborhoods simply barricade themselves with walls and security guards.

The squatters are frequently enterprising people who work hard to improve their communities. They often organize to press governments for social services through movements like Brazil's MTST (see "The Homeless Workers Movement"). Some cities, such as Fortaleza in northeastern Brazil, even contribute building materials so that favela residents can build more permanent structures with basic indoor plumbing. Over time, shacks and lean-tos are transformed through self-help initiatives into still very dense but livable suburbs. The economy of favelas can be quite vibrant, with much activity in the informal sector, and housing may be intermingled with shops, factories, warehouses, and other commercial enterprises. Favelas and their counterparts can become centers of pride and support for their residents, where community work, folk belief systems, crafts, and music (for example, Favela Funk) flourish. Many of the best steel bands of Port of Spain, Trinidad, have their homes in the city's shantytowns.

Figure 3.30 A favela in Rio de Janeiro. A favela in Rio de Janeiro clings to a once-vacant hillside that was considered too steep for apartment buildings. Now home to thousands of people, it is still served mainly by footpaths and narrow alleyways. Water is only available sporadically, so many residents store water in tanks on their roof. [Mario Tama/Getty Images]

VIGNETTE Favelas are everywhere in Fortaleza, Brazil, where they provide a home to about 12 percent of the city's population. Fortaleza grew from about 1 million in the late 1970s to over 2 million by the year 2000. In 2018, there were more than 3.6 million residents, many of whom had fled drought and rural poverty in the interior of the country. During a period of rapid growth in the 1980s, city parks of just a square block or two in residential areas were invaded by squatters. In one upscale neighborhood, 10,000 people occupied a single park. Because of the lack of water and sanitation in the early days of the migration, migrants often had to relieve themselves on the street.

One day, while strolling on the Fortaleza waterfront, Lydia Pulsipher chanced to meet a resident of a beachfront favela who invited her to join him on his porch. There, he explained how he and his wife had come to the city 5 years before, after being forced to leave the drought-plagued interior when a newly built irrigation reservoir flooded the rented land their families had cultivated for generations. With no way to make a living, they set out on foot for the city. In Fortaleza, they constructed the building they used for home and work from objects they collected along the beach. Eventually, they were able to purchase roofing tiles, which gave the building an air of permanency. At the time of the visit, he maintained a small refreshment stand and his wife a beauty parlor that catered to women from the beach favelas. *[Source: Lydia Pulsipher's field notes, updated with the help of John Mueller.]* ■

ELITE LANDSCAPES

In contrast to shantytowns are districts in nearly every city that are modern, planned, and similar to elite urban districts across the world. Examples of such urban landscapes can be found in Rio de Janeiro, São Paulo, and other cities in Brazil's southeastern industrial heartland. These cities have districts that are elegant and futuristic showplaces, resplendent with the very latest technology and buildings and high-end shops that rival those in New York, Singapore, and even Tokyo. Known as **urban growth poles**, these areas attract investment, trade, educated immigrants, and overall economic development—all of which make them stand out from surrounding landscapes.

Planners throughout the region now acknowledge, however, that in the rush to develop modern urban landscapes, municipal governments and developers neglected to underwrite the parallel development of a sufficient urban infrastructure (for an exception, see the discussion of the southern city of Curitiba below).

In short supply are sanitation and water systems, an up-to-date electrical grid, transportation facilities, schools, adequate affordable housing, and medical facilities—all necessary to sustain modern business, industry, and a healthy and educated urban population. In recent decades, there have been massive investments in these systems, especially in the larger and wealthier countries, such as Brazil, Chile, Argentina, and Mexico, that can afford the improvements. These investments are done in recognition that human development through education, health services, and community building, whether privately or publicly funded, is a primary part of creating strong urban economies.

URBAN TRANSPORTATION

In large, rapidly expanding, partially unplanned urban areas with millions of poor migrants, transportation can be a special challenge. Favelas are often on the urban fringes, far from available low-skill jobs and ill-served by roads and public transportation (see Figure 3.30). Workers must make lengthy, time-consuming, and expensive commutes to jobs that pay very little. The entire urban population gets caught up in endless traffic jams that cripple the economy and pollute the environment.

A solution to part of this problem has come from the southern Brazilian city of Curitiba, which has carefully oriented its expansion around a master plan (dating from 1968) featuring bus rapid transit (BRT). Designed as an alternative to more expensive subways, the system uses dedicated bus lanes on the streets of Curitiba, and hundreds of minibuses that stop only at dedicated stations. Making 12,000 trips a day and bringing 1.3 million passengers from often remote neighborhoods to all parts of the city, the system cost 40 times less than a subway system. Being able to get to work quickly and cheaply has helped poor migrants find and keep jobs. The reduced use of cars means that emissions and congestion are lowered and urban living is more pleasant. Curitiba's system has been successfully imitated in 170 cities worldwide, including almost every major city in this region, as well as places like Pittsburg and Las Vegas, serving 33 million people per day.

urban growth poles locations within cities that are attractive to investment, innovative immigrants, and trade, and thus attract economic development like a magnet

CHECK YOUR UNDERSTANDING

1. Why does this region have such densely occupied urban landscapes?
2. Why do cities in this region often lack adequate support services and infrastructure?
3. What are some problems created by primate cities?
4. What factors that encourage rural–urban migration affect women differently than men?
5. How are favelas or shantytowns different from more planned communities?
6. What is one solution to problems related to urban transit that has spread throughout this region?

POPULATION, GENDER, AND CULTURE

3.7 Examine the factors that created a population explosion in this region during the early twentieth century, and reduced population growth in the late twentieth century.

3.8 Describe the major institutions that shape gender roles in this region.

3.9 Evaluate the forces challenging the dominance of the Roman Catholic Church in this region.

The political and economic changes discussed earlier in this chapter are reflected in shifts in the population, gender relations, and culture of this region. The early twentieth century saw a population explosion that trailed off by the late twentieth century for a number of reasons, one of which was changing gender roles. Change has also come to the social practices and structures that guide daily life in this culturally diverse region, including race relations and religious observance.

POPULATION TRENDS

By 2018, about 650 million people were living in Middle and South America, close to ten times the highest estimated population of the region in 1492. This reflects high birth rates and associated low human development that persisted throughout much of the twentieth century. In poor rural agricultural areas, children were seen as sources of wealth because they could do useful farm and household work at a young age and eventually would care for their aging elders. Infant death rates were also high, so some parents had four or more children to be sure of raising at least a few to adulthood. In this

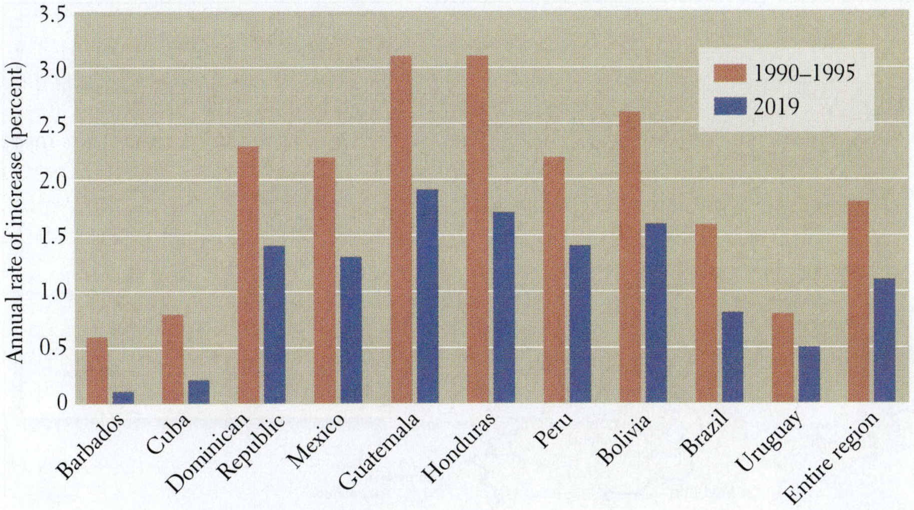

Figure 3.31 Trends in natural population increases, 1990–2018. The orange and blue columns show that rates of natural increase have declined steadily throughout the region. Although they are projected to continue to do so into the future, in many countries, natural population increase remains high enough to outstrip efforts to improve standards of living. Note that the rates of natural increase are for a few selected countries and for the entire region. [Research from: World Population Data Sheet for 2018, at https://www.prb.org/2018-world-population-data-sheet-with-focus-on-changing-age-structures/]

context women also had less access to education and effective birth control methods, and fewer reasons to delay childbirth. Another factor was the Roman Catholic Church's opposition to family planning.

By the 1970s, the region was undergoing a demographic transition (see Figure 1.28). Between 1975 and 2018, the annual rate of natural increase (the birth rate minus the death rate) for the entire region fell from about 1.9 percent to 1.1 percent—a rate of growth lower than the world average (**Figure 3.31**). Some countries are further along this transition, as reflected in the population pyramids for Brazil and Guatemala shown in **Figure 3.32**. Brazil's pyramid is narrower at the bottom than in the middle, indicating an aging population and a sharply declining birthrate. Guatemala's pyramid is widest at the bottom, indicating a younger population with birth rates that are declining more gradually.

The changes outlined in Figures 3.31 and 3.32 reflect gains in human development. Better access to food, shelter, sanitation, and medical care reduced infant mortality, encouraging couples to have only two or three children instead of five or more. As populations urbanized and economies grew, women gained better access to education and employment outside the home, and so delayed childbearing and had smaller families. This trend is continuing as women move into high-level jobs, such as management positions in corporations and government, that traditionally have gone to men. Perhaps the most visible example of this is the women who have become presidents of Chile, Argentina, and Brazil in the last decade.

Because 25 percent of the region's population is under the age of 15, even if couples have only one or two children, the population will continue to grow as this large group reaches the age of reproduction. Population projections for the year 2050 are for 783 million people, and supplying this extra 133 million people with food, water, homes, schools, and hospitals will be a challenge.

POPULATION DISTRIBUTION

The population density map of this region reveals a very unequal distribution of people. Comparing the population density map

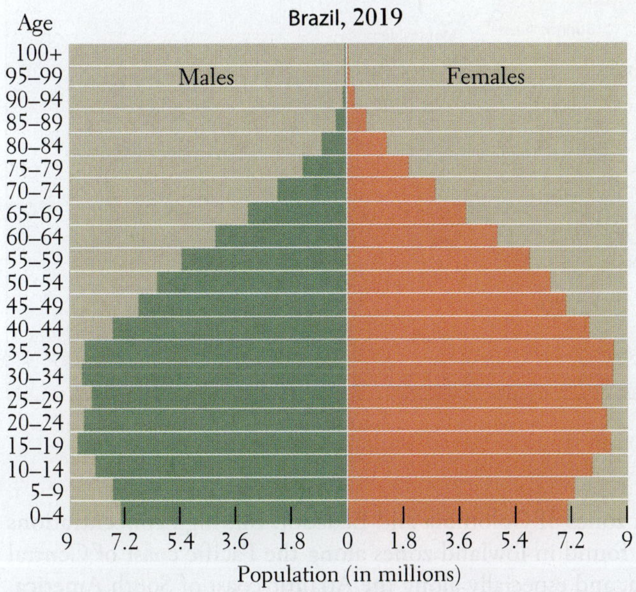

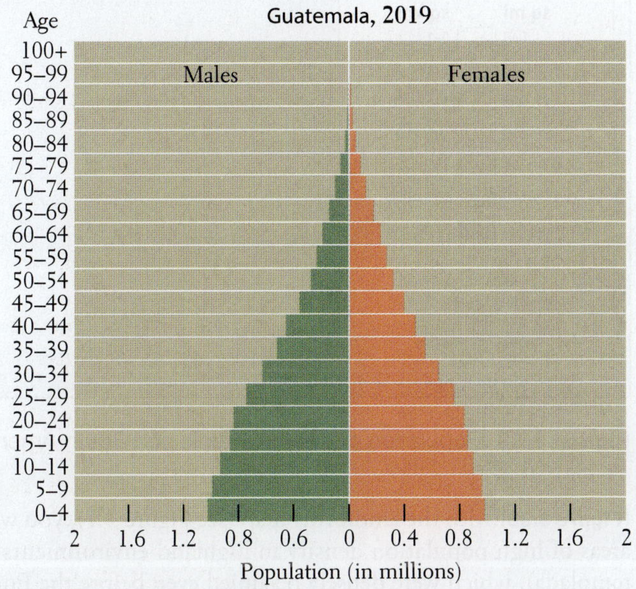

Figure 3.32 Population pyramids for Brazil and Guatemala, 2019. Note that the pyramids for these countries are in two different scales.

Figure 3.33 Population density in Middle and South America.

(**Figure 3.33**) with the landforms map (see Figure 3.1), you will find areas of high population density in highland environments (tierra templada), which were densely occupied even before the European conquest, such as Mexico City, highlands in Guatemala, and the Andean zones of Colombia and Ecuador. But high concentrations are also found in lowland zones along the Pacific coast of Central America, and especially along the Atlantic coast of South America. Most coastal lowland concentrations, in the tierra caliente, are near seaports with vibrant economies and a cosmopolitan social life that attracts people.

CHECK YOUR UNDERSTANDING

1. What factors created a population explosion in this region during the early twentieth century?
2. What factors reduced population growth in the late twentieth century?
3. Why will populations continue to grow in this region for the foreseeable future?

GENDER

Gender roles are strongly influenced by family and religion, both of which are conservative influences in this rapidly changing region. The basic social institution is the *extended family*, which includes cousins, aunts, uncles, grandparents, and more distant relatives. It is via more traditional notions of this institution that gender roles are assigned and enforced. For example, it is generally accepted that the individual should sacrifice many of his or her personal interests to those of the extended family and community and that individual well-being is best secured by doing so. For women this often means pursuing less career-oriented part-time or informal employment so as to have more time to be a homemaker. For men this can mean going out into society and making a living by whatever means necessary, even if it involves lengthy travel and long periods away from family and friends to whom money is sent.

The traditional arrangement of domestic spaces and patterns of socializing illustrates strong family ties that often affect women disproportionately. Extended families frequently live together in domestic compounds of several houses surrounded by walls or hedges. Among other things this facilitates the supervision and socialization of younger generations, with females more involved in cooking and domestic maintenance, and males given a wider range of activity outside of the home. The supervision of the extended family extends into public spaces, where social groups are likely to be family members of several generations rather than unrelated groups of single young adults or married couples, as is often the case in the United States. A woman's best friends are likely to be her female relatives. A man's social or business circles will include male family members or long-standing family friends.

Gender roles in the region are also strongly influenced by the Roman Catholic Church. The patron saint of almost every country in the region is a form of the Virgin Mary, who is held up as the model for women to follow through a set of values known as ***marianismo***, which emphasizes virginity, motherhood, chastity, and service to the family (**Figure 3.34A**). In this still-widespread tradition, the ideal woman is the day-to-day manager of the house and of the family's well-being. She trains her sons to enter the wider world and her daughters to serve within the home. Over the course of her life, a woman's power increases as her skills and sacrifices for the good of all are recognized and enshrined in family lore.

Her husband, the official head of the family, is expected to work and to give most of his income to his family. Still, men have much more autonomy and freedom to shape their lives than women because they are expected to move about in the larger community and establish relationships, both economic and personal. A man's social network is considered just as essential to the family's prosperity and status in the community as is his work.

Very different sexual standards for males and females have evolved in this region. While all expect strict fidelity from a wife to a husband, a man is much more free to engage in multiple simultaneous relationships with the opposite sex. Males measure themselves by the model of **machismo**, in which manliness is considered to consist of honor, respectability, fatherhood, household leadership, attractiveness to women, and the ability to be a charming storyteller. Traditionally, the ability to acquire money was secondary to other symbols of maleness. Increasingly, however, a new, market-oriented culture prizes visible affluence as a desirable male attribute.

LGBTQ Issues

While a general inclination to keep nonheterosexual relationships "in the closet" persists, over the last decade attitudes have begun to change. Respect for the rights of gays and lesbians has grown significantly in recent years: Argentina legalized same-sex marriage in 2010, as did Brazil and Uruguay in 2013, and Mexico in 2015, but homosexuality is still illegal in Guyana and Belize. Even in more accepting countries, social attitudes toward LGBTQ people are often complex, contradictory, and ultimately disapproving. In Brazil, state legislator and pastor Sargento Isidório, who describes himself as an "ex-homosexual," was reelected in a landslide and gained national notoriety in 2014 when he blamed the recent drought in São Paolo on the city's LGBTQ "Pride Parade," which, not incidentally, is also the world's largest such event, drawing more than 3 million people annually. Throughout the region, violence against LGBTQ people is common and is rarely addressed by law enforcement. Brazil's president Jair Bolsonaro boasts of being a "proud homophobe" who would rather his son "die in a car crash" than be gay. Transgender people are seen as running particularly afoul of the ideals of machismo. Officials of the Catholic Church are often among the most outspoken in their disapproval of LGBTQ people. For example, Chile's Cardinal Jorge Medina stated in 2010 that "if a person has a homosexual tendency it is a defect, like missing an eye, a hand, a foot."

CULTURE

The region of Middle and South America is culturally complex because numerous distinct indigenous groups were already present when the Europeans arrived, and many other cultures were introduced during and after the colonial period.

From 1500 to the early 1800s, some 10 million people from many parts of Africa were brought to plantations on the islands and in the coastal zones of Middle and South America. After the emancipation of enslaved peoples of African descent in the

marianismo a set of values based on the life of the Virgin Mary, the mother of Jesus, that defines the proper social roles for women in Middle and South America

machismo a set of values that defines manliness in Middle and South America

Population, Gender, and Culture in Middle and South America

The political and economic changes this region has seen are reflected in shifts in population growth, gender relations, and culture. Population growth has trailed off largely due to higher human development combined with changing gender roles. The map of the Gender Development Index (GDI) shows disparities in the Human Development Index (HDI) by gender. (HDI values are estimated separately for women and men, and the ratio of these two values is the GDI.) The closer the ratio is to 1, the smaller the gap between women and men. The map shows much of South America ranking fairly high on GDI comparable to North America and Europe, while some Andean countries, most of Central America, and Mexico stand out as the least equitable across genders.

THINKING GEOGRAPHICALLY

A How might the ideals of *marianismo* contribute to a lower GDI for El Salvador?

B What ideals of machismo might this man exemplify?

C How does the story of Marielle Franco suggest both rejection and acceptance of women and LGBTQ people in positions of power?

A Marianismo in El Salvador. A girl in El Salvador poses for a photo in front of the Virgin of Guadalupe, who is revered as a symbol of *marianismo* in Central America and Mexico. [JOSE CABEZAS/AFP/Getty Images]

B Machismo in Mexico. A cowboy in Mexico exemplifies the values of machismo. [Marc DEVILLE/Gamma-Rapho via Getty Images]

C Gender and violence in Brazil. A woman holds a photo of Marielle Franco, a black, bisexual councilwoman from a favela in Rio de Janeiro, who was assassinated after giving a speech about killings by the police. [MAURO PIMENTEL/AFP/Getty Images]

British-controlled Caribbean islands and Guyana in the 1830s, more than half a million Asians were brought there from India, Pakistan, and China as indentured agricultural workers. Their cultural impact remains most visible in Trinidad and Tobago, Jamaica, and the Guianas. In some parts of Mexico, Central America, the Amazon Basin, and the Andean Highlands, indigenous people have remained numerous. To the unpracticed eye, they may appear little affected by colonization, but this is not the case.

In the Caribbean and along the east coast of Central America as well as the Atlantic coast of Brazil, mestizos (people who have a mixture of African, European, and some indigenous ancestry) make up the majority of the population. Some of the most colorful festivals in the Americas, such as versions of Mardi Gras, known here as Carnival, are based on a melding of cultural practices from these three heritages (**Figure 3.35**).

In some areas, such as Argentina, Chile, and southern Brazil, people of Central European descent are also numerous. The Japanese, though a tiny minority everywhere in the region, increasingly influence agriculture and industry, especially in Brazil, the Caribbean, and Peru. Alberto Fujimori, a Peruvian of Japanese descent, campaigned on his ethnic "outsider" status to become president of Peru during a time of political upheaval.

In some ways, diversity is increasing as the media and trade introduce new influences from abroad. At the same time, the processes of **acculturation** and **assimilation** are also accelerating, thus erasing diversity to some extent as people adopt new ways. This is especially true in the biggest cities, where people of widely different backgrounds live in close proximity to one another.

acculturation adaptation of a minority culture to the host culture enough to function effectively and be self-supporting; cultural borrowing

assimilation the loss of old ways of life and the adoption of the lifestyle of another culture

Race and Skin Color

People from Middle and South America, especially those from Brazil, often proudly claim that race and color are of less consequence in this region than in North America. They are only partially right.

Figure 3.35 LOCAL LIVES: Festivals in Middle and South America

(A) A member of a *bateria*, or percussion band, during Rio de Janeiro's Carnival, the largest in the world. Carnival, which is celebrated in much of the United States as Mardi Gras, takes place in the days and weeks before Lent, the 40-day period in Christian traditions of fasting and abstinence that precedes Easter. Elaborate celebrations take place in Barranquilla in Colombia, Port-au-Prince in Haiti, Port of Spain in Trinidad, and Montevideo in Uruguay. [Viviane Ponti/Lonely Planet Images/Getty Images]

(B) A procession commemorating Jesus walks on an *alfombra*, or carpet, made of colored sawdust, painstakingly drawn by hand on the streets of Antigua, Guatemala, during *Semana Santa*, or Holy Week, the last week of Lent. The processions, which will destroy the alfombras, originate from the Mayan practice of making elaborate designs with flowers and feathers for royalty to walk on as they made their way to important ceremonies. [Danita Delimont/Gallo Images/Getty Images]

(C) A couple celebrates Día de los Muertos—Day of the Dead—at a cemetery in Metepec, Mexico. Throughout Middle America on November 1, families gather in cemeteries to build private altars dedicated to deceased loved ones. The holiday originates from celebrations honoring the Aztec goddess Mictecacihuatl, Queen of the Underworld, whose job is to watch over the bones of the dead. [MARIO VAZQUEZ/AFP/Getty Images]

In all of the Americas, skin color is now less associated with status than in the past. By acquiring an education, a good job, a substantial income, the right accent, and a high-status mate, a person of any skin color may become recognized as upper class.

Nevertheless, the ability to reduce the significance of skin color through one's actions is not quite the same as race having no significance at all. Overall, those who are poor, less educated, and of lower social standing tend to have darker skin than those who are educated and wealthy. And while there are poor people of European descent throughout the region, most light-skinned people are middle and upper class. In Brazil, where it is popular to downplay the significance of race, the average black worker earns 60 percent of what the average white worker earns. By contrast, in the United States, the average black worker earns 65 percent of what the average white worker earns. Indeed, race and skin color have not disappeared as social factors in this region any more than in the United States. In some countries—Cuba, for example, where overt racist comments are socially unacceptable—it is common for a speaker to use a gesture (tapping his or her forearm with two fingers) to indicate that the person referred to in the conversation is of African descent.

Religion in Contemporary Life

Judaism, Islam, Hinduism, and indigenous beliefs are found across the region, but the majority of the people are at least nominal Christians, most of them Roman Catholic. While the Roman Catholic Church remains highly influential and relevant to the lives of believers, it has had to contend with popular efforts to reform it, as well as with rising competition from other religious movements. From the beginning of the colonial era, the church was the major partner of the Spanish and Portuguese colonial governments. It received extensive lands and resources from colonial governments, and in return built massive cathedrals and churches throughout the region, sending thousands of missionary priests to convert indigenous people. The Roman Catholic Church encouraged these people to accept their low status in colonial society, to obey authority, and to postpone rewards for their hard work until heaven.

People throughout the region converted to the faith, some willingly and some only after the threats, intimidation, and torture of the Spanish Inquisition were brought to the Americas. Many indigenous people put their own spin on Catholicism, creating multiple folk versions of the Mass with folk music, more participation by women in worship services, and interpretations of Scripture that vary greatly from European versions. A range of African-based belief systems (Candomblé, Umbanda, Santería, Obeah, and Voodoo) combined with Catholic beliefs can be found in Brazil, northern South America, Middle America, and the Caribbean—wherever the descendants of Africans have settled. These African-based religions have attracted adherents of European or indigenous backgrounds as well, especially in urban areas.

liberation theology a movement within the Roman Catholic Church that uses the teachings of Jesus to encourage the poor to organize to change their own lives and to encourage the rich to promote social and economic equity

The power of the Roman Catholic Church began to erode in the nineteenth century in places such as Mexico, where social movements aimed at addressing the needs of the poor masses seized and redistributed some church lands. They also canceled the high fees the clergy had been charging for simple rites of passage such as baptisms, weddings, and funerals. Over the years, the Catholic Church became less obviously connected to the elite and more attentive to the needs of poor and non-European people. By the mid-twentieth century, the church was abandoning many of its earlier racist policies, for example by ordaining clergy of African and indigenous descent. Women were also given more of a role in religious ceremonies.

In the 1970s, a Catholic movement known as **liberation theology** was begun by a small group of priests and activists wanting to combat the extreme inequalities in wealth and power common in the region. The movement portrayed Jesus Christ as a social revolutionary who symbolically spoke out for the redistribution of wealth when he divided the loaves and fish among the multitude. The perpetuation of gross economic inequality and political repression was viewed as sinful, and social reform as liberation from evil.

The legacy of liberation theology is now fading. At its height in the 1970s and early 1980s, liberation theology was the most articulate movement for region-wide social change. It had more than 3 million adherents in Brazil alone. But the Vatican objected to this popularized version of Catholicism, and its influence diminished. In countries such as Guatemala, El Salvador, and Argentina, governments targeted liberation theologists, vilifying them as communist collaborators; hundreds were tortured and assassinated. Liberation theology has also had to compete with newly emerging evangelical Protestant movements.

The Americas' First Pope The election of Argentina's Cardinal Jorge Bergoglio to become the leader of the Catholic Church in 2013 reflected realities both in this region and outside of it. Pope Francis, as Bergoglio is now known, is the first non-European pope in almost 13 centuries, and the first ever from the Americas. His election reflects the dramatic decline of Europe's share of the world's Catholics, down from 65 percent in 1910 to only 24 percent in 2010, and the rise of Middle and South America's share of the world's Catholics, up from 24 percent in 1910 to 39 percent in 2010. Pope Francis was also elected by a group of Catholic cardinals who want to reform the church and the Vatican bureaucracy in light of the recent financial corruption and sexual abuse scandals that have damaged the Church's image around the world.

While Francis's election was widely celebrated in Argentina, he was also criticized by some for his complicity in Argentina's Dirty War. As the head of the Jesuit order in Argentina, Francis—then known as Father Bergoglio—was accused in 1985, 2005, and again in 2010 of handing over to the military Catholic priests and laypeople engaged in liberation theology, some of whom were never seen again, and of doing comparatively little to shield others from the dictatorship's death squads.

Bergoglio denied these accusations, but such behavior would not have been out of the ordinary for someone in his position, as most of Argentina's Catholic clergy publicly supported the dictatorship and supplied it with information about supposed terrorists and political activists who opposed the regime. In this they were joined by most of Argentina's elite, including wealthy landowners, industrialists, and the managers of most newspapers. The Argentine Catholic clergy's collaboration with the military, which it formally apologized for in 1996, stands in contrast to the actions of Brazil and Chile's Catholic clergy, who were often highly critical of military governments that perpetrated violence similar to Argentina's Dirty War during the 1960s, 1970s, and 1980s.

Evangelical Protestantism **Evangelical Protestantism** has spread from North America into Middle and South America, and is now the region's fastest-growing religious movement. About 10 percent of the population, or at least 50 million people, are adherents. The movement is growing rapidly in Brazil, Chile, and Middle America, especially among poor and middle classes in both rural and urban settings. In contrast to liberation theology's emphasis on combating the region's extreme inequalities, evangelical Protestants often teach a *gospel of success*, stressing that those true believers who give themselves to a new life of hard work and clean living will experience prosperity of the body (wealth) as well as of the soul.

With its focus on personal salvation and empowerment of the individual through miraculous healing and psychological transformation, evangelical Protestantism is less hierarchical than the Roman Catholic Church. Instead of a central authority to the movement there are a host of small, independent congregations led by entrepreneurial individuals who may be either male or female.

Perhaps two of the most important contributions of evangelical Protestantism are the focus (as in Umbanda and other African-based religions) on helping people, especially urban migrants, cope with the strains of modern life; and the emphasis placed on involving women in leadership roles in church activities and services.

CHECK YOUR UNDERSTANDING

1. What makes this region so culturally complex?
2. What is the basic social institution in the region?
3. How have the ideals of *marianismo* and *machismo* influenced gender identities in this region?
4. What is the fastest-growing religious movement in this region?

SUBREGIONS OF MIDDLE AND SOUTH AMERICA

3.10 Compare how the subregions of Middle and South America have been influenced differently by foreign powers.

This tour of the subregions of Middle and South America focuses primarily on the themes of cultural diversity, economic disparity, and environmental deterioration. These themes are reflected somewhat differently from place to place. For each subregion, examples of connections to the global economy are given.

THE CARIBBEAN

There are many sides to the Caribbean. There are the large island nations of Haiti and the Dominican Republic that are plagued by poverty, instability, and low human development. There is Cuba, which has charted its own course to achieve medium levels of human development. And there are the smaller islands, which also have a strong record of fostering human development. Visitors to Cuba or the smaller islands are often struck by the ramshackle, lived-in landscapes of the islands that don't match the tourist ads. Often isolated in resorts or cruise ships, tourists rarely get to learn that the humble houses, garden plots, and quaint, rutted, narrow streets mask social and economic conditions supportive of a modest but healthy and productive way of life.

Political, Social, and Economic Change

Over the past half-century, most islands have become independent, self-governing states (**Figure 3.36**). Outside of Haiti and parts of the Dominican Republic, these islands are no longer the poverty-stricken places they were 40 years ago. Children go to school, and literacy rates for all but the elderly average close to 95 percent. There is basic health care: mothers receive prenatal care, nearly all babies are born in hospitals, infant mortality rates are low, and diseases of aging are competently treated. Life expectancy is in the 70s, the overall rate of population increase for the Caribbean is the lowest in this region, and some islands do particularly well in making opportunities available to women. Returned emigrants often say that the quality of life on their home Caribbean islands actually exceeds that of wealthier societies because life is enhanced by strong community and family support. And, of course, there is the beautiful natural environment.

Island governments continually search for ways to turn former plantation economies, once managed from Europe and North America, into more self-directed, self-sufficient, and flexible entities that can adapt quickly to the perpetually changing markets of the global economy. High-earning sectors such as tourism or specialty agriculture are tempting, but island economists (such as Sir Arthur Lewis, who won the Nobel Prize in Economics in 1979) are wary of the dependence and vulnerability that too much specialization can bring. When plantation cultivation of sugar, cotton, and *copra* (dried coconut meat) died out in the 1960s, some islands turned to producing "breakfast" crops, such as bananas and coffee, which they sold for high prices under special agreements with the countries that once held them as colonies. Now these protections are disappearing as free market agreements in the European Union make such special arrangements illegal. Other island countries turned to the processing of their special resources (petroleum in Trinidad and Tobago, bauxite in Jamaica), the assembly of such high-tech products as computer chips and pharmaceuticals (in St. Kitts and Nevis), or the processing of computerized data (in Barbados). Most islands combine a changing array of one or more of these strategies with tourism development.

In 2016, tourism and related activities contributed at least 60 percent of the gross national product in island countries such as Antigua and Barbuda, Barbados, the Bahamas, and St. Martin. In most years, Antigua and Barbuda (population 91,000) hosts more than four times its population in tourists, and St. Martin (population 71,000) hosts 47 times its population. Heavy borrowing to import food as local agriculture declines and to build hotels, airports, water systems, and shopping

evangelical Protestantism a Christian movement that focuses on personal salvation and empowerment of the individual through miraculous healing and transformation

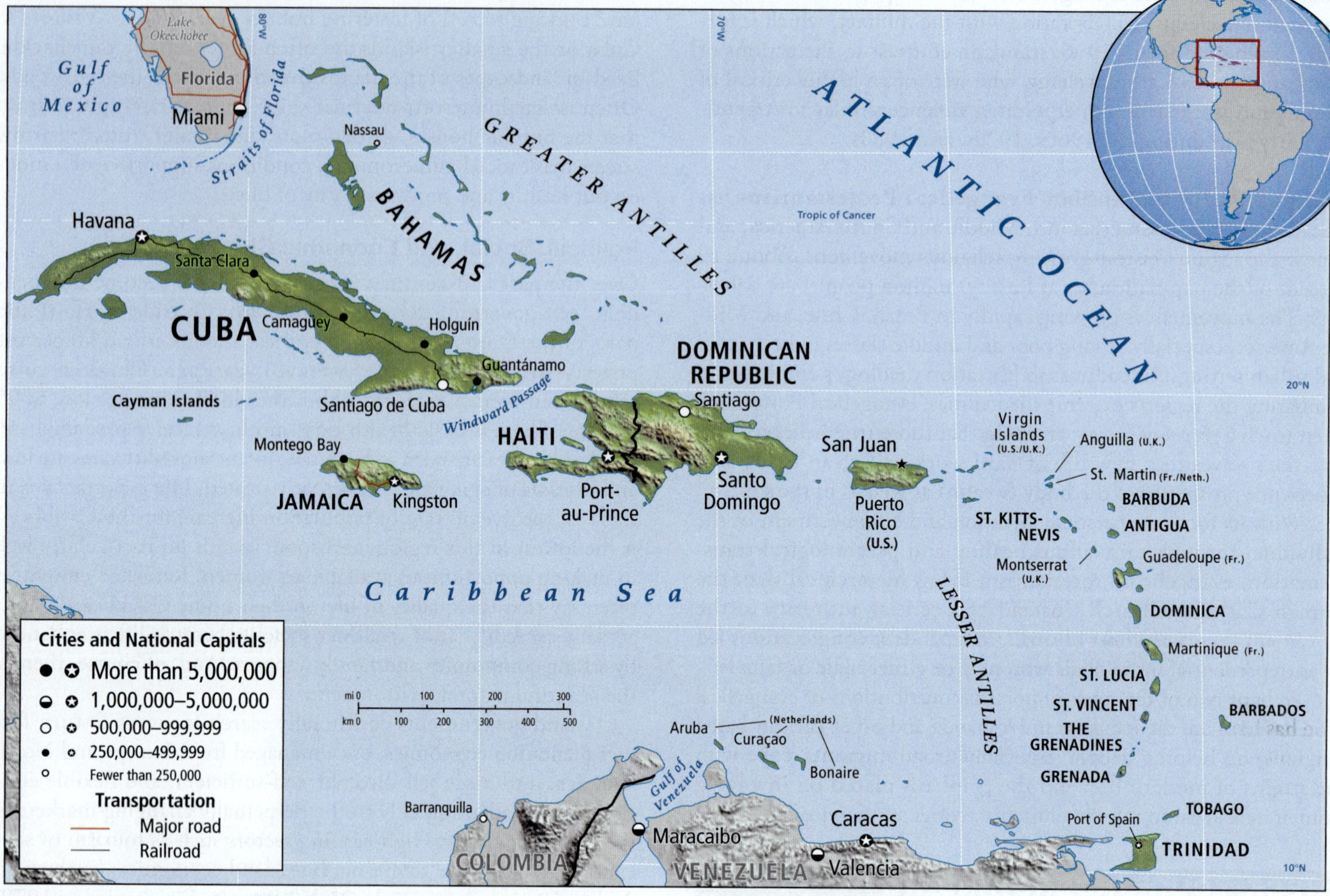

Figure 3.36 The Caribbean subregion.

centers for tourists has left some islands with debts many times their income. Then there is the stress that comes from dealing perpetually with hordes of strangers ignorant of local customs and making erroneous assumptions about the state of human development (see map of HDI in Figure 3.21).

Tourism is often difficult to regulate, in part because it is a complex multinational industry controlled by North American and European travel agencies, airlines, cruise ship lines, and resort owners. Large cruise ships now routinely bring several thousand passengers into small port cities such as Castries or Soufrière in St. Lucia, often overwhelming them with garbage and sewage. Many islands look for tourists who will stay on shore for a week or more and develop a deeper interest in island societies. Ecotourism, sport tourism (such as small-boat sailing), and special-interest seminar tourism (such as cooking, art, yoga, and physical fitness retreats) are being pursued as low-density alternatives to the "sand, sea, and sun" mass tourism that has brought the majority of visitors to the region.

Cuba and Puerto Rico Compared

Cuba and Puerto Rico shared a common history until the 1950s and then diverged starkly. U.S. investors dominated both islands after the end of Spanish rule around 1900, buying plantations and developing resorts and other businesses on both islands. These investors and the U.S. government pushed local dictatorial regimes to keep labor organization at a minimum and popular political movements under tight control. By the 1950s, poverty was widespread and human development low, with most people working as agricultural laborers.

Cuba In 1959, Cuba went through a socialist revolution (see Figure 3.26C) that dramatically improved human development, despite persistently rather low GNI per capita (see Figure 1.19). To many in the region Cuba proved that, contrary to the dictates of neoliberalism, large investments in human development could successfully be made by relatively poor countries. Others were horrified by the Cuban revolution, especially the confiscation of property from the former aristocracy and jailing of political dissidents.

Intense hostility from the United States encouraged Cuba to seek help from the Soviet Union to achieve its social revolution. It provided cheap fuel, significant technical and military assistance, and it bought Cuba's main export crop, sugar, at artificially high prices. With the demise of the Soviet Union in 1991, Cuba's economy declined sharply.

To survive this economic crisis, Cuba sought European investors to help redevelop its tourism industry. Medical tourism has developed because early on in the revolution effective investments were made in high-quality medical and scientific education. More than 5000 patients, mostly from South America, come each year for cosmetic and other procedures. Canada has invested in joint biotech research on hepatitis and cancer drugs with Cuban medical scientists. In exchange for Cuban pharmaceuticals, high-tech medical equipment, and the services of medical social workers and doctors, for years Venezuela furnished Cuba with around 50,000 barrels of oil a day (2018) in a similar exchange. Oil was discovered off Cuba's north shore in 2012, and oil and gas production could become a major component of Cuba's economy in the near future, as numerous oil exploration firms from around the world are already vying for development rights.

Leisure tourism remains Cuba's biggest source of income, earning U.S.$3 billion in 2018 from 4.1 million visitors, mostly Canadians and Europeans, who usually spend a week or more at luxury beach resorts. While a growing number are making a decent living working in the tourism industry (nearly 500,000 jobs are connected in some way to tourism), incomes remain low for most Cubans. For them, consumer goods and even food remain in short supply.

Political relations with the United States remained icy until 2009, when President Obama loosened the rules on travel from the United States, and opened formal diplomatic relations. Since then little has happened, with a trade embargo still in effect. Meanwhile, governments in Europe and the Americas continue to trade with and invest in Cuba.

Puerto Rico As a commonwealth within the United States, Puerto Rico underwent a gradual capitalist metamorphosis from a sugar plantation economy to a manufacturing center. This began in the 1950s with Operation Bootstrap, a government-aided program to transform the island's economy with tax breaks and subsidies offered to international companies to locate on the island. By 1965, assembly plants, petroleum processing, and pharmaceutical manufacturing had transformed the economy, but also heavily polluted the island's coast.

In the 1990s, as part of NAFTA, the U.S. Congress began to eliminate the tax advantages of doing business in Puerto Rico, resulting in many industries leaving. This caused a recession and substantial migration to the United States, where there are now one-and-a-half times as many Puerto Ricans as there are on the island. The remittances they send home, as well as the remaining manufacturing jobs and poverty alleviation programs of the U.S. federal government, have maintained a level of human development comparable to Cuba. However, government expenses remain high for social services and pension obligations and so the island sinks ever deeper into debt—U.S.$133 billion as of 2018. Outside of San Juan, with its skyscraper hotels, Puerto Rico's landscape reflects a shortage of well-paying jobs, an inadequate transportation system, mediocre schools, and few opportunities for advanced training, all of which encourage migration to the mainland.

Some Puerto Ricans support independence from the United States, arguing that this would open some financial possibilities (like the right to declare bankruptcy), and bring closer ties to Middle and South America. Others advocate for statehood for Puerto Rico, which would mean higher status in the United States and perhaps some advantages to its tourism industry.

Haiti and Barbados Compared

Haiti and Barbados present another study in contrasts (**Figure 3.37**). During the colonial era, both were European possessions with plantation economies—Haiti a colony of France and Barbados a colony of Britain. Today they are far apart, with Haiti having the lowest human development in the region, and Barbados among the highest, even though it has much less space, a higher population density, and few resources other than its limestone soil, beaches, and people.

Figure 3.37 Contrast in the Caribbean: Haiti and Barbados. While both islands had similar beginnings as European colonies, their present situations are very different.

(A) Haiti has the lowest human development in the region and suffers political instability that has brought repeated military intervention by outsiders. Shown here are UN troops from Brazil distributing water in a slum in Port-au-Prince. [VANDERLEI ALMEIDA/AFP/Getty Images]

(B) Barbados, on the other hand, is politically stable and has a much more prosperous economy based on services (especially tourism) and export-oriented light manufacturing. Barbadians are generally highly educated and often have time to enjoy spectator sports like horse racing. [John Warburton-Lee/AWL Images/Getty Images]

Haiti At the end of the eighteenth century, Haiti had the richest plantation economy in the Caribbean. When Haitian slaves revolted against the brutality of the French planters in 1804, Haiti became the first colony in Middle and South America to become independent. However, Haiti's early promise was lost as violent and corrupt leaders neither reformed the exploitative plantation economy nor sought a new economic base. Under a long series of incompetent and avaricious authoritarian governments, the people sank into abject poverty, while the land was badly damaged by particularly wasteful, unprofitable plantation cultivation. In the middle of the twentieth century, a class-based reign of terror under François "Papa Doc" Duvalier (1957–1971) and his son Jean-Claude "Baby Doc" Duvalier (1971–1986) pitted the mulatto elite against the black poor. They were followed by a series of weak leaders who promised social reform but delivered little.

Today, Haiti's economy is weak but improving. In the 1980s, multinational corporations opened maquiladora-like assembly plants, employing primarily young women. A violent coup in 1991 brought human rights abuses and a UN trade embargo that forced the closure of many plants. Manufacturing is only now recovering, with many clothing plants and a few high-tech assembly plants. Minerals such as bauxite, copper, and tin exist in Haiti, but infrastructure is so underdeveloped that profitable mining is not yet possible. Haiti's lands are largely deforested and eroded and subject to disastrous flooding (see Figure 3.14C) that constrains agriculture.

Political stability is low and corruption rampant, with efforts to hold fair elections often devolving into violence (see the Figure 3.26 map). Since the early 1990s the UN has maintained peacekeeping troops in Haiti, and several humanitarian aid organizations, in addition to the U.S. government, run programs there. The United States and other foreign powers increased their presence after four hurricanes struck the island in 2008 and a massively devastating earthquake killed 160,000 in 2010 and left 1.5 million homeless. A cholera epidemic then struck, due to damaged infrastructure, killing another 10,000. $9.5 billion in emergency aid from across the globe began to make a difference, in part because the scope of the disaster weakened sources of resistance to change. However, corrupt local officials demanded large fees just to allow in medicine and other relief aid, and most of this aid was administered by foreigners, so a lot of money went to pay foreign consultants.

Barbados Tiny but far more prosperous, Barbados has fewer natural resources and is more than twice as crowded as Haiti. Although both Haiti and Barbados entered the twentieth century with largely illiterate, agricultural populations, their development paths have diverged sharply. In 2018, only 60 percent of Haitians were literate, while Barbados had 100 percent literacy and a diversified economy that includes tourism, sugar production, remittances from migrants, information processing, offshore financial services, and modern industries that sell products throughout the Caribbean. Barbados's prosperity is explained in part by the fact that its citizens successfully pressured the British government to invest in the people and infrastructure of its colony before giving it independence in 1966. Barbadians hold jobs requiring sophisticated skills; they are well educated and well fed, and most are homeowners. Furthermore, the Barbadian government and private businesspeople constantly seek new employment options for the citizens and occupy a central role in Caribbean economic and social development.

CHECK YOUR UNDERSTANDING

1. What are some upsides and downsides of tourism for Caribbean countries?
2. How did Cuba and Puerto Rico's development diverge after 1959?
3. In what way is Haiti's dysfunction atypical of the Caribbean?

MEXICO

Mexico is transitioning into a reasonably well-managed, middle-income democracy (**Figure 3.38**), although corruption and a drug war threaten the country's future (see Figures 3.26B and 3.27). Mexico's economic integration via NAFTA/USMCA with the United States and Canada since 1994 has resulted in both economic growth and a stream of legal and undocumented migrants into the United States (though migration has slowed down and even reversed in recent years).

Formal and Informal Economies

Mexico's formal economy has modernized rapidly in recent decades. Its main components are mechanized, export-oriented agriculture, the region's largest manufacturing sector, petroleum extraction and refining, a large service sector, and the remittances of millions of migrants working in the United States. Mexico's large and varied informal economy employs about half of the total working population of 52 million, and contributes at least 13 percent of the GDP through such activities as subsistence agriculture, street vending, trash recycling, a large black market, and extensive criminal activity.

Geographic Distribution of Economic Sectors

Agriculture employs about 13 percent of Mexico's population, but produces only 4 percent of Mexico's GDP. Only 12 percent of Mexico's land is good for agriculture, and much of that is very hilly by U.S. standards. In the north the old inefficient haciendas and ranches have been replaced by large corporate farms, often run by American companies that employ only a few laborers and technicians. With the aid of government marketing, research, mechanization, and irrigation programs, their high-quality meat and produce goes primarily to the U.S. market. In the tropical coastal lowlands farther south, subsistence cultivation on small plots is much more common, but here, too, tropical fruits and products such as jute, sisal, and tequila are produced for export as well as for internal consumption. In Mexico's convoluted mountains, many small family farms produce a wide variety of crops, including coffee, corn, tomatoes, sesame seeds, hot peppers, and flowers.

The industrial (or secondary) sector employs about 24 percent of Mexico's registered workforce and accounts for 37 percent of the GDP. Mexico's profitable petroleum-refining industry is located along the Gulf coastal plain, while the maquiladoras are concentrated along the U.S. border, where large cities such as Tijuana, Mexicali, Ciudad Juárez, and Reynosa, as well as many smaller ones

Figure 3.38 The Mexico subregion. **A** The U.S.–Mexico border barrier that divides Nogales, Arizona, from Nogales, Mexico. [Scott Olson/Getty Images]

provide an inexpensive labor pool within easy reach of the U.S. market (**Figure 3.39**). In 2018, Mexican factory workers earned an average of about U.S.$20 per day ($2.45 per hour). Compared to the United States, Mexico has far fewer regulations covering worker safety, fringe benefits, and environmental protection, and it charges much lower taxes on industries.

Women have been the employees of choice in maquiladoras because they are more compliant than men, they work on short-term contracts for less pay and, if fired, they leave quietly. Thousands of young women previously sheltered by close-knit families now live in precarious, crowded circumstances where they are routinely sexually harassed by employers, domestic partners, gangs, and drug cartels. Nonetheless, young women continue to migrate to the maquiladoras because their families need the money they send home, which also gives them a higher status in the family than earlier generations of women had.

Today, services employ 60 percent of Mexico's population and produce 62 percent of the GDP. Tourism is a major part of the service sector, with Mexico the sixth most visited country in the world. Tourism is strongly influenced by the United States, which supplies most visitors. Other major service sector areas, such as retail and finance, are similarly tied in with the United States, but mostly through ownership, as with many large banks, and the presence of U.S. companies, such as Walmart. Much service sector employment is also in small operations that cater to everyday needs: shoe repair, internet cafés, cleaning services, beauty parlors—many in the informal economy.

Migration from the Mexican Perspective

The movement of Mexicans to the United States to find work is as controversial in Mexico as it is in the United States because Mexico is losing many citizens (and parents) during their most productive years. Parents leave children to be raised by grandparents or other kin. And there is worry over the safety of family members while crossing the border and when on the job.

Unlike the many Europeans who cut ties with their families and native land when they migrated to North America, Mexicans often undertake their migrations with the express purpose

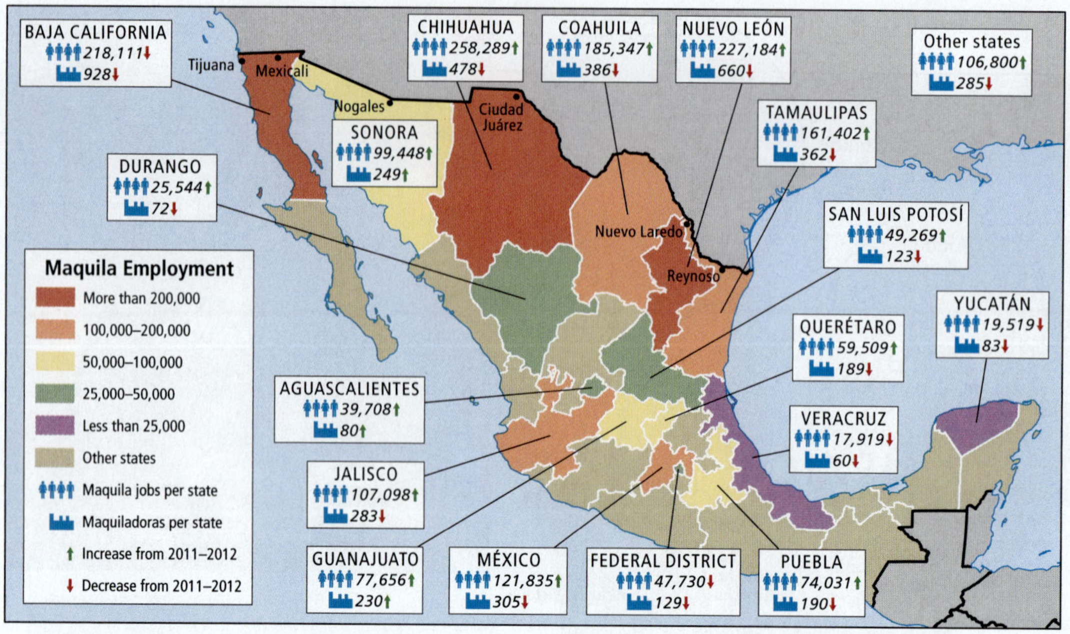

Figure 3.39 Location of maquiladoras in Mexico. The number of facilities and jobs in each state is shown on the map.

of helping out their families and home communities. In recent years, Mexican workers in the United States remitted an estimated U.S.$20 billion annually to their home communities (see "Migration and Remittances"). Most Mexican households receiving remittances are located in relatively better-off states such as Michoacán and Zacatecas, because residents of the poorest states find it difficult to finance a migration to the United States.

Migration to the United States has been declining in recent years, and as more Mexicans receive education it will decline further. Those with more than a high school education are much less likely to migrate than those with just 9 to 12 years of schooling. Since about 49 percent of the Mexican population already uses the internet (see Figure 3.23), government programs are trying to expand use of the internet as a means of getting training and higher education to young adults in order to give them a potentially brighter future at home than they are likely to find as migrants in the United States. Nonetheless, migration is still an important way in which Mexican families improve their well-being.

CHECK YOUR UNDERSTANDING

1. How much of Mexico's workforce is employed in the informal economy?
2. What is unique about Mexico's manufacturing sector?
3. Why is immigration to the United States problematic for Mexico?

CENTRAL AMERICA

It has long been said that Central America's wealth is in its soil. While this may no longer be true, as industry and services have expanded, around one-quarter of the people of the seven countries of Central America remain dependent on agriculture. The majority are indigenous or mestizo (*ladino* is the local term), with most rural people living in small villages surrounded by tiny farms called minifundios (**Figures 3.40** and **3.41**). Some own land but most are sharecroppers and seasonal laborers on large farms and plantations (see the story of Aguilar Busto Rosalino in "Food Production and Development"). Most land is controlled by a tiny minority of wealthy individuals and companies, which contributes to high levels of income disparity. On the small bits of farmland available to the poor for growing their own food and cash crops, local densities may be 1000 people or more per square mile. These circumstances contribute to overall medium levels of human development that are much lower in rural areas (see Figure 3.21).

The Central American isthmus consists of three physical zones that are not well connected with each other: the narrow Pacific coast; the highland interior; and the long, sloping, rain-washed Caribbean coastal region. Along the Pacific coast laborers work on large plantations that grow sugarcane, cotton, and bananas and other tropical fruits; coffee is grown in the hills behind the coast. In the highland interior of Guatemala, Honduras, and Nicaragua, cattle ranching and commercial agriculture have recently displaced indigenous subsistence farmers. Similarly, the humid Caribbean coastal region, for many years sparsely populated with indigenous and African-Caribbean subsistence farmers, has become dominated by commercial agriculture, forestry, tourism development, and resettlement projects for small farmers displaced from the highlands.

Social and Economic Conditions

The people of this region have experienced centuries of hardship, including long hours of labor at low wages and the loss of most of their farmlands to large landholders. In both rural and urban areas, infrastructure development has lagged. Roads are primitive and especially in rural areas education and incomes are low, and population growth rates are high. Most people lack clean water, sanitation, health care, and protection from poisoning by agricultural chemicals.

The Exceptions: Costa Rica and Panama

Two exceptions to these extreme patterns of elite monopoly, mass poverty, and rapid population growth are Costa Rica and, to some

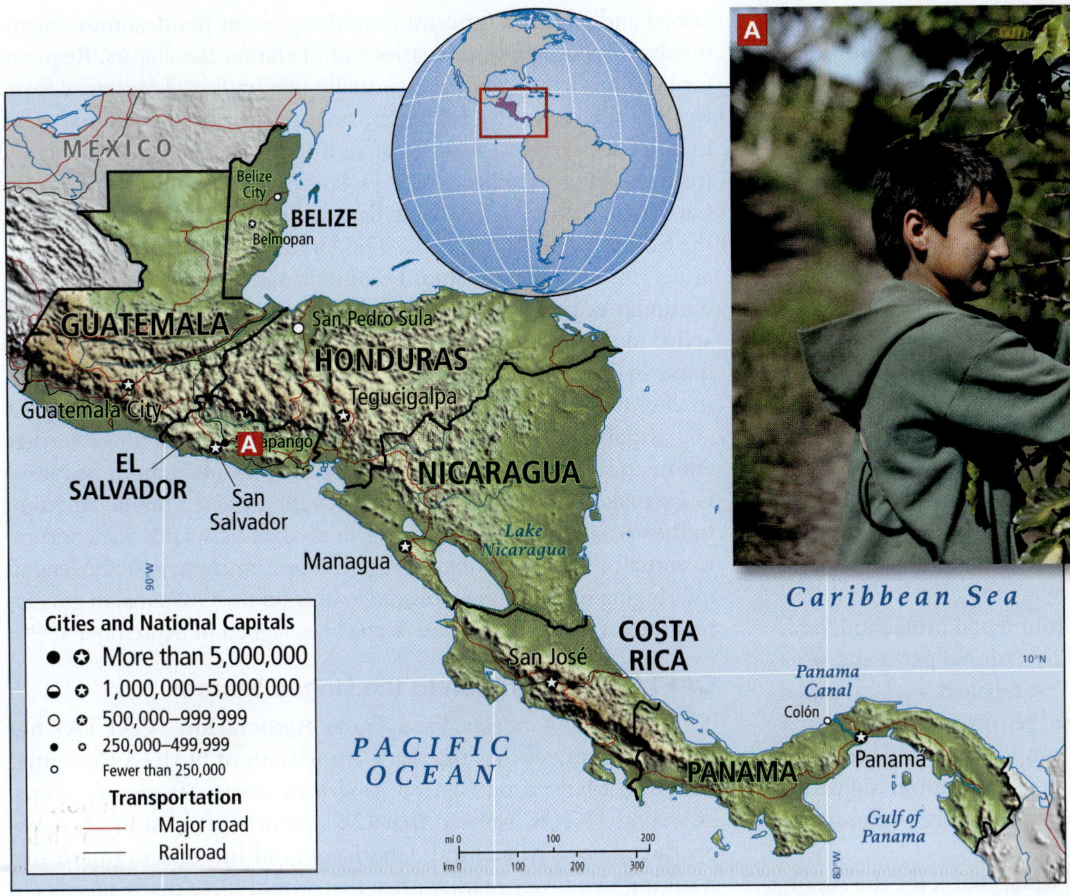

Figure 3.40 The Central America subregion. **A** A boy harvests coffee in El Salvador. [Jose CABEZAS/AFP/Getty Images]

Figure 3.41 Minifundios. The pattern of tiny farm plots (*minifundios*) surrounding a town is visible in this photo of Costa Rica. Some minifundios are the result of government efforts to break up large landholdings and redistribute them among poor farmers. Most minifundios are too small to provide more than supplemental income to their owners. [JodiJacobson/Getty Images]

extent, Panama (see Figure 3.21). In Costa Rica, there were no precious metals to extract and the fairly small native population died out soon after the conquest by Europeans. Without a captive labor supply, the European immigrants to Costa Rica set up small but productive family farms that they worked themselves, not unlike early North American family farms. Costa Rica has strong democratic traditions that began in the nineteenth century, and it has unusually enlightened government policies that are widely seen as examples for future development in Central America. With no standing army, the country has been more free to invest in schools, health care, social services, and infrastructure, contributing to Central America's highest levels of human development. The country also has one of the region's soundest economies, having moved on from dependence on agricultural exports to a more diverse basis in specialized manufacturing, pharmaceuticals, and ecotourism. Nevertheless, while average incomes have risen in recent decades, income inequality, which was already high, has also grown, and for the past two decades poverty has hovered at around 20 to 25 percent. Population growth is low for the region, at 1.3 percent per year.

Costa Rica is a world leader in environmental protection, having set aside 25 percent of its land area in national parks and protected areas. In the 1980s, the country established wetland parks along the Caribbean coast and encouraged sensitive ecotourism at a number of nature preserves. Costa Rica supports scientific research through several international study centers in its central highlands and lowland rain forests, where students from throughout the hemisphere study tropical environments.

Panama is known primarily for the canal that joins the Atlantic Ocean and Caribbean Sea with the Pacific, precluding the need for a long sea voyage around the tip of South America. The United States in fact created the country of Panama in 1903 by fomenting and arming a revolution in what had been a part of Colombia, in order to cement its control over the canal. Opened in 1914, the Panama Canal was built primarily with money from the United States and labor from Afro-Caribbean people, over 27,000 of whom died building the canal, mostly from disease and accidents.

With the United States managing the canal and maintaining a large military presence, Panama remained a virtual colony of the United States. Over the years resentment to the U.S. presence built, resulting in occasional violent protest. Panama's government also accumulated considerable debt owed largely to U.S. banks. As a way of solving both of these problems, negotiations began for the transfer of the canal from the United States to Panama in the 1960s. The final transfer didn't occur until 1999, when the canal was seen as increasingly obsolete due to competition from nearby oil and gas pipelines and potentially from another canal route through Nicaragua. The canal was also a bottleneck for world trade as it was not large enough to accommodate the huge cargo, tanker, and cruise ships of the modern era. However, in 2016, after a decade of expensive renovation, the canal was able to accommodate all but the very largest of ships in current use. The canal is today a major part of the service economy that has given Panama the highest incomes in Central America.

Decades of Civil Conflict

Frustrated with elite-dominated governments unresponsive to the widespread poverty, people in Guatemala, Honduras, Nicaragua, and El Salvador have organized movements for political change, both armed and peaceful, for centuries. Many recent political movements involved socialist revolutionaries and liberation theologists. Requests for even moderate reforms are usually met with stiff resistance from wealthy elites, national militaries, and at times the U.S. military and intelligence services. A series of civil wars flared up intermittently from the 1960s to the 1990s. Violence was particularly intense in Guatemala, where 200,000 civilians were killed or "disappeared" by the military during this time. The United States often backed and armed military dictatorships because it was convinced that the revolutionaries posed a communist/socialist threat to its corporations active in the region. The most recent coup d'état occurred in Honduras in 2009. The United States' role in supporting this coup and its aftermath is complex and still not entirely clear.

Rigoberta Menchú won the 1992 Nobel Peace Prize for her efforts to stop government violence against the indigenous people of Guatemala. Her autobiographical account attracted public attention to the carnage and was important in awakening worldwide concern. Eventually, after a number of regional peacemakers joined Menchú in bringing international pressure to bear on the Guatemalan government, the Guatemalan Peace Accord was signed in September 1996.

CAFTA and Migration to the United States

The Central America Free Trade Association (CAFTA) has extended many of the "free trade" provisions of NAFTA to Central America. Many rural Central American grassroots organizations, as well as NGOs, warned that U.S. exports of subsidized, mass-produced corn to Central America would increase dramatically, resulting in a lower market price that would force many farmers to sell their land and migrate to find work. In the years since CAFTA's implementation, some farmers have been able to adapt to new opportunities, for example by switching crops from corn to fresh vegetables, which they can sell to new international buyers that have appeared in recent years. However, many poorer and less educated farmers have become part of a huge flow of undocumented immigrants to the United States via Mexico, some in highly publicized caravans making their way to the U.S. border.

Nicaragua's Conflicts

Until the late twentieth century, Nicaragua had landownership patterns characteristic of the region: A tiny elite held the usual monopoly on land, while the mass of laborers lived in poverty. By 1910, the U.S. company United Fruit (now Chiquita) had coffee and fruit plantations in the Nicaraguan Pacific uplands and on the Pacific coastal plain. Between 1912 and 1933, the United States kept Marines in the country to quell labor protests that threatened United Fruit's interests, ultimately helping establish the wealthy Somoza family as a dynasty of brutal dictators.

The socialist Sandinista revolution of 1979 ousted the Somoza regime. The Sandinistas, who eventually won several national elections, embarked on a program of land redistribution and agricultural reform and improvement of basic education and health services. However, the country was soon mired in a debilitating war with the Contras, right-wing counterinsurgents backed by local elites and the United States. In what became known as the Iran–Contra affair, the United States secretly supported the Contras (for which Congress had explicitly denied funds) with money from covert arms sales to Iran. A trade embargo imposed by the United

States further contributed to the ruin of the Nicaraguan economy. By the end of the 1980s, Nicaragua was one of the poorest nations in the Western Hemisphere.

In national elections in 1990, an electorate weary of violence voted the Sandinistas out. Starting in 1997, several free elections resulted in moderate governments that failed to bring Nicaragua any measure of prosperity. In 2006, Daniel Ortega, a Sandinista leader and president in the late 1980s, was elected president. Now in his third term, he takes populist, socialist, and anti-U.S. positions similar to those once taken by Chávez in Venezuela. Ortega has since been celebrated for strong economic growth and improvements in human development under his leadership, but also has faced accusations of being a dictator due to his banning of political protests in 2018, censorship of the media, and jailing or killing the political opposition.

CHECK YOUR UNDERSTANDING

1. What factors have contributed to low levels of human development in this region?
2. Why is Costa Rica an exception in Central America?
3. What has the role of the United States been in this region, economically and politically?

THE NORTHERN ANDES AND CARIBBEAN COAST OF SOUTH AMERICA

The five countries in the northernmost part of South America share a Caribbean coastline and extend south into a remote interior of wide river basins and humid uplands (**Figure 3.42**). The Guianas resemble Caribbean countries in that they were once traditional plantation colonies worked by slaves and indentured labor, and today their multicultural societies are made up of descendants of African, South Asian, Southeast Asian, Dutch, French, and English settlers. Venezuela and Colombia share a Spanish colonial past; their populations are largely mestizo, but there is also a small upper class of primarily European heritage and a small population of African origin in the western and Caribbean lowlands. In all the countries of this subregion, small indigenous populations survive, mainly in the interior lowland Orinoco and Amazon basins, where they hunt, gather, and grow subsistence crops.

Figure 3.42 Northern Andes and the Caribbean coast. **A** An aerial view of Caracas, Venezuela, showing slums on the left and high-rise apartments on the right. [LEO RAMIREZ/AFP/Getty Images]

The Guianas

To the north of Brazil lie three small countries known collectively as the Guianas: Guyana, Suriname, and French Guiana. Guyana gained independence from Britain in 1966 and Suriname from the Netherlands in 1975, while French Guiana is still part of France. The common colonial heritage of these three countries is evident in the extractive industries that still dominate their economies: sugar, rice, and banana plantations in the coastal areas, logging and gold, diamond, and bauxite mining in the resource-rich highlands.

The population descends mainly from two major cultural groups who once worked the plantations. Africans were brought in as slaves from 1620 to the early 1800s, and South and Southeast Asians were brought in as indentured servants after the abolition of slavery. The descendants of Asian indentured servants are mostly Hindus and Muslims. Many of them became small-plot rice farmers or owners of small businesses. Those of African descent are primarily Christian; they are both agricultural and urban workers. Politics in the Guianas is complicated by the social and cultural differences between citizens of Asian and African descent. These differences, as well as persistent inequality, have constrained economic growth and human development, which lags the most in Guyana.

Venezuela

Venezuela was once the wealthiest country in South America and could be again due to having the world's largest proven oil deposits. Oil has been the backbone of the country's economy since the mid-twentieth century, and Venezuela is a founding member of the Organization of the Petroleum Exporting Countries (OPEC). Venezuela is among the top suppliers of oil in the world, supplying to the United States, China, and much of this region. Historically profits from oil were limited to a small network of wealthy and powerful people and their families, which led to the government nationalizing the industry in the 1970s, with the idea that the profits would fund public programs to help the poor majority. However, this policy was relaxed, foreign oil firms, such as Chevron, were allowed back in, and oil profits stayed with the elite and the small middle class. By the late 1990s, 63 percent of the country's wealth was controlled by 10 percent of the population, mainly those of European descent, leaving those of mixed native and African descent at the bottom of the income pyramid. Taxes on the wealthy and on foreign investors were kept low, so oil and other assets did not generate enough government revenue to fund badly needed improvements in human development. This failure to invest in its own people is reflected in the capital city of Caracas, where modern freeways separate rows of high-rise buildings from shantytowns that are home to millions living in deep poverty, without access to clean drinking water, sanitation, and adequate education and transportation (see Figure 3.42A).

The landslide election of Hugo Chávez as president in 1998 brought more than two decades of socialist reforms to the country. Chávez emerged at a time when one-third of the population was living on U.S.$2 (PPP) a day, and per capita GNI (PPP) was lower than it had been in 1977. His government invested heavily in health care, education, job creation, and improved access to water and food for the large underclass. Profits from the largely foreign-managed oil industry were redirected to support these social programs, winning him lasting loyalty from the underclass but wiping out support from the small middle class and elites, who feared a turn toward broad government control of the economy. They also earned him intense criticism from the U.S. media and the U.S. government, which he responded to with colorful rhetoric. When addressing the UN in 2006, a day after President George Bush spoke, he said, "The devil came here yesterday, and it stinks of sulfur still today." Supported by historically high prices for Venezuela's oil, Chávez helped lead a region-wide backlash against neoliberalism in the 2000s, frequently condemning the United States' history of intervention in the region. Briefly deposed in a coup d'état in 2001 in which the United States participated covertly, Chavez was quickly reinstated and remained in office, accumulating more power and enacting more reforms, until his death from cancer in 2013.

Chavez's socialist policies depended on high oil prices as well as generous loans from China amounting to $63 billion between 2007 and 2014. The policies lived on under Chavez's hand-picked successor, Nicolas Maduro, but after oil prices fell dramatically in 2014, the strain on Venezuela's economy became intense. Moreover, the loans from China had to be repaid in oil, and with prices now much lower, much more oil had to be supplied. Meanwhile Venezuela's oil industry, which was largely state-controlled by the time Chavez died, suffers from corruption and mismanagement that has sharply reduced its production, sending the country's economy into a tailspin of hyperinflation since 2014. Shortages of food, medical supplies, and other imported goods have led to widespread protests throughout the country, high crime rates, and the departure of 3 to 4 million Venezuelans, mostly to Colombia and other Andean countries. Meanwhile the Maduro government is maintaining its hold on power by selling access to Venezuela's oil fields to Russian and Chinese oil companies, using the money to support the socialist policies that benefit Venezuela's poor majority.

Colombia

A civil war in Colombia has raged on and off for more than 50 years, fueled by economic inequalities, political repression, and the drug trade. The conflict has died down in recent years due to a ceasefire between the government and the largest rebel group, the FARC (Revolutionary Armed Forces of Colombia), which has become a political party with representation in the Colombian Congress. However, ongoing violence has given Colombia the highest total of internally displaced people in the world, at 7.7 million, most of them poor farmers displaced by illegal armed groups.

At the root of the conflict are huge inequalities in wealth and power. Colombia is the world's second-largest exporter of coffee and a major exporter of oil, coal, and illegal drugs. A small proportion of the population, mostly of European descent, has monopolized most wealth by keeping wages low and resisting paying taxes. Via successive neoliberal governments, the wealthy have pushed back strongly against efforts to redistribute some of their extensive landholdings to landless rural people.

Since the 1960s several revolutionary guerrilla groups pushed for reforms for the poor. Opposing them was the Colombian military, as well as private armies funded by wealthy elites, drug traffickers, and several U.S. corporations (including Chiquita and Coca-Cola) that have operations in Colombia. The UN estimates that 80 percent of the fatalities in the conflict were committed by these private armies, or "paramilitaries," and many of those killed were not combatants, but rather leaders of trade unions pushing for better pay for working people.

All warring parties participated in the drug trade, and many still do. This trade is derived from the leaves of the ancient Andean coca plant, traditionally chewed as a fatigue and hunger suppressant, but now processed into the much stronger cocaine and sold internationally (see Figure 3.27). Although coca growers make a somewhat better income from coca than they would from other crops, competing drug-smuggling rings reap most of the profits, intimidating or murdering anyone who opposes them.

Per capita drug use by Colombians is quite low (about one-quarter of the U.S. rate), and there are signs that drug-related violence is decreasing. The Council on Hemispheric Affairs, a nonprofit research organization, suggests strong family ties of most Colombian youth keep them away from drugs, as do community antidrug-violence initiatives such as the *No Mas* (No More) movement. Millions have demonstrated against the drug cartels, and some success in abating the violence has come from rehabilitating drug trade operatives.

Much work is being done to find sources of income for rural people other than coca. For example, in the mountains of southwestern Colombia, 1200 farm families have formed a cooperative that produces a line of 20 products—preserved fruits, sauces, and candies—marketed especially to Latinos in North America and Europe. The cooperative also focuses on child and adult education. Elsewhere in the country, rural people working with the food scientists at the International Center for Tropical Agriculture have developed new varieties of corn that are more productive and nutritious.

Despite the conflict, Colombia has maintained a vibrant tourist economy. Cartagena, for example, is known for its elegant colonial buildings, beautiful beaches, and lively entertainment (**Figure 3.43**).

Figure 3.43 Tourist hotels on the water in Cartagena, Colombia. High-rise hotels for tourists and apartment buildings for wealthy Colombians line the beach in Cartagena. [Jane Sweeney/AWL Images/Getty Images]

CHECK YOUR UNDERSTANDING

1. What are the major extractive industries in the northern Andes and Caribbean coast?
2. Why is the population of the Guianas so diverse?
3. How does the current political system of Colombia differ from that of Venezuela?
4. What conditions are at the root of the conflict in Colombia?

THE CENTRAL ANDES

The central Andes, which includes the countries of Ecuador, Peru, and Bolivia, consists of a high and wide system of parallel mountain ranges and plateaus that carries 70 percent of the world's tropical glaciers, a narrow coastal lowland in Ecuador and Peru (Bolivia has been landlocked since its mineral-rich Atacama Desert coastline was annexed by Chile in 1884), and wide interior lowlands in all three countries that drain east into the Amazon (**Figure 3.44**). On the eve of the Spanish conquest (1532), the coast and mountainous zone was home to the Inca Empire, which was the largest empire in the Americas. The legacy of the Incas is still prominently reflected in the landscape: in the thousands of miles of paved Incan roads and footpaths; in the numerous massive stone ruins, such as those at Machu Picchu in Peru (see Figure 3.16B); and especially in the roughly one-half of the population that is indigenous—the largest proportion in South America.

After the fall of this area to the Spanish, a tiny group of landowners of European descent prospered, while the vast majority of indigenous and mestizo people lived in poverty and servitude, working on large haciendas and in rich mines (copper, lead, and zinc in Peru; tin, bauxite, lead, and zinc in Bolivia). Many attempts at improving human development for the majority of people were made in the twentieth century, most of which failed and some of which brought in horrific violence. However, a recent shift toward growing political involvement by the large indigenous population offers some hope for improving human development.

Settlement Patterns

Settlement has a distinct lowland/highland pattern in this region. The Pacific coast is home to large and modern cosmopolitan cities, including Peru's capital and commercial center, Lima, and Ecuador's leading industrial center, Guayaquil. The lowland people are mainly mestizo, some with African heritage. The majority of indigenous people live in the **Altiplano** (highlands), some in large cities such as Cusco, Peru (the former capital of the Incan empire), La Paz (Bolivia's capital), and El Alto, Bolivia, the highest large city in the world at 13,615 ft (4150 m) and 1.5 million people. The interior Amazonian lowland is less densely settled, mostly with indigenous lowlanders and "colonists" from the highlands coming to farm and work in the mining and oil industries.

The coast is a zone of amazing productivity. In this often-dry area, irrigation has allowed dramatic expansion of plantations and farms in recent years, often funded with IMF loans. Export crops such as bananas, cotton, tobacco, grapes, citrus, apples, and

Altiplano an area of high plains in the central Andes of South America

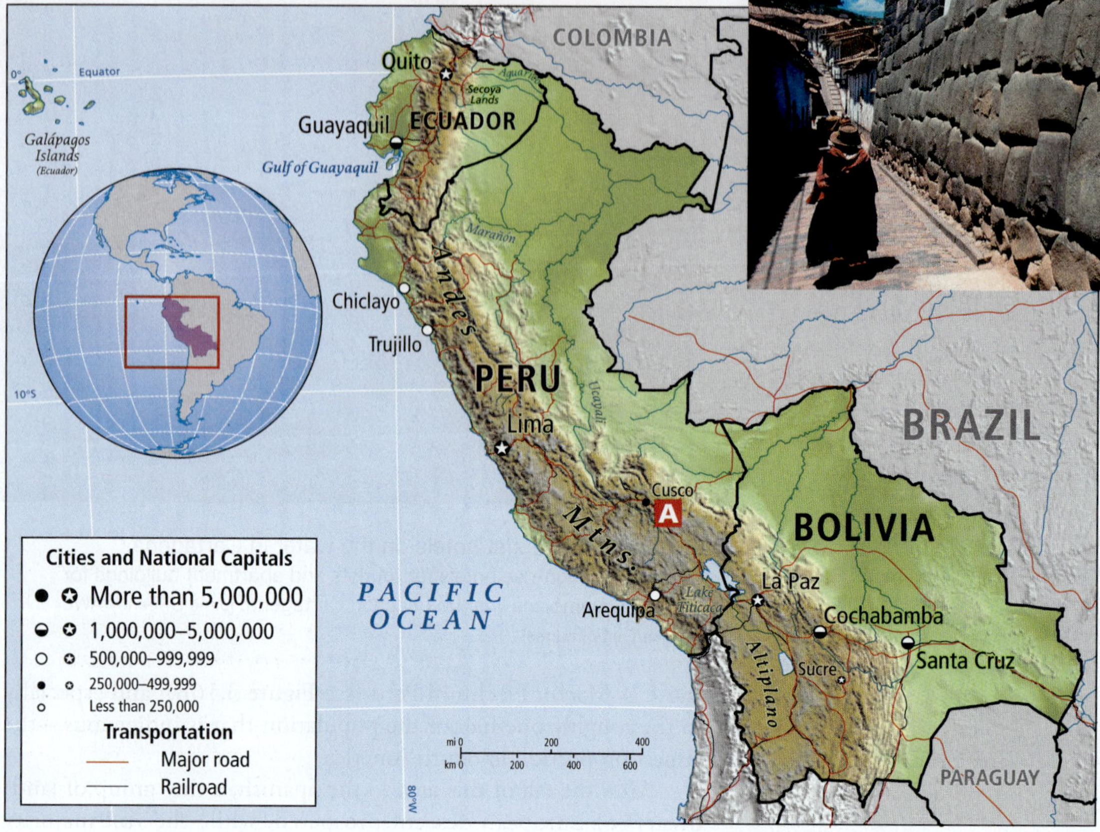

Figure 3.44 Central Andes subregion. **A** Cusco, Peru, was once the capital of the Incan Empire and much Incan stonework has survived since the city fell to the Spanish in 1533. Many Spanish structures were built on top of the Incan walls, such as the street shown above. These walls have withstood the earthquakes that have often hit the Andes much better than the Spanish structures built on top of them. [George Holton/ Science Source/Getty Images]

sugarcane are produced, with most of the profits going to large agribusiness firms. The expansion is bringing new money to the coast, but also straining water resources and displacing smaller-scale farmers. A similar drama is playing out in the ocean off Peru, where nutrient-rich ocean currents support huge numbers of fish. A vibrant export-oriented fishing industry serves European and North American markets and is funded with international loans, displacing smaller scale fishers who once served local markets. Many have been forced to migrate to Lima or Guayaquil.

Climate Change, Water, and Development Issues

Climate change threatens water resources in this subregion, with the glaciers of the central Andes having lost 22 percent of their surface area over the last 35 years. These glaciers feed most of the Andean rivers, which have less volume and hence diminished ability to irrigate and generate hydropower. The water supply of 30 million people is also at risk because highland cities and rural villages get much of their water from glaciers and the rivers they feed. Water shortage is already reducing the productivity of highland crops and curtailing sheep and llamas' production of wool.

Agricultural Restoration in the Peruvian Highlands

In the mountainous interior, the indigenous Aymara and Quechua people are using traditional crops to help create sustainable agricultural change. Support for this movement comes in part from the U.S. government, which wants coca farmers to find profitable alternative crops, but also from Venezuela under Hugo Chavez, who wanted to help alleviate poverty in the Andes. Some funds went to indigenous farmers of traditional crops, such as quinoa (see Table 3.1), an ancient, highly nutritious plant now widely grown and exported to North American and European organic markets.

Neoliberalism and Socialism in Recent Decades

Supporting human development in the central Andean region is important for political and environmental stability. Neoliberal policies imposed by the IMF have periodically forced the privatization of state-run industries and the streamlining of government, resulting in job loss, social turmoil, and only modest economic growth. Dramatic increases in the prices for gasoline, electricity, and transportation have helped raise funds to pay off debts to the IMF, and in the process hurt many poor people, whose wages remain low. In rural areas, neoliberal policies have taken subsidies away from small farms producing local food and instead given them to large export-oriented corporate farms, even though they cause environmental degradation and the loss of livelihood for poor people. Continually low human development in the Andes provided justification for rural revolutionary movements, such as the Sendero Luminoso (Shining Path), which struggled with the government for control of the highlands in the 1980s, killing some 70,000 people in the process.

Many people pushed off their farms by the growth of large-scale export-oriented agriculture have moved to the Amazon lowland regions. Here national governments, eager to repay debts to the IMF or China, have encouraged a wide range of export-oriented resource extraction, including logging, oil development, and the cultivation of African oil palm trees. These activities are shrinking the home territories of indigenous people and degrading rich rain forest ecosystems.

Ecuador's Correa Socialist strategies for lessening gross inequalities have been tried in Ecuador, with varying success. In 2006, Ecuadorians elected a U.S.-educated socialist economist, Rafael Correa, who raised taxes on the wealthy, and used the profits from the oil industry to invest in human development. Part of the Pink Tide, Correa had a reputation for being incorruptible and was reelected in 2009 and again in 2013. He also used Ecuador's military to guard and expand oil development in the Amazon at the expense of many indigenous people. Weak oil prices and protests against some of his policies moved Correa to not seek reelection in 2017. The following government was more neoliberal in orientation, reducing taxes on the oil industry and the wealthy.

Bolivia's Morales Gains in political participation by indigenous people are perhaps most impressive in Bolivia. In the 1980s and 1990s, neoliberal governments privatized state-owned industries, with most sold to foreign interests, to comply with demands by the World Bank and IMF that Bolivia reduce its debt. This brought job losses to the majority and worsened income disparity, setting the stage for massive public protests in the early 2000s. In 2005, Bolivians elected in a landslide the first indigenous head of state in the Americas: Evo Morales, a socialist, former coca farmer, and head of the coca producers' union, who has proven to be innovative, independent, and controversial.

Morales began his presidency advocating against the privatization of water (see "The Water Management Crisis") and for coca farmers, saying that they should not lose the right to grow an age-old crop that produces only a mild high in its natural state just because an outsider had figured out how to make the powerfully addictive drug cocaine from coca. Morales went on to nationalize the oil and gas industries, using the profits to invest in social programs that have boosted human development and reduced income disparity. Morales is currently attempting to create a mining industry focusing on lithium, an essential component in batteries for electric cars and other electronic devices. Bolivia is estimated to hold one-half of the world's lithium but has so far struggled to find investors with the necessary technical capability who are willing to comply with Morales's demands for a larger share of profits than nearby Chile, which also has huge reserves of lithium.

CHECK YOUR UNDERSTANDING

1. How is the Central Andes subregion vulnerable to climate change?
2. What challenges have neoliberal and socialist governments faced in the Central Andes?
3. Where have gains in political participation by indigenous people been the greatest?

THE SOUTHERN CONE

The countries of Chile, Paraguay, Uruguay, and Argentina have diverse physical environments but remarkably similar histories (**Figure 3.45**). The so-called *Southern Cone* of South America had little European settlement during the Spanish empire, but in the late nineteenth and early twentieth centuries, major migrations from Germany, Italy, and Ireland occurred and European immigrants soon outnumbered indigenous and mestizo populations. Only Paraguay has a predominantly mestizo population. Despite a turbulent economic and political history, the Southern Cone countries today have the highest human development in the region.

The Economies of the Southern Cone

Agriculture has been replaced by services as the leading economic sector but remains prominent in the identity of the region and as a source of exports. The primary agricultural zone, the *pampas*, is an area of extensive grasslands and highly fertile soils in northern Argentina and Uruguay (and southern Brazil) that is famous for its grain and cattle. Sheep raising dominates in Argentina's drier, less fertile southern zone, Patagonia. On the Pacific side of the Andes, in Chile's central zone, Mediterranean climates support large-scale fruit production that caters to the winter markets of the Northern Hemisphere. Both Argentina and Chile have greatly expanded their production of wine (see Figure 3.45B). Chile also benefits from considerable mineral wealth, with the world's largest reserves and production of copper.

Although the agricultural and mineral exports of the Southern Cone created considerable wealth in the past, fluctuating prices for raw materials on the global market create economic instability. Substantial poor populations continually press for investments in human development, and each country has experimented with socialist reforms, including state-run industries and subsidies for food, housing, basic health care, and transportation.

Chile's export-oriented industrial, mining, and agricultural sectors grew under neoliberal governments, but inequalities remained high. In 2006 Chile elected a moderate socialist, Michelle Bachelet, who used the profits from government-owned copper mines to invest in human development. She was followed by a more neoliberal president in 2010, reelected in 2014, and again followed by a neoliberal president in 2018.

Argentina elected a moderate socialist, Néstor Kirchner, in 2006, after years of slow economic growth and growing debt under neoliberal governments. Like Chile's Bachelet, Kirchner invested in human development with revenues from Argentina's largely agricultural exports, quelling widespread protests and promoting more equitable economic growth. Kirchner's wife, Cristina Fernández de Kirchner, succeeded him as president, and continued similar policies until lower prices for Argentina's exports produced an economic downturn. She was replaced by a neoliberal president whose policies failed to reverse the downturn or reduce Argentina's debt.

A Itaipu Dam is one of the largest power stations in the world, supplying 90 percent of Paraguay's electricity. [SambaPhoto/Cassio Vasconcellos/Getty Images]

B Food processing is a leading industry throughout the region. Shown here are three workers in a Chilean winery. [Alfredo Maiquez/Lonely Planet Images/Getty Images]

Figure 3.45 The Southern Cone subregion.

Buenos Aires: A Primate City

The primary urban center in the Southern Cone is Buenos Aires, the capital of Argentina and one of the world's largest cities at 15 million (**Figure 3.46**). Over a third of the country's people live in this primate city, which boasts elegant urban landscapes, dozens of international banks, as well as numerous empty factories, pollution, and many areas plagued by low human development. Many past Argentine governments have tried to make Buenos Aires a sophisticated center of skilled labor capable of attracting investors from throughout the world. The reality is that many of the city's residents live in run-down apartments sometimes on wages too low to afford basic nutrition. The investments in human development that Buenos Aires needs may also draw investors, but not necessarily as fast as some would like.

CHECK YOUR UNDERSTANDING

1. What is unique about the populations of the Southern Cone countries?

2. What constrained growth in Chile and Argentina's import substitution industries?

BRAZIL

Brazil's multicultural society can be dazzlingly beautiful, but its landscapes also plainly show the social and environmental effects of poorly planned development. Flamboyant creativity and elegance are seemingly always on display, surrounded by the intense poverty and hardships under which so many labor. Brazil's 210 million people live in a highly stratified society made up of a small, very wealthy elite; a modest but rising middle class; and a large majority of working poor. In Brazil's megacities of São Paulo and Rio de Janeiro, elegant high-rise buildings are surrounded by vast favelas where crime rates are high and homeless street children are seemingly everywhere. According to the latest figures (2017), the richest 10 percent of Brazil's population has 43 percent of the country's wealth, equivalent to 3.5 times what the bottom 40 percent earn. This is one of the widest disparities in Middle and South America and the world, though it is declining.

Figure 3.46 Buenos Aires. Children in a poor part of Buenos Aires simulate snow with polystyrene from a stuffed animal. Behind them looms an unfinished hospital intended to serve their neighborhood; it has instead been left vacant for more than 50 years. Approximately 25 percent of Argentina's population lives in poverty. [JUAN MABROMATA/AFP/Getty Images]

Brazil's Size and Varied Topography

Brazil is slightly larger than the lower 48 contiguous United States. Its three distinctive physical features are the Amazon Basin, the Mato Grosso, and the Brazilian Highlands (**Figure 3.47**). The Amazon Basin, which covers the northern two-thirds of the country, is described in "Landforms" (see also Figures 3.10 and 3.13). The Mato Grosso (meaning great woods) is a seasonally wet/dry interior lowland south of the Amazon with a convoluted surface. Extensively cleared for agriculture in the twentieth century, the drier southern parts were once covered with grasses and scrubby trees adapted to long dry periods, while wetter northern parts had rain forest. The southern third of Brazil is occupied mostly by the Brazilian Highlands, a variegated plateau that rises abruptly just behind the Atlantic seaboard 500 miles (800 kilometers) south of the mouth of the Amazon. The northern portion of the plateau is arid; the southern part receives considerably more rainfall. Ninety percent of Brazil's population lives within 300 miles (482 km) of the coast, but in the temperate zone, near Rio de Janeiro, São Paulo, Curitiba, and Pôrto Alegre, denser settlement extends deeper inland (see Figure 3.33).

Managing Brazil's Large and Varied Economy

Brazil's economy is the largest in Middle and South America and the eighth largest in the world, with natural resources that are the envy of most nations. However, management of those resources has been a challenge, with successive governments failing to recognize the fragility of Amazonian ecosystems (see "Tropical Forests, Climate Change, and Globalization"), and the country's vast mineral wealth mined with insufficient attention to human and environmental consequences.

Brazil's energy resources are diverse and abundant, making it the eighth largest energy producer in the world (larger than Venezuela). Renewable energy supplies about 43 percent of the country's needs, well above the global average of 10 percent renewable energy, with hydroelectric power supplying 15 percent due to the many rivers and natural waterfalls that descend from the highlands. Ethanol, a gasoline substitute derived from sugar cane, supplies another 15 percent. Brazil is the world's second largest producer of ethanol after the United States, and has used this fuel since the 1920s. Oil now provides for 39 percent of Brazil's energy needs, and most of it comes from very large offshore reserves in the Atlantic. Brazil's car industry has developed some of the world's best flexible fuel technology, with cars that can instantly switch from gas to ethanol and get 40 miles to the gallon.

Brazil is the most highly industrialized country in South America, and its global role is expanding. Most of its industries—steel, motor vehicles, aeronautics, appliances, chemicals, textiles, and shoes—are concentrated in a triangle formed by the huge southeastern cities of São Paulo, Rio de Janeiro, and Belo Horizonte.

Agriculture, Activism, and the Amazon

Brazil's agricultural economy is large and varied, but global attention has recently focused on the threat to rain forests posed by

Figure 3.47 The Brazil subregion.

large-scale soybean cultivation. In 2006 environmental activists convinced international soy buyers and producers to stop buying soy grown on former rain forest land. This drastically reduced the amount of rain forest being cleared for soy, and helped cut rates of deforestation to half of what was common before 2006. Government policies also played a role in limiting agricultural expansion in the Amazon, with the socialist Pink Tide governments of 2003–2015 presiding over the dramatic reductions in deforestation. This was largely in response to international pressure, and made easier by the primarily urban base of their political parties. Wealthy farmers and ranchers tend to support the more neoliberal parties that have governed Brazil since 2015, after which deforestation rates have increased somewhat, but still remain well below 2006 levels.

Urbanization

Around 86 percent of Brazil's population is urban, with a number of large cities on the Atlantic perimeter of the country. During the global economic depression of the 1930s, farmworkers migrated to urban areas as world prices for Brazil's agricultural exports fell. The spread of green revolution agriculture has pushed even more people off the land since the 1960s. Jobs in factories created by military governments in the 1960s drew many people to the cities, but since Brazil's competitive edge in the global market was its cheap labor, wages were kept very low. Workers had to organize and protest relentlessly to raise wages above bare survival levels.

Urban poverty is still widespread, with around a third of urban dwellers, many of them underemployed, living in favelas (see Figure 3.30). The poverty in the cities of the northeast rivals that of Haiti, the poorest country in the Americas. However, favela dwellers are famous for their strong community life and support for those in distress. Many turn to music, dance, and religion as a source of strength.

VIGNETTE It's Friday afternoon, and a crowd of white-clad women is gathering outside a house in the Felicidad favela in Fortaleza. Like the surrounding houses, this one is small, but it gleams white; all its surfaces are swathed in marble. Potted palms decorate the porch, and beside them, welcoming the women, is a tall, elegant, middle-aged man also dressed in white. He is a leader in the movement known as Umbanda, one of several belief systems thriving in the coastal cities of Brazil (**Figure 3.48**) that combine African, European, and indigenous spiritual traditions. Each group is led by a male or female who invokes the spirits to help with health problems, remove curses, speak to dead relatives, and give general life advice. Ceremonies can last as long as 8 hours and combine drumming, dancing, spirit possession, healing, and friendly psychological support. Similar movements, such as Voodoo and Santería, are found elsewhere in the Americas, including the Caribbean, the United States, and Canada. *[Source: Lydia Pulsipher's field notes.]* ■

Brasília Brasília, the modern capital of Brazil, is an intriguing example of the effort to lead development with urban growth poles (discussed in "Elite Landscapes"). Built in the state of Goiás in just 3 years beginning in 1957, Brasília lies about 600 miles (1000 kilometers) inland, remote from the coast and the glamorous old administrative center of Rio de Janeiro. The city was promoted as a **forward capital** that would speed the development of the western territories and the Amazon Basin. However, another motive was to trim the badly swollen and highly inefficient government bureaucracy by removing the capital from the centers of Brazilian society.

Symbolism figured more prominently than practicality in the design of Brasília, which was laid out to look from the air like a jet plane. There was to be no central business district, but rather shopping zones in each residential area and one large mall. Pedestrian traffic was limited to a few grand promenades; people were expected to move even short distances in cars and taxis. Public buildings were designed for maximum visual and ceremonial drama, but safety was an afterthought.

Over five decades of actually using this urban landscape, people have made all sorts of interesting changes to the formal design. At the Parliament, legislative staff and messengers created footpaths where they needed them: through flowerbeds and—with little steps notched in the dirt—up and over landscaped banks. Thus they efficiently connected the administration buildings, bypassing the sweeping promenades. Little hints of the informal economy that characterizes life in the old cities of Brazil began to show up—a fruit vendor here, a sidewalk manicurist there. And the shantytowns that the planners had tried hard to eliminate began to rise relentlessly around the perimeter. Overall, in the 50 years since its construction, Brasília's success as a forward capital and growth pole has been limited. The city has drawn 4.3 million people, making it the third largest city in the country, but population and investment remain centered in the coastal cities.

Figure 3.48 African-derived religions in Brazil. Practitioners of Candomblé carry flowers out to a boat during a ritual to honor Yemanja, goddess of the sea, in Amoreiras, Brazil. Candomblé is one of several African-derived religions that is gaining popularity, especially in the urban coastal areas. [Jan Sochor/Latincontent/Getty Images]

Rio and Mega Events In 2014, Rio de Janeiro was the primary host city for the most-watched sporting event on Earth, the FIFA World Cup, and in 2016 the city hosted the world's best-attended sporting event, the Olympics. The preparations for these events were designed to bring in new investments in transportation, tourism, and athletics infrastructure and to project an attractive image of Brazil to the world.

Among the most lasting legacies of the World Cup and the Olympics, which together brought millions of people and billions of dollars of new investment to Rio de Janeiro, is the hardship they place on poorer residents of the city. According to Rio de Janeiro's city government, only 4100 families were removed to the urban periphery to make way for new infrastructure serving the mega events, but local NGOs claim that

forward capital a capital city built in the hinterland to draw migrants and investment for economic development and sometimes for political/strategic reasons

22,000 families comprising 77,000 people were forced to relocate. Most lived near roads, airports, sports facilities, and the city's harbor area, all of which were significantly expanded and renovated. Some lived in favelas, but many legally owned their homes and land. Most were offered some compensation, but found that the move to the urban periphery brought lengthy commutes and transportation costs that ate up 15 to 30 percent of their income. Those who refused had to battle riot police and bulldozers, usually without success.

The scale of these mega events also gave Brazil's notoriously corrupt public officials many chances to steal public funds. In 2015, investigations into corruption at Brazil's state-owned oil company, Petrobras, uncovered millions of dollars in bribes by major Brazilian construction companies to government officials in Rio whose job it was to award construction contracts for the Olympics. The investigation eventually led to the removal from office of Brazil's then president, and the jailing of the previous president in 2018. Anger at corruption fueled many protests of the mega events, which many see as "sanitizing" Rio into a playground for tourists and the rich.

CHECK YOUR UNDERSTANDING

1. Where are most Brazilian cities?
2. Where did pressure to reduce deforestation due to soy cultivation in the Amazon come from?
3. How many urban Brazilians live in favelas?

CRITICAL THINKING QUESTIONS

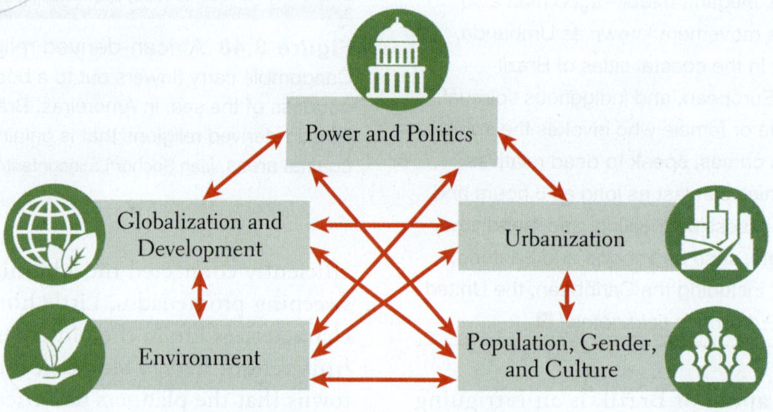

The diagram represents connections among the five geographic themes that structure this book. Listed below are some important questions that have been addressed in this chapter. Answer each question, and indicate where in the diagram you think the topics in each question belong.

1. What are the four main temperature-altitude zones found in Middle and South America, and how have they been occupied and used by humans differently?
2. How are deforestation, climate change, and water scarcity linked in this region?
3. What are some possibly sustainable alternatives to deforestation?
4. How did this region's integration into the global economy result in economic instability and huge inequalities?
5. How have neoliberalism and socialism influenced economic development in this region?
6. What challenges did import substitution face as a development strategy?
7. Why did SAPs fail to achieve stable economic growth?
8. How have foreign interventions shaped the politics of this region?
9. What influence has the international drug trade had in this region?
10. Under what economic conditions have this region's revolutionary movements generally arisen?
11. Why do cities in this region often lack adequate support services and infrastructure?
12. How do primate cities impact the countries they are located in?
13. What kinds of problems do this region's shantytowns have?
14. Why is the Brazilian city of Curitiba unique in this region?
15. What factors created a population explosion in this region during the early twentieth century, and reduced population growth in the late twentieth century?
16. How do major institutions, such as the family and religion, shape gender roles in this region?
17. What forces are challenging the dominance of the Roman Catholic Church?

Key Terms

acculturation 169
Altiplano 181
assimilation 169
austerity 149
Aztecs 143
biodiversity 134
brain drain 161
contested space 157
coup d'état 157
Creoles 144
dictator 157
ecotourism 141
El Niño 135
evangelical Protestantism 171
export processing zones (EPZs) 153
favelas 161
forward capital 187
hacienda 147
import substitution industrialization (ISI) 149
Incas 143
income disparity 152
indigenous 131
isthmus 132
liberation theology 170
machismo 167
maquiladoras 138
marianismo 167
marketization 149
mestizos 144
Middle America 132
nationalize 150
neoliberalism 148
plantation 147
primate city 161
privatization 149
shifting cultivation 143
socialism 148
South America 132
structural adjustment programs (SAPs) 149
subduction zone 133
temperature-altitude zones 135
trade winds 135
urban growth poles 164

More Practice at SaplingPlus

Read the interactive e-text, review key concepts, and check your understanding.

Migrants who have crossed the Croatian border into Slovenia are watched by the police.
[Matej Leskovsek/SIPA/newscom]

4

Europe

Marina Firzik's sewing machine hums to a stop as she reaches for her ringing cell phone. It's a call from Catholic Charities, where she works as a volunteer every Monday. There is an emergency! The panicked voice on the phone recounts how thousands of bedraggled refugees from wars in Syria and elsewhere have reached the southern border of her small Central European country of Slovenia. They are on foot, walking in a steady stream, carrying children, the elderly, and a few mementos from home. After a rough crossing in wooden boats and rubber rafts from Turkey to Greece, most have been on the road for weeks, walking north through the early winter rain and cold of the Balkans. The situation is heartbreaking and chaotic, and Marina is needed to help sort and distribute donated clothing, blankets, and food.

As Marina prepares to leave she is struck by the thought that after years of steady but slow improvements in her way of life since Slovenia declared its independence from communist Yugoslavia in 1991, she may be suddenly faced with disruptive changes. Who are these outsiders, desperate to escape war and find a new life in Europe? Will they try to stay in Slovenia? She is sympathetic and wanting to help, but she is also frightened. She knows about war. Her parents and grandparents were full of tales of hunger and hardship during World War II. She can still see the machine gun pockmarks on many of the older buildings in her town. Now, she is frightened that the new people will bring back that chaos. She hears that most are Muslims, about whom she knows little. Marina rises and turns to see that on her TV screen there is a shocking picture: a stream of refugees stretches into the distant lowlands, monitored from on high by Slovene mounted police. The newscaster says the refugees have walked 1500 kilometers (932 miles) from Greece through the Balkans, carrying children, the infirm, and belongings through rain and mud, often sleeping unprotected in heavy weather (continued in "Migrants, Refugees, and Borders," later in the chapter).

Learning Objectives

Environment: Physical and Human

4.1 Identify Europe's physical features: its peninsulas, landforms, vegetation, and climates.

4.2 Describe the ways in which Europe's vulnerability to climate change varies across the region, and characterize Europe's response to climate change.

4.3 Summarize the range of environmental issues in Europe, including energy use and pollution of air and water.

4.4 Describe Europe's road to global power from ancient cultural complexity and internal and external strife, through the birth of the European Union.

Globalization and Development

4.5 Discuss the features and challenges of the European Union and its efforts to compete globally.

Power and Politics

4.6 Explain how despite the success of the European Union, nationalism still threatens European unity and social cohesion.

4.7 Explain how the European commitment to expanding political freedoms developed from the post–World War II period through the present.

4.8 Analyze how Europe's role as a colonizer continues to affect politics within the EU and in international relations.

Urbanization

4.9 Discuss the evolution of Europe's cities from early trading centers to manufacturing centers to modern high-tech centers.

4.10 Describe modern European cities in terms of density and government housing, and the role of urban infrastructures in transportation and poverty alleviation.

Population, Gender, and Culture

4.11 Explain why Europe's population is aging and how government policies, including immigration, are addressing aging.

4.12 Explain how issues of gender inequality are being differently addressed in various parts of the European Union.

The Subregions of Europe

4.13 Distinguish among the various subregions geographically, economically, and in social protection/active aging policies.

30°W
20°W
10°W
Prime Meridian 0°
10°E
20°E
30°E
40°E
Arctic Circle
ARCTIC OCEAN
Hammerfest
Vladsø
Tromsø
Inari
Ivalo
Kola Peninsula
Narvik
Kiruna
Lapland
White Sea
Reykjavik
Bodo
ICELAND
NORWAY
Lieksa
Umea
Kuopio
Trondheim
SWEDEN
Gulf of Bothnia
Vaasa
Lake Onega
Faroe Islands
Torshavn
Alesund
FINLAND
Lake Ladoga
Svir
Turku
Helsinki
St. Petersburg
ATLANTIC
OCEAN
Gulf of Finland
Bergen
Oslo
Lake Vanern
Tallinn
ESTONIA
Stockholm
Kirkwall
Gulf of Riga
Aluksne
LATVIA
Riga
SCOTLAND
Aberdeen
North Sea
Goteborg
Lake Vattern
Alborg
Glasgow
Edinburgh
DENMARK
Baltic Sea
LITHUANIA
Vilnius
Kaunas
NORTHERN IRELAND
UNITED KINGDOM
Copenhagen
Lund
Malmo
RUSSIA
Kaliningrad
Dnieper
Belfast
Irish Sea
Douglas
Minsk
REPUBLIC of IRELAND
Dublin
Liverpool
Leeds
Manchester
Sheffield
SCHLESWIG-HOLSTEIN
Kolobrzeg
Gdansk
POLAND
Plain
BELARUS
Hamburg
Oder
European
Cork
St. George's Channel
Newcastle
ENGLAND
Birmingham
WALES
Norwich
NETHERLANDS
Amsterdam
Bremen
Elbe
Berlin
Warsaw
Brest
Cambridge
Oxford
Hannover
North
Poznan
Vistula
Land Elevations
meters
feet
4877
16,000
3353
11,000
2134
7000
914
3000
305
1000
152
500
0
0
Thames
London
Southampton
Rotterdam
The Hague
Essen
Dortmund
Duisburg
Leipzig
Warta
Lodz
Wroclaw
Brugge
Antwerp
Dusseldorf
Dresden
Plymouth
English Channel
Calais
Brussels
Cologne
Lille
Bonn
GERMANY
Katowice
Krakow
UKR
BELGIUM
Liege
Rhine
Prague
LUXEMBOURG
Frankfurt
Le Havre
Saarschleife
Nuremberg
CZECHIA
SLOVAKIA
Presov
Luxembourg
Mannheim
Brno
Seine
Paris
Metz
Stuttgart
Danube
Zvolen
Carpathian Mts.
Bratislava
Miskolc
Le Mans
Strasbourg
Munich
Vienna
Iasi
Loire
Budapest
mi 0
100
200
300
km 0
100
200
300
400
500
1:14,800,000
Albers Equal Area Conic Projection
Heldsburg
Salzburg
Nantes
FRANCE
Dijon
Zurich
Vaduz
LIECHTENSTEIN
AUSTRIA
HUNGARY
Kecskemet
Bern
Graz
Kuhtai
ROMANIA
SWITZERLAND
Szeged
Moulins
Geneva
Ljubljana
Arad
Bay of Biscay
Lyon
Alps
SLOVENIA
Zagreb
Pecs
Trieste
Bucharest
Bordeaux
Milan
Venice
Grenoble
Craiova
Turin
Po
CROATIA
BOSNIA & HERZEGOVINA
Belgrade
La Coruña
Gijon
Danube
Pieven
Bilbao
Rhone
Genoa
SAN MARINO
SERBIA
Vigo
Toulouse
Sarajevo
BULGARIA
ITALY
Braga
Nice
MONACO
Viareggio
Florence
Nis
Pyrenees
Marseille
Adriatic Sea
Sofia
Porto
Duero
ANDORRA
Bastia
KOSOVO
Podgorica
MONTENEGRO
Skopje
Zaragoza
Apennines
NORTH MACEDONIA
PORTUGAL
Corsica
VATICAN CITY
Madrid
Rome
Barcelona
Tirane
Thessaloniki
ALBANIA
Bari
Lisbon
SPAIN
Korce
GREECE
Aegean
Tagus
Naples
Valencia
Iberian Peninsula
Sardinia
Kerkira
Balkans
Volos
Palma
Islas Baleares
Tyrrhenian Sea
Ionian Sea
Cordoba
Guadalquivir
Seville
Faro
Cagliari
Athens
Messina
Palermo
Granada
Pirgos
Mediterranean Sea
Strait of Gibraltar
Málaga
Gibraltar
Algiers
Sicily
Tunis
Rabat
MOROCCO
Atlas Mountains
ALGERIA
MALTA

What Makes Europe a Region?

Over the last 500 years, the region of Europe (**Figure 4.1**) has had a central role in world history and geography as a colonizer of distant territories and as a major player in the development of capitalism. In the twentieth century, Europe brought the world to the edge of disaster with two massive wars. Since then it has revived, prospered, and designed a model for regional economic and political integration (the **European Union**, or **EU**) that has been surprisingly successful and has been copied in other world regions. The EU, in 2019 consisting of 28 countries, has been noted for its open markets, emphasis on democracy and basic human well-being, and open internal borders fostering free migration among all countries.

Unfortunately, the EU now faces several grave challenges. First, there are worries that the union might dissolve in the face of growing nationalism and financial and social disparities. Then in 2016 came the shocking vote in the United Kingdom (England, Wales, Scotland, and Northern Ireland) to leave the EU entirely. Popularly known as **Brexit** (Britain's exit), this decision was not well thought out, yet it has far-reaching, but as yet poorly understood, implications for the whole of Europe. Added to Brexit is the unexpected flood, starting in 2015, of millions of immigrants—many of them refugees from failing countries in North Africa and Southwest Asia (especially Syria), sub-Saharan Africa, and South Asia. This influx now threatens to swamp the EU's emerging efforts at multicultural democracy and region-wide freedom of movement. Recent terrorist episodes linked to Muslim immigrants or their Europe-born offspring threaten the amicable resettlement of the newcomers and the dream of a Europe with no internal borders.

The United Kingdom (UK) has repeatedly rescheduled when it will leave the EU, because agreements regarding trade, migration, borders, taxation, social welfare, banking regulations, and more remain unresolved. Because of the uncertainty of this situation, in this eighth edition of this textbook, the UK will usually be counted as one of the now 28 countries in the European Union. A few more hope to join in the next several years. **Figure 4.2** shows all parts of Europe as defined in this book, the 28 members of the EU, and the dates they joined. It also shows the three countries that have opted not to join (Iceland, Norway, and Switzerland) and those that are candidates or potential candidates to join (the six former provinces in what was once Yugoslavia); Turkey, also a candidate, is not shown. Figure 4.2 also shows the four subregions of Europe used here.

Terms in This Chapter

This book divides Europe into four subregions—*North*, *West*, *South*, and *Central Europe* (see Figure 4.2). *Central Europe* refers to all those countries formerly in the Soviet sphere (USSR) that are now in the European Union, as well as the countries that were formerly in Yugoslavia, plus Albania.

European Union a supranational organization that unites most of the countries of Europe

Brexit the 2016 decision by UK voters to leave the European Union; details of how this exit will proceed remain unresolved as of late 2019

◀ **Figure 4.1** Physical map of Europe.

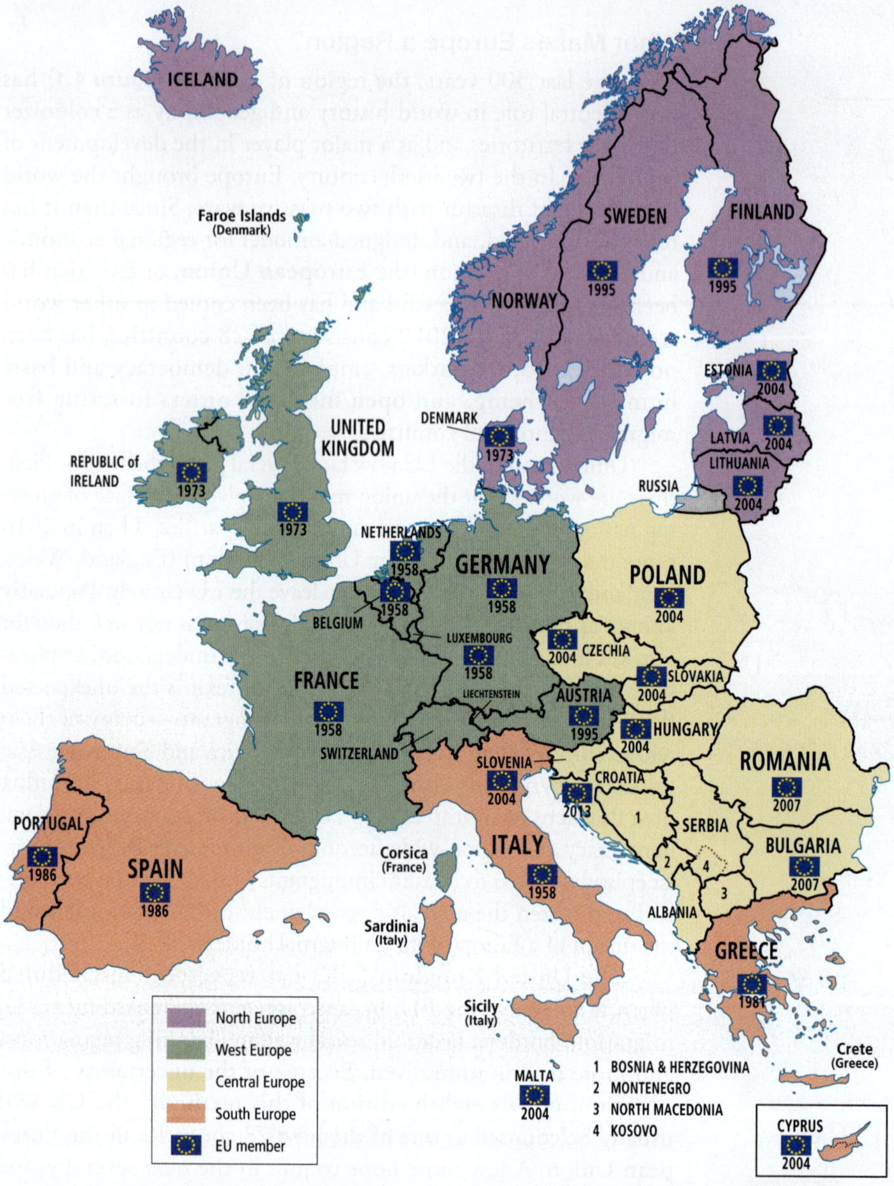

Figure 4.2 Political map of Europe. Current (2019) members of the European Union are shown with an EU symbol along with their dates of joining. European subregional designations are shown for all of Europe.

For convenience, we occasionally use the term *western Europe* to refer to all the countries that were *not* part of the experiment with communism in the Soviet sphere and in Yugoslavia. That is, *western Europe* comprises the combined subregions of North Europe (except Estonia, Latvia, and Lithuania), West Europe (except the former East Germany), and South Europe. When we refer to the countries that were part of the Soviet sphere up to 1989, we use the pre-1989 label *eastern Europe. Central Europe* is now the preferred term for what in the Soviet era was called *eastern Europe.* When we refer to the group of countries collectively known to some as the *Balkans* (Albania, Bosnia and Herzegovina, Bulgaria, Croatia, Macedonia, Montenegro, Romania, Serbia, and Slovenia), we prefer the term *southeastern Europe. EU-28* is the term for just the current 28 member countries of the European Union. *EU-27* will refer to the EU if and when the UK leaves (Brexit). *Czechia* is the now accepted shortened name for the Czech Republic, which will remain its official name.

ENVIRONMENT: PHYSICAL AND HUMAN

4.1 Identify Europe's physical features: its peninsulas, landforms, vegetation, and climates.

4.2 Describe the ways in which Europe's vulnerability to climate change varies across the region, and the character of Europe's response to climate change.

4.3 Summarize the range of environmental issues in Europe, including energy use and pollution of air and water.

4.4 Describe Europe's road to global power from ancient cultural complexity and internal and external strife, through the birth of the European Union.

Europe's environment gives ready access to the world ocean and contains a rich variety of climates and vegetation. Humans and their ancestors have occupied the western end of the Eurasian continent for hundreds of thousands of years and have continuously modified the environment through their activities.

PHYSICAL GEOGRAPHY

Physically, Europe is a peninsula extending off the western end of the huge Eurasian continent. On this giant peninsula of Europe, there are many peninsular appendages, large and small. Norway and Sweden share one of the larger appendages. Other large peninsulas are the Iberian Peninsula (occupied by Portugal and Spain) and those of Italy and Greece. All extend either into the Atlantic Ocean or into the various seas that are adjacent to the Atlantic (see Figure 4.1).

The perimeter of the region of Europe to the north, west, and south is primarily oceanic coastline. This unique access to the world ocean facilitated European exploration and colonization of distant territories. The eastern physical and cultural border of Europe has been difficult to define. Just where Europe ends and Asia begins has been a source of contention for millennia. In this book, the eastern limit of the European region is taken to be the eastern border of the European Union. Other potential parts of Europe—Ukraine, Belarus, western Russia, Moldova, and the Caucasus—are covered in Chapter 5, and Turkey in Chapter 6. Europe's variable landforms, including the many peninsulas that reach into surrounding oceans and seas, all affect the region's climates and vegetation.

Landforms

Although European landforms are fairly complex, the basic pattern is mountains, uplands, and lowlands, all stretching roughly west to

Figure 4.3 The Alps in Austria. [wingmar/Getty Images]

Figure 4.4 Uplands in Germany. [Dierk Boeser/imageBROKER/Shutterstock.com]

Figure 4.5 North European Plain: North German seacoast with windmills. [Karl Johaentges/LOOK-foto/Getty Images]

east in wide bands. As you can see in Figure 4.1, Europe's largest and highest mountain chain, the *Alps*, runs west to east through the middle of the continent, from southern France through Switzerland and Austria (**Figure 4.3**). The alpine mountain formation extends into the Czech Republic and Slovakia, and curves southeast as the Carpathian Mountains into Romania. This network of mountains is mainly the result of pressure from the collision of the northward-moving African Tectonic Plate with the southeasterly moving Eurasian Plate (see Figure 1.6). Europe lies on the westernmost part of the Eurasian Plate. South of the main Alps formation, lower mountains extend into the peninsulas of Iberia and Italy and along the Adriatic Sea through Greece to the southeast. The northernmost mountainous formation is shared by Scotland, Norway, and Sweden. These northern mountains are old (about the age of the Appalachians in North America) and have been worn down by glaciers and millions of years of erosion.

Extending northward from the central Alpine zone is a band of low-lying hills and plateaus curving from Dijon (France) through Frankfurt (Germany) to Krakow (Poland). These uplands (**Figure 4.4**) form a transitional zone between the high mountains and lowlands of the *North European Plain*, which is the most extensive landform in Europe (**Figure 4.5**). The plain begins along the Atlantic coast in western France and covers a wide band around the northern flank of the main European peninsula, reaching across the English Channel and the North Sea to take in southern England, southern Sweden, and most of Finland. The plain continues east through Poland, then broadens to the south and north to include all the land east to the Ural Mountains in Russia.

Crossed by many rivers and holding considerable mineral deposits, the coastal lowland of the North European Plain is an area of large industrial port cities and densely occupied rural areas. Over the past thousand years, people have transformed the natural seaside marshes and vast river deltas into farmland, pastures, and urban areas by building dikes and draining the land with wind-powered pumps. This is especially true in the low-lying Netherlands, where concern over climate change and sea level rise is considerable.

The rivers of Europe link its interior to the surrounding seas. Several of these rivers are navigable well into the upland zone, and Europeans have built industrial cities on their banks. The Rhine carries more traffic than any other European river, and the course it has cut through the Alps and uplands to the North Sea also serves as a route for railways and motorways (**Figure 4.6**). The area where the Rhine flows into the North Sea is considered the economic core of Europe. Rotterdam, Europe's largest port, is located here. The Danube River, larger and much longer than the Rhine, flows southeast from Germany, connecting the center of Europe with the Black Sea. As the European Union expands to the east, the economic and environmental roles of the Danube River basin, including the Black Sea, are getting more attention (**Figure 4.7**).

Vegetation and Climate

Nearly all of Europe's original forests are gone, some for more than a thousand years. Today, forests with very large and old trees exist only in scattered areas, especially on the more rugged Alpine slopes

Figure 4.6 The Rhine River at Basel, Switzerland. [Ingolf Pompe/LOOK-foto/Getty Images]

Figure 4.7 The Danube River at Budapest, Hungary. [_ultraforma_/E+/Getty Images]

midlatitude temperate climate as in south-central North America, China, and Europe, a climate that is moist all year, with relatively mild winters and long, mild-to-hot summers

North Atlantic Drift the easternmost end of the Gulf Stream, a broad, warm-water current that brings large amounts of warm water from the Gulf of Mexico to the coasts of Europe

Mediterranean climate a climate pattern of warm, dry summers and mild, rainy winters

continental climate a climate pattern in which summers are fairly hot and moist and winters become longer and colder the deeper into the interior of the continent one goes

(see Figure 4.3) and in the northernmost parts of Norway, Sweden, and Finland. The dominant vegetation in Europe is crops and pasture grass. Forests are intensively managed and are regenerating on abandoned farmland where agriculture is no longer competitive, so now regrowth forests cover about one-third of Europe.

Europe has three main climate types (**Figure 4.8**): midlatitude temperate, Mediterranean, and continental. The **midlatitude temperate climate**, characterized by year-round moisture, relatively mild winters, and long, mild-to-hot summers, dominates in western Europe, where the Atlantic Ocean moderates temperatures (Figure 4.8A). To minimize the effects of heavy precipitation runoff, people in these areas have developed elaborate drainage systems for their houses and communities. Forests are both evergreen and deciduous.

A broad, warm-water ocean current called the **North Atlantic Drift** brings large amounts of warm water to the coasts of Europe. It is really just the easternmost end of the Gulf Stream, which carries warm waters from the Gulf of Mexico north along the eastern coast of North America, then curves across the North Atlantic to Europe (see Figure 4.8). The air above the North Atlantic Drift is relatively warm and wet, and the eastward-blowing winds that push this air over North and West Europe and the North European Plain bring moderate temperatures and rain deep into the Eurasian continent, creating a climate that, although still fairly cool, is much warmer than elsewhere in the world at similar latitudes. There is some concern that global climate change is weakening the North Atlantic Drift, leading perhaps to a significantly cooler, and drier, Europe.

In the south of Europe, the **Mediterranean climate** prevails—warm, dry summers and mild, rainy winters (Figure 4.8B). In the summer, warm, dry air from North Africa shifts north over the Mediterranean Sea as far as the Alps, bringing high temperatures and clear skies. Crops grown in this climate, such as olives, grapes, citrus, wheat, apples, and other fruits, must be drought-resistant or irrigated. In the fall, this warm, dry air shifts to the south and is replaced by cooler temperatures and rainstorms sweeping in off the Atlantic. Overall, the climate here is mild, and houses along the Mediterranean coast are often open and airy to afford comfort in the long, hot, sunny summers. During the short, mild winters, life moves indoors, where wood fires are used to heat just one or two small rooms even in large houses.

In Central Europe and the far northern Scandinavian peninsulas, all of which are situated away from the moderating influences of the North Atlantic Drift and the Mediterranean Sea, the climate is more extreme. In this region of **continental climate**, summers are fairly short and hot and the winters become longer and colder the farther north or deeper into the interior of the continent one goes (Figure 4.8C). Here, houses tend to be well insulated, with small windows, low ceilings, and steep roofs that can shed snow. Crops that must be adapted to much shorter growing seasons include corn and other grains plus fruit trees and a wide variety of vegetables adapted to the cold, especially root crops and cabbages.

CHECK YOUR UNDERSTANDING

1. What is the significance of Europe's many peninsulas?
2. Describe and locate Europe's three main landform types.
3. Describe and locate Europe's three principal climates.

ENVIRONMENT

Europeans are aware of having dramatically transformed and in many cases degraded their environments over the past several thousand years, and are now taking action on many fronts. The goals of having clean air and water, sustainable development in agriculture, industry, and energy use while maintaining biodiversity are, nevertheless, far from achievement. The European Union (plus nonmember states Norway, Switzerland, and Iceland) has set goals for

4.8 Climates of Europe

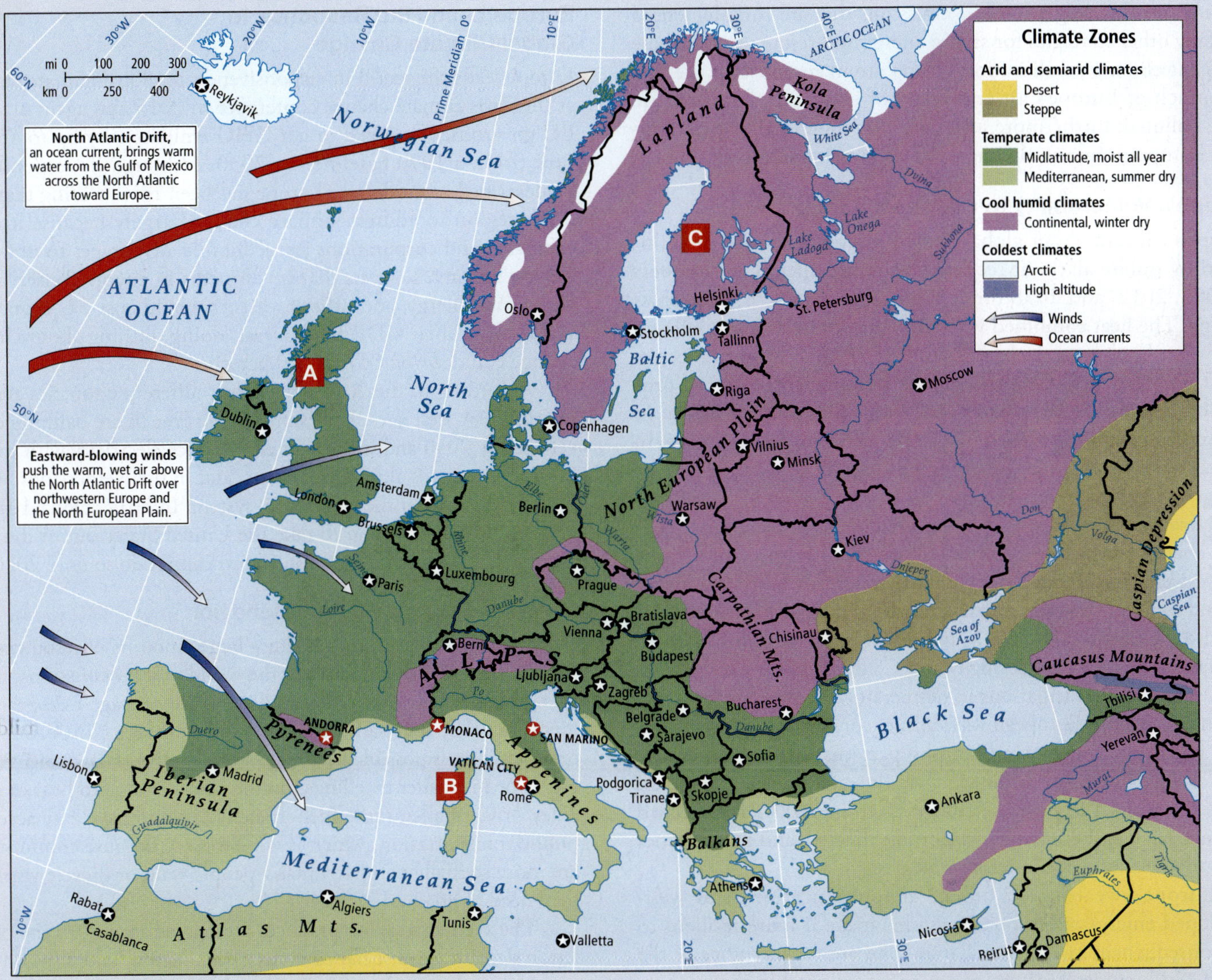

A **Midlatitude climate, moist all year, Scotland.** [David Henderson/Getty Images]

B **Mediterranean climate, summer dry, Corsica.** [Ellen Rooney/robertharding/Getty Images]

C **Continental climate, winter dry, Finland.** [Raimund Linke/Getty Images]

cutting greenhouse gas (GHG) emissions that are complemented by many other strategies for saving energy and resources. But these stringent climate amelioration efforts are in response to the fact that much of Europe's air and many of its seas and streams are quite polluted; furthermore, through their high level of consumption, Europeans negatively impact environments across the globe.

Europe's Vulnerability to Climate Change

Europe's concern about global climate change may also be influenced by public alarm at recent extreme weather. The summers of 2003, 2012, and 2018 broke high-temperature records across Europe. The heat combined with drought caused crops to fail and freshwater levels to sink. In the summer of 2018, wild fires broke out from Sweden to Greece. Nuclear power plants reduced production because the river water used for cooling was too warm. In the Austrian Alps, the Dachstein glacier melted at a record pace. In Switzerland, water for cattle in high mountain pastures had to be brought in by helicopter.

However, there is considerable variation in vulnerability to climate change across Europe. In 2018, northern Europe showed the most extreme record temperatures and droughts, while parts of central and southern Europe experienced record rainfall and snowfall, as was also the case in 2006, 2012, and 2013. By late summer in 2018, the rivers of France and Germany and Central Europe were beyond flood stage. These vacillations in temperature and precipitation are causing changes in European ways of life, constricting budgets, limiting vacations, and causing farming patterns to change. Perhaps most alarming is sea-level rise, which at the rate of three millimeters per year is judged to be irreversible even if the **2017 Paris Accord** is respected. Coastal habitats are already affected, and seaside cities must adopt serious amelioration strategies.

Europe's wealth, technological sophistication, and well-developed emergency response systems make it more resilient to the consequences of climate change than are most regions of the world. Still, some areas are much more vulnerable than others because of their location, dwindling water resources, and rising sea levels, or because of the effects of poverty. **Figure 4.9** illustrates some of these vulnerabilities. The European Environment Agency has created a report that makes it easy to see that climate change is likely to have far-reaching and contradictory effects in different parts of the region. This report, *National Climate Change Vulnerability and Risk Assessments in Europe, 2018*, can be accessed online.

The Mediterranean is likely to have the most consequential changes, including the risk of desertification, loss of biodiversity, water scarcity, more subtropical diseases, and the loss of tourism income. Central and eastern Europe will face decreasing precipitation and risk of forest fires, while northwestern and northern Europe could experience increasing precipitation, flooding, rising seas, and possibly better crop yields, but recent trends suggest increasing heat and drought.

Paris Accord, 2017 a UN-sponsored agreement by 181 countries to lower greenhouse gas emissions

green an adjective indicating a person or group that is environmentally conscious

Green political parties groups that consistently advocate for emission controls, community recycling, grassroots work on local environments, and general humanitarian policies

Europe Leads the Response to Global Climate Change

Europe leads the world in responsiveness to global climate change; at the Paris climate change conference in 2015 and again in 2017, EU governments agreed to cut GHG emissions by up to 40 percent (from a 1990 baseline) by 2030. Although its response can be described as moderate, Europe has been more willing than any other region to address climate change, largely because its governments and corporations see economic advantages to doing so. Recent research suggests that investments in energy conservation, alternative energy, and other measures would cost EU economies 1 percent of their GDP. By contrast, doing nothing about climate change could *shrink* GDP by 20 percent.

According to the 2017 agreement, different countries are taking different strategies. For example, the French are banning diesel vehicles by 2040 and will no longer produce electricity with coal by 2022. Germany is closing coalmines, and in the Netherlands, windmills already power the electric rail system. In 2017, Donald Trump announced his intention to take the United States out of the Paris agreements, but due to legal constraints cannot do so until 2020.

Europe's Everyday Green Behavior

By global standards, Europeans use large amounts of resources and contribute about one-quarter of the world's GHG emissions. However, **green** behavior is a long tradition. The average European resident uses only one-half the energy of the average North American resident. This is possible because Europeans live in apartment complexes or in smaller dwellings that need less energy to heat or cool. They drive smaller, more fuel-efficient cars and have easy access to public transportation, which is widely used. Because communities are more spatially compact, many people walk or bicycle wherever they need to go.

These energy-saving practices are related in part to the population densities and social customs of the region, and also to widespread popular support for ecological principles. **Green political parties** influence national policies in all European countries, and their popularity is increasing in parliamentary elections. Green policies are central to the agenda of the European Union; results include strong regional advocacy for emission controls, well-entrenched community recycling programs, and grassroots work on local environmental concerns. Take for example the case of London's Guerrilla Gardeners. Under cover of darkness, Green activists—who in the daytime are bureaucrats, teachers, stock traders, and computer programmers—sneak into Central London to plant colorful flowers and foliage in traffic islands and roundabouts (**Figure 4.10**). Armed with trowels, spades, mulch, and watering cans, they make quick assaults late at night. The Guerilla Gardeners are part of an international movement in more than 30 countries.

Transportation Impacts on Europe's Environment

Europe has an extraordinarily dense and advanced transportation network. Although Europeans have for many years favored fast rail networks for passengers and cargo rather than private cars and trucks, they have recently been drifting closer to the American model of private cars and trucks driven on sweeping freeways. This change is now happening in even the poorest parts of Europe, where cars have been scarce in the past. However, in response to

PHOTO ESSAY

4.9 Vulnerability to Climate Change in Europe

Despite Europe's wealth, technological sophistication, and well-developed emergency response systems that make it more resilient to the consequences of climate change than most world regions, some areas are vulnerable because of their location, dwindling water resources, susceptibility to rising sea levels and flooding, or the effects of poverty. Here you see three very different situations created by climate change in Europe.

THINKING GEOGRAPHICALLY

A **Other than the church, what clues in this photo indicate the reservoir level is sinking?**

B **Why could one say that these structures are the modern equivalent of seventeenth-century windmills?**

C **How might owning a vehicle decrease a family's vulnerability to flooding?**

A Spain. As the water level in a reservoir in Catalonia, Spain, sinks, a once-submerged eleventh-century church reemerges. Rising temperatures will make Spain's climate drier, its evaporation rate will increase, and its scarce water resources will shrink further, threatening agriculture and drinking water, especially along the Mediterranean coast. [Jasper Juinen/Getty Images]

B The Netherlands. The massive Delta Works, a series of dams, sluices, locks, dykes, levees, and storm surge barriers, protects the Netherlands from rising sea levels, during storms. This massive engineering project renders the Netherlands only slightly less vulnerable to climate change than the rest of Europe, even though 60 percent of its population lives below sea level. [The Asahi Shimbun via Getty Images]

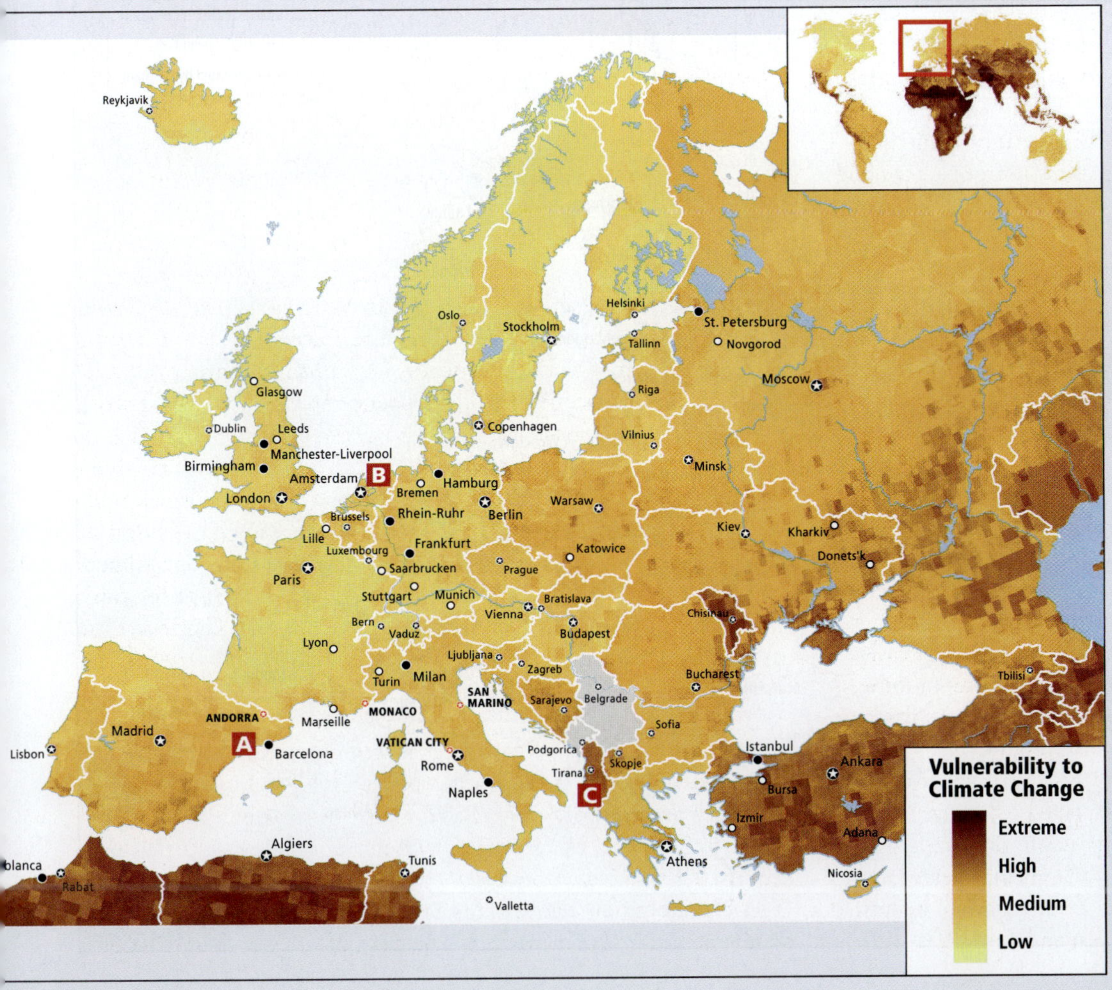

C Albania. Albania's high vulnerability to climate change is related to its low incomes and inadequate infrastructure, including transportation. Therefore, Albania's sensitivity to potential disturbances is high and its resilience low. Many problems stem from misguided government policies. For example, until 1992 it was illegal for Albanians to own private automobiles. Even today, many poorer Albanians depend on horses and small, slow, horse-drawn carts as their main mode of transportation. [ERMAL META/AFP/Getty Images]

Figure 4.10 Guerilla gardeners in London. [Minna Kantonen/Getty Images]

rising fuel costs and CO_2 emissions, the European Union has developed long-term plans that reduce the emphasis on cars and trucks and involve designing *multimodal transport* to link high-speed rail (**Figure 4.11**) to road, air, and water transportation. Highway transport, while convenient, is inefficient and polluting relative to rail or water transport. Private cars used for passenger transportation produce three times more CO_2 emissions than does rail-based public transportation. By truck, 1 gallon (3.7 liters) of gasoline can move 1 ton of cargo a distance of 59 miles (94 kilometers), by rail 202 miles (323 kilometers), and by waterway 514 miles (822 kilometers).

Europe's many navigable rivers, its long, irregular coastline, and the low cost of water transportation have been a boon for the development of links to global trade. Europe has numerous modern ocean ports that cater to container ships: Helsinki, Riga, Hamburg, Copenhagen, Antwerp, Rotterdam, Plymouth, Southampton, Le Havre, Barcelona, Marseille, and Koper. The EU transportation plan, expected to be completed in 2020, includes "Motorways of the Sea," which are upgraded shipping lanes through the Baltic,

Figure 4.11 High-speed rail in Europe, and major seaports. Europe is well served with railroads. This map shows only the high-speed lines. The bright red lines are operated by Germany, often in partnership with other entities, private and government. Deepwater seaports active in global trade also ring the continent. All such ports make use of rail lines to access inland markets.

the North Sea, the English Channel, and along the eastern Atlantic. In the Mediterranean, shipping lanes are being improved from the western Mediterranean to the northern Adriatic, and several ports across North Africa are being added. All of these changes will improve trade not only for European markets, but also for markets of Morocco, Algeria, Tunisia, Malta, Libya, Egypt, Israel, Palestine, Lebanon, and Turkey. One-third of the world's container traffic now goes through the Mediterranean. The container ship industry is keenly attuned to concerns about CO_2 and climate change, and is now using optimal (often slower) speeds that are carefully calculated to minimize fuel consumption and emissions while maximizing profits.

Europe's Energy Issues

In the past, Europe's main energy sources shifted first from wood to coal and then to petroleum and natural gas, and in some countries to nuclear power. However, in response to rising energy costs and commitments to cut GHG emissions, new energy sources are increasingly emphasized. Nonetheless, wood, a major GHG emitter, continues to be a major source of energy for home heating. Wood fuel persists because it is cheap and carries symbolic meaning, evoking traditional home and family values.

The EU-28 are all net importers of energy and get a large portion of their fossil fuel supplies from Russia—30 percent of their crude oil, 33 percent of their natural gas, and 26 percent of their coal, as of 2018. About half of the natural gas now comes from Russia via pipelines through Belarus and Ukraine, while Turkey supplies Russian gas to Europe via the Black Sea and hosts a pipeline that carries gas from the Caspian Sea to Europe. Russia is negotiating for another trans-Turkey pipeline to carry oil and gas to the Mediterranean. Europeans fear this dependency will be used against them by Russia, which periodically withholds flows to Europe through Belarus and Ukraine. But actually, Russia is itself dependent on the EU fuel trade because 70 percent of Russia's fossil fuel exports, which remain state-owned assets, go to the European Union, and these export sales account for half of Russia's government budget.

Another 30 percent of the gas and oil that the European Union consumes comes from various Middle Eastern producers. Large oil and gas deposits in the North Sea, most controlled by Norway (not an EU member), have alleviated Europe's dependence on "foreign" sources of energy, but the production of oil from the North Sea has already peaked and is expected to diminish significantly in the next decade.

The EU is increasingly promoting energy independence and security. The use of nuclear power to generate electricity has been more common in Europe than in North America; Finland, the Czech Republic, Slovakia, Hungary, and Bulgaria have Russian-built nuclear power stations. Russia also supplies 18 percent of the mined uranium used in such plants. But concerns about safety—especially after the 2011 Fukushima reactor disaster in Japan—and the problem of how to safely manage spent nuclear fuel have resulted in deep questioning of the viability of nuclear power. The EU-28 depends on nuclear power for 25 percent of its total needs; France alone uses half of that total. In France, 77 percent of the electricity was generated by nuclear power during 2015 (compared with only 20 percent in the United States), but the use of nuclear power is declining even in France. Germany has stopped building new reactors and is weaning itself off nuclear power as enthusiasm grows across western Europe for renewable energy.

Renewable and Alternative Energy Europe has the highest per capita use of renewable energy on Earth, but remember, the burning of wood, which is constantly renewed in Europe's vast forests, produces much of this renewable energy. Yet, wood use is a problem because it contributes to CO_2 emissions. The EU wants to increase its use of renewable energy in order to reduce fuel imports and thereby strengthen its energy security, stimulate the economy with new energy-related jobs, but also combat climate change by reducing emissions. The EU's goals are to increase use of non-CO_2 renewable energy by 20 percent by 2020; power 60 percent of EU homes with renewably generated electricity by 2020; and reduce its CO_2 emissions by 40 percent by 2030.

A wide array of alternative, non-CO_2–producing energy projects is now attracting significant private investment, allowing government subsidies to decline. Wind power is the favored technology: Europe had 38 percent of the world's installed wind capacity in 2017, enough to supply about 13 percent of the EU's electricity. Solar power, which in 2017 supplied only 6 percent of the EU's electricity, is surging in popularity, in part due to rapidly declining prices for Asian-made solar panels (**Figure 4.12**). Europe has about 75 percent of the world's installed solar power capacity.

Figure 4.12 A huge array of solar panels on a farm in Denmark. Kim Rasmussen is a farmer in Denmark who decided "raising a crop of solar panels" might be a useful addition to his mixed farming strategies—he raises cattle, barley, and wheat, and produces seeds from a variety of grasses, which he sells to seed companies. Set on a sunny hillside, Rasmussen's 12,500 solar panels on two hectares of land power his farm and four nearby villages. [Mac Goodwin]

In Germany, combined output of renewable energy (wind, solar, biomass, and hydroelectric) reached 104 billion kilowatt hours between January and June 2018—enough to power every household in the country. In 2017, 29 percent of the UK's electricity was produced by renewables. Windmills and solar panels are seen across the rural and village landscapes.

Even renewable energy has environmental impacts: windmills kill thousands of birds and spoil pristine seascapes and landscapes; solar panels affect the ecology and beauty of the land they occupy; wood removed from forests changes habitats and reduces CO_2 absorption by trees while producing unwanted CO_2 when burned; and all these sources of energy produce pollution, especially during materials-manufacturing stages.

Air Pollution

There is significant energy-related air pollution in much of Europe, but it is particularly heavy over the North European Plain. This is a region of heavy industry, dense transportation routes, and large and affluent populations. The intense fossil fuel use associated with affluent lifestyles results not only in the usual air pollution but also in *acid rain*, which, due to wind patterns, can fall far from where it was generated.

The former communist countries in Central and North Europe have a special problem with air pollution. Mines in this region produce highly polluting soft coal that is burned in outdated factories and power plants. Central Europe also produces high per capita emissions from burning oil and gas, and it receives air pollution blown eastward from western Europe. In Upper Silesia (Poland's leading coal-producing area), acid rain has destroyed forests, contaminated soils and the crops grown on them, and raised water pollution to deadly levels. Residents have higher rates of birth defects and cancer and lower life expectancies than comparable populations in the rest of Europe. Industrial pollution was one of Poland's biggest obstacles to entry into the European Union, and it only barely met the EU's environmental requirements.

Central Europe's severe environmental problems developed in part because the theories and policies of the Soviet Union portrayed nature as existing only to serve human needs. During the Soviet era, little pollution data was collected and public protests against pollution were prohibited. However, the shift toward more open societies after 1991 has made activism possible in places like Hungary and Bulgaria, and popular protest has resulted in reductions in air pollution as well as public attention to other environmental issues.

The new EU member states in Central Europe (the former Soviet bloc countries) are improving energy efficiency and reducing emissions to some extent. Power plants, factories, and agriculture are polluting less, and the countries with the worst emissions records, such as Poland, have been making the most progress. Nonetheless, the emphasis on market economies also has brought many more ways to generate air pollution, along with other forms of pollution.

Freshwater and Seawater Pollution

Sources of water pollution in Europe include insufficiently treated sewage, chemicals and silt in the runoff from agricultural plots and urban areas, consumer packaging litter, petroleum residues, and industrial effluent (**Figure 4.13**). Most inland waters contain a variety of such pollutants. Any pollutants that enter Europe's inland wetlands, rivers, streams, and canals eventually reach Europe's surrounding coastal waters. The Atlantic Ocean, the Arctic Ocean, and the northern reaches of the North Sea are better able to disperse chemical pollutants dumped into them because they are part of, or closely connected to, the circulating flow of the world ocean. In contrast, the Baltic, Mediterranean, and Black seas, along with the southern North Sea, are nearly landlocked bodies of water that do not have the capacity to flush themselves out quickly and are prone to accumulating chemical pollution and plastic litter.

In the Mediterranean, the effect of the pollution that pours in from rivers, adjacent cities, industries, hotel resorts, and farms is exacerbated by the fact that the sea has just one tiny opening to the world ocean (see Figure 4.1). At the surface, seawater flows in from the Atlantic through the narrow Strait of Gibraltar and moves eastward. At the bottom of the sea, water exits through the same narrow opening, but only after it has been in the Mediterranean for 80 years or more. The natural ecology of the Mediterranean is attuned to this lengthy cycle, but the nearly 460 million people now living in the countries surrounding the sea have upset the balance. Their pollution stays in the Mediterranean for decades. As a result, fish catches have declined, beloved seaside resorts have become unsafe for swimmers, and agricultural workers have become sick.

There are 34 countries with coastlines on Europe's many seas, all with different economies, politics, and cultural traditions. Such diversity makes it difficult to cooperate to minimize pollution or even reduce the risk of severe pollution. Although most of the Mediterranean's pollution is generated by Europe, rapidly growing populations and economic development on the North African and eastern Mediterranean coasts of the sea also pose environmental threats because these countries still lack adequate urban sewage treatment and environmental regulations to control agricultural and industrial wastes. Over the past decades, wars and armed conflict throughout the eastern Mediterranean (Libya, Egypt, Israel, Lebanon, Syria, and Turkey) have added to the pollution and disrupted efforts to stop long-standing polluting customs such as dumping household sewage and industrial waste into the sea and the streams that feed into it.

CHECK YOUR UNDERSTANDING

1. To what extent is Europe both vulnerable to, and responding to, various climate change issues?
2. Describe Europe's strategies for saving energy and resources.
3. What are the several ways Europeans manage to use less energy per capita than people in the United States?
4. What are some regional differences in environmental issues and how do responses vary?

HUMAN PATTERNS OVER TIME

The range of explanations for Europe's powerful and sustained impact on the world varies widely. One old argument, for which there is no evidence, is that Europeans are somehow a superior breed of humans. Another is that Europe's many bays, peninsulas, and navigable rivers have promoted commerce to a larger extent there than elsewhere, a deterministic argument that is inadequate. In fact, much

Figure 4.13 Percentage of coastal zone waters with less-than-good ecological status, 2015. The waters along Europe's coastal zones are to varying extents polluted by agricultural, industrial, and fossil fuel toxins. The nearly landlocked Baltic Sea has the worst pollution, but pollution along the edges of the North Sea is also severe. The white shading around Norway merely means that the government did not provide data (Norway is not in the EU).

of Europe's success is based on technologies and ideas it borrowed from elsewhere. For example, the concept of the peace treaty, so vital to current European and global stability, was first documented not in Europe but in ancient Egypt. In order to understand how Europe gained the leading role around the globe that it still struggles to retain, it is helpful to look at the history of this area.

Sources of European Culture

About 10,000 years ago, the practice of agriculture and animal husbandry gradually spread into Europe from the uplands and plains associated with the Tigris and Euphrates rivers in Southwest Asia (present-day Iraq) and from farther east in Central Asia and beyond. Mining, metalworking, and mathematics also came to Europe from these places and from various parts of Africa. All of these borrowed innovations increased the possibilities for trade and economic development in Europe.

The first European civilizations were ancient Greece (800 to 86 B.C.E.) and Rome (753 B.C.E. to 476 C.E.). Located in southern Europe, both Greece and Rome initially interacted more with the Mediterranean rim, Southwest Asia, and North Africa than with the rest of Europe, which then had only small and relatively impoverished rural populations. Later European traditions of science, art, and literature were heavily based on Greek ideas, which were themselves derived from yet earlier Egyptian and Southwest Asian (Arab and Persian) sources.

The Romans, after first borrowing from Greek culture, also left important legacies in Europe. Many Europeans today speak *Romance* languages, such as Spanish, Portuguese, Italian, French, and Romanian, all of which are largely derived from Latin, the language of the Roman Empire. European laws that determine how individuals own, buy, and sell land originated in Rome. Europeans then spread these legal customs throughout the world through colonization.

The practices that Romans used when colonizing new lands also shaped much of Europe. After a military conquest, the Romans secured dominance in rural areas by establishing large, plantation-like farms on which local people worked and were monitored. Politics and trade were centered in new Roman towns built on a grid pattern that facilitated commercial activity. Commanding structures, like the ceremonial gate in Figure 4.14A, inspired loyalty and facilitated military repression of rebellions. These same systems for taking and holding territory were later used when Europeans colonized the Americas, Asia, and Africa (**Figure 4.14**).

A The Arch of Trajan in the Roman colonial town of Timgad, Algeria, founded around 100 C.E. [Ethel Davies/ robertharding/Getty Images]

B Alhambra, a palace built by Spain's North African Islamic rulers in the thirteenth century. [Izzet Keribar/ Getty Images]

C A holiday festival in a Belgian town, painted by Pieter Bruegel (the elder) in 1559. [De Agostini Picture Library/Getty Images]

10,000 B.C.E. · 1000 B.C.E. · 500 B.C.E. · 0 C.E. · 500 C.E. · 1000 C.E. · 1400 C.E. · 1500 C.E.

8000 B.C.E. Agriculture and animal husbandry introduced via Southwest Asia

753 B.C.E.–476 C.E. Roman civilization

1400 North African Islamic rule of Spain

711–1492

1500s Protestant Reformation

Figure 4.14 A VISUAL HISTORY OF EUROPE

Thinking Geographically

A What about the photo of the ruins of Timgad reflects a particular Roman strategy for organizing space?

B What about the architecture of Alhambra suggests technological sophistication and an astute understanding of climate?

C What about this picture suggests relative prosperity in sixteenth-century Belgium?

The influence of Islamic civilization on Europe is often overlooked. After the fall of Rome, while Europe was in a time known as the *early medieval period* (roughly 450 to 1300 C.E.; sometimes referred to as the *Dark Ages*), Turkish, Persian, Egyptian, and Arab scholars preserved learning from their own ancient traditions and also from Greece and Rome in large libraries, such as those in Alexandria, Egypt, and Constantinople (now Istanbul), Turkey. Also, Muslim Arabs originally from North Africa ruled Spain from 711 to 1492 (see Figure 4.14B), and from the 1400s through the early 1900s, the Ottoman Empire (based in what is now Turkey) dominated much of southeastern Europe and Greece. The Arabs, Persians, and Turks all brought ideas, new technologies, food crops, architectural principles, and textiles to Europe from Arabia, Persia, Anatolia, China, India, and Africa. Arabs also brought Europe its numbering system, mathematics, and significant advances in medicine and engineering, building on ideas they picked up in South Asia.

Beginning 500 years ago, Europe also began to draw on a host of cultural features from the various colonies that Europe established in the Americas, Asia, and Africa. For example, many food crops now popular in Europe came from the Americas (some examples are potatoes, corn, peppers, tomatoes, beans, squash, pineapples, and cacao; see Table 3.1) and Southwest, Central, and East Asia (wheat, leafy greens, garlic, onions, apples, and citrus fruit).

The Inequalities of Feudalism

As the Roman Empire declined, a social system known as *feudalism* evolved during the *medieval period* (450–1500 C.E.). This system originated from the need to defend rural areas against local bandits and raiders from Scandinavia and the Eurasian interior. The objective of feudalism was to have a sufficient number of heavily armed, professional fighting men, or *knights*, to defend a much larger group of *serfs*, who were legally bound to live on and cultivate plots of land for the knights. Over time, some of these knights became a wealthy class of warrior-aristocrats, called the *nobility*, who controlled certain territories. Some nobles gained power over other knights, amassing vast kingdoms.

The often-lavish lifestyles and elaborate castles (**Figure 4.15**) of the wealthier nobility were supported by the labors of the serfs. Most serfs lived in poverty outside castle walls and, much like enslaved people, were legally barred from leaving the lands they cultivated for their protectors.

Early Urbanization Transformed Europe

While rural life followed established feudal patterns, new political and economic institutions were developing in Europe's towns and cities. Here, thick city walls provided defense against raiders, and commerce and crafts supplied livelihoods, allowing the people more independence from feudal knights and kings.

Located along trade routes, people in Europe's early urban areas were exposed to new ideas, technologies, and institutions from Southwest Asia, India, and China. Some institutions, such as banks, insurance companies, and corporations, provided the foundations for Europe's modern economy. Over time, citizens of Europe's urban areas established a pace of social and technological change that left the feudal rural areas far behind.

Urban Europe flourished in part because of laws that granted basic rights to urban residents. With adequate knowledge of these laws set forth in legal documents called *town charters*, people with few resources could protect their rights even if challenged by those who were more wealthy and powerful. Town charters provided a basis for European notions of *civil rights*, which have proved hugely influential throughout the world. With strong protections for their civil rights, some of Europe's townsfolk grew into a small middle class whose prosperity moderated the feudal system's extreme divisions of status and wealth. Occasional fairs and festivals

D Portuguese mercenaries off the coast of India in 1537. [Werner Forman/Universal Images Group/Getty Images]

E The execution of King Louis XVI in 1793 during the French Revolution. [Prisma/Universal Images Group/Getty Images]

F A German iron works around 1900. [Imperial Wharf in Kiel, Germany, c.1900 (colour litho)/German School (20th century)/INDIVISION CHARMET/ Bibliotheque des Arts Decoratifs, Paris, France/Bridgeman Images]

1600 C.E. — 1700 C.E. — 1800 C.E. — 1900 C.E.

1568

1500–1700 Portuguese and Spanish empires expand

1600–1900 Dutch, British, and French empires expand

1682

1780s The Industrial Revolution begins

1793

1789–1799 The French Revolution

1850

D What about this picture suggests how the Portuguese felt about the value of trade with India?

E How is the execution of Louis XVI related to the evolution of democracy in Europe?

F What does this photo indicate about the impact of the Industrial Revolution on employment in Europe?

Figure 4.15 Predjama Castle in Slovenia. Predjama Castle, perched in the mouth of a cave in southwestern Slovenia, exemplifies the feudalism from which modern Europe eventually emerged. First mentioned in the historical record in the thirteenth century, Predjama became legendary as an impregnable fortress in the fifteenth century, when the Austrian Imperial Army laid siege to it for over a year. For months the Austrians were ignorant of a passage through the cave that kept the castle supplied. The Austrians finally succeeded in taking the castle and then the region when Predjama's owner, Erazem Leuger, was betrayed by one of his own men. A cannonball killed Leuger while he sat on a toilet in Predjama's outhouse! [Paul Biris/Getty Images]

provided chances to associate across class divisions (see Figure 4.14C). A related outgrowth of urban Europe was a philosophy known as **humanism**, which emphasizes the dignity and worth of the individual, regardless of wealth or social status.

The liberating influences of European urban life transformed the practice of religion. Since late Roman times, the Catholic Church had dominated not just religion but also politics and daily life throughout much of Europe. Roman Catholic churches were the center of community life and remain prominent in European rural and urban landscapes today. In the 1500s, however, a movement known as the *Protestant Reformation* arose in the urban centers of the North European Plain. Reformers, among them Martin Luther, challenged Catholic practices—such as selling *indulgences*, which supposedly diminished the time a sinner was to suffer in purgatory, and the holding of church services in Latin, a language understood by only a tiny educated minority. The inability to understand Latin services stifled public participation in religious debates. Protestants promoted individual responsibility and more open public debate of social issues, altering the perception of the relationship between the individual and society. These ideas spread faster with the invention of the European version of the printing press (1440s), which enabled widespread literacy, and were the foundation of the **Enlightenment** (1600–1800), a European intellectual movement that emphasized the power of the individual to use reason and logic (science) to understand the world. The central role that cities have come to play in European life is further discussed under the geographic theme "Urbanization."

humanism a philosophy and value system that emphasizes the dignity and worth of the individual, regardless of wealth or social status

Enlightenment the European intellectual movement that emphasized the power of the individual to use reason and logic (science) to understand the world

The Age of Exploration: Europe Goes Global

A direct outgrowth of the greater openness and connectivity of an urbanizing Europe was the exploration and subsequent colonization of much of the world by Europeans. Spain, Portugal, England, and the Netherlands, in competition with one another, explored and then conquered vast overseas territories and created worldwide trade relationships that transformed Europe economically and culturally and laid the foundation for the modern global economy. This commerce and cultural exchange began a period of accelerated globalization that persists today (see the discussion in Chapter 1).

In the fifteenth and sixteenth centuries, Portugal took advantage of advances in navigation, shipbuilding, and commerce to set up a trading empire in Asia (see Figure 4.14D) and a colony in Brazil. Spain soon followed, founding a vast and profitable empire in the Americas and the Philippines. By the seventeenth century, however, England, the Netherlands, and France had seized the initiative from Spain and Portugal. All European powers implemented **mercantilism**, a strategy for increasing a country's power and wealth by acquiring colonies and managing all aspects of their production, transport, and trade for the colonizer's benefit. **Figure 4.16** illustrates the transfers of wealth from the far corners of the Earth to Europe and the types and locations of colonial enterprises that generated the wealth.

mercantilism a strategy for increasing a country's power and wealth by acquiring colonies and managing all aspects of their production, transport, and trade for the colonizer's benefit

Mercantilism brought wealth that supported the Industrial Revolution in Europe (see the next section) by supplying cheap resources from around the globe for Europe's new, mostly urban factories. Industrial labor was provided by those fleeing the feudal countryside, and the colonies became important markets for Europe's manufactured goods.

By the mid-1700s, wealthy merchants in London, Amsterdam, Paris, Berlin, and other western European cities were investing in new industries. Workers from rural areas poured into urban centers in England, the Netherlands, Belgium, France, and Germany to work in the manufacturing industries (see Figure 4.14F) and mining. Wealth and raw materials flowed in from colonial ports in the Americas, Asia, and Africa. Some cities, such as Paris and London,

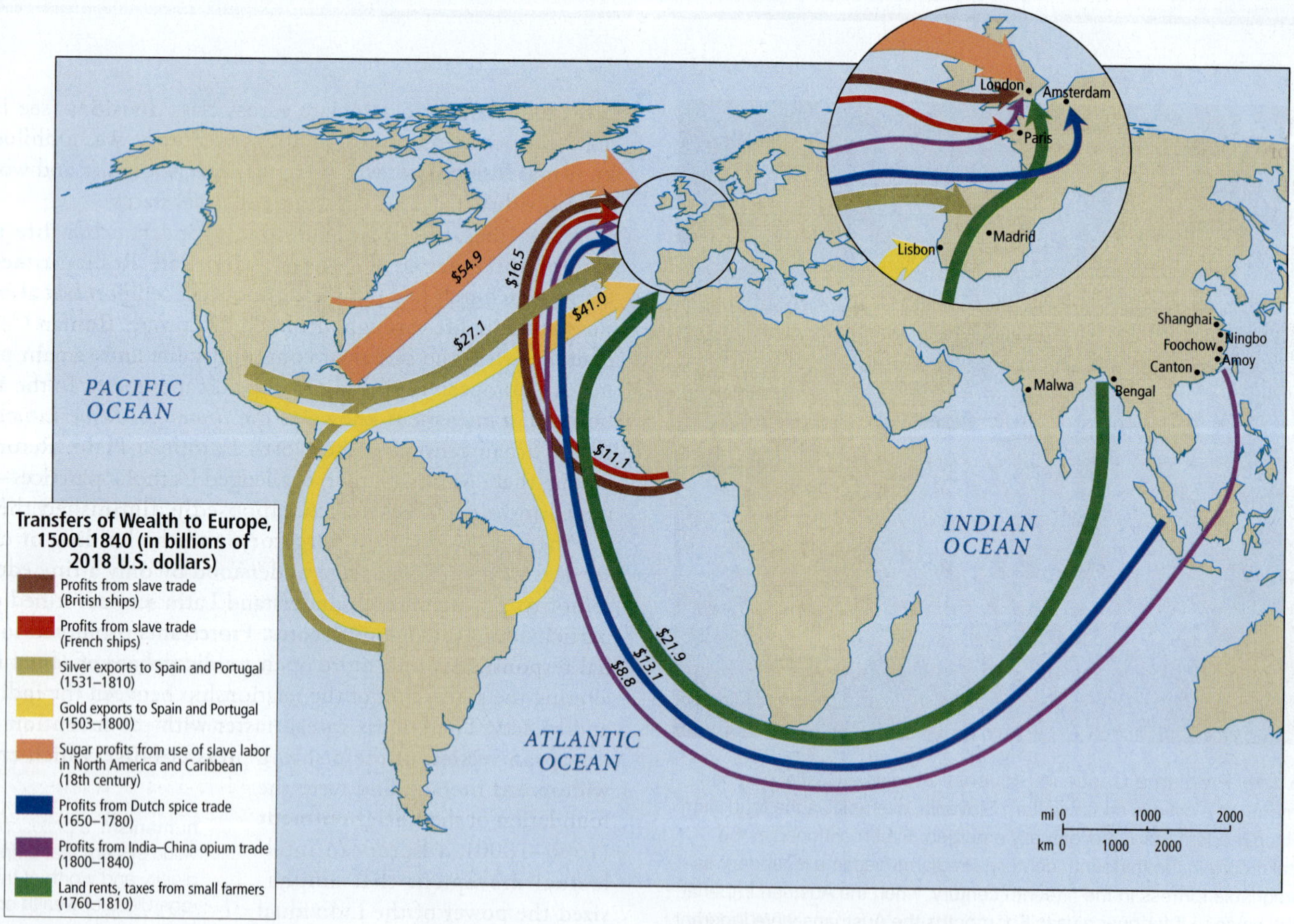

Figure 4.16 Transfers of wealth from the colonies to Europe. During the period of mercantilism, Europe received billions of dollars of income from its overseas colonies. This one-way flow of minerals, agricultural products, and profits of the slave trade fueled the Industrial Revolution in Europe, the construction of magnificent urban structures in Europe's capital cities, and subsequent infrastructure development across the region. In the colonies, mercantilism left a wide range of depleted *landscapes of imperialism*, meaning the landscapes reflected the fact that the wealth created did not remain, but was sent to Europe. [Research from: Alan Thomas, *Third World Atlas* (Washington, DC: Taylor & Francis, 1994), p. 29]

were elaborately rebuilt in the 1800s to reflect their roles as centers of global empires.

By 1800, London and Paris, each of which had a million inhabitants, were Europe's largest cities, and eventually became *world cities* (cities of worldwide economic or cultural influence). London remains a global center of finance and is host to millions of highly educated EU migrants who go there to work; Paris is a cultural center that has influence over global consumption patterns, from food to fashion to tourism, and is also a magnet for migrants.

By the twentieth century, European colonial systems had strongly influenced nearly every part of the world. Within Europe, the overseas empires of England, the Netherlands, and eventually France shifted wealth, investment, and general economic development toward western Europe and away from southern Europe and the Mediterranean. Over time, the British profited the most from their colonies.

Revolutions in Industry and Politics

The wealth derived from Europe's colonialism helped fund two of the most dramatic transformations in a region already characterized by rebirth and innovation: the industrial and democratic revolutions.

Europe's Industrial Revolution—particularly Britain's ascendancy as the leading industrial power of the nineteenth century—was intimately connected with colonial expansion and sugar production. In the seventeenth century, Britain developed a small but influential trading empire in the Caribbean, North America, and South Asia, which provided it with access to a wide range of raw materials and to markets for British goods.

Sugar, produced by British colonies in the Caribbean, was an especially important trade crop (Figure 4.16). Sugar production is a complex process that requires major investments in equipment, and for which a great deal of labor was once needed. Enslaved people were forcibly brought from Africa to help grow and process sugar. The skilled management and large-scale organization needed for the production and distribution of sugar later provided a production model for the Industrial Revolution. The mass production of sugar was facilitated by ever more efficient mechanization, which, in turn, generated enormous wealth and trading opportunities that encouraged industrialization across Europe.

By the late eighteenth century, Britain was introducing mechanization into all its industries, first in textile weaving and then in the production of coal and steel. By the nineteenth century, Britain was the world's greatest economic power, with a huge and growing empire, expanding industrial capabilities, and deploying the world's most powerful navy. Industrial technologies developed in Britain were spread throughout continental Europe (see Figure 4.14F), North America, and elsewhere, transforming millions of lives in the process.

Industrialization led to massive growth in urban areas in the eighteenth and nineteenth centuries. However, the concentration of people and the extremely low living standards in Europe's cities created tremendous pressures for change in the political order, ultimately leading to democratization. Most early industrial jobs were dangerous and unhealthy, demanding long hours and offering little pay. Poor illiterate people were packed into tiny dwellings that they shared with many others. Water was often contaminated, and urban streets were clogged with sewage. Children were often sickened by inadequate sanitation, nutrition, and health care; by industrial pollution; and from being overworked as child laborers (**Figure 4.17**).

Figure 4.17 An engraving from 1871 of barefoot children working in a brickyard in England. Notice that both boys and girls of about age 11 to 13 are carrying heavy loads of clay on their heads. There are no apparent efforts to keep them safe from hazards. [Ann Ronan Pictures/Print Collector/Hulton Archive/Getty Images]

Political Awakening

Opportunities for advancement through education were restricted to those who could afford it, and the ideas and information gained by the few workers who could read gave them the incentive to organize and protest for change. In 1789, the French Revolution led to the first major inclusion of common people in the political process in Europe. Angered by the extreme disparities of wealth in French society and inspired by the popular revolution in North America, the poor rebelled against the monarchy and the elite-dominated power structure that controlled Europe (see Figure 4.14E). Eventually, after the *Reign of Terror* (1793–1794) in France—when more than 40,000 people were executed as rival political factions battled for control—some of the general populace, especially in urban areas, became involved in governing through democratically elected representatives. The political freedoms that grew out of the very violent French Revolution proved short-lived, as governments dominated by the elite soon regained control in France. Nevertheless, over time the French Revolution provided crucial inspiration to later urban democratic political movements in France and many other parts of the world.

The Impact of Communism

During the struggles of the nineteenth century, popular discontent took the form of new revolutionary political movements that proposed radical changes to the established economic and political order. The political philosopher and social revolutionary Karl Marx framed the mounting social unrest in Europe's cities as a struggle between socioeconomic classes. His treatise *The Communist Manifesto* (1848) helped social reformers across Europe articulate ideas about how wealth could be more equitably distributed. East of Europe in Russia, Marx's ideas inspired the creation of a revolutionary communist state in 1917, the Union of Soviet Socialist Republics, often referred to as the *USSR* or the *Soviet Union*.

In West, North, and South Europe (collectively known during this era as *western Europe*), communism (defined and discussed further in "The Cold War") gained some popularity, but this tended to lead to the formation of communist political parties that operated peacefully within the context of democratic political systems. In countries such as France, Italy, Spain, and even Germany, communist political parties are still somewhat influential, occasionally forming governing coalitions with other parties.

By the early twentieth century, after lengthy and often violent struggles against the rich and politically powerful, Europe's huge working class won greater political freedoms, such as the right to vote in elections. These struggles also gained urban industrial workers the right to form unions that could bargain with employers and the government for higher wages and better working conditions. Eventually, political power, wealth, and opportunity were distributed more evenly throughout society. However, it is important to note that the road to these sorts of changes in Europe was rocky and often violent, just as now democratizing processes are halting and disruptive in many other parts of the world.

Two World Wars Affect Europe's Stance Globally

Despite Europe's many advances in industry and politics, at the beginning of the twentieth century, the region still lacked a system of collective security that could prevent war among its rival nations. **Populist** political movements purporting to support "the common people" arose in several parts of Europe, aiming to achieve solidarity by appealing to national or ethnic purity and to the exclusion of "outsiders." Between 1914 and 1945, two extremely destructive world wars left Europe in ruins, no longer the dominant region of the world. At least 20 million people died in World War I (1914–1918) (**Figure 4.18**) and 70 million in World War II (1939–1945). During World War II, Germany's Nazi government killed 15 million civilians in its failed attempt to conquer the Soviet Union. Eleven million civilians died at the hands of the Nazis during the **Holocaust**, a massive execution of 6 million Jews and 5 million gentiles (non-Jews), including ethnic Poles and other Slavs, **Roma** (Gypsies), disabled and mentally ill people, gays, lesbians, transgendered people, and political dissidents. The Italian fascist military under Benito Mussolini was responsible for many thousands of deaths.

populist movement appealing to the interests and prejudices of ordinary people as distinct from the rich and powerful

Holocaust during World War II, a massive execution by the Nazis of 6 million Jews and 5 million gentiles (non-Jews)

Roma the now-preferred term in Europe for Gypsies; some prefer to be called Gypsies as a point of ethnic identity

Iron Curtain a fortified border zone that separated western Europe from eastern Europe during the Cold War

Cold War a period of conflict, tension, and competition between the United States and the Soviet Union from 1945 to 1991

capitalism an economic system characterized by privately owned businesses and industrial firms that adjust prices and output to match the demands of the market

communism a political ideology and economic system, based largely on the writings of the German revolutionary Karl Marx, in which the state owns all farms, industry, land, and buildings and provides for the needs of the people (a version of socialism)

Figure 4.18 Church of the Holy Spirit in Javorca. Memorials to the soldiers on the losing side who fell during World War I are unusual in Europe, but high in the Slovene Alps, an ecumenical church memorializes 2564 Austro-Hungarian soldiers, many teenage boys, who died in nearby Italian Front battlefields. The church was financed and built by surviving soldiers shortly after the war. A Catholic church and a Muslim mosque were also constructed by soldiers in nearby locations. Javorca is now designated a European Heritage Site. [Dr. Jurij Fikfak]

World War II divided Europe and much of the world into two battling camps. While a long list of countries participated, the main combatants were the *Axis* powers of Germany, Italy, and Japan; and the *Allies*—Russia, the United Kingdom, the United States, and France (except during the German occupation of 1940–1944). After World War II ended in 1945 with an Allied victory, a number of changes took place that influence the relationships of European countries even today. Germany was divided into two parts. West Germany became an independent democracy allied with the rest of western Europe—especially Britain and France—and the United States. Russia controlled East Germany and the rest of Central Europe (Latvia, Lithuania, Estonia, Poland, Czechoslovakia, eastern Austria, Hungary, Romania, Bulgaria, Ukraine, Moldova, and Belarus). After the war, under Russia's dominant leadership, these war-devastated countries solidified ideologically into a communist alliance known as the *Soviet bloc*. The line between East and West Germany was part of what was called the **Iron Curtain**, a long, fortified border zone that separated western Europe from what was then called *eastern Europe* (**Figure 4.19**). Yugoslavia in southeastern Europe did not become part of the Soviet bloc but also kept itself insulated from the West, partly to protect its socialized economy and to validate its leadership of countries not aligned with either side of the Cold War.

The Cold War

The division of Europe created a period of conflict, tension, and competition between the United States and the Soviet Union known as the **Cold War**, which lasted from 1945 to 1991. During this time, once-dominant Europe, and indeed the entire world, became a stage on which the United States and the Soviet Union competed for supremacy. The central issue was the competition between **capitalism**—characterized by privately owned businesses and industrial firms that adjust prices and output to match the demands of the market—and **communism** (actually a version of

Figure 4.19 The Iron Curtain, circa 1968. **A** In Czechia (formerly part of Czechoslovakia), the remains of a patrol tower and a stretch of fence built by the Soviet bloc are reminders of the Cold War. [PHB.cz (Richard Semik)/Shutterstock.com]

socialism), in which the state owns all farms, industry, land, and buildings (the so-called *means of production*).

After 1945, in most of what we now call Central Europe, the Soviet Union forcibly implemented a communist-inspired economic model known as **central planning**, in which a central bureaucracy dictated prices and output, with the stated aim of allocating goods and services equitably across society according to need. This system was successful in alleviating long-standing poverty and illiteracy among the working classes and for women, but it was rife with waste, corruption, and bureaucratic bungling. Ultimately, the Soviet economic model collapsed in the early 1990s due to inefficiency, insolvency, high levels of environmental pollution, and public demands for more political freedoms.

During the postwar period, the rest of Europe, especially western Europe, developed capitalist economies with financial support from the United States under the *Marshall Plan* and from the governments of the involved countries. Basic infrastructure, such as roads, housing, and schools, were rebuilt. Economic reconstruction proceeded rapidly in the decades after World War II and included special attention to the social welfare needs of the general public.

Decolonization

Europe's decline during the two world wars also led to the loss of its colonial empires. Many European colonies had participated in the wars, with some (India, the Dutch East Indies, Burma, and Algeria) suffering extensive casualties; almost all emerged economically devastated. After World War II, Europe was less able to assert control over its colonies as their demands for independence grew. By the 1960s, most former European colonies had gained independence, often after bloody wars fought against European powers and their local allies.

The Birth of the European Union

At the end of World War II, European leaders concluded that the best way to prevent the kinds of hostilities that had led to two world wars would be to forge closer economic ties among the European nation-states. The first step toward developing an economic union in Europe began in 1951 with the creation of a common market for coal and steel—two resources essential for war industries, but also for the rebuilding of Europe. The Common Market, as this agreement came to be known, enabled the monitoring of Germany's industrial rehabilitation to be sure it stayed aimed at peace, not rearmament. The next major step took place in 1958, when Belgium, Luxembourg, the Netherlands, France, Italy, and West Germany formed the *European Economic Community* (*EEC*). The members of the EEC agreed to eliminate certain tariffs against one another and to promote mutual trade and cooperation. Denmark, Ireland, and the United Kingdom joined in 1973; Greece, Spain, and Portugal in the 1980s; and Austria, Finland, and Sweden in 1995, bringing the total number of member countries to 15. In 1992, the EEC became the European Union, with the signing of the **Maastricht Treaty**. The 15 member countries agreed to work toward a higher level of economic, social, and political integration that would make possible the free flow of goods and people across national borders.

socialism primarily an economic system in which all people have access to basic goods and services and all large-scale industries are collectively owned, with any profits benefitting society as a whole

central planning a communist economic model in which a central bureaucracy dictates prices and output, with the stated aim of allocating goods equitably across society according to need

decolonization the process of dissolving or ending colonial relationships, institutions, and mind-sets

Maastricht Treaty the 1992 treaty that established the European Union, signed by 15 countries of the European Community in the city of Maastricht, Netherlands

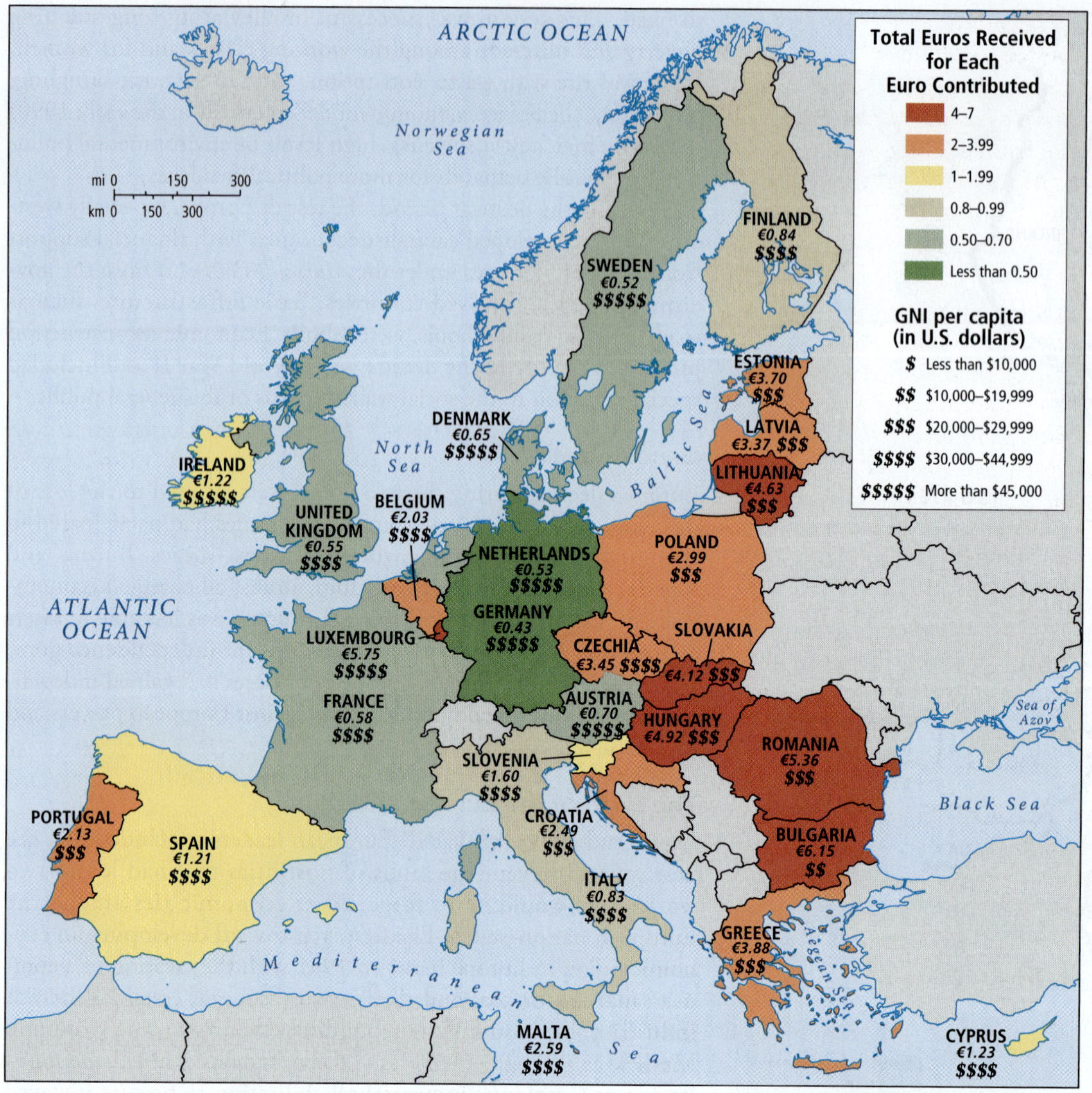

Figure 4.20 The EU budget. The budget of the European Union is remarkable in that some countries, in the interest of the common good, pay far more than they receive in benefits, while others receive much more in benefits than they contribute. GNI figures are symbolized with dollar signs; the euro amounts are the total number of euros received back in services for every euro contributed. The UK with 4 $ signs is a moderately wealthy country, and for every euro it contributes to the budget, it gets only 55 cents back. Other well-off countries also receive less than they contribute. The intention is that eventually the less affluent countries will be brought up to the same level as the other countries and will be able to contribute equally or more generously than others. [Data from: Official website of the European Union, https://europa.eu/european-union/about-eu/funding-grants_en]

Since 1992 the European Union has nearly doubled in size to 28 countries. Figure 4.2 is a map showing all EU members and the dates they joined the union, plus candidate countries, potential candidates, and those who have declined membership. **Figure 4.20** shows the relative wealth (GNI) of these countries, as well as EU budget allocations. The EU budget, €157 billion in 2017, is a tiny fraction (2 percent) of the combined national budgets of the EU-28 (€7,022 billion). The EU budget money is used for a wide range of programs that promote cohesion among the many varied EU member states. These programs include regional and urban development; employment and social inclusion; agriculture and rural development; maritime and fisheries policies; research and innovation; and humanitarian aid.

Nearly all the countries in North and West Europe (plus Spain and Italy) are notably richer than those in Central and South Europe. Many EU budget policies are aimed at addressing this wealth disparity. Some 11 countries receive less from the EU budget than they put in. And while many Europeans understand that the EU budget is quite small and appreciate the rationale behind these budget allocations, others, such as those who voted for Brexit, see this as a major political issue driving the EU apart.

CHECK YOUR UNDERSTANDING

1. Why is it inaccurate to say that Europe's rise to power and wealth was the result of fortuitous geographic location and resources or due to unusually talented people?
2. Explain how the age of exploration and colonialism enhanced Europe's power to a global level and facilitated the Industrial Revolution.
3. How did the two world wars affect Europe's relationships to the wider world?
4. Explain the social rationale behind the EU budget contributions and allocations.

GLOBALIZATION AND DEVELOPMENT

4.5 Discuss the features and challenges of the European Union and its efforts to compete globally.

By the 1960s Europe was shedding its colonial dependencies, participating with the United States as a leader in global affairs, and beginning to implement economic unification. At first as individual countries and then increasingly as a unit, Europe began to use a number of economic development strategies designed to address disparities of wealth and to ensure its ability to compete in the global economy, especially with the United States, Japan, and with industrializing economies of Asia, Africa, and South America. To increase its export potential, Europe has focused on lowering the cost of producing goods in Europe. One strategy has been to shift labor-intensive industries from the wealthiest countries in West Europe, where wages are high, to the relatively poorer, lower-wage states of Central Europe. Generally, this has worked well, helping poorer European countries prosper while keeping the costs of doing business low enough to reduce the incentive for European companies to move away from Europe to places where labor and resource costs are lower still. However, while the economies of Central Europe have grown, in Europe's wealthiest countries, the resultant reduction of industrial capacity (*deindustrialization*) has led to higher unemployment rates, resulting in some resentment among idle workers. And despite these efforts to keep jobs in Europe, some firms have moved abroad to cut costs.

Another way that the EU attempts to address economic and development disparities across the region is to promote social integration via the EU budget. One third of the EU-28 budget is devoted to what is called the **Cohesion Policy** (often called *social cohesion* and discussed further in "Social Protection Systems") aimed at improving overall social integration by removing disparities between countries. The entire budget is made up of funds that EU members have agreed to pool for the good of Europe as a whole. The money is used to improve transportation, energy use, and communications between members; to support the use of sustainable growth to protect the environment; to make Europe competitive globally; and to encourage cooperation in science and learning. The idea is that the EU will be more economically healthy and better able to compete in the global economy if stark disparities in standards of living and well-being are addressed. Therefore, contributions to the EU budget and allocations from the budget are not equitably distributed. In the interest of the common good, affluent countries contribute much more than they receive, and some countries receive much more than they contribute. Figure 4.20 illustrates how budgetary allocations are made to address economic inequities.

ECONOMIC COMPARISONS OF THE EU AND THE UNITED STATES

The EU economy now encompasses more than 512 million people (out of a total of 746 million in the whole of Europe)—roughly 200 million more than live in the United States. Collectively, the EU countries are wealthy, with a joint economy slightly larger than that of the United States. The EU, China, and the United States constantly vie for being the largest economy in the world (recently China has been the largest and the United States, third), though these rankings change from year to year as global markets expand and contract. The European Union exerts a powerful influence on the global trading system (**Figure 4.21**). Nonetheless, it is noteworthy that trade between countries within the EU amounts to about twice the monetary value of trade between the EU as a whole and the outside world.

In contrast to the United States, which usually imports far more than it exports—resulting in a trade deficit—the EU usually maintains a trade balance in which the values of imports and exports are roughly equal. However, there are annual vacillations, which indicate that EU industries may be losing some of their global competitiveness; for example, Europe's clothing stores, once full of rather expensive garments made in Europe, now sell many less expensive products made in Asia. Upscale clothing factories have closed in Italy, Slovenia, and other EU-28 countries. Unlike the United States, EU regulations do not require country-of-origin designations on garments, and this lack of information means EU shoppers are less aware that they are buying non-EU-made items.

Trade between the United States and the EU is now under renegotiation because President Donald Trump asserts that Europe has been unfair during the era when there were free trade agreements. In the United States, tariffs are being imposed on a number of EU products, such as medium-priced and luxury cars. As of this writing, the new agreements have not been finalized.

Income statistics also show contrasts between the EU and the United States. According to the World Bank, the average GNI per capita (PPP) in 2017 for the European Union (U.S.$41,207) was significantly less than that of the United States (U.S.$60,200). China's was U.S.$16,760. Notably, despite the lower GNI per capita (PPP) in the EU, poverty levels are lower in the EU (9.8 percent) than in the United States (16 percent) because the disparity of wealth in the EU is about one-third that of the United States. Over the last 20 years, in the United States wages have stagnated, while a few in the United States have become very wealthy. In Europe, social cohesion policies, discussed above, have kept income more equitable. An important result of these policies is that many EU countries outrank the United States on the United Nations Human Development Index. Nonetheless, there are still wide disparities of wealth in Europe and for the most part it is the former communist countries that are the poorest, with the lowest HDI rankings. **Figure 4.22** addresses these issues.

ECONOMIC INTEGRATION AND A COMMON CURRENCY

Economic integration has addressed a number of problems in Europe. Individual European countries have relatively small populations, which means that they have smaller internal markets for their products. As a result, their companies earn lower profits than those in large countries. The European Union tackled this problem by joining European national economies into a common market and by removing many of the controls on the movement of people and products posed by country borders. The **Schengen Agreement** (signed in 1985 and implemented in 1995, and repeatedly modified) gradually removed border checks, making it possible to travel throughout the EU with few if any border stops. This has been a boon to students, vacationers, and laborers seeking better jobs and pay. Furthermore,

Cohesion Policies (often referred to as *social cohesion*) EU guidelines for programs aimed at removing social and economic disparities across Europe

Schengen Agreement an agreement first signed in 1985 and implemented in 1995 that allows for free movement across common borders in the European Union

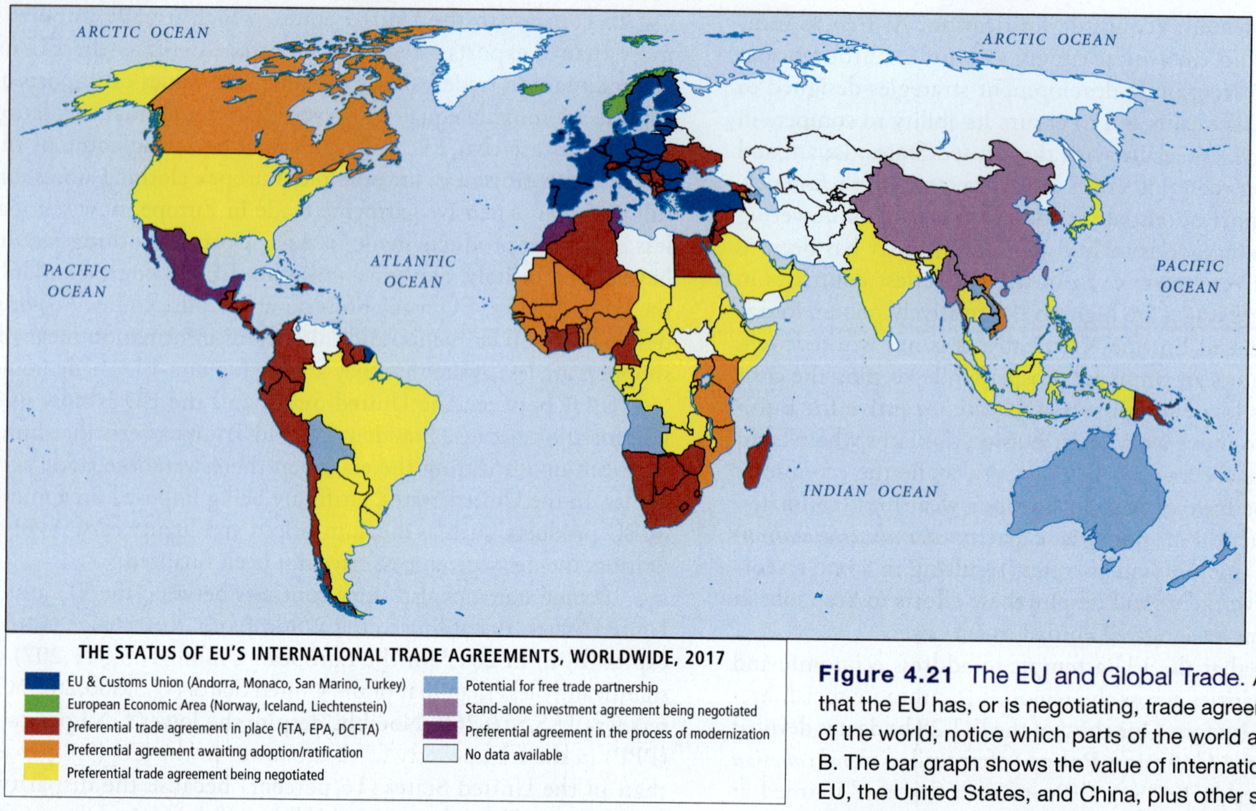

A. The status of EU's international trade agreements, worldwide, 2017.

Figure 4.21 The EU and Global Trade. A. The map shows that the EU has, or is negotiating, trade agreements with most of the world; notice which parts of the world are left out. B. The bar graph shows the value of international trade of the EU, the United States, and China, plus other selected individual countries. C. The graph shows that the EU is dominant globally in the export of manufactured products. The top ten, including the EU, account for 84 percent of the world total in 2017. [Data from European Commission, World Trade Organization estimates]

B. Value of world trade in goods and services for selected countries, 2016.

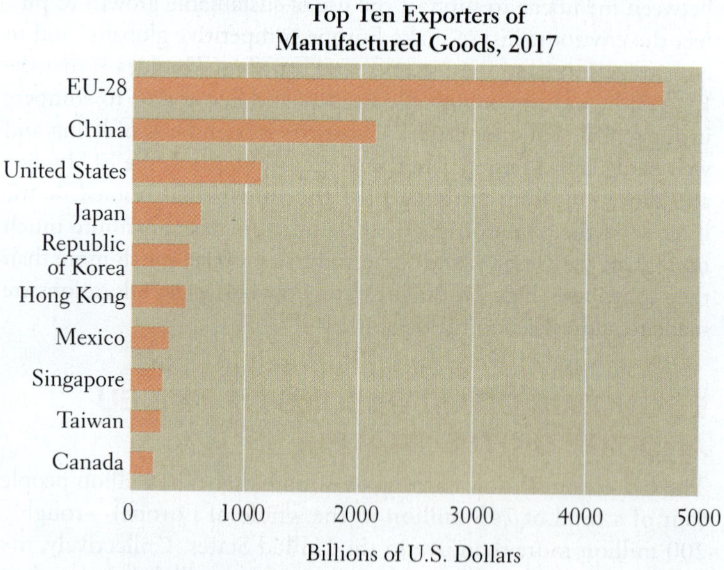

C. Top ten exporters of manufactured goods, 2017.

open borders gave companies in any EU country access to a much larger market and allowed them to take advantage of **economies of scale**—reductions in the unit costs of production that occur when goods or services are efficiently mass-produced, resulting in increased profits per unit. Before the Schengen Agreement, when businesses sold their products to neighboring countries, their earnings were diminished by tariffs and other regulations, as well as by fees for currency exchanges.

economies of scale reductions in the unit cost of production that occur when goods or services are efficiently mass-produced, resulting in increased profits per unit

The Schengen Agreement also meant that countries relinquished a crucial portion of national sovereignty—that related to control of national territory. This has long been a sensitive

PHOTO ESSAY

4.22 Globalization and Development in Europe

The UN Human Development Index (HDI) aims at showing relative well-being (not merely income), and to do so it addresses how countries compare to each other globally in providing the circumstances for people to live long, healthy, and creative lives; gain education, with continuing access to information and ideas; and have access to the resources needed for a decent standard of living. This map confirms that most Europeans live near the top of the global range of human development. Still, there are marked disparities, as shown by each country's numerical rank. The photos show urban landscapes in three countries that rank eleventh, twenty-seventh, and fifty-second on the global HDI.

THINKING GEOGRAPHICALLY

A B C Which of the situations pictured do you think best illustrates the effectiveness of EU social cohesion policies in upgrading HDI?

A B C In which of these situations do you think immigrants would have the most difficulty assimilating? Do immigrants necessarily choose easy assimilation?

A Slow progress in Romania. Human development in Romania, also a communist country until recently, has progressed less rapidly than in the Czech Republic. Incomes have risen and access to modern amenities has improved, but urban and rural landscapes appear "down at the heels." Romania ranks at the lower end of HDI in the European Union. [Courtesy of Mac Goodwin]

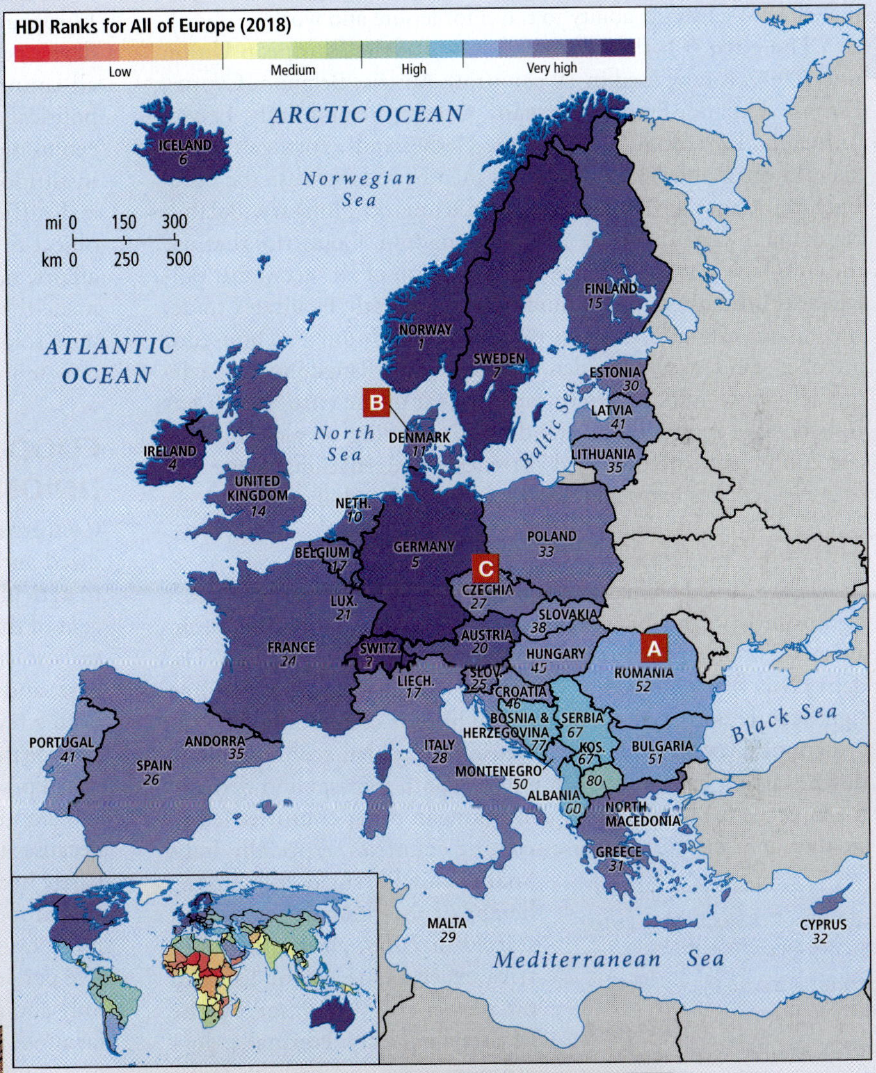

B High human development in Denmark. In Denmark, any signs of poverty are hard to spot and the country is full of affluent villages where people, while enjoying rural life, participate, often remotely, in high-tech employment. [loneroc/Shutterstock.com]

C Modest to affluent lifestyles in the Czech Republic. Until the 1990s the Czech Republic (Czechia) was part of a communist country (Czechoslovakia). Today, access to modern amenities is available to most and the country ranks very high in human development. [Glenn van der Knijff/Getty Images]

issue in Europe, and loss of control over national sovereignty was supposedly a chief rationale behind the Brexit vote in the United Kingdom. There, worries over losing border control were linked to immigration, competition for jobs, fair wages, and control over cheap imports. For the entire EU, the recent influx of non-European immigrants has weakened the resolve to maintain open borders, and in 2017 temporary border controls were erected at certain locations. Nonetheless, the open borders feature of the Schengen Agreement remains very popular among students and the highly skilled, who relish the ability to travel for leisure and work.

The **euro (€)**, the official currency of the European Union since 1999, is now used in 19 countries: Austria, Belgium, Cyprus, Estonia, Finland, France, Germany, Greece, Ireland, Italy, Latvia, Lithuania, Luxembourg, Malta, the Netherlands, Portugal, Slovakia, Slovenia, and Spain. Nine EU members do *not* use the euro: Bulgaria, Croatia, Czech Republic, Denmark, Hungary, Poland, Romania, Sweden, and the United Kingdom. Countries that use the euro have a stronger voice in the creation of EU economic policies, and the use of a common currency greatly facilitates trade, travel, and migration within the European Union. All non-euro member states, except Sweden and the United Kingdom, have currencies whose value is determined by that of the euro, which has considerable international standing. Depending on global financial conditions, the preferred currency of international trade and finance vacillates between the euro and the U.S. dollar.

THE EURO AND DEBT CRISES

In recent years, there have been major challenges to the euro, most significantly in the form of a debt crisis that has tested the mechanisms that governments use to maintain economic stability. The debt crisis first emerged in Greece, where government spending significantly outpaced tax revenues. This was compounded by the growth of Greece's trade deficit during the global economic slowdown starting in 2008, which reduced both tourism and demand for Greece's exports. Similar conditions in other countries led to smaller debt crises in other **eurozone** countries, especially Italy, Spain, and Portugal. But at the height of the financial crisis that started in 2008, almost all countries exceeded the debt limit of 60 percent of GDP set by the Maastricht Treaty. Normally, governments respond to high levels of debt by borrowing from other countries and then reducing the value of their own currency, relative to that of the lenders, which makes these debts easier to pay. This is not an option for any member of the eurozone because no country can by itself manipulate the value of the euro. The EU has the **European Central Bank (ECB)**, which controls the value of the euro, but it is strongly influenced by the largest eurozone economies, such as those of Germany and France. Banks in these countries have loaned large amounts of money to Greece and the other indebted governments of the eurozone, and they would lose money if the value of the euro declined. Because all the economies of the eurozone were threatened by the debt crises, Germany and France had to continue to prop up Greece and other deeply indebted countries with yet more loans.

euro (€) the official (but not required) currency of the European Union

eurozone all the countries using the euro currency

European Central Bank (ECB) directs monetary policy and ensures the soundness of the banking system among eurozone countries

subsidies monetary assistance granted by a government to an individual or group in support of an activity, such as farming, that is viewed as being in the public interest

Common Agricultural Program (CAP) an EU program meant to guarantee secure and safe food at affordable prices, produced sustainably by farmers who earn a fair income

In addition to making large, inexpensive short-term loans to Greece and the other indebted governments through the ECB, the European Union, with crucial support by Germany and France, developed longer-term mechanisms to limit government debt in all countries that use the euro. This involved strengthening the political and financial ties between EU countries so that common economic policies could be better developed and enforced by EU institutions. While many eurozone countries—even those in financial difficulty, such as Greece, Spain, and Cyprus—strenuously object to giving EU institutions more control over their financial affairs, nonetheless, their urgent need for loans from the ECB has made them reconsider these objections. Although short- and long-term solutions have been implemented and countries like Spain are now more stable, the eurozone crisis is still ongoing.

FOOD PRODUCTION FOR THE EUROPEAN UNION AND THE WORLD

Concerns about food security in Europe have led to heavily subsidized and regulated agricultural systems. In the 1980s, **subsidies** to agriculture-related enterprises accounted for more than 70 percent of the EU budget. By 2018, that figure was cut in half. But the early subsidies encouraged large-scale food production by agribusiness, and although agribusiness remains dominant, there has also been a revival of organic and small-scale sustainable approaches by European farmers, and there is a growing market for organically raised produce (**Figure 4.23**).

Food security is especially important to Europeans, partly because stories of World War II food shortages are still so vivid and partly because the EU sees itself as ensuring food security globally. In Europe, most food is now produced on large, efficient mechanized farms. These farms require less labor and are more productive per acre than were farms before the 1970s. One result is that only about 1.6 percent of Europeans are now engaged in full-time farming. A second result of the efficiencies of agricultural mechanization is that the percentage of land in crops has declined since the mid-1990s, while forestlands have increased. These forests are sustainably managed to provide fuel, building materials, wilderness habitat, and places for recreation.

THE COMMON AGRICULTURAL PROGRAM (CAP)

The drastic decline of labor and land in farming has had an emotional effect on Europeans, who see it as endangering their cultural heritage and their goal of food self-sufficiency and food leadership in a globalizing world. To address these worries, the European Union established its wide-ranging **Common Agricultural Program (CAP)**, meant to guarantee secure and safe food supplies at affordable prices, produced sustainably by farmers who

Figure 4.23 LOCAL LIVES: Foodways in Europe

(A) *Tortellini*, stuffed navel-shaped pasta from the region around Modena and Bologna, Italy. One legend claims that the gods Venus and Jupiter arrived at a tavern in Bologna one night, weary from an ongoing battle between Bologna and Modena. When they retired to their room, the innkeeper peeked through the keyhole and saw only Venus's navel. Inspired, he rushed off to the kitchen to create the first tortellini. [Jonathan Gelber/Getty Images]

(B) Swedish *gravlax*—raw salmon cured in salt and dill—is cut into thin slices. Gravlax was invented by fishermen who preserved salmon by salting it and then burying it in the sand on a beach above the high-tide line. This process gave the dish its name: *grav* means grave and *lax* means salmon. [Dorling Kindersley/Getty Images]

(C) In French, *confit* of duck legs refers to meat that is salted, seasoned, and slowly cooked in its own fat. The meat is then stored in the fat, which can preserve it for several weeks. Confit is considered a regional specialty of southwestern France. [Image Source/Getty Images]

earn a fair income. The CAP is also meant to help balance territorial development across the EU and to preserve the quaint rural landscapes everyone associates with Europe, but where few now live.

Until it was drastically revised in 2013, the CAP was criticized for undermining its own goals with its use of tariffs on agricultural goods imported from North America and many developing countries, and for giving subsidies to farmers to underwrite the costs of food production. Tariffs kept out competition and effectively raised food costs for millions of EU consumers. The subsidies tended to favor large, often corporate-owned farms because payments were based on the amount of land under cultivation. CAP policies often resulted in overproduction of food, which lowered market prices and discouraged small family farmers.

On the global scale, the CAP actually undermined worldwide food security because protective agricultural policies like tariffs and subsidies—also used in North America, Japan, and elsewhere—hurt farmers in the developing world. Tariffs lock farmers from poorer countries out of major markets. And subsidies encourage overproduction in rich countries; the resulting occasional gluts of farm products such as butter, eggs, and corn are then dumped—that is, sold cheaply on the world market. Dumping by big producers can lower global prices to the point that farmers in developing countries are driven out of business.

The CAP is now under continual revision. Subsidies to produce certain crops have been eliminated and instead needy farmers are given direct payments to raise their standard of living to an acceptable level. The EU instituted market reforms to increase efficiency and competitiveness, and put more emphasis on improving environmental sustainability and equalizing agricultural development across the EU, including the maintenance of family farms considered important to European identity. Part of the reforms include shrinking the CAP budget, which in the mid-2000s was 58 percent of the total EU budget. By 2018, it was down to 38 percent and is scheduled to be no more than 28.5 percent by 2027.

THE GROWTH OF CORPORATE AGRICULTURE AND FOOD MARKETING

Despite the fact that family farms are valued in the European Union, they continue to decline in number—just as they did several decades ago in the United States. Large farms are created by consolidating small family plots into more profitable operations run by corporations. These farms tend to employ very few laborers and use machinery and chemical inputs.

The move toward corporate agriculture is strongest in Central Europe. When communist governments gained power in the mid-twentieth century, they consolidated many small, privately owned farms into large collectives (except in Slovenia; see the Vignette below). After the breakup of the Soviet Union, these farms were rented to large corporations, which further mechanized them and laid off all but a few laborers. Rural poverty rose. Small towns shrank as farmworkers and young people left for the cities and western Europe. With EU expansion, the CAP has provided even more incentives for large-scale mechanized agriculture in Central Europe. Efforts to counter this trend include green food production and farm tourism, both of which have received CAP subsidies.

During the communist era, Slovenia was unlike most of the rest of Central Europe in that the farms were not collectivized. As a result, the average farm size is just 17 acres (6.8 hectares), plus

Figure 4.24 Vera Kuzmic in her market stall in Ljubljana. [Lydia Pulsipher]

5.6 hectares of woodland. Nearly every farm family has at least one member who commutes to work in a nearby city. Although Slovenia has plenty of rich farmland, as standards of living have risen it has become a net importer of food. Nonetheless, Slovenia's new emphasis on private entrepreneurship, combined with a demand throughout Europe for organic foods, has encouraged some Slovene farmers to carve out a niche for themselves in organic farming, first in local markets and eventually as exporters. The case of Vera Kuzmic is illustrative (**Figure 4.24**).

VIGNETTE Vera Kuzmic (a pseudonym) lives 2 hours by car south of Ljubljana, Slovenia's capital. For generations, her family has farmed 12.5 acres (5 hectares) of fruit trees near the Croatian border. In the economic restructuring that took place after Slovenia became independent in 1991, Vera and her husband lost their government jobs. The Kuzmic family decided to try earning its living in vegetable-market gardening because vegetable farming could be more responsive to market changes than fruit tree cultivation. By 2000, the adult children and Mr. Kuzmic were working on the land, and Vera was in charge of marketing their produce and that of neighbors whom she had also convinced to grow vegetables.

Vera secured market space in a suburban shopping center in Ljubljana, where she and one employee maintained a small vegetable and fruit stall (see Figure 4.24). Her produce had to compete with less expensive, Italian-grown produce sold in the same shopping center—all of it trucked in daily from large corporate farms in northeastern Italy. But Vera gained market share by bringing her customers special orders and by guaranteeing that only animal manure and no pesticides or herbicides were used on the fields. For a while, these strategies kept her in business. But when Slovenia joined the European Union in 2004, she had to do more to compete with produce growers and marketers from across Europe who now had access to Slovene customers.

virtual water the volume of water required to produce, process, and deliver a good or service that a person consumes

Anticipating the challenges to come, the Kuzmics' daughter Lili completed a marketing degree at the University of Ljubljana. The family incorporated their business and Lili is now its Ljubljana-based director, while Vera manages the farm. The Kuzmics continue to focus on selling their produce to Ljubljana's expanding professional population, who are willing to pay extra for fine organic vegetables and fruits. Since Lili's market research showed the wisdom of diversification, now, in a banquet facility on the farm built with CAP funds, Vera also prepares special dinners for tour groups interested in witnessing traditional farm life and in tasting Slovene dishes made from homegrown organic crops. ■

THE WIDE REACH OF EUROPE'S ENVIRONMENTAL IMPACT

As Europe's development grew after World War II, Europeans increasingly became global consumers. Today, their food, clothing, building materials, and industrial products are all based on at least some resources from abroad. As a result of this resource use, Europeans have a wide-reaching environmental impact. One way to grasp the enormity of this impact is to look at **virtual water**. Introduced in Chapter 1, *virtual water* is the volume of water required to produce, process, and deliver a good or service that a person consumes. Table 1.1 shows the virtual water content of a variety of products. The average amount of water required to meet a person's basic needs is about 5 to 13 gallons (20 to 50 liters) per day. But, to arrive at a person's actual annual total water footprint, one must add the water required for basic needs to a person's virtual water usage. Although Europeans do not import as many of their consumer goods as do Americans, they still consume one-fifth of the world's imports; many of these goods have a high *virtual water component* (the water consumed in the production process). Nearly all EU countries import more virtual water than they export, and Europe as a region imports more virtual water than any other. The foreign countries that produce Europe's imported virtual water often have very little water themselves, and the costs of this water loss are not being adequately figured into the price of the products exported to Europe. In all fairness, then, the environmental impacts of Europe's virtual water consumption should be counted against Europe's total impact on the biosphere (the world's environment). Other resource extraction, beyond water, that goes into producing goods sold to Europeans should also be counted in Europe's environmental impact.

CHECK YOUR UNDERSTANDING

1. How is Europe both embracing and wary of globalization?
2. Why are the expensive Common Agricultural Program and food security so important to Europeans' sense of place and safety?
3. How does the influx of refugees and immigrants threaten Europeans' desire for open borders, security, and social cohesion?
4. Why is it fair to say that Europe's environmental impact extends far beyond Europe?

POWER AND POLITICS

4.6 Explain how despite the success of the European Union, nationalism still threatens European unity and social cohesion.

4.7 Explain how the European commitment to expanding political freedoms developed from the post–World War II period through the present.

4.8 Analyze how Europe's role as a colonizer continues to affect politics within the EU and in international relations.

The 28 countries of the European Union and their neighbors are continually renegotiating political and power relationships with each other and with the wider world. In fact, one could say that this continual negotiating stance is the basic *raison d'être* (the ultimate purpose) for the European Union.

POPULAR DEMOCRACY AND NATIONALISM

During the nineteenth and twentieth centuries, the evolution of the concept of democracy in Europe was linked to the idea of **nationalism**, or allegiance to the state. The notion spread that all the people who lived in a certain area formed a nation and that loyalty to that nation should supersede loyalties to family, clan, or individual monarchs. Eventually, the whole map of Europe was reconfigured, and the mosaic of kingdoms gave way to a collection of *nation-states.* All of these new nations were, at varying paces, transformed into democracies by the political movements arising in Europe's industrial cities. But nationalism encourages continual competition with one's neighbors. An overemphasis on nationalism was a major component of both of the two world wars, so in the period after the wars—called the *postwar era*—as the European Union was constructed, nationalism was recognized as a problem and deemphasized. Nonetheless, the urge for a strong national identity lives on in European countries, especially in times of political or economic troubles, and it plays a role today, for example in the case of the Brexit vote in 2016. Britain's surprise vote to leave the European Union had a strong nationalist component.

THE POLITICS OF EU EXPANSION

After World War II, the commitment to political freedoms was most pronounced in Germany. In an effort to make the authoritarianism of Nazi Germany impossible in the future, a new postwar constitution strongly protected political rights. A similar postwar constitution was drawn up in Italy. It is worth noting that in the name of political freedom, both of these postwar constitutions outlawed the public expression of far-right-wing political views: those that espoused white racial superiority or Christian superiority or opposition to certain ethnic minorities. This contradiction of protecting political freedoms by prohibiting discriminatory views plagues a number of left-leaning EU governments, most notably Germany and France, where, despite the prohibition of far-right nationalist parties, these parties have recently won a notable number of Parliamentary seats.

In several parts of Europe, the expansion of political freedoms came late or was compromised by violence (**Figure 4.25**). Spain remained a dictatorship until 1975, and Portugal formally democratized only after a revolution related to decolonization in Africa in 1974 (Figure 4.25D). Until the late 1990s, Northern Ireland had sectarian violence between Catholics and Protestants so severe that free and fair elections were not possible (see Figure 4.25C). Central Europe did not significantly expand political freedoms until the Soviet Union dissolved in 1991 as its former satellites declared independence, one by one. Yugoslavia, a large communist republic in southern Central Europe, always firmly outside the Soviet bloc, also dissolved in the early 1990s. There the growth of political freedoms was hampered by a powerful wave of ethnic xenophobia and violence, which is now muted but continuing (see Figure 4.25A).

As early as the 1980s, just as the European Union was becoming well established, Soviet control over Central Europe began to falter in the face of a workers' rebellion in Poland known as *Solidarity.* Discontent spread even into Russia, where rebellion against authoritarianism led briefly to a softening of control and an opening known as *Glasnost.* By 1990, East Germany had reunited with West Germany. As the Soviet Union dissolved, many economic and political relationships in Central Europe collapsed. Thousands of workers lost their jobs in state-owned factories. In some of the poorest countries, including Romania, Bulgaria, and Hungary, social turmoil and organized crime threatened stability. Membership in the European Union became especially attractive to Central European political leaders and citizens who thought, overly optimistically, it would spur economic development for them and would bring investment by the wealthier EU member countries from West and North Europe.

Standards for EU membership, however, are rather demanding and specific. A country must have both political stability and a democratically elected government. Each country has to adjust its constitution to EU standards that guarantee the rule of law, human rights, and respect for minorities. Each must also have a functioning market economy that is open to investment by foreign-owned companies and that has well-controlled banks. Finally, farms and industries must comply with strict regulations governing the finest details of their products and the health of environments. Meeting these requirements has been a challenge for members such as Romania, Bulgaria, and Hungary.

Two very wealthy, stable democracies chose not to join the European Union: Switzerland and Norway. They have long treasured their neutral role in world politics, and were also concerned about losing control over their domestic affairs. After the 2008 economic recession, Iceland, long a candidate to join the EU, lost enormous wealth because of banking speculation. Iceland has now chosen not to join the European Union, but it maintains a close relationship with the EU through the European Economic Area (EEA) and border pacts in the Schengen Agreement. Iceland participates in the EU open market, with most exports going to the EU; it has adopted many EU laws, and contributes money to EU social cohesion efforts, but it does not use the euro.

Several countries on the perimeter of Europe are candidate EU countries. They include Turkey, North Macedonia, Montenegro, and Serbia. The prospect of Turkey joining the EU has always been in doubt and now seems more remote. Turkey has strained relations with the island country of Cyprus (which was admitted to the

nationalism strong devotion and loyalty to the interests or culture of a particular country, nation, or cultural group

Power and Politics in Europe

Patterns of political power and conflict in Europe and elsewhere today tend to reflect the extent to which people enjoy political freedoms. The most violent recent conflicts in Europe took place in Central Europe (Hungary, Romania, Greece, the Balkans) during the transition away from communism and in Northern Ireland during efforts to break away from British control. The denial of political freedoms allowed extreme ethnic conflict to erupt. The map shows levels of democracy globally, and for Europe, in particular. Armed conflicts and genocides with high death tolls are shown in Europe since 1990 and in postcolonial states globally.

Outside Europe, long and brutal wars were fought to gain independence from European colonial economic exploitation. The denial of local political freedoms by colonial administrations was a major impetus for independence movements. The map of Africa and Asia shows how low levels of democratization persist long after the formal end of colonial rule. European imperialism helped set many of these countries on an authoritarian trajectory that has only recently started to change.

THINKING GEOGRAPHICALLY

A Conflict has long been a part of European history. In the case of Photo A, what seems to have been the primary instigation for violence in Bosnia?

B When Macron apologized for French colonial policies in Algeria, some Algerians felt used and insulted rather than grateful. Why might that be?

C Ireland has long been a site of sectarian violence. What are the basic facts behind the disputes?

D Oddly, Portugal was freed from dictatorship just as it lost its colonies in Africa. Is there a connection between these two occurrences?

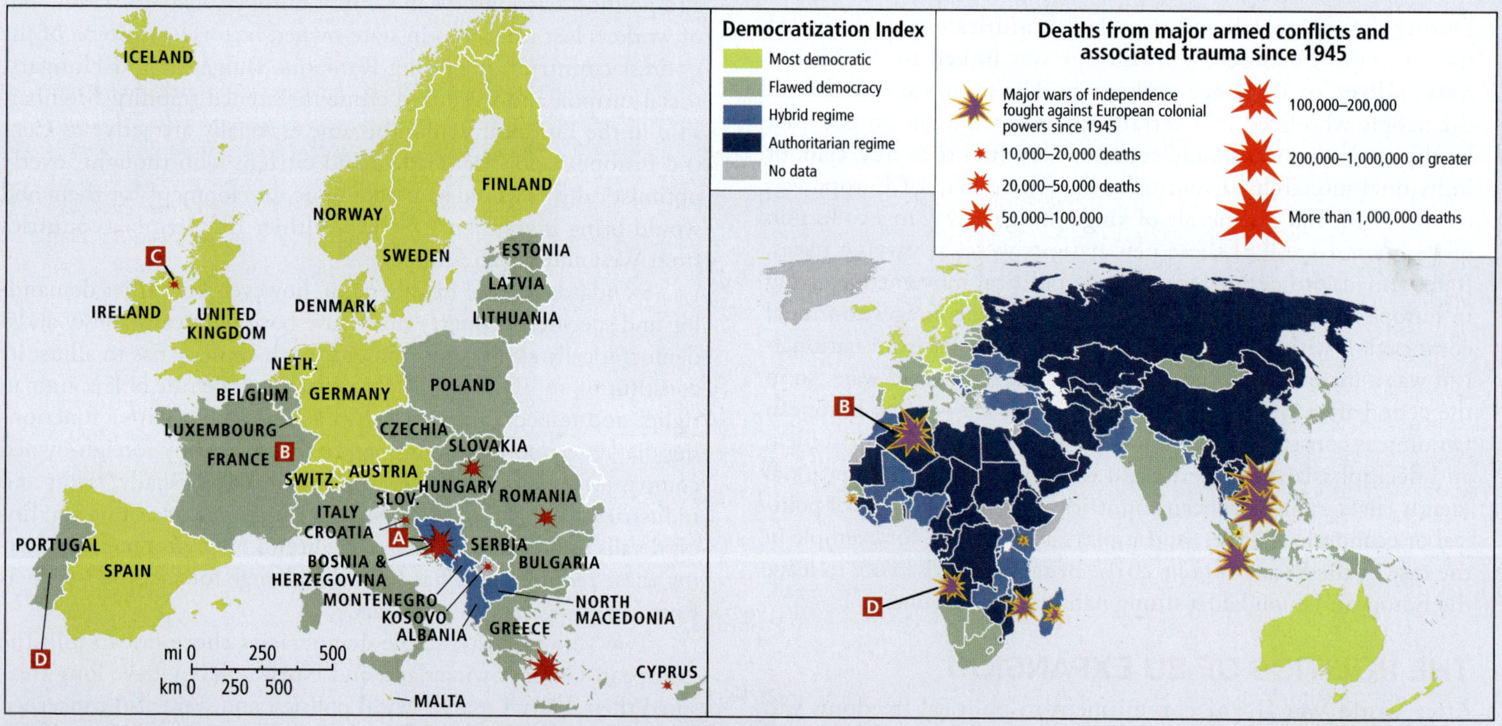

A Bosnia. In the aftermath of Bosnia's civil war in a suburb of Sarajevo in 1996, a man, on crutches having been wounded by a shell, stands deep in thought outside the destroyed Oslobodenje newspaper building in Sarajevo. Attempts by Serbia to pit ethnicities against each other and to limit democracy in the former Yugoslavia sparked several wars during the 1990s. In Bosnia, a civil war resulted in 175,000 deaths. [Tom Stoddart Archive/Premium Archive/Getty Images]

B France and Algeria. In 2018 President Emmanuel Macron of France apologized for France's treatment of Algerians, a move taken to lighten the load of French Algerians who suffer exclusion in France and also a move to improve relations with Algeria [NYT online, October 15, 2018, "What to Do When Your Colonizer Apologizes," by Kamel Daoud]. [LUDOVIC MARIN/AFP/Getty Images]

C Northern Ireland. For many decades Northern Ireland was the site of repeated terrorist attacks between Protestants and Catholics. Two murals in Derry, Northern Ireland, mark an entrance to a "nationalist" neighborhood. The mural on the right depicts a boy wearing a gas mask and holding a homemade petrol bomb. Both murals were created during "the Troubles," a time of violent conflict from 1969 to 1998. Competing claims about Northern Ireland's democratic legitimacy still flourish, though the violence has subsided since peace was agreed to in 1998. [Oli Scarff/Getty Images]

D Portugal and Angola. A group of Angolan rebels captured during Angola's war of independence from Portugal. To hold on to its African colonies, Portugal fought three major wars during the 1960s and 1970s. At the time Portugal itself was an undemocratic authoritarian dictatorship, but the human and financial costs of these wars eventually brought about a revolution in Portugal in 1974. The revolution resulted in independence for the colonies and the establishment of democracy in Portugal. Because of the authoritarian governing structures put in place by the Portuguese and because of the lengthy civil wars that followed decolonization, none of Portugal's former colonies has well-protected political freedoms. [Weber/Hulton Archive/Getty Images]

European Union in 2004) and has a history of human rights violations against minorities (especially against its large Kurdish population). Women in Turkey have experienced a serious contraction of civil liberties in recent years. There are also issues regarding the separation of religion and state. Turkey would be the first majority Muslim country to join the European Union, and recent policies indicate that Turkey is abandoning its official secular posture. Turkey itself has some reservations. Until recently, its economy had grown faster than the EU average, leading some Turks to question the need to join the EU at all. Moreover, in recent years, Turkey appeared to view its geopolitical advantage as being less with Europe and more as a leader in the Middle East. As the attempted revolutions collectively referred to as the *Arab Spring* (beginning in 2010) developed, Turkey subtly changed its behavior toward Europe, going from being a voice of calm and reason regarding political developments in the Middle East to adopting heavy-handed responses to a wave of Arab Spring–type protests arising within Turkey itself. By 2016, as the lengthy crisis in Syria persisted, accompanied by the flight of many Syrian refugees through Turkey and increasing terrorist activity within Turkey, the behavior of Recep Tayyip Erdoğan (first as prime minister and then as president of Turkey) became increasingly autocratic. The trajectory of Turkey's relationships with Europe and the Middle East became ever more difficult to discern.

A few other countries—Ukraine, Moldova, and perhaps even the Caucasian republics (Armenia, Azerbaijan, and Georgia)—have expressed some interest in joining the EU. However, there is strong opposition to this within Europe, and Europe's huge and potentially powerful neighbor, Russia, opposes quite actively the expansion of the European Union to the east into what it considers its *sphere of influence* (see Chapter 5).

POSTCOLONIAL TRIBULATIONS IN EUROPE

Despite Europe's own rapid and peaceful development after World War II, and its success in quieting disastrous ethnic hostilities in southeastern Europe during the 1990s, by 2014, Europe was encountering a rash of political problems related to its centuries of colonialist activities around the world. Many trace the persistence of dictatorships in the Middle East, for example, to the colonialist manipulation by Europe of the politics and economies of Middle Eastern countries. Europe's interference was especially prominent after World War I and has extended into the present in such countries as Algeria, Morocco, Egypt, and Syria. France is a particular case in point, especially with regard to its former colony of Algeria, which became independent after an 8-year war in 1962 (Figure 4.25B). Since the nineteenth century, hundreds of thousands of French had migrated to Algeria to run farms and industries there, exporting most products back to France. French rule was harsh and Algerians were underpaid and powerless. After independence, political turmoil and killings continued, so both French colonists and many Algerian workers migrated to France.

Over time, the Algerian French became concentrated and socially excluded in suburban Paris housing blocks (***banlieues***), far from jobs, training, and other services. They were not admitted into normal French society unless they completely assimilated as French, a process with many hurdles. Most were stigmatized by their appearance, names, and Muslim faith. Some disillusioned youth, even those born and raised in France, were attracted to **Islamic jihadism**, which promised a robust identity and a mission to claim power through terrorism. This is the backstory to the terrorist attacks in France in 2015 and Belgium in 2016, but similar scenarios can be described for the UK's relationships with former colonies in the Middle East and Pakistan and India, as well as for Portugal's and Belgium's relationships with parts of Africa (see map in Figure 4.25). The case of Algeria is only one variant of the difficult **postcolonial** struggles of Muslims to find acceptance and to create a life for themselves in their native-born or adopted countries.

banlieues the troubled suburbs of Paris where a high proportion of recent immigrants live, excluded from decent employment, higher education, and many other advantages of urban life

Islamic jihadism a personal or group struggle to promote Islamic revivalism, often with the threat of or actual use of force

postcolonial the conditions and attitudes that persist in a society after colonization is over

EU GOVERNING INSTITUTIONS

Somewhat similar to the United States, the European Union has legislative, executive, and judicial components of government; but these components have relationships to each other that vary from the U.S. case (**Figure 4.26**). The 512 million citizens of the EU directly elect the *European Parliament*; each country gets a certain portion of seats in Parliament, based on its population, much like the U.S. House of Representatives. Laws must be passed in Parliament by 55 percent of the member states, which must contain 65 percent of the EU total population. In other words, a simple majority does not rule. The European Parliament employs about 6000 civil servants and shares power over the EU budget and legislation with the *Council of the European Union*. The Council of the European Union is similar to the U.S. Senate in that it is the more powerful of the two legislative bodies. However, its members are not elected but consist of one minister of government from each EU country. The Council of the European Union employs about 3500 people and is responsible for implementing the broad range of EU social cohesion policies.

The *European Commission*, made up of one commissioner from each country appointed for a 5-year term, acts like an executive branch of government, proposing and implementing new laws and monitoring the treaties and running the day-to-day business of the EU. European commissioners are expected to uphold common EU-28 interests and not those of their own countries. The head of the European Commission is elected by the European Parliament and serves a 5-year term in what amounts to being a prime minister or head of state for the whole of the EU. Citizen control is maintained because the entire commission must resign if censured by the popularly elected Parliament. The European Commission also includes 33,000 civil servants who work in Brussels to administer the European Union on a day-to-day basis. The *European Council* (not to be confused with the Council of the European Union) is made up of all 28 heads of state who strategize through debate the EU's political direction.

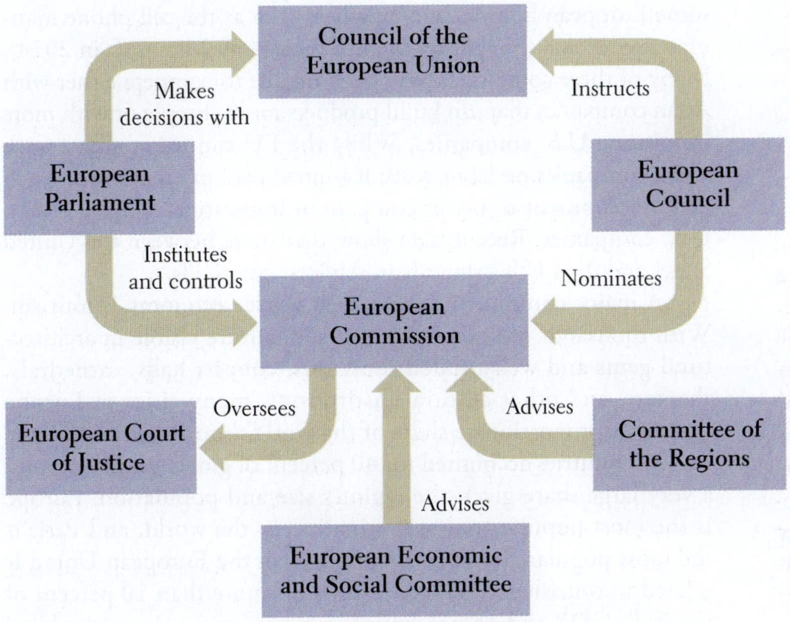

Figure 4.26 EU governing institutions and how they interact. Explanations of each component of this graph are in the text.

The judiciary branch is called the *Court of Justice of the European Union (CJEU)* and consists of two major court systems that interpret EU law and enforce compliance. The *Court of Justice (ECJ)* is the high court and has one judge from each of the 28 member countries. The ECJ deals with EU-wide issues between members; the *General Court (GC)* hears cases dealing with individuals and companies that relate to such things as business law and trademarks. Two advisory groups interact with all the EU governing institutions. They are the *Committee of the Regions* (local and regional representatives from the 28 member states) and the *European Economic and Social Committee (EESC)*, 350 representatives from unions and civil society organizations.

NATO AND THE EUROPEAN UNION AS GLOBAL PEACEMAKERS

The European Union acts as a global peacemaker and peacekeeper through the **North Atlantic Treaty Organization (NATO)**, which is based in Europe. During the Cold War, European and North American countries cooperated militarily through NATO to counter the influence of the Soviet Union. NATO originally included the United States, Canada, the countries of western Europe, and Turkey; it now includes almost all the EU countries as well. Russia views this expansion as a threat.

Since the breakup of the Soviet Union, NATO has focused mainly on providing the international security and cooperation needed to expand the European Union. When the United States invaded Iraq in 2003, most EU members opposed the war. As worldwide opposition to the United States built, the global status of the European Union rose. Thereafter, with the United States preoccupied in Iraq and Afghanistan, NATO and the EU assumed more of a role as a global peacekeeper. For example, NATO forces fought off attempts by pirates to seize merchant ships during the 2009 Somali pirate crisis off the northeast coast of Africa, and undertook a number of operations in the Mediterranean related to the Arab Spring in Algeria, Tunisia, Libya, and Syria. After the November 2015 terrorist attacks in Paris (see the discussion above), there was talk of NATO coming to the aid of France in retaliating against ISIS in Syria. However, the phrasing of the mutual defense provision (Article V of the North Atlantic Treaty) allows all allies to make their own decisions regarding such aid, so NATO, as such, is unlikely to aid in any swift retaliation.

THE FUTURE OF THE EUROPEAN UNION

After much discussion of the complexities of uniting Europe, it is useful to ask how Europeans themselves are feeling about the EU. The rise of populist parties and the success of some right-wing politicians, the Brexit vote in the United Kingdom, and the seeming constant chorus of opposition from the populace all seem to suggest that overall EU consensus is breaking down; but, no, there is plenty of evidence that this is not true. Recent polls show that 60 to 70 percent of EU citizens are pro EU and would not vote to leave the union. The main reasons given are that the EU contributes to economic strength and it improves relationships between countries. Even in Italy and Hungary, where leaders often take anti-EU postures, there is overwhelming citizen support for the EU. In fact, all the problems and uncertainties that the UK has encountered with Brexit seem to have awakened an awareness of the value of the EU. World leaders are also confidently optimistic about the future of the European Union, viewing it as an important model for other world regions.

CHECK YOUR UNDERSTANDING

1. Give some examples of how nationalism instigated political conflict in Europe during the first half of the twentieth century.
2. Why was/is nationalism rejected as a useful political perspective in the European Union?
3. How did the ambition to join the EU and the requirements for doing so affect the development of democratic institutions in post-Soviet Central European countries?
4. Give some examples of how Europe's colonialist past still influences social relationships, political stances, and the acceptance of immigrants today.
5. What are some ways that the design of EU governing institutions reveals efforts to address the varied needs of the 28 EU countries?
6. How has the role of NATO changed over the course of its history?
7. How is recent terrorism in Europe related to nationalism, colonialism, political unrest in former colonies, the rise of the right in Europe, and the current migration crisis in Europe?

North Atlantic Treaty Organization (NATO) a military alliance of European (including Turkey) and North American countries, created during the Cold War to counter the influence of the Soviet Union and to provide international security and cooperation

URBANIZATION

4.9 Discuss the evolution of Europe's cities from early trading centers to manufacturing centers to modern high-tech centers.

4.10 Describe modern European cities in terms of density and government housing, and the role of urban infrastructures in transportation and poverty alleviation.

As noted previously, urbanization in Europe began with trading centers and was part of a general transformation of European life. Industrialization, which was partly funded by profits from the colonies, drew people from the countryside into cities where they then encountered new ideas, such as civil rights, religious reform, and public education. Most importantly, people were drawn to cities because there they could find options for improving their economic circumstances.

Today, Europe's cities, with old town centers surrounded by modern high-rise suburbs and supported by world-class urban infrastructures, are the heart of Europe's society (**Figure 4.27**). More than 70 percent of the EU population lives and works in urban areas. Yet, spatially, much of Europe remains idyllic and productive rural countryside, important symbolically and economically, and as a place for regular recreation.

The economies of most European cities are now primarily service-oriented. Education and research have been important in urban Europe since medieval times, and now Europe has numerous hubs of innovation in the knowledge-intensive service sector so crucial to the modern European economy.

EUROPE'S GROWING SERVICE ECONOMIES

Industrial and agricultural jobs have declined across the region, due to mechanization and to the fact that manufacturing has moved either to a different region in Europe or abroad. Most Europeans (about 73 percent) now find jobs in the service economy. Services, such as the provision of health care, education, finance, tourism, commerce, and information technology, are now the base of Europe's integrated economy, drawing hundreds of thousands of employees to the main European cities. For example, financial corporations that are located in London and serve the entire world play a huge role in the British economy. Many multinational companies are headquartered in London in part to take advantage of the city's financial sector; but they are also attracted by London's lively multicultural urban life. Such multicultural centers are now widely recognized as being fertile zones of innovation. The highly trained specialists who work in these companies come from across the EU and beyond. But for the more than 20 percent of UK workers who lack the special training and skills for the new job market, underemployment became chronic and a source of resentment. Underqualified British workers were a major factor in the anti-immigrant component of the Brexit vote in 2016.

A special category of service jobs in the EU is *knowledge-intensive services*, which now account for 40 percent of all employment. These jobs require considerable formal education, such as graduate degrees, high-end technical training or college, and usually extensive on-the-job training. Despite a fine record of success, there has been concern within the EU about the declining profitability of some European high-tech companies, such as the cell phone manufacturer Nokia (bought by the U.S. company Microsoft in 2014). Many of these companies have been unable to compete either with Asian companies that can build products more cheaply or with more innovative U.S. companies. While the EU cannot compete with Asian companies on labor costs, it is now investing more in research and development to better compete in innovation categories with U.S. companies. Recent data show that trade between the United States and the EU is primarily in services, not goods.

A major component of Europe's service economy is tourism. With thousands of years of history and culture visible in architectural gems and well-funded museums, concert halls, cathedrals, theaters, and other cultural institutions, many cities in Europe draw a disproportionate share of the world's tourists. In 2018, the EU-28 countries accounted for 40 percent of global tourist arrivals, a very large share given the region's size and population. Europe is the most popular tourist destination in the world, and Paris is the most popular city. One job in eight in the European Union is related to tourism, which generates a bit more than 10 percent of the EU's GDP and 15 percent of its taxes (this varies with global economic conditions). Europeans are themselves enthusiastic travelers, frequently visiting one another's countries to attend local festivals (**Figure 4.28**) as well as traveling to many distant locations across the world. This travel is made possible by the long paid annual vacations—usually 4 to 6 weeks—that Europeans are granted by employers. Vacation days can be used a few at a time so that people can take numerous short trips. The most popular holiday destinations among EU members in 2018 were Spain, France, Italy, Germany, and Austria; the first three were favored for their sunny weather and beaches.

In such cities as Budapest, Copenhagen, Vienna, Ljubljana, Venice, Ravenna, Trieste, and Dubrovnik, the relics of ancient urban life are a main attraction for visitors. Having begun as trading centers more than a thousand years ago, these cities and many more still bear the architectural marks of medieval life in their historic centers (Figure 4.27B). These old cities are located either on navigable rivers in the interior or along the seacoasts because water transportation figured prominently (as it still does) in Europe's trading patterns.

After World War II, nearly all the cities in Europe expanded around their perimeters in concentric circles of apartment blocks (see Figure 4.27A). In most cities, well-developed rail and bus lines link the blocks to one another, to the old central city, and to the surrounding countryside. Land is scarce and expensive in Europe, so only a small percentage of Europeans live in single-family homes, although the number is growing. Even single-family homes tend to be attached or densely arranged on small lots. Except in public parks, which are common, one rarely sees the sweeping lawns familiar to many North Americans. Because publicly funded transportation is widely available, many people live without cars, in apartments as close to the city center as they can afford (see Figure 4.27C). People walk more in Europe and when necessary they commute by bus or train (and increasingly by car) from suburbs or ancestral villages to work in nearby cities.

Although deteriorating housing and slums do exist (especially in some of the poorer cities and towns of Central Europe and in very large cities like Paris, Brussels, and London), substantial

4.27 Urbanization in Europe

Europe's cities are famous throughout the world for their architecture, economic dynamics, and cultural variety. Many are quite ancient, with quaint city centers that are now surrounded by a ring of eighteenth- and nineteenth-century buildings and then large, more modern apartment blocks. The various parts of some cities are connected by public transportation, but in others (e.g., Paris) transportation outside of the center is difficult, time consuming, or expensive. Despite their global draw, many European cities will shrink in the next several decades due to declining national populations (as in Italy and parts of Central Europe) as well as due to deliberate efforts to shift growth to other urban centers (as in the case of London).

THINKING GEOGRAPHICALLY

A In what ways does this photo of Berlin illustrate the statement that "Berlin does a better job of linking its high-rise residents to urban life than do Paris or Brussels." What about the history of Berlin could account for so many high-rise apartments near the old city?

B Given how the free transport in Tallinn was financed, what would the attitudes toward taxation need to be?

C Explain why Barcelona would be such a tourist attraction. What type of tourist would go there, and from where are most of the visitors?

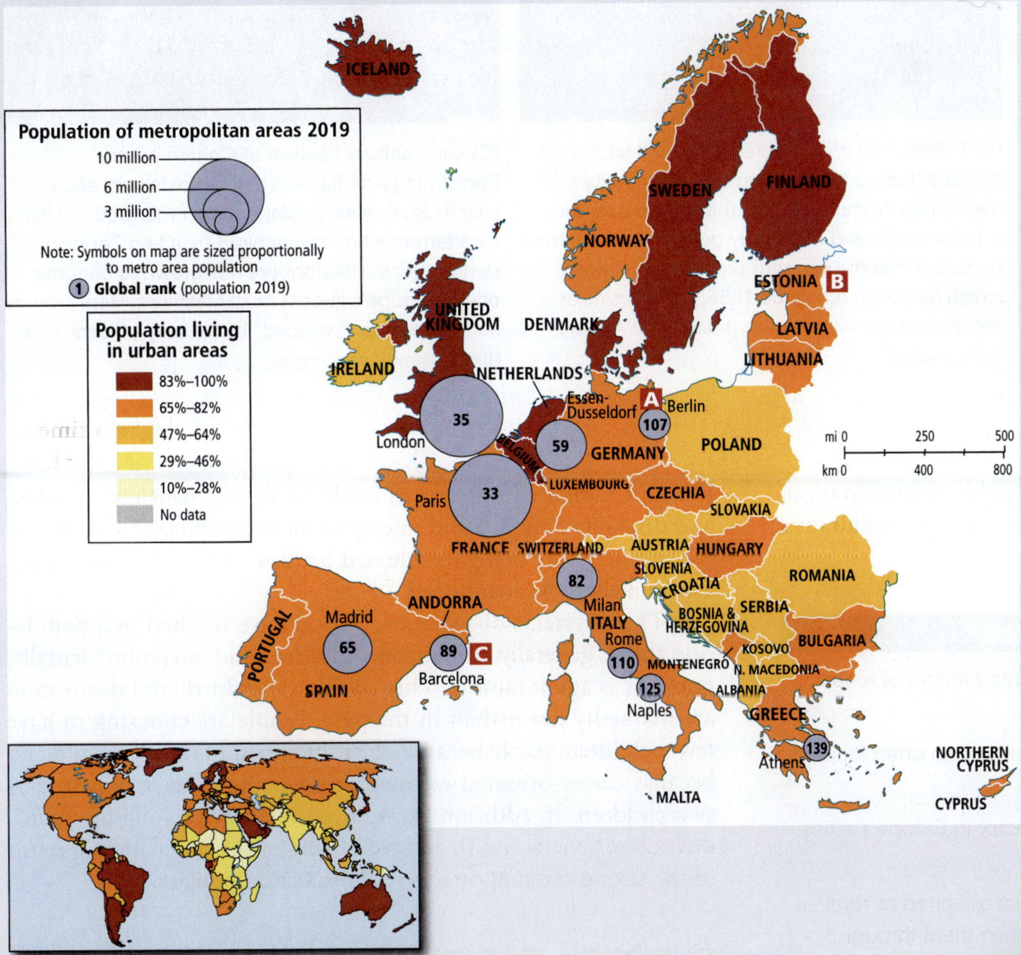

A Berlin. High-rise apartment blocks dot the periphery of Berlin's old urban core. Berlin does a better job of linking its high-rise suburbs to urban life than do Paris or Brussels, where the poor and new immigrants are often isolated and excluded from urban advantages. [ODD ANDERSEN/AFP/Getty Images]

B Tallinn. Tallinn, Estonia (population 441,000), is experimenting with free public transport. To get the free rides you register as a resident, which enables the municipality to get €1,000 from your income tax every year. Then you only need to pay €2 for a "green card" and all your trips are free. [left: Maremagnum/Photographer's Choice/Getty Images; right: Tancredi J. Bavosi/Getty Images]

C Barcelona. Barcelona in Spain is an elegant city with outstanding architecture and upscale shops. La Rambla, a tree-lined pedestrian mall, is a popular strolling place among locals and tourists. [Samuel Aranda/Getty Images]

Figure 4.28 LOCAL LIVES: Festivals in Europe

(A) Every year in Buñol, Spain, participants in *La Tomatina* throw tomatoes at each other. Supposedly, this festival began in 1945 when a group of young men, not allowed in a parade, grabbed tomatoes from a nearby vegetable stand and hurled them. The young men returned the next year and repeated the brawl, though they brought their own tomatoes. Today, truckloads of tomatoes are brought in, as shown here. [Denis Doyle/Getty Images]

(B) Riders in *Il Palio*, a horse race held each year in Siena, Italy. Jockeys from each of the city's 17 major neighborhoods circle the main plaza; they ride bareback and frequently get thrown from their horses. A colorful pageant precedes the race, which has been run since 1656, though similar races were run even earlier. [Mike Hewitt/Allsport/Getty Images]

(C) Glastonbury Festival of Contemporary Performing Arts (known as *Glasto*) is held each year in Somerset, England. First organized in 1970 by a farmer who was inspired by a Led Zeppelin performance, Glastonbury has grown to become one of Europe's largest music festivals. Here crowd-surfing fans are "rescued" by security officers. [Peter Still/Redferns/Getty Images]

public and private investments in housing, public transportation, sanitation, water, and utilities mean that most people maintain a generally high standard of urban living.

CHECK YOUR UNDERSTANDING

1. Explain why early trading towns evolved into centers of ideas and civil rights and then industrialization.
2. Explain several reasons why Europe's cities are so often located on bodies of water.
3. Describe the importance of the service sector in Europe's cities. To what extent is it globally competitive?
4. Why do European urban dwellers remain so attached to rural landscapes that they willingly pay to maintain them through agricultural subsidies?
5. Explain the role of cities in Europe's tourism industry and the role of tourism in Europe's economy.

POPULATION, GENDER, AND CULTURE

4.11 Explain why Europe's population is aging and how government policies, including immigration, are addressing aging.

4.12 Explain how issues of gender inequality are being differently addressed in various parts of the European Union.

In nearly all European countries, tax-supported systems of social protection provide all citizens with basic health care; free or low-cost higher education; affordable housing; pension, survivor, and disability benefits; and unemployment benefits. Across the EU these benefits are being reevaluated because of their high costs, but are unlikely to be abandoned.

These systems of social protection have resulted in a population that is generally affluent and educated and, not coincidentally, one that is aging rapidly. This is so because birth and death rates are markedly lower than in the past. People are choosing to have fewer children, partly because of the high cost of raising a child and because career-oriented women are choosing to have only one or two children. In addition Europeans are enjoying long life spans, due to the general health and well-being of the population, a result of the strong tax-supported social protection policies.

EUROPE'S RAPIDLY AGING POPULATION

Between 1960 and 2018, the proportion of the European population 15 years and under declined from 27 percent to 16 percent (well below the global average of 26 percent), while those over 65 increased from 9 percent to 20 percent (well above the global average of 9 percent). This means that those over 65 now outnumber those under 15, and half the region's population is aged 43 or over, an unusual circumstance in human history. Longer life expectancies have influenced this trend; at birth, the average life expectancy of a EU citizen is 78 for men and 83 for women. However, the major contributor to the aging of the population is the declining fertility rate: fewer babies are being born.

Overall, Europe now has the lowest rate of natural increase in the world. In all of the EU-28, fertility rates are below the replacement rate for developed countries (2.1 children per woman aged 15–49).

The average fertility rate for EU countries is 1.6. The one-child family is becoming common throughout Europe. France has a slightly higher fertility rate of 1.9, because its large immigrant population tends to have slightly bigger families. That pattern will last only until that immigrant population assimilates to French culture. Ireland has a rate of 1.8 because of the strength of Roman Catholic rules against birth control (recently lifted by popular vote), but other Roman Catholic countries (Spain, Italy, Portugal) have rates below 1.5.

The biggest worry is that these population changes could slow economic growth because low birth rates mean fewer consumers are being born. With fewer people, economies and the tax bases of countries will contract over time. Demand for new workers, especially highly skilled ones, may go unmet. Furthermore, the number of younger people available to provide expensive and time-consuming health care for the elderly, either personally or through tax payments, will decline. Currently, for example, there are just two German workers for every retiree. In the United States there are 3.5 workers for every retiree, in India 15, in Indonesia 11, in China 7.

Europe's work force is shrinking, and though the rates vary, this can be seen in the landscape. In rural areas of Spain, for example, more than 1500 villages have been abandoned and fields are left to regrow forests. If present trends hold, every future generation of Europeans will be noticeably smaller than the previous one. The declining birth rate is illustrated in the narrow bases of population pyramids for European countries, which look more like lumpy towers than pyramids. **Figure 4.29** shows the pyramids for

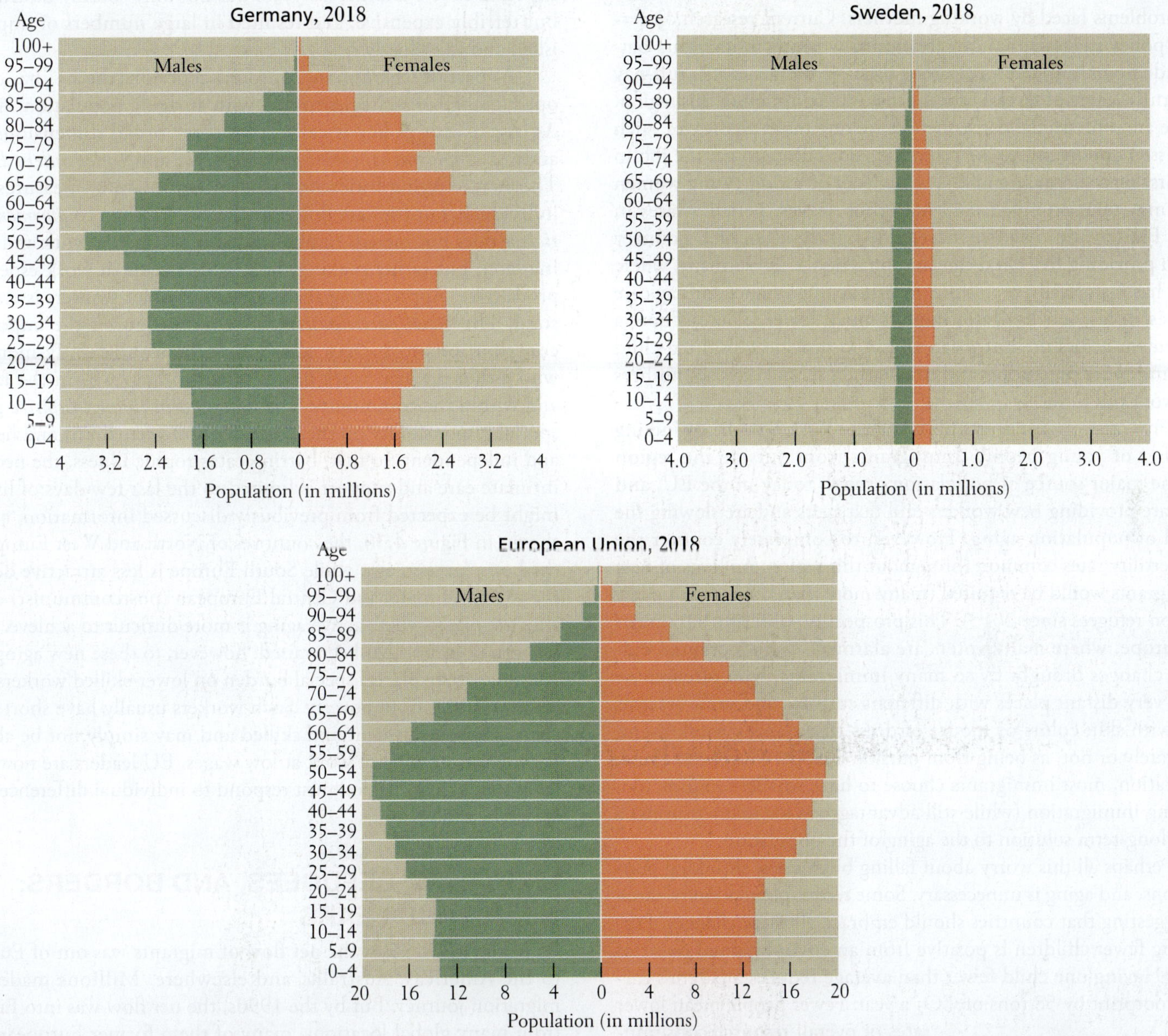

Figure 4.29 Population pyramids for Germany, Sweden, and the EU, 2018. These population pyramids have quite different shapes, but all have a narrow base, indicating that birth rates have declined over time. Notice the details of the pyramid shapes, which indicate important past trends in birth and death patterns. [Data from: "Population Pyramid of Germany," "Population Pyramid of Sweden," and "Population Pyramid of the European Union," International Data Base, U.S. Census Bureau, 2018, at https://www.census.gov/data-tools/demo/idb /informationGateway.php]

Germany, Sweden, and the whole European Union. The narrowing base of each pyramid indicates that by the mid-1970s, the postwar baby boom of the 1950s and 1960s had ended across Europe. By 2000, 25 percent of adult Europeans were having no children at all.

Why are Europeans choosing to not have children? Urbanization and strong economic development play a role, as does the fact that more and more women want professional careers. In Germany, for example, educated women have the lowest fertility rates; and this educated group is growing in size every year. Twenty-five percent of all German women are now choosing to remain unmarried well into their thirties, which means that they will likely have only one child, if any.

European governments have tried to boost fertility rates with a variety of policies, the most successful of which have addressed the problems faced by working mothers. Current research on fertility policy suggests that career-minded working mothers are better aided by child-care provisions than by the generous, full-pay maternity leaves of several months to a year common in this region. There are professional consequences to long maternity leaves, such as missed opportunities for promotions or failure to get the job in the first place because employers are leery of hiring young women who may take extended paid maternity leaves. Fertility rates in both France and Sweden increased slightly after each country began providing free child care and preschools. In Germany, where there has historically been little government assistance for day care or preschools, fertility rates remain much lower. There is now a movement across the European Union to provide subsidized child care and more preschools and to lengthen school days so mothers can work full time.

Immigration could provide a partial solution to the dwindling number of young people. Immigrants from outside the region are the major source of population growth today in the EU, and they are providing new workers and taxpayers and are slowing the trend of population aging. However, to completely counter the low fertility rates common throughout this region, millions of new immigrants would be required (many more than the surge of a few million refugees since 2015). This prospect is politically unpopular in Europe, where many voters are alarmed by the potential cultural changes brought by so many immigrants, who often come from very distant places with different customs and value systems and with skin colors or modes of dress that quickly mark them, accurately or not, as being from outside Europe. Moreover, after a generation, most immigrants choose to have smaller families, too, making immigration (while still advantageous for many reasons) a poor long-term solution to the aging of the population.

Perhaps all this worry about falling birth rates, shrinking populations, and aging is unnecessary. Some recent population research is suggesting that countries should embrace all these changes. Producing fewer children is positive from an environmental point of view. Having one child fewer than average reduces a parent's carbon footprint by 58 tons of CO_2 a year. Fewer people mean lower rates of overall resource consumption and reduced expenses for education. Also it appears that lower fertility addresses gender equity by freeing up women to make career choices that were closed to them when procreation was their primary role. Automation will help in addressing shrinking workforces; and even the aging of populations can be viewed as having advantages, once attitudes are changed.

Active Aging a three-pronged EU policy aimed at changing society's attitudes toward the abilities and needs of people as they age

ACTIVE AGING

Concern over the negative effects that might result from the aging of Europe's population for a time reinforced stereotypes about older citizens. They were described as obsolete, underachieving, unable to learn new skills, a drag on the economy, and expensive to care for. These descriptors were used to push people in their late fifties into early retirement. Anyone over 40 was thought to be not worth an investment in training. But it rapidly became clear that sending middle-aged workers into retirement when they could be expected to live another 30 years was not only socially destructive, but terribly expensive, and resulted in large numbers of impoverished and idle elderly.

As part of its emphasis on social cohesion, the EU has developed a range of strategies to deal with an aging population. **Active Aging**, as this policy is called, has three broad dimensions that address *employment*, *social participation*, and *independent living*. The Active Aging Index, mapped in **Figure 4.30**, measures how individual countries are doing in these areas. The *employment initiative* does not merely emphasize staying longer at work and retiring later, but rather making those added years of work especially productive. Appreciating the potential of older workers is emphasized. The *social participation initiative* recognizes the important contributions to society of older workers and family members who either do paid work or volunteer to help within the family and within the community. The *independent living initiative* places special responsibility on the aging person to remain fit and healthy and independent, so that, barring catastrophic illness, the need for intimate care and support is limited to the last few days of life. As might be expected from previously discussed information, and as shown in Figure 4.30, the countries of North and West Europe are the best places to age, while South Europe is less attractive despite the milder climate, and Central European (postcommunist) countries are places where active aging is more difficult to achieve.

An objection has been raised, however, to these new aging policies: they may place a special burden on lower-skilled workers who so often suffer from poverty. Such workers usually have shorter life spans than the affluent and skilled and may simply not be able to endure longer years of work at low wages. EU leaders are now realizing that aging policies must respond to individual differences and societal contexts.

MIGRANTS, REFUGEES, AND BORDERS: NEEDS AND FEARS

Until the mid-1950s, the net flow of migrants was out of Europe, to the Americas, Australia, and elsewhere. Millions made that migration journey. But by the 1990s, the net flow was into Europe from many global locations, many of them former European colonies (**Figure 4.31**). By 1995, most of the European Union (plus Iceland, Norway, and Switzerland) had implemented the Schengen Agreement allowing the free movement of people and goods across much of Europe, with the exception of the UK (which opted out of the accord), as well as Ireland, Romania, Bulgaria, Croatia,

Figure 4.30 Aging in Europe: The north–south and east–west divide. The new EU policy called Active Aging has three components aimed at easing for everyone the stress of an ever larger elder population: *employment*, *social participation*, and *independent living*. An Active Aging Index calculated for the EU-28 reveals that countries in the north and west are better prepared in all three categories for the increase of older people than are those in the south and east; differences are linked to factors such as levels of education, health, and wealth. [Data from UNECE] **A** A very fit elderly man runs up a mountain in Slovenia. [tomazl/E+/Getty Images]

and non-EU states in southeastern Europe (see thumbnail map in Figure 4.31).

This opening of Europe's borders facilitated trade, employment, tourism, and, somewhat controversially, the unfettered migration of EU citizens. However, the Schengen Agreement also indirectly increased both access for immigrants from outside the European Union and their mobility once inside the European Union. Greece and Italy, both with long Schengen Area coastlines along the Mediterranean, have been major entry points for undocumented migrants, but both are also in serious financial straits and are unable to police their shoreline borders effectively. During the rapid influx of several million immigrants in 2015–2017, despite the Schengen Agreement, a few countries, in the name of national security, reinstated border controls; but not until France did so after the November 2015 Paris attacks were borders closed for any more than a few days. The repeated terrorist attacks in 2015 and 2016 and the simultaneous arrival of more than a million migrants (many of them refugees from Syria seeking asylum) caused much speculation that the Schengen Agreement was dead, an eventuality that would drastically change life and economies in Europe. It is not dead, but it is certainly under threat.

VIGNETTE (continued from the beginning of the chapter) After the call from Catholic Charities, Marina packed a small bag and left to volunteer in assisting the refugees as they came across the Croatian border. There she was shocked to see the pitiful condition of the people and their children (**Figure 4.32**). They were impossibly tired and muddy. To her surprise, she could talk with most of them because they were obviously educated and used English as a second language, like she does. They were, indeed, mostly Muslim, but they didn't seem alien at all and were

Figure 4.31 The EU Schengen Area and Asylum Applicants into Europe, 2018. Countries within the Schengen Border are indicated to the west and north of two heavy blue lines on the main map and as purple territory in the map insert. Note the countries that are not in Schengen. A burst of migration by asylum seekers began in 2015, declined somewhat in 2016, and remained significant but much lower by 2018. Shown on the map are best estimates of the number of asylum applicants (migrants) into each country by 2018 (the latest year for which data are available). Many of these people were fleeing from conflict in North Africa and Southwest Asia; others, called economic migrants, were escaping from poverty and the absence of jobs in their home countries. The distinction between "economic migrants" and "asylum seekers" is flawed, as discussed in "The Rules for Assimilation in Europe." [Data from European Commission EUROSTAT websites at http://ec.europa.eu/eurostat/web/asylum-and-managed-migration/data/main-tables; https://ec.europa.eu/eurostat/web/asylum-and-managed-migration/statistics-illustrated; and http://www.economist.com/blogs/economist-explains/2015/08/economist-explains-18]

so grateful for kind words and a change of clothes or a carriage for a toddler. After a week on the border, the chaos began to clear. Dozens of buses came to pick up the thousands of travelers to drive them to the Austrian border, from where they would proceed to Germany. Marina asked them why they didn't stay in Slovenia, which they had the right to do. Most refugees had Germany as a goal because there they felt they could find work as the doctors, teachers, and businesspeople they had been back home. Opportunities in Slovenia were completely unknown to them. In fact, no one she spoke with had ever even heard of Slovenia until now. ■

Like some citizens of the United States, some Europeans are ambivalent about migrants from anywhere, even though many are themselves descendants of migrants. Until recently, the internal flow of migration has been mostly from Central Europe into North, West, and South Europe. These Central European migrants were generally treated fairly, although prejudices against the supposed backwardness of Central Europe remain evident. Immigrants from outside Europe, so-called *international immigrants*, have met with varying levels of acceptance. Attitudes reached a new low during the Muslim refugee influx of 2015–2018. Perhaps because

Figure 4.32 Discarded items left along the route of the walking refugees. A sympathetic Slovene public brought food, clothing, tents, and even baby carriages to the exhausted refugees as they trudged through the countryside. These gifts were useful for a time, but tired walkers could hardly carry all this with them. Dirty clothes, worn-out shoes, and broken carriages were discarded; and much more had to be jettisoned during the long journey. Some Slovene donors, having never experienced fleeing their homes on foot, felt insulted by the litter. [Michele Amoruso/Pacific Press/Alamy Live News]

these days few Europeans identify themselves as belonging to any religious faith, they were put off by shows of religious fervor by Muslim immigrants; some went way beyond expressing discomfort to outwardly demonstrating hate.

THE RULES FOR ASSIMILATION IN EUROPE

In Europe, culture plays as much of a role in defining differences between people as race and skin color do. **Assimilation** in Europe usually means giving up the home culture and adopting the ways of the new country, including skill in its language and conformity in dress. An immigrant from Asia or Africa may be accepted into the community if he or she has gone through a comprehensive change of lifestyle. For example, in the coastal tourist town of Piran in Slovenia an immigrant medical doctor from West Africa was twice elected mayor. On the other hand, if minority groups—such as the Roma (formerly called Gypsies), who have been in Europe for more than a thousand years—maintain their traditional ways (food, dress, gender attitudes, child rearing), it is nearly impossible for them to blend into mainstream society.

Since World War II, immigrants have come both legally and illegally from Europe's many former colonies and protectorates across the globe. Turks and North Africans often came as **guest workers** and were expected to stay for only a few years, fulfilling Europe's need for temporary workers in certain sectors. They were Muslim and often lived in segregated communities. Other immigrants were refugees from the world's trouble spots, such as Syria, Afghanistan, Iraq, Haiti, Central and South America, and Central Africa. Many came illegally from all of these areas.

Now, Europe's small but growing Muslim immigrant population (**Figure 4.33**) is the focus of assimilation issues in the European Union. Coming from a wide range of places and cultural traditions, these new immigrants from southeastern Europe (the former Yugoslavia), Syria, Lebanon, Jordon, Pakistan, India, and Bangladesh have joined remaining Muslim guest workers from Turkey and North Africa. Some have maintained traditional dress, gender roles, and religious values, while others have quickly assimilated into European culture. Those who do not assimilate, especially in dress and language, run the risk of **social exclusion**.

Studies of public attitudes in Europe show that immigration is least tolerated in areas where levels of income and education are lowest. Central and South Europe are the least tolerant of new immigrants, possibly because of fears that immigrants may drive down wages that are already relatively low. North and West Europe, with higher incomes and generally more stable economies, are the most tolerant. The European Union, through its *social cohesion policies*, is working on curbing illegal immigration from outside Europe while encouraging EU citizens to be more tolerant of legal migrants. The migrants themselves, whether from inside or outside Europe, fear that forced **cultural homogenization**—the tendency toward uniformity—awaits them.

Here should also be mentioned the vocal but increasingly problematic debate about the differences between **economic migrants** and **refugees**. Economic migrants—those who come in search of better job opportunities than they had in their home country—may have immigrated legally or illegally. Refugees (people who have fled from their homes because of war or violent social unrest or discrimination) often become **asylum seekers**—those who seek international protection because they fear war and persecution at home—and hence are classed as legal immigrants. Economic migrants are often thought to be unworthy of a welcome because they are coming out of ambition rather than necessity. There are several things wrong with this designation of "unworthy." Economic migrants are often themselves fleeing conflicts or discrimination of one sort or another in their home countries and therefore may deserve to be viewed as refugees. Also, Europe badly needs more workers who are ambitious, so why should these economic migrants be unwelcome? It may be because so many of them are black- or brown-skinned and/or Muslim, or from former colonies of Europe. Meanwhile, the refugees are for the most part self-identified and often carry no papers, so confirming their claims to be fleeing terrorism and war is very difficult.

It should also be noted that alienation and anger has boiled over in recent years among some Muslim immigrants and their children—resulting in protests, riots, and sometimes even terrorism, as in Paris in 2015. This anger relates primarily to the systematic social exclusion of these less

assimilation the shedding of old ways of life and the adoption of the lifestyle of another culture

guest workers legal workers from outside a country who help fulfill the need for temporary workers but who are expected to return home when they are no longer needed

social exclusion the condition of being systematically left out of important opportunities for participation in society

cultural homogenization the tendency toward uniformity of ideas, values, technologies, and institutions among associated culture groups; the loss of ethnic distinctiveness

economic migrants those who seek employment opportunities better than what they have at home

refugees those who have fled from their homes because of war or violent social unrest or discrimination

asylum seekers refugees who left their homes because of violent persecution and seek new legal status in an adopted country

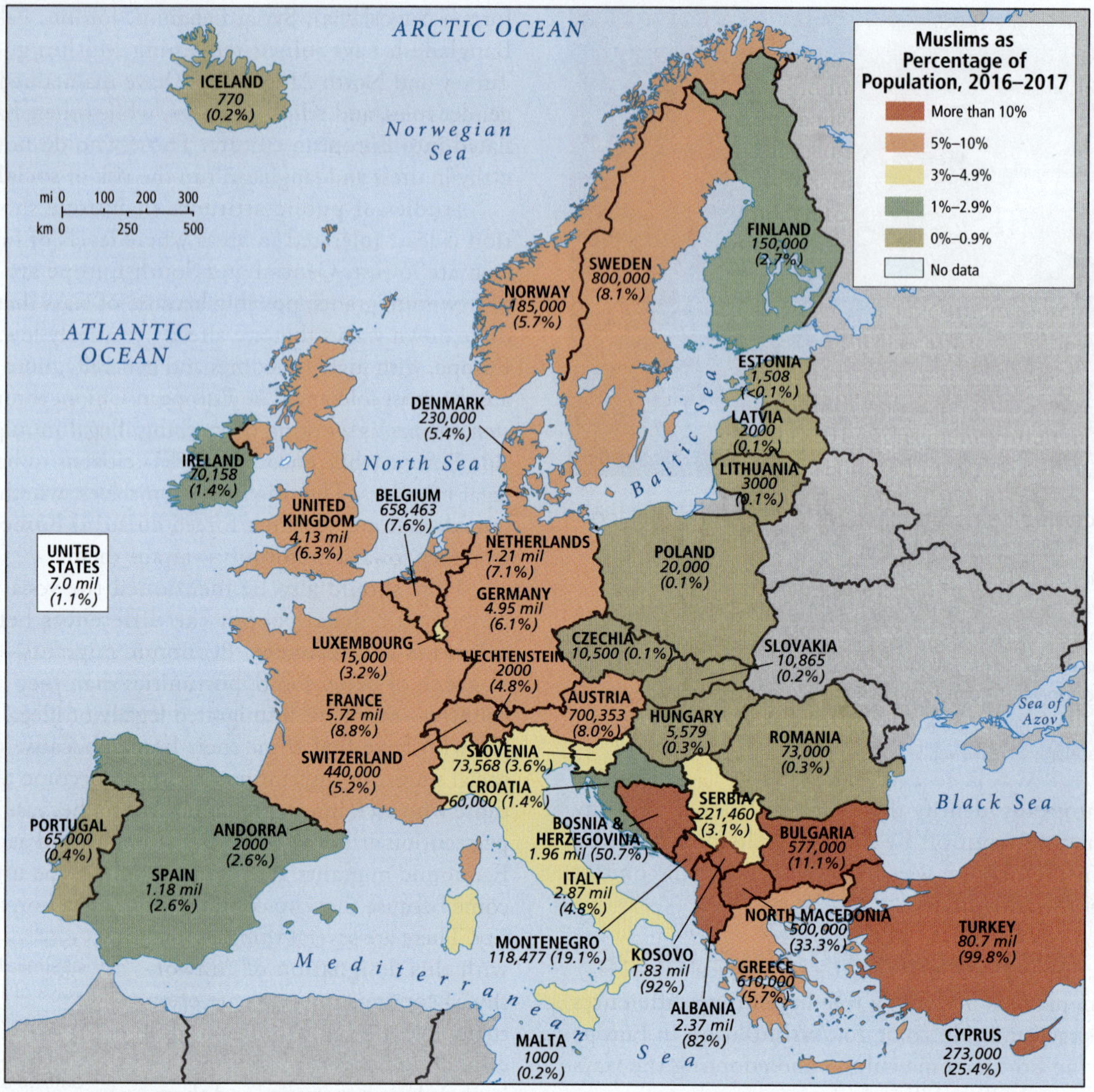

Figure 4.33 Muslims in Europe. Muslims are a small minority in most European countries, but in the EU-28 plus Norway and Switzerland their population has grown from a total of 19.5 million in 1990 to 25.8 million in 2016. Still, they amount to just 4.9 percent of a total population of 526 million. If migration rates remain as low as they were in 2018, the Muslim component is expected to rise to about 11.2 percent of Europe's population by 2050, still vastly lower than the percentage of Christians, other religions, and nonbelievers. Many Muslim immigrants intend to make Europe their permanent home; but just how, or if, the assimilation of Muslims into generally secular Europe will proceed is a major topic of public speculation. [New data (2016) from Pew Research Center on Muslims in EU-28 plus Norway and Switzerland and as a percent of total population of each country, at http://www.pewforum.org/2017/11/29/europes-growing-muslim-population/; and Power from Statistics: Data Information and Knowledge, Outlook Report, EuroStat 2018, at https://ec.europa.eu/eurostat/documents/7870049/9087487/KS-FT-18-005-EN-N.pdf/8c143a88-a67c-46ca-a24d-c34046fc4a49.]

assimilated Muslims from meaningful employment, from adequate social services, and from higher education. When the French media investigated protests by Muslim residents in 2015, they found that the protestors' complaints were indeed legitimate. Young Muslims born in Europe, who have never known life in any other place and who have been schooled in the lofty ideals of the European Union, harbor the greatest resentment about the constricted opportunities they face. However, many Europeans, unaware of the extent to which their own societies discriminate against and exclude Muslims, have come to view Muslim protesters simply as malcontents.

CHANGING GENDER ROLES IN EUROPE

Gender roles have changed significantly from the days when most women married young and worked in the home or on the family farm. Growing numbers of European women are working outside the home, and the percentage of women in professional and technical fields is rising rapidly. Nevertheless, European public opinion among both women and men largely holds that women are less able than men to perform the types of work typically done by men and that men are less skilled at domestic, caregiving, and nurturing duties. In most of Europe, men still have greater social status, hold more managerial positions, and have more autonomy in daily life than women—more freedom of movement, for example. These advantages for men are stronger in Central and South Europe than they are in West and North Europe.

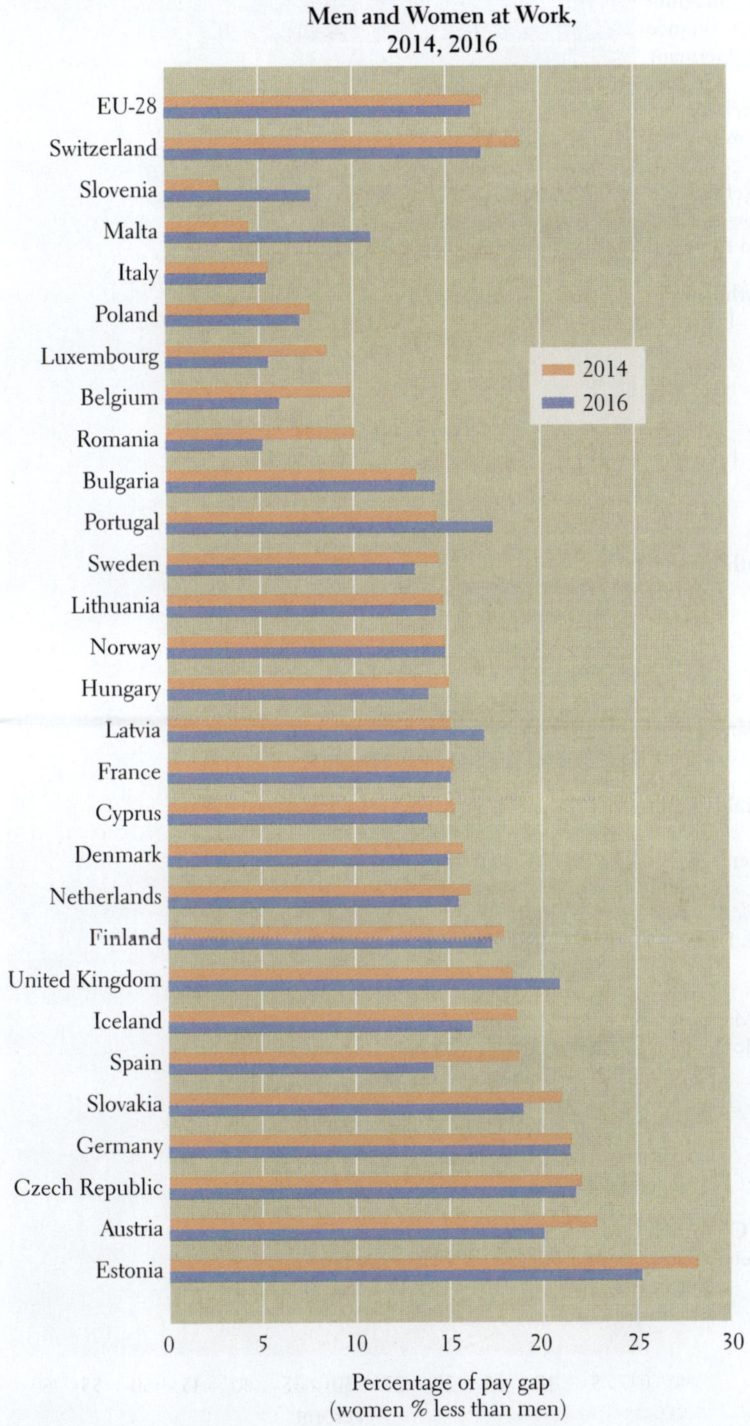

Figure 4.34 The gender pay gap in the EU-28 in 2016, percent less than men earned by women. Nowhere in the EU are men and women paid equally. Furthermore, in those cases when 2016 (blue line) is longer than 2014 (orange line), gender pay differentials have worsened. In countries where the blue line is shorter than the orange line (for example, Poland, Romania, Spain), the pay gap has narrowed. There are no clear regional patterns, and in some of the richest countries women suffer the greatest pay disadvantages when compared to men.

The Gender Pay Gap and Double Day

Perhaps most illustrative of male advantage is the gender pay gap. The gender pay gap, although widespread in Europe, is narrowing and is notably less for younger women, who tend to be more educated and willing to break with traditions in employment. **Figure 4.34** compares 2014 data to 2016 data (the latest available) on the average gross hourly earnings of men and women in the EU. Every employment category is included. If education and experience are equalized, men still earn considerably more than women. Education and health professions tend to pay both genders more equally, but salaries in these professions are also lower on average than in manufacturing and finance, where discriminatory pay is most prevalent. Of course, the gender pay gap, wherever it is found, negatively impacts the financial well-being of European women and their families.

The **double day**—the unpaid domestic work that must be done daily in addition to paid employment—continues to fall mostly to women. Certainly, younger European men now assume more shopping, cooking, housework, and child-care duties than did their fathers, but women who work outside the home are still expected to do most of this unpaid domestic work. UN research shows that in most of Europe, women's domestic duties average 4 hours and 20 minutes each day, while men put in 2 hours and 16 minutes.

Women burdened by the double day generally operate with somewhat less efficiency in a paying job than do men. Where children are involved, women also tend to choose employment that is closer to home and that offers more flexibility in the hours and skills required. These more flexible jobs (often erroneously classified as part time) almost always offer lower pay and less opportunity for advancement, though not necessarily fewer work hours.

Influence of Women on Policy Formation

Despite the persistent gender inequality, many EU policies officially encourage gender equality. Well over half the university graduates in Europe are now women, and the number of managerial posts in the EU bureaucracy held by women is growing. In Parliaments, where women can influence policy making, membership is increasing, but women still constitute less than one-third of elected representatives in 14 of the EU-28 countries in 2018. Only in Finland (42 percent), Sweden (44), and Norway (41) do women come close to filling 50 percent of the seats in the legislature (**Figure 4.35**). Despite their higher rates of university and professional education, the political influence and economic well-being of European women lag behind those of European men.

Women as heads of state are also few. Several women have held high offices in the United Kingdom, but elsewhere in West Europe, this trend is only beginning. In 2005, Germany elected Angela Merkel as its first woman chancellor (prime minister) and elected her to a third term in 2014; in 2018 Chancellor Merkel announced her retirement; Iceland in 1980 elected its first female head of state; others have followed. Slovenia chose Alenka Bratušek as prime minister in 2013. In 2014, Croatia elected a female president, Kolinda Grabar-Kitarović. In March 2019, Slovakia elected

double day the longer workday of women with jobs outside the home who also work as caretakers, housekeepers, and/or cooks for their families

President Zuzana Čaputova, a pro-EU, anticorruption, and environmental activist. Still, across Europe, women generally serve only in the lower ranks of government bureaucracies, where they implement policies but have limited power to formulate them.

The failure of women to gain national elective offices, either as members of parliaments or as heads of state, means that their influence on policy development is diluted. The map of the Gender Development Index (**Figure 4.36**) shows marked division, with the UK, the Netherlands, Germany, Austria, Italy, Greece, Malta, Lithuania, and southeastern Europe having a GDI ranking that is notably lower than Scandinavia, Portugal, Spain, France, and most of the postcommunist states of Central Europe. Why would this be? Scandinavia long ago noticed gender disparities that hurt females; it now takes pride in its gender-neutral policies. Spain, Portugal, France, and Belgium have recently undertaken gender reforms, and the postcommunist states still bear the positive effects of basic gender-equalizing socialist reforms during the communist era. On the other hand, many, even very wealthy countries (those colored red), have remained partially under the sway of patriarchal customs. Nonetheless, when compared to global patterns (see inset map), all of Europe shows increasing gender parity.

Norway is a recognized global leader in redefining gender in society. It does this by directing much of its most innovative work on gender equality toward advancing men's activities in traditional women's arenas. For example, men and women are allowed to share the year of paid parental leave with a newborn or adopted child. This policy, of recognizing a father's responsibility in child rearing and the need for father and child to bond early in life, is gaining popularity among celebrities in the United States under the name *family leave*. In 2015, Mark Zuckerberg, the Facebook cofounder and chief executive, announced that he was taking a 2-month leave to be with his newborn daughter, Max.

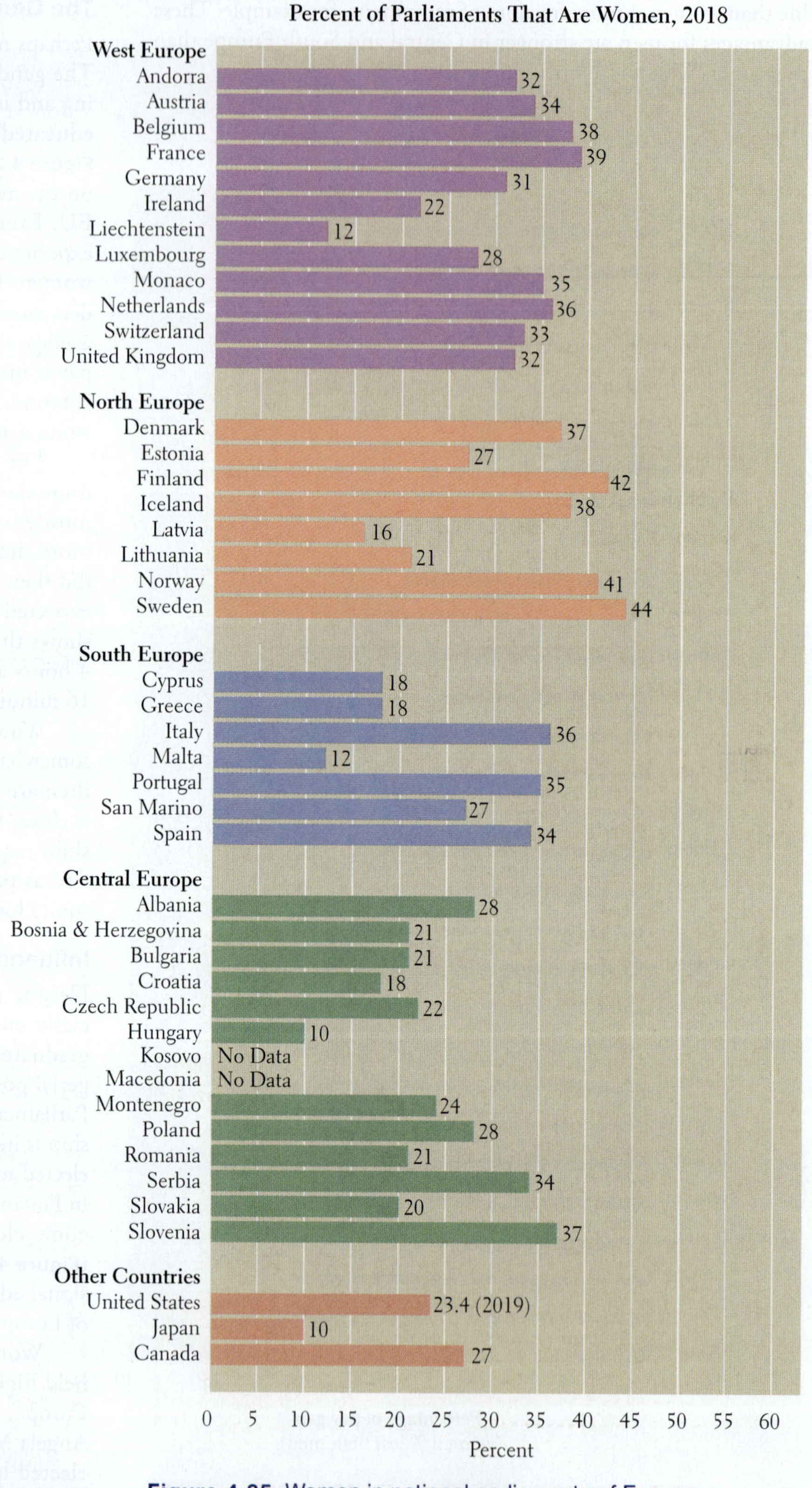

Figure 4.35 Women in national parliaments of Europe, 2018. Women comprise about half the adult population in all EU countries, but only in a very few countries in North Europe do they make up close to half of European legislatures; thus, their influence on policies and legislation is much restricted.

SOCIAL PROTECTION SYSTEMS: GOALS AND OUTCOMES

In the post–World War II era, the European Union was conceived primarily to promote economic cooperation and free trade, and the results have been overall quite positive. But by the mid-twentieth century, public pressure for improved living standards, channeled through the democratic process, had moved most European governments toward becoming **welfare states**. In welfare states, the government accepts responsibility for the well-being of citizens, guaranteeing basic necessities such as education, affordable housing and food, unemployment insurance, and

welfare states government accepts responsibility for the well-being of citizens, guaranteeing basic rights to education, affordable housing and food, unemployment insurance, and health care for all

Population, Gender, and Culture in Europe

The Gender Development Index (GDI) reveals that while European countries generally treat males and females fairly equally, some countries (such as the UK, the Netherlands, Germany, Austria, Italy, Greece, Malta, Cyprus, two of the Baltic states, and much of southeastern Europe) have a GDI ranking that is markedly lower than that of the rest of Europe. This raises concerns about gender equity and the well-being of women in these countries. Notice also that Bosnia-Herzegovina, North Macedonia, Turkey, the Middle East, and North Africa (see world inset map) have notably lower rankings. These are predominantly Muslim areas. It is data like this, revealing deep cultural differences, that worries Europeans about the prospects of assimilation of immigrants.

THINKING GEOGRAPHICALLY

Thinking about Europe as a whole, what about the gender development map is most surprising to you?

A **If you were a woman with only a high school education, where in Europe would you aspire to settle for a well-paying job and why?**

B C D **If a country ranks high on the gender development scale (GDI), is it necessarily an affluent place? Why or why not?**

D **Sweden is a popular country for immigrants to settle in. What factors are immigrants likely to be weighing when they decide on Sweden?**

A A businessman in Venice, Italy. Italy, like a number of other affluent EU countries, ranks lower on gender development than poorer, newer EU members because of tenacious patriarchal attitudes toward women. Women are paid less for equal work and are rarely found in executive positions. There is not yet a national consensus that the genders should aim for equality. [Sam Edwards/Getty Images]

B A scientist in Poland. A biologist collects plant specimens from the forest floor in Bialowieza National Park. Poland ranks high in gender development primarily because during the communist era, women were encouraged to get educated and move into male-dominated occupations, especially science. They were still underrepresented in supervisory positions. [Raymond Gehman/Getty Images]

C An apprentice in Romania. Romania ranks high in gender development. Under former communist policies, women were urged to take even hard manual labor jobs previously filled by males. Here, an ethnic minority woman is learning a trade in a program specifically aimed at the previously underemployed. [Stephen Bisgrove/Alamy Stock Photo]

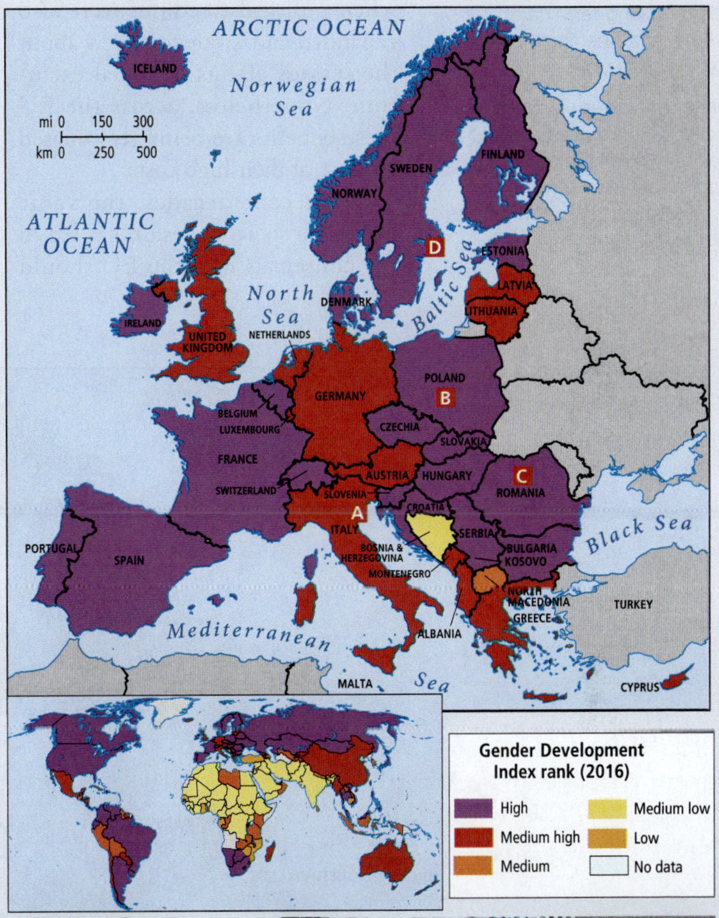

D A father and his children in Stockholm, Sweden. Sweden, which ranks high in gender development, has emphasized equalizing gender roles, including encouraging men to take major roles in child rearing and other domestic duties. The Swedish government plays an active role in influencing how children are raised. [Maskot/DigitalVision/Getty Images]

health care for all. Now, all European countries have some level of tax-supported systems of *social welfare*, or **social protection** (the EU term). The stated goal is to use taxes to fight poverty and promote **social cohesion** by providing all EU citizens with basic health care; assistance with parenting responsibilities (free or low-cost primary, secondary, and higher education); affordable housing; and pension, survivor, disability, and unemployment benefits. The ideas behind these systems developed during the nineteenth and twentieth centuries, and were broadly expanded in western Europe in the last 50 years, when political freedoms were also expanded and awareness grew about the serious effects of social exclusion. Nonetheless, across the EU these benefits are being reevaluated because of their high costs.

social protection the tax-supported systems that provide citizens with benefits such as health care, long-term care, pensions, and relief from poverty and social exclusion

social cohesion an expressed goal of the EU-28 involving balanced and sustainable development, reduced disparities between regions and countries, and equal opportunities for all people

In recent decades, the value placed on social cohesion has led to the consensus that the EU should become a technically advanced, sustainable, and inclusive economy. This, in turn, has reinforced the role of social protection in building a sense of loyalty to the EU. But the costs of social protection systems have become a political issue as economic growth has slowed, as unemployment has increased, and as populations have shrunk and aged (thus needing more protection). The unexpected sudden huge increase in refugee immigration added to political contentions regarding costs, because the migrants are so obviously in need of immediate protection. At the heart of the political debates about social protection costs is the fact that there is a relatively wide disparity of income and well-being among EU countries. Because individual countries give social protection benefits based on an agreed-upon percentage of the country's GDP, benefits vary widely across the EU. This variability tends to encourage migration toward the more generous countries, thereby undermining social cohesion.

The map in **Figure 4.37** shows the geographic distribution of social protection (welfare) systems in the EU (plus Norway and Switzerland). The most important categories of protection in nearly all EU-28 (plus 2) are *old age assistance, sickness and*

Figure 4.37 European expenditures for social protections. In 2016 (the latest data available from the EU), the expenditures for social protection in the EU-28 averaged about 19.1 percent of total EU GDP, and social protection is regarded as the most important function of government. The map shows the percent of GDP each country in the EU-28 spent on social protection in 2016. Some non-EU members are included for comparison. [Data from: http://ec.europa.eu/eurostat/web/social-protection/data/main-tables; and https://ec.europa.eu/eurostat/statistics-explained/index.php?title=File:Total_general_government_expenditure_on_social_protection,_2016_(%25_of_GDP).png]

disability support, with *protection for families and children* coming third. Comparative levels of protection can be classified into three categories, from those regions with the most social protection coverage to those with the least. North and West Europe currently have the most countries with well-funded social protection (dark green on the map). North Europe (Scandinavia), spending 20 to 25 percent of GDP on social protection, attempts to create equality across gender and class lines by providing extensive health care, education, housing, and child- and elder-care benefits to all citizens, from cradle to grave. In all of North Europe, child care is state-supported and widely available, in part to help women enter the labor market but also to give all children a certain common set of values. The hope is that this will ensure that in adulthood every citizen will be able to contribute to the best of his or her capability, and that citizens will not develop criminal behavior or abuse drugs. While finding comparable data is very difficult, surveys of crime victims across the European Union show that the crime rate in Scandinavia is generally lower than in other parts of Europe.

Countries in West Europe have somewhat less lofty goals than North Europe. Austria, Belgium, France, Greece, and Italy (also dark green on the Figure 4.37 map) and those colored medium green (such as Germany, Luxembourg, Poland, Portugal, Slovakia, Slovenia, Spain, and the UK) spend 15 to 19.9 percent of their GDP to assist those in need but do not try to promote upward mobility. State-supported health-care and retirement pensions are available to all, but for the most part social protection systems still reinforce the traditional "housewife contract" by assuming that women will stay home and take care of children, the sick, and the elderly. In this group, Germany probably has the most generous system and is presently liberalizing its policies to aid working women.

Most of the former Soviet bloc countries—the Baltic Republics, Bulgaria, Czech Republic, Hungary, and Romania—spend the least of their GDP on social protection (9.9 percent to 14.9 percent). And because these countries all have fairly low GDPs, it is clear that citizens in the countries in Central Europe have the least protection. In these countries, during the communist era, social protection was comprehensive, resembling the current cradle-to-grave social democratic system in Scandinavia, except that women were pressured to work outside the home. Benefits, usually tied to one's job, often extended to nearly free apartments, health care, state-supported pensions, subsidized food, fuel, vacations, and early retirement. However, in the postcommunist era, state funding has collapsed, support programs died, and jobs disappeared, forcing many people to do without basic necessities. In many postcommunist countries, welfare systems are now being revised, but usually with the goal of reducing benefits drastically and raising the retirement age to over 65.

Europeans generally pay much higher taxes than North Americans. The overall sum of taxes and compulsory contributions for the EU-28 countries averages about 40 percent of GDP. (For the United States it averages 25 percent; and for Canada, less than 30 percent.) In return for paying high taxes, Europeans expect more in services. In some cases, European governments are able to deliver services more cheaply than the so-called *free market* does elsewhere. For example, the European Union spends on average between U.S.$3000 and U.S.$4000 per person for health care, while the United States spends more than U.S.$8000. Even at this much lower cost, the European Union has more doctors, more acute-care hospital beds per citizen, and better outcomes than the United States in terms of life expectancy and infant mortality.

While some argue that Europe can no longer afford high taxes if it is to remain competitive in the global market, others say that Europe's economic success and high standards of living are the direct result of the social contract to take care of basic human needs for all. The debate has been resolved differently in different parts of Europe, and the resulting regional differences have become a source of concern in the European Union, because, with open borders, unequal benefits can encourage those in need to flock to a country with a generous welfare system and overburden the taxpayers there.

Social Protection, Social Cohesion, and Migrants in Europe

When the refugee crisis erupted in the EU in 2015, it looked like the abrupt arrival of well over a million migrants, many of whom were Muslim, might spell the end of social cohesion and the consensus on growth—even, perhaps, the end of the EU open-border policies. Many countries seemed to be vying over which one could be the least welcoming. Then, slowly and spasmodically, EU citizens were awakened to the fact that many parts of the EU could use a boost of young, educated workers who would soon become taxpayers. In early 2016, when quotas on the number of migrants each country would be asked to accept were announced (based on the host country's population), many countries first objected strenuously, and then within a few days changed their minds. At least some began to see that in the medium and longer term, the migrants could have a positive impact on the labor supply, economic growth, and the tax base, provided the right integration and assimilation policies could be developed.

HUMAN WELL-BEING IN EUROPE

To a significant extent, the social protection policies described above are responsible for the high level of well-being in Europe when compared to global patterns. But even in Europe the patterns are uneven. As we have observed in Figure 4.22, which shows rankings on the United Nations HDI, although Europe is one of the richest regions on Earth, there is still considerable disparity between countries in wealth and well-being. Figure 4.20, which maps GNI for Europe, supplies a partial explanation for the disparities: Europeans earn markedly different incomes; and as we have learned, poorer countries invest less in social protection, so the patterns become solidified.

CHECK YOUR UNDERSTANDING

1. What are the chief components of population change in Europe today?
2. Assess the ways in which European landscapes reveal population change.
3. Explain why immigration could address some of Europe's perceived population issues.
4. Explain why worry about Europe's low birth rates and aging populations could be quite unnecessary.
5. Describe the EU's new policies regarding Active Aging.

6. Why might Europe's rules for assimilation be especially hard for Muslim immigrants to follow?
7. What are some of the features of Europe's welfare states that affect daily lives of Europeans?
8. Discuss the status of females in Europe. What are some problems; what are some signs of change?

THE SUBREGIONS OF EUROPE

4.13 Distinguish among the various subregions geographically, economically, and in social protection/active aging policies.

The subregions of Europe take on added significance these days because of the social and financial crises faced by the European Union. The refugee crisis, general immigration, unemployment, aging populations, global competitiveness, and recovery from economic recession are perceived, fairly or not, to fall along subregional divisions. Countries in South and Central Europe are seen as mismanaging the problems, and countries in West and North Europe think of themselves as the more stable and financially responsible entities that must bail out the others.

WEST EUROPE

The countries of West Europe include the United Kingdom, the Republic of Ireland, France, Germany, Belgium, Luxembourg, the Netherlands, Austria, and Switzerland (**Figure 4.38**). Despite their economic success, these countries are contending with several issues that will affect their development and overall well-being:

- Challenges, such as Brexit, to what had been a continual trend toward an open economic union with each other and with neighboring countries to the north, east, and south
- Social tensions, such as those generated by the influx of immigrants from inside and outside Europe, many of them refugees, and failures to incorporate migrants into societies
- Worries about security in light of recent terrorist attacks, and confusion about how to promote the dream of free movement through Europe yet secure borders against terrorists
- The persistence of high unemployment despite general economic growth
- The need to increase the global competitiveness of the subregion's agricultural, industrial, and service sectors while maintaining costly but highly valued social welfare programs

Benelux

Belgium, the Netherlands, and Luxembourg—often collectively called the *Low Countries* or *Benelux*—are densely populated countries that have very high standards of living (see Figure 4.22). These three countries are well located for trade: they lie close to North Europe and the British Isles and are adjacent to the commercially active Rhine Delta and the industrial heart of Europe. The coastal and river-side locations of Benelux, with the great port cities of Antwerp, Rotterdam, and Amsterdam, give these countries easy access to the global marketplace. In addition, the Benelux countries have played a central role in the European Union for many years: Brussels, Belgium, is considered the EU capital, with most EU headquarters, along with the NATO headquarters, located there, all of which draw other business to Brussels.

International trade has long been at the heart of the economies of Belgium and the Netherlands. Both were active colonizers of tropical zones: Belgium in Africa and the Netherlands in the Caribbean, Africa, and Southeast Asia. Their economies benefited from the wealth extracted from the colonies and from global trade in tropical products such as sugar, spices, cacao beans, fruit, wood, and minerals. Private companies based in Benelux still maintain advantageous relationships with the former colonies, which supply raw materials for European industries. Cacao is an example of this: It is exported from West Africa in a raw state to Benelux and other European countries, where it is refined into high-end elegant products by the upscale chocolatiers there. Some of these products are now part of the *fair trade movement* in Europe and America that channels an increased share of chocolate profits to cacao farmers in Africa and elsewhere. Getting some of these profits to the actual field laborers is still problematic.

The largest Benelux nation, the Netherlands, is noted particularly for having reclaimed land that was previously under the sea and for creating attractive living spaces in wetlands (**Figure 4.39**). Today, its landscape is almost entirely a human construct. As its population grew during and after the medieval period, people created more living space by filling in a large, natural coastal wetland. To protect themselves from devastating North Sea surges, they built dikes, dug drainage canals, pumped water with windmills, and constructed artificial dunes along the ocean. Today, a train trip through the Netherlands between Amsterdam and Rotterdam takes one past the port of Amsterdam with lots of ships, fuel storage facilities, and raised rectangular fields crisply edged with narrow drainage ditches and wider transport canals. One is likely to see fields filled with commercial flowerbeds, vegetable gardens, and grazing cattle, all of which serve the needs of the primarily urban population.

Although the Netherlands has a high standard of living, with 17.2 million people, it is one of the most densely settled countries in Europe, and this density has consequences. There is no land left for the kind of high-quality suburban expansion preferred in the Netherlands that wouldn't intrude on agricultural space. And there is not nearly enough space for recreation; bucolic as they appear, transport canals and carefully controlled raised fields are not venues for picnics and soccer games. People now travel great distances to reach their jobs, and these long commutes sap time and patience and add to air pollution and traffic jams. Even for weekend getaways, people usually leave the Netherlands for neighboring countries that have more natural areas. The choice to maintain agricultural space is largely a psychological and environmental one, as agriculture accounts for only 1.6 percent of GDP and 1.2 percent of employment.

France

France, shaped like an irregular hexagon, is bound by the Atlantic on the north and west, by mountains on the southwest and southeast, by Mediterranean beaches on the south, and by lowlands on the northeast. The capital city of Paris, in the north of the hexagon, is the heart of the cultural and economic life of France. As one of

Figure 4.38 West Europe subregion. **A** Rising amid the chaos of London's skyline is an unusual skyscraper known by locals as "the Gherkin" due to its shape, which resembles that of a gherkin pickle. Like many other European cities, London is having a boom in new architectural styles. [View Pictures/Universal Images Group/Getty Images]

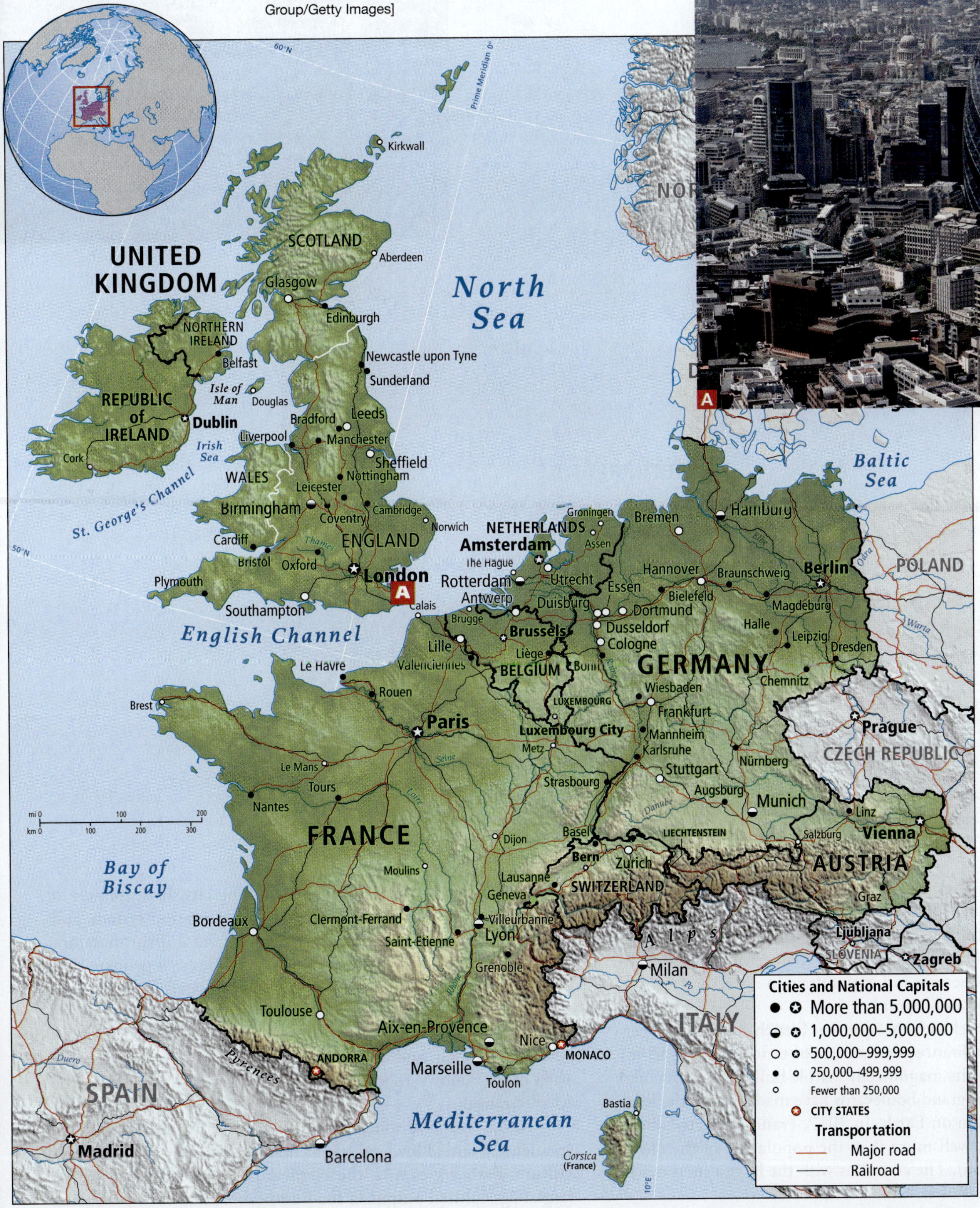

Figure 4.39 Land reclamation in the Netherlands. **A** The Netherlands Land Reclamation map shows the various levels of land in this low-lying coastal country. The tan-colored land was all originally below sea level and has been turned into port facilities, agricultural land, and industrial sites. Massive port facilities outside Rotterdam were created on recently reclaimed land. [Frans Lemmens/Getty Images]

Europe's two world cities (London is the other), Paris has an economy tied to providing highly specialized services—traditionally, fine food, food processing, fashion, and tourism, but increasingly, accounting, advertising, design, and financial, scientific, and technical services. These service industries draw in a steady stream of workers from Europe and beyond.

Few would quarrel with the idea that French culture has a certain cachet and that France is an arbiter of taste and a model for relaxed urban living. Its magnificent marble buildings, manicured parks, museums, and grand boulevards have made France the leading tourist destination on Earth. In 2017, France attracted about 90 million tourists, well more than the population of the entire country (65.1 million). The countries with the largest increases in visitors to France are Russia, China, and Brazil.

Paris, though elegant, is also a working city. It is the hub of a well-integrated water, rail, and road transportation system, and its central location and transport links attract a disproportionate share of trade and businesses, so it qualifies as a primate city. Together with its suburbs (which are administered separately), Paris has almost 11 million inhabitants—roughly one-sixth of the country's total. In the 1970s, French planners decided that Paris was large enough, and they began diverting development to other parts of the country, especially to Toulouse and farther south to the Mediterranean.

To the west and southwest of Paris (toward the Atlantic) are less densely settled lowland basins that are used primarily for agriculture. France vies with the Netherlands and Germany for the largest agricultural output in the European Union, each producing about $70 billion per year. France's climate is mild and humid, though drier to the south. Throughout the country, farmers have found profitable specialties, such as wheat (France ranks fifth in world production), grapes (France and Italy compete for first place in world wine production and wine exports), and cheese (France is second in world production). Despite France's agricultural leadership, agriculture accounts for only 2 percent of the French national GDP (2017) and employs only 2.8 percent of the population. Modernization has made possible ever-higher agricultural production with ever-lower labor inputs. The manufacturing sector, which expanded quickly after World War II, absorbed many former farmworkers; but after 1960, the high-end service sector grew even more rapidly. French industry now accounts for 20 percent of France's GDP, and services account for nearly 77.2 percent.

The Mediterranean coast is the site of the French Riviera and France's leading port, Marseille. Development here is booming. In part because of all of the tourism to the area, so many marinas, condominiums, and parks have been built that they threaten to occupy every inch of waterfront. The downside of this wealthy, densely populated region is that it produces large amounts of the pollutants that contribute to Mediterranean environmental problems (see the discussion in "Freshwater and Seawater Pollution" earlier in the chapter).

France derived considerable benefit and wealth from its large overseas empire, which it ruled until the middle of the twentieth century. Many of the citizens of former French colonies in the Caribbean, North America, North Africa, sub-Saharan Africa, Southeast Asia, and the Pacific (and their descendants) now live in France and bring to it their skills and a multicultural flavor. Paradoxically, while the European French are proud of being cosmopolitan, they are also very protective of their distinctive culture and wish to guard what they consider to be its purity and uniqueness. This has led to a marginalization, and even exclusion, of some (especially Muslim) immigrant minorities, who tend to live on the fringes of the big cities in high-rise apartment ghettos (*banlieues*), where they are poorly served by public transportation, have trouble accessing higher education and technical training, and have unemployment rates that are twice the national average.

Many people in France are deeply concerned about what they perceive to be a decrease in France's world prestige as its industrial competitiveness declines, unemployment rises, social welfare benefits decrease, and traditional French culture is diluted by an influx of migrants from former colonies. Such feelings have resulted in occasional swelling of popular support for the National Front, France's xenophobic right-wing party. So far, these tendencies have been countered by France's deep pride in its sophisticated multicultural identity and contribution to an envisioned egalitarian global community. But during the current influx of Muslim immigrants, xenophobia has increased, strengthening the representation of the National Front (now renamed National Rally) in Parliament. From a purely practical point of view, because it has a low birth rate, France needs immigrants to keep its economy humming, its technology on the cutting edge, and its tax coffers full.

France's status as a leading member of the European Union rivaled that of Germany for many years. France's trade volume (imports and exports) remains the sixth largest in the world, and it produces at least 25 percent of the EU agricultural products. Recently, though, its economic productivity has begun to suffer and its national debt has increased, due in part to expensive social programs and lenient workday rules (the French work only a 35-hour week and have especially generous unemployment benefits). During the EU economic crisis, France paid for an ever-smaller portion of the bailout funds. Some have predicted that France will itself need a bailout because of constrictions in the manufacturing and service sectors.

Germany

The famous image of a happy crowd in 1989 dismantling the Berlin Wall, as the symbolic end of Soviet influence in Central Europe, had particular significance for Germany. For 40 years, the country had been divided into two unequal parts. This is because, at the end of World War II, Russian troops, instead of going home, occupied the eastern third of Germany, and by 1949, the Soviets and their East German counterparts had turned it into a communist state known as East Germany. From the beginning, many East Germans tried to flee to what was then West Germany. To retain the remaining population, the Soviets literally walled off the border in the summer of 1961 (the Berlin Wall was part of the Iron Curtain; see Figure 4.19). East Germany's rich resources of skilled labor, minerals, and industrial capacities were used to buttress the socialist economies of the Soviet bloc and to support the military aims of the Soviet Union.

When the Berlin Wall fell and the two Germanys were reunited, East Germany came home with enormous troubles that proved expensive to fix. Compared to those in the west, its industries were outdated, inefficient, polluting, and in large part irredeemable. Much of its decaying or substandard infrastructure (bridges, dams, power plants, and housing) did not meet the standards of western Europe and had to be remodeled or dismantled and replaced. Its industrial products were not competitive in world markets; East German workers, though considered highly competent in the Soviet sphere, were undereducated and underskilled by western European standards. Reunified Germany is Europe's most populous country (83 million people), the world's fifth-largest economy, and a global leader in industry and trade. But the costs of absorbing the poor eastern zone dragged Germany down in many rankings for years. Unemployment rose sharply (especially among women) in the 1990s and again during the global recession beginning in Europe in 2008. Not only were there layoffs in the east (where unemployment is nearly double the national average), but workers throughout Germany, accustomed to high wages and generous benefits, lost jobs as firms mechanized or moved overseas, some to the United States.

During periodic global economic recessions, Germany's many multinational corporations have an especially difficult time because they are so dependent on global trade in industrial products. Germany long had the largest export economy in the world ($1.4 trillion in 2017), but China—of course a far larger country in area, resources, and population—recently overtook Germany, and in 2017 China had exports worth U.S.$2.157 trillion. The automobile industry, centered in Stuttgart, is an example of German commercial ventures that are moving operations out of Germany to countries, including the United States, where the markets for German cars lie. Expansion into Asia came not only because factory costs are lower there, but also because the market for luxury brands such as Mercedes-Benz is soaring, especially in China. Daimler Trucks, with home offices in Stuttgart, is the world's largest truck

Figure 4.40 Berlin's "bear pit" karaoke. Every Sunday afternoon in Berlin, an event that many visitors describe as magical and heartwarming is held. In the *Mauerpark*, or "Wall park"—a green space established after the dismantling of the Berlin Wall in 1989—a crowd of thousands gathers in a stone amphitheater to cheer on total strangers performing karaoke. Singers almost always get a huge round of applause—whether they're good at karaoke or just come across as a nice person. Gareth Lennon, an Irish bike messenger and karaoke enthusiast, started the event in 2009 as he rode around Berlin with a bike-mounted, battery-powered karaoke system and filmed people as they performed. He happened on the amphitheater one Sunday afternoon; within months, hundreds and then thousands of people began to show up every weekend. [Carsten Koall/Getty Images]

manufacturing company, with plants across Europe and in Africa, Asia, and Brazil. Daimler Trucks North America, now headquartered in Portland, Oregon, is North America's largest manufacturer of heavy-duty trucks.

Because Germany was regarded as the instigator of two world wars, for years West Germany had to tread a careful path. With some help from the U.S.-funded Marshall Plan, it had to see to its own economic and social reconstruction and to rebuilding a prosperous industrial base without appearing to become too powerful economically, politically, or militarily. For the most part, West Germany played this complex role successfully. Since the early 1980s, it has been a leader in building the European Union; with the EU's largest economy, it has borne the greatest financial burden of the European unification process. This burden has only increased during the 2008 economic crisis, the Greek and Italian debt crises, and the refugee crisis beginning in 2015. It is the German people who have taken in most of the refugees (close to 2 million since 2014) and provided a majority of the funds used to assist Ireland, Greece, Spain, Portugal, and Italy with the migrants.

The British Isles

The British Isles, located off the northwestern coast of the main European peninsula, are occupied by two countries: (1) the United Kingdom of Great Britain (England, Scotland, and Wales) and Northern Ireland—often called simply Britain or the United Kingdom (UK); and (2) the Republic of Ireland (not to be confused with Northern Ireland; see Figure 4.38).

The Republic of Ireland The main physical resources in the Republic of Ireland (usually called simply *Ireland*) are its soil, abundant rain, and beautiful landscapes. Ireland's considerable human capital remained underdeveloped for centuries during and after the time it served as a test-case colony where Britain could hone the technique of mercantilism that it then used in the Caribbean, North America, Asia, and Africa. The British confiscated land for their own use and for the use of their local sympathizers; thus, large numbers of Irish people were turned into landless and starving paupers. When they protested, they were imprisoned or sent as indentured labor to the new Caribbean colonies of the 1600s. Into the modern era, the Irish depended on small-plot agriculture and eventually on tourism, while industrialization lagged. As a result, until very recently, the Irish people were the poorest in West Europe. Over the last two centuries, some 7 million Irish emigrated to find a better life or were exiled as prisoners to Australia. In the 1990s, however, a remarkable turnaround began. Ireland attracted foreign manufacturing companies by offering cheap, educated labor; accessible air transport; access to EU markets; low taxes; and other financial incentives. The economy grew so quickly that for a time Irish labor was in short supply and the unemployment rate was 1 percent.

To supply the workers required, Ireland invited back those who had emigrated, and recruited other foreign workers from many locales, especially the former Soviet satellites (for example, Latvia and Poland) now in the European Union. In 2000, approximately 42,000 people migrated into Ireland (18,000 were returning emigrants), including some information technology (IT) workers from the United States. Czech workers packed Irish meat; Filipino nurses worked in most Irish hospitals. The cost of living rose sharply, however, and this rise plus the global recession meant that Ireland was no longer able to attract or even retain foreign investment. Ireland originally had not favored the admission of Central European countries to the European Union because it feared those countries would attract investment away from Ireland by offering skilled but even cheaper labor pools. This was beginning to happen by 2007.

Nevertheless, until the global recession of 2008, Ireland had one of Europe's fastest-growing (though still small) economies, and industry in Ireland played nearly as important a role as services in the country's GNI. In 2007, Ireland had Europe's second-highest per capita GNI, but by the end of 2011, its unemployment rate was 15 percent, affecting 400,000 workers across the skills spectrum. Thousands of guest workers returned to Central Europe, and tens of thousands of native Irish left to other parts of the European Union, to North America, Australia, and New Zealand, leaving behind mortgages they could no longer afford. A banking crisis and rising government debt left Ireland in need of a bailout that amounted to €85 billion (U.S.$108 billion) from the European Union and the IMF, including more than 12 billion pounds (U.S.$19.3 billion) from the United Kingdom. After a stringent austerity period, foreign investors attracted by low taxes built up Ireland's export capacity and by 2015 Ireland was rebounding; its belt-tightening was being recommended as a model for Greece, Spain, and others. Ireland in 2018 ranked fourth on the global

Human Development Index and has a per capita GNI of $62,000. Ireland has prospered as a member of the EU and benefited from EU financial support during the budget crisis, so it will not join the UK if it ends up leaving the EU (Brexit).

Northern Ireland Northern Ireland consists of six counties in the northeastern corner of the island; it is distinct from the Republic of Ireland and is administratively part of the United Kingdom. Northern Ireland began to emerge as a political entity in the 1600s, when Protestant England conquered the whole of Catholic Ireland. England removed or killed many of the indigenous people and settled (primarily Protestant) lowland Scots and English farmers on the vacated land. The remaining Irish resisted English rule with guerrilla warfare for nearly 300 years, until the Republic of Ireland gained independence from the United Kingdom in 1921. Northern Ireland remained part of the United Kingdom. Protestant majorities held political control in Northern Ireland, and the minority Catholics were subject to severe economic and social discrimination. Catholic nationalists unsuccessfully lobbied using constitutional (peaceful) means for a united Ireland. Other Catholic groups, the most radical among them being the Irish Republican Army (IRA), resorted to violence, including terrorist bombings, often against civilians and British peacekeeping forces, who were seen as supporting the Protestants. The Protestants reciprocated with more violence. By 1995, more than 3000 people had been killed in the Northern Ireland conflict or by violence that spread to the United Kingdom.

In 1998, tired of seeing the development of the entire island—and especially of Northern Ireland—blighted by the persistent violence, the opposing groups finally reached a peace accord known as the *Good Friday Agreement*, which was overwhelmingly approved by voters in both Northern Ireland and the Republic of Ireland. The plan for Northern Ireland provides for more self-government shared between Catholics and Protestants, the creation of human rights commissions, the early release of convicted terrorists, the decommissioning of paramilitary forces (such as the IRA), and the reform of the criminal justice system. Since 1998, violence has periodically risen and then abated, and the society has slowly begun to accept peaceful coexistence. The city of Belfast remains mostly divided along religious lines, but some neighborhoods are beginning to integrate.

A major sticking point in the Brexit negotiations has become the border between Northern Ireland (part of the UK) and the Republic of Ireland (an independent country that will stay in the EU no matter what the UK does). Northern Ireland was, in fact a major advocate of Brexit, but like so many, it did not consider all the ramifications of Brexit. Northern Ireland now wants the border with the Republic of Ireland to remain completely open to trade, people, and services, a seeming impossibility if the UK leaves the EU single market and customs union.

The United Kingdom The United Kingdom consists of England, Scotland, Wales, Northern Ireland, and 14 of what are known as British Overseas Territories—remnants of the British colonial empire. In 2014, Scotland, one of the four main parts of the UK, considered separating from the UK administratively, but Scottish voters (by 55 percent) did not accept this separation.

The United Kingdom has a mild, wet climate (thanks to the North Atlantic Drift) and a robust agricultural sector that is based on grazing animals. Its usable land is extensive, and its mountains contain mineral resources, particularly coal and iron. In contrast to Ireland, the United Kingdom has been operating from a position of power for many hundreds of years. Beginning in the seventeenth century, Britain acquired resources from its colonies, first in Ireland and then in the Americas, Africa, and Asia. Together, these resources were sufficient to make Britain the leader of the Industrial Revolution in the late eighteenth century. By the nineteenth century, the British Empire covered nearly a quarter of the world's land surface. As a result, British culture was diffused far and wide, and English became the *lingua franca* (the language of trade between speakers of different languages) of the world.

Britain's widespread international affiliations, set up during the colonial era, positioned it to become a center of international finance as the global economy evolved. Today, London is the leading global financial center, customarily handling more value in financial transactions than New York City, its closest rival. This remains true despite the UK's rejection of the euro (€). That the UK continues to use the pound (£) may have helped its economy during the EU economic crisis; and in fact, anti-EU sentiment in the UK, fed by the UK's loss of empire, by declining political and economic status, and by contentions over the influx of immigrants, of which the UK has a great number, resulted in an unexpected vote in June 2016 to leave the European Union (Brexit). Immediately there were regrets because no real plans existed on how such an important country would leave the EU or what the implications might be. In mid-2019, tensions are rising because required schedules for leaving have been set and reset. The most recent date given for a final decision is October 31, 2019.

Table 4.1 provides a brief synopsis of the main Brexit issues remaining to be decided. There is considerable evidence that many voters had little idea of what they were voting for; others who favored the EU did not bother to vote. Polls have shown that many

Table 4.1 Major Unresolved Issues in the Brexit Debate, 2019
• Can the UK split from the EU and still remain in the customs union?
• If not, will all free trade be stopped and will border controls between the UK and EU be reinstated?
• Since Ireland will stay in the EU, and Northern Ireland will not, how will the border between Ireland and Northern Ireland be handled?
• Will visas be necessary to enter the UK from the EU or to leave the UK for the EU?
• What will be the status of EU migrant workers in the UK (many are highly trained specialists, others are skilled and irreplaceable, and several hundred thousand have petitioned to remain in the EU)?
• What will be the status of UK migrant workers in the EU?
• Will UK retirees living in Spain or elsewhere in the EU be allowed to stay?
• How will banking regulations change?
• Would it not be advisable to have a second referendum?

voters did not understand what the EU is or how leaving would affect them personally. There is also evidence that Russia may have attempted to influence the vote with false reports. Given all the confusion, there is growing pressure for a second referendum, but political maneuvering may prevent that. Given the continuing confusion and the importance of this matter to the future of the European Union, a blog updating the subject of Brexit will be available online in Sapling.

The United Kingdom is no longer Europe's industrial leader. At least since World War II, and some say earlier, the United Kingdom has been sliding down from its high rank in the world economy toward the position of an average European nation. The discovery of oil and gas reserves in the North Sea gave the United Kingdom a cheaper and cleaner source of energy for industry than the coal on which it had long depended, but this did not stop the economic decline. Cities such as Liverpool and Manchester in the old industrial heartland and Belfast in Northern Ireland have had long depressions.

Britain has successfully established technology industries in regions that previously were not industrial—for example, near Cambridge and west of London, in a locale called Silicon Vale. The service sector currently dominates the economy (accounting for 80 percent of the GDP and 84 percent of the labor force). In the last decade, new, though not high-paying, jobs have been created in health care, food services, sales, financial management, insurance, communications, tourism, and entertainment. Even the movie industry has discovered that England is a relatively cheap and pleasant place for film production and editing. Britain's past experience with a worldwide colonial empire has left it well positioned to provide superior financial and business advisory services in a globalizing economy. Such service jobs, however, tend to go to young, educated, multilingual city dwellers, many of them from elsewhere in the European Union, not to the middle-aged, unemployed ironworkers and miners left in the old industrial UK heartland. Because of the EU's open borders, there are now more than 400,000 foreign nationals working in Greater London alone; what will happen to the economy if they leave?

The Jamaican poet and humorist Louise Bennett used to joke that Britain was being "colonized in reverse." She was referring to the many immigrants from the former colonies who now live in the United Kingdom. Indeed, it would be hard to overemphasize the international spirit of the UK. London, for example, has become an intellectual center for debate in the Muslim world. Salman Rushdie, an Indian Muslim writer, lived under British protection after his controversial book *The Satanic Verses* was published and he received death threats; many bookstores carry Muslim literature and political treatises in a variety of languages; Israelis and Palestinians talk privately about peace; descendants of Indian and Bangladeshi immigrants run popular restaurants; and Saudi Arabians have a strong presence. A leisurely stroll through London's Kensington Gardens reveals thousands of people from all over the British Commonwealth—Indian, Pakistani, Malaysian, Nigerian, and West Indian families—pushing baby carriages and playing games with older children. Impromptu soccer games may have team members from a dozen or more countries.

CHECK YOUR UNDERSTANDING

1. How did colonialism transform West Europe economically and culturally? Why is Europe now struggling to remain globally competitive? What is the role of immigrants from former colonies?
2. Deep divisions threaten the stability of West Europe. Show how changing job requirements, rising unemployment for some, the influx of immigrants, and costly social protection programs are precipitating discontent, such as the Brexit vote, as the need to increase the region's global competitiveness becomes clear.
3. Why has it fallen to Germany to bear the greatest burden of supporting other EU countries with failing economies?

SOUTH EUROPE

The Mediterranean subregion of South Europe (**Figure 4.41**) was once unrivaled in wealth and power. The Greek and Roman empires in the ancient period were followed in the medieval period by the Italian trading cities of Venice, Florence, and Genoa, with contacts stretching across the Indian Ocean and by land through Central Asia to eastern China. During the sixteenth century, Spain and Portugal developed large colonial empires, primarily in the Americas, but also they were beginning to venture into Africa and Southeast Asia.

After about 1600, times changed, as elsewhere in Europe economies expanded to a global scale with England, France, and the Netherlands developing distant colonies in the Americas, Africa, and Asia. The Mediterranean ceased to be a center of trade. South Europe declined, and the Industrial Revolution reached this subregion relatively late. Even today, Portugal and Greece and parts of Italy and Spain remain poor in comparison to the rest of the continent; only some of the countries of Central Europe are poorer.

This section pays particular attention to Spain, Portugal, and Italy, contrasting the parts of each country that are now prospering with the parts that have remained poor. Then Greece, an early member of the European Union, is discussed in relation to the effect of its recent financial problems on its status as an EU member and how its financial problems have recently been alleviated by the fact that it is a main entry point for refugees/asylum seekers from the war-torn regions of Southwest Asia and North Africa, and hence is the recipient of some assistance.

In all six countries of South Europe—Portugal, Spain, Italy, Malta, Cyprus, and Greece—agriculture was the predominant occupation through most of the twentieth century. Farmers often lived in poverty, producing crops for local consumption. The lack of industrialization meant emigration (primarily to North and South America, Australia, and New Zealand) was the only route to monetary success. In the 1970s, however, agriculture was modernized and mechanized and refocused to grow commercial crops on large irrigated holdings for export to the rest of Europe and beyond. The abundance of produce flowing from South Europe's agribusinesses is not without problems. The use of pesticides and chemical fertilizers has negatively affected producers and consumers. The surpluses have driven down prices, hurting smaller family farmers across Europe. Agriculture (measured as a percentage of GDP and as a percentage of the labor force) has declined in

Figure 4.41 The South Europe subregion map.

importance in South Europe as manufacturing, tourism, and other services have increased. Middle-class tourist money has brought new life to the economy of some parts of the region, but it has done little to elevate the lives of poor, rural people. In Greece and Italy, for example, income from tourism has been confined to cities and to scenic coastal and island locations, while interior rural areas remain poor and isolated. Ironically, in much of South Europe, the very qualities that attract tourists are being threatened by the sheer number of visitors and the negative effects they have on coastal environments and on rural culture.

Spain and Portugal

In the last quarter of the twentieth century, both Spain and Portugal emerged from centuries of underdevelopment. Just 40 years ago, Spain was a poor and underdeveloped country wracked by years of civil war and then a military dictatorship led by Francisco Franco. Portugal was similarly authoritarian and was still trying to hold on to its African colonies. Even the physical location of these two countries was no longer the advantage it had been during the days of empire (**Figure 4.42**). As commercial activity shifted away from the Mediterranean to West and North Europe and the wider world, poverty and isolation took over. The Pyrenees kept Spain separated from the more prosperous parts of Europe, making land travel and commerce difficult.

In 1975, Spain made a cautious transition to democracy; then, in 1986, it joined the European Union under a cloud of suspicion that it would not measure up. Portugal, after a bloodless coup in 1974, underwent democratic reforms and granted

Figure 4.42 **The Tower of Belem in Lisbon, Portugal.** In order to protect the wealth brought by his empire in Asia, Africa, and the Americas, King Manuel I of Portugal constructed a system of defense around the capital of Lisbon. Built from 1513 to 1519, during an era of great prosperity, the Tower of Belem was both a central part of this defense and a ceremonial gateway to and from the city. For many of the Portuguese expeditions of discovery and colonization, including most of the colonization of Brazil, the Tower of Belem was the last architectural symbol of home the travelers saw as they left on their long voyages. [Mauricio Abreu/Getty Images]

independence to its overseas colonies. Since then, growth has accelerated in both countries with the help of EU funds aimed at bringing these countries into economic and social harmony with the rest of the European Union through investment in infrastructure and in human resources. Spain especially has had success as a result of these improvements. Foreign businesses invested in industry, and factories were modernized. The cities of Barcelona and Madrid became centers of population, wealth, and industry. Both are now linked to the rest of Europe by a network of roads, airports, and high-speed rail lines (see Figure 4.11), and both have become favorite retirement locations for northern and western Europeans.

Some segments of Spanish and Portuguese society have not been able to participate in the new prosperity. Many small farmers remain poor throughout Portugal and in the northwestern and southwestern provinces of Spain, which have few resources. Recently, even the relatively prosperous zones of Barcelona, Madrid, and Lisbon have suffered in the global recession with many workers being laid off. In 2008, unemployment in Spain was 13.9 percent; by 2017, it was even worse at 16 percent.

In view of the high unemployment figures, Spanish and Portuguese workers have been willing to work for significantly lower wages than those paid in North and West Europe, and these lower wage scales have helped draw foreign investment from around the world. This is especially true for Spain, where other attractions include an educated and skilled workforce (partly the result of EU funds for training); somewhat lenient environmental and workplace regulations, and Spain's now-solid democratic institutions, which enable the country to withstand crises (such as the terrorist bombing in Madrid in March 2004, and the 2008 economic recession just mentioned). Spain's economic base includes agribusiness, food and beverage processing, and the production of chemicals, metals, machine tools, textiles, and automobiles and auto parts. Ford, Daimler, Citroën, Opel, Renault, MDI, and Volkswagen all have factories in Spain.

Regional disparities in wealth within Spain and between Spain and Portugal contribute to persistent levels of social discord. Two of Spain's most industrialized and now wealthiest regions—the Basque country in the central north (adjacent to France) and Catalonia in the east (also bordering France)—have especially strong ethnic identities. Their inhabitants often speak of secession from Spain as a response to years of perceived repression by the Spanish government. The Catalan effort at secession has been repeatedly declared unconstitutional, but it persists. Meanwhile, the Basques, once poor but now quite well off, object to being taxed to support poorer districts in Spain. They have organized a separatist movement that periodically resorts to violence. Galicia, in the far northwestern corner of Spain, also has a strong cultural identity and a proclivity for separatist sentiments. All three of these ethnic enclaves emphasize their distinctive languages as markers of identity and chafe against the use of Castilian Spanish as Spain's official language.

In Portugal, although unemployment rates are lower than in Spain, concern over financial futures is higher than in any other South Europe country, and people have cut tax-based expenditures (social protection) in nearly all categories except education. In an effort to ameliorate various divisive cultural and economic disparities, the Spanish and Portuguese governments had borrowed so heavily to boost development that, like Greece, their budget deficits exceeded (by three times) the eurozone limit. By 2012, Spain and Portugal were in danger of defaulting on loans and requiring a bailout. EU-enforced belt tightening then brought on social unrest: miners protesting in Madrid were fired upon with rubber bullets; a range of middle-class citizens joined raucous demonstrations; and across Spain, individual provinces sought emergency federal financial aid.

In Spain, immigration is another important part of the country's current situation. Spain's connections to North Africa date back more than 1300 years; Moorish Muslim conquerors came in 700 C.E. and stayed until they were expelled in the late 1400s, deeply influencing Spanish culture, language, architecture, cuisine, and attitudes toward gender roles. Today, more than 600,000 immigrants to Spain from North Africa (many in Spain illegally) are working in cleaning and maintenance or in factories. They are joined by several million people from sub-Saharan Africa and from Spain's former colonies in Middle and South America. A significant number of these are professionally trained people who despaired of making a decent living in their home countries.

It is important to note, however, that Spain can absorb immigrant workers because the net migration rate is *out* of Spain. Since the 1970s, several million Spanish workers have migrated to work in other European countries, creating employment niches for those coming in. The brain drain to other EU countries may actually benefit Spain eventually as the emigrant Spaniards return with new skills and ideas.

Italy

Italy has the largest economy in South Europe and a high GNI (PPP) per capita ($40,000 in 2017); and on the UN HDI, Italy (28) ranks just below Spain (26) (see Figure 4.22). The north of Italy is industrialized and prosperous, while the south is agricultural, has a high unemployment rate, and is dependent on government payments to alleviate poverty.

Over the millennia, Italian traders have contributed greatly to European culture and prosperity. During the late 1200s, Marco Polo, from a wealthy trading family in Venice, took a trip through Central Asia to China and returned after some 20 years to give medieval Europeans their first impressions of life and commerce in Asia, particularly China. The broader view of the world that Polo brought home contributed to the European Renaissance in northern Italy. Wealthy families such as the Medici were patrons of the arts and literature and invested in beautifying public and private spaces. Italy's enlightened ideas about philosophy, art, and architecture spread throughout Europe.

Italy's prominence faded, however, as first Spain and Portugal and then England, France, and the Netherlands grew rich on their colonies in the Americas, Asia, and Africa. During the 1700s and 1800s, Italy was caught up in a long series of wars between its then-richer and more powerful neighbors, France, Spain, and Austria. During this time, progress was rocky for Italy's ordinary people, buffeted by the political ambitions of politicians and religious leaders (the Pope, the leader of the Catholic

Church, remains headquartered at the sovereign Vatican City in Rome). Wide disparities in wealth and power left whole sections of the country lawless and in decline, especially the southern half of the peninsula and the islands of Sardinia and Sicily, which were collectively known as "the South." There, absentee landlords and central government tax collectors from northern Italy hired *mafiosi* to collect fees from those who worked the land. The locals were given no chance for self-government. Economically, the south remains woefully behind the north, due in part to the pervasive corruption that in some places has a stranglehold on civic institutions.

Northern Italy, by contrast, has one of the most vibrant economies in the world, producing products renowned for their quality and design: Ferrari automobiles, Olivetti (now a subsidiary of Telecom Italia Group) office equipment, and high-quality musical instruments. In Milan, one of the largest cities in Italy, fashion designers such as Giorgio Armani have surpassed even their rivals in Paris. However, the mainstay of the Italian economy has traditionally been the exporting power of thousands of small mom-and-pop factories that make everything from machine valves to buttons, shoes, and leather clothing—all of which can now be made more cheaply in China. The small factories are closing, Italy's share of global trade is dropping, and its budget deficits have exceeded the limits that the European Union places on its member countries (Germany and France also have large budget shortfalls but are more solvent). Like all countries in the eurozone, Italy can no longer simply devalue its money, formerly the *lira*, to make its products less expensive on the world market.

For all of Italy's stylish success as an industrial and services leader in Europe and the world, its ability to continue attracting investment has been damaged by an inefficient bureaucracy, high tax rates, dense tax rules, inadequate infrastructure, and corruption. In 2018, although improvements were instigated by a new, more financially savvy prime minister, Italy had a global competitiveness rank of 43 out of 140, below that of many countries in Asia, Southeast Asia, and Southwest Asia.

Italy has been a democracy since World War II and is a charter member of the European Union. Because of its domestic politics, though, Italy was long known as the "bad boy" of Europe, a designation usually meant to be taken humorously. The label comes in large part from unfair stereotypes of Italians as clever operators on the fringes of legality. But it also derives from the fact that Italians vote governments in and out in rapid succession, and they frequently reelect leaders who have been mixed up in corruption. Italians have devised ways to live with official indiscretions without descending into national crises. During the dozens of governmental emergencies since 1950, a quasi-government based on informal relationships—the *sottogoverno*, or *undergovernment*—has taken over whenever a predicament has developed, allowing daily life to proceed apparently unimpaired.

Italy, like Spain, has for a decade or more dealt with large numbers of immigrants, principally from North and sub-Saharan Africa. As globalization spreads, agricultural and industrial policies that favor development in Europe and other rich areas have had the side effect of jeopardizing farmers, fishers, and small artisans across Africa. Hundreds of thousands of young Africans, usually those with some education, have, in desperation, migrated to Europe to look for work so that they can support families back home. Many have arrived along Italy's southern shores in leaky boats via North Africa, which has few protections for refugees. This poorest part of Italy has not welcomed the Africans, who are stigmatized as public enemies and immediately arrested or victimized by unscrupulous labor and sex worker scouts. Since 2014, the sea route to Italy has accounted for 88 percent of all migrant deaths by drowning in the Mediterranean.

Greece

The Kingdom of Greece became independent from the Ottoman Empire after World War I and for a while was a republic, but in 1935 the Greek monarchy was reestablished. Greece was occupied by both Italy and Germany during World War II; and after the war, it entered a protracted era of communist/anticommunist civic unrest. Greece joined NATO in 1952, a move that seemed to strengthen ties with postwar Europe; but in the mid-1960s, a military coup established a dictatorship that lasted until 1974. Democratic elections followed, and the Greek monarchy was abolished.

In 1981, after years of instability, Greece joined the emerging European Economic Community. When Greece signed on to the 1992 EU Treaty of Maastricht, it became obligated to limit deficit spending and debt levels. Always one of the poorer countries in Europe and one that habitually spent more than it took in, Greece entered the eurozone in 2001, at a time when low interest rates and the strong euro made excessive borrowing possible. Greece borrowed for infrastructure projects and to fund state jobs and benefits, including pensions. The quality of the infrastructure projects was compromised by a culture of corruption and tax evasion that put public funds in the pockets of officials and other elites. Many of the intended recipients of benefits were cheated, and the Greek deficit increased. Greece did not report these problems to the EU banks and regulators. The interest rate on Greece's debt shot up when the debt—then €360 billion (U.S.$384 billion)—was revealed to be so large. Already short of cash, Greece found it impossible to pay even the interest charges and entered a long period of being near default while the debt mounted quickly. Despite many threats to cut Greece loose, the eurozone countries and the IMF continued to negotiate repeated bailouts, which by December of 2015 totaled another €340 billion (U.S.$363 billion). In return, the EU and the IMF demanded stern austerity measures, the brunt of which fell on the elderly and the poorest in Greece, as social protection and pensions were cut.

Greece's situation became more difficult as a result of the refugee crisis. Several Greek islands lie close to the Turkish port city of Izmir, where many thousands of Syrian refugees and other migrants congregated and paid smugglers to take them to the Greek islands. Because Greece is a member of the European Union and within the Schengen border zone, the migrants assumed they would not be turned back. Several hundred thousand migrants illegally entered Europe through Greece during 2015. Oddly, the migrant crisis bought Greece some time and money with which to work on easing the debt crisis, because in order to defend the Schengen border, in 2016 the EU postponed some of the more dire consequences of the debt and began helping Greece to care for the migrants and fend off further immigrant arrivals. By early 2019 the number of migrants was down but Greece and Italy were still taking in the majority—some 54,000 in 2018.

CHECK YOUR UNDERSTANDING

1. Describe the economic evolution of Spain and Italy: periods of growth and prosperity, and regional disparities in wealth and development.
2. What types of financial difficulties have both countries faced?
3. How have Spain, Italy, and Greece's economic woes been exacerbated by migrants?
4. Where in Italy are small firms most numerous and high-tech firms most dominant? Which have been hit hardest by globalization?
5. What economic activities have replaced small farm agriculture in South Europe?
6. Describe how Spain and Portugal are now struggling to find a way to absorb immigrants from their former colonies.
7. How did Greece manage to become so indebted to EU banks?
8. Describe at least three present strains on democratic institutions in South Europe.

NORTH EUROPE

North Europe traditionally has been defined as the countries of Scandinavia—Iceland, Denmark, Norway, Sweden, and Finland—and their various dependencies (**Figure 4.43**). The

Figure 4.43 The North Europe subregion. **A** Denmark is at the forefront of Europe's renewable energy advances. Figure 4.12 and the surrounding text discuss a large installation of solar panels on a Danish farm. Blessed with excellent wind conditions, Denmark also saw a chance to drastically reduce its dependence on oil from the Middle East, so it began to install wind farms, even in offshore areas. By 2015, more than 40 percent of Denmark's energy was coming from wind power. Denmark is now a worldwide leader in wind-power engineering and marketing, but this has come with environmental impacts. Wind turbines have a visual impact and can interfere with wildlife, especially birds and sea creatures. [UniversalImagesGroup/Getty Images]

Faroe Islands (and Greenland) in the North Atlantic are territories of Denmark; the island of Svalbard in the Arctic Ocean is part of Norway. North Europeans are now including the three Baltic states of Estonia, Latvia, and Lithuania in their region, as we do here. All three Baltic states were part of the Soviet Union until September 1991, and these small countries have many remaining links to Russia, Belarus, and other parts of the former Soviet Union. Although their cultures are not Scandinavian, the three countries are trying to reorient their economies and societies in varying degrees to North Europe and the West.

The countries of North Europe are linked by their locations on the North Atlantic, the North Sea, and the Baltic Sea. Most citizens can drive to a coast within an hour or two. The main cities of the region—Copenhagen, Oslo, Stockholm, Riga, and Helsinki—are vibrant ports that have long been centers of shipbuilding, fishing, and the transshipment and warehousing of goods. They are also home to legal and financial institutions related to maritime trade.

Figure 4.44 The Swedish village of Gamleby. This village in Västervik, Sweden, is known for its colorfully painted wooden houses along the old naval port. Many Swedish towns feature such bright paint jobs. A favorite color is Falun red, a copper-based paint originating in Falun in Dalarna County. [Sara Thaw]

Scandinavia

Most people in North Europe live in the southern parts of Scandinavia, where economic activity is concentrated. The landscapes of these warmer lowlands of southern Scandinavia are agricultural, but the economies of these countries are based on industries and services. For example, even though the small country of Denmark produces most of North Europe's poultry, pork, dairy products, wheat, sugar beets, barley, and rye, 76 percent of the Danish economy consists of services (finance, education, design, tourism), high-tech manufacturing, construction and building trades, and fisheries. Sweden is the most industrialized nation in North Europe. The Swedes, who also like to underscore their quaint rural roots (**Figure 4.44**), produce most of North Europe's transportation equipment and two highly esteemed brands of automobiles, the Saab and the Volvo (though some Volvo products are now built outside Europe, and Saab is owned by a consortium based in China and Japan). IKEA, founded in Sweden, is the world's largest furniture manufacturer. Through 375 stores in 50 countries, IKEA markets furniture and housewares famed for their affordability, their spare, elegant design, and their disconcerting necessity to be assembled at home. IKEA is a major consumer of woods logged in the world's tropical zones (see Figure 10.11); China is the largest producer of IKEA products. Helsinki, Finland, is home to some of the world's most successful IT companies, such as Tieto, Nokia, and Novo Group.

The northern part of Sweden and almost all of Norway are covered by mountainous terrain, while Finland is a low-lying land dotted with glacial lakes, much like northern Canada. Sweden and Finland contain most of Europe's remaining forests. These well-managed forests produce timber and wood pulp. Norway, which has less usable forestland, had been considered poorly endowed with natural resources compared with the rest of Scandinavia, but the discovery of gas and oil under the North Sea in the 1960s and 1970s has been a windfall to the country. The exploitation of these resources dwarfs all other sectors of Norway's economy. Norway is able to supply its own energy needs through hydropower, and it exports oil and gas to the European Union. It is now one of Europe's wealthiest countries and ranks highest in the world in human well-being. In anticipation of the North Sea oil running out, Norway has invested a large proportion of its oil profits for use by future generations.

The fishing grounds of the North Sea, the North Atlantic, and the southern Arctic Ocean are also an important resource for the countries of North Europe (as well as for several other countries on the Atlantic). In recent decades, however, overfishing has severely reduced fish stocks, and by 2010 most cod eaten in Europe was imported. Still, fishing remains culturally important and rights of access to the fishing grounds are a cause of dispute. In 1994, the European Union created a joint 200-mile (320-kilometer) coastal exclusive economic zone that ensures equal access and sets fishing quotas for member states. In January 2001, in order to give North Sea fish populations a chance to rebound, EU members plus Norway and Iceland agreed to an annual 3-month hiatus in cod fishing, and a ban on using juvenile fish to feed salmon in commercial farms. By 2006, however, it was clear that these short respites from fishing had not increased the cod population; so a hiatus of as long as 12 years was called for over a much bigger area of the North Sea. A small recovery from 2009 to 2014 gave cause for hope, and recovery for most species continued in 2018. Predictions by the EU are that sustainable fishing in the North Sea will be achieved by 2020. While conservation measures have been crucial in the rebound, current thinking remains that global warming accounts for a significant portion of the drastic slump in catches.

Two important characteristics distinguish the Scandinavian countries from other European nations: their strong social protection systems (see "Social Protection Systems" earlier in the chapter) and the extent to which they have equal levels of participation and well-being for men and women (see Figure 4.36: Population, Gender and Culture in Europe). Sweden's comprehensive social protection system provides stability and a safety net for every one of the country's 10.2 million people. This system is founded on three values. First is the idea of security: that all people are entitled to a safe, secure, and predictable way of life with as little discomfort as possible. This concern extends to interior design in subsidized housing, which is modern, elegant, and practical. Second, the appropriate life is ordered, self-sufficient, and quiet, not marred by efforts to stand out above others. Third, when the first two concepts are practiced properly, the ideal society, *folkhem* ("people's home"), is the result.

These three values help explain why Swedes are willing to pay for a social protection system that provides child care, parental leave, health care, sick leave, elder care, housing subsidies, and other benefits. By 2013, the system amounted to close to one-third of the government's annual budget and was not confined only to citizens; immigrants were also covered, though some Swedes expressed their dissatisfaction with this. In the early 2000s Sweden began to experiment with privatization of education and services under a center-right government; but by 2014, as student performance fell, economic disparities increased, and the poor (many of them recent Iraqi refugees) took to the streets, Swedes began to lobby for a return to high levels of tax-supported social protection. The immigrant crisis of 2015 shook Sweden's commitment to social welfare for all because the wave of migrants was so huge; 100,000 newcomers were expected, and more than twice that number came in (21,560 of whom were asylum seekers; see Figure 4.41). Because of the open-border rules of the Schengen zone, at first Sweden kept no records. When the country's security and its ability to assimilate the newcomers was questioned, Sweden closed its borders for a few days with the intention of reopening them as soon as possible; but this caused worries that the Schengen agreement was now dead.

In September 2018, the right-leaning anti-immigrant party took 17.6 percent of the vote; this was too low to give them a formal role in the government, which will likely be formed by left, center-left, and center-right parties. A poll of voters showed that most hoped that the new government would reaffirm traditional Swedish social values.

For all their emphasis on an orderly peaceful life that is fair for all, Scandinavians are noted for occasional outbursts of exuberant enthusiasm for over-the-top adventures. One such example is the Scandinavian love for heavy metal music. **Figure 4.45** shows the Finnish band Apocalyptica, known for its inventive orchestral embellishments of metal music.

Figure 4.45 Scandinavia and heavy metal music. Riding a decades-long Scandinavian obsession with heavy metal music, the Finnish band Apocalyptica, composed of three cellists and a drummer, formed in Helsinki in 1993. Originally a tribute band that played covers of the U.S. band Metallica, Apocalyptica has become a pioneer of a more symphonic- and folk-influenced style of metal. The tradition continues today as metal bands, clubs, and even metal-based church services grow in popularity. [Neil Lupin/Redferns/Getty Images]

The Baltic States

The three small Baltic states of Estonia, Latvia, and Lithuania are situated together on the east coast of the Baltic Sea. Culturally, these countries are distinct from one another, with different languages, myths, histories, and music. After World War II, all three were forced to become Soviet republics. When the Soviet Union collapsed in 1991, they regained their independence. Despite their recent connection to Russia, their cultural ties traditionally lean toward West and North Europe. Ethnic Estonians, who make up 60 percent of the people in Estonia, are nominally Protestant, with strong links to Finland; they view themselves as Scandinavians. Ethnic Latvians, who make up just 50 percent of the population of Latvia, are mostly Lutheran; they see themselves as part of the old German maritime trade tradition. Lithuanians are primarily Roman Catholic and also see themselves as a part of West Europe.

Populations in all three countries are decreasing and aging. Birth rates are very low and young people now have the option to migrate to North and West Europe. Ethnic Russians, who today make up about a third of the population in both Estonia and Latvia, are having more children than the indigenous populations are. They could become the dominant culture group within a few decades—a crucial development if these ethnic Russians continue to cultivate strong political ties with Russia. Ironically, although the Russian minority was placed in the Baltic states by Russia to counteract Westernization and maintain the influence of the Soviet Union, under EU rules (all three countries are now EU members), these Russians are minorities whose rights must be protected by the EU.

Of all the countries of North Europe, the Baltic States have the most precarious economic outlook. Under Soviet rule, the Russians expropriated their agricultural and industrial facilities and used them primarily for Russia's benefit. Until 1991, 90 percent of the Baltic States' trade was with other Soviet republics. Since then, Estonia has undertaken the most radical economic change, moving toward a market economy and increasing its trade with the West. Estonia is the only one of the three to have had real economic growth, and its accomplishments have attracted foreign aid and private investors. Heavily industrialized Latvia and Lithuania now trade more with the UK, Germany, and the West than with Russia. They have increased their standards of quality, but many of their factories are out of date and still pollute heavily.

The Baltic States see national security as their major problem because their strategic position along the Baltic Sea is coveted by Russia, which has few easy outlets to the world's oceans. In this regard, the status of the Russian exclave of Kaliningrad is crucial. Kaliningrad is called an **exclave** because, although it is politically part of Russia, it lies far from the main part of Russia. Kaliningrad is situated along the Baltic Sea, between two EU states, Lithuania and Poland (see Figure 4.43). The Russian Baltic fleet is headquartered in Kaliningrad, which has a relatively ice-free port (see the discussion in Chapter 5). Russia's strategic interest in Lithuania is unlikely to diminish, and Lithuania fears that political instability in Russia could result in Russia reinvading Lithuania, which is not a member of NATO. The countries of western Europe, unwilling to commit to supporting Lithuania militarily, hope to diplomatically resolve rising differences with the Russians.

CHECK YOUR UNDERSTANDING

1. Why, other than their location, are the Baltic States now included in North Europe?
2. Describe the modern economies of Scandinavia, including the present role of agriculture.
3. How has Norway planned for the future of North Sea oil?
4. Describe the fishing situation in the ocean waters around North Europe.
5. Sweden has a distinctive national consensus on how life should be lived. Describe it.
6. What has been North Europe's record regarding the support of social protection for citizens and new immigrants?
7. Why are the economies of the three Baltic States precarious? Which of the three is doing the best?

CENTRAL EUROPE

Central Europe has undergone a series of profound changes since World War II (**Figure 4.46**). First, the region experienced harsh economic and political conditions under communist governments. Then in 1989, in the euphoria at the end of the Cold War, many believed that democracy and market forces would quickly turn around the region's formerly centrally planned, sluggish, and highly polluting economies. This did not happen. Instead, the southern parts of Central Europe (traditionally called the Balkans) suffered violent political turmoil, and virtually all saw temporary drops in their levels of well-being.

Over time two tiers of countries emerged in Central Europe. In the 1990s, those countries physically closest to western Europe—Poland, the Czech Republic, Slovakia, Hungary, and Slovenia—elected new democratic governments, began more careful environmental policies, made deliberate progress toward market economies, and began to attract foreign investment. Access to consumer goods improved markedly. In 2004, all of these countries joined the European Union, and Croatia joined in 2013. Although conforming to the wide range of economic, environmental, and social requirements set forth by the European Union has not been easy, for many the challenge has been invigorating.

The countries with the greatest economic and social difficulties—the southern tier—remain Albania, Bulgaria, and Romania, and some of the countries that were in the former Yugoslavia (Bosnia and Herzegovina, Montenegro, Serbia—including the region of Kosovo, and North Macedonia). Romania and Bulgaria joined the EU in 2007; but for all these countries, adjusting to democracy and privatization has been difficult. Plagued by corruption and economic chaos, many governments relaxed the pace of reforms as they struggled merely to maintain civil peace. In the 1990s, all of the former Yugoslavia republics except Slovenia became mired in a series of terribly costly genocidal ethnic conflicts that belatedly drew the military intervention of Europe (including NATO) and the United States.

A Time of Troubles in Southeastern Europe

Shortly after the exhilaration of the breakup of the Soviet Union, the socialist counties of Yugoslavia and Albania went through a rather unexpected but brutal war. This region between Austria and Greece, formerly known as the Balkans, is home to many ethnic groups of differing religious faiths (**Figure 4.47**). Too often the assumption has been that it was this diversity that led to the conflict; but over many generations, the various ethnic groups managed to live together peacefully by intermarrying, blending, and realigning into new groups. Often it was outsider conquerors who encouraged ethnic rivalry as a method of control, as was the case with the Hapsburg Empire (governed from Vienna in Austria) and later the Nazis during World War II.

Marshal Josep Broz Tito, Yugoslavia's founder and leader until his death in 1980,

exclave a portion of a country that is separated from the main part

Figure 4.46 The Central Europe subregion. **A** Gyula sausage is cured in a smokehouse in Gyula, Hungary, which enjoys *protected geographical indication*, or PGI. This means that, within the European Union, only sausages from the town of Gyula and those that are made in a certain way may be labeled "Gyula." PGI labeling is one way that the EU has tried to preserve local agricultural economies and traditional ways of life even as industrialization, urbanization, and cultural homogenization have become more dominant. [FERENC ISZA/ AFP/Getty Images]

recognized the ethnic complexity of the countries he was governing. He tried through the power of his personality to foster pan-Yugoslav nationalism. There was ethnic peace during Tito's rule, but he did not lead a national dialogue about the nature and history of multiculturalism or of the benefits of ethnic diversity. When Tito died, Serbs—the most populous ethnic group in the military and the federal government bureaucracy—tried to assert control. In response to the deteriorating economy and burgeoning Serbian nationalism, Slovenia and Croatia, Yugoslavia's northernmost and wealthiest provinces, voted to declare independence in late 1990. Slovenes and Croats themselves became more nationalistic, and a spiral into competing nationalisms threatened.

Given the scale of the interethnic violence that occurred in the following years, it is not surprising that many people, especially outsiders, viewed ethnic hatred as the overriding characteristic of the region and the cause of the conflict. However, more likely causes include an overall lack of ethical leadership, including little guidance by Marshal Tito on multiethnic affairs, invented myths about ethnically pure nation-states, and Serbian geopolitics.

In June of 1991, the Serb-dominated government and military of Yugoslavia allowed the province of Slovenia to separate after just a short skirmish—this happened in part because of Slovenia's location close to the heart of Europe and far from Serbia and in part because there were no significant supportive enclaves of ethnic Serbs in Slovenia. After this loss, the Yugoslav government feared that unless it made a strong show of force in dealing with the Croats and the Bosnians, the rest of the non-Serb populations of Yugoslavia would also move to secede and take with them valuable territory,

Figure 4.47 Ethnic groups in southeastern Central Europe. The patchwork of cultural and religious groups in southeastern Central Europe developed over the last several thousand years as different ethnic groups migrated to the area. They had to contend with each other's differences and with conquerors from the north and the south. Normally, relations were amicable and intermarriage was common, but occasionally hostilities arose, often as a result of outside pressures.

resources, and skilled people. This fear became a reality when Croatia and Bosnia and Herzegovina attempted to secede from Yugoslavia. The Serbs reacted brutally. While suffering ethnic cleansing, Croats and Bosnians themselves retaliated with brutality and ethnic cleansing, as did ethnic Albanians (also known as *Kosovars*).

Conflict sprang up again in the late 1990s in the province of Kosovo, the southern area of Serbia, dominated by ethnic Albanians. In the late 1980s, in order to divert public attention away from the country's economic deterioration, Yugoslav President Slobodan Milosevič, a Serb, aggressively claimed that Serbs were being threatened by Kosovar Albanians. In response, the guerrilla Kosovo Liberation Army was formed, supported by expatriates of Albanian ethnicity, some from the United States.

The various conflicts in and around Serbia devastated the economies of southeastern Europe. Dozens of bridges were bombed; the Danube River, a major transportation conduit through Europe to the Black Sea, was blocked with wreckage. In 2000, an election in Serbia resoundingly removed Slobodan Milosevič from power, and in 2002 he was brought to trial for war crimes in the International Court of Justice (World Court) in The Hague (where he died of natural causes in 2006). To forestall any future troubles, the European Union and the countries of southeastern Europe had earlier agreed that, once Milosevič was gone, they would enact the Balkan Stability Pact. Funded by the European Union, this pact promotes the recovery of the region through public and private investment as well as through social training to enhance the public acceptance of multicultural societies and the strengthening of informal democratic institutions. In May 2006, the people of Montenegro voted to declare independence from Serbia; in 2008 Kosovo declared independence, but its status remains in contention.

Transitions in Agriculture, Industry, and Society

After all this turmoil, not surprisingly, by 2000 many people in all 13 countries in Central Europe were looking back nostalgically to the communist era, when there was a strong authoritarian government, jobs were stable, and health care and education were free. Still, even the seven southern, poorest, and most politically unstable parts of Central Europe have useful natural resources for both agriculture and industry, and all have large, skilled, but inexpensive workforces that have slowly begun to attract foreign investment. The most important trading partners for Central Europe are Germany, France, the Netherlands, Italy, the United Kingdom, Russia, and the United States, but the combination changes from country to country (Serbia and those further south have just recently started developing wider trade networks). Venture capitalists from Hungary, the Czech Republic, Slovenia, and Croatia are beginning to invest in their neighbors to the south.

The Czech Republic (Czechia), in the northern tier of prosperity in Central Europe, can be taken as an example of transitions in agriculture. There, land was expropriated after World War II and turned into state farms and cooperatives. In the 1990s these large state farms were reorganized into private owner-operated cooperatives or restored to the original owners and their descendants. By that time, though, the original owners had mostly become city dwellers so, rather than farming the land themselves, they usually leased their regained land, either in small parcels to those who wished to become family farmers or in large parcels to the new private cooperatives. Eventually, 1.25 million acres (500,000 hectares), representing 30 percent of the Czech Republic's agricultural land, went to individual private farmers, large and small.

The EU economic advisers who readied the Czech Republic for entry into the European Union preferred larger, commercial farms because overproduction, which could contribute to falling commodity prices across Europe, is easier to control on large, corporately managed farms. EU advisers then suggested that the smaller family farms, which tend to be owned by independent-minded individuals, find more lucrative uses for their land, such as farm tourism. "Dude farms"—similar to dude ranches in the western United States—which provide living history demonstrations of traditional food cultivation and preparation techniques, now attract affluent urban families from all over Europe and beyond (see Figure 4.24).

Membership in the European Union has brought both benefits and challenges to wine producers of Central Europe. Their wines now have a wider market—Central Europe, Germany, Austria, Italy, Japan, and the United States—but vintners must adhere to strict EU agricultural policies. For example, Europeans are actually producing too much wine, lowering wine prices for all producers. The European Union wants to reduce the total vineyard acreage across Europe by some 400,000 hectares. Fortunately, Slovenia, as one of the tiniest EU members, will not be grossly affected by this reduction. Currently more than 90 percent of the 100 million liters of wine produced annually in Slovenia is consumed in Slovenia.

Industrial production was a specialty of Central Europe when these countries were part either of the Soviet bloc or Yugoslavia. So it would seem that manufacturing would be the best hope for competing in the EU marketplace. Unfortunately, much of Central Europe's considerable industrial base is still burdened by legacies from the communist era; it is polluting and energy intensive, with an emphasis on coal and steel production, which can't compete in the global market. In south Central Europe, the turn to producing much needed consumer goods to stimulate the domestic economy is just emerging.

Despite industry in Central Europe being slow to succeed, high-end product design is improving in places like Slovenia and Croatia; there, regional trade shows are full of attractive, if still mostly locally sold, goods. Expensive artisanal products—such as handmade designer clothes and shoes, crystal glassware, luxury baby clothes, and super-modern kitchens—are sold throughout Europe and the United States. Meticulously constructed prefabricated energy-efficient houses are being exported to Europe, Africa, and now the United States. Pan-Les, in southern Slovenia, produces a charming 500-square-foot house with a total annual heating bill of U.S.$300 a year. Beautiful, handcrafted Skrabl pipe organs from the same region are catching the attention of American churches.

In both Romania and Bulgaria, corruption has inhibited postcommunist industrial reform. Both countries narrowly avoided censure by the European Union for failing to rein in corruption. In some countries, the involvement of organized crime has plagued the sale of formerly state-owned industries, especially in military sectors. It has been difficult to develop financial regulations and freedom of the press—both requirements for EU membership—because of the amount of suspicion and wariness on the part of some political leaders. However, because the countries in the region want to remain in the good graces of the European Union, most have been working to control risky banking practices, to limit unscrupulous deals, and to reduce organized crime.

Progress in Market Reform

The governments of Central Europe are taking "free market–friendly" steps to help their economies grow more quickly. One strategy is to encourage foreign investment. But the question is: How will foreign investment help those displaced by the changes? After some discouragement over the endless paperwork, firms from western Europe, the United States, and China have been coming to the region with investment projects. The Chinese firm Hisense recently became the dominant investor in the upscale Slovene appliance firm Gorenje; and in Romania, hog farming first attracted the U.S. Smithfield firm, which was then bought out by a Chinese firm (**Figure 4.48**). China's industrial-scale hog farming so thoroughly displaced local hog farmers that many quit farming and migrated to western Europe to work as laborers.

VIGNETTE Grigore Chivu (a pseudonym) wanders through the empty hog pens on his farm near Lugoj in western Romania. For generations, his family made a meager but rewarding living by raising hogs and then processing and selling the meat. A few years ago, just before Romania joined the European Union (EU), he had more than 250 hogs. At Christmas, he and thousands of hog farmers across the country would slaughter a certain number and preserve the meat using time-honored methods. Hog slaughtering was a time of high spirits, celebration, and cooperation as the farm families contemplated the coming feast and

Figure 4.48 Hog farms in Romania and the United States.

(A) Hogs on a small animal farm in Romania. [JoenStock/E+/Getty Images]

(B) An industrial hog farm in the United States, where thousands of hogs are raised at a time. [Andy Sacks/Getty Images]

their profits. Flavorful sausages, which they smoked and hung high in the rafters of their kitchens for further drying, were slowly sold to select customers, providing a steady income well into summer.

Romania's entrance into the European Union required all farmers to conform to EU standards for processing meat. The old methods were no longer allowed, and for some the new standards were prohibitively expensive. Before Chivu and his farmer friends could organize a butchering cooperative that would conform to EU standards and for which development funds were available, an American company, Smithfield, based in Virginia, stepped into the breach. Smithfield, the world's largest pork producer, planned to produce pork products on an industrial scale and market them globally and especially in the EU. Smithfield enlisted the help of Romanian politicians and got permission (and even EU subsidies) to establish a conglomerate that included feed production, hog breeding, modernized sanitary barns for fattening thousands of hogs in small cages, and slaughterhouses.

Then, in September of 2013, Smithfield was bought by a Chinese company, Shuanghui, which planned to redirect much of the company's global production toward supplying the growing demand for pork in China's new middle class. Shortly after Shuanghui purchased Smithfield, Romanian officials signed agreements with their Chinese counterparts to export more than 3 million hogs to China each year.

As the old, picturesque Romanian agricultural landscape is transformed into huge factory farms, the number of independent hog farmers has been reduced by more than 90 percent. Unable to compete with the lower prices industrial farms can charge, Grigore Chivu, like thousands of his fellow hog farmers, prepared to migrate to western Europe, where, because of his age and farming skills, he will obtain only a low-wage job to support his family. There is no work for him in Romania, and he is too young for a pension. ■

CHECK YOUR UNDERSTANDING

1. Describe the two tiers of Central European countries; in what ways do they differ?
2. Why is mere ethnic diversity an insufficient explanation for the Balkan wars of the 1990s?
3. What are the likely real causes of the interethnic violence in the Balkans?
4. What enabled Slovenia's successful separation from Yugoslavia in 1991?
5. Describe the role of Slobodan Milosevič in the Balkan troubles.
6. To what extent is the EU involved in the recovery efforts in south Central Europe (the Balkans)?
7. What are the various EU strategies for reinvigorating rural areas in Central Europe?

CRITICAL THINKING QUESTIONS

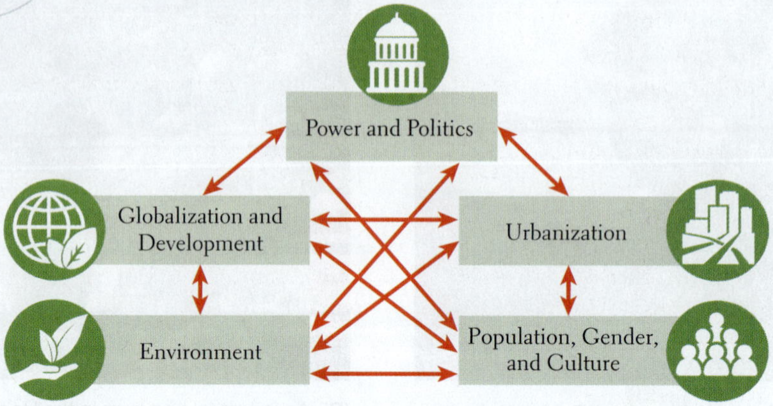

This diagram represents connections among the five geographic themes around which this book is structured. Listed below are some important questions that have been addressed in this chapter. As you answer each question, decide which of the five themes are connected by your response.

1. What aspects of their physical environment have Europeans exploited in order to gain global prestige and wealth?
2. Which subregions are most active in climate-change amelioration? What technologies are they using?
3. Explain how Europe's high level of consumption negatively impacts environments across the globe.
4. Describe some of the negative impacts on other world regions of Europe's long, bumpy road to global power.
5. How did the adoption of a common market and then a common currency affect economic development and banking in Europe?
6. What impact do Europe's heavily subsidized and regulated agricultural systems have on the rest of the world?
7. Describe the role of nationalism in the two world wars. Then discuss the role of nationalism today and how disparity of wealth and the unexpected arrival of thousands of refugees have fueled nationalist feelings.
8. Discuss why EU countries are often described as welfare states. How do the different subregions vary in their support of tax-supported systems of social protection (or social welfare)?
9. State the argument some give for why former colonies of Europe have spawned Islamic jihadism, resulting in terrorist attacks.
10. If you were a poor, undocumented immigrant searching for a way to support your family, would you choose the United States or Europe as a possible destination? Why?
11. Explain the contradiction of Europe's image as a place of idyllic rural farm and village landscapes, yet the reality that it is a highly urban place.
12. Has urbanization inevitably led to democratization in Europe? How yes, and how no? Which groups have typically gotten left out of urban democratic advantages?
13. Why do only a small percentage of Europeans live in single-family homes?
14. Are there some potential advantages to Europe of aging populations and zero or negative growth rates?
15. How do improvements in education and work opportunities for European women affect population growth? Which policies have been most successful in countering Europe's low birth rates?
16. Which of Europe's subregions has initially received the most refugees and immigrants? To where have most migrants moved on?
17. What are at least three of the unresolved issues that are stopping the resolution of a Brexit agreement?
18. Why might a pig farmer in West Africa criticize the EU's generous support for European agribusiness?

Key Terms

Active Aging 226
assimilation 229
asylum seekers 229
banlieues 220
Brexit 193
capitalism 208
central planning 209
Cohesion Policies 211
Cold War 208
Common Agricultural Program (CAP) 214
communism 208
continental climate 196
cultural homogenization 229
decolonization 209
double day 231
economic migrants 229
economies of scale 212
Enlightenment 205
euro (€) 214
European Central Bank (ECB) 214
European Union (EU) 193
eurozone 214
green 198
Green political parties 198
guest workers 229
Holocaust 208
humanism 205
Iron Curtain 208
Islamic jihadism 220
Maastricht Treaty 209
Mediterranean climate 196
mercantilism 206
midlatitude temperate climate 196
nationalism 217
North Atlantic Drift 196
North Atlantic Treaty Organization (NATO) 221
Paris Accord, 2017 198
populist 208
postcolonial 220
refugees 229
Roma 208
Schengen Agreement 211
social cohesion 234
social exclusion 229
socialism 209
social protection 234
subsidies 214
virtual water 216
welfare state 232

More Practice at SaplingPlus

Read the interactive e-text, review key concepts, and check your understanding.

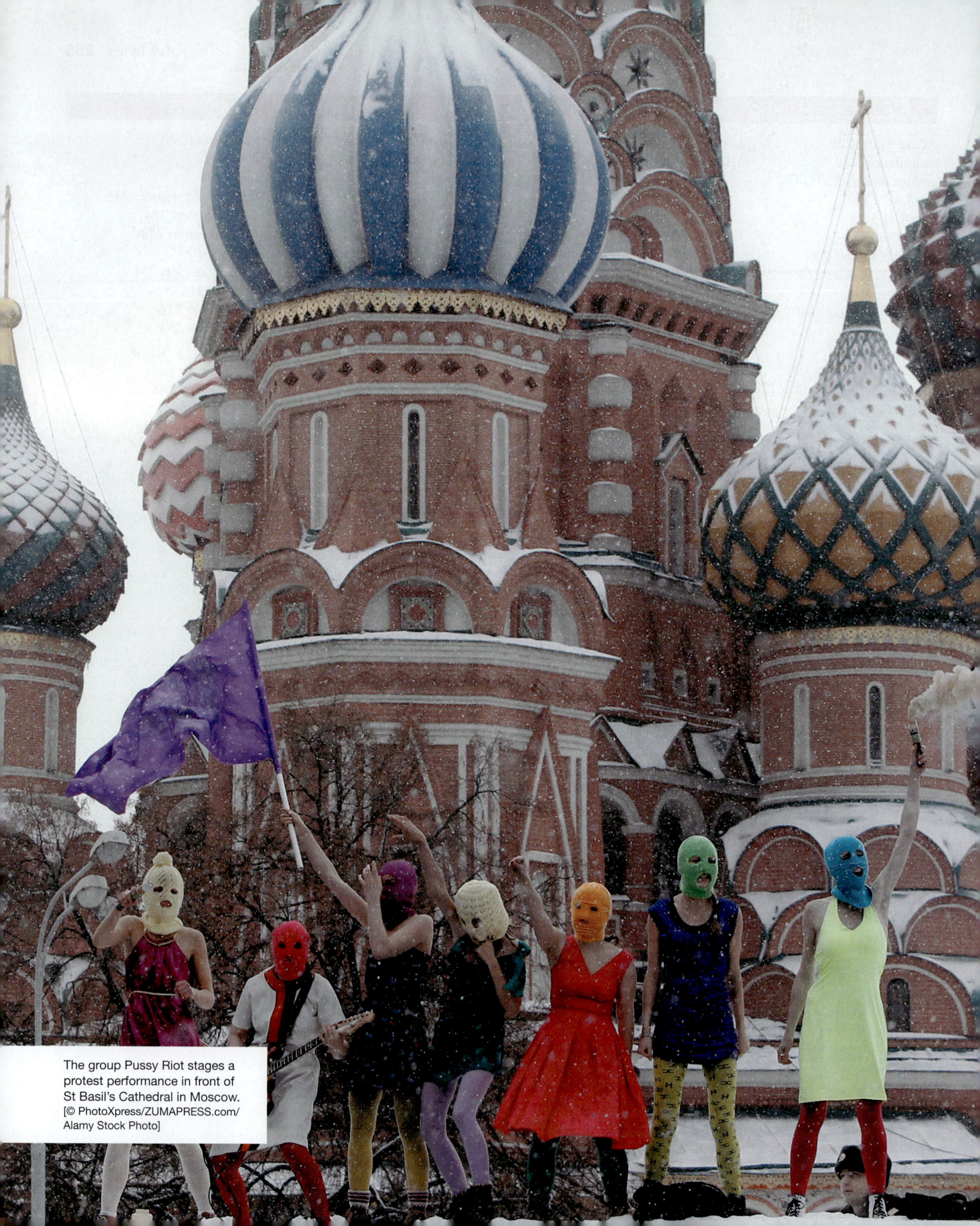

The group Pussy Riot stages a protest performance in front of St Basil's Cathedral in Moscow. [© PhotoXpress/ZUMAPRESS.com/ Alamy Stock Photo]

5

Russia and the Post-Soviet States

Pussy Riot is a group of Russian political activists—mostly women in their twenties and thirties—known for unauthorized but high-profile protests in public spaces. They became a worldwide media sensation in 2012 when they staged a "guerilla performance" at a Moscow cathedral, in which they decried the policies of Vladimir Putin, the country's president. Their aim is to spotlight the lack of democracy in Russia and show how the country has drifted toward authoritarianism under Putin's control. However, being a dissenter in Russia is increasingly difficult as the government has passed new laws aimed at cracking down on protesters. Members of Pussy Riot were put on trial for the cathedral protest and were imprisoned at a remote gulag (labor camp) for "hooliganism," which has reinforced the notion that political dissent in Russia will be repressed by the authorities. Subsequently, the group has continued to stage protests, including during the World Cup soccer final in 2018. In their work, which ranges from music to poetry to performance art, Pussy Riot supports LGBTQ and free speech rights, while they oppose not only the current Russian government but also the antifeminist traditions of the Russian Orthodox church. Western artists and political groups, such as Madonna and Amnesty International, have endorsed Pussy Riot, but in Russia, understanding of their goals and methods is limited. So far, Vladimir Putin has the support of the majority of Russians, who perceive him as a strong leader.

Learning Objectives

Environment: Physical and Human

5.1 Describe the natural landscapes, climate, and vegetation of Russia and the post-Soviet states.

5.2 Analyze how the environment in Russia and the post-Soviet states has been affected by human activity and climate change.

5.3 Summarize how early societies emerged in the region, and describe its communist and postcommunist history.

Globalization and Development

5.4 Explain how the post-Soviet economy has affected people differently in Russia and the post-Soviet states.

5.5 Discuss how Russia and the post-Soviet states are integrated into the global economy.

Power and Politics

5.6 Describe how authoritarianism challenges democracy in the region.

5.7 Analyze past and current military conflicts in Russia and the post-Soviet states.

Urbanization

5.8 Discuss how the region's geography and communist-era planning affected urbanization.

5.9 Describe the primate city phenomenon in Russia and the post-Soviet states.

Population, Gender, and Culture

5.10 Explain changes to Russia's and the post-Soviet states' population, health, and life expectancy.

5.11 Describe how gender affects work and politics in Russia and the post-Soviet states.

5.12 Summarize how religious practices and national identity have evolved in the post-Soviet era.

Subregions

5.13 Identify key characteristics of the subregions of Russia and the post-Soviet states.

ARCTIC OCEAN
Norwegian Sea
North Sea
Barents Sea
Kara Sea
Laptev Sea
Baltic Sea
Gulf of Bothnia
White Sea
Black Sea
Sea of Azov
Caspian Sea
Aral Sea
Persian Gulf
Greenland
Svalbard
Franz Josef Land
North Land
Novaya Zemlya
North Pole 90°N
Arctic Circle
UNITED KINGDOM
Manchester
Birmingham
London
FRANCE
BELGIUM
Brussels
Amsterdam
NETHERLANDS
LUX.
Luxembourg
Hamburg
DENMARK
GERMANY
Copenhagen
Berlin
NORWAY
Oslo
SWEDEN
Stockholm
FINLAND
Helsinki
Tallinn
ESTONIA
Riga
LATVIA
LITHUANIA
Vilnius
Kaliningrad
POLAND
Warsaw
Prague
CZECHIA
Vienna
Bratislava
SLOVAKIA
Budapest
HUNGARY
ROMANIA
MOLDOVA
Chisinau
BELARUS
Minsk
Brest
UKRAINE
Kiev
Lviv
Pripyat
Chernobyl
Homyel
Odessa
Mykolayiv
Kherson
Dnipropetrovsk
Kharkiv
Donetsk
Sevastopol
Carpathian Mts.
Kola Peninsula
Murmansk
Kirovsk
Arkhangelsk
St. Petersburg
Lake Ladoga
Lake Onega
Northern Dvina Canal
North European Plain
Smolensk
Tver
Moscow
Moscow Canal
Yaroslavl
Ivanovo
Bryansk
Tula
Kursk
Lipetsk
Voronezh
Nizhniy Novgorod
Kirov
Kazan
Penza
Saratov
Volga-Don Canal
Tolyatti
Samara
Rostov-on-Don
Volgograd
Astrakhan
Caspian Depression
Orenburg
Orsk
Ufa
Magnitogorsk
Chelyabinsk
Perm
Yekaterinburg
Nizhniy Tagil
Tyumen
Ural Mountains
Pechora
Vorkuta
Salekhard
Yamal Peninsula
Kara Strait
Pechora Basin
West Siberian Plain
Urengoy
Surgut
Dikson
Taymyr Peninsula
North Siberian Lowland
Norilsk
Central Siberian Plateau
Siberia
RUSSIAN FEDERATION (RUSSIA)
Tunguska Basin
Seversk
Tomsk
Omsk
Novosibirsk
Krasnoyarsk
Novokuznetsk
Gladkaya
Bratsk
Bratsk Res.
Tayshet
Irkutsk Basin
Lake Baikal
Angarsk
Irkutsk
Ulan-Ude
Kyzyl
E. Sayan Mts.
W. Sayan Mts.
Vilyuy Res.
Lena
Olenek
Kotuy
Angara
Ob
Irtysh
Yenisey
Kama
Volga
Don
Dnieper
Dvina
Steppes
KAZAKHSTAN
Astana
Aktogay
Qaraghandy
Leninsk
Lake Balkhash
Syr Darya
Amu Darya
Shymkent
Almaty
Bishkek
KYRGYZSTAN
Lake Ysyk
Tashkent
Samarkand
Dushanbe
TAJIKISTAN
Pamirs
TURKMENISTAN
Ashkhabad
UZBEKISTAN
Tien Shan
Altai Mts.
Junggar Basin
Urumqi
Tarim Basin
Taklimakan Desert
Gobi
MONGOLIA
Ulan Bator
Baotou
Lanzhou
Xian
CH
GEORGIA
Batumi
Tbilisi
Grozny
ARMENIA
Yerevan
AZER.
AZERBAIJAN
Baku
Caucasus Mts.
TURKEY
IRAQ
Lake Urmia
Lake Van
Tehran
Elburz Mts.
Zagros Mts.
Esfahan
Mashhad
IRAN
AFGHANISTAN
Kabul
Hindu Kush
Islamabad
Faisalabad
Lahore
PAKISTAN
INDIA
New Delhi
QATAR
Ad Dawhah (Doha)
Abu Zaby
Dubayy
UAE
Muscat
OMAN

Pussy Riot is an example of politics in modern Russia, which emerged from the tumultuous and complete change of the region's political and economic systems. In fact, Russia is the largest country in a region that has entirely changed its political and economic systems in a short period of time (**Figure 5.1**). Three decades ago, the **Union of Soviet Socialist Republics (USSR)**, more commonly known as the **Soviet Union**, was the largest country in the world, stretching from Central Europe to the Pacific Ocean. It covered one-sixth of Earth's land surface. In 1991, the Soviet Union broke apart, ending a 70-year era of nearly complete governmental control of the economy, society, and politics. Over the course of a few years, attempts were made to substitute the Communist Party's economic control with capitalist systems similar to those of Western countries, which are based on competition among private businesses. The transition is still ongoing and has proven difficult.

After 70 years of authoritarian rule, elections took place throughout this region. At first, there was a great degree of optimism about democracy taking hold. Today, however, it is highly debatable whether elections are free and fair, as opposition candidates are marginalized by lack of access to both print and broadcast media. Crime and corruption have been a further setback to the democratization of the region.

Politically, the Soviet Union has been replaced by Russia and 14 independent post-Soviet states. This chapter covers, in addition to Russia, the European states of Ukraine, Belarus, and Moldova; the Caucasian states of Georgia, Armenia, and Azerbaijan; and the Central Asian states of Kazakhstan, Kyrgyzstan, Tajikistan, Turkmenistan, and Uzbekistan (**Figure 5.2**). Three former Soviet republics, the Baltic states of Lithuania, Latvia, and Estonia, are now part of the European Union, which is why they are included in Chapter 4. Russia, which was always the core of the Soviet Union, remains dominant in the region and is influential in the world because of its size (at roughly three-quarters the size of the former Soviet Union, it is still the largest country in the world), population, military, and huge oil and gas reserves.

Geopolitically, this region is still sorting out its relationships. The Cold War between the Soviet Union and the United States and its allies was replaced by a temporary thaw, although the relationship has grown chilly again. Most former Soviet allies in Central Europe have already joined the European Union, and some other countries in the region may eventually do the same. Trade with Europe, especially in oil and gas, is important, yet future trade is uncertain, as Europe has imposed sanctions on Russia due to its military involvement in Ukraine (see the section "Crisis in Ukraine"). Russian interference in European elections is also a source of contention. In the future, the Central Asian states currently allied with Russia may forge stronger connections with their neighbors in Southwest Asia or South Asia. The far eastern parts of Russia—and Russia as a whole—are already finding common trading ground with East Asia and Oceania (**Figure 5.3**).

Union of Soviet Socialist Republics (USSR) the multinational state formed from the Russian empire in 1922 and dissolved in 1991; commonly known as the *Soviet Union*

Soviet Union see *Union of Soviet Socialist Republics*

◀ **Figure 5.1** Regional map of Russia and the post-Soviet states.

Figure 5.2 Political map of Russia and the post-Soviet states. Note that we use the borders that are recognized by most of the international community. Specifically, the Crimean Peninsula is mapped as part of Ukraine, but in reality it is controlled by Russia.

What Makes Russia and the Post-Soviet States a Region?

Caucasia the mountainous region between the Black Sea and the Caspian Sea

Russia and the post-Soviet states make up a large portion of Eurasia, the world's largest continent. It is a vast region but is sparsely populated and with a climate that, in many places, is unforgiving. Russia and the post-Soviet states are today associated as a region primarily because of their history from the nineteenth century onward. Imperial Russia underwent a convulsive revolution in 1917 that resulted in a new political and economic system that lasted until 1991. The leaders of this revolution continued with the imperial practice of territorial expansion that ultimately enveloped countries from the Baltic Sea to the Pacific and from the Arctic Sea to the mountains of **Caucasia** and southern Central Asia. Control of the economic and political systems of the Soviet Union was exercised primarily from the Russian capital, Moscow, through authoritarianism reinforced with military power. After the collapse of the Soviet Union in 1991, Russia's dominance diminished somewhat as western parts of the region were free to orient more toward Europe and the Central Asian republics toward other parts of Asia (see Figure 5.2). Nevertheless, Russia remains the leading power in this region and is currently reasserting its dominance yet again.

Terms in This Chapter

There is no entirely satisfactory name for the countries of the former Soviet Union. In this chapter we use *Russia and the post-Soviet states*. One reason we do so is that economic, political, and social developments from the Soviet days still shape the region today. The post-Soviet states continue to be closely associated with Russia economically, but they are independent countries separate from Russia.

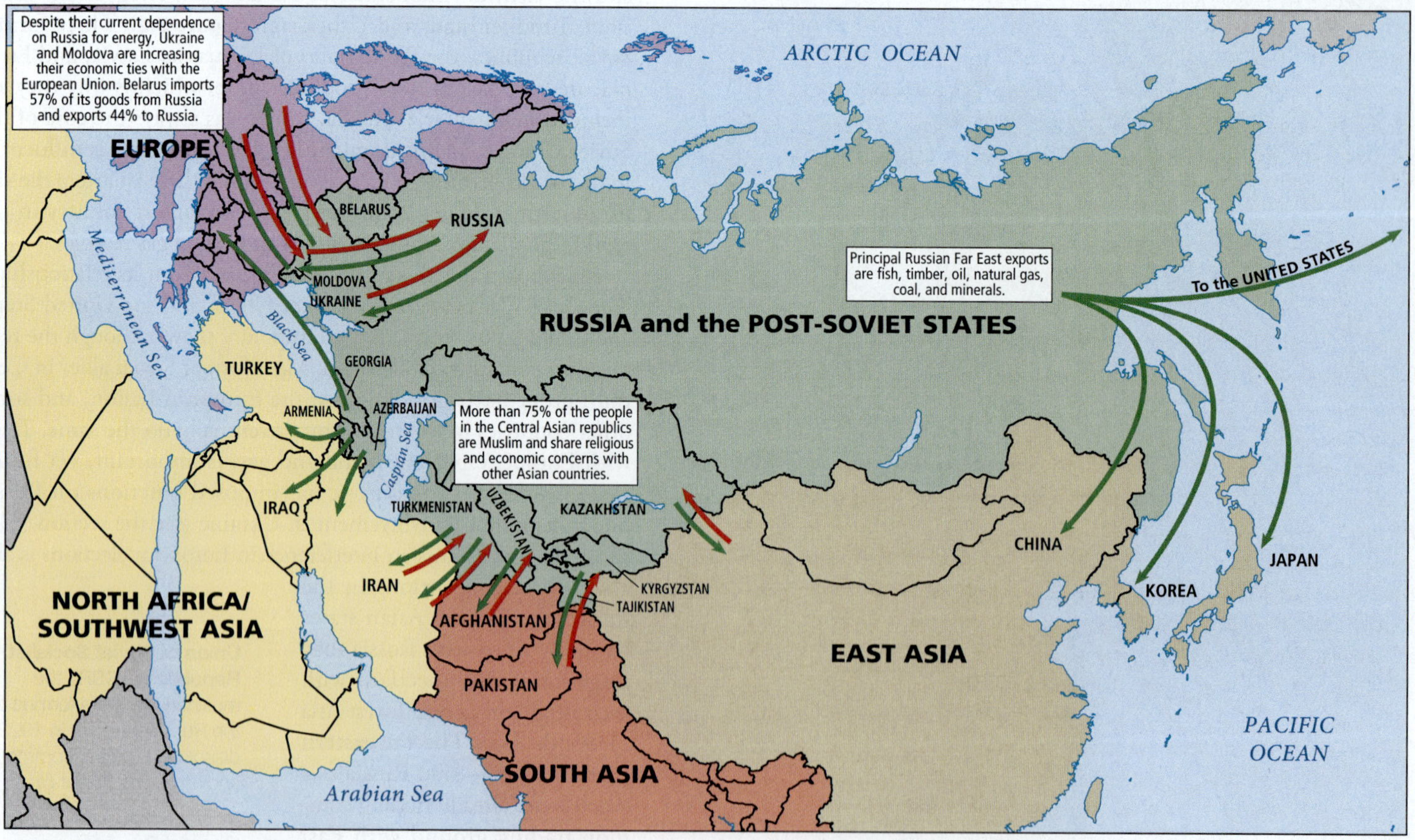

Figure 5.3 Russia and the post-Soviet states: Contacts with other world regions. The region of Russia and the post-Soviet states is in flux as all of the component countries rethink their geopolitical positions relative to one another and to adjacent regions. The arrows indicate various types of contact, from economic to religious and cultural.

Russia itself is formally known as the **Russian Federation**, which includes a large number of provinces, as well as 21 mostly ethnic *internal republics*—such as Chechnya, Karelia, and Tatarstan—which all together constitute about one-tenth of the Federation's territory and one-sixth of its population. The internal republics are sometimes labeled "autonomous," but in reality they do not share power equally with the central Russian government in Moscow.

ENVIRONMENT: PHYSICAL AND HUMAN

5.1 Describe the natural landscapes, climate, and vegetation of Russia and the post-Soviet states.

5.2 Analyze how the environment in Russia and the post-Soviet states has been affected by human activity and climate change.

5.3 Summarize how early societies emerged in the region, and describe its communist and postcommunist history.

The physical and human features of Russia and the post-Soviet states vary greatly over the huge territory they encompass. The physical vastness of the region (especially Russia) is and has always been a major challenge for effective governance, military activities, transportation, resource extraction, and trade.

LANDFORMS

Because the region is relatively complex physically, a brief summary of its landforms is useful. We start our description of the landforms of Russia and the post-Soviet states in the west (Figure 5.1). Moving west to east, there is first the eastern extension of the North European Plain, including the plain of the Volga River, followed by the Ural Mountains, which are often considered the dividing line between European Russia and **Siberia**. East of the Urals are the West Siberian Plain, followed by an upland zone called the Central Siberian Plateau, and finally, in the Far East, a series of mountain ranges bordering the Pacific. To the south there are mountains and uplands (the Caucasus) as well as, in western Central Asia, semiarid grasslands, or **steppes**.

The eastern extension of the North European Plain stretches 1200 miles (about 2000 kilometers), from Central Europe to the Ural Mountains. The part of Russia west of the Urals is often called *European Russia* because the Ural Mountains are traditionally considered part of the indistinct border between Europe and Asia. European Russia is the most densely settled part of the entire region (see Figure 5.24 later in this chapter) and is its agricultural and industrial core. Its most important river is the Volga, which flows into the Caspian Sea. The Volga River and its tributaries form a major transportation route that connects many parts of the North European Plain, including Moscow, St. Petersburg, and the Baltic and White seas in the north, and the Black and Caspian seas in the south.

The Ural Mountains extend north to south from the Arctic Ocean into Kazakhstan (see Figure 5.1). A low-lying range similar in elevation to the Appalachians, the Urals are not much of a barrier to humans and are only partially a barrier to nature (some European tree species do not extend east of the Urals). There are several easy passes across the mountains, and winds carry moisture all the way from the Atlantic into Siberia. Much of the Urals' once-dense forest has been felled to build and fuel new industrial cities.

The West Siberian Plain, east of the Urals, is the largest plain in the world. A vast, mostly marshy lowland about the size of the eastern United States drains toward the north into the Arctic Ocean. The northward-flowing rivers of Siberia have done little to improve Russia's need for east–west communication. Long, bitter winters mean that in the northern half of this area, a layer of permanently frozen soil (**permafrost**) lies just a few feet beneath the surface. Permafrost is formed when the ground warms up during the short summer but the surface layer of organic material and soil insulates against this warming effect, leaving the subsurface always frozen. In the far north, the permafrost comes to within a few inches of the surface. Because water does not percolate down through this frozen layer, surface water accumulates to create wetlands that form above the permafrost. In the summer months these wetlands provide habitats for many migratory birds. Due to climate change, some permafrost in Siberia is melting, which releases carbon into the atmosphere through the decomposition of thawing vegetation. Another potent greenhouse gas—methane—that was previously trapped in the frozen layers is also released into the atmosphere.

In the far north lies a treeless area called the **tundra**, where mostly only mosses and lichens can grow because of the extreme cold, the shallow soils, and the permafrost. The West Siberian Plain has one of the world's largest oil and natural gas deposits, although the harsh climate and permafrost make extraction difficult.

The Central Siberian Plateau and the Pacific Mountain Zone, farther to the east, together equal the size of the United States. Permafrost prevails except along the Pacific coast. There, the ocean moderates temperatures; the many active volcanoes created as the Pacific Plate sinks under the Eurasian Plate supply additional heat. Lightly populated places like the Kamchatka Peninsula and Sakhalin Island on Russia's Pacific coast are havens for wildlife.

To the south of the West Siberian Plain, steppes and deserts stretch from the Caspian Sea to the Chinese border. To the west of these grasslands are the Caucasus Mountains, and facing China and Mongolia are a series of other mountain chains, including the Pamir and Tien Shan. The rugged terrain has not deterred people from crossing these mountains. For tens of thousands of years, people have exchanged plants (apples, onions, citrus, rhubarb, wheat), animals (horses, sheep, goats, cattle), technologies (cultivation, animal breeding, portable shelter construction, rug and tapestry weaving), and religious belief systems (principally Islam and Buddhism).

CLIMATE AND VEGETATION

The climates and associated vegetation in this large region are varied but less so than in other regions because so much territory here is taken up by expanses of midlatitude grasslands and by northern forest and tundra that are cold much of the year. No inhabited place on Earth has as harsh a climate as the northern part of the Eurasian landmass occupied by Siberian Russia (shown in **Figure 5.4C** and the figure map). Winters are long and cold, with only brief hours of daylight. The

Russian Federation Russia and its political subunits, which include 21 internal republics

Siberia the territories of Russia located east of the Ural Mountains

steppes semiarid grass-covered plains

permafrost permanently frozen soil that lies just beneath the surface

tundra a cold, treeless area, between the ice cap and the continental climate forest, where the subsoil is permanently frozen

5.4 Climates of Russia and the Post-Soviet States

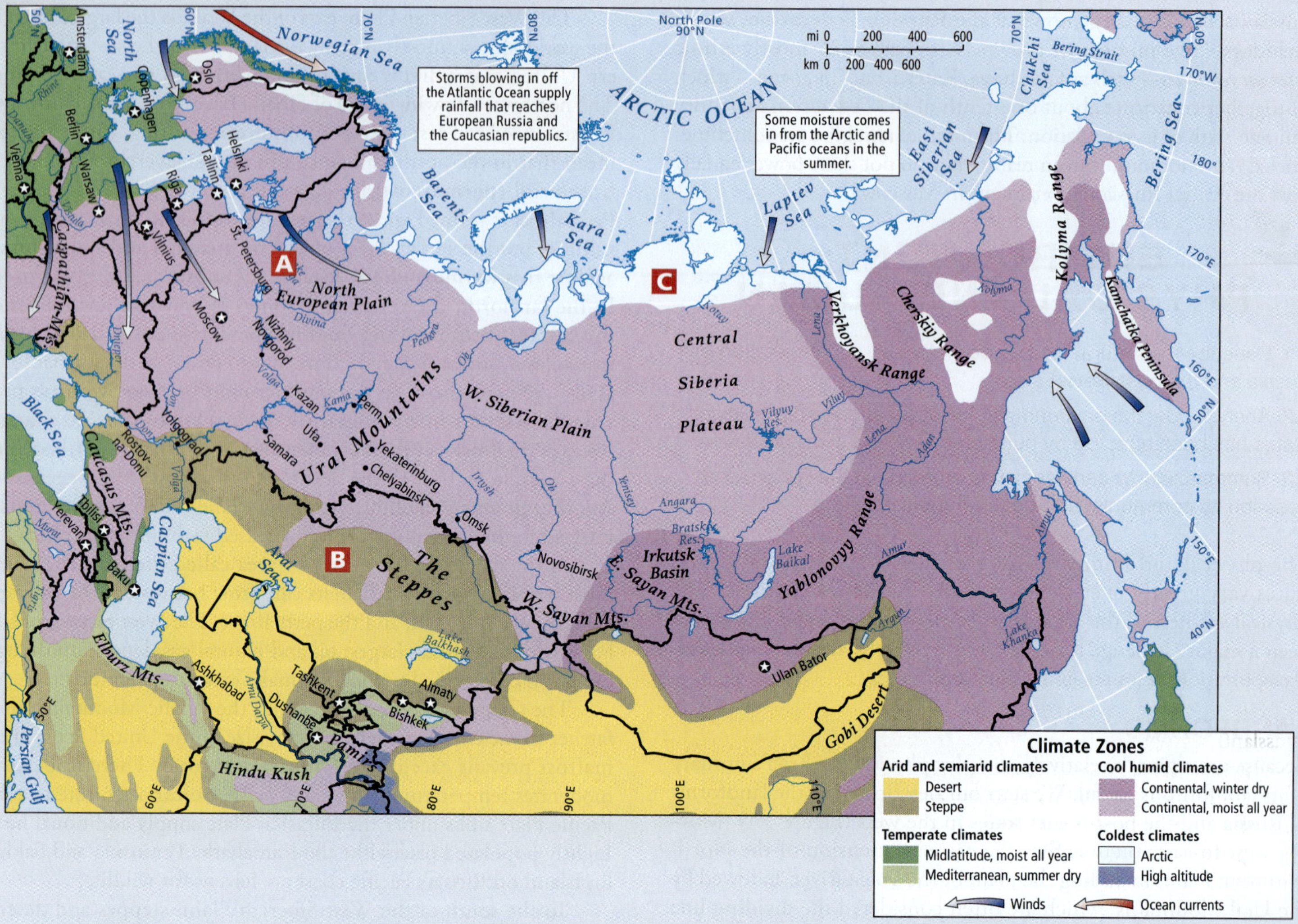

A **Continental, moist all year, Vologda region** [vicsa/iStock/Getty Images]

B **Steppe, Kazakhstan** [Panorama Media/AGE Fotostock]

C **Arctic, Siberia** [vladimir zakharov/Getty Images]

Earth's coldest temperature outside Antarctica has been recorded in western Siberia, at −98°F (−71°C). Summers are short but with daylight much of the time. Precipitation is moderate, coming primarily from the west. In the northernmost areas, the natural vegetation is tundra grasslands and other hardy ground cover. The major economic activities here are the extraction of oil, gas, and some minerals, as well as reindeer herding by the local indigenous population. Just south of the tundra is a vast, cold-adapted coniferous forest known as **taiga** that ranges from northern European Russia to the Pacific (and from a global perspective, this coniferous belt also includes much of Alaska, Canada, and Scandinavia). The largest portion of taiga lies east of the Urals, and here forestry—often unrestrained by ecological concerns—is a dominant economic activity. The short growing season and the large areas of permafrost generally limit crop agriculture, except in the southern West Siberian Plain, where people grow grain.

Because massive mountain ranges to the south block access to warm, wet air from the Indian Ocean, most rainfall in the entire region comes from storms that blow in from the Atlantic Ocean far to the west (see the Figure 5.4 map). But by the time these air masses arrive, most of their moisture has been squeezed out over Europe. A fair amount of rain does reach Ukraine, Belarus, European Russia (see Figure 5.4A), and the Caucasian republics; these regions are especially important areas for food production. The natural vegetation in these western zones is open woodlands and grassland, though in ancient times, forests were common.

East of the Caucasus Mountains, the lands of Central Asia have semiarid to arid climates influenced by their location in the middle of a very large continent (see Figure 5.4B). The summers are scorching but short and the winters are intense. In the desert zones, daytime-to-nighttime temperatures can vary by 50°F (28°C) or more. Northern Kazakhstan produces grain and grazing animals. The more southern areas (southern Kazakhstan, Uzbekistan, and Turkmenistan) have grasslands (steppes), which are also used for herding. Some of the land has been converted to irrigated commercial agriculture using water from glacially fed rivers, but most of it is not useful for farming. The climates in the more mountainous area where Kazakhstan, Uzbekistan, Tajikistan, and Kyrgyzstan meet are varied and support a number of small-scale agricultural activities, some of them commercial.

The construction of land transportation systems has been held back by the climate of the region. Especially to the north, it is difficult to build during the long, harsh winters, which eventually give way to a spring period called the *rasputitsa,* or the "quagmire season," when melting snow and ice (or, to a lesser extent, autumn rain) turn many roads and construction sites into impassable mud pits. Huge distances between populated places also complicate the situation. As a result, few roads or railroads have ever been built beyond western Russia. One exception is the Trans-Siberian Railway, which crosses the continent through the southern reaches of Siberia. The railway was a major infrastructure achievement in the early twentieth century.

CHECK YOUR UNDERSTANDING

1. What parts of Russia are located west and east of the Ural Mountains?
2. What physical features characterize the tundra?
3. Where is the taiga located, and what are its characteristics?
4. Describe how landforms and movements of air masses affect the climate of the region.
5. How have climate and distance affected the development of transportation systems in the region?

ENVIRONMENT

In Soviet ideology, nature was considered the servant of industrial and agricultural progress, and humans were meant to dominate nature on a grand scale. While this sentiment was common throughout much of the world at the time of the formation of the USSR, it does seem to have been taken further here than elsewhere. During the leadership of Joseph Stalin from 1922 to 1953, Soviet resource policy followed the maxim, "We cannot expect charity from nature. We must tear it from her." During the Soviet years, huge dams, factories, and other industrial facilities were built without regard for their effect on the environment or on public health. Cities expanded quickly—with workers flooding in from the countryside—to accommodate the new industries. Some cities were built around industries that produce harmful by-products, creating problems that still linger. For example, the former chemical weapons–manufacturing center of Dzerzhinsk is listed by the nonprofit organization Pure Earth as one of the ten most polluted cities in the world.

Other sources of contamination are multiple and difficult to trace. Such **nonpoint sources of pollution** include automobile exhaust, raw sewage, and agricultural chemicals that drain from fields into water supplies. In all urban areas of the region, air pollution is the result of burning of fossil fuels as more people purchase cars and as the industrial and transport sectors of the economy continue to grow.

After the collapse of the Soviet Union, the region's governments, beset with myriad economic problems, were either reluctant to address environmental issues or incapable of doing so. As one Russian environmentalist put it, "When people become more involved with their stomachs, they forget about ecology." This dilemma is exemplified by the oil and gas industry. On the one hand it is a major source of income for Russia and several of the post-Soviet states, but on the other hand relatively little attention is being paid to the environmental impact of the extraction of fossil fuels. However, recent surveys in Russia suggest that people are becoming more aware of environmental issues and more willing to act to protect the environment. The types of environmental activism that people are engaged in often address problems that affect ordinary Russians—the types of air, water, and soil pollution for which there is an obvious source of pollution as well as a remedy.

People's awareness about more indirect environmental issues, such as climate change, is still lagging. One reason why environmentalism is only slowly gaining ground is related to the vastness of the land, which reinforces the deeply held public attitude that resources are unlimited. If Russia progresses economically over time,

taiga subarctic coniferous forests

nonpoint sources of pollution diffuse sources of environmental contamination, such as automobile exhaust, raw sewage, and agricultural chemicals

Figure 5.5 The aftermath of the Chernobyl disaster. The abandoned city of Pripyat in Ukraine, once home to 50,000 people, was evacuated after the explosion of the Chernobyl nuclear power plant, visible on the horizon. [SERGEI SUPINSKY/AFP/Getty Images]

it is possible that Russians' environmental consciousness will continue to grow, mirroring the historical trend in, for example, the large and resource-rich North American continent. A recent environmental performance index maintained by Yale University ranks Russia as having the 52nd best environment among the countries of the world. It classifies Russia's environmental quality to be on a global medium, which means that it is similar to countries on the same economic development level. In Central Asia, though, Uzbekistan and Tajikistan score very poorly due to problems with air quality and habitat protection.

Nuclear Power and Pollution

Russia and the post-Soviet states are also home to extensive nuclear pollution, the effects of which have spread globally. The world's worst nuclear disaster occurred in Ukraine in 1986, when the Chernobyl nuclear power plant exploded (**Figure 5.5**). The explosion severely contaminated a vast area in northern Ukraine, southern Belarus, and Russia. It spread a cloud of radiation over much of Central and North Europe. In the area surrounding Chernobyl, more than 300,000 people were evacuated from their homes, and Pripyat, which used to have a population of 50,000 people, has been an abandoned ghost town since the disaster. The ultimate health effects are impossible to assess exactly. The United Nations estimates that 6000 people developed cancer due to the accident, although the number may be much higher.

When the Soviet Union collapsed, many of the post-Soviet states inherited nuclear facilities and even nuclear weapons. The concern at the time was that nuclear weapons from the former Soviet arsenal would end up in the hands of terrorists. Due to cuts to military funding, weapons and military supplies were being sold on the black market. Recent developments are more encouraging. In an effort to discourage the temptation to sell nuclear materials, the Russian government has given impoverished military personnel long-overdue pay raises. In addition, all countries in the region now cooperate with the International Atomic Energy Agency in controlling nuclear material.

One such country is Kazakhstan (see Figure 5.2). During the Soviet era, remote areas of eastern Kazakhstan were used as a testing ground for nuclear devices. The residents were sparsely distributed and no one took the trouble to protect them from nuclear radiation. On the bright side, Kazakhstan's government has disarmed the nuclear warheads that it inherited from the Soviet Union and safeguards its remaining nuclear weapons material.

Kazakhstan plans to use its nuclear technology to become a global leader in the nuclear energy sector. It is already the single biggest source of uranium in the world, generating 36 percent of the world's total. In addition to mining and exporting the raw material, a "bank" for nuclear fuel, which is intended to serve the needs of nuclear power plants around the world, opened in 2017 in Kazakhstan. However, global nuclear expansion remains uncertain for several reasons. Following the 2011 Fukushima accident in Japan (see Chapter 9), resistance to nuclear power has increased. Also, other sources of energy, such as solar power and natural gas, are now cheaper than nuclear. And finally, no reliably safe system has yet been found for storing nuclear waste until it is no longer radioactive. Kazakhstan's population may also be suspicious of nuclear power because they have lived for so long with land contaminated by radioactive waste.

Resource Use and the Environment

Russia and the post-Soviet states have considerable natural and mineral resources (see Figure 5.14 later in the chapter). Russia itself has the world's largest natural gas reserves, major oil and coal deposits, and forests that stretch across the northern reaches of the continent. The Central Asian states share substantial deposits of oil and gas, which are centered on the Caspian Sea and extend east toward China. The resources underneath the Caspian Sea have been disputed because when the Soviet Union dissolved, no agreement existed among the newly created states on how to divide control of the seabed. Such a deal, which is predicted to spur offshore oil and gas exploration as well as the building of pipelines below the Caspian Sea, was reached in 2018.

Russia also has major deposits of industrial minerals. For example, the city of Norilsk is rich in nickel—used in steel and other industrial products—and other minerals, such as copper (**Figure 5.6**). Norilsk is located in northern Siberia and was founded during the early Soviet era as a place of resource extraction and industry. In fact, with approximately 200,000 people, it is the largest city in the world at such northern latitudes. But mining and mineral processing comes with a price. Much of the vegetation has been killed off

Figure 5.6 Norilsk. In 2016, the Daldykan River near Norilsk turned red due to an industrial spill from a Norilsk Nickel facility. [© Liza Udilova/Greenpeace]

around the city's metal-smelting complex—the largest facility of its kind in the world. The smelter emits about 1 percent of all global emissions of sulfur dioxide, which helps form acid rain. Nevertheless, Norilsk continues to attract investment crucial to the new Russian economy. Norilsk Nickel, a company that was privatized at the end of the period of communism and is now the largest producer of nickel in the world, dominates the city. With an extensive base of mineral resources, Norilsk Nickel has used its enormous profits to become a global player in the mining industry. Despite its poor environmental record, it owns mining operations in Australia, Finland, and southern Africa and employs almost 100,000 people worldwide. At home in Norilsk, the company is now subject to new regulations that, it is hoped, will improve the local environment in the near future.

Forestry in Russia Russia has 20 percent of the world's forests, which makes it a highly valuable resource. What was used domestically under communism is now part of a global flow of raw material. Russian forestry today is a highly export-oriented sector as most of the timber is shipped to China, where it is manufactured into products like furniture and flooring. Much logging in Russia follows the pattern of clearing large swaths of forests and then moving on to other areas. If forest management were to adopt sustainable techniques involving replanting and continuous maintenance, both economic productivity and forest health could improve. Historically, the most accessible forests were logged; now, timber extraction has been pushed deep into Siberia and Russia's European north. Harvesting and shipping timber from what is often roadless terrain is cumbersome and reduces profitability.

The Loss of the Aral Sea In the Central Asian states, the Aral Sea, once the fourth-largest lake in the world, was fed by the Syr Darya and Amu Darya rivers for millions of years. These rivers brought water from melting glaciers and snow in the mountains of the region.

In the 1960s, the Soviet leadership ordered that the two rivers be diverted to irrigate millions of acres of cotton in naturally arid Kazakhstan and Uzbekistan. So much water was consumed by these projects that within a few years, the Aral Sea had shrunk measurably (**Figure 5.7**). By the early 1980s, no water at all from the

(B) 2000
[NASA]

(C) 2009
[NASA]

(D) 2017
[NASA]

(A)

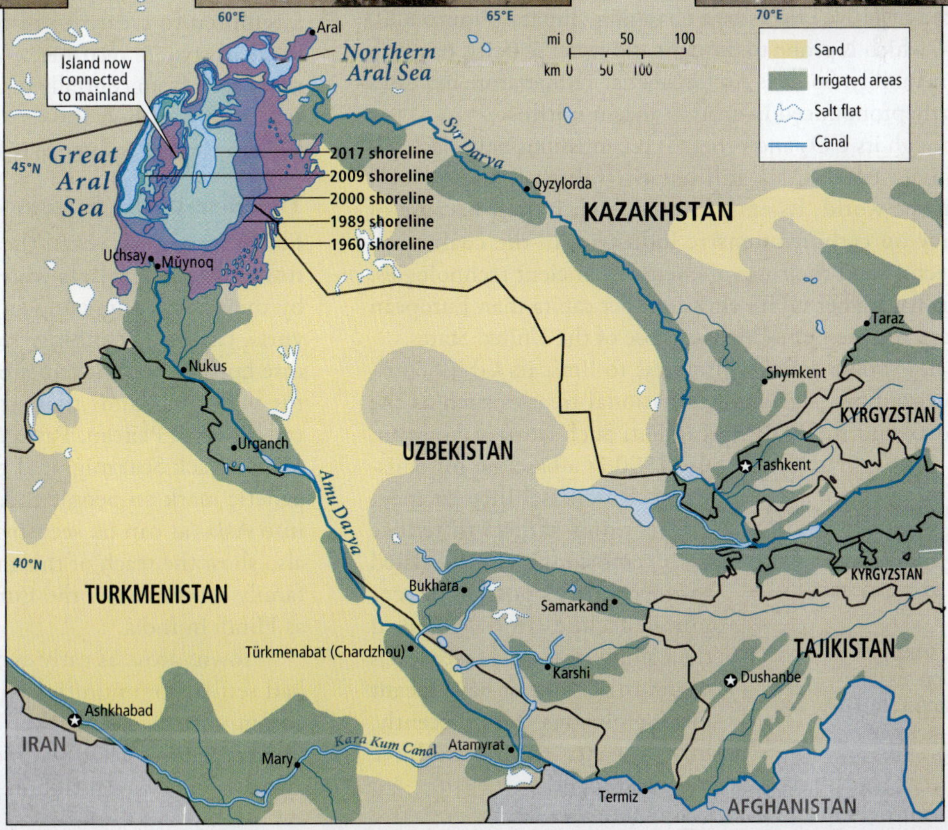

Figure 5.7 The decline and disappearance of the Aral Sea, 1960–2017. (A) Once the fourth-largest lake in the world, the Aral Sea is disappearing as a result of large-scale irrigation projects in Central Asia. Since 2005 when Kazakhstan built a dam, the northern Aral Sea is slowly increasing in size and depth. However, as (B), (C), and (D) show, the southern sections of the sea continue to decline dramatically. In some dry years, the eastern lobe disappears altogether, which is evident in (C), from 2009. The scale of all three photos is the same. [Research from: NASA Earth Observatory, at https://earthobservatory.nasa.gov/WorldOfChange/AralSea]

two rivers was reaching the Aral Sea, and eventually it lost most of its volume and shrank into three smaller lakes. The region's huge fishing industry, which accounted for one-sixth of the fish catch in the entire USSR, died out because of increased salinity in the water. Once-active ports were marooned many kilometers from the water (see Figure 5.8C). The decline and disappearance of the Aral Sea has been described as the largest human-made ecological disaster in history.

Efforts to increase water flows to the Aral Sea have been hampered because the now-independent countries of Uzbekistan and Kazakhstan share the sea and they have different objectives. Uzbekistan wants to continue its massive irrigation programs, which have made it the world's eighth-largest cotton grower (see Figure 5.8B). Therefore, the inflow to the Aral Sea from the south remains marginal. Nevertheless, some restorative actions have been taken. Kazakhstan, which relies on oil more than agriculture, built a dam in 2005 to keep water in the northern Aral Sea from spilling into the southern Aral Sea. The water level recovered, the fishing industry was revived, as were the Kazakh communities around the lake. In the southern Aral Sea, however, water levels remain very low (Figure 5.7C, D). In Uzbekistan, where 26 percent of the workforce is employed in agriculture, the limitation of irrigation has forced economic diversification, and cotton production now makes up just 9 percent of the country's exports, far less than the 45 percent it represented in 1990. Gold mining is now the leading export industry.

CLIMATE CHANGE

Climate change is an issue in this region for many reasons. Although high levels of CO_2 and other greenhouse gases (GHGs) are produced here, Russia's forests play a major role in absorbing global CO_2. And while much of the region is only moderately vulnerable to the negative effects of a changing climate (**Figure 5.8**), Central Asia, which is prone to drought and water scarcity, has high levels of vulnerability. So does Arctic Siberia. Temperature increases are particularly pronounced in areas in the far north.

Even though its CO_2 and other GHG emissions have declined since the Soviet era, Russia still has the fourth-highest GHG emissions in the world. Its emissions are high in part because of a dependency on carbon-intensive industries in the coal, steel, and metal sectors and their use of energy-inefficient technologies. While Russia has higher GHG emissions per capita than European countries, its levels are not as high as those of the United States.

Russia has shown some willingness to limit its GHG emissions by, for example, signing international treaties such as the Kyoto Protocol and the Paris Agreement. Such treaties, however, use 1990 as a baseline for emissions. Russia's ability to meet international targets to reduce GHG emissions is largely related to the country's post-1990 economic decline after the end of the communist era. Lower levels of industrial output have meant lower emissions. More recently, however, its emissions have increased slightly.

nomadic pastoralists people whose culture and economy are centered on tending grazing animals who are moved seasonally to gain access to the best pastures

Silk Road historic trading route between China and the Mediterranean; named after the importance of silk as a desired commodity of trade

Central Asia is vulnerable to the impacts of climate change because of its dependence on water from rivers that are fed by glaciers in the Pamirs and other mountains in Kyrgyzstan and Tajikistan (see Figure 5.8A). In recent decades, the glaciers have shrunk because of higher temperatures and decreases in rain and snowfall. In this area, most precipitation falls in the mountains in the winter and spring. The glaciers act as water-storage systems, supplying melted ice flow to the rivers during the summer, when irrigation is most needed for growing crops. If current trends continue, the glaciers will melt earlier in the year, resulting in lower river flows in the summer. Either Central Asia's agricultural systems would have to adapt to a spring growing season, or farmers would have to store water for use in the summer. Either situation demands complex and expensive changes.

CHECK YOUR UNDERSTANDING

1. How did Soviet authorities regard the problem of urban and industrial pollution?
2. How is this region contaminated by pollution, and what are those sources of pollution?
3. What happened at Chernobyl in 1986?
4. What factors led to the shrinking of the Aral Sea?
5. Why is Central Asia especially vulnerable to climate change?

HUMAN PATTERNS OVER TIME

The core of the entire region has long been European Russia, the homeland of the ethnic Russians. Expanding gradually from this center, the Russians conquered a large area inhabited by a variety of other ethnic groups. These conquered territories remained under Russian control as part of the Soviet Union (1917–1991), which attempted to create an integrated social and economic unit out of the disparate territories. The Soviet Union's collapse brought a rapid shift from centrally planned economies to market economies. It also diminished Russian domination throughout this region, though Russia's influence is still considerable.

The Rise of the Russian Empire

For thousands of years, the dominant people in the region were **nomadic pastoralists** who lived on the meat and milk provided by their herds of grazing animals, and used animal fiber to make yurts, rugs, and clothing. As possibly the first people to domesticate horses, their movements followed the changing seasons across the wide grasslands that stretch from the Black Sea to the Central Siberian Plateau. Pastoralists, possibly from the steppes north of the Black Sea, migrated over long distances and have left their genetic mark on people today living anywhere from Europe deep into Asia, as can be seen by DNA analysis. Studies of linguistics also show the reach of this migration: the Indo-European language family links most of the European languages with languages such as Hindi in India.

Towns arose as early as 5000 years ago in Central Asia, which had settled communities supported by irrigated croplands. These communities were enriched by their locations, which were central for trade along what became known as the **Silk Road**, that vast, ancient, interwoven ribbon of trading routes between China and

PHOTO ESSAY

5.8 Vulnerability to Climate Change in Russia and the Post-Soviet States

The region ranges from low to high levels of vulnerability to climate change. In Central Asia, vulnerability will increase as temperatures rise and glaciers in the Pamirs and other mountains melt. The dependence of so many Central Asians on agriculture makes them very sensitive to drought and water scarcity. Widespread poverty, combined with poorly developed disaster response systems, reduces the overall resilience to these and other disturbances.

THINKING GEOGRAPHICALLY

A Why would climate change cause rivers like the one in this photo to receive less water over the long term?

B How would a decline in Uzbekistan's cotton industry affect the country's economy?

C What major resource did the Aral Sea once provide in abundance?

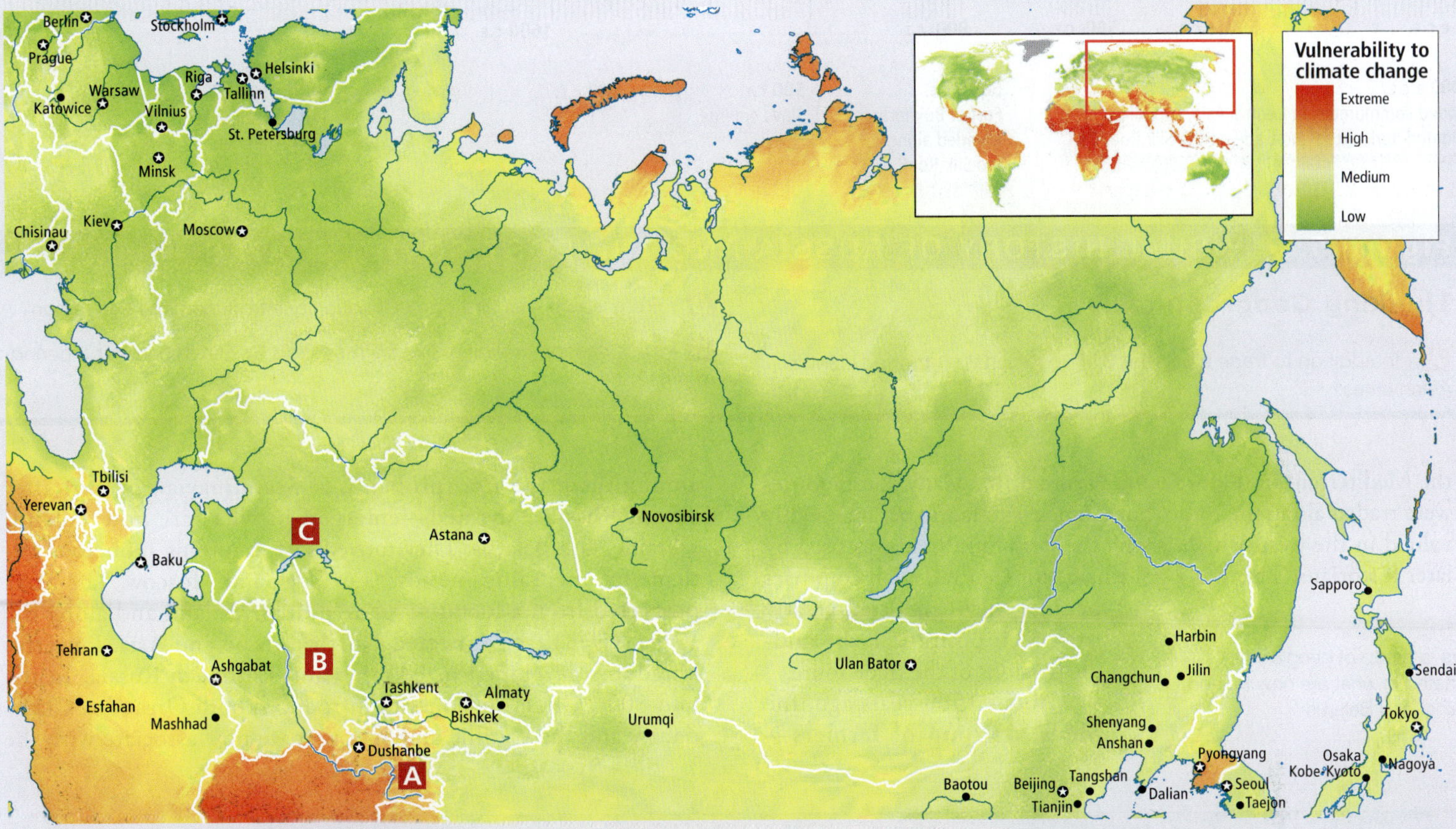

A A river fed by glacial melting in Tajikistan. Most of Central Asia depends on rivers for household water use and irrigation water. If these rivers receive less water from glaciers, there will not be enough water to supply people and their crops. [Martin Moos/Getty Images]

B An irrigated cotton field in Uzbekistan. Twenty-six percent of Uzbeks work in agriculture, and cotton is the country's leading farm export. If irrigation water were to become less available, the livelihoods of millions of Uzbeks would be threatened. [Andrew Peacock/Getty Images]

C A fishing boat in the Aral Sea. The boat is stranded in what used to be part of the Aral Sea, which has shrunk dramatically because the rivers that feed it have been diverted to irrigate crops. Even less water will reach the Aral Sea as glaciers continue melting. [VYACHESLAV OSELEDKO/AFP/Getty Images]

A The fort at Bukhara, Uzbekistan, founded in 500 B.C.E. along the Silk Road. [placchic/Moment/Getty Images]

B The Cyrillic alphabet, named after the Byzantine missionary St. Cyril (left), was introduced around 900 C.E. [Saints Cyril and Methodius, fresco © De Agostini Editore/Bridgeman Images]

C The summer palace of Czar Peter the Great in St. Pete[rsburg] built from 1714 to 1755. [Martin Child/Getty Images]

4000 B.C.E. — 500 B.C.E. — 500 C.E. — 900 C.E. — 1600 C.E. — 1700 C.E.

3000 B.C.E. Settled communities in Central Asia, supported by irrigated agriculture and trade on the Silk Road

500 B.C.E. Fort at Bukhara founded along the Silk Road

500 C.E. Slavs emerge

900 C.E.

1598–1689 Russian annexation of western Siberia

Figure 5.9 VISUAL HISTORY OF RUSSIA AND THE POST-SOVIET STATES

Thinking Geographically

A In addition to trade along the Silk Road, what supported Central Asian cities?

B From where was Orthodox Christianity introduced into this region?

C Why was the capital of Russia moved to St. Petersburg for a period of time?

the Mediterranean (**Figure 5.9A, Figure 5.10**). Many commodities were traded along the route, but Chinese silk fabric became highly valued in the west, which is why the term Silk Road was used by later historians. This was a golden age for Central Asian countries because they benefitted greatly from their position in the middle portions of the trade route.

Slavs a group of people who originated in what are now Poland, Ukraine, and Belarus

About 1500 years ago, the **Slavs**, a group of farmers—among them the *Rus* (probably of Scandinavian origin) after whom Russia is named—emerged in what are now Poland, Ukraine, and Belarus. They moved east, founding numerous settlements, including the towns of Kiev and Moscow. The Slavs prospered from a lucrative trade route over land and along the Volga River that connected North Europe and Asia. Powerful kingdoms developed in Ukraine and European Russia. In the mid-800s, Greek missionaries introduced both Orthodox Christianity and the Cyrillic alphabet that we now associate with the

Figure 5.10 The ancient Silk Road and related trade routes. Merchants who worked the Silk Road rarely traversed the entire distance. Instead, they moved back and forth along part of the road, trading with merchants to the east and west. [Research from: "Maps of the Silk Road," the Silk Road Project and the Stanford Program on International and Cross-Cultural Education, at http://virtuallabs.stanford.edu/silkroad/SilkRoad.html]

Poor men and women hauling freight along the Volga River in 3, as depicted by the Russian painter Ilya Repin. [Sovfoto/UIG Getty Images]

E A World War II–era poster of a Soviet soldier. [Galerie Bilderwelt/Getty Images]

F A journey to the International Space Station begins at Baikonur Cosmodrome in 2012. [Bill Ingalls/NASA/Getty Images]

1800 C.E. | 1900 C.E. | 2000 C.E.

1714–1755

1796–1914 Russian annexation of Central Asia and Caucasia

1873

1917 Bolshevik Revolution

1922–1953 Stalin era

1941–1945

1945–1991 Cold War

2012

D How does this painting convey the poverty in nineteenth-century Russia?

E How many Soviets were killed during World War II?

F How did the Cold War strain the USSR?

Russian language (Figure 5.9B). The Cyrillic alphabet is still used in six of the region's countries.

In the twelfth century, the Mongol armies of Genghis Khan conquered Ukraine and Russia. The **Mongols** were a nomadic pastoral people centered in East and Central Asia. Moscow's rulers became tax gatherers for the Mongols, dominating neighboring kingdoms and eventually growing powerful enough to challenge Mongol rule. In 1552, the Slavic ruler Ivan IV ("Ivan the Terrible") defeated the Mongols, marking the beginning of the Russian empire. St. Basil's Cathedral, a major landmark in Moscow's Red Square, commemorates the victory (**Figure 5.11**).

By 1600, Russians centered in Moscow had conquered many former Mongol territories, integrating them into the growing empire and extending it to the east, as shown in **Figure 5.12**. The first major non-Russian area to be annexed was western Siberia (1598–1689). Russian expansion into Siberia (and even into Pacific North America) resembled the spread of European colonialism around the world, as well as the future westward spread of the United States into contiguous but sparsely populated areas inhabited by indigenous people. Russian colonists took land and resources from the Siberian populations and treated the people of those cultures as inferior. So many laborers migrated from Russia to Siberia that by the eighteenth century they outnumbered the indigenous Siberians. By the mid-nineteenth century, Russia had also expanded its reach to the south in order to gain control of Central Asia's cotton crop, its major export. The building of railroads played an important role in the movement of people and goods.

Figure 5.11 Red Square, Moscow. Red Square is located in the center of Moscow. It is the site of one of the most recognized buildings in Russia, St. Basil's Cathedral (left), which was built between 1555 and 1561 to commemorate the defeat of the Mongols by Ivan the Terrible, the first Russian czar. Each of the colorful "onion domes" tops a chapel dedicated to an Orthodox saint. On the right is the Kremlin. Now the center of the Russian government, the Kremlin was originally built between 1482 and 1495 as a fortress, and it has been the residence of many czars. [MarinaDa/Shutterstock.com]

To the west, the Russian empire was intertwined with the politics of European royalties and empires. In the early nineteenth century, the French emperor Napoleon attempted to conquer Russia but underestimated its size and harsh winter climate. The failed attempt by Napoleon to conquer Russia actually enabled Russia to expand its empire into Poland and Finland (see Figure 5.12).

The Russian empire was ruled by a powerful monarch, the **czar**, who—along with a tiny aristocracy—lived in splendor, while the vast majority of the people lived short, brutal lives in poverty (see Figure 5.9C, D). Many Russians were *serfs*, who were legally bound to live on and farm land owned by an aristocrat. If the land was sold, the serfs were transferred with it. Serfdom was legally ended in the mid-nineteenth century, but the inequities of Russian society persisted into the twentieth century, fueling opposition to the czar.

The Communist Revolution and Its Aftermath

In 1917, Czar Nicholas II was overthrown in a revolution. What followed was a civil war between

Mongols nomadic pastoral people centered in East and Central Asia, who by the thirteenth century had established an empire that stretched from Europe to the Pacific

czar the ruler of the Russian empire

Figure 5.12 Russian imperial expansion, 1300–1945. A series of powerful rulers expanded Russia's holdings across Eurasia to the west and east. Expansion was particularly vigorous after 1700, when Russia acquired Siberia. Note that the centuries-long imperial expansion was reversed in 1991, when the Soviet Union was disbanded and the new post-Soviet states were created. [Research from: Robin Milner-Gulland with Nikolai Dejevsky, *Cultural Atlas of Russia and the Former Soviet Union*, rev. ed. (New York: Checkmark Books, 1998), pp. 56, 74, 128–129, 177]

rival factions with different ideas about how the revolution should proceed. Of those disparate groups, the **Bolsheviks** succeeded in gaining control. The Bolsheviks were inspired by the German revolutionary philosopher Karl Marx, whose writings explained the principles of **communism**. Marx criticized the societies of Europe as inherently flawed because they were dominated by **capitalists**—the wealthy minority who owned the factories, farms, businesses, and other means of production. Marx pointed out that the wealth of capitalists was actually created in large part by workers, who nonetheless remained poor because the capitalists underpaid them for their work. Under communism, workers were called upon to unite to overthrow governments that supported the capitalist system, take over the means of production, and establish an egalitarian society.

The Bolshevik leader Vladimir Lenin argued that the people of the former Russian empire needed a transition period in which to realize the ideals of communism. Accordingly, Lenin's Bolsheviks formed the **Communist Party**, which set up a powerful government based in Moscow. Lenin believed that the government should run the economy, taking land and resources from the wealthy and using them to benefit the common good.

Bolsheviks a faction of communists who came to power during the Russian Revolution

communism an ideology, based largely on the writings of the German Karl Marx, that calls on workers to overthrow capitalism and establish an egalitarian society where workers share what they produce

capitalists a wealthy minority that owns the majority of factories, farms, businesses, and other means of production

Communist Party the political organization that ruled the USSR from 1917 to 1991; other communist countries, such as China, North Korea, and Cuba, also have ruling communist parties

centrally planned economy an economic system in which the state owns all land and means of production, while government officials direct all economic activity, including the locating of factories, residences, and transportation infrastructure

The Stalin Era

In 1922, Lenin's successor Joseph Stalin reorganized the society and economy in a highly authoritarian style. His 31-year rule of the Soviet Union brought a mixture of brutality and revolutionary change. He tried to increase general prosperity through rapid industrialization made possible by a **centrally planned economy**. All economic activity was run by government officials in Moscow, and the state owned all real estate and means of production. The state determined where all factories, apartments, and transportation infrastructure would be located and managed all production, distribution, and pricing of products throughout the country. The idea was that under this system, the economy would grow more quickly, which would hasten the transition to the idealized communist state in which everyone shared equally.

The centerpiece of Stalin's vision was massive government investment in gigantic development projects, such as factories, dams, and chemical plants, some of which are still the largest of their kind in the world. To increase agricultural production, farmers had to join large, government-run collectives.

The Soviet strategy of government control initially met many of the expectations. It brought rapid economic development and higher standards of living. Schools were provided for previously uneducated poor and rural children, and education contributed to major technological and social advances. During the Great Depression of the 1930s, the Soviet Union's industrial productivity grew steadily, allowing for the development of a large military, even while the economies of capitalist countries stagnated.

However, the communist economic development model also had deep flaws. With production geared largely toward heavy industry (the manufacture of machines, transportation equipment, and armaments), the Soviet planners paid less attention to consumer goods and services that could have further improved living standards. Ultimately, the Soviet Union proved to be less innovative than noncommunist countries in western Europe and less able to develop advanced technologies to sustain economic growth over time.

The most destructive aspect of Stalin's rule, however, was his ruthless use of the secret police and mass executions to silence anyone who opposed him. Those who resisted were forced to work in prison labor camps, the so-called *gulags*, many of which were located in Siberia. Prisoners there mined for minerals and also built and worked in newly constructed industrial cities. Millions were executed or died from overwork under harsh conditions. Estimates vary greatly, but between 20 and 60 million people died as a result of Stalin's policies.

World War II and the Cold War

During World War II, the Soviet Union did more than any other country to defeat the armies of Nazi Germany (see Figure 5.9E). Hitler exhausted his war machine in his failed attempt to conquer the Soviet Union. In the process of defeating Germany on the Eastern Front, 23 million Soviets were killed, more than all the other European casualties combined. After the war, the Soviet Union was determined to erect a buffer of allied communist states in Central Europe that would be the battleground of a potential future war with Europe. Because they were unwilling to risk another war, the western allies ceded control of much of Central Europe to the Soviet Union while they busied themselves with reconstructing western Europe. The result was a politically and economically divided Europe—separated by what was called the "Iron Curtain"—and the **Cold War**, a global geopolitical rivalry that pitted the Soviet Union and its allies against the United States and its allies around the world for nearly 50 years (**Figure 5.13**).

A wide variety of economic and political problems led to the eventual collapse of the Soviet Union. Fundamental flaws in Soviet central planning led to innumerable inefficiencies and chronic mismanagement compared to most market economies. Mikhail Gorbachev, the last president of the USSR from 1985 to 1991, addressed these problems by beginning to permit a limited private market economy. Under an overall policy of **perestroika**, or "restructuring," government decision making was decentralized and private entrepreneurs were allowed to legally sell goods and services. Gorbachev also introduced policies known as **glasnost**, or "openness," which encouraged more transparency in the way government operated and more openness in society in general, which allowed for criticism of the government. However, many problems remained, such as spending on the military, including the Soviet space program, and support to Cold War allies around the world (see Figure 5.9F). Scarce

Cold War the geopolitical rivalry that pitted the United States and western Europe, who were espousing free market capitalism and democracy, against the USSR and its allies, who were promoting a centrally planned economy and a communist state

perestroika literally, "restructuring"; the restructuring of the Soviet economic system done in the 1980s in an attempt to revitalize the economy

glasnost literally, "openness"; the policies instituted in the 1980s under Mikhail Gorbachev that encouraged more transparency and openness in Soviet government and society

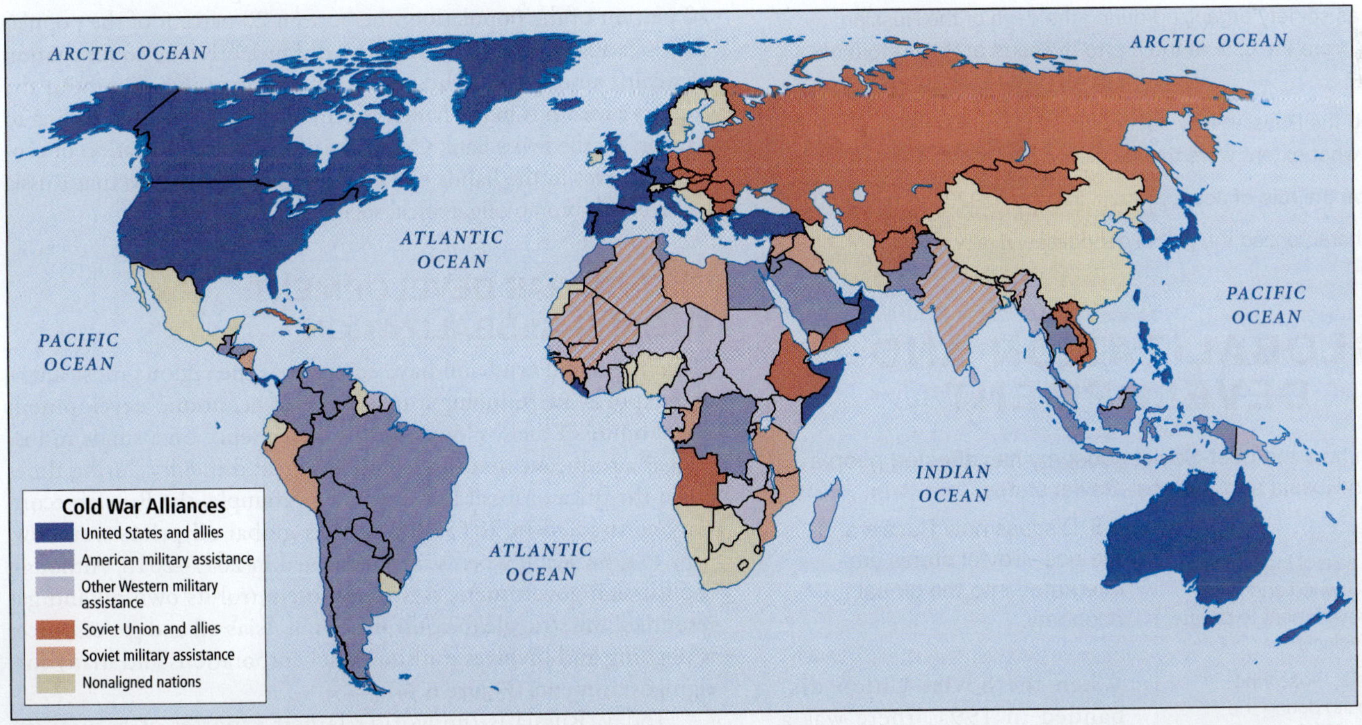

Figure 5.13 The Cold War in 1980. In the post–World War II contest between the Soviet Union and the United States, both sides enticed allies through economic and military aid. The group of countries militarily allied with the Soviet Union was known as the *Warsaw Pact*. Some countries remained nonaligned. [Research from: Clevelander, at http://en.wikipedia.org/wiki/Image:New_Cold_War_Map_1980.png]

resources were diverted away from much-needed economic and social development.

Soviet efforts to promote communism in less developed countries were also expensive, as was extensive political, economic, and military coercion in Central Europe and Central Asia. As problems multiplied, so did resistance to Soviet control, especially in Central Europe in the late 1980s. The Soviet policies of perestroika and glasnost also meant that Central European countries felt more comfortable enacting their own reforms and distancing themselves from the Soviet Union. Because Gorbachev, unlike his predecessors, did not resort to violence to control the countries of Central Europe, they were able to abandon communism and instead establish democracy. Eventually, this dissension spread to the republics that made up the Soviet Union. Communist hardliners attempted a coup against Gorbachev and his reforms in order to save the Soviet Union. The coup failed during a few tense and confusing days in 1991, which paved the way for the dissolution of the USSR.

A war in Afghanistan also helped to stretch the USSR past the breaking point. The Soviets feared that a strongly anticommunist Muslim movement in Afghanistan would influence nearby Central Asian Soviet republics to unite under Islam and rebel against the USSR. In response, the USSR backed an unpopular communist government in Afghanistan. This brought strong resistance from the Muslim fundamentalist mujahedeen movement, and in 1979, Russia launched a war to support the Afghan government. A bloody, decade-long conflict ensued, concluding with the Soviets being defeated by Afghan guerillas (see Chapter 8).

CHECK YOUR UNDERSTANDING

1. Discuss the location and importance of the Silk Road.
2. How was society organized during the days of the Russian Empire, from the czar at the top to the serfs at the bottom of society?
3. Who led the Russian Revolution of 1917, what were their goals, and to what extent were they achieved?
4. Describe the rule of Joseph Stalin from the 1920s to the 1950s.
5. What characterized the global geopolitical rivalry after World War II?

GLOBALIZATION AND DEVELOPMENT

5.4 Explain how the post-Soviet economy has affected people differently in Russia and the post-Soviet states.

5.5 Discuss how Russia and the post-Soviet states are integrated into the global economy.

privatization the sale of industries that were formerly owned and operated by the government to private companies or individuals

oligarchs in Russia, those who acquired great wealth during the privatization of Russia's resources and who use that wealth to exercise power

When the Soviet Union disbanded in 1991, there was a disorderly transition to a market economy that gave the advantage to a small number of well-placed individuals. The Soviet economy consisted almost entirely of industries owned and operated by the government. In the early 1990s, these industries were sold at extremely low prices, often to the high-level Soviet officials who had run them in the old centrally planned economy. The hope was that through this **privatization**, industries would become more competitive and efficient. Almost three decades later, there have been some gains in efficiency and productivity, mostly in a few key export-oriented industries. The most widespread effect of privatization was to make a small number of people fabulously rich very quickly. Many became what is known as **oligarchs**—wealthy and politically connected people who wield enormous, often clandestine, power. These oligarchs continue to exercise power in government and private enterprise.

The Russian government instituted a wide range of policies to try to revitalize the sluggish economy, moving away from the old centrally planned economy as quickly as possible. This transition to a market economy took place so rapidly that it was known as *shock therapy*. Today, about 65 percent of Russia's economy is held privately, a momentous change from 1991, when virtually 100 percent of the economy was run by the state.

Unfortunately, shock therapy came at a steep cost for many Russians. A major part of the reforms was the abandonment of government price controls that once kept goods affordable. Instead, in the new market economy, prices skyrocketed for the many goods that were in high demand but short supply. While a few people became rich, many lost their jobs and could barely pay for basic necessities such as food. Toward the end of the 1990s, as opportunities opened and competition developed, the supply of goods increased and prices fell. But even today, the effects of shock therapy are still being felt. For example, average incomes have grown by 45 percent since 1991, but almost all of this gain has gone to the wealthiest 20 percent of the population; the bottom 20 percent of the population is earning about half what they did in 1991. The concentration of wealth among the richest is extreme. Thirty-five percent of the country's wealth is in the hands of only 110 billionaires, according to a report by the Swiss bank Credit Suisse. Such concentration of economic power in the hands of a very few people confirms that Russia has evolved into an oligarchical society.

OIL AND GAS DEVELOPMENT: FUELING GLOBALIZATION

Natural gas and crude oil have emerged as the region's most lucrative exports, introducing a new wave of economic development based on fossil fuels. However, this dependence on a single industry has also meant that the region has great difficulty during times when the price of fossil fuel drops. For example, the Russian economy contracted in 2015–2016 when global oil prices were low, only to grow again when prices recovered in 2017–2018. Although the Russian government continues to control its own oil and gas resources, the struggle to control Central Asia's oil and gas resources is ongoing and involves multinational corporations and many foreign governments (**Figure 5.14**).

Today, Russia is the world's largest exporter of natural gas and the second-largest exporter of oil. Taxes on energy companies fund more than one-third of the federal budget. While this does guarantee that some of the revenues from the fossil fuel–based

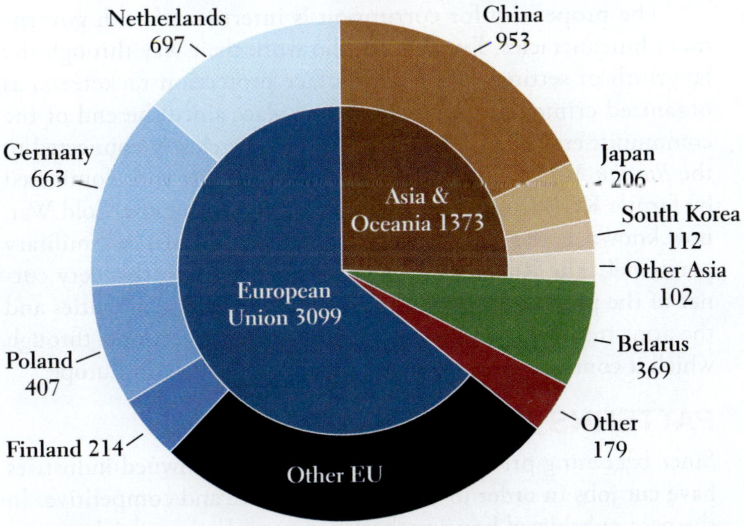

Russia's crude oil and condensate (liquefied natural gas) exports by destination, 2016 (in thousands of barrels per day)

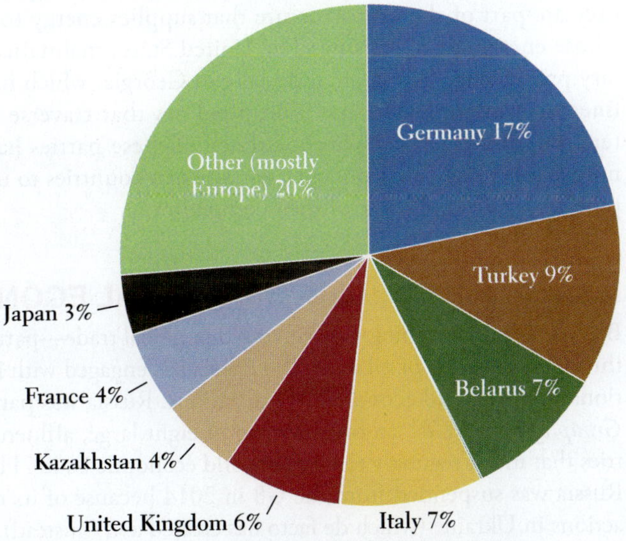

Share of Russia's natural gas exports by destination, 2016

Figure 5.14 Oil and natural gas in Russia and the post-Soviet states. [Research from: U.S. Department of Energy, Energy Information Administration, "Russia Country Analysis Brief," May 2008; and "Country Analysis Brief: Russia," October 31, 2017, at https://www.eia.gov/beta/international/analysis_includes/countries_long/Russia/russia.pdf]

industries benefit Russia's population, the remaining profits are concentrated in the hands of a few oligarchs.

As the charts in Figure 5.14 show, Russia's economic relationship with Europe is largely based on the oil and gas trade. The European Union receives about 45 percent of its natural gas from Russia, with Central Europe much more dependent than other parts of the EU. Russia is also very reliant on this trade; almost 90 percent of its exports go to Europe.

The trade is dominated by Gazprom, which is Russia's largest company as well as the world's largest gas company. Because Gazprom is state-controlled, Russia can use the company to promote its geopolitical interests. Russia occasionally tries to use its gas exports to influence the politics and economies of Ukraine, Belarus, and countries in Central Europe. Even before the war in Ukraine erupted in 2014 (see the section "Crisis in Ukraine"), Russia curtailed Ukraine's access to Gazprom's gas in a dispute over prices and as a response to Ukraine moving closer to the EU rather than being a Russian ally. Because gas pipelines that supply Europe run through Ukrainian territory, this caused gas shortages in parts of Europe. Another controversial pipeline across the Baltic Sea is currently under development, which would connect the German market to Russian gas. However, many in Europe are concerned about increased reliance on Russian energy and argue in favor of other sources of fossil fuels and, especially, renewable energy sources.

In Central Asia, a tug-of-war has evolved between Russia and foreign multinational energy corporations over the right to develop, transport, and sell Central Asian oil and gas resources. The Central Asian states do not have enough capital to develop the resources themselves, but they nevertheless benefit economically from selling the rights to exploit oil and gas to foreign companies.

Because Central Asia is landlocked, the only way to get oil and gas to world markets is via pipelines (see Figure 5.14), the routes of which have become a major source of dispute. Russia, the United States, the EU, Turkey, China, and India have all proposed or developed various routes. The oil pipelines from Kazakhstan and the gas pipelines from Turkmenistan are especially important, as they are part of the infrastructure that supplies energy to China, whose energy usage is rising. The United States maintains a military presence in the region, especially in Georgia, which has pipelines to Europe. Russia has built pipelines that traverse its own territory and reach into Europe. Each of these parties has made numerous efforts to encourage Central Asian countries to use their pipelines instead of those of their competitors.

RUSSIA, BRICS, AND THE GLOBAL ECONOMY

Because it was becoming more involved in global trade—particularly through its exports of oil and gas—Russia has engaged with international political and economic organizations. Russia was part of the *Group of Eight (G8)*, an organization of eight large, affluent countries that meets regularly to discuss world economic issues. However, Russia was suspended from the G8 in 2014 because of its military actions in Ukraine (which de facto has created a G7 instead). Russia also joined the World Trade Organization (WTO) in 2012 in order to promote greater trade with other countries.

Looking to form an economic counterweight to the United States and the EU, Russia has recently been meeting with Brazil, India, China, and South Africa as an association known as BRICS (see Chapter 1). In part because all five countries are large and populous and their economies have strengthened over time, the alliance is a symbol of the emerging power of non-Western countries. In fact, the BRICS countries now represent almost a third of the global economy. Of the five, Russia's economy may be the least diverse in the group—because it is dependent on fossil fuel exports—but it is also the wealthiest on a per capita basis.

The economic and political diversity of the BRICS countries has made it difficult for them to develop a united bloc that would challenge Western power. Some BRICS countries are also geographically isolated from each other, except for Russia and China, who share a border, as do China and India. This geographic connectedness is one reason why Russia recently became the largest oil supplier to China. Russia and China are working together to improve Russia's railroad infrastructure, for both passenger and cargo traffic.

Russia's pivot toward China, which is part of the BRICS goal of further collaboration, is also driven by Western economic sanctions against Russia and the realization that Russia and the West are no longer post–Cold War military strategic partners, but in competition with each other over the old Soviet-dominated territories (see the "Power and Politics" section). This has forced Russia to look for new markets and alliances, such as China. Russia has increased its military budget to support its political and economic objectives after reducing its military spending following the collapse of the Soviet Union.

INSTITUTIONALIZED CORRUPTION

The cost of doing business in Russia is affected by Russia's relatively high level of corruption. In 2017 the organization Transparency International rated Russia the most corrupt of the BRICS countries and the 38th most corrupt country in the world. Overall, the region does not score well in terms of corruption. Most Central Asian states are considered highly corrupt. Civil servants and the police often expect businesses to offer bribes to receive licenses and to avoid other regulatory hurdles. This affects both large corporations and small entrepreneurs.

The propensity for corruption is intertwined with government bureaucracies. Even those who work their way through the labyrinth of setting up a business face protection racketeers, as organized crime has become commonplace since the end of the communist era. Many oligarchs have become closely connected to the *Russian Mafia*, a highly organized criminal network dominated by former KGB (Russia's intelligence agency during the Cold War, now known as the Federal Security Service, or FSB) and military personnel. The Russian Mafia has influence in nearly every corner of the post-Soviet economy, especially in illegal activities and the arms trade, and has developed international networks through which it controls illicit activities abroad, particularly in Europe.

PATTERNS OF WORK

Since becoming privatized, most formerly state-owned industries have cut jobs in order to be more productive and competitive. In the past, subsidized housing, health care, and other social services were often provided along with the job. The new companies that have emerged rarely offer full benefits to employees. There is little job security because many small private firms appear quickly and often fail. Discrimination is also a problem, given the absence of equal opportunity laws. Job ads often contain wording such as "only attractive, skilled people under 30 need apply." The vignette below illustrates gender discrimination in the workplace.

VIGNETTE Consider the following employment situations of two Russian women. One sought work in a traditionally male occupation, and the other worked in a traditionally female occupation. But both recently faced gender stereotyping on the job.

Thirty-one-year-old Svetlana Medvedeva decided to become a boat pilot. After having completed the required education, she applied for a job but was turned down because of her gender. The company cited a 1974 law that bars women from 456 professions, such as miners and train engineers, which are deemed unsafe for women. Similar laws are on the books in other post-Soviet states. Rooted in gender stereotypes, these professional bans legally determine what is "men's work" and what isn't. After a lengthy legal battle, a court decided that Medvedeva was in fact discriminated against, and she eventually got another job in her profession. However, the law itself was not repealed.

Evgenia Magurina had worked seven years for the Russian airline Aeroflot. Then she and other crew members were measured and photographed by their employer. It had decided to link flight attendants' salaries to dress size. Magurina was deemed too big. Her pay was cut by 30 percent and she was scheduled for less desirable flights. She, too, sued. Incredibly, the airline argued in court that Magurina's "extra" weight cost more in fuel. They lost. So while Medvedeva, Magurina, and other women face discriminatory attitudes in the workplace, they have at least some recourse through the legal system. ■

Unemployment figures for the region vary widely. Official unemployment rates for 2017 averaged 5.5 percent in Russia but ranged from less than 2 percent in major urban centers such as Moscow and St. Petersburg, to over 10 percent in remote areas of Siberia and Caucasia. In Belarus, unemployment is officially only 1 percent; however, the old Soviet-style system of employing workers whether they are needed or not is still practiced there. Also, many people do not bother to register as unemployed because of the absence of unemployment benefits. The rate of **underemployment**, which measures the number of people who are working too few hours to make a living wage or who are highly trained but are working at low-paying jobs, is considerable in all countries.

The Informal Economy

Much of the economic activity in the region takes place in the informal sector (**Figure 5.15**). Workers in the informal economy tend to be young, unskilled adults; retirees on tiny pensions; and those with only a low level of education. Much informal work is in the area of personal services. This could be working on automobiles, selling consumer goods in a market stall, or driving an unlicensed taxi. What they all have in common is that they are not officially registered as a business. Increasingly, cell phones and the internet allow buyers and sellers of informal goods and services to be connected. The United Nations estimates that the informal sector makes up 36 percent of Russian employment. Russian men are more likely to work in the informal sector than women. In the Caucasian and Central Asian countries, the number employed in the informal sector is often above 50 percent. On the one hand, such large percentages indicate that people may actually be better off financially than official GDI per capita figures suggest. On the other hand, a large informal economy is commonly considered a sign of economic weakness and underdevelopment.

Despite the fact that the informal economy helps people survive, it is not popular with governments. Unregistered enterprises do not pay business and sales taxes and usually are so underfinanced that they tend not to grow into job-creating formal sector businesses. In many cases, informal businesspeople must pay protection money to local gangsters (the so-called Mafia tax) to keep from being reported to the authorities. On top of that, informal workers are not eligible for social security and public pensions.

Figure 5.15 Street vendor in Moscow, Russia. [NATALIA KOLESNIKOVA/AFP/Getty Images]

GEOGRAPHIC PATTERNS OF HUMAN WELL-BEING

As we have observed elsewhere, gross national income (GNI) per capita (cost of living adjusted) can be used as a general measure of well-being because it shows in a broad way how much people in a country have to spend on necessities. However, because GNI per capita is an average, it does not indicate how income is distributed or how accessible health care, education, and other social services are to most people. In this particular region, the intermediate Human Development Index (HDI) (**Figure 5.16**) ranks are in large part a holdover from the Soviet era, when communism kept wages relatively equal (though they were usually lower outside European Russia) and the strong social safety net provided basic health care, food, and housing for all. Since 1991, disparities in wealth have steadily increased within the Russian Federation and each of the post-Soviet states. As we have observed, in Russia a few people have become fabulously wealthy; their wealth has undoubtedly skewed the average figure. For many Russians, income is actually well below the country's $24,900 GNI per capita (PPP) because Russia has

underemployment the condition in which people are working too few hours to make a living wage or are highly trained but working at menial jobs

PHOTO ESSAY

5.16 Globalization and Development in Russia and the Post-Soviet States

The map depicts ranks on the Human Development Index (HDI), which is a calculation (based on adjusted real income, life expectancy, and educational attainment) of how adequately a country provides for the well-being of its citizens. It is possible to have a low income and a higher HDI rank if social services are sufficient or if there are no wide disparities of wealth. Russia, Belarus, and Kazakhstan rank very high. Two countries in Central Asia only rank in the medium category. But from a global perspective, Russia and the post-Soviet states rank slightly higher on HDI relative to their income level, which may be a leftover result of the Soviet emphasis on broad education and basic social services that improved human well-being during most of the twentieth century.

THINKING GEOGRAPHICALLY

A Where are many of the markets for Russian natural gas located?

B How is small-scale agriculture in Russia organized?

C How would you describe the urban–rural dimension of poverty and wealth?

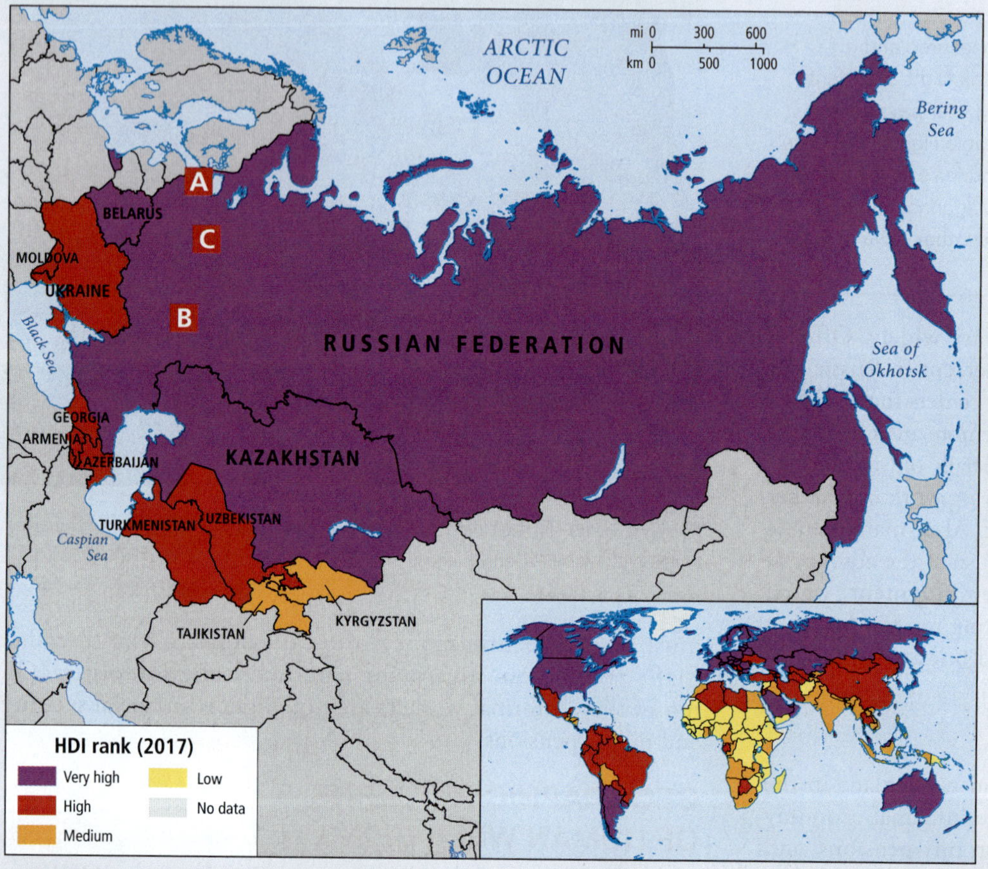

A Gazprom's new headquarters. The well-being of the Russian population is tied to the success of the country's fossil fuel industry. Gazprom, Russia's largest corporation, built a new headquarters in St. Petersburg. Its relocation from Moscow symbolizes the importance of the European market for Gazprom. So as not to disrupt the low-rise historic built environment of St. Petersburg, the new skyscraper is a few miles outside the city center. [Nikita_Maru/Shutterstock.com]

B Locally produced food. Many Russians supplement their diets, as well as income, by growing their own food. Many kitchen gardens are located at dachas, which are rural, seasonal homes (ranging from grand to very modest) within commuting distance of their urban dwellings. Growing garden crops is part of the dacha lifestyle. [Alan Gignoux/Alamy Stock Photo]

C The wealthy in Moscow. Many oligarchs who have become wealthy in the post-Soviet era live in Moscow and shop at the upscale GUM mall. In fact, Moscow recently ranked as having the second most billionaires of any city, after New York City. Contrast this lifestyle with the elderly women in Figure 5.15. [Anton Gvozdikov/Alamy Stock Photo]

changed from a society based on economic equality to one where the gap between poverty and wealth is similar to that of the United States. The difference between the lives of the economic elite in Moscow and those of people left behind in rural villages is extreme. Significant differences in income among the new states has become more apparent in the post-Soviet era. GNI per capita (PPP) in Belarus is $18,100 and in Kazakhstan is $23,400, while in Kyrgyzstan and Tajikistan it hovers around $3600.

FOOD PRODUCTION IN THE POST-SOVIET ERA

Across most of the region, agriculture is precarious at best, hampered by a short growing season and boggy soils. Because of Russia's harsh climates and rugged landforms, only 10 percent of its vast expanse is suitable for agriculture. In the Caucasus mountain zones, rainfall adequate for agriculture coincides with a relatively warm climate and long growing seasons. Together with Ukraine and European Russia, this area is the agricultural backbone of the region (**Figure 5.17**). The best soils are in an area stretching from Moscow south toward the Black and Caspian seas, and extending west to include much of Ukraine and Moldova. In Central Asia, irrigated agriculture is common where water is available and long growing seasons support cotton, fruit, and vegetables.

Changing Agricultural Production

During the 1990s, agriculture went into a general decline across the entire region. In most of the former Soviet Union, yields dropped by 30 to 40 percent compared to previous production levels. This was due mostly to the collapse of the subsidies and trade arrangements of the Soviet Union and its allies. Many large and highly mechanized collective farms were suddenly without access to equipment, fuel, or fertilizers. Since that time, agricultural output has rebounded. With the most productive soil of all the post-Soviet states, Ukraine now exports significant amounts of grain and sunflower-seed oil. Agriculture is one of the country's leading export sectors.

Russia's commercial agriculture is still characterized by many large and inefficient farms, most of which are now run by corporations. In the wake of its conflict with Ukraine, Russia stopped importing food from EU and other Western countries to retaliate for sanctions that those countries imposed on Russia. This has

Figure 5.17 Agriculture in Russia and the post-Soviet states. Agriculture in this part of the world has always been difficult, due to the cold climate, the short growing seasons, low soil fertility, and limited rainfall in all but a few places (Ukraine, Moldova, and Caucasia). [Research from: Robin Milner-Gulland with Nikolai Dejevsky, *Cultural Atlas of Russia and the Former Soviet Union*, rev. ed. (New York: Checkmark Books, 1998), pp. 186–187, 198–199, 204–205, 216–217]

redirected investment into the agricultural sector in Russia and recently increased domestic commercial production. At the same time, Russians are heavily dependent on "homegrown" food—small household plots, gardens, and family farms that produce about half of the total agricultural output on only 20 percent of the arable land (Figure 5.16B). Such small-scale production is often located on the outskirts of urban areas to be close to consumers. In Figure 5.17, the areas labeled "suburban cultivation" correspond to this type of farming. Growing and selling homegrown food is an important component of the informal sector, as discussed above. This production is typically done without agricultural chemicals and therefore is similar to organic, locally grown food popular among consumers in United States and Europe, although more out of economic necessity than a desire for a healthy lifestyle. Long distances across Russia also complicate the delivery of fresh produce, which is another reason for the reliance upon home gardening.

In Central Asia, agriculture has been reorganized with more emphasis on smaller, food-producing family farms. Grain and livestock farming (for meat, eggs, and wool) are important. Farmers in Georgia, favored with warm temperatures and abundant moisture from the Black and Caspian seas, grow citrus fruits and even bananas. Before 1991, most of the Soviet Union's citrus and tea came from Georgia, as did most of its grapes and wine. Georgia still exports some food to Russia, but because of political tensions between the two countries, Georgia is increasing the amount of food and wine it sells to other markets.

CHECK YOUR UNDERSTANDING

1. Who has gained the most from recent economic growth in Russia?
2. Discuss how Russia's natural resources are transforming the role of Russia in the global economy and in regional politics.
3. How does Central Asia's geographic situation affect the export of its energy resources?
4. Why is the informal sector of the economy important in this region?
5. Where does most of the agricultural output of the region come from? Compare that to the population distribution of the region. What are the resulting patterns of food exports and distribution?

POWER AND POLITICS

5.6 Describe how authoritarianism challenges democracy in the region.

5.7 Analyze past and current military conflicts in Russia and the post-Soviet states.

Since 1991, the politics in this region have remained very authoritarian and largely dominated by Russia, which sees itself as the successor to the Soviet Union. For almost two decades, Russia's political scene has been controlled by one person, Vladimir Putin, a former high official in the KGB. Vladimir Putin was first elected president in 2000. Prevented to run again by term limits, Putin assumed the office of prime minister in 2008 but continued to hold de facto power in Russia. Now back as president, he will most likely be in charge until at least 2024.

The outcome of the last presidential election in 2018 was hardly in doubt, due to the lack of other serious contenders for the presidency—some had been jailed—and Putin's dominance in state-controlled media reporting. People in hardscrabble industrial cities, small towns, and rural areas, along with the elderly, view Putin as a force for stability, which is favored by many, considering the political turbulence of Russia during the last three decades. Putin's reelection also has larger global geopolitical ramifications, as his annexation of Crimea and aggressive policies toward Ukraine have stirred up nationalist sentiment in Russia, which has solidified his popularity and grip on power, while the relationship between Russia and the West has been frosty.

While criticism of the government is formally allowed, most people (correctly) fear that such criticism would be met with retribution from the *siloviki* (literally "men of force," many of whom have military or intelligence backgrounds) who control many governing institutions. Political analysts agree that the siloviki have become more prominent within the Russian political structure since the start of the Ukrainian war in 2014. Journalists and others who document corruption or speak out against the government have been killed and jailed. Even Russian critics living abroad have been targeted, including poison attacks in England that have been traced back to the Russian government.

Authoritarianism will likely hold sway in the future as well. One reason is that Russia's current constitution, written in 1993, gives sweeping powers to the president, which will mean that a strong, and likely authoritarian, leader will succeed Putin. Many scholars of this region have also noted that Russia's geography, featuring a vast territory and few mountain or other barriers to prevent invasion by outsiders, favors a single powerful authoritarian ruler with a strong military. Others argue that the presence of authoritarian rulers throughout Russian history has created a political culture in which strong and sometimes brutal leaders thrive.

Elsewhere in the region, authoritarianism and limited political freedoms are also the norm. As evident on the map in **Figure 5.18**, most countries rank as authoritarian and have leaders who are unwilling to share power or allow criticism. Elections often involve systematically intimidating voters and arresting the political opposition. Not only is much of the region authoritarian, but its ratings on the democracy index have worsened over the last decade.

A few of the post-Soviet states have transitioned to somewhat greater political freedoms. A series of so-called color revolutions took place between 2003 and 2005 in Georgia (the Rose Revolution), Ukraine (the Orange Revolution), and Kyrgyzstan (the Tulip Revolution). While these countries hold elections and tolerate some freedom of expression, they retain many forms of authoritarian control and practices that constrain political freedoms. All three are labeled "hybrid regimes" in Figure 5.18. Much like the outcome of the Arab Spring (see Chapter 6), establishing greater democracy by popular protest has proven to be hard.

CRISIS IN UKRAINE

In 2014 in Ukraine, one of the most challenging crises in this region since the fall of the Soviet Union emerged. The crisis has positioned Russia against the EU and the United States in a seeming throwback to the days of the Cold War.

PHOTO ESSAY

5.18 Power and Politics in Russia and the Post-Soviet States

Since World War II, most conflicts in this region have been in Caucasia and Central Asia. Most took place just before or shortly after the breakup of the Soviet Union. Since then, popular movements pushing for more political freedoms have emerged in several countries, but only with modest success. Violent conflict continues to be a major source of political instability in parts of the region.

THINKING GEOGRAPHICALLY

A Why has Abkhazia been a conflict zone from the time of Soviet demise until today?

B How do proximity to Russia and the geography of ethnicity and language in Ukraine correlate with conflict?

C How does Russia's unwillingness to let Chechnya secede relate to fossil fuel resources?

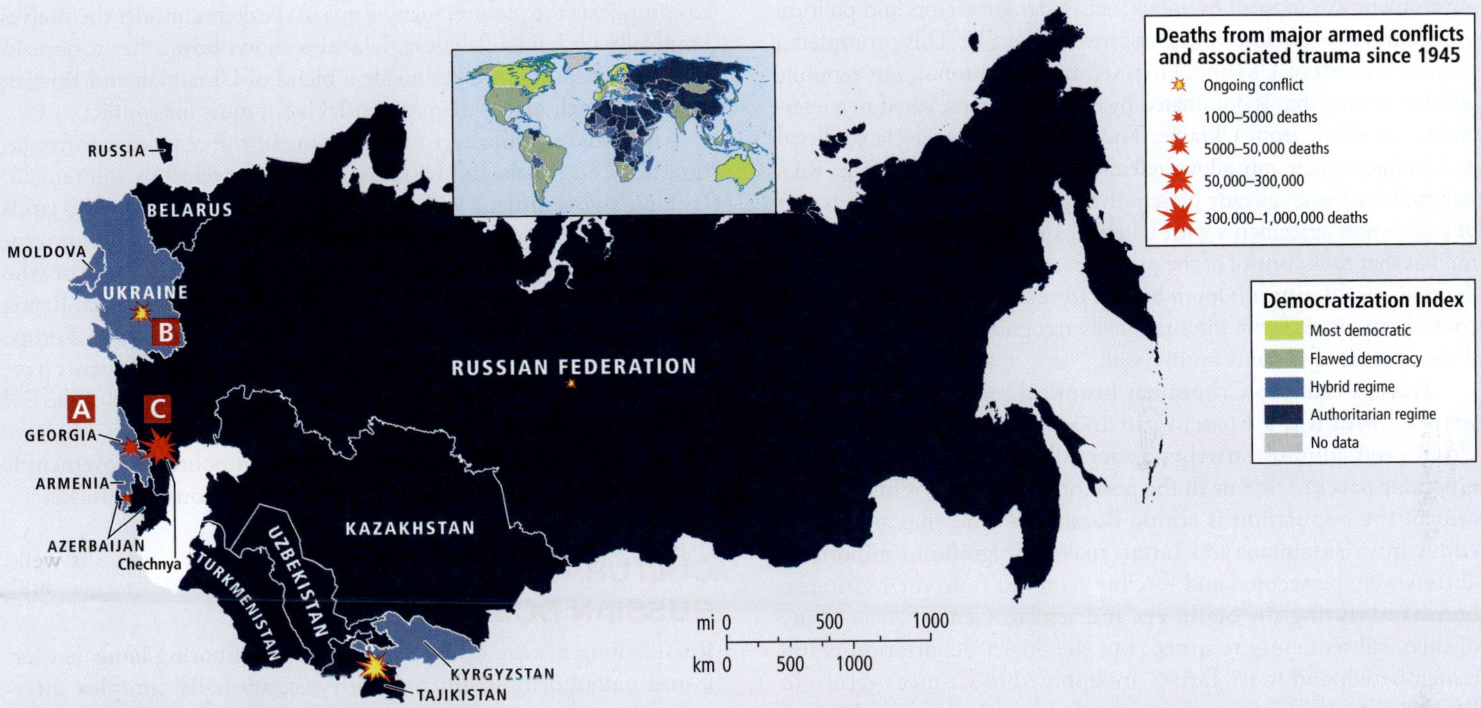

A Abkhazian elections. Abkhazia is a small territory in northwestern Caucasus that broke away from Georgia during the 1990s. However, few countries except Russia recognize it as an independent state. In the picture, Abkhazians fly their national flag during an election campaign; the Abkhazian government has close ties with Russia. [ITAR-TASS News Agency/Alamy Stock Photo]

B Pro-Russian separatist rebels on a road near Donetsk in eastern Ukraine. Pro-Russian rebels and the Ukrainian army have been engaged in conflict since 2014. The rebels fly a flag that represents the Russian Orthodox religion, which indicates that ethnic and religious identity are closely intertwined. [VASILY MAXIMOV/AFP/Getty Images]

C Chechen police and military in a show of force in Grozny, Chechnya. Chechen rebels have fought two unsuccessful wars for independence from Russia and have conducted terror campaigns against Russia since the fall of the Soviet Union. Political instability in Caucasia is one of the major obstacles to the expansion of political freedoms in this region. [STR/AFP/Getty Images]

Ukraine is divided in terms of its geographic orientation. Areas to the west, which border Central Europe, tend to favor closer ties with Europe, while areas to the east, where many of Ukraine's Russian-speaking minority reside, lean toward Russia. The government at the time decided against strengthening the economic ties between Ukraine and the EU, resulting in a wave of protests in Ukraine's capital of Kiev. Like the color revolutions, these protests were supported by U.S. and European governments and pro-democracy NGOs. The government was toppled by an alliance of demonstrators and political groups in favor of closer integration with the EU. This prompted a reaction in eastern Ukraine. First, Crimea, an autonomous republic within Ukraine that is dominated by ethnic Russians, voted in a referendum to secede from Ukraine. The referendum was declared illegal by Ukraine, which only allows referendums at the national scale. Russian military forces, already present in large numbers in Crimea as part of post-Soviet agreements with Ukraine, supported local pro-Russian militias that took control of the province, and the Russian government then annexed Crimea (**Figure 5.19**). However, this de facto Russian control of Crimea is not internationally recognized, which is why it is shown as part of Ukraine in this book.

Russia's claim to Crimea has historical roots. For almost 200 years Crimea was a Russian province. During the Soviet days, Crimea was administratively transferred to Ukraine, which is why it became part of Ukraine in the postcommunist era. While 58 percent of the population is ethnic Russian, Crimea has a complex ethnic mix; Ukrainians and Tartars make up significant minorities. Tartars were persecuted and forcibly removed from their Crimean homeland during the Stalin era and sent to Central Asia. Many of them subsequently returned, but the Soviet deportation is still remembered, and most Tartars are opposed to Crimea's return to Russian control. Russia's annexation of Crimea is, in part, also geopolitical. Control of the large naval base in the city of Sevastopol gives Russia better access to the Black Sea.

In addition to events in Crimea, administrative and police buildings in eastern Ukraine were taken over by local pro-Russian forces. Although Russia denies having done so, most observers agree that the Russian military covertly assisted the rebels. Figure 5.19 shows the territories in eastern Ukraine that are held by the rebels as self-proclaimed "people's republics" independent from Ukraine. The areas have large ethnic Russian populations that support closer ties with Russia. While the rebel-held territories, called the Donetsk Basin by geographers, make up only a small part of Ukraine, it is a populous and heavily industrialized area. Its economy is oriented toward Russia and most people's first language is Russian. (Although to complicate the picture, many Russian speakers consider themselves ethnically Ukrainian!) Figure 5.19 also shows how other regions in eastern and southern Ukraine are a blend of Ukrainian and Russian cultures, which makes them potential flashpoints for conflict.

Regardless of the eventual outcome of the ongoing confrontation between Russia and Ukraine, underlying tensions will remain. Ukraine's government will be challenged to balance the demands of citizens in the western parts of the country, many of whom want closer ties with the EU, with those of ethnic Russians in the east who want closer ties and possibly even political union with Russia. Russia will have to weigh the economic benefits of its gas trade with Europe and the cost of economic sanctions that Western governments have imposed, against its long-term opposition to the expansion of the EU into Ukraine. Additionally, the EU and the United States will have to decide whether supporting an anti-Russian political movement is worth risking a further deterioration of their relations with Russia.

CULTURAL DIVERSITY AND RUSSIAN DOMINATION

Russia's long history of expansion into neighboring lands has left it and neighboring countries with exceptionally complex internal political geographies. As the Russian czars and then the Soviets pushed the borders of Russia eastward toward the Pacific Ocean over the past 500 years, they conquered a number of non-Russian areas. Russians are now the main inhabitants of many of those areas, including today's Ukraine and Kazakhstan. Other territories have been designated semiautonomous *internal republics*

Figure 5.19 The geography of conflict and ethnicity in Ukraine. Language and ethnicity by *oblast* (Ukraine's main administrative regions). Pro-Russian forces currently control the eastern portion of Ukraine and call it the Luhansk and Donetsk People's Republics. Along with the Crimean Peninsula, now annexed by Russia, this is an area where the Russian-speaking population of Ukraine is most numerous.

(**Figure 5.20**) within Russia and have significant ethnic minority populations that are descended from indigenous people or trace their origins to long-ago migrations from Germany, Turkey, or Persia.

Both the czars and the USSR had a policy of **Russification**, whereby large numbers of ethnic Russian migrants were settled in non-Russian ethnic areas and given the best jobs and most powerful positions in regional governments. The goal was to force potentially rebellious regional minorities to conform to the state's goals. However, minorities have resisted Russification as a way to enhance local ethnic identities. Today, the reversed process—de-Russification—is happening especially in Central Asia, where the Russian language and general Russian influence are discouraged (see the vignette in "Urban Central Asia and Caucasia"). For example, Russians made up approximately 37 percent of the population in Kazakhstan at the end of the Soviet era. Over time, many have migrated to Russia. Today, they number only 20 percent.

THE CONFLICT IN CHECHNYA

Shortly after the breakup of the Soviet Union, several internal republics demanded more autonomy, and two of them, Tatarstan and Chechnya, declared outright independence. Tatarstan has since been placated with more economic and political autonomy, and as an enclave surrounded by other parts of the Russian Federation, its independence is unrealistic (see Figure 5.20). However, Chechnya's stronger resistance to Moscow's authority led to the worst bloodshed of the post-Soviet era.

Chechnya, located on the fertile northern flanks of the Caucasus Mountains (see the inset map in Figure 5.20), is home to 1.3 million people. Partially in response to Russian oppression, the Chechens converted from Orthodox Christianity to the Sunni branch of Islam in the 1700s. Since then, Islam has served as an important symbol of Chechen identity and as an emblem of resistance to Russia, which annexed Chechnya in the nineteenth century.

In 1942, during World War II, as the Germans invaded Russia, a group of Chechen rebels simultaneously waged a guerilla war against the Soviets. Near the end of the war, Stalin exacted his revenge by deporting the majority of the Chechen population (as many as 500,000 people) to Kazakhstan and Siberia. There they were held in concentration camps, and many died of starvation. The Chechens were finally allowed to return to their villages in 1957, but they were portrayed as traitors to the Soviet Union. More than a dozen ethnic groups suffered such deportations during Stalin's rule, including the Crimean Tatars as discussed earlier. They were

Russification the czarist and Soviet policy of encouraging ethnic Russians to settle in non-Russian areas as a way to assert political control

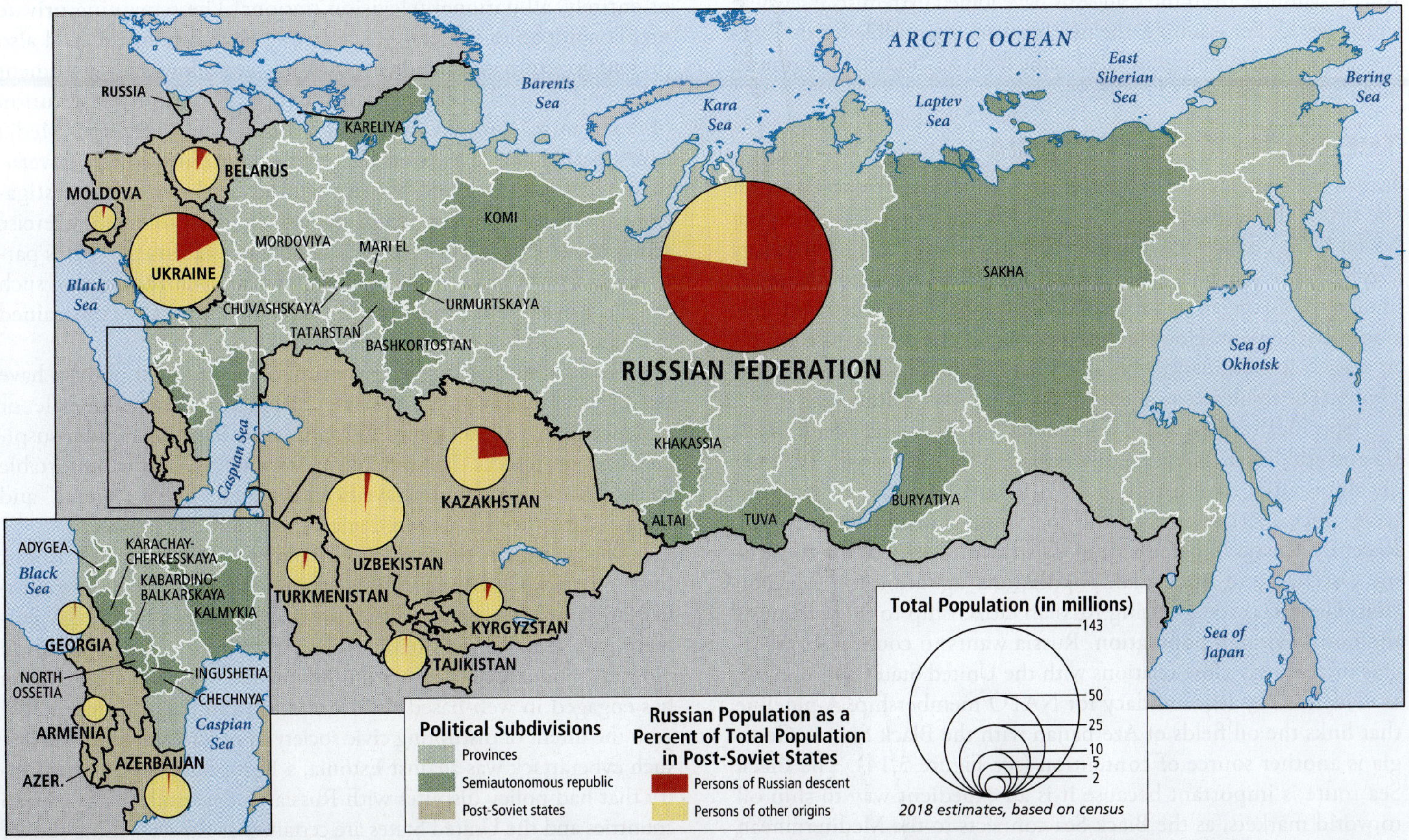

Figure 5.20 Ethnic character of Russia and percentage of Russians in the post-Soviet states. Russia, with all of its internal republics and administrative regions, is formally called the Russian Federation. The ethnic character of many internal republics was changed during the Soviet era, so Russians now form significant minorities there. The pie charts for each country indicate the percentages of Russians in the post-Soviet states. According to the latest census (2010), the ethnic makeup of Russia is 77.7 percent Russian, 3.7 percent Tatar, 1.4 percent Ukrainian, 1.0 percent Chuvash, 1.0 percent Chechen, and 10.2 percent other. [Research from: "Ethnic Groups," *The World Factbook*, Central Intelligence Agency, at https://www.cia.gov/library/publications/the-world-factbook/geos/rs.html]

perceived as a potential threat to the state, and forced migration was also a useful tool to colonize remote and sparsely populated territories across the Soviet Union.

In 1991, as the Soviet Union was dissolving, Chechnya declared itself an independent state. Russia saw this as a dangerous precedent that could spark similar demands by other cultural enclaves throughout its territory. Russia also wished to retain the agricultural and oil resources of the Caucasus; Chechnya is a potential route to move oil and gas to Europe from Central Asia. Russia responded to the Chechen insurgency with bombing raids and other military operations that killed tens of thousands of people and created 250,000 refugees.

Chechen guerrillas have largely given up, most Russian combat troops have been pulled out of Chechnya, and Russia has begun making substantial investments in rebuilding the capital city of Grozny. While this shift is a welcome change, some Chechen rebels have continued their struggle by carrying out brutal terrorist attacks, many of which now take place in Moscow and other areas outside Chechnya. Chechnya remains highly militarized (see Figure 5.18C), and the ongoing conflict there highlights Russia's inability to peacefully address internal political dissent. The terrorist attacks have fueled Russian xenophobia against people from the Caucasus region, which has made a compromise about the future status of Chechnya hard to achieve. The conflict has even been linked to global terrorism, as Muslims from the Caucasus have joined extremists elsewhere in the world. For example, the two brothers responsible for the Boston Marathon bombings in 2013 came from a Chechen background and had been influenced by Islamic radicalism.

THE CONFLICT IN GEORGIA

Just to the south of Chechnya lies Georgia. The difference between the two is that Georgia became an independent country when the Soviet Union dissolved. After a long history of Russian dominance, Georgia now wants to distance itself from Russia. There is support among the Georgian population and government for greater integration with the West. However, the geopolitical goal of Putin's Russia is to reassert its dominance over areas that once were part of the Soviet Union. The result is a local conflict with global implications.

Specifically, Russia and Georgia have sparred over South Ossetia and Abkhazia. These are two regions within Georgia, but they are dominated by ethnic groups different from Georgians who have, since 1991, expressed the goal of separation from Georgia. Recently Russia has stepped up its strategic support for the ethnic Ossetian and Abkhazian populations' agitation for secession from Georgia, even granting Russian citizenship to 90 percent of the non-Georgian population. Russia wants to counteract Georgia's increasingly close relations with the United States and the EU, as evidenced by its candidacy for NATO membership. A pipeline that links the oil fields of Azerbaijan with the Black Sea via Georgia is another source of contention (see Figure 5.14). The Black Sea route is important because it is an expedient way to ship oil to world markets, as the Black Sea connects to the Mediterranean Sea and, ultimately, to the Atlantic Ocean. Russia would like to enhance its control over the oil resources of the Caspian Sea region by routing all pipelines through Russian-controlled territory.

In 2008, Georgia's military, attempting to gain control over rebelling parts of South Ossetia, engaged with the Russian army, which supported the rebels, saying that they had the right to protect (recent) Russian citizens. Today, the Georgian government does not control Abkhazia and South Ossetia and it is not clear whether the two will break away and become independent countries, become provinces within the Russian Federation, or be satisfied by offers of more autonomy within Georgia's federal structure. This ambiguous geopolitical situation has been ongoing for a decade. Therefore it is referred to as a *frozen conflict*. It is also similar to the Ukrainian/Crimean conflict, as Russia directly or indirectly controls territories just beyond its internationally recognized borders.

THE MEDIA LANDSCAPE

During the Soviet era, all media were under government control. There was no free press, and public criticism of the government was a punishable offense. Between 1991 and the early 2000s, the communications industry was a center of privatization, and several media tycoons emerged to challenge the authorities. Privately owned newspapers and television stations regularly criticized the policies of various leaders of Russia and the other states. It appeared that a free press was developing.

Since Vladimir Putin's rise to power, however, critical analysis of the government has become rare throughout Russia. The reasons are manifold. Virtually all television channels present viewers with state-approved information. The government has used direct ownership to do this. The Russian government owns, partially or entirely, all national television stations. The remaining private media companies typically have strong government ties and also present government-friendly news. New laws allow the government to extend control over media through, for example, prosecution of "extremist" content, which is a loosely defined term. Media investigating corruption may be charged with insulting a governmental official. To avoid politically motivated criminal investigations, nominally independent media have been forced to exercise self-censorship in their coverage of controversial issues. This is particularly true of reporting about geopolitically sensitive places, such as Chechnya and Crimea, where media are especially constrained and must conform to the government's position.

Russian journalists openly critical of government policies have been treated to various forms of punishment: censorship, exile, or violence. Since 2000, about 200 journalists have died under suspicious circumstances. Foreign journalists and NGOs uncomfortable to the Putin government have been declared "foreign agents" and deported, or they have been denied entry in the first place.

Today, much information circulates on the internet. About three-quarters of the Russian population has internet access (the numbers are similar or lower in other post-Soviet states), but the government uses selective website blocking and charges against opposition bloggers much like it does with other media. Most famously, Russia has engaged in web-based disinformation campaigns, presumably with the intent of disrupting civic society in other countries. The first such cyberattack was against Estonia, a European post-Soviet country that had policy disputes with Russia. Today, multiple European countries and the United States are certain that Russia is using similar tactics against their political systems.

CHECK YOUR UNDERSTANDING

1. How is this region's long history of authoritarianism still manifested today?

2. How is the conflict in Ukraine related to the internal ethnic geography of the country?
3. How has Russia's history of expansion into neighboring lands influenced the political geography of the region? What are the social, political, and economic impacts of Russification policies?
4. What impact did Vladimir Putin's rise to power have on the media in Russia?
5. What are the causes of conflict in Caucasia (e.g., Georgia and Chechnya)?

URBANIZATION

5.8 Discuss how the region's geography and communist-era planning affected urbanization.

5.9 Describe the primate city phenomenon in Russia and the post-Soviet states.

Even though Russia and the post-Soviet states have the largest landmass of any world region, including immense areas of lightly populated or uninhabited land, this is still largely a region of cities. Especially Ukraine, Belarus, and Russia are highly urbanized. There, 70 percent or more of the population lives in urban areas. The Caucasian republics of Georgia, Armenia, and Azerbaijan are somewhat less urbanized (around 60 percent), and Central Asia is only 48 percent urban, which is below the world average of 55 percent.

Russia's capital city of Moscow, with approximately 17 million people in the urban region, is a primate city (see Chapter 3 for the definition of primate city), as are most capital cities in this region (**Figure 5.21A**). The dominance of primate cities reflects the importance of politics and government in determining where economic development took place during the Soviet and post-Soviet eras.

Moscow is shaped by both its history and its current status as a fast-growing metropolis. More than ever, there is a need for effective transportation to move people around its 1000 square miles. Communist-era plans for the city included a network of wide avenues and roads. Over time, a series of ring roads have been added further and further out from the city center. The city's subway system is the fourth most extensive in the world, with about 200 stations. But even so, traffic congestion is severe. Another outcome of Moscow's rapid growth is housing shortages. Due to high demand, the cost of housing has increased so dramatically that Moscow is now on par with famously expensive cities like New York and Tokyo. Shortages are also exacerbated by the ability of wealthy people to buy apartments and refurbish them into one luxurious dwelling. This is a form of gentrification similar to the process taking place in North American cities (Chapter 2). Moscow is also modernizing with new residential high rises and hotels constantly adding to the city's skyline. A large number of historic buildings have disappeared recently. Even if they are nominally protected landmarks, political corruption allows for development to go unchecked.

Urban life for most people is very different though. Russia's inequality is evident in its urban landscape. Giant apartment blocks built for industrial workers are a legacy of communist-era central planning. They dominate most cities, especially on the fringes of the older, pre-Soviet urban cores (Figure 5.21B).

Outside Moscow, St. Petersburg, and a few other cities that are growing due to energy revenues (see Figure 5.21C), many urban areas are deteriorating. Cramped, drab apartments, badly in need of repair, with shared kitchens and bathrooms, are common. They suffer from the absence of funds for maintenance because economic and population growth has been slow or nonexistent. In some cases, population decline is a cause of the deteriorating quality of housing (see Figure 5.21B).

URBAN LEGACIES OF SOVIET REGIONAL ECONOMIC DEVELOPMENT

One of the more difficult legacies of the Soviet era involves the huge industrial cities in the lightly inhabited expanses of Siberia and the Pacific coast. Founded originally as trading posts and fortress towns during the pre-Soviet Russian empire, for several reasons these towns became the focus of Soviet planners. The interior was home to important mineral resources, so when the Soviet Union industrialized, it was logical to plan for urban growth near those resources. Also, the Soviet Union was concerned about potential conflict with the West; therefore, strategic industries should be developed in protected locations. Finally, the Soviet state was also eager to solidify its control of the vast interior east of Moscow. These places became home to major extractive industries and defense-related infrastructure, but further growth was always hampered by the vast distances and challenging physical geography that separate them from the main population centers in the west. Today, nearly 90 percent of Siberia's people are concentrated in the few large urban areas. Life in these cities is very different from the more traditional rural lifestyles that characterized this region only a few generations ago. The map in **Figure 5.22** shows a string of important cities eastward from Moscow and also shows the connection between mineral resources and industrialization.

Because the region's rivers run mainly north–south, there is a need for land transportation systems, such as railroads and highways, that run east to west to connect these cities with the west. The harsh climate of the region makes this infrastructure exceedingly expensive to build and maintain, and much has fallen into disrepair in the post-Soviet era. Only one poorly paved road and one main rail line—the Trans-Siberian Railway (Figure 5.22)—run the full east–west length through the southern tier of Siberia, connecting Moscow with Vladivostok, the main port city on the Russian Pacific coast.

URBAN CENTRAL ASIA AND CAUCASIA

While Central Asia has historically had a relatively low rate of urbanization, the recent expansion of the energy industries has fueled new urban growth. Cities here have also grown because the region's population growth is significantly higher than Russia's. Tashkent, the capital of Uzbekistan, has 2.3 million people, making it the fourth-largest city in the region. Almaty, the former capital of Kazakhstan, is the next-largest city in Central Asia, at 1.5 million people, followed by Astana, the new capital of Kazakhstan, which has 770,000 people. The largest city of Caucasia is Baku, the capital of Azerbaijan, which is located on the shores of the Caspian Sea. With 2.9 million residents, it is a boom town that combines an ancient urban heritage with energy-driven modernization.

5.21 Urbanization in Russia and the Post-Soviet States

An uneven pattern of urbanization is developing in this region. In several countries, the largest cities are growing rapidly, as they are centers of new development and globalization. Elsewhere, however, many cities are struggling economically and even shrinking in terms of population because they are less able to attract new investment.

THINKING GEOGRAPHICALLY

A Why is Moscow growing so rapidly?

B Where in the post-Soviet city would you find large-scale apartment blocks?

C What is responsible for the growth of Astana?

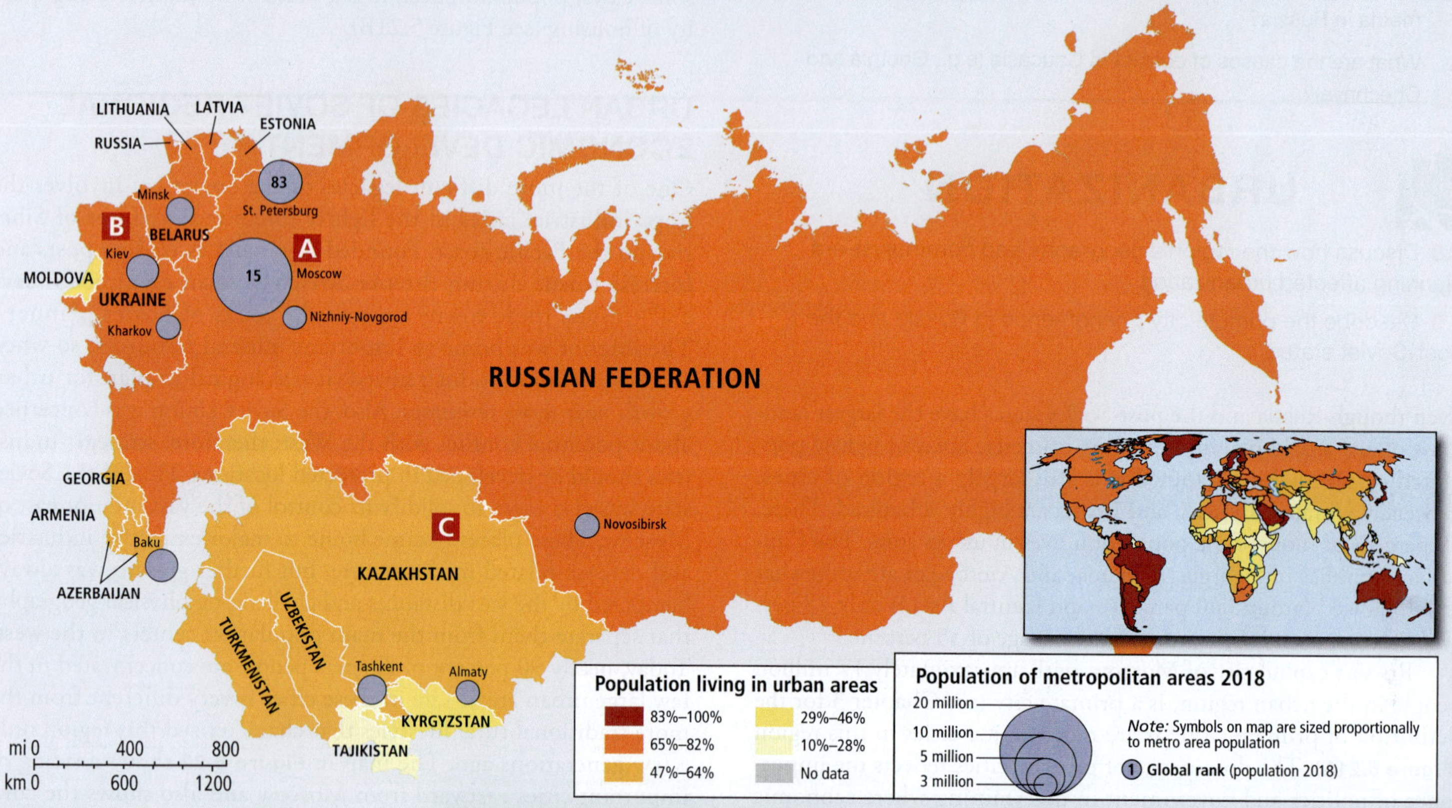

A The region's primate city—Moscow. Moscow's population of 17 million people is larger than the next six largest cities combined. It produces more than 20 percent of Russia's GDP, a percentage that continues to grow because of the city's role as the headquarters of multinational corporations operating in Russia. Moscow's population is likely to increase further, and there are many undocumented migrants in the city who are undercounted by the Russian census. [Mordolff/iStock/Getty Images]

B An apartment building in Kiev, Ukraine. Large-scale homogeneous apartment blocks were built during the Soviet era, usually located outside the city center. Such neighborhoods are typically home to a few thousand people living in a cluster of 10–20-story apartment buildings. Basic retail and government services are also provided in the neighborhood. However, many of the properties are run down today and are considered undesirable. [Giuseppe Anello/Alamy Stock Photo]

C Astana, capital of Kazakhstan. Astana has added massive amounts of new development, which has largely been supported by the increased exploitation of Kazakhstan's oil resources by local and international companies. The population of Astana has more than doubled after it was designated the new capital in 1997 (the old capital was Almaty). [Michael Runkel/robertharding/Getty Images]

Figure 5.22 Principal industrial areas and land transport routes of Russia and the post-Soviet states. The region's industrial, mining, and transportation infrastructure is concentrated in European Russia and adjacent areas. The main trunk of the Trans-Siberian Railway and its spurs link industrial and mining centers all the way to the Pacific, but the frequency of these centers decreases with distance from European Russia. [Research from: Robin Milner-Gulland with Nikolai Dejevsky, *Cultural Atlas of Russia and the Former Soviet Union*, rev. ed. (New York: Checkmark Books, 1998), pp. 186–187, 198–199, 204–205, 216–217; and http://www.travelcenter.com.au/russia/images/trans-sib-map-v3.jpg]

VIGNETTE Yernar Zharkeshov is a 24-year-old university-educated man who is moving home to Kazakhstan after having spent time in Britain and Singapore. He is back in his homeland seeking new opportunities, which are increasing because of its extensive oil exports. The natural choice for him is Astana, a city that was designated the capital of Kazakhstan in 1997. Yernar quickly found a job as a government economist; he is just one of thousands of young people who have migrated to the growing capital.

The idea of Astana came from Nursultan Nazarbayev, who was president of Kazakhstan since its independence from the Soviet Union in 1991 until his retirement in 2019. Without much public debate, he moved the capital from Almaty in the south to the remote and sparsely populated steppe in the north that is known for its harsh climate. Here he realized his vision for a grand and impressive new capital, built more or less from scratch. Astana is meant to symbolize a new, forward-looking Kazakhstan, and Nazarbayev has spared no expense. The city is full of stunning buildings, either designed by world-renowned architects or by Nazarbayev himself (see Figure 5.21C).

Like many other migrants to Astana, Yernar Zharkeshov is an ethnic Kazakh. President Nazarbayev's unspoken motive for moving the capital is to consolidate the country's northern territories, which are currently dominated by ethnic Russians. The presence of these Russians is the result of the practice of Russification during the days of the Soviet Union. Astana has been designed to promote the opposite—the "Kazakhification" of the country's north. ■

CHECK YOUR UNDERSTANDING

1. How are the reasons for housing shortages today different from those of the Soviet era?
2. Which capitals of this region are primate cities? How does the primate city phenomenon affect these countries?
3. How is communist-era central planning still evident in cities in the region?
4. What factors hold back economic growth in the industrial cities located in Siberia and along the Pacific coast?
5. What has fueled recent urbanization in Central Asia?

POPULATION, GENDER, AND CULTURE

5.10 Explain changes to Russia's and the post-Soviet states' population, health, and life expectancy.

5.11 Describe how gender affects work and politics in Russia and the post-Soviet states.

5.12 Summarize how religious practices and national identity have evolved in the post-Soviet era.

With high death rates and low birth rates, Russia, Ukraine, and Belarus are undergoing a late-stage variant of the demographic transition (see Chapter 1). Similar to southern and central Europe, the rate of natural increase is negative. The population in Caucasia grows slowly, while the Central Asian states exhibit relatively high population growth. During the Soviet era, women's opportunities to become educated and work outside the home curtailed population growth. Free health care and adequate retirement pensions also helped lower incentives for large families, as people didn't have to depend on their children for support in old age. Severe housing shortages were an additional disincentive to having children, and many families chose to have only one or two children. All of these factors led to slower population growth during the last years of the Soviet Union, but the population didn't actually decline until the economic collapse and the "shock therapy" of the 1990s.

By the mid-2000s, the rate of shrinkage had slowed down considerably across the region, and by 2012 Russia's population was holding steady. Today, the death rate is about the same as the birth rate, resulting in a natural increase rate of just below zero. Russia is attracting some migrants, mostly from neighboring countries such as the Central Asian states, and so currently has a small population increase. But this growth may not last; Russia is predicted to lose about 10 million people by 2050. The population decline of the 1990s means that fewer people will be in their reproductive years in the coming decades. These population trends are geographically uneven. Western Russia is declining, especially rural areas where many villages are emptying out; only the elderly remain.

Russia has made considerable efforts to fight population decline. All couples were offered a cash bonus for having two or more children. However, similar policies in Europe have been less successful than efforts to address the concerns of career-minded working mothers, who often value access to inexpensive, high-quality day care and longer school days more than they do money. In recognition of the financial burden of having children, the most recent policy specifically targets poor Russians who already receive monthly financial support for the first child. A one-time payment equivalent to more than $5000 is awarded for every subsequent child. Parents of seven or more children are awarded the Order of Parental Glory at a ceremony at the Kremlin; the medal is presented personally by the president. Much like in Europe, immigration could help offset potential population decline, especially among the working-age population. The Russian government is aware of the need for boosting the number of workers and has encouraged immigration. At the same time, anti-immigrant sentiment continues to be a problem in Russia. According to the UN, Russia is home to 11 million migrants, predominantly low-wage workers from other post-Soviet states. That is the largest immigrant population in the world after that of the United States. Workers from Kazakhstan, Kyrgyzstan, and Armenia can freely work in Russia thanks to a customs union agreement among the countries. Others need work visas, which are commonly granted for a hefty fee.

LIFE EXPECTANCY AND POPULATION STRUCTURE

Much of the reason for the population decline, as well as the recent slowing of this decline, relates to life expectancy. Life expectancy differs dramatically according to gender, with women living nine to ten years longer than men in Russia, Ukraine, Georgia, and Belarus. By comparison, around the world, women live on average only four years longer than men. The other countries in the region also have a gender discrepancy above the global norm. Life expectancy for women declined only marginally after the collapse of the Soviet Union, but life expectancy for men dropped considerably in many places. In Russia, for example, between 1987 and 1994, male life expectancy dropped from 64.9 to 57.6 years. Since then, male life expectancy has recovered but only recently did it surpass Soviet-era levels. At 68 years, it is still very low for a developed country.

Major causes of low life expectancies in the region are the loss of health care, which was usually tied to employment, and the physical and mental distress caused by lost jobs and social disruption. These factors, though, might not explain much of the gender difference in life expectancy since the fall of the Soviet Union. Other possible causes for low male life expectancy are alcohol abuse and alcohol-related accidents and suicides, which are much more common among men. Data from the World Health Organization show that residents of Belarus are the heaviest drinkers in the world, consuming 17.5 liters of alcohol on average each year. Russians and Ukrainians engage in the most risky drinking patterns. For example, more than half of all deaths among the working-age population in Russia are alcohol-related. On the positive side, recent increases in male life expectancy may be a result of declining rates of alcohol abuse. This change is likely an outcome of government and private campaigns to reduce alcohol consumption and improved treatment of alcoholism, as well as general economic advancements. On the other hand, during the post-Soviet era, heroin started to flow from Afghanistan, through the porous borders of Central Asia, and into Russia. The UN reports that Russia is the world's largest heroin consumer.

Population pyramids for several countries (**Figure 5.23**) show the overall population trends in the region and reflect geographic differences in life expectancy and fertility. The pyramids for Belarus and Russia somewhat resemble those of European countries (for example, Germany, as shown in Figure 4.29), but there are also differences. First, they are more irregular at the bottom, indicating that birth rates declined sharply for a period, but rebounded during the last decade. Because so few children were born 20 years ago, there are few prospective parents now and into the near future, which means that low birth rates and population stagnation or decline will probably continue for some time. Also, the narrower point at the top shows their much shorter average life span compared to European countries. Also note the gender difference among the elderly as women live longer than men.

The pyramids of Kazakhstan and Kyrgystan also show rebounds in birth rates in the below-20 age groups, although the effect is more

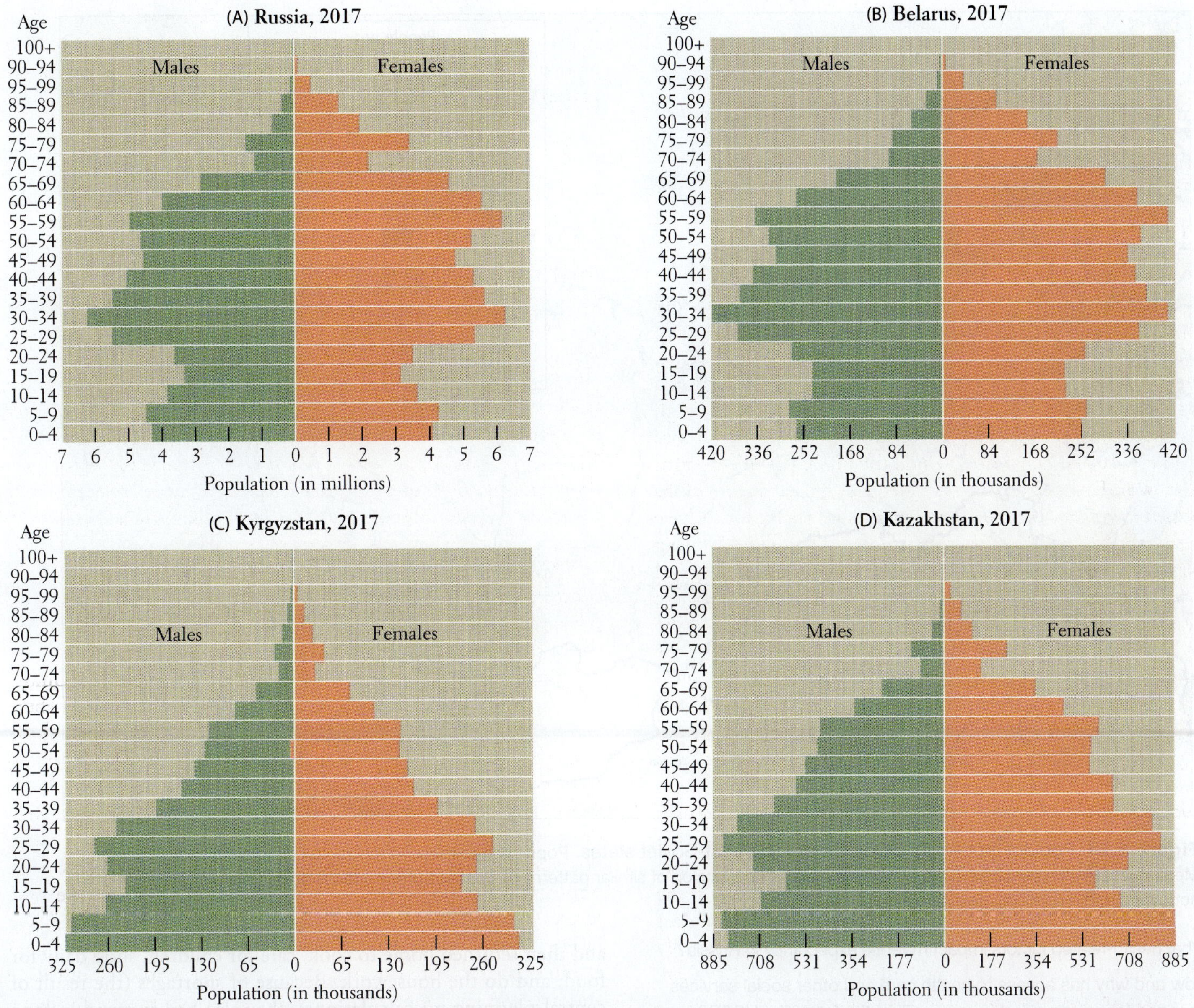

Figure 5.23 Population pyramids for Russia, Belarus, Kyrgyzstan, and Kazakhstan. Note that the pyramid for Russia is at a different scale (millions) than the other three pyramids (thousands) because of Russia's larger total population. This difference does not affect the pyramid's shape. [Data from: "Population Pyramids of Russia," "Population Pyramids of Belarus," "Population Pyramids of Kyrgyzstan," and "Population Pyramids of Kazakhstan," *International Data Base*, U.S. Census Bureau, 2017, at https://www.census.gov/data-tools/demo/idb/informationGateway.php]

pronounced than in Russia. Especially Kyrgyzstan's pyramid indicates a younger population structure—wide at the bottom but narrow at the top—which is common in less affluent and rural economies. In the Central Asian countries, the population is currently growing and will expand during the next decades. The rate of natural increase in Kyrgyzstan and Tajikistan is over 2 percent, which is higher than in any world region outside sub-Saharan Africa.

POPULATION DISTRIBUTION

A broad area of moderately dense population stretches from Ukraine on the Black Sea north to St. Petersburg on the Baltic Sea and east to Novosibirsk, the largest city in Siberia. Here, settlement is highly urbanized, but the cities are widely dispersed (**Figure 5.24**).

A secondary area of dense settlement extends from southern Russia into Caucasia, the mountainous region between the Black Sea and the Caspian Sea, where there are several primate cities of well over 1 million people each. In Central Asia, another patch of relatively dense settlement is found in eastern Uzbekistan. The subregion's largest city, Tashkent, is located there. Nearby irrigated cotton farming and mineral extraction along major Central Asian rivers have resulted in very high rural density.

CHECK YOUR UNDERSTANDING

1. What Soviet-era developments helped curtail population growth? What is being done in the post-Soviet era to encourage population growth?
2. How does life expectancy vary according to gender in this region?

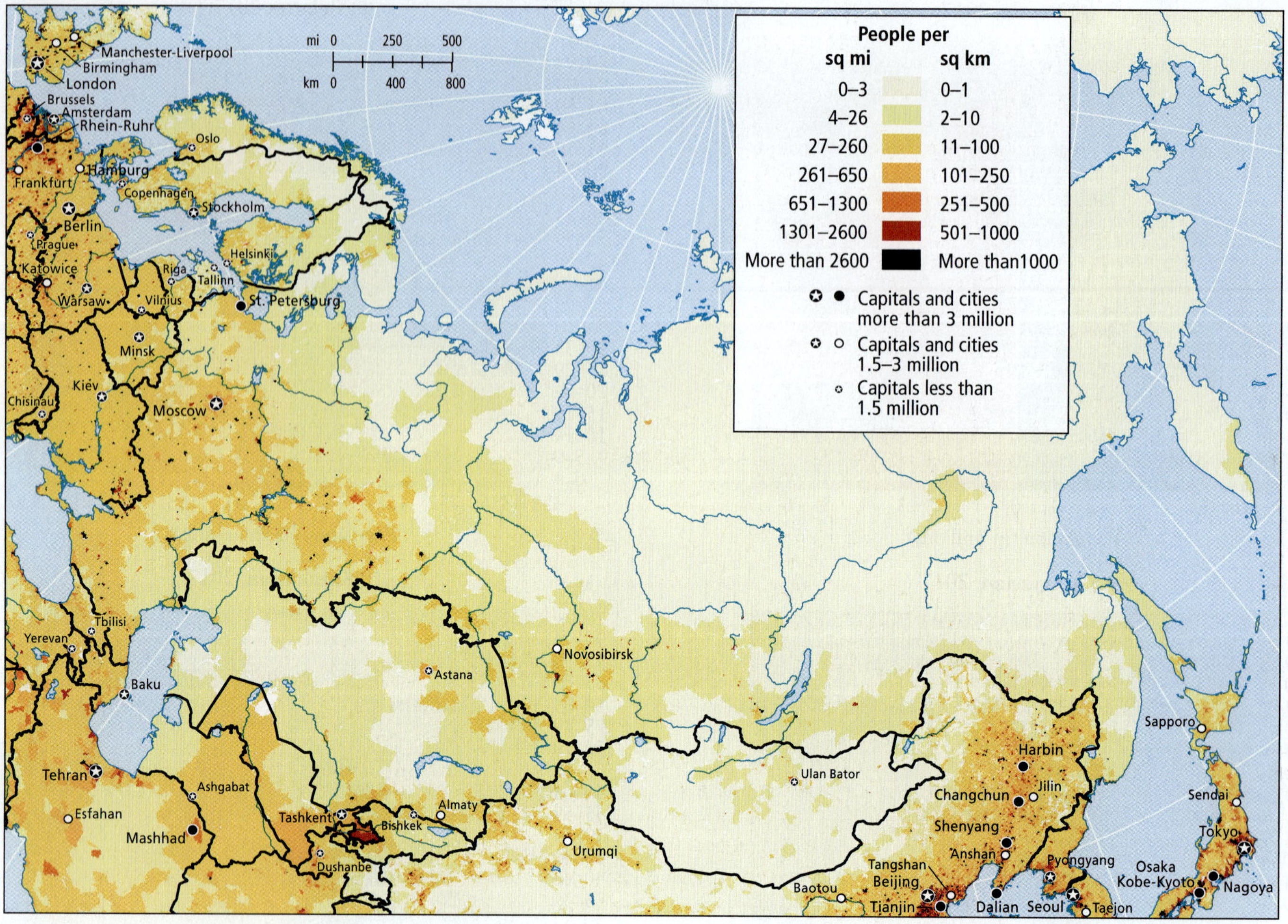

Figure 5.24 Population density in Russia and the post-Soviet states. Population patterns in this region are quite uneven. Medium population densities in the western parts are an extension of similar patterns in Central Europe. However, densities decline northward and, especially, eastward.

3. What health-related factors impact male life expectancy in Russia?
4. How and why has access to health care and other social services changed for so many in this region since the fall of the USSR?

SOCIOCULTURAL ISSUES

New freedoms in the post-Soviet era have encouraged self-expression, individual initiative, and cultural and religious revivals (**Figure 5.25**); on the other hand, this era has brought very hard times to many, as access to jobs and the social safety net has become more difficult.

Gender: Challenges and Opportunities

Men and women have been differently affected by the massive changes in this region. Soviet policies that encouraged all women to work for wages outside the home have transformed this region. By the 1970s, 90 percent of able-bodied women of working age had full-time jobs, giving the Soviet Union the highest rate of female paid employment in the world. However, the traditional attitude that women are the keepers of the home persisted. The result was the *double day* for women. Unlike men, most women worked in a factory or office or on a farm for eight or more hours and then returned home to cook, care for children, shop daily for food, and do the housework. Because of shortages (the result of central-planning miscalculations), they also had to stand in long lines to procure necessities for their families.

These pressures on working women affected diets because there was so little time or space for cooking. The everyday cuisine in most of the region was limited during the communist era; only recently have market forces made a wider range of food available. **Figure 5.26** shows several distinct dishes of the region that are prepared at home or in restaurants. The popularity of hearty soups and stews and the common use of root vegetables and grains (bread is an essential part of every meal) reflect the region's climate, soil, and agricultural systems.

When the first market reforms in the 1980s reduced the number of jobs available to all citizens, President Gorbachev encouraged women to go home and leave the increasingly scarce jobs to men. By the late 1990s, women made up 70 percent of the registered unemployed, despite the fact that because of illness, death, or divorce, many if not most were the sole supporters of their families. Consequently, many had to find new jobs. Today, unemployment for women is slightly lower than for men, although four in ten women don't participate in the workforce.

Figure 5.25 LOCAL LIVES: Festivals in Russia and the Post-Soviet States

(A) A man performs stunts on horseback during the festival of Nauryz in Kyrgyzstan. Nauryz is a Zoroastrian (an ancient religion with roots in Iran) festival that marks the New Year and the first day of spring on March 21. Forbidden during the Soviet era, Nauryz celebrations are now a source of pride throughout Central Asia. [VYACHESLAV OSELEDKO/AFP/Getty Images]

(B) Victory Day celebrates the end of World War II, or the Great Patriotic War as it is called in Russia. Because the country suffered great devastation and many Russians lost their lives, the day holds special meaning. Large public displays celebrate the nation, most notably a military parade on Red Square in Moscow. On a smaller scale, there are historical lectures, ceremonies, fireworks, and patriotic songs. One tradition is to offer flowers to veterans passing by in the street, although with time there are fewer and fewer living World War II veterans. [VIKTOR DRACHEV/AFP/Getty Images]

(C) Celebrated seven weeks before the Eastern Orthodox Christian Easter, Maslenitsa is a week-long festival that is the last chance to revel before the onset of Lent. Maslenitsa has counterparts in Catholic areas, such as the carnivals in Venice and Rio, and Mardi Gras in New Orleans. At the end of the week, as in this picture from Russia, an effigy called Lady Maslenitsa is burned. She represents winter, and the ritual burning is a way to celebrate the coming of spring. [Anton Vergun/Getty Images]

On average, the female labor force in Russia is now more educated than the male labor force. The same pattern is emerging in Belarus, Ukraine, Moldova, and parts of Muslim Central Asia. In fact, the same is true in the United States and western countries as well. In Russia, the best-educated women commonly hold professional jobs, but they are unlikely to hold senior supervisory positions and are paid less than their male counterparts. In 2017, the wages of women—both professional and low-skilled workers—averaged 26 percent less than those of men. While significant, the income gap has narrowed recently and the region as a whole ranks higher in gender income equity than many others (**Figure 5.27**).

The Trade in Women During the postcommunist era, the "marketing" of women became one of the less savory entrepreneurial activities. One part of this has been the internet-based bride services aimed at men in Western countries. A woman in her late teens or early twenties, usually trying to escape economic hardship, pays a small fee to be included in an agency's catalog of pictures and descriptions, which may feature tens of thousands of women. She is then assessed via e-mail or Facebook by the prospective groom, who then travels—usually to Russia or Ukraine—to meet women he has selected from the catalog and to potentially have one of them accompany him to the West to marry.

Sex work has also increased in recent years. The region of the post-Soviet states is a major source for Europe's sex workers. Most of the women—coming from countries like Ukraine, Moldova, and Russia—are supplied by members of the Russian Mafia, who have been known to deceive women who are looking for jobs as maids or waitresses in Europe, and then force them into sex work. This is a form of **trafficking**, which is defined by the UN as the recruiting, transporting, and harboring of people through coercion for the purpose of exploiting them. Within Russia, there are also about 3 million sex workers, some of whom are trafficked from countries in Africa and Asia. According to the U.S. State Department, forced labor trafficking is also prevalent in Russia outside the sex industry.

The Political Status of Women One way for women to address institutionalized discrimination is to increase their political

trafficking the recruiting, transporting, and harboring of people through coercion for the purpose of exploiting them

Figure 5.26 LOCAL LIVES: Foodways in Russia and the Post-Soviet States

(A) A bowl of borscht, or beet soup, a Ukrainian dish with many variants that is popular throughout this region. Beets are an extremely nutritious root vegetable that can tolerate cold weather well, enabling farmers to make the most out of the short growing season found throughout much of this region. Borscht is served warm in the winter and chilled in the summertime. [Alexandra Grablewski/Getty Images]

(B) Caviar is the roe (fish eggs) of the sturgeon and has long been an essential part of Russian cuisine. Once a staple food rather than a luxury, it is now relished more sparingly by ordinary Russians, such as on special holidays like New Year's Eve. Russians often enjoy their caviar spread on toast. High-quality caviar can cost in excess of $100 for a can. Especially the beluga sturgeon harvested from the Caspian Sea is greatly valued, but the species is endangered, so much caviar comes from other species and places, even outside Russia. [CSMaster/Shutterstock.com]

(C) A girl prepares *non*—a kind of bread cooked throughout Central, South, and Southwest Asia (where it is also known as *naan*). Non is traditionally served with great reverence. In Uzbekistan, for example, it is never cut with a knife, but rather is broken by hand and put near each place setting at a table, always with the flat side down (serving it flat side up is considered insulting). One tradition holds that when an Uzbek person leaves a house, he or she should bite off a piece of bread, which will be kept for that person to eat upon return. [VICTOR DRACHEV/AFP/Getty Images]

power. Although women were granted equal rights in the Soviet constitution, they never held much power. The very few in Communist Party leadership often held these positions at the behest of male relatives.

Since the fall of the Soviet Union, the political empowerment of women has grown somewhat, but long-standing cultural biases against women in positions of power remain. In many countries, the number of women in legislatures has risen (**Figure 5.28**), but often this is because male leaders want to appear more progressive. Frequently they promote women who are the least likely to work for change. On average, 18 percent of the legislators in the region's parliaments are women, which is similar to the United States, but below the global norm. Support among women themselves for women's political movements is not widespread, as many fear being seen as anti-male or against traditional feminine roles. To be a feminist is to occupy a tenuous position in society.

Religious and Nationalist Resurgence

The official Soviet ideology was atheism, and religious practice and beliefs were seen as obstacles to revolutionary change. Orthodox Christianity, which was the official religion of the Russian empire, was tolerated, but few people went to church, in part because the open practice of religion could be harmful to one's career. Now, religion is a major component of the general cultural resurgence across the former Soviet Union.

The overall level of religiosity is moderate in Russia by global standards, but a number of people, especially those from indigenous ethnic minority groups, are turning back to ancient religious traditions. In Russia, Ukraine, Moldova, Belarus, Georgia, and Armenia, most people have some ancestral connection to Orthodox Christianity. Today, about 54 percent of people in Russia claim affiliation with the Orthodox church. Those with Jewish heritage form an ancient minority, mostly in the western parts of Russia and Caucasia. Religious observances by both groups increased markedly in the 1990s, and many sanctuaries that had been destroyed or used for nonreligious purposes by the Soviets were rebuilt and restored.

Even President Putin in Russia commonly appears in public with Orthodox priests and attends religious services. This signals a support for the traditional values that the church represents, as well as an emphasis on Russian culture in general. Putin is tapping into growing nationalist sentiment. Russia has been defining itself in opposition to Western culture, which is seen by Russian traditionalists as too secular, multicultural, and accepting of homosexuality. In fact, the Russian parliament passed a law in 2013 against homosexual "propaganda." This vague law makes it possible for the government to target and persecute the country's LGBTQ community. Linking the LGBTQ community with the West also conveniently positions Russia as having a heterosexual, Slavic, and Orthodox culture that needs to remain "pure" of outside influences. By appealing to the sense of national identity among many Russians, it is easier for the government to rally support for policies that are intended to advance Russia's influence outside its borders.

In Central Asia, Islam was repressed by the Soviets, who feared Islamic fundamentalist movements would cause rebellion against the dominance of Russia. Today, Islam is openly practiced and is increasingly important politically across Central Asia, Azerbaijan, and some of the Russian Federation's internal ethnic republics, such as Chechnya and Tatarstan. Especially in the Central Asian states, however, the return to religious practices is often contentious.

5.27 Population, Gender, and Culture in Russia and the Post-Soviet States

The map ranks countries into five categories (from very high to very low) on gender equity in terms of longevity, education, and income. Russia and the post-Soviet states perform a little bit better with regard to gender equality than they do in other measures such as income and health. All countries are in the very high to medium Gender Development Index (GDI) categories. The Soviet system, which encouraged education and careers for women, is partly responsible for this record. The country with the lowest gender equality is only somewhat more unequal than the country with the highest gender equality rank. The GDI differences in the region seem to be partially related to economic differences; that is, countries such as Russia and Kazakhstan with the highest income per capita are in the very high GDI group, while poorer states in Central Asia and Caucasia are ranked medium.

THINKING GEOGRAPHICALLY

A How have sex workers been trafficked in and out of the region?

B How is population decline manifested geographically across Russia?

C Why does religion play an increasingly important role in the post-Soviet era?

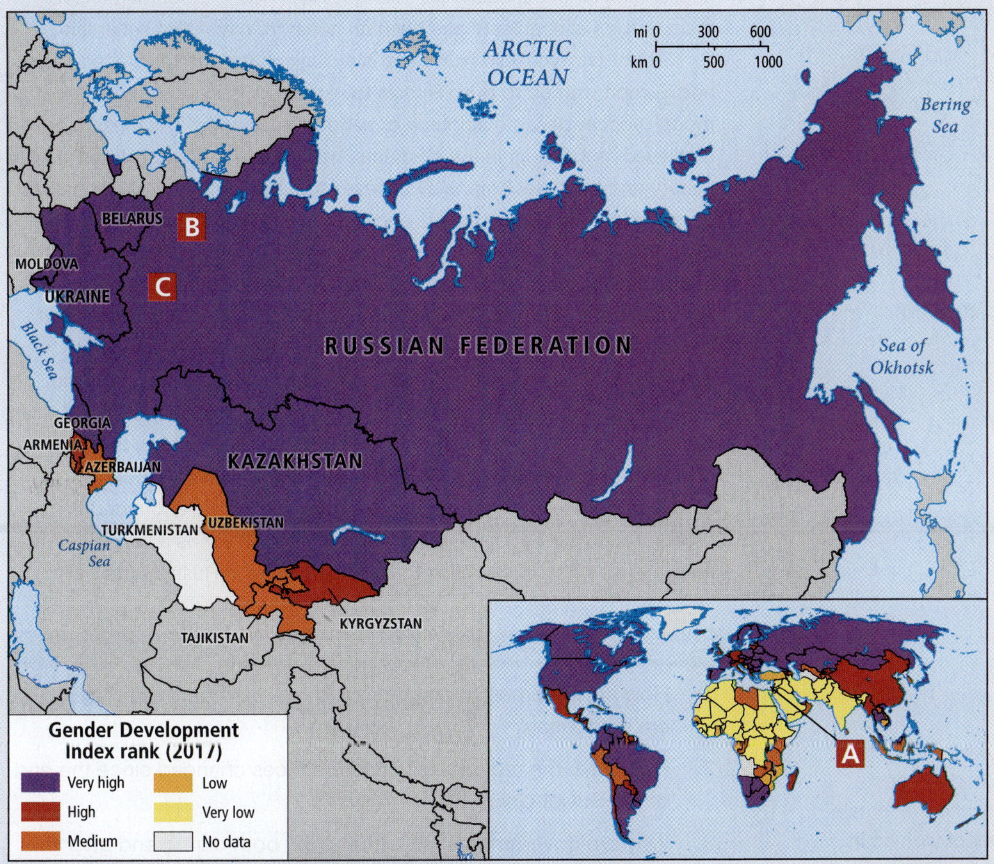

A A Russian sex worker outside a brothel in Tel Aviv, Israel. In Southwest Asia, there is a demand for "white" sex workers, and the lack of economic opportunities in the post-Soviet states has driven young women to fulfill that demand. [Eddie Gerald/Alamy Stock Photo]

B Abandoned houses in the Novgorod region of western Russia. Russia's population decline is geographically uneven. Villages far from major cities are the most likely to decline. Only the elderly remain or the settlements empty out completely. [FotograFFF/Shutterstock.com]

C Russian President Vladimir Putin with the Patriarch of the Russian Orthodox church. Much like elsewhere in the world, religion has become an important element of national identity and a potent element of political nationalism. [ALEXEY NIKOLSKY/AFP/Getty Images]

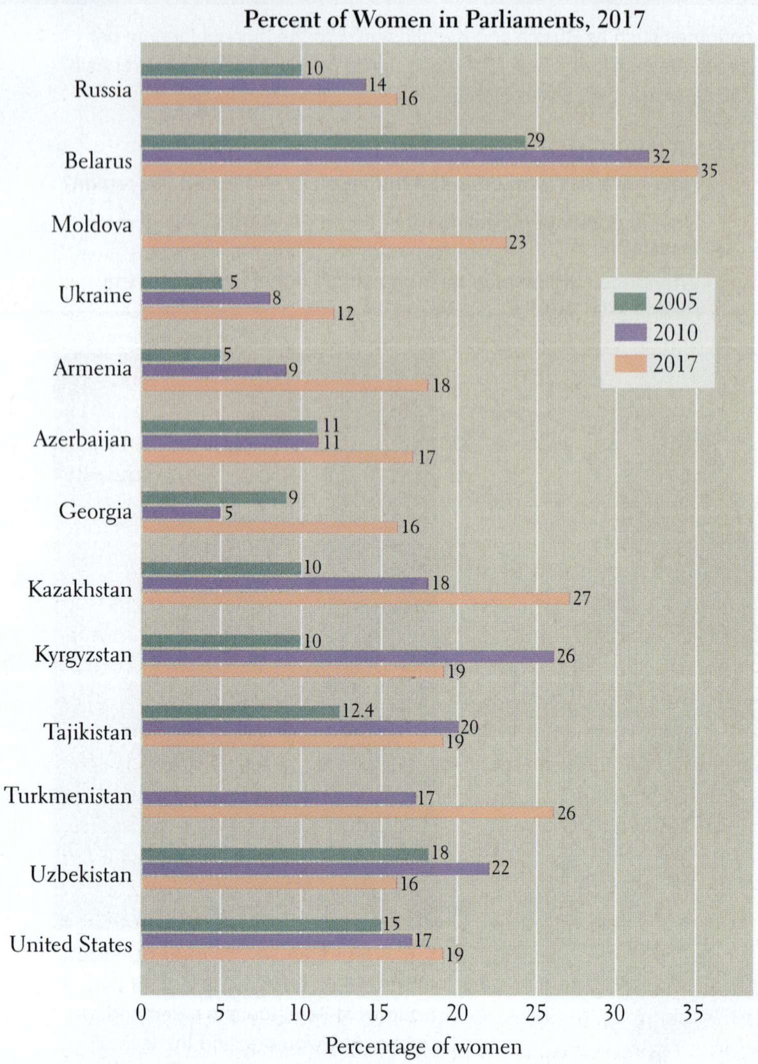

Figure 5.28 Women legislators, 2005, 2010, and 2017. This graph shows the percentage of legislators in Russia and the post-Soviet states (with the United States for comparison) who are female. The general trend has been an increase in the number of women in parliament. However, the countries with the largest percentage of female legislators are not necessarily those with the most open democratic participation. [Research from: World Bank, "Proportion of Seats Held by Women in National Parliaments," at https://data.worldbank.org/indicator/SG.GEN.PARL.ZS]

Some local leaders still view traditional Muslim religious practices as obstacles to social and economic reform.

Militant Islamic movements have often been violently repressed by Central Asian governments, in part out of fear of influences from nearby countries, where Islam plays a greater role in politics. In some cases, these influences are clear—as in Tajikistan's civil war (1992–1997), which was fought between a government made up of former Soviet officials and Islamic rebels who received support from Taliban fighters from across the border in Afghanistan; and the conflict in Chechnya in which militants from Saudi Arabia and other Muslim countries fought. At times, government repression has backfired, as when Uzbekistan and Kyrgyzstan persecuted ordinary devout Muslims, which actually helped the recruitment efforts of Islamic militants. Human Rights Watch, an organization that monitors human rights abuses worldwide, reports that Uzbekistan's government jails and tortures believers who worship independently outside the strict supervision of the state. A new government in Uzbekistan has promised reforms but the future of religious freedoms remains cloudy.

VIGNETTE After Russia's problematic experience with market reforms during the 1990s, the country searched for a new identity that was neither a return to communism nor affirmation of Western-style democracy. The answer has been a rebirth of Russian nationalism. Nowhere is the newfangled celebration of the nation more overt than in the stands of soccer stadiums. Especially for young Russian men, supporting a club team or the national team has been an outlet for nationalist sentiment.

Thirty-two-year-old Ivan Katanaiev says that when the United States and Europe wanted to push Russia toward their "decadent" societal model, soccer became a source of national resistance. The messages displayed on banners in the stadiums, he continues, hark back to Russian history and heritage. Katanaiev believes that the form of ultranationalism displayed by the soccer fans is endorsed at the highest political level. Even President Putin paid tribute to the leader of a soccer fan club who had been killed in a confrontation with a Muslim from the Caucasus. The ethnocentrist nationalism in the soccer community has also meant that it has been fertile ground for recruiters who are looking for people to volunteer to fight on the side of ethnic Russians in the Ukraine conflict.

However, not all soccer fans are comfortable with the mixing of nationalist politics and soccer. Robert Ustian, who hails from the province of Abkhazia in Georgia, is a fan of the club CSKA Moscow, but he has been dismayed by racist chants among the fans. He founded the organization "CSKA Fans Against Racism," which sends a very different message about what it means to be a soccer fan in Russia today. ■

CHECK YOUR UNDERSTANDING

1. How have women participated in the workforce, from the Soviet era until today?
2. How have the region's religious practices changed since the end of the Soviet Union?
3. Why are governments in Central Asia, both during and after the Soviet era, fearful of Islamic movements?
4. How can Russian nationalism be understood in relation to Western culture?

SUBREGIONS OF RUSSIA AND THE POST-SOVIET STATES

5.13 Identify key characteristics of the subregions of Russia and the post-Soviet states.

This section begins with Russia, the largest country in the region and the one that has dominated the entire region for many hundreds of years. The post-Soviet states of Belarus, Moldova, and Ukraine are discussed next because of their location and their close social, cultural, and economic associations with European Russia. The Caucasian states of Georgia, Armenia, and Azerbaijan are then covered. Finally, the countries of Central Asia—Kazakhstan, Kyrgyzstan, Tajikistan, Turkmenistan, and Uzbekistan—are discussed.

RUSSIA

In the post-Soviet era, Russian influence in the region remains strong, even as the post-Soviet states have begun to construct their own economies and political identities. Russia is still the largest country on Earth in terms of area—nearly twice the size of either Canada, the United States, or China—and is rich in natural resources. With 144 million people, it ranks ninth in the world in population. For convenience, we have divided Russia into three parts: European Russia; Central Siberia, which is composed of about half the landmass east of the Urals; and the Russian Far East. There is no clear boundary between Central Siberia and the Far East; in fact, the term *Siberia* is commonly used for Russian territories all the way to the Pacific.

European Russia

European Russia is the area of Russia that shares the eastern part of the North European Plain with Latvia, Lithuania, Estonia, Belarus, Ukraine, and Moldova (**Figure 5.29**). It is usually considered the heart of Russia because it is here that early Slavic peoples established what became the Russian empire, with its center in Moscow. Although it occupies only about one-fifth of the total territory of present-day Russia, European Russia has most of the industry, the best agricultural land, and about 70 percent (100 million people) of the population of the Russian Federation.

A tiny part of European Russia is Kaliningrad, located on the Baltic Sea between Lithuania and Poland. In geopolitical terms, it is an *exclave*, which is when a smaller part of a country is geographically separated from the rest of the country by another territory. This historically German city (then called Königsberg) became part of the Soviet Union after World War II, and then Russia after 1991. The Kaliningrad port is important to Russia because it serves as an outlet to the Atlantic Ocean. Militarily, Kaliningrad is the home to Russia's Baltic naval fleet, which provides the country with strategic access to northern Europe. The port of Murmansk, which faces the Arctic Ocean, also connects to the Atlantic. Because of warm ocean currents, it is actually ice-free year-round despite its location north of the Arctic Circle. However, Murmansk is located far from main commercial routes, which means that until now it has mostly been important for military purposes. After the Cold War, Russia reduced its investment in Murmansk, but is now investing there and in other military bases that face the Arctic Ocean. Russia's interest in the Arctic is driven in part by climate change. Fossil fuel gas resources can be exploited and maritime trade routes are opening up, which may increase Murmansk's role as a commercial port.

The vast majority of European Russians live in cities, their parents and grandparents having left rural areas when Stalin collectivized agriculture and established heavy industry. Most of these people went to work in one of four major industrial regions that were chosen by central planners for their accessibility and their location near crucial mineral resources. One industrial region is centered on Moscow; another in the foothills of the Ural Mountains (part of this zone lies in Siberian Russia); a third along the Volga River from Kazan to Volgograd; and the fourth just north of the Black Sea, extending into Ukraine around Donetsk (see Figure 5.22).

The most densely occupied part of European Russia—stretching from St. Petersburg on the Baltic, south to Caucasia, and east to the Urals (see Figure 5.24)—coincides with that part of the continental interior that has the most favorable climate (see Figure 5.4). Even so, because the region lies so far to the north—Moscow is at a latitude 100 miles (160 kilometers) north of Edmonton, Canada—and in a continental interior, the winters are long and harsh and the summers short.

Urban Life and Patterns in European Russia Moscow remains at the heart of Russian life. The city has grown in the post-Soviet era; about 13 million people live within the city limits and another 4 million live in the metropolitan area. Always a center of Russian culture, Moscow has so changed since 1991 that a common saying is "Moscow and Russia are two different countries." The availability of goods—imported specialty food, electronics, brand clothing, automobiles, and especially luxury products and entertainment of all types—has exploded. Prices are much higher here than in most Russian cities. Entrepreneurs are especially active. While the city is full of affluent, educated young people, the level of inequality is dramatic. Even many educated individuals are struggling to make ends meet in this high-cost city.

The privatization of real estate in Moscow gives an interesting insight into pre- and post-Soviet circumstances. During the communist era, all housing was state-owned and there was a general housing shortage; thousands of families were always waiting for housing, some for more than ten years. Most apartments were built after the mid-1950s in large, shoddily constructed gray concrete blocks; the units included one to three rooms that housed an entire family. In 1991, the government allowed residents of Moscow to acquire, free, about 250 square feet (23 square meters) of usable living space per person—about the size of a typical American living room. Most families simply took ownership of the apartments in which they had been living, so that by 1995, a majority of Muscovite families owned their apartments. A lively real estate market quickly emerged, with a rising wealthy elite willing to pay for an apartment in the city center. Purchase prices in excess of U.S.$1,000,000 are not unusual today. Rents also skyrocketed. Those who could find alternative housing for themselves made tidy sums simply by renting out or selling their city-center apartments (see Figure 5.29A). The demand for commercial space for private businesses reduced the number of dwelling units, exacerbating the housing shortage.

St. Petersburg, called Leningrad during most of the communist period (1924–1991), is Russia's second-largest city, with about 5 million people. Located at the eastern end of the Gulf of Finland, on the Baltic Sea, it has a diverse economy, including finance, shipbuilding, and mixed manufacturing. Czar Peter the Great ordered the city built in 1703 as part of his effort to make Russia more European. From 1713 to 1918, it served as the Russian capital, and it has been one of Russia's cultural centers ever since (see Figure 5.9C). The city has a rich architectural heritage—including many palaces, public parks, and canals—which has drawn comparisons with the ambience of Venice. Since 1991, it has been undergoing a renaissance; the transportation system is being refurbished, urban malls are proliferating, and historic sites are being renovated. The czars' Winter Palace, now called the Hermitage Museum, houses one of the world's most important art collections. Therefore, tourism is now a major part of the economy of St. Petersburg. An obstacle to tourism is political animosity between Russia and the West; for example, all visitors from Europe and the United States must acquire a visa before traveling to Russia.

Figure 5.29 European Russia subregion. **A** The Kotelnicheskaya was completed in 1952 as an apartment building with multiple families sharing a kitchen and bathroom. It is one of seven such grand buildings—nicknamed the Seven Sisters—all of which are famous Moscow landmarks. Today, the building has been extensively remodeled and apartments are owned or rented by an affluent clientele. [Yadid Levy/robertharding/Getty Images]

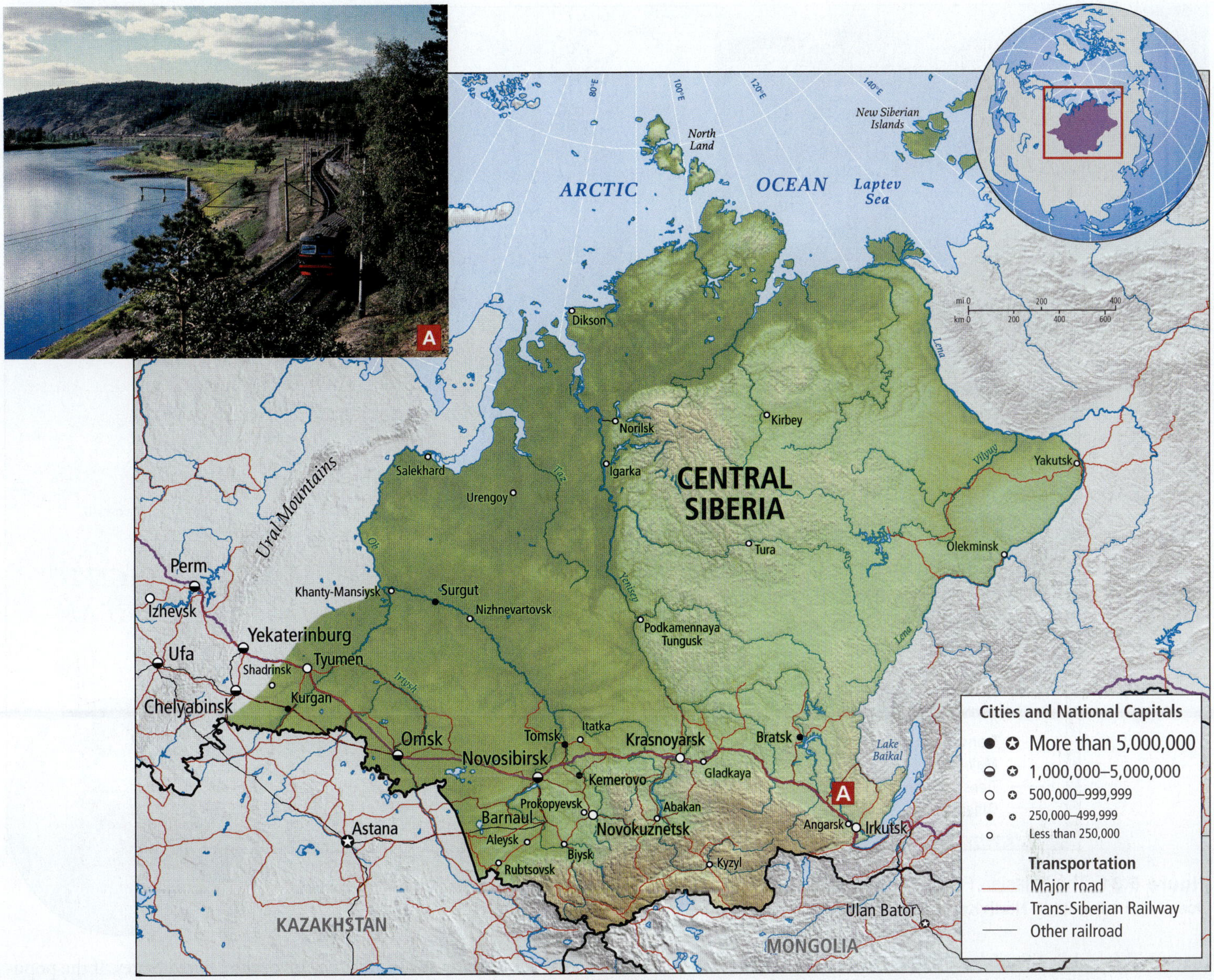

Figure 5.30 Central Siberia subregion. **A** The Trans-Siberian Railway is the main land connection between Siberia and western Russia. [DEA PICTURE LIBRARY/De Agostini/Getty Images]

Central Siberia

Central Siberia encompasses the West Siberian Plain beyond the Urals and about half the Central Siberian Plateau (**Figure 5.30**). It is so cold that 60 percent of the land's subsurface is permafrost, some of which is melting due to climate change (see the "Climate Change" section). This vast central portion of Russia is home to just 31 million people, many of whom moved into the region from European Russia. Settlement is concentrated in cities in the somewhat warmer southern tier.

For millennia, Siberia has been a land of wetlands and quiet, majestic forests. The Siberian taiga is one of the largest forests in the world and sequesters a great deal of atmospheric carbon. Like all economic activities, forest harvesting declined during the post-communist period, but as the economy has improved recently, more clear-cutting is taking place, which lowers the rate of *carbon sequestration*—a process whereby plants absorb carbon though photosynthesis and store it in their biomass. In the northern zones closer to the Arctic, climate change melts the permafrost. Collapsing land is a growing problem in the few cities of Siberia, like Norilsk, where buildings crack and are made structurally unsound by the disappearance of permafrost. Another effect is that organic material that used to be frozen is now exposed to the air, which results in the release of methane and other GHGs. Basically, global warming generates more GHG emissions, which further heats the climate through this ecological feedback mechanism.

Although the entire area of Russia east of the Urals continues to be occupied by a scattering of indigenous groups—some still practicing nomadic herding, hunting, and gathering—Central Siberia is dotted with bleak urban landscapes and industrial squalor (see the discussion of Norilsk in the "Resource Use and the Environment" section). This is the legacy of the Soviet effort to claim territory from the indigenous peoples and to establish industries

Figure 5.31 The Russian Far East subregion. **A** Military vessels in Vladivostok harbor, located on the Pacific Ocean and adjacent to North Korea. [Iain Masterton/Getty Images]

that would exploit Siberia's valuable mineral, fish, game, and timber resources. Novosibirsk, Siberia's de facto capital and financial center, is an example of the result of this policy.

Novosibirsk is the largest Russian city east of the Urals (1.7 million people) and the third-largest city in Russia after Moscow and St. Petersburg. It is also extremely isolated; it takes about 50 hours by train to get from Novosibirsk to Moscow, which lies more than 1600 miles to the west. As a major stop along the Trans-Siberian Railway (see Figure 5.30A), Novosibirsk has the highest concentration of industry between the Urals and the Pacific. During the Cold War, it was a center for strategic industries such as chemicals and weaponry; it contained more than 200 heavy industry plants. After a post-Soviet slump, Novosibirsk's banks and financial services are drawing outside investors interested in acquiring factories in the vicinity, and the city's population is currently increasing slightly.

The Russian Far East

Here in this extensive territory of mountain plateaus with long coastlines on the Pacific and Arctic oceans (**Figure 5.31**), almost 90 percent of the land is permafrost. The coastal volcanic mountains stop the warmer Pacific air from moderating the Arctic cold that spreads deep into the continent. Making up over one-third of the entire country of Russia, this area is nearly two-thirds the size of the United States. If the population of 6 million were evenly distributed across the land, there would be just 2.6 persons per square mile (1 per square kilometer), but 75 percent of the people live in just a few cities along the Trans-Siberian Railway and on the Pacific coast. At a rift in the Earth's crust, the deepest and most voluminous freshwater lake in the world—Lake Baikal—is also located in the subregion's southern tier.

The residents of the Russian Far East are indigenous people or immigrants and exiles from Russia. Indigenous groups include the Yakuts, which is the largest ethnicity of the vast Sakha internal republic to the north. The Yakuts historically hunt, fish, and keep reindeer. In southern Siberia near Mongolia, the Buryats share language and customs with other Mongol groups, such as nomadic herding. Some of the ethnic Russians are descendants of those sentenced to hard labor in the region's labor camps for political crimes in the Soviet era (some made famous in Aleksandr Solzhenitsyn's book *The Gulag Archipelago* were along the Kolyma River). These Far Easterners worked in timber and mineral extraction enterprises and in isolated industrial enclaves. In the Soviet days, many immigrants headed for cities along the Pacific coast, especially Nakhodka and Vladivostok, near North Korea. Vladivostok is strategically important as a naval base and is also Russia's largest Pacific port (see Figure 5.31A).

The interior of the Russian Far East has many resources, but it has not been developed because of its distance from European Russia and its difficult physical environments. Only 1 percent of the territory is suitable for agriculture, mostly along the Pacific coast and in the Amur River basin. Relatively unexploited stands of timber cover much of the area. The wilderness supports many endangered species, such as the Siberian tiger, the Amur leopard, and the Kamchatka snow sheep. The region's timber, coal, natural gas, oil, and minerals first attracted Soviet central planners and now attract private investors. The proximity to Chinese markets has increased timber harvesting. A consortium of multinational companies produces oil and gas off Sakhalin Island that is destined for Asian markets. During the last decade, liquefied natural gas has been shipped to Japan, which is just 30 miles south of Sakhalin. With no pipelines in place and with few domestic energy sources, Japan is willing to pay the extra price for natural gas that has to be liquefied before it can be transported on oceangoing tankers. At the same time, areas in Siberia and the Far East that are not part of the growing natural resource economy have suffered from economic and population decline.

The region has further potential for linkages with the greater Pacific Rim. So far, fossil fuel is delivered via pipelines to the Chinese interior, but port facilities on the Pacific coast can make the region more accessible to Asia. An oil pipeline from Siberia is expected to terminate at Nakhodka on the Russian Pacific coast, which will give Russia access to petroleum markets in Japan and the western Pacific, and a spur to China will supply that country as well. Because of its rich resource reserves and its location so far from the Russian core, the Russian Far East is likely to integrate economically with other Pacific Rim nations, and might eventually seek further independence from European Russia. The Far East has a strategic position connecting the BRICS partners Russia and China, which means that Russia is likely to pay close attention to the development of the region.

CHECK YOUR UNDERSTANDING

1. What are the three principal geographic divisions of Russia, and how do their economies and cultural traditions differ?
2. What characterizes Russia's fast-growing primate city and capital?
3. Describe the settlement patterns of Central Siberia and the Far East.

BELARUS, UKRAINE, AND MOLDOVA

Sandwiched between Russia and the Central European countries that were once part of the Soviet Union's sphere of influence are the post-Soviet states of Belarus, Ukraine, and Moldova (**Figure 5.32**). Each of these countries is the home of a distinct ethnic group, but Russian residents form significant and influential minorities in all three. Due to their location, these countries have been tied to Russia, though Moldova and Ukraine are developing closer associations with Europe. This has caused serious geopolitical tension with Russia, most notably the conflict in Ukraine.

Figure 5.32 The Belarus, Moldova, and Ukraine subregion. **A** Ukrainian farm fields. [Creative Travel Projects/Shutterstock.com]

Belarus

In size and terrain, Belarus resembles Minnesota; its flat, glaciated landscape is strewn with forests and dotted with thousands of small lakes, streams, and marshes that are replenished by abundant rainfall. Much of the land has been cleared and drained for collective farm agriculture. The stony soils are not particularly rich; nor, other than a little oil, are there many known useful resources or minerals beneath their surface. Belarus absorbed 70 percent of the radiation contamination after the Chernobyl nuclear accident in Ukraine in 1986 because of its proximity to the accident site. Twenty percent of Belarus's agricultural land and 23 percent of its forestland were contaminated. Scientific and agricultural methods have been developed to reduce the level of radiation in the soil, but many people still rely on food with elevated levels of radioactivity.

During the early twentieth century western Belarus and Ukraine were part of Poland and had a large number of Jewish residents. After border changes, population resettlement programs, and the Holocaust, Belarus and Ukraine emerged in the post–World War II era as republics within the Soviet Union. Especially Belarus was thoroughly Russified during this time. Although 84 percent of the population of 9.5 million is ethnic Belarusian (a Slavic group) and only 8 percent is Russian, the Russian language predominates, and Belarusian culture survives primarily in museums and historical festivals. The drab urban concrete landscape, such as in the capital Minsk, looks and feels like Soviet Russia. The Belarusian economy remains dominated by state firms that sell goods domestically or to Russia; only a few industries are privatized. Belarus's GDP per capita is significantly higher than in Ukraine and Moldova, in part because Belarus receives subsidies from Russia, such as oil at below market price, which is economically beneficial. The economy, however, has struggled with slow growth for a number of years. A sluggish economy has renewed interest in privatization, but so far few reforms have been implemented.

Belarus became an independent state when the Soviet Union collapsed in 1991, but it remains tightly controlled by its autocratic leader, Alexander Lukashenko, who has been in power since 1994 and allows little meaningful democracy and freedom of speech. Lukashenko does not accede to all Russian demands, but because Belarus relies heavily on Russia—especially for cheap energy—it is unlikely to seriously rebel. Russia is also able to manipulate the flow of oil and gas to Europe that is transported via pipelines across Belarus (see Figure 5.14).

Russia sees Belarus as a useful **buffer state** against the influence of the EU and NATO. A buffer state is situated between two rival and more powerful political entities, in this case Russia and the European Union, and serves as neutral territory that limits direct contact and potential conflict between the two powers.

Moldova and Ukraine

Moldova has a mix of cultures—Romanian in the west and Slavic in the east—while Ukraine is primarily Slavic in its traditions, including language, cuisine, religious customs, and architectural forms. The distinctive ethnic artistry of these two countries (elaborate holiday breads and pastries, intricate embroidery and lace, and in Ukraine, ornately painted Easter eggs) is prominently displayed in homes and marketed at home and abroad. Both countries seem to be tending toward closer association with the EU, but in Ukraine, long-standing ties to Russia complicate that propensity. Moldova is also affected by internal divisions that have developed as Slavic minorities in the eastern section of the country have formed a breakaway state called *Transnistria*, whose future political status is uncertain, although it currently receives political and economic support from Russia. Transnistria has the potential to be another political and military flashpoint, just like Ukraine. For now, it remains yet another frozen conflict in post-Soviet geopolitics.

buffer state a country that is situated between two rival and more powerful political entities and serves as a neutral territory, limiting direct contact and potential conflict between the two powers

Moldova and Ukraine have warmer climates than do Belarus and Russia, and with their agricultural resources, they have the potential to increase their agricultural production significantly. Figures 5.17 and 5.32A show that intensive crop cultivation is far more typical here than elsewhere in the post-Soviet states. Not only is the climate warmer, but the area's open farming landscape is endowed with rich, black-colored soil (called *chernozem*). Moldova, which borders Romania to the west, is much smaller (3.5 million people) than the more economically diverse Ukraine (42 million people) but produces many of the same products—grain, sugar, milk, and vegetables, for example—as well as manufactured goods related to agriculture. Moldova is increasing the amount of higher-value commodities, such as wine, nuts, and fruits. Both Ukraine and Moldova rely less and less on Russia, and trade with Russia is now about 10–15 percent of all trade. Instead, exports and imports have been reoriented toward Europe. Moldova's trade with Romania, which is geographically and culturally close, is especially important. Odessa on the Black Sea, once the main port for the entire Russian empire, is still the main conduit for goods leaving and entering Ukraine. The future of Ukraine's industrial production is uncertain because the industrial heartland is located in the eastern part of the country, which is where the military struggle between government and separatist forces is taking place (see the section "Crisis in Ukraine").

The European Union has a strong interest in Ukraine and Moldova. The two countries are potential future members of the EU, but their membership is dependent on economic and political progress toward greater democracy, and is affected by the ebb and flow of EU relations with Russia, which does not want either country to be reoriented toward Europe. The relationship is highly strained at the moment, as Russia supports separatists in eastern Ukraine and has annexed Crimea. Another issue is border security. The EU's long eastern border faces Ukraine and Moldova. The EU cooperates with Ukraine and Moldova to tighten border security in order to reduce the number of undocumented immigrants and the illegal drugs that follow a route from Asia, through Ukraine and Moldova, and on to western Europe. Finally, oil and gas pipelines from Russia cross Moldova and Ukraine on their way to Europe (see Figure 5.14), and the region is also a potential route for future pipelines from Caucasia to EU countries.

CHECK YOUR UNDERSTANDING

1. In what ways is Belarus still similar to the old Soviet Union?
2. What are the agricultural resources of Moldova and Ukraine?
3. How are Moldova and Ukraine increasingly reoriented toward Europe rather than Russia?

CAUCASIA: GEORGIA, ARMENIA, AND AZERBAIJAN

Caucasia is located around the rugged spine of the Caucasus Mountains, which stretch from the Black Sea to the Caspian Sea. (The language map in **Figure 5.33** serves as the map for this subregion.) Culturally and physically, this region includes a piece of the Russian Federation on the northern flank of the mountains as well as the three independent states of Georgia, Armenia, and Azerbaijan, which occupy the southern flank of the mountains in what is known as *Transcaucasia*. Transcaucasia is a band of subtropical intermountain valleys and high volcanic plateaus that drop to low coastal plains near the Black and Caspian seas (see Figure 5.4).

In this mountainous space the size of California, live more than 50 ethnic groups—Armenians, Chechens, Ossetians, Abkhazians, Georgians, and Tatars, to name but a few—most speaking separate languages. The groups vary widely in size, from a few hundred (for example, the Ginukh in Dagestan) to more than 6 million (the Turkic Azerbaijanis). Some are Orthodox Christians, many are Muslims, a few are Jews, and some retain ancient elements of local animistic religions. All these groups, including Chechens and others who live in Russian Caucasia (see the discussion on political conflict in Caucasia), are remnants of ancient migrations. For thousands of years, Caucasia was a stopping point for nomadic peoples moving between the Central Asian steppes, the Mediterranean, and Europe. Other ethnic groups were affected more recently by the Soviet-instigated relocation and subsequent return, such as the Chechens. The region's history as a global meeting place, together with the mountainous topography that has fostered isolated self-reliance, has resulted in the ethno-linguistic patchwork that is Caucasia. Today, many Caucasians maintain ties to Europe, Russia, and North America, where hundreds of thousands of emigrants from the region live. The trend of emigration intensified after the communist era, which has led to the depletion of an able-bodied workforce and brain drain. Armenia in particular has a high outmigration rate, resulting in recent population losses. Migrants typically cite unemployment and poor economic prospects as reasons to migrate. An open labor market among some post-Soviet states has also enabled Armenians to migrate. But the growth of a Caucasian diaspora—people living away from their ancestral homeland—has also had some positive effects, as those living abroad send remittances back to people in the Caucasian countries. Remittances into Georgia and Armenia make up 12–13 percent of their respective economies.

As a result of external political maneuvering and boundary making, an ethnic group may now have members in several different Caucasian states or in the internal republics of Russia. Because of their strong ethnic loyalties, this pattern can result in persistent conflict (**Figure 5.34**). After the Soviet collapse, the three culturally distinct states of Georgia, Azerbaijan, and Armenia each gained independence. More so than in other parts of the former Soviet Union, the move to independent status resulted in ethnic warfare. Very quickly, several ethnic groups within these already tiny states took up arms to obtain their own independent territories. The Abkhazians and the South Ossetians fought against Georgia. The Christian Armenians in

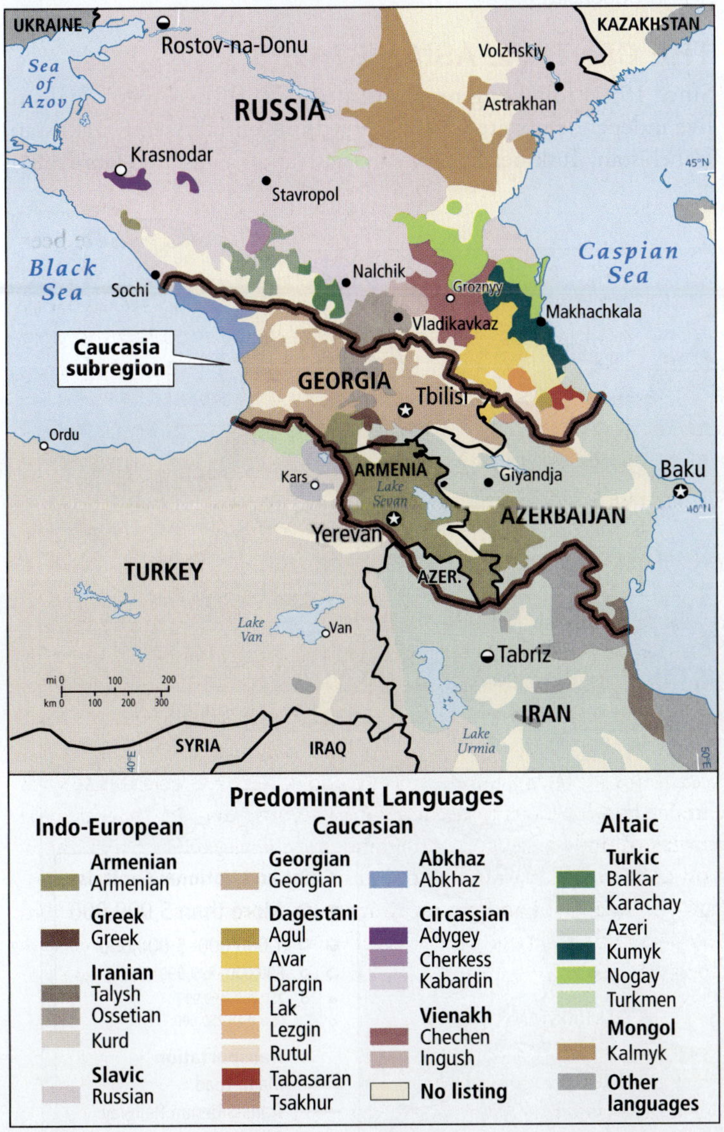

Figure 5.33 The Caucasia subregion. This map illustrates the diversity of ethnolinguistic groups in Caucasia and its adjacent, culturally related neighbors.

Figure 5.34 Areas of contention in Caucasia.

Nagorno Karabakh (an enclave of Armenia located inside Azerbaijan) fought against the Muslim majority of Azerbaijan; 30 years later, Armenian military still controls the area and sporadic fighting still breaks out between the two countries. On the Russian side of the border, separatists in Chechnya took up arms for political independence. Although the ethnic strife across Caucasia has some local causes, it has often been instigated by the greater powers that surround Caucasia—Turkey, Russia, and Iran—who are focused on their own national agendas: access to strategic military installations or to agricultural and mineral resources. The historic animosity between the nations of Armenia and Turkey is particularly prominent; it dates back to the mass killings of Armenians in Turkey during World War I, a tragic event that Armenia wants Turkey to accept, unsuccessfully so far, as genocide.

Access to the region's significant oil and gas reserves has recently been another source of conflict in Caucasia. Especially getting the oil and gas safely into the world market is a major problem (see Figure 5.14). Pipelines, trucks, and ocean tankers all have to pass through contested territory or across difficult terrain. The routing of necessary infrastructure involves considering which countries are at odds with each other and which are allies. Regional governments and Western oil companies have built a pipeline from Azerbaijan (where most of the oil is located) to the Black Sea, where the oil can easily be shipped to Europe. Transit through Chechnya is problematic because of Chechnya's conflict with Russia. There are now pipelines through Caucasian Russia, and new ones are under construction. A pipeline also carries oil from Azerbaijan through Georgia to Turkey's Mediterranean coast.

CHECK YOUR UNDERSTANDING

1. What landforms dominate in Caucasia?
2. Discuss ethnic and cultural diversity in Caucasia.
3. What political conflicts have emerged in Caucasia during the post-Soviet era?
4. How is transportation of petroleum a politically sensitive challenge in Caucasia?

THE CENTRAL ASIAN STATES

Since 1991, following nearly 12 decades of Russian colonialism, five independent states have emerged in Central Asia: Kazakhstan, Uzbekistan, Turkmenistan, Kyrgyzstan, and Tajikistan (**Figure 5.35**).

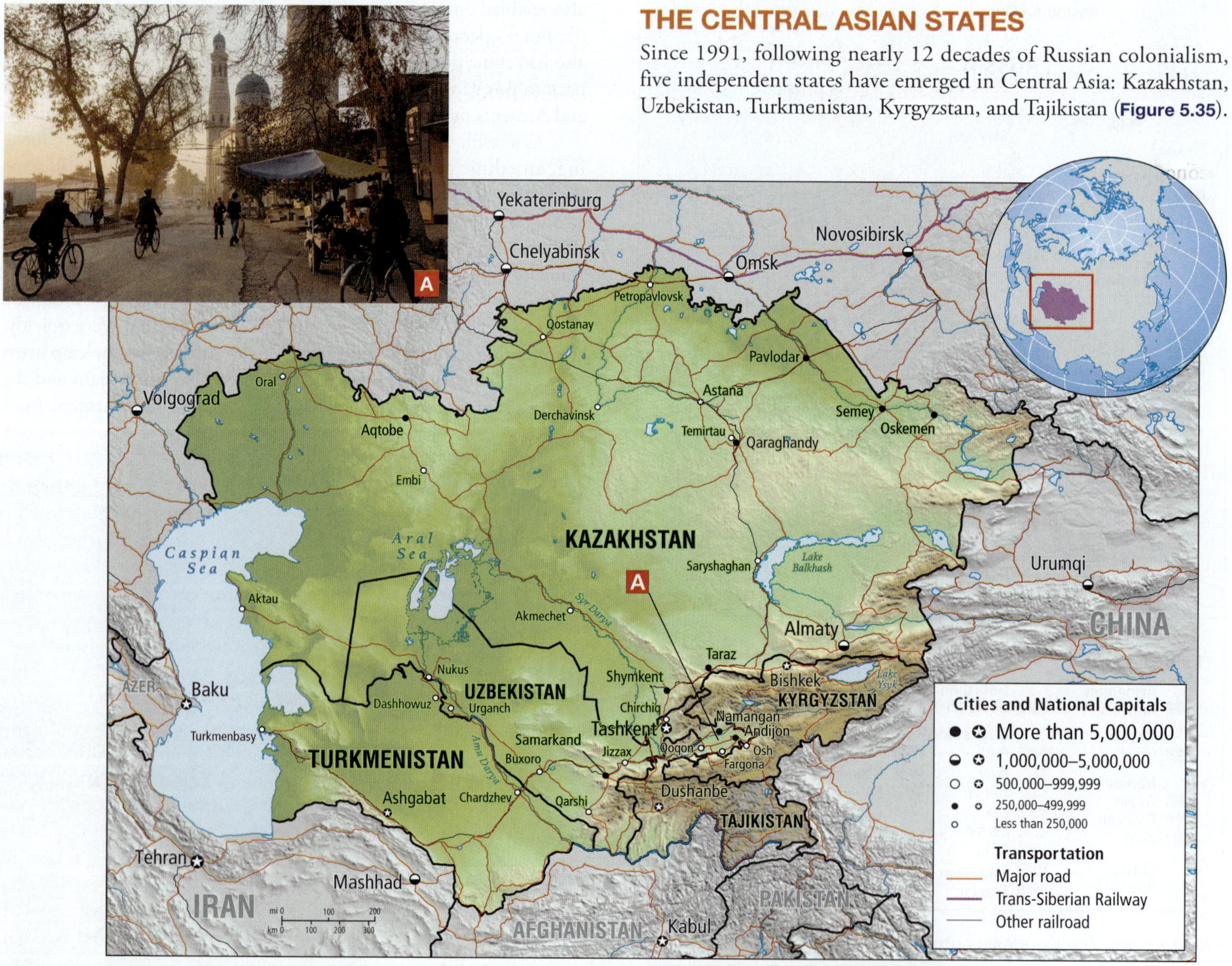

Figure 5.35 The Central Asian States subregion. **A** A bazaar in the densely settled Andijan province in Uzbekistan. [Theodore Kaye/Alamy Stock Photo]

Each of these new nations has traditions that are recognizably different from the others, yet all draw on deep, common traditions.

Physical Setting

Central Asia lies in the center of the Eurasian continent, in the rain shadow of the lofty mountains that lie to the south in Iran, Afghanistan, and Pakistan (see Figures 5.3 and 5.4). Its dry continental climate is a reflection of this location. What rain there is falls mainly in the north, on the Kazakh steppes (see Figure 5.4B), where wide grasslands support limited grain agriculture and large herds of sheep, goats, and horses. In the south, deserts crossed by rivers carry glacial meltwater from nearby high mountain peaks. In Soviet times, these rivers were tapped to irrigate fields of cotton and grains (see "The Loss of the Aral Sea" section). In southern settlements, houses have food gardens walled to protect against drying winds. These gardens nurture many types of melons, herbs, onions and garlic, and tree crops—plums, pistachios, apricots, and apples—all of which were first domesticated in this region in ancient times.

Uzbekistan and Turkmenistan are largely low-lying plains. Kazakhstan has plains in the north and uplands in the southeast that grade into high mountains. Kyrgyzstan and Tajikistan lie high in the Hindu Kush, the Pamir, and Tien Shan mountains. These two countries have exceedingly rugged landscapes: Tajikistan's elevations go from near sea level to more than 22,000 feet (6700 meters), and those in Kyrgyzstan are similar. Both offer little in economic potential; Tajikistan is the poorest country in the entire region and has undergone numerous changes in government and a 1990s civil war involving many political and religious factions, from which the country has not yet recovered.

Past and Present in Central Asia

Civilization flourished in Central Asia long before it did in lands to the north. The ancient Silk Road, a ribbon of trade routes connecting China with the fringes of Europe, operated for thousands of years (see Figure 5.10). Vestiges of nomadic life, common in the days of the Silk Road, can still be seen from Uzbekistan to Siberia. The *yurt*, a portable dwelling of felt on a collapsible wood frame, once provided mobile and durable shelter to millions of nomadic people throughout Central Asia and parts of Russia, Mongolia, and western China (**Figure 5.36**). In a grassland environment where wood and stone material for dwellings was in short supply, the yurt was a sensible solution. It is still being used today, but the decline of nomadism has meant that yurts mainly provide supplementary living space in rural areas. Its symbolic meaning is significant, though; for example, a stylized yurt sits at the center of Kyrgyzstan's flag. In urban Central Asia, modern versions of ancient trading cities still dot the land; Samarkand, which has been dubbed a "crossroad of cultures" by the United Nations, is one such city. At other locations there are remnants of the past, such as the oasis city of Merv (in today's Turkmenistan), which was one of the largest cities in the world around 1200 before it was sacked by the Mongol armies of Genghis Khan. Commerce along the Silk Road diminished as trade shifted to sea lanes after 1500. Central Asia then entered a long period of stagnation until, in the mid-nineteenth century, czarist Russia developed an interest in the region's major export crop, cotton.

Russian imperial agents built modern mechanized textile mills to replace small-scale textile firms that had employed people to do traditional cotton, wool, and silk weaving. As part of the Russification process, the new mills employed imported Russians and were often located in newly built Russian towns alongside railroad lines. In rural areas, cotton fields often replaced the wheat fields that had fed local people.

Figure 5.36 The yurt: A celebrated but fading way of life. A Kazakh herding family and its yurt, the traditional dwelling of the steppes of Central Asia. [Education Images/Universal Images Group/Getty Images]

Kazakhstan, with a long border with Russia, is the most Russified of the Central Asian states today; over 20 percent of the population is ethnically Russian, the Russian language has an official although secondary status in the country, and Orthodox Christianity is the second most popular religion after Islam. Islam—dominant in the other Central Asian states—was officially tolerated during the Soviet era but was undermined by the communist government's promotion of atheism. Fearing that the open practice of Islam would encourage pan-Islamism and weaken Russia's hold on the area, Russia restricted Islamic practices, such as access to mosques, pilgrimages to Makkah (Mecca), and Muslim-influenced dress.

For Central Asia, the transition since independence in 1991 has been rocky. Some conditions of daily life deteriorated. Trade patterns were interrupted; factories closed; transportation and services, never well developed, declined; and poverty spread as health standards fell. Although elections are held and free markets are opening up opportunities for entrepreneurship, governments remain authoritarian and patriarchal, and the heads of state tend to be autocratic. Elections are routinely hijacked by fraudulent vote counts. Most Central Asian governments, though nominally Muslim themselves, fear Islamic fundamentalism and repress even moderate Muslims. Three of the countries—Turkmenistan, Uzbekistan, and Tajikistan—border volatile Afghanistan. The conflict in Afghanistan has the potential to create a "spillover" effect of religious strife into Central Asia.

Figure 5.37 A market in Central Asia. The *bazaar* (marketplace) in Osh, Kyrgyzstan, bustles with the activity of produce sellers and shoppers, much as have similar places throughout Central Asia for thousands of years. [James Strachan/Getty Images]

Russia has tried to maintain a strong influence on Central Asia, but international interest in the region's oil and gas resources is challenging that influence. In Kazakhstan especially, large oil and gas reserves are driving rapid economic growth. GDP per capita (PPP) exploded from U.S.$7800 in 2000 to $26,400 in 2017, and growth may continue in the future, based on the development of new oilfields in western Kazakhstan near the Caspian Sea and the construction of new pipelines (see Figure 5.14). Since the Soviet days, Kazakhstan's existing pipeline network has connected with Russia, but a new pipeline also links the country to China. Part of Kazakhstan's new wealth is being invested in constructing a lavish new capital, Astana (see the vignette in "Urban Central Asia and Caucasia"). Oil wealth is also beginning to make a difference in human well-being, as is reflected in Kazakhstan's relatively high HDI rank, which is similar to that of Russia and significantly above those of the other Central Asian states. Kazakhstan now sees itself as an emerging leader in Central Asia and sells its fossil fuel within the subregion. This is an example of *Eurasianism*, which is a geopolitical idea that emphasizes Central Asia as a strategic region in between Europe and Asia. Kazakhstan has also asserted more independence vis-à-vis Russia, which is symbolized by a recent move away from using the Cyrillic script associated with Russian language and culture.

As communism has faded, Central Asia is reconnecting with its ancient Silk Road days, when trade and long-distance travel were the heart of the economy. The Chinese government is pursuing a new "Silk Road strategy" that is intended to reignite these economic connections (see "Belt and Road Initiative" in Chapter 9). A largely contraband trade in consumer goods (clothes, electronics, cars) and illegal substances (from drugs to guns) with Iran, Afghanistan, and western China is also blossoming. Opportunities to render services such as food vending, transportation, and accommodations to tourists and traveling businesspeople are opening up (**Figure 5.37**).

CHECK YOUR UNDERSTANDING

1. How has economic development proceeded since the Central Asian states became independent in 1991?
2. How has democratization proceeded since the Central Asian states became independent in 1991?
3. How are the Central Asian states reconnecting with nearby Asian countries and to their ancient Silk Road days?

■ CRITICAL THINKING QUESTIONS ■

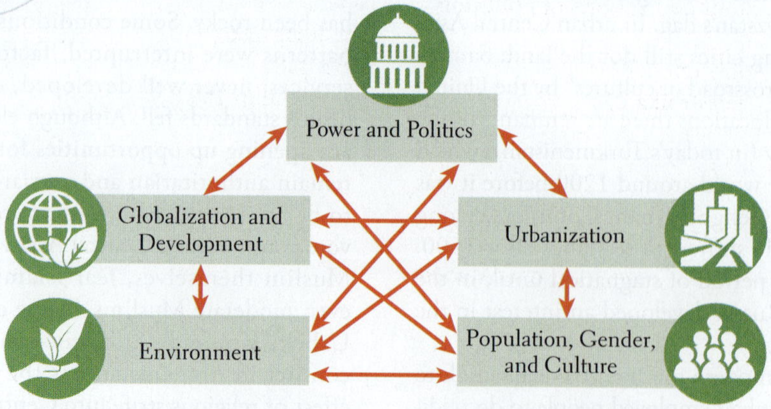

The diagram represents connections among the five geographic themes that structure this book. Listed below are some important questions that have been addressed in this chapter. Answer each question, and indicate where in the diagram you think the topics in each question belong.

1. How would you describe the environmental philosophy of the Soviet Union? Is it different today in Russia and the post-Soviet states? Why or why not?
2. The shrinking of the Aral Sea has long-term historical causes. How is the management of this environmental disaster different now than during the Soviet era?
3. There are many different sources of greenhouse gas emissions in Russia. How are they affected by various forms of economic activities?
4. Considering the importance of fossil fuel extraction for some countries in the region, discuss the pros and cons of fossil fuel from an economic development perspective. What are the economic and geopolitical implications of fluctuating oil prices?
5. In general terms, discuss how the change from a centrally planned economy to a more market-based economy has affected career options and standards of living for the elderly, for young professionals, for unskilled laborers, for members of the military, for women, and for former government officials.
6. Russia exports large amounts of natural gas to Europe. Does that mean Europe is dependent on Russia, or vice versa? Discuss both sides to this argument.
7. Ethnic and national identities are affecting how the internal republics of Russia envision their futures. How might the geography of Russia change if these feelings were to intensify? In what ways might the global economy be affected?
8. Many Russians are dependent on homegrown food. Should this be considered a strength or a weakness from the perspective of economic development and/or environmental sustainability?
9. Why might the conflict in Ukraine be seen as a revival of Cold War tensions? In what ways is it different from Cold War conflicts?
10. The oligarchs occupy a dominant position in Russian society. How do they affect the development of Russia and its relationship with the rest of the world, both politically and economically?
11. One can make the argument that a geopolitical realignment of the post-Soviet states is taking place. Give examples of how this process is unfolding.
12. Urbanization is influenced by both economic forces and government policy. How would you describe that interaction in the context of the Soviet Union?
13. The region exhibits an uneven pattern of urban growth and decline. What is causing this and how does it play out geographically?
14. Discuss the changing birth rate in Russia and other countries in the region. What are the major causes of change? Where are similar declines happening elsewhere?
15. Discuss the religious affiliations of the different post-Soviet states and how those have impacted geopolitics in the region.
16. The ideology of nationalism is experiencing a resurgence in Russia. Compare that to similar trends in other world regions.
17. Immigrants come from post-Soviet states to Russia. What are the forces behind this migration? Contrast the economic advantages with the political opposition to such migrations.
18. To different degrees, there are Russian minorities in all post-Soviet states. What are the historical reasons for that, and why has the number of Russians declined there in the post-Soviet era?

Key Terms

Bolsheviks 270
buffer state 298
capitalists 270
Caucasia 260
centrally planned economy 270
Cold War 271
communism 270
Communist Party 270
czar 269
glasnost 271
Mongols 269
nomadic pastoralists 266
nonpoint sources of pollution 263
oligarchs 272
perestroika 271
permafrost 261
privatization 272
Russian Federation 261
Russification 281
Siberia 261
Silk Road 266
Slavs 268
Soviet Union 259
steppes 261
taiga 263
trafficking 289
tundra 261
underemployment 275
Union of Soviet Socialist Republics (USSR) 259

More Practice at SaplingPlus

Read the interactive e-text, review key concepts, and check your understanding.

During the Arab Spring protests in Cairo, Egypt, a woman checks her laptop while the military patrols the city. [Daniel Berehulak/Getty Images]

6

North Africa and Southwest Asia

On December 17, 2010, 26-year-old Mohamed Bouazizi sold fruits and vegetables from a hand-drawn cart on the streets of Sidi Bouzid, a town in central Tunisia. The problem was that he lacked a permit to do so. Spotted by the police, his cart was promptly confiscated. Reports also suggest that the police were verbally and physically abusive toward Bouazizi. After the incident, he marched up to a government building and set himself on fire; he died in a hospital shortly thereafter.

What would cause a man to take such drastic measures? Discontent among the youthful populace across North Africa and Southwest Asia regarding corruption, repressive policing, and the lack of economic opportunities had been building for some time. The action of one man immediately resonated with other people, first through protests in Sidi Bouzid, then elsewhere in Tunisia. Protests were captured on cell phone cameras, and those images inspired millions to take action.

In Tunisia, increasingly uncontrollable protests called upon President Ben Ali to resign, after 23 years of dictatorial rule. When the military refused the president's orders to crack down on the demonstrations, the regime fell, the president fled, and similar uprisings spread throughout the region, as shown by the photo of the young Egyptian woman in Cairo. The Arab Spring was in full force.

Today, the consequences of Mohamed Bouazizi's act of desperation include a new, more democratic constitution in Tunisia; two election cycles in which different political parties have replaced each other in power; but also uncertainty regarding the role of religion in public life and whether or not greater economic prosperity is on the horizon.

Learning Objectives

Environment: Physical and Human

6.1 Describe natural landscapes and climate in North Africa and Southwest Asia.

6.2 Analyze how the relationship between society and environment in North Africa and Southwest Asia is dominated by access to water.

6.3 Discuss how major religions originated in the region, the emergence of Islamic empires, and eventual Western dominance.

Globalization and Development

6.4 Explain how fossil fuel resources are distributed across North Africa and Southwest Asia.

6.5 Discuss the effect of fossil fuel resources on economic development in the region.

Power and Politics

6.6 Describe how democratization and political conflict have evolved after the Arab Spring.

6.7 Analyze how religion, particularly Islamism, has affected politics in North Africa and Southwest Asia.

Urbanization

6.8 Discuss how and where new wealth has been invested in urban development.

6.9 Describe domestic and international migration patterns into the region's cities and the quality of life of those migrants.

Population, Gender, and Culture

6.10 Describe different demographic structures among the countries of North Africa and Southwest Asia.

6.11 Explain gendered spaces in North Africa and Southwest Asia and what that means for women's lives.

6.12 Discuss the limited rights of women in North Africa and Southwest Asia.

Subregions

6.13 Identify key characteristics of the subregions of North Africa and Southwest Asia.

ATLANTIC OCEAN
Mediterranean Sea
Black Sea
Gulf of Guinea
Bight of Benin
Gold Coast
Ivory Coast
Tyrrhenian Sea
Ionian Sea
Adriatic Sea
Aegean Sea
Sea of Azov
Balearic Islands
Corsica
Sardinia
Sicily
Crete
Madeira Islands (Portugal)
Canary Islands (Spain)
Azores
Iberian Peninsula
Pyrenees
Alps
Apennines
Atlas Mountains
Ahaggar Mountains
Anatolian Plateau
Sahara
Sahel
Libyan Desert
Nubian Desert
Bosporus Strait
Suez Canal
Lake Nasser
Lake Chad
Lake Volta
Lac de Kossou
Duero
Guadalquivir
Danube
Dnieper
Don
Nile
Blue Nile
White Nile
Niger
Benue
Gambia
Black Volta
White Volta
Oti
Ubangi
Lake Tuz
UNITED KINGDOM
London
NETHERLANDS
Rotterdam
Amsterdam
BELGIUM
Brussels
GERMANY
Berlin
Munich
POLAND
Warsaw
Vilnius
Minsk
BELARUS
RUSSIA
Kiev
Kharkiv
UKRAINE
Dnipropetrovsk
Donetsk
Rostov-na-Donu
Odesa
MOLDOVA
Chisinau
ROMANIA
Bucharest
Paris
Luxembourg
FRANCE
Prague
CZECH. REP.
SLOVAKIA
Bratislava
Vienna
AUSTRIA
Budapest
HUNGARY
Bern
SWITZERLAND
Vaduz
Ljubljana
SLOVENIA
Zagreb
CROATIA
BOSNIA AND HERZEGOVINA
Sarajevo
Belgrade
SERBIA
Podgorica
MONTENEGRO
Tiranë
ALBANIA
Skopje
MACEDONIA
Sofia
BULGARIA
GREECE
Athens
Turin
Milan
MONACO
SAN MARINO
VATICAN CITY
Rome
ITALY
Naples
ANDORRA
PORTUGAL
Lisbon
Madrid
Toledo
SPAIN
Barcelona
Istanbul
Ankara
TURKEY
Izmir
Adana
Ordu
Aleppo
Al Ladhiqiyah
Nicosia
CYPRUS
LEBANON
Beirut
Damascus
ISRAEL
Tel Aviv
Jerusalem
Gaza
OPT
Amman
Valletta
MALTA
Tanger
Rabat
Casablanca
Fes
Marrakech
MOROCCO
Erfoud
El Aaiún
Western Sahara (Morocco)
Oran
Algiers
Constantine
Beskra
El Oued
Touggourt
Ouargla
Ain Sefra
ALGERIA
Tamanrasset
Tunis
Gafsa
Sfax
TUNISIA
Tripoli
Banghazi
Tobruk
LIBYA
Zillah
Waw An Namus
Al Jawf
Sidi Barrani
As Sallum
Alexandria
Port Said
Cairo
Al Jizah
Suez
Al Aqabah
EGYPT
Al Uqsur (Luxor)
Aswan
Aswan High Dam
Bir Misahah
Dunqulah
SUDAN
Khartoum
Al Fashir
SOUTH SUDAN
Shambe
Juba
Nouakchott
MAURITANIA
MALI
Tombouctou
NIGER
Niamey
CHAD
N'Djamena
SENEGAL
THE GAMBIA
GUINEA-BISSAU
GUINEA
Conakry
Bamako
Freetown
SIERRA LEONE
Monrovia
LIBERIA
CÔTE D'IVOIRE
Yamoussoukro
Abidjan
Ouagadougou
BURKINA FASO
GHANA
Accra
TOGO
Lomé
BENIN
Porto Novo
NIGERIA
Abuja
Ibadan
Lagos
CAMEROON
Douala
Yaoundé
Malabo
EQUATORIAL GUINEA
SÃO TOMÉ and PRÍNCIPE
São Tomé
Libreville
GABON
CONGO
Brazzaville
Kinshasa
Cabinda (Angola)
ANGOLA
CENTRAL AFRICAN REPUBLIC
Bangui
DEMOCRATIC REPUBLIC of THE CONGO
UGANDA
Kampala
RWANDA
Kigali
BURUNDI
Bujumbura
TANZ.
Las Palmas
50°N
40°N
30°N
20°N
10°N
0° Equator
30°W
20°W
10°W
0° Prime Meridian
Land Elevations
meters feet
4877 16,000
3353 11,000
2134 7000
914 3000
305 1000
152 500
0 0
mi 0 100 200 300 400 500
km 0 200 400 600 800
1:24,500,000
Lambert Azimuthal Equal Area Projection

As the story of Mohamed Bouazizi suggests, the region is no stranger to political conflict. For different reasons, many of which we will explore in this chapter, conflicts have been local, regional, and even global in scale. Since 2010, a series of events called the Arab Spring was brought on by deep and structural problems. Large numbers of the poor, the uneducated, the educated unemployed, minorities, and women have been pushed further and further into poverty and powerlessness by political and economic systems that have privileged the wealthy and politically connected. Movements like the Arab Spring arise when enough people decide they can no longer tolerate these systems.

The region of North Africa and Southwest Asia contains 21 countries plus the occupied Palestinian Territories. Physically, the region is part of two continents, Africa and Eurasia (**Figure 6.1**). The North African countries of this region stretch from Western Sahara and Morocco on the Atlantic Ocean, through Algeria, Tunisia, Libya, and Egypt along the Mediterranean Sea, plus Sudan, which lies south of Egypt. Eurasia begins on the eastern side of the Red Sea with the Arabian Peninsula, which consists of Saudi Arabia, Yemen, Oman, the United Arab Emirates, Qatar, Bahrain, and Kuwait; the countries of the Eastern Mediterranean littoral (shoreline), including Jordan, Israel, the occupied Palestinian Territories, Lebanon, and Syria; and Turkey, Iraq, and Iran (**Figure 6.2**). This region is often referred to as the Middle East, a term we do not use in this book for two reasons. First, it arose during the colonial era and describes the region from a limited Euro-American geographic perspective. To someone in China or Japan, the region lies to the far west, and to someone in Russia, it lies to the south. Second, the term Middle East, while geographically imprecise, typically refers to the area east of the Mediterranean but does not include all the countries in North Africa and Southwest Asia.

What Makes North Africa and Southwest Asia a Region?

To most outsiders, North Africa and Southwest Asia is a region characterized by five qualities: It is the center of the religion of Islam; it holds a great deal of Earth's petroleum reserves; water is generally scarce here; Arab culture dominates most countries; and women are discriminated against more intensely than in most of the rest of the world. While these five features are all present, they alone provide a far too simplistic picture of the region.

The vast majority of people practice **Islam**, a monotheistic religion that emerged between 601 and 632 C.E., which is when the founder of the religion, the Prophet Muhammad, received revelations from God, according to believers. Islam is a faith that is interpreted in many different ways. Most Muslims are moderate in their thinking and accept the validity of other beliefs, especially that of Christianity; for example, Jesus (the central figure in Christianity) plays an important role in Islam. A few Muslims are drawn to ultrafundamentalist versions of Islam known as **Islamism**. The term *Islamism* refers to a Muslim

Islam a monotheistic religion that emerged on the Arabian Peninsula in the seventh century C.E.; its tenets are based on the writings of the Prophet Muhammad

Islamism a grassroots religious revival in Islam that seeks to curb secular influences in society and to replace secular governments and civil laws with governments and laws guided by Islamic principles

◀ **Figure 6.1** Physical features of North Africa and Southwest Asia.

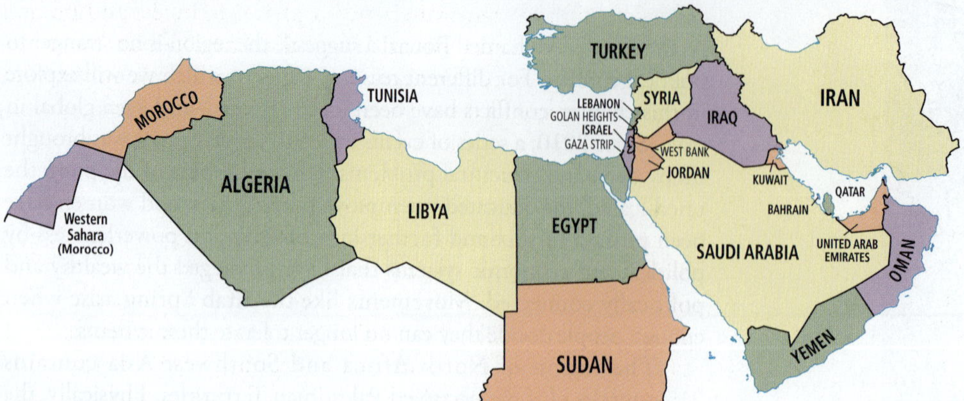

Figure 6.2 Political map of North Africa and Southwest Asia.

religious activist movement that seeks to curb secular influences that are spreading as a result of globalization. Some Islamists think that Western culture has a corrupting influence on Islamic countries, while others are more open to a diversity of ideas. The global public opinion institute Pew Research Center reports that people in many Muslim countries strongly support religious pluralism and the freedom to worship, while also embracing Islamic principles as a basis for legislation, somewhat contradictory findings. However, when it comes to violent, radical groups such as Al Qaeda, the Taliban, and the Islamic State, 13 to 15 percent of Muslims surveyed around the world support them.

Fossil fuel deposits—made up of oil, coal, and natural gas formed over millions of years from the fossilized remains of dead plants and animals—are highly uneven in their global distribution. Fossil fuel *resources* include all existing fossil fuels, but only those that can be extracted under current technological and economic conditions are considered *reserves* (see Figures 6.19 and 6.20 later in the chapter). In North Africa and Southwest Asia, oil and natural gas reserves are mainly located and extracted around the Persian Gulf. These fossil fuels are exported throughout the world at tremendous profit, but the revenues are not equitably distributed to the people in the countries from which they are generated. For example, in the oil-rich areas around the Persian Gulf, profits have traditionally gone to the members of a few large, privileged families, and only modest amounts are spread to the rest of the citizens. More recently, however, a number of these **Gulf states** have invested in public services, education, and infrastructure to promote a higher level of well-being among most people. Countries without significant fossil resources remain poor, which has resulted in a region characterized by uneven economic development.

This region is generally arid, but the degree of aridity and its impact on development vary widely. Newly revealed water resources that are found deep underground may change development options.

fossil fuel a source of energy formed from the remains of dead plants and animals; fossil fuel includes oil, coal, and natural gas

Gulf states Arab countries that border the Persian Gulf: Saudi Arabia, Kuwait, Bahrain, Oman, Qatar, and the United Arab Emirates

occupied Palestinian Territories (oPT) Palestinian lands occupied by Israel since 1967

Arab culture and language are widespread, but many people in the region are not Arab. The populations of Turkey and Iran, the second and third most populous countries in the region, are of non-Arab ethnicities, as are many minority populations, such as the Kurds and Berbers. Some Christians in the region are ethnically Arabs, while others are not.

Finally, the role and status of women are in transition. The experiences of women vary widely from country to country. In Turkey, women have many more options for education and work than in Saudi Arabia or Yemen. Gender issues also vary between more rural and urban areas. Urban women tend to be more educated, outspoken, and active in commerce, public life, and government, while rural women tend to lead more secluded domestic lives with few educational opportunities. The movement toward greater gender equity is generally strongest in urban areas.

Terms in This Chapter

In this book, as mentioned earlier, we choose not to use the common term *Middle East*. The term *Arab world* is used only where it applies, since many people in the region are not of Arab ethnicity.

We use the term **occupied Palestinian Territories (oPT)** to refer to Gaza and the West Bank, which are areas under the control of Israel. The word “occupied” is lowercase to show the supposed temporary quality of the occupation. The U.S. Department of State uses the term *Palestinian Territories*. The United Nations, upon recently agreeing to give the territory observer status, now refers to it as the *State of Palestine*. Many, but not all, countries in the world recognize it as an independent state.

The Arabic language uses a different alphabet than English, which means that Arabic words and place names have to be represented with English letters. The goal is to use a spelling that is understandable to English speakers and as close as possible to the original pronunciation. However, such transliteration—the representation of the sounds of another language—is not a straightforward process. In fact, different spellings of the same word are not unusual. For example, in this book we use Qur’an (instead of Koran) and Makkah (instead of Mecca).

ENVIRONMENT: PHYSICAL AND HUMAN

6.1 Describe natural landscapes and climate in North Africa and Southwest Asia.

6.2 Analyze how the relationship between society and environment in North Africa and Southwest Asia is dominated by access to water.

6.3 Discuss how major religions originated in the region, the emergence of Islamic empires, and eventual Western dominance.

Landforms and climates are particularly closely related in this region. The climate is dry and hot in the vast stretches of the relatively low, flat land; it is somewhat moister where mountains capture orographic rainfall. The lack of vegetation perpetuates aridity. Without plants to absorb and hold moisture, the rare but occasionally copious rainfall simply runs off, evaporates, or sinks rapidly into underground aquifers, of which there are many across the region.

CLIMATE

No other region in the world is as dry as North Africa and Southwest Asia (**Figure 6.3**). A belt of dry air that circles the planet between roughly 20° N and 30° N creates desert climates in the Sahara of North Africa, the Eastern Mediterranean, the Arabian Peninsula, Kuwait, Iraq, and Iran.

The Sahara's size and location under this high-pressure belt of dry air make it a particularly hot desert region. In some places, temperatures can reach 130°F (54°C) in the shade at midday. With little cloud cover, water, or moisture-holding vegetation to retain heat, nighttime temperatures can drop quickly to below freezing. Nevertheless, in even the driest zones, humans survive as traders and nomadic herders at scattered oases, where they maintain groves of drought-resistant plants such as date palms. Desert inhabitants often wear light-colored, loose, flowing robes that reflect the sunlight, retain body moisture during the day, and provide warmth at night.

In the uplands and at the desert margins, enough rain falls to nurture grass, some trees, and limited agriculture. Such is the case of a belt from northern Morocco along the Mediterranean coast to Tunisia; in the highlands of Sudan, Yemen, and Turkey; and in the northern parts of Iraq and Iran (see Figure 6.3B–D). The rest of the region, generally too dry for cultivation, has for generations been the prime herding lands for nomads, such as the Kurds of Southwest Asia, the Berbers and Tuareg in North Africa, and the Bedouin of the steppes and deserts on the Arabian Peninsula. Recently, most nomads and their descendants have settled in farming communities or urban places. Some of their lands are now irrigated for commercial agriculture, but the general aridity of the region means that sources of irrigation water (rivers and aquifers) are scarce.

LANDFORMS AND VEGETATION

The rolling landscapes of rocky and gravelly deserts and steppes cover most of the region (see Figure 6.1). In a few places, mountains capture sufficient moisture to allow plants, animals, and humans to flourish. In northwestern Africa, the Atlas Mountains, which stretch from Morocco on the Atlantic coast to Tunisia on the Mediterranean coast, lift damp winds from the Atlantic Ocean, creating orographic rainfall of more than 50 inches (127 centimeters) per year on windward slopes. In some Atlas Mountain locations, there is enough snowfall to support a modest ski industry.

A rift that formed between two tectonic plates—the African Plate and the Arabian Plate—separates Africa from Southwest Asia (see Figure 1.6). The rift, which began to form about 12 million years ago, is now filled by the Red Sea. The Arabian Peninsula lies to the east of this rift. Mountains bordering the rift in Yemen rise to over 12,000 feet (3660 meters). They capture enough rainfall to sustain agriculture, which has been practiced there for thousands of years.

Behind these mountains to the northeast lies the great desert region of the Rub'al Khali (which means "the empty quarter"). Like the Sahara, it has virtually no vegetation. The desert is generally flat although sand dunes of the Rub'al Khali, which are constantly moved by strong winds, are among the world's largest, some as high as 2000 feet (610 meters; see **Figure 6.4**).

The landforms of Southwest Asia are more complex than those of North Africa. The Arabian Plate is colliding with the Eurasian Plate and pushing up the mountains and plateaus of Turkey and Iran (see Figure 1.6). Turkey's mountains lift damp air that passes over Europe and the Mediterranean from the Atlantic, resulting in considerable precipitation (**Figure 6.5**). The rain makes it to the mountains of western Iran, but the farther east one goes, the drier it gets. The tectonic movements that create mountains also create earthquakes, especially in Turkey and Iran.

There are only three major river systems in the entire region, and all have attracted human settlement for thousands of years. The Nile flows north from the moist central East African highlands. It crosses arid Sudan and desert Egypt and forms a large delta on the Mediterranean. The Euphrates and Tigris rivers both begin in the mountains of Turkey. The rivers then flow southeast to the Persian Gulf. A much smaller river, the Jordan, starts as snowmelt in the uplands of southern Lebanon and flows through the Sea of Galilee to the Dead Sea. Most other streams are for most of the year dry riverbeds, or *wadis*, carrying water only after the generally light rains that fall between November and April (**Figure 6.6**).

North Africa and Southwest Asia were home to some of the very earliest agricultural societies (see the "Human Patterns over Time" section). Today, rain-fed agriculture is practiced primarily in the highlands and along parts of the Mediterranean coast, where there is enough precipitation to grow citrus fruits, grapes, olives, and many vegetables, though supplemental irrigation is often needed. While irrigated agriculture is an age-old type of farming, it has become increasingly significant in modern times, both in the valleys of the major rivers, where seasonal flooding fills irrigation channels, and where aquifers are tapped by wells.

ENVIRONMENT

Environmental concerns are only beginning to become an overt focus in this region. In part, this is because for thousands of years

6.3 Climates of North Africa and Southwest Asia

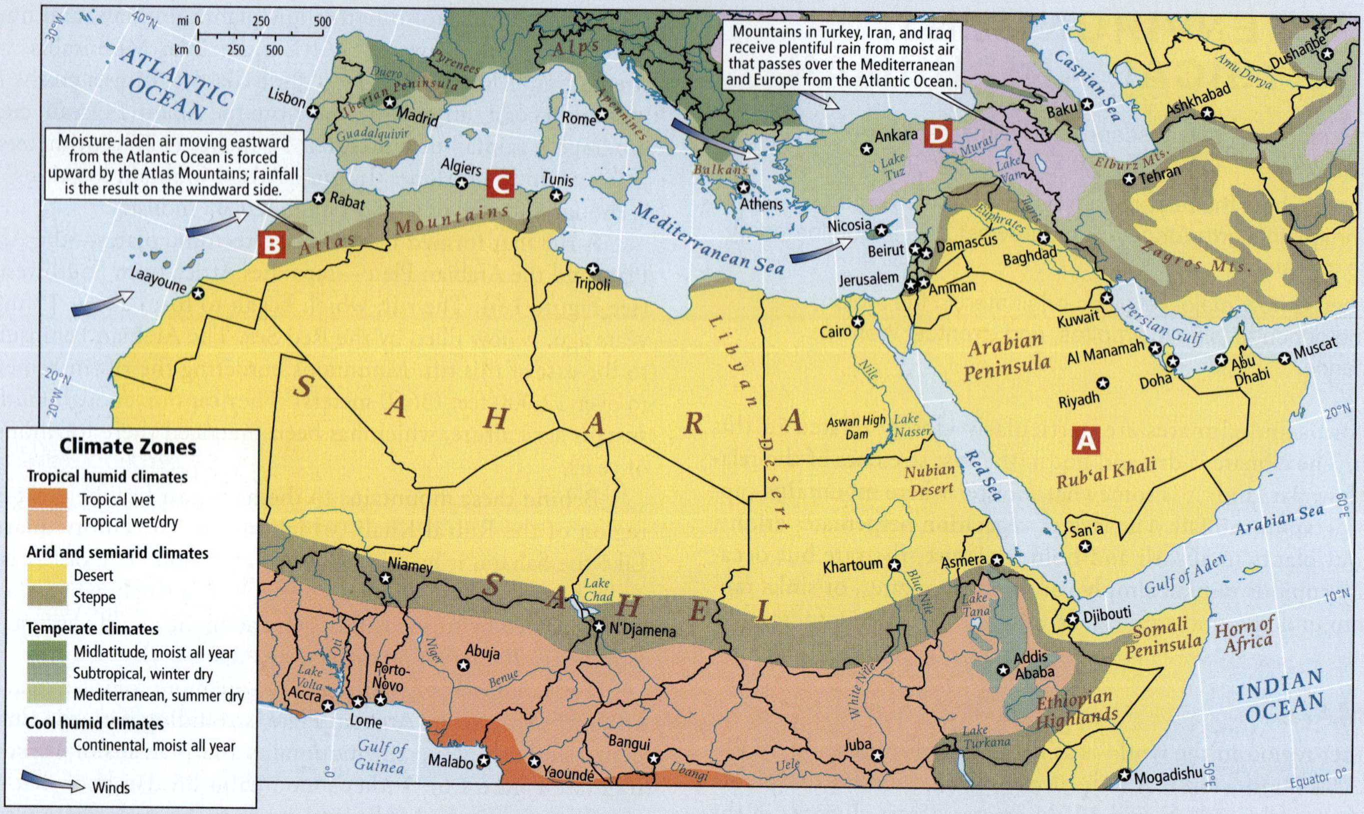

A **Desert, Saudi Arabia** [Saleh AlRashaid/Getty Images]

B **Steppe, Morocco** [J Boyer/The Image Bank/Getty Images]

C **Mediterranean, summer dry, Algeria** [Frederic Soreau/Photononstop/Getty Images]

D **Continental, moist all year, Turkey** [Ozcan MALKOCER/Getty Images]

Figure 6.4 Sand dunes in the Oman section of Rub'al Khali. [Malcolm MacGregor/Getty Images]

Figure 6.5 Mount Ararat, Turkey. At 16,854 feet (5137 meters), Mount Ararat is a snow-capped peak of volcanic origin. It is located in the far east of Turkey near the Armenian border. In the Bible's Book of Genesis, Noah builds a boat to escape a flood that encompasses much of Earth. Mount Ararat is commonly assumed by biblical scholars to be the higher ground on which Noah's Ark came to rest. [Franz Aberham/Getty Images]

people here have confronted the challenges of a naturally arid environment and have been quite attuned to its scarcities (**Figure 6.7**).

An Ancient Heritage of Water Conservation

The **Qur'an** (or **Koran**), the holy book of Islam, guides believers to avoid degrading the environment and to share resources, especially water. In actual practice, the residents of this region conserve water better than most people in the world. Daily bathing is a religious requirement and is often done in a *hammam*, or public bath, where water use is minimized. For millennia, groundwater at higher elevations has been captured and moved to lower-lying dry fields and villages via constructed underground water tunnels, called *qanats*,

Figure 6.6 A wadi in the Atlas Mountains of northern Algeria. Due to intermittent rainfall, riverbeds that are dry for much of the year are a common physical feature of the region. [DEA/De Agostini/C. SAPPA/Getty Images]

that minimize evaporation. The qanats allow for irrigation during periods of the year when surface water is scarce. Likewise, traditional architectural designs that maximize shade and airflow are used to create buildings that stay cool. This includes thick stone walls and small windows that keep out intense daytime heat, and wind towers that funnel breezes above the ground level into the first floor of buildings.

Despite their history of water-conserving technologies and practices, however, this region's 540 million residents now have such limited water resources that even clever combinations of ancient and modern measures are no longer sufficient to ensure an adequate supply of water. Growing populations, water pollution, and unwise modern usages of water virtually guarantee that water shortages will be more extreme in the future, especially with the likelihood of climate change–related drier conditions. Turkey, Iraq, and Iran have the most plentiful water resources. However, all other countries in the region suffer from physical freshwater scarcity, meaning that they have less water than the minimum the United Nations considers necessary to support basic human development—1000 cubic meters per person per year. The poorer countries in the region also suffer from economic water scarcity, which is when people or countries do not have the financial means to utilize existing water resources.

Research on groundwater storage in North Africa reveals that deep aquifers may hold up to 100 times the water available through renewable freshwater resources (**Figure 6.8**). This *fossil water*, deposited thousands of years ago, is unevenly distributed and so deep that special pumps are required to bring it to the surface. There probably isn't enough to meet the expected increases in demand for agricultural irrigation and drinking water for rapidly growing urban populations, but this stored groundwater may help alleviate scarcities in the short term.

Water and Food Production

The predominant use of water in North Africa and Southwest Asia

Qur'an (or Koran) the holy book of Islam, believed by Muslims to contain the words God revealed to Muhammad

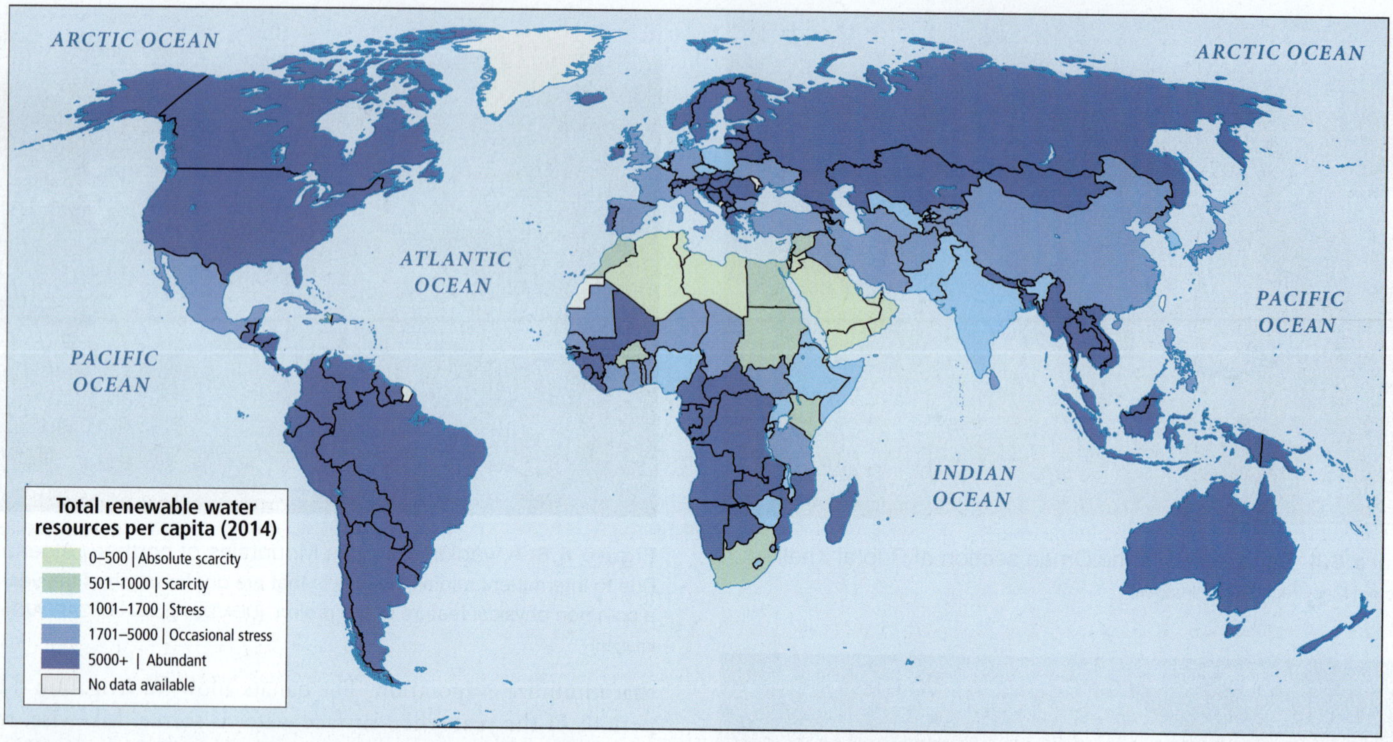

Figure 6.7 Total renewable water resources per capita in cubic meters, 2014. Most of the region experiences water stress or scarcity, which occur when so much water is withdrawn from rivers, lakes, and aquifers that not enough water remains to meet growing human and ecosystem requirements for sustainability. [Research from: *The United Nations World Water Development Report 2016: Water and Jobs*, p. 16, at http://www.unesco.org/new/en/natural-sciences/environment/water/wwap/wwdr/2016-water-and-jobs/.]

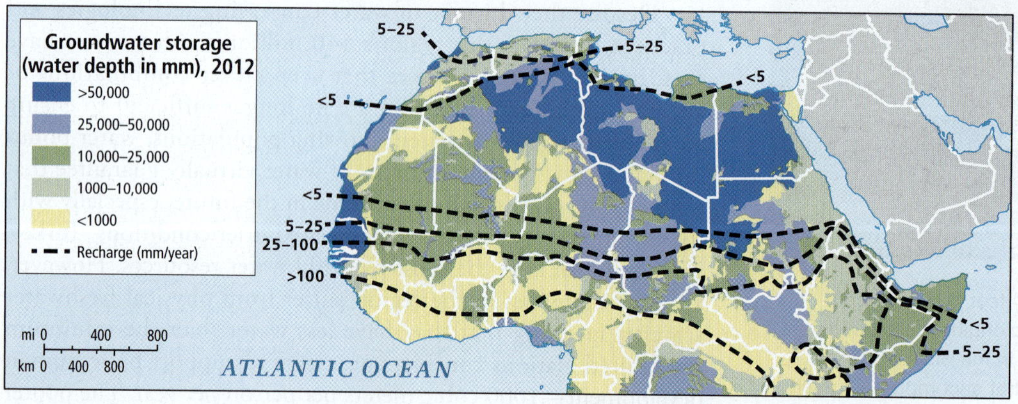

Figure 6.8 Water for the future? A map of fossil groundwater deep below the deserts in North Africa. This ancient water deposit, already tapped in Libya, is expensive to extract and is unlikely to meet future needs because it is not being replenished. [Research from: A. M. MacDonald, H. C. Bonsor, B. E. O. Dochartaigh, and R. G. Taylor, "Quantitative Maps of Groundwater Resources in Africa," *Environmental Research Letters*, April 19, 2012, at http://iopscience.iop.org /1748-9326/7/2/024009/pdf/1748-9326_7_2_024009.pdf.]

is for irrigated agriculture, even though agriculture does not contribute significantly to national economies. In Tunisia, for example, agriculture accounts for 10 percent of GDP but 83 percent of all the water used. Only a small amount of available water is used for household and industrial purposes. Most national governments actually subsidize irrigated agriculture because they are worried about depending completely on imported food. Irrigated agriculture also supports jobs and family economies that are crucial to rural communities. Some of this region's food specialties appear in **Figure 6.9**.

salinization a process where evaporation of water leaves salty minerals behind in the soil; may occur naturally or as a result of irrigation

Until the twentieth century, agriculture was confined to a few coastal and upland zones where rain could support cultivation, and to river valleys (such as the Nile, Tigris, and Euphrates valleys) where farms could be irrigated with simple gravity-flow technology. Some agriculture has also been possible in *oases*—small areas in the desert where vegetation can survive because groundwater is found near the land surface. However, to accommodate population growth and development, most countries in the region now have ambitious mechanized irrigation schemes that have expanded agriculture deep into formerly uncultivable desert environments.

Over time, irrigation projects damage soil fertility through **salinization**. When irrigation is used in hot, dry environments, pure water evaporates, leaving behind a salty residue of the minerals that exist in the water. When too much residue accumulates, the plants are unable to grow or even survive. This human-induced loss

Figure 6.9 LOCAL LIVES: Foodways in North Africa and Southwest Asia

(A) A Moroccan *tagine* stew is prepared in a heavy clay pot (also called a tagine) with two parts: a round bottom and a cone-shaped lid with a hole in the top. The bottom holds the food, while the hole in the lid allows excess steam to escape, preventing the food from overcooking. Tagines are placed directly on hot coals or a grill and left to cook for hours. [Gary Yeowell/Getty Images]

(B) Falafel is fried in Gaza, Palestine. Either fava beans or chickpeas are mixed with parsley, scallions, garlic, and spices and then formed into a ball or patty that is deep-fried. Falafel is often made into a flatbread sandwich and served with tomatoes, cucumbers, lettuce, and sauces. The origin of this popular vegetarian street food, which has spread throughout this region and much of the world, is unclear. Egyptian Coptic Christians say they invented it as a substitute for meat during Lent. Others claim it originated in ancient Egypt, or even in South Asia. [Abid Katib/Getty Images]

(C) *Khubz*, or flatbread, is made in a *tannur*, a wood- or charcoal-fired oven. Khubz consists of a dough that is rolled flat and then pressed onto the interior side walls of the tannur, where the radiant heat from the fire rapidly bakes it. Khubz is so central to diets in this region that in recent decades most governments have subsidized it, keeping prices artificially low. Policies to remove subsidies have resulted in food riots among the urban poor, which can escalate to political uprisings. [SABAH ARAR/AFP/Getty Images]

of fertility is one of the largest environmental water issues that the world faces today in arid and semiarid regions. Note that the process of salinization occurs naturally as well. Some of the deserts in the region are salt flats where, at some point in time, surface water has evaporated, resulting in a top layer of salty minerals.

Israel has developed relatively efficient techniques of drip irrigation that uses hoses and pipes that deliver the precise amount of water, drop by drop, for each plant to grow, with minimal runoff. This dramatically reduces the amount of water used, limits salinization, and frees up water for other uses. Until very recently, however, poorer states have been unable to afford this somewhat complex technology.

Vulnerability to Climate Change

North Africa and Southwest Asia are especially vulnerable to the multiple effects of climate change (**Figure 6.10**). Higher temperatures are already increasing evaporation rates and water scarcity (Figure 6.10B, C). Also, as the climate warms, a sea level rise of a few feet could severely impact the Mediterranean coast, especially the Nile Delta, one of the poorest and most densely populated lowland areas in the world (Figure 6.10A), and the Persian Gulf coasts, where vast new high-tech cities that lie at sea level could be flooded (see Figure 6.28B later in the chapter).

Independent of any environmental effects of climate change, global efforts to reduce dependencies on fossil fuel consumption could devastate oil- and gas-based economies and transform the region's geopolitics, leaving countries vulnerable to losing some of their global power and income levels. Despite enormous wealth from fossil fuel sales, few countries have undertaken the significant economic diversification needed to prepare for the reduced global consumption of the fossil fuels they sell. The countries in the region also contribute to climate change by emitting greenhouse gases. In fact, the affluent Gulf states have the highest emissions in the world per capita. Most recent data showed that Qatar had the world's highest emissions, with Bahrain and Kuwait also in the top five. These are small countries with overall emissions that are dwarfed by China and the United States, but their development is nevertheless based on wasteful energy consumption.

War and insurgency have also been major sources of environmental degradation in this region. The largest oil spill in history occurred during the first Gulf War (1991), when the Iraqi government released 300 million gallons of oil into the Persian Gulf in order to thwart a land invasion by the United States. During the Gulf War, there were also oil fires that burned for months, and they had measurable impact on global greenhouse gas emissions.

Climate Change and Desertification Climate change could accelerate **desertification**, the conversion of nondesert lands into deserts. As temperatures increase, soil moisture decreases because of evaporation. As a result, plant cover is reduced, which causes less protection for arid soil. Bare patches of soil can become badly eroded by wind, and eventually sand dunes can blow onto formerly vegetated land.

In the grasslands (steppes) that often border deserts, a wide array of land use changes contributes to the general drying. For example, as groundwater levels fall because

desertification a set of ecological changes where a dry area with some vegetation is turned into desert

PHOTO ESSAY

6.10 Vulnerability to Climate Change in North Africa and Southwest Asia

Water scarcity, sea level rise, desertification, food scarcity, and political instability are some of the factors that make parts of this region highly vulnerable to climate change. The map shows how the poorest (such as Sudan and Yemen) and conflict-prone (such as Syria and Iraq) countries have the highest degree of vulnerability. Affluent Gulf states can cope better with the effects of climate change.

THINKING GEOGRAPHICALLY

A **What about Alexandria's location makes it particularly vulnerable to sea level rise?**

B **What are some of the factors in the region that make it vulnerable to climate change and complicate emergency responses to climate change, thus limiting resilience?**

C **How do conflict and climate change in combination create vulnerable populations?**

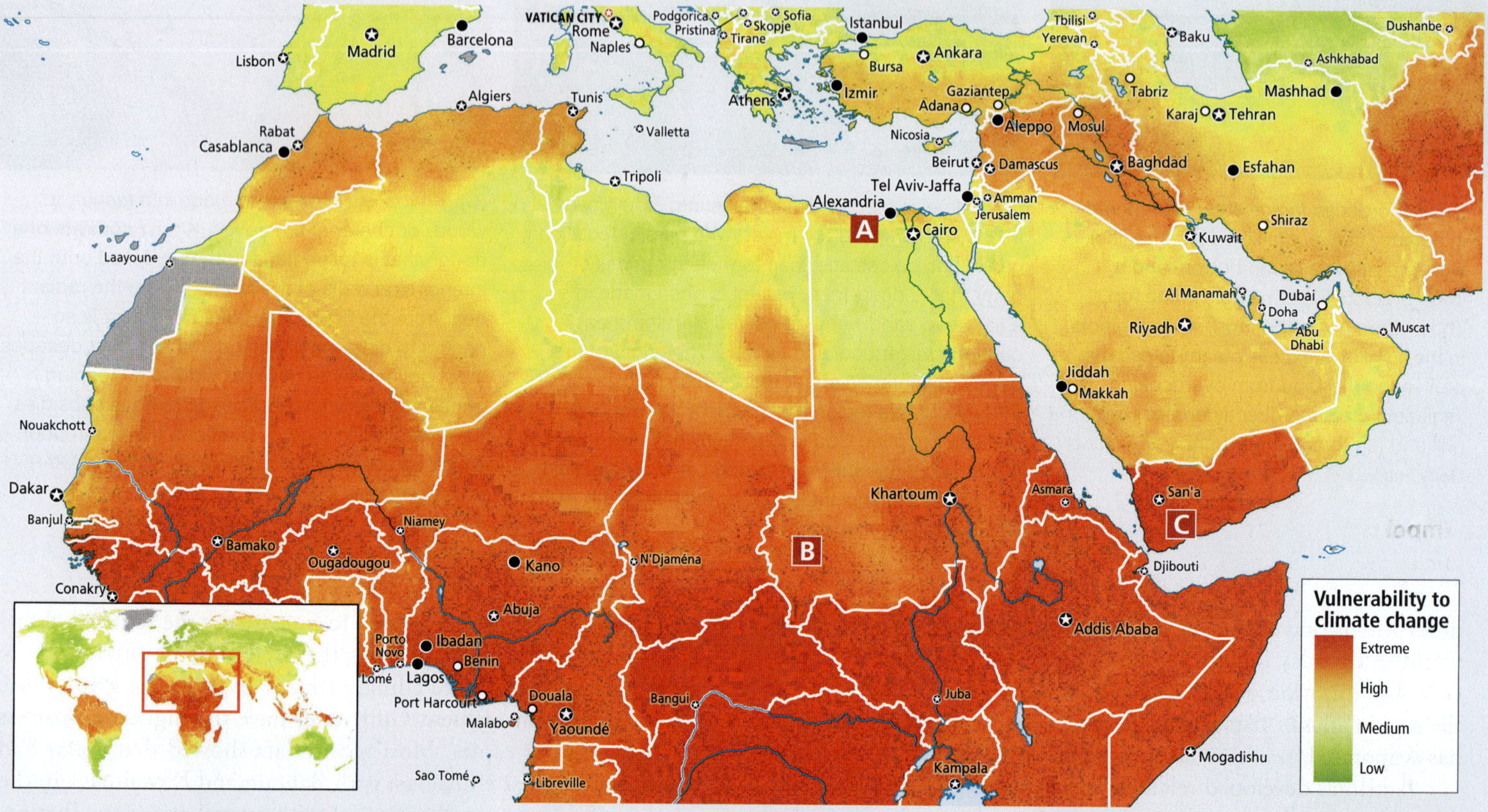

A Alexandria, Egypt. Located on the Mediterranean coast of the Nile Delta region, Alexandria is a low-lying city exposed to rising sea levels. Efforts to improve the sea wall around the city, visible at the right of the photo, may be outpaced by rising sea levels. Eighty percent of the country's imports and exports run through Alexandria. [James Baigrie/Getty Images]

B Sudan. Refugees from Darfur, Sudan, line up for food and water in neighboring Chad. Sudan is exposed to drought, and poverty has left the population very sensitive to any disruption in food or water supplies. Meanwhile, political instability is uprooting people from their homes and complicating emergency responses. [Scott Nelson/Getty Images]

C Wadi Mur, Yemen. Residents of Wadi Mur, Yemen, ride out a sandstorm. The same factors that make Sudan so vulnerable to climate change also affect Yemen. Additionally, a conflict within the country has resulted in even more internally displaced people. These populations further stretch Yemen's resources. [C. Boisvieux/ZUMA Press/Newscom] water is being pumped out for the needs of urban populations and irrigated agriculture, some plant roots no longer reach sources of moisture, causing the plants to die. In other cases, nomadic herders have taken up settled cattle ranching, which also contributes to desertification. Governments encourage settlement because the mobility of nomads is viewed as complicating management for modern states where records must be kept, taxes collected, and children sent to school with regularity. However, ranching on fragile grasslands may lead to overgrazing, which can kill the grasses and increase evaporation. If irrigation is used to support ranching, such projects can also deplete groundwater resources, resulting in long-term desertification. Incidentally, encouraging nomads to settle may actually increase their vulnerability to climate change, because it is their very mobility and skills at cycling seasonally through multiple environments that have helped nomads adapt to climate variability over many generations.

Around the perimeter of the Saharan desert, there have been several attempts at preventing and reducing the risk of desertification. In North Africa, Algeria has maintained a "green barrier" since the 1970s, which stretches the entire east-to-west extent of the country, in between the Mediterranean north and the desert south. There, afforestation (establishing forests on land not previously forested) efforts have helped to maintain vegetation and soil quality in the face of desertification.

Imported Food and Virtual Water Because agriculture is so difficult due to the aridity of this region, the diets of nearly all people here include imported food. The total per capita water use of this region must thus include the water used to produce this imported food. Virtual water is the volume of water used to produce all that a person consumes in a year (see Chapter 1). For example, 1 kilogram (2.2 pounds) of beef requires 15,500 liters (4094 gallons) of water to produce, while 1 kilogram of goat meat requires just 4000 liters (1056 gallons). Goat meat, which is popular in this region and produced locally, has a much less significant water component than beef, but beef consumption is on the rise. One kilogram of corn requires 900 liters (238 gallons) of water, and 1 kilogram of wheat, 1350 liters (357 gallons). Beef, corn, and wheat are common imports; in fact, North Africa and Southwest Asia import more wheat than any other world region, which is ironic because this is where the wheat plant was first domesticated.

Strategies for Increasing Access to Water Some strategies have been developed for increasing supplies of fresh water, but each presents a set of difficulties. All of them are expensive and some have enormous potential to cause more wasting of water.

The fossil fuel–rich countries of the Persian Gulf have invested heavily in seawater **desalination** technologies that remove the salt from seawater, making it suitable for drinking or irrigation. For example, desalination plants supply 70 percent of Saudi Arabia's drinking water and some of its wheat field irrigation. Many more desalination plants are under construction in Saudi Arabia. Desalination is an industrial process that separates salt from fresh water; however, it uses huge amounts of energy, burns fossil fuels, and results in emissions that contribute to global warming. If all the costs of producing food with desalinated water were counted—the fuel burned, the cost of irrigation and other equipment, and the fact that irrigated soil inevitably loses productivity due to soil salinization—the wheat produced by this method would be far too expensive for anyone to buy. Governments have not charged agricultural users the true production cost of water, which has kept the wheat artificially cheap, though there are now plans to remove these subsidies. In recognition of some of these problems Saudi Arabia has turned to solar power to fuel desalination plants. Not only is solar power increasingly cost efficient, but it is also an abundant resource in the region.

Many countries pump groundwater from underground aquifers to the surface for irrigation or drinking water. Starting in the 1980s, Libya invested some of its earnings from fossil fuels into one of the world's largest groundwater pumping projects, known as the *Great Man-Made River*. This project includes a series of underground pipelines that draws on the ancient fossil water deposit under the Saharan desert mentioned earlier to supply water to Libya's coastal cities and agricultural fields. With this irrigated agriculture, Libya could reduce its dependency on food imports and even export some foods to the European Union. However, due to political turmoil in Libya, no such exports are currently feasible. Another problem is that the rate of natural replenishment of the aquifer is far slower than the rate of extraction, making this use of groundwater ultimately unsustainable.

Dams and reservoirs built on the region's major river systems increase water supplies. In Egypt, for example, the Aswan dams on the Nile River generate much of the country's electricity, reduce flood risk, and improve navigation, but the natural cycles of the river have been altered by the construction of the dams. Downstream of these two dams, water flows have been radically reduced, making it necessary to use supplemental irrigation, and the lack of flooding means that fertility-enhancing silt is no longer deposited on the land. As a result, expensive fertilizers are now necessary. Moreover, with less silt coming downstream, parts of the Nile Delta are sinking into the sea or are experiencing saltwater intrusion because the seawater is flowing into aquifers and wells in coastal zones, rendering once-fresh water unusable. Upstream, the artificial reservoir created by the dams has flooded villages, fields, wildlife habitats, and historic sites.

Dams and Cross-Border Relations The need to share the water of rivers that flow at or across national boundaries often complicates the political relationships among the countries involved. The Aswan High Dam sits on Egypt's border with Sudan, an area that has been a source of contention for many years. The colonial-era Nile Waters Agreement created by the British gives most of the river's flow to Egypt, some to Sudan, while nine other upstream countries in sub-Saharan Africa were not included in the treaty. Therefore, when Ethiopia began to build a hydroelectric dam on the Blue Nile tributary in 2011, Egypt expressed serious concerns. In the short term, the dam will be able to hold back one year's flow of the Blue Nile that won't reach Egypt. Also, the Ethiopian dam may allow Sudan to increase the amount of water it uses for irrigation because the dam will send water year-round downstream to Sudan rather than the current

desalination the removal of salt from seawater—accomplished through the use of expensive and energy-intensive technologies—to make the water suitable for drinking or irrigating uneven seasonal flow. The three countries are now negotiating how to share water in the context of the new dam.

In the case of Turkey's Southeastern Anatolia Project, which involves the construction of several large dams on the upper reaches of the Euphrates River for hydropower and irrigation purposes, the dams have reduced the flow of water to the downstream countries of Syria and Iraq (**Figure 6.11**). In negotiations over who should get Euphrates water, for example, Turkey argues that it should be allowed to keep more water behind its dams because the river starts in Turkey and most of its water originates there as mountain rainfall. Meanwhile, Iraq's claim to the water is linked to the fact that the Euphrates travels the longest distance in Iraq.

CHECK YOUR UNDERSTANDING

1. What climates cover most of the land in this region?
2. What areas in the region have access to land and water for agriculture?
3. How is the scarce commodity of water a source of conflict?
4. What new technologies offer solutions to water scarcity? To what degree are they sustainable?
5. How will people get enough water to grow food in the future, especially if climate change makes this dry region even drier?
6. How might a dam like the Aswan on the upper part of the Nile River in Egypt contribute to the salinity and lowered fertility in lowland regions that were once naturally flooded but are now irrigated by pumped water?
7. What countries do the Euphrates/Tigris and the Nile rivers connect? What are the geopolitical implications of dams on these rivers?
8. Why does a dam create soil moisture problems for downstream users of river water once the dam is operational?

HUMAN PATTERNS OVER TIME

Often heralded as a "cradle of civilization," important developments in agriculture, societal organization, and urbanization took place long ago in this part of the world. Also, three of the world's great religions were born here: Judaism, Christianity, and Islam.

More recently, North Africa and Southwest Asia have struggled with the impacts of outsiders. In the current era, social and political change has lagged behind economic development, which itself varies widely around the region.

Agriculture and the Development of Civilization

Between 8000 and 10,000 years ago, formerly nomadic peoples founded some of the earliest known agricultural communities in the world. These communities were located in an arc formed by the uplands of the Tigris and Euphrates river systems (in modern Turkey and Iraq) and the Zagros Mountains of modern Iran. This zone, which stretches to the eastern shore of the Mediterranean, is often called the **Fertile Crescent** because of its once-plentiful fresh

Fertile Crescent an arc of relatively lush, fertile land, where nomadic peoples began the earliest known agricultural communities and where the world's first cities emerged

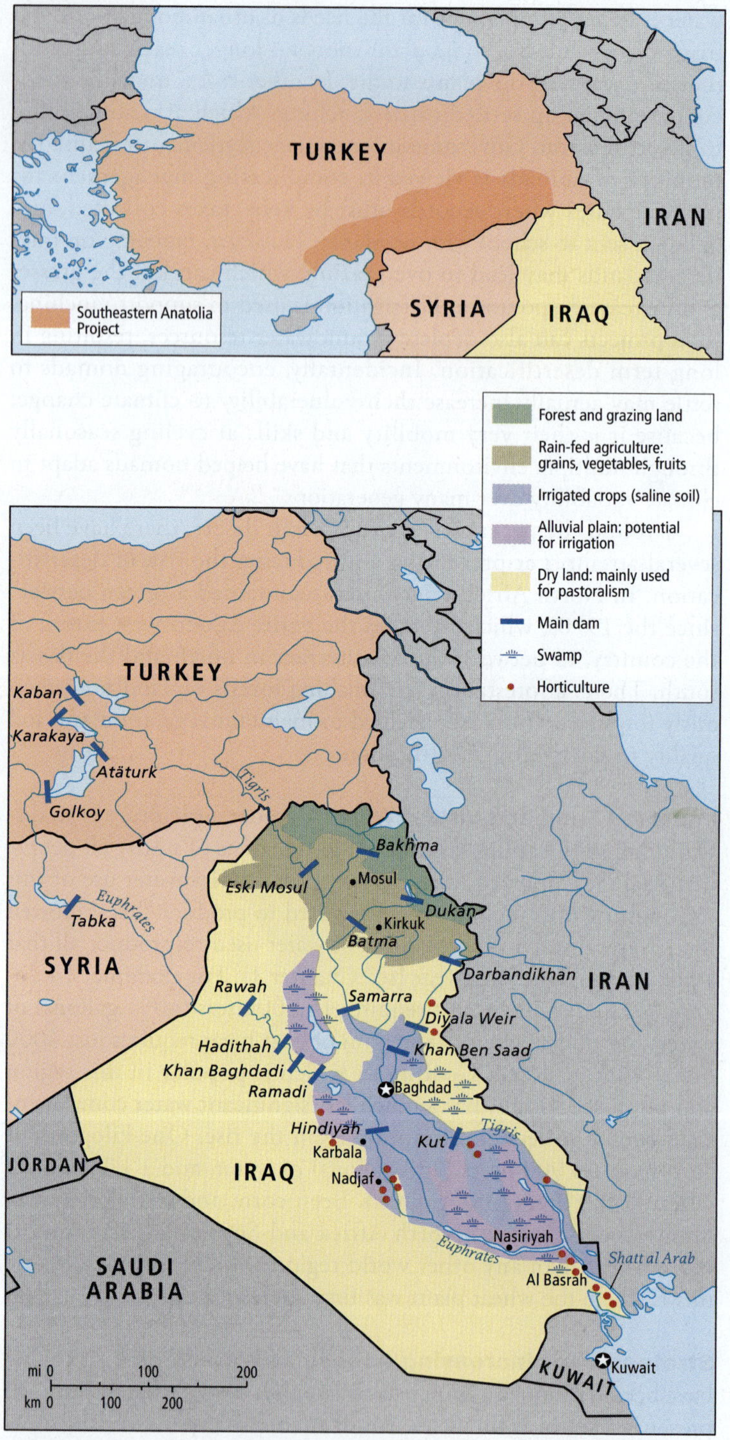

Figure 6.11 Dams on the Tigris and Euphrates drainage basins. Turkey's projects to manage the Tigris and Euphrates river basins through dam construction have international implications. Water that is retained in Turkey does not reach its neighbors. The main map shows Turkey's dams on the headwaters of the two rivers, as well as dams built in Syria, Iraq, and Iran. The smaller map shows the full extent of Turkey's Southeastern Anatolia Project. [Research from: United Nations Environmental Programme, "Turning the Tides" map, in *Vital Water Graphics: Problems Related to Freshwater Resources*, at http://www.unep.org/dewa/assessments/ecosystems/water/vitalwater/22.htm.]

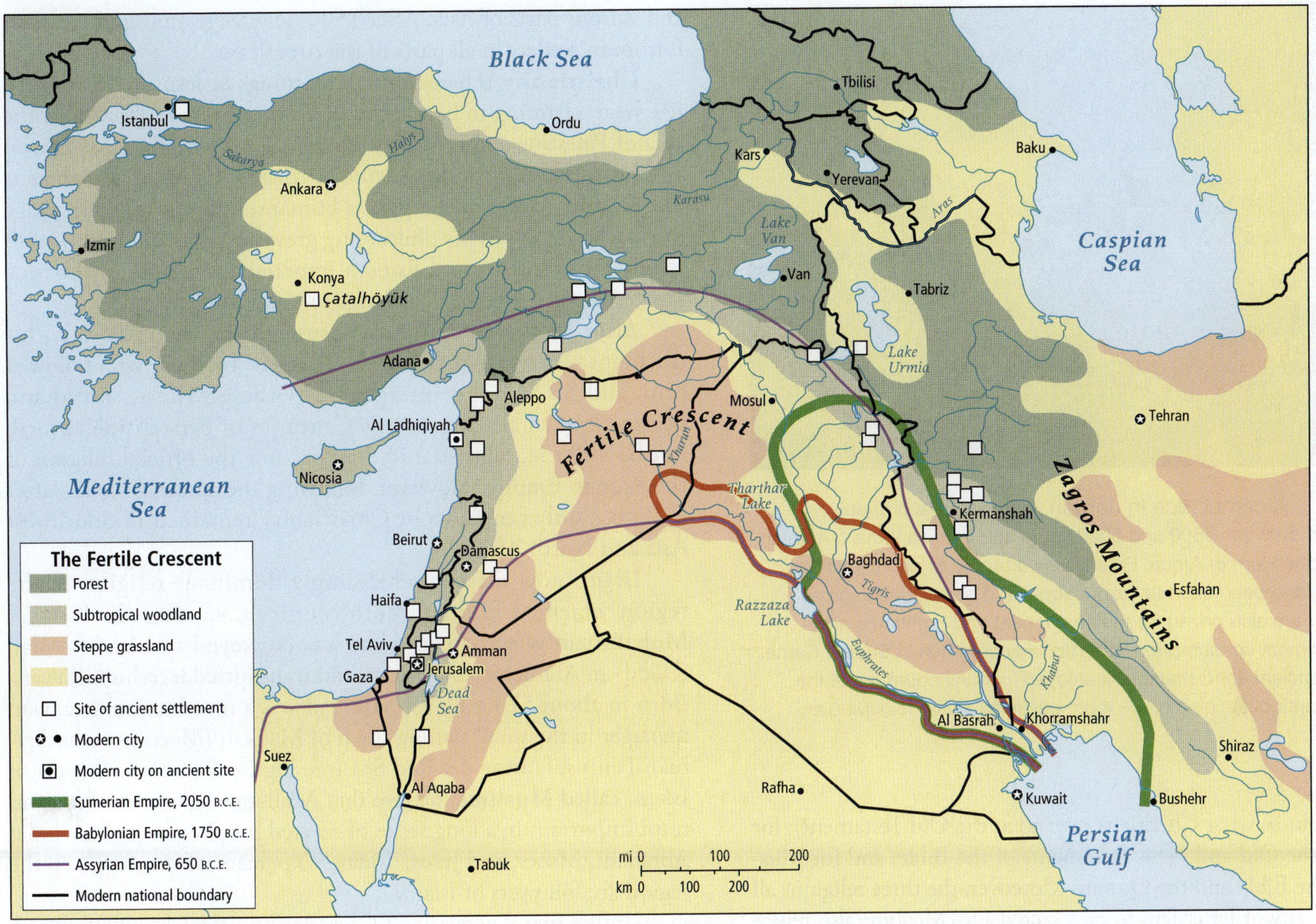

Figure 6.12 The Fertile Crescent. About 10,000 years ago, people in the Fertile Crescent began domesticating cereal grains, legumes, and animals, especially sheep and goats. The use of domesticated animals spread to Europe and Africa as agricultural peoples traded their surpluses for other goods or moved into other regions. Major empires developed successively in the eastern part of the Fertile Crescent, such as the Sumerian, the Babylonian, and the Assyrian empires. [Research from: Bruce Smith, *The Emergence of Agriculture* (New York: Scientific American Library, 1995), p. 50.]

water; its fertile soil that is seasonally replenished by flooding; its open forests and grasslands; and its abundant wild grains and animals (**Figures 6.12** and **6.13**). The climate of the Fertile Crescent is thought to have been more conducive to agriculture in ancient times.

Early settlements eventually grew into societies based on widespread irrigated agriculture, especially along the Tigris and Euphrates. Over several thousand years, agriculture spread to the Nile Valley and across North Africa, northwest into Europe, and east to the mountains of Persia (modern Iran). Ultimately, other cultivation systems across the world were influenced by developments in the Fertile Crescent.

The Fertile Crescent is also considered the birthplace of urbanization. Eventually, small settlements associated with agriculture took on urban qualities: dense populations, specialized occupations, concentrations of wealth, and centralized government and bureaucracies. City dwellers, who did not engage in agriculture, thrived because surrounding rural areas were capable of producing a surplus of food. The center of early urbanization was Sumer (in modern southern Iraq), where a number of cities existed 5000 years ago. By contemporary standards Sumerian cities were not very large—at the most 50,000 residents—but these city-states extended their influence over the surrounding territory. Today, these places only exist as archeological sites, but cities in the western Fertile Crescent that emerged a little bit later are still inhabited. Jericho (in modern occupied Palestinian Territories) and Damascus (in modern Syria) are among the cities that claim to be the "oldest in the world."

The Coming of Monotheism: Judaism, Christianity, and Islam

The very early religions of this region were founded on polytheism, a belief in many gods who controlled natural phenomena; such was the case through the Greek era and into the Roman period. But starting several thousand years ago, **monotheism**—the belief system based on the idea that there is only one god—began to emerge. The three major monotheistic world religions—Judaism, Christianity, and Islam—all have connections to interrelated sacred texts. For Jews, it is the Torah (the

monotheism the belief system based on the idea that there is only one god

Figure 6.13 A camel race in Jordan. Camels were probably domesticated between 4500 and 5000 years ago in the southern Arabian Peninsula or the Horn of Africa. Extensively adapted to arid climates, camels have long been used to carry people and heavy loads across deserts. They are also prized for their lean meat and nutritious milk. Today, camel meat consumption is particularly common in Sudan. Camel racing is an ancient sport that is still enjoyed in many countries of the region, and betting has also made it big business. [Salah Malkawi/Getty Images]

first five books in what Christians refer to as the Old Testament); for Christians, the Old and New Testaments of the Bible; and for Muslims, both the Bible and the Qur'an. Moreover, the three religions all consider the city of Jerusalem to be a sacred site. Muslims also revere Makkah (Mecca) and Al Madinah (Medina) in Saudi Arabia.

Judaism was founded approximately 4000 years ago. According to tradition, the patriarch Abraham led his followers from Mesopotamia (modern Iraq) to the shores of the Eastern Mediterranean (modern Israel and the occupied Palestinian Territories), where he founded Judaism. Judaism is characterized by the belief in one god, an ethical code summarized in the Ten Commandments, and an enduring ethnic identity reinforced by dietary and religious laws.

After rebelling against the Roman Empire, which culminated in their expulsion in 73 C.E. from the Eastern Mediterranean, some Jews were enslaved by the Romans and most migrated to other lands in a movement known as the **diaspora**. The term *diaspora* can mean the dispersion of any group of people from their original homeland, but it is especially associated with the historical movement of Jews from Palestine across North Africa, Europe, and various parts of Asia. After 1500, Jews were among the earliest European settlers in all parts of the Americas.

Judaism a monotheistic religion characterized by the belief in one god, a strong ethical code summarized in the Ten Commandments, and an enduring ethnic identity

diaspora the dispersion of Jews around the globe after they were expelled from the Eastern Mediterranean by the Roman Empire beginning in 73 C.E.; the term can also refer to other geographically dispersed culture groups

Christianity a monotheistic religion based on the belief in the teachings of Jesus of Nazareth, a Jew who described God's relationship to humans as primarily one of love and support, as exemplified by the Ten Commandments

Muslims followers of Islam

Christianity is based on the teachings of Jesus of Nazareth, a Jew who, claiming to be the son of God, gathered followers in the area of Palestine about 2000 years ago. Jesus, who became known as Christ (meaning anointed one or Messiah), taught that there is one God who loves and supports humans, but who will also judge those who do evil. This philosophy grew popular, and both Jewish religious authorities and Roman imperial authorities of the time saw Jesus as a dangerous challenge to their power.

After Jesus's execution in Jerusalem in about 32 C.E., his teachings were written down (the Gospels) by those who followed him, and his ideas, as interpreted by these writers, spread and became known as Christianity. Centuries of persecution ensued, but by 400 C.E., Christianity had become the official religion of the Roman Empire. However, following the spread of Islam after 622 C.E., only remnants of Christianity remained in Southwest Asia and North Africa.

Islam, now the overwhelmingly dominant religion in the region, emerged in the seventh century C.E., after the Prophet Muhammad wrote down what was conveyed to him by Allah ("God" in Arabic) in the Qur'an and transmitted it to his followers. Born in about 570 C.E., Muhammad was a merchant and caravan manager in the small trading town of Makkah (Mecca) on the Arabian Peninsula near the Red Sea (see Figure 6.16B). Followers of Islam, called **Muslims**, believe that Muhammad was the final and most important in a long series of revered prophets, which includes Abraham, Moses, and Jesus. Today, 93 percent of the people in the region are followers of Islam.

Unlike many versions of Christianity, Islam has virtually no central administration, though there are numerous clerical leaders who help their followers interpret the Qur'an. (The Shi'ite version of Islam tends to have a stronger religious hierarchy; see the "Islamic Religious Law and the Sunni–Shi'ite Divide" section). An important effect of the lack of a central authority is that the interpretation of Islam varies widely within and among countries, from group to group, and from individual to individual. There are 1.8 billion Muslims around the world today.

The Pillars of Islam The Five Pillars of Islam embody the central teachings of the religion. Not all Muslims are fully observant, but the pillars have an enormous impact on daily life, including on festivals and religious holidays (**Figure 6.14B**).

1. The declaration of faith among the believers—a phrase that states Allah is the only God and Muhammad is his messenger.
2. Daily prayer at five designated times (daybreak, noon, midafternoon, sunset, and evening). Although prayer is an individual activity, Muslims are encouraged to pray in groups and in mosques.
3. Fasting (no food, drink, or smoking) during the daylight hours of the month of Ramadan, followed by a celebratory meal after sundown (Figure 6.14B). Ramadan falls in the ninth month of the Islamic calendar, which is a lunar calendar, meaning that Ramadan will take place during a different time every year. Ramadan celebrates the revelation of the Qur'an to Muhammad.

Figure 6.14 LOCAL LIVES: Festivals of North Africa and Southwest Asia

(A) Passover is a holiday that is celebrated by Jews in Israel and around the world. It commemorates the biblical event of the Jewish population escaping enslavement in Egypt. Passover is celebrated every spring and involves a ritual meal called a *Seder*. In the picture, members of a tiny ethno-religious group in the Jewish tradition, the Samaritans, engage in a Passover pilgrimage at Mount Gerizim, which is located on the West Bank. [GALI TIBBON/AFP/Getty Images]

(B) A banquet is set in Gaza, Palestine, for *Iftar*, the sunset meal that Muslims eat to break the fast each night in the holy month of Ramadan. During Ramadan, observant Muslims refrain from eating, drinking, quarreling, or having sex between sunrise and sunset. They offer extra prayers each day as a demonstration of their submission to Allah. Wealthy Muslims often sponsor public banquets, such as the one shown here. [MOHAMMED ABED/AFP/Getty Images]

(C) An Iranian woman celebrates *Nauryz*, an ancient holiday of Zoroastrian origin celebrated throughout Iran, parts of Turkey, and much of Central Asia (see Figure 5.25). It occurs at the March equinox and celebrates the coming of spring. People jump over bonfires while singing a verse of purification that is meant to remove sickness and problems, replacing them with warmth and energy. [BEHROUZ MEHRI/AFP/Getty Images]

4. Obligatory almsgiving (*zakat*) in the form of a "tax" of 2.5 percent. The alms are given to Muslims in need. Zakat is based on the recognition of the injustice of economic inequity. Although it is usually an individual act, the practice of government-enforced zakat is returning in some Islamic countries. The tradition of almsgiving means that charitable institutions are numerous and important in Islamic life. It is doubtful, however, that this pillar has significantly reduced inequality in the region.
5. Pilgrimage (**hajj**) at least once in a lifetime, for those who are able financially and physically to make the journey, to the Islamic holy places in or near Makkah (Mecca) during the twelfth month of the Islamic calendar. The most important hajj sites are the Masjid Al-Haram (the largest mosque in the world) and the Kaaba, a holy structure that sits in the middle of the courtyard of the mosque.

Saudi Arabia occupies a prestigious position in Islam, as it is the site of two of Islam's three holy shrines: Makkah, the birthplace of the Prophet Muhammad and of Islam, and Al Madinah (Medina), the site of a mosque that is located on the site of Muhammad's home and also contains his burial place. (The third holy shrine is in Jerusalem.) The fifth pillar of Islam has placed Makkah and Al Madinah at the heart of Muslim religious geography for more than 1400 years. Today, a large private sector service industry, owned and managed by members of the huge Saud family that rules Saudi Arabia, organizes and oversees the 5- to 7-day hajj for up to 3 million devout visitors. The number of people descending on Makkah every year has increased as there are more Muslims than ever (by natural increase or religious conversion), and the level of affluence in the Islamic world has increased so that more can afford the trip. This presents numerous problems for the city of Makkah, from providing housing for all pilgrims to managing huge crowds at sacred sites. During the 2015 hajj, as many as 2000 people were crushed to death in a stampede as too many worshippers congregated at the same site outside Makkah. Unfortunately, such events have been recurrent over the years during the hajj.

Islamic Religious Law and the Sunni–Shi'ite Divide Beyond the Five Pillars, Islamic religious law, called **shari'a**, or "the correct path," guides daily life according to the principles of the Qur'an. But there are many interpretations of the Qur'an, several renderings of shari'a, and a wide variety of versions of observant Muslim life. Some Muslims believe that no other legal code is necessary in an Islamic society, because shari'a provides guidance in all matters of life, including worship, finance, politics, marriage, sex, diet, hygiene, war, and crime. Other Muslims think that secular law is more useful in modern societies that are increasingly multicultural because secular law makes allowances for different religious sensibilities. The debate about whether shari'a or secular law is better has raged for hundreds of years, and it continues to divide traditionalists and moderates in Islamic countries.

Insofar as interpretations of the Qur'an and shari'a are concerned, the Muslim community is split into two major groups that formed after the death of Muhammad, when divisions arose over who should succeed the Prophet and have the right to interpret the

hajj the pilgrimage to the city of Makkah (Mecca) that all Muslims are encouraged to undertake at least once in a lifetime

shari'a Islamic religious law that guides daily life according to the interpretations of the Qur'an

Qur'an for all Muslims: **Sunni** Muslims, who today account for 85 percent of the world community of Islam, and **Shi'ite** (or sometimes called **Shi'a**) Muslims, who live primarily in Iran. A majority of Muslims in Iraq and Bahrain are also Shi'ite, while large Shi'ite minorities exist in Lebanon. Elsewhere, Sunnis dominate. While Sunnis have a relatively decentralized religious system for interpreting shari'a, Shi'ites recognize an authoritative priestly class, whom they call imams, ayatollahs, or mullahs.

This division continues today with different ceremonies and disagreements about theology and the interpretation of shari'a. These religious differences are exacerbated by countless local disputes over land, resources, and political power. In Saudi Arabia, for example, where the Shi'ites are a minority, the group lives in segregated places that are marginalized and underfunded by the Sunni-dominated government. In another case, Iraq, the long-standing conflict between Sunnis and Shi'ites intensified after the U.S. invasion in 2003, as rivalries arose over which group should control government affairs and fossil fuel resources. As the main Shi'ite country, Iran plays an influential role in Shi'ite communities elsewhere. One aspect of the Shi'ite tradition is that the clerics also tend to be political leaders. In Sunni countries, there is a clearer divide between religious leadership and the secular affairs of the government, even if government policy is influenced by religion. The political role of Shi'ite clerics may come from the minority status of Shi'ite communities around the region, and as they tended to be shut out of national political power, the mosque became the community center for both spiritual and political matters. As conflict in the region has intensified, the Sunni–Shi'ite divide has also become more pronounced in many countries. What has also emerged is a geopolitical struggle between the regional powers Saudi Arabia and Iran, which represent two opposing visions of Islam.

The Diffusion of Islam Among the first converts to Islam were the Bedouin—nomads of the Arabian Peninsula. By the time of Muhammad's death in 632 C.E., they were already spreading the faith and creating a vast Islamic sphere of influence. Over the next century, Muslim armies built an Arab–Islamic empire over most of Southwest Asia, North Africa, and the Iberian Peninsula of Europe (**Figure 6.15**).

The spread of Islam is an example of **diffusion**—the process by which ideas, things, and people move across space and time. The diffusion pattern of Islam was dependent on multiple political, cultural, and geographic factors. The Arab world adopted Islam first in a contiguous pattern of expansion farther away from the core region around Makkah and Al Madinah. In other places, there were natural barriers (the massive Saharan desert) or political and cultural barriers (established Christian kingdoms in Europe) that delayed or limited the spread.

Sunni the larger of two major groups of Muslims

Shi'ite (or Shi'a) the smaller of two major groups of Muslims; Shi'ites are found primarily in Iran and southern Iraq

diffusion the process by which ideas, things, and people move across space and time

Ottoman Empire influential Islamic empire centered in today's Turkey that lasted from the 1200s to the early 1900s

Not only was Islam spread by person-to-person interactions, but many other forms of Arab culture spread this way as well. While most of Europe was stagnating during the medieval period (450–1500 C.E.), the Arab–Islamic empire nurtured learning and economic development. Muslim scholars traveled throughout Asia and Africa, advancing the fields of architecture, history, mathematics, geography, and medicine. Centers of learning flourished from Baghdad (Iraq) to Toledo (Spain). The early Arab–Islamic era also fostered vibrant economies and wide-ranging trade. The Islamic world was centrally located between the Mediterranean and Asia, which allowed it to play a dominant role in long-distance trade connecting Asia, Africa, and Europe. The traders founded settlements and introduced new forms of urban and living spaces, such as the courtyard design (see Figure 6.32C later in the chapter). The architectural legacy of Arabs and Muslims not only diffused to, but lives on in, Spain, India, Central Asia, the Americas, and countless places across the world.

By the end of the tenth century, the Arab–Islamic empire had begun to break apart. From the eleventh to the fifteenth centuries, Mongols from eastern Central Asia (eventually converting to Islam by 1330 C.E.) conquered parts of the Arab-controlled territory, including the destruction of Baghdad, which was the largest and wealthiest city during this golden age of Islam. Around the same time, nomadic Turkic herders from Central Asia began to converge in western Anatolia (Turkey) where they eventually forged the **Ottoman Empire**, which became the most influential Islamic empire the world has ever known, and lasted until the end of World War I in 1918.

By the 1300s, the Ottomans had become Muslim, and by the 1400s, they had defeated the Christian Byzantine Empire centered in Constantinople, which was the successor to the Roman Empire. The Ottomans took over Constantinople, renamed it Istanbul, and eventually ruled territories from North Africa to southeastern Europe and Southwest Asia. Previously, in the 1490s, the Arab Muslims had lost their control of the Iberian Peninsula to Christian kingdoms. Islam currently still dominates in a huge area that stretches from Morocco to western China and includes northern South Asia, as well as parts of Southeast Asia (see the inset of Figure 6.15). The presence of Islam in Southeast Asia is a reminder of the historic Arab-controlled trade around the Indian Ocean and a form of noncontiguous diffusion.

Once a location was completely conquered, the Ottoman Empire, like the Arab–Islamic empire before it, encouraged religious tolerance toward the conquered peoples so long as they adhered to a religion with a sacred text. Jews, Christians, Buddhists, and Zoroastrians (adherents of a religion that predated Islam in Iran and surrounding areas) were allowed to practice their religions, although there were attractive economic and social advantages to converting to Islam. Multicultural urban life in Ottoman cities facilitated vast trading networks spanning the known world, and Istanbul became a cosmopolitan capital with elaborate buildings and lavish public spaces that outshined anything in Europe until the nineteenth century (**Figure 6.16C**).

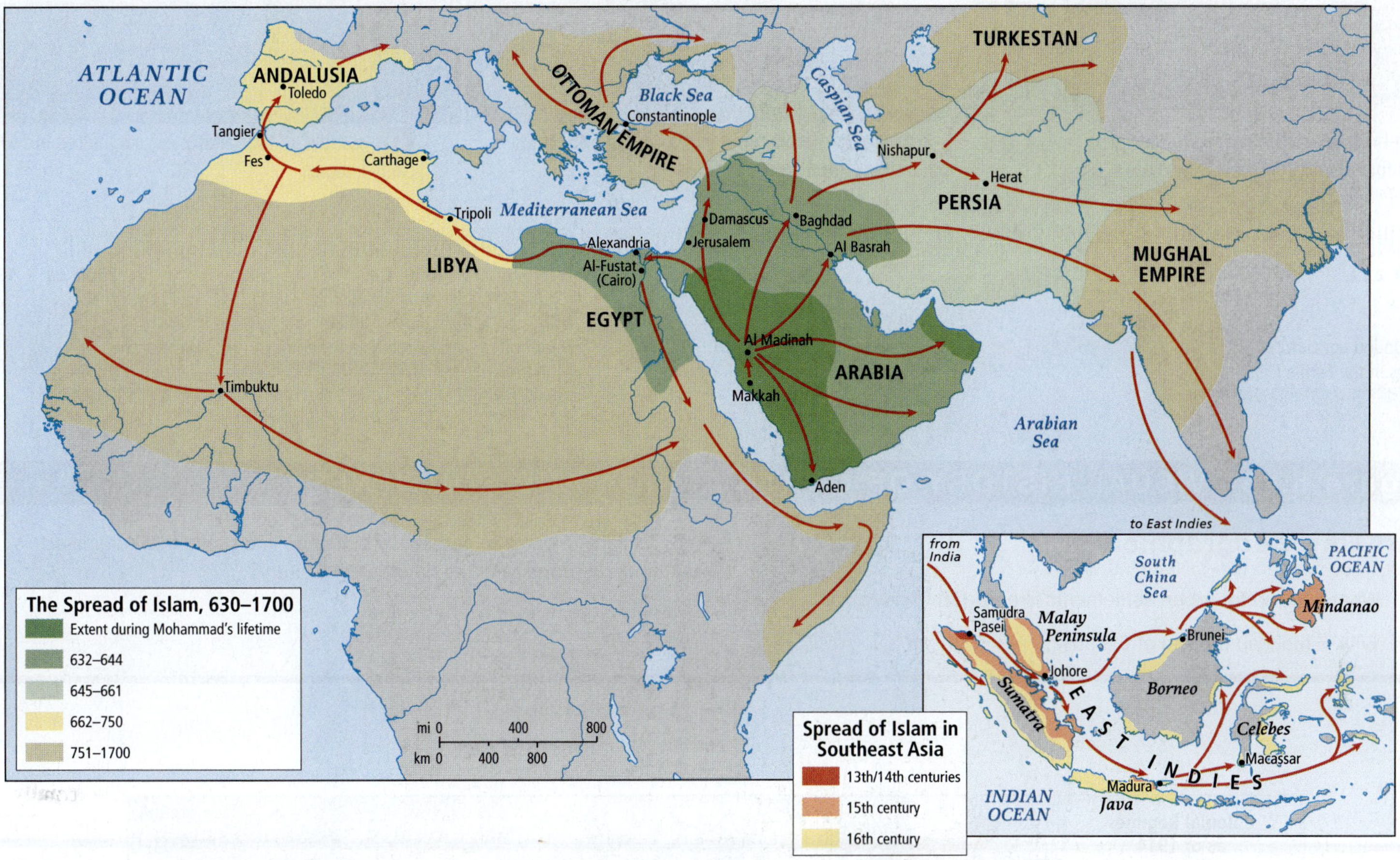

Figure 6.15 The diffusion of Islam, 630–1700. In the first 120 years after the death of the Prophet Muhammad in 632, Islam spread, primarily by conquest. Over the next several centuries, Islam was carried to distant lands by both traders and armies. [Research from: Richard Overy, ed., *The Times History of the World* (London: Times Books, 1999), pp. 98–99.]

Western Domination and State Formation

The Ottoman Empire ultimately withered in the face of a Europe made powerful by colonialism and the Industrial Revolution. Throughout the nineteenth century, North Africa provided raw materials for Europe. By 1830, France was exercising direct control over parts of the North African territory of Algeria (see Figure 6.16D). France took control of Tunisia in 1881 and Morocco in 1912, Spain controlled a bit of the Mediterranean coast and what is now called Western Sahara, Britain gained control of Egypt in 1882 and Sudan in 1898, and Italy took control of Libya in 1912 (**Figure 6.17A**).

The modern borders of North Africa and Southwest Asia are in many cases the product of European colonialism. The two most important colonizers—Great Britain and France—negotiated among themselves how to control the region as the power of the Ottoman Empire waned. This is evident today through the straight lines that were devised by the colonial mapmakers. For example, the borders that separate Syria, Jordan, and Iraq today were originally intended to create two spheres of influence by separating British and French-governed territories.

World War I (1914–1918) brought the fall of the Ottoman Empire, which had allied itself with Germany. At the end of the war, the victorious powers dismantled the Ottoman Empire and all of the former Ottoman territories; only Turkey was recognized as an independent country. The rest of the formerly Ottoman-controlled territory was allotted to France and Britain as protectorates (see Figure 6.17B). On the Arabian Peninsula, Bedouin tribes were consolidated under Sheikh Ibn Saud in 1932, and Saudi Arabia began to emerge as an independent country.

World War II (1939–1945) further affected the political development of North Africa and Southwest Asia. Most significantly, in the aftermath of the Holocaust in Europe, the Jewish state of Israel was created in the Eastern Mediterranean on land inhabited by Arab farmers and nomadic herders as well as by some Jews (see "Situation 1: Israel and Palestine"; see also Figure 6.16E).

By the 1950s, European and U.S. energy companies played a key role in influencing who ruled Iran and Saudi Arabia—countries where vast oil deposits were to become especially lucrative. In Egypt, a major cotton producer, European textile companies played a similar role. Officials in the governments of these countries received financial benefits from foreign companies and showed their loyalty to these companies with low taxes on oil and cotton exports and easy access to land. While a tiny ruling elite

A Mud-brick houses in Harran, Turkey, were inhabited for at least 5000 years. [Wu Swee Ong/ Getty Images]

B Masjid Al-Haram, built in Makkah in 630. [Rabi Karim Photography/Getty Images]

C Istanbul's "Blue Mosque," completed in 1616. [Ayh... Getty Images]

10,000 B.C.E. | 3000 B.C.E. | 0 C.E. | 500 C.E. | 1000 C.E. | 1500 C.E. | 1600 C.E.

- **8000 B.C.E.** Nomads founded agricultural communities in the Fertile Crescent
- **3000 B.C.E.**
- **400 C.E.** Roman Empire, officially Christian
- **630**
- **632** Death of Muhammad
- **632–1500s** Arab–Islamic empires
- **1300s** Rise of the Ottoman Empire
- **1616**

Figure 6.16 VISUAL HISTORY OF NORTH AFRICA AND SOUTHWEST ASIA

Thinking Geographically

A What allowed for urban settlements like Harran to develop?

B Why is Makkah the site of the hajj?

C Which Turkish city had lavish buildings and parks that outshined anything in Europe until the nineteenth century?

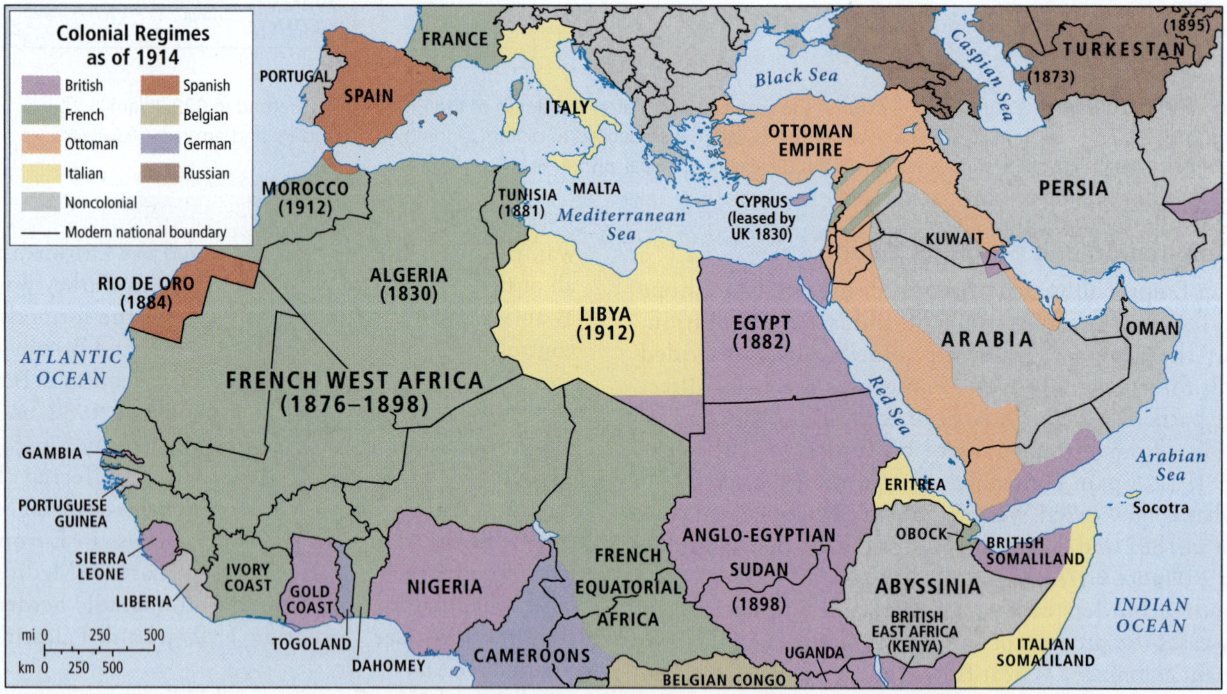

Figure 6.17 Colonial regimes in North Africa and Southwest Asia. (A) European powers began influencing the affairs of the region in the nineteenth century and expanded their control by 1914 at the beginning of World War I. The dates on the map indicate when the Europeans took control of each country. [Research from: *Hammond Times Concise Atlas of World History* (Maplewood, NJ: Hammond, 1994), pp. 100–101.]

grew fabulously wealthy, even the modest revenues generated by oil and other taxes were not invested in creating opportunities for the vast majority of poor people. Over time, ever more political power accrued, especially to foreign energy companies. The United States and Western Europe supported those autocratic local leaders who, in turn, were sympathetic to the interests of Western companies

D France's wars in Algeria in the 1830s. [DEA/ . PIROZZI/Getty Images]

E Jewish, Arab, and UN representatives negotiate a cease-fire during the 1948 war. [Keystone-France/ Getty Images]

F Shia Muslims celebrate a religious festival in the city of Karbala in Iraq. [AHMAD AL-RUBAYE/ AFP/Getty Images]

1700 C.E. | **1800 C.E.** | **1900 C.E.** | **2000 C.E.**

- **1830s**
- **1830–1918** European colonial expansion
- **1918** Fall of the Ottoman Empire
- **1948** War resulting in the creation of the state of Israel
- **1950s–present** Strong European and U.S. influence on politics
- **2018**

D When did France first assert control over the territory in what is now Algeria?

E The state of Israel was created in the aftermath of what global event?

F What is the religious composition of Iraq?

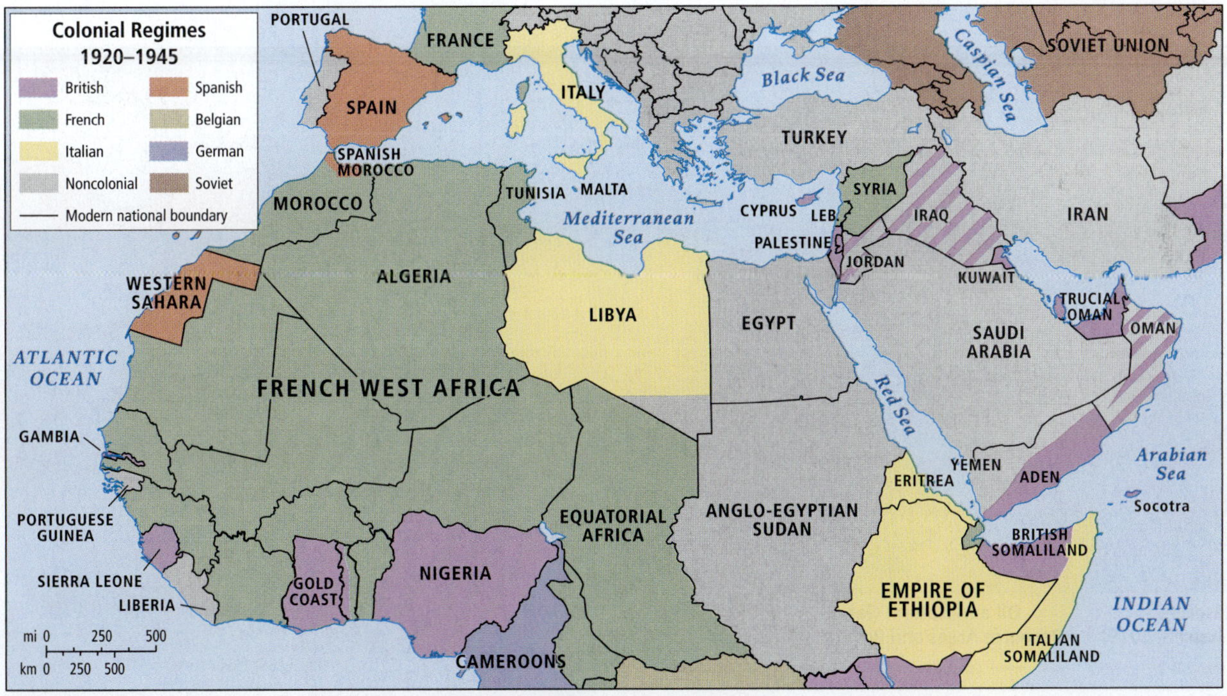

(B) Between 1920 and 1945, what was left of the Ottoman Empire in the Eastern Mediterranean became *protectorates* (territories that have their security and foreign affairs managed by other countries) administered by the British and French. The striped areas reflect the British colonial practice of allowing local rulers to be in charge of daily administrative activities while the British retained control of many of the colonies' policies and actions. [Research from: *Rand McNally Historical Atlas of the World* (Chicago: Rand McNally, 1965), pp. 36–37; and *Cultural Atlas of Africa* (New York: Checkmark Books, 1988), p. 59.]

and Cold War strategies vis-à-vis the Soviet Union. As a result of these concerns, both Cold War adversaries supported undemocratic governments. The Assad regime in Syria is such an example, and current conflict there cannot be fully understood without taking into account how international geopolitics have shaped the region for decades.

CHECK YOUR UNDERSTANDING

1. What are the factors that made the Fertile Crescent the earliest civilization in history?
2. How are the three major monotheistic world religions—Judaism, Christianity, and Islam—interconnected?
3. How do the Five Pillars of Islamic Practice influence the lives of Muslims in the region?
4. How did European colonial powers rule or control most countries in the region?
5. Why did the United States and Western Europe support autocratic local leaders in the region?

GLOBALIZATION AND DEVELOPMENT

6.4 Explain how fossil fuel resources are distributed across North Africa and Southwest Asia.

6.5 Discuss the effect of fossil fuel resources on economic development in the region.

There are major economic barriers to peace and prosperity within North Africa and Southwest Asia. In some countries, the wealth from fossil fuel exports remains in the hands of an elite few, while most people work at low-wage urban jobs or are relatively poor farmers or herders. Persian Gulf states have started to make significant investments in education and social services, but even there a large underclass of immigrant workers exists. The economic base is unstable because the main resources are fossil fuels and agricultural commodities, both of which are subject to wide price fluctuations on world markets.

FOSSIL FUEL EXPORTS AND DEVELOPMENT

This region, which has close to two-thirds of Earth's known reserves of oil, is the major oil supplier to the world. Natural gas reserves and exports are significant as well. Most oil and gas production is located around the Persian Gulf (**Figures 6.18** and **6.19**). The North African countries of Algeria and Libya also have oil and gas reserves. It is important to note, however, that several countries in this region—Morocco, Tunisia, Syria, Lebanon, Jordan, Israel, and Turkey—have minimal oil and gas resources and are net importers of fossil fuels. Meanwhile, Egypt and Sudan produce enough to supply most of their own needs, but export little. Regardless of their own petroleum resources, all countries in the region are affected by the geopolitics of oil and gas.

European and North American companies were the first to exploit the region's fossil fuel reserves early in the twentieth century. These companies paid governments a small royalty for the right to extract oil (natural gas extraction has grown more recently). Oil was processed at on-site refineries owned by the foreign companies

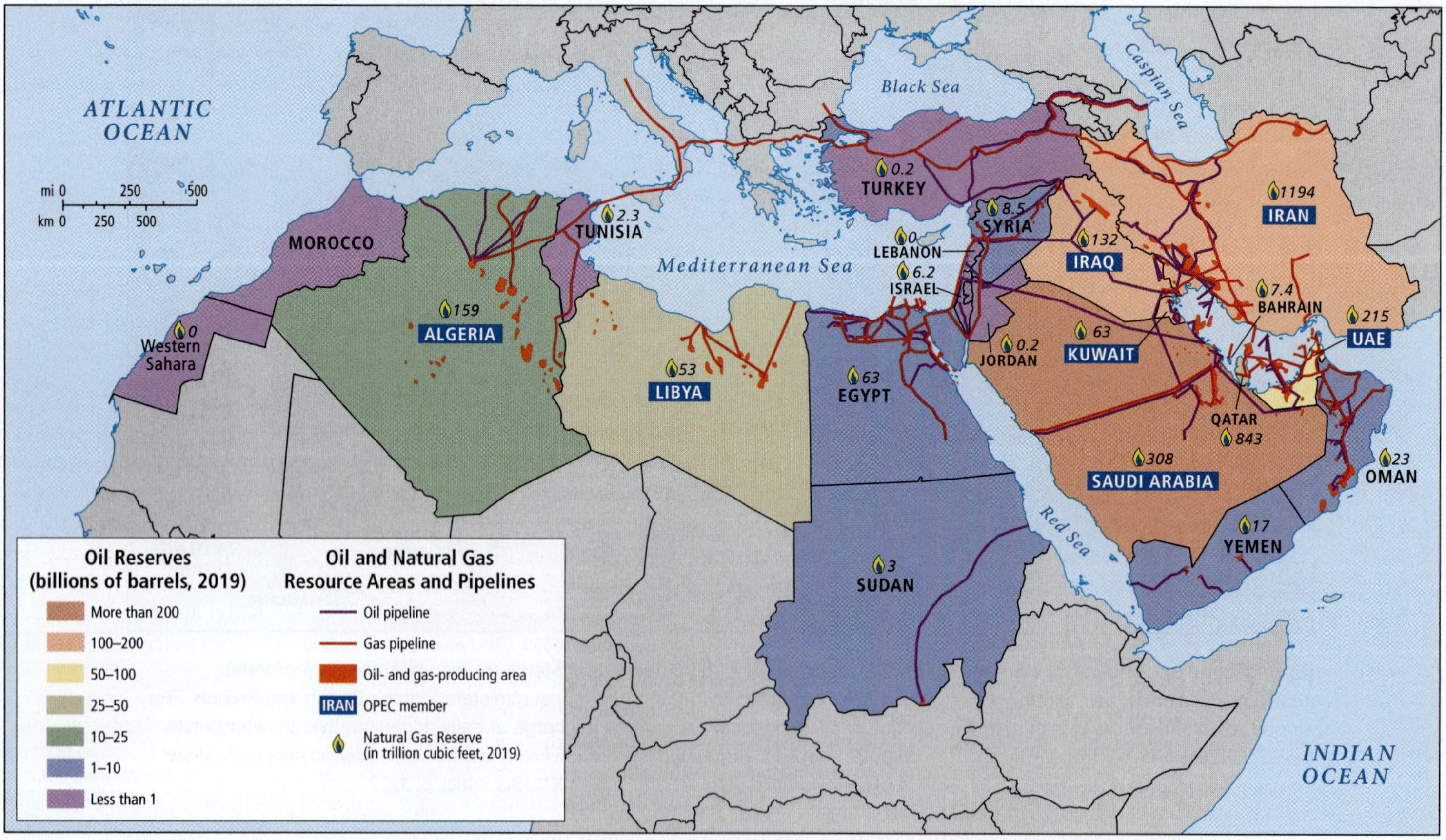

Figure 6.18 Oil and gas reserves and resources in North Africa and Southwest Asia. The map shows the oil reserves, oil and gas resource areas, and pipelines in the region. [Data from: Energy Information Administration, International Energy Statistics, at https://www.eia.gov/beta/international/data/browser; and "Selected Oil and Gas Pipeline Infrastructure in the Middle East," at http://www.eia.doe.gov/cabs/Saudi_Arabia/images/Oil%20and%20Gas%20Infrastructure%20Persian%20Gulf%20(large)%20(2).gif.]

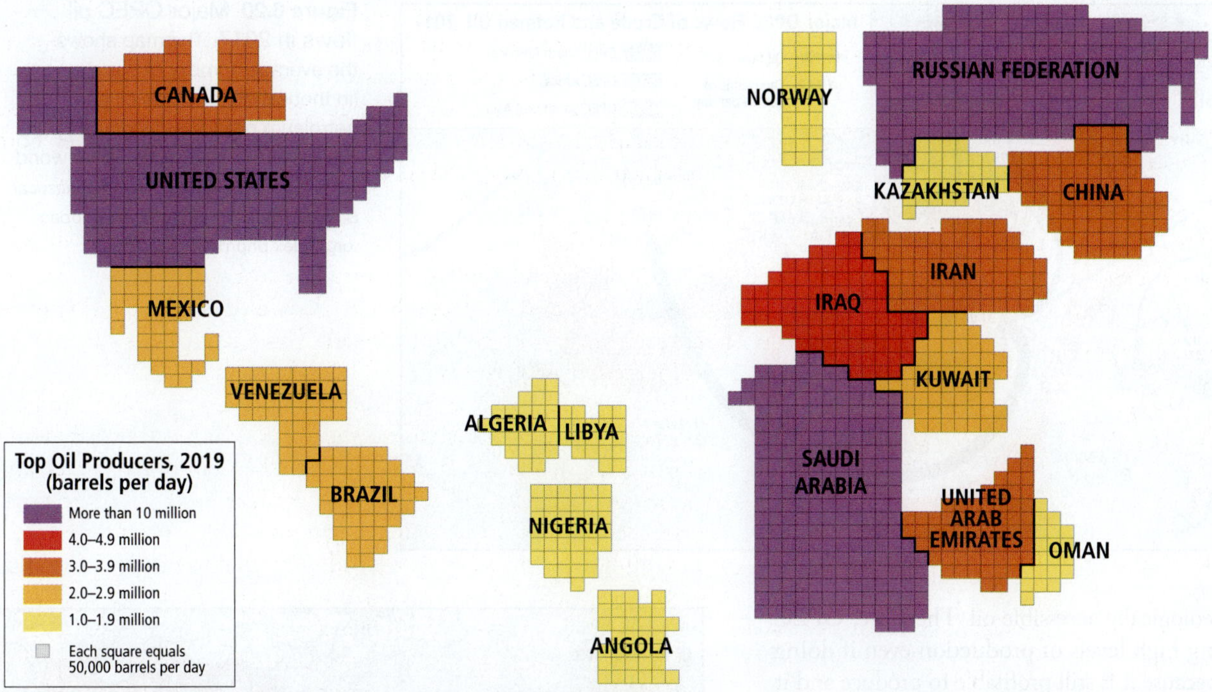

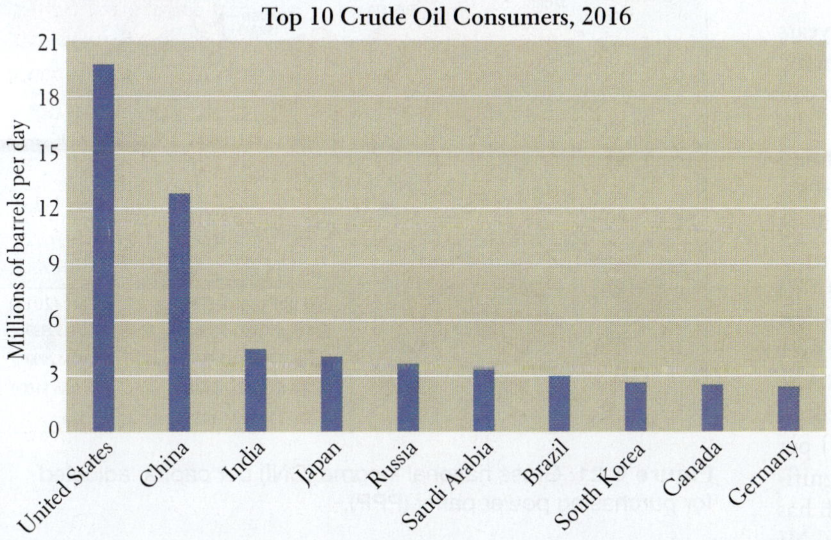

Figure 6.19 Oil production and consumption by country. In the map, the size of countries corresponds to the amount of oil production. This can be contrasted with the leading oil consumers of the world in the lower bar graph. [Data from: Energy Information Administration, International Energy Statistics, at https://www.eia.gov/beta/international/data/browser; and U.S. Energy Information Administration, "What countries are the top producers and consumers of oil?" at https://www.eia.gov/tools/faqs/faq.php?id=709&t=6.]

and sold at very low prices, primarily to Europe and the United States and eventually to other countries, such as Japan. Since then, most of the countries in the region have established national oil companies that are majority government owned. This enables them to control their own resources rather than being overly dependent on multinational corporations from Europe or the United States. It is estimated that 90 percent of global oil reserves are now controlled by these national oil companies.

Global oil and gas prices may rise and fall dramatically. This happens for both economic and political reasons. For example, in 1973 governments in the Gulf states raised the price of oil and gas as a penalty on the United States for its support for Israel in the Yom Kippur War between Israel and neighboring Arab countries (discussed further below). The oil price increase in 1973 was made possible by the establishment of national oil companies, and importantly, the founding in 1960 of a **cartel**—a group of producers that collectively control production levels and set prices for products—known as **OPEC (Organization of the Petroleum Exporting Countries)**. For a few months after the Yom Kippur War, there was a total halt of oil shipments from the Gulf states to the United States and to a few other countries that supported Israel. When the Gulf states resumed their shipments, they raised the price of oil. OPEC remains the main organization of oil-producing states; membership, which changes from time to time, now includes all of the states indicated in **Figure 6.20**. OPEC members cooperate to periodically restrict or increase oil production, thereby significantly influencing the price of oil on world markets. For example, since 2015 the price of oil has been somewhat low, resulting in lower revenues for oil producers around the world and triggering cutbacks in production. However, many OPEC countries, especially the Gulf states, have low

cartel a group of producers that controls production levels and sets prices for products

OPEC (Organization of the Petroleum Exporting Countries) a cartel of oil-producing countries—currently, Algeria, Angola, Congo, Ecuador, Equatorial Guinea, Gabon, Iran, Iraq, Kuwait, Libya, Nigeria, Saudi Arabia, the United Arab Emirates, and Venezuela—that was established to regulate the production and price of oil

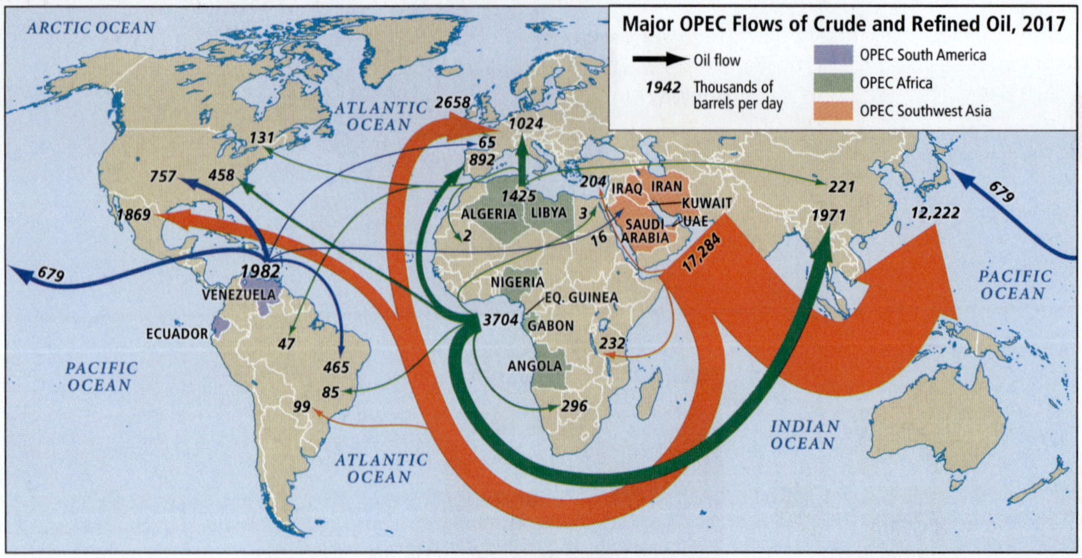

Figure 6.20 Major OPEC oil flows in 2017. The map shows the average number of barrels (in thousands) of crude oil and petroleum products distributed per day by OPEC to regions of the world. [Data from: *2018 OPEC Annual Statistical Bulletin*, Table 5.1, at https://asb.opec.org/index.php/data-download.]

production costs due to geologically accessible oil. Therefore, OPEC may decide on maintaining high levels of production even if doing so perpetuates low prices because it is still profitable to produce and it enables OPEC to gain a larger share of global oil sales.

Many OPEC countries, which were exceedingly poor 50 years ago, have become much wealthier since 1973. As we have observed in previous chapters, gross national income (GNI) per capita can be used as a general measure of well-being because it shows broadly how much money people in a country have to spend on necessities. The term GNI is often followed by purchasing power parity (PPP), which indicates that all figures have been adjusted for cost of living and so are comparable.

Figure 6.21 points to two situations regarding GNI in this region. First, from country to country, there is wide variation—less than U.S.$4000 per person per year in Syria and Yemen and over U.S.$40,000 per person per year in Bahrain, Kuwait, Oman, the United Arab Emirates (UAE), Saudi Arabia, and Qatar. Second, oil and gas wealth does not necessarily translate into high GNI per capita (compare Figure 6.21 with Figure 6.18)—Iraq has significant oil wealth but only modest GNI per capita. Turkey, which has a diversified economy and very little oil, nonetheless has a GNI rank higher than that of Iran (in the middle range), which has significant oil and gas reserves.

GNI per capita is an annual snapshot of an economy. However, oil dependency means that the region is susceptible to economic downturns. For example, oil income in Saudi Arabia shot up from U.S.$2.7 billion in 1971 to U.S.$117 billion in 2017. At first, Saudi Arabia and the other Gulf states were slow to invest their fossil fuel earnings in basic human resources at home, and they only gradually began to explore other economic strategies in case oil and gas ran out. Like their poorer neighbors, OPEC countries remain very dependent on outside countries for their technology, manufactured goods, skilled labor, and expertise.

Recently, the Gulf states have begun to invest heavily in roads, airports, urban expansion, irrigated agriculture, and petrochemical industries. Saudi Arabia is investing in new cities in different parts of the country to spur investment and advance economic and social development. Saudi Arabia has observed the economic success of its smaller Gulf neighbors, such as UAE and Qatar, and wants to replicate their approach to economic development based on foreign investment, reduced government bureaucracy, and more liberal social rules.

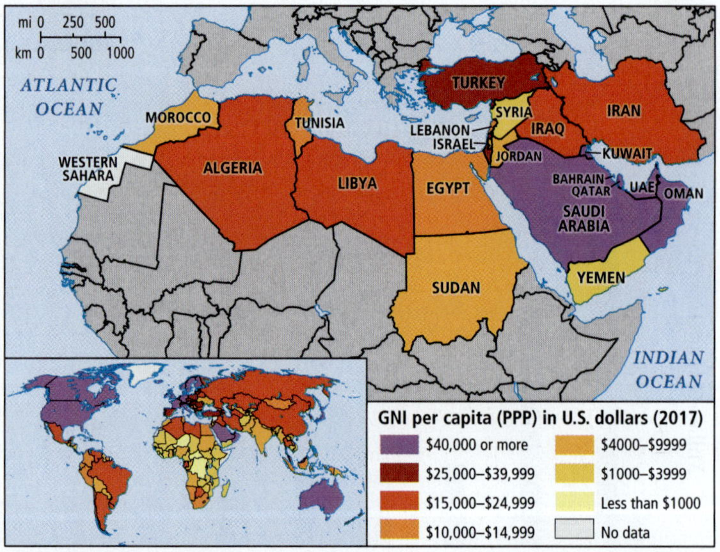

Figure 6.21 Gross national income (GNI) per capita, adjusted for purchasing power parity (PPP).

As a first round of investment, resources poured primarily into visible physical development. More recently, oil wealth has also been invested in education, social services, public housing, and health care. Libya, before the Arab Spring, and Kuwait developed ambitious plans for addressing all social service needs earlier than other countries, with especially heavy investment in higher education. Increasingly, Gulf states like UAE and Qatar are pouring money into education as well to develop domestic human resources. Meanwhile, the non-OPEC countries (those that do not produce fossil fuels) do not share directly in the wealth of the Persian Gulf states and cannot afford comprehensive public services. For the most part, the oil-rich countries have not helped their poorer neighbors develop.

Many factors beyond OPEC's control strongly influence world oil prices. After the 1973 OPEC oil embargo, other countries made

PHOTO ESSAY

6.22 Globalization and Development in North Africa and Southwest Asia

The Human Development Index (HDI) is a calculation (based on income, life expectancy, and educational attainment) of how adequately a country provides for the well-being of its citizens. In this region, HDI varies greatly from the low to the very high range, with Israel ranking the highest (22nd in the world). Iraq ranks lower on HDI than it does on the GNI scale, indicating that the revenue generated from oil is not reaching many people in the country. Jordan and Tunisia, on the other hand, rank high on the HDI scale but somewhat low on GNI per capita. This probably signifies that the government is investing in social services that will enhance human well-being.

THINKING GEOGRAPHICALLY

A What types of jobs are available to labor migrants in the Gulf states?

B What flows of money, resources, and people characterize globalization in the Gulf states?

C Which countries in the region are the most economically diverse?

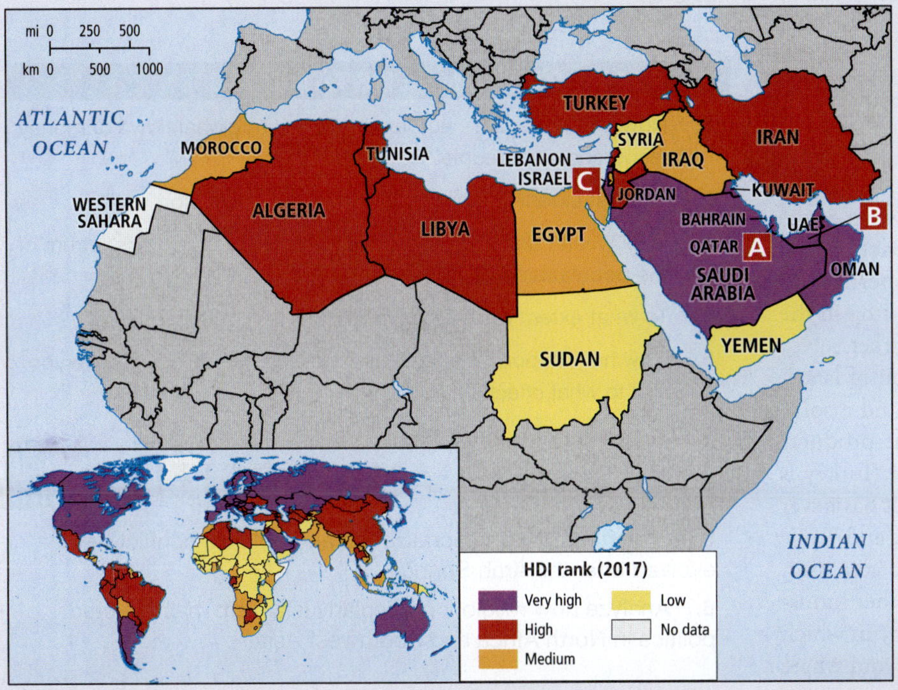

A Construction workers in Qatar. Qatar has the highest percentage of migrant workers in the world. [ionel sorin furcoi/Getty Images]

B Burj al Arab hotel in Dubai. This luxury hotel is shaped as a dhow, the traditional sailing vessel of the Persian Gulf. [RABIH MOGHRABI/AFP/Getty Images]

C Silicon Wadi. Israel spends more than any country in the world on research and development as a share of GDP. Silicon Wadi is a nickname for a concentration of high-tech companies around Tel Aviv. [Ariel Jerozolimski/Bloomberg via Getty Images]

successful efforts to increase their domestic supply. More recently, rapid industrialization in China and India has driven up their use of fuel and other petroleum products, creating long-term pressures that are raising the price of oil. Employing a theory known as "peak oil," some experts argue that in an era of diminished conventional oil reserves—which these experts say has already begun—prices will be driven higher. On the other hand, new oil exploration technologies, such as "fracking," have increased supply, which put pressure downward on oil prices (see "Oil Extraction" in Chapter 2). Efforts to combat climate change with renewable energy sources have been increasingly successful and could also reduce the demand for oil. However, so long as gasoline-fueled cars remain a major mode of transportation, demand for oil will remain robust. Current global oil flows are summarized in Figure 6.20.

ECONOMIC DIVERSIFICATION AND GROWTH

Greater **economic diversification**—expansion into a wider range of economic activities—could have a significant impact on the region, bringing economic growth and broader prosperity and thereby cushioning against diminishing oil reserves or a drop in the prices of oil, gas, or other commodities on the world market.

By far the most diverse economy in the region is that of Israel, which has a large knowledge-based service economy and a solid manufacturing base. Israel's goods and services and the products of its modern agricultural sector are exported worldwide. Turkey is the next most diversified, in part because—like Israel—it has never had oil income to fall back on. Egypt, Morocco, Libya, and Tunisia are also starting to move into many new economic activities. Some fossil fuel–rich countries tried to diversify into other industries. The government in the UAE developed an economy in which only 20 percent of GDP is based directly on fossil fuels, and where trade, tourism, manufacturing, and financial services are now dominant. Even so, most Gulf states are still very dependent on fossil fuel exports and the spin-off industries they generate.

Diversification has also been limited by economic development policies that favored *import substitution* (see Chapter 3). Beginning in the 1950s, many governments, such as those of Turkey, Egypt, Iraq, Israel, Syria, Jordan, Tunisia, and Libya, established state-owned enterprises to produce goods for local consumption. These enterprises were protected from foreign competition by tariffs and other trade barriers. With only small local markets to cater to, profitability was low and the goods were relatively expensive. Many of the import substitution enterprises have been closed or privatized.

Economic diversification also is also limited by the lack of investment. For example, rather than invest in local enterprises, members of the Saudi royal family generally place their wealth in more profitable firms abroad. Even when private and public investment has stayed in the region, it has gone into lavish projects that are meant to display wealth, which may not necessarily prove to be economically productive. Only recently have governments recognized that they need to invest locally in order to plan wisely for a time when oil and gas run out.

economic diversification the expansion of an economy to include a wider array of activities

Finally, both international and domestic military conflicts and the ensuing political tensions have stymied economic diversification because they have resulted in some of the highest levels (proportionate to GDP) of military spending in the world. Military spending diverts funds from other types of development, such as health care and education (**Figure 6.23**). The top spenders in the world are Saudi Arabia, Libya, and Oman, which dedicate more than 10 percent of their GDP to the military. In fact, eight of the top ten per capita spenders are in North Africa and Southwest Asia. The aftermath of the Arab Spring has made the region even more insecure, and the perceived need for a strong military is as marked as ever.

CHECK YOUR UNDERSTANDING

1. How are the region's economies linked to global flows of money, resources, and people?
2. To what extent has oil wealth led to modernization?
3. Is oil wealth benefitting those in power or a broader spectrum of the population?
4. To what extent does the region have a diversified economy?
5. How have import substitution policies been applied in the region and to what effect?

POWER AND POLITICS

6.6 Describe how democratization and political conflict have evolved after the Arab Spring.

6.7 Analyze how religion, particularly Islamism, has affected politics in North Africa and Southwest Asia.

Political and economic cooperation in the region has been thwarted by a complex tangle of hostilities between neighboring countries and within countries. Some are instigated by antagonism between Sunnis and Shi'ites, but many of these hostilities are the legacy of outside interference by Europe and the United States in regional politics, including colonial intrusions, and by the activities of global oil and gas corporations. The Israeli–Palestinian conflict, which has profoundly affected politics throughout the region, is rooted in the persecution of the Jews in nineteenth-century Europe, culminating in the Holocaust during World War II. The Iran–Iraq war of 1980–1988 and the Gulf War of 1990–1991 were instigated, in part, by pressures on petroleum resources, again from Europe and the United States. The violence in Iraq, though begun by a homegrown dictator, came to a head when the United States invaded and occupied Iraq, beginning in 2003. Though the United States has now formally withdrawn, violence continues to plague Iraq and neighboring Syria.

THE ARAB SPRING AND BEYOND

The political protests that swept across North Africa in 2010 and into Southwest Asia by 2011 profoundly changed the region. Politically, it resulted in the toppling of long-standing dictatorships in Tunisia, Libya, and Yemen. A protest also overthrew the

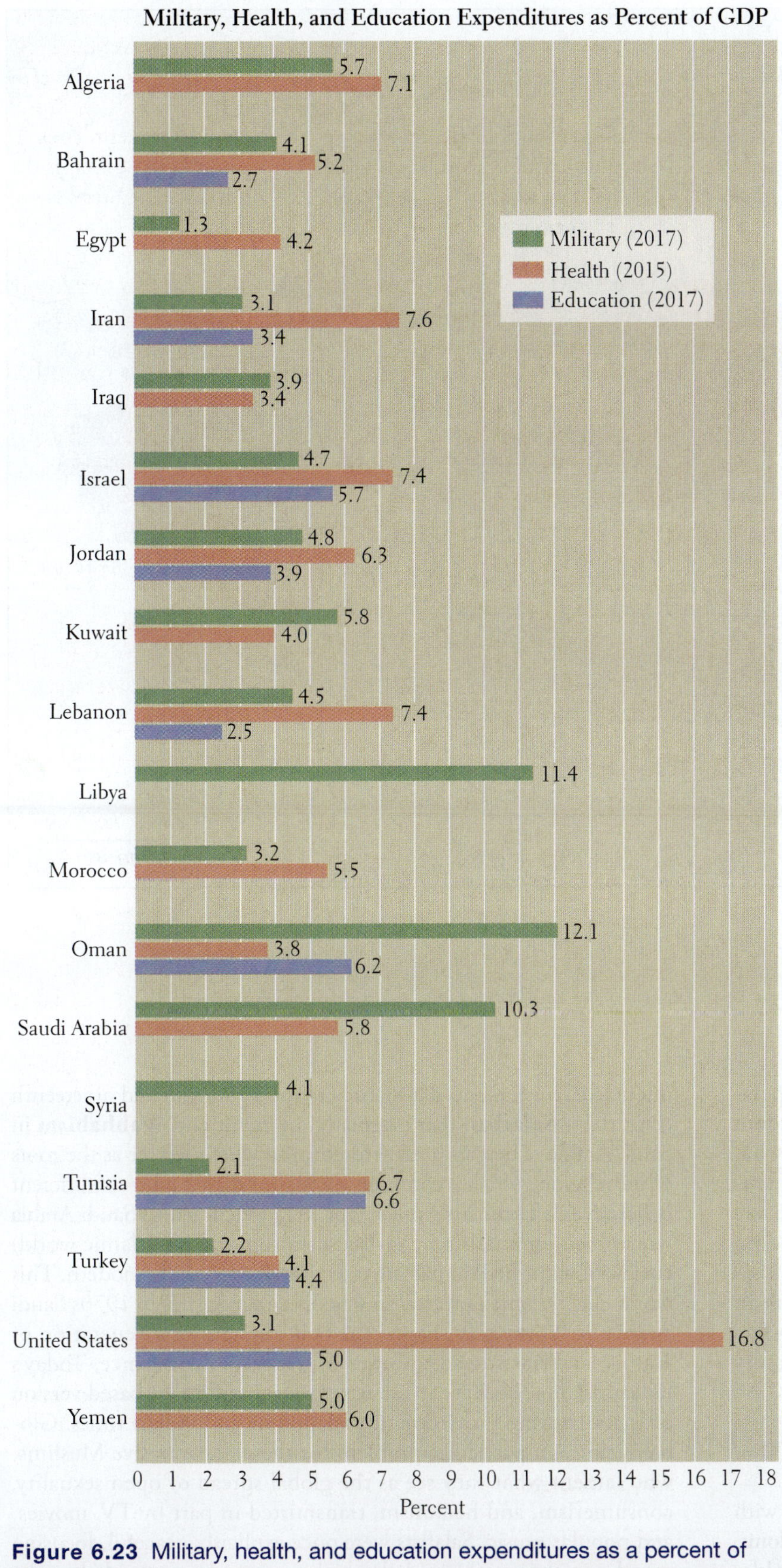

Figure 6.23 Military, health, and education expenditures as a percent of GDP, 2015–2017. The graph shows those countries in the region that have data for each variable. The United States is included for comparative purposes. [Data from: World Bank, "Military expenditure (% of GDP)," 2017, at https://data.worldbank.org/indicator/ms.mil.xpnd.gd.zs; and Table 8-9, UN Human Development Report, 2018, at http://hdr.undp.org/sites/default/files/2018_human_developme nt_statistical_update.pdf.]

military regime in Egypt, although the military returned to power after only a couple of years. In Syria, the outcome of political protest has been most problematic where initial opposition to President Assad morphed into a civil war involving multiple groups. **Figure 6.24** summarizes the outcome of the Arab Spring country by country. The early demonstrations across the region were precipitated by high unemployment and rising food prices caused by the global recession that began in 2008, poor living conditions, government corruption, and the absence of freedom of speech and other political freedoms. An undercurrent of dissent among women was also palpable in every country.

The political disquiet reflected the fact that, for much of their history, most governments in this region gave citizens very little ability to influence how decisions were made. Laws were simply interpreted to suit the factions that held power. Elections were either nonexistent or were rigged to reelect those already in office or their chosen successors. Some Gulf states are absolute monarchies even without the pretense of democracy. Meanwhile, freedom of speech and of the press was strictly limited, especially if it involved criticism of the government. As a result, most people across the region saw their governments as unresponsive and corrupt and viewed themselves as powerless to influence government in any way other than by massive protests. The role that Islam should play in society has been an especially contentious issue, and it relates directly to determining which system of laws should be adopted as well as which rights women should be afforded.

Different Paths Toward Reform

The Arab Spring movement began in Tunisia (see the chapter opening page) and spread to Egypt, then to Libya, and eventually to most countries in the region. The size and focus of the protests varied—see **Figure 6.25**—but they all raised hopes for real political reforms. In Tunisia, a compromise between secular and Islamic parties led to a new constitution, although many women are worried about the influence of conservative Islamists. Another challenge is posed by Islamist militants who have launched terrorist attacks in the country. In Egypt, an Islamist-oriented government was elected that proceeded to adopt a constitution that did not adequately protect the rights of more secular political groups, minorities, or women. Political turmoil and violence resulted as the Islamists governed in ways that outraged their numerous political opponents. Subsequently the Islamists were removed from power by Egypt's military, who tolerate little dissent from either democracy activists or Islamists.

More incremental approaches to reform were chosen by the kings of Jordan and Morocco, resulting in less chaos and violence. In both cases, the respective kings negotiated with reformists to keep some of their royal powers while agreeing to give up others. When elections were held, candidates from a wide range of political perspectives won positions. In Morocco, for example, an Islamist party has been successful but far from gaining a

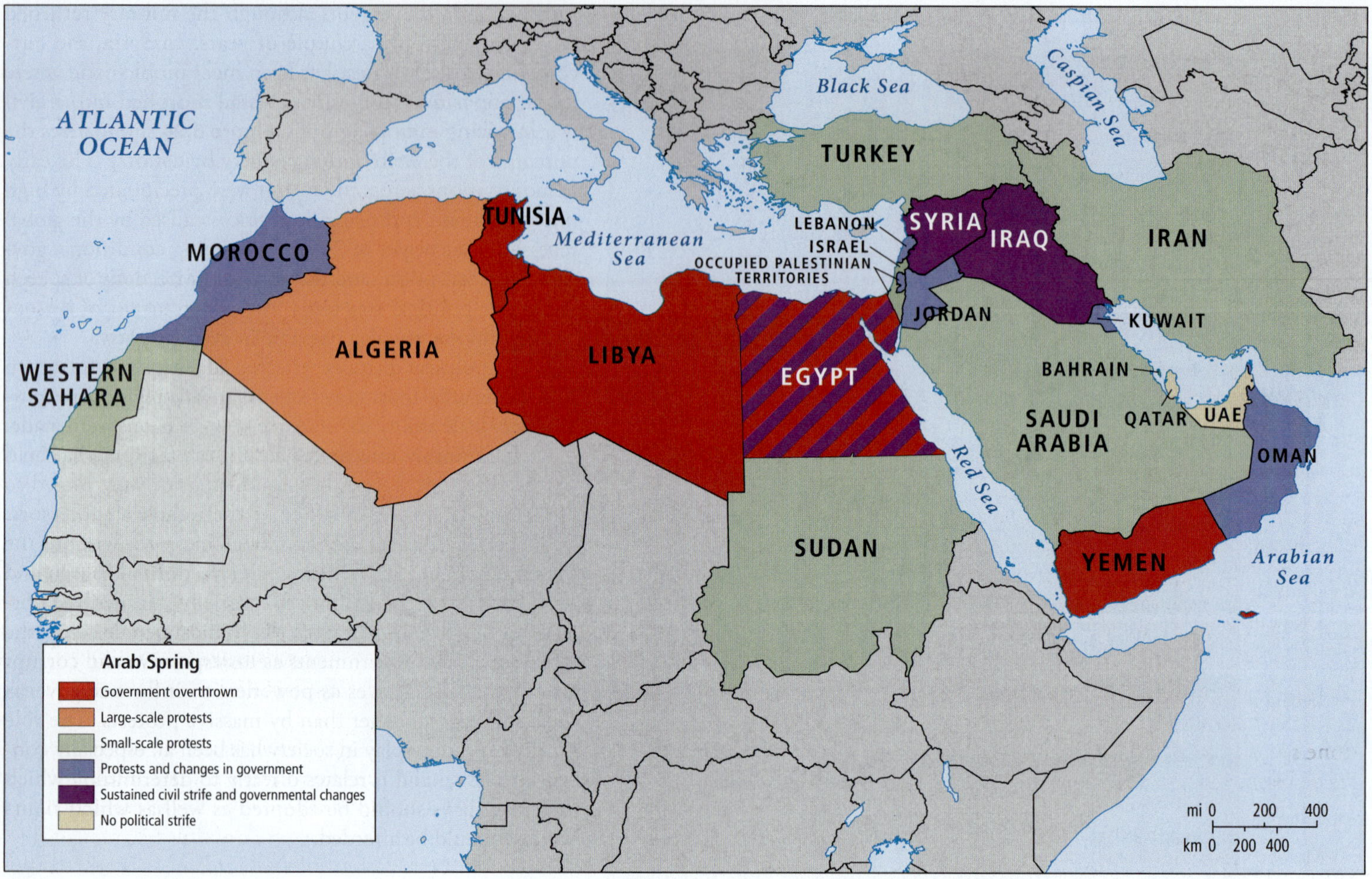

Figure 6.24 Political outcomes of the Arab Spring.

majority of the votes; thus, the Islamists had to accept cooperation and compromise as a necessity, an important precedent for the future. Jordan's path toward reform has been somewhat slower. The monarch has appointed reform-minded governments rather than declaring open elections. Women and minorities are guaranteed some seats in the parliament. However, freedom of the press is also more limited in Jordan than it is in Tunisia and Morocco, which suggests a more open political culture in North Africa compared to Jordan, where the ruler is rarely criticized publicly.

Salafism a religious movement that started in Egypt and that has influenced the ideology of radical Islamism

Wahhabism a religious movement that started in Saudi Arabia and that has influenced the ideology of radical Islamism

jihad in Arabic, struggle; either understood as a personal struggle to be a good Muslim, or against enemies of Islam

Islamism in a Globalizing World

Ironically, the Arab Spring, with its emphasis on political freedoms, actually opened up space for Islamist groups who took advantage of the new openness to assert their own agenda. Contemporary radical Islamism has its roots in two interrelated movements dating back to the eighteenth and nineteenth centuries—**Salafism** that originated in Egypt and **Wahhabism** in Saudi Arabia. They both stress a return to what they see as the roots of the religion, which is common among fundamentalists of different beliefs. Salafist scholars from Egypt were welcomed to Saudi Arabia where their emphasis on pan-Islamism (uniting the Islamic world) coalesced with the Wahhabists' rejection of everything modern. This worldview became powerful in schools and mosques in 1970s Saudi Arabia; it thoroughly shaped the belief system of a generation of Saudis, and has spread to other Muslim countries since. Today's radical Islamists believe in an extreme, purist Qur'an-based version of Islam that has little room for adaptation to modern times. Globalization is a particular problem for these conservative Muslims, who lament what they see as the global spread of open sexuality, consumerism, and hedonism, transmitted in part by TV, movies, and popular music. Salafists were once explicitly peaceful, focusing mostly on the much-needed social services they provided to the poor. But since the 1990s, the idea of **jihad** (a struggle) against the West has gained popularity among younger Islamists. The concept of jihad is multifaceted—it can mean a personal struggle to be a good Muslim or to do peaceful missionary work, but has in public discourse increasingly been associated with religiously inspired violence by jihadists against invaders of their homeland, or what they perceive as enemies of Islam. Many conservative Muslims—radical Islamists or not—object to what they see as the Western emphasis on the liberalization of women's roles, especially the idea that women should be educated and active outside the home (see Figure 6.34 later in the chapter).

Historically, Islamism is rooted in the interaction between religious and governmental authority. Governments in Saudi Arabia, Yemen, the UAE, Oman, and Iran are **theocratic states** in which Islam is the official religion and political leaders are considered to be divinely guided by both Allah and the teachings of the Qur'an. Elsewhere, such as in Turkey as well as in Tunisia, Libya, and Syria (at least before the 2011 rebellion began), governments are officially **secular states**, where religious parties are not allowed and the law is neutral on matters of religion. In practice, even in this region's secular states, Islamic ideas influence government policies. Combined with an authoritarian political culture, this has meant that political freedoms, such as free speech and the right to hold public meetings, were severely limited outside of Islam. In countries such as Egypt, for decades the only public spaces in which people were allowed to gather were mosques, and the only public discussions free of censorship by the government were religious discourses. In this context, many political movements became rooted in Islam that might not have done so otherwise. In Egypt, the Muslim Brotherhood, which is an old organization with Salafist roots, won the first post–Arab Spring election because it was highly organized and could mobilize its supporters. The Brotherhood developed for decades as a religious social movement, and even though it was banned on and off by the Egyptian authorities, it emerged as a potent political force during the Arab Spring. In a similar fashion, Shi'ite minorities around the region were shut out of political power; instead, they often rallied around their local religious leaders who then became de facto political leaders.

The Iranian Theocracy

Iran's recent political history and its status as a theocratic state based largely on a Shi'ite version of Islamism complicate its role in the modern world. The theocracy was formed under the leadership of the Islamic fundamentalist Ayatollah Khomeini, a Shi'ite spiritual leader who led a revolution in 1979 against the last Shah (the title of the monarch in Iran). Today, the official head of state is the Supreme Leader, a religious cleric. At the same time, the Supreme Leader shares power with an elected president. In the recent past, some presidents have been modernizing reformers, while others have been conservative traditionalists. The president in Iran has power over day-to-day affairs of the government, but ultimate power rests with the Supreme Leader, which is a lifetime appointment. This system is backed up by the Revolutionary Guard, a militia that cracks down on any threat to Iran's Islamic system of government.

When the Arab Spring erupted across North Africa, many wondered if Iran (which is not Arab) would join in the protests. The reason was that discontent had been simmering in Iran for some time. However, there were no major protests during post–Arab Spring Iranian elections, partly because the leaders of the protests were in jail or in exile, but also because a moderate presidential candidate, who eventually ended up winning the election, was backed by reformists. Nevertheless, significant disturbances have occurred in Iran as recently as 2018, due to the lack of economic opportunities as well as political repression.

THE ROLE OF MEDIA AND THE INTERNET IN POLITICAL CHANGE

The flow of information about politics and protests, such as those described above, is increasingly complex. Traditional media is, at times, restricted by government censorship, although the internet and social media are harder to control. In some countries—Tunisia, Israel, and Lebanon, for example—the press is reasonably free and opposition newspapers can be aggressive in their criticism of the government. But in much of the region, journalism can be a risky career, often leading to imprisonment. In Saudi Arabia, a dozen newspapers are available every morning, but all are owned or controlled by the royal Saud family, and all journalists are constrained by the fact that they may not print anything critical of Islam or of the Saud family, which numbers in the tens of thousands. When accidents happen or some malfeasance by a public official is revealed, the story is blandly reported with little effort to explore the causes of events or their effects, or to hold responsible officials accountable.

An example of journalistic change is the broadcasting network Al Jazeera, founded by the ruling emir of Qatar, but now privately owned and independent of the Qatar government. Al Jazeera is credited with changing not only the climate for public discourse across the region but also public opinion outside of the region through Al Jazeera English. The various versions of Al Jazeera now openly cover controversy with no apparent censorship.

The roles of the internet and cell phone technology as forces for political change emerged first during Iranian protests in 2009, when the fact that so many Iranians had video-capable cell phones meant that via the internet, the world saw instantly the brutality of the Iranian police and the Revolutionary Guard. Use of the internet (e-mail, Facebook, crowdsourcing, and Twitter) became commonplace during the Arab Spring, as ordinary people with nothing but an inexpensive cell phone were able to give real-time reports about activities in the streets of Cairo, Tripoli, and Benghazi. Twitter became the chief medium because it is free, mobile, and quick. Crowdsourced data could then be compiled and mapped, which allows protesters to see how and where a crisis unfolds and coordinate their actions accordingly, as the chapter-opening photo shows.

While new technology can connect people in ways not possible in the past and diffusion of information is quicker and less dependent on geographic proximity than ever, political action still takes place in concrete locations. For example, the epicenter of the Arab Spring was Tahrir Square in Cairo. That is where much of the Egyptian protests against the autocratic military

theocratic states countries that require all government leaders to subscribe to a state religion and all citizens to follow rules decreed by that religion

secular states countries that have no state religion and in which religion has no direct influence on affairs of state or civil law

PHOTO ESSAY

6.25 Power and Politics in North Africa and Southwest Asia

Political freedoms are relatively few in this region, which is one of the least democratic in the world. Many countries suffer from a violent political culture where wars and terrorism have given authoritarian regimes an excuse to forcibly repress legitimate political opposition groups.

THINKING GEOGRAPHICALLY

A **What circumstances made many suspicious of the U.S. claim that the 2003 war with Iraq was undertaken to bring democracy to Iraq?**

B **Why is the 1973 war between Israel and Egypt cited as one of the causes of political repression within countries of the region?**

C **How has the religious composition in Yemen affected its role in the geopolitical conflict of the region?**

D **How has the geopolitical situation changed for Kurds?**

E **What accounts for the higher death toll in Libya's Arab Spring than Egypt's?**

A U.S. Navy warships patrol the Persian Gulf. U.S. and European militaries have had a large presence in this region for a number of years, especially because of the importance of the region's oil exports to the global economy. The U.S.-led invasion of Iraq in 2003 was justified as an effort to rid Iraq of weapons of mass destruction and to ensure greater political freedoms for the people of Iraq. However, safeguarding foreign access to the region's oil reserves was a larger motivation. [Frank Rossoto Stocktrek/DigitalVision/Getty Images]

B An Egyptian man displays a Star of David on his shoe. An act of symbolic desecration takes place during an anti-Israel rally. Ongoing tensions between Israel and its Arab neighbors have emboldened radical groups on both sides. Most Arab governments don't recognize the state of Israel. Meanwhile, Israel bans its citizens from visiting "enemy countries" (a number of Arab states and Iran) without special permission. [KHALED DESOUKI/AFP/Getty Images]

C San'a, the capital of Yemen, is bombed. The Arab Spring resulted in the removal of the long-term dictator of the country. However, since then different factions have been vying for power. Yemen has become a proxy war between Sunni countries, first and foremost Saudi Arabia, and Shi'ite Iran. Saudi air strikes are considered human rights violations by many. [Mohammed Hamoud/Anadolu Agency/Getty Images]

D Kurds live in many nations. Kurds form a majority of the population in southeastern Turkey and in neighboring parts of Syria, Iraq, and Iran. Kurds have long been denied political freedoms and in some cases even the right to speak their own language. The relationship between the Turkish government and the Kurds remains conflictual, while in northern Iraq the Kurds have greater autonomy than they previously had. Kurdish troops in Syria—including many women soldiers—were instrumental in defeating the Islamic State. [BULENT KILIC/AFP/Getty Images]

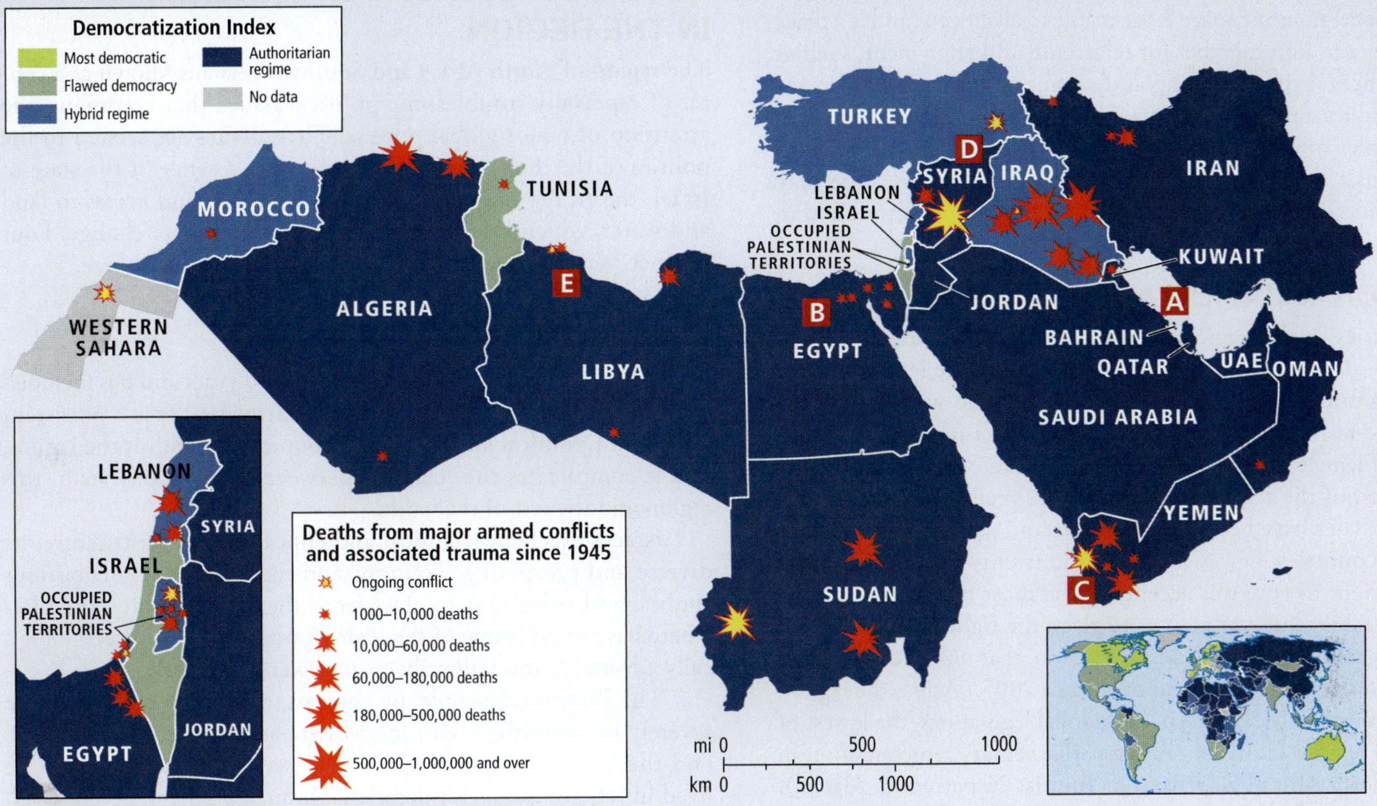

E Thousands celebrate in Tripoli, Libya, on an anniversary of Libya's revolution of 2011. Like the war in Syria, this conflict, in which 25,000–30,000 people were killed, was ignited by the government's brutal repression of Arab Spring protests. Civil strife has not ended either; the central government is weak, and local militias—and even competing governments—wield power in Libya. [MAHMUD TURKIA/AFP/Getty Images]

happened. Tahrir Square is therefore an important example of a **public space**. A public space is accessible to all citizens and is a place where they can join together for relaxation and enjoyment, such as performances or just strolling and people-watching, but also for political rallies and even protests. Urban squares, plazas, and parks are examples of public spaces, and without them it can be hard for citizens to channel legitimate political grievances. One may even say that democracy is dependent on public spaces to function properly.

DEMOCRATIZATION AND WOMEN

All countries in this region now allow women to vote, and two countries—Israel and Turkey—have elected female leaders (prime ministers) in the past. Nevertheless, women who want to actively participate in politics still face many barriers. Turkey was an early adopter of female suffrage and allowed women to vote in the 1930s. For the rest of the region, the process has been slow and uneven. The Gulf states have been the most resistant, and in the most conservative countries, it was not until the twenty-first century that women's right to vote was accepted. The most notorious example is Saudi Arabia where women were given the right to vote and run in local elections as recently as 2015. (Note that only local elections are allowed in Saudi Arabia and only since 2005.) Across the region, women average only 16 percent of national legislatures, the lowest of any world region. However, female parliamentary representation in parts of North Africa (31 percent in Tunisia, 26 percent in Algeria), Israel (28 percent), and Iraq (25 percent) is higher than in the United States (19 percent). Women lost ground in Egypt when the post–Arab Spring elections brought fewer women to office than had been the case beforehand. Gulf states like Qatar and Oman have less than 1 percent females in parliament, and those political bodies do not have much power over legislative affairs anyway.

CHECK YOUR UNDERSTANDING

1. Why did people protest during the Arab Spring and how has the outcome of the Arab Spring varied within the region?
2. To what extent is media free in the region?
3. What is the status of women's participation in politics?
4. What are the goals of Islamism? Has the influence of Islamism increased or decreased recently?

public space a place, such as a square or a park, that is open for all citizens to join together for personal enjoyment or political participation

FOUR GEOPOLITICAL SITUATIONS IN THE REGION

The region of North Africa and Southwest Asia is known as a center of especially troublesome political issues that command the attention of major global powers. These issues are related to the politics of the international oil trade, the presence of the state of Israel, the rivalry between Sunnis and Shi'ites, and access to land and water, which has been complicated by climate change. Four distinct, yet often interrelated, situations are presented here.

Situation 1: Israel and Palestine

The Israeli–Palestinian conflict has lasted 70 years and has included several major wars and innumerable skirmishes. It is a persistent obstacle to political and economic cooperation within the region, and it complicates the relations between many countries in this region and the rest of the world.

Israel's excellent technical and educational infrastructure, its diverse and prospering economy, and the large aid contributions (public and private) it receives from the United States and elsewhere have made it one of the region's wealthiest, most technologically advanced, and militarily most powerful countries.

The Palestinian people, by contrast, after years of conflict, are severely impoverished with inadequate access to education, jobs, and the basics of human well-being (see **Table 6.1**). Many have lived in refugee camps for decades. Through a series of events, Palestinians have lost most of the land on which they used to live. They now reside on the West Bank (home to 2.9 million Palestinians) and the Gaza Strip (1.9 million), with many also living in Israel and neighboring Jordan. The Jordanian census does not distinguish between people of Palestinian origin and Jordanian natives, but the two groups are probably equal in size. The West Bank and Gaza are both highly dependent on Israel's economy, where about 120,000 work in the construction, manufacturing, and service industries (**Figure 6.26B**). Israel often takes military action in the West Bank and Gaza in retaliation for Palestinian suicide bombings and rocket fire, primarily launched from Gaza. Many parts of the West Bank occupied Palestinian Territories continue to be encircled by security walls built by Israel to discourage potential Palestinian attackers from entering Israel. In effect, the wall also curtails Palestinian access to places they need to go on a day-to-day basis (Figure 6.26A).

On the bright side, economic cooperation is a fact of life for Israelis and Palestinians. Economic ties between Palestine and Israel

Table 6.1 Circumstances and human well-being among Palestinians and Israelis, 2017–2018

	Population in millions, mid-2018	GNI PPP in U.S.$, 2017	Life expectancy at birth, 2018		HDI rank, 2018	Infant mortality rate, 2018	Overall life satisfaction, 2018 index*
			Males	Females			
Palestinians	4.8	5,560	72	75	119	18	4.6
Israelis	8.5	38,060	81	84	22	3	7.3

*Ranked on a scale of 1 to 10, with 0 representing least satisfied and 10 representing most satisfied.
Source: Data from *2018 Population Data Sheet* (Population Reference Bureau) and *2018 Human Development Indices and Indicators* (United Nations).

Figure 6.26 Conflict and cooperation between Israelis and Palestinians.

(A) Palestinians on their way to Jerusalem to celebrate Ramadan pass through a heavily guarded checkpoint at the West Bank barrier in Bethlehem, Palestine. The barrier has hindered the flow of people and goods between Israel and the West Bank, especially damaging the latter's economy. [MUSA AL-SHAER/AFP/Getty Images]

(B) Palestinian workers at a SodaStream factory in the West Bank. Israeli companies like SodaStream have access to the expertise, equipment, and investment capital that make the growth of manufacturing industries possible, while Palestinians supply cheap labor. Such Israeli operations are controversial, and a boycott movement of Israeli companies making products in the occupied Palestinian Territories is gaining traction in Europe and the United States. [MENAHEM KAHANA/AFP/Getty Images]

have been essential to both for many years. Israel is the largest trading partner of the West Bank and the Gaza Strip, and provides Palestinians with a currency (the Israeli shekel), electricity, and most imports. Israel needs the labor of tens of thousands of Palestinian workers in fields, factories, and homes. During periods of calm, Israelis and Palestinians quietly reach agreements to expand Palestinian access to jobs in, and trade with, Israel. Outbreaks of violence impede the ease of movement of people through roadblocks, but the two sides continue to remain economically interdependent.

The Creation of the State of Israel In the late nineteenth century, as a response to centuries of discrimination and persecution in Europe and Russia, the **Zionist** movement emerged with the aim of establishing a homeland for the Jewish diaspora. Small groups of European Jews began to purchase land from absentee landowners in a part of the Ottoman Empire known as Palestine. Some of the purchased land was uncultivated, but landowners had also leased their lands for many years to Arab Palestinian tenant farmers and herders or had granted them the right to use the land freely. Such rights were no longer recognized by the Zionists who acquired the land to be farmed and occupied by Jews. Historians still debate whether or not the Zionists or the former landlords adequately compensated the displaced indigenous inhabitants.

On their newly purchased lands, the Zionists established communal farming settlements called *kibbutzim*, into which a small flow of Jews came from Europe and Russia. While Jewish and Palestinian populations intermingled in the early years, Zionist land purchases displaced more and more Palestinians, and violence became increasingly common in the 1920s and 1930s.

In 1917, the British government adopted policies that favored the establishment of a national home for Jews in the Palestine territory. In 1923 after World War I, the United Kingdom received what is called a "mandate" from the League of Nations—the precursor to the United Nations—to administer the territory of Palestine (**Figure 6.27A**). British control of Palestine made it easier for Jewish people to migrate there. By 1946, following the genocide of 6 million Jews in Nazi Germany's death camps during World War II, strong sentiment had grown throughout the world in favor of a Jewish homeland in the space in the Eastern Mediterranean. Jewish migration to Palestine continued to accelerate.

Conflict and the Two-State Solution In November 1947, after an intense debate in the UN General Assembly, the UN adopted the "Plan of Partition with an Economic Union" that ended the earlier mandate and called for the creation of both Arab and Jewish states (the *two-state solution* still under discussion today) with special international status for the city of Jerusalem (see Figure 6.27B).

The Palestinians and neighboring Arab countries fiercely objected to the establishment of a Jewish state and feared that they would continue to lose land and other resources. Then, as now, the conflict between Jews and Palestinian Arabs was less about religion than control of land, settlements, and access to water. The sequence of the changes in allotments of land to Israel and to the Palestinians over the last 95 years can be followed in Figure 6.27.

On the same day that the British ended the mandate and withdrew their forces, May 14, 1948, the political entity representing the Jewish population (the Jewish Agency for Palestine) declared the state of Israel on the land designated to them by the Plan of Partition. Warfare between the Jews and Palestinians began immediately. Neighboring Arab countries—Lebanon, Syria, Iraq, Egypt, and Jordan—supporting the Palestinian Arabs invaded Israel the next day. Fighting continued for several months, during which Israel prevailed militarily. An armistice was reached in 1949, and as a result, the Palestinians' land shrank still further, with

Zionism a movement that started in nineteenth-century Europe to create a Jewish homeland in Palestine

Figure 6.27 Israel and Palestine, 1923–today.

(A) (B) (C)

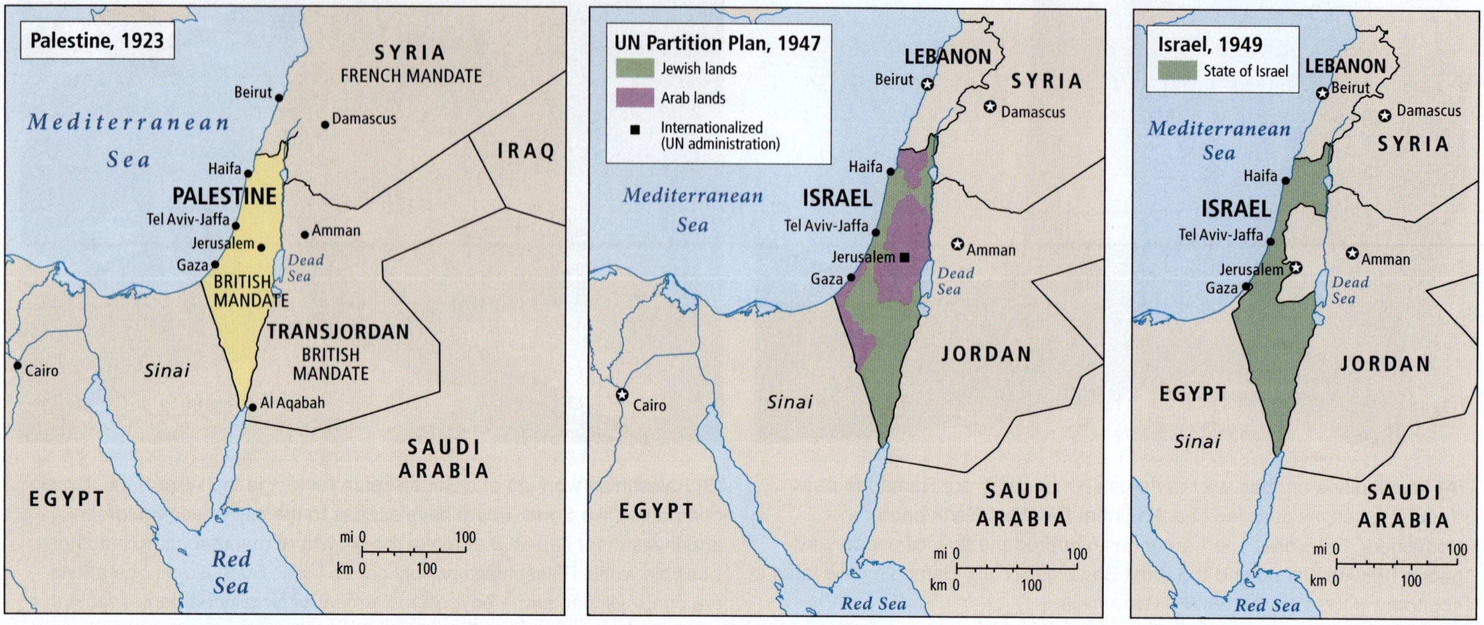

Israel and Palestine, 1923–1949. (A) Palestine, 1923. Following World War I, Britain controlled Palestine and Trans-Jordan (the precursor to Jordan). **(B)** The UN Partition Plan, 1947. After World War II, the United Nations developed a plan for separate Jewish and Arab-Palestinian states. **(C)** Israel, 1949. The Jewish settlers won a war against Arab forces, creating the state of Israel, which was larger than the original UN-proposed Jewish state. [Research from: Colbert C. Held, *Middle East Patterns—Places, Peoples, and Politics* (Boulder, CO: Westview Press, 1994), p. 184; and Foundation for Middle East Peace, *The Israeli Settlements in the Occupied Territories*, 2002, at http://www.firstpr.com.au/nations.]

(D) (E) (F)

Israel and the Palestinian Territory after 1949. When the state of Israel was created in 1949, its Arab neighbors were opposed to a Jewish state. **(D)** In 1967, Israel soundly defeated combined Arab forces and took control of Sinai, the Gaza Strip, the Golan Heights, and the West Bank. **(E)** In subsequent peace accords, Sinai was returned to Egypt, but Israel maintained control of the Golan Heights, the Gaza Strip, and the West Bank, claiming that they were essential to Israeli security. **(F)** Although the Palestinians were granted some autonomy in the Gaza Strip and West Bank, during the 1990s, the Israelis continued to build Jewish settlements in the West Bank and Golan Heights. This map shows West Bank settlements where 380,000 Jewish settlers live. [Research from: "Map: Golan Heights," at http://www.fmep.org/reports/archive/vol.-22/no.-6/map-golan-heights; and Yotam Berger, "How Many Settlers Really Live in the West Bank? Haaretz Investigation Reveals," *Haaretz*, at https://www.haaretz.com/israel-news/.premium.MAGAZINE-revealed-how-many-settlers-really-live-in-the-west-bank-1.5482213.]

the remnants (Gaza and the West Bank) incorporated into Egypt and Jordan, respectively (see Figure 6.27C). In essence, the Palestinians ended up as a nation without a state.

In the repeated conflicts over the next decades—especially the Six-Day War in 1967—Israel again defeated its larger Arab neighbors, expanding into territories formerly controlled by Egypt (Gaza and Sinai), Syria (Golan Heights), and Jordan (West Bank of the Jordan River; see Figure 6.27D). Since 1948, hundreds of thousands of Palestinians have fled the war zones, many ending up in refugee camps in nearby countries. Some Palestinians stayed inside Israel, especially to the north near Lebanon, and became Israeli citizens. Approximately 20 percent of Israelis are Palestinians, but they have not been treated by the authorities as equal to Jewish Israelis. De facto discrimination has resulted in fewer social services in Palestinian neighborhoods and villages, less access to government jobs, and restrictive building permits that limit residential choice. In 2018, Israel's parliament passed a controversial law declaring the country as a Jewish state, which has furthered Arab-Israelis' sense of marginalization.

On the occupied territories, the Palestinians have mounted prolonged uprisings, known as the **intifada**. The first ran from 1987 to 1993, when an agreement called the Oslo Accords provided that Israel withdraw from parts of the Gaza Strip and the West Bank, and that the Palestinian Authority be the entity that would enable Palestinians to govern themselves in their own state. For a period during the 1990s, the situation improved as Palestinians seemed to have achieved something close to statehood. However, a second intifada from 2000 to 2005 was fueled by extremists on both sides, failed peace negotiations, and ongoing tensions over the expansion of Israeli settlements in Gaza, the West Bank, and the Golan Heights (see Figure 6.27F). Shorter intifadas have occurred since then. Both sides have suffered substantial trauma and casualties. According to B'Tselem, an Israeli human rights group, far more Palestinians have died in this violence (from 2009 to 2018, there have been 3390 Palestinian deaths, primarily in the Gaza Strip, versus 175 Israeli deaths).

Over the years since 1947, the vision of a two-state solution has not died. In 2012, the Palestinians were awarded nonmember observer state status at the United Nations, and thus the Palestinians gained official recognition that they had never before achieved, even if they don't have an official vote in the UN as an independent country. Israeli allies in the UN, primarily the United States, have blocked the full recognition of Palestine as a state. From the perspective of international law, to consider Palestine as an independent state is also problematic, because it does not have full control over its territories; the Israeli military is, in practice, in charge of much of the West Bank. Moreover, the occupied Palestinian Territories have two separate and antagonistic governments, one in Gaza and one in the West Bank.

Territorial Disputes When Israel occupied Palestinian lands in 1967, the UN Security Council passed a resolution requiring Israel to return those lands in exchange for peaceful relations between Israel and neighboring Arab states. This *land-for-peace* formula, which set the stage for an independent Palestinian state, has been only partially fulfilled, as noted above.

Despite the land-for-peace agreement, Israel secured ever more control over the land and water resources of the occupied territories. Israel took what appeared to be a major step toward peace in 2005 when it removed all Jewish settlements and military on-the-ground presence from the Gaza Strip, but this progress was negated by a blockade of Gaza's economy. Israel, and to some extent Egypt, have control over almost all the goods and people that cross in and out of the land and sea borders of Gaza. Israel also controls the supply of water, electricity, and communications infrastructure in Gaza. Frustrated by being sealed off from the outside world, thousands of Palestinians gathered by the border wall between Gaza and Israel in 2018 and clashed with Israeli security forces. In response to attacks from Gaza, Israel has additionally engaged in air strikes on Gaza that have devastated the territory. The problem is that the Palestinian Authority is divided between the Islamist Hamas movement, which is in control of Gaza, while the more moderate Fatah controls the West Bank. This situation with two geographically separate territories governed by two different administrations has been a reality since 2007. Hamas gained legitimacy among many Palestinians because, modeled after the Muslim Brotherhood in Egypt, it provided much-needed social services. But periodically there are violent protests by Palestinians against Hamas' inefficient administration of Gaza. At the same time, because of its continuing attacks on Israel, Hamas is considered a terrorist organization by the United States and the EU. Egypt is also concerned about Hamas, which is why the single border crossing between Egypt and Gaza is rarely open.

The territorial dispute is also affected by continuous Israeli settlements in the West Bank. Some 380,000 Israeli settlers have colonized the West Bank and thousands of Palestinians have been displaced as a result. The **West Bank barrier** (Figure 6.26a) encircles the Jewish settlements on the West Bank and separates approximately 25,000 Palestinians from their neighbors and their farm fields. It also blocks roads that once were busy with small businesses, effectively annexes about 9 percent of the West Bank to Israel, and severely limits Palestinian access to much of the city of Jerusalem, most of which is now on the Israeli side of the barrier. The barrier, declared illegal by the UN's International Court of Justice, is nonetheless very popular among Israelis because it has reduced the number of Palestinian attacks.

The most intractable of territorial disputes may be over the city of Jerusalem. The city is situated right at the boundary between Israel and the West Bank. West Jerusalem is heavily Jewish, while East Jerusalem is predominantly populated by Palestinians, although a large number of Jewish settlers have moved there over time. In the middle sits the old city of Jerusalem. This is where three world religions have converged for centuries, and all three now claim the right to sacred spaces in the city. Since the days of the Crusades, many Christian churches have been built in the old city and pilgrims from around the world visit there. The Western Wall is a remnant of the Jewish temple from biblical times and the most important site of prayer in contemporary Judaism. Only a few hundred feet from there is the Dome of the Rock, a Muslim shrine at the location of Mohammad's ascent to heaven according to Islamic thought. Next to it sits the Al Aqsa Mosque, commonly

intifada a prolonged Palestinian uprising against Israel

West Bank barrier an Israeli-built wall or fence that now surrounds much of the West Bank and encompasses many of the Jewish settlements there

considered the third holiest site in Islam (after Makkah and Al Madinah). A potential two-state solution will have to resolve who should control and govern the city, and how different faiths would have access to Jerusalem. Today, Israel claims Jerusalem as its capital and most government affairs are conducted from there, including the *Knesset*, Israel's parliament. But because most countries do not recognize Jerusalem's status as the capital, almost all foreign embassies are located in Tel Aviv. The United States made a controversial move of its embassy to Jerusalem in 2018.

Situation 2: Iraq

Iraq has considerable oil reserves, the fifth largest in the world. For decades, the United States and Western countries had an amicable relationship with the Iraqi government, driven in large part by their demand for oil. In fact, the United States backed a 1963 coup that installed a pro-U.S. government that evolved into the regime of Saddam Hussein. Iraq's antagonistic relationship with Iran, including a 1980–1988 war, was another U.S. reason for supporting Iraq. But relations with Iraq took a dramatic turn for the worse in 1990 when Saddam Hussein, Iraq's dictator, invaded Kuwait. The United States and its allies forced Iraq's military out of Kuwait in the Gulf War of 1990–1991.

A second U.S. invasion of Iraq in 2003, to depose Saddam Hussein, initially met little resistance; however, the subsequent U.S. occupation encountered insurgent attacks. The death toll for Iraqis is estimated in the hundreds of thousands and possibly over 1 million, when counting Iraqi deaths caused by the harsh conditions created by the war and its aftermath.

The failure to understand the simmering tensions between Iraq's major religious and ethnic groups crippled the U.S. intervention. Sunni Arabs in central Iraq had dominated the country under Saddam Hussein, although they constituted only about 20 percent of the population. Shi'ites in the south, with over 60 percent of the population, have dominated politics since the fall of Saddam. This has given more influence to neighboring Iran, whose mostly Shi'ite population has an affinity with Iraqi Shi'ites. Most observers agree that the new Shi'ite-dominated government has not been very inclusive; instead, Iraqi Sunnis increasingly feel alienated from political decision making. Meanwhile, Kurds in the northeast, once brutally suppressed under Saddam Hussein, are allied with Kurdish populations in Turkey, Iran, and Syria. They have resisted cooperation with the Iraqi national government in Baghdad by forming an autonomous and self-governing region.

The military force of Iraq proved to be ineffectual in maintaining the security of the country. In this weak security space, a militia that we know as the Islamic State quickly gained ground in the territories in central Iraq, where most of the population was Sunni. The roots of the Islamic State can be traced to Al Qaeda insurgents who fought against the U.S. occupation and later the Shi'ite-led government of Iraq. Now its own organization, the Islamic State expanded its control into parts of central and northern Iraq, as well as into bordering Syria.

The objective of the Islamic State was to spark an Islamist rebellion across the region rather than just overthrowing the governments in Iraq or Syria. The Islamic State was successful in using social media to spread its vision and was able to attract foreign fighters. Tens of thousands of young men, and some women, came from other Muslim countries, and a few from the West, to join the fight. The threat of the Islamic State united a wide range of actors against the group—including the United States, Russia, the Kurds, and Iran—and at current writing, the Islamic State is virtually defeated on the ground, although it continues to be a source of inspiration for Islamists around the world to carry out terrorist attacks against perceived enemies. When the Iraqi government now has to focus on other domestic issues, the question is if the Sunni–Shi'ite divide can be overcome.

Situation 3: Syria

As noted above, the conflict in Iraq spilled over into Syria. The reason is that Syria was already embroiled in a civil war, and Islamic State forces could relatively easily move across the border and establish a presence in Syria.

The background to this development is the Arab Spring protests that spread around Syria in 2011. Already before the Arab Spring, Syria was unusually fragile due to periods of drought that have threatened the livelihood of many. Climate change and water scarcity are now thought to be among the stressors that helped to bring on the Syrian rebellion.

The political causes of rebellion in Syria are many. The existence of Syria is a legacy of French colonialism, which brought together a large number of ethnic and religious groups. Several factions of Muslims, as well as Christians and even a few Jews, lived together. But after the Assad family swept to power in a coup in 1970, Syria slowly morphed into an authoritarian state. By the time Bashar al-Assad took over from his father in 2000, Syria had gained a reputation for backing assassinations in Lebanon and bombings in Iraq, and for supporting Hezbollah—a Lebanese-based radical Shi'ite group that has close links to Iran, opposes Israel, and uses terrorist tactics.

Under Assad, open debate, elections, and creativity were stifled by an authoritarian government and a society rife with corruption. Dictatorial rule for 40 years and ethno-religious divisions set up Syria for failure during the Arab Spring. The Assad family belongs to the minority Alawite sect of Shi'ites, while the majority of Syrians are Sunni Muslims. To keep political equilibrium, the Assads favored a secular state similar to Turkey's. However, Sunnis were kept out of governing circles. Sunni antagonism against Alawites grew because Alawites led particularly privileged lives, with many advantages coming to them from the Assad family.

Arab Spring protests were met with brutal repression in Syria. In time, it became apparent that the rebel factions came from opposing points of view—some favoring pro-democratic secular changes, others wanting more Islamist conservative reforms. The outcome of the conflict in Syria has been tragic. The UN estimates the number of deaths to be approximately 400,000. Moreover, 5 million refugees have left the country, while 6 million are internally displaced within Syria. Such numbers are staggering for a country of 18 million. Most Syrian refugees remain in camps in bordering countries. The refugee stream to Europe is widely written about in media, but it should be noted that only 20 percent of Syrian refugees have made it to Europe; most remain in Jordan, Turkey, Lebanon, or are internally displaced.

Western governments, while expressing sympathy for the secular factions of the war, have been reluctant to intervene for fear

of becoming bogged down in a conflict without a clear end point. Russia, on the other hand, supports the Assad regime, which is an old ally. The reasons are twofold: Russia's only military base in the region is in Syria and it seeks to maintain that military presence. Also, there are Islamists along the southern border of Russia in Caucasia and Central Asia; thus, Islamist ideologies are as much of a threat to Russia as they are to the West. Another country that would like to see a resolution to the conflict is Turkey. It wants to limit the refugee movement into its country; ethnic Kurds live on both sides of the border and such connections can be destabilizing; and terrorist attacks have occurred recently in Turkey, some of which may be attributed to the Islamic State.

In 2017, the Islamic State's informal capital of Raqqa fell. It is nearly defeated in the rest of Syria as well. Particularly Kurdish forces were central to the military demise of the Islamic State. Most Kurds are committed to a secular state, which placed them in opposition to the Islamic State. Kurdish forces were known to include many female soldiers, which is also an indicator that its ideology is far different than that of Islamists (see Figure 6.25D).

Situation 4: The Sunni–Shi'ite Rivalry

One underlying factor in the conflicts in Iraq and Syria is the geopolitical rivalry between the Sunni and the Shi'ite branches of Islam. The Shi'ite leader is undoubtedly Iran, and among the leading Sunni countries, Saudi Arabia is particularly interested in keeping Iran at bay. This divide is played out in multiple locations throughout the region.

In Syria, Iran has supported the Assad regime for decades. The Assad family are, as mentioned above, members of the Alawite sect, which is a sub-branch of Shi'ite Islam. But perhaps more than for reasons of religious kinship, Iran supports Assad because any other potential government would be anti-Iran Sunni. Iran has provided Syria with subsidized oil, military support, and training of security forces to help suppress popular unrest in Syria.

Equally important for Iran is that Syria provides a geographic connection to Lebanon. There, Iran has a strategic ally in the Lebanese Shi'ite militia Hezbollah. Through Syria, Iran provides Hezbollah with military support, which is often directed at Israel. This is the main reason that Israel views Iran—a country that is almost 1000 miles (1600 kilometers) away—as its leading security threat.

On the Arabian Peninsula, the Saudi Arabia versus Iran conflict is more explicit. After the Arab Spring, Yemen has struggled to form a stable government. The biggest cause of instability in Yemen is the Houthi, a Shi'ite ethnic group with roots in the northwest of Yemen. The Houthi's military advances across much of Yemen have prompted air strikes by Saudi Arabia and some of its allies. The Shi'ite Houthi are reputedly funded and armed by Iran. Yemen is both becoming the site of a proxy war between the leading Sunni and Shi'ite countries in the region and drifting toward becoming a **failed state**, meaning that the government has lost control and is no longer able to fulfill its basic responsibilities as a defender of a sovereign state.

Finally, Saudi Arabia has initiated economic sanctions against its small but wealthy neighbor Qatar. The reasons include Qatar's support for the Muslim Brotherhood during the Arab Spring, Qatari-based Al Jazeera's frank media critique of Saudi and other Arab governments, and Qatar's willingness to have good relations with Iran. The latter is arguably the key geopolitical issue; Saudi Arabia will crack down on any country in the region that is pursuing an independent foreign policy toward Iran.

CHECK YOUR UNDERSTANDING

1. What are the competing claims of territorial control between Israel and the Palestinians?
2. How has the United States' long and complicated relationship with Iraq changed over time?
3. What led to the rise and demise of the Islamic State?
4. What caused the civil war in Syria and what has been the outcome?
5. Why are Iran and Saudi Arabia enemies?

URBANIZATION

6.8 Discuss how and where new wealth has been invested in urban development.

6.9 Describe domestic and international migration patterns into the region's cities and the quality of life of those migrants.

The rate of urbanization in Southwest Asia and North Africa varies greatly—from 100 percent in the small Gulf city-states Kuwait and Qatar to 36 to 37 percent in Sudan and Yemen. On average, the region is more urban than the world as a whole. In part, that is an outcome of a long historical process whereby people have clustered in cities as agricultural opportunities in the desert landscape were marginal and as cities of the region played an important role in far-flung trade. However, contemporary forms of urbanization are also transforming this region. Two globalized patterns of urbanization with distinct geographic signatures are now apparent: one pattern in the newly oil-rich countries, and another in those countries where limited development has primarily resulted in rural-to-urban migration, with its associated crowded slum housing.

Determining the size of cities in historical times is not always easy, but estimates indicate that five (six if counting Córdoba, Spain, which was Muslim at the time) of the ten largest cities in the world in 1000 C.E. were in Southwest Asia and North Africa, and four of the top ten in the year 1500. This is not surprising considering the importance of the region in economic, cultural, and political affairs at the time. Today, there are still numerous cities with a medieval core that is dominated by narrow, pedestrian-only streets where commerce is taking place. Behind the bustle of these public spaces, the traditional living arrangement was that of a courtyard surrounded by residential spaces, often for extended families. These ancient city centers are referred to as the *medina*, or simply the old city (see Figure 6.32A later in the chapter). Under the influence of European colonizers, this form of indigenous urban design was abandoned and new neighborhoods were built with Western-style apartment blocks and wide streets in a grid network. In effect, two different types of cities sit adjacent to each other.

failed state a country where the government has lost control and is no longer able to fulfill its basic responsibilities as a defender of a sovereign state

In addition to the distinct urban form, other cultural features also shape the urban experience. The call to prayer five times per day, which is one of the five pillars of Islam, defines the aural character of cities and towns in all parts of the region. The call to prayer typically comes from minarets that tower above the urban landscape. In the past, the call was done in person; today, it is more likely to be a voice recording. Nevertheless, the chant is quite audible and one of the most distinct features of Muslim cities. As it is proscribed that prayer should take place five times per day, in order for Muslims to have easy access to a nearby mosque, the urban landscape is dense with mosques.

Until the 1970s, most of the population still lived in small rural settlements. Then, people began to migrate in significant numbers from villages to urban areas in response to economic forces driven by oil wealth and globalization, and by agricultural modernization that displaced small farmers. In locations where agriculture is feasible, the shift toward green revolution export-oriented agriculture reduced the amount of labor required to produce crops. In North Africa, drought also instigated rural-to-urban migration (**Figure 6.28C**). Poor migrants ended up in emerging slums on the periphery or in the historic old city where dwellings didn't have modern amenities. By 2018, 62 percent of the region's people lived in urban areas (see the Figure 6.28 map) and 44 cities could claim a population greater than 1 million people. The largest metropolitan area in the region, and one of the largest in the world, is Cairo, Egypt, with 16.5 million residents in 2018 (Figure 6.28A). Istanbul in Turkey is home to 14 million people. Both were centers of historic empires that have continued to grow in modern times. Tehran, the capital of Iran, also has 14 million people.

The petroleum-rich Gulf states are now highly urbanized. Between 70 and 100 percent of the (still small) population now live in urban areas, which are extravagant in design, sprawling and auto-dependent, and very modern in appearance. Some of these cities have grown from almost nothing to several million inhabitants in just a few decades. These modern Gulf cities draw investment in high-tech ventures attracted by a low rate of taxation or no taxes, and high-end tourism. The new ventures and the construction of office and living space require a wide range of workers from all over the world.

The most high-profile example of new development is probably Dubai in the UAE. There, massive projects like several artificial islands in the shape of palm trees (see, for example, Palm Jumeirah in Figure 6.28B), the iconic hotel Burj al Arab (Figure 6.22B), and the tallest building in the world, Burj Khalifa, are all intended to increase the profile of Dubai as the premier tourism and business center of the region. The neighboring city of Abu Dhabi, in addition to being an oil industry center, tries to brand itself as a cultural destination. Its branch of the famous Paris Louvre is now the largest art museum on the Arabian Peninsula. At the same time, poor migrants who provide the necessary construction labor for such high-profile luxury projects are concentrated in overcrowded apartments or temporary labor camps on the outskirts of the city. However, the recession that started in 2008 badly damaged the economies of the Gulf states. Most of the massive building projects that were underway in the UAE screeched to a halt. While the economy has rebounded, the recession showed that Dubai's real estate and tourism-driven economy is highly cyclical and vulnerable to economic downturns, just like the region's oil industry. The recession also affected remittances (payments) sent back to home countries because, with drastically fewer jobs available in the recession, those countries that have many temporary workers in the Gulf states experienced a steep decline in remittances.

Outside the Gulf states, there have been far less planning and financing for urban growth. For example, in 1950, Cairo had about 2.4 million residents, while today it is home to 16.5 million people. Some of these millions of new residents live in huge makeshift slums, both on the outskirts of the city and in the medieval interiors of the old city, where ancient houses are crumbling and plumbing and other services are chronically dysfunctional. In Egypt, 17 percent of city dwellers live in what the UN defines as slum conditions. In the poorest countries in the region, conditions are much worse. In Yemen, 67 percent of urban residents live in slums; in Sudan, an astounding 94 percent do.

The planning that has taken place in Egypt to deal with overcrowding has largely failed. Everybody agrees that Cairo, Egypt's primate city, has simply grown too large, and in order to spur decentralization, urban planners in Egypt have for decades laid out satellite developments on undeveloped land at some distance from Cairo. However, these "new towns" have not attracted many residents, and some homes have remained vacant for years and even decades. The problem is not that they are unattractive places, but that most Egyptians trying to escape overcrowded Cairo can't move there because of the lack of employment opportunities and transportation services. There are not enough local jobs or public transit options in these new towns for those without cars to get to places where job prospects are better.

URBAN MIGRATION

The prospect of better educational opportunities, jobs, and living conditions draws rural internal migrants to Cairo and other cities outside the Gulf states. However, because stable, well-paying jobs are scarce, many migrants end up working in the informal economy (see Chapter 3), sometimes as street vendors, as casual laborers, or as menial service providers. Unfortunately, education also does little to ensure employment; for decades, this region has been noted for its large numbers of unemployed and underemployed university graduates.

In the cities of the Gulf states, on the other hand, there has been a deficit of trained native young people. Instead, immigrants come from all over the world to function as temporary guest workers. Some work as laborers on construction sites and as low-wage workers throughout the service economy, but many work in shops and professions. In some countries, such as the UAE and Qatar, guest workers make up 80 to 90 percent of the population. Most employers prefer Muslim guest workers, and over the last two decades, several hundred thousand Muslim workers have arrived from the occupied Palestinian Territories, Egypt, Pakistan, and India. Some female domestic and clerical workers come from Muslim countries in Southeast Asia, including the Philippines. Overwhelmingly, these immigrant workers are temporary residents with no job security and no right to become citizens. Their visas are contingent upon an employer who is willing to hire them, and their passports are typically confiscated by their employers; lose your job

Urbanization in North Africa and Southwest Asia

Globalization has brought two distinct patterns of urbanization to this region. In fossil fuel–rich countries, populations are more urbanized and city governments have undertaken lavish building booms, drawing in laborers and highly skilled workers from across the globe. In countries without fossil fuel wealth, cities are receiving massive flows of poor, rural migrants. This is partially a result of economic reforms (emphasizing export agriculture and industrialization) aimed at making rural and urban economies more productive.

THINKING GEOGRAPHICALLY

A What has driven the migration of rural people to the region's old established cities, such as Cairo?

B What circumstances dealt a blow to high-tech and tourism development in the Gulf states, such as Dubai?

C By 2018, what percent of the population in this region lived in urban areas?

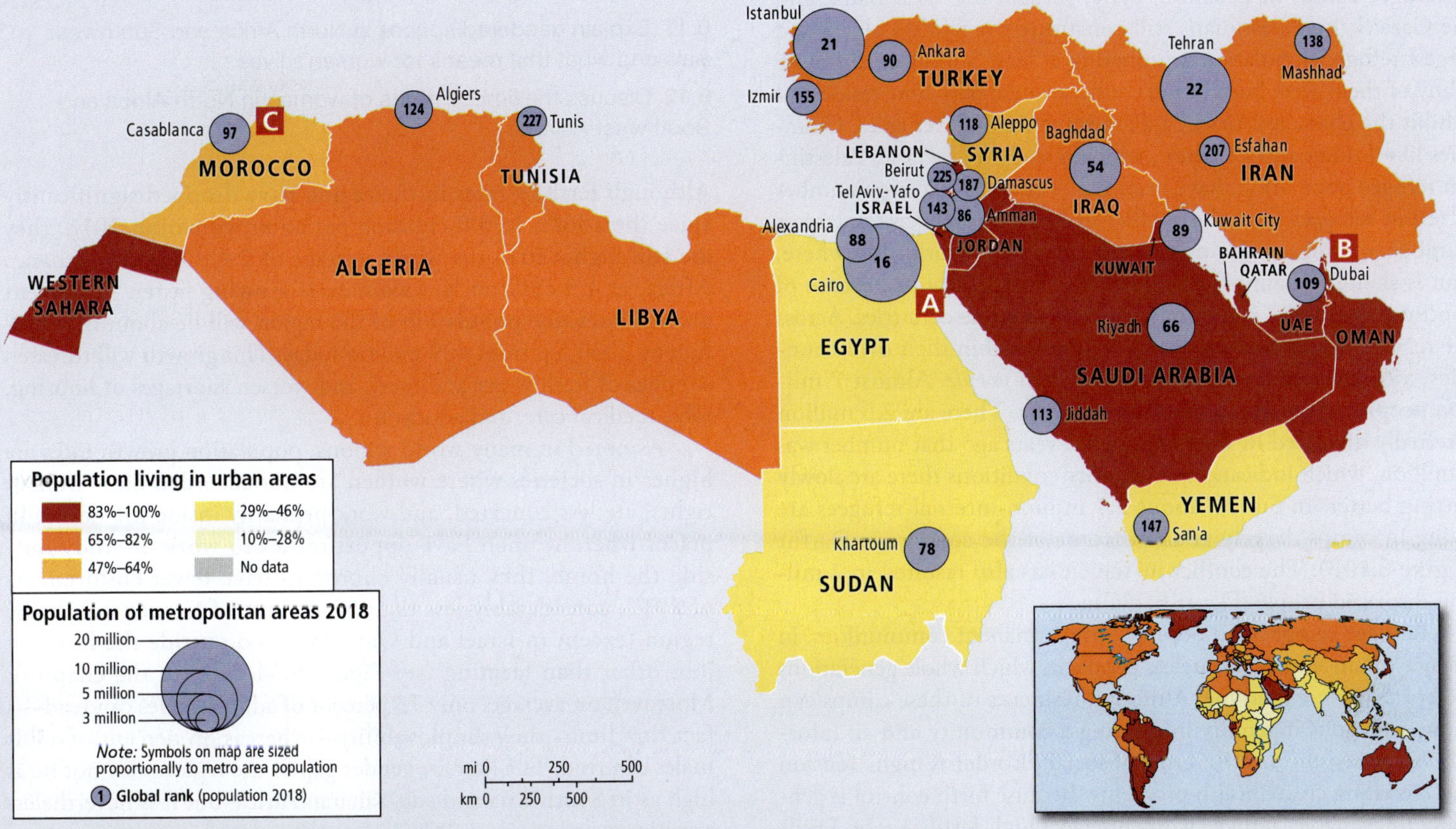

A A *bazaar* (market) located in the dense, central parts of Cairo with narrow streets. After decades of migration from rural areas, Cairo is the region's largest city; it is severely overcrowded. [Phillip Hayson/Getty Images]

B Palm Jumeirah is an artificial palm-shaped island with luxury residences in Dubai, UAE. Palm Jumeirah and several similar developments are central to Dubai's efforts to build a globalized tourism-based economy that will prosper long after the region's fossil fuel resources are exhausted. [Motivate Publishing/Gallo Images/Getty Images]

C A Moroccan man searches for valuables in a garbage dump outside Casablanca. The city's population is growing quickly due largely to migration from Morocco's drought-stricken interior, but the lack of jobs in the city is resulting in escalating unemployment and crime. [ABDELHAK SENNA/AFP/Getty Images]

and you will be deported. The migrants send most of their income to their families at home and often live in stark conditions alongside the opulence of those enriched by oil and gas. The wealth of a country like Qatar, the country with the highest GDP per capita in the world in 2017, is built on extreme labor exploitation.

Refugees comprise a major category of migrants, as this region has the largest number of refugees in the world. Usually, they are escaping human conflict, but environmental disasters such as earthquakes or long-term drought also displace many people. When Israel was created in 1949, many Palestinians were placed in refugee camps in Lebanon, Syria, Jordan, the West Bank, and the Gaza Strip. Palestinians still constitute the world's oldest and largest refugee population, numbering at least 5 million, although many of them were born in the country where they now reside and exhibit the characteristics of a diaspora more than refugees. Countries like Jordan and Lebanon, which already had a large Palestinian refugee population, have been the recipients of a large number of recent refugees from the conflict in Syria. Turkey has the largest number of refugees, 2.7 million people, in the world. Elsewhere, Iran is sheltering almost a million Afghans and Iraqis because of continuing violence and instability in their home countries. Across the region, even more people are refugees within their own countries, a category called *internally displaced people*. Almost 7 million people are internally displaced in Syria. There are 2.6 million internally displaced in Iraq, although 2 years ago that number was 4 million, which indicates that security conditions there are slowly getting better. In Sudan, about 2.1 million internal refugees are living in camps, largely as an outcome of the conflict in Darfur (Figure 6.10B). The conflict in Yemen has also resulted in 2 million displaced people (Figure 6.25C).

Refugee camps often become semipermanent communities, in essence urban slums, of stateless people in which whole generations are born, mature, and die. Although residents of these camps can show enormous ingenuity in creating a community and an informal economy, the cost in terms of social disorder is high. Tension and crowding create health problems. Because birth control is generally unavailable, refugee women have a high fertility rate. Disillusionment is widespread. Years of hopelessness, extreme hardship, lack of education, or lack of well-paid employment take their toll on youth and adults alike. Because of these factors, it is easy for Islamists to find new recruits, which they do by providing basic services to camp residents. Moreover, even though international organizations also provide for refugees, the refugees constitute a huge strain on the resources of their host countries. In Jordan, for example, native Jordanians are a minority in their own country because more than 3 million Palestinian refugees and their children, and the new influx of Syrians, account for well over half the total population of the country, and their presence has changed life for all Jordanians.

CHECK YOUR UNDERSTANDING

1. What is the pattern of urbanization in the oil-rich countries?
2. How has globalization shaped different patterns of urbanization throughout the region?
3. What is the pattern of urbanization in the oil-poor countries?
4. How have economic reforms aimed at improving global competitiveness, especially in export-oriented agriculture, influenced urbanization outside the Gulf states?
5. Where are refugee populations in the region located, and why?

POPULATION, GENDER, AND CULTURE

6.10 Describe different demographic structures among the countries of North Africa and Southwest Asia.

6.11 Explain gendered spaces in North Africa and Southwest Asia and what that means for women's lives.

6.12 Discuss the limited rights of women in North Africa and Southwest Asia.

Although fertility rates in the region have dropped significantly since the 1960s, to 3.0 children per adult woman in 2018, they are still higher than the world average of 2.4. Only sub-Saharan Africa, at 4.9 children per woman, is growing faster. At current growth rates, the population of the region will be about 600 million by 2030, up from 489 million today. This growth will threaten supplies of fresh water and food, and worsen shortages of housing, jobs, medical care, and education.

As noted in many world regions, population growth rates are higher in societies where women are not accorded basic human rights, are less educated, and work primarily inside the home. In places where women have opportunities to work or study outside the home, they usually choose to have fewer children. As of 2017, considerably less than 50 percent of women across the region (except in Israel and Qatar) worked outside the home at jobs other than farming (see Figure 6.34, later in the chapter). Moreover, on average, only 73 percent of adult females can read—a fact that limits their employability—whereas 86 percent of adult males can read. The literacy gender gap in the region may not be as high as in South Asia and sub-Saharan Africa, but it is nevertheless significant.

The deeply entrenched cultural preference for sons in this region is both a cause and result of women's lower social and economic standing. It also contributes to population growth, as families sometimes continue having children until a desired number of sons is reached. At the same time, these effects vary from country to country, which is evident in the different shapes of the population pyramids in **Figure 6.29** (see also the discussion in Chapter 1). Iran had high population growth until approximately 25 years ago. Since then birth rates have dropped dramatically, which is represented by the narrower bottom of the pyramid. This is not happening in Israel, where women still have as many as 3.2 children, on average (compared to Iran's 2.0). At such a high rate, Israel is an anomaly among affluent countries. The traditionalist ultra-Orthodox community in Israel has high birth rates; however, the smallness of that group does not explain the high fertility rate for the country as a whole. Instead, a cultural norm of many children is widespread in society, possibly associated with a strong sense of nationhood in a conflictual region (see Figure 6.31B later in the chapter). In Qatar and the UAE, gender imbalance is extreme (see Figure 6.29C). The unusually large numbers of males over the age

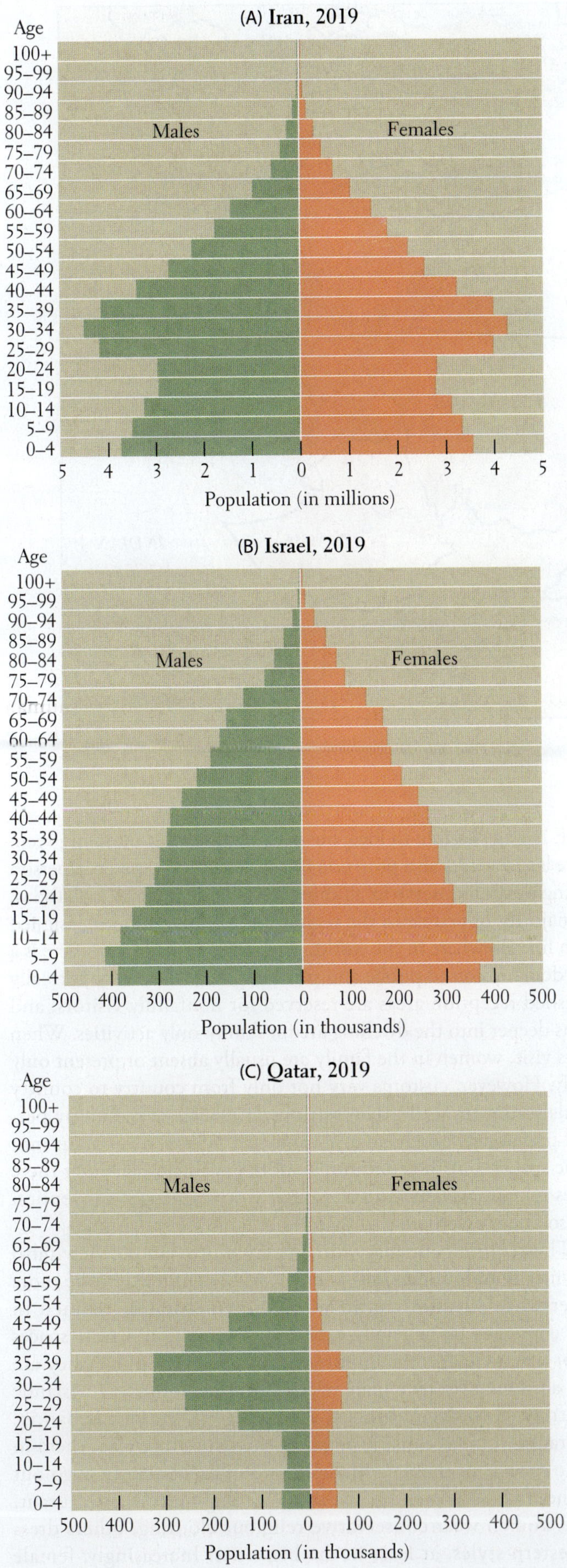

of 15 are the result of the presence of numerous male guest workers (Figure 6.25A). Overall, opportunities for women are opening and this change is likely to reduce women's interest in having large families. Slower population growth will mean that fewer resources will have to be invested in housing, schools, and services.

POPULATION DISTRIBUTION

Although the region as a whole is nearly twice as large as the United States, most of the population is concentrated in the few areas that support agriculture. Vast tracts of desert are virtually uninhabited, while most of the region's population is packed into coastal zones, river valleys, and mountainous areas that capture orographic rainfall (compare **Figure 6.30** with Figure 6.3). Population densities in these areas can be quite high. For example, the well-watered Nile delta is home to a very dense rural population, while some of Egypt's urban neighborhoods have more than 260,000 people per square mile (100,000 people per square kilometer), a density four times higher than that of New York City, the densest city in the United States.

SOCIOCULTURAL ISSUES

This section explores the broad trends in this region with regard to families, gender, and space. Like most institutions in this region, the family remains predominantly patriarchal. Related to this are divisions between the public spaces of society, which are usually occupied by men, and the private spaces of the home, where women spend most of their time. More legal limits are imposed on women here than in any other region, with various countries restricting women's political freedoms, how they dress, and their ability to occupy public spaces. As the vignette below suggests, such limits are most restrictive in Saudi Arabia, but exist to different degrees throughout the region. Change on these fronts is happening, as women push for more equitable treatment, which is evident in an improved gender development index (GDI) measurement for some countries. **Figure 6.31** shows the regional GDI in North Africa and Southwest Asia. Countries are ranked according to the degree to which women and men are equal with regard to longevity, education, and income.

Families and Gender

The family is the most important institution in this region. Although the role of the family is changing, a person's family is still such a large component of personal identity that the idea of individuality is almost a foreign concept. Each person is first and foremost part of a family, and the defining parameter of one's role in the family is gender identity (see Figure 6.31).

The head of the family is nearly always a man; even when a woman is widowed or divorced, she is under the tacit supervi-

Figure 6.29 Population pyramids for Iran, Israel, and Qatar. The population pyramid for Iran **(A)** is at a different scale (millions) from those for Israel **(B)** and Qatar **(C)** (thousands). [Data from: U.S. Census Bureau, "Population Pyramids for Iran, Israel, Qatar," in *International Data Base*, 2019.]

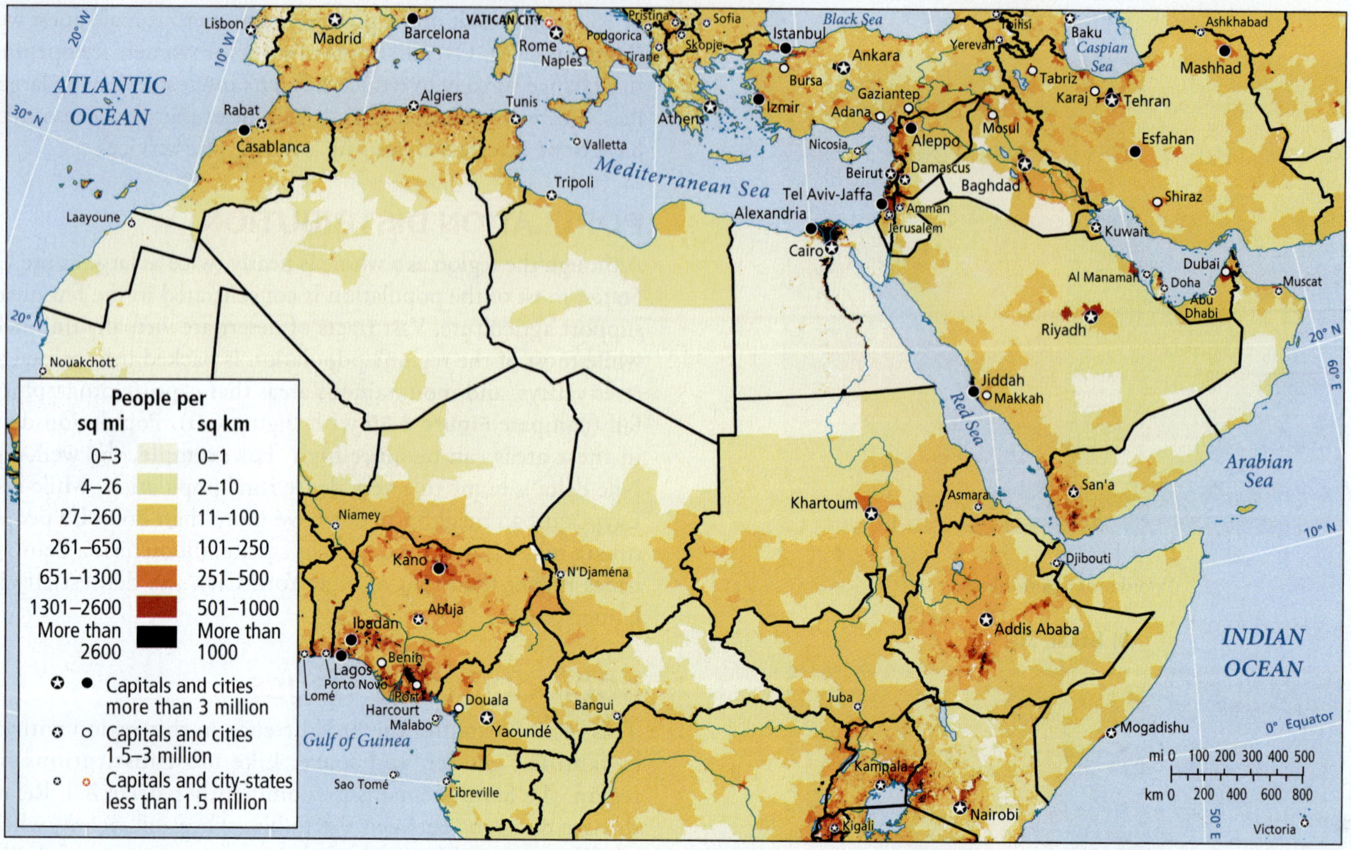

Figure 6.30 Population density in North Africa and Southwest Asia.

sion of a male, perhaps her father, her adult son, or another male relative. An educated unmarried woman with a career outside the home is likely to live in the home of her parents or a brother, and to defer to them out of respect in decision making. Such **patriarchal** ideas are beginning to change as the result of modernization.

Gender Roles and Gendered Spaces

Carefully specified gender roles are common in many cultures, and there is often a spatial component to these roles. In the region of North Africa and Southwest Asia, in both rural and urban settings, the ideal is for men and boys to go forth into *public spaces*—the town square, shops, the market (**Figure 6.32A**). Women are expected to inhabit primarily *private spaces*. But there are many possible exceptions to those ideals and they vary greatly from country to country.

To facilitate this ideal of public/private spaces for the sexes, traditional family compounds included a courtyard that was primarily a private, female space within the home (Figure 6.32C); the only men who could enter it were relatives. For others, a separate room to entertain guests was maintained. For the urban upper classes, female space was an upstairs set of rooms with latticework or shutters at the windows, which increased the interior ventilation and from which it was possible to look out at street life without being seen. These latticeworks also protrude from the façade so that the person looking out can observe in all directions (see Figure 6.32B). Today, the majority of people in the region live in urban homes designed much like elsewhere in the world, yet even in these places, there is a clear demarcation of public and private space. One or two formally furnished reception areas are reserved for nonfamily visitors, and rooms deeper into the dwelling are for family-only activities. When guests visit, women in the family are usually absent or present only briefly. However, customs vary not only from country to country but also from rural to urban settings, by social class, and by personal preference. Today, many women as well as men go out into public spaces, but the conditions under which women enter these spaces remain an issue and a woman who challenges convention does so at her own peril and that of her family's reputation.

patriarchal relating to a social organization in which the father is supreme in the clan or family

female seclusion the requirement that women stay out of public view

The requirement that women stay out of public view (also known as **female seclusion**) is most strictly enforced in the more conservative Muslim countries of the Gulf states. Women in these countries are generally expected to remain in private spaces except when engaged in important business; even then, they are to be accompanied by a male relative. In the more secular Islamic countries—Morocco, Tunisia, Libya, Egypt, Turkey, Lebanon, and Iraq—women regularly engage in activities that place them in public spaces. In these countries, a group of women may go out together for social events, such as an evening meal at a restaurant. Some women wear conservative religious clothing; others dress in Western styles, at least in the big cities. Increasingly, female

PHOTO ESSAY

6.31 Population, Gender, and Culture in North Africa and Southwest Asia

The map shows the regional Gender Development Index (GDI). Countries are ranked according to the degree to which women and men are equal with regard to longevity, education, and income. Of the 21 countries, only Kuwait ranks in the highest category. The small but wealthy country has made rapid progress in providing opportunities for women. Two other Gulf states, Qatar and UAE, rank in the high category together with Israel. On the other hand, 13 countries rank in the lowest category, which shows that much of the region ranks lower on gender development than do most other parts of the world.

THINKING GEOGRAPHICALLY

A How are women participating in public life?

B Relative to other regions, how high is this region's population growth rate?

C Where in the region do we find the highest population densities?

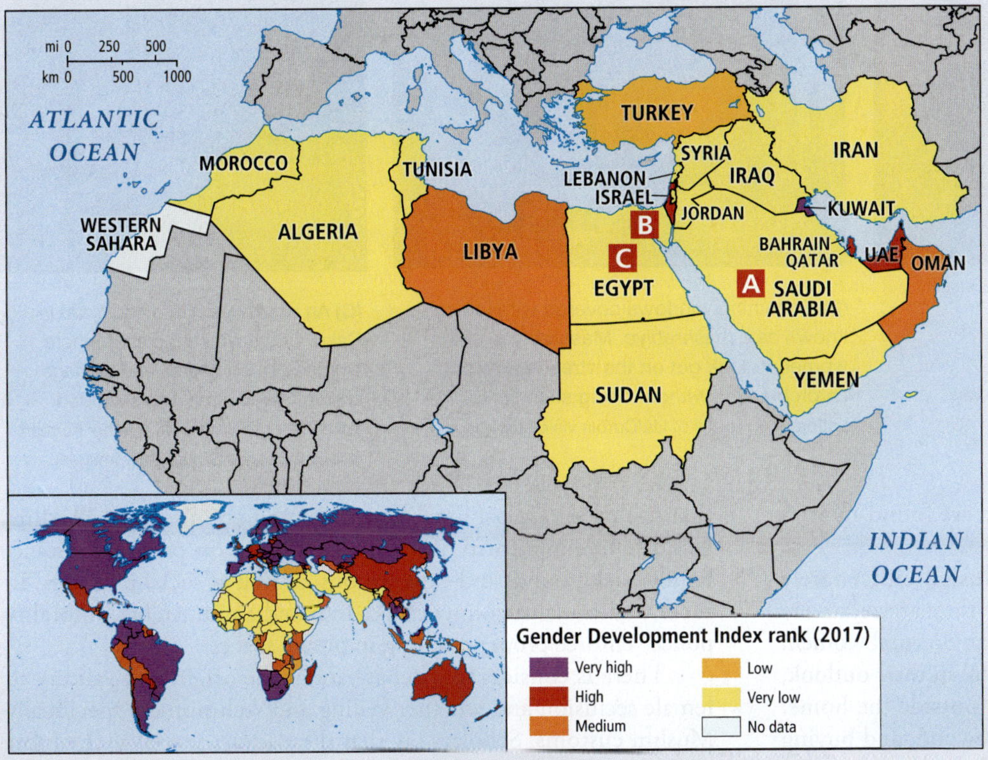

A Saudi woman driving a car. Lifting of the ban against women drivers is seen as a step toward modernization. [AMER HILABI/Getty Images]

B Orthodox Jewish Israeli family with six children. For a predominantly affluent and urban country, Israel has an unusually high fertility rate, among both religious and secular families. [MENAHEM KAHANA/AFP/Getty Images]

C High population densities along the Nile River in Egypt. With an urbanization rate of only 43 percent, many Egyptians are farmers. Cities also cluster along the Nile transportation corridor. [Radius Images/Getty Images]

Figure 6.32 Public and domestic spaces. Many of the older cities and buildings in this region reflect the division between public space and private or domestic space.

(A) The ancient city center of Marrakech, Morocco, is known as the medina. Narrow, irregular streets and keyhole-shaped arches that span the street are common architectural features of public spaces. [TDway/ Shutterstock.com]

(B) Projecting windows covered with lattice, known as a *mashrabiya*. Mashrabiyas allow women to look out on the street below and catch breezes without being seen. [Eric Lafforgue/Art in All of Us/Corbis via Getty Images]

(C) An interior view of a courtyard in Cairo. Courtyards are places where women can do chores and manage children while remaining secluded from the outside world. [Werner Forman/ Universal Images Group/Getty Images]

doctors, lawyers, teachers, and businesspeople are found in even the most conservative societies.

Affluent urban women may observe seclusion either rarely (especially if they are highly educated), or, if they are relatively affluent but less educated, even more strictly than do rural women. Although rural women are often more traditional in their outlook, they have many tasks that they must perform outside the home: agricultural work, carrying water, gathering firewood, and buying or selling food in markets. In North Africa, women of the rural Berber ethnicity often travel to sell their goods at markets and have more spatial freedom than most Arab women. Meanwhile, upper-class women who can afford servants to perform daily tasks in public spaces can more easily stay secluded and afford the amenities (TV, movies, cell phones) that relieve the boredom and isolation of seclusion.

Many women in this region use clothing as a way to create private space (**Figure 6.33**). This is done with the many varieties of the **veil**, which may be a garment that totally covers the woman's body and face or just a scarf that covers her hair. There are many different names for such clothing depending on its design and the country. A *hijab* is a headscarf, a *niqab* covers much of the face, while a *chador* or *burka* is a full-body garment. In some cultures, even prepubescent girls wear the veil; in others, they go unveiled until their transition to adulthood is observed. The veil allows a devout Muslim woman to preserve a measure of seclusion when she enters a public space, thus increasing the territory she may occupy with her honor preserved. A modern young woman may choose to wear a headscarf with makeup, jewelry, jeans, and a T-shirt to signal to the public that she is both a fashionable woman and an observant Muslim. The use of some form of veil is dependent on personal choice, family background, and expectations, as well as social pressures. In more traditionalist countries like Iran and Saudi Arabia, "morality police" enforce proper clothing in public spaces.

veil a piece of clothing that covers a woman's hair, face, or much of her body

There is considerable debate about the origin and validity of female seclusion and whether veiling and seclusion are specifically Muslim customs. Scholars say that these ideas, as well as the (now unusual) custom of having more than one wife, actually predate Islam by thousands of years and do not derive from the teachings of the Prophet Muhammad. In fact, Muhammad may have been reacting against such customs when he advocated equal treatment of males and females. Muhammad's first wife Khadija did not practice seclusion and worked as an independent businesswoman whose counsel Muhammad often sought.

The Rights of Women

This region has notably more restrictive customary and legal limits on women than any other. In Saudi Arabia, and to some extent other Gulf states, women cannot travel independently without male supervision. Yet even in these contexts, changes have recently occurred. Some women in the Gulf states have become more active in public life, education, and business. Qatar has implemented remarkable changes for women. They can now drive, attend a university, be elected to political office, and work side by side with men. An important impetus for change in the region is the increasing number of women who are becoming educated and employed outside the home. Some of the Gulf states, together with Israel, have the highest percentage of the region's women in the labor force (**Figure 6.34**). However, most women in this region still do not work outside the home; when they do, they are paid, on average, only about

Figure 6.33 Variations on the veil as portable seclusion. There is an almost infinite variety of interpretations of the veil.

(A) Women in Dubai now work as police officers, although they follow the tradition of wearing a headscarf. [Artur Widak/NurPhoto via Getty Images]

(B) Schoolgirls in Iran wear a uniform that covers most of their hair and a suit that covers most of their body. [John Borthwick/Getty Images]

(C) These Tunisian women are covered except for their eyes. [FETHI BELAID/AFP/Getty Images]

40 percent of what men earn for comparable work. Only South Asian countries have a similarly large wage gap based on gender.

In a number of countries, women outnumber men in universities; most notably, in Saudi Arabia women make up 70 percent of university students (but only a small fraction of the workforce). More broadly, unleashing the full economic potential of this increasingly well-educated segment of the population would have tremendous positive impacts on the region.

Saudi Arabia remains the most restrictive country, but even there reform is underway. It is now possible for a woman to register a business without first hiring a male manager. The right to drive a car has long been a demand from Saudi women. A few young Saudi women even staged mini-demonstrations by posting YouTube videos of themselves driving. Although they were eventually punished by the state for this behavior, they inspired further challenges to convention. In 2018, the laws were finally changed and women are

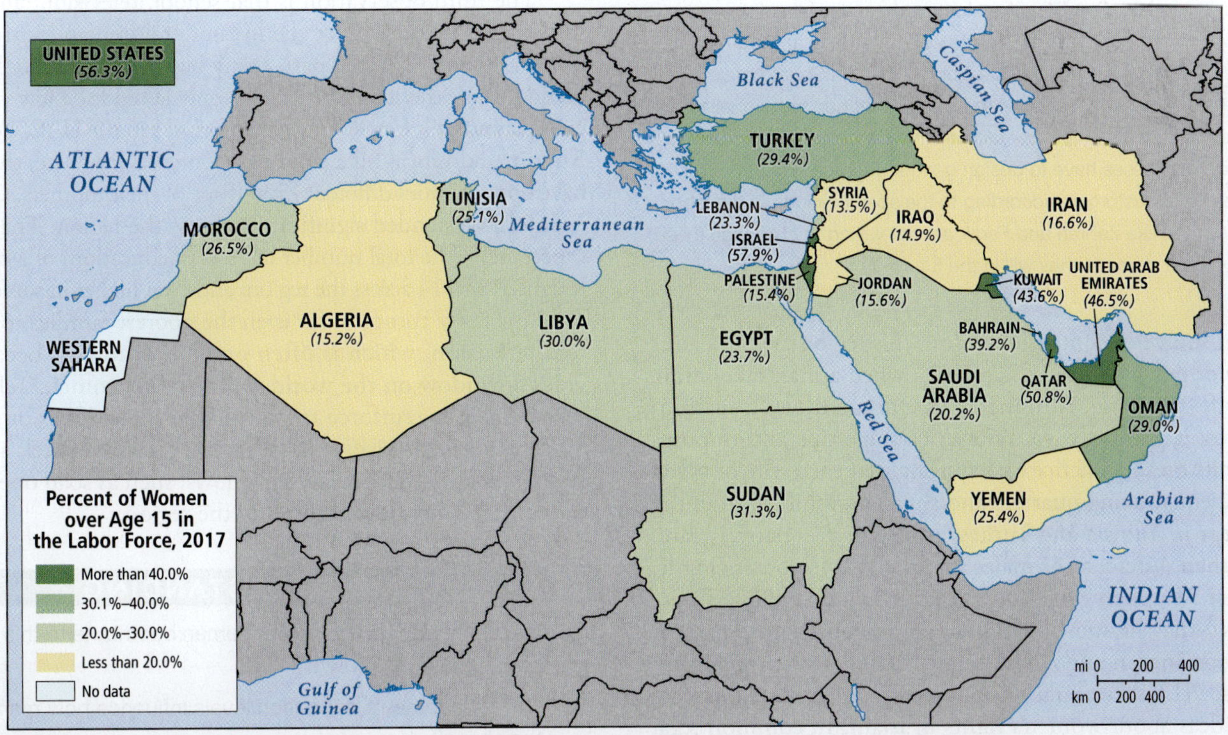

Figure 6.34 Percentage of the region's women who are in the labor force (2017). The female participation rate in the labor force has, for the most part, increased over time. However, the percentages in the map should also be compared to higher numbers for working males, which range from 64 percent (Jordan) to 95 percent (Qatar) in the region. [Data from: *2018 UN Human Development Report*, Table 5.]

indeed allowed to drive. This is more than a symbolic issue because the ability to drive provides women with greater spatial freedom as they no longer have to wait for a male family member to drive them around. Rules have also been relaxed on men and women attending public events such as concerts and soccer matches. The person behind many of the reforms is the new Crown Prince Mohammed bin Salman, who appears to be interested in modest social reform in addition to modernizing the kingdom's economy. Nevertheless, legal restrictions of Saudi women are still profound, which is explored in the following vignette.

VIGNETTE Because Saudi Arabia is an absolute monarchy, the legal system consists of a limited set of actual laws and a large number of royal decrees; some of the latter may be inconsistent and contradictory. The legal rights of citizens are therefore not always easy to decipher. Recently, a small number of women started to challenge discriminatory practices. Granted, Islamic shari'a law is the basis for limited women's rights in family law that governs marriage and divorce, as well as in courtroom proceedings that stipulate a woman's testimony is worth less than a man's. However, legal practices are also a matter of interpretation by a rigidly patriarchal society. Twenty-two-year-old law student Mohra Ferak, concerned about women's rights, organized a series of popular seminars for women to inform them about their legal rights and how to maneuver in the existing legal system. Such information is not always publicly available in a country where the desire for privacy in family matters usually trumps public debate. Another young and educated female, the attorney Bayan Mahmoud Zahran, spends time at a shelter for divorced and widowed women—two vulnerable groups in Saudi Arabia—offering them legal advice.

Many aspects of the Saudi legal system are based on gender segregation, which has some eerie parallels with the American South before the civil rights era and with apartheid-era South Africa. In Saudi Arabia, schooling for girls was originally met with protests that had to be quelled by military troops, restaurants only serve women in a separate secluded area, and all types of businesses have to design gender-segregated spaces for both employees and customers according to the specifications of the labor department. Women like Zahran and Ferak are not willing to challenge these conventions yet, but perhaps they will in the future. ■

Another source of contention within this region and abroad is the practice of polygyny (see Chapter 7): when a man takes more than one wife at a time. Although the Qur'an allows a man up to four wives, it generally does not encourage this and imposes financial limits on the practice by requiring that each wife be given separate and equal living quarters and support. While legal in the region (except in Tunisia and Turkey), polygyny is relatively rare, with fewer than 4 percent of males in North Africa practicing it. In Southwest Asia, polygyny—not covered in any reliable statistical surveys—may be somewhat more common. About 5 percent of marriages in Jordan and 8 to 12 percent in Kuwait are polygamous. The social justifications given for polygyny further illustrate traditional ideas about women's rights in Islam. In common legal practice in many countries, an unmarried woman under 40 must be regarded as a minor requiring protection. Therefore, if a man takes her as a second or third wife, she gains support and safety. Similarly, a widow is saved from disgrace and poverty if her dead husband's brother takes her as an additional wife.

Female genital mutilation (FGM; see Chapter 7) is a practice that intends to control female sexuality. However, it is not widely practiced in this region except in Sudan and Egypt, although it is illegal. As of 2016, according to UNICEF (United Nations Children's Fund), about 70 percent of girls and young women in Egypt have undergone the practice. The prevalence of FGM in Sudan and Egypt seems to be culturally connected to nearby sub-Saharan countries where the practice is also very common. On the positive side, the practice seems to be declining; during the 1980s, 97 percent of girls in Egypt underwent the procedure.

The Lives of Children

Three observations can be made about the lives of children in the Islamic cultures of North Africa and Southwest Asia. First, in most families children contribute to the welfare of the family starting at a very young age. In cities, they run errands, clean the family compound, and care for younger siblings. In rural areas, they also tend grazing animals, fetch water, and tend gardens. Second, their daily lives take place overwhelmingly within the family circle. Their companions are adult female relatives and siblings and cousins of both sexes. Even teenage boys in most parts of the region identify more with family than with age peers.

In rural areas, prepubescent girls can move around in public spaces as they go about their chores in the village. The U.S. geographer Cindi Katz found that until puberty, rural Sudanese Muslim girls have considerably more spatial freedom than do girls of similar ages in the United States, who are rarely allowed full access to their own neighborhoods. After puberty, rural girls may be restricted to the family compound and required to wear the veil.

The third observation is that school, television, and the internet increasingly influence the lives of children and introduce them to a wider world. In the past, many boys went to school for up to a decade, while the schooling of girls only lasted for a few years. Now, a girl's education is longer than that of a boy's in the UAE, Qatar, Oman, Kuwait, and Libya. Like educated women everywhere, these girls will have opportunities different from those of their mothers. Overall, education has expanded significantly across the region. The "school life expectancy" (the total number of years of education) of a child today is roughly 12 to 14 across the region, and even higher in some countries.

It is fairly common for even the poorest families to have access to a television, which is often on all day, in part because it provides a window on the world for secluded women. Television can serve either to reinforce traditional cultural values or as a vehicle for secular perspectives, depending on which channels are watched. In particular, Turkish television shows, such as soap operas and historic dramas, are popular around the region.

CHECK YOUR UNDERSTANDING

1. How does the low status of women contribute to this region's high population growth?
2. How do women's education levels influence how many children they have?
3. In what ways is the family the most important societal institution in the region?
4. What do we mean by saying that the families in this region are patriarchal in structure?

SUBREGIONS OF NORTH AFRICA AND SOUTHWEST ASIA

6.13 Identify key characteristics of the subregions of North Africa and Southwest Asia.

The subregions of North Africa and Southwest Asia present a mosaic of the issues that have been discussed so far in this chapter. Although all countries in the region except Israel share a strong tradition of Islam, they vary in how Islam is interpreted in national life and the extent to which Westernization is accepted. They also vary in prosperity and in the evenness of the distribution of wealth among the general population. All subregions have a dry landscape, but to different degrees pockets of fertile land exist where there is access to water.

THE MAGHREB

North Africa has often been *exoticized* in old American movies, meaning portrayed as different and distinctly foreign. People may conjure up intriguing images of Berber or Tuareg camel caravans transporting exotic goods across the Sahara from Tombouctou (Timbuktu) to Tripoli; smugglers and Cold War spies in the Moroccan city of Tangier; Barbary Coast pirates; the World War II battles waged in the desert of North Africa; or the classic lovers played by Humphrey Bogart and Ingrid Bergman in *Casablanca*. These images—some real, some fantasy, some merely exaggerated—are of the western part of North Africa, what Arabs call the Maghreb ("the place of the sunset," or more loosely translated, the westernmost part of the Arab world). The reality of the Maghreb, however, is much more complex than popular Western images of it.

The countries of the Maghreb stretch along the North African coast from Western Sahara through Libya (**Figure 6.35**). A low-lying coastal zone is backed by the Atlas Mountains, except in Western Sahara and Libya. Despite the region's overall aridity, these mountains trigger sufficient rainfall in the coastal zone to have a Mediterranean climate that supports export-oriented agriculture. Algeria, Tunisia, and Libya have interiors to the south that reach into the huge expanse of the Sahara.

The cultural landscapes of the Maghreb reflect the long historic relationships all the countries have had with the broader Mediterranean region. Tunisia, for example, is where Carthage emerged as a rival to the Roman Empire. Later Tunisia was influenced by Berber people and cultures (see the discussion below), was incorporated into the Ottoman Empire, and eventually became a French colony. Throughout the subregion, European domination lasted from the mid-nineteenth century well into the twentieth century, during which time the people of North Africa took on many European ways: consumerism; mechanized market agriculture; and manners of dress, language, and popular culture. Especially, French colonialism had a strong impact on the subregion (see Figure 6.17). The cities, beaches, and numerous historic sites of the

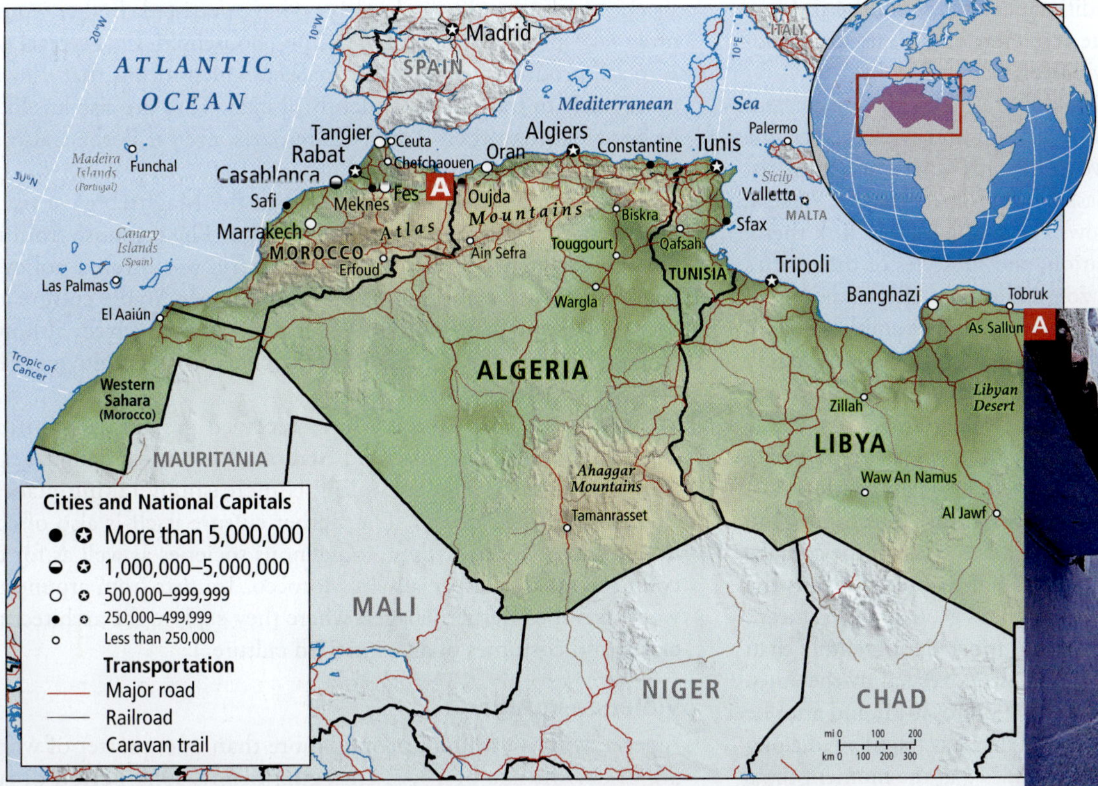

Figure 6.35 The Maghreb subregion (Northwest Africa). **A** A street in Chefchaouen, Morocco, a city built in the fifteenth century to defend northern Morocco against Portuguese invasion. The city is known for its blue-washed houses. It is a popular tourist destination for Europeans, given its proximity to Tangier and Ceuta, the port cities where the ferries from Spain arrive. [Sally Walton/Getty Images]

Maghreb, such as Carthage and the ancient Moroccan cities of Fes and Marrakech, continue to attract millions of European tourists every year who come to buy North African products and to enjoy a culture that, despite Westernization, seems exotic.

The agricultural lands in the coastal zones of the Maghreb are strategically located close to Europe, where there is a strong demand for Mediterranean food crops: olives, olive oil, citrus fruits, dates, grains, and fish. Large amounts of marijuana are also grown in the Rif Mountains of northern Morocco and smuggled into Europe. Oil and gas production is also very important in two of the countries in the subregion—Algeria and Libya. They supply a large amount of oil and gas to the European Union. Fossil fuel sales account for 30 percent of Algeria's GDP and 95 percent of its exports. With a small population and large reserves, oil and gas are even more important for Libya. However, the country's fossil fuel sales collapsed after the Arab Spring as its energy infrastructure—pipelines, refineries, and oil ports—have been seriously disrupted by continuous political instability. The current output is only about half of what it was before 2011.

Europe is also a source of jobs for people from the Maghreb. Local firms are supported by European investment and tourism, and millions of guest workers have migrated to Europe. There are millions of North African migrants particularly in Spain, France, and Italy, many of them undocumented. Libya has also emerged as an important transit country for refugees from countries like Eritrea and Somalia who try to make it to Europe. Because of the absence of a functioning government in Libya during the post–Arab Spring period, smugglers can operate with ease in Libya and transport refugees in small vessels across the Mediterranean to the southern shores of Italy or Spain. These voyages are extremely dangerous, and thousands of refugees have lost their lives (see Chapter 4).

In North Africa, settlement and economic activities are concentrated along the narrow coastal zone where water is more available (see Figure 6.30). Most people live in the cities that line the Atlantic and Mediterranean shores—Casablanca, Rabat, Tangier, Algiers, Tunis, and Tripoli—and in the towns and villages that link them. The architecture, spatial organization, and lifestyles of these cities mark them as cultural transition zones between African and Arab lands and Europe. The city centers retain an ancient ambience, with narrow walkways and huge heavy doors leading to secluded family compounds, but most people live in modern apartment complexes around the peripheries of these old cities (see Figure 6.35A). These developments coexist with a large number of people living in urban slums and hardscrabble rural villages. In Libya, the population cluster of the Maghreb tapers off toward the desert east. The capital and largest city Tripoli is located in the west, while the rest of the country is sparsely populated (with about 6.6 million people, Libya is the least populous of the Maghreb countries). This geography has meant that the country is organized more around tribal identity rather than a unified national identity. The country was governed by the autocratic leader Muammar al-Qaddafi from 1969 to 2011, and after he was overthrown during the Arab Spring, Libya has fragmented along tribal lines—at the moment there are competing governments in Tripoli and the eastern city of Benghazi. Libya is increasingly becoming a failed state.

Berber an indigenous group of people in North Africa that are most numerous in Morocco and Algeria

Many North Africans seek an identity that is less European and that retains their own distinctive heritage. Historians note that for those people who were Westernized during the era of European domination and who now make up the educated middle class, independence from Europe meant the right to establish their own secular countries, with constitutions influenced by European models. But by the 1990s, Islamist movements were challenging these independent secular states, linking them negatively with ongoing European and North American influences. The Arab Spring–related shifts toward greater political freedoms have, ironically, tended to strengthen these Islamist political movements, which are opposed to wider political freedoms for all. Now those North Africans who prefer secular governments are fearful of the outcome if Islamists win elections.

The Berber–Arab Cultural Mix

Part of the Maghreb is a mix between **Berber** and Arab cultures. Berbers are the indigenous people of North Africa and they are most numerous in Morocco and Algeria. Census statistics here don't record language and ethnic identity, but there may be around 20 million speakers of Berber languages and many more who are of Berber descent. Ethnic mixing and the lack of a distinct ethnic identity among people of Berber background who live in urban areas also complicate the picture, as does the fact that the Berbers themselves are divided into separate groups with different languages. Among rural Berbers, primarily living in the Atlas Mountains, the ethnic identification is stronger. There, language, traditional clothing, and small-scale agriculture—sometimes subsistence and nomadic in character—have persisted. Berber women often take an active role in the farming household. One typical task is to sell produce at the nearest market, which means that they are more likely to travel independently of men. They are also less likely to be veiled than their Arab counterparts, even if Berbers also are Muslims (**Figure 6.36**).

The independence of countries in the Maghreb gave rise to greater Berber awareness and nationalism. The response from the new governments of Morocco and Algeria was often a policy of Arabization—integration of Berber into larger Arabic culture and society. Despite these efforts, Berber society has survived, although emigration from the mountains in search of urban employment has created greater assimilation.

More recently, Berbers have received greater recognition. Their languages are considered "national" or official languages in Morocco and Algeria. Schooling is increasingly taking place in Berber languages as well. The Berber culture itself is also of economic value, both for these indigenous societies as well as for the countries at large, especially in Morocco. Tourists from around the world travel to Berber villages, where they explore the architecture, crafts, and costumes of an exoticized culture.

Violence in Algeria

Algeria, with 43 million people, more than one-quarter of whom are under the age of 15, is emerging from a long period of turmoil. Inhabited since antiquity by groups that eventually formed the Berber culture, Algeria has also long been home to outsiders who have often dominated the coastal areas. Many Phoenician, Carthaginian, and Roman coastal settlements flourished until the

Figure 6.36 Berber women in Chefchaouen, Morocco. Berbers in the Maghreb often live in rural areas. The task of women is often to travel to a nearby city to sell homemade farm goods and crafts to urban residents and tourists. [Charles O. Cecil/Alamy Stock Photo]

Arab–Muslim empires swept through the Maghreb in the seventh century C.E. By the 1500s, Algeria was controlled by the Ottoman Empire, during which time the coast became a major center of piracy. This "Barbary Coast" of Algeria was eventually subdued by British and later U.S. forces, and in the 1830s, France invaded Algeria and took control of the country for 130 years. French rule provoked deep resentment among Algerians that is still felt to this day. Tens of thousands of French colonists immigrated to Algeria, and many were given the most fertile land that had been forcibly taken from Algerians.

Agitation against French rule increased dramatically after World War II. The violent protests and repression by the French military that resulted paved the way for a war of independence. This brutal war, which lasted from 1954 to 1962, resulted in the deaths of between 400,000 and 1,500,000 people. It brought full independence from France, but also an enduring legacy of violence and authoritarian rule to Algeria.

After independence, Algerian politics were dominated by the military and secular socialists. Such was common around the Maghreb and other countries at this time, and the ideology of *Arab nationalism* took hold. That ideology stressed anticolonialism, unity among Arab states, and economic and cultural modernization based on socialism. The long-time leader of neighboring Libya, Muammar al-Qaddafi, was also a forceful proponent of Arab nationalism. The strong role of the military didn't leave much room for democracy, though. Islamists, who wanted an explicitly Islamic government above all else, became the main opposition in Algeria. When elections were allowed, the government, then controlled by the secular socialists, intervened to nullify elections that would have given the Islamists control of the government. Another war, this time between Islamist militias and the Algerian military, resulted in more than 100,000 deaths. A military-backed secularist president stayed in power from 1999 to 2019 because of his ability to bring stability to the overall political and economic system of Algeria. The protests of the Arab Spring, which (temporarily) brought Islamists to power in neighboring Tunisia and Egypt, reminded many Algerians of the bloody civil war of the 1990s. Because of this, Islamist militants have gained little momentum in current Algerian politics.

Trouble in Western Sahara

Western Sahara is a sparsely populated (600,000 inhabitants) desert country just south of Morocco. The country's importance in the subregion has to do with its status as a disputed territory. Much of Western Sahara is under the control of Morocco, but an independence movement, the Polisario Front, representing the local population, the mainly nomadic Saharawis, has been fighting Morocco's dominance since the 1970s. A long sand wall built by Morocco runs the length of Western Sahara (1,700 miles, or 2,700 kilometers) in an approximately north-to-south direction (**Figure 6.37**). The sand wall separates Moroccan-controlled areas to the west from areas to the east controlled by the independence movement, which is headquartered across the border in Algeria. The latter has caused political friction between Morocco and Algeria.

The roots of the conflict date back to the colonial era. Spain controlled the territory from the 1800s to the 1970s, when a transition toward independence was planned. Morocco, which borders Western Sahara to the north, then moved hundreds of thousands of Moroccans into Western Sahara as a way to lay claim to the area. One reason is that Western Sahara has phosphate deposits (phosphate is used to produce industrial chemicals and agricultural fertilizers) and there may be oil deposits offshore. United Nations–sponsored peace talks have for decades been unable to resolve the status of Western Sahara. Possible outcomes could be full independence for Western Sahara or some form of semiautonomous status within Morocco.

CHECK YOUR UNDERSTANDING

1. What is the historic and contemporary relationship between the Maghreb and Europe?
2. How does Berber culture differ from the majority Arab culture?
3. How has Algeria's experience of war between the secularist military-backed government and Islamists in the 1990s shaped its response to the protests of the Arab Spring?
4. Why is the territory of Western Sahara occupied by Morocco?

Figure 6.37 Sand wall in Western Sahara near the border of Mauritania. [PATRICK HERTZOG/AFP/Getty Images]

THE NILE: SUDAN AND EGYPT

The Nile River begins its trip north to the Mediterranean in the hills of Uganda and Ethiopia in central East Africa. The countries of Sudan and Egypt share the main part of the Nile system (**Figure 6.38**), which is their chief source of water. Although Sudan and Egypt have in common the Nile River, an arid climate, and Islam (over 90 percent of the population are Muslims in both countries), they differ culturally and physically. Egypt, despite troubling environmental problems and major civil disruptions associated with the Arab Spring, has a sizeable economy and plays an influential role in global affairs, whereas Sudan struggles with civil war, famine, and deep poverty.

Sudan

Sudan is a large country with a moderately sized population—42 million people, less than half the 97 million in Egypt. The country has two distinct environmental zones. The dry steppes and hills of the south are home to animal herders. The lower, even drier Saharan north stretches to the border with Egypt. Although Sudan mostly consists of dry grassland and desert, it does contain the main stem of the Nile and its two chief tributaries, the White Nile and the Blue Nile. This river system brings water from the upland south that is used to irrigate fields of cotton for export. Most Sudanese live in a narrow strip of rural villages along these rivers (see Figure 6.30; see also Figure 6.38). Most of Sudan's cities are clustered around the capital, Khartoum, where the White and Blue Niles join.

South Sudan gained independence from Sudan in 2011 after decades of bloody civil war (discussed further in Chapter 7). The historical roots of animosity between these two states can be found in the politics of oil, religion, and race. The two countries share a large oil field. Islam has been the main religion of Sudan for centuries, but the southern Sudanese people, for the most part, are either Christian or hold various indigenous beliefs, and most are

Figure 6.38 The Sudan and Egypt subregion.

African ethnicities. In western Sudan (Darfur), there are also Muslim, Arabic-speaking Africans. Despite the religion and language they share with the north, they are antagonistic toward the Muslim, Arabic-speaking, lighter-skinned Sudanese of the north, who for thousands of years raided the south for slaves. In 1983, the Islamist-dominated government in Khartoum decided to make Sudan a completely Islamic state and imposed shari'a on the millions of non-Muslim southerners. As a result, Sudan endured multiple civil wars.

The war that led to South Sudan's independence was fought partially over the imposition of shari'a, and also over control of South Sudan's modest oil reserves. From 1983 to the present, this north–south civil war claimed 2 million lives and forced 4 million people to abandon their homes. Since a 2005 peace agreement, many who had fled the southern Sudan conflict have moved back into the region. As a result, the already inadequate resources are being strained. Since then, the fragile and ethnically diverse new country of South Sudan has experienced civil war. Sudan and South Sudan have also struggled over an oil-rich border region.

A second civil war occurred in Darfur, located in western Sudan. This is an area that is part of a cultural and environmental border zone: an ethnically Arab population to the desert north and multiple African ethnicities in the steppe landscape to the south (see discussion of the Sahel in Chapter 7). With limited water resources, persistent droughts, and a population divided between pastoralists (see Chapter 7) and farmers, Darfur is an example of how conflict can be driven by environmental change and geopolitics. Possibly the result of climate change, Darfur had experienced droughts and water shortages for years. Arab herders from north Darfur were forced southward to access water for their farm animals. When doing so, they encroached on farmland used by the local population, mostly of African ethnicities. Some of the herders gave up agriculture altogether and resorted to banditry, forming the Janjaweed militia. To complicate matters, some Darfur groups want greater local autonomy from the central government in Khartoum and a larger share of Darfur's oil reserves. The response from the Khartoum government was to arm the Janjaweed militia. There is strong evidence that the Khartoum government gave the Janjaweed free rein to rape, rob, and kill its opponents.

Omar al-Bashir, when president of Sudan, became the first sitting head of state to be indicted by the International Criminal Court in the Hague for his role in the killing of thousands of Darfuri villagers by the Janjaweed, although as of this writing, he remains at large. A peacekeeping force under the banner of the African Union reduced some of the violence against civilians, but only after the conflict claimed the lives of more than 400,000 Darfurians and more than 2 million were displaced, many to the neighboring country of Chad. Most of these people remain refugees and dependent on UN humanitarian aid.

Egypt

The Nile flows through Egypt in a somewhat meandering track, from Egypt's southern border with Sudan north to its massive Mediterranean delta. Egypt is so dry that the Nile Valley and the Nile Delta are virtually the only habitable parts of the country (see Figures 6.31C and 6.38). Agriculture along the banks of the Nile has always fed the country and provided high-quality cotton for textiles.

With 97 million people, Egypt is the most populous of the Arab countries, and it has been the most politically influential. Its geographic location, bridging Africa and Asia, gives it strategic importance, and the country plays an influential role in such global issues as world trade and in the peace process between the occupied Palestinian Territories and Israel. Egypt has long been a recipient of U.S. aid (it was the fourth largest recipient in 2016, after Iraq, Afghanistan, and Israel), which limits its ability to take independent positions. At the same time, Egypt remains an ally against Islamists; U.S. money flows despite the Egyptian government's dubious human rights record.

Agricultural Development Although cities, especially the capital Cairo, are growing most rapidly, Egypt is only 43 percent urban (Figure 6.28A). In fact, the number of people still living and working on the land has also increased, and Egypt remains one of the most rural countries in the region. Those employed in agriculture constitute 26 percent of Egypt's workforce. The Nile fields are no longer sufficient to feed Egypt's people, and food must be imported. To afford this imported food for their families, rural men often migrate to neighboring countries, where they work to supplement the family income, leaving the women in charge of farming and village life.

Reorganization of Egypt's agricultural system along the lines of green revolution strategies was meant to boost Egypt's food security (its ability to feed its own people) and its ability to export water-intensive crops such as cotton and rice. Not all of these intended benefits materialized, and the result of the modernization of agriculture was that millions of farmers were forced off their land and into the cities.

The Nile Delta Egypt's two main cities, Cairo and Alexandria, are both in the Nile Delta region. With 16.5 million people in its urban region, Cairo, at the head of the delta, is one of the most densely populated cities on Earth. Alexandria, on the Mediterranean, has a population of 5 million people (see the Figure 6.28 map). Outside the cities, the Nile Delta is intensively cultivated farmland with a large rural population. Egypt's crowded delta region faces a crisis of clean water availability and the threat of waterborne disease and pollution. Many Egyptians get their drinking water from sewage-polluted streams, and farmland is often irrigated by the same sewage-infested water. In the absence of effective waste disposal, many Egyptians throw waste directly into the river. Animal excrement also ends up there, which results in the spread of diseases. In Chapter 4, we discuss the problem of pollution around the Mediterranean. The Nile is the single largest source of pollution into the Mediterranean Sea. To address the problem, Egypt has attempted to treat industrial, agricultural, and urban solid and liquid wastes. Nevertheless, hundreds of thousands of tons of sewage, phosphates from farms, and harmful heavy metals enter the Mediterranean every year.

Economic Development Egypt's economy has for many years been plagued by stagnation, inflation, and unemployment. It was dealt a double blow by the global recession that began in 2008 and by the political and economic turmoil that followed the Arab Spring. The economy actually contracted during the Arab Spring because of precipitous declines in tourism, manufacturing, and construction. Unemployment now stands at 12 percent, and inflation continues to be a problem.

One important industry that has been affected by the Arab Spring turmoil is tourism (**Figure 6.39**). The Egyptian government and the private sector have invested in a tourism infrastructure since the 1970s. The most obvious tourist destinations are the remnants of the ancient Egyptian civilizations. The famous and impressive Giza pyramid complex, including the famous sphinx, is located within comfortable reach from Cairo. These structures were built as tombs for ancient pharaohs. Along the Nile further to the south, the Valley of the Kings at Luxor is a secondary center of ancient tombs and temples. For religiously inspired tourists, Mount Sinai is where Moses, according to scripture, received the Ten Commandments. Nearby, all-inclusive beach resorts in the Sinai Peninsula have also attracted large numbers of visitors looking for a sun-and-sand vacation at a reasonable price. However, the number of tourists visiting Egypt has dropped from a peak of almost 15 million in 2010 to 5.4 million in 2017. Safety concerns are the major reason for this decline. There have been a number of terrorist attacks on tourists, and Sinai is a center of Islamist activity in Egypt. In 2015, a Russian airplane full of tourists was brought down by terrorists, which has many wondering whether Egyptian authorities can adequately protect visitors to the country.

A feature of Egypt that boosts its economic development and places the country in a central position of global trade is the Suez Canal. The canal connects the Mediterranean with the Red Sea, and since it was opened in 1869, the canal has shortened shipping routes between Europe and Asia enormously compared to the alternative route around Africa. Today, the canal handles 7 percent of global oceangoing trade. The economic benefits of the canal include shipping fees, employment in ports and logistics, and manufacturing based on material that flows through the canal. The canal has been improved and expanded many times; as recently as 2015, Egypt deepened the canal and built new parallel channels to enhance its capacity, which can generate new revenue for the country. So far the results have been disappointing; the traffic volume has remained flat during the last few years. The Suez Canal is largely dependent on the European economy and the oceangoing traffic it generates between Europe and Asia.

sheikhs traditionally, patriarchal leaders over ancestral tribal groups; today, the title is still used to denote someone in a leadership position

Figure 6.39 Tourists visiting ancient ruins in Egypt. The famous Sphinx is a large statue, located in the Giza pyramid complex near Cairo, of a mythical human-lion creature. It was built around 2500 B.C.E. [JOSEPH EID/AFP/Getty Images]

CHECK YOUR UNDERSTANDING

1. How has Sudan struggled with dissension in the Darfur region and southern Sudan?
2. How has the Nile River affected population distribution in Egypt?
3. What are the environmental problems associated with the Nile River?
4. How has the political turmoil of the Arab Spring affected Egypt's important tourism industry?

THE ARABIAN PENINSULA

The desert peninsula of Arabia has few natural attributes to encourage human settlement, and today large areas remain virtually uninhabited (see Figure 6.30). The land is persistently dry and has large areas that are barren of vegetation. Streams flow only after sporadic rainstorms that may not come again for years. The peninsula's one significant resource—oil—amounts to about 30 percent of the world's proven reserves. Saudi Arabia has by far the largest portion, with approximately 16 percent of the world's reserves.

Traditionally, control of the land was divided among several ancestral tribal groups led by patriarchal leaders called **sheikhs** (the title is still used today, broadly denoting someone in a leadership position). The sheikhs were based in desert oasis towns and in the uplands and mountains bordering the Red Sea. The tribespeople they ruled were either nomadic herders who followed a seasonal migration path over wide areas in search of pasture and water for their flocks, or poor farmers who settled where rainfall or groundwater supported crops. Additional income was earned by trading with camel caravans that crossed the desert (once the main mode of transportation) and by providing lodging and sustenance to those on religious pilgrimages to Makkah.

Wealth in Saudi Arabia and the Gulf States

Saudi Arabia is not only at the heart of this region culturally because it is the home of Makkah and other Islamic holy sites, but it is politically and economically central as well. Saudi Arabia's leaders, with their stranglehold on the country's vast oil wealth, have sought flashy symbols of modernity (**Figure 6.40**) but have managed to resist the political changes sweeping through much of this region. However, many other powerful changes are afoot.

In the early twentieth century, the sheikhs of the Saud family, in cooperation with conservative religious leaders of the Wahhabi sect, united the tribal groups to form an absolutist monarchy called Saudi Arabia, which occupies much of the peninsula. The Saud family consolidated its power just as reserves of oil and gas were becoming exportable resources. The resulting wealth has added greatly to the power

Figure 6.40 The Arabian Peninsula subregion. **A** Al-Faisaliah Tower in Riyadh, Saudi Arabia, is named after King Faisal (1903–1975) and is the second-tallest building in the kingdom. [micheldenijs/E+/Getty Images]

and prestige of the Saud family. Their tight control over Saudi Arabia has inhibited the development of opportunities for young people, despite the considerable material wealth from petroleum resources.

Discontent, especially among young Saudi adults, is widespread. Sometimes people express this discontent by embracing fundamentalist Islam—see the discussion of Wahhabism and the rise of Islamism, above. A majority of the suicide hijackers involved in the 9/11 attacks in the United States were Saudi, as was the founder of the Al Qaeda movement, Osama bin Laden, a member of a Saudi family that became rich in the construction business. Analysts familiar with Saudi Arabia suggest that Islamist violence has an appeal because there is no civil forum for airing discontent. A transition toward younger leadership is on the horizon, although the extent to which that will lead to political reforms is uncertain.

Despite the overall conservatism of Arabian society, oil money has changed landscapes, populations, material culture, and social relationships across the peninsula. Where there were once mud-brick towns and camel herds, there are now large, modern cities. The population of the peninsula has burgeoned to 85 million people, Saudi Arabia being the largest of those with 33 million. Irrigated agriculture has made the peninsula nearly self-sufficient in food, though only in the short term, since the irrigation is unsustainable.

The six smaller nations on the perimeter of the Arabian Peninsula, now ruled by ancestral clans that resisted the Saudi family expansion in the 1930s, have varying profiles. Kuwait and the United Arab Emirates, each with about 6 percent of the world's oil reserves, are rapidly modernizing and are extremely affluent as the oil incomes only have to be shared among a small population. Kuwait was a forerunner in the Gulf region in terms of providing its citizens with social services like education and health care; now, the UAE and others are following that same path. All Gulf states have created government-owned so-called *sovereign wealth funds* where profits from the oil and gas industry are placed. These funds are then invested in industries around the world. The primary goal is to generate further revenue, although investments are also made to advance the political interests of respective countries. The size of the sovereign wealth funds—some 3 trillion U.S. dollars—means that they are major players in the global economy.

As a way to exhibit its new wealth, Qatar will host the soccer World Cup in 2022, which will cost hundreds of billions of dollars for new stadiums and infrastructure—a staggering expenditure for a country with 2.7 million residents. Bahrain and Dubai (a semi-independent emirate of the UAE) generate income not so much from oil as from services related to oil production and transport, banking, and by providing entertainment and shopping to

Figure 6.41 Ski Dubai in the Mall of the Emirates. Malls in Dubai are noted for their opulence. In a climate-controlled section of this mall, the region's first indoor ski resort has been created. While the temperature outside may be above 110°F (43°C), skiers wear winter clothing to stay warm while diners enjoy a meal. [Gabriela Marj/Bloomberg via Getty Images]

neighboring wealthy Saudis and other tourists. In Dubai, where daytime summer temperatures average 110°F (43°C), a shopping mall features a frigid indoor ski slope where skiers can rent skis and winter clothing (**Figure 6.41**). Bahrain was the only one of the small Gulf states that experienced political turmoil during the Arab Spring. There, discontent has been brewing for a long time because the majority Shi'ite population is ruled by a Sunni monarchy. During the Arab Spring, military forces from other Sunni countries in the region intervened on behalf of the Sunni rulers. The situation remains tense.

The final Gulf state country is Oman, which has moderate oil reserves but enough to sustain a high level of well-being. As the gateway to the Persian Gulf, Oman occupied an important position in the historic trade around the Indian Ocean. Contemporary Oman is ruled by a sultan who has directed the country toward a path of modernization. Oman is unique in the region as most people adhere to Ibadism, which is an Islamic sect that is neither Sunni nor Shi'ite.

Changes in Attitudes Toward Work in the Gulf States

The actual day-to-day work necessary to keep the Arabian Peninsula societies running has been performed by contract laborers from South and Southeast Asia at very low wages. Scientific and engineering expertise has been supplied from Europe and the Americas at much higher rates of compensation. Workers and expertise have been imported because a shortage of skills exists in the Gulf states. The causes are multiple: Women tend to not work outside the home even though they may be educated; young men have been discouraged from doing manual labor because of the low status of such work; institutions of higher learning have not been available within the region until recently; and the motivation to undertake demanding jobs day after day has been squelched by the substantial stipends paid to most citizens from oil revenue.

Times are now changing and regional leaders have concluded that youthful idleness is not healthy. Furthermore, when surveys showed that expatriate laborers from India and Southeast Asia were increasingly dissatisfied with their jobs, it became clear that it was time to reorder society. Education, job training for local youth, and the reform of social attitudes toward work are now being emphasized. Expansion of higher education is taking place, although it is dependent on university lecturers from elsewhere in the world. The dignity and emotional rewards of work are stressed, and former rules against the two genders working together are being relaxed. The goals are to instill pride in personal self-sufficiency and in on-the-job accomplishments and to lessen dependence on workers from outside the region. Meaningful lives for the population may also prove an antidote against political radicalization.

Yemen: History, Poverty, and Conflict

Yemen occupies an important juncture in the world's shipping lanes, at the southern end of the Red Sea on the Gulf of Aden. This strategic location served Yemen well in the past as a crossroads of trade, from which it grew prosperous. For example, the early global coffee trade was controlled from Yemen. Its rich history is also represented in the capital of San'a and its distinct architecture of upright "tower houses" with intricate brick and stone patterns. Yemen's cultural connection with places across the Red Sea—such as Djibouti and Somalia—is exemplified by the common consumption of *khat*, which is a leafy plant that is chewed by people throughout much of the day. Khat is a stimulant that is considered an illegal drug in many countries outside Yemen and the Horn of Africa.

Yemen is about one-fourth as large as Saudi Arabia, its neighbor to the north, but the two countries have nearly the same-size population (29 million compared with Saudi Arabia's 33 million). The mountains of western Yemen capture a modest amount of orographic rainfall, and the area can therefore sustain agriculture and a relatively dense population compared to the rest of the Arabian Peninsula. However, Yemen is not well endowed with oil although it exports some oil. Production, however, has already passed its peak and current production levels are low. Yemen's standards of education and of living remain by far the lowest on the peninsula.

The Arab Spring brought political change to Yemen in 2012, when a new government was elected, ending a 33-year regime notorious for corruption and authoritarianism. However, the new government struggled to assert its control across the country. Disgruntled sheikhs withdrew support from the government, apparently because the patronage system failed them, Al Qaeda trained new recruits in Yemen's uncharted territory that borders the Saudi Arabian desert, and most recently the Shi'ite Houthi rebel group took over portions of the country, which prompted an ongoing military intervention from Saudi Arabia (see the discussion on the Sunni–Shi'ite rivalry).

CHECK YOUR UNDERSTANDING

1. How has oil money changed landscapes, populations, material culture, and social relationships across the Arabian Peninsula?
2. Why is today's Yemen a politically unstable state with internal conflicts?

THE EASTERN MEDITERRANEAN

Jordan, Lebanon, Syria, Israel, and the occupied Palestinian Territories have all been preoccupied over the last 60 years with political and armed conflict (**Figure 6.42**). In the 1970s, the Arab–Israeli conflict spilled over into Lebanon as the country was home to many Palestinian refugees. Lebanon is a diverse but fragmented country with Sunni Muslim, Shi'ite Muslim, and Christian populations about equal in size, with a smattering of other groups, such as the small Druze ethno-religious group that is loosely connected to Shi'ite Islam. Lebanon has an unusual governance structure where parliamentary seats and public office holders are apportioned according to the different sects of the country. Lebanon, and especially its capital Beirut, have historically been centers of commerce in the subregion, but the complex internal ethno-religious politics as well as influences from nearby countries—both Israel and Syria have at times occupied parts of Lebanon—have made it politically unstable. In the recent past, armed conflict between Hezbollah (an anti-Israeli Shi'ite militia) and Israel destroyed much of southern Lebanon. Syria has had conflicts with most of its neighbors for many years (the Golan Heights territory of southern Syria has been occupied by Israel for decades), and Jordan has taken in large numbers of Palestinian refugees over many decades.

Figure 6.42 The Eastern Mediterranean subregion.

The countries of this subregion are strategically located adjacent to the rich markets of Europe and the potentially lucrative markets of Central Asia and the Persian Gulf states. The strategic location dates back in history. The famous Silk Road that connected Asia with the Mediterranean had its terminus point near Aleppo, Syria, which is also reflected in that city's ancient quarter. Unfortunately, Aleppo has been a flashpoint in the Syrian civil war and much of its heritage was consequently destroyed. Under a peacetime scenario, the potential for tourism in this subregion would be strong. As part of the Fertile Crescent, the cities located here are among the most historic in the world; there are also impressive Roman ruins, and an abundance of important religious sites that would attract Christian, Jewish, and Muslim pilgrims from around the world. Places like the unique Dead Sea and the dramatic rock city of Petra in Jordan are also major attractions. At the moment, the subregion is a long way from peace, and in Syrian territories recently controlled by the Islamic State, the group has destroyed historic monuments that reflect the cultural and religious past.

The climate makes the subregion suitable for Mediterranean export agriculture. Jordan already exports vegetables, citrus fruits, bananas, and olive products. Until the recent civil unrest, private investors, especially from the Gulf states, helped Syria expand its industrial base to include pharmaceuticals, food processing, and textiles, in addition to gas and oil production. Natural gas discoveries in the waters off Israel, Lebanon, and Syria could make the Eastern Mediterranean subregion more energy independent and interconnected.

Israel is a model of development. Despite its rather meager natural resources, Israel has managed to develop economically and now has one of the most educated, prosperous, and healthy populations in the region (see the GNI and HDI information in Figures 6.21 and 6.22). Israeli engineering is globally competitive, and Israeli innovations in cultivating arid land have spread to the Americas, Africa, and Asia. A range of sophisticated industries is located along the coast between the large port cities of Tel Aviv and Haifa. Moreover, Israel benefits from qualities that other countries lack. As the homeland for the world's Jews, it has a pool of unusually devoted immigrants and financial resources contributed by the worldwide Jewish community and a number of foreign governments, particularly the United States. Regardless of its advantages, Israel's continual conflicts with its neighbors have turned it into an armed fortress and have constrained economic growth because it can't trade with surrounding countries in a normal way. Nevertheless, young and highly trained people are more likely to immigrate to Israel rather than to leave. Thus Israel has a positive net migration rate—more immigrants than emigrants.

Israel also has to battle internal divisions constantly; it is not a homogenous Jewish state. Arabs make up close to 20 percent of the population. Moreover, there is a divide in Israeli society between the majority secular-modern Jewish population, many of whom are moderately religious or nonreligious, and the ultra-orthodox community whose members live traditional lives according to their interpretation of religious texts. The latter group makes up only 10 percent of the Israeli population, but its birth rates are very high and thus it constitutes a growing demographic share. On religious grounds, the ultra-orthodox are also exempt from Israeli military service of up to 3 years, which is frowned upon by other members of society. Another division has to do with where one is born; a quarter of the Jewish population comes from somewhere other than Israel. Many are immigrants from the United States, Russia, Africa, or Asia, and they bring with them cultural traditions and attitudes from their places of origin.

CHECK YOUR UNDERSTANDING

1. What is the ethno-religious makeup of the Eastern Mediterranean?
2. How could the resolution of hostility in the subregion provide a stimulus for economic growth?

THE NORTHEAST

Turkey, Iran, and Iraq (**Figure 6.43**) are culturally and historically distinct from one another. For example, a different language is spoken in each country: Farsi in Iran, Turkish in Turkey, and Arabic in Iraq. Yet the three countries have certain similarities. At various times in the past, each was the seat of a great empire, and each was deeply influenced by Islam. Each occupies the attention of Europe and the United States because of its location, resources, or potential threats. All three countries share common concerns, such as how to allocate scarce water and how to treat the large Kurdish population that occupies a zone overlapping all three countries (see Chapter 1).

Turkey

Over the last century, Turkey has been more closely affiliated with Europe and North America than with any country in its home region. Turkey has been a member of NATO (see Chapter 4) ever since 1952, shortly after this mutual defense organization was created (1949) to address the perceived Cold War threat from the Soviet Union. Greece also joined at that time, and the two countries formed a strategic eastern Mediterranean military pivot point for NATO. After this long association with Europe, a faction in Turkey wants to join the European Union. Many Turks have spent years in Europe as guest workers, and Europe–Turkey business relations are significant.

VIGNETTE Veli-Çetin Avci and his wife Elif are an elegant young couple, both mechanical engineers and officers in Mekanik, a family-held firm that imports computer-driven tool-making machines from the European Union (**Figure 6.44**). They sell to Turkish firms that produce parts used in the global automobile industry. When asked about Turkey's chances of joining the EU (discussed in Chapter 4) and to what extent membership in the organization would help or hurt Turkey, their answer

Figure 6.43 The Northeast subregion and the Kurds. The area of Kurdish concentration overlaps five countries. The shaded areas with a yellow boundary show the concentration of Arab Sunni Muslims in Iraq.

Figure 6.44 Elif Avci and her husband Veli-Çetin Avci enjoy a rare evening out with friends. [Mac Goodwin]

had two parts. They responded immediately that joining the EU would be good for Turkey because its requirements for membership would force the adoption of higher standards in all aspects of Turkish life and would increase democratic participation. But, they hastened to add, their business would probably fail as a result because, given EU membership requirements for open markets, their middleman position between European and Turkish firms would be quickly taken over by the powerful European firms from which they now buy machines. Nevertheless, these business professionals believe that being part of the EU would give them many opportunities to modify their present business in order to take advantage of freer access to European markets and to global trade networks. ■

In recent years, enthusiasm for joining the EU has begun to wane as Islamic identity is on the rise in Turkey and the EU is increasingly wary of Turkish membership. The EU has criticized Turkey on human rights grounds, including the lack of freedom of religion and the press, crackdown on political dissent, and weak protections for minorities such as the Kurds. Less publicly stated are European worries about absorbing a predominantly Muslim state into an at least nominally Christian Europe. Turkish membership would also mean that the EU would directly border the conflict zone of Syria and Iraq, which is an undesirable prospect from the European perspective. And, finally, Turkey has an antagonistic relationship with the EU member state of Greece, which would likely try to block potential Turkish membership. Meanwhile, Turks worry that their culture will not be sufficiently respected in Europe, but more importantly, Turkey has assumed an increasingly influential role in its own nearby neighborhood as the implications of the Arab Spring become more discernible.

As discussed earlier, Turkey, once the core of the Ottoman Empire, was dismantled after World War I. After independence in 1923, Turkey undertook a path of radical Europeanization, led by a military officer, Mustafa Kemal Atatürk, who is revered as the father of modern Turkey. Atatürk and his followers declared Turkey a secular state; encouraged women to discard the veil and seclusion and men to do the same with the fez, the traditional Turkish head garment; and modernized the laws and state bureaucracy. Atatürk also promoted state-sponsored industrialization and actively sought to establish connections with Europe. Today, laws that prohibit people from wearing religious head coverings in government offices or in schools and universities have been relaxed. While industrialization and modernization have proceeded in western Turkey, much of eastern Turkey remains agricultural and relatively poor. Islam, long deemphasized as a matter of state policy, is nevertheless an overriding influence on daily life, and there is a resurgence of Islamic fundamentalism. A moderate Islamist party has been governing Turkey on and off for many years now. The president's wife is often seen wearing a head scarf in public, which indicates that Turkey is nudging in a conservative direction. The military has traditionally viewed itself as the guardian against Islamism in Turkish society. A military coup was attempted against the sitting government as recently as 2016, but it failed and has left Turkish politics in turmoil. Current president Erdoğan has used the coup as a pretext for concentrating power in the office of the president and purging civil servants who are not sufficiently loyal to him.

Turkey straddles the Bosporus Straits, a narrow passage from the Black Sea to the Mediterranean that often is described (with exaggeration) as the division between Europe and Asia. This location gives Turkey potential advantages as an intermediary if it can successfully increase economic links with Europe and Asia. Istanbul, a booming city of 14 million (see the Figure 6.30 map), is located at the Bosporus and is the regional headquarters for hundreds of international companies. Of Turkey's 81 million people, 75 percent are city dwellers. Many of them once lived in rural areas of Turkey, then traveled to Europe to work in factories and on farms; they have returned eager to pursue a standard of urban living similar to what they experienced in Europe. Remittances from Turkish guest workers in Europe have financed innumerable large homes that now dot the landscape of western Turkey. Today, as the Turkish economy has expanded and Turkish migrants have lived in Europe for decades and their connections with the homeland have diminished, the value of remittances has declined.

Turkey's progress is also intertwined with its management of relatively abundant water resources. With its mountainous topography, Turkey receives the most rainfall of any country in the region, and the headwaters of the economically and politically important Tigris and Euphrates rivers are in the mountains of (Kurdish) southeastern Turkey. Agriculture employs 18 percent of Turkey's workforce and the country exports Mediterranean fruits, nuts, and vegetables. As new irrigation and power generation systems on the Euphrates come on line, agricultural and industrial production should increase dramatically. Despite insufficient electrical power, Turkey's diversified economy manages to produce a wide range of goods, including apparel, food, textiles, transport equipment, and leather goods, all of which are exported to Europe and Southwest Asia.

A Future Kurdistan? In addition to the recent civil war in Syria, just to the south, a long-standing obstacle to Turkey's stability is its ongoing conflict with its Kurdish minority (see Chapter 1). The Kurds have lived in the mountainous borderlands of Iran, Iraq, Syria, Turkey, and Caucasia (see Figure 6.43) for at least 3000 years. The division of the Ottoman Empire after World War I by France and Britain divided the Kurds among lands held by Turkey, Iran, Iraq, and Syria, leaving the Kurds without a state of their own. All four countries have maintained hostile relations with their Kurdish minorities. The conflict springs from complex disputes over control of territory and resources, but it is exacerbated by cultural and language differences as well as the Kurds' resistance to state control. The total number of Kurds in the region is somewhere around 30 million people, with the highest concentration in Turkey. There is a sizable Kurdish diaspora in Europe as well, especially in Germany. The Kurds are predominantly Sunni Muslims and their language is related to Farsi, which is spoken in Iran.

The war in Iraq and now the civil conflict in Syria have complicated the Kurds' situation, in part because the loyalties of the transnational Kurdish community to the states in which they are located are now even more in question and also because Kurdish influence in Iraq and Syria has increased in recent years. Kurdish forces have been instrumental in subduing the Islamic State in both Syria and Iraq. After the U.S. invasion of Iraq in 2003, the Kurds developed great autonomy and self-governance in northern Iraq, which also allowed them to build their military capacities. This is the closest the Kurds have come to an actual state of their own—a Kurdistan. The situation in Syria is fluid, but Kurdish forces now control areas of northern Syria, too. This is a concern to the surrounding countries who do not want to cede territory to a potential future Kurdistan.

Kurdish relations with Turkey have been especially problematic in the past. To diminish the sense of identity among the Kurdish people, it was common practice in Turkey to refer to Kurds as "mountain Turks." Various Kurdish rebel factions fought Turkish forces for decades. The Kurdish struggle within Turkey seemed to be abating for a while; however, the civil war in Syria has reignited the Kurdish conflict in Turkey. On the one hand, the Turkish government has adopted an increasingly nationalistic tone, while on the other hand, Kurds in Turkey are supportive of their ethnic brethren on the Syrian side and their quest for territorial control.

Iran

Iran occupies a transitional geographic position in this subregion. It is predominantly a non-Arab country, which was historically referred to as Persia by the outside world. There are minority groups of Arab, Turkish, Kurdish, and Caucasian heritage, which occupy western Iran. Its eastern parts are occupied by people with language and ethnic roots in South Asia, especially Afghanistan and Pakistan. Because its northern flank borders the Caspian Sea, Iran shares in the debate over the future of this inland sea and its resources, especially water and oil, with Caucasia and Central Asia.

Like Turkey, Iran has abundant natural resources and a large, educated population—it has 82 million people and is growing at a modest rate—that could make it a regional economic power. Iran's large petroleum reserves give it influence in the global debate on energy use and pricing, but social and economic turmoil have held the country back for many years.

Iran had been on a path of secular and economic reforms patterned after those instituted by Kemal Atatürk in Turkey since the 1920s. In 1953, to keep power in the hands of Iran's monarch, the shah, the United States orchestrated a coup d'état against the winner of the national election. This helped the United States retain access to Iran's oil and allowed the shah to hold onto power for 26 more years, during which he continued his efforts at Westernization. But his emphasis on military might, political repression, and royal grandeur paid for with oil money overshadowed any genuine efforts at economic progress and social equality.

The 1979 Islamic Revolution was initially welcomed by many Iranians as an antidote to Western influence. However, those who had hoped for a more democratic Iran were disappointed by the new theocratic state. Resisters were imprisoned, and many were even executed. Among those most affected were women. Many upper-class women had lived emancipated lives under the shah's reforms, studying abroad and returning to serve in important government posts. After the revolution all women past puberty had to wear long black chadors, and they could no longer drive, travel in public alone, or work at most jobs. Despite such restrictions, some women supported the return to seclusion, seeing it as a way to counter the unwelcome effects of Western influences. The turmoil of the revolution was characterized by several episodes in which Westerners were taken hostage by the government and held for many months, and by general resentment against the West, which in turn led to decades of isolation and conflict. During the 1980s, Iran engaged in a devastating war with Iraq.

Since the 1979 Iranian Revolution, there have been improvements in basic human well-being, and some of the radical restrictions on women's role in public life have been liberalized, but only limited economic growth and hesitant moves toward greater political freedom have transpired (see the discussion of the Iranian political system). Iran has a GNI per capita commensurate with that of its neighbors; it ranks in the second best category on the UN's Human Development Index, but poorly on gender equality (see Figures 6.21, 6.22, and 6.31).

The Revolutionary Guard, whose main role is as a security force protecting the theocratic state from internal dissent, controls a large share of the Iranian economy, which is detrimental to the development of a well-functioning market economy. On the positive side, Iran's oil refinery capacity is being expanded so that the country can benefit further from its oil wealth. Despite its large oil resources, Iran had to import gasoline due to a lack of domestic refinery capacity, but a new refinery will bring Iran closer to self-sufficiency of gasoline and other crucial petroleum products. The agriculture sector remains relatively weak. Even though agriculture employs 16 percent of the population, Iran has to import almost twice the amount of food it exports. The dominant agricultural export is pistachio nuts, but water shortage and soil salinity are persistent problems for producers.

Moves toward political reform have been constrained by the conservative Islamic clerics who wield immense power through a wide range of governmental bodies. For example, Iran's Council of Guardians, made up of Islamic clerics, decides who can run for election and how elections are administered; it can veto any law passed by the *Majlis* (the Iranian parliament). During the last two decades, Iran has vacillated between electing reformers and Islamist hardliners as presidents. Sometimes elections have been highly contentious and street protesters who suspect rigged elections were brutally suppressed. However, this engagement in politics shows that Iran has a well-informed and sophisticated internet-savvy electorate, skilled at social networking and using Facebook and Twitter. Young people in Iran have lived under an Islamist regime all their lives and many of them don't particularly like it. Therefore, they are less swayed by the type of jihadism that has found fertile ground among discontented Muslim youth in many other countries. In fact, Iran is a relatively young society with large numbers of people under 30 years of age. Since the 1990s, the Iranian government has promoted family planning measures to limit population growth, and has been successful in doing so.

Iran also has developed a nuclear energy program that some suspect will be used to create nuclear weapons. Although many Iranians may not be supporters of the cleric-led regime, they view a domestic nuclear energy favorably as a matter of national pride and the right to self-determination. Under the current centrist president, Hassan Rouhani, Iran agreed to permit UN nuclear inspections inside Iran. In return, international sanctions were lifted so that Iran could participate in global trade. A recent rejection of the Iran agreement by the United States has made this prospect uncertain.

Iraq

Iraq, home to one of the earliest civilizations on Earth and to the Babylonian Empire of biblical times, was carved out of the Ottoman Empire after World War I. Most of Iraq's 40 million people live in the area of productive farmland in the country's eastern half, on the floodplains of the Tigris and Euphrates rivers. Although the recent chaos in Iraq makes it difficult to obtain accurate information about population numbers and distributions, Sunni Arabs (about 11 million people) generally reside in central Iraq around the capital of Baghdad and northward from there. About 7 million ethnic Kurds, who are also Sunni Muslims, live mostly in the northern mountains in the regions bordering Turkey, Syria, and Iran. The southern third of the country is occupied by at least 19 million Shi'ite Muslims concentrated around Al Basrah at the head of the Persian Gulf. Iraq's main oil fields are also in the south. Minority pockets of other groups live throughout the country. These groups lived in more ethnically integrated communities before the Iraq war in 2003.

Foreign activity in Iraq is not new. In the aftermath of World War I, Iraq was taken over by the British until 1932, when it became an independent monarchy. Following independence, the Iraqi monarchy, like many governments in the region, maintained strong alliances with Britain and the United States. A tiny elite monopolized the wealth that was generated by increasing oil production, but they did not invest in social or economic development. A powerful minority of Arab Sunni Muslims controlled the government and most of the wealth. The Shi'ites living in the south, where the oil is located, and the Kurds in the north benefited little from oil earnings.

In 1958, the military overthrew the monarchy and created a secular socialist republic—much like the Arab nationalist movement discussed in the Maghreb region. Large proven oil reserves (the fifth largest in the world) became the basis of a state-owned oil industry, and the profits from oil financed a growing industrial and agricultural base. The proceeds from oil and agriculture produced a decent standard of living for most Iraqis, who also benefited from government-sponsored education and healthcare systems, but corruption and political repression were ever present.

Like a number of other countries in the region, Iraq has been in a state of war and crisis for several decades. Two major events—the Iran–Iraq war (1980–1988) and the Gulf War (1990–1991), precipitated by Iraq's invasion of Kuwait—and years of severe economic sanctions that followed crippled the country's economy. In 2003, the United States launched a war against Iraq that removed then-dictator Saddam Hussein. Iraq held its first truly democratic elections in 2010. Allowing democratic elections has fundamentally changed power relations in Iraq because the Shi'ites are the largest sectarian group in Iraq. During Saddam's long reign, Arab Sunnis dominated the country. The new Shi'ite-led government has ruled in a divisive fashion, favoring Shi'ite interests.

CHECK YOUR UNDERSTANDING

1. In which ways is Turkey at a crossroads between Europe and Asia?
2. What forced the ethnic Kurds into living in many different countries? How have the Kurds been able to recently gain control of some territories?
3. What happened in Iran in 1979? How did those events shape contemporary Iran?
4. How did the fall of Saddam Hussein change the power relations between the country's majority Shi'ites and minority Sunni?

CRITICAL THINKING QUESTIONS

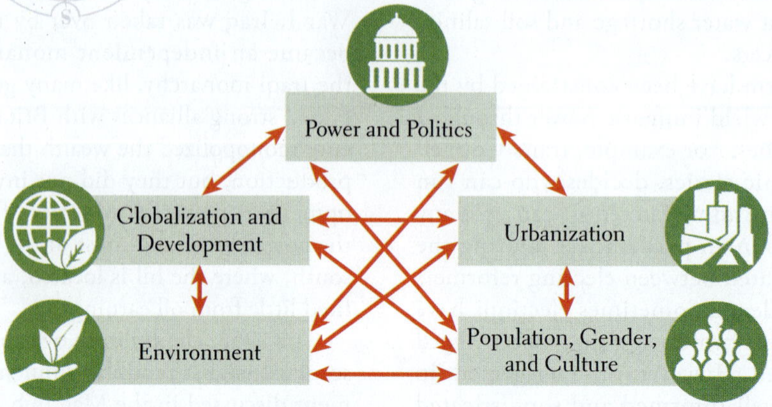

The diagram represents connections among the five geographic themes around which this book is structured. Listed below are some important questions that have been addressed in this chapter. Answer each question, and indicate which themes are involved in your response.

1. Landforms, climate, and rainfall are interrelated and play a dominant role in the development of North Africa and Southwest Asia. Discuss how societies have been shaped by and adjusted to these forces of physical geography.
2. In what ways has climate change affected agriculture in North Africa and Southwest Asia?
3. Observe the global distribution of Islam. What economic and political factors of the past can explain the pattern?
4. High levels of fossil fuel reserves can be thought of as both a blessing and a curse. What would we mean by that?
5. What does it mean if in some countries of this region agriculture may contribute only a small amount to the GDP, yet employ 40 percent or more of the people? What public policies would be appropriate in these circumstances—for example, should agriculture be deemphasized? How might this deemphasis affect food security?
6. In this naturally dry region, countries have different degrees of access to water depending on their economic development level. How are water access and economic development connected?
7. Discuss the possibility that scarcity of water is or will become a cause of violence in the region. What is the evidence against this happening?
8. Describe the circumstances that led to support in Europe and the United States for the formation of the state of Israel. Why has the two-state solution (which would include a Palestinian state) not come to fruition?
9. The Arab Spring protests are attributed to public discontent with the lack of democracy in North Africa and Southwest Asia; however, there are other factors that may have played a role. What other themes in the diagram can you point to that are related to the Arab Spring?
10. Compare and contrast the public debate over the proper role of religion in public life in your own country and in one country from this region (for example, Turkey, Morocco, Egypt, or Saudi Arabia). Contrast the roles of religious fundamentalism in the debates in your country and in the country you chose from this region.
11. How have globalization and immigration influenced the recent development of urban landscapes in the Gulf states?
12. The region has a fertility rate above the global average, although it is trending downward. What are some economic, cultural, and political forces responsible for such changes?
13. Gender is a complex subject in this region. Choose a rural, traditional location somewhere in the region and a modern, urban location elsewhere in the region, and make a list of the forces in each that would affect the future of a 20-year-old woman. Describe those hypothetical futures objectively; that is, without using any judgmental terminology.
14. How is the influence of religion in North Africa and Southwest Asia affected by historical factors, economic conditions, and the ethnic composition of different countries?
15. Explain why the Shi'ite–Sunni divide is not only religious-theological in character, but also plays an outsized role in the geopolitics of the region.
16. There are significant differences in terms of ethnic and religious conflict, political governance, and economic development across the subregions of North Africa and Southwest Asia. What are the underlying factors that can explain those differences?

Key Terms

Berber 350
cartel 325
Christianity 318
desalination 315
desertification 313
diaspora 318
diffusion 320
economic diversification 328
failed state 339
female seclusion 344
Fertile Crescent 316
fossil fuel 308
Gulf states 308
hajj 319
intifada 337
Islam 307
Islamism 307
jihad 330
Judaism 318
monotheism 317
Muslims 318
occupied Palestinian Territories 308
OPEC (Organization of the Petroleum Exporting Countries) 325
Ottoman Empire 320
patriarchal 344
public space 334
Qur'an (or Koran) 311
Salafism 330
salinization 312
secular states 331
shari'a 319
sheikhs 354
Shi'ite (or Shi'a) 320
Sunni 320
theocratic states 331
veil 346
Wahhabism 330
West Bank barrier 337
Zionism 335

More Practice at SaplingPlus

Read the interactive e-text, review key concepts, and check your understanding.

Supporters of the opposition party clash with police at the airport in Nairobi, Kenya, after the 2017 elections. [YASUYOSHI CHIBA/AFP/Getty Images]

7

Sub-Saharan Africa

Ory Okolloh grew up in a poor family in Kenya that struggled to pay school fees and medical bills. When she was 22 her father died from HIV/AIDS because he could not afford treatment. Through great personal effort she attended the University of Pittsburgh and later Harvard Law School. In 2006, she left a job at a major law firm in the United States and returned to Kenya to help create Mzalendo ("patriotism" in Swahili), a website that uses information from citizen journalists and bloggers to keep track of activity in Kenya's parliament, often highlighting corruption and encouraging better governance.

When widespread violence broke out during Kenya's 2007 presidential election, resulting in over 1000 deaths and forcing 600,000 from their homes, Okolloh suggested that there was a need for a Google map that could receive text messages from victims and witnesses of the violence. After two days of hard work by Kenyan programmers, the map was providing better information about the crisis than any news source or government agency. Now a widely used software application called Ushahidi ("testimony" in Swahili), it maps text messages and other communications to, for example, help rescuers locate people buried by earthquakes in Haiti, control disease outbreaks in West Africa, inventory wildlife in Kenya, aid Syrian refugees fleeing to Europe, and monitor hate crimes in the United States. Ushahidi was also used to collect data on voting irregularities and violence during the 2017 election, although it was notably more peaceful than the 2007 election.

Learning Objectives

Environment: Physical and Human

7.1 Describe how sub-Saharan Africa's unique arrangement of landforms developed.

7.2 Explain the advantages and disadvantages of this region's agriculture and irrigation systems in the face of climate change.

7.3 Interpret the impacts of European colonialism in this region.

Globalization and Development

7.4 Explain how sub-Saharan Africa's position in the global economy leaves it struggling with economic instability and low human development.

Power and Politics

7.5 Evaluate the factors that have come together to create weak governments in this region.

Urbanization

7.6 Evaluate the vast unplanned slums found in this region and the forces that have created them.

Population, Gender, and Culture

7.7 Explain the factors slowing population growth in this region.

7.8 Describe how educated career women are having an impact on this region.

7.9 Contrast the various religious traditions that have influenced this region.

Subregions of Sub-Saharan Africa

7.10 Identify key characteristics of the subregions of sub-Saharan Africa.

ALGERIA
LIBYA
Libyan Desert
EGYPT
Canary Islands (Spain)
El Aaiún
Western Sahara
Waw An Namus
Al Jawf
Tamanrasset
Bir Misahah
S a h a r a
MAURITANIA
Nouakchott
MALI
Tombouctou
NIGER
CHAD
SUDAN
SENEGAL
Dakar
Praia
Banjul
THE GAMBIA
GUINEA-BISSAU
Bissau
GUINEA
Bamako
Djenné
Niamey
Ouagadougou
BURKINA FASO
Lake Chad
N'Djamena
Al Fashir
S a h e l
Kano
Kaduna
Abuja
NIGERIA
Gaoua
BENIN
Conakry
SIERRA LEONE
Freetown
CÔTE D'IVOIRE
Bouaké
Yamoussoukro
GHANA
TOGO
Porto Novo
Oshogbo
Ibadan
Lagos
Lomé
Accra
Kumasi
LIBERIA
Monrovia
Abidjan
Grain Coast
Ivory Coast
Gold Coast
Bight of Benin
Port Harcourt
CAMEROON
Douala
Yaoundé
Malabo
CENTRAL AFRICAN REPUBLIC
Bangui
SOUTH SUDAN
Land Elevations
meters feet
4877 16,000
3353 11,000
2134 7000
914 3000
305 1000
152 500
0 0
1:26,500,000
Lambert Azimuthal Equal Area Projection
Gulf of Guinea
EQUATORIAL GUINEA
SÃO TOMÉ & PRÍNCIPE
São Tomé
Libreville
GABON
CONGO
Congo Basin
Kisangani
Stanley Falls
RWANDA
ATLANTIC OCEAN
DEMOCRATIC REPUBLIC OF THE CONGO
Bukavu
Bujumbura
Brazzaville
Kinshasa
Pointe-Noire
Cabinda (Angola)
Kananga
Mbuji-Mayi
Kigoma
Lake Tanganyika
Luanda
Ascension (U.K.)
Katanga Plateau
Lubumbashi
ANGOLA
ZAMBIA
Lusaka
Lake Kariba
Victoria Falls
St. Helena (U.K.)
Ondangwa
Tsumeb
ZIMBABWE
NAMIBIA
Skeleton Coast
Namib Desert
BOTSWANA
Kalahari Desert
Gobabis
Windhoek
Mariental
Gaborone
Pretoria
Johannesburg
Keetmanshoop
Luderitz
Karasburg
Orange
Bloemfontein
Maseru
LESOTHO
Durban
SOUTH AFRICA
Cape Town
Port Elizabeth
Cape of Good Hope

People across the region of sub-Saharan Africa (**Figure 7.1**) are designing and implementing solutions uniquely suited to Africa's needs. For example, Mzelenda and Ushahidi are part of a cell phone revolution that has been transforming sub-Saharan Africa for decades (**Figure 7.2**). Nine out of ten Kenyans have cell phones, mostly "flip phones," which are put to extensive use. More than 20 percent of the Kenyan gross domestic product (GDP) flows through mobile phone–based banking systems that enable rural small business owners to pay bills and employees without making time-consuming trips to distant banks. Farmers with crops to sell use their phones to find markets with the highest prices, bypassing middlemen who charge high transport and marketing fees. This chapter will show that people across the region are designing and implementing solutions uniquely suited to Africa's needs.

Sub-Saharan Africa is where modern humans originated, and it has some of Earth's most spectacular wildlife and a broad variety of landscapes and climates. Its 48 countries (**Figure 7.3**) occupy a space bigger than North America and Europe combined. While its population is now growing faster than that of any other region, it is not particularly densely populated. Neither is it so poverty-stricken as to be incapable of helping itself.

During the era of European colonialism (the 1500s to the 1950s), massive amounts of people and resources were taken from Africa and brought to Europe and the Americas. Even after African countries became politically independent in the 1950s, 1960s, and 1970s, wealth continued to flow out of Africa as it does to this day. There are hopeful signs that wealth is increasingly being created in Africa for Africans, by Africans, but much of the region remains impoverished and mired in armed conflict. While some foreign powers are taking advantage of this instability, important problems such as corruption are being addressed, encouraging more African migrants to return home with the skills needed to innovate for Africa.

Today Africa is one of the fastest-growing regions in the world economically, but many countries remain dependent on the export of raw materials, which leaves them vulnerable to rapid changes in global prices for these goods. China's growing trade with Africa has been a source of economic stability in recent years, but this has only increased the region's dependence on commodity exports, and China's demand for resources also has its cycles of expansion and contraction.

What Makes Sub-Saharan Africa a Region?

Sub-Saharan Africa is a region distinct from North Africa for physical, cultural, and historical reasons. The Sahara Desert has long presented a major physical obstacle, though in prehistory it had a wetter climate and trading caravans have traversed it for thousands of years. Even so, sub-Saharan Africa developed largely separate from North Africa due to the Sahara.

◀**Figure 7.1** Regional map of sub-Saharan Africa.

Figure 7.2 Ushahidi platform users worldwide. The map shows a selection of the locations in which Ushahidi has been deployed since 2008. What kind of use can you imagine for Ushahidi where you live?

Sub-Saharan Africa is defined by the countries shown in Figure 7.3. For cultural and historical reasons, the new country of South Sudan is included in this region. For similar reasons, the country of Sudan is not included because its location on the Nile River and the strong Arab influence in the capital of Khartoum have brought Sudan into closer association with North Africa and Southwest Asia. Sudan is instead discussed in Chapter 6.

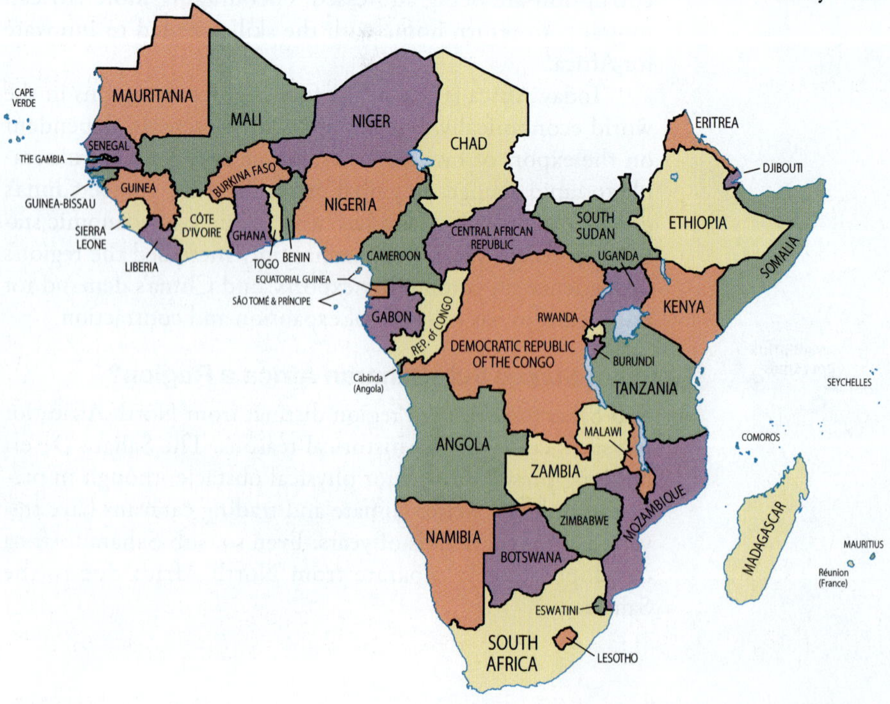

Figure 7.3 Political map of sub-Saharan Africa.

Terms in This Chapter

In this chapter, we refer only occasionally to the whole continent, and then we refer to it simply as Africa.

Two neighboring countries called Congo—the Democratic Republic of the Congo and the Republic of Congo—are abbreviated in this text. The Democratic Republic of the Congo (formerly Zaire) carries the name of its capital in parentheses: Congo (Kinshasa). The Republic of Congo is called Congo (Brazzaville).

ENVIRONMENT: PHYSICAL AND HUMAN

7.1 Describe how sub-Saharan Africa's unique arrangement of landforms developed.

7.2 Explain the advantages and disadvantages of this region's agriculture and irrigation systems in the face of climate change.

7.3 Interpret the impacts of European colonialism in this region.

The African continent is big—the second largest after Asia and about three times the size of the United States. But Africa's great size is not matched by its surface complexity. Sub-Saharan Africa has no major mountain ranges, but it does have several high volcanic peaks. Africa's climates are mostly tropical and subtropical, with only a tiny band of temperate climate in South Africa.

Today many African environments are straining under human pressure, and the region as a whole is more vulnerable than any other to climate change, largely due to its poverty. This poverty is in part a result of the disruptions and exploitation that came with European colonialism. And yet despite these challenges, Africa continues to make enormous contributions to the world, as it has throughout its history.

LANDFORMS AND RIVERS

The surface of the continent of Africa can be envisioned as a raised platform, or plateau, sloping downward to the north and bordered in the east, south, and west by fairly narrow coastal lowlands (see Figure 7.1). Geologists explain this by placing Africa at the center of the ancient supercontinent Pangaea (see Figure 1.8). As landmasses broke off from Africa and moved away—North America to the northwest, South America to the west, and India to the northeast—Africa readjusted its position only slightly. Because its shift was so small, it did not pile up long, linear mountain ranges, as did the other continents when their plates collided with one another (see Chapter 1). However, this process did leave steep escarpments (long, high cliffs) that define the transition between plateau and coast. These have long obstructed transportation and hindered connections to the outside world. Africa's lengthy, uniform coastlines provide few natural harbors, with Cape Town in South Africa as an exception (**Figure 7.4**).

Africa continues to break apart along its eastern flank, where the Arabian Plate has already split away and drifted to the northeast, leaving the Red Sea separating Africa and Asia. Another split, known as the Great Rift Valley, extends south from the Red Sea more than 2000 miles (3200 kilometers) (**Figure 7.5**). The Great Rift Valley dominates Africa's eastern interior in an irregular series of tectonic fractures that curve inland from the Red Sea and extend southwest through Ethiopia and into the Great Lakes region where Lake Victoria forms the headwaters of the White Nile. Africa's highest peaks are volcanoes created by thinner continental crust in the still splitting Great Rift Valley. These include Mount Kilimanjaro (19,324 feet [5890 meters] high) and Mount Kenya (17,057 feet [5199 meters] high). At their peaks, both have permanent snow and ice, though these features have shrunk dramatically due to global climate change and associated local environmental changes. The Great Rift Valley contains numerous archaeological sites that have shed light on the early evolution of humans and other large mammals.

Figure 7.5 Mount Kilimanjaro, Great Rift Valley, Tanzania. [oversnap/E+/Getty Images]

Of the many rivers in sub-Saharan Africa, three major ones originate in the Great Rift Valley from streams feeding the region's great freshwater lakes. The Nile is the longest river in the world (though some say the Amazon is longer), running from Burundi in the Great Rift Valley, north to Lake Victoria, and on to the Mediterranean, being joined along the way by the Blue Nile, which drains much of Ethiopia. The Zambezi River originates in the headwaters of Lake Malawi flowing south to the Indian Ocean in Mozambique, flowing over part of the Great Escarpment to create Victoria Falls (**Figure 7.6**), which is by some measures the world's largest waterfall. The Congo River has the second largest flow of any river in the world after the Amazon, due to its location in Africa's

Figure 7.4 Escarpment at Cape Town, South Africa. [Twist and Shout/Westend61/Getty Images]

Figure 7.6 Victoria Falls/Zambezi River, Zambia. [DEA/V. GIANNELLA/Getty Images]

Figure 7.7 Niger River Delta, Nigeria. [PIUS UTOMI EKPEI/AFP/Getty Images]

zone of heaviest precipitation. The deepest river in the world, its mouth is on the Atlantic and its headwaters feed Lake Tanganyika, which is the second largest freshwater lake in the world by volume (after Lake Baikal in Russia).

inland delta a place where a river reaches a flat inland area and splits into multiple channels

intertropical convergence zone (ITCZ) a band of warm winds that converge from both the north and the south at the equator, pushing air upward and causing copious rainfall

Sahel a band of arid grassland that runs east–west along the southern edge of the Sahara

The Niger River gives Africa its largest river delta (**Figure 7.7**), where it flows into the Gulf of Guinea, as well as one of the region's two inland deltas. An **inland delta** occurs when a river reaches a flat inland area and splits into multiple channels, losing much of its water to evaporation and nurturing lush habitat that supports major wildlife populations. Originating in the Guinea Highlands, the Niger flows northeast through Mali, where its inner delta is located, and then on to Niger and Nigeria. A larger inland delta is the Okavango, which evaporates completely into the Kalahari Desert in Botswana (**Figure 7.8**).

Figure 7.8 Okavango Delta, Botswana. [Kelly Cheng Travel Photography/Getty Images]

Many rivers in sub-Saharan Africa have large dams, primarily to aid irrigation and to produce power. There are an estimated 1270 such dams in the region and, like dams everywhere on Earth, they are being reevaluated due to the ecological changes they cause and the lack of equity in the distribution of irrigation water and electricity.

CLIMATE AND VEGETATION

The absence of high mountain ranges gives this region a banded pattern of climate zones shaped mostly by global wind patterns, though it is modified by elevation (see **Figure 7.9** map). Most areas are tropical or subtropical with temperatures generally above 64°F (18°C) year-round. More temperate zones are found at the southern tip of the continent and in the cooler upland zones (hills, mountains, high plateaus). Seasonal climates in Africa differ more by the amount of rainfall than by temperature.

Most rainfall comes to Africa by way of the **intertropical convergence zone (ITCZ)**, a band of warm winds that converge from both the north and the south at the equator, pushing air upward and causing copious rainfall (see photo D and inset map E in Figure 7.9). The rainfall produced by the ITCZ is most abundant in the Congo Basin, where dense tropical rain forests flourish (see Figure 7.9A).

The ITCZ shifts north and south seasonally, generally following the area of Earth's surface that has the highest average temperature at any given time. Thus, during the height of summer in the Southern Hemisphere in January, the ITCZ brings rain south to water the dry grasslands, or steppes, of Botswana. During the height of summer in the Northern Hemisphere in August, the ITCZ brings rain north to the **Sahel**, a grassy transition zone 200 to 400 miles (320 to 640 kilometers) wide that runs east–west along the southern edge of the Sahara (see the Figure 7.9 map).

Africa also shows the influence of global bands of dry air at about 30 degrees north and south of the equator. Here air that has converged and risen at the ITCZ and then traveled away from the equator falls back to the surface of earth, having dried out along the way. This forms a subtropical high-pressure zone that shuts out moister air, creating the Sahara at 30° N latitude and the Namib and Kalahari deserts at 30° S latitude. Other deserts near this band of dry air, which varies due to the effects of "continental breathing" (see the Chapter 1 "Temperature and Air Pressure" section), include the deserts of Australia, South America, Mexico, the Arabian Peninsula, and Iran.

The tropical wet climates that support equatorial rain forests are bordered on the north, east, and south by seasonally wet/dry subtropical woodlands (see Figure 7.9B). These give way to moist tropical savannas or steppes, where tall grasses and trees intermingle in a semiarid environment. These tropical wet, wet/dry, and steppe climates have supported agriculture and herding for thousands of years, but the amount of rain varies in cycles that are decades long.

7.9 Climates of Sub-Saharan Africa

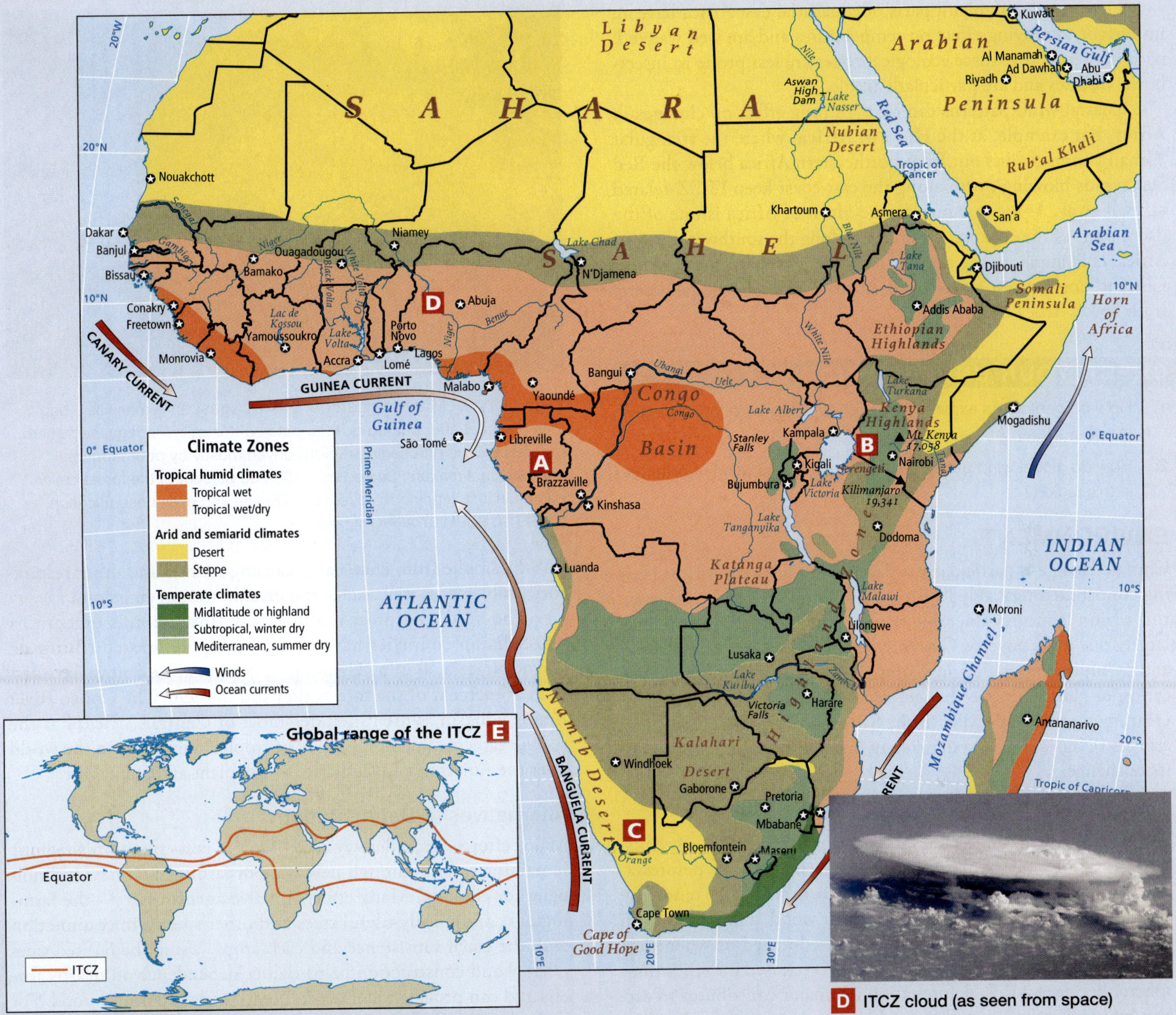

D ITCZ cloud (as seen from space) [NASA-JSC]

A Tropical wet, Gabon [Jacques Jangoux/Alamy]

B Subtropical, winter dry, Kenya [Nigel Hicks/Dorling Kindersley/Getty Images]

C Desert, Namibia [David Yarrow Photography/Getty Images]

The volcanic highland parts of equatorial east Africa have an important modifying effect on temperatures and are well watered, with the result that these ecological zones are less prone to insect-borne diseases and are particularly fertile.

Coastal wind patterns can have a strong effect on climates in Africa. For example, at the **Horn of Africa**, where the triangular Somali peninsula juts out from northeastern Africa below the Red Sea, winds blowing north along the east coast keep ITCZ-related rainfall away. As a consequence, the Horn of Africa is one of the driest parts of the continent. Similarly, along the southwest coast of Africa, cold air above the northward-flowing water of the Benguela Current blocks moist air from the Atlantic, creating coastal deserts from South Africa to Angola.

CHECK YOUR UNDERSTANDING

1. How do geologists explain the absence of high mountain ranges in sub-Saharan Africa?
2. How do global wind patterns influence the distribution of climates in this region?

Figure 7.10 Making charcoal in Madagascar. A man fills a bag with charcoal after cutting down a patch of forest in Madagascar, where deforestation is proceeding rapidly, driven mostly by poor farmers and charcoal makers. Since 1950, 40 to 50 percent of the forest cover has been lost, leading to severe soil erosion and even desertification. [ROBERTO SCHMIDT/AFP/Getty Images]

ENVIRONMENT

Climate change is particularly unfair for sub-Saharan Africa. Having contributed relatively little to the buildup of greenhouse gases, this region is more vulnerable than any other to climate change, largely due to its poverty. Currently, deforestation by rural Africans and by multinational logging companies is adding to the intensification of global climate change, while increasing water scarcity is creating conflict, especially in the Sahel. However, new and older aspects of agricultural systems may help this region adapt to climate change.

Deforestation and Climate Change

Sub-Saharan Africa contributes to an increase in global CO_2 emissions and potential climate change primarily through deforestation. Trees absorb CO_2 as they photosynthesize, thus removing carbon from the air and storing it in their wood, a process known as **carbon sequestration**. When trees are burned or when they decompose after they die, the stored CO_2 is released into the atmosphere. Because of this, deforestation is a major contributor to carbon buildup in the atmosphere and ultimately to climate change.

Most of Africa's deforestation is driven by the growing demand for farmland and fuelwood, although logging by international timber companies is also increasing. Africans use wood (or charcoal made from wood) to supply nearly all their domestic energy (**Figure 7.10**). Wood remains the cheapest fuel available, in part because of African traditions that consider forests to be a free resource held in common. Even in Nigeria, a major oil producer, most people use fuelwood because they cannot afford petroleum products. Not only are charcoal prices rising across the region as forests disappear, but the smoke from charcoal is causing asthma and other respiratory problems, as well as creating greenhouse gas emissions.

Horn of Africa the triangular Somali peninsula that juts out from northeastern Africa below the Red Sea and wraps around the Arabian Peninsula

carbon sequestration the removal and storage of carbon taken from the atmosphere

agroforestry growing economically useful crops of trees on farms, in conjunction with the usual plants and crops

The need to protect the forests of this region is urgent. Six sub-Saharan countries have been severely deforested (Burundi, Togo, Nigeria, Benin, Uganda, and Ghana) and now have less than 16 percent of the forests they did in 1990. The Congo River drainage basin, which includes most of Central Africa, has the second largest remaining stands of tropical rainforest in the world after the Amazon, and similarly diverse plant and animal life.

Alternatives to Deforestation

Major efforts are underway to relieve the pressure on this region's remaining forests through new ways of farming and new technologies for cooking. Many efforts involve **agroforestry**—the farming of economically useful trees and other species in conjunction with the usual subsistence and cash crops. Using the farmed trees for fuel and construction helps reduce dependence on nearby forests and can provide income through the sale of farmed wood. For example, a family in Mali can potentially double their income from raising traditional crops and animal herding by using agroforestry to produce fuelwood, fencing and building materials, medicinal products, and food. However, critics caution that agroforestry sometimes introduces invasive species, nonnative plants that threaten many African ecosystems by outcompeting indigenous species.

Agroforestry has long been promoted by the Green Belt Movement, founded by Dr. Wangari Maathai of Kenya, which helps rural women plant trees to use as fuelwood and to reduce soil erosion (**Figure 7.11**). The movement grew from Maathai's belief that a healthy environment is essential for democracy to flourish. Thirty years and 30 million trees after she commenced her work, Maathai was awarded the 2004 Nobel Peace Prize for her contributions to sustainable development, democracy, and peace.

Bamboo is increasingly being used in agroforestry projects, due to its fast growth rate, its ability to thrive on erosion-prone

Figure 7.11 Scientists at the World Agroforestry Centre in Nairobi, Kenya. The World Agroforestry Centre has worked closely with the Green Belt Movement and Dr. Wangari Maathai. [Wendy Stone/Corbis via Getty Images]

deforested lands, and its potential as a source of charcoal that burns longer and produces less smoke than wood. China has financed projects that have had successes in substituting bamboo for forest wood in Ethiopia and Ghana, and with processing bamboo into high-value flooring, plywood, and fabric. Most of this region's bamboo plantations use species native to Africa that can provide habitat for wildlife, such as gorillas, elephants, and lemurs. There is also significant global demand for bamboo, mostly in Asia where 8–10 million people make a living from bamboo in China alone.

The use of improved cook stoves that burn either less wood or alternative clean fuels can reduce pressure on forests and reduce indoor air pollution, which is a major health problem throughout this region and is caused by the inefficient charcoal cook stoves in wide use today. So far, more efficient charcoal stoves have had more success than stoves that use cleaner fuels, mostly because they are cheaper to operate and the fuel is more available. Cleaner fuels, such as ethanol produced from existing crops, have proven difficult to produce and hard to distribute widely.

Agriculture and Climate Change

Agricultural systems in Africa are undergoing a rapid transition from subsistence farming to commercial farming. Unfortunately, this change is happening at a time when the advantages of traditional agriculture in adapting to climate change are just beginning to be appreciated. Meanwhile, some aspects of commercial production increase vulnerability to climate change while others decrease it.

Traditional Agriculture Most sub-Saharan Africans practice **subsistence agriculture**—farming that provides food for only the farmer's family—usually on small farms of about 2 to 10 acres (1 to 4 hectares). Most subsistence farmers are female, though males may assist in the initial preparation of a plot. These farmers carefully attend to the needs of individual plants and understand the intricate ecological relationships between plants in a garden. They raise a diverse array of crops and a few animals as livestock. Family members may also fish, hunt, herd, and gather some of their food from forest or grassland areas. The wide variety of food sources stabilizes access to nutrition, or **food security**, and also means that the cuisine of sub-Saharan Africa is highly varied (**Figure 7.12**).

To maintain soil quality in a tropical wet or wet/dry climate, subsistence farmers have for generations used **shifting cultivation**, in which small patches of forest are cleared, with the detritus either burned or left to decay (**Figure 7.13**). The clearings are cultivated with a wide variety of plants for 2 or 3 years and then abandoned for a decade or longer and left to reforest; meanwhile, new cleared patches are planted. After a few decades, the soil in the abandoned plots naturally replenishes its organic matter and nutrients and is ready to be cleared and cultivated again. However, when the population increases greatly, demand for farmland means that **fallow periods** are shortened and the soil becomes degraded.

These traditional techniques for growing food—subsistence, mixed, and shifting cultivation—have advantages and disadvantages in the present era of climate change. They can provide a diverse array of strategies for coping with the changes in temperature and rainfall that climate change may bring. For example, traditional farmers in Nigeria (usually women) grow complex tropical gardens, often with 50 or more species of plants. Some of the plants can handle drought, while others can withstand intense rain or heat, meaning that some will thrive, no matter what the weather brings.

However, the subsistence nature of traditional African agriculture can also leave families without much cash. In years when harvests are too low to provide surplus that can be sold, and hunting and gathering fail to provide supplementary food for the family, there may not be enough money to buy food. While such situations can lead to famine, it is important to note that the most serious famines in Africa have occurred not because of low harvests but rather because of political instability that disrupts economies and the systems that distribute food.

Commercial Agriculture Much of Africa is now shifting over to **commercial agriculture**, in which crops are grown for cash rather than solely as food for the farm family. This type of production also has advantages and disadvantages with respect to climate change. If harvests are good and prices for crops are adequate, farmers can earn enough cash to get them through a year or two of poor harvests. Having some cash income can also allow poor families to invest in other means of earning a living, such as opening a grinding mill to help process other farmers' harvests or building a stone oven to bake neighbors' bread for a fee. In addition, they will have money for school fees so that their children can prepare for careers other than farming.

subsistence agriculture farming that provides food for only the farmer's family and is usually done on small farms

food security the condition of stable access to sufficient, safe, and nutritious food

shifting cultivation a system of agriculture in which small plots of forest are cleared, the dried brush is burned to release nutrients, and the clearings are planted; each plot is used for 2 or 3 years and then abandoned for many years of fallowing

fallow period the time (ideally, a decade or more) during which an agricultural plot is allowed to rest and regrow a natural vegetation cover

commercial agriculture farming in which crops are grown deliberately for cash rather than solely as food for the farm family

Figure 7.12 LOCAL LIVES: Foodways in Sub-Saharan Africa

(A) Ethiopian and Eritrean cuisine features spicy vegetable and meat stews and salads served with injera, a type of savory pancake made from fermented grains. Pieces of the injera are torn off and used to scoop up the rest of the meal. Injera is traditionally cooked on a clay plate over an open fire. Because cooking injera on an open fire requires a lot of wood, people also use more efficient stoves, including electric models that are popular in cities. [Photostock Israel/Oxford Scientific/Getty Images]

(B) Recently harvested cassava root is transported in Nigeria. Originally domesticated in Brazil, cassava was brought to Africa in the seventeenth century as cheap transportable food in the slave trade. Because it grows abundantly in a wide variety of soils and environments and is easy to cultivate, it has become a popular crop, despite its relatively low nutritional value. Sub-Saharan Africa produces more cassava than any world region, with Nigeria as the world's largest producer. [Christian SAPPA/Gamma-Rapho/Getty Images]

(C) A potjiekos, or "small pot stew," is prepared in South Africa. Cast iron pots for cooking were brought to South Africa by migrants from the Netherlands, who developed the potjiekos during their expansion north from Cape Town into the interior. These "trekkers" made stews of wild game, vegetables, and spices each evening, keeping the leftovers in the pot for the next day's journey. The trekkers also added fresh ingredients as they traveled. Making a potjiekos is a social activity, with fireside conversation throughout the 3 to 6 hours that it takes to cook the stew. [Graeme Williams/Getty Images]

At whatever scale commercial agriculture is practiced—whether small scale and locally managed, or large scale and owned and operated by international corporations (agribusiness)—problems often develop that are out of the control of the farmer. Some of the most common commercial crops, such as corn, peanuts, cacao beans (used for making chocolate), rice, tea, and coffee, are often less adapted to environments outside their native range (see Figure 7.14A in the next section). If temperatures increase or water becomes scarce or overabundant, these crops are likely to produce poorly. Moreover, to maximize profits, these crops are often **mono-cropped**, that is, grown in large fields containing only a single plant species. This can leave crops vulnerable to pests and diseases. And since only one cash crop is grown, should that crop fail, the farmer may have no other garden foods to rely on. Because the prices for most commercial crops are linked to global markets, they can be highly unstable. Overproduction in distant places can cause local prices to fall dramatically, while crop failures elsewhere can do the opposite.

Commercial agriculture is often much less finely tuned to local conditions than traditional agriculture. It usually requires permanently cleared fields, and because soil fertility in the tropics declines rapidly once the trees are removed, crops are more likely to fail even under ideal climatic conditions. Chemical fertilizers can compensate for soil infertility, but are often too expensive for ordinary farmers, and may lose their effectiveness when used repeatedly. Moreover, rains often wash much of the fertilizer into nearby waterways, thus polluting them. This may ultimately hurt farmers and fishers by

mono-crop agriculture farming in which a single crop species is grown in a large field

Figure 7.13 Shifting cultivation in Madagascar. Farmers burn a field before planting it. [Christian VAISSE/Gamma-Rapho via Getty Images]

spoiling drinking water and reducing the quantity of available fish—important sources of protein for rural people.

These aspects of commercial agricultural systems make them risky even under the most ideal climatic conditions, let alone in the hotter and more drought-prone environments that climate change is producing. Foreign agricultural "experts" who often push commercial systems can be woefully ignorant of local conditions and the value of local cultivation techniques. For example, throughout this region women grow most of the food for family consumption and are guardians of the knowledge that makes traditional farming systems work. However, despite a growing awareness of women's importance in African agriculture, women are frequently not included or even consulted by the foreign experts who promote commercial agricultural development projects. At best, women are employed as field laborers, rarely as policy planners or project supervisors or even equipment operators. This ignorance of women's contributions has facilitated the replacement of diverse traditional agricultural systems, based on numerous plant species and deep knowledge, with less stable commercial systems that depend on a single plant species and knowledge that is often not suited to local conditions.

Some agricultural scientists have begun to incorporate the traditional knowledge of African farmers, female and male, into commercial agriculture systems. For example, scientists at Nigeria's International Institute for Tropical Agriculture are developing cultivation systems that, like traditional systems, use many species of plants that help each other cope with varying climatic conditions. Most of these systems are designed to include both subsistence and commercial agriculture, and to give families both a stable food supply *and* the cash to help them ride out crop failures and pay school tuition for their children.

Water and Climate Change

Freshwater shortages are increasing as climate change brings higher temperatures, lower soil moisture, and more erratic weather, including both floods and drought. Although sub-Saharan Africa has a large wet tropical zone, much of the region is seasonally dry, and in these zones conflicts are increasing rapidly as livelihoods are strained by drought (**Figure 7.14A**).

Conflict, Desertification, and Climate Change in the Sahel Conflict between nomadic animal herders and farmers is as old as human history. In the lands just south of the Sahel recent decades of drought and desertification, partially related to climate change, are forcing herders to move south into agricultural areas, bringing a new intensity to age-old conflicts.

Herding, or **pastoralism**, is practiced by millions of Africans, primarily in savannas, on desert margins, and in the mixture of grass and shrubs called open bush. Herders live off the milk, meat, and hides of their animals, and must develop a finely tuned understanding of environmental geography and seasonal cycles to survive. Over the course of a year, they typically circulate through wide areas, taking their animals to available pasturelands and watering holes, trading with settled farmers along the way for grain, vegetables, and other necessities.

Many dry areas that herders use in Africa are now undergoing desertification, the process by which arid conditions spread to areas that were previously moist (see "Climate Change and Desertification" in Chapter 6). This drying out is often the result of the loss of native vegetation. Poorly managed animal herds may be partially to blame, along with agricultural practices that leave soil exposed to wind erosion. Climate change is also bringing higher temperatures that reduce soil moisture, making it harder for many plants to grow. Economic development schemes that encourage commercial cattle raising also play a role. Cattle need more water and forage than do traditional herding animals and therefore place greater stress on native grasslands than do goats or camels.

Over the last century desertification has shifted the Sahel to the south. For example, the *World Geographic Atlas* in 1953 showed Lake Chad situated in a forest well south of the southern edge of the Sahel. By 1998, Lake Chad sat in the northern fringe of the Sahel and the Sahara itself was encroaching (see the Figure 7.9 map). By 2018, the lake had shrunk to a tenth of the area it occupied in 1953.

With less water, herders have been pushed further south, sparking violent conflicts with farmers that have killed tens of thousands across the southern fringes of the Sahel in recent years. Conflict has been especially intense in Nigeria, home to the largest population of nomadic herders in the world, the Fulani, who number around 8 million (another 16 million are city dwellers). Although these bloody conflicts have a religious aspect (the herders are generally Muslim and the farmers Christian), ideology is less of a factor than access to well-watered land that can support pastures. Governments in Nigeria and Ghana are trying to resolve the conflict by setting up grazing reserves and ranches for the herders, though similar "sedentarization" schemes are generally unpopular with the herders, as it means giving up their nomadic way of life.

New Large- and Small-Scale Irrigation Projects Irrigation is a key strategy for increasing any farmer's resilience to increasing drought and other climate change impacts. Many governments favor large-scale irrigation projects to address the need for increased food supplies in the face of current and future water scarcity. Often these projects transform major rivers, such as the Niger, which we focus on here.

The Niger rises in far western Africa, in the tropical wet Guinea Highlands, and carries summer floodwaters northeast into the normally arid lowlands of first Mali and then Niger (see the Figure 7.9 map and **Figure 7.15**). There, the waters spread out into an inland delta with extensive wetlands. For a few months of the year (June through September), the wetlands ensure the livelihoods of millions of fishers, farmers, and pastoralists. These people share the territory in carefully synchronized patterns of land use that have survived for millennia. Wetlands along the Niger produce eight times more plant matter per acre than the average wheat field does. They provide seasonal pasture for millions of domesticated animals, and serve as an important habitat for birds and other wildlife.

The governments of Mali and Niger now want more dams on the Niger River so they can channel its water into irrigated agribusiness projects that they hope will help feed the more than 26 million

pastoralism a way of life based on herding; practiced primarily on savannas, on desert margins, and in the mixture of grass and shrubs called open bush

7.14 Vulnerability to Climate Change in Sub-Saharan Africa

Much of sub-Saharan Africa is highly exposed and sensitive to drought, flooding, and other climatic disturbances. Poverty increases sensitivity to climate change. However, traditional techniques for growing food can provide diverse strategies for coping with the changes in temperature and rainfall that boost resilience. Many farmers are shifting to commercial agriculture, which is more expensive, polluting, and sensitive to shifting weather patterns, but the cash it provides can increase a family's resilience to climate change. Meanwhile, shifting rainfall patterns are straining ecosystems and increasing conflicts. Governments and managers of protected areas are developing new methods of improving resilience to climate change, but in this region of rapid population growth and political and economic instability the future remains uncertain.

THINKING GEOGRAPHICALLY

A How might climate change in Africa or elsewhere cause problems for tea producers?

B Since 10,000 wildebeests drowned after heavy rains in 2007 (instead of the average loss of 1,000 animals), should they be prevented from migrating?

C How is agricultural diversity illustrated in this photo?

D How is poverty evident in this photo?

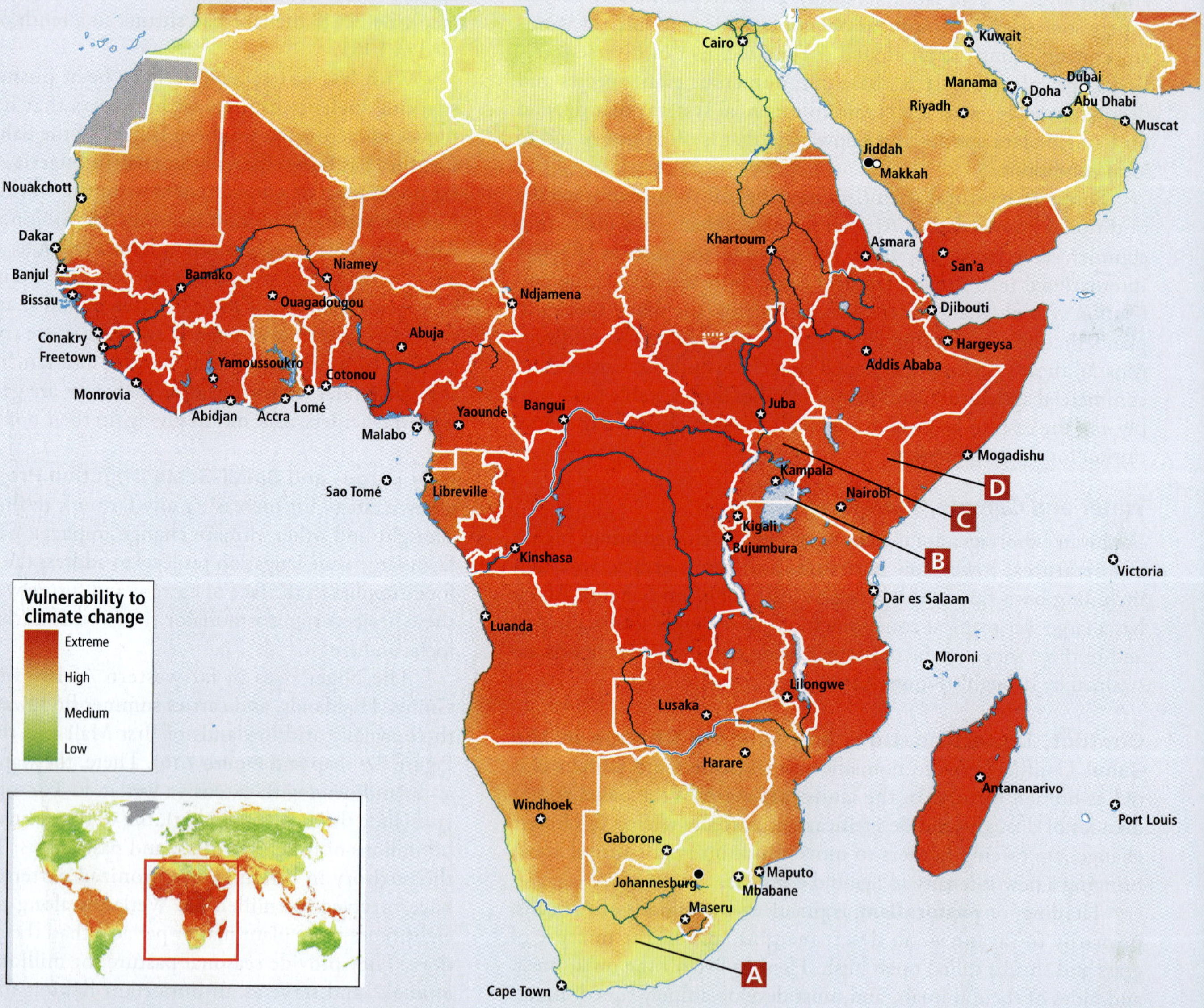

A Tea harvested for export in South Africa. Commercial crops like tea can increase farmers' resilience to climate change by providing them with cash to buy food. But tea, like many commercial crops, requires expensive and polluting fertilizers, and prices for tea are unstable. [Hoberman Collection/UIG via Getty Images]

B The annual wildebeest migration in Tanzania and Kenya. More than a million wildebeest and other wildlife take part in the annual migration through the Maasai game reserve in Kenya. Flooding and shifting rainfall patterns now threaten the migration. [Nigel Pavitt/Getty Images]

C A landscape of subsistence and mixed agriculture in Uganda. These diversified systems make farmers somewhat more resilient to climatic disturbances, but provide little cash to help farmers buy other food they might need. [Ivan Vdovin/Getty Images]

D Somali refugee children fetching water in Dadaab, Kenya. Nearly 500,000 people live in the refugee camp in Dadaab, making it one of the largest camps in the world. Drought, poverty, and political instability come together in the Horn of Africa to create extremely high vulnerability to the effects of climate change. [TONY KARUMBA/AFP/Getty Images]

Figure 7.15 A fisherman casts his net in the Niger River wetlands, where carefully synchronized resource-use patterns have been developed over millennia. What aspect of this fishing method suggests that catches are fairly small? [Wade Davis/Getty Images]

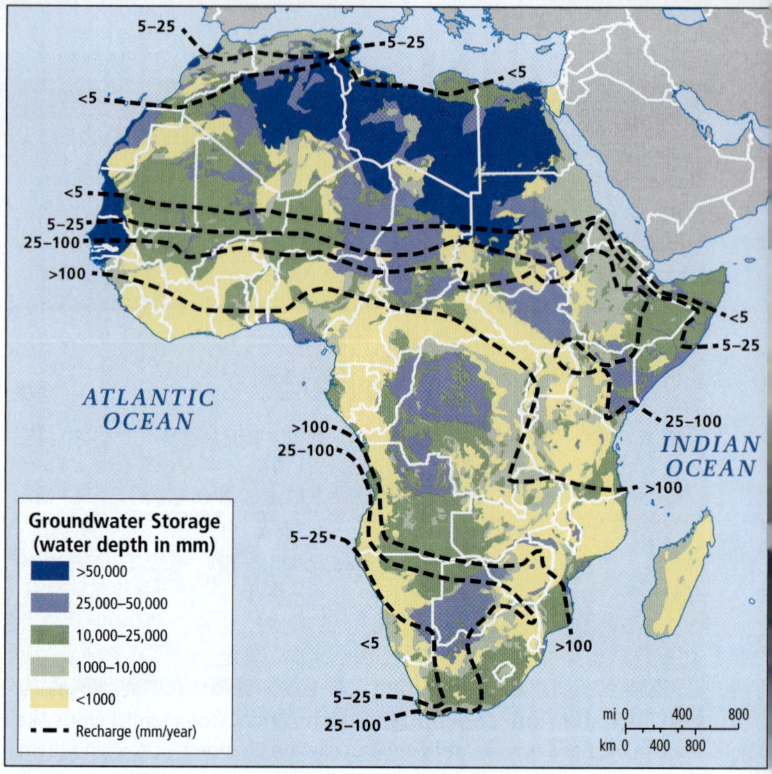

Figure 7.16 Newly exposed water resources. The quantity of water stored in aquifers in Africa is nearly 100 times that of previous estimates. However, because much is fossil water deposited as many as 5000 years ago, it cannot be recharged (replenished) quickly under present drier climate conditions. Also, many densely populated areas have only small deposits. Compare this map of the depth of groundwater deposits and number of years needed to recharge with the Figure 7.31 population map. [Research from: A. M. MacDonald, H. C. Bonsor, B. E. O. Dochartaigh, and R. G. Taylor, 2012, "Quantitative Maps of Groundwater Resources in Africa," *Environmental Research Letters 7*, doi:10.1088/1748-9326/7/2/024009; data for boundaries of surficial geology of Africa from U.S. Geological Survey; and data for country boundaries from ArcWorld, 1995–2011 ESRI.]

people in both countries. However, the dams would forever change the seasonal rise and fall of the river, and the irrigation systems pose a threat to human health due to the standing pools of water they often create. These breed mosquitoes that spread tropical diseases such as malaria and zika virus, and harbor the snails that host schistosomiasis, a debilitating parasite that enters the skin of humans who spend time standing in still water to fish or do other chores.

Many smaller-scale irrigation technologies are available that are cheaper and less disruptive. In some parts of Senegal, for example, farmers are using hand- or foot-powered pumps to bring water from rivers or ponds directly to the individual plants that need it. This is in some ways a more modern version of traditional African irrigation practices whereby water is delivered directly to the roots of plants by human, often female, water brigades. Smaller-scale projects like these are often better at conserving water and are much cheaper and simpler for small farmers to operate. They also avoid the social dislocation and ecological disruption (erosion and flooding) of larger projects involving dams and reservoirs. Already used successfully for 15 years, these low-tech pumps may help farmers adjust to the drier conditions that may come with climate change.

New Groundwater Discoveries A recent comprehensive review and quantitative mapping of all available data on groundwater for the African continent (**Figure 7.16**) revealed that the quantity of **groundwater** (water naturally stored in aquifers over periods measuring thousands of years) was nearly 100 times that of previous estimates. In many cases, the water was close enough to the surface to be accessed via inexpensive pumps.

groundwater water naturally stored in aquifers as many as 5000 years ago during wetter climate conditions

This does not mean that Africa suddenly has an inexhaustible supply of water. Most water was deposited long ago under much wetter conditions and is not being replenished as fast as it will be pumped out. Groundwater is also endangered by commercial agriculture because of its reliance on chemical fertilizers and pesticides that eventually work their way into the aquifers. Nor is it easy to get the groundwater to the people who need it. The largest reservoirs of groundwater are in lightly populated areas (such as North Africa and Botswana; see Figure 7.31 later in the chapter). The most heavily populated country, Nigeria, has only a relatively small supply of groundwater. Nonetheless, this news on groundwater resources is welcome because climate change is likely to reduce precipitation in irregular and unpredictable patterns across the continent, and because per capita water use is likely to expand exponentially as this region develops economically (see Chapter 1).

Wildlife and Climate Change

Sub-Saharan Africa's world-renowned wildlife faces multiple threats from both human and natural forces, all of which could become more severe as the climate changes. The Intergovernmental Panel on Climate Change (IPCC), a body of scientists tasked by the United Nations with assessing scientific information relevant to understanding climate change, estimates that 25 to 40 percent of the species in Africa's national parks may become endangered as a result of climate change.

Wildlife managers are developing new management techniques to help animals survive. For example, one of the greatest wildlife spectacles on the planet took a tragic turn in 2007. The annual 1800-mile-long (2900-kilometer-long) natural migration of more than a million wildebeests, zebras, and gazelles in Kenya's Maasai Mara game reserve requires animals to traverse the Mara River. In the best of times, this is a difficult migration that usually results in roughly a thousand animals drowning. However, recent years have seen extremely heavy rains, possibly related to climate change, swelling the Mara River to record levels and drowning up to 10,000 animals during the migration (see Figure 7.14B). Park managers are now taking a more active role in helping the migrating animals cope with unusual climatic conditions that may worsen with climate change. This may involve stopping animals from attempting a river crossing or directing them to a safer crossing. Shifting rainfall patterns could also force the migration out of the Maasai Mara game reserve and onto private lands where animals could be hunted or become trapped by fencing.

Farmers' dependence on hunting wild game (bushmeat) for part of their food and income is already a major threat to wildlife in much of sub-Saharan Africa. For example, farmers who need food or extra income are killing endangered species such as gorillas, chimpanzees, and especially elephants in record numbers. If crop harvests are diminished by global climate change, many farmers will become even more dependent on bush meat and on income from selling contraband, such as ivory. The threat to wild populations of various species has led to calls to expand and establish new protected areas for wildlife.

Africa's national parks constitute one-third of the world's preserved national parkland, and they have long struggled to limit poaching (illegal hunting) within the park boundaries by members of surrounding communities. Poaching is often fueled by local economic hardships and by demand outside Africa for exotic animal parts (tusks, hooves, penises) as medicines and aphrodisiacs, especially in Asia, where efforts to educate consumers about the damage done by their purchases are only beginning. In 2011, the western black rhino was declared extinct, due largely to poaching driven by demand for its horns by practitioners of Chinese medicine.

Some parks are now using profits from ecotourism to fund development in nearby communities that previously depended, in part, on poaching. For example, Zambia's Kasanka National Park was plagued by poaching until 1985, when the managers decided to generate employment for nearby villages through tourism-related activities. They built tourist lodges and wildlife-viewing infrastructure and started cottage industries to make products to sell to tourists. Funding also goes to local clinics and schools, and students are included in research projects within the park. Today, poaching in Kasanka is rare, its wildlife populations are booming, tourism is growing, and local communities have an ongoing stake in the park's success.

CHECK YOUR UNDERSTANDING

1. What is the primary way in which sub-Saharan Africa contributes to CO_2 emissions?
2. What are some of the advantages and disadvantages of traditional agricultural systems in the face of climate change?
3. How can commercial agriculture both help and hinder efforts to adapt to climate change?
4. What are some challenges in using newly discovered groundwater resources to help alleviate water scarcities in parts of this region?
5. What threats to wildlife is climate change bringing to this region? How are some of sub-Saharan Africa's parks responding?

HUMAN PATTERNS OVER TIME

Africa's rich and ancient history is often overshadowed by the dramatic changes that came with European colonialism, beginning about 600 years ago. These changes were so powerful that even today the lingering effects of European domination are felt throughout the region.

Africa has long been misunderstood and dismissed by outsiders as a place where little of significance has occurred. European slave traders and colonizers were quick to dehumanize Africa as the "Dark Continent," ignoring the substantial and elegantly planned cities of Benin in western Africa, Djenné in the Niger River basin (**Figure 7.17C**), and Loango in the Congo Basin, which European explorers encountered in the 1500s. Even today, most people outside Africa are unaware of its contributions to world civilization.

The Peopling of Africa and Beyond

Africa is the original home of humans (Figure 7.17A). We now know that there were a number of extremely ancient hominid species that lived simultaneously in Africa and that moved out of the continent into Asia and later Europe. It was probably in eastern Africa (in what are today the highlands of Ethiopia, Kenya, and Uganda) that one of the first human species (*Homo erectus*) evolved more than 2 million years ago. Differing anatomically from present-day humans, these early, tool-making humans ventured out of Africa, reaching north of the Caspian Sea and beyond into Asia as early as 1.8 million years ago. Anatomically, modern humans (*Homo sapiens*) evolved about 200,000 years ago from earlier hominoids in eastern Africa and migrated toward the eastern Mediterranean about 90,000 years ago. They then spread across mainland and island Asia, only later turning west into Europe. After coexisting in Eurasia and Europe with earlier dispersions of *Homo erectus*, *Homo sapiens* eventually outcompeted them.

Early Agriculture, Industry, and Trade in Africa

Agriculture dates back 7000 years in the Sahel and the highlands of present-day Sudan and Ethiopia, and was brought south to equatorial Africa 2500 years ago and to southern Africa about 1500 years ago. Trade routes spanned the continent, extending north to Egypt and Rome and east to India and China. Gold, elephant tusks, and timber from tropical Africa were exchanged for salt, textiles, beads, and a wide variety of other goods.

About 3400 years ago, people in the vicinity of Lake Victoria (now adjacent to Uganda, Kenya, and Tanzania) learned how to smelt iron and make steel. By 700 C.E., when Europe was still recovering from the collapse of the Roman Empire, a remarkable civilization with advanced agriculture, iron production, and

A A museum diorama of early humans (*Homo erectus*) in southern Africa [Peter V. Bianchi/National Geographic/Getty Images]

B The ruins of Great Zimbabwe [DESMOND KWANDE/AFP/Getty Images]

C The Great Mosque, Djenné, Mali, built 1200-rebuilt in 1905 [Education Images/UIG via Getty Imag

Figure 7.17 VISUAL HISTORY OF SUB-SAHARAN AFRICA

Thinking Geographically

A What are the clues in the picture that food may have been a source of conflict among early humans and between humans and other animals?

B What about this structure challenges views about sub-Saharan Africa's past?

C What does the existence of a mosque in Mali in 1200 indicate about the geography of trading patterns in this part of Africa?

gold-mining technology had developed in the highlands of southeastern Africa in what is now Zimbabwe. This empire, now known as the Great Zimbabwe Empire, traded with merchants from Arabia, India, Southeast Asia, and China, exchanging the products of its mines and foundries for silk, fine porcelain, and exotic jewelry. The Great Zimbabwe Empire collapsed around 1500 for reasons not yet understood (see Figure 7.17B).

Complex and varied social and economic systems existed in many parts of Africa well before the modern era. Several influential centers, made up of dozens of linked communities, developed in the forest and the savanna of the western Sahel. There, powerful kingdoms and empires rose and fell, such as Djenné (see Figure 7.17C) in Mali (200 B.C.–1300 C.E.); Ghana (700–1000 C.E.), centered not in modern Ghana, but in what is now Mauritania and southwestern Mali; and the Mali Empire (1250–1600 C.E.), centered a bit to the east on the Niger River in what became the famous Muslim trading and religious center of Tombouctou (Timbuktu). Originally founded about 1100 C.E. by Tuareg nomads as a seasonal camp, African rulers periodically sent large trade caravans carrying gold and salt through Tombouctou on to Makkah (Mecca), where their opulence was a source of wonder.

Africans also traded enslaved people. Long-standing customs of enslaving people captured during war fueled this trade, as was the case in many ancient cultures beyond Africa. The treatment of enslaved people within Africa was sometimes brutal and sometimes reasonably humane. Long before the arrival of Islam, a slave trade developed with Arab and Asian lands to the east. After the spread of Islam began, around 700 C.E., the slave trade continued, and Muslim traders exported close to 9 million African slaves to parts of Southwest, South, and Southeast Asia (**Figure 7.18**). Whenever enslaved Africans were traded to non-Africans, the deeply embedded indigenous checks on brutality were often abandoned. For example, to ensure sterility and promote passivity, Muslim traders often preferred to buy castrated males.

The course of African history shifted dramatically in the mid-1400s, when Portuguese traders came to Africa's west coast. The names given to stretches of this coast by the Portuguese and other early European maritime trading powers reflected their interest in Africa's resources: the Gold Coast, the Ivory Coast, the Pepper Coast, and the Slave Coast.

By the 1530s, the Portuguese had organized a slave trade with the Americas that was more widespread and brutal than any trade of African enslaved people that preceded it. The trading of enslaved people was eventually dominated by British companies, but French, Dutch, and Portuguese companies played a major role in creating the elaborate production systems, based on slave labor, that fueled Europe's Industrial Revolution.

European trading companies established forts on Africa's west coast and paid nearby African kingdoms with weapons, trade goods, and money to conduct slave raids into the interior. Some enslaved people were taken from enemy kingdoms in battle while many more were kidnapped from their homes and villages in the

D European slave trading in 1814 [Bibliothèque de l'Arsenal, Paris, France/Archives Charmet/Bridgeman Images]

E African troops under British control in colonial Nigeria in 1900 [Reinhold Thiele/Thiele/Getty Images]

F A soccer stadium in Cape Town, South Africa, built for the 2010 World Cup [F1online/Getty Images]

1600 C.E. | 1700 C.E. | 1800 C.E. | 1900 C.E. | 2000 C.E.

1400–1900
European slave trade

1840–1916
European colonial powers, meeting in Europe, establish borders of modern Africa

1948–1994
Official apartheid in South Africa

2020

D What is the evidence of European involvement in the slave trade in this image?

E What about this photo suggests hierarchies of status and power in British colonial Africa?

F What about this stadium conveys the wealth and Westernization of South Africa?

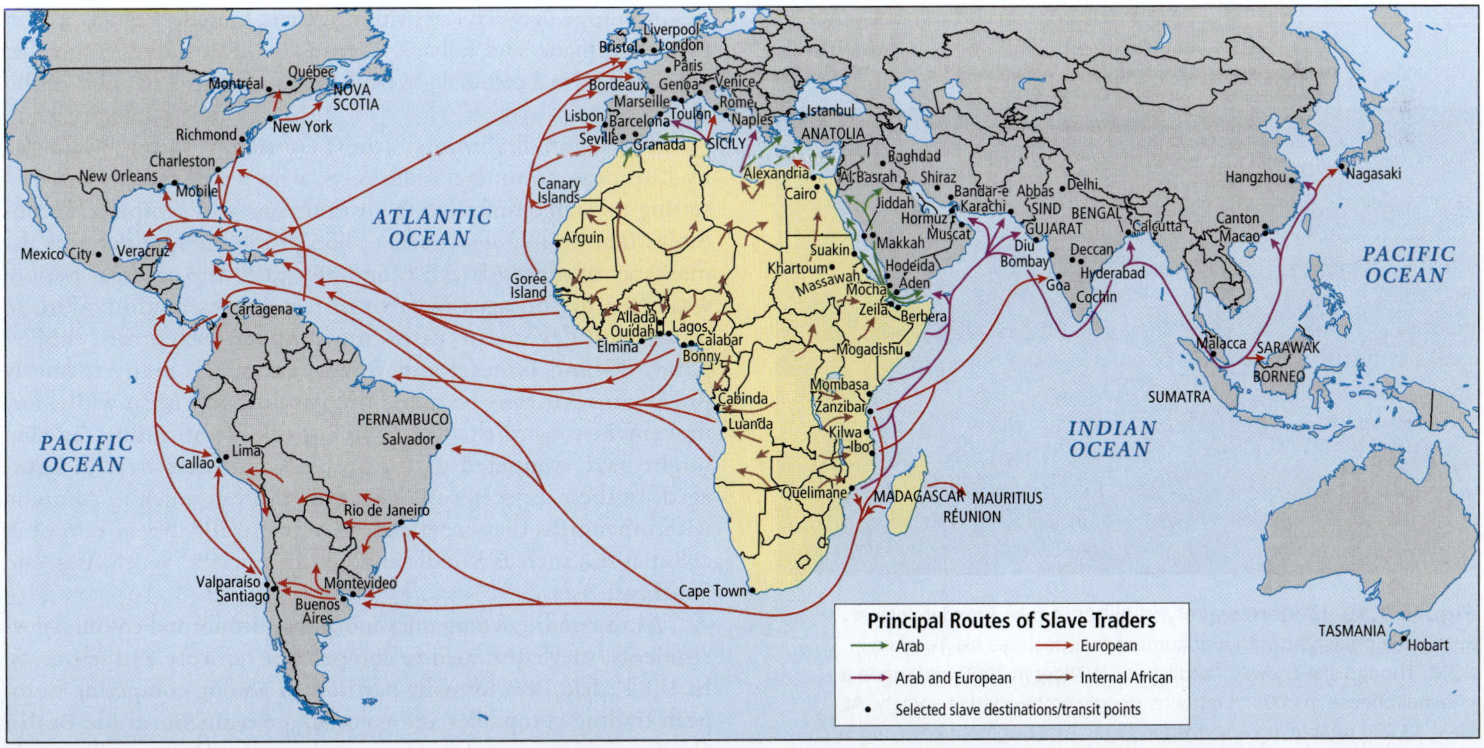

Figure 7.18 Map of the African slave trade. The origin of the African slave trade is indigenous and ancient, predating Islam by thousands of years. Outside of Africa, many slave markets were established around the Mediterranean and the Indian Ocean and eventually, after 1500 C.E., millions of enslaved people were brought to the Americas. [Research from: Joseph H. Harris, in Monica B. Visona et al., *A History of Art in Africa* (New York: Harry N. Abrams, 2001), pp. 502–503.]

forests and savannas. Most enslaved people traded to Europeans were male because they brought the highest prices and the raiding kingdoms preferred to keep captured women for their reproductive capabilities. Between 1600 and 1865, about 12 million captives were packed aboard cramped and filthy ships and sent to the Americas. One-quarter or more of them died at sea. Of those who arrived in the Americas, about 90 percent went to plantations in South America and the Caribbean. Between 6 and 10 percent were sent to North America (see Figure 7.18).

The European slave trade severely drained parts of Africa of human resources and set in motion a host of damaging social responses within Africa that even today are not completely understood (see Figure 7.17D). The trade enriched the African kingdoms that could successfully conquer their neighbors and impoverished or destroyed more peaceful or less powerful kingdoms. It also encouraged the slave-trading kingdoms to become dependent on European trade goods and technologies, especially guns.

Modern slavery, also known as human trafficking, persists in this region today to a greater extent than anywhere else on Earth, and it is a growing problem that some argue exceeds the transatlantic slave trade of the past. Slavery is today most common in the Sahel region, where several countries have made the practice officially illegal only in the past few years (**Figure 7.19**). Today, people may become enslaved during war. They may be sold by their parents or relatives to pay off debts or forced into slavery when migrating to find a job in a city. Some migrants are even enslaved by gangsters in Europe, where they become street vendors, domestic servants, or prostitutes.

1884 Berlin Conference A meeting that divided Africa among European trading companies and colonial powers, establishing many boundaries that persist to this day

Figure 7.19 Modern slavery in Niger. A man born into slavery in Niger is presented with a document that declares his freedom in 2008. Though slavery was "abolished" in Niger in 1960 and made a criminal offense in 2003, it remains deeply embedded in society. At least 43,000 people are enslaved in Niger, most of them born into such status. Enslaved people have no rights and little access to education, and spend most of their lives herding cattle, tending fields, or working in their "master's" house. What does this photo suggest about the social relationships of the man being freed from slavery in Niger? [Kasia Wandycz/ Paris Match via Getty Images]

In Côte d'Ivoire, Burkina Faso, and Nigeria, forced child labor is used increasingly in commercial agriculture such as cacao (chocolate) production. In Liberia, Sierra Leone, Cameroon, and Congo, young boys have been enslaved as miners and soldiers, and in Nigeria, the Islamic militant group Boko Haram has kidnapped at least 2000 young girls since 2014. They were seized apparently to provide wives for Boko Haram soldiers, many of whom were, themselves, captured and radicalized as young boys. It is hard to know exactly how many Africans are currently enslaved, but the Global Slavery Index estimates there were 9.2 million slaves in sub-Saharan Africa in 2018.

CHECK YOUR UNDERSTANDING

1. When and where did anatomically modern humans (*Homo sapiens*) evolve from earlier hominids?
2. When and where did agriculture first develop in sub-Saharan Africa?
3. What are some differences among the slave trades that have existed in this region?

From European Companies to African Colonies

As the trans-Atlantic slave trade wound down by about the mid-nineteenth century, thanks largely to abolitionist movements in Europe, many European trading companies in Africa switched to buying raw materials to supply Europe's growing industries. Most companies were sponsored by European monarchs and wealthy investors, and competition between them mirrored rivalries in Europe between the British, French, Dutch, Belgians, Portuguese, Germans, and Italians. Eventually these trading companies grew into formal colonial governments that controlled most of this region.

The colonial administrations continued to be dominated by European commercial interests, who were less interested in setting up stable functioning governments and prosperous economies than in making profits. Food production, which was the mainstay of most African economies until the colonial period, was often discouraged in favor of resource extraction. With so much land devoted to cash-crop production (cotton, rubber, palm oil) and mineral and wood extraction, many formerly prosperous Africans became hungry and poor. Meanwhile, any movement toward the development of African industries that might have competed with European industries was discouraged. In these aspects, sub-Saharan Africa has much in common with other areas that experienced commercially driven European colonialism, such as Middle and South America, South Asia, and Southeast Asia.

As the trading companies morphed into formal colonial governments, they continued to compete for territory and resources. In 1884 Africa was formally partitioned among competing European trading companies and colonial governments at the **Berlin Conference**, convened by German chancellor Otto von Bismarck, who declared, "My map of Africa lies in Europe." With some notable exceptions, the boundaries of most African countries today derive from the boundaries established between 1840 and 1916 by European traders and administrators. By World War I, only two

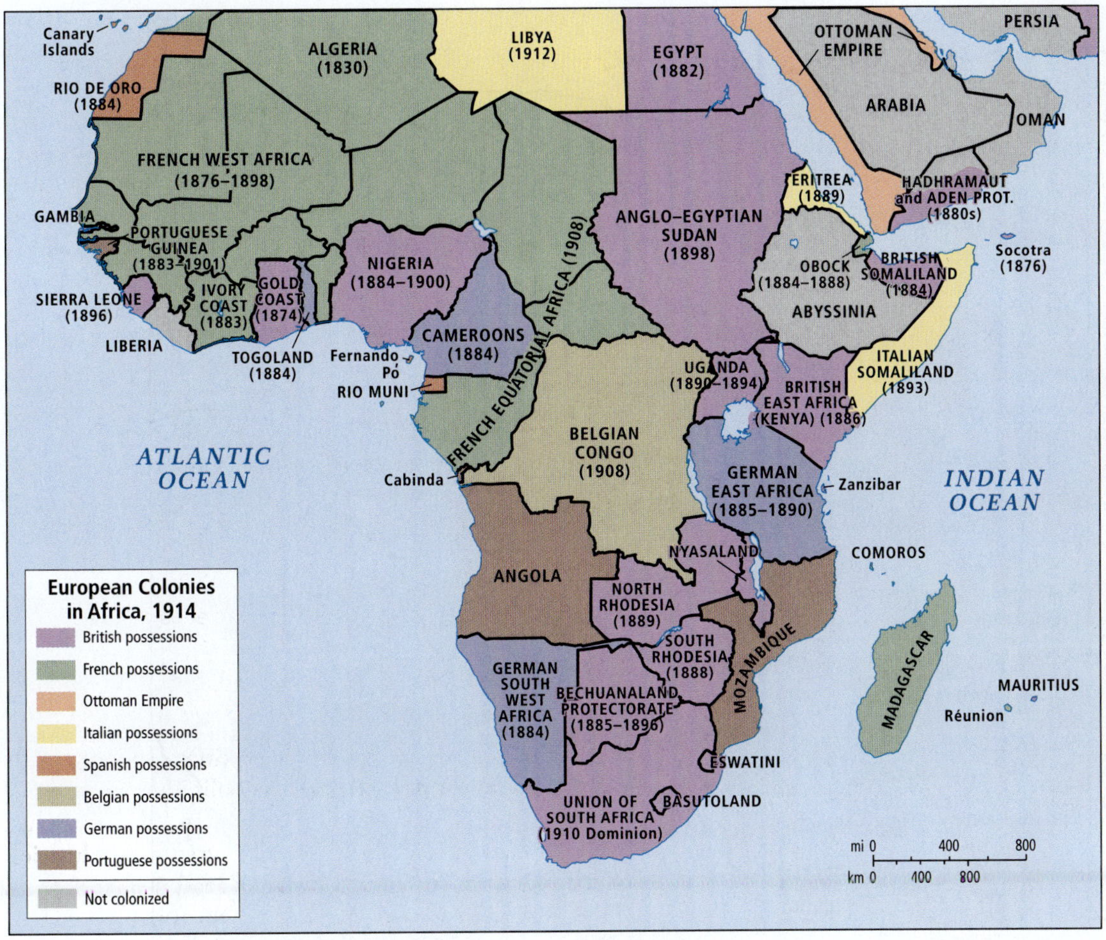

Figure 7.20 The European colonies in Africa in 1914. The dates on the map indicate the beginning of officially recognized control by the European colonizing powers. Countries without dates were informally occupied by colonial powers for a few centuries. [Research from: Alan Thomas, *Third World Atlas* (Washington, DC: Taylor and Francis, 1994), p. 43.]

areas in Africa were still independent (**Figure 7.20**): Liberia on the west coast (populated by former enslaved people from the United States) and Ethiopia (then called Abyssinia) in East Africa, which managed to defeat early Italian attempts to colonize it.

Divide and Rule The territorial divisions drawn by Europeans lie at the root of many of Africa's current problems. In some cases, the boundaries were purposely drawn to divide tribal groups and thus weaken them. In other cases, nomadic groups who used a wide range of environments over the course of a year were forced into smaller, less diverse lands or forced to settle down entirely, thus losing their traditional livelihoods. As access to resources shrank, hostilities between competing groups developed and were even encouraged by colonial officials so that they could manipulate the outcome through a set of policies known collectively as "**divide and rule**."

After independence, when Africans assumed control of their own countries, governance was complicated because officials inevitably belonged to one local ethnic group or another and were not seen as impartial in their attempts to resolve conflicts. Many older indigenous traditions for ensuring ethical behavior, discouraging greed on the part of leaders, and resolving ethnic conflict had been weakened during the colonial era, and the political institutions that replaced them were often weak, less open in their discussions of issues and not adept at conflict resolution. This amplified hostilities between ethnic groups, some of which have grown into civil wars (Figure 7.29C). The colonial and independence histories of sub-Saharan Africa's two largest economies helps further illustrate the complex legacy of European imperialism in this region.

Companies, Colonies, and Conflict in Nigeria The state of Nigeria is a creation of British imperialism, with disparate groups—speaking 395 indigenous languages—joined into one unusually diverse country whose borders were set by European trading interests (**Figure 7.21**). To facilitate ruling such a diverse territory, the British reinforced a north–south dichotomy that mirrored the physical and cultural patterns: dry and Muslim in the north, wet and Christian in the south.

After centuries in which the slave trade dominated European activity along the coast of the Niger delta, British and French abolition of slavery in the early 1800s meant traders switched to selling European cotton cloth and buying African oil palm, which provided lubricants for the machinery of the Industrial Revolution. The trade became so lucrative and widespread that Yoruba and Igbo ethnic groups along the coast came to be ruled by British trading companies that grew into formal British colonies by the 1860s. Throughout this period British missionaries encouraged the Yoruba-Igbo to attend Christian schools that were open to the public.

North of the coast in the Sahel areas, the Royal Niger Company,

divide and rule the deliberate intensification of divisions and conflicts among "native" groups by European colonial powers

Figure 7.21 Nigeria's ethnic geography. There are thousands of ethnic groups distributed across sub-Saharan Africa. Nigeria alone has more than 400. When Europeans set up political units, they either ignored ethnic boundaries or purposely divided large ethnic groups in order to control them more easily. The black dotted lines are district political boundaries; the colors represent the country's various ethnic groups. [Research from: Ethnic Map of Nigeria, Online Nigeria: Community Portal of Nigeria, http://www.onlinenigeria/com/mapethnic.asp#.]

a British trading company, was using its administrative and military power to monopolize trade along the Niger river and keep out similar French and German companies. Here the company ruled local ethnic groups, such as the Hausa and Fulani, by offering their leaders bribes or other financial inducements to cooperate. In these predominantly Muslim areas, Christian missionary schools were not allowed and local leaders did not encourage public education. However the renowned military skill of the Hausa-Fulani led the Royal Niger and other British trading companies to recruit them to serve in the colonial militias that conquered much of what became Nigeria. In 1914 the Sahel territories controlled by the Royal Niger Company were merged with existing British colonies along the coast to form the Colony and Protectorate of Nigeria.

These two very different areas struggled to coexist under British rule, and when Nigeria became independent in 1960, tensions increased. The south had more than ten times as many primary and secondary school students as the north and was more prosperous, and southerners also held most government civil service positions. Yet the northern Hausa-Fulani dominated the top political posts and the military, in part because of their long association with British colonial administrators in the north. Over the years, bitter and often violent disputes erupted between the southern Yoruba-Igbo and northern Hausa-Fulani regarding the distribution of increasingly scarce clean water, jobs, and oil revenues. To an outsider today it might seem that these groups are fighting each other because of ancient grudges, but in fact they were thrown together just over a century ago by British trading interests into a political system that amplified their differences.

The Colonization of South Africa South Africa is unique in this region as the only country with a substantial population of European origin (around 4 million). Efforts by Europeans to maintain political and economic control over the much larger African population led to an infamous system of racial segregation.

The Cape of Good Hope is a rocky peninsula with a Mediterranean climate that shelters a harbor on the southwestern coast of South Africa. Portuguese navigators seeking a sea route to Asia first rounded the Cape in 1488 and traded in the area, though the Dutch East India Company took possession to support their trade in Asia by 1650. Dutch farmers, called Boers, grew food to resupply Dutch trading ships, often with the help of slaves brought from Indonesia and Madagascar. Many Boers grew resentful of the corrupt and authoritarian rule of the Dutch East India Company and moved inland. Some became nomadic cattle herders and subsistence farmers on the high, rolling plateaus of the interior, which are reminiscent of the Great Plains in North America, but with a much more temperate climate. Temperatures here range between 60 and 70°F (16 and 20°C).

The British captured the Dutch colony in 1795 during the Napoleonic Wars. When slavery was outlawed throughout the British Empire in 1834, large numbers of Boers moved to the northeast in order to keep their enslaved people. There, in what became known as the Orange Free State and the Transvaal, the Boers often came into violent conflict with African groups, especially the Zulu, who resisted Boer expansion with large armies.

In the 1860s, extremely rich deposits of diamonds and gold were unearthed in these areas, attracting huge numbers of British prospectors and leading to a rapidly growing mining industry. While white "skilled" miners were well paid, African miners were forced to work for minimal wages under extremely unsafe conditions, living in compounds that travelers of the time compared to large cages.

Britain, eager to claim the wealth of the mines, invaded the Boer states unsuccessfully in 1880–1881, and again during 1899–1902, prevailing only after a massive, brutal, and expensive war. The terms of peace promised self-government, which was granted in 1910 to the Union of South Africa, which included the cape colony and all the Boer states in a country roughly twice the size of Texas.

The right to vote and other political freedoms were granted only to the 20 percent of the population that was white. Not until 1994 would full political rights (voting, freedom of expression, freedom of assembly, freedom to live where one chose) be extended to black South Africans who by then made up more than 80 percent of the population—and to Asians and mixed-race or "coloured" people (2 percent and 8 percent of the population, respectively).

In 1948, the long-standing customary practice of segregation in South African society was formalized by **apartheid**, a system of laws that required everyone except whites to carry identification papers at all times, to live in racially segregated areas, and to use segregated transportation and other infrastructure (**Figure 7.22**). Eighty percent of the land was reserved for the use of white South Africans, who at that time made up just 10 percent of the population. Black, Asian, and "coloured" people were assigned to ten ethnically based "homelands" (such as Transkei, KwaZulu, and Swazi). The South African government considered the homelands to be independent enclaves within the borders of, but not legally part of, South Africa. Nevertheless, the South African government exerted strong influence in the homelands. Democracy theoretically existed throughout South Africa, but nonwhites were allowed to vote only in the homelands.

The African National Congress (ANC) was the first and most important organization that fought to end racial discrimination in South Africa. Formed in 1912 to work nonviolently for civil rights for all South Africans, the ANC grew into a movement with millions of supporters, most of them black. Its members endured decades of brutal repression by the white minority, and its most prominent leader, Nelson Mandela, was jailed from 1962 to 1990.

Figure 7.22 Apartheid in South Africa. A South African woman protests apartheid by using a "whites only" car on a train in 1952. South Africa's long history of European domination officially ended only in 1994 with the first national elections in which black South Africans were allowed to vote. [Keystone-France/Gamma-Keystone via Getty Images]

Resistance to minority white rule increased until the late 1980s, when ongoing violence verging on civil war combined with international political and economic pressure forced the white-dominated South African government to initiate reforms that would end apartheid. The homelands were dismantled and, in 1994, the first national elections took place in which black South Africans could participate, bringing Nelson Mandela to power. His efforts at reconciliation between the country's different racial groups had earned him a Nobel Peace Prize in 1993.

Today in South Africa, the thorny process of dismantling systems of racial discrimination and corruption continues, with varying amounts of success. Meanwhile there is an ongoing outflow of young, educated, mostly white South Africans to Australia, the United Kingdom, the United States, and New Zealand.

Power and Wealth in the Aftermath of Independence

While the era of formal European colonialism in Africa was relatively short, lasting roughly 80 years from the 1880s to the 1960s, its impacts are still felt today in the political and economic instability that plague this region. Few states have achieved broad prosperity, and political freedoms have begun to expand only recently. Most countries have limped along inside borders created to facilitate resource extraction and weaken African resistance to foreign control, with tensions between rival groups often breaking out into violence.

Sub-Saharan Africa emerged from the exploitation of the colonial era just in time to get sucked into the vortex of globalization, where small or weak countries are often left with little control over their own fates. Out of colonies created to facilitate resource extraction for the benefit of outsiders, countries have emerged that

apartheid a system of laws mandating racial segregation in South Africa from 1948 until 1994 remain dependent on the export of cheap raw materials, the prices of which are highly unstable in the global economy. Meanwhile notoriously corrupt colonial governments, many of which began as private trading companies with monopolies on the trade of key exports in their areas of operation, have morphed into independent governments plagued by bribery and embezzlement. Many are known as **kleptocracies**, where officials steal the resources of their countries for themselves. Economic instability combined with rampant corruption has led to huge gaps between rich and poor.

Even today African economic elites and policy makers, like their colonial predecessors, are often insensitive to the widespread poverty and hardship around them. This was demonstrated during the summer of 2010 when South Africa hosted the 2010 World Cup, the world's largest sporting event and a first for this region, which up to that point had never hosted a global sporting event of similar scale. The preparations involved demolishing the homes of tens of thousands of poor, urban residents who were relocated to make way for the elegant new soccer stadiums (see Figure 7.17F). Nonetheless, South Africa, with its highly profitable manufacturing, service, and mining economy, and its recent political reconciliation, has brought higher levels of human development to more Africans than any other country in the region.

CHECK YOUR UNDERSTANDING

1. What were the main objectives of European colonial administrations in Africa?
2. What policies could be considered "divide and rule"?
3. How is South Africa unique among sub-Saharan African countries?
4. With what world region does sub-Saharan Africa share a dependence on raw materials exports?

GLOBALIZATION AND DEVELOPMENT

7.4 Explain how sub-Saharan Africa's position in the global economy leaves it struggling with economic instability and low human development.

For many centuries, sub-Saharan Africa has contributed human labor and raw materials to the global economy, but the profits from turning these human resources and raw materials into higher-value manufactured and processed products have gone to wealthier countries. Even today, decades after the end of formal European colonialism and resource extraction, most economies in this region are still dependent on the export of raw materials. This results in economic instability because prices for raw materials are generally low, compared to more finished goods, and can vary widely from year to year. While a few countries are diversifying and industrializing, most still sell raw materials and all rely on expensive imports of food and manufactured goods. As a result, most countries are relatively poor, and investments in human development, while increasing in recent years, have remained generally low.

kleptocracy A government in which officials exploit and steal the resources of their country to enrich themselves and extend their political influence

commodities raw materials that are traded, usually to other countries, for processing or manufacturing into more valuable goods

Raw materials that are traded on the world markets for processing or manufacturing elsewhere into more valuable goods are called **commodities**. Sub-Saharan Africa produces a wide array of commodities, such as cotton; cacao beans; coffee; timber; palm oil; unrefined petroleum; gas; precious stones; and metals, such as copper, gold, and iron ore. However the profits from commodity production are usually too low to lift poor countries out of poverty, in part because competition from other poor countries producing the same commodities often drives down prices.

The wide fluctuations of commodity prices on global markets create economic instability for producers. Commodities in their raw state are of more or less uniform quality; therefore, on the world market, they bring about the same price. For example, on the major commodities exchanges, a pound of copper may sell for U.S.$3 to $5, regardless of whether it is mined in South Africa, Chile, or Papua New Guinea. Because of instant communication and the global scale of commodity trading, any change that influences the global supply or demand for a particular commodity can cause immediate instability in the pricing of that commodity. If an earthquake in Chile damages a major copper mine, the temporary restriction in the supply of copper will drive up the price on world markets. In response, governments that make money from sales of raw copper may overestimate their future revenues and commit to expensive infrastructure projects. But the demand for copper can quickly decrease for many reasons; for example, a slowing of housing construction or innovations in building industry materials can reduce the use of copper for wiring and plumbing, resulting in a rapid drop in the world market price of copper. In response to the loss of revenues from copper, governments may suddenly halt infrastructure projects and cut essential services such as electricity, road maintenance, or health care.

Since 2000, there has been a noticeable shift in many African countries away from reliance on just one or two commodities, but no overall movement away from commodity dependence. Economic instability and low incomes remain the norm throughout the region (**Figure 7.23**).

SUCCESSIVE ERAS OF GLOBALIZATION

Successive eras of globalization have transformed Africa over the past several centuries. They include the era of colonial exploitation of commodities; the transition from colonial status to political independence during which large debts were incurred, bringing on the era of structural adjustment and the rise of informal economies; and the current era, in which some African countries are moving away from commodity dependence and toward new types of economic development and regional integration.

The Colonial Exploitation Era

Colonial governments, having evolved out of trading companies based on the export of commodities of agricultural or mineral raw materials, were guided by the idea that colonies existed to benefit their colonizers. As a result laborers and farmers were paid poorly

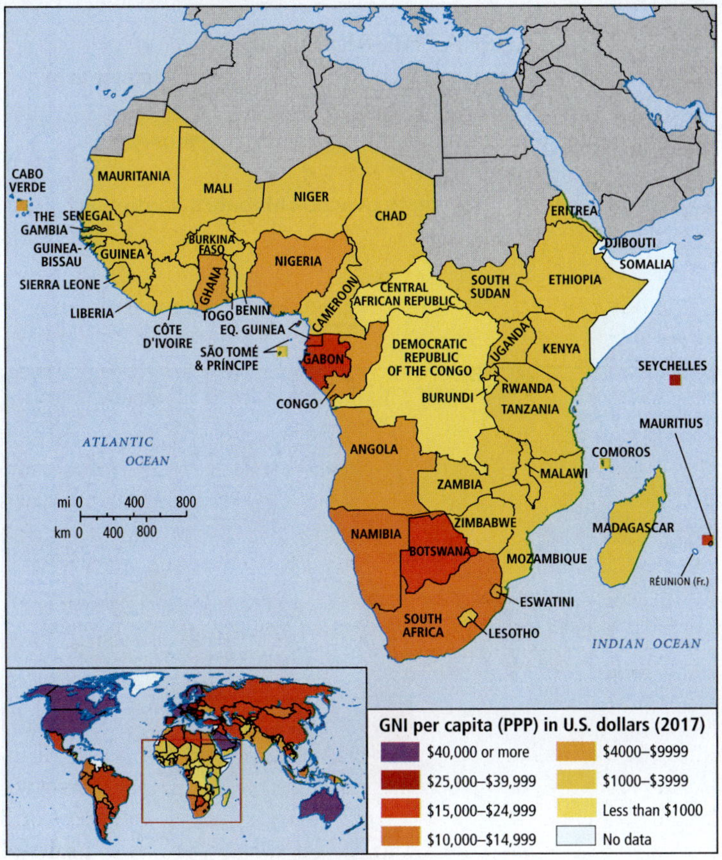

Figure 7.23 Gross national income (GNI) per capita purchasing power parity (PPP), 2018. Most middle-income economies in this region are based on oil exports with the exception of South Africa, Botswana, and Namibia, which export mainly minerals.

and any economic activities that might compete with those of Europe, such as more profitable manufacturing-based industries, were discouraged. Human development remained low, with education and health care generally neglected. For years, **commodity dependence** and widespread poverty characterized all sub-Saharan African economies.

The only sub-Saharan African country to develop a manufacturing base at this time was South Africa, where profits from commodity exports (mainly minerals) were reinvested in the manufacturing of mining and railway equipment. In the late twentieth century, South Africa developed a large service sector with particular strengths in finance and communications that support the mining and manufacturing industries. Today, with only 6 percent of sub-Saharan Africa's population, South Africa produces close to 30 percent of the region's economic output and has brought increases in human development to the largest number of people.

The Era of Debt Accumulation and Structural Adjustment

During the early independence era most countries in this region accumulated substantial debts that had the effect of keeping them dependent on resource extraction. Massive loans were offered to African governments by a network of U.S. and European corporations and banks, as well as by international development agencies such as the World Bank, the International Monetary Fund (IMF), or the foreign aid agencies of the governments of wealthy countries, such as the United States Agency for International Development (USAID). The projects they proposed were mostly large-scale development projects, such as massive hydroelectric dams, industrial parks, and other infrastructure that under ideal circumstances could have supported a new era of industrialization. However, most African governments were small and staffed with people who didn't have the training to manage these giant projects, so private corporations from the United States or Europe were brought in to build them. This meant that African countries remained dependent on outsiders for technical and organizational expertise, and most of the money for the projects went to foreigners.

The potential of many projects was also diminished by the corruption of high-level politicians and bureaucrats, such as Mobutu Sese Seko of Congo (Kinshasa), who embezzled billions of dollars and deposited them in secret bank accounts in Switzerland and the Caribbean (**Figure 7.24C**). Many lenders were willing to turn a blind eye to this corruption because of Cold War efforts to sway African countries away from the socialist development models promoted by the Soviet Union and toward more capitalist "free market" systems. Other lenders knew that even if the development projects failed, the African governments that borrowed the money could still be held accountable for repayment of the loans. Repayment could be financed by raw materials exports to North America and Europe.

A breaking point came in the early 1980s, when an economic crisis swept through the region and much of the rest of the developing world, collapsing the prices of key raw materials exports, and leaving most countries unable to repay their debts at all. In response, the IMF and the World Bank stepped in to ensure that the loans were repaid. Structural adjustment programs (SAPs; see Chapter 3) were imposed that required governments to sell off inefficient government-owned enterprises, often at very low prices. In addition, funding was cut for government agencies providing education, health care, agricultural assistance, and aid to poor people so that tax revenues could be devoted to loan repayment. Meanwhile economies were redirected back to their commodity export sectors because these were the only activities that could even begin to make enough money to repay the loans. If countries refused to implement SAP requirements, they would be cut off from any future lending for economic development.

SAPs did have some useful results. They tightened bookkeeping procedures and thereby curtailed corruption and waste in bureaucracies. They closed some corrupt state-owned enterprises, opened some sectors of the economy to medium- and small-scale business entrepreneurs, and made tax collection more efficient. But overall, SAPs had many unintended consequences and failed at their primary objective—reducing debt (**Figure 7.25**).

As unemployment rose, so did political instability. As key infrastructure, such as roads and electrical power grids, deteriorated for lack of new investment, the cost of doing business increased. This held back economic development and scared away potential investors. SAPs also reduced food security because agricultural resources were shifted toward the production of cash crops for export. Between 1961 and 2009, per capita food produc-

commodity dependence economic dependence on exports of agricultural and mineral raw materials

7.24 Globalization and Development in Sub-Saharan Africa

This region has struggled to free itself from dependence on commodity exports, only to be saddled with massive debts accumulated by corrupt rulers borrowing from foreign interests and international institutions. Policies implemented to repay these loans tended to reinforce commodity dependence, as has Asian interest in the resources of this region. Meanwhile, the needs of millions of poor people are often addressed only by underfunded grassroots development projects. As a result, the Human Development Index (HDI) in this region is the lowest in the world, reaching middle levels only in the countries that have enough money to invest in better schools, health-care systems, and housing.

THINKING GEOGRAPHICALLY

A Why might China be particularly interested in investments in Africa's transportation infrastructure?

B What type of development organization has focused on footpaths?

C Why was the IMF willing to overlook Mobutu's corruption?

A One of the estimated 20,000 Chinese workers in Angola. China has given loans and aid to Angola in excess of U.S.$20 billion since 2015 in return for guarantees of Angola's future oil production. While human development is rising in Angola, it remains low. [Per-Anders Pettersson/Getty Images]

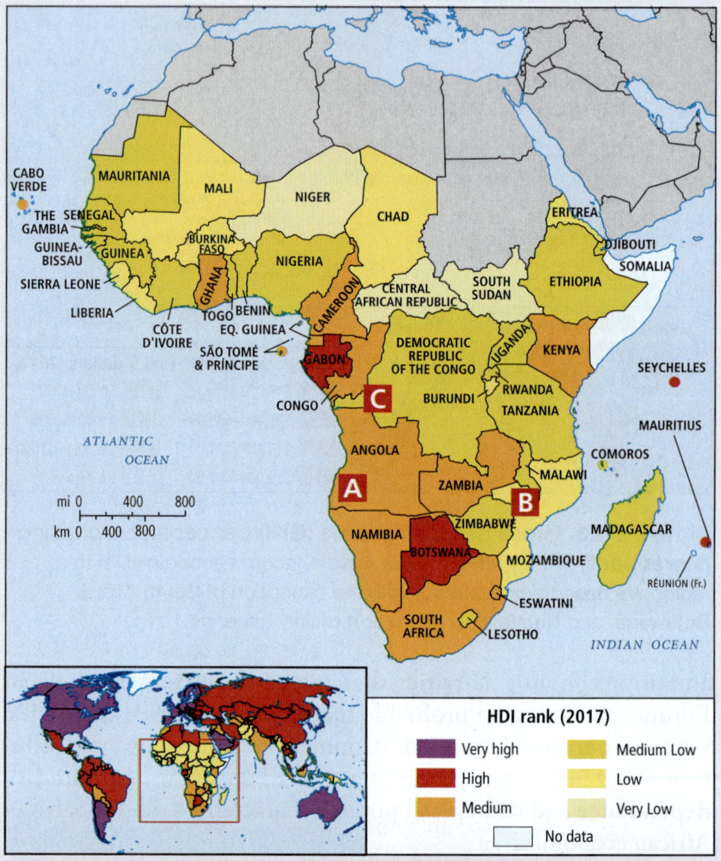

B Rural transport in Malawi. Women and girls in Malawi carry water home from a well along miles of unpaved footpaths. In rural areas, as much as 87 percent of goods are transported this way, imposing grueling hours of labor on millions and constraining human development. [Helen H. Richardson/The Denver Post via Getty Images]

C Corruption in Congo (Kinshasa). A soldier walks through an abandoned palace of former president Mobutu Sese Seko, who embezzled between \$4 and \$15 billion from development loans with full knowledge of lenders such as the IMF. During his 32-year reign, human development remained low despite the country's huge mineral resources. [Tyler Hicks/Liaison/Getty Images]

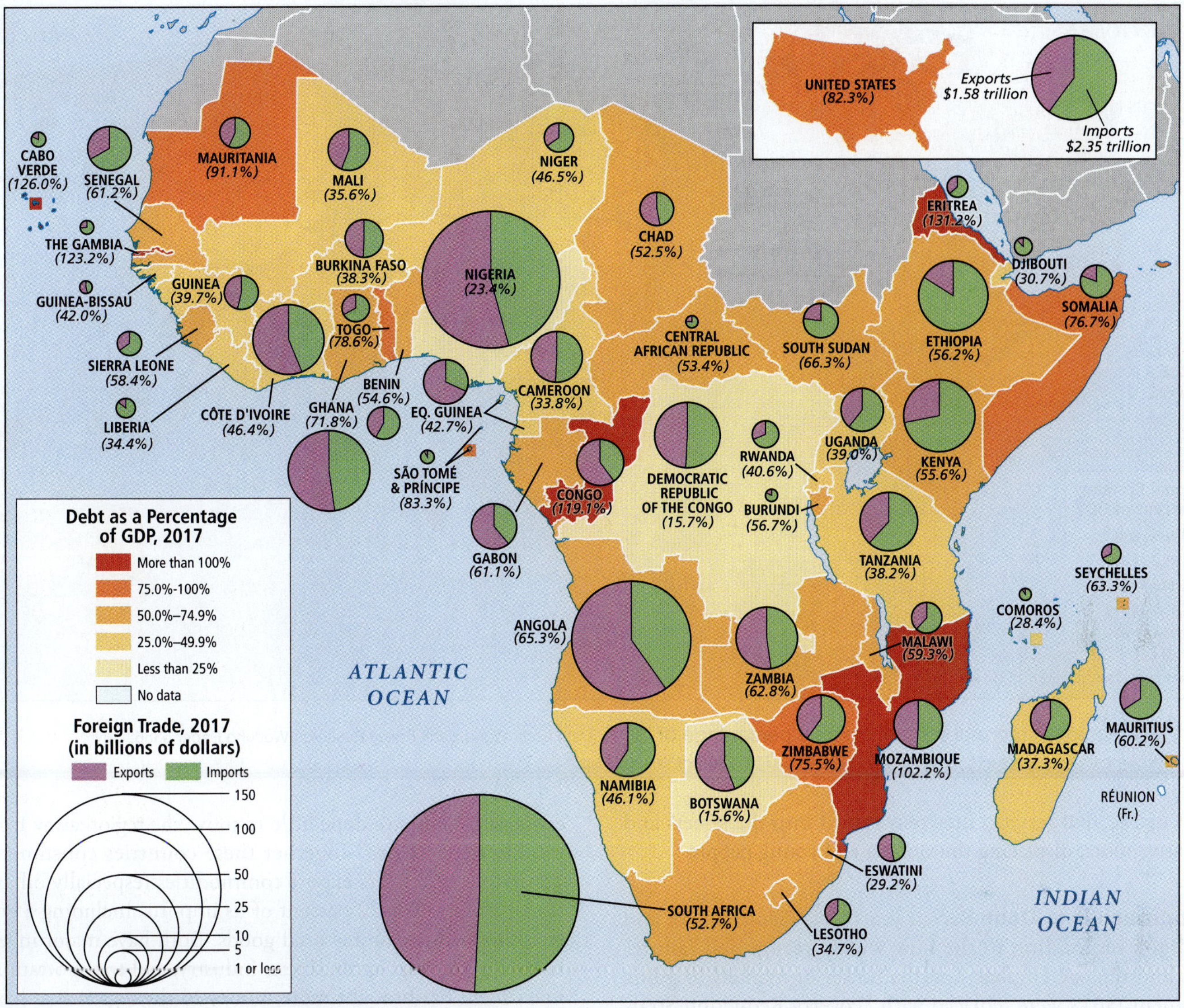

Figure 7.25 Economic issues: Public debt, imports, and exports. Public debt is increasing across Africa, partly as a result of borrowing to fund development projects (Note that the U.S. public debt—see inset—far exceeds that of most sub-Saharan countries). In many sub-Saharan countries, imports exceed exports. [Date from: CIA World Fact Book.]

tion in sub-Saharan Africa actually decreased by 14 percent, making it the only region on Earth where people were not eating as well as they had in the past. This nearly 50-year period of structural adjustment policies, from which the region is still recovering, was a painful time that constrained and at times even lowered human development.

Informal Economies A variety of small-scale, unregulated, officially "off the books," and sometimes risky and illegal activities gave some relief from the hardships created by SAPs. A varied mix of ancient and new methods of providing services and products helps many Africans make ends meet, for example through growing and selling garden produce, preparing food, vending cheap goods on the street, selling time cards (or credits) for mobile phones, making craft items and utensils, or providing child and elder care. More risky and illegal work includes sex work, distilling illegal liquor, and smuggling and selling drugs, weapons, endangered animals, bushmeat, and ivory. Many of these latter activities take place "under the radar," and so may involve criminal acts, wildly unsafe activities, and hazardous substances.

The role of the informal economy is especially important in sub-Saharan cities, where it has grown from one-third of all employment to more than two-thirds. Among women, it is thought to produce more than 90 percent of all job opportunities. Although it is difficult to assess the situation precisely, for sub-Saharan Africa as a whole, the informal economy is estimated to account for about 38 percent of the GDP—the highest in the world (**Figure 7.26**). Informal economy jobs are a godsend to many poor people, but they create problems for governments because the informal economy typically goes untaxed, so less money is available to pay for government services or to repay debts. Moreover, as the informal sector grows, profits to individual entrepreneurs decline as more people compete to sell goods and services to those with little disposable income. And although women typically dominate informal economies, when large numbers of men lose their jobs in

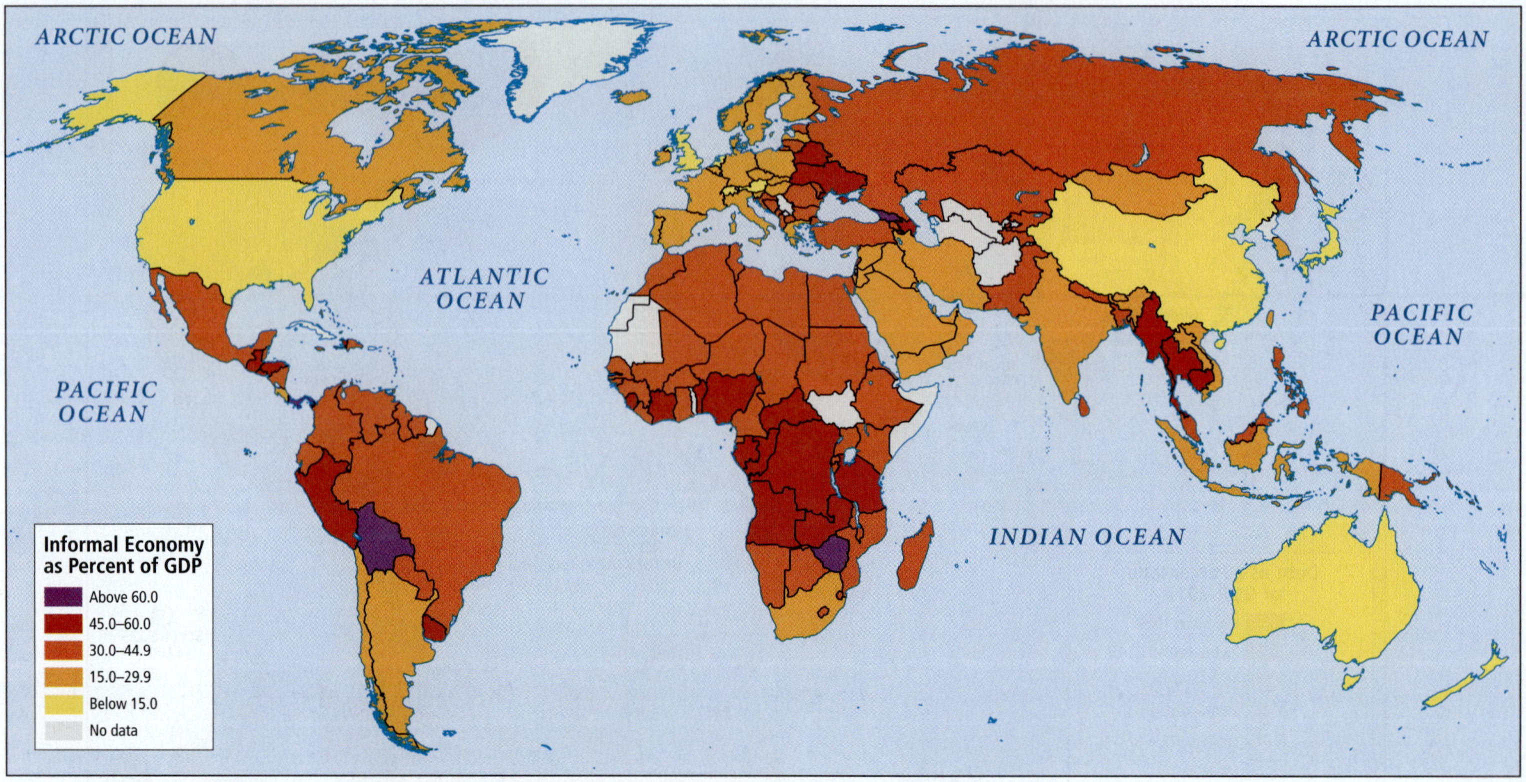

Figure 7.26 Regional informal economies as a percentage of GDP. [Data from: World Bank Policy Research Working Paper 5356, July 2010, http://documents.worldbank.org/curated/en/311991468037132740/pdf/WPS5356.pdf.]

factories or the civil service, they may crowd into the streets and bazaars as vendors, displacing the women and young people.

New Approaches to Debt Recent years have seen the IMF and World Bank responding to the now widely recognized failures of SAPs and the overemphasis on the power of markets to guide development. SAPs were replaced with "Poverty Reduction Strategy Papers" (PRSPs) that focus more on reducing poverty, encouraging manufacturing, and promoting more democratic reforms that reduce corruption. Compliance with PRSPs also involved debt "forgiveness," meaning that debts were paid off by the IMF, World Bank, or African Development Bank. Most sub-Saharan governments have had their debt forgiven, and while some have accumulated new debts, poverty has been significantly reduced and human development increased in most of the region in recent years. Nevertheless, overall levels of human development remain quite low.

The Current Era of Diverse Globalization

The current wave of globalization has brought new sources of investment, jobs, and better infrastructure to much of Africa. While Europe and the United States are still the largest sources of investment in sub-Saharan Africa, China is financing many new projects. Many African companies are also earning sufficient profits to reinvest in other enterprises in the region, and Africans who have migrated abroad are also contributing significantly.

Asian Investments in Africa China and, to a lesser extent, India, view sub-Saharan Africa as a new frontier for their large and growing economies and have done little to move the region away from commodity dependence. Together these countries consume roughly 18 percent of Africa's export commodities (especially oil and minerals), and provide 22 percent of its imports, including a wide variety of low-cost manufactured goods. Both have major investments in energy, mining, agribusiness, industry, and transportation. Asian investment has brought more money to the region that has helped raise human development, but economic instability remains a problem because commodity dependence is still the norm.

China's massive investments in Africa have both similarities and differences to those of U.S. and European interests that created the era of debt and SAPs. Large loans for infrastructure projects, such as new railways, dams, oil refineries, and business parks, are being made to governments, even those notorious for their corruption, with repayment of the loans guaranteed by Africa's natural resources. Most of the work on the projects is being done by Chinese companies, with Africa gaining little expertise in the design, construction, or maintenance of these projects. Similarly, these investments have resulted in large debts that must now be paid in the form of the raw materials exports.

There are major differences, too. China's loans to Africa are made without the conditions of SAPs, such as privatization of state-run businesses and cutting government spending on education and health care. China's willingness to utilize its massive population of relatively low-income people is another difference. Hundreds of thousands of Chinese have come to Africa as part of its development projects. Some have been brought in as low-skilled laborers, which has outraged many, given Africa's desperate need for employment (see Figure 7.24A). Some Chinese have also come in posing as wealthy investors, only to set up shop as small-scale traders in African cities, provoking similar outrage.

As in other world regions, Chinese imports have displaced many locally produced goods. For example Nigeria once maintained an age-old textile industry based on hand-dyed fabrics with graphic indigenous patterns. Once valued across the world for their authenticity, Chinese manufacturers took the traditional designs home and reproduced them in Chinese factories far more cheaply, and exported them back into Nigeria. Since the 1980s when this process began, roughly 97 percent of Nigeria's textile jobs (some 250,000) have been eliminated as Nigerians are now buying Chinese-made textiles. Meanwhile Nigeria's trade deficit with China is growing ever larger.

India's trade with Africa is similar to China's, in that it is based on raw materials exports to India, and imports to Africa of Indian manufactured goods. India has however been somewhat more willing to invest in sectors with more potential to benefit the region, such as manufacturing and information technology. For example, the India-based cell phone company Airtel operates a borderless network that now covers 18 countries: Burkina Faso, Chad, Congo (Kinshasa), Congo (Brazzaville), Gabon, Ghana, Kenya, Madagascar, Malawi, Niger, Nigeria, Rwanda, Seychelles, Sierra Leone, Tanzania, South Sudan, Uganda, and Zambia. Airtel's customers can make calls across the network at local rates without incurring surcharges, and recently, the company added internet access. The borderless aspect of Airtel's network, and others like it, is helping to increase trade and interaction between African countries, which have long been goals of African regional economic development efforts.

Regardless of its overall similarities to the loans offered by U.S. and European interests that led to the era of debt and SAPs, Asia's interest in Africa's resources does at least give the region more bargaining power with foreign interests. For example, in order to counter China's growing interest in this region, the United States has recently increased its funding for new development projects in Africa, with many new projects focusing on investments in human development, including better educational and health-care infrastructure.

Africans Investing in Africa The role of African professionals in globalization is growing. As was noted in the opening vignette, perhaps the most dramatic sign of change for African economies is that educated Africans who are skilled in IT and other high-tech fields are investing their time and efforts at home instead of migrating to richer regions. Their incomes, along with remittances (money sent home by Africans working and living abroad, primarily in Europe and North America), are funding small businesses, schools, and new economic transformations, such as the rapid spread of cell phones.

Figure 7.27 is a map of mobile phone subscriptions by country as of 2018. Between 1998 and 2018, mobile phone subscribers in sub-Saharan Africa increased by more than 100 times, from about 4 million to about 444 million. Because few have smartphones (or computers), access to the web is limited. Smartphone connections are increasing access to the internet, but still stand at only 33 percent. In 2017, for the region as a whole, the mobile industry already directly employed 3 million, contributed around 7.1 percent of GDP through $110 billion of value added, and around $14 billion to government budgets through tax payments.

A number of African innovations have made cell phone service available to the very poor. Prepayment credits are now sold in very small units, and many people now make a living selling such small credits. Other Africans have created ways to charge mobile phones in remote villages where there is no electricity, as the Malawian William Kamkwamba figured out how to do (for more on Kamkwamba's story, see the vignette in the "Energy Needs" section).

REGIONAL ECONOMIC DEVELOPMENT

Many Africans see great potential in regional economic integration similar to that of the European Union (EU). According to studies by the World Trade Organization, if sub-Saharan countries could increase trade with each other by just 1 percent, the region would show a total added income of U.S.$200 billion per year—a substantial internal contribution toward the alleviation of poverty. However, in most years less than 20 percent of the total trade of sub-Saharan Africa is conducted between African countries. This is partly because so many countries produce the same raw materials for export, and everywhere except South Africa industrial capacity is so low that the raw materials cannot be absorbed within the continent. So African countries compete with each other and with all other global producers to sell to the main customers, which are currently in Europe, North America, China, and India. This failure to trade with each other can be traced to divisive colonial policies, lack of transportation and communications grids, and arcane bureaucratic regulations that impede the flow of goods, people, and ideas.

Over the last several decades, a number of regional trading blocs have been formed to encourage neighboring countries to trade with each other and cooperate in the production of manufactured (value-added) exports (**Figure 7.28**). By combining the markets of several countries, much as the EU does, regional trade blocs can create a market size sufficient to foster industrialization and entrepreneurialism. The goals of Africa's many different regional trade blocs (nine distinct blocs as of 2018) include reducing tariffs between members, forming common currencies, reestablishing peace in war-torn areas, cooperating to upgrade transportation and communications infrastructure, and building regional industrial capacity. Building a full-scale, continent-wide economic union along the lines of the EU is a long-term goal.

The goals of regional and local economic integration are being boosted by the creation of **value chains**. These link parts of the production chain of a final product in order to maximize regional or local economic impact. For example, farmers in Endau, Kenya, were able to increase their profits fivefold by switching from growing corn to producing sorghum after a local nongovernmental organization (NGO) negotiated a deal with a brewery in Nairobi to use the sorghum in its beer. Some farmers have used their boosted income to increase their profits further by buying milling equipment and grinding the sorghum of other local farmers into sorghum flour, which is more valuable than unprocessed sorghum. Countries and regions with strong value chains are better able to keep the profits their industries create from

value chains links between various aspects of a production line to maximize efficiency and profits

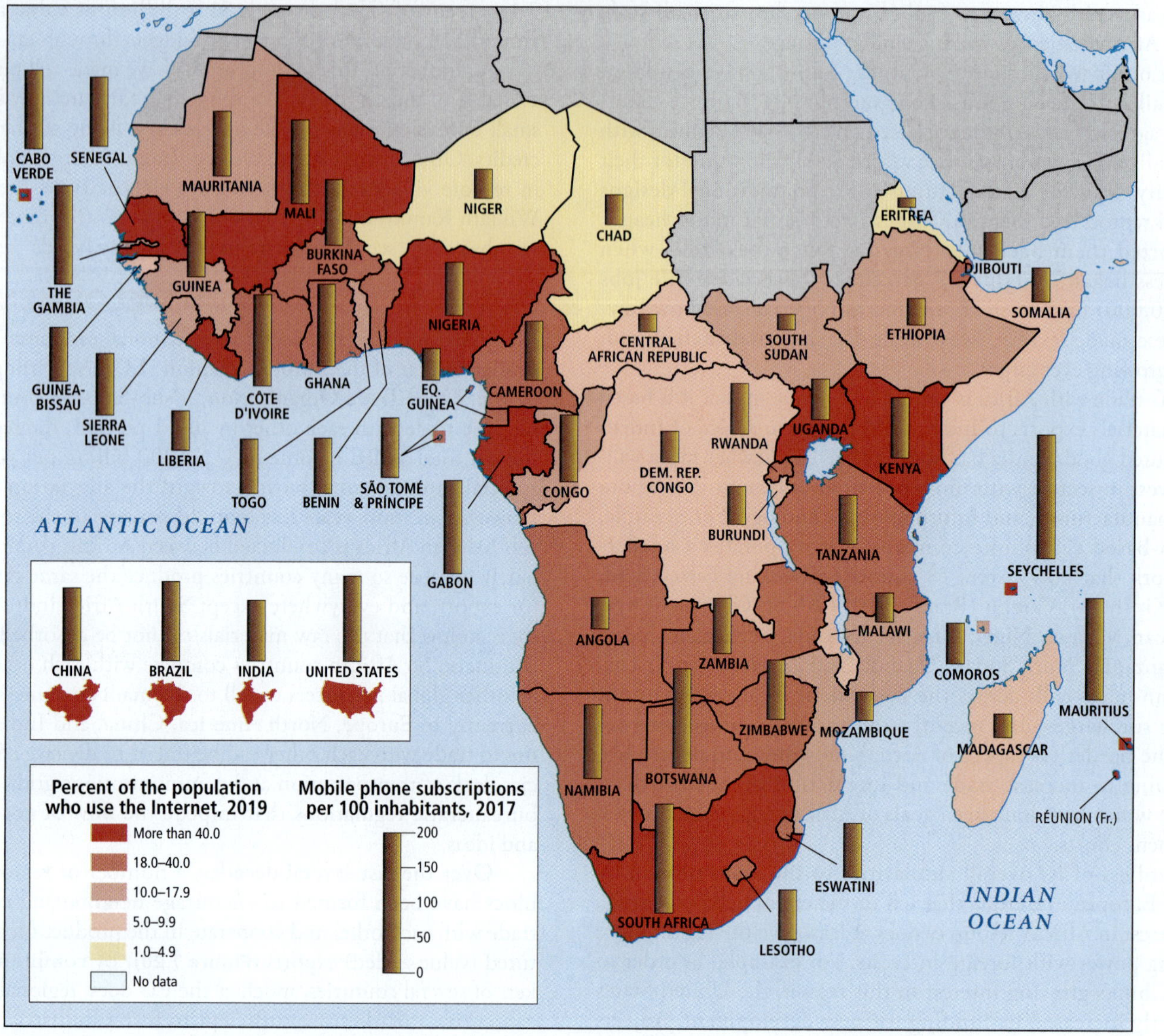

Figure 7.27 Mobile phone users in sub-Saharan Africa. Over the past decade, mobile phone use in sub-Saharan Africa has boomed. It is estimated that as of 2018, there were more than a billion mobile phones in the region. Because of innovations that make it possible to buy service access in tiny amounts, many of the new users will be the very poorest citizens. One of the most popular uses of phones in Africa is to move money from one bank account to another.

going elsewhere. Such value chains are one of the fastest-growing aspects of regional economic development worldwide.

LOCAL ECONOMIC DEVELOPMENT

Hoping to utilize existing talent and consumer demand for African-made products, local agencies and public and private donors have long pursued development designed to foster very basic innovation at the local level that can then be marketed across the region. Often called **grassroots economic development**, projects are designed to use local skills to create products or services for local consumption with simple technology that requires minimal or no investment in imported materials.

grassroots economic development economic development projects designed to use local skills to create products or services for local consumption with simple technology

Control of the projects remains in local hands so that participants retain a sense of ownership and commitment in difficult economic times. Often these are funded with microcredit loans, which makes them self-sustaining (see "Microcredit: A South Asian Innovation for the Poor" in Chapter 8). Despite their potential to increase human development, funding for these projects remains tiny compared to large-scale infrastructure projects that have brought debt and maintained this region's commodity dependence.

Transportation Needs

The issue of improving rural transportation illustrates how a focus on local African needs can generate unique solutions. When non-Africans learn that transportation facilities in Africa are in need of development, they usually imagine building and repairing roads for cars and trucks. But a recent study that analyzed village

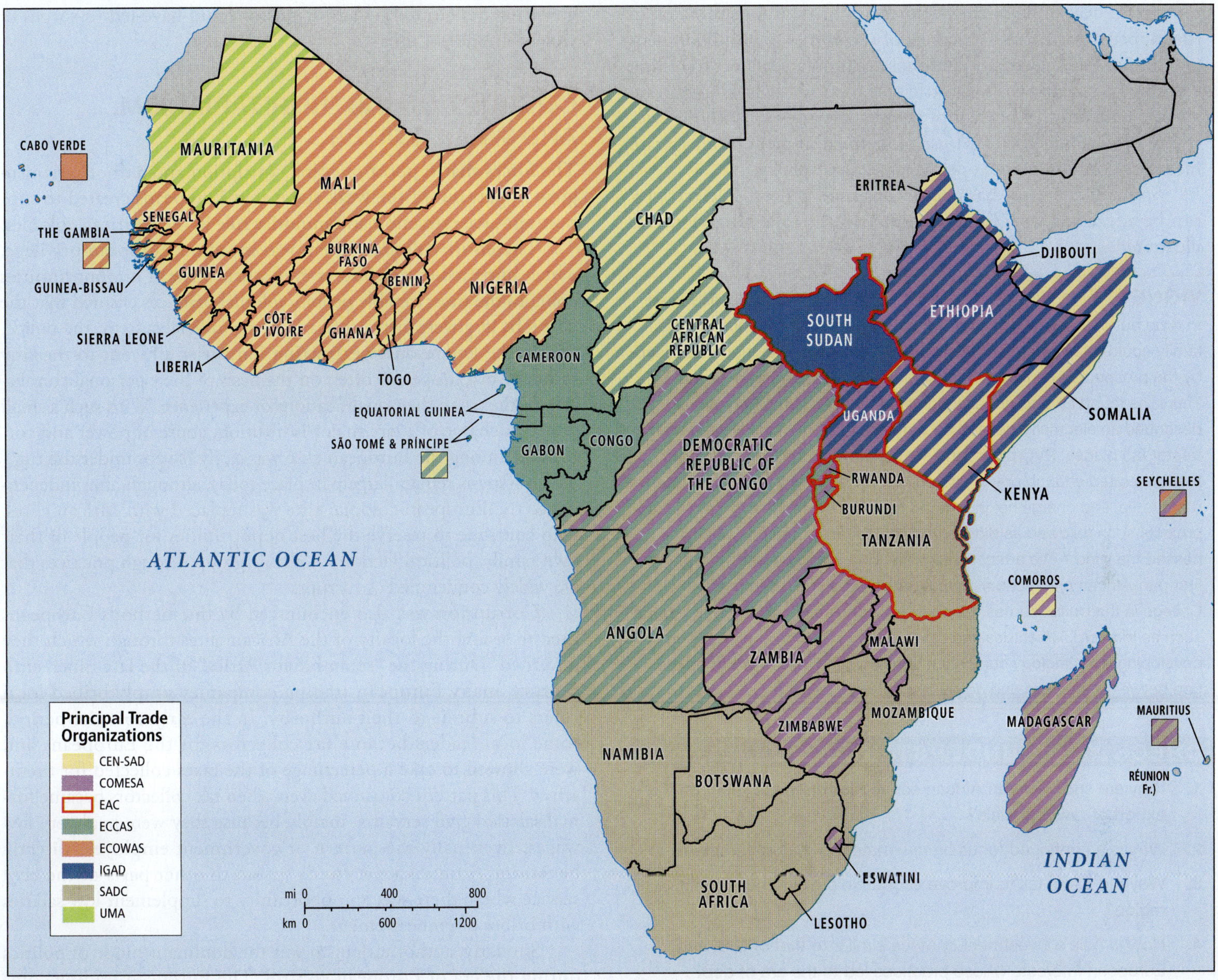

Figure 7.28 Principal trade organizations in sub-Saharan Africa. The confusing pattern of regional trade organizations now operating in sub-Saharan Africa is indicative of the emerging status of internal African trade. These economic communities in sub-Saharan Africa continually change and some countries belong to two or three such organizations. Usually, one country is dominant in an organization. Notice that the Central African Republic and Congo (Kinshasa) belong to several trade groups. Because political stability is essential to trade, some of the strongest trade organizations, such as the Economic Community of West African States (ECOWAS), have become involved in resolving conflicts within and between countries. [Research from: UN Conference on Trade and Development, Economic Development in Africa Report 2013, pp. 9–10, http://unctad.org/en/PublicationsLibrary/aldcafrica2013_en.pdf.]

transportation on a local level found that 87 percent of the goods moved are carried via narrow footpaths on the heads of women. Women "head up" (their term) firewood from the forests, crops from the fields, and water from wells, with the average adult woman spending about 1.5 hours each day moving the equivalent of 44 pounds (20 kilograms) more than 1.25 miles (2 kilometers).

Unfortunately, the often-treacherous footpaths trodden by Africa's load-bearing women have been virtually ignored by African governments and international development agencies, which tend to focus solely on roads for motorized vehicles (which are also badly needed). Grassroots-oriented NGOs are now making far less expensive but equally necessary improvements to Africa's footpaths. Some are installing drainage ditches along footpaths and rehabilitating difficult to traverse sections, while others focus on providing women with bicycles, donkeys, and even motorcycles that can travel on the footpaths. This saves time and energy for the women, who can direct more of their efforts to becoming educated and generating income.

Energy Needs

Africa exports oil, but its own energy needs remain largely unmet. Wood is the most common fuel for cooking, and this accounts for a great deal of damaging deforestation. In all of sub-Saharan Africa, between 50 and 75 percent of the homes (620 million people) do not have electricity, and those that do are subject to rolling blackouts on a daily basis. Electricity is particularly lacking in rural areas, where only 14 percent of people have power. Without electricity medical facilities can't use modern equipment or refrigerate medicines, and industries operate far below capacity.

Some of the most basic levels of home and office electricity can be addressed with local solutions, as the following vignettes illustrate.

VIGNETTE In Malawi, 14-year-old William Kamkwamba was forced to drop out of school when a famine struck his country in 2001 and his family could no longer afford the U.S.$80 school fee. Although depressed by the prospect of no future, he visited the local library when he was able. There, he found an English-language book entitled *Using Energy* that described an electricity-generating windmill. With an old bicycle frame, tractor fan blades, PVC pipes, and scraps of wood, he built a windmill that generated enough power (stored in a car battery) to light his home, run a radio, and charge neighborhood cell phones. More elaborate energy projects soon followed as did international fame as "the boy who harnessed the wind." The attention allowed William to graduate from the first pan-African prep school in South Africa and in 2014 from Dartmouth College in the United States. Since then he has been working with groups such as Ideo.org, which designs development projects, and WiderNet developing a technology curriculum to help young inventors bridge the gap between "knowing and doing." ■

CHECK YOUR UNDERSTANDING

1. Why are sub-Saharan African economies still dependent on the export of raw materials?
2. Why is a continued focus on raw materials exports problematic?
3. Why did SAPs fail to improve economic development in this region?
4. How is Asia's investment in Africa similar to that of U.S. and European interests whose practices led to the era of debt accumulation and SAPs?
5. How have informal economies provided some relief from the hardships created by SAPs?
6. Why does regional economic integration along the lines of the EU hold promise for this region?
7. What aspects of life in this region make grassroots economic development attractive?

POWER AND POLITICS

7.5 Evaluate the factors that have come together to create weak governments in this region.

Despite a shift in sub-Saharan Africa toward more political freedoms, many governments remain authoritarian, nontransparent, and corrupt. Public participation is growing, often due to electronic media, and free and fair elections have brought about dramatic changes in some countries. In other countries, elections and governments tainted by suspicions of fraud have led to surges of violence (**Figure 7.29**).

CORRUPTION, AUTHORITARIANISM, AND WEAK STATES

The forces that support ongoing corruption and authoritarianism grew in part out of the weak states that Europeans created to support resource extraction and minority rule in their African colonies.

The entire colonial system was based on European privilege, which nurtured corruption at every turn. Instead of opportunities going to the most capable people, colonial practices ensured that the best jobs in government, education, the judicial system, the police, and in the key resource extracting industries all went to the tiny minority of Europeans, often on the basis of their personal connections rather than their qualifications or experience. With such a small group in charge of every major institution, abuse of power and corruption among the European elite was easily "swept under the rug." These systems remain largely in place today, although after independence the European colonial elite was replaced with African elites, who continue to reserve the best opportunities for people of their own family, political faction, or ethnic group through practices that are widely condemned as corrupt.

Corruption was also encouraged by the methods Europeans used to secure the loyalty of the African elites through which they governed. During the "scramble for Africa" of the late nineteenth century, many European trading companies simply bribed local rulers to submit to their authority. In the early colonial period, some local leaders became tax collectors for the Europeans and were allowed to take a percentage of the taxes collected for themselves. This pattern continued even when tax collectors became formal salaried civil servants, mainly because they were paid very low wages. Eventually the pattern of government employees directly benefiting from taxpayer funds spread to other parts of the civil service where there was an opportunity to supplement low salaries with bribes or embezzlement.

Similarly, authoritarianism was the dominant mode of politics during the era of European control. The European trading companies were interested much less in investing in human development or providing public services than in extracting resources, and this pattern continued with the colonial governments that took over from them in the early twentieth century. Authoritarianism was the main way that colonial governments functioned, as their power came from distant European countries, not from a body of citizens who willingly submitted to their authority. Faced with the difficult task of governing highly diverse African populations that were thrown together within borders created by the European trading companies, colonial governments regularly denied basic rights to Africans, such as the right to freedom of speech and the press, and the right to protest. Many African governments perpetuate these authoritarian practices, operating as they do within the same borders set down by the Europeans, and having maintained many basic institutions of governments from the colonial period.

Much of the political instability and violence of contemporary Africa relates to the weakness of governments, which tend to be small, underfunded, and less able to provide quality public services, such as roads, utilities, education, and health care. Many are unable to counter periodic challenges to their authority that arise in response to corruption, authoritarianism, and low-quality public services. These weaknesses are in part an outgrowth of the colonial era, when governments were deliberately kept to the bare minimum that was needed for resource extraction. In the independence era, most governments have struggled with internal rebellions that have often lead to civil war, much like their colonial predecessors did. Wars *between* African countries are much less common, in large part because this region's weak states can't afford the cost.

The weakness of African governments has hindered moves away from corruption and authoritarianism and toward greater **transparency** and more political freedoms. Change is often blocked by powerful elites in government, and even when badly needed reforms are enacted, governments often don't have the capacity to implement the reforms. More fair and frequent elections have helped bring more accountability, but they have also brought violence during political campaigns as governments often order police to repress protests with brutal force, or allow rival ethnic and political groups to battle each other in ways reminiscent of the "divide and rule" tactics of the colonial era.

Foreign influences that are often described as *neo-colonial* have been an ongoing force in favor of authoritarianism, corruption, and weak governments in the era of African independence. Many foreign governments, especially the United States, the UK, France, the USSR (before its fall in 1991), and China, will support authoritarian African regimes if it boosts their political influence in a country or facilitates resource extraction. International banks in Europe, North America, and Oceania have been willing to make generous loans to African governments, asking no questions when senior officials ask for help hiding millions or even billions of dollars stolen directly from these loans. This type of corruption effectively counters well-intended efforts by NGOs to strengthen African governing institutions in ways that promote democracy, human development, or better provision of public services.

The Case of Congo (Kinshasa)

The Democratic Republic of Congo is a huge, diverse country with a weak, notoriously corrupt government that has been unable to assert control over large parts of its territory for much of its history. About the size of Europe, its population of 84 million is divided into over 250 ethnic groups, the largest of which is only about 14 percent of the population. Exports of Congo's huge mineral wealth fueled the brutal dictatorship of Mobutu Sese Seko, as well as a civil war that broke out after Mobutu's overthrow, resulting in over 5 million deaths. Congo's weak state has been unable to create the transport infrastructure needed to maintain control over the country, or an educational and health-care system that might have forged a sense of national unity. Even the military is chronically underfunded and at times left to plunder the country. Mobutu famously told his troops, "You have guns, you don't need a salary." Plagued by multiple ongoing conflicts that often break along ethnic lines, Congo has the largest population of internally displaced people in this region at 2.8 million.

Congo took shape in the 1880s as the personal property of Belgium's King Leopold II, who gave concessions to a number of trading companies interested in the area's valuable rubber and ivory resources. Over two decades of legendary brutality followed, with company militias and Leopold's own forces threatening torture, imprisonment, or execution if the population of the Congo Basin failed to pay taxes in the form of rubber, ivory, and other resources. Leopold's state was eventually taken over by the Belgian government, who abandoned the most brutal practices but invested relatively little in the country. When Belgian colonial officials withdrew in 1960, there were only 16 Congolese with a university education in the entire country.

The new head of government, Patrice Lumumba, tried to reduce the huge inequalities of the colonial era by having the government take control of foreign-owned companies (also known as *nationalizing*) in order to keep their profits in the country. This along with the support he received from the Soviet Union provoked the U.S. and Belgian governments, fearful of the growth of communism in Africa, to arrange for Lumumba to be killed and replaced by Mobutu Sese Seko.

Mobutu went on to become one of Africa's most corrupt leaders, taking control of Congo's institutions, businesses, and industries for the benefit of himself and his supporters. As a result, trained technicians and capital left the country, and Congo's economy, physical infrastructure, educational system, health-care system, and food supply rapidly declined. However, due to his willingness to support the United States over the Soviet Union, Mobutu was granted enormous loans by the IMF, even above the protests of the IMF officers in Congo who claimed, correctly as it turned out, that Mobutu would embezzle the money and the loans would never be repaid. By the 1990s, Mobutu had sent between $4 billion and $15 billion abroad to secret bank accounts while the vast majority of the Congolese people were poor, uneducated, and anxious for change.

After years of growing resistance to Mobutu's rule, Laurent Kabila launched a successful coup in 1997 with support from Uganda and Rwanda, who coveted Congo's mineral wealth. Hampered by the decline in Congo's infrastructure and human resources, Kabila was unable to assert control over resource-rich areas in Congo's east, and by 1999, Angola, Namibia, Zimbabwe, Chad, Rwanda, and Uganda all had troops in Congo aggressively competing for parts of its territory and resources. More than 5 million Congolese, including Kabila, died as a result of the conflict, and 1.6 million were displaced.

Ongoing violence, including rape, kidnapping, and mass killings, is linked to the exploitation of mineral resources in border areas far from the capital, Kinshasa, that the Congolese government cannot control. Meanwhile, the Chinese government and the American mining firm Freeport-McMoRan are competing to exploit these resources. Political transitions in Congo are often violent, though this is currently confined to the larger cities that the government controls.

THE ROLE OF GEOPOLITICS

Cold War geopolitics between the United States and the former Soviet Union deepened and prolonged many African conflicts. After independence, some sub-Saharan African governments turned to socialist models of economic development, often receiving economic and military aid from the Soviet Union. Other governments

transparency in politics, the state of being open to observation and participation by the public

Power and Politics in Sub-Saharan Africa

The system of African states that exists today was created to extract resources, not to provide basic services or invest in human development. This focus has led to authoritarian governments that struggle to control large disparate populations with repression and brutality. Meanwhile, corruption has flourished in public institutions that grew out of colonial governments designed to serve the interests of a tiny elite. In the context of these problems, the weakness of these governments, which are generally small, underfunded, and ineffective, has encouraged rebellion and civil war. Greater public participation in politics is beginning to create more accountable governments that are less authoritarian and corrupt; however, elections are often tainted by fraud, and many have led to surges in violence.

THINKING GEOGRAPHICALLY

A In what ways does this picture show both the strengths and limits of political freedoms in Kenya?

B Why might it be significant that Ellen Johnson-Sirleaf was elected as Liberia's president without the help of quotas that reserve certain elected positions for women?

C What does the photo suggest about the strength of this rebel militia and its ability to acquire uniforms and arms?

D What does this photo convey about the general well-being of the people depicted?

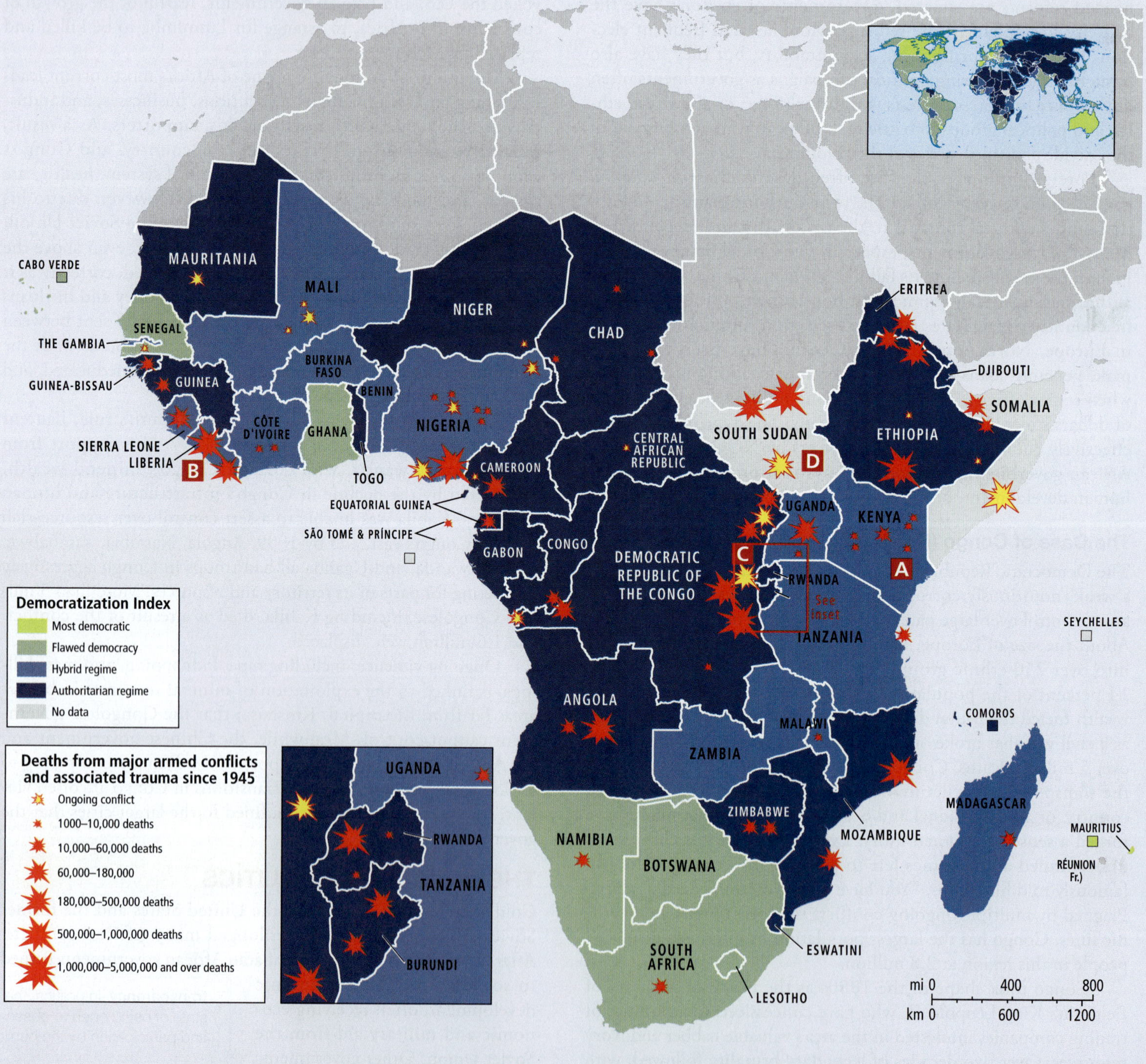

A Kenya. A man in Nairobi, Kenya, runs from police at a demonstration where he has been protesting demands by legislators for an increase in their salaries. That this demonstration could take place at all is a step forward, as is the presence of the media. However, the fact that participants in this nonviolent protest were beaten, tear-gassed, and arrested shows how limited the political freedoms are in Kenya. [TONY KARUMBA/AFP/Getty Images]

B Liberia. Ellen Johnson-Sirleaf of Liberia is the first woman to be elected head of state in this region. Her election in 2005 was considered free and fair by international observers and helped Liberia begin to resolve tensions that had produced two devastating civil wars. She has introduced policies that encourage public participation in government policy making. [Chris Hondros/Getty Images]

C Congo. Members of a rebel militia in eastern Congo that is battling the government for control of the area's metal resources: gold, tin, and coltan (a mineral used in cell phones). [Leon Sadiki/City Press/Gallo Images/Getty Images]

D South Sudan. Refugees from Sudan relocate to South Sudan after a referendum in 2011 established the latter's independence from the former. South Sudan is plagued by political corruption and is involved in continuing hostilities with Sudan; both of these problems have constrained the export of its main resource, oil. [Paula Bronstein/Getty Images]

became allies of the United States, receiving similar aid (see Figure 5.13). Both the United States and the USSR, rather than truly aiding newly independent nations, often tried to undermine each other's African allies by arming and financing rebel groups.

In the 1970s and 1980s, southern sub-Saharan Africa became a major area of East–West tension. The United States aided South Africa's apartheid government in military interventions against Soviet-allied governments in Angola and Mozambique. Another area of Cold War tension was the Horn of Africa, where Ethiopia and Somalia fought intermittently throughout the 1960s, 1970s, and 1980s. At different times, the Soviets and Americans funded one side or the other. The failure by both sides of the Cold War to anchor the aid they dispensed to any requirements that it be used for human development enabled massive corruption. Once this source of cash was removed at the end of the Cold War, many corrupt officials turned to selling off natural resources and commodities.

RADICAL ISLAMISM IN THE SAHEL

Over the last 15 years radical Islamism has grown in the Sahel as a part of conflicts that seemingly have little to do with religion. The conflicts in Sudan and South Sudan between Christian and radical Islamists revolve around control of oil resources. Further east in the Horn of Africa, in northern Uganda, Ethiopia, and Somalia, Al Qaeda–linked Islamists operate economically as organized criminals, but also bomb hotels to scare away foreigners and disrupt civil society. As mentioned briefly in the section "Early Agriculture, Industry, and Trade," above, in religiously mixed Nigeria, a militant Islamist group known as Boko Haram, that occasionally claims links with ISIS or other Muslim militants in Southwest Asia, has attempted to establish itself as an Islamic caliphate in the northern Sahel areas of the country. Boko Haram is largely made up of those who were captured as children a decade or more ago in Sudan, Liberia, Sierra Leone, and elsewhere and forced to become soldiers. Boko Haram continues to steal children, especially schoolgirls, who are then married off to Boko Haram fighters. The newly elected President Buhari of Nigeria, a Muslim from the north, swore to reign in Boko Haram, and in early March 2016, evidence of his success was revealed when dozens of emaciated Boko Haram fighters surrendered to the Nigerian military police.

CONFLICT AND REFUGEES

With only 11 percent of the world's population, this region contains about 19 percent of the world's refugees. If people displaced within their home countries (also known as *internally displaced people*) are also counted, the region has about 28 percent of the world's refugee population (Figure 7.29D). Refugees pour back and forth across borders and within countries, often trying to escape **genocide**, the deliberate destruction of an ethnic, racial, religious, or political group. Women and children constitute 75 percent of Africa's refugees because many adult men who would be refugees are either combatants, jailed, or dead.

As difficult as life is for these refugees, they also place a severe burden on the often poverty-stricken and politically unstable areas that host them. Even with help from international agencies, the host areas find their own development plans deferred with the arrival of so many distressed people who must be fed, sheltered, and given health care. Large portions of economic aid meant to address development needs have been diverted to deal with the emergency needs of refugees.

genocide the deliberate destruction of an ethnic, racial, religious, or political group

SUCCESSES AND FAILURES OF DEMOCRATIZATION

In sub-Saharan Africa, efforts to expand political freedoms have produced mixed results. The number of elections held in the region has increased dramatically. In 1970, only 11 states had held elections postindependence. By 2006, 25 out of the then 44 sub-Saharan African states had held open, multiparty, secret-ballot elections, with universal suffrage. By 2014, every country had held some form of election, though not all could be considered "free and fair" (see the map in Figure 7.29).

While the implementation of democratic elections has increased (see Figure 7.29B), transparency in everyday government actions is often absent. The growth of political freedoms—such as freedom of speech, the freedom to assemble in public, and the ability to participate in policy formation at the local and national levels—has been irregular and uneven. Public frustration with suspicious election outcomes, authoritarian policies, and violent repression of peaceful protests (see Figure 7.29A) has often led to rebellion (see Figure 7.29C).

At times, flawed elections have brought about massive violence. In the Congo (Kinshasa) in 2006, the first elections held there in 46 years brought violence that forced 1 million Congolese people to flee their homes. However, more peaceful elections followed in 2012. Uganda has been the site of repeated elections assessed by outside observers as rigged and fraudulent, with opposition candidates arrested and their supporters intimidated.

Zimbabwe has for decades been a tragic circus of fraudulent elections. In the 1970s, Robert Mugabe became a hero to many for his successful guerilla campaign in what was then called Rhodesia (now Zimbabwe) against a white minority government allied with apartheid South Africa. Mugabe was elected president in 1980, following relatively free multiparty elections. Over the years his policies became more authoritarian and corruption flourished, with many Zimbabweans impoverished while Mugabe became fabulously rich. In the 1990s, he implemented a highly controversial land-redistribution program that took land from white Zimbabweans and redistributed it to his supporters, resulting in food shortages, economic crisis, and political violence that created 3 to 4 million refugees. Mugabe held on to his office through the rigged elections in 2002 and 2008, but was then forced to share power in 2009, after which he returned to power with more rigged elections until he was removed in a coup by the Zimbabwean army in 2017.

In other places, elections have helped end civil wars as the possibility of becoming respected elected leaders has induced former combatants to lay down their arms. In Sierra Leone and Liberia, public outrage against wars created by corrupt ruling elites resulted in elections that brought peace and a change of leadership. Over the long term, fair and regular elections and transparency force governments to be more responsive to the needs of their citizens.

GENDER, POWER, AND POLITICS

One major characteristic of the expansion of political freedoms has been the increase in the number of African women in positions of political power. In Liberia—a country devastated by logging fraud, child-soldier recruitment, and the general brutality of its dictator Charles Taylor—President Ellen Johnson-Sirleaf, a former World Bank economist, began to work on democratic and environmental reforms immediately after she took office (see Figure 7.29B). In Rwanda, racked by genocide and mass rapes in the 1990s, women currently make up 64 percent of the national legislature, the highest percentage in the world. And Rwandan women are also leaders at the local level, where they represent 40 percent of the mayors. In Mozambique, 40 percent of the parliament is female; in Burundi, 31 percent; in Senegal, 43 percent; and in South Africa, 42 percent. Altogether, there are 23 African countries where the percentage of women in national legislatures is well above the world average of 20 percent (the U.S. figure is 23.7 percent). Some of this is a result of quotas that require a larger number of female members of parliament. These policies often reflect commitments to the empowerment of women in the wake of conflicts where women suffered disproportionately. Nevertheless, many female African leaders, including Ellen Johnson-Sirleaf and at least half of Rwanda's female parliamentarians, were elected without the aid of quotas.

CHECK YOUR UNDERSTANDING

1. Why are so many governments in this region authoritarian, nontransparent, and corrupt?
2. What are some consequences of the weakness of African states?
3. What territories has Congo (Kinshasa)'s government found it hardest to assert control over?
4. How do elections contribute to better government?
5. How did the Cold War contribute to corruption?
6. What factors have led to more women in positions of political power?

URBANIZATION

7.6 Evaluate the vast unplanned slums found in this region and the forces that have created them.

Sub-Saharan Africa is undergoing a massive wave of urban growth, much of it unplanned. Cities are growing at around 5 percent per year, the fastest rate of urbanization in the world. Between 50 and 70 percent of the urban population now lives in impoverished slums characterized by crowded fire-prone housing, inadequate sanitation, and poor access to jobs, clean water, and food.

RECENT RAPID URBAN GROWTH

Sub-Saharan Africa is still majority rural (60 percent), and its trend toward urbanization is recent. In the 1960s, only 15 percent lived in cities, compared to 40 percent today (see **Figure 7.30** map) and a projected 50 percent by 2030. In 1960, just one sub-Saharan African city—Johannesburg, South Africa—had more than 1 million people; in 2018, 50 cities did. The largest sub-Saharan African city is Lagos, Nigeria, where in 2018 various estimates put the population at between 14 and 21 million; by 2020 the Lagos state government projects it will have 25 million people. Much growth is taking place in primate cities (see Chapter 3). For example, Kinshasa, Congo, with 12.3 million people, is almost five times the size of Congo's next-largest city, Mbuji-Mayi; Luanda, Angola (7.5 million), is 11 times the size of Lubango; Lagos is twice the size of Onitsha.

Sub-Saharan African cities are unique in that natural increase, not rural-to-urban migration, accounts for the majority of growth. Normally urban life decreases birth rates as women delay childbirth due to better access to education and jobs, and as the cost of educating children makes them more of a financial burden than they are in rural areas, where they are often put to work in the fields at a young age. Urban birth rates are lower than in the rural areas of this region, but they are still unusually high, accounting for over half of urban growth, with migration contributing much less than in other world regions.

Africa's high urban birth rates are related to widespread poverty, a lack of educational and employment opportunities, and high incidence of disease due to inadequate access to health care, clean water, and sanitation (see Figure 7.30A, D). With jobs and schooling hard to come by even in the cities, urban women have fewer reasons to delay childbirth. As in rural areas, poor access to health care in cities means that parents have more children to ensure that some survive into adulthood. All of these factors relate to a lack of planning in the cities of this region.

CAUSES AND CONSEQUENCES OF UNPLANNED URBAN GROWTH

Governments and private investors have not addressed the need for urban jobs or affordable housing for a variety of reasons. Colonial era companies and governments focused on the extraction of raw materials had a more rural focus, and potentially profitable urban manufacturing industries were deliberately discouraged so as not to compete with European manufacturing. As a result, urban employment remained limited and cities stayed fairly small. Colonial governments also feared larger urban populations would support more vigorous political movements for independence. Similarly, many independence era African governments have discouraged urban development for political reasons. With populations still largely rural, many ruling political parties have a rural political base whose needs compete with those of the cities. Because urbanites often support political parties that oppose the rural ruling parties, national governments often resist funding urban planning or the creation and maintenance of urban infrastructure. International lending agencies have also favored rural development programs in order to support raw materials exports that were the basis of loan repayment strategies. For example, from 1992 to 2005 the World Bank lent $14.3 billion to support rural and agricultural investments in this region while spending only $81 million on upgrading slum conditions.

These anti-urban biases, combined with the weakness and corruption of African governments, has produced vast, unplanned, one-story slums that surround older urban centers. Fifty to 70 percent of Africa's urban population now lives in these extremely crowded spaces that they themselves have constructed from found materials.

PHOTO ESSAY

7.30 Urbanization in Sub-Saharan Africa

Urban populations are exploding because of high birth rates in cities and migration from rural areas. By 2030, half of the region's population will be urban, most living in vast, unplanned, one-story slums that are plagued by violence and inadequate access to water, food, sanitation, and education. A lack of attention to urban issues results from governments focused on the extraction of raw materials, which are usually in rural areas. Potentially profitable urban manufacturing industries were discouraged during the colonial era so as not to compete with European manufacturing. This continued during the independence era with international lending agencies favoring rural development, instead of urban development, in order to facilitate resource extraction and loan repayment. These anti-urban biases have produced chaotic and violent cities where many people survive through the informal economy. [Data from: 2018 Population Data Sheet and Demographia World Urban Areas 2018:01.]

THINKING GEOGRAPHICALLY

A What do the types of houses, the water, and the boats suggest about the location of this slum?

B What are some advantages of urban gardening?

C How do gangs influence politics?

D How might the lack of clean water and sanitation create an incentive for people to have large families?

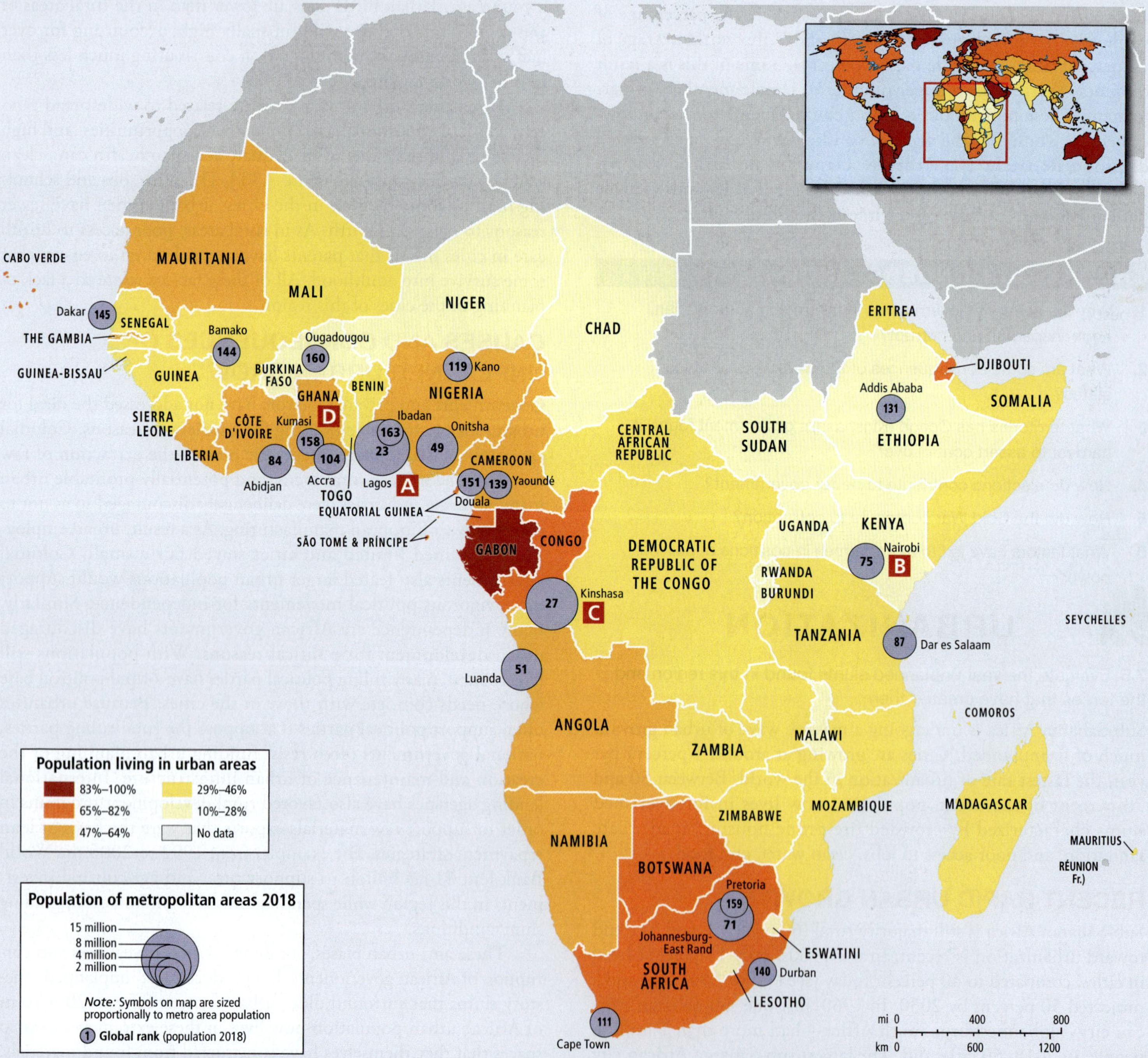

A Lagos, Nigeria. A mother navigates through the Makoko slum in Lagos, Nigeria, the largest city in this region and one of the fastest growing in the world. Makoko is home to more than 100,000 people. Waterborne diseases such as malaria are common, as is flooding, which kills and displaces residents on a regular basis. Because of these hazards and the fact that the community sits on potentially valuable waterfront property, the municipal government of Lagos periodically evicts some of Makoko's residents and demolishes their homes. [PIUS UTOMI EKPEI/AFP/Getty Images]

B Nairobi, Kenya. A "sack garden" is one of many innovative self-help projects found in the Kibera slum in Nairobi, Kenya. High food costs and high unemployment rates make such urban gardening attractive to many African city dwellers. [TONY KARUMBA/AFP/Getty Images]

C Kinshasa, Congo. Opposing gangs battle for control of territory in a slum in Kinshasa, Congo. With police often unwilling to enter slums, gangs are free to engage in robbery, theft, human trafficking, the drug trade, and other illegal activity. [Junior D. Kannah/AFP/Getty Images]

D Accra, Ghana. A young man drinks "sachet water" during a political protest in Ghana. Bagged water is popular among urban dwellers who do not have easy access to clean water. [PIUS UTOMI EKPEI/AFP/Getty Images[

Basic sanitation is absent, as is access to clean water and nourishing food. Transportation is provided by a jumble of buses, vans, motorcycles, and boats in waterfront locales (see Figure 7.30A). Parents frequently have to travel long hours through extremely congested traffic to reach distant jobs, getting much of their sleep while sitting on a crowded bus and leaving their children unsupervised and susceptible to the influence of gangs (see Figure 7.30C).

Uncontrolled urbanization has many destabilizing effects in this region. With manufacturing industries stunted and commodity export economies unstable, many slum dwellers seek employment in the informal economy, which is itself undependable and sometimes dangerous. Gangs control many aspects of the informal economy in slum areas, such as the provision of electricity stolen from powerlines and distributed to individual homes. Gang members and other slum dwellers have also become a volatile political force responsible for much of the violence that accompanies hotly contested elections.

Slums also encourage what geographer Deborah Potts describes as *circular migration*, or back-and-forth migration between rural and urban areas. People generally move from the countryside to cities for jobs and educational opportunities. However, the volatility of many African commodity-based economies, combined with the low living standards found in many slums, means that many urban migrants move back to the countryside at least temporarily. This contributes to overall instability in the region, with millions of lives in constant flux.

The Scarcity of Clean Water in Sub-Saharan Cities

Most water distribution systems in African cities are underdeveloped, serving only public spigots in a few areas and individual homes only in the wealthier neighborhoods. These systems are susceptible to contamination by harmful bacteria from untreated sewage and garbage. Only the largest sub-Saharan African cities have sewage treatment plants, and few of these extend to the slums that surround them. The result is frequent outbreaks of waterborne diseases such as cholera, dysentery, and typhoid. It is estimated that unsafe water and sanitation facilities result in an annual loss of U.S.$28.4 billion from illness and premature death in Africa alone.

Sachet Water: Effective Urban Adaptation or Plastic Plague?

Half-liter bags of chilled purified water have become part of the urban landscape across Africa (see Figure 7.30D). Conceived of by a dentist from the United States visiting El Salvador, the idea of distributing water in heat-sealed plastic bags filled at a reliably clean source has become a widespread practice. Geographer Sara Beth Keough and anthropologist Scott Youngstedt studied how sachet water is impacting Niamey, Niger, finding that people buy the water because of its perceived purity, but that sachets pose a greater health risk once they are discarded. Niamey lacks consistent waste collection, so wealthy neighborhoods pay private workers to collect sachets and other garbage, which they then relocate to open areas in poorer neighborhoods, or on the urban periphery. In the rest of Niamey sachets tend to clog sewer drains, increasing the risk of the very waterborne diseases that the sachet water is meant to reduce. Recycling the sachet bags requires specialized equipment lacking in Niger and much of Africa.

Food Security in Urban Areas

Maintaining adequate food supplies in Africa is complicated by the lack of refrigerated storage facilities, transportation that is unreliable and expensive, and supplies that are unpredictable. Increasingly, food is shipped into sub-Saharan cities from distant parts of the continent or other world regions, but such food is often expensive. Some urban Africans are harking back to their agricultural roots and beginning to produce their own food in the tiny spaces between houses and in the wastelands that surround urban shantytowns.

The growing of produce in and around cities, on even the tiniest scraps of land, has an advantage over both rural market gardening and imported food in supplying urban people with safe, nutritious food. Fruits and vegetables can be highly perishable, so if urban consumers can produce their own food, they are likely to eat better, to create less waste, and to experience lower food transport costs. Also, food waste can be immediately recycled as compost, reducing or eliminating the need for fertilizers; the greenbelts created by urban gardens can reduce urban air and water pollution.

If urban gardening is to become a viable food security solution, however, some adjustments will be necessary. At present, urban farmers often use brownfields—land where the soil is contaminated by past industrial activities. Moreover, urban farmers frequently use wastewater that may not be safe to irrigate their crops. But these problems can be corrected. Raised cultivation beds can be sealed off from the contaminated soil of brownfields, and for irrigation, rainwater can be harvested off nearby roofs, or, alternatively, wastewater can be treated before it is used.

Because few urban gardeners own the land they cultivate, their gardens can be confiscated for development without warning or compensation. One adaptation to this uncertainty is "sack gardens" (see Figure 7.30B) that can be relocated if necessary. Finally, to avoid expensive fertilizers and pesticides, urban farmers often compost food waste and use organic techniques to control pests.

CHECK YOUR UNDERSTANDING

1. What is unique about urban growth in this region?
2. What factors have come together to create the vast slums found throughout Africa?
3. What are some destabilizing effects of unplanned urbanization in this region?
4. What is circular migration?
5. In what ways is sachet water both a solution and a problem for urban public health?
6. How do some urban residents increase their food security?

POPULATION, GENDER, AND CULTURE

7.7 Explain the factors slowing population growth in this region.

7.8 Describe how educated career women are having an impact on this region.

7.9 Contrast the various religious traditions that have influenced this region.

Africa is undergoing massive and rapid change, as populations grow, gender norms change, and cultures transform. Of all the world regions, populations are growing fastest in sub-Saharan Africa, yet growth rates are slowing as the countries proceed through stage two of the demographic transition (see Figure 1.28). Investments in human development are reducing incentives for large families, as are greater economic and educational opportunities for women. This last shift also challenges cultural traditions upholding male dominance throughout the region. The rapid pace of change is also transforming many other aspects of the region's cultures, including religion, language, and ethnic identity.

POPULATION GROWTH

In a little over 50 years, sub-Saharan Africa's population has multiplied by five, growing from about 200 million in 1960 to 1.05 billion in 2018. By 2050, the population of this region could reach just over 2 billion. Africa as a whole is much less densely populated than most of Europe and Asia (Figure 1.26), but places that are relatively uncrowded now may change dramatically over the next few decades (**Figure 7.31**). In some rural areas, the population density may far exceed the carrying capacity of the land, which could lead to widespread migration to urban areas.

Sub-Saharan Africa's average fertility rate (five births per adult woman) is the highest in the world in part because many view children as both an economic advantage and a spiritual link between the past and the future. Children and young adults can perform important work on family farms, and in this region of generally poor health care parents often have extra children in the hope of raising at least a few of them to adulthood.

Five countries in this region have gone through, or are approaching, stage three of the *demographic transition*—the sharp decline in births and deaths that accompanies higher human development (see Figure 1.28)—South Africa, Botswana, Seychelles, Réunion, and Mauritius (the last three are small island countries off Africa's east coast). Circumstances have changed enough to make smaller families desirable and attainable: In all five countries human development has risen to high or medium-high levels and per capita incomes (GNI) are three to ten times the sub-Saharan

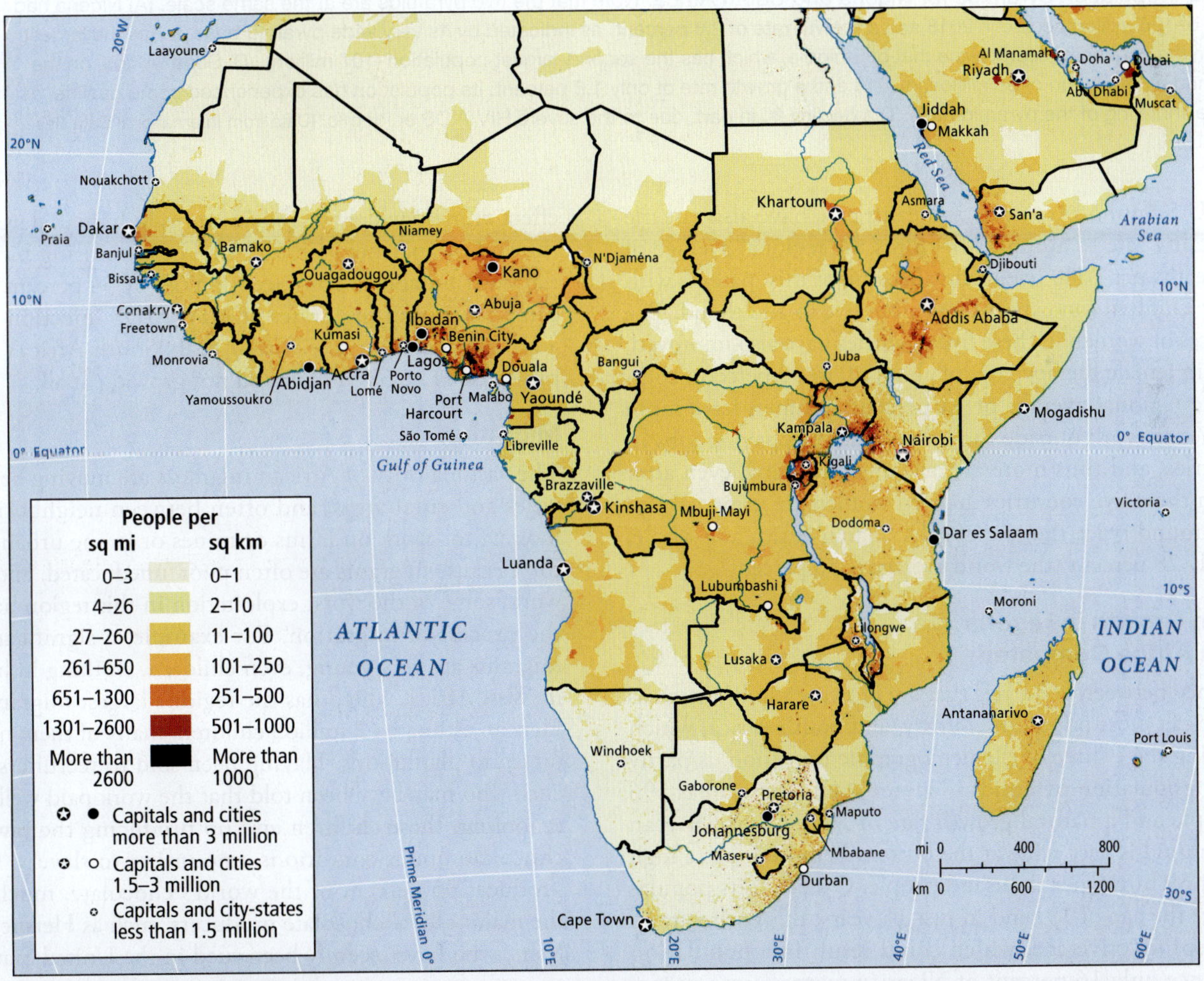

Figure 7.31 Population density in sub-Saharan Africa. Of all the world regions, populations are growing fastest in sub-Saharan Africa. Major concentrations include the Niger Delta and the highlands around the Great Rift Valley. Most countries are in stage two of the demographic transition, though a few of the more prosperous countries have entered stage three (see Figure 1.28). Investments made in human development and more opportunities for women are beginning to slow population growth throughout the region. [Research from: Deborah Balk, Gregory Yetman, et al., Center for International Earth Science Information Network, Columbia University, http://www.ciesin.columbia.edu.]

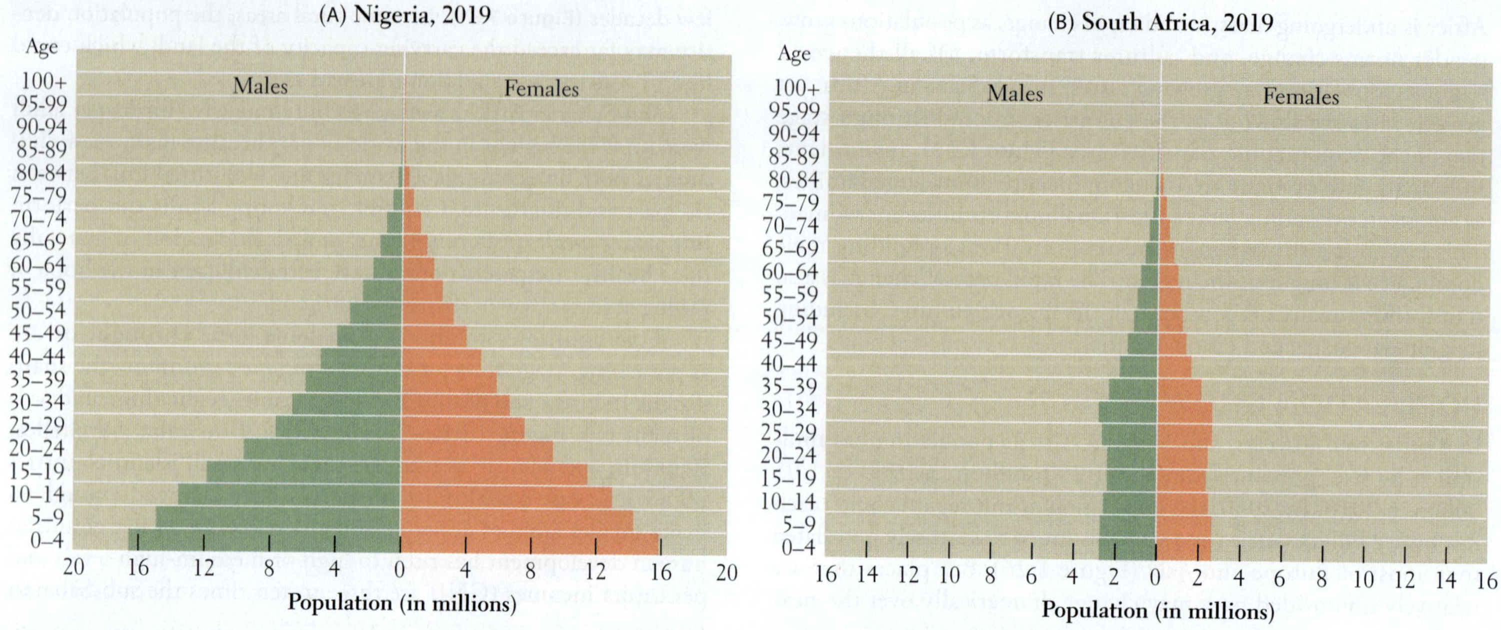

Figure 7.32 Population pyramids for Nigeria and South Africa. Note that the two pyramids are at the same scale. (A) Nigeria had a population of 196 million people in 2018 and a growth rate of 2.6 percent, as indicated by its very wide pyramid base. It has the largest population on the continent, nearly twice that of Ethiopia, which has the second-largest population (107 million). (B) South Africa, on the other hand, had a population of 58 million in 2018 and a growth rate of only 1.6 percent; its population has experienced some decline, as shown by the shrinking of the pyramid base. The decline is, in part, due to the severe HIV/AIDS epidemic. [Data from International Data Base, U.S. Census Bureau.]

average of U.S.$3683. Advances in health care have also cut the infant mortality rate to around two-thirds the regional average of 54 infant deaths per 1000 live births. Because of this, parents with two or three children can expect them all to live to adulthood. The circumstances of women in these five countries have also improved, as reflected in female literacy rates of about 80 to 90 percent, compared to the regional average of 54 percent. Women have more opportunities to work outside the home at decent-paying jobs in these countries, and thus more are delaying childbirth even after marriage. In these five countries married women are using contraception at around twice the rate of sub-Saharan Africa as a whole, which is only 28 percent (the world average is 56 percent).

Population Growth in Nigeria and South Africa Compared

The difference between countries that are growing rapidly, such as Nigeria, Africa's most populous country, and South Africa, which is approaching stage three of the demographic transition, is partly captured by population pyramids in **Figure 7.32**. Nigeria has 196 million people and a natural growth rate of 2.6 percent per year. Nigeria's pyramid is very wide at the bottom because it has a high birth rate of 39 births per thousand people. Over half its population is under the age of 19, and in just 15 years, this entire group will be of reproductive age, which could result in a population explosion since only 16 percent of Nigerian women use any sort of birth control. In 40 years Nigeria's population is projected to be almost half a billion and still growing rapidly. By contrast South Africa has 55 million people and a natural growth rate of 1.6 percent. In contrast to Nigeria, South Africa's pyramid has contracted at the bottom because its birth rate has dropped from 35 per 1000 to 21 per 1000 over the last 20 years. This decrease is primarily an effect of more widely shared advances in human development since the end of apartheid (in 1994). The decreasing birth rate is partially attributable to contraception use by 54 percent of South African women, but high rates of HIV/AIDS infection among young adults also plays a role. In 40 years South Africa's population is projected to be 67 million and not growing at all.

Migration

The vast majority of African migrants are moving between African cities and rural areas, and often between neighboring countries, looking for work on farms or mines or in the urban informal sector. Because migrants are often poor, uneducated, and desperate for work, some of the worst exploitation in this region happens during the process of migration. For example, a significant number of migrants are very young, even children, working in hazardous jobs in West Africa, which has the region's largest migrant population. Here an estimated 2 million children, many of them migrants, work on cacao plantations. Having been sold to recruiters by their parents, who may have been told that the work paid well and involved schooling, these children end up producing the raw material for chocolate under conditions amounting to slavery. Côte d'Ivoire produces 35 percent of the world's chocolate, much of it sold to the major global chocolate producers, such as Hershey, Nestle, and Mars, who have recently been sued in the United States for buying cacao from plantations known to use enslaved migrant children.

Those sub-Saharan migrants who leave the continent to make their way to North Africa and then Europe number in the hundreds of thousands. Overwhelmingly young males who are ambitious, educated, and physically fit, they represent a loss of human capital to their home communities. Those who do find work usually send money home to their families, boosting this region's overall economic stability. However, these migrants are also subject to danger and exploitation. Hundreds die each year trying to cross the Mediterranean while thousands more are held in a form of debt slavery by gangs and organized crime once they reach Europe. Happily, an increasing number of young migrants who do find well-paying jobs in Europe eventually return home to start development projects.

POPULATION AND PUBLIC HEALTH

Diseases have long had transformative effects on this region, shaping human interactions with the environment in complex ways. For example, people living between the 15th parallels north and south of the equator are exposed to "sleeping sickness" (trypanosomiasis), which is spread among people and cattle through the bites of tsetse flies. The disease attacks the central nervous system and results in death if untreated. Several hundred thousand Africans suffer from sleeping sickness, and most of them cannot afford the expensive treatment. But the disease also affects cows, making animal herding unproductive in much of the region. Conservationists, who often struggle to keep herders from encroaching on the habitat of endangered wildlife populations, have at times praised the tsetse fly and trypanomiasis as the best "game warden" Africa has. This perspective however diminishes the human tragedy that sleeping sickness takes as well as the role that rural Africans can play in maintaining wildlife populations.

Infectious diseases are by far the largest killers in sub-Saharan Africa, responsible for over 50 percent of all deaths. Infections of the lungs kill around 1 million people a year, mostly due to smoke inhalation from fires used for cooking and heating in the homes of poor people. HIV/AIDS is the next largest killer at around 700,000 people a year (see discussion below), followed by diarrheal diseases (600,000 a year), which are caused primarily by lack of access to clean water and adequate sanitation. Tuberculosis, a lung infection that many HIV-positive people contract, kills 400,000, as does malaria, a tropical mosquito-borne disease that is particularly deadly for children under the age of 5. Because mosquitoes lay their eggs in standing fresh water, the incidence of malaria has increased with the construction of dams and irrigation projects.

Until recently, relatively little funding was devoted to controlling these diseases. Now, however, major international donors are funding research into alternatives to wood-burning cook stoves (which can reduce lung infections), affordable HIV medications, and a vaccine that will prevent malaria in most people.

HIV/AIDS in Sub-Saharan Africa

The leading cause of death for African women of reproductive age is acquired immunodeficiency syndrome (AIDS), caused by the human immunodeficiency virus (HIV) (**Figure 7.33**). Sub-Saharan Africa, with 14 percent of the world's population, has 70 percent of all people living with HIV, a total of around 25.7 million people. The epidemic has led to more than 15 million orphaned children, many of whom have no family left to care for them or to pass on vital knowledge and life skills (see Figure 7.36A). Millions of parents, teachers, skilled farmers, craftspeople, health-care workers, and other trained professionals have been lost. As a result of systematic local, national, and international efforts the epidemic has declined significantly in the past two decades with new HIV infections cut in half since 2001 largely due to increased education about prevention. Treatment has also become more affordable with antiretroviral therapy drugs that cost $10,000 per year in 2000 now costing around $75 per year. This is due to a combination of subsidies for treatment funded by the UN, governments in the region, and private philanthropists, as well as cheap generic drugs from pharmaceutical companies in India and China. As a result, 60 to 80 percent of Africans with HIV are now living much longer, healthier lives, and fewer children are orphaned. So far, however, the goal of providing treatment to everyone with HIV has proved elusive, and around a million new infections occur every year.

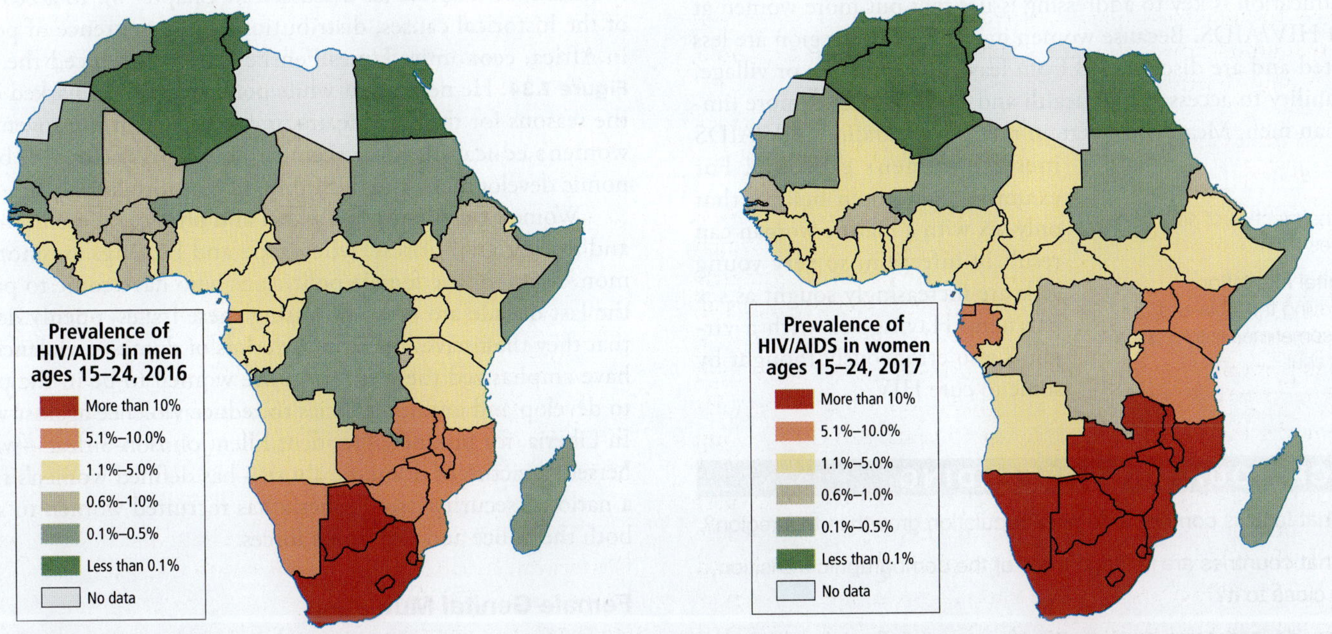

Figure 7.33 HIV/AIDS: male versus female prevalence. HIV/AIDS is more prevalent among women than men, especially in Southern Africa. [Data from: UNAIDS, Global Report 2018 Fact Sheet, http://www.unaids.org/sites/default/files/media_asset/unaids-data-2018_en.pdf.]

HIV and Gender

Women in this region are at greater risk of HIV infection for a number of cultural and economic reasons. Worldwide, women account for half of all people living with HIV, but in sub-Saharan Africa, 59 percent of all people living with HIV/AIDS are women. In Botswana, the richest country in continental Sub-Saharan Africa, nearly 26 percent of the adult female population is living with HIV (compared to 17 percent of men).

Many women are exposed in cities, where HIV is more prevalent, or as a result of circular migration between urban and rural areas. Low-income female urban dwellers and migrants are at a high risk of infection, especially if they become employed in the sex industry in order to survive economically. In some cities, virtually all sex workers are infected, in part because they are frequently subject to sexual violence, which increases the risk of HIV infection. Sexual violence also contributes to higher levels of HIV among women throughout the region.

Circular migration has hastened the spread of HIV, as many urban men, especially those with families back in rural villages, visit sex workers on a regular basis, bringing HIV back to their unwitting wives. As transportation between cities and the countryside has improved, bus and truck drivers have also become major carriers of HIV to rural villages.

HIV/AIDS Education

The most effective means of prevention are massive education programs that lower rates of infection among those who can read and understand explanations about how HIV is spread. For example, Senegal started HIV/AIDS education in the 1980s, and the levels of infection there have so far remained low. Major education campaigns in Uganda lowered the incidence of new HIV infections from 15 percent to just 4.1 percent between 1990 and 2004. By contrast, infection rates soared in places such as South Africa, where a few top politicians put forth untenable theories about the causes of HIV infection or denied that HIV/AIDS was a problem at all.

Education is key to addressing issues that put more women at risk of HIV/AIDS. Because women in much of this region are less educated and are discouraged from leaving their home or village, their ability to access sexual health and HIV services is more limited than men. Meanwhile, certain myths surrounding HIV/AIDS increase women's exposure. For example, some men believe that only sex with a mature woman can result in infection, so very young girls are increasingly sought as sex partners. Having sex with a virgin is also erroneously thought by some to cure HIV.

polygyny the practice of having multiple wives

female genital mutilation (FGM) removing the labia and clitoris and sometimes stitching the vulva nearly shut

CHECK YOUR UNDERSTANDING

1. What factors contribute to high population growth in this region?
2. What countries are in stage three of the demographic transition, or close to it?
3. How do increasing opportunities for women influence population growth?
4. Why are migrants in this region particularly vulnerable to exploitation?
5. What type of disease is the greatest killer in this region?
6. What factors make women in this region more likely than men to contract HIV?

GENDER ISSUES

Practices that guide how males and females relate to each other come from a variety of sources in Africa, all of which are undergoing cultural change and transformation. Especially in rural areas gender relationships are more socially controlled, with most marriages based on alliances between families, not "romantic" or personal attraction. Husbands and wives spend most of their time doing their tasks with family members of their own sex rather than with each other, thanks to a gendered division of labor and responsibilities. In most cases women are responsible for domestic activities, including raising the children, attending to the sick and elderly, and maintaining the house. Women collect water and firewood and prepare nearly all the food. Men are usually responsible for preparing land for cultivation. In the fields intended to produce food for family use, women sow, weed, and tend the crops, as well as process them for storage. In fields where cash crops are grown, men perform most of the work and retain control of the money earned. When husbands in search of cash income for the family migrate to work in mines or in urban jobs, within Africa or beyond, women take over nearly all of the agricultural work, often using simple hand tools, not machines or even draft animals. When there are small agricultural surpluses or handcrafted items to trade, it is women who transport and sell them in the market. Throughout Africa, married couples often keep separate accounts and manage their earnings as individuals, but when a wife sells her husband's produce at the market, she usually gives the proceeds to him.

It is important to note that the practice of men having multiple wives—**polygyny**—is more common in sub-Saharan Africa (where it has ancient pre-Muslim, pre-Christian roots) than in Muslim North Africa (as discussed in Chapter 6). In a 2012 study of the historical causes, distribution, and persistence of polygyny in Africa, economist James Fenske (2013) produced the map in **Figure 7.34**. He noted that while polygyny is on a marked decline, the reasons for this decline are unclear. Current improvements in women's education don't seem to reduce its incidence, but economic development and declining infant mortality do.

Women face significant sexual and physical abuse in the home and during civil unrest, when rape and beatings are more common. Many of the female politicians who have come to power in the last decade are now addressing these issues, openly declaring that they themselves have been victims of domestic violence. Most have emphasized the need for more women to be in the position to develop and enforce policies to reduce violence against women. In Liberia, for instance, President Ellen Johnson-Sirleaf—who was, herself, a victim of domestic abuse—has defined women's rights as a national security issue. Liberia has recruited women to serve in both the police and the armed forces.

Female Genital Mutilation

A practice known as **female genital mutilation (FGM)** (formerly called female circumcision) is common in parts of this region. It

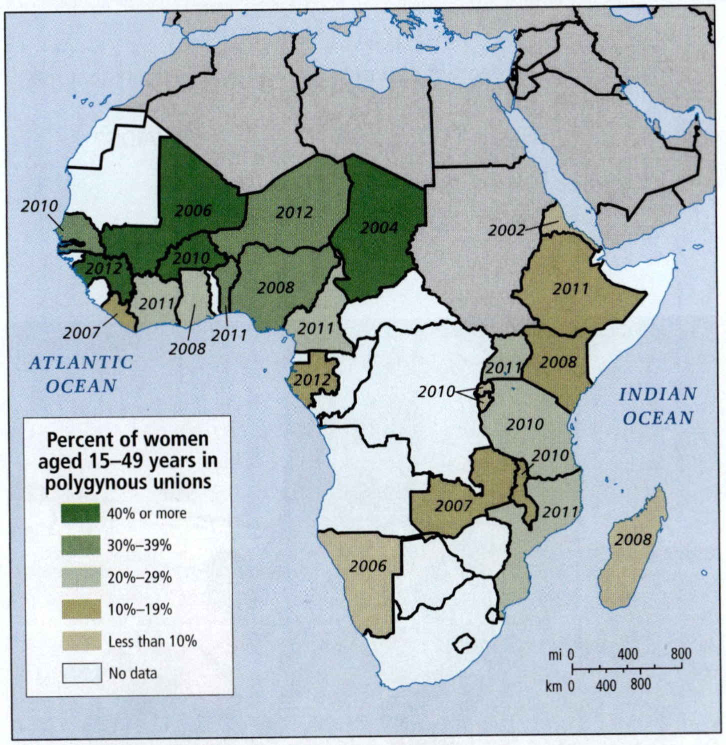

Figure 7.34 Polygyny in Africa. The map depicts the spatial distribution of polygyny in sub-Saharan Africa for various years from 2004 to 2012. Region-wide, 72 percent of sub-Saharan women indicated they are their husband's only wife but that their spouses have had relations with other women, 19 percent are in two-wife marriages, 7 percent are in three-wife marriages, and the remaining 2 percent are in marriages with four or more wives. Polygynous unions are more common in rural areas, where women are less educated and child survival is lower. Often polygynous husbands are much older, better educated, and of a higher social status than their wives (see Figure 7.36C). [Research from: James Fenske, "African Polygamy: Past and Present," Editorial Express, February 15, 2012, https://editorialexpress.com/cgi-bin/conference/download.cgi?db_name =CSAE2012&paper_id=115; data from the Demographic and Health Surveys (DHS) Program STAT compiler (DHS, 2014), published by the UN in The World's Women 2015, p. 17, Figure 1.14.]

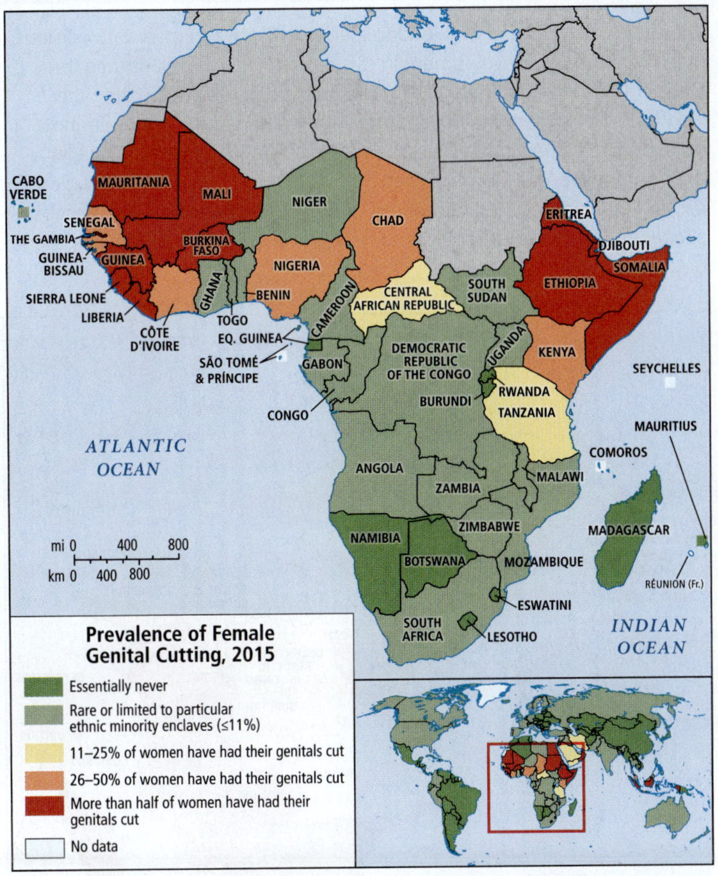

Figure 7.35 Percent of girls and women living in African countries who have undergone FGM. The World Health Organization estimates that there are 91.5 million girls and women living in Africa today who have undergone some form of FGM. Of those, 12.4 million are girls between 10 and 14 years of age. [Data from: Prevalence of female genital mutilation/cutting (for relevant countries only) at http://www.womanstats.org/substatics/femalegenitalcutting_2015_2correctstatic.png.]

involves the removal of a young girl's entire clitoris and parts of her labia, usually without anesthesia. In the most extreme cases (called infibulation), her vulva is stitched nearly shut. This mutilation far exceeds that of male circumcision, eliminating any possibility of sexual stimulation for the female and making urination and menstruation difficult. It has no known medical benefit; leaves women more susceptible to infection, especially HIV; and makes intercourse painful and childbirth dangerous because the flesh scarred by the mutilation is inelastic. A 2006 medical study conducted with the help of 28,000 women in six African countries showed that women who had undergone FGM were 50 percent more likely to die during childbirth.

Documented mostly in the Sahel but also elsewhere (**Figure 7.35**), FGM predates Islam and Christianity and is performed among all social classes in Christian, Muslim, and animist (see the definition in "Indigenous Belief Systems") religious traditions. Traditionally, the practice is viewed as an essential part of a girl's passage into womanhood and probably originated to ensure that a female was a virgin at marriage and that she would be faithful to her husband due to a low interest in intercourse beyond procreation.

A growing number of African leaders consider FGM an extreme human rights abuse. It is outlawed in 16 countries and in decline almost everywhere. The most successful campaigns to eradicate FGM are those that emphasize the threats it poses to a woman's health, thus making it socially acceptable to not undergo this ritual. Among the Kikuyu of Kenya, nearly all females would have had FGM 40 years ago, but today most opt instead for right-of-passage ceremonies that joyously mark a girl's transition to puberty. Similarly, the Maasai now perform a rite of passage that includes dressing in elaborate beaded dresses and participating in ritual bathing instead of FGM (**Figure 7.36B**).

Lesbian, Gay, Bisexual, and Transgender Rights

The rights of lesbian, gay, bisexual, and transgender (LGBT) people are not well protected in any country in this region. Violent homophobia is relatively common, and most LGBT people choose to stay "closeted." Many countries have laws against homosexual

7.36 Population, Gender, and Culture in Sub-Saharan Africa

There are many factors that lead to inequality between men and women in terms of human development in this region. The map illustrates the Gender Development Index (GDI), which is created by estimating the Human Development Index separately for women and men. The ratio of these two values is the GDI. The closer the ratio is to 1, the smaller the gap between women and men. The map shows most of this region ranking fairly low on GDI compared to the rest of the world. The photos show some factors that contribute to higher gaps in human development between men and women.

THINKING GEOGRAPHICALLY

A What are some factors that put African women at a higher risk of contracting HIV/AIDS?

B How does FGM affect the risk of contracting HIV/AIDS?

C Where is polygyny most common in this region?

A HIV/AIDS. In South Africa, boys and girls play at a day-care center for children who have lost their parents to HIV. Women in this region are more likely than men to have HIV, which reduces female life expectancy, leading to a greater gap in human development between men and women. [Melanie Stetson Freeman/The Christian Science Monitor via Getty Images]

B Alternatives to FGM. Maasai women in Kenya gather to discuss alternative rites of passage that can replace female genital mutilation. FGM leads to complications in childbirth that lower the life expectancy of women. Less-educated mothers are more likely to encourage FGM for their daughters, who are also more likely to be taken out of school and married after enduring the practice. [Jonathan Torgovnik for The Hewlett Foundation/Reportage by Getty Images]

C Polygyny. The wives of a chief in Cameroon perform a traditional dance. Polygyny is associated with gender gaps in human development because less-educated women are more likely to be in polygynous marriages, as are better-educated men. [Eric TRAVERS/Gamma-Rapho via Getty Images]

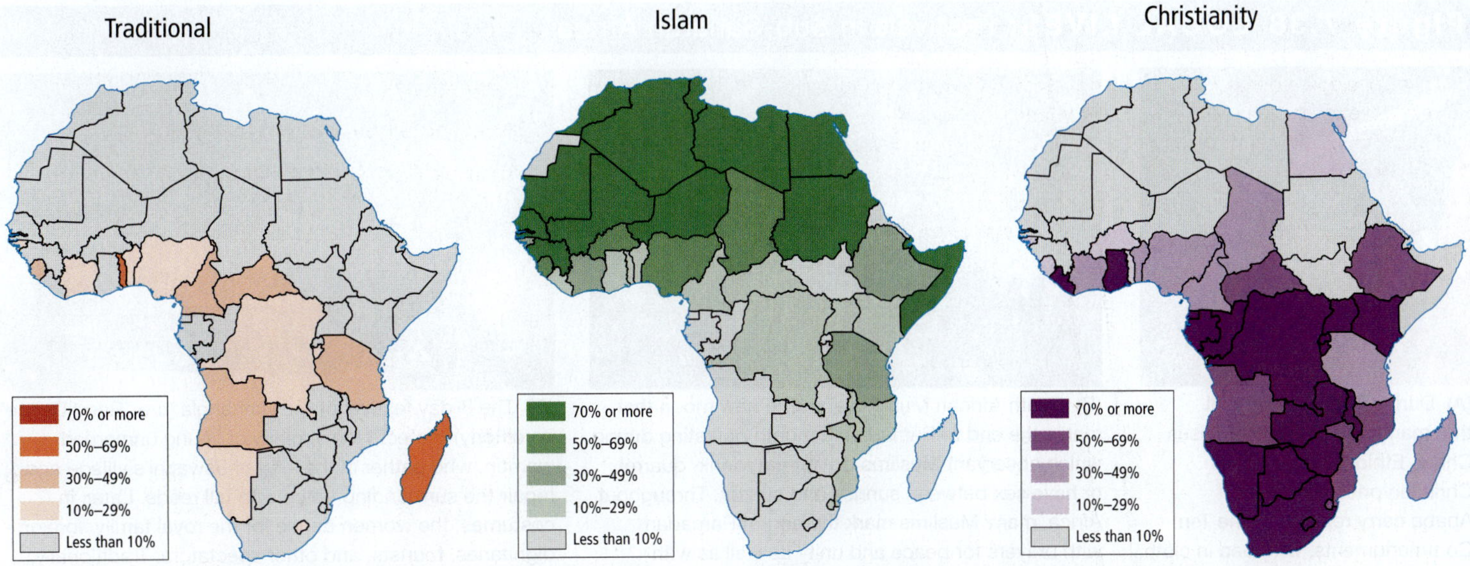

Figure 7.37 Religions in Africa. Notice that the various religions in Africa overlap in distribution; in many countries, one is dominant but others are present. [Research from: Matthew White, "Religion in Africa," in *Historical Atlas of the Twentieth Century*, October 1998, http://users/erols.com/mwhite28/aforelg.htm; revised with new data from the CIA, The World Factbook, 2009, https://www.cia.gov/cia/publications/factbook/index.html.]

behavior or same-sex marriage, and political and religious leaders play a role in perpetuating the idea that all deviations from heterosexuality are "un-African" and should be punished harshly. Anti-LGBT stances have taken on a geopolitical aspect in that LGBT rights are commonly characterized as a Western value that is polluting the non-Western world. Nonetheless, there are a few outspoken supporters of LGBT rights, and South Africa, with its unique experience with anti-apartheid activism, is the one country that has a constitutional ban on discrimination based on sexual orientation.

CULTURE: RELIGION

Africa's rich and complex religious traditions derive from three main sources: traditional or indigenous African belief systems (many of them animist), Islam, and Christianity.

Indigenous Belief Systems

Traditional African religions, found in every part of the continent, may be the most ancient on Earth, since this is where human beings first evolved. **Figure 7.37** highlights the countries in which traditional beliefs remain particularly strong. Traditional beliefs and rituals often are used to bring departed ancestors into contact with living people, who in turn are the connecting links in a timeless spiritual community that stretches into the future. The future is reached only if family members procreate and perpetuate the family heritage through storytelling.

Most traditional African beliefs can be considered **animist** in that spirits, including those of the deceased, are thought to exist everywhere—for example, in trees, streams, hills, and art. In return for respect (expressed through ritual), these spirits offer protection from sickness, accidents, and the ill will of others. African religions tend to be fluid and adaptable to changing circumstances. For example, in West Africa, Oshun, the goddess of water traditionally credited with healing powers, is now also invoked for those suffering economic woes.

Religious beliefs in Africa, as elsewhere, continually evolve as new influences are encountered. If Africans convert to Islam or Christianity, they commonly retain aspects of their indigenous religious heritage. The three maps in Figure 7.37 show a spatial overlap of belief systems, but they do not convey the philosophical blending of two or more faiths, which is widespread. In the Americas, the African diaspora has influenced the creation of new belief systems developed from the fusion of Roman Catholicism and African beliefs. Voodoo in Haiti, Santería in Cuba, and Candomblé in Brazil (see Chapter 3) are examples of this fusion.

Islam and Christianity

Islam began to extend south of the Sahara in the first centuries after Muhammad's death in 632 C.E., and today, according to a recent Pew Research Center study (2010), about 15 percent of sub-Saharan Africans are Muslim (**Figure 7.38B**). Islam is now the predominant religion in the north, throughout the Sahel, and in parts of East Africa, where Muslim traders from North Africa and Southwest Asia brought the religion. Powerful Islamic empires have arisen here since the ninth century. The latest of these empires challenged European domination of the region in the late nineteenth century.

Christianity first came to the region through Egypt and Ethiopia shortly after the time of Jesus Christ, well before it spread throughout Europe (see Figure 7.38A). Christianity did not come to the rest of Africa until nineteenth-century missionaries from Europe and North America became active along the west coast. Many Christian missionaries provided the education and health services that colonial administrators had neglected. In the twentieth century, old-line established churches began to gain adherents in Africa. The Anglican Church (Church of England) grew so rapidly that by 2000, Anglicans in Kenya, Uganda, and Nigeria outnumbered those in the United Kingdom. The Anglican Church in Africa attracts the educated urban middle class, and because of

animism a belief system in which spirits, including those of the deceased, are thought to exist everywhere and to offer protection to those who pay their respects

Figure 7.38 LOCAL LIVES: Festivals in Sub-Saharan Africa

(A) During Timkat, a festival that marks the baptism of Jesus Christ, Ethiopian Orthodox Christian priests in Addis Ababa carry replicas of the Ten Commandments, wrapped in cloth, on their heads. [Matjaz Krivic/Getty Images]

(B) South African Muslims spot the new moon that marks the end of Ramadan, a month of fasting during which observant Muslims do not eat, drink, quarrel, or have sex between sunrise and sunset. Throughout Africa, many Muslims mark the end of Ramadan with prayers for peace and unity as well as with drumming, singing, dancing, and general celebration. [GIANLUIGI GUERCIA/AFP/Getty Images]

(C) The 8-day festival of the Umhangla (the "Reed Dance" ceremony) protects the virginity of young unmarried women, who gather in the king of Eswatini's village and repair the surrounding fence with tall reeds. Later, in costumes, the women dance for the royal family, foreign dignitaries, tourists, and other spectators. Traditionally, the king also chooses a new wife. The women who wear red feathers in their hair are the daughters of the king. [Denny Allen/Getty Images]

their strength in numbers, African Anglicans are able to influence central policy decisions in the wider church, often expressing and enforcing a more conservative point of view than that held by Europeans or American Anglicans. Sub-Saharan Africa is now home to one-fifth of the Christians in the world, and these Christians constitute 57 percent of the sub-Saharan population.

Today, both Muslims and Christians express strong support for democracy and religious freedom, but on the other hand, there is also strong support for government based on the Bible or shari'a law. The very conservative perspectives on LGBT rights discussed above are held by both Christians and Muslims, while Christians tend to be somewhat more supportive of increasing women's rights.

At present, Islam's political role is increasingly contentious because of the activities of radical Islamist sects (for example, Boko Haram in Nigeria and Al Shabaab in the Horn of Africa) that have challenged governments in the Sahel countries, as well as South Sudan, Ethiopia, Somalia, Uganda, and Kenya. These groups appeal to many young unemployed youth struggling against economic marginalization and cultural destabilization related to globalization. However, the origins and financing of these radical movements remain mysterious.

Across the continent, the variety of African religious traditions are celebrated with festivals. As is the case with festivals everywhere, such events offer an opportunity to represent ideal versions of culture, a time to revel in joy and renewed religious dedication, and a chance to make important social contacts, reinforce identity, and perform community service (see Figure 7.38).

Evangelical Christianity and the Gospel of Success In contrast to the Anglican Church, modern evangelical versions of Christianity appeal to the less-educated, more recent urban migrants—the most rapidly growing segment of the population. These independent, evangelical sects interpret Christianity as being aligned with traditional beliefs in the importance they attach to: the need to make sacrificial gifts to spiritual leaders and ancestors, the power of miracles, and the idea that devotion brings wealth. The result is one of the world's fastest-growing Christian movements.

VIGNETTE Preachers at the Miracle Center in Kinshasa describe the gospel of success forcefully: "The Bible says that God will materially aid those who give to Him. We are not only a church, we are an enterprise. In our traditional culture you have to make a sacrifice to powerful forces if you want to get results. It is the same here."

Generous gifts to churches are promoted as a way to bring divine intervention to alleviate miseries, whether physical or spiritual. Practitioners donate food, television sets, clothing, and money. One woman gave 3 months' salary in the hope that God would find her a new husband.

Like all religious belief systems, the gospel of success is best understood within its cultural context. Many of the believers, new to the city, feel isolated and are looking for a supportive community to replace the one they left behind. In return for material contributions to the church and volunteer services, members receive social acceptance, companionship, and community assistance in times of need. ■

CULTURE: ETHNICITY AND LANGUAGE

Ethnicity, as we have seen throughout this book, refers to the shared language, cultural traditions, and political and economic institutions of a group. The map of Africa in **Figure 7.39** shows a rich and complex mosaic of languages but still fails to adequately depict Africa's cultural diversity (compare it with Figure 7.21 of Nigeria).

Most ethnic groups have a core territory in which they have traditionally lived, but very rarely do groups occupy discrete and exclusive spaces. Often, several groups share territory, practicing different but complementary ways of life and using different resources. For example, one ethnic group might be subsistence cultivators, another might herd animals on adjacent grasslands, and a third might be craft specialists working as weavers or metalsmiths.

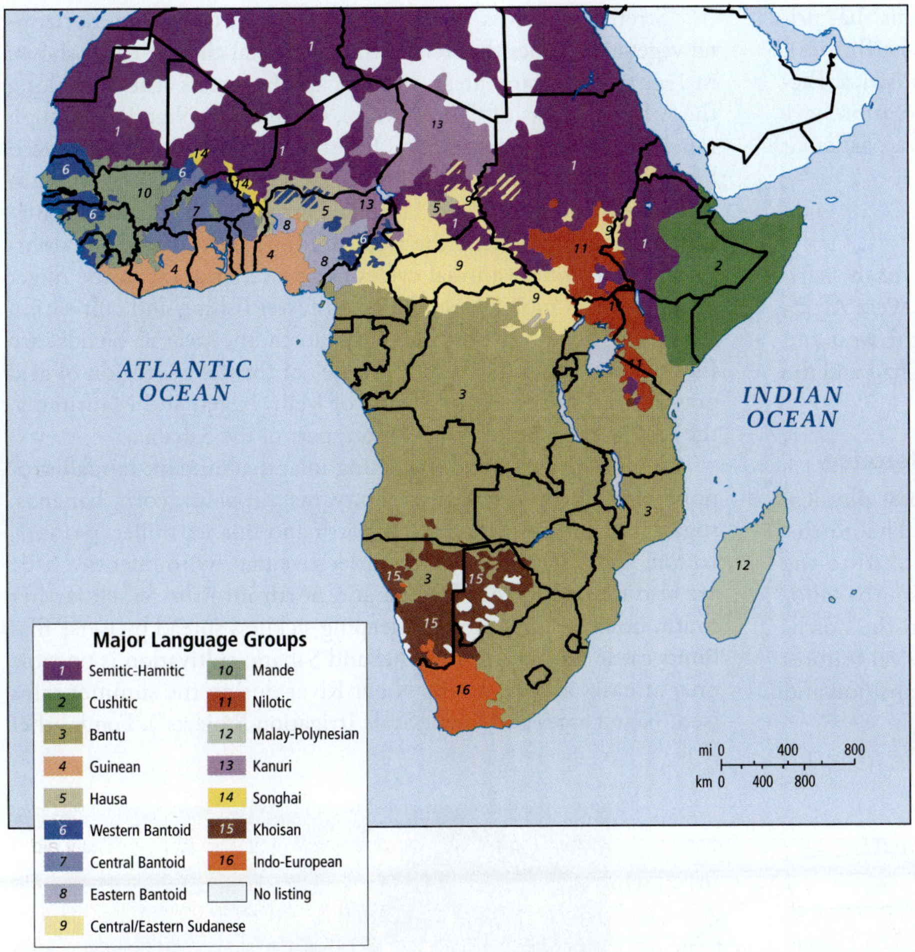

Figure 7.39 Major language groups of Africa. Notice the variety of languages in such countries as Nigeria and Kenya; also note how country borders often split language groups into several parts. [Research from: Edward F. Bergman and William H. Renwick, *Introduction to Geography—People, Places, and Environment* (Englewood Cliffs, NJ: Prentice-Hall, 1999), p. 256; and Jost Gippert, TITUS Didactica, http://titus.uni-frankfurt.de/didact/karten/afr/afrikam.htm.]

People may also be very similar culturally and occupy overlapping spaces, but identify themselves as being from different ethnic groups. For example, Hutu farmers and Tutsi cattle raisers in Rwanda share similar occupations, languages, and ways of life. However, European colonial policies purposely exaggerated ethnic differences as a method of control, assigning a higher status to the Tutsi. Hutu and Tutsi now think of themselves as having very different ethnicities and abilities. In the 1990s and again in 2004, the Hutu-run government encouraged attacks on the minority Tutsi people. Hutu were also killed, but those killed tended to be people who opposed the genocide. The best estimates indicate that more than 1 million people, most of them Tutsi, died during the two genocide catastrophes.

Some African countries have only a few ethnic groups; others have hundreds. Cameroon, sometimes referred to as a microcosm of Africa because of its ethnic complexity, has 250 different ethnic groups. Different groups often have extremely different values and practices, making the development of cohesive national policies difficult. Nonetheless, the vast majority of African ethnic groups have peaceful and supportive relationships with one another.

To a large extent, language correlates with ethnicity. More than a thousand languages, falling into more than a hundred language groups, are spoken in Africa; Bantu is the largest such group (see Figure 7.39). Most Africans speak their native tongue and a **lingua franca** (language of trade). Some languages are spoken by only a few dozen people and are dying out, while other languages, such as Hausa, are spoken by millions of people from Côte d'Ivoire to Cameroon. Lingua francas—such as Hausa, Arabic, and Swahili—are taking over because they better suit people's needs or have become politically dominant. Former colonial languages such as English, French, and Portuguese (all classed as Indo-European languages in Figure 7.39) are also widely used in commerce, politics, education, and on the internet.

lingua franca a language of trade

CHECK YOUR UNDERSTANDING

1. What do marriages in rural Africa tend to be based on?
2. What are the roots of polygyny in this region?
3. How does female genital mutilation impact human development?
4. Why might LGBT people in this region choose to stay closeted?
5. What are most traditional African religions based on?
6. How did Christianity first come to this region?

SUBREGIONS OF SUB-SAHARAN AFRICA

7.10 Identify key characteristics of the subregions of sub-Saharan Africa.

Sub-Saharan Africa can be divided into four subregions that primarily reflect geographic location (rather than other affinities): West Africa, Central Africa, East Africa, and Southern Africa. Factors such as traditions and the current geopolitical situation have also guided the placement of a particular country in a particular subregion.

WEST AFRICA

West Africa, which occupies the bulge on the west side of the African continent, includes 15 countries (**Figure 7.40**). West Africa is physically framed on the north by the Sahara, on the west and south by the Atlantic Ocean, and on the east by Lake Chad and the mountains of Cameroon.

Horizontal Zones of Physical and Cultural Difference

West Africa can be thought of as a series of horizontal physical zones, from dry in the north to moist in the south. The north–south division also applies to economic activities linked to the environment—herding in the north, farming in the south—and, to some extent, to religions and cultures—Muslim in the north, Christian in the south. Most of these cultural and physical features do not have distinct boundaries, but rather zones of transition and exchange.

Stretching across West Africa from west to east are horizontal vegetation zones that reflect the horizontal climate zones shown in Figure 7.9. Many areas of tropical rain forests still exist along the Atlantic coasts of Côte d'Ivoire, Ghana, and Nigeria, although much of this once-great coastal forest has been logged and replaced by agriculture. To the north, this moist environment changes into drier woodland mixed with savanna. Farther north still, as the environment becomes drier, the trees thin out and the savanna dominates. Only where annual cycles create wetlands, as in the Niger River basin, is moisture sufficient to foster fishing and cultivation as well as grazing. Still further to the north, the savanna blends into the yet drier Sahel (Arabic for "shore" of the desert) region of arid grasslands. In the northern parts of Mali, Niger, and Mauritania, the arid land becomes actual desert, part of the Sahara.

Along the coast, and stretching inland, abundant rainfall supports crops such as coffee, cacao, yams, oil palm, corn, bananas, sugarcane, and cassava. Farther north and inland, millet, peanuts, cotton, and sesame grow in the drier savanna environments. Cattle are also tended in the savanna and north into the Sahel; farther south, however, the threat of sleeping sickness spread by tsetse flies limits cattle raising. In the Sahel and Sahara, cultivation is possible only at oases and along the Niger River during the summer rains (see "New Large- and Small-Scale Irrigation Projects"). Food, other

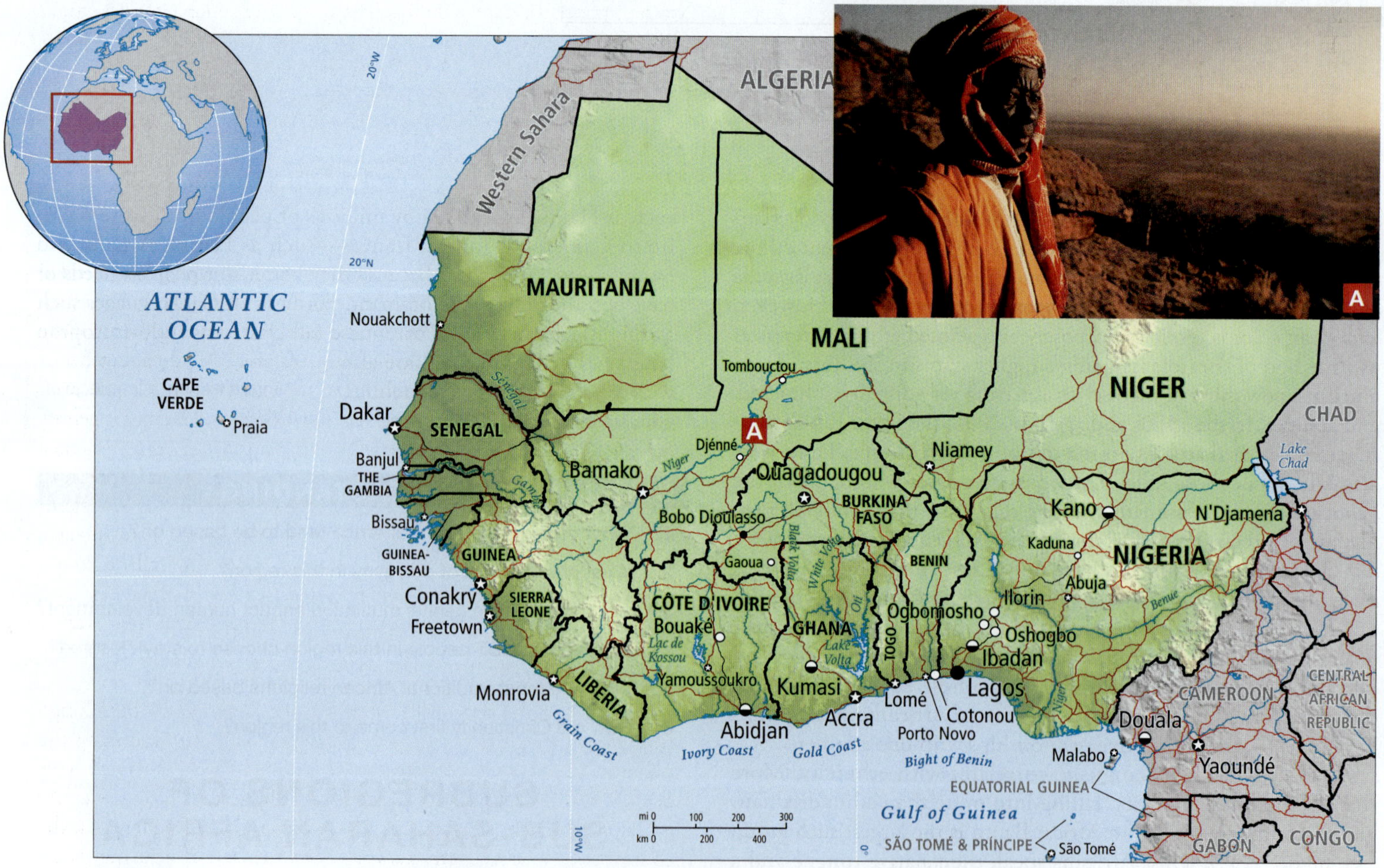

Figure 7.40 The West Africa subregion. **A** An elder of the Dogon people in Mali looks out onto the lands of his neighbors, who are Fulani herders. The Dogon are farmers renowned for their sculpture, masked dances, and religious traditions. [Wade Davis/Getty Images]

than that provided by animals and the fields along the riverbanks, must come from trade with the coastal zones.

West African environments have dried out as more and more forests and woodlands are cleared by international loggers or by local people for cultivation, grazing, and fuelwood. When the forest cover is gone, the land is more vulnerable to erosion, and sunlight evaporates moisture. Desiccating desert winds now blow south all the way to the coast, and this drying out of the land is exacerbated by the recurrence of natural cycles of drought. The ribbons of differing vegetation running west to east are not as distinct as they once were, and even coastal zones occasionally experience dry, dusty conditions reminiscent of the Sahel.

Regional Conflict in Coastal West Africa

The 1990s and early 2000s saw intertwined wars spreading from Liberia to Sierra Leone and Côte d'Ivoire, based on local disputes related to poverty, exploitation, and corruption. In the 1820s, Liberia was colonized by freed slaves from the United States and the Caribbean, on land occupied by indigenous West African groups for at least a thousand years. A highly stratified plantation-based society developed, with the colonists becoming an elite class supported by the labor of indigenous groups who were essentially enslaved. Rubber plantations were established by German and U.S. interests in the twentieth century and were later joined by oil palm plantations to become the backbone of Liberia's largely agricultural economy.

In 1989, Libya funded Charles Taylor to lead a rebellion in Liberia against a government that had itself come to power as a rebellion against the elite class that developed out of the colonists. Taylor commandeered Liberia's diamond and gold mines, as well as its forests, home to many endangered species, including forest elephants and chimpanzees, using the money to fund his personal armies. Made up largely of kidnapped and enslaved children, Taylor's armies fought those of other Liberian warlords in a 14-year civil war that took the lives of 150,000 civilians. The war also spilled over into neighboring Sierra Leone, where another 75,000 people died. Meanwhile foreign investors such as U.S. televangelist Pat Robertson, as well as logging companies based in Europe, China, and Southeast Asia, made profits from corrupt deals made with Taylor.

In 2002, the conflict in Liberia also spilled into Côte d'Ivoire, which was split between a rebel-held north and a government-held south. Abidjan, the former capital on the southern coast, had fine roads, skyscrapers, fancy French restaurants, and a modern seaport. Meanwhile the north was much poorer, dominated by cacao plantations that employ immigrant agricultural workers at very low wages and, in some cases, use enslaved children from nearby Burkina Faso and southern Mali. The country was eventually reunited, but periodic conflicts still break out based on the north–south dichotomy.

The conflicts in Liberia and neighboring countries came to a close as Taylor lost control of the country to rebel groups in 2003. Exiled to Nigeria and then turned over to authorities, Taylor was convicted of war crimes by the International Criminal Court in The Hague, Netherlands, for crimes against humanity, for which he is serving a 50-year sentence. In 2006, a highly respected, reform-minded woman, Ellen Johnson-Sirleaf, was elected president of Liberia and went about rebuilding the country, earning a Nobel Peace Prize for her work in 2011.

Human rights groups and chocolate lovers in Europe and elsewhere (Africans consume very little chocolate or cacao products) successfully pressured international chocolate producers to address the use of enslaved children and adults as laborers on cacao plantations. Other reform efforts focus on the creation of value chains (see "Regional Economic Development") for the benefit of African cacao growers so that they can earn more money and use less exploitive labor practices.

CHECK YOUR UNDERSTANDING

1. What are some aspects of the north–south division in West Africa?
2. What was the basis of Liberia's economy until the civil war that began in 1989?
3. How did Charles Taylor fund his armies?
4. What labor practices have become an issue for chocolate producers in West Africa in recent years?

CENTRAL AFRICA

Looking at a map of Africa with no prior knowledge, you might reasonably expect the countries of Central Africa—Cameroon, the Central African Republic, Chad, Congo (Brazzaville), Equatorial Guinea, Gabon, São Tomé and Príncipe, and Congo (Kinshasa)—to be the hub of continental activity, the true center around which life on the continent revolves (**Figure 7.41**). Paradoxically, Central Africa plays only a peripheral role on the continent. Its lack of infrastructure, combined with dense tropical forests and difficult terrain, make it the least accessible part of the continent, and recent armed conflict over resources and political power has left it in a deteriorating state that discourages potential investors.

Wet Forests and Dry Savannas

Central Africa consists of a core of wet tropical forestlands surrounded on the north, south, and east by bands of drier forest and then savanna (see the Figure 7.9 map). The core is the drainage basin of the Congo River, which flows through this area and is fed by a number of tributaries. Although rich in resources, the forests form a barrier that has isolated Central Africa from the rest of the continent for centuries and has impeded the development of transportation, commercial agriculture, and, until recently, even mining. Moreover, as is true in all tropical zones, the soils of these forests are not suitable for long-term cash-crop agriculture. Most Central Africans are rural subsistence farmers who live along the rivers, roads, and the few railroad lines. The forests themselves are home to nomadic hunter-gatherers, such as the Mbuti and Pygmy people, but these people and their way of life are highly endangered as outsiders scramble for control of Central Africa's forest and mining resources (see Figure 7.41A).

Although the sixteenth-century city of Loango on the Congo coast was a model of urban development for its time, during the colonial period and until recently, cities in Central Africa have been small in size, with little urban infrastructure. Now cities are growing rapidly as refugees stream in, escaping civil violence in the countryside. The new arrivals find a serious lack of housing, clean water, adequate education systems, and urban employment in the formal sector.

Figure 7.41 The Central Africa subregion.

A Traditional dancers in Cameroon depict the animals of the rain forest in a production designed to increase awareness of the damage created by deforestation. [Eco Images/Universal Images Group/Getty Images]

A Colonial Heritage of Violence

The region's colonial heritage has contributed to the difficulties described above. In what is now known as the Democratic Republic of the Congo (Kinshasa), for example, the Belgian colonists left a terrible legacy of brutality. In 1885, Belgium's king, Leopold II, with a personal fortune to invest, obtained international approval for his own personal colony, the Congo Free State, with the promise that he would end slavery, protect the native people, and guarantee free trade. He did none of these, and instead forced millions to extract Congo's resources for his own profit.

The young writer Joseph Conrad witnessed the cruel behavior of Belgian colonial officials when he obtained a job on a steamer headed up the Congo River in 1890. He saw a campaign of terror that featured enslavement, whipping, beheadings, and lopping off the hands of those who didn't meet their production quotas for rubber and other commodities. Conrad's novella *Heart of Darkness* (published in 1902), and a pamphlet by Mark Twain, "King Leopold's Soliloquy" (published in 1905), awakened Europeans and Americans to Leopold's atrocities, and efforts began to end his regime.

Postcolonial Exploitation

When the countries of Central Africa were granted their independence in the 1960s, they were left with little money to develop infrastructure, little experience with self-government, and diverse

Figure 7.42 A music video being filmed in Kinshasa. [JOHN WESSELS/AFP/Getty Images]

populations of multiple ethnic groups that had little in common. Most countries, such as Congo (Kinshasa) (see "The Case of Congo (Kinshasa)"), fell into the hands of corrupt leaders. Political upheavals resulted when some ethnic groups were excluded from representation in government, with corruption and violence pushing economies into a downward spiral.

As with much of this region, Central Africa saw solid economic growth and rising incomes from the 2000s to about 2013, based largely on China's strong demand for raw materials exports, especially oil and minerals. However, as China's growth slowed down in 2014, growth has slowed in Central Africa, jeopardizing investments in infrastructure and human development that have kept the peace in recent years.

VIGNETTE At a shabby bar in Kinshasa, Sam sings a jazzy solo in the muggy evening air to a sharply dressed crowd. Dressing stylishly has become a bit of an obsession among Kinshasa's young professionals. The recent spate of peace in Congo has sparked a resurgence in Kinshasa, where luxury apartments now attract talented local entrepreneurs and returning migrants fresh from Paris seeking to support their country, or perhaps make a small fortune in a hoped-for boom-time. Trendy restaurants serve plentiful river shrimp, musicians enliven the nightlife, and film crews shoot music videos (**Figure 7.42**). Money flows at least for the moment.

But Sam, like most Kinois, lives outside this bubble of prosperity in a crowded slum without clean water or electricity. Almost 90 percent are underemployed. This great divide of wealth leaves Congo politically unstable and hinders continued economic growth. The main sign of progress in the slum is that everyone has a cell phone. That coupled with the return of educated affluent migrants could mean some hope for an alternative future. ■

CHECK YOUR UNDERSTANDING

1. What are some factors that hold back economic development in Central Africa?
2. What has been the basis of economic growth in Central Africa in recent decades?
3. What are returning migrants bringing back to countries like Congo (Kinshasa)?

EAST AFRICA

The subregion of East Africa (**Figure 7.43**) occupies the Horn of Africa, which reaches along the southern shore of the Red Sea and the Gulf of Aden, and includes the countries of Djibouti, Eritrea, Ethiopia, Somalia, and the newly independent country of South Sudan. To the south of the Horn, the subregion also includes the coastal countries of Kenya and Tanzania; the interior highland countries of Uganda, Rwanda, and Burundi in the Great Rift Valley; and the islands of Madagascar, Comoros, Seychelles, Mauritius, and Réunion in the Indian Ocean across the Mozambique Channel. Across the subregion, the relatively dry coastal zones contrast with the moister interior uplands; further to the west, the drying effects of the Sahara are apparent.

Agriculture still dominates employment, with around 60–90 percent of the population engaged in farming in every country. Services are actually the largest sector of the economy everywhere but Uganda, where agriculture is still the highest earner. Tourism is a major component of the service sector except where political instability and terrorism have scared away visitors, as in the case of Rwanda and Burindi during the 1990s and Uganda during the 1970s. The national parks of Kenya and Tanzania are global attractions (see Figure 7.43B). Manufacturing is only beginning to be developed, and is found primarily in Ethiopia, Kenya, and Tanzania, where food processing and textiles dominate. On the coast, where port cities have a thousand-year history of trade across the Arabian Sea and Indian Ocean, trade-related services are central to economies.

Conflict in the Horn of Africa

The entire Horn of Africa has suffered periodic war, state failure, and associated famines since the mid-1980s. While the immediate causes appear to be governments and political officials who have manipulated innumerable situations for personal gain, many conflicts have been inflamed by outsiders, including former colonial powers and Cold War adversaries, the United States and USSR. The result in Somalia, Ethiopia, and now South Sudan, all often referred to as failing states, has been an environment of chaos and lawlessness for ordinary citizens. Meanwhile growing populations and related increases in consumption are straining the region's resource base.

South Sudan's Decades of War South Sudan occupies what was once the southern third of the oil-rich country of Sudan (covered in Chapter 6) (see Figure 7.43A). With a population mostly made up of darker-skinned Christians who have sub-Saharan origins, South Sudan has always been culturally quite different from the more Arabic north, which often dismissed the south as backward and refused to grant the south autonomy that had been promised in the years leading up to Sudan's independence from the British in 1956. A civil war resulted, lasting from 1955 to 1972 and causing half a million deaths. A second civil war followed from 1983 to 2005, in which 2 million died, mostly civilians who starved to death. The second war was in part about profits from South Sudan's oil resources, which the northern-dominated government of Sudan had been keeping for itself. South Sudan finally won its independence in 2011, but the border between the two countries remains in contention, and Sudan is now arming militias that oppose the South Sudanese government.

Figure 7.43 The East Africa subregion.

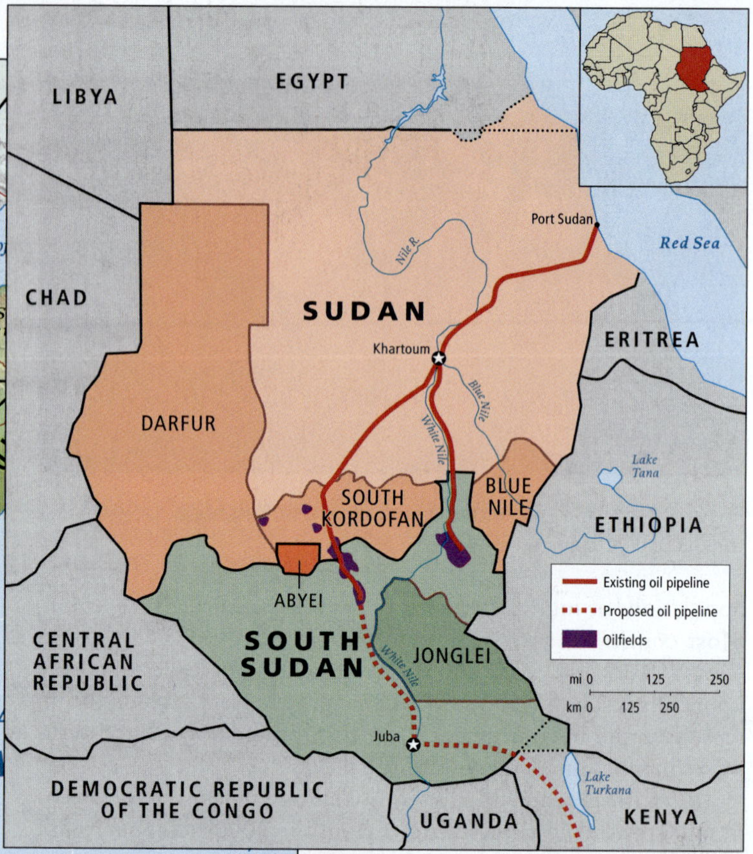

A Map of South Sudan, a new but already war-torn country. Because it is landlocked, it lacks easy access to the world ocean for its oil pipelines.

B Lions in Kenya's Maasai Mara game reserve are increasingly being targeted by poachers for their skins. [Gunter Ziesler/Getty Images]

South Sudan is landlocked and thus has no easy way to get its oil to the world market. Private oil companies based in China and Malaysia have brokered a settlement with the intention of acquiring the oil and transporting it to a Red Sea port in pipelines through Sudan.

A civil war broke out in South Sudan in 2013 due to tensions between the president and his vice president as well as age old ethnic rivalries, resulting in over 400,000 deaths so far and over 4 million displaced from their homes. The country's oil wealth is being used to arm pro-government militias responsible for much of the violence, and to enrich a small network of government officials.

Somalia's Civil War The ongoing civil war in Somalia arose out of tensions between Somalis that were inflamed by outsiders. Somalia has for thousands of years been inhabited mainly by ethnic Somali seminomadic herders who traded their livestock as well as tree resins (such as frankincense and myrrh) with neighboring areas and across the Arabian peninsula. Long governed by local Islamic sultanates, British and Italian trading interests arrived along the coast in the late 1880s, but their control of the interior was resisted until the 1920s and even then their presence was minimal. Independence was granted in 1960, but at the insistence of Ethiopia and its Cold War ally, the United States, borders were drawn that gave Ethiopia Ogaden, a large western interior area that was populated by ethnic Somalis. The government of independent Somalia then allied with the Soviet Union, which gave it extensive economic development and military assistance until 1977, when Somalia invaded Ogaden.

The aftermath of the Ogaden War set the stage for Somalia's ongoing civil war. Somalia's army rapidly conquered Ogaden, with the backing of Somali separatists, only to find the Soviet Union aiding Ethiopia to retake the territory. By 1978, the Soviets had brought in massive amounts of weaponry and 18,000 Cuban troops, who had been fighting against U.S.-backed forces in Angola's civil war. Somalia soon lost control of Ogaden, leading to a revolt against the Somali government by leaders of the Somali army. During the 1980s the Somali government, which was then an ally of the United States, faced ongoing opposition to its authority, to which it responded with brutality that impoverished and alienated large parts of the civilian population. Tensions rose until open civil war broke out in the 1990s, eventually leading to the collapse of the Somali government.

In this context of chaos militant Islamists—most recently, Al Shabaab—have attempted to impose order and conformity through the concept of an Islamic state ruling via shari'a law (see "Islamic Religious Law and the Sunni–Shi'ite Divide" in Chapter 6). Al Shabaab (meaning the youth) have imposed an oppressive interpretation of Islamic shari'a law, for example mandating the stoning to death of women for adultery, which has provoked resistance and foreign intervention. Unable to hold territory in the face of interventions by peacekeepers from other East African nations, Al Shabaab has responded with terrorist attacks mainly in Kenya and in Somalia's capital of Mogadishu.

Early in the conflict the United States sided with corrupt warlords against a diverse array of Islamist militants. This resulted in the debacle in Mogadishu, Somalia, depicted in the 2001 film *Black Hawk Down*. Since then the United States has switched to using drone strikes to target Islamists in Somalia, though this tactic has also resulted in civilian deaths and damage to agricultural infrastructure that has alienated many Somalis.

Somali Piracy Piracy has been practiced for thousands of years off the coast of East Africa, feeding off of seaborne trade with the Arabian Peninsula, South Asia, and beyond, but a recent rash of piracy has been driven by the lawlessness of the Somali civil war. In 2008, pirates from Somalia became a focus of international concern when 40 commercial ships from multiple nations were attacked, seized, and their crews held hostage for ransom. This piracy developed out of the response of Somali fishers who noticed that their catches from the rich fishing grounds off the coast of Somalia were declining due to large commercial fishing fleets (from Italy, France, Spain, Greece, Russia, the Arabian Peninsula, Iran, and China), who took advantage of the absence of government in Somalia to fish its waters without paying any fees or following any regulations that would keep the fishery healthy. Many Somalis also claim that criminals from Europe and Russia are disposing of toxic waste off the coast, resulting in suspicious ailments. What began as efforts to discourage illegal fishing and waste disposal, turned into piracy when the Somalis learned that they could extort money from the relatively unarmed oil tankers and other ships passing through the Gulf of Aden. Western warships and private security companies hired by the oil tankers created a lull in piracy after 2012, but subsequent lax security led to a resurgence of piracy by 2017. Recently the Somali government has given licenses to Chinese companies to fish most of its waters. Ideally the licenses will provide money to regulate foreign fishing and monitor waste dumping, while reserving enough fishing areas to let Somali fishers make a living.

Mixed Farming in the Interior Valleys and Uplands

The Great Rift Valley is home to a form of mixed farming that combines herding and crop cultivation. Animal manure is used to fertilize a wide variety of crops that are closely adapted to particularly fragile semiarid environments, and pastures and fields are rotated systematically. These ancient and specialized mixed-agriculture systems were ignored and disdained by European colonizers, and agronomists have only recently begun to appreciate the precision of African upland mixed agriculture, which is now threatened by civil disorder, development schemes, and population pressure.

Coastal Lowlands

East Africa's shores from southern Somalia southward (known regionally as the Swahili Coast) have long attracted traders from around the Indian Ocean—from China, Indonesia, Malaysia, India, Persia (Iran), Arabia, and elsewhere in Africa. The result has been a grand blending of peoples, cultures, languages, plants, and animals. In the sixth century, Arab traders brought Islam to East Africa, where it remains an important influence in the north and along the coast. Traders, speaking the lingua franca of Swahili, established networks that linked the coast with the interior. These networks penetrated deeply into the farming and herding areas of the savanna and into the semifeudal cultures of the highland lake country.

Today, trade across the Indian Ocean continues, with Arab countries still major destinations of East African raw materials exports, and China and India now the top sources of imported manufactured goods in most countries. East Africa exports mostly agricultural products (live animals, hides, bananas, coffee, cashews, processed food, tobacco, and cotton) and imports machinery, transportation and communication equipment, industrial raw materials, and consumer goods.

Contested Space in Kenya and Tanzania

Kenya and Tanzania are modestly industrialized countries in East Africa that remain largely dependent on agricultural exports, like cacao, and some finished products: tea, coffee, beer, horticultural products, and processed food, as well as cement and petroleum products. Both countries are regional hubs of trade, yet both are among the poorest countries in the world (see Figure 7.23).

Beginning in ancient times, traders dealt in slaves, ivory, and rare animals, items that can no longer be legally traded. British colonists took over in the early nineteenth century, relocating highland farmers in Kenya, such as the Kikuyu, to make way for coffee plantations. In the savannas of Kenya and Tanzania, they displaced pastoralists, such as the Maasai, some of whose lands became game reserves designed to protect the savanna wildlife for the use of European hunters and tourists.

East African Wildlife The national parks of East Africa are globally significant holders of biodiversity, playing an important economic role in their countries, and looming large in the rest of the world's image of Africa. The former colonial game reserves of Kenya and Tanzania are today the finest national parks in Africa, offering protection to wild elephants, giraffes, zebras, and lions, as well as to many less spectacular but no less significant species (see Figure 7.43B). The future of the parks is precarious, however, because of competing demands for the land by herders and cultivators, and due to poaching.

Tourism, mostly in the form of visits to these parks, accounts for up to 20 percent of foreign currency earnings in these countries. This is crucial for the repayment of debts to owed foreign lenders, such as the IMF and the World Bank, who demand repayment in foreign currencies. In 2018, tourism brought Kenya and Tanzania each around 9 percent of their GDP ($9.7 billion and $4.7 billion, respectively), and provided around 9 percent of total employment.

Poaching is a continual threat to the overall economy in these countries. In Kenya, a living elephant is worth close to $15,000 a year in income from tourists who come to see it. Nonetheless, some still view hunting elephants for their ivory as a profitable activity, even though a hunter earns no more than $1000 per dead animal. Recent years have seen a highly organized slaughter of elephants in the zone where Congo, South Sudan, Uganda, Kenya, and Tanzania come together. Thousands have been killed, their ivory stolen, and their bodies left to rot by poachers catering to demand for animal body parts from traditional Chinese medicine practitioners and patients. The perpetrators involve current and former soldiers militants from all the countries in this zone.

Recent growth in the wildlife-based tourism industry throughout Africa is good news for wildlife populations, as cash-strapped governments are more likely to take steps to protect animals that earn them money. Meanwhile, thanks to conservation work some critically endangered wildlife populations, such as mountain gorillas, have been growing in recent years.

The Islands

Madagascar, lying about 310 miles (500 kilometers) across the Mozambique Channel from East Africa (see the Figure 7.43 map), is home to 22 million people, half of whom live on less than $1 a day. Most are poor farmers whose labor supports a small wealthy elite that manages an economy based on the processing and export of the country's commodities. It is also home to an amazing array of plant and animal life unique to the island, having evolved after Madagascar split from Africa 88 million years ago (**Figure 7.44**). Although the island has undergone much change over the last few centuries, it remains an amazing evolutionary time capsule that is highly prized by biologists. Humans arrived around 2000 years ago from Africa and Southeast Asia and have removed 80–90 percent of the islands forests, and now threaten many plants and animals with extinction (see Figure 7.10).

Figure 7.44 Madagascar's unique animal life. Silky sifakas are one of several critically endangered species of lemur, a primate found only in Madagascar. Approximately 250 silky sifakas remain in the wild, all of them in a few protected areas in the rain forests of northeastern Madagascar. [ajlber/iStock/Getty Images]

The other East African island nations—the Comoros, the Seychelles, and Mauritius—are much smaller, wealthier, and less physically complex than Madagascar. All of the islands have a cosmopolitan ethnic makeup, the result of thousands of years of trade across the Mozambique Channel and the Indian Ocean. Further cultural mixing in the islands took place during European (primarily French and British) colonization. During the colonial era, these islands supported large European-owned plantations worked by laborers brought in from Asia and the African mainland. More recently, tourism has been added to the economic and biological diffusion mix. Most visitors come from the African mainland, from Asian countries surrounding the Indian Ocean, and from Europe.

Madagascar underwent IMF structural adjustment policies in the 1990s that failed to alleviate poverty. The island's political leaders tend to be wealthy business people with little sense of the biological treasure that Madagascar is or of the necessity to address poverty and social welfare issues. Many of Madagascar's 100 protected areas receive no funding from the government, and must solicit donations from Europe and North America to operate. Conservationists are working with local populations living outside protected areas to provide them with alternatives to illegal wood gathering and farming; however, these efforts are undercut by large illegal logging operations that have the blessing of corrupt high-level politicians.

Figure 7.45 The Southern Africa subregion.

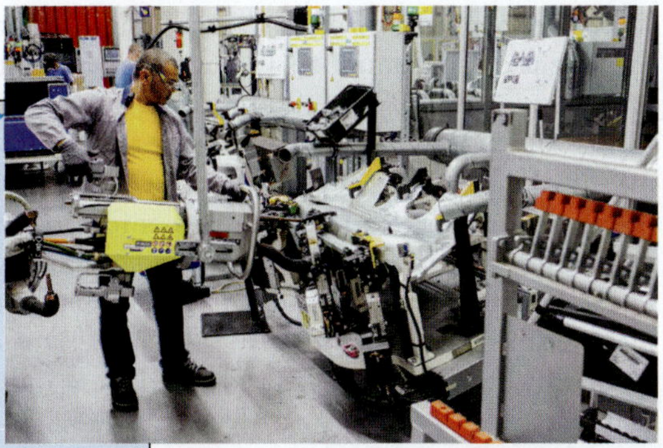

A A worker builds cars under license to Volkswagen at a plant in Uitenhage, South Africa. [Waldo Swiegers /Bloomberg via Getty Images]

CHECK YOUR UNDERSTANDING

1. What sector dominates employment in East Africa?
2. How is South Sudan's oil wealth being used?
3. What foreign powers played a role in the Ogaden War?
4. How did Somali piracy develop?
5. What does mixed farming in the interior valleys and uplands combine?
6. What are the main threats to East Africa's national parks?
7. Why does Madagascar have so many unique plants and animals?

SOUTHERN AFRICA

Southern Africa is unique from the rest of this region due in part to the legacies of large-scale European settlement, which include a bitter history of racial segregation and war, high levels of inequality, but also the region's strongest manufacturing base. Southern Africa also has much in common with the rest of the region, including dependence on commodity exports, and economies that are increasingly dependent on China's demand for commodities.

Large-Scale European Settlement

Southern Africa (**Figure 7.45**) was the only area where Europeans composed more than 1 percent of the population, due largely to the temperate climate that reduced the incidence of malaria, sleeping sickness, and other diseases. It also has extensive lands suitable to European-style cultivation and ranching techniques. The first Europeans were Portuguese and Dutch traders who stayed mostly along the coasts, followed by Dutch farmers, the Boers, who eventually spread throughout what became South Africa, Namibia, Zimbabwe, and Zambia. Large-scale migration from Britain occurred with the development of mining in South Africa. By the 1880s most of Southern Africa was laid claim to by British and Portuguese trading companies, that soon grew into formal European colonial administrations.

Southern Africa's European populations never came close to a majority, composing only 20 percent of the population in South

Africa, 14 percent in Namibia, and under 7 percent elsewhere at the height of their numbers in the early twentieth century. However, they exerted a powerful influence over the colonial governments that amounted to "white minority rule." While the strict regime of racial segregation of apartheid South Africa spread only to Namibia, white dominance was enforced through extensive racial discrimination in the British colonies, and through labor relations approximating slavery in Mozambique and Angola. The main legacies of white minority rule were extremely high income inequality, lengthy wars for independence, and ongoing "white flight."

As pressure for decolonization and the end of white minority rule grew throughout the first half of the twentieth century, white farmers, mine workers, and urbanites resisted. Zambia was the first to gain independence in 1962, which prompted an exodus of white farmers and miners to the nearby colony of Rhodesia (current Zimbabwe), where a white-dominated government declared independence from British rule in 1965. Here a lengthy civil war between the white-dominated Rhodesian government and African nationalists lasted until 1979, when the nation of Zimbabwe was declared and majority democratic rule was instated. A subsequent exodus of whites to South Africa and Europe followed, though most developed farmland remained in the hands of white farmers. Lengthy wars of independence fought by African nationalists in Angola and Mozambique against white majority rule were fought from 1961 until 1974, when independence from Portugal was finally gained. These wars prompted further white emigration to Europe and South Africa.

By 1974, South Africa and its colony of South West Africa (later Namibia) were the last remaining white minority-rule countries in Africa. Tensions between them and their African majority-rule neighbors, as well as between ruling whites and the majority African population within South Africa and Namibia, were increasingly shaped by the Cold War. South Africa was an ally of the United States, supporting rebellions against Soviet-supported governments in Angola and Mozambique. These wars, lasting until 2002 and 1992 respectively, resulted in over 1.5 million deaths, mostly from starvation due to the interruption of commerce in food. The Soviet Union also provided training and weaponry to African nationalists in South Africa and Namibia, making possible many of the attacks by the armed wing of the ANC on South Africa's security apparatus during the 1980s that convinced South Africa's government to negotiate an end to apartheid.

High Economic Inequality Efforts to address Southern Africa's high levels of economic inequality have met with limited success. South Africa and Namibia consistently have the highest levels of economic inequality in the world, and the situation has worsened since the end of apartheid in 1994. In these now majority-rule countries inequality is increasingly less about race, and more about high unemployment rates (27 percent for South Africa, 34 percent for Namibia). Notably, nearby Botswana also has high inequality and high unemployment (18 percent), having never had apartheid. In all these countries, mining provides the bulk of exports while employing relatively few people. Meanwhile manufacturing industries, which could provide much more employment, are small or employing fewer people.

South Africa's manufacturing sector is the most well developed in all of Africa but is now struggling to provide employment as it adopts globally competitive processes that require more machinery and less labor. During the apartheid era, the development of manufacturing industries was a central strategy for providing full employment to white South Africans and stable overall economic growth that would reduce challenges to white minority rule. The government invested profits from mineral exports into sophisticated manufacturing techniques for mining equipment, food processing, and textiles. These industries were never designed to employ the large nonwhite population, and in the post-apartheid era they have not grown to provide enough employment for the country's large and growing population. Textile manufacturing in South Africa has had a particularly difficult time competing with cheap imported Chinese textiles, and has been shedding jobs in recent years. Meanwhile South Africa's growing export-oriented automobile industry has followed trends in European and North American manufacturing that maintain cost competitiveness against lower wage manufacturers by replacing human labor with automation (see Figure 7.45A). On this trajectory, auto manufacturing in South Africa may be globally competitive, but unable to provide large-scale employment to a country that desperately needs it.

High levels of inequality in land ownership, and controversial attempts at land redistribution, are an economic legacy of white majority rule. The best agricultural lands were owned by whites throughout much of this region until major efforts at land redistribution to Africans were made after the transition to majority rule. These efforts have met with different levels of success, resulting in reductions in poverty in Mozambique, but also in food shortages and declining agricultural production in Zimbabwe after chaotic, corrupt, and mostly illegal land redistribution. Commitment to more orderly land redistribution in South Africa has meant that over 70 percent of land remains in the hands of whites, with only 1 percent redistributed since the end of apartheid. Similar policies in Namibia mean that around half of the best farmland remains white-owned.

Commodity Dependence

Southern African economies remain largely dependent on commodity exports, and several countries are dependent on a single commodity for 60–90 percent of their exports. Botswana's main export is diamonds (69 percent), Zambia's is copper (80 percent), and Angola's is petroleum (90 percent). This is particularly dangerous because the prices for these commodities fluctuate widely from year to year, resulting in both economic and political instability because these governments depend on taxes collected from exports for much of their operating budget. Even in the most diversified economy, South Africa, commodities make up over 65 percent of exports, while in Namibia, Malawi, and Mozambique almost all exports are commodities. While these countries have a more diverse array of commodities exports that shelters them from some price variations, they are still highly exposed to general worldwide slumps in the prices for commodities, as has happened in recent years in response to China's weaker demand for commodities due to its slower economic growth.

China in Zambia Zambia's rich copper resources have drawn investment and management by the Chinese, whose need to bring electricity to China's rapidly industrializing cities has created a large demand for copper wire. The Chinese-run mines are efficiently managed, and production has soared, but very little of the profit reaches Zambian citizens. Despite the low wages of local miners, who earn just U.S.$50 a month, the Chinese mine managers insist on importing Chinese mine workers (see Figure 7.24). Working conditions for both Zambian and Chinese laborers can be deadly. Zambian activists note that, as in colonial times, the extraction of Zambia's resources is primarily benefiting foreign investors. Nonetheless, mine development has fueled enough economic growth to support some new urban shopping malls, and most young Zambian adults can now afford mobile phones.

Food Insecurity in Southern Africa

Inadequate access to nutritious food affects much of the population in Zambia, Zimbabwe, Malawi, and Mozambique. In all of these countries between 50 and 80 percent of people are employed in agriculture, most of them subsistence farmers living in poverty, and between 30 and 40 percent of children suffer stunted growth due to malnutrition. With so many people practicing subsistence cultivation there is often not enough surplus production to supply cities or even rural areas when crops fail. This necessitates expensive food imports.

Government policies are often at the root of food insecurity. For example, Zimbabwe's haphazard and corrupt redistribution of land from white commercial farmers to black Zimbabweans beginning in 2000 was disastrous for the country's food security. Because so many farms were seized illegally, with encouragement of the government, the new occupants could not secure loans to keep the often highly mechanized farms running. As a result many formerly productive farms are now lying fallow and the government is trying to entice the evicted farmers back, so far with limited success.

Food security is often a complex issue, involving not just government policies but also environmental and demographic issues, as the case of Malawi illustrates.

Food Crisis in Malawi Malawi is persistently the most food insecure country in Southern Africa, in part due to misguided government policies, but also due to soil depletion, rapid population growth, and agricultural systems that are vulnerable to climate variability.

In 1980, there were about 6 million Malawians and food production was sufficient to feed the people, with surpluses left over for export. By 2018, there were over 19 million Malawians, many facing periodic starvation due to declining farm yields. In many areas fragile tropical soils, which need years of rest between plantings, were forced into continuous plantation cultivation of commercial export crops. This was encouraged by the Malawian government, which depends on taxes collected from agricultural export crops, such as tobacco and tea, for most of its operating budget. To accommodate this, smallholders, the primary local food producers, lost land to large export producers. Their plots became too small to allow for the necessary fallow periods, and their productivity subsequently fell. By the 1990s, 86 percent of Malawi's rural households had less than 5 acres (2 hectares) of land, yet they were responsible for producing nearly 70 percent of food for internal consumption.

As Malawi became more and more dependent on imported food during the 2000s, periodic increases in the price of maize and other imported foods combined with occasional drought to produce high levels of food insecurity. Starvation became common in a place that had once been food secure.

Foreign aid for subsistence farmers combined with heavier rainfall have helped reduce food insecurity in recent years. Training in soil and moisture conservation and greater exposure of Malawian farmers to scientific research have improved some livelihoods; however, the use of so much of Malawi's land for export agriculture continues to place food security at risk.

Reasons for Optimism in Southern Africa

Despite its many challenges there are a number of reasons for optimism in Southern Africa (**Figure 7.46**). Treatment for HIV is now affordable for all but the poorest, which is a major transformation for Southern Africa, which was hit harder by the HIV/AIDS epidemic than anywhere else in the world. The region has seen relative peace since the end of the wars in Mozambique (1975–1992) and the Angolan civil war in (1975–2002), followed by nearly a decade of economic growth and rising living standards in most countries due in part to China's demand for raw materials exports.

Figure 7.46 Durban, South Africa. The beachfront and harbor of Durban, South Africa's third-largest city (after Johannesburg and Cape Town). Durban has been a hub of manufacturing, trade, and transportation, and because of its coastal location, it is becoming a center for export-oriented industries. Durban's tropical climate and beautiful beaches make it a popular destination for tourists. [Franz Marc Frei/LOOK-foto/Getty Images]

After decades of Chinese investment, Angola now rivals Nigeria as the largest exporter of oil in Africa. Meanwhile South Africa has in recent decades developed a globally competitive export-oriented automobile-manufacturing industry that, while limited in its employment potential, is still a major earner and an inspiration to the rest of this region (Figure 7.45A).

South Africa's peaceful recovery from apartheid is still an inspiration to many parts of the world torn apart by conflict. When Nelson Mandela was elected president in 1994, after more than 300 years of European colonization and 50 years of apartheid, he and Bishop Desmond Tutu, a prominent leader in the South African Anglican Church (and a Nobel Peace Prize winner himself in 1984) insisted that in order to heal the racial divide, the country needed to face the reality of what had happened under apartheid (see Figure 7.22). Truth and Reconciliation Commissions asked those who had been brutal to step forward and admit what they had done in detail; those who spoke honestly were formally forgiven for their misdeeds.

Since then, under majority African rule, there has been a noticeable absence of the disasters predicted by some whites, many of whom continue to leave the country. While corruption at the highest levels has increased, and economic inequality is by some measures actually worse today than under apartheid, there has been no collapse of the economy. On the contrary, South Africa still accounts for a third of sub-Saharan Africa's production, and other Southern African countries continue to receive many of their imports from South Africa. Its cities are still the most prosperous, well managed, and best planned on the continent.

CHECK YOUR UNDERSTANDING

1. What aspect of European colonialism sets Southern Africa apart from the rest of sub-Saharan Africa?
2. Where were lengthy wars of independence fought in Southern Africa?
3. How did the Cold War affect this region?
4. Why do some Southern African countries have such high economic inequality?
5. What type of exports are most Southern African economies dependent on?
6. What are some factors that contribute to food insecurity?
7. What aspect of South Africa's recent history is an inspiration to many parts of the world torn apart by conflict?

CRITICAL THINKING QUESTIONS

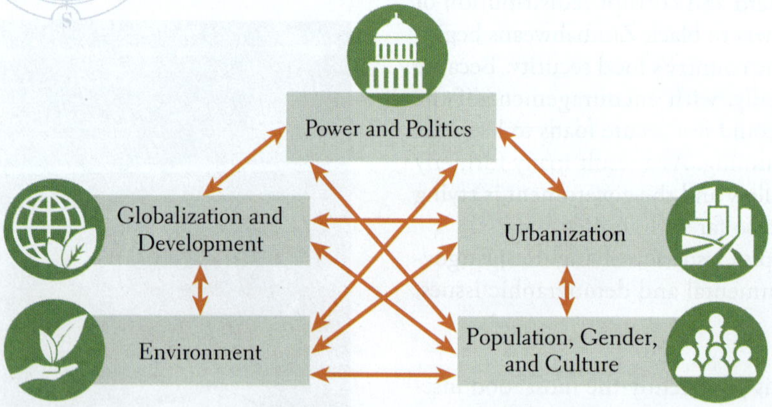

The diagram represents connections among the five geographic themes that structure this book. Listed below are some important questions that have been addressed in this chapter. Answer each question, and indicate where in the diagram you think the topics in each question belong.

1. How did Africa's unique arrangement of landforms develop?
2. What are some advantages and disadvantages of this region's agricultural systems in the face of climate change?
3. What are some of the main forces that lead to deforestation in this region?
4. What are some of the legacies of the early European trading companies in this region?
5. Why do so many sub-Saharan African countries struggle with economic instability?
6. How is Asia's investment in Africa similar to that of the U.S. and European interests whose practices led to the era of debt accumulation and SAPs?
7. Why are governments generally weak in this region?
8. How did the Cold War contribute to corruption and debt accumulation in Congo (Kinshasa)?
9. Why are large unplanned slums so common in this region?
10. How is urbanization influencing population growth in sub-Saharan Africa?

11. Why are populations growing so fast in this region?
12. What forces are slowing population growth?
13. How are better-educated women who are able to pursue careers having an impact on this region?
14. What factors make women in this region more likely than men to contract HIV?
15. How does female genital mutilation affect human development?
16. What are most traditional African religions based on?
17. Why did Europeans manipulate ethnic identities in this region?
18. What distinguishes Southern Africa from the rest of this region?

Key Terms

1884 Berlin Conference 382
agroforestry 372
animism 409
apartheid 385
carbon sequestration 372
commercial agriculture 373
commodities 386
commodity dependence 387
divide and rule 383
fallow period 373
female genital mutilation (FGM) 406
food security 373
genocide 398
grassroots economic development 392
groundwater 378
Horn of Africa 372
inland delta 370
intertropical convergence zone (ITCZ) 370
kleptocracy 386
lingua franca 411
mono-crop agriculture 374
pastoralism 375
polygyny 406
Sahel 370
shifting cultivation 373
subsistence agriculture 373
transparency 395
value chains 391

More Practice at SaplingPlus

Read the interactive e-text, review key concepts, and check your understanding.

Medha Patkar addresses displaced people now living in a slum outside Mumbai. [Sebastian D'Souza/AFP/Getty Images]

8

South Asia

In 2018, Medha Patkar stood on the banks of the Narmada River, in the tiny village of Chota Barda in Madya Pradesh, India. She was speaking to several thousand farmers who had marched there to do "jal satyagraha," or "water truth-insistence," which involves submersion in the Narmada as an act of nonviolent political resistance. Part of Chota Barda will soon be flooded by the reservoir created by the Sardar Sarovar Dam.

This was not the outcome that Patkar, the leader of the "Save the Narmada River" movement, had hoped for. She first arrived here in 1985 to survey the villages that would be submerged by the dam, and then took part in decades of battles to stop its construction. However, to date, over 320,000 farmers and fishers have been forced to move. Only a few have received adequate compensation, and many have ended up in crowded urban slums.

While some have benefited from the dam, others have not. In Gujarat, the nearby state where the dam is located, 30 million people were supposed to have more access to water, primarily for irrigation. Some farmers are getting water, but many complain that smaller distribution canals have yet to be built. In the downstream unflooded sections of the Narmada River, poorly managed releases of water have caused massive die-offs of aquatic life, which have put many fisher people out of work.

So, Medha Patkar speaks to a partially submerged crowd at Chota Barda: "Those who have everything, they should think about those who have nothing. There should be no divide between the rich and the poor, between men and women, between castes, between this religion or that. In this struggle, we are all together and united. And we stand for non-violence."

Learning Objectives

Environment: Physical and Human

8.1. Analyze how landforms and the intertropical convergence zone interact to shape seasonal rainfall patterns in South Asia.

8.2. Evaluate the factors that put lives at risk due to climate change in this region.

8.3. Compare the many impacts and legacies of British rule in South Asia.

Globalization and Development

8.4 Evaluate how South Asia's recent economic development and openness to globalization have affected society.

Power and Politics

8.5 Evaluate the impact of religious nationalism on the politics of this region.

Urbanization

8.6 Assess the forces creating divergent patterns of urbanization in South Asia.

Population, Gender, and Culture

8.7 Identify the factors slowing population growth in this most populous and most densely populated of world regions.

8.8 Explain why men significantly outnumber women in South Asia.

8.9 Describe current social and political transitions that challenge traditional cultural practices related to gender.

Subregions of South Asia

8.10 Identify key characteristics of the subregions of South Asia.

NORTHERN AFGHAN PROVINCES
1 JOWZJAN
2 BALKH
3 SAMANGAN
4 KONDUZ
5 BAGHLAN
6 PARVAN
7 VARDAK
8 TAKHAR
9 KAPISA
10 KABUL
11 LOWGAR
12 BADAKHSHAN
13 LAGHMAN
14 KONARHA
15 NANGARHAR
Caspian Sea
Ashkhabad
TURKMENISTAN
Mashhad
Dushanbe
TAJIKISTAN
Pamirs
Muztagata elev. 24,757
Taklimakan Desert
Karakoram Mts.
K2 elev. 28,250
Hindu Kush
NORTHERN AREAS
NORTH-WEST FRONTIER
Khyber Pass
C H I
Ladakh
Plateau
Himal
Mazar-e-Sharif
FARYAB
BADGHIS
Herat
HERAT
GHOWR
BAMIAN
Kabul
ORUZGAN
GHAZNI
PAKTIA
FARAH
AFGHANISTAN
PAKTIKA
ZABOL
Kandahar
NIMRUZ
HELMAND
KANDAHAR
Zagros Mts.
IRAN
FED. ADM. TRIBAL AREAS
Quetta
PAKISTAN
BALOCHISTAN
Central Makran Range
Peshawar
Islamabad
Rawalpindi
Srinagar
JAMMU & KASHMIR
Jammu
AZAD KASHMIR
Gujanwala
Lahore
Amritsar
HIMACHAL PRADESH
Faisalabad
PUNJAB
Multan
Ludhiana
PUNJAB
Chandigarh
Indo-Gangetic Plain
UTTARANCHAL
Bahawalpur
HARYANA
Meerut
Delhi
DELHI
New Delhi
NEPAL
Kathmandu
Indus
SINDH
Hyderabad
Karachi
RAJASTHAN
Jaipur
Agra
UTTAR PRADESH
Lucknow
Ganga (Ganges)
Yamuna
Jodhpur
Thar Desert
Kanpur
Ayodhya
Ghaghra
Allahabad
Varanasi (Benares)
Patna
Ganga
Kochi
INDIA
Gulf of Oman
Muscat
OMAN
AE
Indus River Delta
Rann of Kutch
Bhuj
Gulf of Kutch
GUJARAT
Ahmadabad
MADHYA PRADESH
Bhopal
Jabalpur
JHARKHAND
Indore
Vindhaya Range
Rajkot
Vadodara
Kathiawar Peninsula
Palitana
Narmada
Satpura Range
Surat
CHHATTISGARH
Jamshedpur
DAMAN & DIU
Aurangabad
Nagpur
ORISSA
Gulf of Cambay
DADRA & NAGAR HAVELI
Nashik
MAHARASHTRA
Arabian Sea
Mumbai (Bombay)
Pune
Bhima
Godavari
Hyderabad
ANDHRA PRADESH
Visakhapatnam
Eastern Ghats
Krishna
Vijayawada
Yanam
Western Ghats
Panaji
GOA
KARNATAKA
Karwar
Deccan Plateau
Bangalore
Chennai (Madras)
Mysore
Laccadive Sea
Mahe
Nilgiri Hills
Pondicherry
Malabar Coast
Coimbatore
KERALA
TAMIL NADU
Karikal
Kavaratti
Madurai
Palk Strait
LAKSHADWEEP
Cochin
Alleppey
Jaffna
Trivandrum
SRI LANKA
Gulf of Mannar
INDIAN
Colombo
Central Highlands
MALDIVES

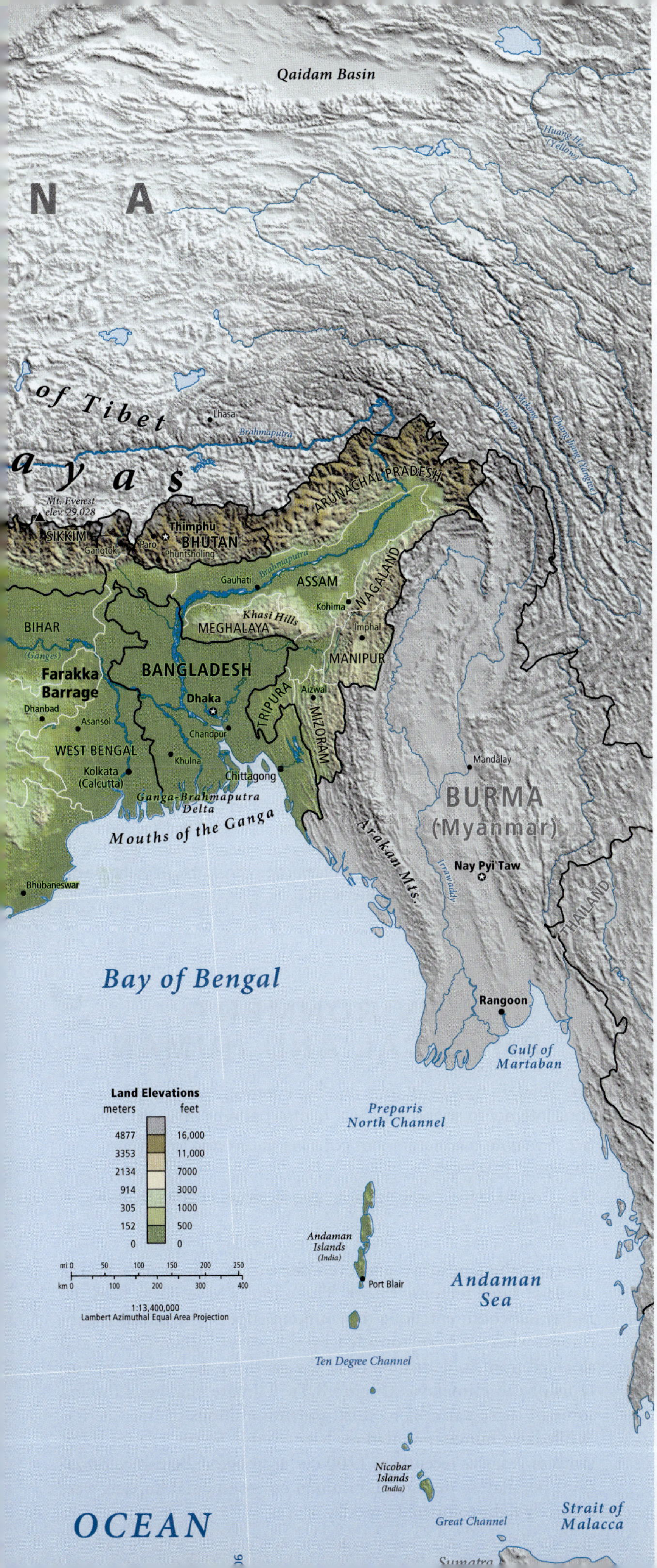

Conflicts such as the one over the Sardar Sarovar Dam are common in throughout South Asia, which is still a poor and densely populated region (**Figure 8.1**), and they are becoming more common as inequalities increase between those who gain the most from economic development and those who benefit less.

India, the giant of South Asia, has ambitions to become "the next China," with recent economic growth making it the third-largest economy in terms of purchasing power parity (PPP). Long mired in deep poverty and saddled with a corrupt political culture, India now projects a dynamic image to the world with its large, highly skilled English-speaking middle class, a state-of-the-art technology sector, giant infrastructure projects like the dam (**Figure 8.2**), and a government, led by Prime Minister Narenda Modi, that is eager to remove obstacles to economic growth. And yet many "obstacles" to economic growth stem from issues of fairness in a society where so many people barely survive.

So far this region's well-established, democratic decision-making structures have helped resolve many conflicts. South Asia's ability to maintain a hold on democracy is thus more important than ever as it tries to balance the ambitions of leaders like Modi with the rights of poor people who find themselves "in the way" of economic development.

What Makes South Asia a Region?

The countries of South Asia are shown in **Figure 8.3**. Physically, they are part of the *Indian subcontinent*—the portion of a tectonic plate that joined the Eurasian continent nearly 60 million years ago (discussed below). Historically and culturally, they share a history linked to conquerors from Central Asia, explorers and traders from the Arabian Peninsula and Southeast Asia, and more recent colonizers from Europe.

Terms in This Chapter

South Asians have adopted new place-names to replace the names given to them during British colonial rule. The city of Bombay, for example, is now officially *Mumbai*, Madras is *Chennai*, Calcutta is *Kolkata*, Benares is *Varanasi*, and the Ganges River is the *Ganga River*.

◀ **Figure 8.1** Physical features of South Asia.

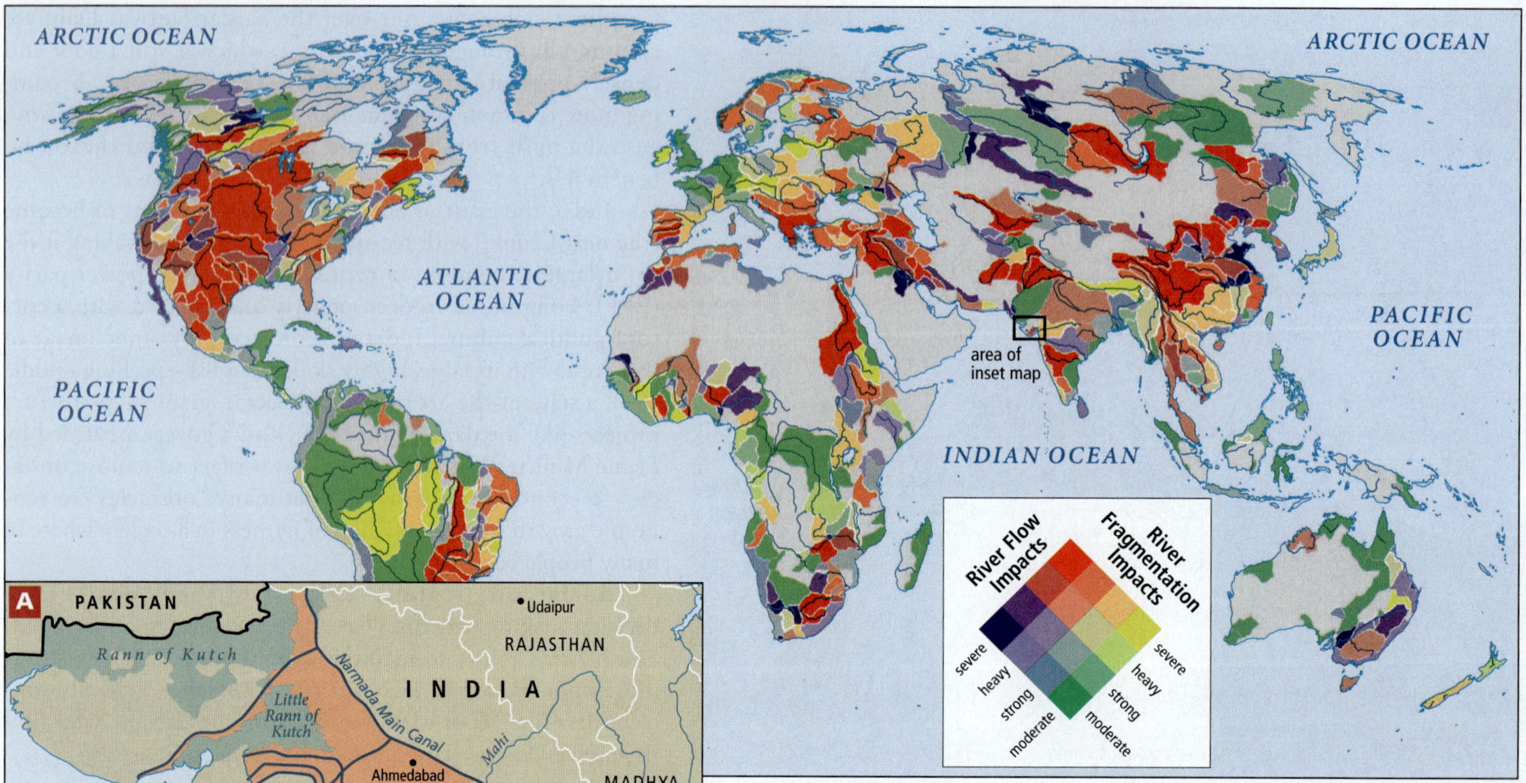

Figure 8.2 The Sardar Sarovar Dam on the Narmada and ecological impacts of major dams around the world. The Sardar Sarovar Dam is only one of hundreds of thousands of dam projects throughout the world that together have displaced between 40 and 80 million people. The map shows some of the impacts of dams around the world on water flow regulation and fragmentation of rivers. **A** The submergence zone (red) and water distribution canals (blue) for the Sardar Sarovar Dam. [McGill University HydroLab]

Figure 8.3 Political map of South Asia.

ENVIRONMENT: PHYSICAL AND HUMAN

8.1. Analyze how landforms and the intertropical convergence zone interact to shape seasonal rainfall patterns in South Asia.

8.2. Evaluate the factors that put lives at risk due to climate change in this region.

8.3. Compare the many impacts and legacies of British rule in South Asia.

Many of the landforms and even climates of South Asia are the result of huge tectonic forces. These forces have positioned the Indian subcontinent along the southern edge of the Eurasian continent, where it is surrounded by the warm Indian Ocean and shielded from cold airflows from the north by the massive mountains of the Himalayas (Figure 8.1). Climate change is forcing some of these patterns to shift, putting millions of lives at risk. While large human populations have lived in South Asia for thousands of years, as recently as 1700 C.E. (just before British colonization) population density and human environmental impacts were relatively light compared to today.

Figure 8.4 Himalaya Mountains, Nepal, as seen from space. [Courtesy NASA]

Figure 8.5 Indus River Valley, northern Pakistan. [Eco Images/Universal Images Group/Getty Images]

LANDFORMS

The most distinctive feature of the Indian subcontinent originates from the collision of two tectonic plates, the Indian-Australian Plate and the Eurasian Plate, with edges made of continental crust (see Figure 1.6), which is thicker and less dense than oceanic crust. The result is the world's highest mountains, the Himalayas, which rise more than 29,000 feet (8800 meters), formed by the crumpled and buckled edges of these two plates, as well as other mountain ranges that curve away from the central impact zone. In the west are the Hindu Kush and Karakoram range, and in the east, the Mizoram range and others that border China and Burma (Figure 8.1 and **Figure 8.4**). Over time the continuous compression has lifted the Plateau of Tibet, which rose up behind the Himalayas, to an elevation of more than 15,000 feet (4500 meters) in some places. To the south of this mountain zone is a large peninsula made up of land that broke off of Africa 60 million years ago (Figure 1.6).

The three great rivers of South Asia—the Indus, the Ganga, and the Brahmaputra—all begin within 100 miles (160 kilometers) of one another in the Himalayan highlands near the Tibet, Nepal, and India borders (**Figure 8.5**), where they are fed by the seasonal melting of glaciers. The main river basins of the Indus and Ganga lie to the southwest and south of the Himalayas in what is called the Indo-Gangetic Plain (**Figure 8.6**). The Indus flows southwest through Pakistan and ultimately joins the Arabian Sea. The Ganga flows southeast to the Bay of Bengal. The Brahmaputra flows east through Tibet and then turns south through Arunachal Pradesh and Assam into Bangladesh, where the mouths of the Ganga and Brahmaputra are blended together in a giant, flood-prone, ever-changing delta.

Lying south of the Indo-Gangetic Plain and the Narmada River Valley is the Deccan Plateau, an area of modest uplands (1000–2000 feet [300–600 meters] in elevation) interspersed with river valleys (**Figure 8.7**). This upland region is bounded on the east and west by two moderately high mountain ranges, the Eastern and Western Ghats (**Figure 8.8**). The Ghat mountain ranges descend to long, narrow, and flat coastlines made of river deltas and floodplains (**Figure 8.9**). The river valleys and coastal zones are densely occupied; the uplands are only slightly less so.

Figure 8.6 Indo-Gangetic Plain, north India. [Adrian Pope/Getty Images]

Figure 8.7 Deccan Plateau, Karnataka, India. [Michele Falzone/Getty Images]

Because of its continual high degree of tectonic activity and deep crustal fractures, South Asia is prone to devastating earthquakes, such as the magnitude 7.7 quake that shook the state of Gujarat in western India in 2001, which killed over 14,000; the magnitude 7.6 quake that hit the India–Pakistan border region in

Figure 8.8 Western Ghats, Kerala, India. [Michele Falzone/Getty Images]

Figure 8.9 Coastal lowlands, South India. [John Lund/Getty Images]

2005, which killed over 86,000; and the magnitude 7.8 quake in Nepal in 2015, which killed over 8000. Coastal areas are also vulnerable to tidal waves, or *tsunamis*, that are caused by undersea earthquakes. In 2004, a massive tsunami originating off Sumatra in Southeast Asia wrecked much of coastal Sri Lanka and southern India, killing over 54,000 people in South Asia and another 230,000 in Southeast Asia.

monsoon a seasonal period of precipitation or dryness: the wet summer monsoon begins with the arrival of rain in June; the dry winter monsoon begins by November

CLIMATE AND VEGETATION

The climate of South Asia is shaped by a seasonal reversal of winds known as **monsoons** (**Figure 8.10**). These monsoon winds are

Figure 8.10 Summer and winter monsoons in South Asia.

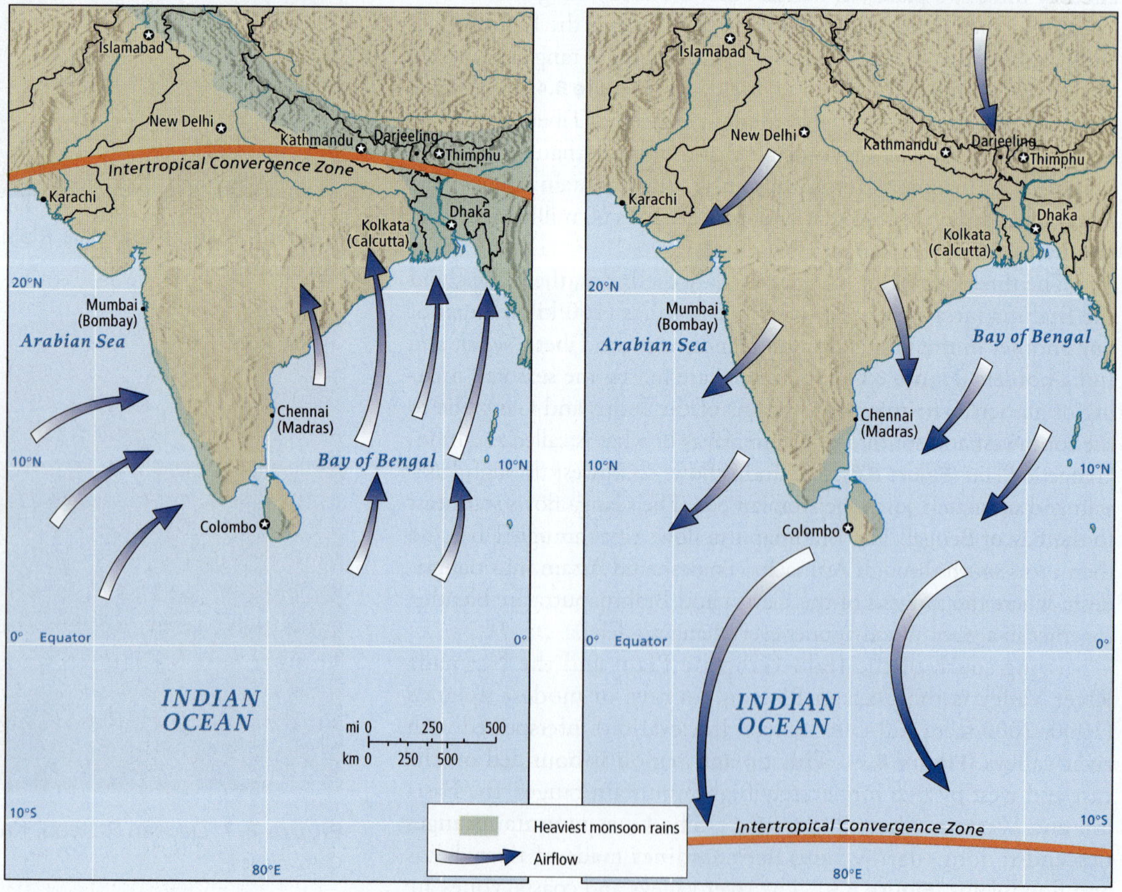

(A) In the summer, the ITCZ is drawn north, picking up huge amounts of moisture from the ocean, which are then deposited over India, Sri Lanka, and Bangladesh.

(B) In the winter, cool dry air blows from the Eurasian continent south across India, pushing the ITCZ south.

affected by the *intertropical convergence zone (ITCZ)*, which, as the name suggests, is a zone created when air masses moving south from the Northern Hemisphere and north from the Southern Hemisphere converge near the equator. As the warm air rises and cools, it drops copious amounts of precipitation. As described in Chapter 7, the ITCZ shifts north and south seasonally. The intense rains of South Asia's summer monsoons are likely caused by the ITCZ being sucked onto the land by a vacuum created when huge air masses over the Eurasian continent heat up and rise into the upper troposphere.

The summer monsoon (see Figure 8.10A) begins in early June, when the warm, moist ITCZ air first reaches the mountainous Western Ghats. The rising air mass cools as it moves over the mountains, releasing rain that nurtures patches of dense tropical rain forests and tropical crops on the upper slopes of the Western and Eastern Ghats and in the central uplands. Once on the other side of India, the monsoon gathers additional moisture and power in its northward sweep up the Bay of Bengal, sometimes turning into tropical cyclones.

As the monsoon system reaches the hot plains of Bangladesh and the Indian state of West Bengal in late June, columns of warm, rising air create massive, thunderous cumulonimbus clouds that drench the parched countryside. Monsoon rains sweep east to west, parallel to the Himalaya Mountains, moving across northern India to Pakistan and finally petering out over the Kabul Valley in eastern Afghanistan by July. Rainfall is especially intense in the east, north of the Bay of Bengal, where the East Khasi Hills hold the world record for annual rainfall—about 35 feet, even though no rain falls for half the year. These patterns of rainfall are reflected in the varying climate zones (**Figure 8.11**) and agricultural zones of South Asia (see Figure 8.17 later in the chapter). Although sufficient rain falls in central India to support forests, most land has long been cleared of forest (see "Deforestation" section) and planted in crops. Much of the remaining forests are small and so separated from each other that they no longer provide suitable habitat for India's wildlife. Forest vegetation is more common in the Himalayan highlands and foothills, but even in the mountains there is widespread deforestation, often leaving only patchy areas of formerly richly forested land.

Periodically, the monsoon seasonal pattern is interrupted and serious drought ensues. This happened in July and August of 2009, when the worst drought in 40 years struck much of South Asia (see the discussion below). Then in September 2009, heavy rains came to central south India, causing crop-damaging floods that killed hundreds of people. Increasingly, scientists are concluding that the extreme droughts and floods of recent years are not an anomaly but instead are part of global climate change.

The winter monsoon (see Figure 8.11B) is underway by November each year, when the cooling Eurasian landmass sends cooler, drier, heavier air over South Asia. This cool, dry air from the north sinks down to the lower elevations and pushes the warm, wet air back south to the Indian Ocean. Very little rain falls in most of the region during this winter monsoon. However, as the ITCZ retreats southward across the Bay of Bengal, it picks up moisture that is then released as early winter rains over parts of southeastern India and Sri Lanka.

The monsoon rains deposit large amounts of moisture over the Himalayas, much of it in the form of snow and ice that add to the existing mass of glaciers (see Chapter 1). Meltwater from these glaciers feeds the headwaters of the Indus, the Ganga, and the Brahmaputra rivers, which carry enormous loads of sediment. When the rivers reach the lowlands, their velocity slows and much of the sediment settles out as silt, which is distributed by successive floods. As illustrated in the diagram of the Brahmaputra River in **Figure 8.12**, the movement of silt constantly rearranges the floodplain landscape, complicating human settlement and agricultural efforts. However, the seasonal deposit of silt nourishes much of the agricultural production in the densely occupied plains of Bangladesh. The same is true on the Ganga and Indus plains.

CHECK YOUR UNDERSTANDING

1. Why is South Asia prone to earthquakes?
2. What are the three major rivers of this region and where do they originate?
3. What is a potential result of interruptions in monsoon rainfall patterns?
4. How does the movement of silt in rivers impact the plains of Bangladesh?

ENVIRONMENT

Climate change puts more lives at risk in South Asia than in any other region in the world, primarily due to water-related issues. Over the short term, droughts, floods, and the increased severity of storms imperil many urban and agricultural areas. Over the longer term, sea level rise may profoundly affect coastal areas and glacial melting poses a threat to rivers and aquifers. South Asia has also emerged as the third-largest contributor of greenhouse gasses, mostly due to its growing use of energy.

Vulnerability and Resilience to Climate Change

All across South Asia, people face a wide range of challenges made more serious by climate change (**Figure 8.13**). Here, we focus on six of these challenges, all of which relate to water in some way, followed by some of the responses to climate change that can increase this region's resilience to climate change.

1. Drought and Flooding Significant sections of South Asia have lived with water shortages for millennia, mostly due to low average rainfall, and also because of extreme seasonal variability in rainfall. The Indus Valley of Pakistan, for example, has many ancient structures that captured water during wet periods for use during dry times (see Figure 8.18A later in the chapter). Now, shifting rainfall patterns and rising temperatures linked to climate change are creating drier conditions in much of South Asia, and occasionally causing abnormally heavy rainfall that can result in widespread flooding, as was the case in Pakistan in 2010 when 20 million peoples were affected by flooding, and over 2000 were killed.

Droughts and floods can easily destroy harvests. Some South Asian methods of water conservation improve resilience to both flooding and drought. Since ancient times, many South Asians have used artificial ponds to increase the rate at which water deposited during the summer monsoon percolates through the soil and into underground aquifers rather than evaporating. This practice

8.11 Climates of South Asia

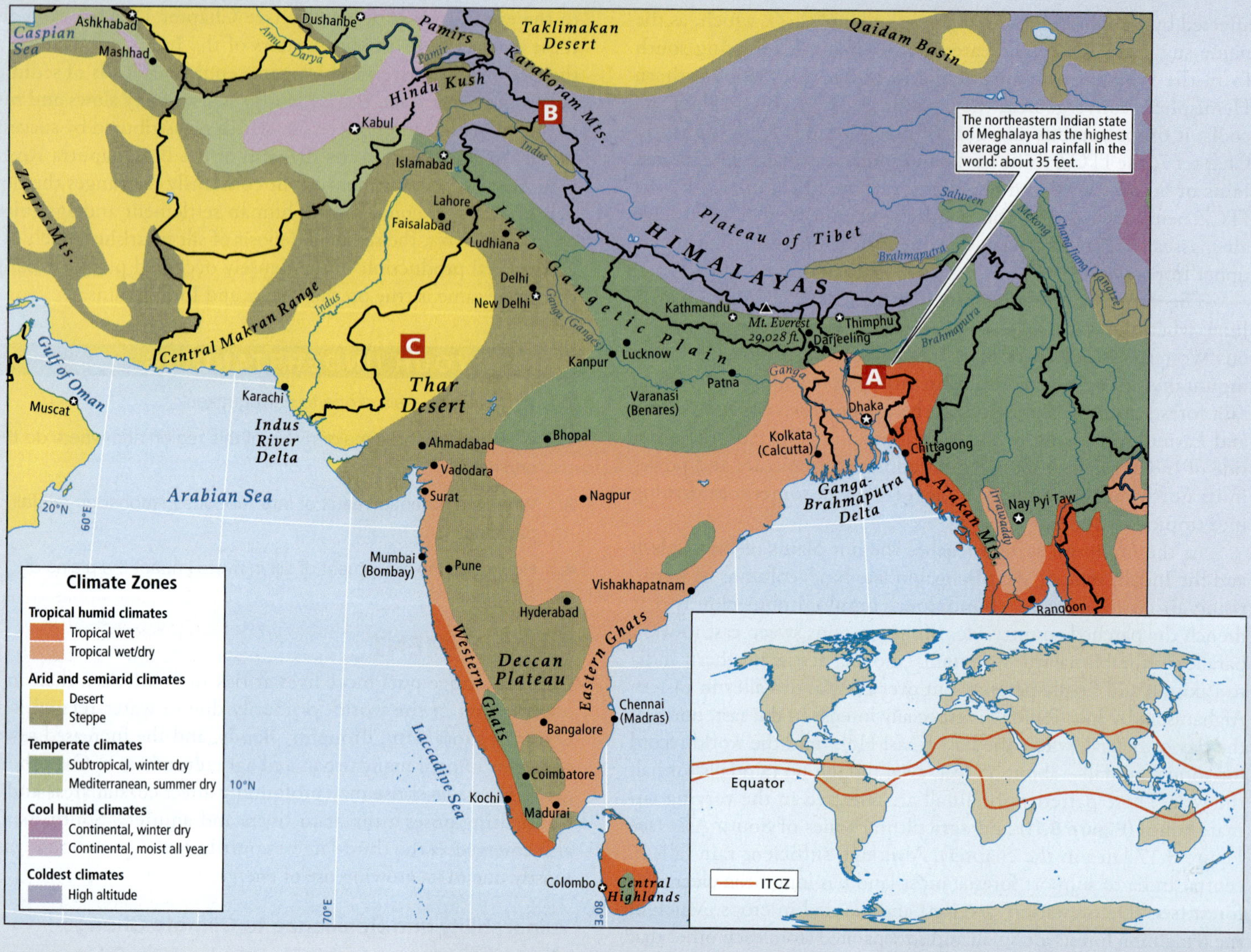

A **Tropical wet, Meghalaya, India** [Tushar S. Chowdhury/Moment Open/Getty Images]

B **High altitude, Kashmir, India** [TAUSEEF MUSTAFA/AFP/Getty Images]

C **Desert, Jaisalmer, India** [Shantanu Bedarkar/Moment Open/Getty Images]

(A) Pre-monsoon stage: The river normally flows in multiple channels across the flat plain.

(B) Peak flood stage: During peak flood times, the great volume of water overflows the banks and spreads across fields, towns, and roads. The force of the water carves new channels, leaving some places cut off from the mainland.

(C) Post-monsoon stage: The river returns to its banks, but some of the new channels persist, changing the lay of the land and access for those who use the land. As the river recedes, it leaves behind silt and algae that nourish the soil. New ponds and lakes form and fill with fish.

Figure 8.12 The Brahmaputra River in Bangladesh at various seasonal stages. People who live along the river have learned to adapt their farms to a changing landscape, and along much of the river, farmers are able to produce rice and vegetables nearly year-round. [Research from: *National Geographic*, June 1993, p. 125.]

makes more water available for irrigation during the dry season via wells. More recent methods involve constructing small berms (raised strips of land) on fields and check dams in gullies to hold rainwater for longer so that it can drain into the soil. Because these "artificial recharge" techniques reduce the amount of water that remains on the surface during the monsoon, they help limit flood damage.

The most common method of boosting resilience to drought in South Asia is irrigation, with many ancient and new methods accounting for as much as 80 percent of water use (see Figure 8.13B). Most irrigation techniques in this region use ditches to spread water through fields, but this requires large amounts of water and can result in the buildup of salt in the soil. The Indian government now subsidizes about half of the cost of drip irrigation technology, which uses up to 90 percent less water. Even so, drip irrigation is still too expensive for the vast majority of farmers, who often cultivate just three or four acres and don't earn enough to cover the cost of even subsidized drip systems. A lower-cost adaptation to drought involves switching from water-intensive crops, such as rice, cotton, and sugarcane, to less-water-intensive crops such as lentils, chickpeas, and vegetables. The government of India is beginning to encourage this by extending subsidy programs that support prices for the water-intensive crops to include the less intensive crops.

2. Glacial Melting Because South Asia's three largest river systems are fed, in part, by glaciers high in the Himalayas, the issue of glacial melting is of particular concern. Unless effective action against climate change is taken, Himalayan glaciers could eventually disappear, causing many large South Asian rivers to run nearly dry during the winter dry season. Around half of the region's population depends on these glacially fed rivers for drinking water, domestic and industrial uses, and irrigation.

While the immediate effect of glacial melting may be flooding, the real threat is that over the long term, smaller glaciers will provide less water each year to rivers and to recharge the ancient aquifers beneath the heavily populated Indo-Gangetic Plain south of the Himalayan Range.

Both water conservation and increased water storage will be needed for supplies to last through the dry winter monsoon. High Himalayan communities in Ladakh, on the India–China border, are experimenting with retaining autumn glacial melt in stone catchments and other structures where the melt refreezes over the winter and is available for spring irrigation. While the efficacy of these "artificial glaciers" is unclear, they do represent a reproducible grassroots response to climate change.

3. Sea Level Rise Tens of millions of impoverished farmers and fishers live near sea level in South Asia, most of them in Bangladesh, which has more people vulnerable to sea level rise (see Chapter 1) than any other country in the world (Figure 8.13A). With 166 million people already squeezed into a country the size of Iowa, over 10 percent of Bangladeshis might have to find new homes if sea levels rise by 3 to 5 feet (0.9 to 1.5 meters). The biggest economic impacts of sea level rise would result from the submergence of parts of South Asia's largest cities, such as Dhaka in Bangladesh,

PHOTO ESSAY

8.13 Vulnerability to Climate Change in South Asia

Climate change is putting more lives at risk in South Asia than in any other region. Many poor people live in areas that are exposed to climate-related hazards, such as floods, droughts, and sea level rise. Meanwhile recovery from these events is complicated by political instability and governments that often lack the capacity to plan for and respond to large-scale disasters. More prosperous areas often have better developed emergency response systems, but even there disparities in wealth can leave large populations highly vulnerable. Fortunately responses to climate change are developing as individuals, companies, and governments struggle to reduce their vulnerability. Some build on age-old strategies for coping with climate variability, while others are entirely new.

THINKING GEOGRAPHICALLY

A How might such embankments help reduce sensitivity to storm surge and sea level rise?

B Why is the use of simple irrigation technology a sensible strategy in a country like Afghanistan?

C What suggests that this girl comes from a very poor family?

D List all the indications in this photo that Mumbai has drainage problems.

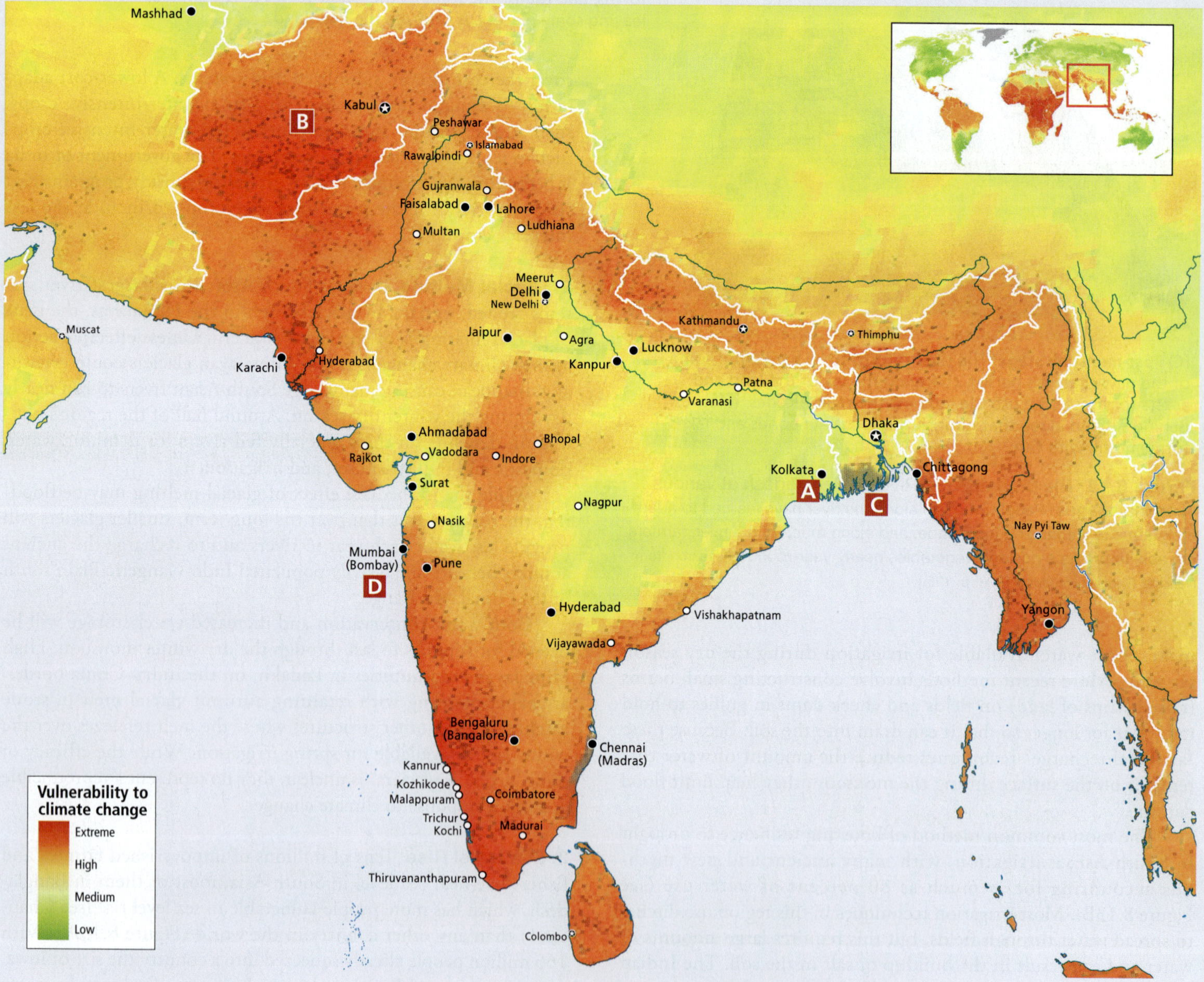

A Reducing sensitivity to sea level rise in Kolkata. Workers add boulders to reinforce an embankment that guards against erosion south of Kolkata (formerly Calcutta) on an island that is very exposed to sea level rise and storm surge. Such embankments can reduce sensitivity to water level changes, but they are expensive to build and require maintenance. [DESHAKALYAN CHOWDHURY/AFP/Getty Images]

B Reducing sensitivity to water shortage in Afghanistan. In Afghanistan, a man makes improvements to irrigation infrastructure that helps reduce the sensitivity of nearby areas to water shortage. Widespread poverty and political instability contribute to Afghanistan's extremely high vulnerability to climate hazards. [Robert Nickelsberg/The LIFE Images Collection/Getty Images]

C Reducing vulnerability in Bangladesh. A girl in coastal Bangladesh washes fish outside her home, which was flooded and damaged after a cyclone destroyed the river embankment that once protected her village from high tides. Bangladesh's large poor population is very exposed and sensitive to sea level rise; however, government capacity to plan for and respond to climate-related hazards has significantly reduced its overall vulnerability. [MUNIR UZ ZAMAN/AFP/Getty Images]

D Increasing resilience in Mumbai. A high tide during monsoon rains floods part of Mumbai's rail system. Low-lying Mumbai (formerly Bombay) is at significant risk for sea level rise; while the city's wealth and well-developed emergency response systems have increased its resilience in the face of climate hazards, much of Mumbai remains quite sensitive. [Kunal Patil/Hindustan Times via Getty Images]

Mumbai (see Figure 8.13D and Chapter 1) and Kolkata in India, and Karachi in Pakistan.

Also threatened by continuing sea level rise are the Maldives Islands in the Indian Ocean, 80 percent of which lie 3 feet (or 0.9 meter) or less above the sea. Beach erosion is so severe that homes built only a few years ago in this richest of South Asia's countries are falling into the sea.

All these places are developing responses to climate change, though the resources available to do so are often very limited (see Figure 8.13A). Over the short term, cities that can afford the expense, such as Mumbai or other large coastal cities, are building dikes, storm walls, and other "defensive" structures, or are improving their drainage systems. Smaller cities and towns and rural areas are using cheaper "green" approaches, such as restoring coastal ecosystems, such as mangrove swamps, that can help keep coastlines from eroding. In southern Bangladesh, coastal farmers are responding to increased soil salinity, due to intruding seawater, by switching to aquaculture (fish farming). The Bangladesh Rice Research Institute has also developed varieties of rice that are more tolerant of salty soils. However, many poor farmers are unable to adapt in these or other ways, and hundreds of thousands are forced to migrate each year, moving to other parts of Bangladesh but also to India. Over the long term, tens of millions of Bangladeshis will likely have to relocate, a scenario that is sure to bring conflict with India, which is already home to several million illegal Bangladeshi migrants.

4. Conflicts over Rivers With more than 20 percent of the world's population and only 4 percent of its fresh water, this region has a high potential for conflict over water resources, especially rivers, many of which travel through multiple countries. The three main river systems all originate partially in Tibet, where the Chinese government is building numerous dams that could reduce water flows to South Asia. The most water-scarce areas are in Afghanistan, Pakistan, and northwest India, but water disputes may be found everywhere.

The Indus Waters Treaty covers the six rivers that start in India and flow into Pakistan. Sponsored by the World Bank and signed in 1960, the internationally recognized treaty has allowed India to use a set amount of water from the portions of the rivers in its territory without sparking war with Pakistan. The treaty remained intact despite three wars between India and Pakistan, though India has threatened to review the treaty in response to terrorist attacks in Kashmir that it suspected Pakistan of supporting.

In contrast, of the 54 rivers that flow through India into Bangladesh, only one has a treaty. This reflects both the much larger water resources in eastern South Asia and the lower potential for war between India and Bangladesh. India uses the absence of treaties to its advantage, leveraging its size and power to dominate Bangladesh, which, unlike Pakistan, cannot appeal to internationally recognized treaties with well-developed conflict-resolution mechanisms. Instead, Bangladesh often has to wage elaborate diplomatic and public relations campaigns merely to get a response from India. Usually, poor people and river ecosystems end up suffering.

Perhaps the most serious water conflict in South Asia surrounds India's diversion of as much as 60 percent of the Ganga River flow to Kolkata to flush out navigation channels where silt is accumulating and hampering river traffic. These diversions, via a dam across the Ganga called the Farakka Barrage (see Figure 8.1), deprive Bangladesh of normal freshwater flow to the Ganga–Brahmaputra delta, allowing salt water from the Bay of Bengal to penetrate inland, ruining agricultural fields and damaging its fishing industry. In the late 1990s, India signed a treaty with Bangladesh promising a fairer distribution of water, but Bangladesh still receives a considerably reduced flow. Massive protests in Bangladesh have been met with indifference in India, where the government is unapologetic about putting the interests of Kolkata's 16 million people ahead of the 40 million rural Bangladeshis affected by its water diversions. One positive development in recent years is India's willingness to let more water flow into Bangladesh to raise the level of the Ganga, facilitating the migration of fish up the river from Bangladesh. This will directly benefit Indian and Bangladeshi fishers while helping to sustain biodiversity along the river.

Without strong global or regional agreements to help countries adapt to climate change, the potential for conflict over water will increase. Any new global agreements to mitigate climate change impacts could weaken India's dominance by recognizing the rights of Bangladesh to access water resources at current levels. In the absence of such agreements, India could even more assertively dominate a Bangladesh weakened by sea level rise.

5. Water Use and Pollution As living standards rise in South Asia, so does the use of water and its pollution. As we learned in Chapter 1, humans require an average of 5 to 13 gallons (20 to 50 liters) of clean water per day for basic domestic needs: drinking, cooking, and bathing/cleaning. In South Asia's poorest urban and rural areas, per capita domestic water consumption is about 5 to 6 gallons per day. Consumption increases as incomes rise and people acquire conveniences that consume large amounts of water, such as a toilet that can use several gallons in a single flush. The middle and upper classes consume closer to 13 gallons per day, which is still low compared to the 80–100 gallons used by the average U.S. citizen.

Wastewater reuse is already widely practiced in South Asia, where it is often used for irrigation by vegetable farmers around cities who value it as a reliably available water resource that has a high nutrient content. This use helps explain why less than 30 percent of this region's wastewater is treated, as cities can often sell it for a small profit, instead of spending money to treat it. While health risks to consumers of wastewater-irrigated vegetables are poorly understood, health risks to farmers are clear and significant, with those using untreated wastewater reporting a higher incidence of skin conditions and allergies. Moreover, as South Asia's cities grow in size and wealth, the amount of wastewater generated, as well as the presence of more pollutants, will necessitate more treatment and better collection systems before wastewater can be safely reused.

As is often the case in South Asia, religion plays a role in the resolution of serious problems such as water pollution in the Ganga, the holiest river for Hindus, and also the source of water for 500 million people, more than any river in the world. Although it has long been polluted, urban and industrial growth along the Ganga in India now make it one of the five most polluted rivers in the world. Meanwhile climate change is contributing to erratic

Figure 8.14. Pollution in the Ganga River. Members of India's National Cadet Corps clean the banks of the Ganga River. Plastic pollution is a huge and growing problem in South Asia, as is pollution of rivers with untreated wastewater, which is a major cause of diarrheal diseases. In India 5 to 12 percent of premature deaths are caused by diarrheal diseases. [Ritesh Shukla/NurPhoto/Getty Images]

changes in the Ganga's flow, leading to both floods and very low flow volume. In 2014 Narendra Modi became prime minister of India as the leader of a Hindu nationalist political party that made many promises to clean up the Ganga (**Figure 8.14**). However, noticeable improvements have been elusive, with many wastewater treatment projects stalled by corruption and hampered by inadequate supervision of private contractors. Modi's government has also been unwilling to curtail development projects that negatively impact the Ganga, such as the numerous hydroelectric dams on the river's upper reaches, which threaten to further reduce the river's at times dangerously low volume.

6. Virtual Water and Exports Much of South Asia's water is used to create agricultural exports, such as rice, cotton, or clothing made from locally produced cotton. This makes South Asia a major exporter of virtual water (Chapter 1), which is problematic in a region where water is increasingly scarce. Afghanistan, Pakistan, and northwest India are all naturally dry regions that have been made more arid by thousands of years of human use, and all are heavily dependent on agricultural exports that require irrigation. Currently, India is the world's largest exporter of rice, and the second-largest exporter of cotton, both of which require enormous amounts of irrigation water. Pakistan's exports are dominated by cotton clothing, which uses primarily locally grown cotton, and Afghanistan is almost entirely dependent on agricultural exports. The costs of the water resources depleted in the production of these goods are often not sufficiently accounted for in their prices, due to government subsidies for rice and cotton.

As climates become hotter and dryer, developing South Asian countries will have to find less water-intensive items to export, or else risk pushing already scarce water resources beyond their limits. This will mean moving more of the population out of the agriculture sector, which currently supplies jobs for close to 50 percent of the region's workforce. In this shift, India is leading the way with its growing high-tech service and manufacturing industries, which contribute much more to the country's gross national income (GNI) than its vast agricultural sector, using a fraction of the water.

Contributions to Climate Change

Although per capita emissions of greenhouse gases (GHGs) for South Asia are some of the lowest in the world, with so many people its emissions are significant. India is the world's third-largest contributor of GHGs, mostly due to coal-powered electricity generation but also due to agriculture, where large amounts of methane are produced by rice cultivation and cattle. Deforestation is a minor contributor to South Asia's GHG emissions, but efforts are underway across the region to reduce pressure on forests and promote reforestation as a way of taking carbon out of the atmosphere. Meanwhile total GHG emissions may grow more slowly in coming years as the growth of solar energy installations has already outpaced growth in new coal power plants.

Transitions to Lower GHG Emissions While GHG emissions from South Asia are high and growing, major efforts are being made by governments to reduce emissions, motivated both by a sense of responsibility for maintaining the biosphere, but also to remedy many current problems related to air pollution.

Air Pollution

Of all the world regions, South Asia has the worst air pollution. According to recent studies compiled by the World Health Organization (WHO), of the ten cities with the worst air pollution (as measured by particulate matter suspended in the air), nine are in South Asia (six in India, three in Pakistan). In Delhi, the region's largest city, air pollution is among the world's worst, with more than half of children suffering permanently impaired lung capacities due to the persistent haze of smog and smoke. On average in India life expectancy is cut short by 5 years due to air pollution–related illnesses.

In urban areas, much of the blame falls on vehicles, industries, and power plants, but rural South Asia with its massive population may be a larger contributor on a regional and global scale. South Asia burns more agricultural waste, fuelwood, and animal dung than any world region. Up to a quarter of the region's air pollution is linked to the many agricultural fields, especially those growing rice, that are burnt every year after the harvest. In addition, fuelwood and animal dung are burned on millions of small, inefficient cookstoves and open fires found throughout rural and urban areas. For decades this burning has produced a brown cloud of smoke so large that it is visible from space. Known as the *Asian brown cloud*, it is large enough to influence monsoon precipitation, which is reduced in central India and intensified in northwest India and Pakistan.

Numerous efforts are underway to reduce air pollution, and most of them also reduce GHG emissions, but so far none have managed to put a dent in the problem due in part to its massive scale. A wide variety of organizations are trying to prevent burning by buying crop residues to use as packing material, in furniture, or to convert them into clean-burning fuels via gasification technologies. However, with one state in India, Punjab, producing 20 million tons of agricultural waste each year from rice alone, it may be a while before these projects can have an impact.

Some cities, such as Delhi, have made major attempts at improving air quality, such as by closing nearby coal-fired power plants, enforcing stricter emissions controls on vehicles, and limiting the burning of solid fuels throughout the city. These efforts have brought modest results, with Delhi no longer the world leader in air pollution, having fallen to sixth place behind five other north Indian cities (one of which, Faridabad, is in Delhi's metro area). At the national scale, the government of India has been wary of limiting emissions at the cost of economic growth, instead relying heavily on new technologies such as solar power generation and electric vehicles.

Solar and Wind Power Generation

South Asia, led by India, has ambitious plans to reduce both urban air pollution and GHG emissions with some of the world's largest existing solar power–generating installations (**Figure 8.15**). This represents a massive transition away from coal and new hydroelectric projects, neither of which can now compete with the low cost of solar and wind power generation. India's government has committed to building 100 gigawatts of solar power and 75 gigawatts of wind power by 2022, which is more than was installed worldwide for either source in 2017. While some are skeptical of this goal, in recent years India has been fourth in the world in adding new wind capacity and second only to China in adding new solar capacity.

Figure 8.15 A solar power–generating facility in Gujarat. Located atop the main irrigation canal from the Sardar Sarovar Dam, the solar panels help lower rates of water evaporation from the canal, and avoid the need to use scarce land for power generation. Villages and towns along the canal can also access electricity. Similar canal-top solar projects across India are planned to add several gigawatts (GW) to the national power grid in coming years. If the full length of Gujarat's canal network were covered with solar panels, roughly 60 GW could be generated, equivalent to 16 percent of India's current electricity demand. [SAM PANTHAKY/AFP/Getty Images]

Electric Vehicles

Vehicles are a major source of GHG emissions and air pollution, especially in South Asian cities, where vehicle ownership is expanding rapidly. India's national government has set an ambitious goal of 30 percent of new vehicles being electric by 2030, but India is already a world leader in electric vehicle use due to the growing popularity of electric scooters, motorcycles, and rickshaws (**Figure 8.16**). Rickshaws are small open-air three-wheeled taxis found throughout many Asian cities. Some are similar to three-wheeled bicycles, while others are motorized. Both kinds of rickshaw are now rapidly being converted to electric power throughout South Asia, with India's fleet of electric rickshaws estimated at around 2 million and rapidly growing. Most drivers are adopting the vehicles on their own without government incentives because they are cheaper to operate and maintain. With over a hundred million South Asians traveling by rickshaw every day, and the government encouraging expansion as a way to reduce air pollution and GHG emissions, many expect the electric vehicle sector to grow rapidly in the coming years. Electric motor scooters and motorcycles are spreading rapidly for similar reasons of cost effectiveness and ease of maintenance, which is a huge development given that sales of these vehicles in South Asia are roughly six times those of cars.

Deforestation Deforestation has been occurring in South Asia since the first agricultural civilizations developed between 5000 and 10,000 years ago. Ecological historians have shown that as the forests vanished, the northwestern regions of the subcontinent (from India

Figure 8.16 An electric rickshaw being manufactured near Ahmedabad in Gujarat. Most of the engines and batteries are imported from China; however, Indian suppliers are starting to appear. [SAM PANTHAKY/AFP/Getty Images]

to Afghanistan) became increasingly drier. The pace of deforestation has increased dramatically over the past 200 years. By the mid-nineteenth century, perhaps a million trees a year were felled for use in building railroads alone. As part of their plans to reduce GHG emissions, South Asian countries have plans to protect their remaining forests and to expand tree plantations. However, in the face of ongoing deforestation these can be considered *plans* only.

South Asia's forests are still shrinking because of commercial logging and expanding village populations that use wood for building and for fuel. The highest rate of deforestation is in Nepal, where 25 percent of forest cover has been lost since 1990, leaving the country with 22 percent of its original forests. Pakistan also has a high rate of deforestation, and has addressed its lack of forests outside of the mountainous north by creating tree plantations. The Indian government often gives misleading information about the extent of deforestation in the country by counting tree plantations of one or two species as forest land, even though these lands tend to be much less biodiverse than natural forest. Hence India officially reports an increase in forest area, even though on average an area the size of Connecticut is deforested each year.

Many of South Asia's remaining forests are in mountainous or hilly areas, where forest clearing dramatically increases erosion during the rainy season. In addition to the loss of CO_2-absorbing forests, massive landslides can destroy villages and close roads. With fewer trees and less soil to retain the water, rivers and streams become clogged with runoff, mud, and debris. The effects can reach so far downstream that increased flooding in the plains of Bangladesh is now linked to deforestation in the Himalayas.

Resistance to Deforestation

South Asian countries have a healthy and vibrant culture of environmental activism that has alerted the public to the consequences of deforestation. In 1973, for example, in the Himalayan district of Uttarakhand (then known as Uttaranchal), India, a sporting-goods manufacturer planned to cut down a grove of ash trees so that their factory, in the distant city of Allahabad, could use the wood to make tennis racquets. The trees were sacred to nearby villagers, however, and when their protests were ignored, a group of local women took dramatic action. When the loggers came, they found the women hugging the trees and refusing to let go until the manufacturer decided to find another grove.

The women's action grew into the *Chipko movement* (literally, "hugging"), which is also known as the *social forestry movement* (**Figure 8.17**). The movement has spread to other forest areas and has been responsible for slowing deforestation and increasing ecological awareness. Proponents of the movement argue that the management of forest resources should be turned over to local communities. They say that people living at the edges of forests possess complex local knowledge of those ecosystems that has been gained over generations—knowledge about which plants are useful for food, medicines, and fuel, and as building materials. Those who live in forested areas are more likely to manage the forests carefully because they want their descendants to benefit from them for generations to come.

Protected Areas

South Asia has a widespread network of areas that are officially protected from human development, and a tradition of setting aside

Figure 8.17 The Chipko Movement. In 1973, a grove of ash trees in Uttarakhand, India, was threatened by a sporting-goods manufacturer who wanted to make them into tennis racquets. Women from a nearby village "hugged" the trees until they were no longer threatened, sparking the Chipko, or "tree hugging," movement that spread to other threatened forests. [Bhawan Singh/The India Today Group/Getty Images]

areas for wildlife that is at least 2000 years old. However, outside of the Himalayas the total area protected is quite small (between 4 and 6 percent of the land area), and even these areas are often impacted by humans. One example is the Mudumalai Wildlife Sanctuary and neighboring national parks in the Nilgiri Hills (part of the Western Ghats). This area harbors some of the last remaining forests in southern India. Here, in an area of about 600 square miles (1554 square kilometers), live a few of India's last wild tigers and a dozen or more other rare species, such as sloth bears and barking deer. Even much smaller forest reserves play an important role in conservation. At 287 acres (116 hectares), Longwood Shola is a tiny remnant of the ancient tropical forests that once covered the Nilgiris.

Phillip Mulley, a naturalist, Christian minister, and leader of the Badaga ethnic group, points out that the indigenous peoples of the Nilgiris must now compete for space with a growing tourist industry (2.8 million visitors in 2018). In addition, huge tea plantations were cut out of forestlands by the Tamil Nadu

A A water storage tank at Mohenjo-daro, in Pakistan. [Luca Tettoni/robertharding/Getty Images]

B The Taj Mahal, a Mughal mausoleum in Agra, India. [Mlenny/E+/Getty Images]

2000,000 B.C.E. — 5000 B.C.E. — 0 C.E. — 1500 C.E. — 1600 C.E. — 1700 C.E.

200,000 B.C.E. Early humans enter South Asia

2500 B.C.E. Mohenjo-daro

1526–1857 C.E. Mughal Empire

1643–1653 Taj Mahal

Figure 8.18 VISUAL HISTORY OF SOUTH ASIA

Thinking Geographically

A This structure was designed to store water from what source?

B Suggest some principal features of the structure and grounds of the Taj Mahal.

state government to provide employment for Tamil refugees from the conflict in Sri Lanka. So while the forestry department and citizen naturalists are trying to preserve forestlands, the social welfare department, faced with a huge refugee population, is cutting them down.

CHECK YOUR UNDERSTANDING

1. What are some of the climate change impacts that put so many lives at risk in South Asia?
2. How would the disappearance of Himalayan glaciers impact South Asia's rivers?
3. Why is India able to dominate Bangladesh much more than Pakistan in disputes over water from rivers?
4. What water-intensive crops form the basis of important exports from South Asia?
5. What are some solutions to chronic air pollution in this region?
6. What are the primary threats to forests in this region?

Indus Valley civilization the first substantial settled agricultural communities in South Asia, which appeared about 4500 years ago along the Indus River in modern-day Pakistan and northwest India

HUMAN PATTERNS OVER TIME

Large human populations have lived in South Asia for at least 50,000 years, with a population estimated at 75 million as early as 2000 years ago. Home to some of Earth's oldest and most influential civilizations, South Asia faces the world's great transitions with an enviable ability to innovate while maintaining ancient traditions. Tens of thousands of years of continuous human occupation, and the integration of powerful influences from outside the region, have given South Asia astounding cultural diversity to draw on in this time of change.

The Indus Valley Civilization

There are indications of early humans in South Asia as far back as 200,000 years ago, but the first real evidence of modern humans in the region is about 38,000 years old. The first large agricultural communities, known as the **Indus Valley civilization** (or *Harappa culture*), appeared about 4500 years ago along the Indus River in what is modern-day Pakistan and northwest India. The architecture and urban design of this civilization were quite advanced for the time, with homes featuring piped sewage disposal. Towns were well planned, and laid out in grid patterns. Evidence of a trade network that extended to Mesopotamia and eastern Africa has also been found. Vestiges of the Indus Valley civilization's agricultural system survive to this day in parts of the valley, including infrastructure for storing monsoon rainfall to be used for irrigation in dry times (**Figure 8.18A**).

Scholars have long debated the reasons for the decline of the Indus Valley civilization. Some believe that complex geologic (seismic) and ecological changes (drier climate) brought about a gradual demise. Others argue that foreign invaders brought a swift collapse.

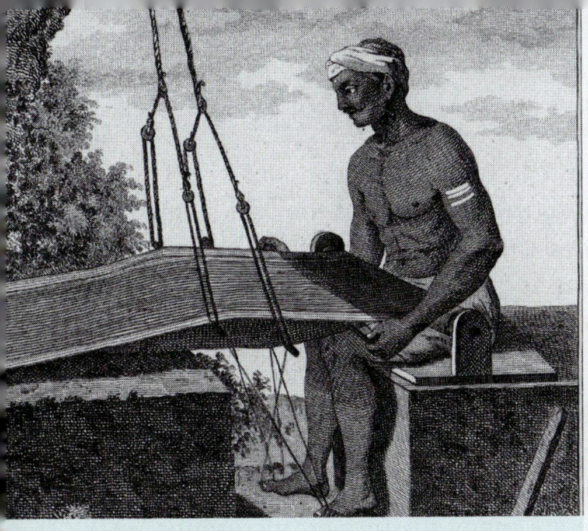

C The traditional textile-weaving industry in Bengal, India. [Science & Society Picture Library/Getty Images]

D Gandhi leading the Salt March to the sea. [Mansell/The LIFE Picture Collection/Getty Images]

E A shopping and office district near New Delhi, which is a center of employment for India's growing middle class. [Kuni Takahashi/Bloomberg via Getty Images]

1800 C.E. — 1850 C.E. — 1900 C.E. — 1950 C.E. — 2000 C.E.

1857 "Mutiny" against the British East India Company

1885–1947 Pro-democracy independence movements

1781

1930

1947 Partition of India

1960s Green revolution and mass migration to urban areas

2008

C What is significant about the fact that in 1750, before the onset of British colonialism, South Asia produced 12 to 14 times more cotton than Britain and the rest of Europe combined?

D What was the immediate result of the Salt March?

E What does this photo suggest about India's modern economy?

Wealth, Invasions, and Trade

As agriculture spread throughout the Indo-Gangetic Plain, large and prosperous economies developed, supporting wealthy cities and powerful rulers whose technology and sophistication became legendary worldwide. Supported by the world's largest agricultural population, accounting for around a quarter of the world's population by the year 1 C.E., the kingdoms, empires, and city-states of South Asia produced one-third of global economic output (GDP) for around a thousand years.

This wealth attracted numerous invaders over the course of South Asian history, the first recorded ones coming from Central and Southwest Asia into the rich Indus Valley and Punjab about 3500 years ago. Many scholars believe that these people, referred to as Indo-European (a linguistic term; see the next section), in conjunction with those of the Harappa and other indigenous cultures, instituted some of the early elements of classical Hinduism, the predominant religion of India today. In the millennia that followed, wave after wave of other invaders arrived, including the Persians, the armies of the Greek general Alexander the Great, and numerous Turkic and Mongolian peoples. Defensive structures against these invaders can still be found across northwest South Asia.

Much interaction also came via sea trade with other areas of advanced civilization, such as Mesopotamia, Egypt, the Roman and Greek empires, and Southeast Asia. This trade was large and often imbalanced in favor of South Asia, which produced highly valued manufactures, such as fine silk and cotton fabric, as well as spices that traders had to pay for with gold and silver, which were some of the only imported items that South Asians wanted. Cultural influences from South Asia were especially powerful in Southeast Asia, where many areas adopted South Asian written languages such as Sanskrit, Hindu and later Buddhist belief systems, aspects of the caste system, and South Asian administrative systems.

Starting about 1000 years ago, Arab traders and religious mystics introduced Islam via land-based trade routes to what are now Afghanistan, Pakistan, and northwest India, and by sea to the coasts of southwestern India and Sri Lanka.

In 1526, the **Mughals**, a group of Turkic Persian people from Central Asia, invaded from the north, intensifying the growth of Islam. The Mughals reached the height of their power and influence in the seventeenth century, controlling the north-central plains of South Asia and becoming the wealthiest empire in the world at the time.

Mughal rule declined throughout the 1700s due to weak emperors, wars of succession, and lack of funds to pay its local administrators, who then rebelled, leading to the secession of several regional states and kingdoms (**Figure 8.19**). This created an opening for a devastating invasion from Persia in the west that further weakened Mughal rule. Meanwhile European trading companies were gaining more control of trade in the region. Of these, the British East India Company was the most successful, gaining control of the wealthiest province of the former Mughal empire, Bengal (modern-day Bangladesh and Indian West Bengal), in 1757.

Mughals a dynasty of Central Asian origin that ruled India from the sixteenth century to the nineteenth century

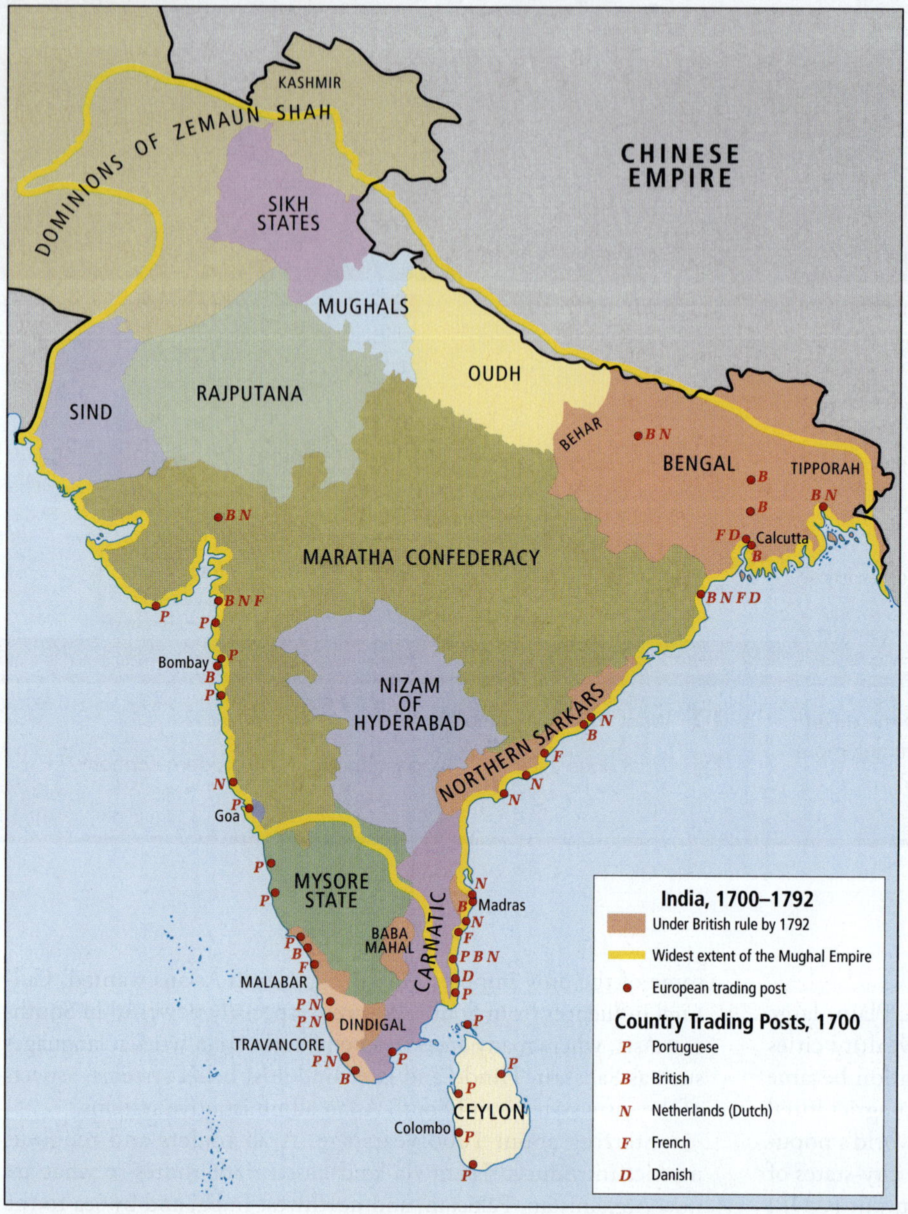

Figure 8.19 Precolonial South Asia. By 1700, several European nations had established trading posts along the coast of India and Ceylon (now Sri Lanka). After the death of the Mughal ruler Aurangzeb in 1707, the Mughals' ability to assert strong central rule throughout South Asia declined and parts of the empire rebelled, resulting in a number of regional states that competed with one another for territory and power. Poor administration and constant wars between these South Asian states paved the way for British conquest of most of the region by 1818. [Research from: William R. Shepherd, *The Historical Atlas* (New York: Henry Holt, 1923–1926), p. 137; and Gordon Johnson, *Cultural Atlas of India* (New York: Facts on File, 1996), p. 111.]

The cultural legacy of the Mughals remained even as the power and range of the dynasty declined. One aspect of this legacy is the 520 million Muslims now living in South Asia, the largest Muslim population of any world region. The Mughals also left a unique cultural heritage that includes the Taj Mahal (Figure 8.18B), miniature painting, and a rich tradition of lyric poetry. The Mughals contributed to the evolution of the Hindi language, which became the language of trade of the northern subcontinent and which is still used by more than 400 million people.

Hinduism a major world religion practiced by approximately 900 million people, 800 million of whom live in India

Language and Ethnicity

One result of the long history of trade and invasions of South Asia over the millennia is that today there are many distinct ethnic groups, each with its own language or dialect. In India alone, 18 languages are officially recognized, but there are actually hundreds of distinct languages. This complexity results partly from strong cultural traditions that allow groups to maintain distinct identities in the face of invasion or migration. As shown in **Figure 8.20**, the Dravidian language-culture group, represented by numbers 14 through 19, is an ancient group, thought to have originated in Pakistan, that predates the Indo-European invasions by a thousand years or more. Today, because of ancient migrations from Pakistan, Dravidian languages are found mostly in southern India, but a remnant of the extensive Dravidian past can still be found in the Indus Valley in south-central Pakistan.

By the seventeenth century, Hindi—a blend of Sanskrit-based and Persian-based Indo-European languages—was the dominant language throughout northern India and what is now Pakistan, where it is known as Urdu. Today, variants of Hindi serve as national languages for both India and Pakistan, though it is the first, or native, language of only a minority. English is a common second language throughout the region. As the language of the British colonial bureaucracy, English remains a language used at work by professional people of all categories. Between 10 and 15 percent of South Asians speak, read, and write English. Many others use a version of spoken English.

Religious Traditions

The main religious traditions of South Asia are Hinduism, Buddhism, Sikhism, Jainism, Islam, and Christianity (**Figure 8.21**). (For a discussion of Islam, Christianity, and Judaism, see Chapter 6.)

Hinduism is a major world religion practiced by approximately 900 million people, 800 million of whom live in India. It is a complex belief system, with roots in both ancient literary texts (known as the *Great Tradition*) and highly localized folk traditions (known as the *Little Tradition*).

The Great Tradition is based on 4000-year-old scriptures called the *Vedas*, which argue that all gods are merely illusory manifestations of the ultimate divinity, which is formless and infinite. Some devout Hindus worship no gods at all, and instead engage in meditation, yoga, and other spiritual practices designed to bring people to a state described as "infinite consciousness." The average

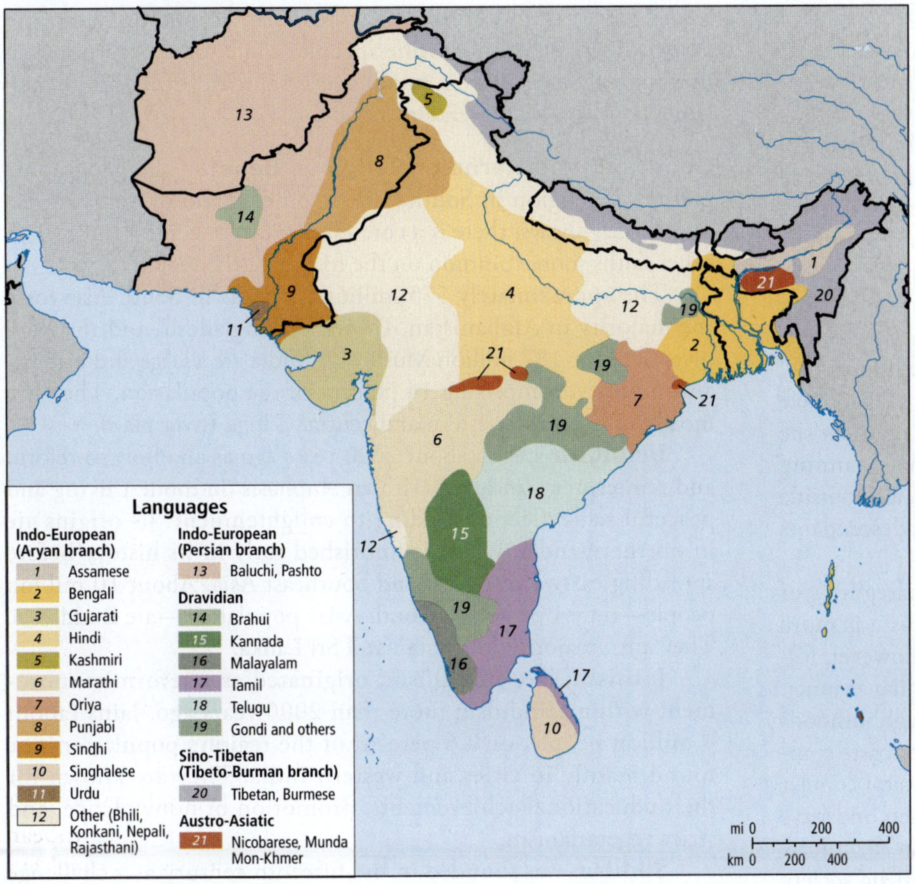

Figure 8.20 Major language groups of South Asia. The modern pattern of language distribution in South Asia is testimony to the fact that this region has long been a cultural crossroads. [Research from: Alisdair Rogers, ed., *Peoples and Cultures* (New York: Oxford University Press, 1992), p. 204.]

person, however, is thought to need the help of personified divinities in the form of gods and goddesses (the Little Tradition). While all Hindus recognize some deities (such as Vishnu, Shiva, Ganesh, and Krishna), many deities are found only in one region, one village, or even one family.

Some beliefs are held in common by nearly all Hindus. One is the belief in reincarnation, the idea that any living thing that desires the illusory pleasures (and pains) of life will be reborn after it dies. A reverence for cows, which are seen as only slightly less spiritually advanced than humans, also binds Hindus together. This attitude, along with the Hindu prohibition on eating beef, may stem from the fact that cattle have been tremendously valuable in rural economies as the primary source of transport, field labor, dairy products, fertilizer, and fuel (animal dung is often burned).

Caste Hinduism includes the **caste system**, a complex and evolving way of dividing society into hereditary hierarchical categories (see Figure 8.30 later in the chapter). In the present form of caste, one is born into a given subcaste, or community (called a ***jati***), that traditionally defines much of one's life experience—where one will live, where and what one can eat and drink, with whom one can associate, one's marriage partner, and often one's livelihood. There are

caste system a complex, ancient Hindu system for dividing society into hereditary hierarchical classes

jati in Hindu India, the subcaste into which a person is born, which traditionally defines the individual's experience for a lifetime

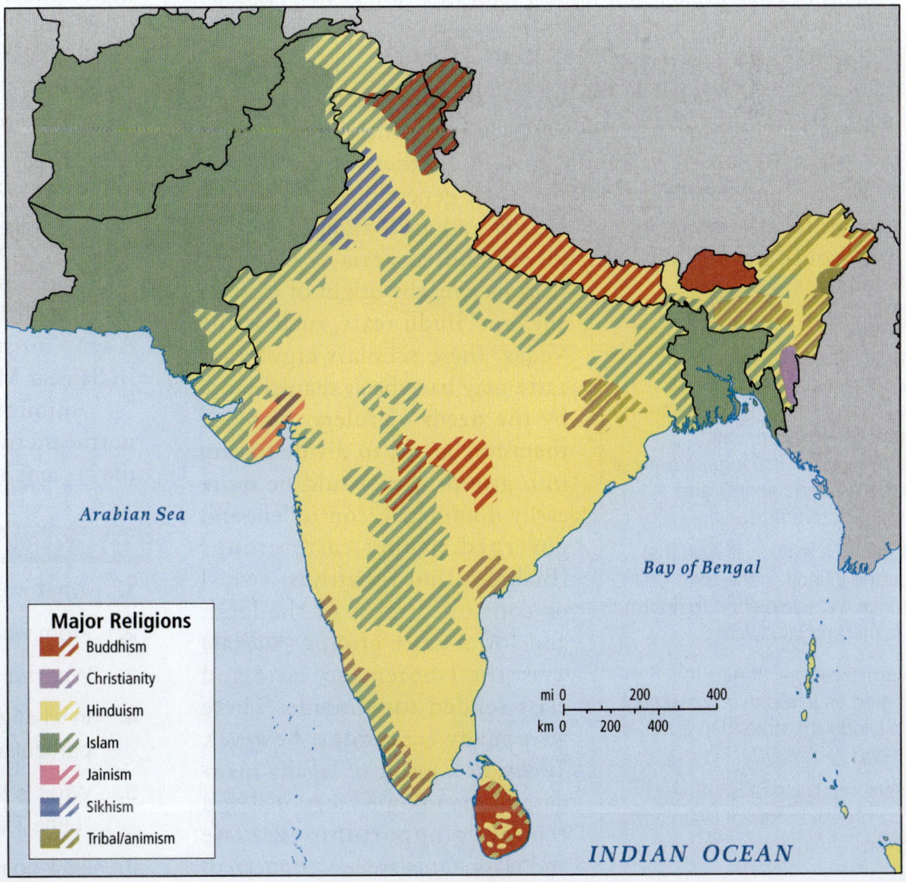

Figure 8.21 Major religions in South Asia. Notice that while Hinduism dominates in India, there are variable and overlapping (striped) patterns of other religious traditions in India, Nepal, and Sri Lanka. On the other hand, Bhutan, Bangladesh, Pakistan, and Afghanistan have very little religious diversity. [Research from: Gordon Johnson, *Cultural Atlas of India* (New York: Facts on File, 1996), p. 56.]

four main divisions or tiers of *jati*, called ***varna***, which are organized hierarchically.

Brahmins, members of the priestly caste, are the most advantaged in caste hierarchy. Thus, they must conform to those behaviors that are considered most ritually pure (for example, strict vegetarianism and abstention from alcohol). As is the case with many castes, Brahmins are found in numerous occupations outside their place as priests in the *varna* system (as barbers and hairdressers, for example). Below Brahmins, in descending rank, are *Kshatriyas*, who are warriors and rulers; *Vaishyas*, who are landowning (small-plot) farmers and merchants; and *Sudras*, who are low-status laborers and artisans. A fifth group, the *Dalits*—"the oppressed," or untouchables—is actually considered to be so lowly as to have no caste. Dalits perform those tasks that caste Hindus consider the most despicable and ritually polluting: killing animals, tanning hides, cleaning, and disposing of refuse. A sixth group, also outside the caste system, is the *Adivasis*, who are thought to be descendants of the region's ancient aboriginal inhabitants.

Although *jatis* are associated with specific subcategories of occupations, in modern economies, this aspect of caste is more symbolic than real. Members of a particular *jati* do, however, follow the same social and cultural customs, dress in a similar manner, speak the same dialect, and tend to live in particular neighborhoods or villages. This spatial separation arises from the higher-caste communities' fears of ritual pollution, such as through physical contact or the sharing of water or food with lower castes. When one stays in the familiar space of one's own *jati*, one is enclosed in a comfortable circle of families and friends that becomes a mutual aid society in times of trouble. The social and spatial cohesion of *jatis* helps explain the persistence and respect paid to a system that, to outsiders, seems to put a burden of shame and poverty on the lower ranks. However, it is important to note that caste and class are not the same thing. Class refers to economic status, and there are class differences within caste groups because of differences in wealth.

Most historians, anthropologists, and other social scientists who have examined the origins of caste generally agree that the current form of caste is much stricter and more rigid than what one would have found even three hundred years ago. Rather than placing the origin of caste in ancient Hindu texts, such as the Vedas, these scholars argue that caste may have been shaped more by the needs of rulers, many of them foreigners, to divide society into groups that would be more easily dominated, controlled, and governed. Upper-caste groups (Brahmins and Kshatriyas) owned or controlled most of the land, and lower-caste groups (Sudras) were the laborers, so caste and class tended to coincide. There were many exceptions, however. Today, as a result of legally mandated expanding educational and economic opportunities, caste and class status are less connected. Some Vaishyas and Sudras have become large landowners and extraordinarily wealthy businesspeople, while some Brahmin families struggle to achieve a middle-class standard of living. By and large, however, Dalits remain very poor.

varna the four hierarchically ordered divisions of society in Hindu India underlying the caste system: Brahmins (priests), Kshatriyas (warriors/kings), Vaishyas (merchants/landowners), and Sudras (laborers/artisans)

Buddhism a religion of Asia that originated in India in the sixth century B.C.E. as an effort to reform and reinterpret Hinduism

Jainism a religion of Asia that originated as a reformist movement within Hinduism more than 2000 years ago

Sikhism a religion of South Asia that combines beliefs of Islam and Hinduism

Geographic Patterns of Religious Beliefs The geographic pattern of religion in South Asia is complex and overlapping. As Figure 8.21 shows, there is a core Hindu area in central India, with other faiths more common on the fringes of the region.

The approximately 575 million Muslims in South Asia form the majority in Afghanistan, Pakistan, Bangladesh, and the Maldives; and the 187 million Muslims in India are a large and important minority, comprising 14 percent of the population. They live mostly in the northwestern and central Ganga River plain.

Buddhism began about 2600 years ago as an effort to reform and reinterpret Hinduism with an emphasis on modest living and peaceful self-reflection leading to enlightenment. Its origins are in northern India, where it flourished early in its history before spreading eastward to East and Southeast Asia. About 10 million people—only 1 percent of South Asia's population—are Buddhists. They are a majority in Bhutan and Sri Lanka.

Jainism, like Buddhism, originated as a reformist movement within Hinduism more than 2000 years ago. Jains (about 6 million people, or 0.6 percent of the region's population) are found mainly in cities and western India. They are known for their educational achievements, promotion of nonviolence, and strict vegetarianism.

Sikhism was founded in the fifteenth century as a challenge to both Hindu and Islamic systems. Sikhs believe in one god, hold high ethical standards, and practice meditation. Philosophically, Sikhism accepts the Hindu idea of reincarnation but rejects the idea of caste, though in everyday life caste plays a role in Sikh identity. The 23 million Sikhs in the region live mainly in Punjab, in northwestern India. Their influence is greater than their numbers because many Sikhs hold positions in the government, the military, and police forces throughout India.

The first Christians in the region are thought to have arrived in the far southern Indian state of Kerala with St. Thomas, the apostle of Jesus, in the first century C.E. Today, Christians and Jews are influential but tiny minorities along the west coast of India. A few Christians live on the Deccan Plateau and in northeastern India near Myanmar (formerly called Burma).

Animism (see Chapter 7) is practiced primarily in central and northeastern India by the Adivasi or "tribal" communities that are the ancient aboriginal inhabitants of South Asia.

CHECK YOUR UNDERSTANDING

1. What was a remarkable feature of the Indus Valley civilization?
2. From where have most invasions of South Asia come?
3. What were some of the civilizations that traded with South Asia?
4. In what world region did South Asian civilization have a particularly strong impact?
5. What is the difference between the "Great Tradition" and the "Little Tradition" in Hinduism?
6. How does "caste" differ from "class"?

British Colonial Rule and Its Legacies

Acting increasingly as both a regional power and as an extension of the British government, the British East India Company (here after referred to as "the Company") conquered most of South Asia by 1818, taking advantage of divisions among South Asian states and utilizing superior military technologies and tactics. The rule of the Company, and the direct rule by the British government that followed it, is today widely criticized in both South Asia and the UK for its economic impact on South Asia, taking it from a producer of 22 percent of global income in 1700 to a producer of just over 4 percent of global income shortly after independence was achieved in 1947. This happened in part as a result of British policies that discouraged industrial development in South Asia and encouraged the production of agricultural raw materials to feed industries in Britain.

By making South Asia part of the British Empire, the British accelerated the process of globalization, transforming the region politically, socially, and economically. Even areas not directly ruled by the British felt the influence of their empire (**Figure 8.22**). Afghanistan repelled British attempts at military conquest, but the British continued to intervene there, trying to make Afghanistan a "buffer state" between British India and Russia's expanding empire. Nepal remained only nominally independent during the colonial period, and Bhutan became a protectorate of the British Indian government.

The Deindustrialization of South Asia As in their other colonies, the British extracted huge amounts of raw materials to feed their industries back home. Most British investments in South Asia were made in the export-oriented agricultural sector and in related

Figure 8.22 The British Conquest of India, 1753–1914. After winning control of much of South Asia, Britain controlled lands from Baluchistan (now Pakistan) to Burma (now Myanmar), including Ceylon (now Sri Lanka) and the islands between India and Burma. The princely states and protectorates maintained limited control of their internal affairs in return for annual payments and political loyalty to the British. [The Map Archive.]

transportation infrastructure, such as railroads, that also facilitated military control. Industries that could potentially compete with those developing in Britain were discouraged through tariffs. Areas that already had established industries, such as Bengal, were transformed under Company rule into suppliers of much less lucrative raw materials to British industries. These policies persisted even after a massive rebellion against the British in 1857, after which Company rule was replaced by direct rule by the British government.

In 1750, Bengal produced 12 to 14 times more cotton cloth than Britain alone and more than all of Europe combined. Bengali weavers, long known for their high-quality muslin cotton cloth, initially benefited from the increased access that the Company gave them to overseas markets in Asia, the Americas, and Europe. However, over the course of the eighteenth and nineteenth centuries, Britain's own highly mechanized textile industry—based on cotton grown in India and the American South—developed cloth that the Company used to replace Bengali muslin. High tariffs were placed on the sale of Bengali fabrics throughout the British Empire, while fabrics made in Britain had relatively low taxes. By the mid-nineteenth century, the Bengali textile industry survived in only a few places (see Figure 8.18C). As one British colonial official put it, "While the mills of Yorkshire prospered, the bones of Bengali weavers bleached on the plains of India."

Further damage to Bengal's economy came from extremely high levels of taxation, which increased the risk of famine. Within South Asian society, the Company acted as a tax collector, first for the Mughals, and then for itself. It mixed this role with its trade-related activities, often requiring taxes be paid in the form of crops it exported, such as cotton and spices. Company demands for taxes increased over time, which meant that less food could be grown. In 1770 the combined effects of drought and high taxes resulted in a famine that killed more than 10 million Bengalis, roughly a quarter of the population.

Famines occurred throughout South Asia at a greater frequency and with more severity than occurred before or after British rule. In part this was due to British reluctance to offer relief to famine-stricken areas, instead allowing exports of food from India to continue even during famines. In the ensuing chaos, agricultural output lagged for years in the famine-struck areas as millions died or were pushed off of their farms, becoming landless laborers, migrants to emerging urban centers, or indentured servants sent overseas to work on plantations in British colonies in the Americas, Africa, Asia, and the Pacific. The last great famine of British India occurred shortly before independence, claiming an estimated 4 million lives in Bengal during World War II, largely due to wartime policies that severely disrupted the economy. The government never officially recognized a state of famine, and relief efforts were extremely inadequate.

civil disobedience protesting of laws or policies by peaceful direct action

Partition the breakup following Indian independence that resulted in the establishment of Hindu India and Muslim Pakistan

Democratic and Authoritarian Legacies Contemporary South Asian governments retain institutions and practices put in place by the British that tend to resist change via authoritarian bureaucracies. While the governments have proved functional over time, there have been numerous riots, rebellions, and other disturbances in virtually every country. Democratic governments were not instituted on a large scale until the final days of the empire, but since independence in 1947 (see below), people have been able to use the political freedoms afforded by democracy to voice their concerns, allowing many peaceful transitions between elected governments. Still, the struggle to retain and build democratic institutions continues.

Independence and Partition

The tremendous changes brought about by the British inspired many resistance movements among South Asians. Some of these were militant movements intent on pushing the British out by force, such as the unsuccessful rebellion of South Asian soldiers employed by the British East India Company in 1857. Other movements, such as the Indian National Congress (founded in 1885), used peaceful political means to agitate for more democracy, which they saw as the route to political independence. Both militant and peaceful political actions were brutally repressed over decades of struggle.

Nonviolence as a Political Strategy In the early twentieth century, Mohandas Gandhi, a young lawyer from Gujarat—who would later become known by the honorific "Mahatma," meaning "great soul"—emerged as a central political leader in South Asia's independence movement. He used tactics of **civil disobedience** to nonviolently defy laws imposed by the British that discriminated against South Asians. Gathering a large group of peaceful protesters, he would notify the government that the group was about to break a discriminatory law. If the authorities ignored the act, the demonstrators would have made their point and the law consequently weakened. If the government instead used force against the peaceful demonstrators, it would lose the respect of the masses. Throughout the 1930s and 1940s, this technique was used to undermine British authority across South Asia with events like the Salt March of 1930. There Gandhi led tens of thousands of people on a march to the sea, where they made salt by evaporating seawater, thus breaking a law that made it illegal for South Asians to produce salt (see Figure 8.18D). The purpose of the law had been to facilitate British rule by controlling a vital human nutritional necessity. Breaking the law on such a massive scale created an international media frenzy that catapulted Gandhi to global notoriety, moving millions of South Asians to support independence from Britain. Gandhi's tactics of civil disobedience became central to South Asian political culture and were used in the United States during the civil rights movement of the 1950s and 1960s.

The Partition of India in 1947 When India achieved independence in 1947 it was divided into two independent countries: India, which was predominantly Hindu; and West and East Pakistan, which were predominantly Muslim (**Figure 8.23**). (Afghanistan, Bhutan, and Nepal were never officially British colonies; Ceylon [now Sri Lanka] became independent in 1948.) This division—called **Partition**—which Gandhi greatly lamented, was perhaps the most enduring and damaging outcome of colonial rule.

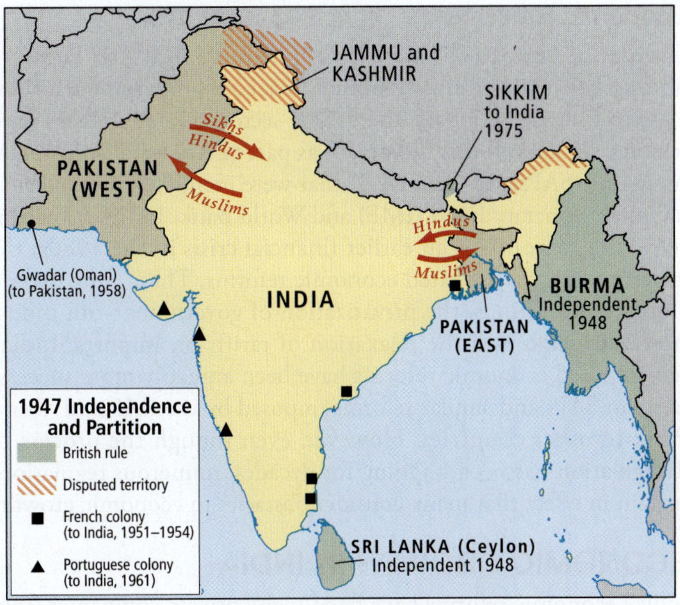

Figure 8.23 Independence and Partition. India became independent of Britain in 1947, and by 1948, the old territory of British India was partitioned into the independent states of India and East and West Pakistan. The Jammu and Kashmir region was contested space, and remains so today. Sikkim went to India, and both Burma and Sri Lanka became independent. Following a brutal civil war, East Pakistan became the independent country of Bangladesh in 1971. [Research from: *National Geographic*, May 1997, p. 18.]

Muslim political leaders first suggested the idea of two nations out of concern for the fate of a minority Muslim population in a united India with a Hindu majority. Though highly controversial, it became part of the independence agreement between the British and the Indian National Congress (India's principal nationalist party). Northwestern and northeastern India, two Muslim majority areas with very different cultures and separated by over 1300 miles (2200 km) of India, became a single country consisting of two parts known as West and East Pakistan. Although both India and Pakistan maintained secular constitutions with no official religious affiliation, more than 7 million Hindus and Sikhs migrated to India from Pakistan, and a similar number of Muslims left India for Pakistan. In the process, civil society broke down: families and communities became divided, looting and rape were widespread, and between 1 and 3.4 million people were killed in numerous riots.

Partition was the tragic culmination of the divide-and-rule approach the British used throughout their empire (see Chapter 7). This approach heightened tensions between groups, for example South Asian Muslims and Hindus, thus creating a role for the British as seemingly indispensable and benevolent mediators. Instead of relieving tensions, the partition of India and Pakistan laid the groundwork for wars, skirmishes, terrorism, and an arms race between India and Pakistan. In 1971, after a bloody civil war, Pakistan was divided into Bangladesh (formerly East Pakistan) and Pakistan (formerly West Pakistan).

The Postindependence Period

In the more than 70 years since the departure of the British, South Asians have experienced both progress and setbacks. Democracy has expanded steadily, albeit somewhat slowly. India is now the world's most populous democracy. It is gradually dismantling age-old traditions that hold back women, poor low-caste Hindus, and other disadvantaged groups, and a vibrant, if still small, middle class is emerging (see Figure 8.18E). Pakistan has held a number of elections, and while some of its governments have been effective, more often they have been corrupt, militaristic, authoritarian regimes. The long feud between Pakistan and India about the border between the two countries (discussed in the "Conflict in Kashmir" section), meanwhile, has sapped resources and ruined innumerable lives. Bangladesh, despite the damage it sustained in its war of independence from Pakistan in 1971, has had a more stable, responsive, and less militaristic (though not trouble-free) government than Pakistan.

Industrialization became a main goal after independence, partly in response to the dismantling of the textile industry during the colonial period. An emphasis on technical training has produced several generations of highly skilled engineers whose talents are in demand around the world. In most countries in the region, urban-based industrial and service economies now constitute a far larger share of GNI than agriculture (which nonetheless continues to employ 60 percent or more of the workforce). The information technology (IT) sector is growing especially rapidly in India, which is now the world's third-largest economy in terms of PPP.

CHECK YOUR UNDERSTANDING

1. What factors led to the decline of the Mughal Empire?
2. How did British rule affect the South Asian textile industry?
3. What was the impact of famines in South Asia under British rule?
4. What tactics did Mohandas "Mahatma" Gandhi use to undermine British authority in South Asia?
5. How did Partition affect tensions in South Asia after independence?

GLOBALIZATION AND DEVELOPMENT

8.4 Evaluate how South Asia's recent economic development and openness to globalization have affected society.

South Asia is a region of startling economic contrasts. India is a good example: it is home to hundreds of millions of desperately poor people, but it is also a global leader in the computer software industry and in engineering innovation. It is celebrated for being the world's third-largest economy in terms of purchasing power, with a large and growing middle class and even a robust space program. And yet its poor often see little improvement in their own lives from this economic development.

Not surprisingly, globalization benefits some South Asians more than others. Educated and skilled South Asians with jobs in export-connected and technology-based industries and services are paid more and generally enjoy better working conditions. Less skilled workers in both urban and rural areas are left with demanding but very low-paying jobs.

FROM SELF-SUFFICIENCY TO GLOBALIZATION

After independence, national self-sufficiency was the main goal for all parts of the region, but after decades of slow economic growth and persistent poverty, most governments began to embrace globalization as a route to greater prosperity. Starting in the 1980s in India, strong growth occurred in the IT and other high-tech industries, as well as in manufacturing. Meanwhile, changes in agriculture drastically increased the amount of food available but failed to improve incomes for most farmers.

Agriculture still employs almost 50 percent of the workers in South Asia, but the contribution of agriculture to most national economies is much lower, averaging less than 20 percent of the GNI for the region. The industrial sector employs far fewer people but produces between one-quarter to one-third of GNI in all countries except Nepal and Afghanistan. The service sector has expanded more rapidly than either agriculture or industry and now produces more than half of the GNI in every country except Bhutan.

Central Planning and Self-Sufficiency

After independence from Britain in 1947 and wary of further domination by outsiders, the new leaders in India, Pakistan, Bangladesh, and Sri Lanka turned to socialist ideas and models of industrial growth inspired by the Soviet Union, believing that central planning (see Chapter 5) by the government would speed industrialization and reduce poverty. They also wanted their respective countries to become self-sufficient, avoiding dependence on manufactured goods imported from the industrialized world (see Chapter 3 for a discussion of import substitution industrialization). To reach their goal of self-sufficiency, governments took over the industries they believed to be the "commanding heights" of a strong economy: steel, coal, transportation, communications, and a wide range of manufacturing and processing industries.

South Asian central planning in the decades after independence generally failed to meet goals for a variety of reasons. One problem was that policies intended to boost employment often contributed to inefficiency by encouraging industries to employ as many people as possible, even if they were not needed, and requiring government approval before an industrial or manufacturing worker could be fired. For example, during the central planning era, it took more than 30 Indian workers to produce the same amount of steel as 1 Japanese worker, making Indian steel uncompetitive in the world market. In addition, as in the former Soviet Union, decisions about which products should be produced were made by ill-informed government bureaucrats and not driven by consumer demand. Until the 1980s, items that would improve daily life for the poor majority, such as cheap cooking pots or simple tools, were produced only in small quantities and were of inferior quality. At the same time, there was a relative abundance of large kitchen appliances and large cars that only a tiny minority could afford. Meanwhile the vast majority of the population remained poor farmers in an agricultural sector that received much less investment and attention from the government. Because their incomes remained low they couldn't buy many of the manufactured goods being produced.

Economic Reforms

During the late 1980s and 1990s, much of South Asia began to undergo economic reforms aimed at increasing competitiveness and creating secure jobs in the private sector. In many other world regions, similar reforms took place as part of structural adjustment programs (SAPs; see Chapter 3) that were mandated by the International Monetary Fund (IMF) and World Bank. In India, by contrast, as a response to an earlier financial crisis in the 1980s, the government itself initiated economic reforms. These consisted of, among other things, the privatization of government-run industries and banks and the relaxation of tariffs on imports. India's self-imposed economic reforms have been arguably more successful than SAPs and similar reforms imposed by the IMF and World Bank in other countries. However, even though the process of deregulation has been ongoing for decades, numerous regulations remain in effect that many consider obstacles to economic growth.

ECONOMIC GROWTH IN INDIA

India's economic reforms have freed many private companies from a maze of regulations, spurring economic growth rates of 7–9 percent over the past two decades, some of the highest in the world. Growth has been led by increases in services provided in the IT sector, and in manufacturing. So far wealthy Indians have reaped most of the rewards from this system, but most incomes have gone up noticeably.

The reforms have enabled both foreign and Indian companies to invest heavily in manufacturing and other industries (**Figure 8.24**). Vehicle manufacturing has grown dramatically in recent years, making India the world's sixth-largest vehicle manufacturer and inspiring claims that India is "the next China." Vehicle manufacturing is significant given the central role this industry has played in the developing of the broader manufacturing sectors in other countries. Vehicle manufacturing produces many "spin-off" industries (for example in the steel industry, in the manufacturing of parts, maintenance of vehicles, and sales and finance of vehicles) that can employ up to seven times the number of people directly employed in the manufacture of whole vehicles.

Many Indian companies are building innovative electric two- and three-wheeled vehicles that are popular for their low operating costs. These vehicles may help reduce air pollution and GHG emissions even without significant government incentives for their purchase. Meanwhile nearly every major global automobile company is setting up a factory in India, drawn by its large and cheap workforce, its excellent educational infrastructure (for the middle and upper classes), its growing middle class's demand for vehicles, and its proximity to large markets in the rest of Asia.

One possible outcome of all this growth is that as manufacturing jobs increase, many of India's rural and urban poor will see their incomes rise. They will then increase their consumption of Indian-made products, fueling further growth. By 2018, GDP per capita (PPP) had risen in all Indian states (see Figure 8.24) and averaged U.S.$6980, nearly six times higher than it was in 1992. So far, most of the benefits from India's self-implemented economic reforms seem to be going to the highly skilled and educated urban upper class, resulting in wider urban–rural income disparities.

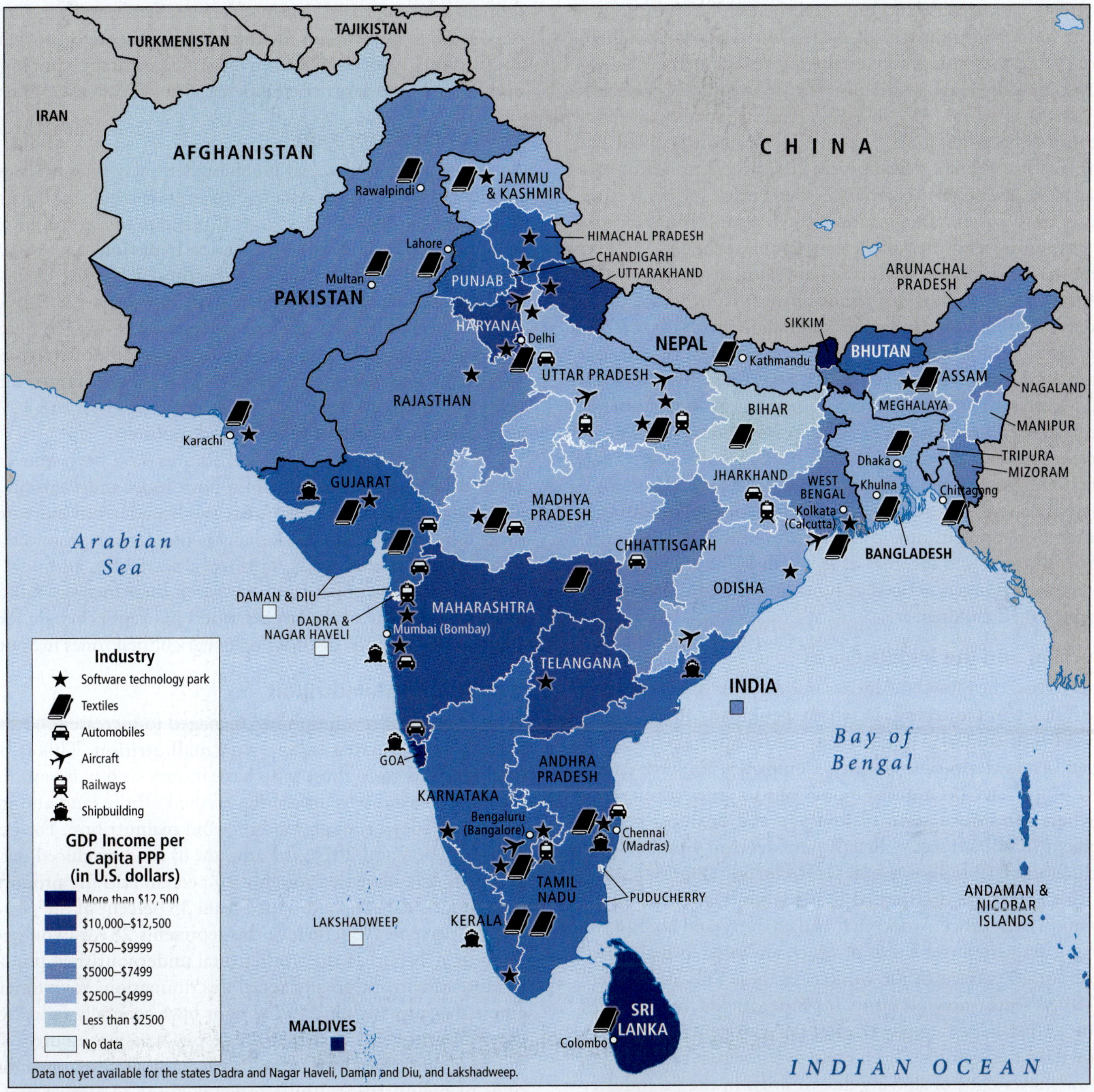

Figure 8.24 GDP income per capita (PPP) and industrial and IT service centers in India. Incomes have grown rapidly throughout South Asia in recent years, as new manufacturing industries and IT service centers have expanded, especially in India. Incomes are highest in southern India and in the states around Delhi. [Research from: https://data.worldbank.org/indicator/NY.GNP.PCAP.PP.CD?view=map; and "Comparing Indian States and Territories with Countries: An Indian Summary," *The Economist*, http://www.economist.com/content/indiansummary.]

Incomes and Human Development Across Indian Society

While economic growth in India has led to a larger base of consumers, high levels of income inequality mean that fewer people can afford expensive items. The wealthiest 1 percent of households, roughly 13 million people, enjoy high levels of human development, owning 58 percent of India's wealth, earning over $17,500 ($70,000 in PPP terms), and enjoying excellent access to health care and education. In this population, the lowest-earning families can afford to own a two-bedroom apartment, two economy cars, send their children to private schools and public universities in India, and vacation outside of India in Asia. The wealthiest 2 million of this group can own a house, one or two mid-size European or Japanese cars, send their children to private schools and universities in India, and vacation in Europe or Australia. The wealthiest few hundred thousand earn over $1 million per year, own large houses and expensive cars, and can send their children to private schools in India and universities in the United States or United Kingdom. Outside of this group, buying power is much lower.

India's "middle class," which can be defined as a household earning $8400–$17,500 ($33,000–$70,000 PPP), makes up about 17 percent of the population (roughly 215 million people), has medium levels of human development, less access to health care and education, and owns less than 22 percent of India's wealth. They can afford to rent an apartment, own a motor scooter and maybe lease a car, send their children to public schools, and take a vacation in India every few years. The wealthiest quarter of this group can send their children to public universities in India.

Around 900 million people are "working poor" with low levels of human development due to poor access to health care and education, and incomes between $700–$6300 ($2900–$26,000 PPP). Collectively owning less than 20 percent of India's wealth, they mostly live in urban slums or rural areas without sanitation or health care, they can afford a bicycle or used motor scooter, their children attend public schools, and they have almost no luxuries.

There are about 100 million desperately poor people with very low human development due to inadequate health care, no sanitation or education, and incomes of less than $700 per year. They endure frequent hunger, periods of homelessness, and little to no schooling for their children.

Outsourcing and the Middle Class

Since the 1990s, the growth of India's middle class has accelerated due to **offshore outsourcing**, in which a company in the more-developed world contracts to have some of its business functions performed in a lower-income country. Companies in North America and Europe have been outsourcing jobs to cities throughout India, where the information technology and business process management (IT-BPM) sector directly employs over 4 million people, with another 12 million employed indirectly (**Figure 8.25A**). Jobs in this sector are dominated by telephone-based technical support and "back office" work such as data entry and accounting for foreign companies, two-thirds of which are based in the United States. About 40 percent of the workers in these jobs are women. The IT-BPM sector now accounts for 8 percent of India's GDP and is the second-largest employer, after the government.

Growth of the sector has been rapid over the past two decades, but it now shows signs of slowing due to shifts in the IT industry toward replacing some IT-BPM jobs with software-based automation and artificial intelligence. However, many are optimistic that the more technologically sophisticated workers in this sector can move into more creative work in the broader IT sector, such as software development, web design, engineering, and pharmaceutical research, that are less dependent on managing business processes for foreign companies. With 68 million college-educated Indians, most of whom are willing to work for under $10,000 a year, more employment in the highly globalized IT sector is likely.

offshore outsourcing the contracting of business or production functions to foreign companies where labor and other costs are lower

RURAL ECONOMIC DEVELOPMENT

Due to changes in food production systems in South Asia, food supplies and the incomes of wealthy farmers have increased. Meanwhile, smaller farmers and impoverished agricultural workers, discouraged by low prices for their crops and in some cases displaced by new agricultural technology, have been forced to seek other employment in rural areas or in cities.

The Green Revolution

Agricultural productivity is much higher today than it was at independence, when South Asia had to import large amounts of food to prevent famines. In large part this is due to the spread of green revolution technologies (see Chapter 1) during the 1960s. These boosted grain harvests dramatically through the use of new seeds bred for high yield and resistance to disease and wind damage, fertilizers, mechanized equipment, irrigation, pesticides, herbicides, and double-cropping (producing two crops consecutively per year). Where the new techniques were used, yield per unit of farmland improved by more than 30 percent between 1947 and 1979, and both India and Pakistan became food exporters.

The benefits of the green revolution have been uneven, with some areas, such as the Punjab in both India and Pakistan, seeing much higher incomes. These areas also benefit from more extensive irrigation networks. However, many poorer farmers who were unable to afford the special seeds, fertilizers, pesticides, and new equipment could not compete and have seen their incomes fall further, as the productivity of the larger farmers has kept prices for food low. **Figure 8.26** shows the distribution of agricultural zones in South Asia.

Hunger and Malnutrition

While the green revolution has managed to increase food supplies, it has not eliminated hunger and malnutrition. This is because food tends to go to those who have money to buy it, and many of the farm workers who have been pushed off the land and into cities cannot afford enough food to avoid malnutrition. For example, between 1964 and 2015, the amount of food produced per capita in South Asia increased roughly 20 percent and the proportion of undernourished people dropped from 33 percent of the population to 15.7 percent. Nonetheless, this represents 281 million people—more than a third of the world's total undernourished population. Because of corruption and social discrimination, government programs that provide food to the poor have generally failed to reach those most in need. For example, despite massive resources devoted to improving child nutrition, 43 percent of children in India show signs of malnutrition. India today is home to almost half of the world's malnourished children.

Farm Size and Income

As in much of the developing world, agriculture across South Asia is based largely on traditional small-scale systems, with average farms around 3 acres in size. These small farms are highly productive per acre, but so small that they can't lift the farmers out of poverty, especially given the low prices paid for most crops. Farms tend to be larger (9–12 acres) where agriculture is more profitable, such as in the Punjab and the Indian state of Haryana, where wealthier farmers are buying out their neighbors. Elsewhere farms remain small and are getting smaller as fields are subdivided among the children of farmers over time.

Persistently low prices for agricultural produce mean that farm incomes remain low, relative to other forms of employment,

PHOTO ESSAY

8.25 Globalization and Development in South Asia

Reform and innovation have contributed to greater economic growth and higher human development over the past several decades in South Asia. India's economic reforms have freed many private companies from a maze of regulations, nurturing its unique IT service sector and some manufacturing industries. Meanwhile, an innovative banking system for very poor people has developed in Bangladesh and spread throughout the developing world. Throughout South Asia, many new enterprises are choosing to locate in rural areas where large rural populations will work for lower wages than in cities.

THINKING GEOGRAPHICALLY

A What factors may constrain growth in the information technology and business process management (IT-BPM) sector in the future?

B Why is the growth of rural nonfarm jobs significant?

C What are some criticisms of microcredit programs?

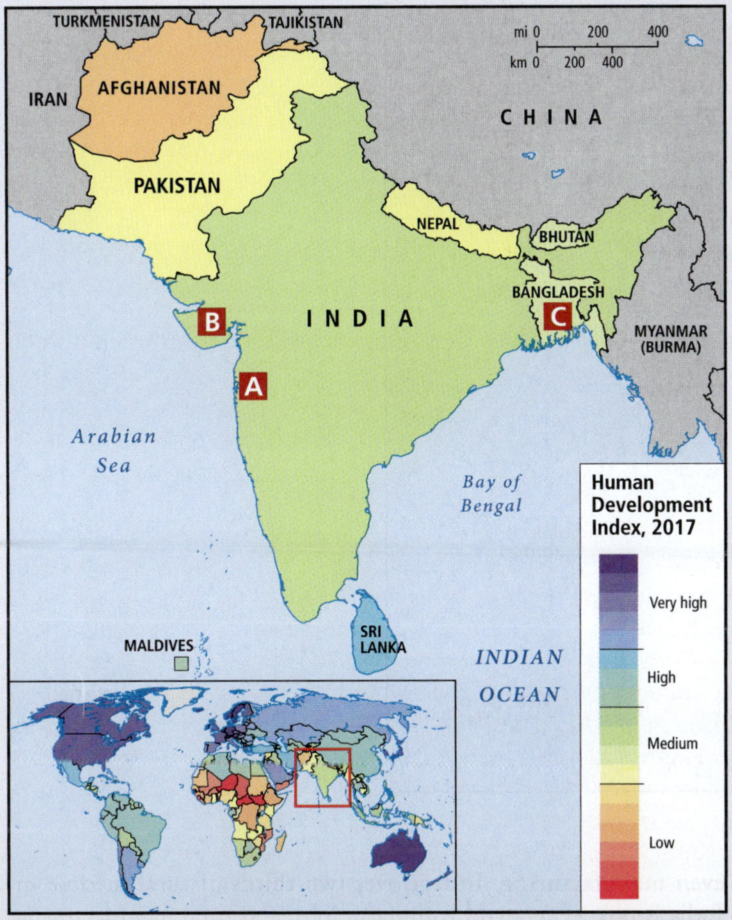

A IT-BPM sector. A software development team in Mumbai, India, develops iPhone apps. Companies in North America and Europe have been outsourcing jobs to cities throughout India, where the IT-BPM sector directly employs over 4 million people. [Kainaz Amaria/Bloomberg via Getty Images]

B Rural manufacturing jobs. Electric scooters are produced in rural Gujarat. Over two-thirds of rural income in India is generated by nonfarm jobs, and rural manufacturing now adds more to the country's GDP than urban manufacturing. [SAM PANTHAKY/AFP/GettyImages]

C Microcredit. Borrowers of small loans from the Grameen Bank operate a business manufacturing and repairing fishing nets in rural Bangladesh. [Majority World/UIG/Getty Images]

Figure 8.26 Major farming systems of South Asia. [Research from: John Dixon and Aidan Gulliver with David Gibbon, *Farming Systems and Poverty: Improving Farmers' Livelihoods in a Changing World* (Rome and Washington, DC: FAO and World Bank, 2001), http://www.fao.org/farmingsystems/FarmingMaps/SAS/01/FS/index.html.]

and agriculture remains the least efficient economic sector, meaning that it has the lowest return on investments of land, labor, and cash. In response to low incomes many have left farming, and agricultural employment has declined dramatically over the past two decades, dropping from 60 percent of total employment in 2000 to around 40 percent of total employment today.

Growth in the Rural Nonfarm Economy

Some people who have left farming have moved to the cities, but many have stayed in rural areas due to a dramatic growth in rural nonfarm employment. This has major implications for a region with such enormous rural populations. Many have long assumed that declines in farm employment would lead to a massive rush to the cities, which would be overwhelmed by vast slums, as has happened in parts of sub-Saharan Africa. While cities have grown, many farmers have stayed in rural areas due to growth in rural nonfarm jobs in services such as transport, trading, construction, and even manufacturing. In fact over two-thirds of rural income in India is now generated by nonfarm jobs, and rural manufacturing now adds more to GDP than urban manufacturing (Figure 8.25B). Rural nonfarm employment has boosted both incomes and food security for rural households, many of which can maintain their small farms and still find jobs nearby in manufacturing or services. With the ability to both grow their own food, and purchase it if need be with income from a diverse array of rural jobs, South Asia's large rural populations are emerging as a major source of economic stability after years of being dismissed as "backward" and economically insignificant.

microcredit a program based on peer support that makes very small loans available to very-low-income entrepreneurs

MICROCREDIT: A SOUTH ASIAN INNOVATION FOR THE POOR

Over the past four decades, a highly effective strategy for lifting people out of extreme poverty has been pioneered in South Asia. **Microcredit** makes very small loans (generally under U.S.$100) to very-low-income people. Throughout the world most banks have not bothered lending in small amounts because the profits are so low. This forces poor people to rely on small-scale moneylenders

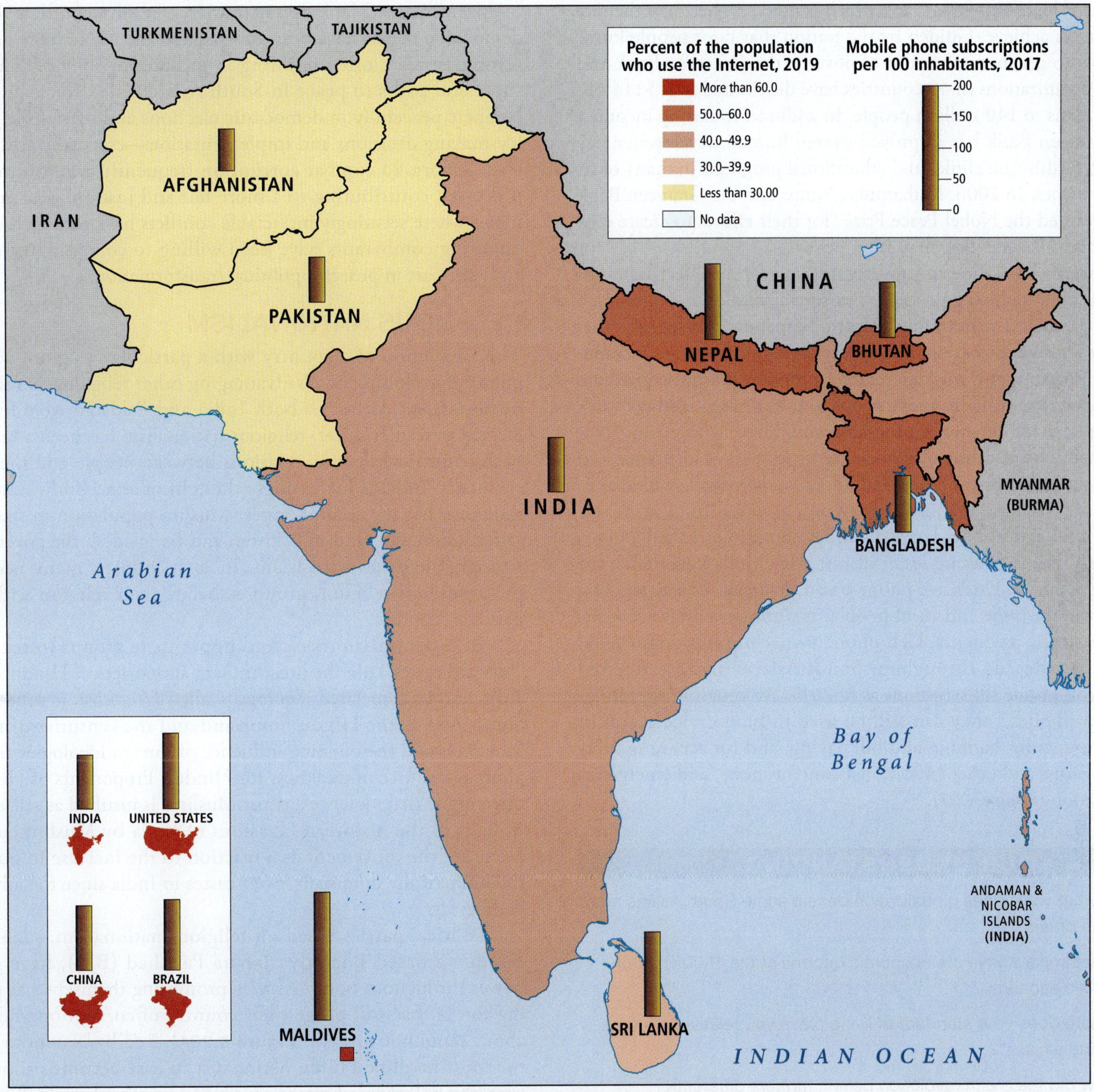

Figure 8.27 Cell phone and internet penetration in South Asia. The map shows the number of internet users in each country and the percentage of the country's population that uses the internet. By 2018, more than 600 million people, or one-third of South Asians, were using the internet. Nepal and Bangladesh have the highest percentage of users (55 and 53 percent) and Afghanistan the lowest (17 percent). [Data from: https://www.internetworldstats.com/stats3.htm; and https://data.worldbank.org/indicator/IT.CEL.SETS.P2?locations=IN-PK-BD-NP-LK-BT-MV.]

who charge interest rates as high as 30 percent or more *per month.* In 1976, Muhammad Yunus, an economics professor in Bangladesh, started the Grameen Bank, or "Village Bank," which makes small loans, mostly to people in rural villages who wish to start businesses (Figure 8.25C). So far, the Grameen Bank has grown dramatically, loaning over U.S.$11 billion to more than 8 million borrowers. Similar microcredit projects have been established in India and Pakistan and throughout Africa, Middle and South America, North America, and Europe.

Microcredit loans often pay for the start-up costs of small enterprises, which include such wide-ranging ventures as cell phone–based services, chicken raising, small-scale egg production, the construction of pit toilets, and the distribution of home water purification systems. Potential borrowers (more than 90 percent of whom are women) are organized into small groups that are collectively responsible for repaying any loans to group members. If one member fails to repay a loan, then everyone in the group is denied loans until the loan is repaid. This system, reinforced with weekly meetings, yields a

98 percent loan repayment rate, which is much higher than the rate most banks achieve. Guided by the notion that poor people know best how to get themselves out of poverty, the Grameen Bank and similar organizations in 40 countries have distributed over $114 billion in loans to 140 million people. In addition to raising incomes, the Grameen Bank has improved overall human development by creating health-care clinics and educational programs in many of its local branches. In 2006, Muhammad Yunus and the Grameen Bank were awarded the Nobel Peace Prize "for their efforts to create economic and social development from below."

Nevertheless, some controversy lingers over the effectiveness of microcredit at reducing poverty, with several studies showing much less success than is commonly claimed. Some research suggests that broader regional or national-scale economic growth raises incomes more noticeably, and that microcredit programs do not contribute enough at these scales to move people out of poverty. Others studies suggest that microcredit programs would have a larger impact if combined with additional financial services such as insurance and savings accounts, as well as legal education and emergency food aid.

Some of these criticisms may be answered by new developments in cell phone–based lending, sometimes called "nanofinance," that is, making loans of just a few dollars workable for lending agencies. Extensive phone-based financial services are now affordable for poor and rural people, including savings accounts and electronic payments. Cell phone ownership has skyrocketed in the past decade throughout South Asia, with an estimated 1.5 billion phone subscriptions as of 2017. Although most phones are "flip phones," they can still be used to buy service access in tiny amounts for banking and bill paying, and for accessing literacy programs and other educational content, news, and emergency information (**Figure 8.27**).

CHECK YOUR UNDERSTANDING

1. In what ways does globalization benefit some South Asians more than others?
2. What impact have the economic reforms of the 1990s had on India's economy?
3. Where does your standard of living place you relative to India's "middle class"?
4. How did the green revolution benefit farmers differently?
5. Why is nonfarm employment important in South Asia?
6. How has microfinance achieved so much success in this region?

POWER AND POLITICS

8.5 Evaluate the impact of religious nationalism on the politics of this region.

religious nationalism the association of a country with a particular religion, with the intent of excluding or disadvantaging other religions

Conflict is common in South Asia, and the causes are often complex and layered. Religious differences often mask underlying forces, such as poverty, economic rivalry between groups, and authoritarianism and corruption in governments. Many opportunities to resolve conflicts through democratic means have been missed, and the use of force has been favored instead, often resulting in prolonged violence. The most successful paths to peace in South Asia enable diverse groups to compete peacefully in democratic elections and participate in policy-making decisions and implementation—especially at the local level. Efforts to combat corruption frequently arise out of these processes, contributing to a more fair and just political environment, where seemingly intractable conflicts have been defused and opposing combatants have been willing to compete in elections and take part in peaceful political transformation.

RELIGIOUS NATIONALISM

The association of a country with a particular religion, with the intent of excluding or disadvantaging other religions, is **religious nationalism**. Although both India and Pakistan were formally created as secular states, religious nationalism has been a reality in both countries, shaping relations between people and their governments. India is increasingly thought of as a Hindu state, even though it has the second-largest Muslim population in the world (after Indonesia), and in Pakistan and Bangladesh the government strongly identifies with Islam. In each country, many people in the dominant religious group associate their religion with their national identity.

In India, urban men from upper-caste groups (often called "forward castes") are the predominant supporters of Hindu nationalism and its associated ideology. Called *Hindutva*, it conceives of South Asia as the Hindu homeland and the centuries of rule by foreigners and the ongoing influence of foreign ideologies and religions as a source of weakness for Hindus. Proponents of Hindutva often argue that violence against Muslims is justified as self-defense in light of the historical invasions of India by Muslim empires. Some see the movement as a reaction to the increase in power of the numerically dominant lower castes in India since the advent of democracy.

Political parties based on religious nationalism, such as the Hindutva-based Bharatiya Janata Parishad (BJP), have gained power throughout South Asia by promoting themselves as purifying forces that will purge their country of corruption and bring about economic growth (**Figure 8.28A**). Led by Narendra Modi, the most hardline Hindu nationalist to ever become prime minister, the BJP focused attention on Modi's record as chief minister in Gujarat during a period of booming economic growth. Nevertheless, Modi was shamed throughout the campaign for not intervening during three days of organized violence in Gujarat during 2002, in which roughly 2000 Muslims were murdered by largely Hindu mobs organized by officials in the local government, the police, and BJP-affiliated religious organizations. Until he became prime minister, Modi was banned from entering the United States due to his role in the riots.

Since the BJP came to power in 2014 there has been a rise in hate speech and violence against Muslims, low-caste Hindus, and untouchables (Dalits). India has strong laws on hate speech directed at specific religious or caste-based groups, and since the BJP came to power in 2014, there has been a 500 percent rise in hate speech by high-level politicians, 78 percent of which came from BJP lawmakers. Organized mob violence and lynchings have

PHOTO ESSAY

8.28 Power and Politics in South Asia

Conflict is common in South Asia, with religious differences often masking underlying causes such as poverty, economic rivalry between groups, authoritarianism, and corruption in governments. Recent efforts to combat corruption are part of a general shift away from authoritarianism and toward greater political freedoms. At times this movement has paved the way for peaceful reconciliation between former combatants, who have been willing to compete in elections and take part in peaceful political transformation.

THINKING GEOGRAPHICALLY

A Why might Modi choose to speak before an image of a god going into battle?

B How can a biometric ID card reduce corruption?

C What aspects of this photo suggest that the convoy is part of a government-funded army?

A Hindu nationalism. India's Prime Minister Narendra Modi addresses supporters of the Hindu nationalist Bharatiya Janata Party (BJP) in Dehradun, India, before a banner depicting the Hindu god Krishna blowing into a conch shell before he rides into battle. Violence against Muslims has increased under BJP rule in India. [Rishi Ballabh/ Hindustan Times via Getty Images]

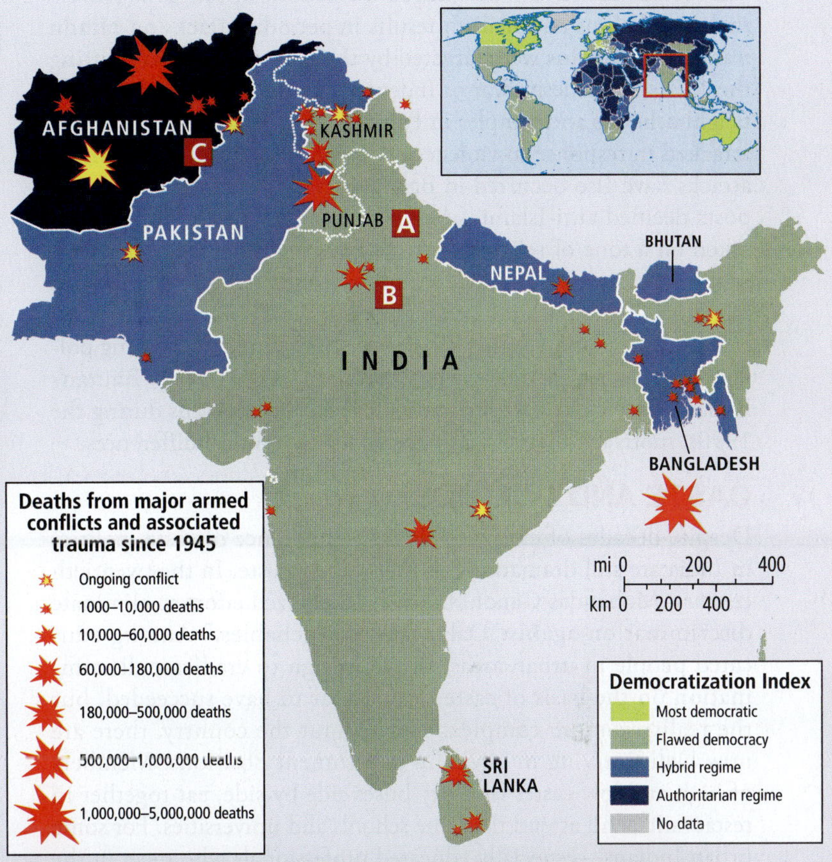

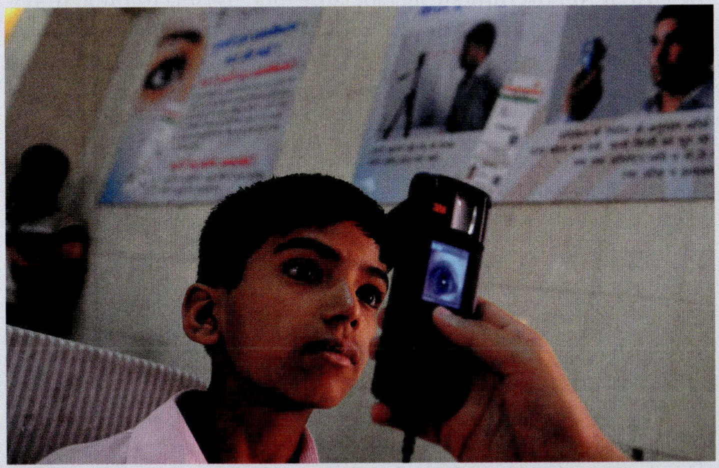

B Anticorruption measures. A boy in New Delhi has his retina scanned for his Aadhaar project biometric ID card, which aims to reduce corruption in the distribution of government aid to India's poor. [Priyanka Parashar/Mint via Getty Images]

C Armed conflict. Local residents watch a burning supply convoy in Pakistan that was attacked by Taliban militants as it attempted to carry fuel to U.S.-led NATO forces in Afghanistan. Pakistan no longer openly supports U.S. and NATO efforts in Afghanistan. [BANARAS KHAN/AFP/ Getty Images]

also increased since 2014, often centering around accusations of slaughtering cows and consumption of beef. These are activities that both Muslims and low-caste Hindus and Dalits have traditionally been allowed, but that are now banned or highly regulated in most Indian states. Conflict with Pakistan has also increased, with border skirmishes in 2019 boosting the image of Modi and the BJP as defenders of "Hindu India" against a Muslim Pakistan during national elections in India that year.

In Pakistan and Bangladesh, Islam is the official state religion but secular constitutions safeguard freedom of religion. Nevertheless, religious nationalism results in periodic attacks on Hindu minorities. Hindus were targeted by the Pakistani military during the 1971 Bangladeshi War of Independence. More recently Hindu neighborhoods and temples in Pakistan and Bangladesh have been attacked in response to violence against Muslims in India. Similar attacks have also occurred in Bangladesh in response to Facebook posts deemed anti-Islamic. Afghanistan's ongoing strife has often taken on a tone of religious nationalism, with efforts to eradicate its Buddhist past. The civil war in Sri Lanka, while not religious in nature, had a religious element with the Singhalese Buddhist majority population dominating the government and enacting policies that discriminated against Tamil Hindus. Similarly, Bhutan's Buddhist monarchy expelled 100,000 Nepalese Hindus during the 1990s, motivated by religious, cultural, and political differences.

CASTE AND POLITICS

Despite decades of efforts to fight the influence of caste, politics in India are still dramatically influenced by caste. In the twentieth century, Mohandas Gandhi began an organized effort to eliminate discrimination against Dalits or "untouchables." Among educated people in urban areas, the campaign to eradicate discrimination on the basis of caste may appear to have succeeded, but the reality is more complex. Throughout the country, there are now Dalits serving in powerful government positions. Members of high and low castes ride city buses side by side, eat together in restaurants, and attend the same schools and universities. For some urban Indians—especially educated professionals who meet in the workplace—caste is deemphasized as the crucial factor in finding a marriage partner. However, studies suggest that fewer than 5 percent of marriages cross *jati* lines, let alone the broader gulf of *varna.* Nearly everyone notices the tiny social clues that reveal whether one is of a higher or lower caste, and in rural areas, where the majority of Indians still reside, the divisions of caste remain prevalent.

One reason caste continues to be relevant is that at the local level, most political parties design their vote-getting strategies to appeal to caste loyalties. They often secure the votes of entire *jati* communities with political favors such as new roads, schools, or development projects. These arrangements fly in the face of the official ideologies of the major political parties and of Indian government policies, which actively work to eliminate discrimination on the basis of caste.

Some of the policies meant to counter discrimination have had the unintended effect of encouraging stronger identification with caste. In the 1920s, after centuries of governing with and promoting the interests of the forward castes, British India instituted policies that reserved a certain number of school admissions and government jobs for three disadvantaged groups: people of lower castes (often referred to as OBC for "other backward class"), Dalits (untouchables), and Adivasis (the ancient aboriginal peoples of South Asia). After independence, India extended such policies to the point where now 49.5 percent of government jobs are reserved for people from these three groups, who together constitute approximately 64 percent of the Indian population. Critics argue that with so many sought-after scholarships and jobs now dependent on one's caste status, people are encouraged to identify with caste now more than ever. Indeed, several new political parties that explicitly support the interests of OBC, Dalits, and Adivasis have gained considerable power in recent decades. Meanwhile students and faculty at many elite institutions of higher education have successfully protested against the established quotas for lower-caste applicants. Meanwhile, many forward castes such as the Patels and a number of Brahmin and Kshatriya castes have agitated for their own special treatment, arguing that some of their less-wealthy members are unfairly shut out of opportunities by their caste status. Politically, these groups form the basis of support for the BJP.

MOVEMENTS AGAINST GOVERNMENT INEFFICIENCY AND CORRUPTION

As in the many world regions where most states grew out of European colonial governments, and before them European trading companies, corruption has a long history in South Asia. The British East India Company was notoriously corrupt, with its directors and staff widely criticized in London for their plunder of India's wealth and their practice of paying large bribes to Indian rulers, as well as British judges and parliamentarians, to extend their political influence. When the British government took over after 1857, they continued many corrupt company practices, such as paying both British and Indian civil servants very modest salaries that were meant to be supplemented by generous "fees" and "commissions" usually paid by private British contractors hired by the government to execute projects, such as the building of roads and dams. By today's standards, these practices would be considered bribery. This and other types of corruption grew after independence, with the switch to government-led socialist models of economic development bringing extensive opportunities for bribery among civil servants and politicians. After the reforms of the 1990s, corruption seems to have actually increased as a result of laws that made it easier for corrupt politicians, civil servants, and businesspeople to move money out of India and into secret bank accounts abroad.

In recent years anticorruption movements have been a powerful force on the political scene. The rise of the BJP in India can be partially attributed to the image that many of its politicians, such as Prime Minister Modi, successfully cultivate of being so piously devoted to the nation that they would never take a bribe. Modi's government has launched ambitious anticorruption drives, but most evidence suggests that corruption is just as bad under the BJP as under previous governments.

Some technologies make it more difficult for bureaucrats and elected officials to demand bribes. For example, cell phones are increasingly being used by governments to pay civil servants and contractors. Because transactions over a cell phone are more traceable, it is now harder for upper-level government officials to demand bribes from the people they supervise in return for issuing their paychecks. Another example is the Aadhaar project, an ambitious effort to provide all 1.3 billion Indians with a unique photo and biometric digitized ID card (Figure 8.28B). The Aadhaar project is designed to facilitate the distribution of government benefits of all types, much of which currently fail to reach those most in need because there are few formal records that can be used to identify them. Currently, much of the money intended to reach such people ends up in the hands of corrupt officials. By 2018, a total of 1.22 billion cards had been issued, achieving almost complete coverage of India, with the government claiming savings of more than $2 billion due to reduced corruption and avoided errors.

Citizens' movements are also using technology to address corruption. Confronted by a bureaucrat who asked for nearly $200 to issue a legitimate income tax refund, one Indian couple in Bangalore launched the website I Paid a Bribe, aimed at collecting information about crooked officials. The idea caught on quickly, and similar sites now exist in more than 17 countries.

REGIONAL CONFLICTS

The most intense armed conflicts in South Asia today are regional conflicts in which nations dispute territorial boundaries or a minority actively resists the authority of a national or state government. Most of these hostilities arise from the authoritarian tendencies of governments that at times work against the growth of political freedoms that might otherwise defuse political tensions.

Conflict in Kashmir

Since 1947 and the postindependence division of India and Pakistan, between 60,000 and 100,000 people have been killed in violence in Kashmir. At the root of the violence is a struggle for territory between India and Pakistan, neither of which is willing to let the people of Kashmir resolve the dispute democratically (Figure 8.28 map).

Kashmir has been a Muslim-dominated area since the 1400s, and in 1947, some Kashmiris believed that it should be turned over to Pakistan for this reason. Although the Hindu maharaja (king) of Kashmir wished for his country to remain independent at the time, the most popular Kashmiri political leader and significant portions of the populace favored joining India. When Pakistan-sponsored raiders invaded western Kashmir in 1947, the maharaja quickly agreed to his nation becoming part of India. A brief war between Pakistan and India resulted in a cease-fire line that became a tenuous boundary.

Pakistan attempted to invade Kashmir again in 1965 but was defeated. India and Pakistan are technically still waiting for a United Nations (UN) decision about the final location of the border (**Figure 8.29**). The Ladakh region of Kashmir (Figure 8.1) is the site of a more limited border dispute between India and China.

After years of military occupation, most Kashmiris now support independence from both India and Pakistan. However, neither country is willing to hold a vote on the matter. Anti-Indian Kashmiri guerrilla groups equipped with weapons and training from Pakistan have carried out many bombings and assassinations. Blunt counterattacks launched by the Indian government have killed large numbers of civilians and alienated many Kashmiris.

Another complication in the Kashmir dispute is the fact that both India and Pakistan—which came close to war against each other in 1999, 2002, and 2019—have nuclear weapons. Because of the nationalistic fervor of the protagonists, many see the conflict in Kashmir as more likely to result in the use of nuclear weapons than any other conflict in the world.

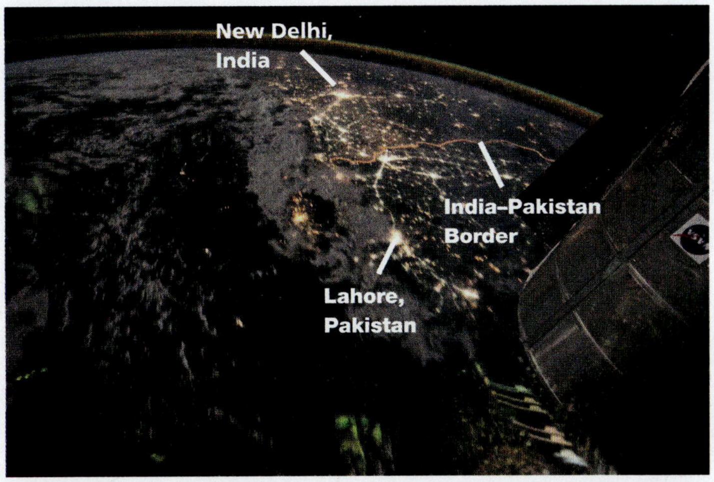

Figure 8.29. Illuminated border. The India–Pakistan border can be seen from space, thanks to the floodlights illuminating a fence built to discourage arms smuggling related to the region's many conflicts, especially the one in Kashmir. A terrorist attack led to a border skirmish. [NASA]

War and Reconstruction in Afghanistan

Afghanistan is home to one of the longest running and currently most lethal conflicts in South Asia. In the 1970s, political debate in Afghanistan became polarized between urban elites, who favored economic modernization and democratic reforms, and conservative religious leaders, whose positions as rural landholders and unelected traditional leaders were threatened by the proposed reforms. Divisions intensified as successive governments, all of which came to power through military coups, became more and more authoritarian. Political opponents were imprisoned, tortured, and killed by the thousands, resulting in a growing rural insurgency.

In 1979 the Soviet Union invaded Afghanistan, fearing that a civil war in Afghanistan would destabilize neighboring Soviet republics in Central Asia. Rural conservative leaders and their followers formed an anti-Soviet resistance group, the *mujahedeen.* As Afghan resistance to Soviet domination increased, the mujahedeen became ever more strongly influenced by militant Islamist thought and by Persian Gulf Arab activists who provided funding and arms. At the time, the United States, still searching for Cold War allies against the Soviet Union, joined with Pakistan in supporting the mujahedeen. Moderate, educated Afghans who favored democratic reforms fled the country during this turbulent time, hoping to go back eventually when peace returned.

In 1989, after heavy losses—14,000 Soviet soldiers killed and billions of dollars wasted—the Soviets gave up and left Afghanistan. Anarchy prevailed for a time as mujahedeen factions fought one another, adding to the 1.5 million civilians and combatants killed in the war with the Soviets.

In the early 1990s, a radical religious, political, and military movement called the **Taliban** emerged from among the mujahedeen. The

Taliban an archconservative Islamist movement that gained control of the government of Afghanistan in the mid-1990s

Taliban wanted to control corruption and crime and minimize Western ways—especially those related to the role, status, and dress of women—that had been introduced in earlier decades by the urban elites and reinforced by the Russian occupation. The Taliban strictly enforced *shari'a*, the Islamic social and penal code (see Chapter 6), greatly restricted women (see "Women and the Taliban in Afghanistan"), and promoted only fundamentalist Islamic education. While publicly banning the production of opium, to which many Afghan men had become addicted, the Taliban privately promoted its production and export to raise funds for their efforts. By 2001, the Taliban controlled 95 percent of the country, including the capital of Kabul.

Following the events of September 11, 2001, the United States and its allies focused on removing the Taliban, who had given shelter to Osama bin Laden and his international Al Qaeda network. By late 2001, the Taliban were overpowered by an alliance of Afghans supported heavily by the United States, the United Kingdom, and eventually NATO (see the Figure 8.28 map).

In 2003, the United States launched the war in Iraq that diverted national attention, troops, and financial resources away from Afghanistan. Almost immediately, the Taliban resurfaced, effectively thwarting the ability of Afghanistan's new government to ensure security and to meet the needs of people outside Kabul. Based in rural areas in both Pakistan and Afghanistan, the Taliban are now aided by widespread distrust of the government in Kabul, which is seen as corrupt (see Figure 8.28C). It appears that most people in Afghanistan favor a democratic government based on Muslim principles, but functionally flawed elections and corruption in government have eroded faith in democracy since the first elections in 2004.

In May 2011, bin Laden was killed in a raid by U.S. forces in the town of Abbottabad, Pakistan. Another raid resulted in the death of the next-highest Al Qaeda commander, and sporadic drone attacks killed both combatants and civilians. With the threat from Al Qaeda seemingly diminished, calls within the United States for a withdrawal from Afghanistan increased. Although 90,000 U.S. troops returned home, the reemergence of the Taliban in recent years ensures a U.S. presence for the indefinite future (see Figure 8.28C).

Nepal's Rebels

After a civil war that ended generations of monarchical rule, Nepal has endured years of political turmoil. An elected legislature and multiparty democracy were introduced in Nepal in 1990, but until 2008, a royal family governed with little respect for the political freedoms of the Nepalese people.

In 1996, inspired by the ideals of the late Chinese leader Mao Zedong (but with no apparent support from China), Maoist revolutionaries took advantage of public discontent and waged a "people's war" against the Nepalese monarchy. Following a decade of civil war, during which 13,000 Nepalese died, the Maoists both took military control of much of the countryside and developed strong political support among most Nepalese. Persistent poverty and lack of the most basic development under the dictatorial rule of then-monarch King Gyanendra led to massive protests that forced Gyanendra to step down in 2006. Soon thereafter, the Maoists declared a cease-fire with the government.

In 2008, the Maoists won sweeping electoral victories that gave them a majority in parliament and made their former rebel leader prime minister. Then, in May 2009, with his many conditions for reforming Nepalese society still unmet, the Maoist prime minister resigned, allowing the Communist Party of Nepal to come to power. The Maoists, Communists, and other parties collaborated in writing a new constitution, which came into effect in 2015, allowing for power sharing between competing political parties, the dominance of forward castes within the government and most political parties, and dividing the country into a federation of states. While there has been some violence in recent years, a return to widespread conflict seems unlikely, especially as both China and India court Nepal's government with new development projects. Both countries are interested in Nepal's water resources and its huge potential to generate hydroelectric power. In 2018 the Maoist and Communist parties merged into the Nepal Communist Party; the merger has oriented the country toward China, which the new party sees as more generous and less prone to interfere in Nepal's politics and economy than India, which has historically dominated Nepal.

CHECK YOUR UNDERSTANDING

1. How has the rise of Hindu nationalism in India affected its political culture?
2. How does caste affect politics in India?
3. What new tools are being used to combat corruption in this region?
4. How did democracy contribute to the resolution of Nepal's civil war?

URBANIZATION

8.6 Assess the forces creating divergent patterns of urbanization in South Asia.

South Asia has two main patterns of urbanization shaped by growing income disparities as wealthy South Asians reap most of the benefits of recent economic growth. The areas that the rich and the middle classes occupy include sleek, modern skyscrapers bearing the logos of powerful global companies, universities, upscale shopping districts, and well-appointed apartment buildings. The areas that the urban poor occupy are chaotic, crowded, and violent, with overstressed infrastructure and menial jobs. These two patterns often coexist in very close proximity (**Figure 8.30**), sometimes taking the form of a third urban pattern, the urban village.

Of the world's top 25 largest metropolitan areas, five are located in South Asia, even though only 35 percent of the region's population live in urban areas. Delhi has the third-largest urban population in the world (after Tokyo, Japan, and Jakarta, Indonesia) at 27 million people, Mumbai has 23 million, Dhaka has 17 million, Kolkata has 15 million, and Karachi has 13 million. This growth is likely to continue, and South Asia's current urban population of about 630 million people could expand to as many as 720 million by 2025. Even so, this region's huge rural population means that South Asia will not be majority urban until 2045 or later.

PHOTO ESSAY

8.30 Urbanization in South Asia

While only 35 percent of South Asia's population is urban, the region is home to some of the world's largest cities, such as Delhi, Mumbai, Dhaka, Kolkata, and Karachi. Throughout the region, between 30 and 60 percent of the urban population live in slums plagued by shoddy construction and inadequate access to water and sanitation.

THINKING GEOGRAPHICALLY

A What security infrastructure can be seen in this photo?

B What circumstances have brought South Asians into crowded urban apartments such as this?

C What about these buildings suggests their temporary nature?

A Upper-middle-class housing. An apartment complex for upper-middle-class civil servants in New Delhi, India. Buildings feature onsite car parking, solar panels, and rainwater harvesting. [Ruhani Kaur/Bloomberg via Getty Images]

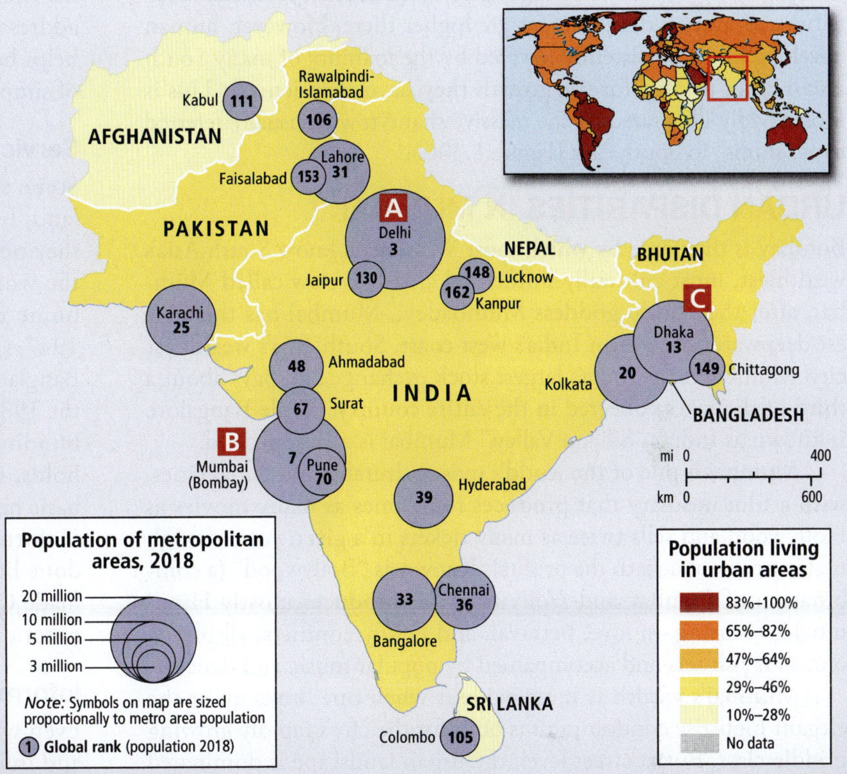

B Lower-middle-class apartment. A family in a lower-middle-class one-room apartment in Mumbai. The young woman to the right has a scholarship to attend Bard College in New York. [INDRANIL MUKHERJEE/AFP/Getty Images]

C Urban slum. A slum in Dhaka, Bangladesh, that grew up in a railroad right-of-way. Many slums develop in areas that have clear risks to human habitation and few safeguards against potentially deadly hazards. [Godong/UIG/Getty Images]

Cities are also growing fast economically, with half of the world's 30 fastest-growing cities in terms of GDP found in this region. Some of this growth is driven by manufacturing and other industries, but many South Asians are moving to cities for high-skill service sector jobs, such as in IT-BPM, education, health care, and engineering (Figure 8.25A). Larger cities tend to have more of these jobs and better services to support them, such as schools and health care; in addition, literacy rates, life expectancy, and other factors that contribute to human development are higher there. However, human development is persistently lowered by the inability of many South Asian cities to plan for the growth they have experienced. This is most vividly illustrated by the massive shantytowns, usually referred to as "slums" in South Asia (Figure 8.30C).

URBAN DISPARITIES IN MUMBAI

Bombay is the name by which most Westerners know South Asia's wealthiest, most culturally influential city. It is now called Mumbai, after the Hindu goddess Mumbadevi. Mumbai has the largest deepwater harbor on India's west coast. South Asia's wealthiest city, Mumbai hosts India's largest stock exchange and pays about a third of the taxes collected in the entire country. While Bangalore is known as India's "Silicon Valley," Mumbai is a close second.

Mumbai is one of the world's most culturally influential cities, with a film industry that produces four times as many movies as Hollywood and sells twice as many tickets in a given year (though making only one-sixth the profits). Known as "Bollywood" (a combination of *B*ombay and *H*ollywood), it produces mostly Hindi movies focusing on love, betrayal, and family conflicts, all played out on lavish sets and accompanied by popular music and dance.

Mumbai's wealth is most evident when one looks up at the elegant high-rise condominiums built for the city's rapidly growing middle class. But at street level, the urban landscape is dominated by great numbers of people living on the sidewalks, in narrow spaces between buildings, and in the large slums served mostly by narrow alleyways that house around 40 percent of Mumbai's population. The largest slum is Dharavi, which houses up to a million people in less than 1 square mile (2.6 square kilometers). Dharavi is said to contain 15,000 one-room factories turning out thousands of products that are sold in the global marketplace, many of them from India's recycled plastic and metal.

SOLUTIONS FOR SOME PROBLEMS FACED BY SLUM DWELLERS

Depending on the city, between 30 and 60 percent of the urban population lives in slums that often lack basic services such as sanitation, health care, education, or even a legal right to the land they are built on. The largest cities have around 40 to 60 percent of their population living in slums, which is one of the reasons South Asian cities have some of the highest population densities in the world. Slums are the result not just of massive migration to the cities, but also a lack of well-funded urban planning. The lack of services provided to slum dwellers means they have to be highly resourceful on a daily basis, and in this region they are developing a number of innovative solutions to improve their lives.

Addresses

Because they are often occupying land illegally, slum dwellers do not have formal addresses, which makes their interaction with governments much more difficult, and likely to result in eviction. On any given day, without notice, bulldozers can show up and destroy an entire neighborhood to make way for a new real estate project that mostly likely won't be of any benefit to the slum dwellers. Addressing the Unaddressed is an NGO based in Kolkata, India, that uses GPS and Google Maps to give postal addresses to slum dwellers, allowing them to receive mail and thus formalize their relationship with local government planners. Having addresses enables slum dwellers to be counted in censuses, which helps bureaucrats secure funds to upgrade services to slums instead of simply clearing them out.

Services

Some slums are legal and residents do have formal titles to their land, but local governments deem them such a low priority that they never receive services such as sanitation. Such is the case with the world's largest slum, Orangi Township in Karachi, Pakistan, home to 2.4 million people. Formed shortly after Partition in 1947, Orangi became home to many migrants from India and later Bangladesh after it gained independence from Pakistan in 1971. By the 1980s it still had no basic sanitation system, so residents started building one themselves. Now serving over 96 percent of households, it is part of the Orangi Pilot Project, which also provides basic health, family planning, and microcredit services to residents. Unfortunately the progress made on providing basic services has done little to resolve other problems, such as violence, which has made Orangi the most murder-prone neighborhood in the already notoriously violent city of Karachi.

Information

Even when services are available in slums, they are of low quality and information about them is hard to find. Dhaka is considered the least "liveable" city in this region, due to the low quality of its basic services such as education and health care, and the generally poor state of infrastructure such as roads and sewers. Residents of Bauniabad, a slum in Dhaka, created an app that provides the locations of schools, health care, banks, and other essential services in their neighborhood, as well as a way for people to provide feedback for improvement of these services. Through the neighborhood's Kolorob (Clamour) app, students at a local school were able to successfully lobby administrators for a library, and clients of a local health clinic got a much-desired area for breast feeding.

RISKS TO CHILDREN IN CITIES

Around 10 million South Asian children between the ages of 5 and 14 are primarily working instead of attending school. Most of these children are employed in rural areas as agricultural laborers with their families, but in urban areas children work in much riskier settings, usually outside the home and without supervision by a family member. Many children labor as domestic servants, in small export-oriented factories, as street vendors, and as scavengers at dumpsites.

Clearly, many of these occupations are entirely unsafe and inappropriate for children. But a more complex image emerges when one considers the many urban craft industries, in which children are often trained in an ancient art and supervised by family members, but nonetheless denied full access to education. For example, carpet weaving is an ancient artistic and economic enterprise in South Asia. Traditionally, it has been a family-run enterprise, with women and children serving as weavers and men as merchants. For thousands of years, young children have learned weaving skills from their parents and become proud members of the family's home-based production unit. But in today's rapidly modernizing society, the role of children as family workers must be balanced with the need for children to learn the skills that will enable them to survive and prosper in a modern economy, such as the ability to read, write, and do basic math.

As global trade has increased the demand for fine, hand-woven carpets, most of the profit has gone not to the weaving families but to South Asian middlemen and foreign traders from Europe and America. The possibility of reaping large profits has led some unscrupulous carpet merchants to set up factories where kidnapped children are forced to produce carpets. In response to public outrage some rug sellers in the United States and Europe have developed certification programs that ensure their merchandise is made with no child labor at all. The question remains: Is there room for cottage industry employment of children within the family circle?

The UN, South Asian governments, and nongovernmental organizations (NGOs) like RugMark have instituted an active program to curb child labor in the hand-woven carpet industry while remaining open to the positive experience of a child learning a skill and becoming part of a family production unit. India has established a national system to certify that exported carpets are made in shops whose child employees go to school, eat an adequate midday meal, and receive basic health care. Such carpets bear the label "Kaleen" or "RugMark."

RISKS TO ADULTS IN CITIES

When rural adults move to cities, the need to find employment, the crowded conditions, the lack of housing, and the welter of new experiences can leave them vulnerable to exploitation. Inexperienced and undereducated rural women and girls are sometimes recruited or forced into **sex work**, the selling of sexual acts for a fee. The vast, urban, slum-based brothels in which sex workers labor are legendary in the cities of South Asia and have been documented in, among other places, *Half the Sky*, a book and video created by Nicholas Kristof and Sheryl WuDunn.

Brothels are notoriously exploitative of women wherever they are found, but the age of the mobile phone is changing the geography of sex work by allowing sex workers to move out of brothels. As a result, they can develop more control over those who will become their clients, and manage their incomes, which in former times were immediately seized by pimps and madams. However, being spatially autonomous also leaves them less likely to have access to condoms or to learn how to avoid exposure to HIV, for which sex workers remain at high risk.

Rural men new to cities are also at risk: on-the-job accidents, exposure to HIV, extreme pressure to support families on low wages by working long hours at physically demanding jobs, and the loss of camaraderie with fellow villagers are just a few of the hardships experienced by such men.

THE URBAN VILLAGE

There is yet a third aspect to urban life in South Asia. Were one to carefully observe places as widely separated and culturally different as Mumbai, Chennai, Kathmandu, and Peshawar, one would discover that beyond the main avenues, wealthy commercial districts, high-rises, and shantytowns, these cities also contain thousands of tightly compacted, reconstituted villages where daily life is intimate and familiar, not anonymous as in Western cities. Koli is one such place.

Figure 8.31 The village of Koli. Harsha Tapke chats with friends in the fish market at the Koli village in downtown Mumbai. In 1534 when Portuguese colonists arrived, Koli was a small fishing settlement on a beautiful bay. The village remains, though real estate developers are constantly scheming to take their land. A few Koli still fish, but declining water quality means catches are smaller. Most residents are educated and work as bureaucrats or in the city's service sector. [Abhijit Bhatlekar/Mint via Getty Images]

VIGNETTE Koli is an ancient fishing village that predates the city of Mumbai and is now squeezed between Mumbai's elegant coastal high-rises and the Bay of Mumbai (**Figure 8.31**). Ringed by fishing boats, Koli is a labyrinth of low-slung, tightly packed homes. Some villagers still fish every day. The screeching of taxis and buses is soon lost in quiet calm as one ducks into a narrow covered passageway that winds through the village and branches in multiple directions. At first, Koli appears impoverished, but inside, well-appointed homes, some with marble floors, TVs, and computers, open onto the dimly lit but pleasant alleys. The visitor soon learns that this is no slum or warren of destitute shanties, but rather a community of educated bureaucrats, tradespeople, and artisans who constitute South Asia's rising urban middle class. ■

CHECK YOUR UNDERSTANDING

1. Describe the three general patterns of urbanization in South Asia.
2. Why are people moving to cities in South Asia?
3. What are some solutions that slum dwellers are developing to improve their lives?
4. What are some risks to children, women, and men in South Asian urban areas?
5. Why is Mumbai so culturally influential?

sex work the provision of sexual acts for a fee

POPULATION, GENDER, AND CULTURE

8.7 Identify the factors slowing population growth in this most populous and most densely populated of world regions.

8.8 Explain why men significantly outnumber women in South Asia.

8.9 Describe current social and political transitions that challenge traditional cultural practices related to gender.

In this most populous of world regions, growth is slowing as the demographic transition takes hold. Birth rates are falling due to rising incomes, urbanization, better access to health care, and the fact that women are finding more opportunities to study and work outside the home, and thus are delaying childbearing and having fewer children. These changes are also altering the culture of this region in ways that both strengthen and weaken patriarchal customs. On the one hand greater opportunities for women mean that age-old customs of male dominance are slowly but surely giving way to greater respect for the rights of women. However, as parents decide to have fewer children, there is more pressure for mothers to have male children, who, in this still highly patriarchal region, are seen as more of an asset compared to females, who are seen as a liability. This has resulted in populations with more males than women in most of South Asia.

POPULATION

Since overtaking East Asia in 2008, South Asia is now the most populous world region with 1.8 billion people. It is also the smallest world region and the most densely populated (**Figure 8.32**; see also Figure 1.26). In fact, densities in the Ganga–Brahmaputra delta are higher than in the most densely populated U.S. metropolitan areas, and India's overall population density is nearly that of the metro areas of Miami or Milwaukee, and much higher than Washington, DC, or San Diego. India will have overtaken China as the most populous nation by 2030, at which point China's population will likely be shrinking at the rate of about 20 million people per decade, while India's will be growing by over 100 million per decade.

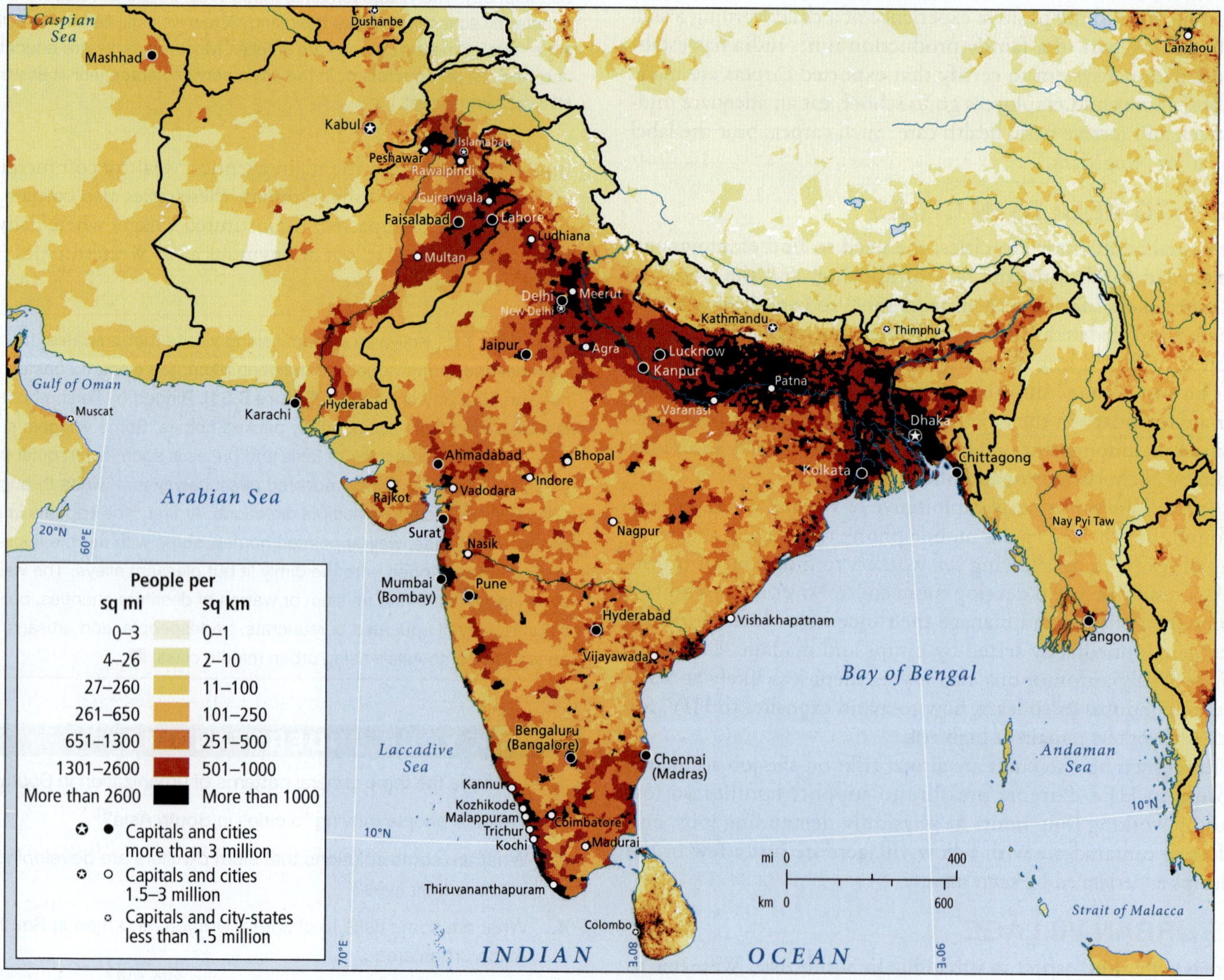

Figure 8.32 Population density of South Asia.

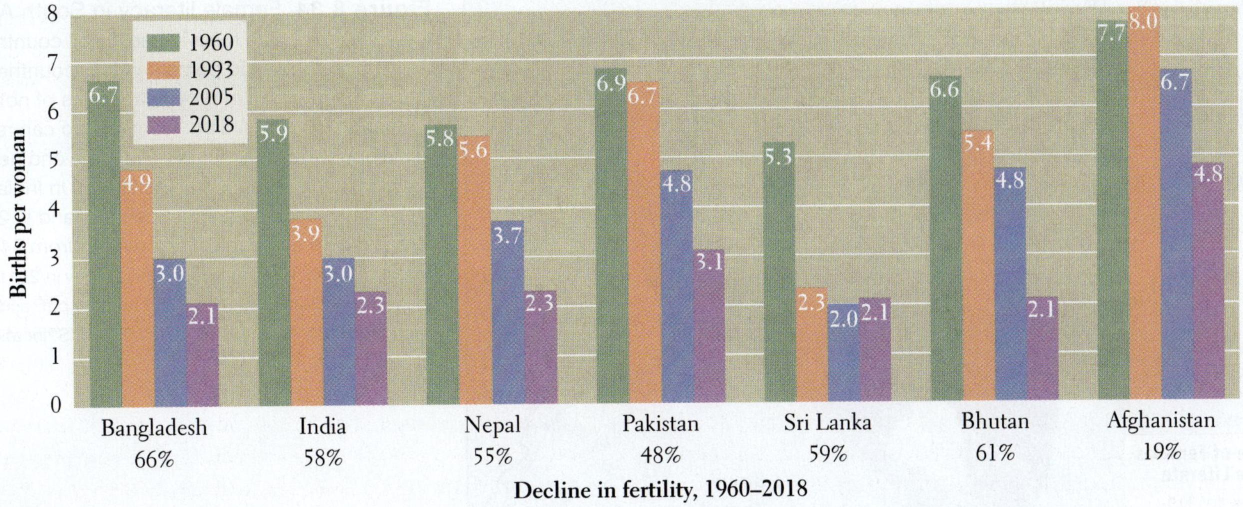

Figure 8.33 Total fertility rates in South Asia, 1960–2018. All South Asian countries except Afghanistan have experienced a substantial decline in fertility since 1960. Afghanistan's rate has declined less than 20 percent over the past 50 years, probably because of almost continuous conflicts, the overwhelmingly rural population (77 percent), and the low literacy rates for both men (43.1 percent) and women (12.6 percent). [Data from: *World Population Data Sheet 2018*, https://data.worldbank.org/indicator/SP.DYN.TFRT.IN?locations=IN-BD-BT-PK-AF-NP-LK.]

However, a notable trend is that current rates of population growth have slowed to approximately 1.5 percent per year and fertility is declining significantly. Only in Afghanistan and Pakistan are fertility rates still above three children per woman (**Figure 8.33**). Families are choosing to have fewer children for a variety of reasons, such as improved educational and employment opportunities for women, better health care, and urbanization.

Population density in this region is highest along river valleys and coasts. The Ganga–Brahmaputra delta in Bangladesh and India has the highest densities (see Figure 8.32). The delta's rural agricultural and fishing areas, inhabited for tens of thousands of years, have population densities associated with suburban and even urban areas in most parts of the world. Other major areas of settlement include the entire Ganga River basin, the Indus River basin in Pakistan, and the area centered on Lahore known as the Punjab, named for the five rivers that flow across the foothills of the Himalayas to the Indus River.

Slowing Population Growth, Health Care, and Urbanization

Efforts to slow population growth through better health care have been under way for more than 50 years. Today, urbanization and the rising status of women are also helping to reduce the birth rate.

With improved health care—such as rehydration to control the effects of diarrhea in infants—far fewer babies are dying in infancy. In 1992, infant mortality rates were 91 per 1000 live births in India, 109 in Pakistan, and 120 in Bangladesh. By 2018, those rates were reduced to 34 in India, 66 in Pakistan, and 28 in Bangladesh. With better assurance that their babies will survive to adulthood and be able to care for their elderly parents, couples often now choose to have just two children.

At the same time that improved access to health care has helped propel the reduction in family size, the partial shift in population from rural to urban areas has reduced the economic incentive for having a large family (see the Figure 8.30 map). Whereas children in rural areas (starting at an early age) contribute their labor to farming, thus increasing the family income, the labor of children in urban areas is less likely to boost the family income. Children in urban areas are more likely to attend school, and thus represent a cost to the family in terms of school fees, books, and uniforms.

Slowing Population Growth and Gender

Improvements in the status of women have also slowed population growth. As in other parts of the world, women in South Asia who have the opportunity to go to school and/or work at a paying job tend to delay childbearing and have fewer children. A major indicator of these opportunities is female literacy (**Figure 8.34**), which is highest in southern India and in Sri Lanka, followed by Bangladesh and eastern and central India, and lowest in the belt that stretches from the northwest in Afghanistan across Pakistan, western India, Nepal, Bhutan, and the Ganga Plain. In the regions where literacy rates are higher, women have better access to education and resources as a result of more equitable employment, marriage, and inheritance practices.

Population Pyramids

A comparison of the population pyramids as well as some other statistics for Sri Lanka and Pakistan helps illustrate the influence of health care and the status of women on birth rates (**Figure 8.35**; see also Figures 8.32 and 8.33). As a country that is far along but not completely through a demographic transition (see Figure 1.28), Sri Lanka has a much higher GNI per capita PPP (U.S.$12,490) than Pakistan (U.S.$5830). Health care is generally better in Sri Lanka, as indicated by its much lower infant mortality rate (in 2018, Sri Lanka had 7 deaths per 1000 live births versus Pakistan's 66 per 1000). Sri Lanka has also worked harder to educate and economically empower its women. Three key indicators of women's education and empowerment that are associated with reduced

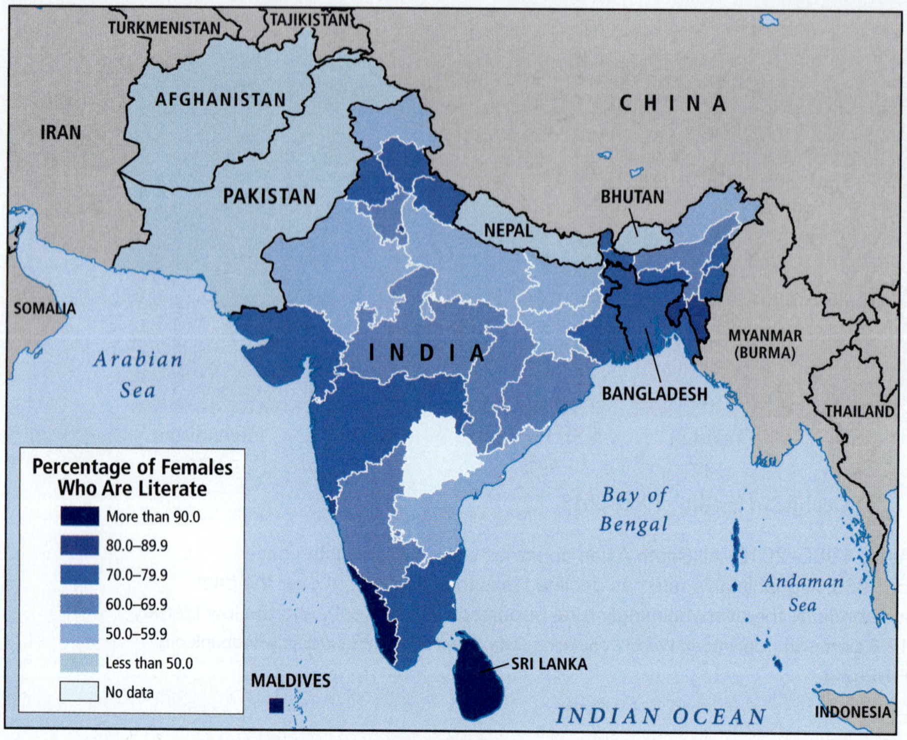

Figure 8.34 Female literacy in South Asia. Female literacy lags behind male literacy in all countries in South Asia, and is below 50 percent in four countries. Female literacy is crucial to improving the lives of not only women, but also children: women who can read often seize opportunities to earn an income and nearly always use this income to help their children. In India, overall literacy has risen in the past decade, and in 2011, female literacy there reached 65 percent, up from 54.5 percent in 2005. [Data from: http://censusindia.gov.in/2011-prov-results/data_files/india/Final_PPT_2011_chapter6.pdf; and https://data.worldbank.org/indicator/SE.ADT.LITR.FE.ZS?locations=AF-BD-IN-PK-NP-LK.]

fertility rates are much higher in Sri Lanka than in Pakistan: female literacy (92 percent versus 44 percent), the percentage of young women who attend high school (73 percent versus 19 percent), and the percentage of women who work outside the home (36 percent versus 24 percent). Surprisingly, Sri Lanka has achieved all this with an urbanization rate of just 18 percent, far lower than Pakistan's 38 percent.

Despite Pakistan's rather dismal statistics compared to those of Sri Lanka, its population pyramid does indicate that birth rates have slowed significantly (notice that the bottom of the pyramid in Figure 8.35B narrows). If this trend continues, Pakistan, as well as all South Asian countries except Afghanistan, will enjoy what is called the *demographic dividend:* a surge of young people entering the workforce and choosing to have fewer children, who will thus have the time and energy to study, work, and contribute to the economy and civil society. South Asian countries that have experienced a demographic dividend will eventually have to deal with an aging population for which there will be fewer younger caregivers and taxpayers to support the elderly. Current population growth trends suggest that this transformation may take 50–100 years in the case of Pakistan, and 20–40 years in the case of Sri Lanka.

Sex Ratio Imbalance

South Asian populations have a significant imbalance in the **sex ratio**, the ratio of females to males in a population, because of cultural customs that make sons more likely than daughters to contribute to a family's wealth. A popular toast to a new bride is "May you be the mother of a hundred sons." Many middle-class couples wishing to have sons go to ultrasound clinics that specialize in identifying the sex of a fetus, with the intention of aborting female fetuses. This practice is now illegal in much of the region but still widespread due to lax enforcement. Poorer South Asians may neglect the health of female children, with some even committing female infanticide (the dowry implications of this are discussed in "Dowry and Status"). India's census of 2011 indicated a sex ratio imbalance of around 115 men for every 100 women, meaning there is a shortage of between 30–70 million women. Similar situations are developing in other South Asian countries.

sex ratio the ratio of females to males in a population

India has the highest sex ratio imbalance in South Asia. While recent data suggest that it may be diminishing, it will likely result in as many as 30 percent more men than women looking to marry by 2050. This situation may already be increasing violence against women, with reports of rape and sex trafficking up in recent years. In the years to come India could see surges in crime, drug abuse, and political violence related to the presence of so many young men with no prospect of having a family. In all cultures, the possibility of building a family is a stabilizing influence for young men.

Notably, the one Indian state where women outnumber men is Kerala, where the government has made women's development a priority by funding education for women well beyond basic levels. The results are reflected in its female literacy rate, which is the highest in India at 92 percent (the average for India as a whole is 63 percent; see Figure 8.34 for the female literacy map), and in the number of women who work outside the home. In Kerala, far from being viewed as an economic liability to the family, daughters are seen as an asset, with many holding high positions in government, education, health care, IT industries, and other professions. Women regularly travel public streets alone and in groups, and commonly work in public places.

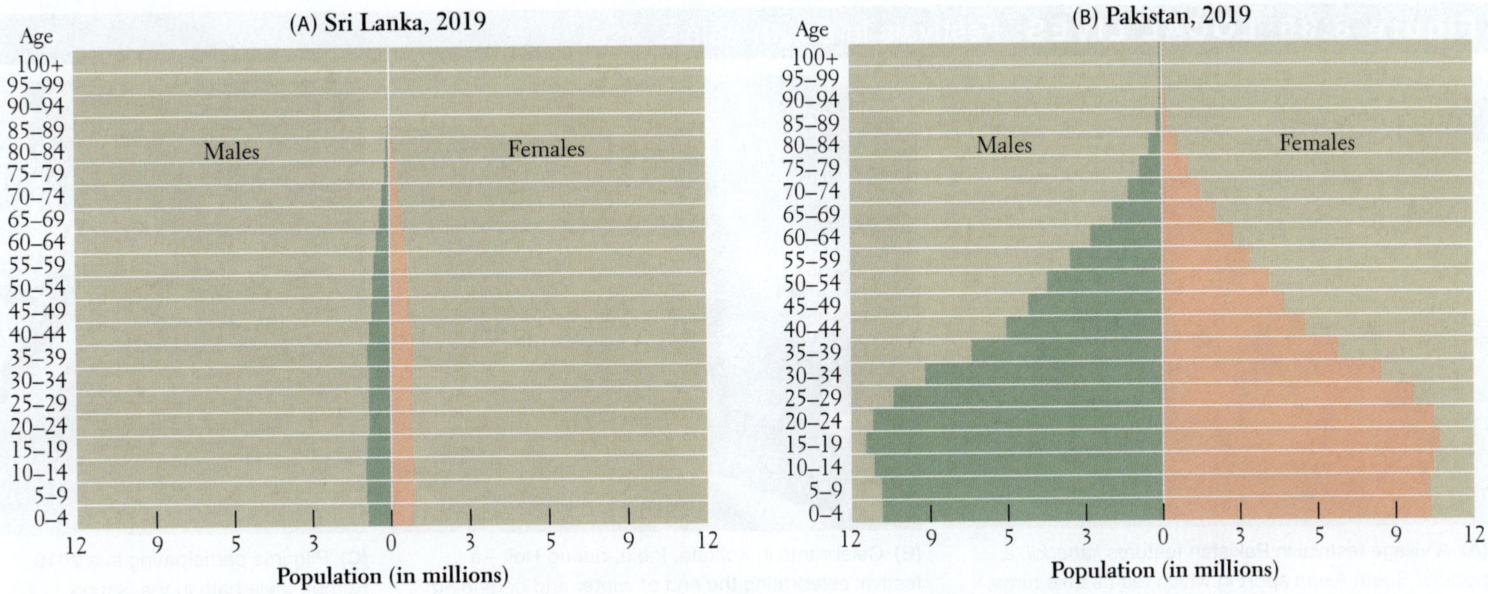

Figure 8.35 Population pyramids for Sri Lanka and Pakistan. [Data from: International Data Base, U.S. Census Bureau, http://www.census.gov/population/international/data/idb/informationGateway.php.]

CHECK YOUR UNDERSTANDING

1. Why is population growth slowing in this region?
2. What are the origins of the region's sex ratio imbalance?
3. Where are women the least restricted in India?

CULTURAL ISSUES

Within the life of one South Asian village or urban neighborhood, there can be considerable cultural variety. Differences based on caste, economic class, ethnic background, gender, religion, and even language are usually accommodated peacefully by long-standing customs, such as religious and ethnic festivals and foodways (**Figures 8.36** and **8.37**), that guide cross-cultural interaction. However, South Asia is also undergoing a number of social and political transitions that challenge traditional cultural practices, especially those related to gender.

The Texture of Village Life

Over 60 percent of South Asians live in rural villages, and many now living in cities were born in a village they occasionally visit, so for most people village life is a common experience.

VIGNETTE The anthropologist Faith D'Aluisio and her colleague Peter Menzel offer a peek into village life as night falls in Ahraura, a village in the state of Uttar Pradesh in north-central India. In the enclosed women's quarters of a walled compound, Mishri is finishing her day by the dying cooking fire as her 1-year-old son tunnels his way into her sari to nurse himself to sleep. Mishri, who is 27, lives in a tiny world bounded by the walls of the courtyard she shares with her husband, five children, and several of her husband's kin. Like many villages in northern India, her village observes the practice of *purdah*, in which women keep themselves apart from men.

That Mishri can observe purdah is a mark of status because it shows she need not help her husband in the fields. Within the compound, she works from sunup to sundown, chatting only briefly with two women who cover their faces and scurry from their own courtyards to hers for the short visit. Mishri is devoted to her husband, who was chosen for her by her family when she was 10; out of respect, she never says his name aloud. ■

VIGNETTE Richard Critchfield studied village life in more than a dozen countries. He found the Bangladeshi village of Joypur in the Ganga–Brahmaputra delta, set in "an unexpectedly beautiful land, with a soft languor and gentle rhythm of its own." In the heat of the day, the village is sleepy: naked children play in the dust, women meet to talk softly in the seclusion of courtyards, and chickens pock for seeds.

In the early evening, mist rises above the rice paddies and hangs there "like steam over a vat." It is then that the village comes to life, at least for the men. The men and boys return from the fields, and after a meal in their home courtyards, the men come "to settle in groups before one of the open pavilions in the village center and talk—rich, warm Bengali talk, argumentative and humorous, fervent and excited in gossip, protest, and indignation" as they discuss their crops, an upcoming marriage, or national politics. ■

Social Patterns in the Status of Women

The status of women in South Asia varies significantly along rural/urban and religious divides. Generally speaking, rural women have far less freedom than do urban women. In rural India, middle- and upper-caste Hindu women are often more restricted in their movements than are lower-caste women because they have a status to maintain. Meanwhile, lower-caste women who go into public spaces may have to contend with sexual harassment and exploitation from upper-caste men. The socioeconomic status of Muslim women in South Asia is lower than that of their Hindu and Christian counterparts. In India, some of this is related to the generally lower incomes and standard of living for Muslims, which also usually means lower educational levels. Muslim women also work

Figure 8.36 LOCAL LIVES: Festivals in South Asia

(A) A village festival in Pakistan features *kabaddi*, a popular South Asian sport in which teams take turns sending a "raider" across a field's center line. That person must tag, or in some cases wrestle to the ground, members of the other team and then return to his or her own side without taking a breath. Kabaddi has been played at the Indian National Games since 1939 and at the Asian Games since 1991. [Amir Mukhtar/Getty Images]

(B) Celebrants in Kolkata, India, during Holi—a festival celebrating the end of winter and beginning of spring. Holi evolved from temple worship practices involving the application of color to statues. In a riotous and celebratory atmosphere, people of different ages, genders, castes, and economic backgrounds temporarily disregard their differences and hurl the colors of the coming spring at each other. [Sanjay Kanojia/AFP/Getty Images]

(C) Pilgrims participating in a 2010 Kumbh Mela bath in the Ganga River at Haridwar. During this event, which is held for 45 days every 3 years in different cities across north and central India, Hindus purify themselves by bathing in the sacred waters of the river. Recent Kumbh Melas have attracted over 120 million participants. [PEDRO UGARTE/AFP/Getty Images]

Figure 8.37 LOCAL LIVES: Foodways in South Asia

(A) Boys in Hampi, Karnataka, eat a south Indian *thali*, or feast, featuring rice, lentils, and various other vegetarian dishes served on a banana leaf. Most South Asian cuisine is eaten with the hands; in South India and Sri Lanka, banana leaves often serve as plates. [Huw Jones/Getty Images]

(B) Chapatis are cooked on a griddle at a wedding in Rajasthan in northwest India. Chapati is a kind of bread made of stone-ground wheat flour and cooked on a griddle or curved iron *tava* located above a fire or oven. The bread is unleavened—it is made of dough that never rises because it doesn't contain yeast. Most popular in northern South Asia, where wheat is more widely grown than rice, chapatis are torn into pieces and used to scoop up vegetable or meat dishes. [Dario Mitidieri/Getty Images]

(C) Street food in a crowded marketplace in Dhaka, Bangladesh, during Ramadan. On offer are whole chickens, lamb, and various vegetable dishes, including potatoes and lentils. Usually, the food is taken home and shared with family members. [MUNIR UZ ZAMAN/AFP/Getty Images]

outside the home less than non-Muslim women in India. Low rates of education and workforce participation for women also prevail in Muslim-dominated countries such as Pakistan and Afghanistan, though women generally have more freedom in Bangladesh.

Purdah The practice of concealing women from the eyes of nonfamily men, especially during women's reproductive years, is known as **purdah**. It is observed in various ways across the region. The practice is strongest in Afghanistan and across the Indo-Gangetic Plain, where within both Muslim and Hindu communities, women are often secluded within structures (**Figure 8.38**) and wear veils or head coverings. Purdah is less strict in central and southern India, but even there, separation between unrelated men and women is maintained in public spaces. In general, low-caste Hindus do not observe this custom, but that is changing as some low-status households with increasing incomes have adopted purdah as a sign of their wealth.

Purdah practices have influenced the architecture of South Asia. Homes are often situated in walled compounds that seclude kitchens and laundries as women's spaces. In grander homes, windows to the street are usually covered with lattice screens, known as *jalee*, that allow in air and light but shield women from the view of outsiders.

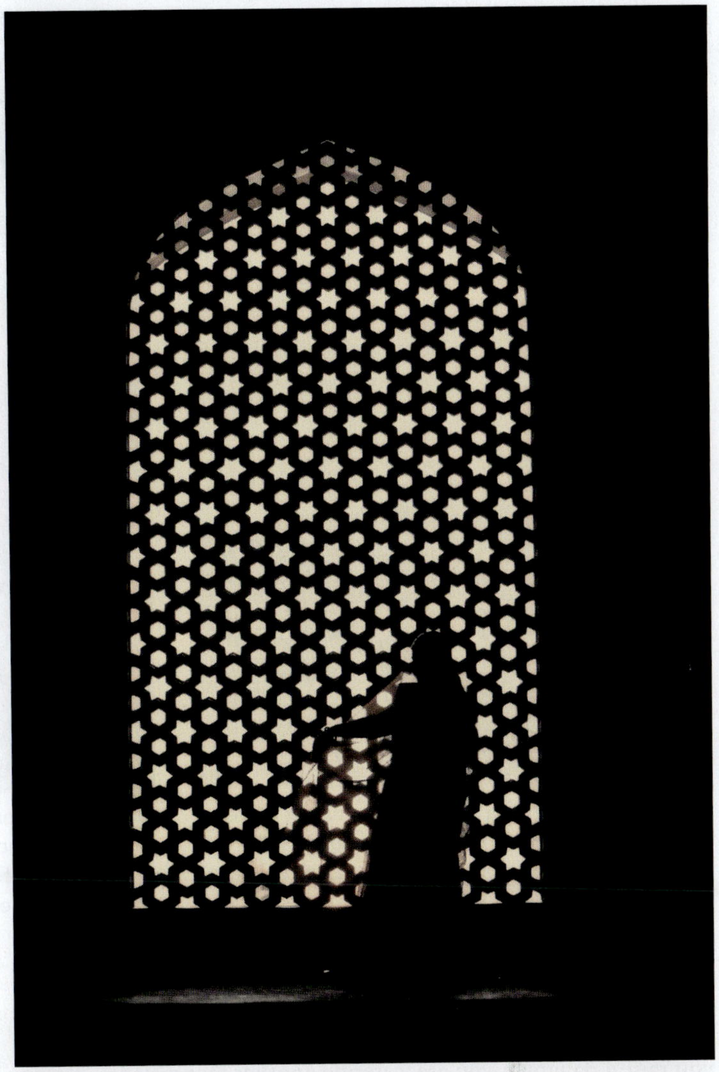

Figure 8.38 Material culture of purdah. A lattice screen in Fatehpur Sikri, India. Known as *jalee*, lattice screens are often found in parts of South Asia where women are secluded. Like the louvers and latticed windows of Southwest Asia (see Figure 6.32B), *jalee* allow ventilation and let in light but shield women from the view of strangers. [1001nights/iStock/Getty Images]

Marriage, Motherhood, and Widowhood Throughout South Asia, most marriages are arranged by the parents of the prospective bride and groom. Particularly in wealthier, better-educated families, the wishes of the bride and groom are considered, but in some cases they are not (**Figure 8.39**). This is especially true of a child marriage, when a young girl (often as young as 12 and in some places even younger) is married off to a much older man.

Usually, a bride goes to live in her husband's family compound, where she becomes a source of domestic labor for her mother-in-law. Most brides work at domestic tasks for many years until they have produced enough children to have their own crew of small helpers, at which point they gain some prestige and a measure of autonomy.

Motherhood in South Asia determines much about a woman's status within her community. A woman's power and mobility increase when she has grown children and becomes a mother-in-law herself. On the other hand, in some communities, the death of a husband, regardless of cause, is considered a disgrace to a woman and can completely deprive her of all support and even her home, children, and reputation. Widows may be ritually scorned and blamed for their husband's death. Widows of the higher castes rarely remarry, and in some areas, they become bound to their in-laws as household labor or may be asked to leave the family home. Most simply become marginalized "aunties" in extended families and help with household duties of all sorts.

Dowry and Status A **dowry** is a sum of money paid by the bride's family to the groom's family at the time of marriage. Dowries may have originated as an exchange of wealth between families that practiced purdah. With her ability to work reduced by purdah, a woman was considered a liability for the family that took her in. Changing dowry customs appear to be a cause of the growing incidence of various kinds of domestic violence against females in Pakistan, India, and Bangladesh. Until the last several decades, only wealthy families gave the groom a dowry—in this case, a substantial sum that symbolized the family's wealth, meant to give a daughter a measure of security in her new family.

As with purdah, increases in affluence and education have reinforced and popularized the custom of dowry. As more men become educated, their families feel that their increased earning potential calls for larger and larger dowries. The practice has also spread to poor lower-caste families wanting to upgrade their status, and the dowries they must pay to get their daughters married can be financially crippling. A village proverb captures this dilemma: "When

purdah the practice of concealing women from the eyes of nonfamily men

dowry a price paid by the family of the bride to the groom (the opposite of bride price); formerly, a custom practiced only by the rich

Figure 8.39 Caste and marriage. Caste remains a powerful force with respect to marriage. Fewer than 5 percent of marriages between Hindus cross subcaste or *jati* lines.

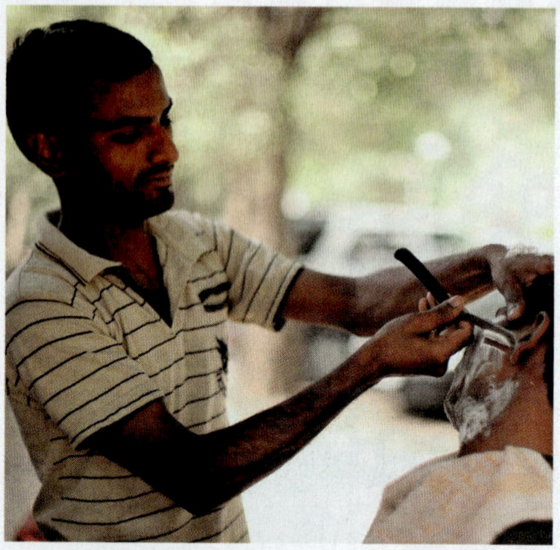

(A) A barber, often a member of a "barber caste" or *jati*, practices his trade in Chandigarh, India. Because barbers come into contact with so many people, they are often sought out by parents seeking a spouse for their children. [Mark Kolbe/Getty Images]

(B) A Brahmin priest officiates at a Hindu wedding ceremony in Varanasi, India. The specifics of the ceremony vary considerably, depending on the caste of the bride and groom. [Religious Images/UIG/Getty Images]

(C) Wedding guests in New Delhi dance to a brass band that accompanies the groom, who rides a white horse. The members of brass bands are often low-caste Hindus or low-status Muslims, in part because the work is part time and often low paid. [ROBERTO SCHMIDT/AFP/Getty Images]

you raise a daughter, you are watering another man's plant." Dowry is thought to contribute to selective abortion and infanticide of females and thus the sex ratio imbalance.

GENDER, POLITICS, AND POWER

As countries in South Asia have moved toward greater respect for political freedoms, the status of women in the region has risen, though slowly (**Figure 8.40**). India, Pakistan, Bangladesh, and Sri Lanka have all had female heads of state (prime ministers) in the past. However, it is important to note that all of these women were either wives or daughters of previous heads of state. Women have been notably less successful in local elections, and at the parliamentary level Indian, Sri Lankan, and Bhutanese women remain very poorly represented (**Table 8.1**).

The very low percentage of women in India's parliament inspired a confederation of Muslim and Hindu women's groups to lobby for legislation that would temporarily (for a 15-year trial period) reserve one-third of the seats in the lower house of parliament and in state assemblies for women. Such one-third quotas are already in place in Pakistan (20.2 percent), Nepal (32.7 percent), and Bangladesh (20.7), but are so far only close to being met in Nepal. If recent voter turnout and political activism trends continue, there is likely to be a major improvement in female representation in the region's national parliaments and local offices.

Table 8.1 Percentages of South Asian Women in Parliament, 2019

Sri Lanka	5.3
Bhutan	14.9
Pakistan	20.2
Nepal	32.7
Bangladesh	20.7
India	12.6
Afghanistan	23.6

Data from: Women in National Parliaments, Inter-Parliamentary Union website, http://archive.ipu.org/wmn-e/classif.htm

Women and the Taliban in Afghanistan

Women in Afghanistan have frequently suffered brutal repression since a conservative Islamist movement, the Taliban, gained control of the government there in the mid-1990s. Prior to that time, rights for women were slowly but steadily improving. Upper-class women in particular had many freedoms; they could dress in Western styles and attend gender-integrated universities. The Taliban supports strict and radical interpretations of Islamic law, forcing females, including urban professional women, to live in seclusion. In regions where the

8.40 Population, Gender, and Culture in South Asia

The map illustrates the Gender Development Index, reflecting inequality in achievements between women and men along three dimensions: reproductive health, empowerment, and participation in the labor market. Most of South Asia has relatively low GDI, with only Nepal and Sri Lanka ranking in the medium range. However, progress is being made due to greater attention to the region's sex ratio imbalance and the political rights of women and LGBTQ people.

THINKING GEOGRAPHICALLY

A Why is there a sex ratio imbalance in this region?

B Why might hijras want more cultural acceptance in South Asia?

C Why might the Taliban be particularly concerned about women voting?

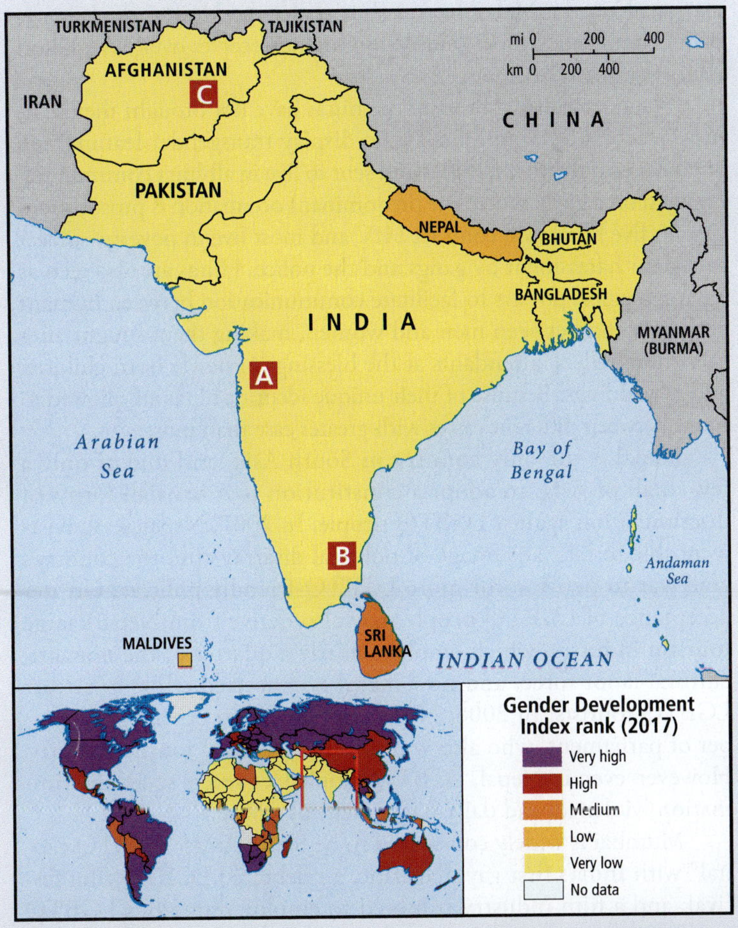

A Violence against women. Participants in the "March for Dignity" walked 6200 miles (10,000 kilometers) from Mumbai to Delhi to raise awareness of sexual assault. Violence against women may be increasing in South Asia due to a growing sex ratio imbalance. [CHANDAN KHANNA/AFP/Getty Images]

B Transgender people. South Asia's largest transgender festival, in Koovagam, India, draws thousands of hijras who reenact the marriage of the god Krishna, who changed from male to female form in order to wed a warrior, Aravan. [ARUN SANKAR/AFP/Getty Images]

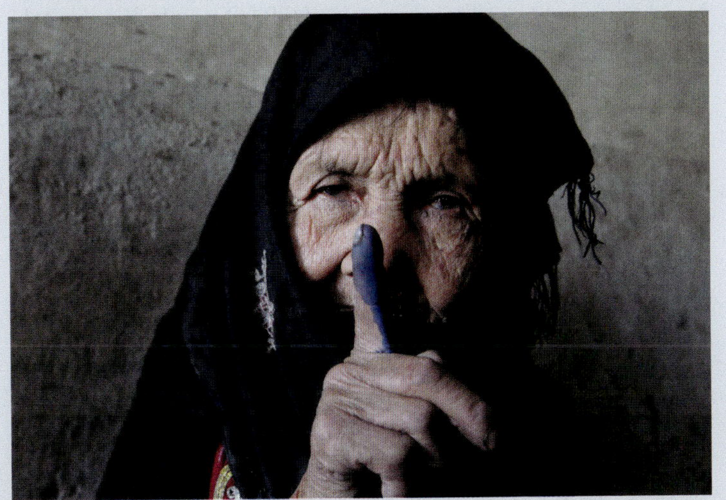

C Political rights of women. A woman displays her inked finger, which was marked after she voted in an election in Kabul, Afghanistan. The Taliban called for a boycott of the election and threatened anyone who participated with violence. [Paula Bronstein/Getty Images]

Taliban retains control, girls and women are not allowed to work outside the home or attend school. In virtually all parts of the country, despite the decline of Taliban control, women must wear a heavy, completely concealing garment, called a *burqa* (or burka), whenever they leave the house. (Men also must follow a dress code, though a less restrictive one.) Although the Taliban were driven from official power in November 2001, they maintain control of many rural areas. Even efforts to provide women and girls with a basic education can run into violent opposition.

VIGNETTE From behind her microphone at Radio Sahar ("Dawn"), Nurbegum Sa'idi speaks on a wide range of topics. Located in the city of Herat, Radio Sahar provides 13 hours of daily programming consisting of educational items that address cultural, social, and humanitarian matters as well as music and entertainment. For example, one recent broadcast aimed at informing women of their legal rights followed the life of a young woman who was physically abused by her husband and his entire family. The woman took the brave step of asking for a divorce. As a result, she was forced into hiding, where she was counseled on the steps she might take next. A reported 600,000 Afghan women and youth listen to Radio Sahar while they do their daily chores. Radio Sahar is one in a network of independent women's community radio stations that has sprung up in Afghanistan since early 2003.

Girls on the Air, a film by Valentina Monti, reveals the diversity of ideas and hopes for the future shared by the young journalists who founded Radio Sahar. A clip can be seen on YouTube. ■

In order to reach women held in deep seclusion, the Afghan government is now making available basic reading and writing lessons on special mobile phones distributed free to Afghan women. The reading/writing software was developed by an Afghan IT firm with U.S. Agency for International Development (USAID) assistance. As a woman achieves literacy, she can add other subjects to her phone through free apps.

LGBTQ Issues in South Asia

South Asia's stance on LGBTQ issues is complex and contradictory. Same-sex sexual activity has only recently been legalized in India, and Nepal has a number of laws protecting LGBTQ people, but throughout South Asia abuse, violence, and discrimination against LGBTQ people are the norm. And yet South Asian countries are global leaders in formally recognizing the existence and rights of transgender people as a "third gender" in constitutions and in government documents, with India now reserving some scholarships and employment opportunities for transgender people. In 2005, India became the first country in the world to allow citizens to self-identify as a third gender on their passports. Nepal followed suit in 2007, Pakistan in 2009, and Bangladesh in 2013, and almost all of South Asia now recognizes a third gender in most types of government documents (Afghanistan and Bhutan are exceptions), with several extending preferential treatment in government hiring, as is offered to other minorities. This formal acceptance is due largely to the deep cultural roots and substantial numbers of transgender people in South Asia.

hijra A trans female identity numbering 6–7 million people, found mostly in north India, Nepal, Pakistan, and Bangladesh

The largest group of transgender people are the **hijra**, a trans female identity numbering 6–7 million people, found mostly in north India, Nepal, Pakistan, and Bangladesh (Figure 8.40B). Hijras are often considered to be the cultural descendants of Mughal-era eunuchs, who were castrated males employed to guard the harems of rulers and other wealthy and powerful people. However, many references to hijras may also be found in pre-Mughal texts, such as the ancient Hindu epics, the Ramayana and Mahabharata. Many hijras identify with the most important Hindu gods, such as Shiva and Vishnu, who have transgender forms (called Ardhanari and Mohini, respectively), both of whom play a role in the Mahabharata. The patron goddess of the hijras is Bahuchara Mata, whose temple in Gujarat receives over 1.5 million visitors annually.

The deep cultural roots of the hijras have not brought them cultural acceptance. Young boys who display transgender leanings are often rejected by their families and sent to live in all-hijra communities centered on a guru, where the predominant occupation is prostitution. One in five hijras will contract HIV and most live in poverty, subject to regular harassment by gangs and the police. Hijras are also seen as having magical powers to facilitate communication between humans and gods and between men and women, making them sought after as fortune-tellers, attendants at the blessings of newly born children and at weddings. Because of their unique identity, hijras are allowed to move between different castes with greater ease than most.

Nepal is the only country in South Asia, and one of only a few in all of Asia, to adopt a constitution that bans all forms of discrimination against LGBTQ people. In 2007, Nepalese activists were able to take advantage of political disarray after the country's civil war to petition for more LGBTQ-friendly policies. Greater acceptance of LGBTQ people may also derive from Nepal's large tourism industry, which employs nearly a quarter of the nonagricultural labor force, and now makes a concerted effort to attract LGBTQ tourists. In 2008, Nepal elected its first openly gay member of parliament, who also works in the LGBTQ tourist industry. However, even in Nepal, LGBTQ people face severe social discrimination, violence, and daily harassment by police.

Mumbai is widely considered to be South Asia's "LGBTQ capital" with India's first gay magazine, a celebrated LGBTQ film festival, and a film industry rumored to employ numerous LGBTQ people. And yet here, too, discrimination is the norm. Virtually no one in Bollywood has "come out," and the abuse of LGBTQ characters in films is usually presented as comedic. As with other major South Asian cities, the anonymity of big-city life is a major draw of Mumbai for LGBTQ people. An increasing number work in the "tech support" call centers serving overseas corporate clients, where one's appearance and ostensible sexual orientation are less important than one's ability to speak English clearly to foreigners. Mumbai's "Pride Parades" are known for their anonymity, with many participants sporting colorful masks to avoid being "outed" to their unknowing families. With anonymity so important, blackmailing LGBTQ people has become common, especially when people arrange to meet using the online dating and LGBTQ chatrooms that have become so popular in recent years.

Throughout South Asia, the recent surge of religious nationalism in politics is opposed by many LGBTQ people who see religion as an obstacle to social acceptance and political participation. Indeed, many Muslim politicians interpret the Qur'an as forbidding homosexuality, and most proponents of Hindutva argue the same with regard to Hinduism. While some officials of the BJP point out the ambiguous gender of many Hindu gods, none has been willing to advocate for LGBTQ rights. Throughout South Asia, most of the expansion of political rights for LGBTQ people has been achieved through the courts, not through legislatures. While a number of hijras have run for office, and gay officials—two mayors and one state legislator—have been elected, their status in society remains low. However, their new official status as a third gender has brought some gains. For example, Bangladesh is now recruiting hijras as traffic police as a way to offer them a livelihood outside of sex work. Although change may come slowly, few are losing hope yet. As one hijra politician running for office in Pakistan put it, "This is the first drop of rain. If we do not have success in this election then next time, next time, next time."

CHECK YOUR UNDERSTANDING

1. How has the practice of purdah spread?
2. How does the dowry custom contribute to sex ratio imbalance?
3. How does motherhood influence a woman's status in South Asia?
4. What do all female heads of state in South Asia have in common?
5. Why is South Asia a world leader in recognizing the rights of transgender people?

SUBREGIONS OF SOUTH ASIA

8.10 Identify key characteristics of the subregions of South Asia.

The subregions of South Asia are grouped primarily according to their physical and cultural similarities, so in several cases, parts of India are grouped with adjacent countries.

AFGHANISTAN AND PAKISTAN

Afghanistan and Pakistan (**Figure 8.41**) share location, landforms, and history, and they have both been involved in recent political disputes over global terrorism. Historically, many cultural influences have passed through these mountainous countries into the rest of South Asia: Indo-European migrations, Alexander the Great and his soldiers, the continual infusions of Turkic and Persian peoples, and the Turkic–Mongol influences that culminated in the Mughal invasion of the subcontinent at the beginning of the sixteenth century. Today, both countries are primarily Muslim and rural, but each country has some large cities. Both countries must cope with arid environments, scarce resources, and the need to find ways to provide

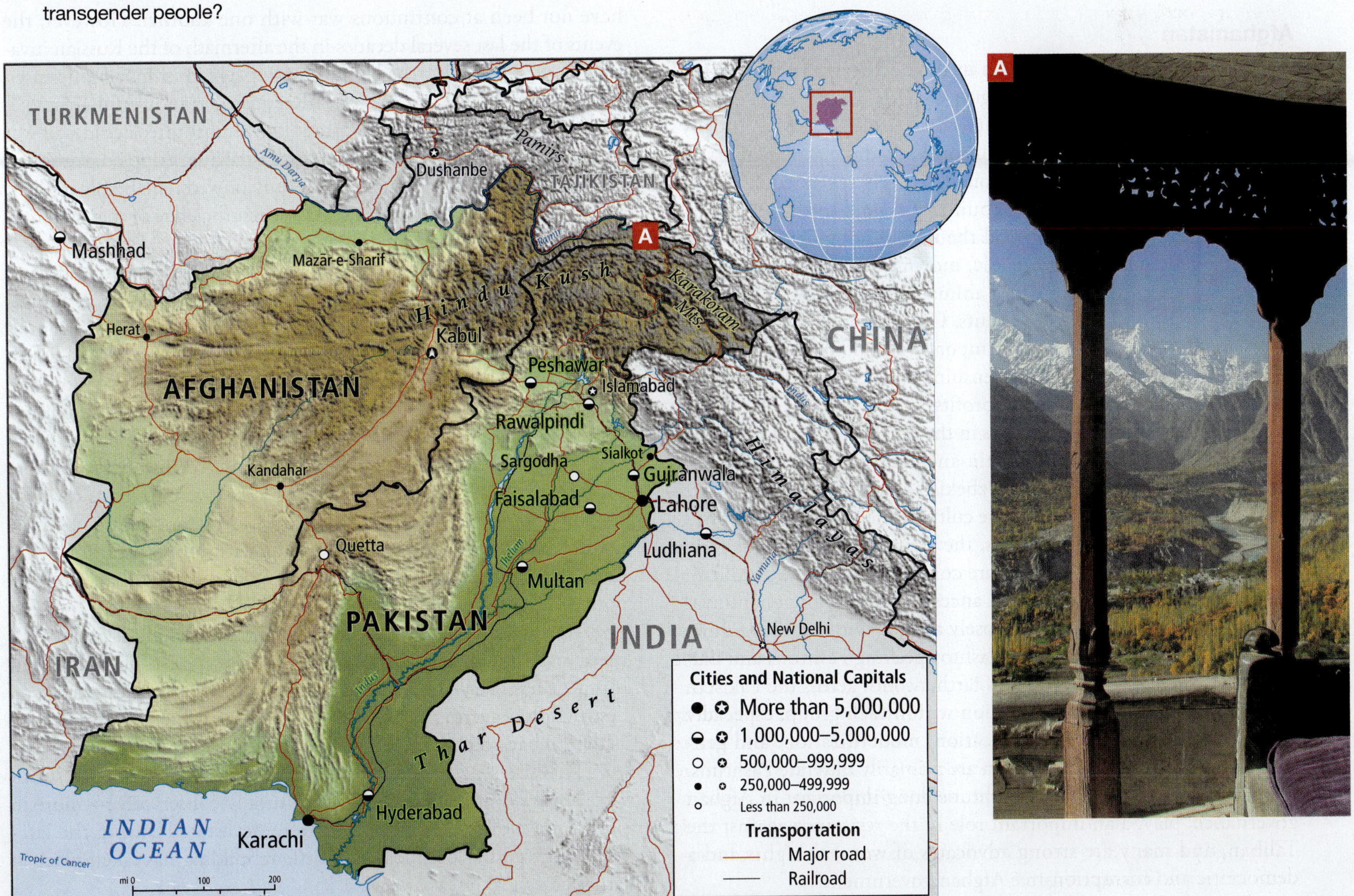

Figure 8.41 The Afghanistan and Pakistan subregion. **A** The Hunza Valley in northern Pakistan is framed by the windows of the 700-year-old Baltit Fort in Karimabad. The people of Hunza Valley are known for their relatively high literacy rate (90 percent), which is due in part to educational programs funded by a wealthy hereditary imam known as the Aga Khan. [Iqbal Khatri/Getty Images]

rapidly growing populations with higher standards of living. Both countries are home to conservative Islamist movements. Pakistan has a weak elected government that, to stay in power, is dependent on a strong military whose loyalties—to the government or to Islamic fundamentalists—are constantly in question. Afghanistan's more extreme poverty is made worse by the armed strife under which it has suffered for more than 30 years (see below).

The landscapes of this subregion are shaped by the ongoing tectonic collision between the landmasses of India and Eurasia. At the western end of the Himalayas, the collision uplifted the lofty Hindu Kush, Pamir, and Karakoram mountains of Afghanistan and Pakistan (see Figure 8.41A). This system of high mountains and intervening valleys swoops away from the Himalayas and bends down to the southwest and toward the Arabian Sea. Landlocked Afghanistan, bounded by Pakistan on the east and south, Iran on the west, and Central Asia on the north, lies entirely within this mountainous system. Pakistan has two contrasting landscapes: the north, west, and southwest are in the mountain and upland zone just described; the central and southeastern sections are arid lowlands watered by the Indus River and its tributaries. In both countries, earthquakes regularly occur due to tectonic compression and active mountain building.

Afghanistan

The Hindu Kush and the Pamirs in the northeast of Afghanistan are rugged and steep-sided (see Figure 8.41). The mountains extend west, then fan out into lower mountains and hills, and eventually into plains to the north, west, and south. In these gentler but still arid landscapes characterized by rough, sparsely vegetated meadows and pasturelands, most in the country struggle to earn a subsistence living from cultivation and the keeping of grazing animals. Rural people remain seminomadic, moving in order to locate forage and water resources for their animals (**Figure 8.42**). The main food crops are wheat, fruit, and nuts. Opium poppies are native to this region, but they were not an important modern cash crop until the Soviet invasion in 1979; the ensuing wars created the environment for opium production and profits.

The less mountainous regions in the north, west, and south are associated with Afghanistan's main ancient ethnic groups. In the north along the Turkmenistan, Uzbekistan, and Tajikistan borders are the associated groups that share cultural and language traditions with the people of these countries, the Turkmen, Uzbek, and Tajik (**Figure 8.43**). The Hazara, who are concentrated in the middle of the country, trace their biological ancestry to the Mongolian invasions of the past, yet today are closely aligned with Iranian culture and languages. In the south, the Pashto-speaking Pashtuns and Baluchis are culturally akin to groups farther south across the Pakistan border. There is significant variation within each group, especially regarding views on religion, education, modernization, and gender roles. For example, the Taliban are primarily illiterate Pashtuns. On the other hand, educated Pashtuns, long important in Afghan governance, played an important role in the resistance against the Taliban, and many are strong advocates of women's rights and a democratic and corruption-free Afghan government.

The ethnic diversity of Afghanistan and its neighbors has thwarted many efforts to unite the country under one government (see Figure 8.43). Although the various ethnic groups have remained separate and competitive, contrary to Western media reports, they have not been at continuous war with one another. However, the events of the last several decades in the aftermath of the Russian invasion of 1979 have disturbed long-standing relationships and increased feelings of distrust (see "War and Reconstruction in Afghanistan"). More important, the sheer devastation of almost three decades of war has left Afghanistan's economy and infrastructure crippled. Although the traditional rural subsistence economy proved remarkably resilient in the face of ongoing civil strife, the self-sufficiency of that system is regularly compromised by drought.

Figure 8.42 Rural Afghanistan. A young Afghan of Turkmen ethnicity moves a caravan through Balkh Province, Afghanistan, near the border with Turkmenistan. Camels and donkeys are a common mode of transportation throughout the drier parts of South Asia. [QAIS USYAN/AFP/Getty Images]

The cities of Afghanistan contrast with the countryside and with one another. Back in the 1970s, the capital Kabul was a modernizing city with utilities and urban services, tree-lined avenues and parks, and lifestyles similar to those in cities elsewhere in South Asia. Most women were not veiled and girls attended schools and wore the latest fashions. Then the Russians invaded and brought to Kabul a version of Western culture that quickly alienated the more conservative countryside. Alcohol consumption and partying were common, some Russian women dressed in a way Afghans thought provocative, respect for religious piety was nil, and all this became negatively associated in Afghan minds with modernization. After the mujahedeen defeated the Russians and the Taliban came to power, modernization became anathema.

Today, Kabul has an official population of 3 million people, but the actual number is probably several million greater. The built environment has been much diminished by wars, with major periods of destruction occurring as the Russians were forced to retreat and then as the Taliban were themselves driven out in 2001.

The cities of Herat in the west and Kandahar in the south were always more traditional and culturally conservative—more in tune with the hinterlands. Both were devastated during the late-twentieth-century conflicts, and little rebuilding has taken place in either city in recent years.

Pakistan

Although not much larger in area than Afghanistan, Pakistan has more than five times as many people (200 million). Some populated

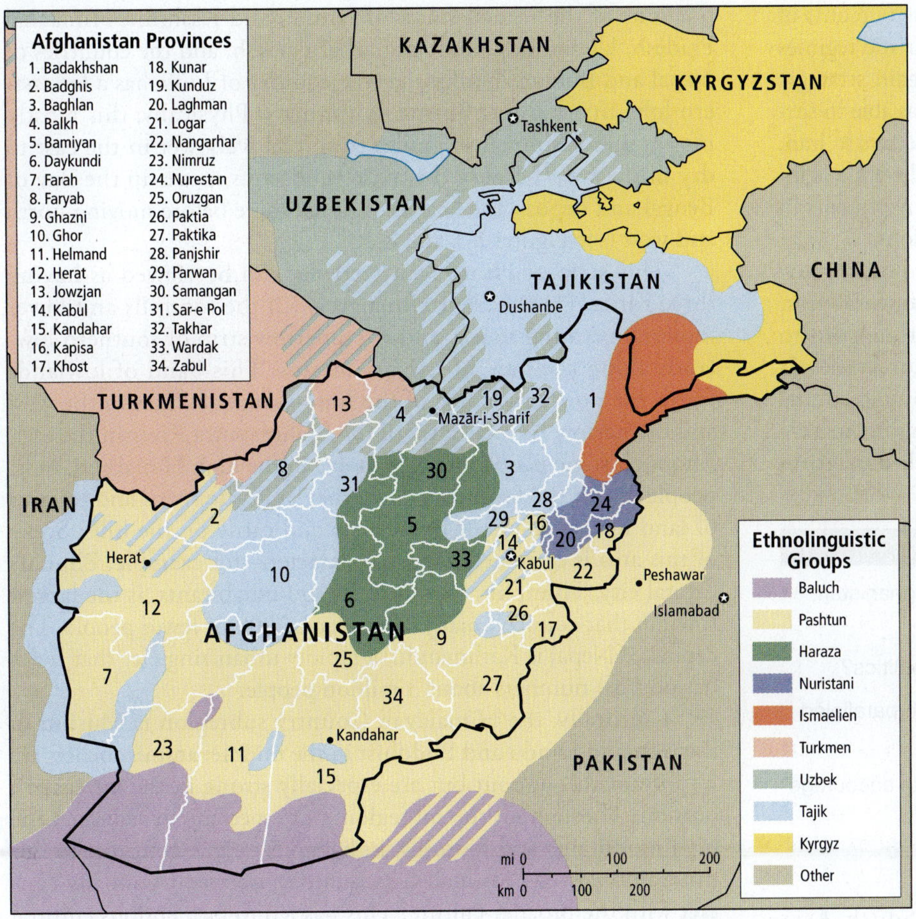

Figure 8.43 Language and ethnicity in Afghanistan. The overlapping of ethnolinguistic groups illustrates how, over the millennia, particular groups have become divided and migrated to different places.

areas are sprinkled throughout the arid mountain districts, depending on herding and subsistence agriculture, but it is the lowlands that have the largest populations. There, the ebb and flow of the Indus River and its tributaries during the wet and dry seasons form the rhythm of agricultural life. The river brings fertilizing silt during floods and provides water to irrigate millions of cultivated acres during the dry season.

Economy Agriculture in the Indus River basin, especially in the Pakistani Punjab, was formerly a primary economic sector. Cash crops such as cotton, wheat, rice, and sugarcane were grown in large irrigated fields, as they had been for many thousands of years. Recent irrigation projects, however, have overstressed the system, resulting in waterlogged and salinized soil. As a result, growth in agricultural productivity has slowed in recent years.

For much of the 1980s to 2007, Pakistan had strong annual economic growth rates of around 6 percent or better. This rate fell to just 2.7 percent because of the global recession in 2007, and persistent civil disruptions linked to Al Qaeda and the war in Afghanistan. Manufacturing and services dominate the economy, with export-oriented textiles, garment-making, and yarn-making industries growing around most major cities, especially Lahore and Karachi.

Having gone through an economic history similar to India of socialism followed by economic reforms leading to more openness to globalization, Pakistan is now attracting foreign investment in manufacturing (especially automobiles and two-wheelers) as well as offshore oil development. As in India, much of the wealth from this new growth has gone to the wealthy, as most manufacturing jobs are fairly low paying. Pakistan's IT sector is much smaller than India's, and is based mostly on individuals working as freelancers via the internet. As in India widespread familiarity with the English language has boosted Pakistan's ability to attract global clients. **Figure 8.44** is a map that shows a variety of centers of economic activity in Pakistan (compare this with the topographic map of Pakistan in Figure 8.41).

Militants and the Conflicts in Afghanistan and Kashmir Pakistan's northern mountainous hinterlands, where government authority is difficult to establish, have become a major base for Islamic militants involved with both the conflict with India over Kashmir and the wars in Afghanistan. Pakistan's military claims that it has no influence over these militants, who have repeatedly attacked the Indian side of Kashmir. Even Pakistan's allies, such as the United States and China, find this hard to believe. Public opinion polls indicate relatively little support among Pakistanis as a whole for Islamic extremist militant groups operating in the country, but the groups operate in remote rural areas where their support base is larger. These polls also indicate widespread opposition to U.S. military operations against Islamic militants in Pakistan.

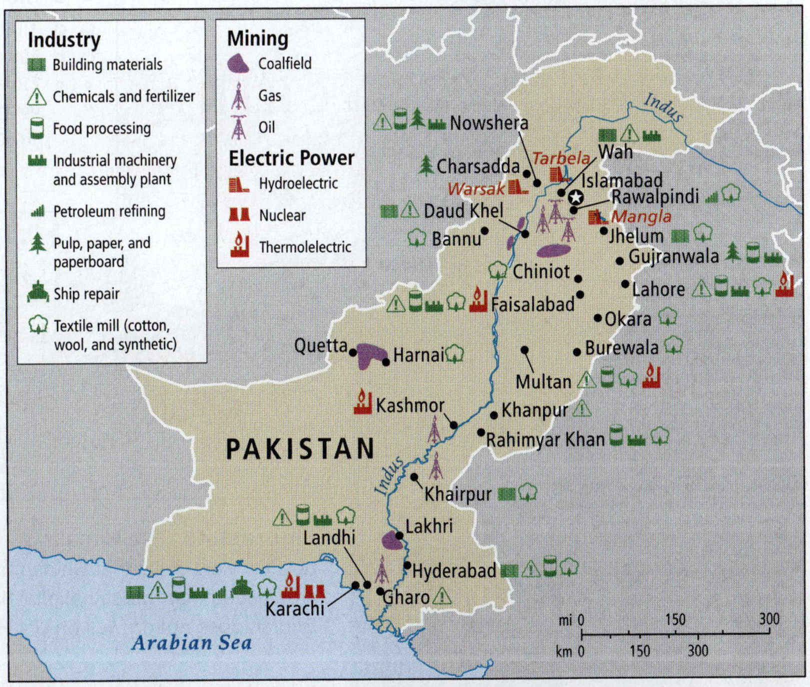

Figure 8.44 Industrial, mining, and power centers in Pakistan.

In many ways, Pakistan's difficulties of today are the outcome of Cold War geopolitics, when antidemocratic and militaristic regimes were propped up and heavily armed by the West to gain strategic advantage over the USSR. During this time, Pakistan was able to surreptitiously acquire nuclear capabilities and sell nuclear secrets to Iran, Libya, and North Korea. Billions of dollars of Western aid went to support the Pakistani military, but relatively little went to investments in human development or to strengthen democratic institutions.

In an effort to address these problems, the United States now provides around a billion or more dollars to Pakistan each year, much of it designated to strengthen civil society through aid to legislative and judicial systems, and investments in human development. Despite improvements in health care and public education systems, many Pakistanis worry that this social aid is merely intended to "Westernize" them, especially through education offered to women and girls.

CHECK YOUR UNDERSTANDING

1. What are some similarities between Pakistan and Afghanistan, physically and culturally?
2. How has Afghanistan's ethnic diversity affected its politics?
3. In what ways has Pakistan's recent economic growth paralleled India's?
4. How might recent public opinion polls in Pakistan both encourage and discourage U.S. policy makers?

HIMALAYAN COUNTRY

The Himalayas form the northern border of South Asia (**Figure 8.45**). This subregion is constructed of the mountainous portions of the Indian states of Jammu and Kashmir, Himachal Pradesh, Uttaranchal, and Arunachal Pradesh, and the countries of Nepal and Bhutan. Notice that the country of India has a far eastern lobe that borders Burma (Myanmar). Physically, this mostly mountainous subregion grades from relatively wet in the east to dry in the west because the main monsoons move up the Bay of Bengal and deposit the heaviest rainfall there before moving north and west (see Figures 8.1 and 8.10).

This strip of Himalayan territory can be viewed as having three zones: (1) the high Himalayas, (2) the foothills and lower mountains to the south, and (3) a narrow strip of southern lowlands along the base of the mountains. This band of lowlands forms the northern fringe of the Indo-Gangetic Plain in the west and the narrow Assam Valley of the Brahmaputra River in the east. Although some people manage to live in the high Himalayas, most people live in the foothills, where they cultivate strips and terraces of land in the valleys or herd sheep and cattle on the hills. Some of this area is extremely rural in character—for example, Bhutan's capital city, Thimphu, with just 98,000 inhabitants, is the largest town in that country; many towns have 2000 or fewer people. The capital of Nepal is Kathmandu, a rapidly urbanizing city that (with its suburbs) numbers about 3 million people.

Culturally, the Himalayan Country subregion is Muslim in the west and Hindu and Buddhist in the middle; animist beliefs are important throughout but are especially strong in the far eastern portion. Throughout the subregion, but especially in valleys in the high mountains and foothills, indigenous people continue to live in traditional ways in small communities, isolated from daily contact with the broader culture. This has led to tremendous cultural diversity. In the Indian state of Arunachal Pradesh, for instance, at

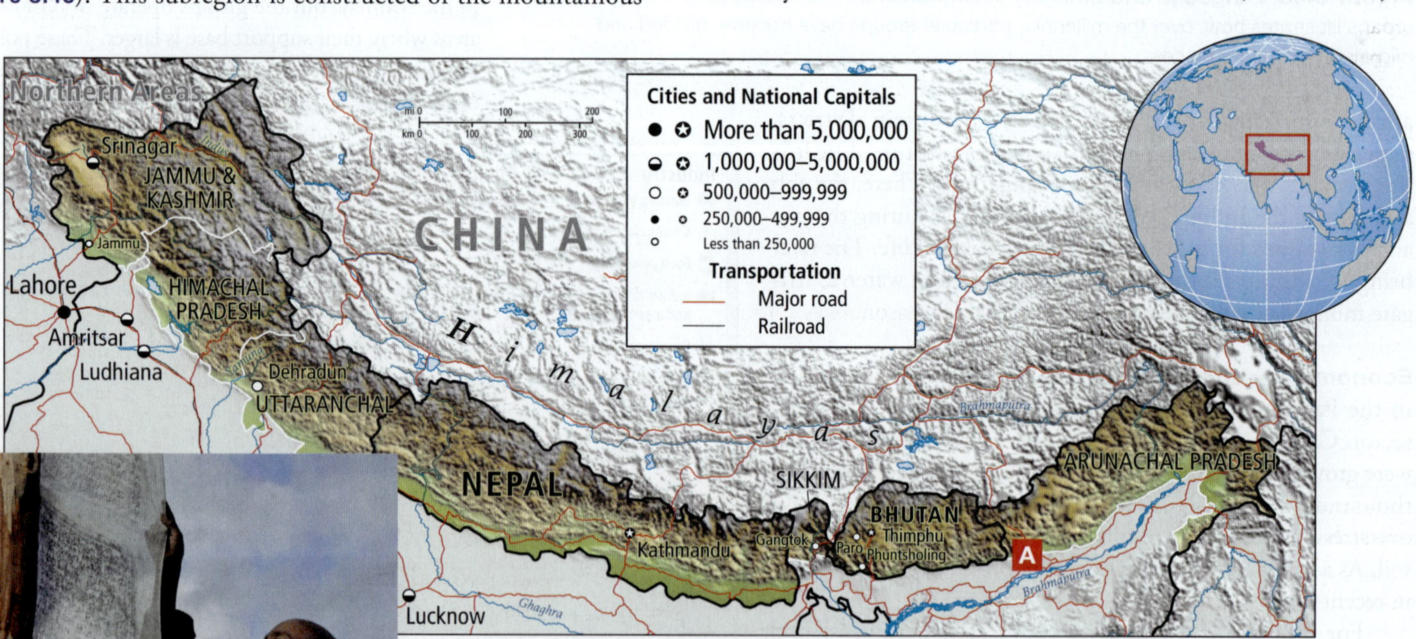

Figure 8.45 The Himalayan Country subregion. **A** A young monk projects the positive view of life that Bhutan has become renowned for since 1972, when the then-king announced that "gross national happiness" was more important than gross national product. Bhutan does not do well on UN Human Development Index indicators. [Keren Su/Getty Images]

the eastern end of the subregion, a population of less than 1 million speaks more than 50 languages.

Despite cultural differences, the Himalayan people have learned to survive in their difficult mountain habitat by relying on one another. An example of this cross-cultural reciprocity comes from central Nepal, where two indigenous groups engage in a complicated seasonal cycle of trade that links them with both Tibet to the north and India to the south.

VIGNETTE The Dolpo-pa people are yak herders and caravan traders who live in the high, arid part of Nepal where they can produce only enough barley and corn to feed themselves for half the year. Through trade, they parley that half-year supply of grain into enough food for a whole year.

At the end of the summer harvest, they load some of their grain onto yaks and head north to Tibet, where they trade grain to Tibetan nomads in return for salt, a commodity in short supply in Nepal. They keep some salt for their own use and carry the rest to the Rong-pa people in the central foothills of Nepal, where they exchange the salt for sufficient grain to last through the winter. The Rong-pa live in a zone where they cultivate wheat and beans but also herd sheep and goats, and salt is a necessary nutrient for them and their animals. If the Dolpo-pa do not bring enough salt to meet all the Rong-pa needs, the Rong-pa load some goats and sheep with bags of red beans and set out for Bhotechaur in western Nepal, where they meet Indian traders at a bazaar. They trade their beans for iodized Indian salt, sell a sheep or two, and buy some cloth and perhaps a copper pan to take home. ■

Most people in the Himalayan subregion are very poor. Statistics for the Indian states in this subregion hover near those for Nepal, which has an annual per capita GNI of just U.S.$2710, adjusted for PPP. In Nepal, the average life expectancy is 70, literacy is about 65 percent (76 percent for men, 48 percent for women), and over one-third of children suffer from stunted growth due to malnutrition. At present, only 18 percent of the land is cultivated, and the country's high altitude and convoluted topography make agricultural expansion unlikely. The government's strategy, therefore, is to improve crop productivity and lower the population growth rate (now 1.8 percent per year).

Bhutan, with a GNI per capita that is roughly four times that of Nepal, is still far from rich. The country is reliant on agriculture in difficult mountainous environments and on trade of mostly cottage industry products with India, which also extends various types of aid to Bhutan for infrastructure development. Bhutan has become famous for its positive view of life and was the first country to promote the idea of measuring human well-being with a "gross national happiness" quotient (see Figure 8.45A). It now promotes this concept to environmentally and spiritually conscious tourists who pay a fee of $250 a day to visit the country. Bhutan is also known for being the last place on Earth to acquire television and the internet, which occurred only over the past decade. A 10-minute video that shows the effect TV has had on Bhutan society can be found on the PBS Frontline website.

The Indian state of Arunachal Pradesh, at the far eastern end of the Himalayan subregion, is one of the more lightly populated and environmentally rich areas of the subregion. Moist air flowing north from the Bay of Bengal brings plentiful rain as it lifts over the mountains. This is one of the most pristine regions in India. Forest cover is abundant, and a dazzling array of flora and fauna occupies habitats at descending elevations: glacial terrain, alpine meadows, subtropical mountain forests, and fertile floodplains. Conditions at different altitudes are so good for cultivation that lemons, oranges, cherries, peaches, and a variety of crops native to South America—pineapples, papayas, guavas, beans, corn (maize), and potatoes—are now grown commercially for shipment to upscale specialty stores, primarily in India.

Over the past 20 years, the Himalayan Country subregion has undergone numerous changes brought about by the increasing numbers of tourists trekking and climbing the mountains and seeking spiritual enlightenment at its numerous holy sites. Tourism, in particular, has been a mixed blessing, creating economic opportunities for some but also having a transforming effect on the culture and the landscape.

CHECK YOUR UNDERSTANDING

1. How do people in the Himalayas use reciprocal trade to survive?
2. What is Bhutan famous for?
3. What gives Arunachal Pradesh its agricultural potential?

NORTHWEST INDIA

Northwest India stretches almost a thousand miles from the states of Punjab and Rajasthan at the border with Pakistan, eastward to encompass the famous Hindu holy city of Varanasi (**Figure 8.46**). It is dry country, yet it contains some of the wealthiest and most fertile areas in India.

A key factor of life in this region is the availability of irrigation water. Rain occurs mainly in the north along the Himalayas, and only during the monsoon from about June to September, so making use of this land's year-round growing season requires irrigation. This is available in the north in the Indian states of Punjab and Haryana, and in most of Uttar Pradesh, but not in Rajasthan. The Indian part of Punjab, which also stretches into Pakistan, is exceptionally well irrigated with water from the five rivers for which this area is named (Punjab means "five waters"). Along with the nearby state of Haryana, Punjab is the most productive and profitable agricultural area in South Asia. To the south, Rajasthan, with only a few fertile valleys, is dominated by the Thar Desert in the west, which covers more than a third of the state. Until recently Rajasthan's tiny agricultural economy made it one of the poorer states in India.

Rajasthan

Rajasthan has been transformed by India's economic development over the past two decades, through new manufacturing and a growing IT sector. Until the early 2000s Rajasthan was dependent largely on agriculture, which, despite less than 1 percent of the land being arable, still produced 50 percent of the state's GDP. A thriving tourist industry, focused on the palaces and fortresses constructed by Hindu warrior princes of the past who fought off invaders from Central Asia, accounted for much of the other half of the GDP.

Major investments in infrastructure and human development during the 2000s began to attract manufacturers and IT companies by the 2010s. As a result, Rajasthan is now wealthier on a per capita basis than Uttar Pradesh, though still poorer than Punjab and Haryana.

Punjab and Haryana

Nearly 85 percent of Punjab's land is cultivated, and in some years Punjab alone has provided nearly two-thirds of India's food reserves. The typical crops are wheat, rice, sugarcane, corn, and barley. Punjab's productive agriculture contributes to its high average annual per capita GDP (PPP) of about U.S.$8000, about a

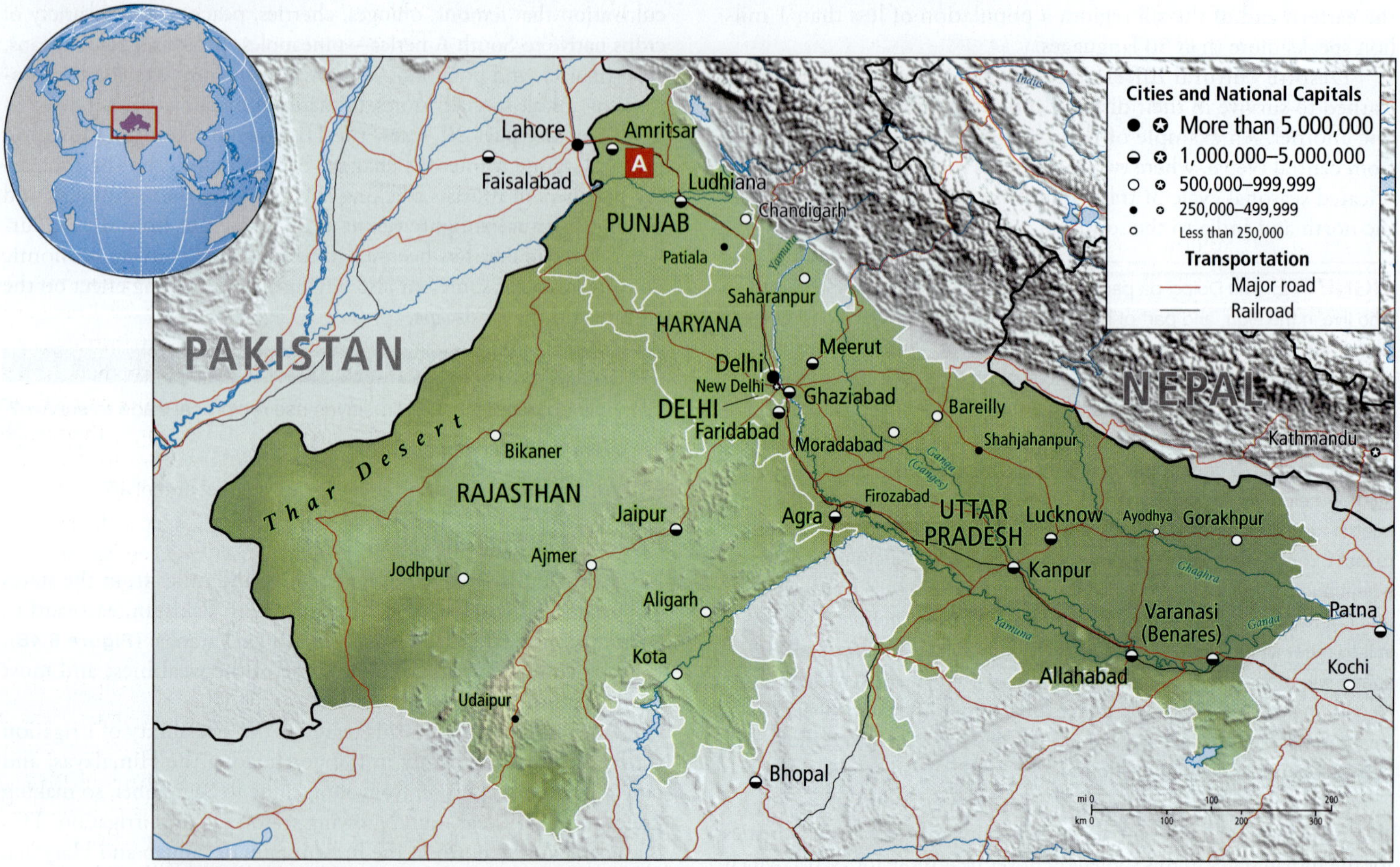

Figure 8.46 The Northwest India subregion. **A** The Golden Temple in Amritsar, Punjab, is the most sacred place for Sikhs. [Adrian Pope/Photographer's Choice RF/Getty Images]

third more than Rajasthan. However, the long-standing dominance of agriculture has meant relatively less development in manufacturing and IT, especially compared to nearby Haryana, which, in addition to being an agricultural powerhouse like Punjab, is also a major center for manufacturing cars, motorcycles, tractors, and household appliances, and a major IT center. Benefiting from its location next to Delhi, which provides access to markets, decision makers, and funding for infrastructure, Haryana's per capita GDP (PPP) of about U.S.$11,000 is one of the highest in India.

Delhi

India's capital, New Delhi, is located approximately in the center of Northwest India. The British built the city center in 1931 just south of the old city of Delhi—an important Mughal city—amid the remains of seven ancient cities. It has all the monumental hallmarks of an imperial capital, and all the problems of a big city where so many are poor. The Delhi metropolitan area (including the old and new cities and outlying suburbs) counts more than 27 million people and attracts a continuing stream of migrants. Many have left nearby states to escape conflict or poverty, or both, such as those who have fled Tibet to escape Chinese oppression and others who have fled the conflicts with Pakistan since the 1980s. Annual per capita GDP (PPP) in Delhi (U.S.$18,600) is roughly three times the average for India, but actual incomes for most people are far lower because a tiny minority of extremely wealthy people pulls up the average. Delhi's rather high literacy rate (86 percent), while considerably higher than India's overall rate (74 percent), is held down by the continual arrival of migrants from poor, rural areas. In fact, Delhi has far fewer schools than it needs for its population because jobs increasingly require a high level of skills, not mere literacy.

Rapid growth has made providing even the most basic services a challenge. Water, power, and sewer facilities are insufficient, and 75 percent of the city's structures violate local building codes. About 20 percent of Delhi's population lives in slums.

Uttar Pradesh

Home to 200 million people living in an area roughly the size of Oregon, Uttar Pradesh is the most populous state in India and the second poorest with a per capita GDP (PPP) of $3100. Over 78 percent of Uttar Pradesh's population is rural, with extensive farmlands fed by irrigation water from the Ganga and Yamuna rivers. Agriculture is the main employer but IT service centers and modern manufacturing industries, such as electronics, are clustered around Delhi. Traditional manufactures such as textiles, instrument making, and metal working are found in ancient cities throughout the state. Together these industries earn as much as the agricultural sector, while employing a fraction of the people and using much less land and water.

Uttar Pradesh is a major source of culture and a political center for South Asia. It is home to ancient cities such as Varanasi that are mentioned in the epic poems of Hinduism, as well as great architectural legacies of the Mughals, such as the Taj Mahal and the city of Fatehpur Sikri. As India's most populous state, Uttar Pradesh is a major political stronghold, providing more seats in parliament than any other and providing 9 of India's 14 prime ministers, more than any other state. As such its needs as a poor, largely rural state have tended to highlight calls for support from the government for agriculture.

CHECK YOUR UNDERSTANDING

1. How has Rajasthan been transformed by recent economic development?
2. What is the major employer in Punjab?
3. Where are most of Haryana's industries located?
4. Why is Delhi's literacy rate relatively low for a city of its size and importance?
5. What are the major rivers providing irrigation water in Uttar Pradesh?

NORTHEASTERN SOUTH ASIA

Northeastern South Asia, in strong contrast to Northwest India, has a wet, tropical climate. This subregion bridges national and state boundaries, encompassing the states of Bihar, Jharkhand, and West Bengal in India; the country of Bangladesh; and the far eastern Indian states of Meghalaya, Assam, Nagaland, Manipur, Mizoram, and Tripura—all clustered at the north end of the Bay of Bengal (**Figure 8.47**). This area is viewed as a subregion because of its dominant physical and human features. The Ganga and Brahmaputra rivers and the giant drainage basin and delta region created by those rivers, as well as the wet climate and fertile land, have nourished a population that is now among the most densely populated in the world (see the Figure 8.32 map).

The Ganga–Brahmaputra Delta

The largest delta on Earth is that formed by the Ganga and Brahmaputra rivers. Every year, the two rivers deposit enormous quantities of silt, building up the delta so that it extends farther and farther out into the Bay of Bengal. The rivulets of the delta change course repeatedly, and then periodically the bay is flushed out by a huge tropical cyclone (see Figure 8.12). The people of the delta have learned never to regard their land as permanent, and they have adapted their dwellings, means of transportation (see Figure 8.47A), and livelihoods to the drastic seasonal changes in water level and shifting deposits of silt. Villages sit on river terraces or, in the lowlands, are raised on stilts above the high-water line. Moving about in small boats, people fish during the wet season, but when the land emerges from the floods, they return to farming. Called *nodi bhanga lok* ("people of the broken river"), those who occupy the constantly shifting silt of the delta region are looked down upon by more permanent settlers on slightly higher ground. Because they must often flee rising floodwaters, they are less secure financially and are thought by their neighbors to lack the qualities of thrift and good citizenship that come from living in one place for a lifetime. Currently, their livelihoods (and at times their very lives) are threatened by periodic cyclones and climate change–related sea level rise, compounded by the loss of nourishing sediment caused by river diversions upstream.

West Bengal

With a population of 80 million packed into an area slightly larger than the U.S. state of Maine, West Bengal is India's most densely populated state (see Figure 8.32). Refugees have dramatically increased the population of West Bengal, which took in roughly 3 million people from Bangladesh during Partition in 1947, hundreds of thousands of migrants fleeing the violence of Bangladesh's War of Independence in 1971, and refugees from the Chinese takeover of Tibet (1959–1976). Today, better employment opportunities draw a continual flow of Bangladeshi migrants (most of them migrating illegally) across the border into India. The result is a mix of cultures that are frequently at odds, with Hindus and Muslims often disputing with indigenous animist people over land occupied by new migrants.

Most people in this crowded state earn a living in agriculture, but agriculture accounts for only 19 percent of West Bengal's GDP. Many farmers also work as rice and jute cultivators or as tea pickers—all labor-intensive but low-paying jobs. One consequence of the dense population's dependence on agriculture is that the land has become overstressed and soil fertility has declined. In addition, impoverished farm laborers who must gather firewood because they cannot afford other fuels deplete the surrounding woodlands.

Kolkata

Kolkata (formerly known as Calcutta), a giant, vibrant city of 16 million people, is famous in the West for Mother Teresa's ministrations to its poor at Nirmal Hriday (Home for Dying Destitutes). Bengalis, however, are also proud of their two native Nobel laureates—Rabindranath Tagore (literature) and Amartya Sen (economics)—and of their Academy Award–winning filmmaker Satyajit Ray, who received a lifetime achievement Oscar.

Built on a swampy riverbank in 1690 as a trading post for the British East India Company, Kolkata served as the capital of British India until 1911 when the government was moved to New Delhi, in part to escape the growing Bengali agitation for Indian independence. Much of Kolkata's colonial built environment has since become lost in the squatters' settlements that invaded the city's parks

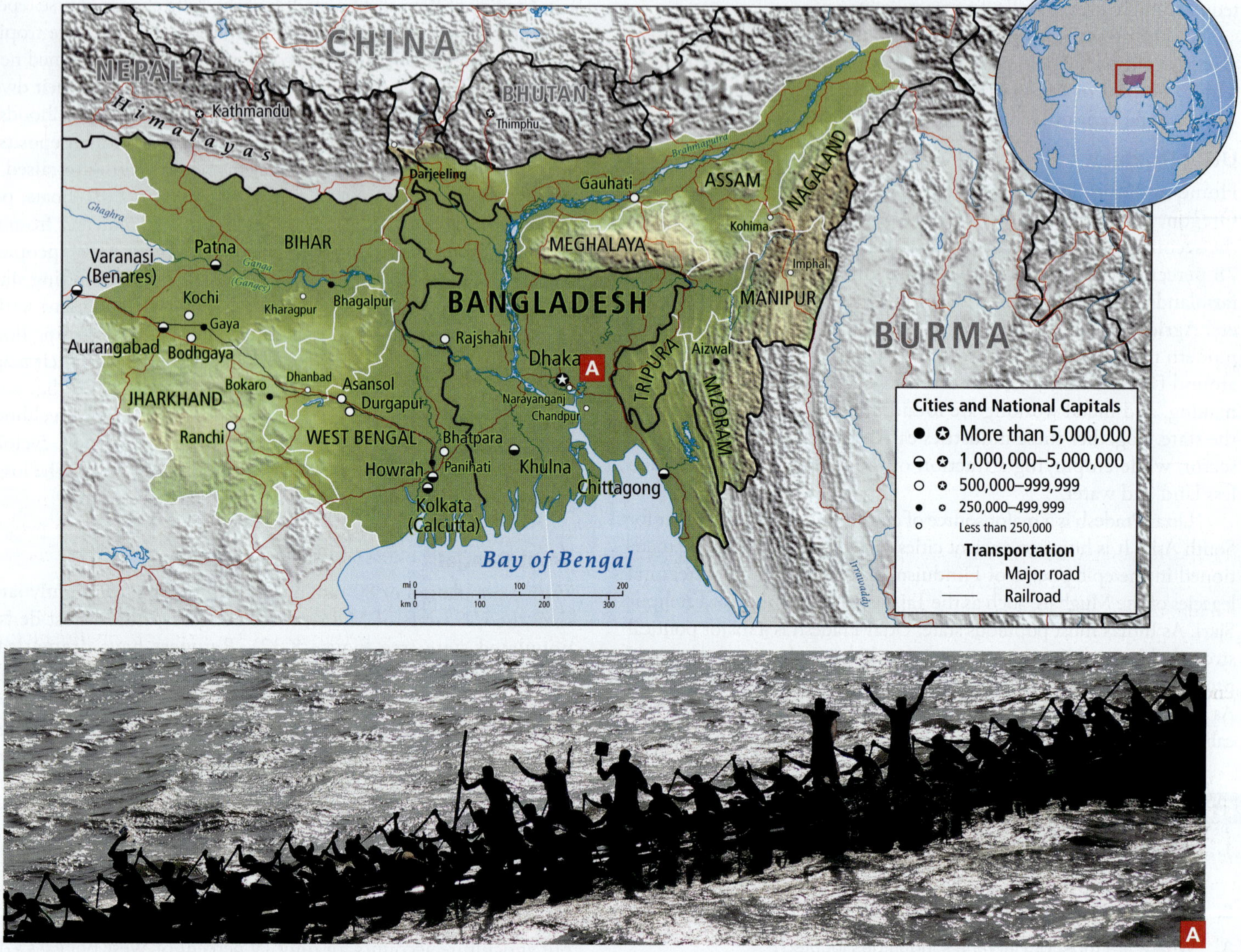

Figure 8.47 The Northeastern South Asia subregion. **A** Men row in a traditional boat race on the Buriganga River near the center of Dhaka, Bangladesh. Boat races are an ancient sport in this land of rivers, and boats almost always have a leader who energizes his crew with songs, rhymes, and gestures. [MUNIR UZ ZAMAN/AFP/Getty Images]

and periphery after Partition and agricultural collapse brought millions of refugees and migrants. Outmoded regulations limited incentives to start businesses and to improve private property. In the late 1990s, the physical decline, economic inertia, and lack of opportunity in Kolkata sent young professionals fleeing to other parts of the world.

By the 2000s, as India's economic growth began to pick up, those who had fled Kolkata were returning and new graduates stayed to partake in an IT-fueled economic revival. At least 50 large construction projects are under way that are expected to eventually attract another 7 or 8 million people to this lowland metropolitan area. Many of these are being built on the wetlands south of the city (in areas exposed to future sea level rise). Development is also happening in the city center, with many once lavish but presently run-down buildings from the colonial era threatened by new construction projects (**Figure 8.48**). A small minority of residents are organizing to preserve Kolkata's historic architecture, but in a city with so much poverty the cause has failed to become a popular rallying cry, and municipal officials are at times unsympathetic.

Far Eastern India

Far Eastern India has two physically distinct regions: the river valley of the Brahmaputra where the river descends from the Himalayas, and the mountainous uplands stretching south and east of the river between Burma on the east and Bangladesh on the west (see the Figure 8.47 map). Although migrants have recently arrived here from across South Asia, ancient indigenous groups that are related to the hill people of Burma, Tibet, and China have traditionally occupied Far Eastern India.

Assam The Indian state of Assam encompasses the Brahmaputra River valley, eastern India's most populated and most productive area. Hindu Assamese make up two-thirds of the population of 31 million, and indigenous Tibeto-Burmese ethnic groups make

Figure 8.48 **Historic architecture.** The Futnani Chambers building in Kolkata, built over a century ago, is in danger of collapse due to the construction of a subway tunnel beneath it. [Frank Bienewald/LightRocket via Getty Images]

up another 16 percent; the rest of the people are recent migrants. The Indian government has attempted to reduce the proportion and influence of the Assamese people (who have continually objected to being under New Delhi's control) by making large tracts of land available to outsiders, such as Bangladeshi refugees, Nepalese dairy herders, and Sikh merchants. Since the late 1970s a low-level insurgency pitting separatist Assamese groups against India's national government and the new settlers has resulted in about 30,000 deaths.

Just under half the people in Assam work in agriculture, where tea is a major cash crop accounting for over half of India's total production. The other main industry is oil, with Assam being the second place in the world where petroleum was discovered (in 1889). Until the 1990s Assam's oil and natural gas accounted for more than half that produced in all of India, but because of India's expanded development of offshore oil and gas fields in the 2000s, Assam's share has fallen to less than a quarter. However, with India importing more than 80 percent of its petroleum, and the government eager to become more energy independent, keeping Assam a part of India is a major priority.

Forestry is also a major land use, with forests covering about 25 percent of the land area of Assam. Encroachment from agriculturalists is a constant threat, and large forests remain primarily in the hilly areas that have government protection. Efforts to relieve pressure on forests through bamboo cultivation are particularly relevant in Assam, where bamboo is a major part of the culture, having long been used as a building material and as a food. Globally India is second only to China in production of bamboo, but almost none is exported. Responding to greater use of bamboo flooring in the United States and Europe, Assam's state government is creating incentives for export-oriented bamboo cultivation and processing.

Colorful names such as "Land of Jewels" and "Abode of the Clouds" convey the exotic beauty of the emerald valleys, blue lakes, dense forests, carpets of flowers, and undulating azure hills in the mountainous sections of eastern India that surround Assam. In these uplands, occupied largely by indigenous ethnic groups, people produce primarily for their own consumption and devise ingenious ways to make use of local natural resources. The state of Mizoram ranks second in India in literacy (88 percent) due to the influence of Christian missionary schools.

Bangladesh

Bangladesh is one of the world's poorest and most densely populated countries, and one that is making some of the most progress in controlling population growth and improving all aspects of human development. Population density is extreme, with more than 166 million people living in an area slightly smaller than Alabama. This is a slightly higher density than the most dense metropolitan area in the United States, Los Angeles. However, fertility rates have dropped dramatically from 7 children per woman in 1974 to 2.1 in 2018, due in large part to the 62 percent of women now using some form of contraception. One-quarter of Bangladeshis live on less than $2 a day, but this is down from nearly half in 2000. One-third of Bangladeshi children under the age of 5 suffer from stunted growth, down from three-quarters in 1991. Adult literacy went from 34 percent in 1990 to 73 percent in 2019, and life expectancy jumped from 49 years in 1975 to 73 in 2018.

Much of this improvement is due to the revival of the textile industry, dismantled by the British more than 200 years ago. Bangladesh is now the second-largest exporter of garments after China, with 91 percent of its exports and 14 percent of its GDP coming from this sector alone. While it directly employs roughly 4.5 million people, whose wages have lifted millions of families out of poverty, most of the earnings from this sector go to a tiny group of wealthy elites who actively work to keep wages in the industry as low as possible, and maintain unsafe workplace standards. A garment factory fire in 2012 outside Dhaka killed 117 workers, while a building collapse at a factory in 2014 killed over 1100. The major global fashion brands that Bangladesh supplies, such as Walmart, the Gap, Disney, and Tommy Hilfiger, have forced better working conditions, but not better pay.

Bangladesh's march toward higher human development results from much more than its textile manufacturing sector. Publicly funded family planning (birth control) has empowered women to have far fewer babies and to turn their energy toward gaining education for themselves and their children. Microcredit has put entrepreneurial activities within reach of the ambitious poor. Government-funded research into improved rice varieties has protected against crop failures while social safety-net spending has kept the poorest from starving. Meanwhile millions of Bangladeshis working abroad regularly send money home to their families.

CHECK YOUR UNDERSTANDING

1. Where is the world's largest delta?
2. What part of India has the highest population density?
3. Why does Kolkata have such an unusually rich architectural heritage?
4. The conflict in Assam involves what groups?
5. What are some signs of improved human development in Bangladesh?

CENTRAL INDIA

The Central India subregion stretches across the middle of India, from Gujarat in the west to Orissa in the east (**Figure 8.49**). It contains India's last untouched natural areas as well as much of its industry. The Narmada River, site of many planned hydroelectric and multipurpose dams, flows across the subregion and empties into the Gulf of Cambay.

Figure 8.49 The Central India subregion. **A** A boy and his trained monkey walk along Mumbai's Marine Drive. In the background is Malabar Hill, one of the most expensive neighborhoods in the world, with costs per square foot higher than those of Manhattan in New York City. [INDRANIL MUKHERJEE/AFP/Getty Images]

Some of India's most significant environmental battles are being fought in the central highlands of this subregion; the protests over the damming of the Narmada River, described at the beginning of this chapter, are only one example. Central India has most of India's remaining forest cover and is home to a concentration of national parks and sanctuaries. However, because the forest cover is patchy, not continuous, many wild plants and animals are isolated in such small populations that their extinction is highly likely, even if it is somewhat delayed by the parks' protection. In recognition of this problem, there are now plans to reconstitute forest corridors between parks. Estimates of future human population growth, however, do not bode well for the future of wildlife anywhere in the subregion.

Central India is notable for its several industrial areas, most of which are now rebounding from the global recession. In the state of Gujarat, on the Arabian Sea, the industrial sector accounts for 45 percent of state GDP, and for 34 percent in nearby Maharashtra. These states are also over 45 percent urban, compared to the more rural and agricultural states of Madya Pradesh and Odisha. Still, even these states have pockets of industrial activity, such as Indore (see the Figure 8.49 map), a commercial and industrial hub whose residents think of it as a mini Mumbai, and nearby, the town of Pithampur, known for several automobile plants, a steel plant, a container plant, and small appliance factories. Mumbai (Figure 8.49A), with over 23 million inhabitants, exerts enormous economic and cultural influence throughout Central India.

CHECK YOUR UNDERSTANDING

1. Why are Central India's forest areas important?
2. What part of Central India has the most industrial employment?
3. What is the largest and most influential city in Central India?

SOUTH INDIA AND SRI LANKA

South India and the country of Sri Lanka (**Figure 8.50**) are set apart from the rest of South Asia by their greater prosperity, higher human development, and more urbanized populations. South India has a higher proportion of well-educated people and is India's wealthiest subregion. Human development is particularly high in Kerala, which has a long tradition of elected communist state governments that have invested heavily in education, social services, and women's development. Culture also sets Southern India apart with Dravidian languages dominating, instead of Indo-European languages as in the rest of India.

This subregion's peninsular and island location means that it has two wet seasons, one as the ITCZ passes toward the north in May and June, and another as it is pushed south in October and November. Consistent rainfall and year-round growing seasons provide a stable base for agriculture, which is more diverse here than in much of India. The southwestern coast of India (known as the Malabar Coast) has a narrow coastal plain backed by mountains (the Western Ghats). The sea-facing slopes of these mountains are some of the wettest in India, supporting forests of teak, rosewood, and sandalwood, all highly valued furniture woods. Small parts of the Deccan Plateau, a series of uplands to the east, are also forested. Here, dry deciduous forests yield teak, tree-farmed eucalyptus, cashews, and bamboo. Several large rivers and numerous tributaries flow eastward across this plateau and form rich deltas along the lengthy and fertile coastal plain facing the Bay of Bengal.

Economic Growth in Bangalore and Chennai

These two South Indian cities have emerged as rival economic leaders for all of South Asia, with Bangalore having almost doubled in size over the past decade to now rival the once much larger Chennai in size. Bangalore, Karnataka, has drawn thousands of well-trained technicians from all over India to work in the country's largest IT-BPM service sector, as well the national center for aeronautics, space, and defense industries. Meanwhile Chennai, in Tamil Nadu, is India's largest automobile production center, and manufacturer of a wide range of items from appliances and home furnishings to solar-powered lighting and pharmaceuticals. Located on the Bay of

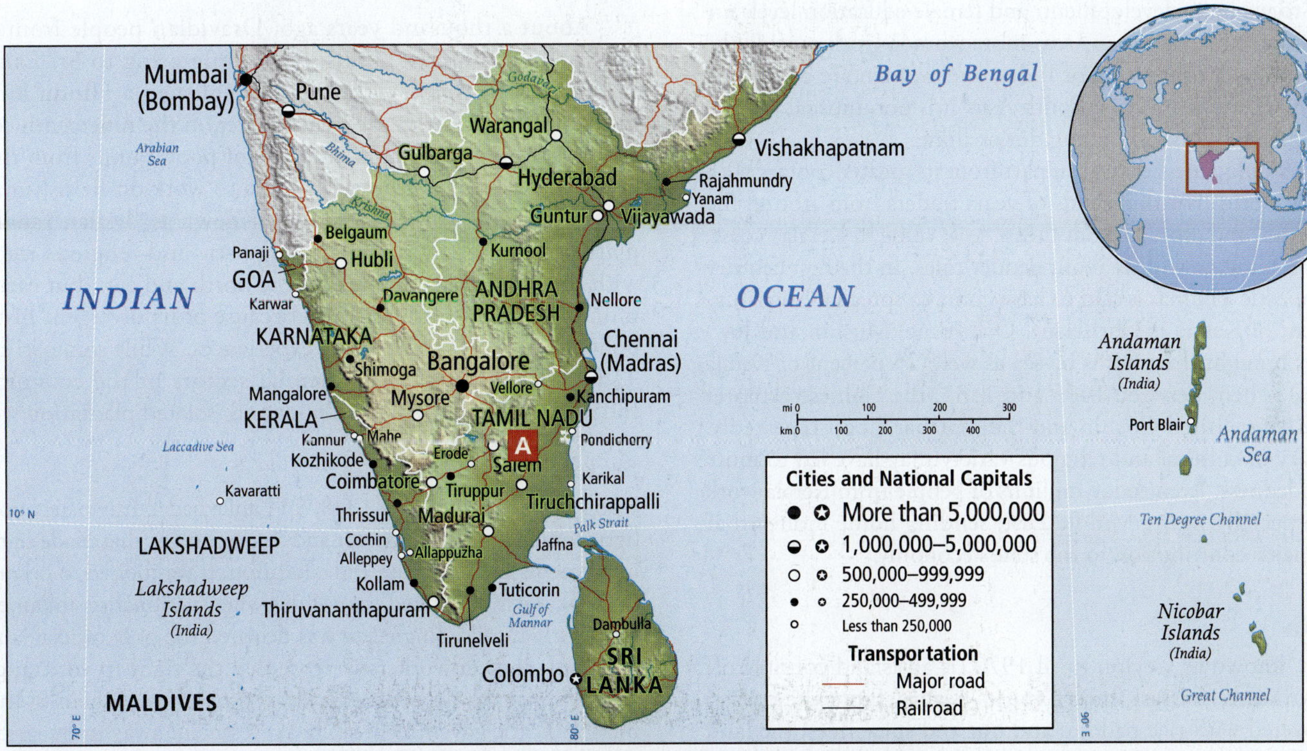

Figure 8.50 The Southern South Asia subregion. A A Hindu temple in Kanchipuram, Tamil Nadu, India. South India is known for its distinctive temple style, which features ornately decorated towers (called *gopurams*) and a water reservoir (called a *kalyani*). [Alinari Archives/Alinari via Getty Images]

Bengal, Chennai's factories benefit from the city's port, one of the largest in India. Like many large Indian cities, Chennai is also now an IT-BPM center that competes with Bangalore for talent.

Kerala

Kerala, a primarily coastal state in far southwestern India, is famous for its high human development, greater attention to women's development, and unique cultural and political history. Since the 1950s, it has had a series of elected communist governments that have strongly supported broad-based social services, especially education for both males and females (see Figure 8.34), giving Kerala consistently the highest levels of human development in all of South Asia.

Culturally, Kerala for centuries had matriarchal systems of inheritance in which property and ancestry were passed on through the women's line. Though partially dismantled by the British, elements of the system live on, with husbands often residing with their wife's family (a pattern also found in Southeast Asia) instead of the reverse, and children often taking their mother's last name instead of their father's. Kerala's communist governments have paid special attention to women's development, and female education levels are the highest in India. In urban areas more women work outside the home than in any other state of India, and women are often seen in public and, unlike much of South Asia, it is not unusual to see a woman going about her shopping duties alone.

Kerala's uniqueness derives in part from its greater contact with the outside world. For thousands of years traders from around the Indian Ocean, and especially Southeast Asia, came to Kerala's coast, possibly bringing new ideas about gender roles. In the first century C.E., the apostle Thomas is said to have come to spread Christianity (20 percent of Kerala is Christian). Over time, Muslim and Jewish traders brought their faiths by sea as well (25 percent of Kerala is Muslim). Then, between 1405 and 1433, the Chinese explorer General Zheng visited Cochin and the Malabar Coast repeatedly. This history of cultural and religious variety may have led to more open-mindedness. Even today millions of people from Kerala work abroad, especially in the Middle East, sending home money that makes a major contribution to the state's economy.

Sri Lanka

Sri Lanka, known as Ceylon until 1972, is an island country off India's southeastern coast known for its beauty. From the coastal plains, covered with rice paddies and nut and spice trees, the land rises to hills where tea and coconut plantations are common. At the center is a mountain massif that reaches to nearly 8200 feet (2500 meters) in elevation. The summer monsoons bring abundant moisture to the mountains, giving rise to lush forests and several unnavigable rivers that produce hydroelectric power. Close to 30 percent of the land is cultivated, and just over 25 percent remains forested.

The original hunter-gatherers and rice cultivators of Sri Lanka, today known as Veddas, now number fewer than 5000. Several thousand years ago, people from northern India came to Sri Lanka and built numerous city-kingdoms. Known as Singhalese, these descendants of northern Indians now constitute about 74 percent of the population of 22 million. The Singhalese brought Buddhism to Sri Lanka, and today 70 percent of Sri Lankans (primarily Singhalese) are Buddhists (**Figure 8.51**).

Figure 8.51 A woman prays at the Golden Temple in Dambulla, Sri Lanka. A gilded statue of the Buddha sits atop a 250-year-old monastery. The statue was added in 1998, using funds donated by Sri Lankans living abroad. [Grant Faint/The Image Bank/Getty Images]

About a thousand years ago, Dravidian people from southern India, known as Tamils, began migrating to Sri Lanka; by the thirteenth century, they had established a Hindu kingdom in the northern part of the island. Later, in the nineteenth century, the British imported large numbers of poor Tamils from the state of Tamil Nadu in southeastern India to work on British-managed tea, coffee, and rubber plantations. Known as "Indian Tamils," the plantation-based Tamils share linguistic and religious traditions with the "Sri Lankan Tamils" of the north and east, but each community considers itself distinct because of its divergent historical, political, economic, and social experiences. While some Sri Lankan Tamils dominate the commercial sectors of the economy, the Indian Tamils have remained poor and isolated plantation workers.

Sri Lanka's Civil War

Upon its independence in 1948, Sri Lanka had a thriving economy led by a vibrant agricultural sector and a government that made significant investments in health care and education. It was poised to become one of Asia's most developed economies when nationalists inflamed conflicts with Tamils. Singhalese was declared the only official language, Tamil plantation workers were denied the right to vote, and there were attempts to deport hundreds of thousands of them to India. In the 1960s, the government shifted investment away from agricultural development and toward urban manufacturing and textile industries, which were dominated by Singhalese. By 1983, the Tamil minority, lacking political power and influence, turned to guerilla warfare against the Singhalese, creating an army known as the Tamil Tigers.

For more than 30 years, the entire island was subjected to repeated terrorist bombings and kidnappings. Peace agreements were attempted several times, but in the end it was an overwhelming military victory by the government, combined with an effective crackdown on international funding for the Tamils, that forced the Tamil surrender in May 2009. However, dissatisfaction with the peace process has resulted in an ongoing flow of Tamil and other refugees from Sri Lanka. Tamils have since asked the UN to investigate war crimes committed during the war, but the Sri Lankan government has insisted on conducting its own lengthy investigations.

Despite many years of violence, economic growth has been surprisingly robust in Sri Lanka. Driven by strong growth in food processing, textiles, and garment making, Sri Lanka is today one of the wealthiest nations in South Asia on a per capita GDP basis. Its government has made major investments in human development that have benefited most of its citizens. Sri Lanka's economy may be able to perform well enough that prosperity will dampen any further ethnic or religious strife.

CHECK YOUR UNDERSTANDING

1. What sets South India apart from the rest of South Asia?
2. What industries have spurred rapid growth in Bangalore?
3. What factors help account for the higher status of women in Kerala?
4. How did Singhalese nationalists inflame tensions in Sri Lanka?

CRITICAL THINKING QUESTIONS

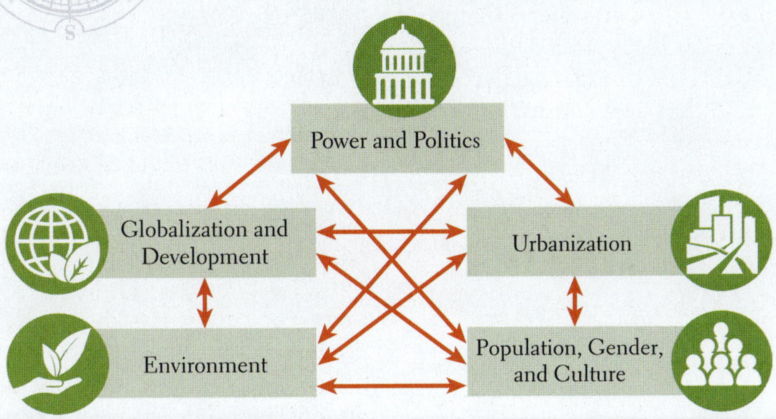

The diagram represents connections among the five geographic themes that structure this book. Listed below are some important questions that have been addressed in this chapter. Answer each question, and indicate where in the diagram you think the topics in each question belong.

1. How do landforms and the ITCZ interact to shape seasonal rainfall patterns in South Asia?
2. Why are so many lives at risk due to climate change in this region?
3. What are some solutions to environmental problems in this region that also help address climate change?
4. What are some legacies of British rule in South Asia?
5. How have South Asia's recent surge in economic development and greater openness to globalization affected skilled urbanites and rural farmers differently?
6. Why is the growth of nonfarm employment in rural India significant?
7. How has religious nationalism influenced the politics of this region?
8. How are efforts to combat corruption influencing the politics of this region?
9. Why is the conflict in Kashmir potentially of global significance?
10. Why are there divergent patterns of urbanization in South Asia?
11. What are some solutions that South Asians have developed for the problems faced by slum dwellers?
12. What are some factors slowing population growth in this most populous and most densely populated of world regions?
13. Why do men significantly outnumber women in South Asia?
14. How has the custom of dowry become both financially draining and popular?
15. Why is South Asia a world leader in recognizing the rights of transgender people?
16. How have the Taliban attempted to change the status of women in Afghanistan?

Key Terms

Buddhism 444
caste system 443
civil disobedience 446
dowry 467
hijra 470
Hinduism 442
Indus Valley civilization 440
Jainism 444
jati 443
microcredit 452
monsoon 430
Mughals 441
offshore outsourcing 450
Partition 446
purdah 467
religious nationalism 454
sex ratio 464
sex work 461
Sikhism 444
Taliban 457
varna 444

More Practice at SaplingPlus

Read the interactive e-text, review key concepts, and check your understanding.

sks to protect against smog
ei/Alamy Stock Photo]

9

East Asia

When Chinese journalist Chai Jing was pregnant, she found out that her soon-to-be-born daughter had a tumor. It turned out to be benign, but Jing was convinced that her baby girl's condition was caused by air pollution. Therefore, she embarked on a year-long journey detailing China's horrendous pollution problem. Spending her own money, U.S.$159,000, she turned her investigation into a documentary film called *Under the Dome*, which can be seen on YouTube. Perhaps inspired by Al Gore's film about climate change, *An Inconvenient Truth,* Jing's film is partly lecture and partly clips that document children who have never seen white, fluffy clouds against a blue sky; officials who bluntly say that economic development is more important than environmental protection; and polluters who callously ignore the fines that they accrue when they exceed pollution limits. When the film was released in 2015, it had more than 100 million views in a 2-day period.

When she was making the film, Jing had support from state-sponsored television, and government censors who reviewed the script gave her the go-ahead. At first the government was apparently willing to allow public exposure of the pollution problem, but when the film became hugely successful, it blocked its distribution. Surveys of attitudes in China show, not surprisingly, that pollution is one of the leading concerns of the Chinese people.

Learning Objectives

Environment: Physical and Human

9.1 Describe natural landscapes and climate in East Asia.

9.2 Discuss how society has reshaped and affected the region's environment with regard to water resources, food production, and air pollution.

9.3 Explain how Chinese empires dominated East Asia and how communism subsequently transformed much of the region.

Globalization and Development

9.4 Explain how economic progress in postwar Japan paved the way for economic development elsewhere in the region.

9.5 Discuss how China's economy has been transformed from the era of early communism to contemporary globalization.

Power and Politics

9.6 Describe the lack of political freedom in China.

9.7 Analyze the geopolitical struggles of the seas surrounding East Asia.

Urbanization

9.8 Compare the characteristics of the major cities in East Asia.

9.9 Explain why Chinese economic growth is concentrated in coastal cities.

Population, Gender, and Culture

9.10 Explain the lingering effects of the one-child policy on population and gender in China.

9.11 Describe how Japan is dealing with the problem of an aging population.

9.12 Analyze the status of ethnic minorities throughout East Asia.

Subregions

9.13 Identify key characteristics of the subregions of East Asia.

RUSSIA
KAZAKHSTAN
Astana
W. Sayan Mts.
Tannu Ola
Lake Baikal
Ulan-Ude
Yablonovyy Range
Lake Balkhash
Saryesik Atyrau Desert
Uus Nuur
Hovsgol Nuur
Hovd
Har Us Nuur
Hentiyn Nuruu
Choybalsan
Hulun Nur
Buir Nur
Ulan Bator
Hangayn Mts.
Altai Mts.
Junggar Basin
Urumqi
Yining
Bishkek
Almaty
KYRGYZSTAN
UZB.
Victory Peak elev. 24,700
MONGOLIA
Dzamin Uud
Greater Khi
NEI MONGOL (INNER MONGOLIA)
Gobi Desert
Tian Shan
Kuqa
Kashi
Turfan Depression
Bosten Hu
Muztagata elev. 24,757
TAJ.
Tarim Basin
Taklimakan Desert
XINJIANG UYGUR
Lop Nor
PAKISTAN
K2 (Godwin Austen) elev. 28,250
Yutian
Yumen
Baotou
Hohhot
Zhangjiakou
Beijing
Tangshan
TIANJIN
Tianjin
Datong
Ordos Desert
Yinchuan
SHANXI
Baoding
HEBEI
Shijiazhuang
Taiyuan
North China Plain
Qaidam Basin
Golmud
QINGHAI
Qinghai Hu
Xining
NINGXIA HUIZU
Loess Plateau
Luliang Shan
Taihang Shan
Handan
Zibo
Jinan
SHANDONG
GANSU
Lanzhou
CHINA
TIBET (XIZANG)
Plateau of Tibet
Xianyang
Xian
Luoyang
Zhengzhou
Xuzhou
HENAN
SHAANXI
Qin Ling
Bengbu
Huainan
Hefei
Nanjing
Delhi
New Delhi
Himalayas
Daba Shan
Xinyang
HUBEI
Dabie Shan
ANHUI
SICHUAN
Chengdu
Sichuan Basin
Chongqing
Qiaotou
Three Gorges Dam
Yichang
Wuhan
NEPAL
Kathmandu
Xigaze
Lhasa
Gyangze
Lucknow
Kanpur
Mt. Everest elev. 29,035
Thimphu
BHUTAN
Gongga Shan elev. 24,790
HUNAN
Changsha
Nanchang
Pingxiang
JIANGXI
Ganga Plain
Patna
Brahmaputra
Dalou Shan
Guiyang
GUIZHOU
INDIA
BANGLADESH
Dhaka
Kolkata (Calcutta)
Chittagong
Mouths of the Ganga
Yunnan-Guizhou Plateau
Kunming
YUNNAN
Wuliang Shan
Nan Ling
FUJIAN
Wuyi Shan
GUANGXI ZHUANGZU
GUANGDONG
Rongcheng
Shantou
Guangzhou (Canton)
Nanning
Hong Kong (China S.A.R.)
Macao (China S.A.R.)
BURMA
Hanoi
Zhanjiang
Haikou
Gulf of Tonkin
HAINAN
LAO
Vientiane
South China Sea
THAILAND
VIETNAM
Bangkok
CAMBODIA

The experience of Chai Jing illustrates some of the negative consequences of China's rapid economic development. Incomes have risen and cities have boomed, but so have air and water pollution and greenhouse gas (GHG) emissions, although there are signs that new, cleaner technologies may mitigate these pollution problems.

East Asia, with 1.6 billion people, is the second most populous world region after South Asia. The region is dominated by China, which has a population of almost 1.4 billion people. China's dominance is due not only to its population, but also to its physical size and its role in the global economy. Just as the striking economic rise of Japan, South Korea, and Taiwan over the past 60 years has been the model for developing countries, the opening of China to the global economy will continue to transform East Asia and the world in the future. The changes underway in this region are immense.

What Makes East Asia a Region?

The vast East Asian territory stretches from the Taklimakan Desert in far western China to Japan's rainy Pacific coastline, and from the frigid mountains of Mongolia in the north to the tropical landscapes of Hainan, China's southernmost province, in the south (**Figure 9.1**). East Asia is comprised of the countries of China, Mongolia, North Korea, South Korea, Japan, and Taiwan (the last has operated as an independent country since World War II but is claimed by China as a province; **Figure 9.2**). These countries are grouped together because of their cultural and historical roots, many of which are in China. Because of China's great size, historical influence, enormous population, and huge economy, it is given particular emphasis in this chapter. Japan, whose large and prosperous economy makes it a major player on the world stage, is also emphasized.

Terms in This Chapter

East Asian place-names can be very confusing to those unfamiliar with East Asian languages. We give place-names in English transliterations of the appropriate Asian language, taking care to avoid redundancies. For example, *he* and *jiang* are both Chinese words for river. Thus the Yellow River is the Huang He, and the Long River is the Chang Jiang in Chinese; it is redundant to add the term *river* to either name. The Chang Jiang is often called the Yangtze River in English, a name that denotes the lower reaches of the river, which is where Westerners first encountered it. The word *shan* appears in many place-names and usually means mountain.

Pinyin (a spelling system based on Chinese sounds) versions of Chinese place-names are now commonplace. For example, China's capital was once called Peking in English but is now Beijing, and the southern province of Canton is Guangzhou.

Although China refers to Tibet as Xizang, people around th world who support the idea of Tibetan self-government avoi using that name. This text uses Tibet for the region (with Xizar in parentheses).

◀ **Figure 9.1** Regional map of East Asia.

Figure 9.2 Political map of East Asia.

ENVIRONMENT: PHYSICAL AND HUMAN

9.1 Describe natural landscapes and climate in East Asia.

9.2 Discuss how society has reshaped and affected the region's environment with regard to water resources, food production, and air pollution.

9.3 Explain how Chinese empires dominated East Asia and how communism subsequently transformed much of the region.

A quick look at the regional map of East Asia (Figure 9.1) reveals that the topography here is perhaps the most rugged of any world region. East Asia's varied climates result from the meeting of huge warm and cool air masses and the dynamic interaction between land and oceans. The region's large human population has affected the variety of ecosystems that have evolved there over the millennia and that still contain many important and unique habitats.

.ANDFORMS

.e complex topography of East Asia is partially the result of the .v-motion collision of the Indian subcontinent with the south- .dge of Eurasia over the past 60 million years. This tremen- .force created the Himalayas and lifted up the Plateau of Tibet .ed in gray and gold in Figure 9.1), which can be considered .est of four descending steps that define the landforms of . East Asia, located roughly west to east.

.econd step down from the Himalayas and the Plateau of .road arc of basins, plateaus, and low mountain ranges . yellowish tan in Figure 9.1). These landforms include .ling highland grasslands and deep, dry basins and des- . China (such as the Taklimakan Desert) and Mongo- .e Sichuan Basin and the rugged Yunnan–Guizhou Plateau to the south, which is dominated by a system of deeply folded mountains and valleys that bend south through the Southeast Asian peninsula.

., usually .ith an .ean

The third step, directly east of this upland zone, consists mainly of broad coastal plains and the deltas of China's great rivers (shown in shades of green in Figure 9.1). Starting from the south, this step is defined by three large lowland river basins: the Zhu Jiang (Pearl River) basin, the massive Chang Jiang basin, and the lowland basin of the Huang He on the North China Plain. Each of these rivers has a large delta. Despite the deltas being subject to periodic flooding, they have historically been used for agriculture. However, coastal cities have now spread into many of these deltas, filling in wetlands and farms with zones of dense population and industrialization (see Figure 9.27). Low mountains and hills (shown in light brown in Figure 9.1) separate these river basins. China's Far Northeast and the Korean Peninsula are also part of this third step.

The fourth step consists of the continental shelf, covered by the waters of the Yellow Sea, the East China Sea, and the South China Sea. Numerous islands—including Hong Kong, Hainan, and Taiwan—are anchored on this continental shelf; all are part of the Asian landmass.

The islands of Japan have a different geological origin: they are volcanic, not part of the continental shelf. They rise out of the waters of the northwestern Pacific in the tectonically active zone where the Pacific and Philippine plates dive beneath the Eurasian plate. Lying along a portion of the Pacific Ring of Fire (see Figure 1.7), the entire Japanese island chain is particularly vulnerable to disastrous volcanic eruptions, earthquakes, and **tsunamis**. A tsunami is a seismic sea wave that is usually triggered by an earthquake with an epicenter underneath the ocean. Volcanic Mount Fuji, the highest peak in the country and a recognizable national symbol (see Figure 9.11F), last erupted in 1707. However, the mountain is still classified as active, and deep internal rumblings have been detected recently. In 2011, the largest earthquake in recorded Japanese history (registering 9.2 on the Richter scale) hit off the coast of Honshu, Japan, near the city of Sendai. The quake and subsequent tsunami killed about 18,000 people and damaged several nuclear reactors located on the coast, resulting in the second-worst nuclear accident ever in the world (discussed further in "Natural Hazards and Energy Vulnerability").

The East Asian landmass has few flat portions, and the flattest land is either very dry or very cold. Consequently, the large numbers of people who occupy the region have had to be particularly inventive in creating spaces for agriculture. They have cleared and terraced entire mountain ranges, until recently using only simple hand tools (see Figure 9.4A). They have irrigated drylands with water from melted snow, drained wetlands using elaborate levees and dams, and applied their complex knowledge of horticulture and animal husbandry to help plants and animals flourish in difficult conditions.

CLIMATE

East Asia has two principal contrasting climate zones, shown in the map in **Figure 9.3**: the dry interior west and the wet (monsoon) east. Recall from Chapter 8 that the term *monsoon* refers to the seasonal reversal of surface winds that flow from the Eurasian continent to the surrounding oceans during winter and from the oceans inland during summer.

The Dry Interior

Because land heats up and cools off more rapidly than water does, the interiors of large landmasses in the midlatitudes tend to be

9.3 Climates of East Asia

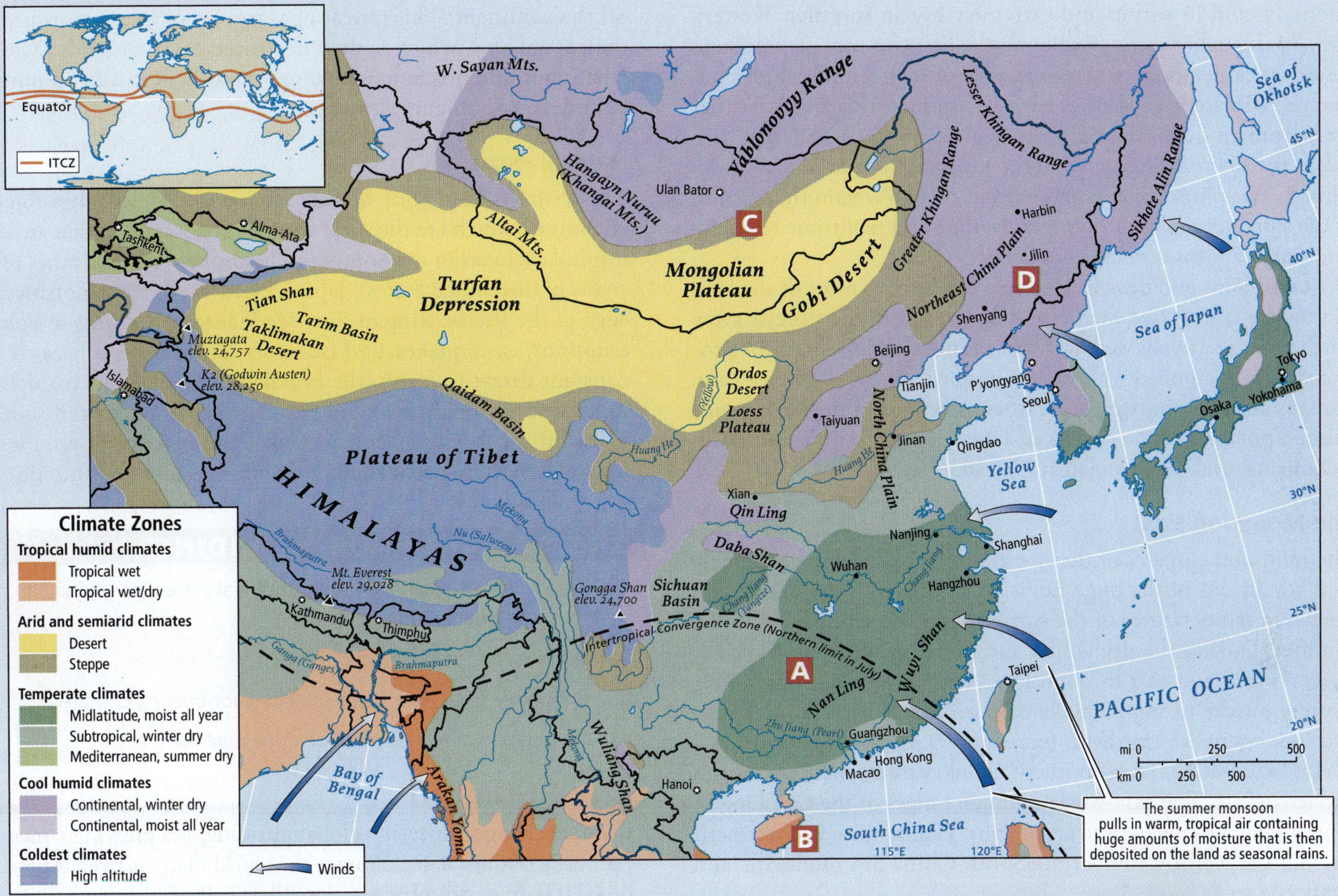

A Midlatitude, moist all year, Guilin, Guangxi Zhuang Autonomous Region, China [BIHAIBO/iStock/Getty Images]

B Tropical wet, Hainan, China [Travelasia/Asia Images/Getty Images]

C Steppe, Mongolia [Tuul & Bruno Morandi/Getty Images]

D Continental, winter dry, northeastern China [HelloRF Zcool/Shutterstock]

intensely cold in winter and extremely hot in summer. Western East Asia, roughly corresponding to the first two topographic steps described earlier, is an extreme example of such a midlatitude continental climate because it is very dry. In fact, this area is farther away from an ocean than any other place on Earth's surface (this *pole of inaccessibility* is located in the Junggar Basin in northwestern China). With little vegetation or cloud cover to retain the warmth of the sun after nightfall, summer daytime and nighttime temperatures may vary by as much as 100°F (55°C).

Grasslands and deserts of several types cover most of the land in this dry region (see Figure 9.3C). Only scattered forests grow on the few relatively well-watered mountain slopes and in protected valleys supplied with water by snowmelt. In all of East Asia, humans and their impacts are least conspicuous in the large, uninhabited portions of the deserts of Tibet (Xizang), the Tarim Basin in Xinjiang, and the Mongolian Plateau.

The Monsoon East

The monsoon climates of the east are influenced by the extremely cold conditions of the huge Eurasian landmass in the winter and the warm temperatures of the surrounding seas and oceans in the summer. During the dry winter monsoon, descending frigid air sweeps south and east through East Asia, producing long, bitter winters on the Mongolian Plateau, on the North China Plain, and in China's Far Northeast (see Figure 9.3D). While occasional freezes may reach as far as southern China, winters there are shorter and less severe. The east–west mountain ranges of the Qin Ling in central China partially protect southern China from cold northern air, and the warm waters of the South China Sea moderate land temperatures.

As the continent warms during the summer monsoon, the air over land rises, which creates low air pressure that pulls in high-pressure wet, tropical air from the adjacent seas. The warm, wet air from the ocean deposits moisture on the land in the form of seasonal rains. This effect is reinforced by the intertropical convergence zone (ITCZ)—where air masses from north and south converge, are forced upward into the atmosphere, and generate rainfall—which is located over southern China in the summer (see Chapter 7). As the summer monsoon moves northwest, it must cross numerous mountain ranges and displace cooler air. Consequently, its effect is weakened toward the northwest. Thus, the far southeast is drenched with rain and has warm weather for most of the year (see Figure 9.3B), whereas central China has about 5 months of summer monsoon weather (Figure 9.3A). The North China Plain, north of the Qin Ling and Dabie Shan ranges, receives even less monsoon rain—about 3 months of monsoon each year. Very little monsoon rain reaches the dry interior.

Korea and Japan have wet climates year-round, similar to those found along the Atlantic Coast of the United States, because of their proximity to the sea. They still have hot summers and cold winters because of their northerly location and exposure to the continental effects of the huge Eurasian landmass. Japan and Taiwan actually receive monsoon rains twice: once in spring, when the main monsoon moves toward the land, and again in autumn, as the winter monsoon forces warm air off the continent. This retreating warm air picks up moisture over the coastal seas, which is then deposited on the islands. Much of Japan's autumn precipitation falls as snow, in particular in northern latitudes and at higher altitudes.

typhoon a tropical storm in the western Pacific Ocean; called a cyclone or hurricane elsewhere

Natural Hazards

The entire coastal zone of East Asia is intermittently subject to **typhoons** (which are the same weather phenomenon that are called tropical cyclones in the Southern Hemisphere or hurricanes off the coasts of the United States). Japan's location along the northwestern edge of the Pacific Ring of Fire (see Figure 1.7) results in volcanic eruptions, earthquakes, and tsunamis. These natural hazards are a constant threat in Japan; the heavily populated zone from Tokyo southwest through the Inland Sea (between Shikoku and southern Honshu) is particularly endangered. Earthquakes are also a serious natural hazard in Taiwan and in China's mountainous interior.

CHECK YOUR UNDERSTANDING

1. Which are the four main topographical zones, or "steps," that form the East Asian continent?
2. What is the Pacific Ring of Fire?
3. Which are East Asia's two principal contrasting climates?
4. What types of natural hazards does East Asia face?

ENVIRONMENT

Climate change, especially global warming, is a growing concern in East Asia (**Figure 9.4**). China is the world's largest overall producer of GHGs (but still produces less than half of U.S. emissions on a per capita basis; see Chapter 1). China has recently been working to limit its emissions for a variety of reasons, but even though it has significantly increased its efficiency through the use of new technologies, its emissions still make up 27 percent of the world's total. This amount will increase as China's urban households use more cars, air conditioning, appliances, and computers. Japan, South Korea, and Taiwan are also major GHG emitters.

East Asia is responding to climate change in multiple ways. Japan has led these efforts for decades. In 1997, the Japanese city of Kyoto hosted the meeting in which countries first committed to reducing their GHG emissions. Japanese automakers such as Toyota were among the first to develop and sell hybrid gas–electric vehicles. China and Japan are now first and second in the world in installed solar power–generating capacity (see Figure 9.4D). China is also the world leader in manufacturing photovoltaic cells; in fact, two-thirds of the world's solar panels are made in China. More and more of this production is being used domestically. Large on-the-ground solar panel plants are being installed in China's sunny, dry interior, where the level of incoming solar radiation is high. The electricity from these is transported via the grid to energy-hungry cities in eastern China. However, mitigating climate change is not the sole reason for the growth of green energy; the high level of air pollution from cars and coal-fired power plants must be reduced by turning to cleaner power sources.

Water Shortages

Glaciers on the Plateau of Tibet capture monsoon moisture and store it as frozen ice, which is then slowly released as meltwater.

China's largest rivers, the Huang He and the Chang Jiang, are partially fed by these glaciers, which are now melting so rapidly due to global warming that scientists predict they will eventually disappear. One effect could be significantly lower flows in these two great rivers during the winter, when little rain falls. Both have already begun to run low during winter, and trade has been affected because riverboats and barges have become stranded on sandbars. Another effect could be a reduction in the amount of irrigation water available for dry-season farming.

Nearly every year, abnormally low rainfall or abnormally high temperatures also create droughts somewhere in China. These droughts often cause more suffering and damage than any other natural hazard.

Droughts can be worsened by human activity on a local or regional scale. When people begin to live or farm in dry environments, as many millions have done in China and Mongolia during the twentieth century, *desertification* (see Chapter 6) can result. People clear vegetation to grow crops, which require more water than the natural vegetation, so the crops must be irrigated with surface water or water pumped from underground aquifers (Figure 9.4A). Many dry areas in China are subject to strong winds that can blow away topsoil once the vegetation is removed. Furthermore, irrigated crops are often less able than natural vegetation to hold the soil. This has resulted in huge dust storms much like those that plagued the central United States in the 1930s Dust Bowl (see Chapter 2). High dunes of dirt and sand have appeared almost overnight in some parts of China that border the desert, threatening crops, roads, and homes. Dust storms can also move over long distances from western China and affect large cities, such as Beijing, near the coast. Particulate matter from these dust storms even circles the entire globe in the upper atmosphere.

VIGNETTE In Ningxia Huizu Autonomous Region on the Loess Plateau in north-central China, Wang Youde squints out at what are now sand-colored low hills barren of vegetation, thinking about how these vast tracts of former farmland have been transformed into deserts by a combination of human error and climate change.

The Loess Plateau was already prone to dust storms during times of drought (*loess* means "wind-deposited soil"). Agricultural expansion into loess areas led to the removal of thick, natural, deep-rooted grasses, which once helped hold down the soil. One can still see the agricultural terraces on the arid slopes where now not even grass grows. Wang Youde's family and 30,000 others fled the area when Youde was 10 years old because one day a sand dune covered their village. From his early childhood, he remembers flowers, birdsong, and occasional snowfalls; all have vanished. Now Youde is back, heading up a project to revegetate thousands of hectares with drought-resistant plants that will hold the soil. By hand, squares of braided straw or stones are laid down to keep water from running off the land and to protect planted seedlings. The seedling survival rate is only 20 to 30 percent. Yet Youde says, "Every time we see an oasis that we have created we are very satisfied . . . [b]ecause we have poured sweat and blood into our work." His adult children are helping him, hoping to remain with the family and escape the hardships of migrating to find urban factory work. ■

Water shortages are particularly intense in the North China Plain, which produces 60 percent of China's wheat, 45 percent of its corn, and 35 percent of its cotton. Here the water table is falling about 3 feet per year, mainly because of the increased use of groundwater for wheat irrigation. Groundwater must be used for agriculture because much of the available surface water is dedicated to urban needs. (The North China Plain is one of the most densely populated and urban regions of China.) Withdrawals of water from the Huang He often make the river's lower sections run completely dry during the winter and spring before it reaches the ocean. A gigantic development called the South–North Water Transfer Project is under way, designed to attempt to rectify this problem (see Figure 9.4B). The project diverts water from the Chang Jiang basin in central China, which has a relative surplus of water, to the Huang He basin in the north, using dams, canals, and pipelines. The upper reaches of the Huang He receive water from the Plateau of Tibet, while more water is transferred to the lower Huang He in the coastal provinces in the east. The first segment started to bring water to Beijing in 2014 and the city now relies on the project for 70 percent of its water. The project is not expected to be fully completed until decades into the future.

China is also trying to avoid water shortages, focusing on water conservation to stretch existing supplies. Already, 30 percent of China's urban water is recycled, and many cities are trying to raise this percentage. China is also making a major effort to remove pollutants from wastewater discharged by industry and farming, which together account for 85 percent of water use.

Japan, the Koreas, and Taiwan have monsoon rainfall patterns, which give them a generally wetter climate and make them less vulnerable to drought than China and Mongolia.

Flooding in China

Flooding has been a hazard in China for hundreds of years. In fact, the six deadliest floods in human history took place in China. The same shifting patterns of rainfall that may worsen droughts can also worsen flooding. Under usual conditions, the huge amounts of rain deposited on eastern China during the summer monsoon periodically can cause catastrophic floods along the major rivers. If global warming leads to even slight changes in rainfall patterns, flooding could be much more severe (see Figure 9.4C). Engineers have constructed elaborate systems of dikes, dams, reservoirs, and artificial lakes to help control flooding. However, these systems failed in 1998, when heavy rains in the Chang Jiang basin caused some of the worst flooding in decades, resulting in the deaths of approximately 4000 people. The pattern is now repeating frequently on the Chang Jiang and other rivers in central and southern China. During the rainy summer of 2010, the new Three Gorges Dam (see "Three Gorges Dam: The Power of Water") could absorb much of the floodwaters, but the amount of water was nevertheless too much to handle and the floodgates of the dam had to be opened, leading to mudslides and a large number of casualties from drowning. The summer of 2018 was yet again a bad year for flooding with hundreds of fatalities.

Recently, urban areas have suffered the most from flooding. The number of Chinese cities affected by annual floods more than doubled from 2008 to 2015. The problem is that urban sprawl has expanded but the drainage infrastructure has not been sufficiently improved. Since the big flood of 1998, the amount of land in China that has been urbanized has more than doubled.

PHOTO ESSAY
9.4 Vulnerability to Climate Change in East Asia

China is particularly vulnerable to drought, desertification, flooding, and other hazards that climate change may intensify. The map shows a high degree of vulnerability in central China, where a large population and intensive agriculture result in pressure on existing water resources. With less access to groundwater and surface water, the region is likely to experience exacerbated water stress. Farther inland, the already dry lands of western China are also vulnerable as rainfall diminishes. Vulnerability is also related to economic development level, as the case of the Korean peninsula shows. Poverty-stricken North Korea is classified as having a high level of vulnerability to climate change, unlike affluent South Korea.

THINKING GEOGRAPHICALLY

A What is one process that can result from people beginning to live or farm in dry environments?

B What is causing the water table to fall each year on the North China Plain?

C Flooding is especially likely in what part of China?

D Why is East Asia among the leaders in the world in renewable energies?

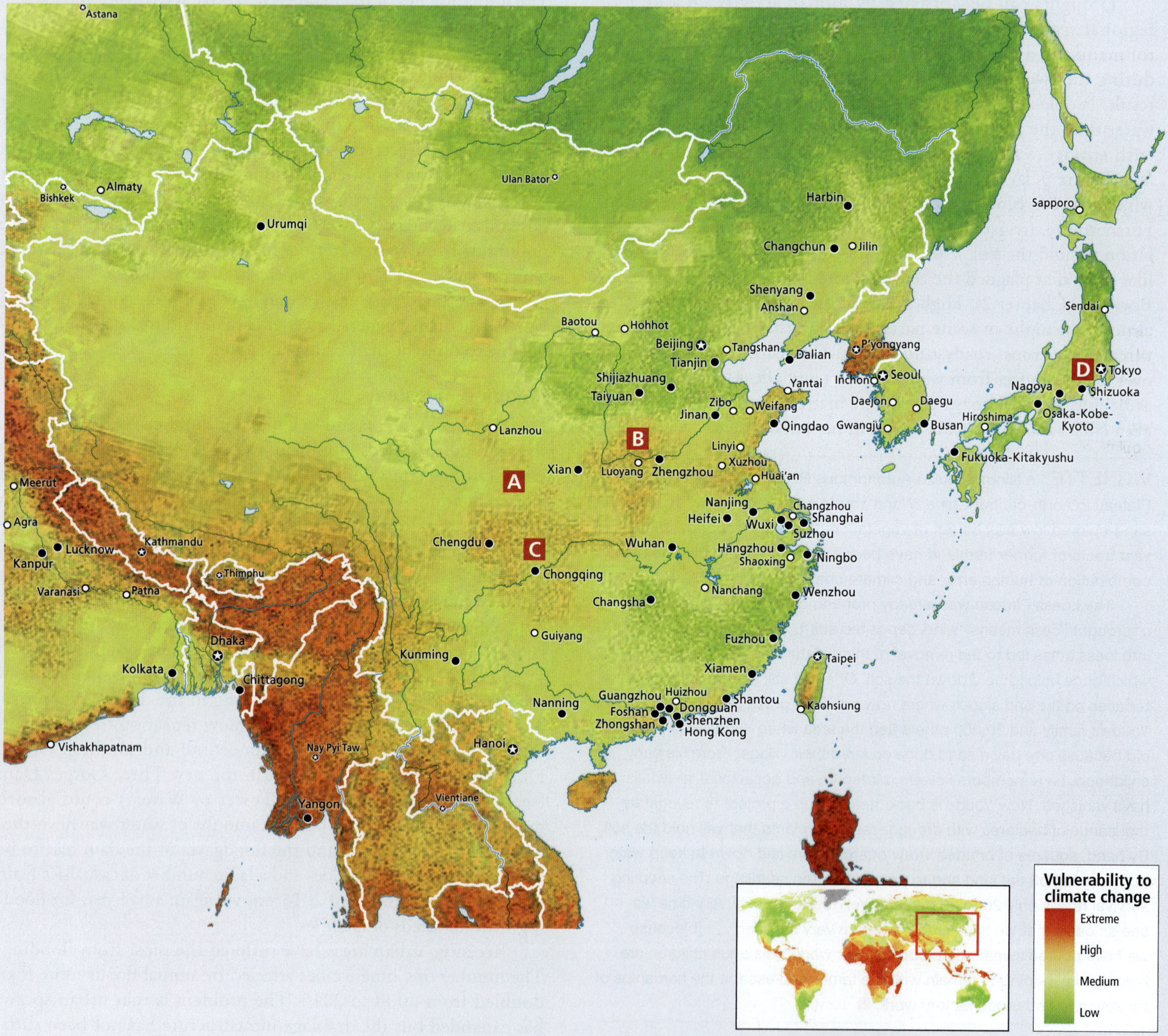

A The terraced hillsides of China's north-central Gansu Province are in a dry upland zone, where agriculture depends on rainfall or irrigation water taken from rivers or underground aquifers. Rainfall is already unpredictable and could become more so with climate change. Melting glaciers on the Plateau of Tibet threaten to reduce river flows, and the overuse of groundwater for irrigation is depleting aquifers. Large rural populations are increasingly being forced to relocate to other parts of China. [Simon Yu/Getty Images]

B A section of the south-to-north water diversion project crossing through the city of Jiaozuo in the Henan Province. Due to severe water shortages and the effects of increases in population, China's national government is building huge canals and other infrastructure to divert water from distant rivers in central and western China to Beijing and surrounding areas. [Li Bo Xinhua News Agency/Newscom]

C Flooding hits the Chang Jiang at Chongqing, China, carrying off boats and barges. Climate change could result in more flooding. This is especially likely in eastern China, where there is much rainfall during the summer monsoon. [Visual China Group/VCG/Getty Images]

D Solar panels installed on the roofs of houses in the city of Ota in central Japan. East Asia has made large investments in solar power. China and Japan have the most installed solar power–generating capacity in the world. More renewable power is needed to offset emissions from coal-fired power plants. [Kyodo News/Getty Images]

Most of this urbanization has taken place on floodplains, creating high amounts of impervious land cover (asphalt and concrete) that leads to serious drainage problems. Natural land surfaces have the capacity to absorb much rainwater while paved surfaces do not, resulting in floods.

Natural Hazards and Energy Vulnerability

Until the Sendai earthquake and tsunami of 2011, Japan, along with many other countries, planned to increase its use of nuclear power as a way to reduce GHG emissions. However, the tsunami washed over the Fukushima nuclear plant and destroyed its cooling systems, which caused a meltdown of several of the plant's nuclear reactors. The severity of the nuclear disaster that followed in the wake of the tsunami halted or curtailed many plans for expanding nuclear power in Japan and across the world. The accident forced the evacuation of more than 300,000 people and temporarily contaminated the water supply of Tokyo. For miles around the reactor, the ground, food, and livestock were contaminated and thousands of gallons of water containing 10,000 times the normal level of iodine-131 (a radioactive substance) were released into the Pacific Ocean. Some places remain ghost towns; nobody has been allowed to resettle them.

Another unfortunate outcome of the Fukushima disaster is that Japan has since increased its GHG emissions from coal energy because of the suspension of the nuclear power plants. On the positive side, the accident is expected to boost reliance on renewable resources, including solar power (see Figure 9.4D). Japan has restarted a few of its nuclear reactors, but whether nuclear power will make a full comeback in Japan is questionable. Before the accident, Japan relied on nuclear power for 30 percent of its electricity; today it is only 2 percent.

FOOD AND SUSTAINABILITY

East Asia's *food security*—the capacity of the people in a geographic area to consistently provide themselves with adequate food—is increasingly linked to the global economy. The wealthiest countries in the region can buy food on the global market, but they are less able to produce sufficient food domestically for two main reasons. Much agricultural land has been lost to urban and industrial expansion; and rising levels of affluence have led to a demand for more meat, which is either imported or requires the importation of animal feed.

About 75 percent of the food consumed in Japan, South Korea, and Taiwan is imported. Japan is the fourth-largest food importer in the world, and with limited amount of grazing land available, it imports more meat than any other country. In China, self-sufficiency in grain production is important to national identity because devastating famines were a recurring problem throughout history. China is nearly self-sufficient with regard to basic necessities, but it relies on imports of grain and soybeans for animal feed and specialty food items. China imports food such as meat, dairy, and seafood from Europe, North America, and other Pacific Rim countries. A growing middle class is driving this consumption and China is now the second-largest food importer in the world (after the United States).

wet rice cultivation a prolific type of rice production that requires the plant roots to be submerged in water for part of the growing season

The rise in demand for agricultural commodities from China and other economically expanding countries has had an impact on global food prices. In 2007–2008, the world market price for grain and other basic food commodities shot up dramatically, although the reasons for the increases in food prices also include high oil prices (because producing food is often energy intensive) and the use of corn to produce ethanol fuel (ethanol is used as a replacement for gasoline in the Americas and elsewhere) rather than as a food source. While global food prices peaked in 2011, they remain relatively high from a historical perspective.

Food Production

Just over half of East Asia's vast territory can support agriculture (**Figure 9.5**). In much of this area, food production has been pushed well beyond what can be sustained over the long term. As a result, East Asia's fertile zones are shrinking. In China, roughly one-fifth of the agricultural land has been lost since the Communist Revolution of 1949, largely because of urban and industrial expansion and agricultural mismanagement that created soil erosion and desertification (see the vignette in "Water Shortages"). This is not surprising; when Japan, South Korea, and Taiwan developed economically, which they did before China, their available farmland declined too.

As urban populations become more affluent, they consume more meat and other animal products that require more land and resources than the plant-based national diet of the past. Soybeans are one example of this change in diet. Soybeans are no longer used mainly for human food but instead for animal feed, for farmed fish food, and especially for high-grade cooking oil. More than 80 percent of the soybeans used in China are now imported, primarily from Brazil, the United States, and Argentina. Due to trade disputes between the United States and China, this and other trade patterns may shift in the near future.

Even with the increase in meat consumption, vegetable dishes with an ancient history remain popular throughout East Asia (**Figure 9.6A**), and meat and dairy dishes are still prepared in traditional ways in places where raising farm animals has been the norm (see Figure 9.6C).

Rice Cultivation

The region's most important grain is rice, and over the millennia its cultivation has transformed landscapes throughout central and southern China, Japan, Korea, and Taiwan. In these areas, rainfall is sufficient to sustain **wet rice cultivation**, which can be highly productive.

Wet rice cultivation requires elaborate systems of water management, as the roots of the plants must be submerged in water early in the growing season. Centuries of painstaking human effort have channeled rivers into intricate irrigation systems, and whole mountainsides have been transformed into descending terraces that evenly distribute the water. Writing about wet rice cultivation in Sichuan Province, geographer Chiao-Min Hsieh describes how "[e]verywhere one can hear water gurgling like music as it brings life and growth to the farms." However, these same cultivation techniques, combined with extensive forestry and mining, have led to the loss of most natural habitats in all but the most mountainous, dry, or remote areas. With so little suitable agricultural land left, further expansion of the area under wet rice cultivation is unlikely.

Figure 9.5 China's agricultural zones. China exhibits regional specialization in agricultural products. This is due to differences in climate and soil, as well as economic reforms that have discouraged regional self-sufficiency within China. [Research from: "World Agriculture," *National Geographic Atlas of the World*, 8th ed. (Washington, DC: National Geographic Society, 2005), p. 19; and "China: Economic, Minerals" map, *Goode's World Atlas*, 21st ed. (New York: Rand McNally, 2005), pp. 39, 207.]

Figure 9.6 LOCAL LIVES: Foodways in East Asia

(A) South Korean women making kimchi at a charity event. Kimchi, a Korean dish made of fermented and often highly seasoned vegetables, was developed at least 3000 years ago as a food storage and preservation technique. Specific kinds of kimchi are made at different times of the year because fermentation happens at variable rates, depending on the particular vegetables available for fermenting at that point in the year and on the temperature at which each vegetable ferments. [Chung Sung-Jun/Getty Images]

(B) A sushi restaurant in Tokyo. Sushi is a combination of cooked, vinegared rice and other ingredients, usually raw seafood. [Jose Fuste Raga/Getty Images]

(C) A woman in rural Mongolia prepares *byaslag*, a cheese made in Mongolia of yak, cow, sheep, or goat milk. Unlike most cheeses originating in Europe, byaslag is not aged but rather is desiccated in the dry air of Mongolia. It is one of many unique products made with animal milk in Mongolia. [Tuul & Bruno Morandi/Getty Images]

Fisheries and Globalization

Many East Asians depend heavily on ocean-caught fish for protein (see Figure 9.6B). People in Japan, South Korea, and China eat more than four times as much fish and shellfish as do Americans. The Japanese in particular have had a huge impact on the seas of not only this region but also the entire world. There are some 4000 coastal fishing villages in Japan, sending out tens of thousands of small crafts to work nearby waters each day.

There are also many large Japanese fishing vessels, complete with onboard canneries and freezers, that harvest oceans around the world. They can do so because waters at a certain distance away from land, typically 200 nautical miles (230 miles, or 370 kilometers), are considered international waters and outside the jurisdiction of any individual country. Environmentalists have criticized Japan for overfishing. For example, Japanese fishing off western Africa has reduced the catches of local fishers so much that many have been forced to migrate to Europe for work. There is little room for Japan to expand its fish imports and exports because the global wild fish catch has been static or in decline since 1990 due to overfishing. Moreover, consumption in Japan has waned, especially among the younger population. Japan's per capita fish consumption in 2016 was 51.4 pounds (23.3 kilograms), compared to 89 pounds (40.2 kilograms) in 2002. In fact, the consumption of meat exceeded fish for the first time in 2006.

Instead, China is today both the largest producer and the largest consumer of fish in the world. Unlike Japan, the world's second-largest consumer of fish, most of China's production is based on aquaculture rather than on wild-caught fish. It is likely that aquaculture will expand in the future, in China and elsewhere, as an important source of protein in an increasingly populous world. In fact, global aquaculture production has exceeded wild-caught fish since 2013.

People and Animals

Animals are used in East Asia not only for food but also for medicinal, aesthetic, and cultural purposes. For example, traditional Chinese medicine involves, among other things, the capturing of wild animals. Then, dried seahorses, lizards, and starfish are found in markets around China because they are believed to have healing properties. Unfortunately, because of this practice, some species are now endangered.

In Japan, aquaculture provides living ornamentation to gardens and other carefully tended landscapes. Carp have been domesticated and bred to have a variety of colors and patterns (**Figure 9.7**) and are a central part of many Japanese gardens. As an adaptive species that does well under a wide variety of aquatic conditions, ornamental carp are now found around the world. Animals are often central to a culture, as is the case in Mongolia, where many species of animals are fundamental to the nomadic culture that historically dominated there. Sheep, goats, yaks (a long-haired Central Asian bovine), horses, and cattle are the most commonly raised animals. A way of life once common in areas of dry grasslands around the world, nomadic living is becoming less common than it was in the past as places like Mongolia become more urbanized.

Figure 9.7 Ornamental carp in Japan. *Nishikigoi* (also known as koi), a variety of carp, in a garden in Himeji, Japan. Carp were domesticated in both ancient China and Rome as a food source. Nishikigoi, which means "brocaded carp," are an ornamental variety developed in Japan in the 1820s. Currently, the trade in ornamental nishikigoi is larger than the trade of carp raised for food. [Alexander Safonov/Getty Images]

CHECK YOUR UNDERSTANDING

1. Why has East Asia suffered for so many years from enormously destructive droughts and devastating floods? What has been done to protect the population from such hazards?
2. How might the melting of China's highest glaciers affect the country's rivers? Which areas might be more affected than others?
3. In which ways is China more vulnerable to global warming than Japan, South Korea, or Taiwan?
4. What are the connections between East Asia's serious environmental problems, high population density, rapid urbanization, and environmentally unsustainable economic development?
5. To what extent is East Asia self-sufficient with regard to food necessities?
6. How has rice cultivation transformed landscapes throughout the region?

THREE GORGES DAM: THE POWER OF WATER

The Three Gorges Dam (**Figure 9.8**) is at this time the largest dam in the world, at 600 feet (183 meters) high and 1.4 miles (2.3 kilometers) wide. It was designed to improve navigation on the Chang Jiang and control flooding, but it is most lauded for its role in generating hydroelectricity for China, a country that uses considerable amounts of energy and is working to reduce its GHG emissions. Supplying about 3 percent of China's electricity needs, the dam replaces the burning of 30 million tons of coal annually. But a dam also comes with a social cost. The Chinese environmental activist Dai Qing notes that China has 22,000 large dams, all of which have displaced people—perhaps as many as 60 million—without any attention to their rights as stakeholders in the projects.

Many critics see serious design and planning flaws in the Three Gorges project. Of greatest concern is the dam's position above a seismic fault. The dam was built at the east end of the Three Gorges because the deep canyons provided a prodigious reservoir for water to power turbines, providing electricity for all of central China, from the sea to the Plateau of Tibet. Unfortunately, the enormous weight of the water and its percolation through geological fissures in the 370-mile-long (600-kilometer-long) reservoir behind the dam could trigger earthquakes. Already at issue are huge landslides along the gorges, lubricated by the rising reservoir water. Meanwhile, cracks in the dam raise doubts about its structural integrity. Similar defects led to the failure of China's much smaller Banqiao Dam during a 1975 typhoon, which caused 150,000 deaths from the flooding and an ensuing famine. Even if the Three Gorges Dam remains structurally sound over the long term, its potential to generate power will probably be reduced by the buildup of eroded silt behind the dam.

Any failure of the dam would be a financial as well as human disaster. The construction of the dam cost approximately U.S.$25 billion, but the ultimate costs may end up being three times this figure, due in part to unforeseen negative environmental and social impacts and to theft from the project by corrupt officials.

Also of concern is the incalculable cost associated with relocating the 1.3 million people who once lived where the dam now forms a reservoir. Thirteen major cities have been submerged, along with 140 large towns, hundreds of small villages, 1600 factories, and 62,000 acres (25,000 hectares) of farmland. In some cases, communities were rebuilt immediately uphill from the submerged location and people only had to relocate a short distance, others had to migrate to distant parts of the country. The reservoir has also destroyed important archaeological sites as well as some of China's most spectacular natural scenery. There are significant environmental costs as well. The giant sturgeon, for example, a fish that can weigh as much as three-quarters of a ton and is as rare as China's giant panda, is critically endangered. Sturgeon used to swim more than 1000 miles (1600 kilometers) up the Chang Jiang, past the location of the dam, to spawn. Now the sturgeon's migration, and consequently their reproductive process, has been irretrievably altered. The rare *baiji* river dolphin was, in fact, declared extinct recently because it has not been sighted for years.

The plan to build the Three Gorges Dam came just as UN development specialists were deciding that the benefits dams could bring do not sufficiently outweigh the many problems they cause. Decades ago, international funding sources such as the World Bank withdrew their support for the Three Gorges Dam because of concerns over the social and environmental costs and other shortcomings of the project. However, Chinese industrialists who need the energy, construction companies that have prospered from building the dam and its many ancillary projects, and government officials eager to impress the world and leave their mark on China continue not only to support the *Da Ba* (the Big Dam), but also to look for other locations around the world where China can gain influence and profit by building dams.

Figure 9.8 Three Gorges Dam. The largest dam in the world is capable of supplying about 3 percent of China's electricity. While the dam generates clean power, it has also caused significant disruptions to the local ecosystem. [Visual China Group/VCG/Getty Images]

AIR POLLUTION: CHOKING ON SUCCESS

Air pollution is often severe throughout East Asia, but the air quality in cities is particularly poor. After South Asia, the pollution in East Asian cities ranks among the worst in the world. China's pollution problems are particularly important because its demand for energy is growing so fast. Globally, coal burning is a major source of air pollution, and China is the world's largest consumer and producer of coal, accounting for 48 percent of all the coal burned in the world each year. Between 1980 and 2017, China's coal consumption grew almost sixfold. During the last few years, however, China's appetite for coal has stagnated in favor of other energy sources.

The combustion of coal releases high levels of pollutants—suspended particulates, smog-causing ozone, and sulfur dioxide—all of which can cause respiratory ailments. In Chinese cities, these emissions are often much higher than the World Health Organization (WHO) defines as safe. On the positive side, WHO also concludes that urban air pollution has dropped significantly during the last few years across China. **Figure 9.9** shows the geography of urban air pollution in China. The worst pollution is often in cities, where homes are in close proximity to industries that depend heavily on coal for fuel. The use of coal to heat homes is still common in China and contributes further to pollution, especially during the winter heating season. Because of that, millions of households are now mandated by the government to switch to less-polluting gas or electrical heat. The cities of northern China suffer from high levels of air pollution, especially in the Hebei Province, where a combination of heavy industry and

Figure 9.9 China's most and least polluted cities in terms of air quality. Based on a 2016 report from China's Ministry of Environmental Protection, the map shows the 10 cities in China that have the most air pollution and the 10 cities with the least air pollution (out of 338 monitored cities). Pollution is measured by the levels of particles (soot), carbon monoxide, and ozone found in the air. [Data from: "Hebei Has 6 of 10 Most Polluted Chinese Cities in 2016," *China Daily*, January 22, 2017, http://www.chinadaily.com.cn/china/2017-01/22/content_28023186.htm.]

atmospheric conditions that trap pollution is problematic. The region tends to be subject to temperature **inversion**—when the ground cools down more than the warm, wet air above does. This creates conditions where the cool air becomes "trapped" near the land surface; and along with it, industrial and vehicle emissions build up, leading to intolerable pollution levels. In southern areas such as the Guangdong Province, manufacturing is less energy intensive and the air is better.

As the capital city, the air quality of Beijing has been considered a problem for the city's international competitiveness (although it is not on the list of the 10 most polluted cities in Figure 9.9). Multinational corporations were wary of conducting business in Beijing and employees didn't want to live there because of the air quality. Today, Beijing's air is significantly better than just a few years ago.

Sulfur dioxide from coal burning also contributes to acid rain, which is displaced to the northeast by prevailing winds that reach coastal East Asia. Acid rain is particularly damaging to freshwater ecosystems and forests. South Korea is particularly afflicted, as the following vignette illustrates. Particulates from China's coal burning are transported globally by high-altitude, west-to-east flowing jet streams, thus affecting air quality in North America and Europe. Locally, toxic mercury from coal plants accumulates in aquatic life and, ultimately, in humans.

inversion the unusual condition when warm air overlies cool air, impeding normal circulation and trapping pollution

VIGNETTE On January 20, 2014, the weather announcer Kim Bo-gyung on the South Korean *Arirang News* warned her viewers that roughly 2 inches of snow were expected overnight. That in itself was not very unusual; South Korea commonly has chilly winter air that comes from continental Asia and picks up moisture when it moves over the Yellow Sea. This day, however, the snow contained more than just water. It was, in fact, acid snow. Pollutants such as nitrogen oxide and sulfur oxide travel from China and combine with water in the atmosphere, and in winter form snow that is deposited in South Korea. The pH during this weather event was as low as 3.8, which is the same level of acidity as wine or orange juice. The weather announcer cautioned people with sensitive skin to stay away from the snow. Pollution that originates in China sometimes moves beyond South Korea across the entire Pacific Ocean. New research indicates that 10 percent of sulfate in the air over the western United States actually comes from China. ■

Air pollution from vehicles is also severe. While per capita automobile ownership is much lower than in the United States, the streets of large Chinese cities are clogged with traffic. For years, China's vehicles have had very high rates of lead and carbon dioxide emissions. The government has begun to work on the problem. Emission standards are tightened every few years, based on the EU's standards.

Air Quality Elsewhere in East Asia

Public health risks related to air pollution are also serious in the largest cities of Japan (Tokyo and Osaka), Taiwan (Taipei), Mongolia (Ulan Bator), and South Korea (Seoul), and in adjacent industrial zones. Even with antipollution legislation and increased enforcement, high population densities and rising expectations about better living standards make it difficult to improve environmental quality. Taiwan is a case in point.

Taiwan's air quality challenges are caused by the island's extreme population density of 1888 people per square mile (729 per square kilometer), its high rate of industrialization, and its close proximity to China. There are 91 motor vehicles (cars or motorcycles) in Taiwan for every 100 residents—more than 21 million exhaust-producing vehicles on this small island. In addition, there are nearly eight registered factories per square mile (three per square kilometer). But just like in China, efforts to improve air quality are in the making. Industries and power plants are switching from coal to cleaner-burning natural gas. The formerly polluted capital Taipei has made especially great strides.

Both North and South Korea depend on hydro- or nuclear-generated electricity to run factories and heat buildings, unlike elsewhere in East Asia. North Korea has relatively few industries and uses very few cars, so its air pollution levels are thought to be low for the region (although few reliable data are available for North Korea). South Korea uses fossil fuels in its numerous industries and has many gasoline-powered cars, which are the sources of most of its internally generated air pollution. Mongolia generally has the region's cleanest air, but in Ulan Bator, the pollution from coal heaters has inspired some imaginative projects. For example, whole sections of Ulan

Bator are now heated by centrally located boilers, which supply hot water to apartment buildings and individual dwellings. This is sorely needed, especially in the frigid wintertime when the need for home heating is great and smog accumulates around the city.

CHECK YOUR UNDERSTANDING

1. What are the pros and cons of the Three Gorges Dam?
2. What are the causes of air pollution in East Asia?

HUMAN PATTERNS OVER TIME

East Asia is home to some of the most ancient civilizations on Earth. Settled agricultural societies have flourished in China for more than 7000 years, which makes China among the oldest continuous civilizations in the world. After centuries of leading the world technologically and economically, East Asia was gradually eclipsed by European powers beginning in the seventeenth century. Outright domination by Europeans occurred only in certain places and not until the nineteenth century. The twentieth century was violent and tumultuous, and after 1945 when World War II ended, the countries of East Asia adopted entirely new economic systems. Three of the countries—China, Mongolia, and North Korea—became communist, and stayed so until the 1980s, when all but North Korea began to allow more capitalist practices. The other three countries—Japan, South Korea, and Taiwan—took a more capitalist route after World War II and today have among the highest standards of living in the world. China, while having a moderate GDP per capita, is now a global economic superpower that rivals the United States and the European Union.

Chinese civilization evolved from several hearths, including the North China Plain, the Sichuan Basin, and the lands of interior Asia that were inhabited by Mongolian nomadic pastoralists. On East Asia's eastern fringe, the Korean Peninsula and the islands of Japan and Taiwan were profoundly influenced by the culture of China, but they were isolated enough that each developed a distinctive culture and maintained political independence most of the time. In the early twentieth century, Japan industrialized rapidly by integrating the European influences that China disdained. As **Figure 9.10** shows, both ancient and contemporary traditions are celebrated in East Asia today.

Imperial China

Although humans have lived in East Asia for tens of thousands of years, the region's earliest complex civilizations appeared in China about 4000 years ago. Written records exist only from the civilization that was located in north-central China. There, a small, militarized, feudal aristocracy (see Chapter 4 for a discussion of feudalism) controlled vast estates on which the majority of the population lived and worked as semi-enslaved farmers and laborers. The landowners usually owed allegiance to one of the petty kingdoms that dotted northern China. These kingdoms were relatively self-sufficient and well defended with private armies.

An important move away from feudalism came with the Qin empire (beginning in 221 B.C.E.), which instituted a trained and

FIGURE 9.10 LOCAL LIVES: Festivals in East Asia

(A) A man dressed as a samurai—a member of Japan's pre-industrial-era warrior class—for a festival in Fukushima, Japan. Thousands of parades, processions, and street festivals are held each year in Japan, often celebrating events in local history or legend. [TORU YAMANAKA/AFP/Getty Images]

(B) The annual, month-long ice festival in Harbin, China, is one of the largest festivals of its kind in the world. Held on and off since 1963, each festival focuses on a different theme, and ice sculpture teams from around the world come to compete. Harbin is the northernmost of China's major cities, and the average January temperature is 0°F (−18°C), which means that ice sculptures are safe from melting. As the image shows, illuminated ice sculptures are especially radiant at night. [Tao Zhang/NurPhoto via Getty Images]

(C) A man participates in *Naadam*, or "Games," which form the basis of festivals held throughout Mongolia in early July to celebrate the country's nomadic history. There are three games practiced: archery, horseracing, and wrestling. Females compete in the archery and horseracing events. [Bruno Morandi/robertharding/Getty Images]

A The oldest Confucian temple, built in Confucius's hometown of Qufu, China, in 478 B.C.E. [IMAGEMORE Co, Ltd./Getty Images]

B Part of a terra cotta army that has been excavated near Xian. The army was buried with the first Qin emperor in 210 B.C.E. [Robert G Brown/Design Pics/Getty Images]

C A section of the Great Wall built in 1570. [Melanie Stetson Freeman/The Christian Science Monito via Getty Images]

5000 B.C.E. | 4000 B.C.E. | 3000 B.C.E. | 2000 B.C.E. | 1000 B.C.E. | 0 C.E. | 1000 C.E. | 1500 C.E.

5000 B.C.E. Agricultural societies develop in China

551–478 B.C.E. Confucian philosophy begins

221–206 B.C.E. Qin empire develops alternatives to feudalism

3000 B.C.E.–1644 C.E. Great Wall of China built (in sections)

FIGURE 9.11 VISUAL HISTORY OF EAST ASIA

Thinking Geographically

A What is the model for Confucian philosophy?

B How did the Qin empire influence subsequent Chinese empires?

C The Great Wall was built in response to what?

D Which major Chinese city and port was a British possession from the time of the Opium Wars until recently?

salaried bureaucracy in combination with a strong military to extend the monarch's authority into the countryside. One part of the legacy of the Qin empire is shown in **Figure 9.11B**.

The Qin system proved more efficient than the old feudal system it replaced. The estates of the aristocracy were divided into small units and sold to farmers. The empire's agricultural output increased because the people more efficiently utilized the land they now owned. In addition, the imperial administration effectively built and maintained levees, reservoirs, and other tax-supported public works that reduced the threats of flood, drought, and other natural disasters. Although the Qin empire was short-lived, subsequent empires maintained Qin methods, which have proved essential in governing a united China.

Confucianism and East Asian Culture

The philosophy of **Confucianism** is closely related to China's ruling tradition. Confucius, who lived from 551 to 479 B.C.E., was an idealist who was interested in reforming government and society. He thought human relationships should involve a set of defined roles and mutual obligations. Confucian values include courtesy; knowledge; integrity; and respect for and loyalty to family lineage, parents, local community, and government officials. These values diffused across the region and are still widely shared throughout East Asia (see Figure 9.11A).

Confucianism a Chinese philosophy that teaches the importance of stability and social order based on traditional institutions such as the patriarchal family, community, and the state

Confucian Gender Roles The model for Confucian philosophy was the patriarchal extended family. The oldest male held the seat of authority and was responsible for the well-being of everyone in the family. Beyond the family, the Confucian patriarchal order held that the emperor was the grand patriarch of all China, charged with ensuring the welfare of society. Imperial bureaucrats were to do his bidding and commoners were to obey the bureaucrats.

Over the centuries, Confucian philosophy penetrated all aspects of East Asian society. Concerning the ideal woman, for example, a student of Confucius wrote: "A woman's duties are to cook the five grains, heat the wine, look after her parents-in-law, make clothes, and that is all! When she is young, she must submit to her parents. After her marriage, she must submit to her husband. When she is widowed, she must submit to her son." These concepts about limited roles for women affected society at large, where the idea developed that sons were the more valuable offspring, with public roles, while daughters were primarily servants within the home.

Cycles of Expansion, Decline, and Recovery Although the Confucian system at times facilitated the expansion of imperial China, its resistance to change also led to periods of decline (**Figure 9.12**). Heavy taxes were periodically levied on farmers, bringing about farmer revolts that weakened imperial control. Threats from outside, particularly invasions by nomadic people from what are today Mongolia and western China, inspired the creation of massive defenses such as the Great Wall (see Figure 9.11C), built along China's northern border. The wall wasn't always successful in repelling invasions, but at the same time, invading nomadic people eventually became indistinguishable from the Chinese, and thus Chinese culture and civilization have absorbed a tremendous mixture of different influences.

D British warships attack near Guangzhou in southern China during the First Opium War in 1840. [Hulton Archive/Getty Images]

E A poster made during the Cultural Revolution in 1969, showing workers reading a book of the thoughts of Mao Zedong. [David Pollack/Corbis via Getty Images]

F A high-speed train passes Mt. Fuji in Japan in 2013. [Vidler Steve/Prisma by Dukas Presseagentur GmbH/Alamy]

1600 C.E. | 1700 C.E. | 1800 C.E. | 1900 C.E. | 2000 C.E.

1500s–1949
European and Japanese imperialism throughout East Asia

1945–Present
Japan rebuilds after WWII

1949–Present
Communist China

1980s–Present
Economic reforms in China

2013

E After the Communist Revolution in China, how did the Chinese government control all aspects of economic and social life?

F How did Japan's political and economic development proceed after World War II?

FIGURE 9.12 The extent of Chinese empires, 221 B.C.E. to 1850 C.E. The Chinese state has expanded and contracted throughout its history. [Research from: *Hammond Times Concise Atlas of World History* (Maplewood, NJ: Hammond, 1994).]

One nomadic invasion did result in important links between China and the rest of the world. In the 1200s, the Mongolian military leader Genghis Khan and his descendants were able to conquer all of China. They also pushed west across Asia as far as what are now Hungary and Poland (see Chapter 5). It was during the time of this Mongol empire that traders such as the Venetian Marco Polo made the first direct contacts between China and Europe. These connections proved much more significant for Europe, which was dazzled by China's wealth and technologies, than for China, which saw Europe as backward and barbaric.

Indeed, from 1100 to 1600, China remained the world's most developed region, despite cycles of imperial expansion, decline, and recovery. It had the largest economy, the highest living standards, and the most magnificent cities. Improved strains of rice allowed dense farming populations to expand throughout southern China and supported large urban industrial populations. Nor was innovation lacking: Chinese inventions included paper making, printing, paper currency, gunpowder, and improved shipbuilding techniques.

Why Did China Not Colonize an Overseas Empire?

During the Ming dynasty (1368–1644), Zheng He, an admiral in the emperor's navy, directed an expedition that could have led to China conquering a vast overseas empire. From 1405 to 1433, Zheng He sailed 250 ships—the biggest and most advanced fleet that the world had ever seen. Zheng He took his fleet throughout established Chinese trade routes to Southeast Asia, across the Indian Ocean, and all the way to the east coast of Africa.

The lavish voyages of Zheng He lasted for almost 30 years, but they never resulted in an overseas empire like those established by European countries a century or two later. These newly explored regions simply lacked much that China needed or wanted. Moreover, back home the empire was continually threatened by the armies of nomads from Mongolia, so any surplus resources were needed for upgrading the Great Wall. Eventually the emperor decided that Zheng He's explorations were not worth the effort. Instead, China reduced its contacts with the rest of the world as the emperor focused on repelling the Mongols. The stance of isolationism was not only defense driven, but also an outcome of a long-standing attitude in China characterized by suspicion of outside cultures. As a result, the pace of technological change slowed, leaving China ill-prepared to respond to growing challenges from Europe after 1600. China did, however, become a regional colonizing power by extending control to include territories in East and Southeast Asia (see Figure 9.12).

European and Japanese Imperialism

By the mid-1500s, during Europe's Age of Exploration, Spanish and Portuguese traders interested in acquiring China's silks, spices, and ceramics found their way to East Asian ports. They brought a number of new food crops from the Americas such as corn, peppers, peanuts, and potatoes. These new food sources contributed to a spurt of economic expansion and population growth, and by the mid-1800s, China's population was more than 400 million; at that same time, Europe had 270 million people.

By the nineteenth century, European merchants gained access to Chinese markets and European influence increased markedly. In exchange for Chinese silks and ceramics, British merchants supplied opium from India, which was one of the few things that Chinese merchants would trade for. The emperor attempted to crack down on this drug trade because of its debilitating effects on Chinese society. The result was the Opium Wars (1839–1860), in which Britain badly defeated China. Hong Kong became a British possession, and British trade, including its opium trade, expanded throughout China (see Figure 9.11D).

The final blow to China's long preeminence in East Asia came in 1895, when a rapidly modernizing Japan won a decisive naval victory over China (and to solidify its emerging regional dominance, Japan also defeated Russia). After this first defeat by the Japanese, the Qing dynasty made only halfhearted attempts at modernization, and in 1912, this last of the Chinese dynasties was overthrown by an internal revolt and collapsed. From the time of the decline of the Qing empire (1895) until China's Communist Party took control in 1949, much of the country was governed by provincial rulers in rural areas and by a mixture of Chinese, Japanese, and European administrative agencies in the major cities.

China's Turbulent Twentieth Century

In response to the absence of a central state authority, two rival political groups arose in China in the early twentieth century. The Nationalist Party, known as the Kuomintang (KMT), was an urban-based movement that appealed to a wide variety of social classes. The Chinese Communist Party, on the other hand, found its base among the rural poor.

By 1937, Japan had control of most major Chinese cities. The KMT did not resist the Japanese effectively and were confined to the few interior cities not under Japanese control. The communists, however, waged a constant guerrilla war against the Japanese throughout rural China, where they gained widespread support. Japan's brutal occupation caused 10 million Chinese deaths, including those of 250,000 to 300,000 civilians in the city of Nanjing, then China's capital, in a massacre that is known as "the rape of Nanjing." Such acts of wartime brutality, as well as the keeping of local "comfort women" (women forced into prostitution) by the Japanese army in the occupied territories during World War II, still complicate the relationship between Japan and its East Asian neighbors.

When Japan finally withdrew in 1945, defeated at the end of World War II by Allied forces, the more popular communists pushed the KMT into exile in Taiwan. In 1949, the Communist Party, led by Mao Zedong, proclaimed the country the *People's Republic of China,* with Mao as president.

Mao's Communist Revolution Mao Zedong's revolutionary government became extremely powerful, dominating all the outlying areas of China, and launched a brutal occupation of Tibet (Xizang). The People's Republic of China was in many ways similar to past Chinese empires. The Chinese Communist Party replaced the Confucian bureaucracy and Mao Zedong became a sort of emperor with unquestioned authority. China received support from the Soviet Union, but the two communist states remained wary of each other for decades.

Among the early beneficiaries of the revolution were the masses of Chinese farmers and landless laborers. On the eve of the revolution, huge numbers lived in abject poverty. Famines were frequent, infant mortality was high, and life expectancy was low. The vast majority of women and girls held low social status and spent their lives in unrelenting servitude.

The revolution drastically changed this. All aspects of economic and social life became subject to central planning by the Communist Party. Land and wealth were reallocated, often resulting in an improved standard of living for those who needed it most. Major efforts were made to improve agricultural production and to reduce the severity of floods and droughts. Everyone, regardless of age, class, or gender, was mobilized to construct almost entirely by hand huge public works projects—roads, dams, canals, whole mountains terraced into fields for rice and other crops. "Barefoot doctors" with rudimentary medical training dispensed basic medical care, midwife services, and nutritional advice to people in even the most remote locations. Schools were built in the smallest of villages. Opportunities for women became available, and some of the worst abuses against them, such as the crippling binding of feet to make women's feet small and childlike, were stopped.

Change, however, came at enormous human and environmental costs. During the **Great Leap Forward** (a government-sponsored program of massive economic reform initiated in the 1950s), 30 million people died, many from famine brought on by poorly planned development; others because they were persecuted for opposing the reforms. Meanwhile, deforestation, soil degradation, and agricultural mismanagement became widespread. In the aftermath of the Great Leap Forward, some Communist Party leaders tried to correct the inefficiencies of the centrally planned economy, only to be demoted or jailed.

In 1966, partially in response to the failures of the Great Leap Forward, a political movement known as the **Cultural Revolution** enforced support for Mao and punished dissenters. Everyone was required to study the "Little Red Book" of Mao's sayings (see Figure 9.11E). Educated people and intellectuals were a main target of the Cultural Revolution because they were thought to instigate dangerously critical evaluations of Mao and the central planning of the Communist Party. Tens of millions of Chinese scientists, scholars, and students were sent out of the cities to labor on farms, in mines and industries, or to jail, where as many as 1 million died. Children were encouraged to turn in their parents. Petty traders were punished for being capitalists, as were those who adhered to any type of organized religion. The Cultural Revolution so disrupted Chinese society that by Mao's death in 1976, the communist regime had been seriously discredited.

After Mao's death in 1976, a new leadership instituted limited market reforms, but the Communist Party retained tight political control. In 2009, after more than 30 years of reform and remarkable levels of economic growth, China's economy became the second largest in the world, behind the United States. At that point, China's economy also became the largest in East Asia, surpassing its rival Japan (although Japan's per capita wealth is still three times that of China's). However, the disparity of wealth in China has been increasing, human rights are still often abused, and political activity remains tightly controlled even as discontent occasionally boils over into open protests against the government.

CHECK YOUR UNDERSTANDING

1. Why do we consider China to be among the oldest continuous civilizations in the world?
2. What is Confucianism? How has it been used to govern parts of East Asia?
3. How was China affected by European and Japanese imperialism?
4. Why didn't China acquire an overseas empire like the Europeans did?
5. How did Mao Zedong's revolutionary communist government, which took power in China after World War II, rearrange Chinese society?

Japan Becomes a World Leader

Although China's influence was preeminent in East Asia for thousands of years, Japan, with only one-tenth the population and 5 percent of the land area of China, dominated East Asia economically and politically for much of the twentieth century. Japan's rise as a modern global power resulted largely from its response to challenges from Europe and North America.

Beginning in the mid-sixteenth century, active trade with Portuguese colonists brought new ideas and military technology that strengthened Japan's feudal lords (*shoguns*), allowing them to unify the country with the support of the military. However, the shoguns monopolized contact with the European outsiders, refusing to allow Japanese people to leave the islands, on penalty of death.

A second period of radical change began when a small fleet of U.S. naval vessels arrived in Tokyo Bay in 1853. The foreigners, carrying military technology far more advanced than Japan's, forced the Japanese government to open the economy to international trade and political relations. In response, a group of reformers (the *Meiji*, which means "enlightened rule") seized control of the Japanese government, setting the country on a crash course of modernization and industrial development that became known as the *Meiji Restoration*. During this time, Japanese students were sent abroad and experts were recruited from around the world, especially from Western countries, to teach everything from foreign languages to modern military technology. Investments emphasized infrastructure development, especially in transportation and communication. The improvements that resulted enabled Japan's economy to grow rapidly, surpassing China's size in the early twentieth century.

Between 1895 and 1945, Japan, which was relatively poor in resources, fueled its economy with resources from its own colonial empire. Equipped with imported European and North American military technology, its armies occupied first Korea, then Taiwan, then coastal and eastern China, and eventually Indonesia and much of Southeast Asia, as shown in **Figure 9.13**. Many people in these areas still harbor resentment about the brutality they suffered at Japanese hands. Japan's imperial ambitions ended with its defeat in World War II and its subsequent occupation by U.S. military forces until 1952.

Great Leap Forward a failed economic reform program under Mao Zedong intended to quickly raise China's industrial level

Cultural Revolution a political movement launched in 1966 to force the entire population of China to support the continuing revolution

Figure 9.13 Japan's expansions, 1875–1942. Japan colonized Korea, Taiwan (then known as Formosa), Manchuria, China, parts of Southeast Asia, and several Pacific islands to further its program of economic modernization and to fend off European imperialism in the early twentieth century. [Research from: *Hammond Times Concise Atlas of World History* (Maplewood, NJ: Hammond, 1994).]

Immediately following World War II, the U.S. government imposed many social and economic reforms on Japan. Japan was required to create a democratic constitution and reduce the emperor's role to a symbolic one. Its military was reduced dramatically, making Japan reliant on U.S. forces to protect it from attack. Politically, it was advantageous for the United States to transform Japan into a Cold War ally.

With U.S. support, Japan rebuilt rapidly after World War II, and it eventually became a giant in industry and global business, exporting automobiles, electronic goods, and many other products (see Figure 9.11F). Japan's economy is still among the world's largest, wealthiest, and most technologically advanced, even if growth has been sluggish the last two decades.

Chinese and Japanese Influences on Korea, Taiwan, and Mongolia

The history of the remaining East Asia region has largely been shaped by what transpired in China and Japan.

demilitarized zone (DMZ) a border area between rival states where military activities are prohibited; usually refers to the border between North and South Korea

The Korean War and Its Aftermath Korea was a unified but poverty-stricken country until 1945. At the end of World War II, the United States and the Soviet Union, as victorious allies, agreed to divide the former Japanese colony. The Soviet Union occupied and established a communist regime to the north, while the United States took control of the southern half of the Korean peninsula, where it instituted reforms similar to those in Japan. After the United States withdrew its troops in the late 1940s, North Korea attacked South Korea. The United States returned to defend the south, leading a 3-year war against North Korea and its allies, the Soviet Union and China.

After great loss of life on both sides and the devastation of the peninsula's infrastructure, the Korean War ended in 1953 in a truce. A **demilitarized zone (DMZ)** was established near the 38th parallel; it serves as the de facto border between North and South Korea. A DMZ could be any border area between rival states where military activities are prohibited, but the term usually refers to Korea. This border, which is 2.5 miles (4 kilometers) wide, is called "demilitarized" because it is a buffer zone not claimed by either side, although in the surrounding area, both sides are heavily armed.

North Korea closed itself off from the rest of the world, and to this day it remains isolated and impoverished, occasionally gaining international attention by showing off its military and nuclear capacity (see Figure 9.18C). South Korea, on the other hand, now has a prosperous and technologically advanced market economy. Relations between the two countries remain uneasy, with occasional skirmishes breaking out along the border. On and off, there are signs of rapprochement and the two countries engage in peace negotiations.

Taiwan's Uncertain Status Historically, Taiwan was a poor, agricultural island on the periphery of China. Then, between 1895 and 1945, it became part of Japan's regional empire. In 1949, when the Chinese nationalists (the Kuomintang) were pushed out of mainland China by the Chinese Communist Party, they set up an anticommunist government in Taiwan, officially naming it the Republic of China. For the next 50 years, the United States aided Taiwan just like it did Japan and South Korea. As a result, Taiwan became modern and industrialized. Its economy quickly overshadowed that of China and remained dominant until the 1990s, serving as a prosperous icon of capitalism right next door to massive communist China. Today, Taiwan remains an economic powerhouse. Taiwanese investors have been especially active in Shanghai and the cities of China's southeast coast.

Mainland China has never relinquished its claim to Taiwan. As China's economic and military power has increased, the United Nations, the World Bank, and the United States, along with most other countries, agencies, and institutions, have judiciously tiptoed around the issue of whether Taiwan should continue to be treated as an independent country or as a rebellious province of China. Even if the overwhelming number of people in Taiwan are ethnically Chinese, the self-perceived identity among most people is Taiwanese rather than Chinese-only. Most Taiwanese also feel that their country should hold on to its sovereignty (see Figure 9.18D).

Mongolia Seeks Its Own Way For millennia, Mongolia's nomadic horsemen posed periodic threats to China, so much so that the Great Wall was built, reinforced, and extended to combat them (see Figure 9.11C). China has long been obsessed with both deflecting and controlling its northern neighbor, and China did control Mongolia from 1691 until the 1920s. Revolutionary communism spread to Mongolia soon thereafter, and Mongolia continued as an independent communist country under Soviet, not Chinese, guidance until 1989, shortly before the breakup of the Soviet Union. Communism brought education and basic services. Literacy for both men and women rose above 95 percent. Situated between two enormous countries, Mongolia has been wary of both Russia and China during the difficult road to a market economy that it has been on since 1989. In need of cash to participate in the modern world, many families have elected to abandon nomadic herding and permanently locate their portable *ger* homes near Ulan Bator and search for paid employment. While the economy has developed, the rapid and drastic change in lifestyle has also resulted in the deterioration of the traditional family structure in a society that formerly took pride in an egalitarian if not prosperous standard of living.

CHECK YOUR UNDERSTANDING

1. How did Japan rise up to become a regional power?
2. What was the outcome of the Korean War? How have North and South Korea developed differently?
3. How did Taiwan and South Korea industrialize after World War II?
4. What foreign powers influenced Mongolia during its communist era?

GLOBALIZATION AND DEVELOPMENT

9.4 Explain how economic progress in postwar Japan paved the way for economic development elsewhere in the region.

9.5 Discuss how China's economy has been transformed from the era of early communism to contemporary globalization.

There are different ways to measure how well specific countries enable their citizens to enjoy healthy and rewarding lives, but it is generally agreed that income per capita should not be the sole measure of well-being. **Figure 9.14** depicts countries' rank on the Human Development Index (HDI), which is a calculation of how adequately a country provides for the well-being of its citizens. The region's human development ranks "very high" or "high." This is a reflection of mostly high gross national incomes (GNI) per capita, adjusted for purchasing power parity (PPP), and a healthy and well-educated population. However, it should be noted that China has an economic development level near the world average, which is significantly lower than some of its wealthier neighbors, such as Japan and South Korea. North Korea is an outlier as a severely impoverished country.

How did most of the countries reach this relatively high level of well-being? After World War II, the countries of East Asia established two basic types of economic systems. The communist regimes of China, Mongolia, and North Korea relied on central planning by the government to set production goals and to distribute goods among their citizens. In contrast, Japan and later Taiwan and South Korea established **state-aided market economies** with the assistance and support of the United States and Europe. In this type of economic system, market forces, such as supply and demand and competition for customers, determine many economic decisions. However, the government intervenes strategically, especially in the financial sector, to make sure that certain economic sectors develop in a healthy fashion. Investment in the country by foreigners is also limited so that the government can retain more control over the direction of the economy and so that economic development benefits domestic interests. In the cases of Japan, South Korea, and Taiwan, government intervention was designed to enable **export-led growth**. This economic development strategy relies heavily on the production of manufactured goods destined for sale abroad—primarily to North America and Europe—while limiting imports for local consumers.

More recently, the differences among East Asian countries have diminished as China and Mongolia have set aside strict central planning and adopted many aspects of state-aided market economies and export-led growth. China in particular relies heavily on exports of its manufactured goods to North America and Europe. Japan, South Korea, and Taiwan became less export-oriented over time as their own affluent domestic markets emerged.

state-aided market economy an economic system based on market principles but with strong government guidance; in contrast to the limited government, free market economic system of the United States and, to a lesser degree, Europe

export-led growth an economic development strategy that relies heavily on the production of manufactured goods destined for sale abroad

PHOTO ESSAY

9.14 Globalization and Development

The economic development level and general well-being of the population are very high in Japan, South Korea, Hong Kong, and Taiwan. The map shows the Human Development Index (HDI), which is based on adjusted real income, life expectancy, and educational attainment (note that HDI data are not reported for Taiwan). China and Mongolia rank in the second-best category ("high") for human development. Both have made significant progress over the last 20 years. However, the introduction of a market economy in China has increased inequalities between individuals and between regions within the country. In contrast, both South Korea and Japan have a relatively equitable distribution of economic resources. Data for North Korea are hard to come by, but the country is undoubtedly very poor, due to severe government mismanagement.

THINKING GEOGRAPHICALLY

A Where are the world's largest ports located, and why?

B China's export-oriented production is concentrated in which part of the country?

C What makes Tokyo one of the most important "world cities"?

D What is the objective of China's Belt and Road Initiative?

E How has China's agriculture changed over time?

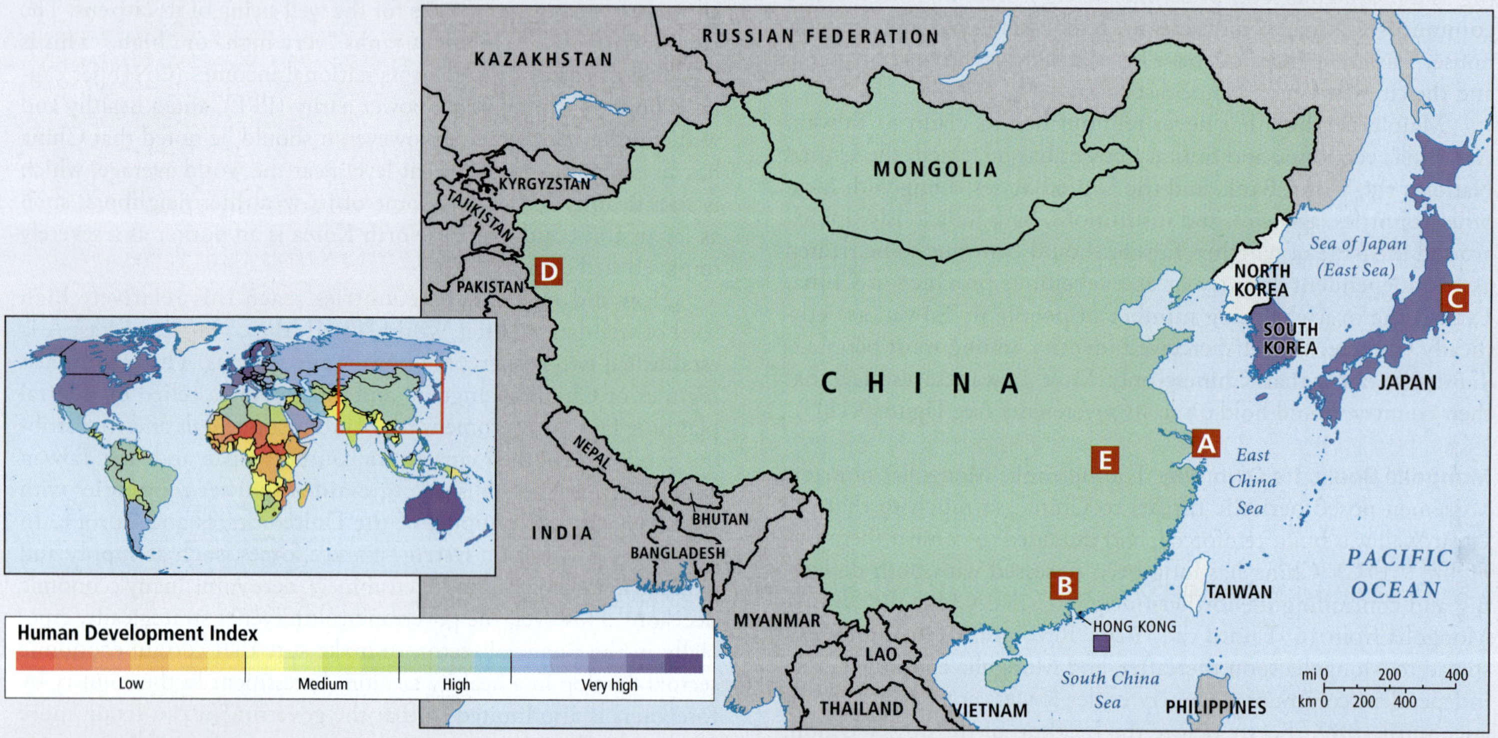

A The Shanghai Yangshan deepwater port with containers and loading cranes. Shipping containers around the world are standardized so they can easily be loaded from ocean-going vessels to port facilities and then onto railcars or trucks to be delivered to their final destination. This system has made global shipping increasingly efficient and inexpensive. [pat138241/iStock/Getty Images]

B Workers in Dongguan, Guangdong Province, check stuffed toys for defects. Many people have migrated from central China to the southern coast, where much manufacturing for export is concentrated. [Feng Li/Getty Images]

C Sony corporate headquarters in Tokyo. Sony is a multinational conglomerate that specializes in electronics, entertainment, and financial services. Tokyo has one of the largest concentrations of multinationals in the world. [Kiyoshi Ota/Bloomberg via Getty Images]

D Belt and Road Initiative. A modern highway links the remote Xinjiang Province in western China with Pakistan. China's Belt and Road Initiative aims to connect China with other countries in Asia and beyond. Improving infrastructure, such as roads, rail, and ports, is part of that plan. [Peter Ydeen/Shutterstock.com]

E Chinese agriculture. China has increased its agricultural mechanization and productivity over time. As this picture of a Chinese farmer harvesting rice illustrates, most farmsteads have access to a level of capital investment that allows for the purchase of small farm machinery. [hanhanpeggy/iStock/Getty Images]

THE JAPANESE MIRACLE

Throughout the nineteenth century, the economies of Japan, Korea, and Taiwan were minuscule compared with China's. During the twentieth century, though, all three grew tremendously, in part because of ideas that originated in Japan.

Japan Rises from the Ashes

Japan's recovery after its crippling defeat at the end of World War II is one of the most remarkable tales in modern history. Except for the imperial capital of Kyoto, which was spared because of its historical and architectural significance, all of Japan's major cities were destroyed by the United States. Most notably, the United States bombed Hiroshima and Nagasaki with nuclear weapons, and leveled Tokyo with incendiary bombs.

Key to Japan's rapid recovery was its state-aided market economy, in which the government guided private investors in creating new manufacturing industries. The strategy was export-led economic growth, with the Japanese government negotiating trade agreements with the United States and Europe. These and other deals ensured that large and wealthy foreign markets would be willing to import Japanese manufactured goods. South Korea, Taiwan, countries in Southeast Asia, and eventually China and many other parts of the developing world have imitated Japan's model of a state-aided and export-oriented economy. Also central to Japan's recovery, but less imitated abroad, were arrangements between the government, major corporations, and labor unions that guaranteed lifetime employment by a single company for most workers in return for relatively modest pay.

The government-engineered trade and labor arrangements produced explosive economic growth of 10 percent or more annually between 1950 and the 1970s. The leading sectors were export-oriented automobile and electronics manufacturing. Japanese brand names such as Sony, Panasonic, Yamaha, and Toyota became household words in North America and Europe. Products made by these companies sold at much higher volumes than would have been possible in the Japanese and nearby Asian economies. However, growth started to slow considerably around 1990, followed by a stagnation phase called the lost decades. Despite this stagnation, Japan's postwar "economic miracle" continues to have an immense worldwide impact as a model, and Japan remains a significant actor in the world economy. Japan purchases resources from all parts of the world for its industries and domestic use. These purchases and investments in various local economies create jobs for millions of people around the globe.

Productivity Innovations in Japan

Over the years, Japan has made major innovations in manufacturing that have boosted its productivity and have been diffused to other industrial economies, changing the spatial arrangements of industries. The **just-in-time system** clusters together companies that are part of the same production process so that they can deliver parts to each other when they are needed (**Figure 9.15**). For example, factories that make automobile parts are clustered around the final assembly plant, delivering parts literally minutes before they will be used. This saves money by making production faster and more efficient, and reducing the need for warehouses.

just-in-time system the system pioneered in Japanese manufacturing that clusters companies that are part of the same production system close together so that they can deliver parts to each other precisely when they are needed

Figure 9.15 Japan's just-in-time system. In this example of the just-in-time system, related industries are clustered together around a Toyota plant in Toyota City, Japan. The facilities supply each other with various automobile parts and other inputs needed for the production process. [Kurita KAKU/Gamma-Rapho via Getty Images]

A related innovation is the *kaizen system* of continuous improvement in manufacturing and business management. In this system, production lines are constantly surveyed for errors, which helps ensure that fewer defective parts are produced. Production lines are also constantly adjusted and improved to save time and energy. Both the just-in-time and the kaizen systems have been imitated by companies around the world and have been taken overseas by Japanese companies that invest abroad. For example, Toyota uses the kaizen and just-in-time systems in its U.S. plants.

MAINLAND ECONOMIES: COMMUNISTS IN COMMAND

After World War II, economic development on East Asia's mainland proceeded on a dramatically different course than it did in Japan. Communist economic systems transformed China, Mongolia, and North Korea. Private property was abolished and the state took full control of the economy, loosely following the example of the Soviet centrally planned economy (see Chapter 5). These sweeping changes transformed life for the poor majority but ultimately proved less resilient and successful than was hoped.

By design, most people in the communist economies could not consume more than the bare necessities. On the other hand, the policy—called the *iron rice bowl* in China—of guaranteeing nearly everyone a job for life, sufficient food, basic health care, and housing was better than what they had before. One drawback was that overall productivity remained low.

The Commune System

When the Communist Party first came to power in China in 1949, its top priority was to make monumental improvements in both agricultural and industrial production. The communist governments in North Korea and Mongolia held similar goals, though these countries had much smaller populations and resource bases to work with.

In the years following World War II, an aggressive agricultural reform program joined small landholders together into cooperatives so that they could pool their labor and resources to increase production. In time, the cooperatives became full-scale communes, with an average of 1600 households each. The communes, at least in theory, took care of all aspects of life. They provided health care and education, and built rural industries to supply items such as clothing, fertilizers, small machinery, and even tractors. The rural communes also had to fulfill the ambitious expectations that the leaders in Beijing had for better flood control, expanded irrigation systems, and especially more food production.

The Chinese commune system met with several difficulties. Rural food shortages developed because farmers had too little time to farm. They were required to spend much of their time building roads, levees, terraces, and drainage ditches, or working in the new rural industries. Local Communist Party administrators often compounded the problem by overstating harvests in their communes to impress their superiors in Beijing. The leaders in Beijing responded by requiring larger food shipments to the cities, which caused even greater food shortages in the countryside. Although the Chinese agricultural communes were inefficient and created devastating food scarcities during the Great Leap Forward, they did eventually result in a stable food supply that kept Chinese people well fed. After years of market reforms, China today produces more grain, fruit, vegetables, meat, and eggs than any other country.

On the Korean Peninsula, North Korea and South Korea took very different economic paths. North Korea invested in military strength at the expense of broader rural economic development. The country not only lacks good farmland, but the government has not sufficiently funded farming inputs, such as seeds, fuel, and fertilizer, to produce enough food for the population. Commune-style agriculture has been so neglected that production is precarious. However, compared to the 1990s when outright starvation was widespread, the situation is much better today. Even so, North Korea has used rocket launchings, nuclear tests, and even the threat of war to intimidate its neighbors, and it is selling its military technology abroad to pay for its food imports. According to the United Nations, roughly 40 percent of North Koreans suffer from undernourishment, which is why the world has allowed food aid into North Korea. About 30 percent of North Korea's food is imported, much of it in the form of food aid. But the intransigence of the North Korean regime can also backfire, as cutbacks in food aid from South Korea, the United States, and the UN suggest. The black market is a source of food imports into North Korea from across the Chinese border.

In Mongolia, communist policy followed the Soviet model of collectivization, though it was tailored to Mongolia's economy, which at the time was largely one of herding and agriculture. There were minimal changes to the nomadic pastoralist way of life, but the role of mining and industry was emphasized. Today, 31 percent of the population works in agriculture, but other sectors of the economy are more productive, as agriculture's contribution to GDP is only 13 percent.

Focus on Heavy Industry

The communist leadership in China, North Korea, and Mongolia believed that investment in heavy industry would dramatically raise living standards. Massive investments were made in the mining of coal and other minerals and in the production of iron and steel. Especially in China, heavy machinery was produced to build roads, railways, dams, and other infrastructure improvements that leaders hoped would increase overall economic productivity. However, the vast majority of the population remained impoverished agricultural laborers who received little benefit from industries that created jobs mainly in urban areas. Not enough attention was paid to producing consumer goods and household items that would have driven modest internal economic growth and improved living standards for the rural poor. Even in the urban areas, growth remained sluggish because, as also was the case in the Soviet command economy, small miscalculations by bureaucrats resulted in shortages and production bottlenecks that constrained economic growth.

Struggles with Regional Disparity

For centuries, China's interior west has been poorer and more rural than its coastal east. The interior west has been locked into agricultural and herding economies, while the economies of the east have benefited from trade and industry. The first effort to address regional disparities, right after the revolution, was an economic policy that focused on a combination of investment from the central government and the development of **regional self-sufficiency**. Each region was encouraged to develop as an independent entity with both agricultural and industrial sectors that would create jobs and produce food and basic necessities. This policy did little to lessen regional disparities and in fact constrained economic growth by inhibiting internal trade between regions.

ECONOMIC REFORMS IN CHINA

In the 1980s, China's leaders enacted economic reforms that changed the country's economy in five ways. First, building on communist-era methods of financial control but also integrating strategies pursued in Japan, China used the state-owned banking system to direct investment into export-oriented manufacturing industries. Second, at the local level, economic decision making was decentralized. Farmers now could make household-level decisions, subject to the approval of the commune, about how to use land and which crops to grow. Third, the reforms allowed existing and newly established businesses to sell their produce and goods in competitive markets. Fourth, **regional specialization**, rather than regional self-sufficiency, was encouraged in order to take advantage of regional variations in climate, natural resources, and location, and thereby encourage national economic integration. Basic trade theory suggests that regions that have the factors of production to produce certain goods efficiently will do so and then trade for other goods that other regions produce. In practice, this has meant that coastal urban areas focus on export-oriented manufacturing

regional self-sufficiency an economic policy in communist China that encouraged each region to develop its own industrial and agricultural resources

regional specialization specialization (rather than self-sufficiency) in the production of goods that takes advantage of regional variations in resources and location and advanced service production, while resource-based industries and production for domestic consumption are fostered in the interior provinces. Finally, the government allowed foreign direct investment in Chinese export-oriented coastal enterprises and the sale of foreign products in China.

While these reforms dramatically improved the efficiency with which food and goods were produced and distributed, regional disparities have widened and rural–urban disparities are sometimes extreme. For example, rural households in some interior provinces have an income that is approximately a third that of their urban counterparts (**Figure 9.16**). Internal migration (see "Economic Transitions") is a direct consequence of this spatially uneven development.

China's economic reforms have transformed not only China's economy but also the economies of wider East Asia and indeed the whole world. Today China is the world's largest producer of manufactured goods, supplying consumers across the globe. Of the other communist-led countries, Mongolia has participated in this transformation only moderately, mainly as a raw material exporter, and North Korea has not participated at all.

EAST ASIA'S ROLE AS MEGA-FINANCIER

As U.S. government debt has mounted in recent decades (up to U.S.$16 trillion in 2018), so has concern about the fact that China and Japan now own some of this debt. Among foreign holders of U.S. debt, China owns about 18.5 percent, and Japan 16 percent (**Figure 9.17**); many people are worried that either country could destabilize the U.S. economy by selling or refusing to buy more U.S. debt. However, such an event is highly unlikely. In order to finance U.S. consumption of foreign goods and to keep the value of the dollar high, the export economies of China and Japan are investing in U.S. debt, and thus U.S. capacity to keep on importing. Dumping their vast holdings of U.S. debt and therefore destabilizing the U.S. economy would mean that both China and Japan would not be able to sell as many of their goods to consumers in the United States, which would hurt their own economies. Hence China and Japan have strong incentives to continue financing U.S. government debt.

East Asians are able to use their reserves to finance development elsewhere in the world, thereby securing privileged trade deals.

Figure 9.16 China's regional GDP and rural–urban income per capita disparities, 2017–2018. Notice the disparity in GDP per capita across China, as indicated by the colors of the provinces, as well as the rural–urban income disparity in each province, as represented by a percentage (the per capita disposable income of rural households as a percentage of per capita disposable income of urban households). [Data from: National Bureau of Statistics of China, at http://data.stats.gov.cn/english/easyquery.htm?cn=E0103.]

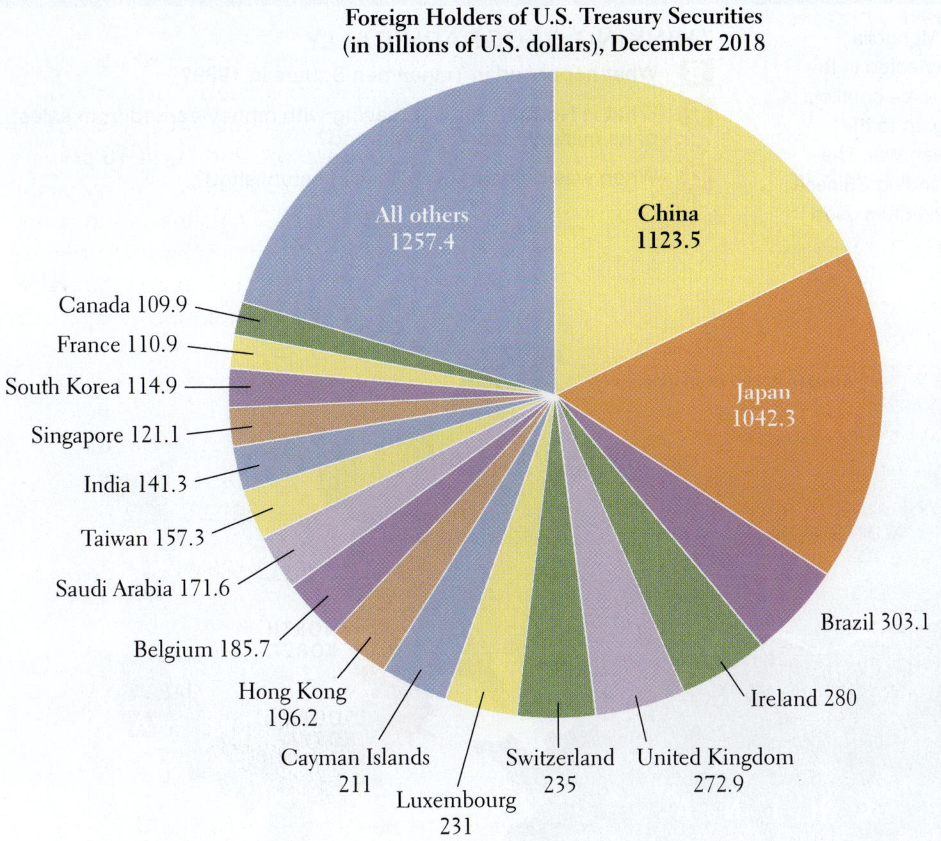

Figure 9.17 Foreign holders of U.S. Treasury securities, December 2018. The total dollar amount of U.S. securities held by foreign countries is more than $6 trillion; note, though, that the United States also holds large amounts of securities from foreign countries. [Data from: U.S. Treasury, "Major Foreign Holders of Treasury Securities (in billions of dollars)," at http://ticdata.treasury.gov/Publish/mfh.txt.]

The current such project is called the *Belt and Road Initiative*, which is primarily focused on transportation infrastructure—roads, rails, and ports—that connects a wide range of countries to China. The overland route of the Belt and Road Initiative goes from China to Central Asia, Southwest Asia, and even Central Europe. It has been compared to the historic Silk Road trade route (see Figure 5.10). The maritime route includes Southeast and South Asia as well as Africa. The arrangement is that participating countries—there are currently 78 of them—receive loans from China for projects built by Chinese firms. In recent years, China's lending to developing countries has outpaced even that of the World Bank. For example, China has financially supported countries that have an adversarial relationship with Western-dominated lending institutions. While most countries are pleased to receive new investment, some voices warn about being swept up in an emerging neocolonial relationship, with China as the new dominant. While China wants to forge new, favorable trade connections where the recipient country will supply China with needed industrial materials, resources, and fossil fuels, the Belt and Road Initiative is also a way for China to invest its financial surplus and to find new markets for Chinese firms (Figure 9.14D).

While China has short-term, concrete economic objectives, some of these investments are meant to project *soft power*—a country's attempt to persuade others to do what it wants without resorting to force. The goal of Chinese soft power is to shape long-term attitudes about China around the world and to ensure friendly allies for decades to come. China is using investments and soft power around the world but without making any demands on these countries to engage in political reforms, such as improving their human rights records. If successful, the ambitious project could reshape the future global economic landscape in China's favor.

CHECK YOUR UNDERSTANDING

1. What characterizes the successful economic development strategy that East Asian countries pioneered?
2. What Japanese industrial management innovations have successfully diffused around the world and created clusters of economic activity?
3. How did communist economic systems transform China, Mongolia, and North Korea?
4. How have the reforms that China's leaders enacted since the 1980s changed the country's economy and affected the global economy?
5. Which countries hold much U.S. public foreign debt? Is that a problem? Why or why not?

POWER AND POLITICS

9.6 Describe the lack of political freedom in China.

9.7 Analyze the geopolitical struggles of the seas surrounding East Asia.

The level of political freedom is highly uneven throughout East Asia. Japan's current democratic political structure was established after World War II, South Korea's in the late 1980s, and Taiwan's in the mid-1990s. Mongolia has dramatically expanded its political freedom since abandoning communism in 1992. China, however, remains under the tight control of an authoritarian regime, and North Korea is even more tightly held. The map in **Figure 9.18** shows each country's level of democracy. The red starbursts indicate places where civil unrest has broken out since 1945.

JAPAN'S POLITICAL SHIFTS

Japan emerged as the first modern democracy in the region. After its defeat at the end of World War II, Japan adopted a new constitution that stripped the emperor of any meaningful political power, embraced pacifism over militarism, and established parliamentary democracy. Significantly, the government that played such a central role during the post–World War II rise of the Japanese economy was controlled from 1955 to 2009 by one political party, the

PHOTO ESSAY

9.18 Power and Politics in East Asia

Political freedoms are better protected in Japan, Taiwan, Mongolia, and South Korea than in China and North Korea. This is reflected in the democratization index depicted in the map below. Large-scale conflicts mostly occurred decades ago, such as the events leading up to the formation of the People's Republic of China, and the Korean War. The ongoing conflict indicator on the map refers to the longstanding Chinese *laogai* forced labor prison system. It is included here as it is often used against political dissenters, such as the Uygurs today.

THINKING GEOGRAPHICALLY

B What happened in Tiananmen Square in 1989?

C What is North Korea purchasing with money earned from sales of its military technology abroad?

D When was democracy in Taiwan established?

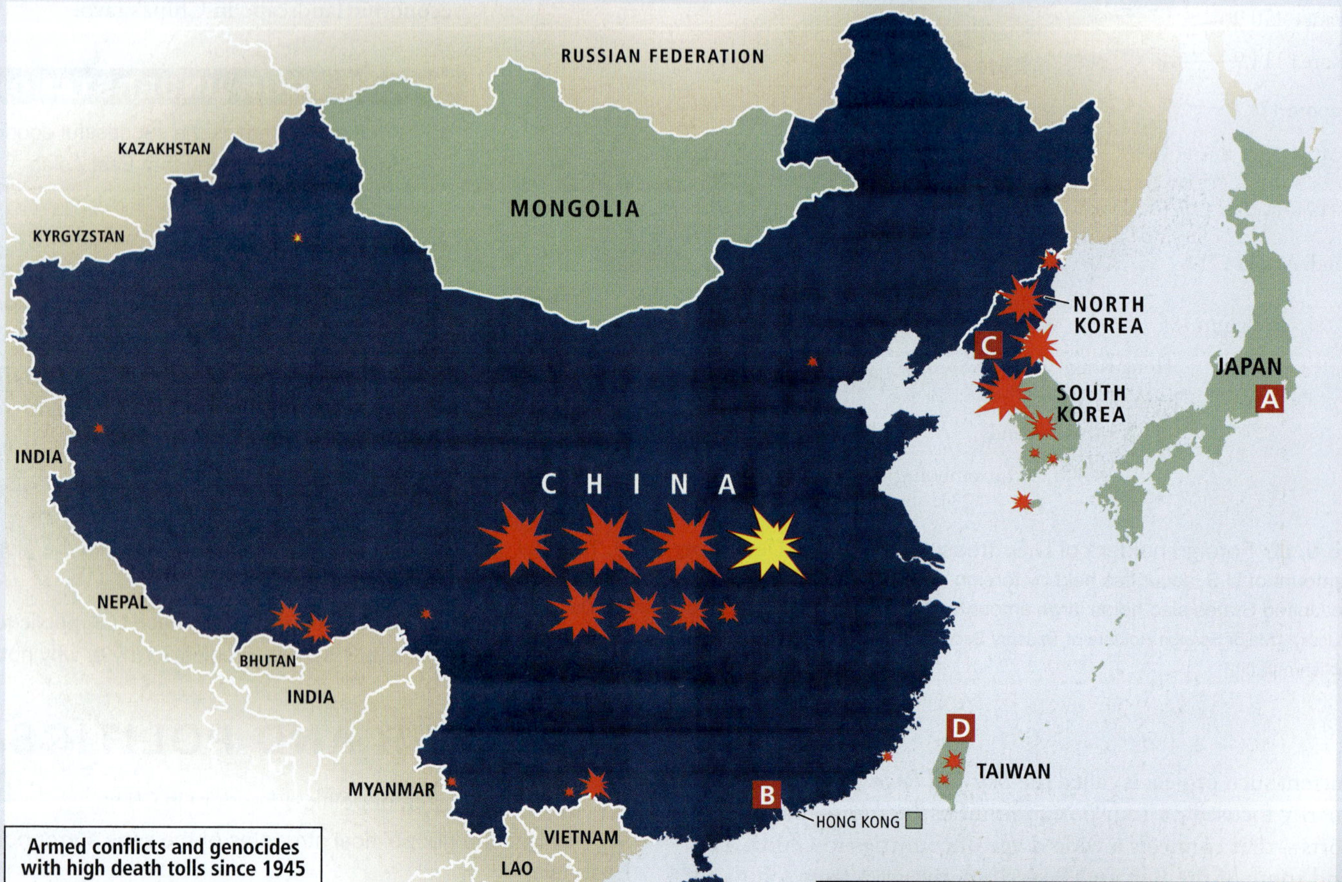

A Japanese war dead honored. Japanese politicians visit the Yasukuni Shrine in Tokyo, which honors Japan's war dead, including convicted war criminals from World War II. Condemned by China, Taiwan, and Korea, these visits spark protests throughout the region. [KAZUHIRO NOGI/AFP/Getty Images]

B **Tiananmen Square protests remembered.** The annual vigil in Hong Kong commemorating those who died in the 1989 Tiananmen Square massacre in Beijing has become a focus for China's pro-democracy movement. As many as 150,000 people have shown up for the vigil in recent years. Hong Kong is the only part of China where the right to engage in political protests is protected by law. [ANDREW ROSS/AFP/Getty Images]

C **A display of military might.** Soldiers march in a parade for the anniversary of North Korea's founding in P'yongyang, North Korea. The United States and North Korea's neighbors all express varying levels of concern about North Korea's military capabilities, especially its nuclear arsenal. [AP Images/Ng Han Guan]

D **Disputed islands.** A Taiwanese activist puts her country's flag on a model of one of the Senkaku/Diaoyu islands, one of several disputed islands in the East China Sea that is claimed by Japan, China, and Taiwan. Many Taiwanese see this and other disputes as ways to reaffirm the country's continued independence from China, which considers Taiwan a rebellious province. [STR/AFP/Getty Images]

Liberal Democratic Party (LDP). This may have been an outcome of the homogenous and consensus-oriented culture of Japan. The LDP remains dominant in Japanese politics, although it may face more challenges in the future. One of the main differences between LDP and the leading opposition parties is that the more nationalistic LDP takes a hawkish stance against China and all of Japan's other neighboring countries (see Figure 9.18A). Having closer relations with China may be advantageous for Japan, though, as China has the potential to be a major market for Japanese exports in the future.

POLITICAL CHANGE IN CHINA?

With China now a globalized economy, many wonder how much longer the Communist Party can remain in control without allowing a significant expansion of political freedom throughout the entire country. The Communist Party officially says that China is a democracy, and indeed for many years some elections have been held at the village level and within the Communist Party. However, representatives of the National People's Congress, the country's highest legislative body, are appointed by the Communist Party elite, who maintain tight control throughout all levels of government. Change might arrive as the population becomes more prosperous, educated, and widely traveled, thereby becoming exposed to places that have more political freedoms. However, the Communist Party remains determined to repress any major reform movements that arise from outside the party's senior leadership. For many years, China pursued a de facto collective leadership model, although recently the country's president, Xi Jinping, has concentrated power in his own hands.

Less than a decade after market reforms began, demands for political change that were inspired by similar events in communist central Europe culminated in a series of pro-democracy protests that drew hundreds of thousands of people to Beijing's Tiananmen Square in 1989 (see Figure 9.18B). These protests were brutally repressed, with thousands (the precise number is uncertain) of students and labor leaders massacred by the military. The Tiananmen Square event made clear that China's integration into the global economy would take place on political terms dictated by the state.

Informed consumers and environmentalists in developed countries have criticized China for its "no holds barred" pursuit of economic growth. Much of China's growth has been dependent on environmentally destructive activities and harsh conditions for workers, both of which effectively reduce production costs so that the prices of Chinese goods are low on world markets.

Information Technology and Political Freedom

Because of the spread of information via the internet, people's push for political freedom from within China has increased. Since the revolution in 1949, China's central government has controlled the news media in the country. By the late 1990s, though, the expanding use of electronic communication devices was loosening central control over information. The expansion of internet access has risen dramatically since then; 802 million people (or 58 percent of the population) were connected by 2018 (**Figure 9.19**). However, internet access is not evenly distributed. A digital divide exists between major cities like Beijing and Shanghai, where about 75 percent of the people have internet access, and the rural interior, where only about 40 percent do. As economic development and communication infrastructure spread over time, it is likely that the divide will diminish. In fact, access to the internet in rural China is significantly higher than it was a decade ago.

Today, millions of Chinese people have access to an international network of information. It is much more difficult for the government to give inaccurate explanations for problems caused by inefficiency and corruption. The availability of the internet to ordinary Chinese citizens supports the expansion of political freedom; however, the backlash from the government in the form of censorship has been fierce.

Social networking technologies have emerged as a powerful tool for collective action in China. In recent years, the use of the multipurpose app WeChat has grown exponentially. With about 1 billion registered users in China, it has outpaced the Twitter-like service Weibo with "only" 400 million users. This type of virtual person-to-person interaction has been instrumental in forcing major media coverage and public discussion of incidents that expose government corruption and mismanagement.

Nevertheless, the internet in China is not the open forum that it is in most Western countries. Chinese internet users are required to register their real names when using an online account. Postings on social media sites like Weibo are constantly screened by the government to monitor anonymous criticisms about party officials. And if people in China use politically sensitive words in social postings or internet searches, they may receive a reminder that they have just used prohibited language. This type of censorship has informally been dubbed the "Great Firewall" and particularly filters content that originates outside China. Google, in a dispute with the Chinese government over state-mandated censorship, moved its operations to Hong Kong, where more freedom is allowed. Meanwhile, a web browser from the Chinese company *Baidu* dominates the search engine market, with about 75 percent of users. It is easier for Chinese authorities to censor a domestic company than to censor other companies whose information may flow from servers around the world.

Protests and Political Freedom

Protests by workers for better pay and living conditions, and by farmers and urban dwellers displaced by new real estate developments, are becoming more common in China. The government has stopped issuing complete statistics on protests, but the nonprofit group Human Rights Watch estimates that 100,000 to 200,000 such protests are being held each year.

Most protests across China are small-scale, local protests that only involve a few dozen people. They are held with some frequency; virtually every day some form of resistance against those in power takes place. The protesters are usually not directly motivated by broader political concerns. The triggering mechanism is often when an individual or small group of people sense that they have been ignored and mistreated by authorities. The international Pew Research Center has conducted attitude surveys that indicate that ordinary Chinese people think corrupt officials are the most important problem that is facing the country. Corruption is followed by environmental pollution and the growing gap between the rich and poor. It is often these concerns that are the underlying factors of dissent, and when combined with individual grievances, will cause people to engage in public protests.

Figure 9.19 Internet use in East Asia, 2016. Japan, South Korea, and Taiwan have very high internet use rates, while China has the most internet users of any country in the world (an estimated 802 million people). Usage in China varies greatly among provinces. For comparison, 89 percent of people in the United States use the internet. [Data from: Statistica, "Internet Penetration Rate in China in 2016, by Region," at https://www.statista.com/statistics/277265/percentage-of-internet-users-in-china-by-province/.]

Since confidence in public officials is weak, as the Pew survey indicates, people don't expect that the legal system will treat them fairly; thus they engage in direct street protests. Most of the time, news about protests doesn't travel far. Internet censorship, as discussed above, has the capacity to spatially contain dissent. For example, for protests that take place in rural areas or small towns far away from the coastal urban centers, attention in Chinese media is limited or nonexistent. People have started to distribute videos of disturbances on social media, but before they "go viral," censors often take note and remove them. At times, though, many people have joined interest groups that pressure the government to take action on particular problems.

More systemic threats to the power monopoly of the Communist Party are dealt with more harshly. The university professor and human rights activist Liu Xiaobo was a veteran of the Tiananmen Square protests in 1989 who has continuously called for democratic elections in China. For that, he was arrested, and sentenced to a long prison sentence. While incarcerated, Liu Xiaobo was awarded the Nobel Peace Prize, though neither he nor his family was allowed to attend the ceremony in Norway. Liu died of cancer in 2017, still under the control of the Chinese authorities. Another group that is perceived as a threat by the Chinese government is Falun Gong, a Buddhist-inspired movement. Falun Gong represents an emergence of new religious groups that operate independent from the state. Many of its members have been arrested, but Falun Gong has also been able to mount a worldwide media campaign for its cause, which is another reason it is disliked by the government.

In an effort to maintain control over China's increasingly articulate protestors, the government has allowed elections to be held for what are called "urban residents' committees." These elections are supposed to serve as an outlet for people to be able to peaceably voice their frustrations and create a limited amount of change at the local level. However, many of these elections are hardly democratic, with candidates selected by the Communist Party. In cities where unemployment is high and where protests have been particularly intense, elections tend to be more free, open, and democratic.

GEOPOLITICS OF THE SEA

Political tensions have been rising between the countries of East Asia even as their economies have become more linked. One especially important issue concerns the territorial control of the waters off the coast of East Asia. While these disagreements are rooted in contemporary economic conditions and political calculations by governments, they are also fueled by the painful legacies of World War II and its aftermath. Japan and Russia are in a long-standing dispute over the Kuril Islands in the Pacific Ocean. Recently, two major disputes between East and Southeast Asian countries have arisen over uninhabited islands that lie over oil and natural gas reserves.

In the East China Sea, the Senkaku Islands (in China, these are known as the Diaoyu Islands) are claimed by China, Japan, and Taiwan. The conflict over the Senkaku/Diaoyu Islands has escalated lately, mainly because of competing claims on them by China and Japan. Because China and Japan are the world's first- and second-largest importers of fossil fuels, respectively, they each have an interest in controlling nearby reserves. The waters around the islands are also rich fishing grounds, in the midst of vital shipping lanes, and militarily strategic. Within China, the conflict often inflames widespread resentment about Japan's brutal occupation of China before and during World War II. Japanese nationalists, on the other hand, eager to counter China's rising power and armed with several treaties that explicitly give it control over the islands, have staged protests and planted Japanese flags on the islands. Chinese ships sail around the islands in what Japan claims are its territorial waters. China also says it should have control over the surrounding air space. Many now see this as East Asia's most risky potential military conflict of the post–World War II period.

The control of the Senkaku/Diaoyu Islands is also important for legal reasons. International law, based on the United Nations Convention on the Law of the Sea (UNCLOS), stipulates that waters extending 12 nautical miles (1 nautical mile is approximately 1.15 miles, or 1.85 kilometers) out from a country's shoreline are its *territorial waters*. Nonmilitary foreign ships have the right to pass through those waters, but otherwise a country can govern and regulate its territorial waters much like it does its land. In addition, an *exclusive economic zone* extends much farther, 200 nautical miles. While this zone is not a country's territory, the country has the exclusive rights to mineral exploitation on the seabed. By claiming the Senkaku/Diaoyu Islands, China and Japan therefore also assert the rights to resources in the East China Sea in the vicinity of the islands. In reality, the exact boundaries of territorial waters and, especially, exclusive economic zones are hard to determine. If no agreement is reached between disputing countries, the UN's International Court of Justice can issue a ruling. That has not been done in this case.

The second dispute concerns the Spratly Islands of the South China Sea (see also Chapter 10). Both China and Taiwan are pursuing claims to these islands, as are several Southeast Asian countries, largely for the same reasons as with the Senkaku/Diaoyu Islands. China's strategy here has been to build an artificial island that now also has an airfield. By actually populating these islands, it is easier for China to invoke the international Law of the Sea in order to claim the rights to oceanic resources. The other countries are worried that China is militarizing its claims over the South China Sea and that it intends to move air force and navy vessels into these waters as a way to protect its interests. None of the other countries in the region are as militarily powerful as China.

conurbation an extended urban area consisting of several nearby cities that have grown together over time

world city one of the most important cities in the global system in terms of economic, political, and cultural dominance

CHECK YOUR UNDERSTANDING

1. How is the internet emerging as a tool to promote the expansion of political freedoms in China?
2. How has China's government responded to protests and other efforts to promote political change by Chinese citizens?
3. Why do so many countries claim control of uninhabited islands in or near the waters of East Asia?
4. Which East Asian countries have high levels of political freedom, and which ones do not?

URBANIZATION

9.8 Compare the characteristics of the major cities in East Asia.

9.9 Explain why Chinese economic growth is concentrated in coastal cities.

Throughout most of the twentieth century, the most dynamic East Asian cities were in Japan, South Korea, and Taiwan. Tokyo, Japan, is the world's largest urban area, with 38 million inhabitants. The reason it is so big is that it is a **conurbation**—a number of cities located so close together that they have grown together over time to form one large urban area. Some of the other urban centers in the area, located around Tokyo Bay, include Yokohama and Kawasaki. As the capital and economic center of one of the largest economies in the world, Tokyo has attracted huge amounts of investment. In fact, it is considered a **world city**, which are the most important cities in a globalized world, serving as control-and-command centers of the global economy with headquarters of multinational corporations and important financial institutions, holding great sway over global political affairs, and being home to renowned cultural institutions and media. In addition to Tokyo, most geographers consider London and New York City to round off the list of the top-tier world cities.

The vast Kobe–Osaka–Kyoto area in Japan may not have world city status, but it is also a conurbation and has 17 million people (see the map in **Figure 9.20**). South Korea is home to Seoul, East Asia's second-largest urban area, at 24 million people, and the fifth-largest urban economy in the world (after Tokyo, New York, Los Angeles, and London). The urban area of Taiwan's capital of Taipei is home to 8.6 million people and has the third-highest per capita income in East Asia, after Hong Kong and Macao. Seoul and Taipei are especially good examples of primate cities that dominate their respective countries (see Chapter 3).

However, for the last 30 years, the strongest urban growth in East Asia, and even in the entire world, has been in China. From

2014 to 2016, 45 of the 50 fastest-growing large cities in the world, as measured by increase in GDP, were in China. Despite the rapid growth, urbanites still represent only 59 percent of China's total population, so more urban growth is likely, especially considering that in every other country in the region—even North Korea—the urbanization rate is higher than in China. Most East Asian cities are extremely crowded, and future growth poses a challenge for planners and residents (see Figure 9.20A, D). Especially in China, developers are targeting rural land on the outskirts of cities. There, villagers and farmers are rarely given fair compensation for the land that they occupy or farm. The reason is that the individual Chinese farmer does not own farmland; he or she only leases it from the local government, which may be inclined to sell the land to high-bidding developers.

SPATIAL DISPARITIES

China's focus on export-oriented manufacturing based in urban areas has led to rural areas lagging behind cities in access to jobs and income, education, and medical care. The map in Figure 9.16 depicts the differences. GDP per capita is significantly lower in China's interior provinces than it is in coastal provinces, and within each province there is a disparity between rural and urban places. Notice that rural–urban GDP disparities, while still significant, are less extreme in northeastern and coastal provinces than in interior and western provinces. Similar rural-to-urban disparity patterns are found in all other countries in East Asia. Even in wealthy countries such as South Korea, circumstances in the remote countryside remain rudimentary compared to a glitzy and high-tech metropolis like Seoul.

INTERNATIONAL TRADE AND INVESTMENT

When China initiated economic reforms in the 1980s, many people were wary of the disruption that could result from abruptly opening the economy to international trade. To ease the transition, China first selected five coastal cities as sites where foreign technology and management could be imported to China and free trade established. These **special economic zones (SEZs)** now operate like export processing zones (EPZs) found in other world regions (**Figure 9.21**; see Chapter 3). Since then, the program has expanded to a large number of cities, many of them in the interior of the country. There are a wide variety of zones with different names and they are administered in different ways, but all provide footholds for international investors and multinational companies eager to establish operations in the country.

Today the SEZs and other zones are China's greatest **growth poles**, meaning that their development, like a magnet, is drawing yet more investment and migration as they have an increasingly better workforce and better infrastructure as a result of the initial investment. The first coastal SEZs were spectacularly successful. In just 25 years, many coastal cities grew from medium-sized towns or even villages into some of the largest urban areas in the world. Foreign direct investment remains concentrated on the eastern coast, which accounts for 85 percent of China's exports. Areas along the southern coast near Hong Kong and the central coast near Shanghai have especially well-developed export economies. As shown in Figure 9.16, GDP per capita (PPP) rates remain noticeably higher on the eastern coast relative to China's interior. The government is mindful of these geographic differences and is trying to develop the inland areas. Generating massive amounts of electricity at the Three Gorges Dam is one way of addressing the divide, as it can attract new industrial development to China's interior and act as a counterweight to the booming coast. Foreign direct investment is also increasingly directed toward interior provinces.

Transportation Improvements in China

Recently, growth poles in the interior have had higher rates of growth than those on the coast. One reason for this is that China spends 9 percent of its GDP on transportation improvements (by comparison, the United States spends 2 to 3 percent), which has made central and western China more accessible. New highways and railroads may eventually enable the interior to catch up with the economic development of the coast. The world's longest high-speed rail now takes a traveler from Beijing to southern China in less than 10 hours, rather than the 22 it used to take, and is routed through central China, not along the coast.

ECONOMIC TRANSITIONS

In 2013, China reached a major milestone that every rapidly developing country eventually reaches: its booming service sector began to create more jobs than did its industrial sector. Many of these jobs offer much better pay and working conditions than industrial jobs do, which is creating a serious shortage of workers in the industrial sector.

Until recently, China's spectacular urban growth was based on hundreds of millions of new urban migrants who were willing to put up with adverse conditions to earn a little cash. Fewer migrants are now willing to work in such jobs, creating periodic labor shortages. Some factory owners have been forced to offer higher pay, better working conditions, and shorter workdays or more time off. The government is also trying to offset labor shortages by making changes to the *hukou* (household registration) system, beginning to give formal urban residency status to many people who have migrated to urban areas illegally.

The background is that until recently, most rural-to-urban migration in China was illegal because the ***hukou* system**, an ancient practice reinforced in the Communist era, effectively tied rural people to the place of their birth. The objective was to maintain social order by limiting migration so cities wouldn't be overwhelmed by people for whom there was inadequate housing and employment. Today the system is being reformed and migrants are being allowed to move to approved cities as long as they obtain permission from the authorities. This is a major change for millions of people who have been ignoring the *hukou* system for decades, becoming part of what is called the **floating population**,

special economic zones (SEZs) free trade zones within China, which are commonly called export processing zones (EPZs) elsewhere

growth poles zones of development whose success draws more investment and migration to a region

***hukou* system** the system in China where each person's permanent residence is registered and any person who wants to migrate must obtain permission from authorities to do so

floating population the Chinese term for people who live and work in a place other than their household registration location; many are people who have left economically depressed rural areas for the cities

9.20 Urbanization in East Asia

Some of the wealthiest and fastest-growing urban areas in the world are in East Asia. China's cities are undergoing particularly rapid growth. The largest urban region in the world is located here—Tokyo, the capital of Japan. Tokyo also holds great sway over global affairs, as most Japanese multinational corporations are headquartered there. China does not have one corresponding principal city, which is evident in the map below. The capital Beijing to the north acts as the political center, while Shanghai, situated at the mouth of the Chang Jiang, is the economic powerhouse of central China. In the global economy, the cities of China's southern coast are increasingly important. The other countries of East Asia conform to the primate city phenomenon with one city that is overwhelmingly dominant.

THINKING GEOGRAPHICALLY

A Migrants to urban areas who ignore the *hukou* system are considered part of what population?

B How many people live in the Tokyo region, the world's largest conurbation?

C Why can Seoul be classified as a primate city?

D What happened to many of the former residents of Pudong, the new financial center of Shanghai?

A Beijing. Migrants from rural areas arrive in Beijing. Most of China's urban growth is based on migration from rural areas. Urbanization is likely to continue in the future as China is less urbanized than its neighboring countries. [PETER PARKS/AFP/Getty Images]

B Tokyo. Ginza is one of Tokyo's leading commercial districts with upscale shopping, dining, and entertainment. Based on its total GDP, Tokyo can be considered the wealthiest city in the world and its lavish shopping opportunities are commensurate with that status. [Education Images/Universal Images Group/Getty Images]

C Seoul. Large expanses of high-rise apartment buildings characterize much of the current urban development throughout East Asia. Shown here are the outskirts of Seoul, South Korea. [geophoto/Multi-bits/The Image Bank/Getty Images]

D Shanghai. A man searches for lumber in the remains of an old neighborhood in Shanghai that is being demolished to make way for high-rise apartments for the city's burgeoning population of wealthy residents. The people of this neighborhood were forcibly evicted by the city government and resettled in cheaper apartments on the urban fringe. In order to modernize, few attempts at renovating historic neighborhoods, called hutongs, are being made in urban China. [Hong Wu/Getty Images]

Figure 9.21 Foreign investment in East Asia. The map shows China's original SEZs and one additional important category, the economic and technology development zones (ETDZs). The colors on the map reflect levels of foreign investment in each of the countries of the region and in each of China's provinces. [Data from: https://data.worldbank.org/indicator/BX.KLT.DINV.CD.WD and http://china-trade-research.hktdc.com/business-news/article/Facts-and-Figures/Mainland-China-Provinces-and-Cities/ff/en/1/1X000000/1X06BOQA.htm.]

a term used in China to describe those who have no rights to subsidized housing, schools, or health care in the place to which they have migrated. These migrants generally work in menial, low-wage jobs and make agonizing sacrifices to send money home to children and spouses living in rural areas. The number of people migrating from rural to urban areas in China has increased. China's migrant workers now number 287 million people, approximately half of the working people in China's cities and a third of the entire workforce (see **Figure 9.22** and Figure 9.20A).

For decades, the Chinese government only weakly enforced the *hukou* system because urban industries were in need of a growing workforce. The lax enforcement enabled the mobility of the labor force while keeping that labor vulnerable and exploitable. The current reforms to the *hukou* system are designed both to encourage more migration to cities, which have had labor shortages in recent years, and to control the places that migrants can relocate to. The goal is to freely allow migration to small and medium-sized cities while still limiting migration to the biggest cities so they are not overwhelmed by new residents. A major benefit for migrants is that they now have better access to housing, schools, and health care. Another new policy reform is that urban migrants no longer have to give up their right to farmland in the village that they left behind, which provides some sense of security if they would one day go back.

The extra costs these changes impose—especially the higher wages and the improved access to subsidized housing, education, and health care for millions of urban migrants—mean that China is no longer the cheapest place to manufacture products. Some factories have already moved to Vietnam, Bangladesh, and countries in Africa with even cheaper labor. With excellent worker productivity in the United States, some American companies that once outsourced to China are now returning home.

Some industrial employers hope that enough new migrants will be drawn to the cities by the *hukou* reforms that the labor shortages will disappear. The experiences of other countries, though, suggest that China's urban service sector will continue to attract more and more migrants. The rapid growth of the service sector is being fueled by the purchasing power and consumption of

Figure 9.22 Migration and urbanization in the new Chinese economy. Millions of people in China are migrant workers who have moved from rural to urban areas to find work, usually without the government's permission and hence without the full rights of citizenship they would be entitled to had they stayed in their rural homes. The most common destination is the Guangdong Province, which some now call the "world factory," and into the lower Chang Jiang region around Shanghai. However, the pool of cheap, rural labor is drying up. As migrations slow, the price of labor goes up, and China's coastal manufacturing economy will have to pay higher wages or relocate.

China's increasingly wealthy urban populations. Moreover, China's population growth is slowing down significantly, which also means that industries cannot rely on the labor surplus that has so far characterized the Chinese economy.

As noted previously, during Japan's economic miracle from the 1950s to 1970s, its economy grew annually at a rate of 10 percent. After that, the growth slowed down as the economy matured. The same transition toward an affluent economy happened in South Korea and Taiwan as well. China appears to be following the same path. For a long time, China's economic growth was around 10 percent, but it has been nudging downward recently—it was 6.9 percent in 2017 and 6.6 percent in 2018. As the country grows wealthier, maintaining the breakneck pace of development becomes unrealistic. As the vignette below shows, the relative slowdown has impacted many people who only recently have been exposed to capitalist institutions such as stock markets.

VIGNETTE Jin is a recently retired woman in Shanghai who has found a new hobby: investing in the stock market. As the economy expands, the stock market typically follows along. Why not, Jin reasons, grow your retirement savings by investing in the stock market? She was inspired by watching talk shows on television that promoted easy ways to get rich. Jin had her nephew install trading software called Big Wisdom on her new computer. She also has the Big Wisdom app on her cell phone so she can trade whenever she feels like it.

The emerging culture of trading is also creating new social spaces. In order to tap into the latest news about which stocks are hot, people like Jin go to brokerage houses, which are places that are open to the public where large screens with real-time changes in stock price can be observed and computer stations are available for buying and selling stocks. Over numerous cups of tea, small-time investors socialize and exchange tips all day long.

But here's the problem. China's stock markets have been extremely volatile lately. Trading in China is dominated by individual investors much like Jin. Economic experts say that the demographics of Chinese investing—the investors are also known as the "pajama traders"—makes the market unpredictable. Small investors who trade frequently and are looking for short-term gains exhibit herd behavior. This creates sudden swings in the stock market, which has taken sudden dives on multiple occasions. The problem has been exacerbated by a recent decline in GDP growth, which has made investors nervous, and the lack of reliable information to base decisions on. The institutions of capitalism are still quite immature in China. ■

HONG KONG'S SPECIAL ROLE

Hong Kong, which was a British possession from the end of the first Opium War in 1842 until 1997, has long had a special role as a link to the global economy for China. Before 1997, some 60 percent of foreign investment in China was funneled through Hong Kong, and since then Hong Kong has remained the financial hub for China's booming southeastern coast (**Figure 9.23**).

Hong Kong is one of the most densely populated cities in the world; its 7.4 million people are packed into only 23 square miles (60 square kilometers), an area roughly the same size as Manhattan in New York City but with four times as many people. Hong Kong is composed of Hong Kong Island, which is the city's center, and a rocky peninsula near the delta of the Zhu Jiang (Pearl River), the most important river in southern China. Skyscraper-dominated Hong Kong is also one of the richest cities in the world: its per capita GNI (adjusted for PPP) in 2017 was above that of the United States, at U.S.$64,100.

In July 1997, Britain's 99-year lease of Hong Kong ran out and Hong Kong became a special administrative region (SAR) of China. Hong Kong still operates under a political and legal system established by the British that allows for more political freedom. Even so, more than a million residents, many of them quite wealthy, left before China took over in 1997, worried that their fortunes and freedoms would diminish. China has allowed more political freedom in Hong Kong, in part because of the island's important role as a link to global trade. However, interference from Beijing in Hong Kong's local elections has led to clashes in the streets between protesters and the police. At least for now, the citizens of Hong Kong are allowed to voice their opinions unlike in the rest of China.

SHANGHAI'S LATEST TRANSFORMATION

Shanghai has historically been a trendsetting city. Its opening to Western trade in the early nineteenth century spawned a period

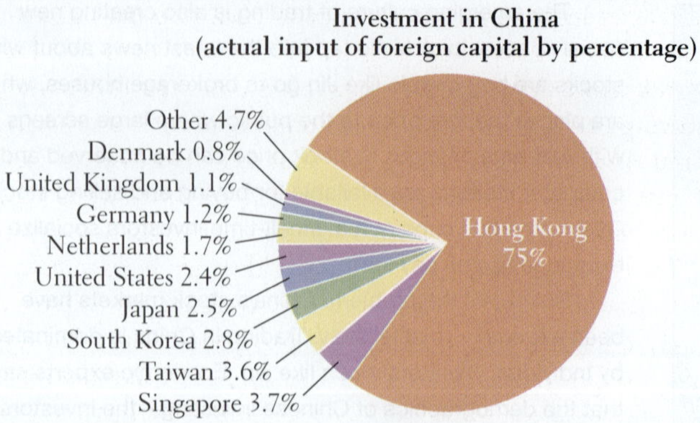

Figure 9.23 Net foreign direct investment in China, 2017. Foreign investment in China reached $136 billion in 2017. Note that investment from Hong Kong, even though it is part of China, is considered "foreign" here. Many investments that actually originate in various countries across the globe are funneled through Hong Kong banks, as was the case when Hong Kong was a British territory. [Data from: "China: Foreign Investment," at https://en.portal.santandertrade.com/establish-overseas/china/foreign-investment.]

of phenomenal economic growth and cultural development that led to it being called the "Paris of the East." As a result of China's reentry into the global economy, Shanghai is undergoing another boom, which has enriched some people and dislocated others (see Figure 9.20D). The Shanghai urban area is the largest in China, with 24 million residents. On average, these residents of Shanghai are now as affluent as people in Central and Southern European countries such as the Czech Republic and Portugal. Shanghai today has the world's single busiest cargo port, and the region around the city is now responsible for as much as a quarter of China's GDP.

In less than a decade, the city's urban landscape has been remade by the construction of more than a thousand business and residential skyscrapers, subway lines and stations, highway overpasses, bridges, and tunnels. Shanghai's shopping district on Nanjing Road is as imposing as any such district around the world. For hundreds of miles into the countryside, suburban development linked to Shanghai's economic boom is gobbling up farmland and displacing farmers.

Pudong, the city's new financial center, sits across the Huangpu River from the Bund—Shanghai's famous, elegant row of big colonial-era brownstone buildings that served as the financial capital of China until half a century ago. Previously, Pudong was a maze of dirt paths and neighborhoods of simple, tile-roofed houses, but its former residents were pushed out to make way for soaring high-rises (see Figure 9.20D). New construction is restricted in the historic section of the Bund, which is another reason why Pudong's central location has been attractive as a new business center. Pudong is home to Shanghai's stock exchange and the tallest skyscrapers in China. Buildings like the Shanghai Tower and the Oriental Pearl Tower have become a visual symbol for the economic vitality of Pudong, the city of Shanghai, and all of China.

CHECK YOUR UNDERSTANDING

1. How has China's focus on export-oriented manufacturing affected rural areas?
2. How are the two concepts of *hukou* and floating population interrelated?
3. What role did SEZs play in China's economic development?
4. What milestone in its economic structure did China reach in 2013?
5. What special role has Hong Kong played in China's transformation?
6. What sets Shanghai apart among China's major cities?

POPULATION, GENDER, AND CULTURE

9.10 Explain the lingering effects of the one-child policy on population and gender in China.

9.11 Describe how Japan is dealing with the problem of an aging population.

9.12 Analyze the status of ethnic minorities throughout East Asia.

From a global perspective, East Asia's rate of natural increase is low, 0.4 percent. This rate is very similar to that of North America (0.3 percent) but higher than in Europe (–0.1 percent). In China, this low population growth is partially due to the legacy of government policies that harshly penalized families for having more than one child. But in China as well as elsewhere, urbanization and changing gender roles are also resulting in smaller families, regardless of official policy. Only in Mongolia (with 3 million people) are women still averaging over two children each, but even there family size is shrinking.

RESPONDING TO AN AGING POPULATION

Low birth rates mean that fewer young people are being added to the population, and improved living conditions mean that people are living longer across East Asia. The overall effect is that the average age of the populations is rising. Put another way, populations are aging. For Mongolia and North Korea, it will be several decades before the financial and social costs of supporting numerous elderly people will have to be addressed. China will eventually face serious future problems with elder care because the one-child policy and urbanization have so drastically reduced family kin-groups. Japan, on the other hand, has already been dealing with the problem of having a large elderly population that requires support and a reduced number of young people to do the job.

Japan's Options

Japan's population has stopped growing altogether and the country is aging rapidly, raising concerns about economic productivity and humane ways to care for dependent people. The demographic transition (see Chapter 1) is in its late stages in this highly developed country, where 92 percent of the population lives in cities. Japan has a negative rate of natural increase (–0.3), the lowest in East Asia and the world, except for a few countries in Central Europe. If this trend continues, Japan's population is projected to plummet from the current 126 million to 111 million by 2040.

At the same time, the Japanese have the world's longest life expectancy, at 84 years. As a result, Japan also has the world's oldest population, 28 percent of which is over the age of 65. By 2050, the average age in Japan is projected to be 53, which can be compared with 41 years, the projection for the United States. Retired people in Japan may then account for 40 percent of the population, and Japan's labor pool could be reduced by more than a third, but it would still need to produce enough to take care of more than twice as many retirees as it does now. Clearly, these demographic changes will have a momentous effect on Japan's economy, and the search for solutions is underway. One possibility is increasing immigration to bring in younger workers who will fill jobs and contribute to the tax rolls, as the United States and Europe have done.

In Japan, recruiting immigrant workers from other countries is a very unpopular solution to the aging crisis. Many Japanese people object to the presence of foreigners, and the few small minority populations with cultural connections to China or Korea have for years faced discrimination. Nonetheless, foreign workers are dribbling into Japan in a multitude of legal and illegal ways. Many are so-called guest workers from South Asia, brought in to fill the most dangerous and lowest-paying jobs with the understanding that they will eventually leave. Others are the descendants of Japanese people who once migrated to South America (Brazil and Peru). Regardless, Japan's foreign population remains tiny, making up only 1.8 percent (2.3 million people) of the total population. A recent UN report estimates that Japan would have to admit more than 640,000 immigrants per year just to maintain its present workforce and avoid a 6.7 percent annual drop in its GDP.

In a novel approach to Japan's demographic changes, government and industry have invested enormous sums of money in robotics over the past decade. Robots are already widespread in Japanese industries such as auto manufacturing, and their industrial use is growing. Now they are also being developed to care for the elderly, guide patients through hospitals, look after children, staff hotels, and even make sushi. By 2025, the government plans to replace up to 15 percent of Japan's workforce with robots (**Figure 9.24**).

China's Options

The proportion of China's population over 65 is now only 10 percent, but this will change rapidly as conditions improve and life expectancies increase by 5 to 10 years to become more like those of China's affluent East Asian neighbors. China's current population pyramid shows that there are many people in their fifties who will be retiring in the not-too-distant future (**Figure 9.25A**). However, a crisis in elder care is already upon China for two other reasons: the high rate of rural-to-urban migration and the shrinkage of family support systems because of the one-child policy (discussed below).

The lingering effects of the one-child policy have transformed Chinese families and Chinese society as a whole. For example, an only child has no siblings, so within two generations, the kinship categories of brother, sister, cousin, aunt, and uncle have disappeared from most families, meaning that any individual has very few, if any, related age peers with whom to share family responsibilities.

Figure 9.24 A humanoid robot serves retail customers in Japan. As the population is aging, which leads to fewer people in the workforce, Japan experiments with innovative ways of solving this vexing problem. [Rodrigo Reyes Marin/Nippon News/AFLO/Alamy Live News]

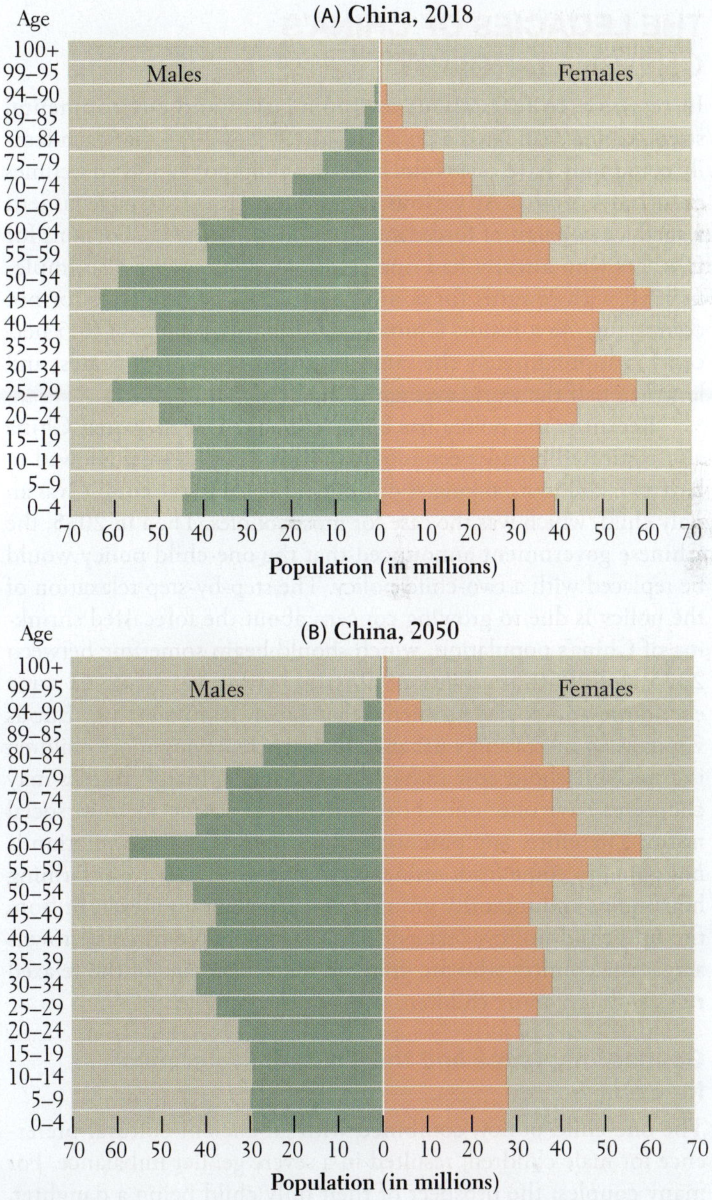

Figure 9.25 Population pyramids for China, 2018 and 2050 (projected). [Data from: U.S. Census Bureau, Population Division, International Data Base, 2018, at https://www.census.gov/data-tools/demo/idb/informationGateway.php.]

Most children are doted on by several adults, and children are not taught by siblings to share. Conscious efforts must be made to instill self-sufficiency in only children.

When hundreds of millions of young Chinese people were lured into cities to work, rural areas benefitted from remittances. However, the one-child family also meant that for every migrant, two aging parents have been left to fend for themselves, often in rural, underdeveloped areas. China's parliament passed a law that requires family members to visit and support their elderly relatives. While such a law may be unenforceable, it shows how worried the government is about the social consequences of the prospect of an aging society, including the increasing spatial mismatch between the elderly who need care and their younger relatives who often have moved elsewhere in search of employment.

THE LEGACIES OF CHINA'S ONE-CHILD POLICY

In response to fears about overpopulation and environmental stress, China mandated a policy from 1979 to 2013 that each family could only have one child. The one-child policy was sometimes enforced brutally. At various times and places, the government waged a campaign of forced sterilizations and forced abortions for mothers who already had one child. The policy was also implemented with rewards for complying and with large fines for not complying. As a result, China's rate of natural increase (0.5 percent) is approximately the same as that of the United States, and less than half the world average (1.2 percent).

Recently, the policy has been relaxed. In 2013, when China also instituted broader economic reforms, couples were allowed to have two children, so long as one of the people in the couple was an only child, which was the case for most couples. Then in 2015, the Chinese government announced that the one-child policy would be replaced with a two-child policy. The step-by-step relaxation of the policy is due to growing concern about the forecasted shrinking of China's population, which should begin sometime between 2025 and 2050 (see Figure 9.25B), creating many of the same economic and social problems that Japan is now facing. But as China is an increasingly urban society where having more children is a net household cost rather than a benefit, many couples may choose to have only one child, which has become the new social norm. Therefore, any potential minor boost in the population is not going to come from residents of large cities, nor from farming households and minority groups who have long been exempt from the one-child policy; instead, it could be people in small towns across the country who are going to take advantage of the relaxed rules and have more children.

Gender Imbalance and the Cultural Preference for Sons

The one-child policy, combined with an ancient cultural preference for male children, resulted in a severe gender imbalance. For many couples, the prospect of their only child being a daughter, without the possibility of having a son in the future, was devastating. The preference for sons relates to patriarchal Confucian values that have prevailed throughout this region for millennia (see "Confucian Gender Roles"). In fact, gender imbalance has emerged in the Koreas and Taiwan even without the one-child policy, which suggests that the influence of Confucian values is strong.

China's more severe gender imbalance is illustrated by its population pyramid (see Figure 9.25A). The average global sex ratio at birth is 107 boys to 100 girls, which evens out to 101 boys to 100 girls by the age of 5. In China, however, there are 116 boys for every 100 girls born. In fact, for nearly every category until age 60, males outnumber females. The census data show that there are already 34 million more men than women.

What happened to the missing girls? There are several possible answers. Given the preference for male children, the births of these girls may simply have gone unreported by families hoping to conceal their daughters as they tried to conceive a male child. There are many anecdotes of girls being raised secretly or even disguised as boys. Also, adoption records indicate that girls are given up for international adoption much more often than boys are. Or the girls may have died in early infancy, either through neglect or infanticide. Finally, some parents conduct medical tests that can identify the sex of a fetus. There is evidence that in China, as elsewhere around the world, many parents choose to abort female fetuses.

There is some evidence that attitudes are changing (**Figure 9.26**). For example, in Japan, South Korea, and Mongolia, the percentage of women receiving secondary education equals or exceeds that of men. In Japan, there is a slight gender imbalance *in favor of females*. In China, social policy makers have tried for decades to eliminate the old Confucian bias toward men by empowering women economically and socially. In some ways, they are succeeding, as Chinese women now participate in the workforce to a large degree. Nevertheless, the preference for sons persists at a weakened level.

A Shortage of Brides

A major side effect of the preference for sons is that there is now a growing shortage of women of marriageable age throughout East Asia. In 2018, China alone had an estimated deficit of 12 million women ages 20 to 35. Females are also effectively "missing" from the marriage rolls because many educated young women are too busy with career success to meet eligible young men.

Research suggests that at least 10 percent of young Chinese men will fail to find a mate; and poor, rural, uneducated men will have the most difficulty. The shortage of women will lower the birth rate yet further, which will contribute to the expected shrinkage of the population over the next century. Without spouses, children, or even siblings, there will be no one to care for single men when they age. Furthermore, China's growing millions of single young men are emerging as a potential threat to civil order, as they may be more prone to drug abuse, violent crime, HIV infection, and sex crimes. Cases of kidnapping and forced prostitution of young girls and women are reported.

POPULATION DISTRIBUTION

In East Asia, people are not evenly distributed across the land (**Figure 9.27**). China, with 1.4 billion people, has almost one-fifth of the world's population. However, 90 percent of these people are clustered on the approximately one-sixth of the total land area that is suitable for agriculture and commerce, and over half of these live in urban areas. The population is concentrated especially densely

PHOTO ESSAY

9.26 Population, Gender, and Culture in East Asia

The map shows how well countries are ensuring gender equality. The Gender Development Index (GDI) is based on the same variables as HDI—longevity, education, and income—but adjusted for gender differences. A high rank indicates that the genders are tending toward equality. Here, the differences among the countries don't align directly with the level of economic development. All countries rank from medium to very high. Japan does well and is in the very high GDI category. Affluent South Korea, however, only ranks in the medium category. Interestingly, Mongolia has a very high ranking, which may be a result of the emphasis on gender equity under communism and traditional cultural attitudes that promote some measures of gender equality.

THINKING GEOGRAPHICALLY

A **Why did the Chinese government recently rescind the one-child policy?**

B **How have places like Singapore benefited from the historic migration of Chinese?**

C **What are the social and economic effects of aging in Japan?**

A A young Chinese couple with their one child at the entrance to the Forbidden City at Tiananmen Square, Beijing, China. Even if they are now allowed to have a second child, they may choose not to do so. [Chederros/Nomad/AGE Fotostock]

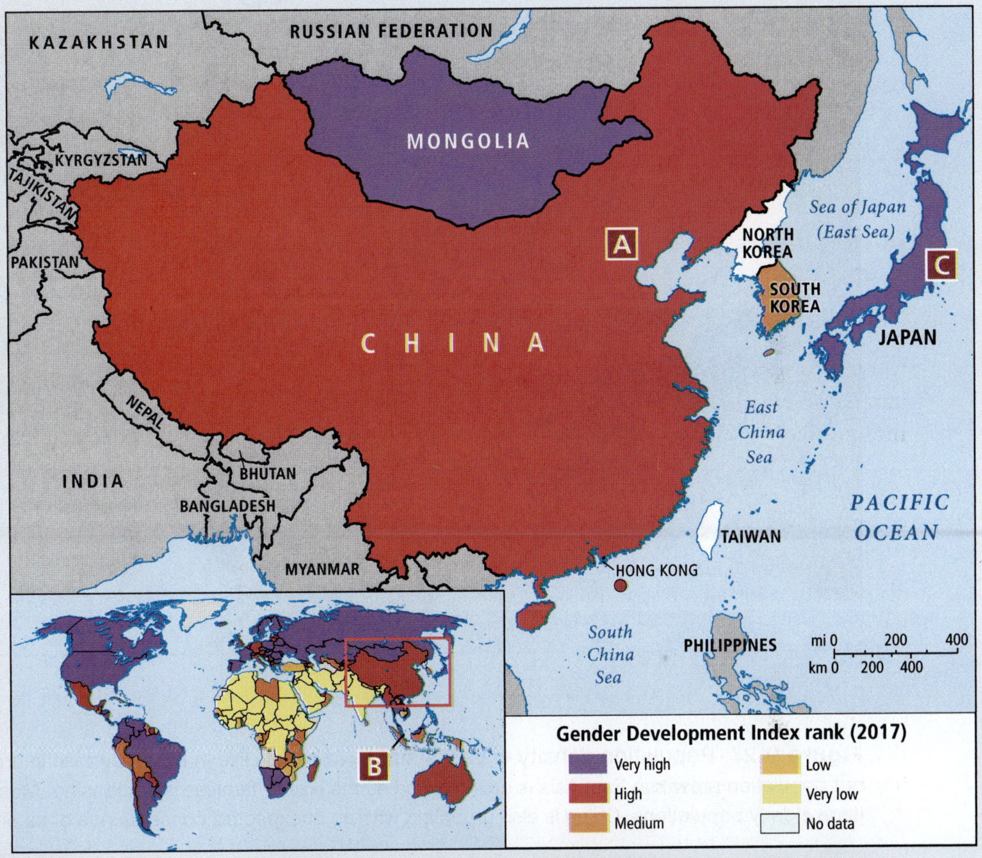

B A streetscape in the Chinatown neighborhood in Singapore. Chinatowns were established by the Overseas Chinese all over the world and they continue to be centers of Chinese cultural and economic life. [mlphotobc/iStock/Getty Images]

C Elderly people in Tokyo, Japan, exercise at the grounds of a temple during Japan's Respect-for-the-Aged Day. [YOSHIKAZU TSUNO/AFP/Getty Images]

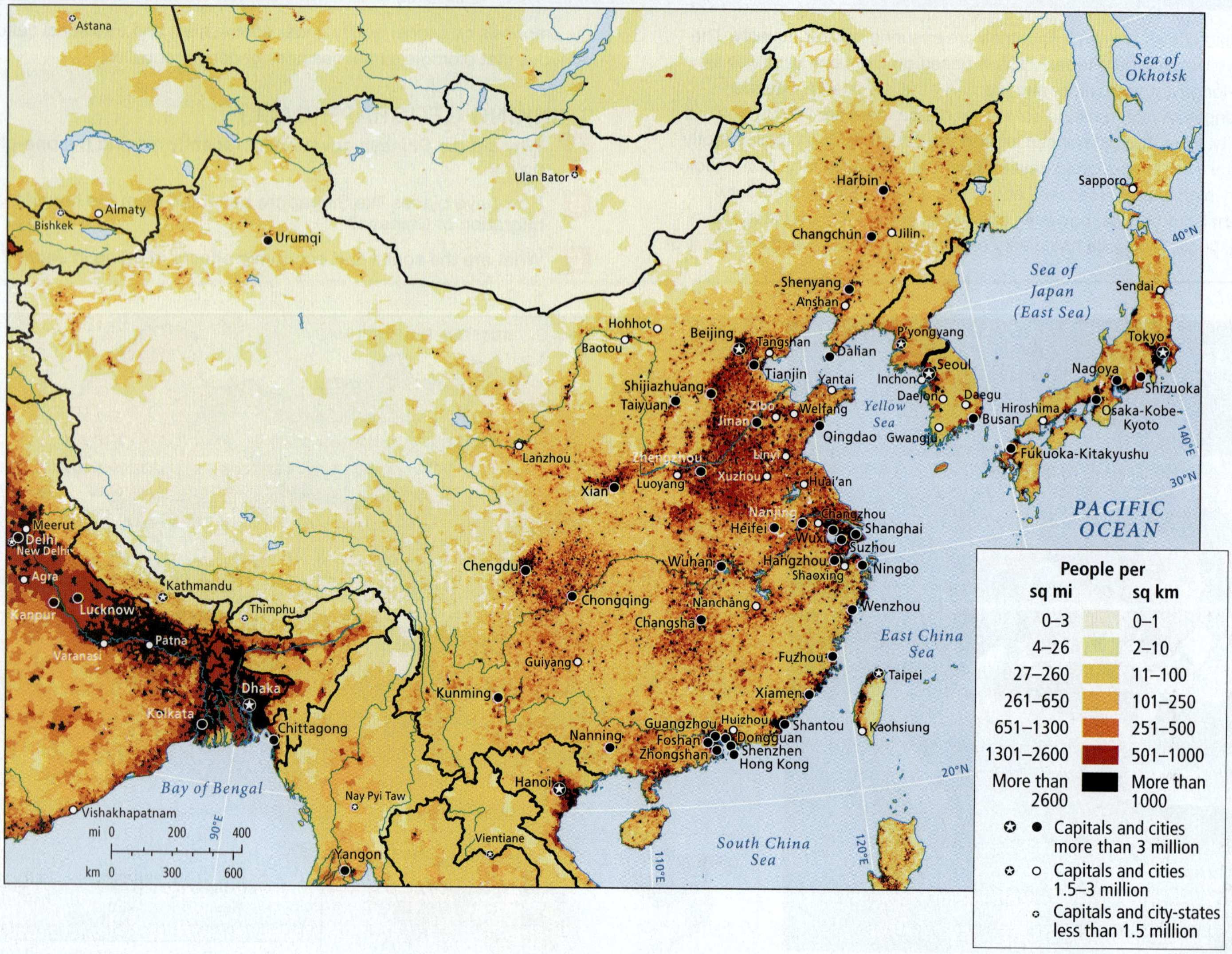

Figure 9.27 Population density in East Asia. More people live in East Asia than in any other world region except for South Asia, but population growth in East Asia is now slowing as the size of families is decreasing. Most of the region now faces the challenge of caring for large elderly populations. China is also grappling with an unexpected consequence of its one-child policy: a shortage of women.

in eastern China, including the North China Plain stretching from Beijing down to Shanghai, the coastal zone from Shanghai to Hong Kong that includes the delta of the Zhu Jiang (Pearl River) in the southeast, the Sichuan Basin, and the middle Chang Jiang (Yangtze) basin.

The west and south of the Korean Peninsula are also densely settled, as are northern and western Taiwan. In Japan, settlement is concentrated in a band that stretches from the cities of Tokyo and Yokohama on the island of Honshu facing the Pacific, south through the coastal zones of the Inland Sea to the islands of Shikoku and Kyushu. This urbanized region is one of the most extensive and heavily populated metropolitan zones in the world, accommodating 86 percent of Japan's total population. The rest of Japan is mountainous and more lightly settled. Mongolia is only lightly settled, with one modest urban area.

CHECK YOUR UNDERSTANDING

1. What are the factors behind gender imbalance—more men than women in many age categories—in China?
2. What problems have been created by the resulting shortage of women in China?
3. How is the large population of East Asia unevenly distributed across space?
4. What are the consequences of aging in Japan?
5. What are the consequences of aging in China?

SOCIOCULTURAL ISSUES

Most countries in East Asia have one dominant ethnic group, but all countries have considerable cultural diversity. In China, for example, 93 percent of Chinese citizens call themselves *people of the Han*. The name harks back about 2000 years to the Han empire but it gained currency only in the early twentieth century, when nationalist leaders were trying to create a mass Chinese identity. Figure 9.13 shows how the Han empire extended to territories that are now populated by ethnically Chinese people. The term *Han* simply connotes people who share a general way of life and pride in Chinese culture. The main language spoken by the Han is Mandarin, which is the type of Chinese spoken in Beijing and much

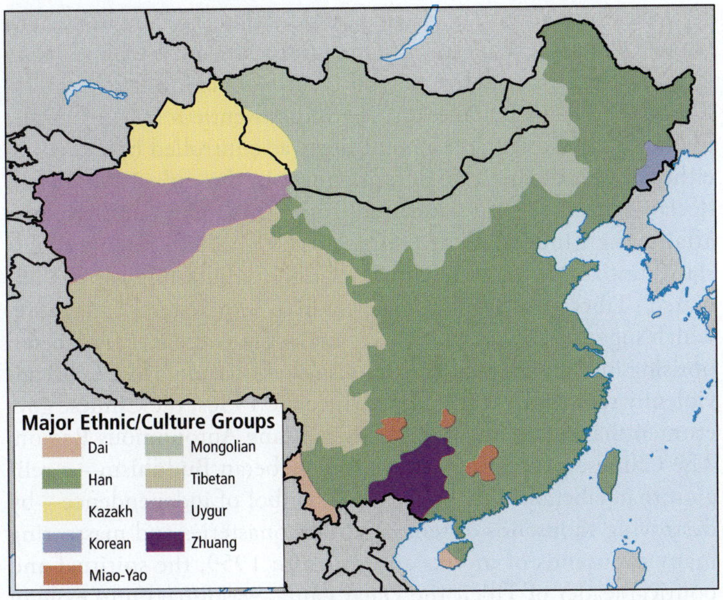

Figure 9.28 Major ethnic groups of China. This map shows the areas traditionally occupied by the Han and by ethnic minorities. It does not show the recent resettling of Han in Xinjiang and Tibet, nor does it show the Hui, people from many ethnic groups whose ancestors converted to Islam and who are found in disparate locations along the old Silk Road and in southeastern China. [Research from: Chiao-min Hsieh and Jean Kan Hsieh, *China: A Provincial Atlas* (New York: Macmillan, 1995), p. 12. The website http://www.index-china.com/minority/minorityenglish.htm includes a comprehensive survey of minorities in China.]

of the north. There are many other Chinese dialects, especially in southern China.

China's non-Han minorities number about 120 million people in more than 55 different ethnic or culture groups scattered across the country. Most live outside the Han heartland of eastern China (**Figure 9.28**). Some of these areas have been designated *autonomous regions,* where minorities theoretically manage their own affairs. In practice, however, the Communist Party in Beijing controls these regions, especially those considered to have potentially rebellious populations and those that have resources of economic value. We profile a few of China's ethnic groups here.

Western China's Muslims

Muslims of various ethnic origins have long been prominent minorities in China. All are originally of Central Asian origin and most are Turkic people who historically have been nomadic herders. Others specialized in trading. They tend to be concentrated in China's northwest and often think of themselves as quite separate from mainstream China.

Uygurs (pronounced WEE-gurs) and Kazakhs, who are Turkic-speaking Muslims, live in the autonomous region of Xinjiang in the far northwest (see Figure 9.1). Though relatively few Uygurs or Kazakhs remain nomadic herders today, contact between them and similar peoples in Central Asia has been revived since China's market reforms began and the Soviet Union dissolved (**Figure 9.29**).

The Beijing government claims this area's oil, other mineral resources, and irrigable agricultural land for national development. Accordingly, it has sent troops and many millions of Han settlers to Xinjiang through what is known as the *Go West* policy. The Han settlers fill most managerial jobs in government administration, the military, and the mineral extraction and power generation industries. An important secondary role of the Han is to dilute the power of Uygurs and Kazakhs within their own lands. In Xinjiang, there are now almost as many Han as Uygurs (9 versus 10 million), plus small numbers of other minorities.

The Beijing government has rushed to develop Xinjiang and its capital Urumqi by establishing economic development zones in the area, but the Uygurs have been left out of most policy-making roles and indeed have been excluded from participating in the economic boom. One young Uygur man in Urumqi writes of his discontent: "I am a strong man, and well-educated. But [Han] Chinese firms won't give me a job. Yet go down to the railroad station and you can see all the [Han] Chinese who've just arrived. They'll get jobs. It's a policy to swamp us."

Until recently, the Uygur people of Xinjiang expressed their resistance to Han dominance merely by reinvigorating their Islamic culture. Islamic prayers were increasingly heard publicly, more Muslim women were wearing Islamic dress, Uygur was spoken rather than Chinese, and Islamic architectural traditions were being

Figure 9.29 The Silk Road market in Kashi. Kashi (also known as Kashgar), a city in China's Xinjiang Uygur Autonomous Region, has been an important trading center along the Silk Road for at least 2000 years. The market is located in the old city of Kashi, which is crisscrossed with walled courtyards and narrow alleys. Currently, it is being demolished or modernized, depending on one's perspective. Many Uygurs view this as a form of cultural genocide orchestrated by the Chinese government. [Paul Springett D/Alamy Stock Photo]

revived. Since 2009, more active resistance groups have formed in the wake of violence erupting in the streets of Urumqi between Uygurs and the Han. Uygur separatists have attacked Chinese targets, while Beijing accuses all Uygur activists of being Islamic fundamentalists bent on terrorism. Most recently, the government has cracked down heavily on Uygurs, whether they are political activists or not. A 2018 UN report suggests that 1 million people in Xinjiang have been sent to "reeducation camps" where they are to be instilled with a sense of allegiance to the Chinese state.

Distinct from the Uygurs and Kazakhs are the Hui, who number about 11 million. The original Hui people were descended from ancient Turkic Muslim traders who traveled the Silk Road from Europe across Central Asia to Kashgar (now Kashi) and on to Xian in the east (see the map of the Silk Road in Figure 5.10; see also Figure 9.29). Subgroups of Hui live in the Ningxia Huizu Autonomous Region and throughout northern, western, and southwestern China. There they continue as traders, farmers, and artisans. The tradition of commercial activity and adoption of the Chinese language among the Hui have facilitated their success in China. Many are active in the market economies of southeastern China as businesspeople, technicians, and financial managers, using their money not just to buy luxury goods but also to revive religious instruction and to fund their mosques, which are now more obvious in the landscape.

The Tibetans

The Tibetans, in contrast to the prosperous and assimilated Hui, are an impoverished ethnic minority group of 5 million people scattered thinly over a huge, high, mountainous region in western China. The history of Tibet's political status vis-à-vis China is long and complex, characterized by both cordial relations and conflict. During China's imperial era (prior to the twentieth century), Tibet maintained its own government but was controlled by China, as either a subordinate state or a province. In the early 1900s, Tibet declared itself separate and free from China and conducted its affairs as an independent country. It was able to maintain this status until 1949–1950, when the Chinese communist army "liberated" Tibet, promising a "one country, two systems" structure, which suggested a great deal of regional self-governance. A Tibetan uprising in 1959 led China to abolish the Tibetan government and violently reorder Tibetan society. Since the 1950s, the Chinese government has referred to Tibet as the Xizang Autonomous Region. The Chinese government suppressed Tibetan Buddhism—a religion that Tibetans rally behind as a symbol of independence—by destroying thousands of temples and monasteries and massacring many thousands of monks and nuns. In 1959, the spiritual and political leader of Tibet, the Dalai Lama, was forced into exile in India along with thousands of his followers, where he still resides.

Since the 1990s, the Beijing government's strategy has been to overwhelm the Tibetans with secular social and economic modernization and with Han Chinese settlers rather than outright military force (though China maintains a military presence in Tibet). To attract trade and quell foreign criticism, China now spends hundreds of millions of dollars on housing and on roads, railroads, and a tourism infrastructure that capitalizes on European and American interest in Tibetan culture. China presents its actions in Tibet as part of its overall strategy to integrate the entire country economically and socially. Schools are being built and jobs opened up to young Tibetans. A new railway link effectively connecting Tibet to the rest of China was completed in 2006 (**Figure 9.30**). The Han in Tibet see the railway as a public service that will promote Tibetan development, but Tibetan activists see it as a conveyor belt for more Han dominance over the Tibetan economy and culture.

Within Tibetan culture, women have held a somewhat higher position than in other cultures of East Asia. Among nomadic herders, women could have

Figure 9.30 The Beijing–Tibet Railway. The western section of this railroad opened in 2006. It is the world's highest railway, with nearly 600 miles (1000 kilometers) of line that are at an altitude above 13,000 feet (4000 meters). The trip from Beijing to Lhasa now takes 41 hours, and it costs much less to transport goods and people via the railway than it does using the existing highway. However, many Tibetans worry that the railway will bring more Han migration to Tibet.

A A view of the train as it passes through the mountains of the Plateau of Tibet. [View Stock/Getty Images]

more than one husband, just as men were free to have more than one wife. At marriage, a husband often joins the wife's family. By comparison, Han Chinese culture has typically regarded the women of Tibet and other western minorities as barbaric precisely because their roles were not circumscribed. They rode horses, they worked alongside the men in herding and agriculture, and they were generally more assertive in daily life than their Chinese counterparts.

Indigenous Diversity in Southern China

In Yunnan Province in southern China, more than 20 groups of ancient native peoples live in remote areas of the deeply folded mountains that stretch into Southeast Asia. These groups speak many different languages, and many have cultural and language connections to the indigenous people of Tibet, Myanmar, Thailand, or Cambodia. Some people follow animist religions based on the worship of natural features and ancestral connections; others are Buddhists. Gender relations are different here than among the Han. A crucial difference may be that among several groups, most notably the Dai, a husband moves in with his wife's family at marriage and provides her family with labor or income. A husband inherits from his wife's family rather than from his birth family.

Taiwan's Many Minorities

In Taiwan and the adjacent islands, the Han account for 95 percent of the population, but Taiwan is also home to 60 indigenous minorities. Some have cultural characteristics—languages, crafts, and agricultural and hunting customs—that indicate a strong connection to ancient cultures in Southeast Asia and the Pacific. The mountain dwellers among these groups have resisted assimilation more than the plains peoples. Both groups may live on tribal land set aside for indigenous minorities if they choose, but most are now being absorbed into mainstream, urbanized, Han-influenced Taiwanese life. The Han are themselves not homogenous in Taiwan. They are divided into subgroups based on their ancestral home in China, and the people (and their descendants) who fled to Taiwan around the time of the establishment of the People's Republic of China in 1949 are often called "mainlanders."

The Ainu in Japan

There are several indigenous minorities in Japan and most have suffered considerable discrimination. A small and distinctive minority group is the **Ainu**, characterized by their light skin, heavy beards, and thick, wavy hair (**Figure 9.31**). The Ainu are a racially and culturally distinct group thought to have migrated many thousands of years ago from the northern Asian steppes. They once occupied Hokkaido and northern Honshu and lived by hunting, fishing, and some cultivation, but they are now being displaced by forestry and other development activities. Few full-blooded Ainu remain because, despite prejudice, they have been steadily assimilated into the mainstream Japanese population. The Ainu officially only number about 25,000, but there most likely are many more who are part Ainu or don't identify themselves as Ainu in the Japanese census due to perceived discrimination. Somewhat belatedly, the Japanese parliament officially recognized the Ainu as an indigenous minority in 2008.

Figure 9.31 The Ainu of Japan. In 2008, Japan's government formally recognized the Ainu as an indigenous people of Japan, more than 500 years after the Ainu began to be marginalized by mainstream Japanese culture. In the picture, a group of Ainu engages in a blessing of food ceremony during a worldwide summit of indigenous people held in Sapporo, Japan. [Franck Robichon/EPA/Shutterstock.com]

East Asia's Most Influential Cultural Export: The Overseas Chinese

China has had an impact on the rest of the world not only through its global trade, but also through the migration of its people to nearly all corners of the world. The first recorded emigration by the Chinese took place more than 2200 years ago. Following that, China's contacts then spread eastward to Korea and Japan, westward into Central and Southwest Asia via the Silk Road, and by the fifteenth century, to Southeast Asia, coastal India, the Arabian Peninsula, and even Africa.

Trade was probably the first impetus for Chinese emigration. The early merchants, artisans, sailors, and laborers came mainly from China's southeastern coastal provinces. Taking their families with them, some settled permanently on the peninsulas and islands of what are now Indonesia, Thailand, Malaysia, and the Philippines. Today they form a prosperous urban commercial class known as the **Overseas Chinese**. The quintessential Overseas Chinese state in Southeast Asia is Singapore, where 75 percent of the population is ethnically Chinese (Figure 9.26B; see also Chapter 10).

In the nineteenth century, economic hardship in China and a growing international demand for labor spawned the migration of as many as 10 million Chinese people to countries all over the world. By the middle of the twentieth century, many others fleeing the repression of China's Communist Revolution joined those from earlier migration. As a result, "Chinatowns" are found in cities all over the world. The term *Overseas Chinese* has been extended to apply to Chinese emigrants and their descendants in all such locations.

Ainu an indigenous minority group in Japan characterized by their light skin and thick, wavy hair, who are thought to have migrated from the northern Asian steppes

Overseas Chinese Chinese emigrants and their descendants, especially those in Southeast Asia

CHECK YOUR UNDERSTANDING

1. Who are the largest minority groups in East Asia and where are they located?
2. Why is there resentment and sometimes open resistance to Han Chinese domination in the various minority homelands?
3. Who are the Overseas Chinese and where do they live?

SUBREGIONS OF EAST ASIA

9.13 Identify key characteristics of the subregions of East Asia.

Because China is so large and diverse, we consider its four major subregions—northeastern, central, southern, and far north and west China (**Figure 9.32**). The other countries have different cultural traditions, both ancient and modern, and each makes up a separate subregion of East Asia.

loess windblown material that forms deep, fertile, but easily erodible soils
alluvium river-borne sediment

CHINA'S NORTHEAST

China's northeast consists of the Loess Plateau, the North China Plain, and the Far Northeast subregion (**Figure 9.33**). The Loess Plateau and the North China Plain are the ancient heartland of China. By the eighth century, the city of Chang'an (now where the modern city of Xian is located) had 2 million inhabitants in the urban region and may have been the largest city in the world. After 900 C.E., the center of Chinese civilization shifted to the North China Plain, but the Loess Plateau remained a crucial part of China. Xian served as the eastern terminus of the Silk Road that connected China with Central Asia and Europe (see Figure 5.10).

The Loess Plateau (see Figure 9.33) and the North China Plain are linked physically as well as culturally. Both are covered by fine yellowish **loess**, or windblown soil particles. Loess soils are typically quite fertile but erode easily. For thousands of years, dust storms have picked up loess from the surface of the Gobi and other deserts to the west and carried it east. Over time, the windblown loess has filled up deep mountain valleys in Shanxi and Shaanxi provinces, creating an undulating plateau. The Huang He, in a second massive earth-moving process, transports millions of tons of **alluvium** (river-borne sediment) from the Loess Plateau to the coastal plain each year. Over the millennia, this river has created the North China Plain by depositing its heavy load of sediment in what was once a much larger Bo Hai Sea.

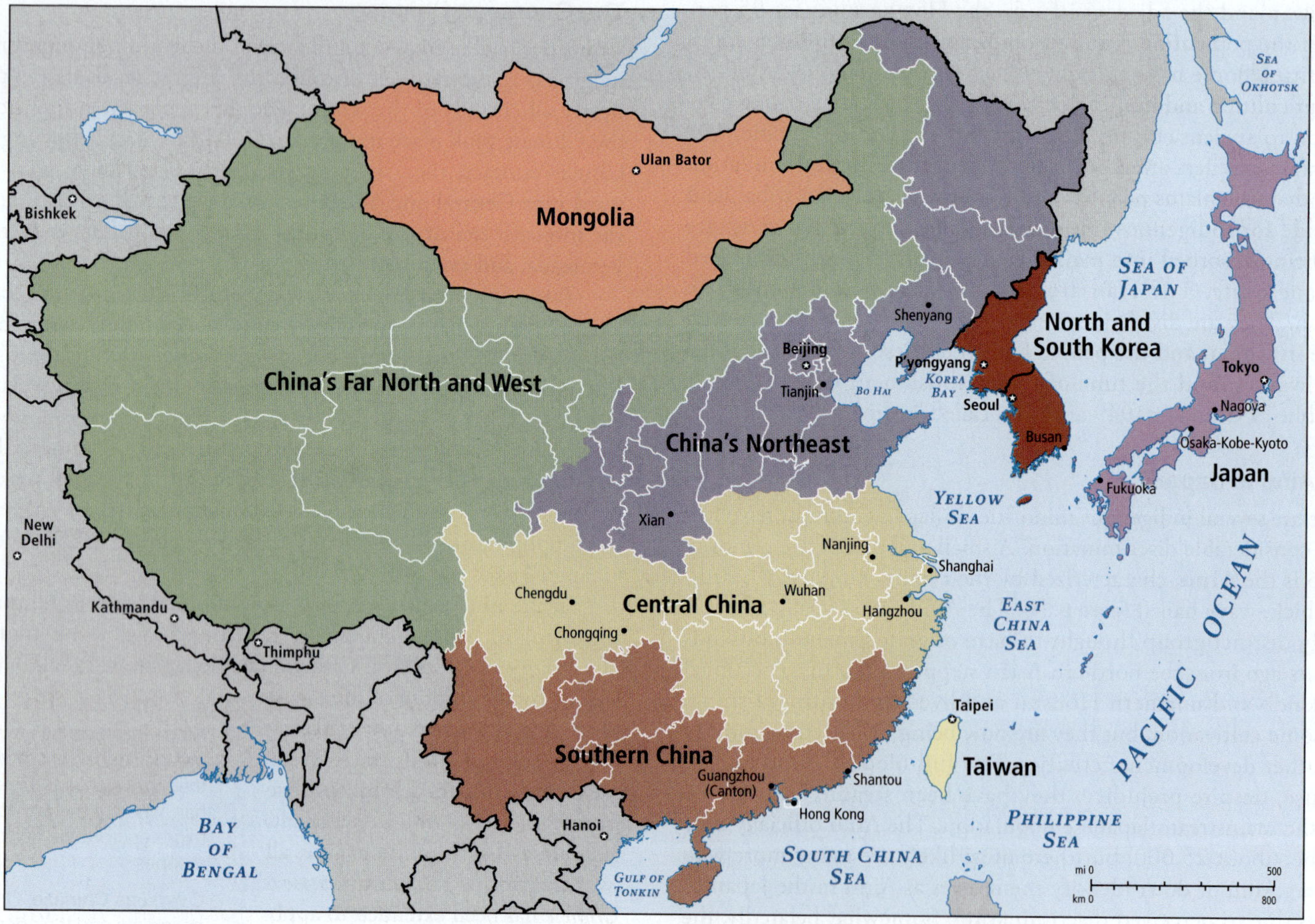

Figure 9.32 Subregions of East Asia. This map shows the four subregions of China as well as the other four subregions of East Asia: Taiwan, Japan, South and North Korea, and Mongolia.

Figure 9.33 China's Northeast subregion.

The Loess Plateau

An unusually diverse mixture of peoples from all over Central Asia and East Asia have found the Loess Plateau a fertile, though challenging, place to farm and herd. Among the many cultural groups that share this densely inhabited and productive plateau are the Hui and the Han of China and, particularly in the western reaches, Mongols, Tibetans, Uygurs, and Kazakhs. Many of the plateau inhabitants farm cotton and millet—a tall-grass grain grown in many parts of Asia and Africa—in irrigated valley bottoms, or raise sheep on the drier grassy uplands. China's largest coal reserves are also found here.

Thousands of years of human occupation have stripped the plateau of its ancient forests, leaving only grasses to anchor the loess. Because the loess is so thick—hundreds of feet deep in many areas—and the winds persistent, some people have carved residences and animal sheds below the surface of the plateau (**Figure 9.34**). Torrential rains cut deep gullies into the loess, and such gullies now cover much of the landscape, causing people to abandon the agricultural terraces. For decades, the government has maintained a revegetation campaign to stabilize slopes (see the vignette in "Water Shortages").

Figure 9.34 "Cave dwellings" of the Loess Plateau, China. In parts of Shanxi and Shaanxi provinces, the loess is so thick and firm that energy-efficient, cave-like houses can be dug out from it. First, a courtyard about 10 meters deep is dug out. Then rooms are excavated off the main courtyard. [Best View Stock/Getty Images]

The North China Plain

The North China Plain is the largest and most populous expanse of flat, arable land in China. Since the sixth century, it has been home to most of the imperial families, and from this region the Han Chinese have dominated the country.

Today the North China Plain is one of the most densely populated spaces in the country: 214 million people live in a space about the size of France. The majority are farmers who work the generally small fields that cover the plain. Most of them produce wheat, which grows well in the relatively dry climate. The forests that once blanketed the plain were cut down thousands of years ago, and now the only trees that survive are those surrounding temples and those planted as windbreaks.

The Huang He, the river that created the North China Plain, remains the plain's most important physical feature, serving as a major transportation artery and a source of irrigation water, but also producing disastrous floods. After the river descends from the Loess Plateau, its speed slows dramatically on the flat surface of the plain, and its load of sediment begins to settle out. This sediment raises the level of the riverbed year after year until, in some places, it lies higher than the plain. Occasionally, during the spring surge, the river breaks out of the levees that have been built to control it and rushes across the surrounding densely occupied plain. The spring surge has forced the Huang He to cut many new channels over time (**Figure 9.35**), and it also endows the plain with a new layer of fertile soil as much as a yard (a meter) thick each year. Because it both enriches and destroys, the river is referred to as both the Mother of China and China's Sorrow. The Huang He is also connected to the watershed of the Chang Jiang to the south by a canal. The Da Yunhe (Grand Canal), considered the oldest and longest canal in the world, was completed about 610 C.E. and originally was built to carry troops and supplies north. Later, in the eighteenth century, it was used to carry taxes (in the form of grains) to the capital, Beijing.

China's Far Northeast

The northeastern corner of China inland of the Korean peninsula was for centuries considered a peripheral region, partly because of its location and partly because of its harsh climate. The winters are long and bitterly cold, the summers short and warm. The growing season is less than 120 days. Nonetheless, China's Far Northeast subregion found an important niche in the country's economy

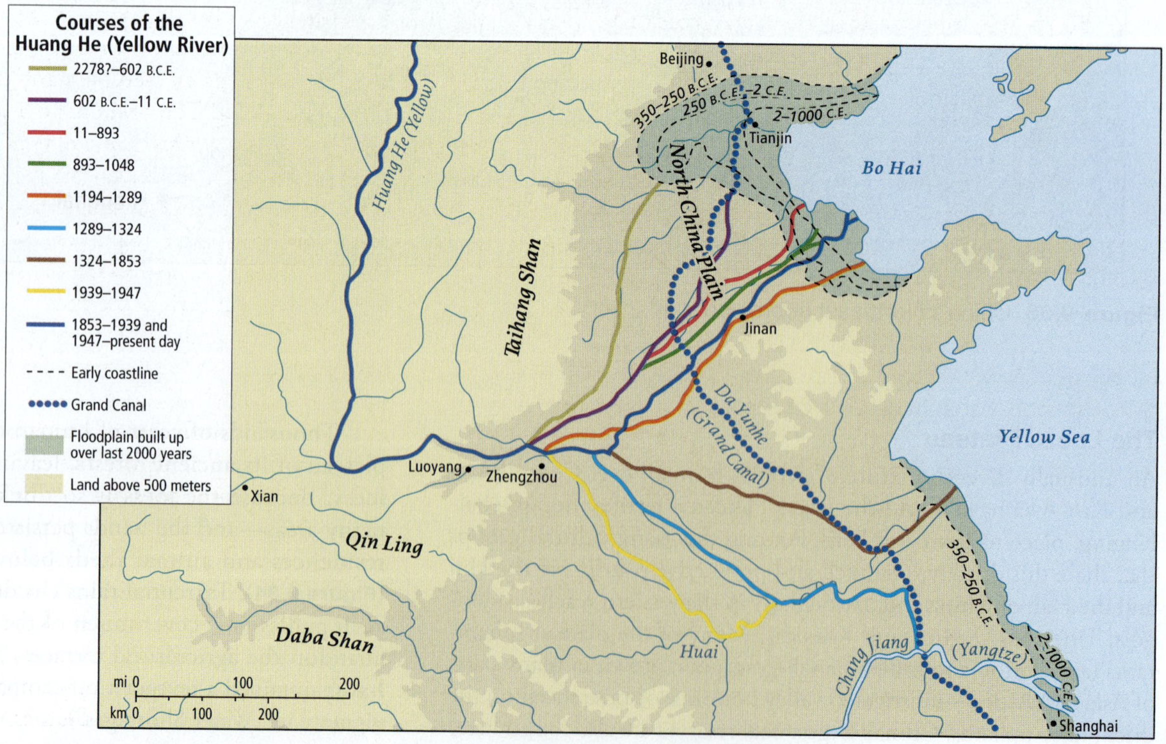

Figure 9.35 The Huang He's changing course. The lower course of the Huang He has changed its direction of flow many times. In 2000 B.C.E., it flowed north and entered the Bo Hai, south of Beijing. Then it repeatedly shifted to the east, then southeast, like the hand of a clock and finally cut south all the way to the Chang Jiang delta at Shanghai. Now it once again flows into the Bo Hai.

in the post-revolution 1950s, when its rich mineral resources—oil, coal, gas, gold, copper, lead, and zinc—made possible major industrialization, and its fertile state farms began to produce wheat, corn, soybeans, sunflowers (for oil), and beets. The Far Northeast was known for its steel, coal, and petroleum production (it is still the country's leading oil-producing area), and for its "iron man" workers who labored heroically for the revolution in return for meager wages and barracks-like housing. The Heilong Jiang (Amur River), which runs more than 1000 miles (1600 kilometers) along the border between Russia and China, became a conduit for trade with Russia, other parts of China, and Japan.

However, in the current era of market economy and globalization, the geographic focus of development has shifted to Guangdong Province in Southern China, to the international banking center of Hong Kong, to the Chang Jiang (Yangtze) delta and Shanghai, and even to Xinjiang in far western China. Today, China's Far Northeast is considered a rust belt. Its famous steel production has been surpassed by facilities in provinces along the central coast. Its outdated industrial facilities are being dismantled to make way for new factories that can compete in the global economy. Even so, the population of the industrial heartland of the northeast is stagnant or even declining slightly.

Much of the revitalization is concentrated in cities on the coast, such as Dalian, and is financed by nearby Korean and Japanese firms that train young people to do telemarketing, information technology (IT), or manufacturing work. The Japanese occupied China's Far Northeast between 1932 and 1945. They called the region Manchuria, which is a term that is still used internationally. The Japanese regime is remembered as brutal, but they were the first to introduce industrial development to the region. Now, because the Japanese are bringing investment cash and a chance for economic renewal, they are welcomed.

Beijing and Tianjin

There are 12 million residents in the urban core of Beijing, China's capital city, and 21 million in its metropolitan region. Beijing lies at the northern edge of the North China Plain. It is the administrative headquarters of the People's Republic of China and has several of the nation's most prestigious universities. Because Beijing is not located right at the coast, the nearby port city of Tianjin, an industrial and transportation center, is important for the regional economy.

Beijing was once a grand imperial city full of architectural masterpieces—notably the Forbidden City, the former imperial headquarters, which is made up of approximately 1000 buildings. The Forbidden City, as well as the rest of central Beijing, is laid out as a square. The square is a traditional shape in Chinese urban design, as it represents Earth in ancient Chinese philosophy, and Chinese cities are planned as a microcosm of Earth. The rectangular city walls of historic Beijing are now gone and instead replaced by ring roads that follow the same pattern. In fact, much of Beijing's historic character was lost under the communists, who tore down many monuments. Today, in the context of China's acceptance of capitalism, much of Beijing is again being reshaped, this time as a center for international commerce, with new neighborhoods of high-rise apartments, office towers, and hotels replacing communist-era apartment blocks and older neighborhoods of nineteenth-century, tile-roofed, extended-family homes. During this development process, Beijing and the cities around the North China Plain suffered from extreme levels of air pollution, although an apparent turnaround is in progress. Investments in modern energy systems and more stringent emission regulations are starting to pay off.

CHECK YOUR UNDERSTANDING

1. What are some of the environmental problems on the Loess Plateau?
2. What is the importance of the Huang He in China's Northeast?
3. How is the Far Northeast economically different than the rest of China?

CENTRAL CHINA

Central China consists of the upper, middle, and lower portions of the Chang Jiang (Yangtze River) basin (**Figure 9.36**). Like many of Eurasia's rivers, the Chang Jiang starts in the Plateau of Tibet. It descends to skirt the south of Sichuan Province, then flows through the Witch Mountains (Wu Shan) and the Three Gorges Dam, leaving the upper basin and flowing into the middle basin in China's densely occupied central plain. It then winds through the coastal plain (the lower basin) and enters the Pacific Ocean in a huge delta, the site of the famous trading city of Shanghai.

Sichuan Province

Sichuan, with 83 million people, has some of China's richest resources: fertile soil, a hospitable climate, inventive cultivators, and sufficient natural raw materials to support diversified industries. For years, Sichuan embodied the ideal of complementary agricultural and industrial sectors, but it could not keep pace with the coastal industries that attracted its young people. Now, old industrial cities like Chongqing are changing and are successfully recruiting foreign investors.

The heart of the province is the Sichuan Basin, also called the Red Basin because of its underlying red sandstone. It is a region of hills and plains crossed by many rivers that drain south toward the Chang Jiang. The basin is on a south-facing slope and is surrounded by mountains that are highest in the north and west. The mountains form a barrier against the arctic blasts of winter and trap the moist, warm air that moves up from the southeast. For these reasons, the Sichuan climate is generally mild and humid, and the basin is so often cloaked in fog or low cloud cover that it is said, "A Sichuan dog will bark at the sun."

Over thousands of years, Sichuan's relatively affluent farmers have cleared the native forests and manicured the landscape, so that only a few patches of old-growth forest are left in the uplands and mountains. The rivers have been channeled into an intricate system of irrigation streams for wet rice cultivation (see "Rice Cultivation") and, more recently, for growing the fine vegetables and fruits that affluent urban dwellers across China want. Also known by its older spelling Szechuan, the region's spicy cuisine is world renowned.

The population density of Sichuan is more than 800 people per square mile (300 per square kilometer). Most of the people are farmers, growing rice, wheat, and corn, and as a sideline, raising animals such as silkworms, pigs, and poultry. In the new market economy, it is often these sidelines that earn them the most cash

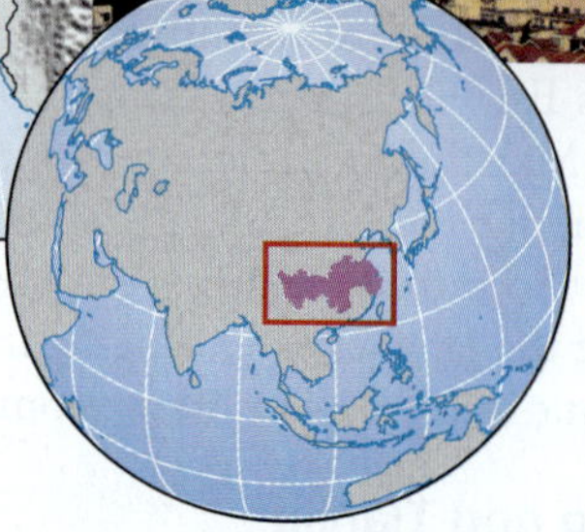

Figure 9.36 The Central China subregion. **A** Looming over Shanghai's Pudong district are three of China's four tallest buildings: the Shanghai World Financial Center, the Jin Mao Tower, and the Shanghai Tower, which is the second-tallest building in the world (after Burj Khalifa in Dubai). [Wei Fang/Getty Images]

income. In the mountain pastures to the west of the basin, Tibetan herders raise cattle, yaks, sheep, and horses. (Sichuan has the largest population of Tibetans outside Tibet—close to 1 million.)

The two main cities of the Sichuan Basin are Chengdu, the provincial capital (with a population of 11 million in the metropolitan area), and Chongqing (with a metro population of 9 million). Chengdu is a transportation hub and is home to light industries, especially food processing and textile and precision instrument manufacturing. Some of its higher-quality products are sold in the global market. Chongqing, with its iron, steel, and machine-building industries, is an old industrial city that suffered from decline. Now, Chongqing is targeted for renewal as Sichuan attracts workers back to the interior from the coastal provinces. As a result of the construction of the Three Gorges Dam (see "Three Gorges Dam: The Power of Water"), Chongqing, which lies at the western end of the dam reservoir, is also a hub for shipping and other new economic activities, especially as it benefits from better navigation and an abundance of electricity from the Three Gorges Dam.

The Central and Coastal Plains

After the Chang Jiang emerges from the Three Gorges Dam, it traverses an ancient, undulating lake bed. This former lake bed is the middle basin of the Chang Jiang. It and the river's lower basin are rich agricultural regions dotted with industrial cities.

The middle basin and the lower basin have been filled with alluvium (river-borne sediments) carried down from the Plateau of Tibet and the Sichuan Basin. Other rivers entering the middle basin from the north and south also bring in loads of silt, which are added to the main river channel. The Chang Jiang carries a huge amount of sediment—as much as 186 million cubic yards (142 million cubic meters) per year—past the large industrial city of Wuhan. This is particularly pronounced during the rainy season in the summer and early fall (the area is affected by a moderate monsoon—see "The Monsoon East"). This sediment, which under natural conditions was deposited on the basin floors during annual floods, enriched agricultural production. But because the floods often destroyed people, property, animals, and crops, the Chang Jiang, like the Huang He, now flows between levees. Still, it occasionally breaches the levees and floods both rural and urban areas. As the river approaches the East China Sea, it deposits the last of its sediment load in a giant delta, at the outer limits of which lies the trading city of Shanghai.

The climate of the middle and lower basins is milder than that of the North China Plain. The Qin Ling range and other, lower hills that extend eastward across the northern limits of the basin block some of the cold northern winter winds. The mountains also trap warm, wet southern breezes, so the basins retain significant moisture during most of the year. The growing season is 9 to 10 months long. The natural forest cover has long since been removed to provide agricultural land for the rural population. Summer crops are rice, cotton, corn, and soybeans; winter crops include barley, wheat, rapeseed (canola), sesame seeds, and broad beans. There are more than 490 million people in the middle and lower basins of the Chang Jiang. In the past, many of them were farmers, but every year more people migrate to the many industrialized cities of the region. Shanghai overshadows all of these cities in importance.

For many centuries, Shanghai's location facing the East China Sea, with the whole of Central China at its back, positioned the city well to participate in whatever international trade was allowed. When the British forced trade on China in the 1800s, Shanghai became their base of operations. Before the Communist Revolution, Shanghai was among the most cosmopolitan cities on Earth, home to well-educated citizens, wealthy traders, and a number of underworld figures as well. After the revolution, the Communist Party designated the city as the nexus of capitalist corruption, and it fell out of favor as a place for government investment.

When China began to open up to world trade in the 1980s, Shanghai entrepreneurs were well situated to take advantage of government incentives, such as reduced taxes, assistance in preparing building sites, and permission to take profits out of the country. These incentives helped motivate developers to build factories, housing, shopping malls, and elegant, extraordinarily tall skyscrapers (see Figure 9.36A). Today, some of the Overseas Chinese (see "East Asia's Most Influential Cultural Export: The Overseas Chinese"; see also Chapter 10)—particularly those who fled the communist system to places such as Taiwan, Singapore, and Malaysia—have returned to invest in everything from media to factories, shopping centers, and entertainment. Europeans and North and South Americans are attracted to Shanghai, too, hoping to find joint ventures with Chinese partners so that they can tap into the huge pool of consumers emerging in China.

CHECK YOUR UNDERSTANDING

1. What are the main physical and agricultural characteristics of Sichuan?
2. How do events upstream on the Chang Jiang affect areas downstream?

SOUTHERN CHINA

Southern China has two distinct sections (**Figure 9.37**). The first is made up of the mountainous and mostly rural provinces of the inland. The second includes the provinces on the southeastern coast, where the booming cities of China's evolving economy are located.

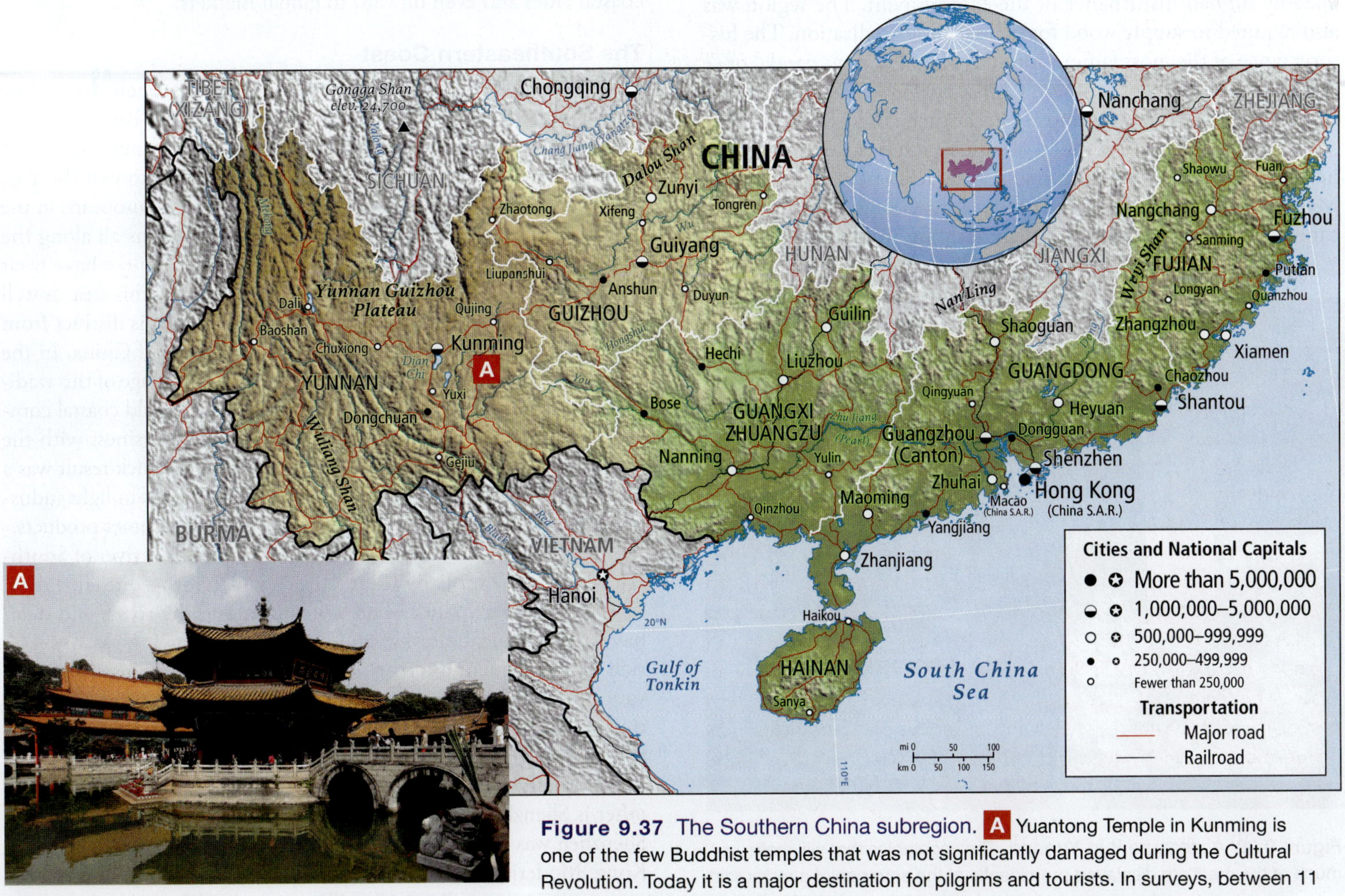

Figure 9.37 The Southern China subregion. **A** Yuantong Temple in Kunming is one of the few Buddhist temples that was not significantly damaged during the Cultural Revolution. Today it is a major destination for pilgrims and tourists. In surveys, between 11 and 16 percent of Chinese profess to be Buddhists, which makes Buddhism the largest organized religion in China (most Chinese, however, do not have a religious affiliation). [Bruce Yuanyue Bi/Getty Images]

The Yunnan–Guizhou Plateau

The provinces of Yunnan and Guizhou share a plateau noted for its natural beauty and mild climate and for being the home of numerous indigenous groups that are culturally distinct from the Han Chinese. Some of them also practice Buddhism, which is a prominent religion in nearby Southeast Asia (see Figure 9.37A). The plateau is a rough land of deeply folded mountains that trend north–south, through which the Nu (Salween) and Mekong rivers flow. The heavily forested valleys are deeper than they are wide. The north–south orientation naturally connects the region more to Southeast Asia than to the Chinese coast. In some places, rope and bamboo bridges have been slung across the chasms. The landforms here are unstable, and earthquakes cause heavy damage to settlements and terraces of rice paddies.

The valuable natural resources of the Yunnan–Guizhou Plateau are primarily biotic. Yunnan Province, with its fertile soil, is an important source of produce and meat for the bustling cities on the southeastern coast (**Figure 9.38**). Yunnan is called "the national botanical garden" because many of China's plant and bird species are native to the area's exotic landscapes. Guilin is noted for its fantastic formations of eroded limestone, known as *karst* (**Figure 9.39**). Until the 1970s, the tropical forests of far southern Yunnan (close to Myanmar, Thailand, and Lao) harbored elephants, bears, porcupines, gibbons, and boa constrictors. However, the flora, fauna, and geologic treasures of these provinces have been severely threatened by human disturbance of the environment. The region was also required to supply wood for China's industrialization. The historic range of the most famous forest dweller, the giant panda, once extended over much of southern China, but the animal is now limited to 40 panda preserves, mostly in the Sichuan Province.

The mountains of Yunnan also lie at the heart of a booming heroin trade that flows from major producers in remote areas of Myanmar and Lao across the border into China. Kunming, the capital of Yunnan Province, is a key location in this heroin trade.

Figure 9.38 A restaurant in Yunnan. The vegetables that will make up much of the meal are displayed prominently at the entrance of a restaurant in Dali, Yunnan. A mild climate and fertile soils make Yunnan an important source of vegetables for much of China. [Christian Kober/Getty Images]

Figure 9.39 The fantastic karst landscapes of Yunnan's Stone Forest. The region is composed of karst (limestone) formations that have been eroded by water and wind to form jagged peaks jutting out from the surrounding land. This phenomenon is found in its most extreme form near Kunming in the Stone Forest (called the *Shilin* in Chinese), which has become a world-renowned tourist attraction. [Tony Waltham/robertharding/Getty Images]

From there, heroin is transported to China's northern and eastern coastal cities and even onward to global markets.

The Southeastern Coast

The southeastern coastal zone of China has been a window to the outside world for centuries. During the Tang Dynasty (618–907 C.E.), its ports began launching ships that journeyed as far as Africa. Around the same time, Arab traders began to visit this part of China. By the fifteenth century, some of the first Europeans in the region described a string of flourishing trading towns all along the coast. The overwhelming majority of Overseas Chinese have their roots along China's southeastern coast. The people of this area, as well as many Overseas Chinese, speak Cantonese, which is distinct from the Mandarin Chinese spoken in most other parts of China. In the 1980s, the central government decided to take advantage of this tradition of outside contact by designating several of the old coastal communities as SEZs, with special rights to conduct business with the outside world and to attract foreign investors. The quick result was a chain of cities attracting millions of migrants to work in light industries that now supply the world with a variety of consumer products.

Much of this development is along the principal river of Southern China, the Zhu Jiang (Pearl River), which forms one large delta in Guangdong Province. The lowlands along the rivers and delta have a subtropical climate and a perpetual growing season. The rich delta sediment and the interior hinterlands are used to cultivate sugarcane, tea, fruit, vegetables, herbs, timber, and mulberry trees for the raising of silkworms, all of which are processed and packaged in the various SEZs and exported to global markets.

The city that epitomizes SEZ development more than any other is Shenzhen. When it was declared China's first SEZ in 1979, Shenzhen was little more than a fishing village situated next to Hong Kong. Modern Shenzhen emerged as a manufacturing hub that has grown to a metropolis of 13 million people. When the electronics companies hire temporary laborers to expand production for the

Christmas gift season, millions more flow into the city. Shenzhen is still an urban landscape of rapid change. During the recession, growth ground to a halt, but like most boom-and-bust cities, it is again experiencing renewed growth. However, reduced competitiveness in low-wage manufacturing due to increasing real estate and labor costs has initiated a shift toward high-tech manufacturing and development. Shenzhen is emerging as a Chinese Silicon Valley. Off the coast lies Hainan Island, which is the only place in China with a tropical climate (see Figure 9.3B). Well-endowed with sandy beaches, it has become a leading tourist destination and a place where wealthy people from chilly northern China keep second residences.

Hong Kong

Densely populated Hong Kong is the economic heart of the southeastern coast. For 99 years, Hong Kong was controlled by Britain and served as a hub of the Asian colonial economy. In 1997, Britain's lease on Hong Kong ran out and the city reverted back to Chinese control. However, Hong Kong's economic importance has not diminished and the city, in itself, has the world's eleventh-largest trading economy. It used to have a large manufacturing base but much of that has migrated to mainland China; Hong Kong is instead a major financial center. The Hong Kong area also has the largest port facilities in the world. Technically the ports of Shanghai and Singapore are larger, but the port in Hong Kong proper, plus two newer and more spacious ports in nearby Shenzhen and Guangzhou, together carry far more oceangoing goods than does any other urban area in the world. Basically, global investment is funneled into southeastern China via Hong Kong banks, and manufactured goods flow out of the region via its world-class ports. Through *containerization*, goods are shipped in containers that are standardized in size so that they can fit on trucks, rail, or ships around the world, which has made global transportation quick and efficient. Given the very obvious success of many cities along China's southeastern coast, from Macao to Shanghai, Hong Kong will probably continue in its role as a financial hub in the development of this very rapidly growing region. As a de facto city-state, Hong Kong's position as a business center resembles that of Singapore in Southeast Asia.

Macao

Macao (also spelled Macau), built on a series of islands across the estuary from Hong Kong, is the oldest permanent European settlement in East Asia. Portuguese traders arrived in about 1516, and by 1557 had established a raucous colonial trading and gambling center. Macao remained a Portuguese colony until the Chinese government regained control of it in 1999, just like it did with Hong Kong in 1997. Today, this small territory has a population of 675,000—primarily workers who earn a living in the tourism industry. Macao remains China's only gambling center; some of its casinos are managed by Las Vegas firms. Macao's GDP per capita is even higher than Hong Kong's and many times that of the mainland.

CHECK YOUR UNDERSTANDING

1. Where in Southern China can we find several indigenous populations and heavily forested mountainous terrain with unique landforms?
2. How are Hong Kong and Macao different from the rest of China?

FAR NORTH AND WEST CHINA

The Far North and West subregion of China (**Figure 9.40**) once occupied a central role in the global economic system. Traders carried Chinese and Central Asian products such as silks, rugs, spices and herbs, and ceramics over the Silk Road to Europe, where they exchanged them for gold and silver. Today, this route is cumbersome for long-term trade, but the subregion is recovering some of its ancient economic vitality as it connects the consumers of eastern China with the resource economies of central Asia.

Despite its trading past, this large interior zone has historically been considered backward by the rulers of eastern China because of its dry, cold climate; its vast grasslands; its history of nomadic herding; and—in the lands of the Uygurs and Kazakhs—its persistent adherence to Islam. Settlements are widely dispersed, and most agriculture requires irrigation. Three of the subregion's four political divisions have been designated autonomous regions (not provinces) because of the high percentage of ethnic minority populations there, but their people do not enjoy real autonomy. The central government in Beijing retains control over political and economic policies because it sees this subregion as crucial to China's future: it has energy and other resources for industrial development, it is close to the emerging oil-rich economies of Central Asia and Russia, and it affords a place to settle Han Chinese people.

Xinjiang

The Xinjiang Uygur Autonomous Region in northwestern China is historically and physically part of Central Asia. Yet it is the largest of China's political divisions, accounting for one-sixth of China's territory.

The Uygur minority and its conflict with the central government in Beijing and resettled Han Chinese are discussed earlier. Xinjiang, which has 22 million inhabitants, consists of Central Asian ethnic groups, most of whom are Muslims, and recent Han immigrants. The main ethnic group in the subregion is the Turkic-speaking Uygurs. They, and other groups, once made their living as nomadic herders and animal traders, moving with their herds in search of water and forage (**Figure 9.41**). The nomadic residence of choice was the **yurt** (or *ger*, in Mongolia). These cozy houses—round, heavy, felt tents stretched over collapsible willow frames—can be folded and carried on horseback, in horse-drawn carts, or today, in trucks (see Figure 9.44A later in the chapter). In the last few decades, many nomadic people have taken city jobs and abandoned yurts in favor of modern apartments.

Xinjiang consists of two dry basins: the Tarim Basin, occupied by the Taklimakan Desert, and the smaller Junggar Basin to the northeast. Both are virtually surrounded by 13,000-foot-high (4000-meter-high) mountains topped with snow and glaciers. In this distant corner of China, far from the world's oceans, rainfall is exceedingly sparse. Snow and glacial meltwater from the high mountain peaks are important sources of moisture. Much of the meltwater makes its way to underground rivers, where it is protected from the high rates of evaporation on the surface. Long ago, people

yurt (or *ger*) round, heavy, felt tent stretched over collapsible willow lattice frames and used by nomadic herders in northwestern China, Mongolia, and Central Asia

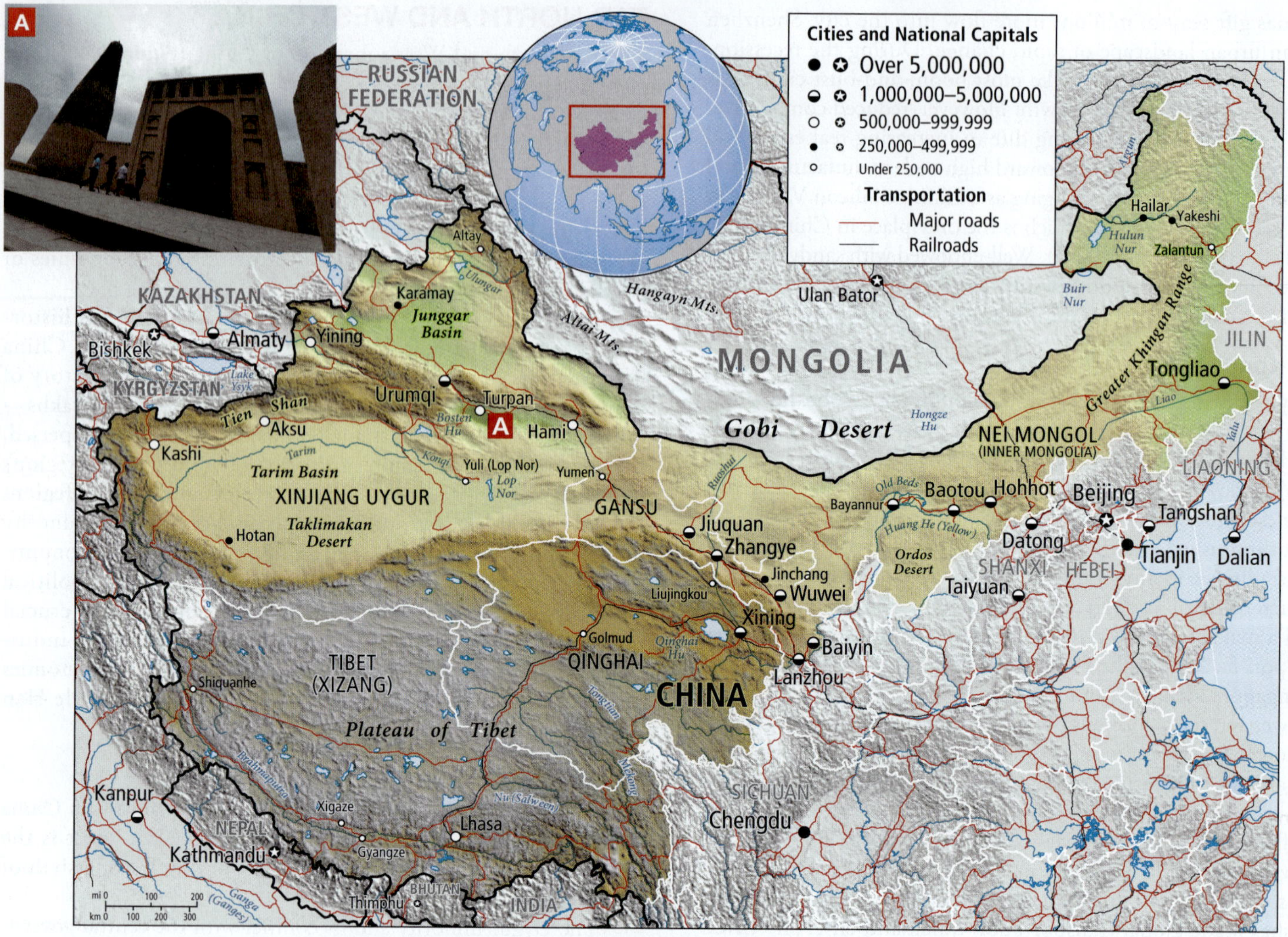

Figure 9.40 China's Far West and North subregion. A The Emin Minaret and Mosque in Turfan, Xinjiang, China. Built in 1778 to honor a Uygur general who had suppressed a local rebellion against the Chinese emperor, the structure is still a place of worship. [MARK RALSTON/AFP/Getty Images]

built tunnels called ***qanats*** deep below the surface to carry groundwater dozens of miles to areas where it was needed. Qanats have made productive some of the hottest and driest places on Earth. The Turfan Depression, situated between the Junggar and the Tarim basins, is a case in point. In this basin that descends 500 feet (150 meters) below sea level, temperatures often reach 104°F (40°C), and evaporation rates are extremely high. But because the qanats bring irrigation water, this area produces some of China's best foods: melons, grapes, apples, and pears.

In Xinjiang today, the local and the global, the very traditional and the very modern, confront each other daily. With the breakup of the Soviet Union, citizens of the new republics of Central Asia have been eager to revive their trading heritage, and they have oil and gas to sell. China is welcoming them and attracting outside investors from Europe and the Americas by establishing free trade zones in cities such as Kashi and Urumqi.

qanats underground tunnels, built by ancient cultures and still used today, that carry groundwater for irrigation in dry regions

Figure 9.41 People of the Xinjiang Uygur Autonomous Region. Nomads in Altay, Xinjiang, herd their livestock along a caravan. Nomadism is increasingly rare in China, as in the rest of the world. [STR/AFP/Getty Images]

The Plateau of Tibet

Situated in far western China, the Plateau of Tibet is the traditional home of the Tibetan people. Administratively it includes the Xizang Autonomous Region and Qinghai Province, which lie from 10,000 feet (3000 meters) to 13,000 feet (4000 meters) above sea level. They are surrounded by mountains that soar thousands of feet higher. They have cold, dry climates (late June can feel like March does on the American Great Plains) because of their high elevation and because the Himalayas to the south block warm, wet air from moving in from the Indian Ocean. Across the plateau, but especially along the northern foothills of the Himalayas, snowmelt and rainfall are sufficient to support a short growing season for barley and vegetables such as peas and broad beans. Meltwater from Himalayan snow and glaciers forms the headwaters of some major rivers: the Indus, Ganga (Ganges), Brahmaputra, Irrawaddy, Mekong, Chang Jiang, and Huang He. Many of the larger settlements in Tibet are located along the valley of the Brahmaputra (called Yarlung Zangbo in Chinese) in the southern part of the province.

Traditionally, the economy of the Plateau of Tibet has been based on the raising of grazing animals. The main draft animal is the yak, which also provides meat, milk, butter, cheese, hides, and fiber, as well as dung and fat for fuel and light. Animal husbandry on the sparse grasses of the plateau has required a mobile way of life so that the animals can be taken to the best available grasses at different times of the year. For several decades, though, the Chinese government has pressured Tibetan herdspeople to settle in permanent locations so that their wealth can be taxed, their children schooled, their sick cared for, and their dissidents curtailed. Still, throughout the plateau, many native peoples continue to live mobile yet solitary lifestyles, as they have for centuries, occasionally adopting some aspects of modern life and adapting to its restrictions. The political history of the Tibetan minority in China and suppression of Tibetan culture (**Figure 9.42**) are discussed in "The Tibetans."

Figure 9.42 The Potala Palace rises over Lhasa, Tibet. Built in the seventeenth century as both a monastery and seat of government for much of Tibet, it was home to the last ten Dalai Lamas, a lineage of Buddhist teachers that goes back to 1391. Since China's invasion of Tibet in 1959, and the subsequent exile of the fourteenth Dalai Lama to India, the Potala Palace has been made into a museum. [Ray Cheung/Getty Images]

CHECK YOUR UNDERSTANDING

1. How is China's Far North and West subregion physically, economically, and culturally different from the rest of China?
2. What does "designated autonomous region" mean?
3. Why does the central government in Beijing retain control over political and economic policies in China's Far North and West?

MONGOLIA

The present country of Mongolia is the north-central part of what was once a much larger cultural area in eastern Central Asia known by the same name (**Figure 9.43**). Between 1206 and 1370, the Mongols, under Genghis Khan and his grandson Kublai Khan, created the largest-ever land-based empire, stretching from the coast of China to central Europe. While in control of China, the Mongols fostered international trade and broke the control of traditional elites by abolishing their automatic access to privilege. They refined the manufacture of textiles, jewelry, and blue and white porcelain, and trade in these wares flourished along the Silk Road. Although the Mongols brought many important reforms to China, they also practiced authoritarian rule and discrimination against ethnic Chinese. Deposed by the Chinese in 1366, the Mongols retreated to the north; over the next 300 years, their control of territory in Central Asia and Europe dwindled. Eventually, the southern part of Mongolia (somewhat confusingly called Inner Mongolia) became part of China.

The Mongolian Plateau lies in the heart of Central Asia, directly north of China and south of Siberia. It is high and dry, with an extreme continental climate. The temperature ranges from very cold in the winter to moderate in the summer, such as in the capital of Ulan Bator, where a summer day is often no warmer than 70°F (21°C). The physical geography of Mongolia can be broken down into four major zones. In the far southeast and extending across the border into China is the Gobi Desert—actually a very dry grassland that grades into true desert in particularly dry years or where it is overgrazed. It is especially prone to increased aridity with global climate change. To the west and northwest of the Gobi is a huge, rolling, somewhat moister grassland. The remaining two zones are Mongolia's two primary mountain ranges: the forested Hangayn Nuruu, in north-central Mongolia, and the grass- and shrub-covered Altai Mountains, which sweep around to the west and south and into the Gobi Desert (see the Figure 9.43 map).

With just over 3 million people, Mongolia occupies a territory so large that its average population density is one of the lowest on Earth, at 4 people per square mile (1.5 per square kilometer). For thousands of years, the economy has been based on the nomadic herding of sheep, goats, camels, horses, and yaks. Nomadic pastoralism dates at least as far back as the domestication of plants (8000 to 10,000 years). Nomadic pastoralists must understand the intricate biological requirements of the animals they breed, and they

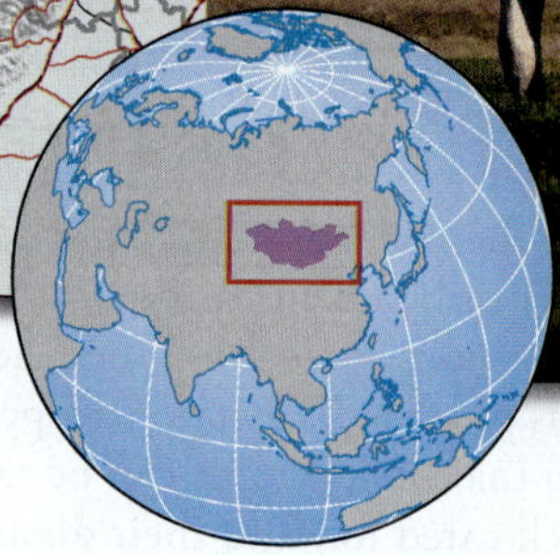

Figure 9.43 The Mongolia subregion. A A Kazakh man practices his hunting skills with a trained golden eagle in western Mongolia. For thousands of years, foxes and rabbits have been hunted by the Kazakhs of western Mongolia, who used bows and arrows—and more recently, guns—to kill or wound their prey. They can pursue a wounded animal for miles on horseback, using trained golden eagles to finish the kill. [Chalermkiat Seedokmai/Getty Images]

must also know the ecology and seasonal cycles of the landscapes they traverse with their herds.

Today, only 32 percent of Mongolians are still engaged in rural activities, often related to this traditional lifestyle (**Figure 9.44A**; see also Figure 9.43A). But as Mongolia has invested heavily in IT and wireless networks, even nomadic pastoralists far out in the steppe use cell phones. The other 68 percent now live in cities, primarily the capital, Ulan Bator (see Figure 9.44B). Urban workers are employed in a wide range of services and in industries related to the processing of minerals from Mongolia's mines and animal products such as hides, fur, and wool.

The Communist Era in Mongolia

Influenced by the Russian Revolution of 1917, and after considerable internal turmoil, Mongolia first declared independence from China in 1921; three years later it became a communist republic under the Soviet sphere of influence. From 1924 to 1989, Mongolia sought guidance from Soviet advisers and technicians. They helped set up a system of central planning that reorganized the nomadic pastoral economy according to socialist policies, but did so without drastically disrupting traditional lifeways. Rural households of extended families became economic collectives. Some households continued to herd and breed animals, others engaged in sedentary livestock production, and still others farmed crops. During this time, Mongolia also began to industrialize and urbanize. The Mongolian economy was supported by the Soviets, who subsidized a wide array of social services. Education and health care was available to all; life expectancy rose dramatically.

Postcommunist Boom and Bust

The collapse of the Soviet Union in 1991, coupled with lower world prices for the minerals that Mongolia produces, led to declines in economic growth, human well-being, and social order. The Soviet advisers left without training Mongolian replacements to run the economy. Layoffs and factory closings caused a severe decline in living standards. Some people who were working in urban jobs were forced to return to the countryside.

Similar to developments in postcommunist Russia, some signs of a better future were eventually emerging. Membership in the WTO brought increased international trade, and foreign investors took an interest in the country, especially in its minerals. China, with its booming economy just to the south, has been particularly interested in Mongolia's copper, coal, and gold. Compared to nearby post-Soviet states and China, Mongolia has also had a positive political development. It is ranked as a flawed democracy, but its elections and media are predominantly free and fair.

The global recession that started in 2008 was much more severe in Mongolia than in any of the other East Asian countries,

Figure 9.44 Tradition and modernity in Mongolia.

(A) The interior of a ger (called a yurt in other parts of central Asia) that belongs to a family in northern Mongolia. A ger is a felt-covered, wood lattice–framed structure traditionally used by the nomadic peoples of Mongolia and Central Asia as their main dwelling. In the past, gers were moved by horses; trucks are now widely used for the seasonal moves that seminomadic herders frequently make to provide adequate pasture for their sheep. [Ira Block/National Geographic Image Collection Magazines/Getty Images]

(B) The outskirts of Ulan Bator, Mongolia, where the gers of recent migrants are overshadowed by large stationary homes. Sixty-eight percent of Mongolians now live in cities. [Ozgur Cagdas/Alamy Stock Photo]

primarily because Mongolia depended on the export of raw materials—fleece from sheep and goats, and mined minerals—for which market prices dropped precipitously. Prices eventually turned upward again, bringing renewed economic growth. But Mongolia, as a landlocked country, is heavily dependent on raw material exports to China and energy imports from Russia, which means that recent economic shocks may be repeated in the future. At the moment, the country struggles with debt and budget deficits due to fluctuations in foreign investment. Mongolia appears to be a boom-and-bust country that is at risk of developing a *resource curse*, which is when a country focuses its development strategy on resource extraction at the expense of other economic sectors. Such countries may then become overly dependent on commodity prices and suffer from economic volatility, and government corruption can be endemic as people in power benefit from the sudden influx of cash.

Gender Roles in Mongolia

Traditionally, Mongolian women enjoyed a status approximately equal to that of men, but outside forces, such as Tibetan Buddhism in the sixteenth century and Chinese rule in the seventeenth and eighteenth centuries, eroded their status. The communist era restored, or even enhanced, egalitarian gender attitudes. The daily work of women herders was valued by the government, and like their husbands, women were eligible for retirement pensions. A woman could leave an unsuccessful marriage because the government provided support for her and her children. A few women became successful professionals and party officials.

In today's modern nomadic pastoral economy, as in the past, women usually choose which stock to breed, tend animals, preserve meat and milk, and produce many essentials from animal hides and hair. Women provide the materials for the construction of gers and also dye and weave furnishings for the interiors. Both boys and girls are taught to ride horses at an early age, and both help with the herding. As they mature, women become more responsible than men for the routines of daily life in the home. Men engage primarily in pasturing the herds and arranging the marketing of the animals and animal products. Men also perform other tasks—such as the occasional planting of barley and wheat—that are carried out beyond the home compound.

Demographically, Mongolian women have a life expectancy that is 9 years longer than men, they participate in the workforce to an extent similar to that of women in the United States, and they have surpassed Mongolian men in educational attainment. As in so many other countries that are moving from communism to a market economy, the informal economy—in which women are particularly active—has been an essential component of the Mongolian economic transition.

CHECK YOUR UNDERSTANDING

1. What are the differences between past and current economies of Mongolia?
2. Why is Mongolia afflicted by booms and busts?
3. How would you describe Mongolia's record on gender equality?

KOREA, NORTH AND SOUTH

Some of the most enduring international tensions in East Asia have been focused on the Korean Peninsula. As we have seen ("The Korean War and Its Aftermath"), after centuries of unity under one government, the peninsula was divided in 1953. The two resulting countries are dramatically different. Communist, poor, and inward-looking North Korea does not participate in the global economy, while more cosmopolitan, populous, democratic, and affluent South Korea has pursued a model of state-aided capitalist development.

Physically, Korea juts out from the Asian continent like a downturned thumb (**Figure 9.45**). Low-lying mountains cover

Figure 9.45 The Korea subregion. **A** The largest shipbuilding facility in the world, Hyundai's Ulsan Shipyard. South Korea, using advanced shipbuilding technologies and highly efficient shipyard management, was for years the leading shipbuilding country in the world. However, Chinese shipyards have recently edged South Korea in terms of tonnage produced, although sometimes these ships were built by South Korean companies. The eight largest shipbuilding corporations in the world are all from East Asia (South Korea, Japan, and China). [SeongJoon Cho/Bloomberg via Getty Images]

much of North Korea and stretch along the eastern side of the peninsula into South Korea, covering nearly 70 percent of the peninsula. There is little level land for settlement in this mountainous zone. The rugged terrain disrupts ground communications from valley to valley. Along the western side of Korea, floodplains slope toward the Yellow Sea, and most people live on these western slopes and plains. Although the peninsula is surrounded by water on three sides, its climate is essentially continental because it lies so close to the huge Asian landmass. The same cyclical monsoons that "inhale" and "exhale" over the Asian continent (see "The Monsoon East") bring hot, wet summers and cold, dry winters.

The Korean Peninsula was a unified country as early as 668 C.E. Scholars believe that present-day Koreans are descended primarily from people who migrated from the Altai Mountains in western Mongolia, because the Korean language appears to be most closely related to languages from that region. Other groups—Chinese, Japanese, and Mongols—invaded the peninsula, sometimes as settlers, other times as conquerors. The Buddhist and Chinese Confucian values brought by some of these groups have influenced Korean society. Korea is also noted for its early advances in mathematics, medicine, and printing.

Contrasting Political Systems

Although an armistice ended the Korean War in 1953, North Korea and South Korea each assumed a strong nationalist stance, which has resulted in more than 65 years of hostile competition between the two. Both governments adopted the Korean concept of *juche*, which means "self-reliance" or "the right to govern yourself in your own way." In North Korea, juche was interpreted as unquestioned loyalty to the Great Leader, Kim Il Sung, and later to his son, Kim Jong Il, and since 2011, his grandson Kim Jong Un, who is now president (**Figure 9.46A**). There is little pretense of political freedom, and the government remains in essence a military dictatorship. Extraordinarily large amounts of money fund national defense, which is a very powerful institution in the

Figure 9.46 Life in North Korea.

(A) Hailed as the largest show of its kind in the world, the Arirang Festival celebrates the nation and its leaders. The celebration is held in P'yongyang in the largest stadium in the world. More than 100,000 people perform, and the festival features elaborately choreographed dances, gymnastics, and music. After having been cancelled for a few years due to lack of funds, the North Korean government reinstated the festival in 2018. [Ayse Topbas/Getty Images]

(B) A man drives an ox cart through a street in P'yongyang that has a picture of former president Kim Il Sung hung in the background. The vast majority of North Koreans who labor in factories and on farms are plagued by outdated and often broken-down equipment. Virtually nobody owns a car, which is why the oversized street in the picture is devoid of traffic. [David Hume Kennerly/Getty Images]

country. North Korea limits its trade and other involvement with its continental neighbors, China and Russia. Its restrictive internal and external economic policies make it one of the poorest nations in the world, with a GDP per capita that is estimated to be as low as many African countries (see Figure 9.46B). Despite its apparent desire for isolation, North Korea has occasionally threatened South Korea and nearby Japan.

North Korea has provoked global concern by producing nuclear material, forcing UN monitors to leave the country, and then withdrawing from the Treaty on the Non-Proliferation of Nuclear Weapons. After years of nuclear testing, by 2017 it was clear that North Korea had developed nuclear weapons and operational missiles to deploy them. North Korea says it must prepare for an attack by the United States or Japan. It is generally agreed that the presence of nuclear arms in North Korea is cause for concern worldwide. The new leader, Kim Jong Un, has pursued a militaristic path similar to those of his predecessors. Sanctions on North Korea have been imposed by the UN and United States, although ongoing negotiations aim to reduce military tensions on the Korean Peninsula.

The nuclear issue has diverted attention from massive mismanagement of the country's resources, especially food. Such mismanagement has resulted in very serious human rights abuses in North Korea. There are vast prisons for those who have deviated from the Communist Party line, such as by engaging in entrepreneurialism (trading). Eyewitness escapees have told the outside world of beatings, torture, and executions.

In South Korea, by contrast, juche has come to mean vigorous individualism, coupled with pride in and loyalty to one's own people and nation. Social criticism and even aggressive protests by labor unions and other social movements are allowed, with the understanding that one's ultimate loyalty is still to South Korea. Shortly after World War II, South Korea allied itself politically with the United States, Japan, and Europe, and sought economic growth through foreign aid and capitalist development. But despite its economic success, South Korea was governed by a series of military dictatorships until the late 1980s, when democratic elections were held.

South Korea's economic success is also transforming the previously homogenous society, albeit slowly. Immigrant male workers and young immigrant women who marry Korean men—both groups typically originating in poorer Asian countries—are gradually altering the ethnic composition of South Korea. For example, every tenth marriage now involves a foreign spouse. This recent trend toward multiculturalism means that the country is inevitably moving toward a more inclusive notion of what it means to be Korean.

Contrasting Economies

The two Koreas differ dramatically in their economic resources. North Korea (with a population of 25 million) has physical resources for industrial development, including forests and deposits of coal and iron ore. The many rivers descending from its mountains have considerable potential to generate hydroelectric power. South Korea (with a population of 52 million) has better resources for agriculture because its flatter terrain and slightly warmer climate make it possible to grow two crops of rice and millet per year. In recent decades, South Korea has surpassed North Korea in agricultural production many times over, and its electronic, automotive, chemical, and shipbuilding industries successfully compete with those in the Americas, Europe, and Japan (see Figure 9.45A). North Korea's industry is rudimentary, but includes some mining and weapons manufacturing.

A major reason for South Korea's economic success has been the formation of huge corporations known as ***chaebol***.

chaebol huge corporations in South Korea that are involved in a variety of economic sectors

These conglomerates—companies that are involved in a wide variety of economic sectors—include such internationally known companies as Samsung and Hyundai. South Korea's government has assisted the chaebol by making credit easily available to them and by helping them purchase foreign patents. Nevertheless, the chaebol have come under increasing criticism in recent years because their close connections with the government have led to corruption and mismanagement.

Despite these concerns, the South Korean economic system has worked well and South Koreans, who were extremely poor at the end of the Korean War, can now afford the products they formerly only exported. For example, the GNI per capita is now close to the European Union average. The use of advanced telecommunication technologies and the modern urban landscapes, especially in the capital Seoul, testify to South Korea's success in the global economy.

North Koreans have benefited from some communist social policies, particularly access to basic health care and education. In general, however, North Korea's industries are inefficient and its workers poorly motivated. Furthermore, through the demise of the Soviet Union, the country lost markets and access to raw materials, fuel, and technical training. Poor harvests on collective farms and recurrent cycles of floods and droughts have brought recurrent food shortages and even the extensive famines that happened from 1995 through 2001. Today the situation is not quite as dire, but a 2017 UN report concluded that about 40 percent of North Koreans are undernourished. The infant mortality rate in 2018 was 12 per 1000 in North Korea, compared to just 2.8 per 1000 in South Korea.

CHECK YOUR UNDERSTANDING

1. How does South Korea's model of state-aided capitalist development work?
2. Why is North Korea a threat to South Korea and other countries in the region?

JAPAN

Japan consists of a chain of four main islands and hundreds of smaller ones (**Figure 9.47**). Prone to severe earthquakes, tsunamis, and typhoons, and so mountainous that only 18 percent of its land can be cultivated, the Japanese archipelago might seem an unlikely place to find 126 million people living in affluent comfort. Yet its citizens have learned to cope with these limitations and have made the most of crowded conditions in cities and countryside alike.

A Wide Geographic Spectrum

Modern Japanese populations are descended from migrants from the Asian mainland, the Korean Peninsula, and the Pacific Islands. Ideas and material culture imported from these places include Buddhism, Confucian organization, architecture, the Chinese system of writing, the arts, and agricultural technology.

Japan stretches over a range of latitudes roughly comparable to that between Canada's province of Nova Scotia and the Georgia coast (and even as far south as the Florida Keys, if one counts Japan's tiny, southernmost islands). Although half a world away, it has a similar range of climates to that of the North American Eastern Seaboard. The climate is cool on the far northern island of Hokkaido; temperate on the islands of Honshu, Kyushu, and Shikoku; and subtropical in the southern Ryukyu Islands. Despite the moderating effect of the surrounding oceans, the great seasonal climate shifts of nearby continental East Asia give Japan a more seasonal variation in temperature than it would have if it lay farther out to sea.

Honshu, the largest and most densely populated of Japan's islands, has the most mountains and forests. Because it has gained access to forest products from Southeast Asia and other parts of the world, Japan has been able to keep much of its own forest land in reserves. Today, the interior mountains of Honshu (and Hokkaido) remain largely forested, although many of the slopes are planted with a monoculture of *sugi* (Japanese cedar) that is harvested commercially as building material. The few flat lowlands and coastal areas of Honshu were once intensively cultivated and supported a dense rural population. While some farmland remains, industrial cities now fill many of these lowlands. Japan's largest flatland is the Kanto Plain, which is where the Tokyo–Yokohama urban agglomeration is located. Thirty percent of Japan's total population lives there. The flatlands that ring the Inland Sea are also heavily urbanized, especially the Kobe–Osaka–Kyoto metropolitan area, which is the second-largest concentration of people in Japan (see the Figure 9.47 map). Many Japanese cities have a modern appearance although Kyoto, the country's ancient imperial capital located in the interior of Honshu, was spared destruction during World War II and is home to numerous historic temples and shrines (**Figure 9.48**).

Japan's other three main islands are Hokkaido, the least populated island and Japan's northern frontier, and Kyushu and Shikoku, two smaller southern islands. Off Kyushu's southern tip, the very small, mountainous Ryukyu Islands stretch out in a 650-mile (1000-kilometer) chain, reaching almost to Taiwan. Although many areas are densely settled and are important for tourism and specialty agriculture, such as the growing of pineapples, the Ryukyu Islands are considered a poor, rural backwater compared to the four big islands.

The Challenge of Living Well on Limited Resources

Japan has among the highest ratios of people to farmland in the world: more than 7000 people depend on each square mile (more than 2700 per square kilometer) of cultivated (arable) land. Thus, Japanese agriculture relies heavily on high-yield varieties of rice and other crops, as well as on irrigation, fertilization, and mechanization. Because flat land is scarce, Japanese farmers have had to make efficient use of hill slopes by planting tea bushes and orchards and by terracing for rice paddies.

Today less than 3 percent of the population fish or farm for a living. Japanese farms are highly productive in terms of output per acre; however, their food is some of the most expensive in the world. Foreign rice costs about half as much as domestically grown rice does, but many Japanese people associate locally grown rice with quality and prefer it. Another farm product, Kobe beef, has a worldwide reputation for quality, but it sells for more than U.S.$100 per pound.

The seas surrounding Japan, where warm and cold ocean currents mix, are an especially important source of food for the

FIGURE 9.47 The Japan subregion.

A During rush hour in the Tokyo railway, "passenger arrangement staff" squeeze as many commuters onto a train as possible. They also make sure passengers can get off crowded trains safely, and signal train conductors when it is safe for the trains to depart. [Ton Koene/AGE Fotostock]

B Dancers at a *yosakoi* festival in Sapporo, Hokkaido. Yosakoi is a modern and highly energetic form of a traditional dance that is performed by large costumed groups at festivals and sporting events. [Shayne Hill Xtreme Visuals/Getty Images]

C A 60-foot-high statue of a *Gundam*, or giant robot, in Odaiba, Tokyo. This statue represents a character from a popular *anime* (Japanese animation) TV series. [John S Lander/LightRocket via Getty Images]

Figure 9.48 The historic urban landscape of Kyoto. Unlike many other cities in Japan, Kyoto has a well-preserved old town with narrow streets and traditional wood-frame structures. [Jon Chica/Shutterstock.com]

country, especially considering the limited amount of farmland. There are some 4000 coastal fishing villages and tens of thousands of small craft that work these waters and bring home great quantities of tuna, halibut, mackerel, salmon, and other fish. In addition, a large aquaculture industry produces oysters, seaweed, and other foods in shallow bays, and freshwater fish in artificial ponds.

Despite all these efforts at food production, Japan is one of the world's largest consumers of imported agricultural products, especially seafood, grains, meat, and processed foods, and it has the lowest levels of food self-sufficiency (the ability to produce its own food supply) among developed countries. As shown in **Figure 9.49**, the average household spends more than 28 percent of its expenditures on food (about $673 per month in 2018). Americans, by contrast, spend about 13 percent.

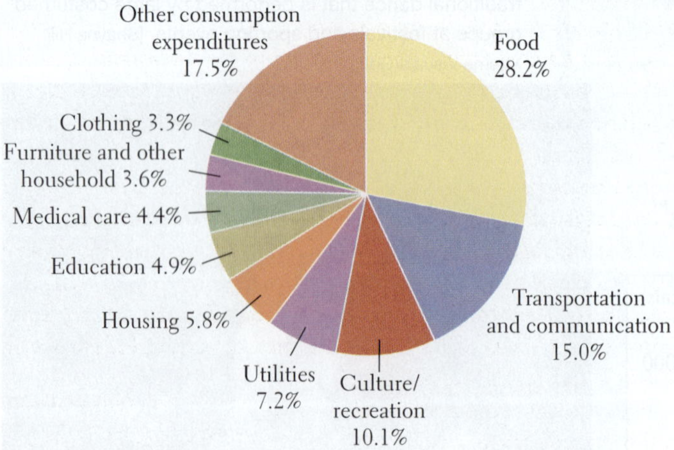

Figure 9.49 Average monthly expenditures for a Japanese household, 2018. [Data from: Statistics Japan, http://www.stat.go.jp/english/data/kakei/156.html.]

Japan's Culture of Contrasts

Perhaps more than any other country in the world, Japan is renowned for its ability to confound foreigners. Here, in an attempt to penetrate some of Japan's cultural complexity, we look at a few of its unique cultural phenomena.

Japan has combined its long-standing traditions with customs from around the world, resulting in unique cultural contrasts. We see the mix in many ways: in the kinds of clothes people wear, in the foods they eat, in the music they listen to, in the sports they enjoy, and in the cultural impact they have abroad. Thus, among other contrasts, Japan is a land of kimonos and blue jeans, sushi and hamburgers, the *koto* (a traditional stringed instrument) and rap music, sumo wrestling and baseball. Japanese couples who marry in traditional wedding kimonos often change their clothing after the ceremony to a Western-style tuxedo and a white bridal dress for the reception. Japanese culture is a blend of modern influences and respect for tradition and authority.

There is often a distinction in Japanese life between what is displayed on the outside, *omote*, and what is private on the inside, *ura*. The former often protects the latter, as in the contrasts between *omote-ji*, the outer layer of cloth in a kimono, and *ura-ji*, the layer closest to the skin. This cultural attitude is also expressed geographically. *Omote-Nippon*, the urban-industrial eastern (Pacific Ocean) side of Japan, trades with the world, while *Ura-Nippon* is the more traditional and secluded western (Sea of Japan) side of the country.

Working in Japan

Japan's distinctive culture is in part reflected in the traditional working lives of its people, as well as in their attitude toward the outsiders who perform many of the country's most menial jobs.

The Japanese are well known for their hard work, high-quality output, and devotion to their jobs. Since before World War II, the norm in Japan has been that an employer provides lifetime employment, regular paid vacations, a pension upon retirement, and even subsidized housing. Job-hopping is considered reprehensible. Personnel management instills loyalty to the employer. The prototypical Japanese corporate employee is a "salaryman"—a male, usually with a family he spends a lot of time away from, who lives to work. In sprawling Tokyo where the commute home is long and the employee is expected to work late, he can stay overnight at a capsule hotel, which is little more than a bed-in-a-tube made available at a low cost (**Figure 9.50**). Back at work, individualism is regarded as selfish. Employees who have good ideas may not present them for several years, or may give credit to someone else. The Japanese language does not even have a word for *entrepreneur*; only recently has the English term become a buzzword in Japanese.

This norm is now receding in the face of larger economic and social changes. Since the slowdown of the Japanese economy, which has been ongoing since the 1990s, many companies had to reduce their workforces and employee benefits. Moreover, some younger male and female graduates in business and technological fields have turned away from the old pattern in favor of the freedoms of short-term employment. Although they may settle for a lower pay and give up job-related benefits, women enjoy less gender discrimination in temporary or contract jobs, and temporary workers do not feel compelled to acquiesce to their superiors or to work overtime. This freelance approach to work allows them time

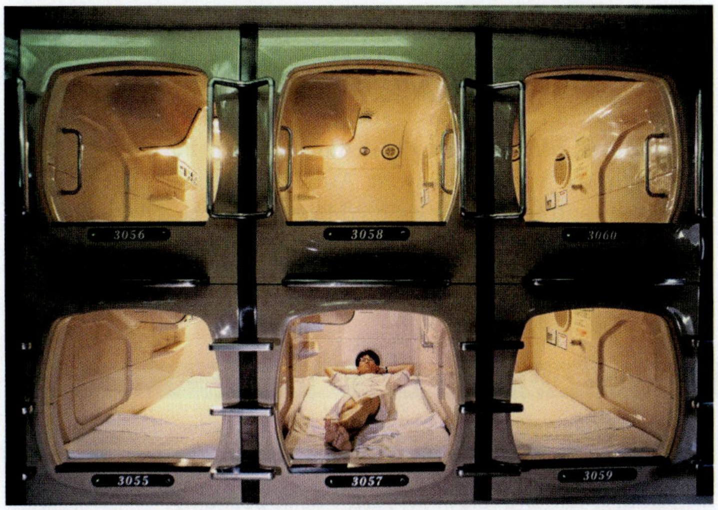

Figure 9.50 Capsule hotel in Tokyo. The capsule concept was invented decades ago, starting out as a place to stay for businessmen who worked late and missed the last commuter train home to the suburbs. Today, capsule hotels also attract budget travelers and tourists who are curious about this unique lodging experience. [Barry Lewis/ Alamy Stock Photo]

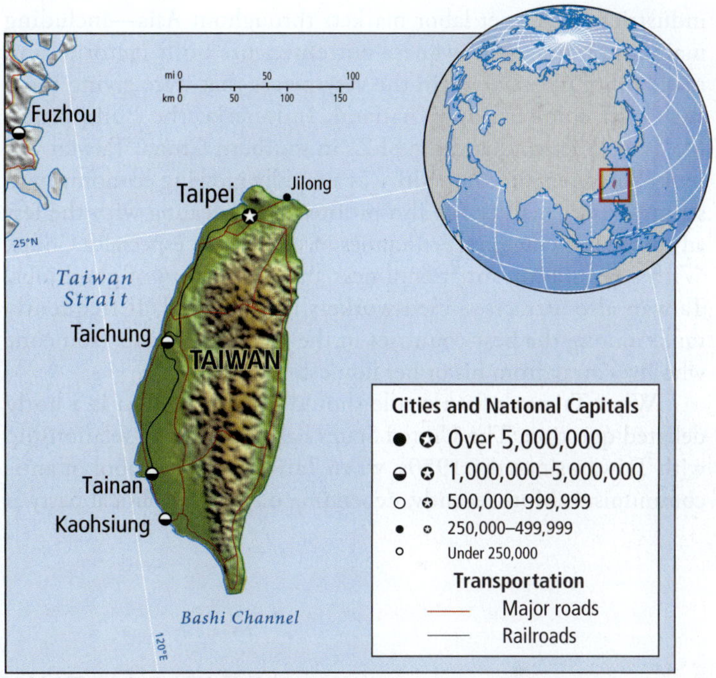

Figure 9.51 The Taiwan subregion.

to start their own businesses and perhaps spend more time with their families. At the same time, the old loyalty system makes it difficult to find a new job, and firms worried about industrial espionage are often unwilling to hire someone who has worked for a competing company.

Foreign Workers Foreign workers from poor countries come to Japan as guest workers to take low-paying jobs that today's Japanese workers avoid. They come from places such as China, South and Southeast Asia, and even Latin America and Africa. Some Latin American immigrants are descendants of Japanese people who went abroad in search of work some generations ago, when Japan was a poorer country. Foreign men often work in construction, in factories, and as dishwashers in restaurants; women work as hotel maids, as cleaners, and in other low-status jobs. Many foreign women are also employed as hostesses in bars and as sex workers. The number of foreign workers in the Japanese economy is growing slowly, although they make up only 1.7 percent of the Japanese population. The Japanese workforce needs to expand in order to remain productive in the future, which most people are acutely aware of.

CHECK YOUR UNDERSTANDING

1. How does Japan differ from north to south?
2. What are the distinct characteristics of food production in Japan?
3. In what ways is Japan a culture of contrasts and complexities?

TAIWAN

Taiwan is located a little over 100 miles (160 kilometers) off China's southeastern coast (**Figure 9.51**). With an area of 14,000 square miles (36,000 square kilometers) and almost 24 million people, Taiwan is a crowded place. A mountainous spine runs from the northeastern corner to the southern tip of the island, with a rather steep escarpment facing east and a long, gentler slope facing west. Most of the population lives on the western side, especially at lower elevations along the coastal plain. As farming has declined, populations have become concentrated in a few urban centers. The greatest concentration is in the far north, in the area surrounding Taipei, the capital.

Despite its small size and small population, Taiwan is one of the most prosperous countries of the **Asia–Pacific region**, a huge trading area that includes East Asia, South Asia, Southeast Asia, and Oceania. It ranks fifteenth globally in the size of its exports, and its GDP per capita in 2017 was U.S.$50,300. After the Communist Revolution in China, Taiwan took advantage of its ardent anticommunist stance, its skilled refugees from China, and its geographic location close to the mainland to draw aid and investment from Europe, America, and eventually Japan. The island's economy quickly changed from overwhelmingly rural and agricultural to mostly urban and industrial. Green revolution technologies (see Chapter 1) allowed a surplus labor force in the countryside to be transferred to urban jobs, and today only 5 percent of Taiwanese people still work in agriculture. Many industries made products for Taiwan's impressive export markets. Then, slowly, the economic emphasis changed again, from labor-intensive industries to high-tech and service industries requiring education and technical skills.

Today, local consumers absorb a major proportion of Taiwan's own production: home appliances, electronics, and vehicles. As Taiwan lost its labor-intensive

Asia–Pacific region a huge trading area that includes East Asia, South Asia, Southeast Asia, and Oceania industries to cheaper labor markets throughout Asia—including mainland China—Taiwanese entrepreneurs built factories and made other investments in the very places that were giving them such stiff competition: Thailand, Indonesia, the Philippines, Malaysia, Vietnam, and the SEZs in southern China. Taiwan was thus able to remain competitive as a rapidly growing economy with strong export markets; it also profited from dealing with the less advanced but emerging economies in the region, especially China. With a reputation for friendliness and availability of good jobs, Taiwan also attracts skilled workers from abroad. It frequently ranks among the best countries in the world for expats (someone who lives away from his or her home country).

What Taiwan's future role should be in East Asia is a hotly debated question. The United States has had a strong relationship with Taiwan since the 1950s, when Taiwan was a symbol of anti-communism. More recently, depending on which political party is in charge, sometimes the Taiwanese government favors relaxed relations with China and at other times the tone is more nationalistic (see Figure 9.18D). China, on the other hand, regards Taiwan as an integral part of the mainland, and Taiwan could easily be militarily subdued, as China periodically demonstrates by flexing its military muscles in the Straits of Taiwan. At the same time, Taiwan is geographically and financially situated to enhance the development of mainland Chinese markets and the integration of the Asia–Pacific economy. China is currently accepting this reality.

CHECK YOUR UNDERSTANDING

1. Taiwan claims sovereignty, while China considers Taiwan one of its provinces. Why is that?
2. What is Taiwan's role in the Asia–Pacific economy?

CRITICAL THINKING QUESTIONS

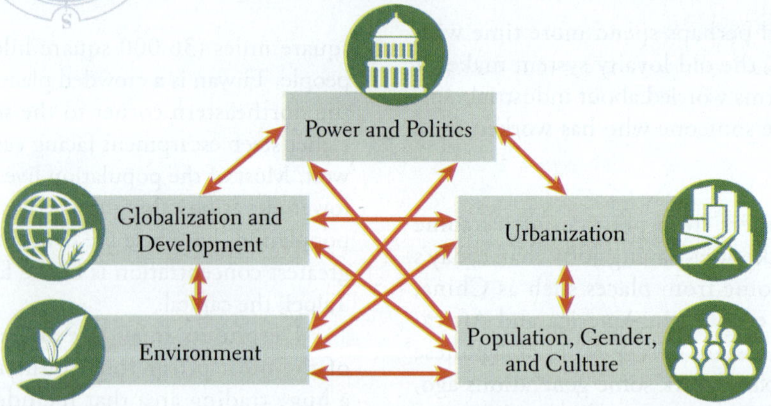

The diagram represents connections among the five geographic themes that structure this book. Listed below are some important questions that have been addressed in this chapter. Answer each question, and indicate where in the diagram you think the topics in each question belong.

1. Why is the interior west of continental East Asia (western China and Mongolia) so dry and subject to extremes in temperature? What are the effects of this climate on local population size and forms of subsistence?
2. Historically, China has been afflicted with recurring famines. While famine is no longer a concern, China's access to food has changed with economic development. How?
3. How might a less authoritarian political system have resulted in changes in the overall conception and design of China's Three Gorges Dam?
4. What is the contradiction between China's communist political system and its adoption of free markets? How do you see democracy and human rights evolving in China in the future?
5. In what ways might globalization have taken a different course if the voyages of Admiral Zheng He had inspired China to conquer a vast colonial empire, like that of Great Britain?
6. Contrast Japan's pre–World War II policies in East Asia with its current role in the region. Describe the principal similarities and differences.
7. In what ways has the one-child policy helped maintain stability in China? Even though the one-child policy is no longer in use, how might it have lingering effects on China in terms of future gender balance and economic growth?
8. How has East Asia's spatial pattern of urbanization been shaped by integration into the global economy?
9. What information technologies are now available to the Chinese and what role can such technologies have in the democratization of China?
10. East Asian countries are able to buy food on the global market but not produce all the food they need at home. What effects might this have for East Asia and the rest of the world?

11. Japan is concerned about its rapidly aging population. What are those concerns and what might be the solutions to such problems?
12. What areas of East Asia are most vulnerable to climate change? How has this vulnerability already been manifested in terms of water shortages?
13. Air pollution is of grave concern in Chinese cities. Make a case for why air pollution will improve during the next decade.
14. How do you think Confucianism has impacted contemporary political and demographic developments in China?
15. Modern economic development in East Asia first took hold in Japan, then South Korea and Taiwan, and eventually China. Why did the region follow this spatiotemporal pattern?
16. How would you describe the political relationship between the central government of China and the country's ethnic minorities?

Key Terms

Ainu 529
alluvium 530
Asia–Pacific region 547
chaebol 543
Confucianism 500
conurbation 516
Cultural Revolution 503
demilitarized zone (DMZ) 504
export-led growth 505
floating population 517
Great Leap Forward 503
growth poles 517
hukou system 517
inversion 498
just-in-time system 508
loess 530
Overseas Chinese 529
qanats 538
regional self-sufficiency 509
regional specialization 509
special economic zones (SEZs) 517
state-aided market economy 505
tsunami 488
typhoon 490
wet rice cultivation 494
world city 516
yurt (or *ger*) 537

More Practice at SaplingPlus

Read the interactive e-text, review key concepts, and check your understanding.

Indigenous person in Sarawak, Malaysia, on the island of Borneo. **Inset:** Oil palm plantations in Sarawak. [Yvan Cohen/LightRocket via Getty Images; inset: Mattias Klum/National Geographic/Getty Images]

10

Southeast Asia

In 2005, indigenous people in the Malaysian state of Sarawak, on the island of Borneo, attended a public meeting wearing orangutan masks. They carried signs informing onlookers that, although the government protects natural areas for orangutans, it ignores the basic rights of indigenous people to live on their own ancestral lands and practice their forest skills.

By the 1980s, 90 percent of Sarawak's lowland forests had been degraded or clear-cut and replaced by oil palm plantations. As a consequence, indigenous people found that even on uncleared land, their hunts declined. Vegetative diversity was lost and streams and rivers became polluted. Virtually none of the profits from palm oil were returned to the communities affected by the conversion of forests to plantations.

In the 1990s, a group in Berkeley, California, concerned about deforestation in Sarawak, organized the Borneo Project. They became a "sister city" to one indigenous group, the Uma Bawang, who began a community-based mapping project. Using rudimentary compass-and-tape techniques, they mapped the extent and biological content of their forest home. Since then, indigenous people in Sarawak have used global positioning systems (GPS), geographic information systems (GIS), and satellite imagery to make more sophisticated maps. In 2001, these maps proving indigenous land use helped win a precedent-setting court case protecting indigenous lands from encroaching oil palm plantations. This court ruling was challenged in 2005, but in 2009 the Malaysian court ruled in favor of the Uma Bawang, stating that in 1939, native land rights had been protected when British colonial officials mapped the boundaries of native lands. Now Malaysian indigenous people have the right to sue the government for past illegal leases to logging companies, and as of 2014, some 8000 such cases were in litigation.

Learning Objectives

Environment: Physical and Human

10.1 Identify Southeast Asia's physical features: its landforms, vegetation, and climates.

10.2 Relate Southeast Asia's most critical environmental issues to climate change.

10.3 Explain how Southeast Asia was settled, dominated by outside groups for centuries, and eventually achieved autonomy.

Globalization and Development

10.4. Describe how the formation of state-aided market economies and export-led production have been key to the economic development of this region.

10.5. Explain why regional economic cooperation is expanding, but only slowly.

Power and Politics

10.6. Discuss the recent expansion of political freedoms in Southeast Asia, and describe how authoritarianism, corruption, and violence have at times reversed these gains.

Urbanization

10.7. Describe urbanization in Southeast Asia and discuss the factors that are fueling urban growth.

10.8. Explain why the largest Southeast Asian cities receive the most migrants, despite having insufficient housing, water, sanitation, and jobs.

Population, Gender, and Culture

10.9. Explain the factors that influence population dynamics in Southeast Asia.

10.10. Discuss how the divergent array of cultural traditions existing in close proximity has the potential to breed social conflict.

Subregions

10.11. Distinguish between the various subregions of Southeast Asia geographically, economically, and sociopolitically.

CHINA
INDIA
BHUTAN
BANGLADESH
Dhaka
Chittagong
Brahmaputra
Putao
Gongga Shan
elev. 24,700
Yunnan–Guizhou Plateau
Guiyang
Kunming
Changsha
Nanchang
Nan Ling
Wuyi Shan
East China Sea
Taipei
Taichung
Kaohsiung
TAIWAN
Taiwan Strait
Ryukyu Is.
Guangzhou (Canton)
Zhu Jiang (Pearl)
Shantou
Hong Kong
Wuliang Shan
Mandalay
MYANMAR (BURMA)
Nay Pyi Taw
Loikow
Bago Mts.
Arakan Yoma
20°N
Bay of Bengal
Rangoon
Gulf of Martaban
Preparis North Channel
Andaman Islands
Andaman Sea
10°N
Ten Degree Channel
Nicobar Islands
Great Channel
Lao Cai
Ha Giang
Fan Si Pan
elev. 10,312
Dien Bien Phu
Phong Saly
Hanoi
Haiphong
Muong Sai
Samneua
Ban Ban
Luang Prabang
Chiang Rai
Chiang Mai
LAO
Vientiane
Thanh Hoa
Gulf of Tonkin
Vinh
Ban Nape
VIETNAM
Dong Hoi
Udon Thani
Khon Kaen
Mekong
Annam Cordillera
Quang Tri
Hue
Savannakhet
Da Nang
Tak
THAILAND
Pakxe
Surin
Dangrek Range
Lop Buri
Saraburi
Bangkok
Kon Tum
Play Ku
Qui Nhon
Siem Reap
Khone Falls
Lomphat
Batdambang
CAMBODIA
Kratie
Da Lat
Nha Trang
Cam Ranh
Phan Rang
Phnom Penh
Ho Chi Minh City (Saigon)
Kompong Som
Can Tho
Mekong Delta
South China Sea
Gulf of Thailand
Surat Thani
Malay Peninsula
Phuket
Strait of Malacca
Songkhla
Pattani
Banda Aceh
Alor Setar
George Town (Penang)
Ipoh
MALAYSIA
Kuantan
Medan
Kuala Lumpur
Melaka (Malacca)
Johor Baharu
SINGAPORE
Riau Archipelago
0° Equator
Barisan Mountains
Padang
Kerinci
elev. 12,484
Sumatra
Palembang
Bandar Lampung
INDIAN OCEAN
Krakatoa
elev. 2667
Jakarta
Cirebon
Tegal
Bandung
Semarang
Surakarta
Mt. Merapi
elev. 9724
Yogyakarta
Malang
Surabaya
Madura
Java
Java Sea
Bali
Denpasar
Lombok
Mataram
Mt. Tambora
elev. 9354
Lesser Sunda Islands
Flores
10°S
INDONE
Luzon Strait
Aparri
Luzon
Sierra Madre
Baguio
Philippine Sea
Mt. Pinatubo
elev. 6683
Olongapo
Quezon City
Pasig
Manila
Mt. Taal
elev. 987
Lucena
Mt. Mayon
Mindoro
PHILIPPINES
Palawan
Panay
Cebu
Negros
Sulu Sea
Zamboanga
Mindanao
Mt. Kinabalu
elev. 13,451
Kota Kinabalu
Sabah
Bandar Seri Begawan
BRUNEI
MALAYSIA
Niah
Sarawak
Kuching
Tarakan
Celebes Sea
Borneo
Pontianak
Kalimantan
Samarinda
Balikpapan
Banjarmasin
Makassar Strait
Sulawesi (Celebes)
Makassar (Ujungpandang)
Manado
100°E
110°E
120°E

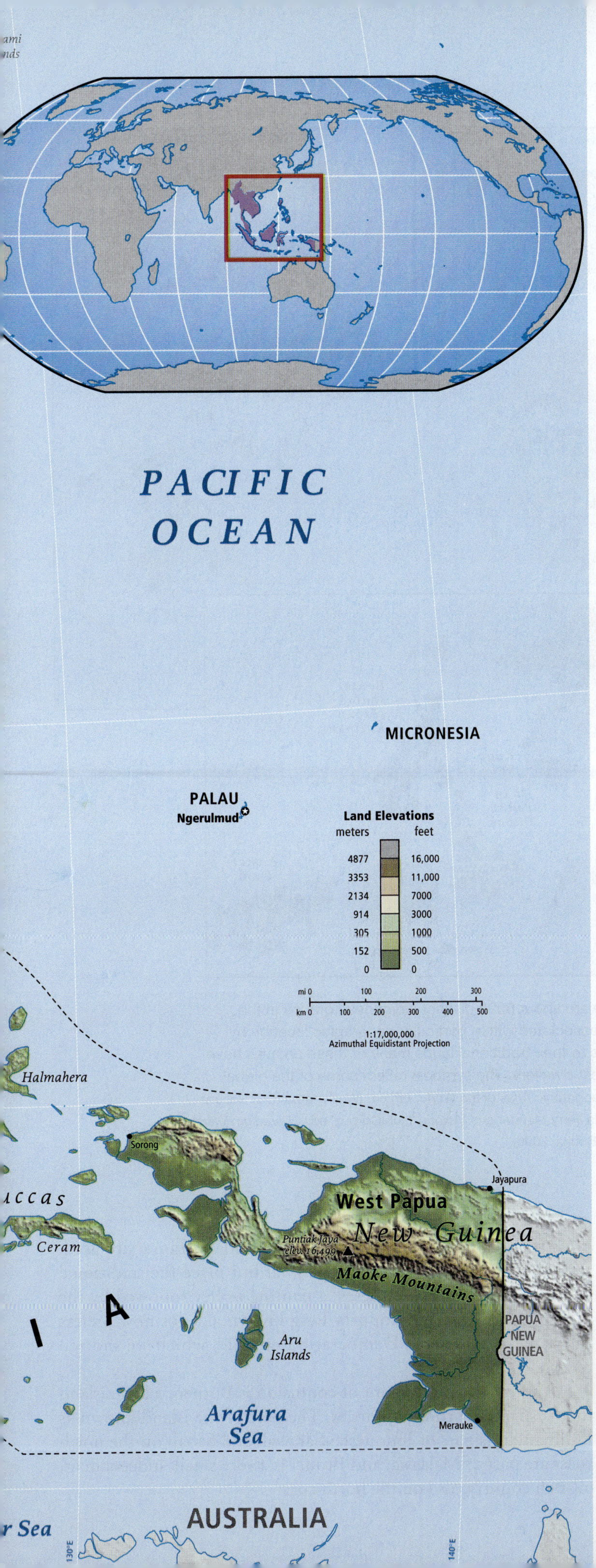

As we saw in the case of Sarawak, it is not unusual for compliant or corrupt governments to help private developers take indigenous lands. The tactics used by the Uma Bawang to regain control of their ancestral lands in Borneo (**Figure 10.1**) mirror the accounts of similar struggles in Middle and South America and highlight a number of themes in this book: the environment, globalization and development, and power and politics.

In particular, this case shows how issues that may seem to have only local significance can be rooted in global trends. The connection of the local to the global presents both challenges and opportunities for the world's indigenous people (**Figure 10.2**). On one hand, the global market provides demand for timber and palm oil—crops that are destroying their rainforests; on the other, support from distant groups like the Borneo Project help tiny communities such as the Uma Bawang have their land claims validated. In fact, the use of high-tech mapping to help indigenous groups secure their legal rights to ancestral lands has now become a global phenomenon. Hundreds of indigenous groups throughout the world are collaborating with mapping specialists, many of them geographers, often resulting in their land rights being formally recognized by governments for the first time. In 2007, the United Nations Department of Economic and Social Affairs passed the Declaration on the Rights of Indigenous People; three reports entitled *State of the World's Indigenous Peoples I, II, III,* which are available online, document the condition and emerging issues of indigenous communities, including mapping efforts.

What Makes Southeast Asia a Region?

Southeast Asia is physically a manifestation of tectonic forces that are described in the next section. Aside from physical commonalities, the peninsula and island countries of Southeast Asia today share a deep cultural past, related to southwestern China, but also to India and the Muslim world. With the exception of Thailand, they all also share more recent experiences with European colonialism. Then, during and after World War II, most countries went through severe political turmoil while gaining their independence, only to embark on rapid journeys through industrialization and urbanization to relative prosperity. There have been ups and downs: Smooth paths to stable democratic processes and institutions have been elusive, urbanization is occurring very rapidly, and development strategies have too often carried grave environmental and social side effects. Still, Southeast Asia's accomplishments, such as rapid industrialization, expansion of the middle class, education for the masses, the empowerment of women, improved public health care and food security, and slower population growth, are admired and emulated by other developing regions.

Terms in This Chapter

The mainland Southeast Asian countries are Myanmar (Burma), Thailand, Lao, Cambodia, Vietnam, Malaysia, and Singapore; the

Figure 10.1 Physical features of Southeast Asia.

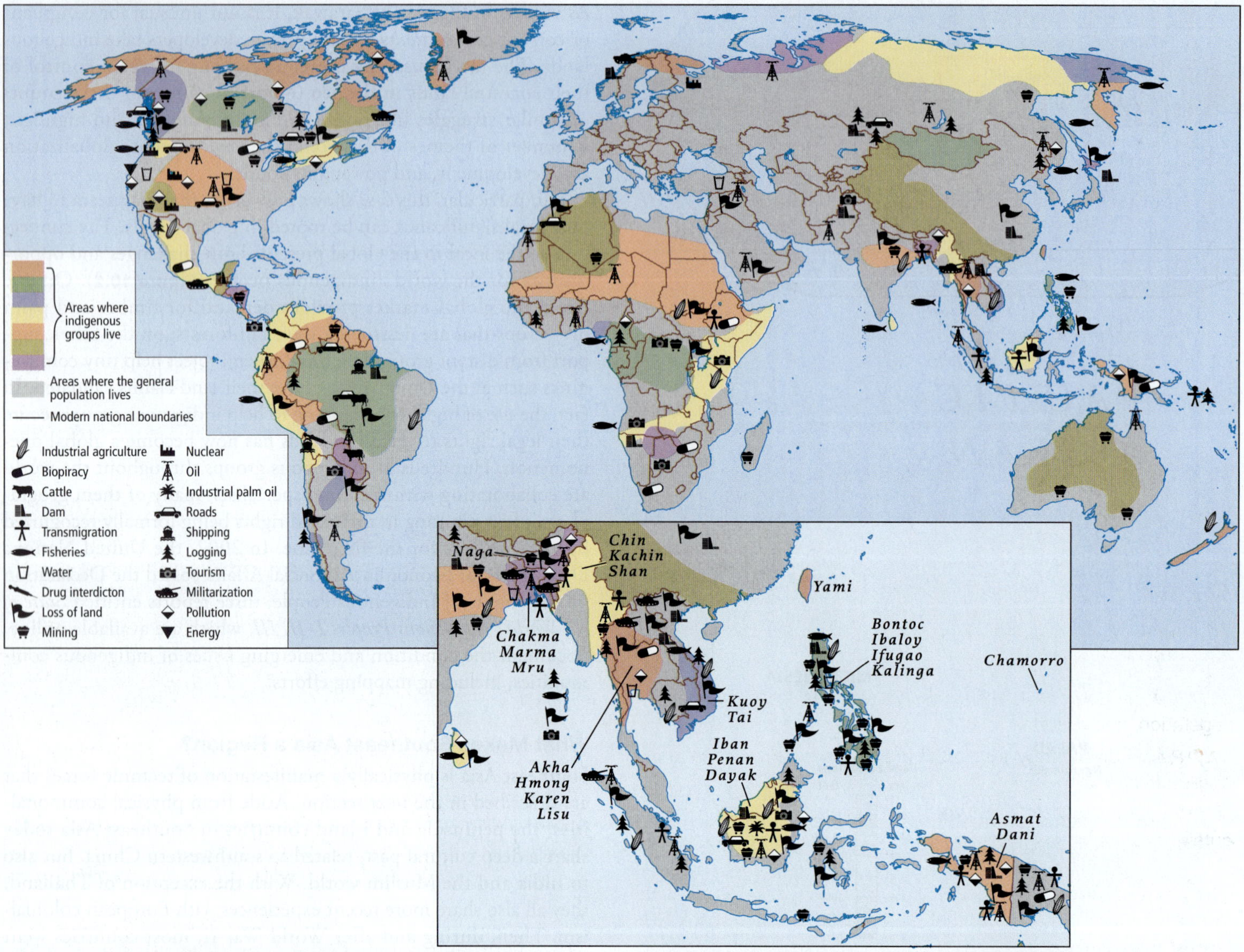

Figure 10.2 Indigenous issues in Southeast Asia and worldwide. There are about 5000 distinct indigenous groups in the world. On this map, each color represents one or more of these groups that are associated with a particular geographic location by their language, culture, and way of life. In order to identify and protect their rights to their traditional lands many of these peoples have participated in community mapping projects, similar to those of the Sarawak forest dwellers. The symbols reflect some of the global issues that affect a particular group. [Research from: "Struggling Cultures," *National Geographic Atlas of the World*, 8th ed. (Washington, DC: National Geographic Society, 2005), p. 15; and "Globalization: Effects on Indigenous Peoples" map, in Jerry Mander and Victoria Tauli-Corpuz, eds., *Paradigm Wars: Indigenous Peoples' Resistance to Economic Globalization* (San Francisco, CA: Sierra Club Books, 2006).]

island countries are Indonesia, Brunei, Timor-Leste (East Timor), and the Philippines. Malaysia occupies part of the peninsula and the northern parts of the island of Borneo (**Figure 10.3**).

Some governments in Southeast Asia choose to dispense with place-names that originated in their colonial past. Governments also make name changes as a political strategy. Such is the case with Burma, where a military government seized control in a coup d'état in 1990, shortly after the election of Aung San Suu Kyi. The military placed her under house arrest and changed the country's name to Myanmar. As discussed later in the chapter, the government of Burma is once again in transition, and in press reports, the names of both Burma and Myanmar are used. In this text, we use the name Myanmar but place the traditional name of Burma in parentheses, a naming policy followed by the U.S. government. The country long known as Laos now prefers the name Lao People's Democratic Republic, which we shorten to Lao.

Another potential point of confusion is Borneo, a large island that is shared by three countries. The part of the island known as Kalimantan is part of Indonesia; Sarawak and Sabah on the north coast are part of Malaysia; and Brunei is a very small, independent, oil-rich country, also on the north coast.

Figure 10.3 Political map of Southeast Asia.

ENVIRONMENT: PHYSICAL AND HUMAN

10.1 Identify Southeast Asia's physical features: its landforms, vegetation, and climates.

10.2 Relate Southeast Asia's most critical environmental issues to climate change.

10.3 Explain how Southeast Asia was settled, dominated by outside groups for centuries, and eventually achieved autonomy.

The physical patterns of Southeast Asia have a continuity that is not immediately obvious on a map. A map of the region shows a unified mainland region that is part of the Eurasian continent and a vast and complex series of islands arranged in chains and groups (Figure 10.1). These landforms are actually related in tectonic origin. Climate is another source of continuity; most of the region is tropical or subtropical.

LANDFORMS

Southeast Asia is a region of peninsulas and islands (see Figure 10.1 and Figure 10.3). Although the region stretches over an area larger than the continental United States, most of that space is ocean; the area of all the region's landmass amounts to less than half that of the contiguous United States. Myanmar (Burma), Thailand, Lao, Cambodia, and Vietnam occupy the large Indochina peninsula that extends south of China. This peninsula itself sprouts the long, thin Malay Peninsula that is shared by outlying parts of Myanmar (Burma) and Thailand, a main part of Malaysia, and the city-state of Singapore, which is built on a series of islands at the southern tip. The **archipelago** (a series of large and small islands) that fans out to the south and east of the mainland is grouped into the countries of Indonesia, Malaysia, Brunei, Timor-Leste (East Timor), and the Philippines. Indonesia alone has some 17,000 islands, and the Philippines, 7000 (**Figure 10.4**).

Figure 10.4 Aerial view of the Philippine archipelago. [Per-Andre Hoffmann/LOOK-foto/Getty Images]

Figure 10.5 The Irrawaddy River gorge. The Irrawaddy flows seaward from the mountains of China through Myanmar (Burma). [CORY RICHARDS/National Geographic Image Collection/Alamy]

The irregular shapes and landforms of the Southeast Asian mainland and archipelago are the result of the same tectonic forces that were unleashed when India split off from the African Plate and gradually collided with Eurasia (see Figure 1.6). As a result of this collision, which is still underway, the mountainous folds of the Plateau of Tibet reach heights of almost 20,000 feet (6100 meters). These folded landforms bend out of the high plateau and turn south into Southeast Asia. There, they descend rapidly and then fan out to become the Indochina peninsula (Figure 10.1). Deep gorges originating in the high mountains (**Figure 10.5**) widen out into valleys that stretch toward the sea. Several rivers flow south from the mountains of the Yunnan–Guizhou Plateau of China to the Andaman Sea, the Gulf of Thailand, and the South China Sea. The major rivers of the Indochina peninsula are the Irrawaddy and the Salween in Myanmar (Burma); the Chao Phraya in Thailand; the Mekong, which flows through Lao, Cambodia, and Vietnam; and the Black and Red rivers of northern Vietnam. Several of these

archipelago a group, often a chain, of islands

Figure 10.6 The Mekong River delta. Ho Chi Minh City (formerly Saigon) is located on the Mekong River delta. [Keren Su/DanitaDelimont/Newscom]

Figure 10.7 Volcano in Indonesia. Anak Krakatau volcano erupts at Sunda strait in South Lampung, Indonesia, on December 23, 2018. Although volcanoes are destructive, they also deposit new materials that ultimately enrich tropical soils for agriculture. [ANTARA FOTO/REUTERS/Newscom]

Figure 10.8 Post-earthquake/tsunami damage in Sumatra, Indonesia. Banda Aceh in northern Sumatra was one of the areas most devastated by the 2004 earthquake and tsunami that claimed 230,000 lives. [KAZUHIRO NOGI/AFP/Getty Images]

rivers have major delta formations—especially the Irrawaddy, Chao Phraya, and the Mekong—that are intensively cultivated and settled (see "Food Production and Climate Change"; see also **Figure 10.6**).

The curve formed by Sumatra, Java, the Lesser Sunda Islands (from Bali to Timor), and New Guinea conforms approximately to the shape of the Eurasian Plate's leading southern edge (see Figure 1.6). Where the Indian–Australian Plate plunges beneath the Eurasian Plate along this curve, hundreds of earthquakes occur and volcanoes abound, especially on the islands of Sumatra and Java (**Figure 10.7**). Volcanoes and earthquakes are also common in the Philippines where the Philippine Plate pushes against the eastern edge of the Eurasian Plate and is in turn pushed by the Pacific Plate. The volcanoes of Southeast Asia are considered part of the Pacific Ring of Fire (see Figure 1.7).

Volcanic eruptions and the mudflows and landslides that follow eruptions complicate and endanger the lives of many Southeast Asians. Earthquakes are also especially problematic because of the tsunamis they can set off (**Figure 10.8**). The tsunami of December 2004, triggered by a giant earthquake (9.1 in magnitude) just north of Sumatra, swept east and west across the Indian Ocean, taking the lives of 230,000 people (170,000 in Aceh Province, Sumatra) and injuring many more. It was one of the deadliest natural disasters in recorded history. Since then, there has been a series of strong earthquakes, with several along coastal Sumatra in August and September of 2009, and April and May of 2010. In July 2018 a 6.4 magnitude earthquake hit a tourist center in Lombok, Indonesia, killing 14 people, and in September 2018 an earthquake and tsunami hit Sulawesi, killing 832 people. Between the 17th and 20th of October 2018, a series of nine earthquakes (magnitude 4.8 to 5.5) struck the region from the Philippines to Papua New Guinea to the Sulawesi islands east of Borneo; they did not generate notable tsunamis.

Sundaland part of the submerged Eurasian continental shelf that, when previously above sea level, served as a bridge between the mainland and islands

The now-submerged shelf of the Eurasian continent, known as **Sundaland** (**Figure 10.9**), extends under the Southeast Asian peninsulas and islands. It was above sea level during the recurring ice ages of the Pleistocene epoch, when much of the world's water was frozen in glaciers. The exposed shelf allowed ancient people (including *Homo erectus*) and Asian land animals (such as elephants, tigers, rhinoceroses, and orangutans) to travel south to what became the islands of Southeast Asia when sea levels rose.

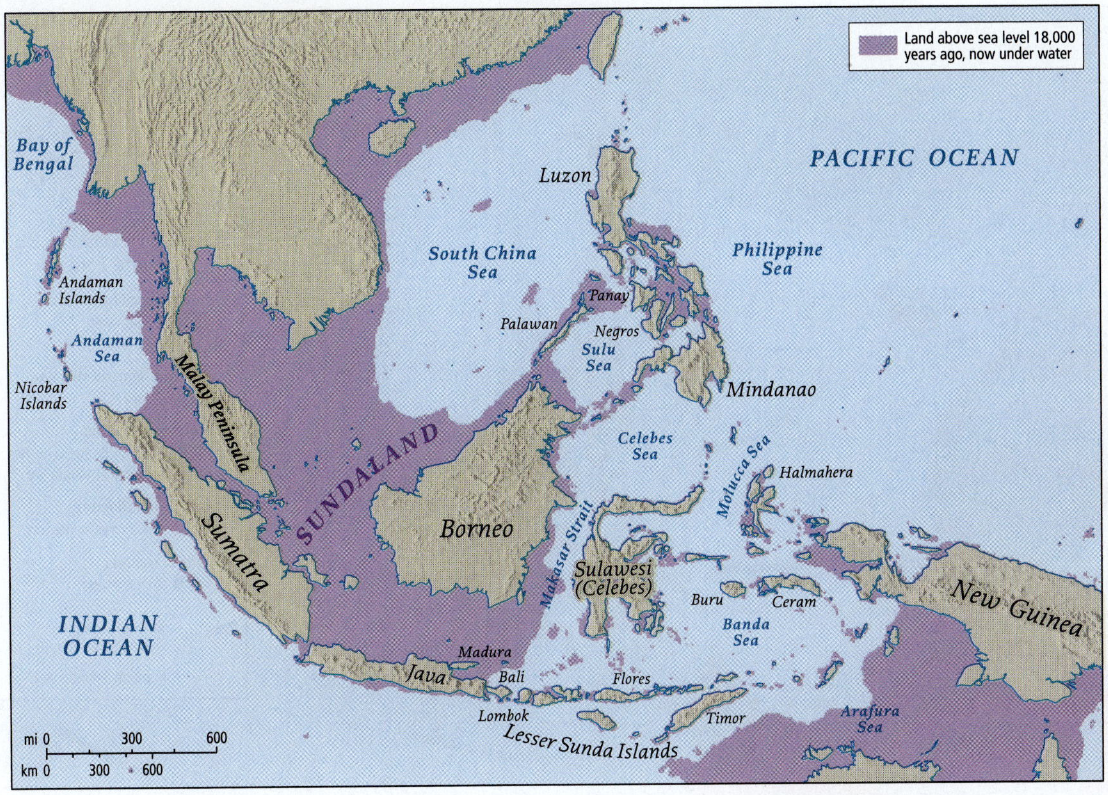

Figure 10.9 Sundaland as it was 18,000 years ago, at the height of the last ice age. The now-submerged shelf of the Eurasian continent that extends under Southeast Asia's peninsulas and islands was exposed during the last ice age and remained above sea level until about 16,000 years ago, when that ice age was ending.

CLIMATE AND VEGETATION

The largely tropical climate of Southeast Asia is distinguished by continuous warm temperatures in the lowlands—consistently above 65°F (18°C)—and heavy rain (**Figure 10.10**). The rainfall is the result of two major processes: the monsoons (seasonally shifting winds) and the movement of the intertropical convergence zone (ITCZ), an area centered roughly at the equator, where moisture-laden surface winds converge from the Northern and Southern Hemispheres and rise upward, resulting in rainfall. The wet summer (monsoon) season extends from May to October, when the warming of the Eurasian landmass sucks in moist air from the surrounding seas and pulls the ITCZ northward (see the Figure 10.10 map). Between November and April, there is a long dry season on the mainland, when the seasonal cooling of Eurasia causes dry, cool air from the interior of the continent to flow out toward the sea, pushing the ITCZ southward (see Figure 11.5). On the many islands, however, the winter can also be wet because the air that flows from the continent warms and picks up moisture as it passes south and east over the seas. The air releases its moisture as rain after ascending high enough over elevated landforms to cool. With rains coming from both the monsoon and the ITCZ, the island parts of Southeast Asia are among the wettest areas of the world.

Irregularly every 2 to 7 years, the normal patterns of rainfall are interrupted, especially in the islands, by the El Niño phenomenon (see Figure 11.6, model of El Nino). In an El Niño event, the usual patterns of air and water circulation in the Pacific are reversed. Ocean temperatures are cooler than usual in the western Pacific near Southeast Asia. Instead of warm, wet air rising and condensing as rainfall, cool, dry air sits on the ocean surface. The result is severe drought, often with catastrophic effects for farmers and for tinder-dry forests that regularly catch fire. The cool El Niño air can trap smoke and other pollutants at Earth's surface, creating unusually toxic smog and such low visibility that planes sometimes crash.

The soils in Southeast Asia are typical of the tropics. Although not particularly fertile, they will support dense and prolific vegetation when left undisturbed for long periods. The warm temperatures and damp conditions promote the rapid decay of **detritus** (dead organic material) and the quick release of useful minerals. These minerals are taken up directly by the roots of the living forest rather than enriching the soil. Because rainfall is usually abundant during the summer wet season (when drought is only episodic), this region has some of the world's most impressive forests in both tropical and subtropical zones (see Figure 10.10). On the mainland, human interference, coupled with the long winter dry season, has reduced the forest cover. The relationships between humans and the environment that affect forests elsewhere in the region are discussed in the opening vignette and in the next section.

CHECK YOUR UNDERSTANDING

1. Describe the tectonic forces that caused the shape and landforms of the Southeast Asian mainland and archipelagos.
2. Describe the temperature and moisture conditions that normally characterize Southeast Asian lowlands and islands. How do these conditions change during irregular El Niño events?

detritus dead organic material (such as plants and insects) that collects on the ground and in the soil

10.10 Climates of Southeast Asia

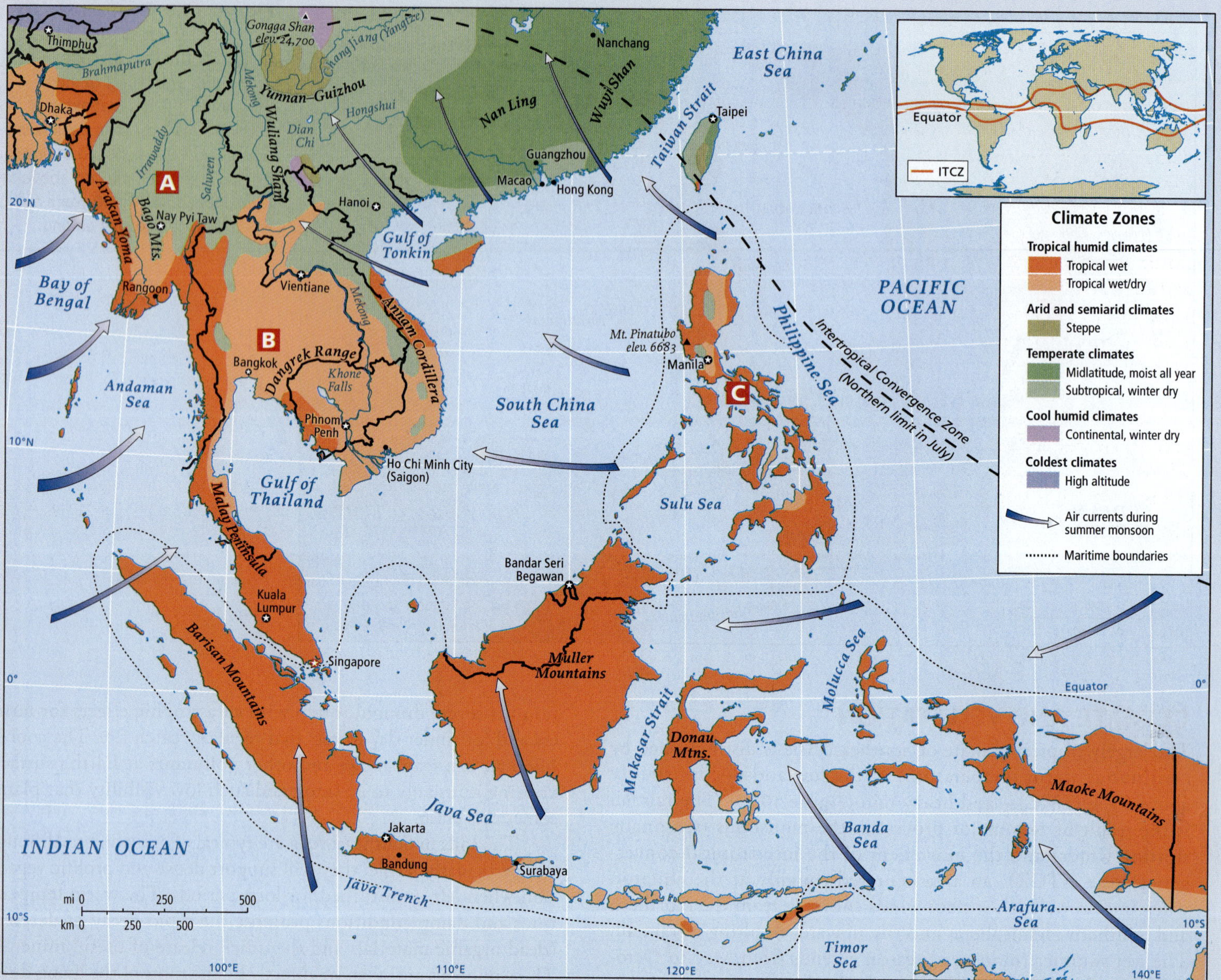

The presence of abundant wet, tropical air, combined with the movements of the intertropical convergence zone (ITCZ) across the region twice a year, makes Southeast Asia one of the wettest regions in the world.

A Subtropical, winter dry, Myanmar (Burma) People riding elephants in the winter in the Bago Mountains of Myanmar (Burma). [Frans Lemmens/Getty Images]

B Tropical, wet/dry, Khao Yai, Thailand An elephant herd grazes in a grassland opening in the tropical forests of Thailand. [Feargus Cooney/Getty Images]

C Tropical, wet, Alaminos, Philippines Filipino tourists bathe in Hidden Valley mineral springs in Alaminos. [Danita Delimont/Gallo Images/Getty Images]

DEFORESTATION AND CLIMATE CHANGE

Southeast Asia has the second-highest rate of deforestation of any world region, after sub-Saharan Africa. With only 5 percent of the world's forests, Southeast Asia has accounted for nearly 25 percent of worldwide deforestation in the last 10 years. The forces driving deforestation in Southeast Asia are global. Map A in **Figure 10.11** shows the supply chain that takes wood from various tropical regions primarily to China, where it is turned into furniture and other products. Map B shows the destinations of tropical wood furniture primarily from China but also from Vietnam, Malaysia, and Indonesia. As you can see, North America and the European Union (EU) nations are the major consumers. Therefore in the global chain of tropical wood consumption, the producers, the processors, and the consumers all play major roles.

Deforestation and the conversion of land to agricultural uses in Southeast Asia are having a profound impact both on regional ecosystems and on the entire global biosphere. Enormous amounts of carbon dioxide (CO_2) are released when forests are removed and the underbrush is burned. Since live trees absorb CO_2, fewer trees also mean that less CO_2 is absorbed from the atmosphere. Every day, 13 to 19 square miles (34 to 50 square kilometers) of Southeast Asia's rain forests are destroyed, much of it illegally. A great deal of this deforestation takes place in Malaysia, Myanmar (Burma), and Indonesia, where the logging of tropical hardwoods for sale on the global market is a major activity (see Figure 10.11). For decades, these activities have made Indonesia among the world's biggest contributors to global climate change. Since 1950, the tropical forest cover on the Indonesian part of the island of Borneo alone has decreased by about 60 percent.

The effects of all this clearing are both local and global. One effect is that indigenous people of the forest lose their living space and resources when the land is given over to export agriculture. A second is that pesticide use increases as agribusinesses try to protect oil palm plantations, which are now driving much of the deforestation in Southeast Asia. Another effect, particular to Southeast Asia, is the loss of habitat for orangutans (**Figure 10.12**), the Sumatran tiger, the Sumatran rhinoceros, and tens of thousands of other lesser known species that are displaced when rain forests are converted to agricultural uses. Finally, deforestation processes (vegetation decay, drying out of the land, and fires) cause the release of greenhouse gases

Figure 10.11A Trade in tropical timber. Large quantities of tropical timber sold to China (Map A) are made into furniture, plywood, and flooring, which are then sold to consumers in the developed world. Most of the consumer demand for tropical timber products (Map B) is in North America, Europe, and Japan. The trade shown in the graphics is mostly legal, but much of the wood that fuels China's wood-processing industries is actually harvested illegally in Southeast Asia. In 2006 the World Wildlife Fund estimated that more than 40 percent of the timber in Indonesia was cut and traded illegally. The data mapped are the latest trade flow data available (2016). [Data from: Frances Maplesden, "Situation and Trends in World Timber Trade and the Tropical Timber Trade in Asia-Pacific," presentation to Asia-Pacific Forestry Week, Clark Freeport Zone, Pampanga, Philippines, February 22–26, 2016; Biennial Review and Assessment of the World Timber Situation, 2015–2016, pp. 19, 20, 34; and International Tropical Timber Organization (ITTO).]

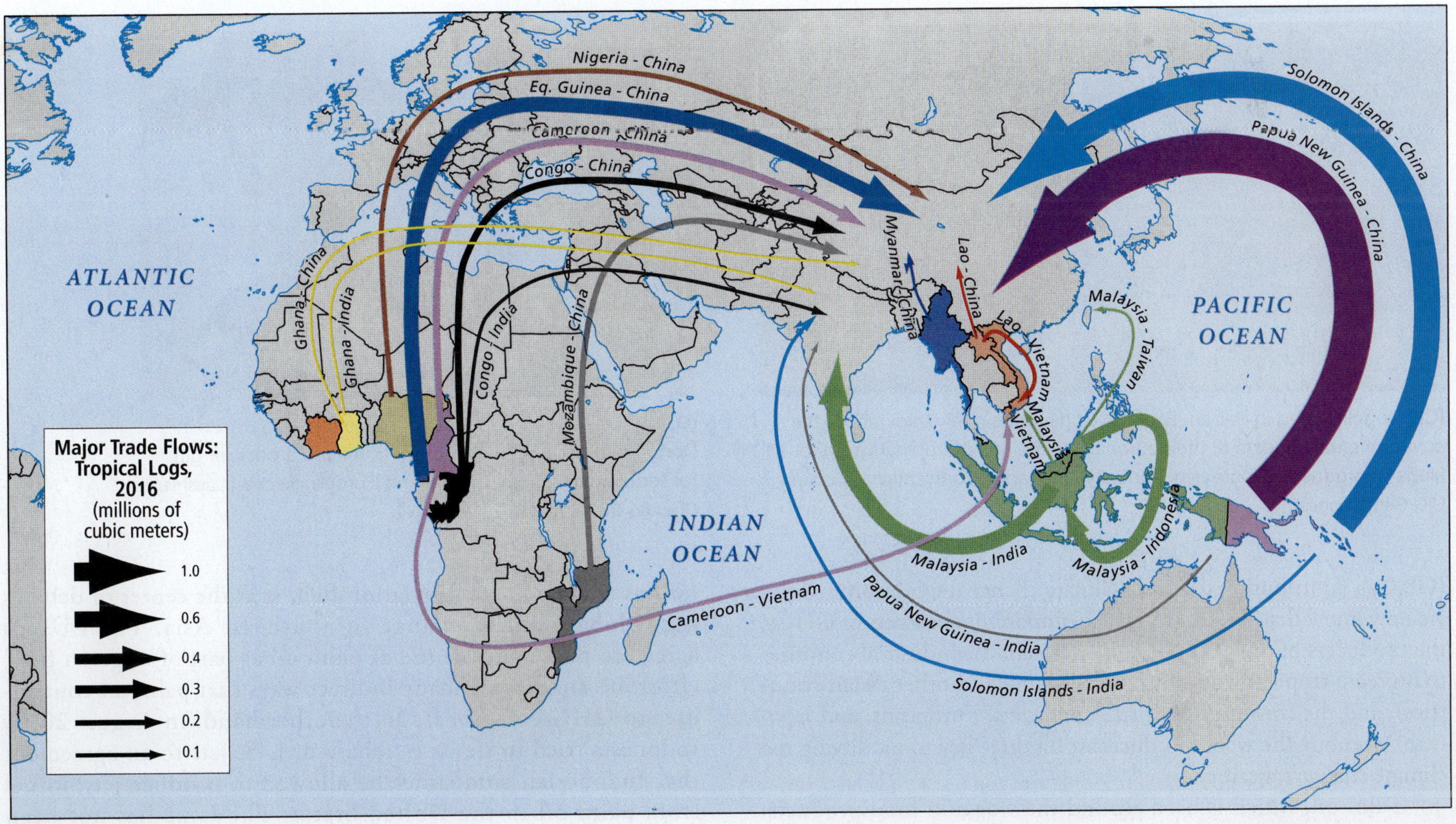

(A) Origins and destinations of industrial tropical timber.

Figure 10.11B, C, D

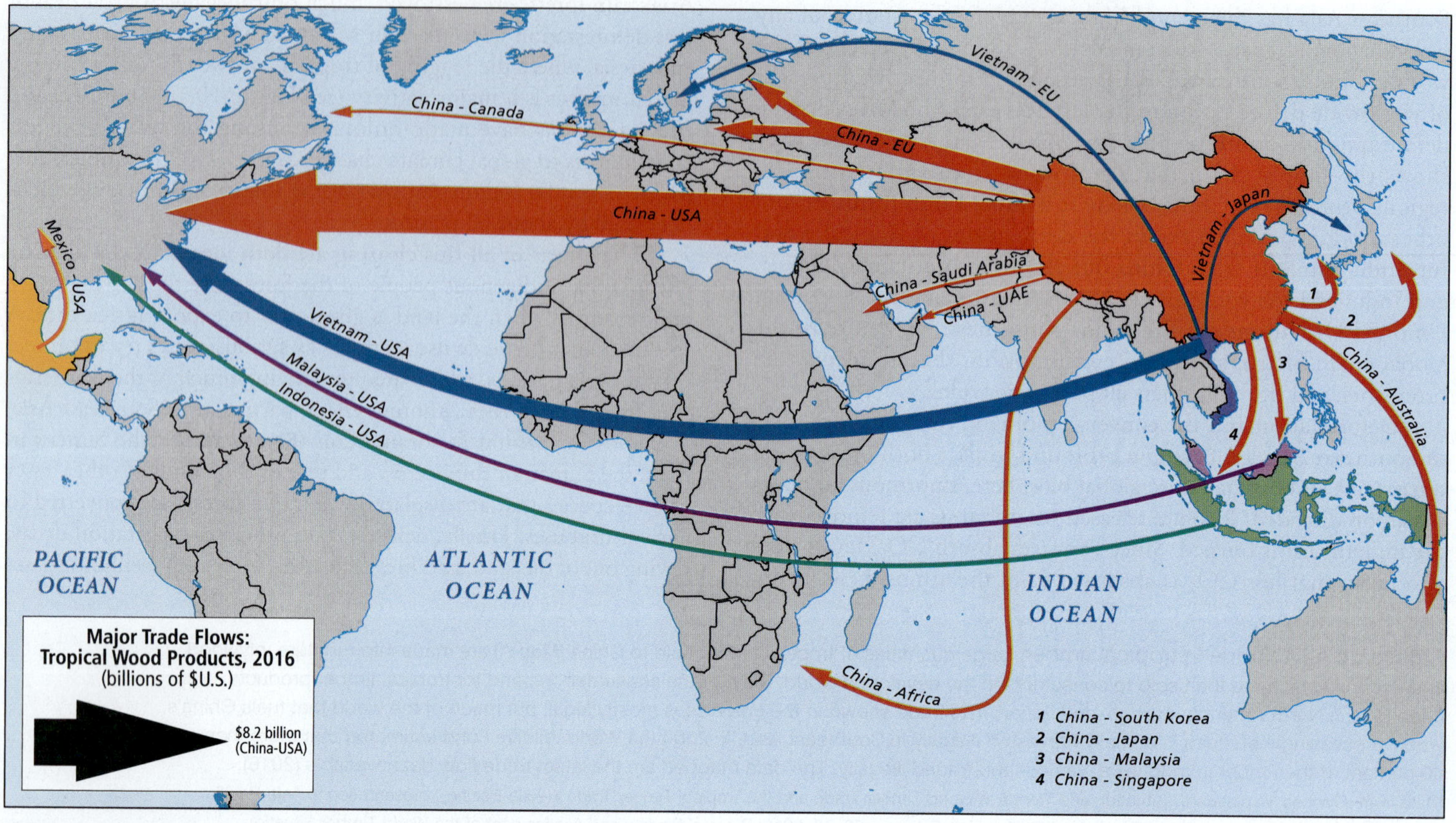

(B) Trade flows of tropical timber products: furniture.

(C) A logging road in Sabah, Borneo. Consumers rarely think about the environmental impacts of their purchases. The use of tropical wood for all kinds of products is increasing globally with rapid urbanization. [Auscape/UIG/Getty Images]

(D) Tropical hardwoods are made into toy train tracks in Dongguan, China. Little wooden trains are a popular gift for toddlers in North America and Europe. [Michael Lassman/Chicago Tribune/MCT/Getty Images]

(GHGs), contributing to global climate change (**Figure 10.13**). Since we now know that tropical forests are important absorbers of GHGs, the producers of the wood (Southeast Asia), the industrial countries that create tropical wood products (China and other Asian countries), and the consumers (North Americans, Europeans, and Japanese) who buy the wood products are all three key to modifying the climate change trajectory.

Palm oil, which is used around the world in food products, as a cooking oil, in soaps and cosmetics, as a machine oil, and now as a component of aviation fuel, is at the center of debates over global climate change in Southeast Asia. The EU has agreed to phase out the use of palm oil as part of a "green fuel" effort, because of the many indirect ways that palm oil contributes to GHG emissions. On the other hand, in August 2018, Indonesia tried to tie its purchase of U.S. jets to an agreement that Indonesian companies be allowed to produce jet biofuel from palm oil in the United States. The issue has not been resolved.

Figure 10.12 A wild female orangutan and her baby eat flowers in Tanjung Puting National Park in Kalimantan, Indonesia. The deforestation and fragmentation of habitats caused by roads have rendered many plants, such as the Kalimantan mango, extinct; loss of such foods contributes to the endangerment of orangutan species (orangutan means *man of the forest* in Malay). [Paula Bronstein/Getty Images News/Getty Images]

Figure 10.13 A satellite image of fires used to clear forests for agriculture in Indonesia and Malaysia. The CO_2 produced contributes to climate change and degrades local air quality. [Jesse Allen/Earth Observatory/MODIS/NASA]

The oil palm plant originated in West Africa and was brought to Southeast Asia early in the colonial era. Today, Indonesia and Malaysia are the world's largest palm oil producers, and the expansion of their oil palm plantations has resulted in extensive deforestation. Much of the deforestation is in lowland peat swamps, which, in a natural state, are capable of storing large amounts of carbon in their soils. To create space for oil palm plantations, the swamps are drained and most of their vegetation is burned (see Figure 10.13). Often the fires spread underground to the peat beneath the forests, smoldering for years after the surface fires have been extinguished. These subsurface fires also release carbon into the atmosphere—the daily carbon emissions of fires in 2015 were so enormous that they surpassed the emission output of the entire United States. In recent years, smoke from burning forests and peat has periodically covered much of the region, dramatically reducing air quality and even making it necessary for people in rural areas to wear gas masks.

Surprisingly, palm oil has been promoted as a potential solution to global climate change because it can be converted into fuel for transport vehicles. The claim is that because the palm trees absorb CO_2 from the atmosphere just as the original forest did, palm oil could be considered a "carbon-neutral" fuel. This is a fallacy because palm trees do not absorb carbon at a rate equivalent to natural rain forests—and if forests are burned to clear land for plantations, it can take decades to offset the initial carbon emissions produced by deforestation. If a peat swamp is cleared for an oil palm plantation, more carbon is released through the burning of subsurface peat than centuries of CO_2 absorption by growing oil palms could counteract.

Even as deforestation proceeds across the region, many people are beginning to understand the costs of removing so many trees. Forests and peat swamps are now known to perform billions of dollars per year in services: well-managed forests provide renewable plant and animal food, medicine and construction materials, and domestic fuel resources. The peat swamps act like giant sponges, soaking up water and thus mitigating floods. Forests and swamps also purify fresh water, preserve biodiversity, and absorb enormous amounts of CO_2.

CHECK YOUR UNDERSTANDING

1. What happens to CO_2 in the atmosphere when there are fewer trees due to legal or illegal logging?
2. How does the expansion of oil palm cultivation lead to excessive increases in the release of carbon into the atmosphere?
3. Judging from the data in Figure 10.11A, B, C, and D, compare the roles of the United States and China in the global chain of tropical wood product consumption.

FOOD PRODUCTION AND CLIMATE CHANGE

Food production contributes to global climate change in two ways. First, it is a major contributor to deforestation, and second, some types of cultivation actually produce significant GHGs.

Shifting Cultivation

Shifting cultivation (also known as *slash-and-burn* or *swidden* cultivation) has been practiced sustainably for thousands of years in the hills and uplands of both mainland and island Southeast Asia (**Figure 10.14 map**). To maintain soil fertility in these warm, wet environments where nutrients are quickly lost to decay, traditional farmers move their fields every 3 years or so, letting old plots lie fallow for 15 years or more. The regrowing forest on once-cleared fields not only regenerates the soil and inhibits runoff, it also absorbs significant amounts of CO_2 from the atmosphere.

A Tapping rubber on a plantation in Cambodia. [Thierry FALISE/Gamma-Rapho/Getty Images]

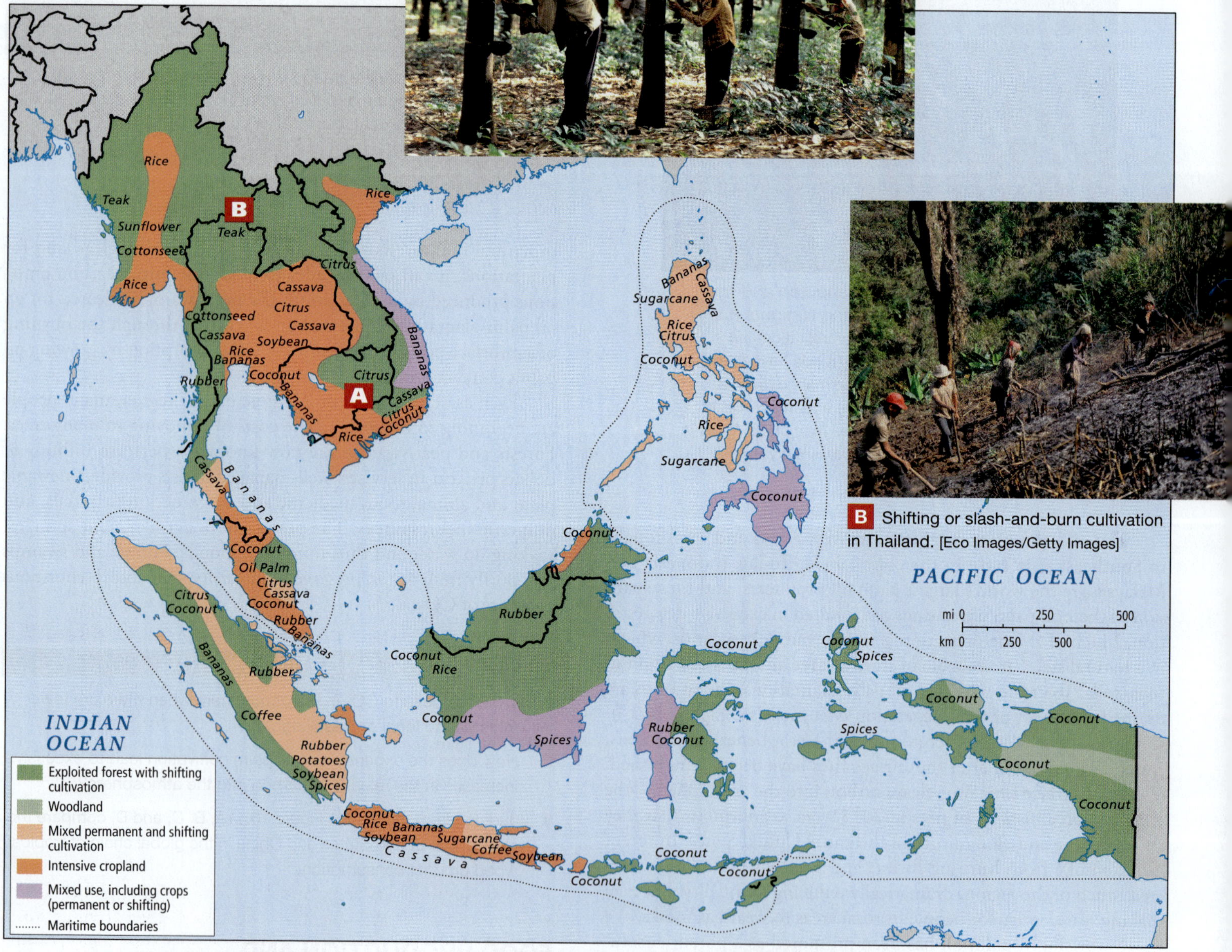

B Shifting or slash-and-burn cultivation in Thailand. [Eco Images/Getty Images]

Figure 10.14 Agricultural patterns in Southeast Asia. Little virgin forest remains in this region. Tropical forestlands that persist are still used for subsistence (food) and small-scale cash crops like coconut and spices. A wide variety of commercial crops, such as rice, oil palm, rubber, bananas, citrus, and coffee, are grown for local consumption and export. [Research from: *Hammond Citation World Atlas* (Maplewood, NJ: Hammond, 1996), pp. 74, 83, 84.]

However, if fallow periods are shortened or are disrupted by logging, soil fertility can collapse, making future cultivation impossible. In some cases, fertility can be temporarily restored through the use of chemical or organic fertilizers, but these can be too expensive for most farmers, and chemical fertilizers may lose effectiveness over time or be too unhealthy for food cultivation. Tropical soils left bare of forest for too long eventually turn into hard, infertile, sunbaked clay called *laterite*.

Where population densities are relatively low, subsistence farmers can practice *sustainable shifting cultivation* indefinitely, but to allow for the long fallow periods necessary to restore fertility, this system requires larger areas than other types of agriculture. As population density increases, farmers are usually forced to shorten fallow periods, thus inhibiting forest regrowth and soil regeneration. As more and more plots are cleared, they become too close to one another. Where rural population densities are high, shifting cultivation now accounts for a significant portion of the region's deforestation. Moreover, because forest debris is usually burned, shifting cultivation can result in wildfires. This is especially true during an El Niño period, when rainfall is low. El Niño–induced wildfires have increased in recent years, particularly in Indonesia, further contributing to deforestation and carbon emissions there.

Wet Rice Cultivation

Another major indirect contributor to global climate change is Southeast Asia's most productive and intensive form of agriculture. *Wet rice cultivation* (sometimes called *paddy rice*) entails planting rice seedlings by hand in flooded and often terraced fields on land that was once forested (**Figure 10.15**). The steep highland fields are cultivated by hand, while lowland fields are cultivated with hand-guided plows pulled by water buffalo or tractors. Wet rice cultivation has transformed landscapes throughout Southeast Asia. It is practiced throughout this generally well-watered region, especially on rich volcanic soils and in places where rivers and streams bring a yearly supply of fresh silt.

Figure 10.15 Highland rice terraces in the Philippines. The terracing of steep slopes has a long history across Asia. It was a strategy to increase cultivable land by making thousands of small flat shelves that would control erosion in high-rainfall areas and hold soil and moisture for plants. [Per-Andre Hoffmann/LOOK-foto/Getty Images]

The flooding of rice fields also results in the production of methane, a powerful GHG that is responsible for about 20 percent of GHG emissions. It is estimated that up to one-third of the world's methane is released from flooded rice fields, where organic matter in soil undergoes fermentation as oxygen supplies are cut off. Wet rice has been cultivated for thousands of years, but the increase in human population in the last 25 years has driven a 17 percent expansion of the area devoted to wet rice. The map in Figure 10.14 shows patterns of field and forest crops across the region.

Commercial Agriculture

In step with a global trend, during the recent decades of relative prosperity in Southeast Asia, small farms once operated by families have been combined into large commercial farms owned by local or multinational corporations. These farms produce cash crops for export, such as rubber, palm oil, bananas, pineapples, tea, and rice (see the inset photo on the chapter opening page and Figures 10.14 and 10.15). Commercial farming on large tracts of deforested land reduces the need for labor by using mechanization, irrigation, and chemicals for fertilizer and pest control. The objectives of commercial farming are to generate big yields and quick profits, not to develop long-term sustainable agriculture. Many commercial farmers have achieved dramatic boosts in harvests (especially of rice) by using high-yield crop varieties, the result of green revolution research that has been applied in many parts of the world (see Chapter 8).

As we have noted in relation to other world regions, large-scale commercial (so-called *green revolution*) farming ultimately has significant negative environmental effects, including the loss of wildlife habitat and hence the loss of biodiversity (see Figure 10.12); increases in soil erosion, flooding, and chemical pollution; and the depletion of groundwater resources. In addition, subsistence farmers unable to afford the new technologies find they cannot compete with large commercial farms and are forced to migrate to cities to look for work.

CHECK YOUR UNDERSTANDING

1. Considering the differences between plantation and slash-and-burn cultivation, which of the two is more likely to allow native plants to survive, thus preserving biodiversity?
2. How might the different goals of the various agricultural systems affect climate change patterns?
3. What are some important ways in which development practices are contributing to food insecurity?

WATER AND CLIMATE CHANGE

Many of Southeast Asia's potential vulnerabilities to global climate change are related to water resources. A number of vulnerabilities rooted in warming temperatures may affect the region's water resources and hence its economy and food supply; they include glacial melting, increased evaporation, loss of freshwater habitats, coral reef bleaching, heavier storms, and storm surge flooding. **Figure 10.16** illustrates four of these vulnerabilities. Additionally,

10.16 Vulnerability to Climate Change in Southeast Asia

Southeast Asia's vulnerability to climate change relates to how nature supplies water to the region and how water ecosystems are affected by climate and human use. Although the region is one of the wettest on Earth, water is not a secure resource. Several rivers begin on the remote Tibetan plateau as glacial melt. However, the Tibetan glaciers are melting at rates faster than they can be replenished; eventually the rivers could dry up. Warmer air is increasing evaporation rates, thus desiccating crops and lowering lake levels. Warming oceans produce more frequent and intense rainstorms that bring floods. Here we explore vulnerabilities related to flooding, high rates of evaporation, coral reef bleaching, and tropical storms.

THINKING GEOGRAPHICALLY

A **The text describes several causes of coral bleaching. Explain the link between coral bleaching and human activity in the case of rising ocean temperatures.**

C **How might flood-resistant cities help people adapt to climate change?**

D **Typhoons are particularly significant for people living on which landforms?**

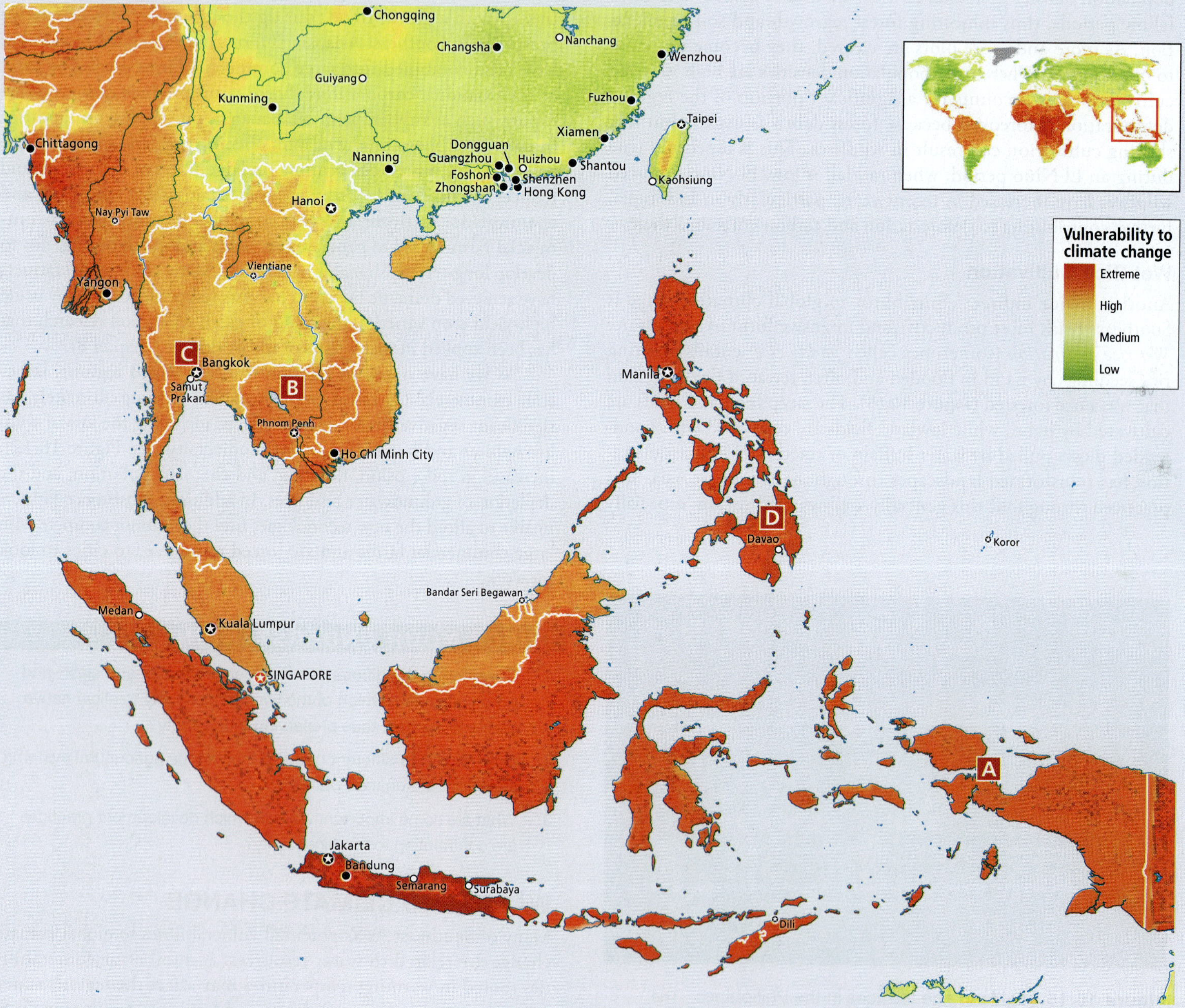

A A partially bleached coral in Cenderawasih Bay, West Papua, Indonesia. Ocean temperatures are rising, resulting in more coral bleaching events that severely degrade reefs, which are nurseries for many aquatic life forms. [Reinhard Dirscherl/Getty Images]

B A husband and wife harvest shrimp and fish on Cambodia's Tonlé Sap Lake. Sixty percent of Cambodia's protein comes from this inland freshwater lake. Higher temperatures may be increasing rates of evaporation from the lake, causing water levels to drop and fish catches to decline. [LEMAIRE Stéphane/hemis.fr/Getty Images]

C Residents of Bangkok, Thailand, paddle along a flooded street. Catastrophic flooding is becoming more common in the city. Climate change could increase flooding along rivers as patterns of rainfall and glacial melting change. [Roland Neveu/LightRocket/Getty Images]

D A father and children on Mindanao, Philippines, sift through the wreckage of Typhoon Bopha in search of coconuts. Typhoons (hurricanes) are projected to increase in frequency as a result of the warmer temperatures created by climate change. [STR/AFP/Getty Images]

human intervention in river systems, especially efforts both to control the water and to use it for electricity generation, further increases vulnerability by interfering with ecological balance.

Melting Glaciers and Increased Evaporation

Like much of Asia, mainland Southeast Asia's largest rivers (the Irrawaddy, Salween, Mekong, and Red Rivers) are fed during the dry season (November through March) by glaciers high in the Himalayas. These glaciers are now melting at a rate that could result in their eventual disappearance. As glacial melting accelerates, the immediate risk is catastrophic flooding. While some areas have long been adapted to dramatic seasonal floods, which have taken place in this region for most of human history, other areas are unprepared for floods (see Figure 10.16C).

A longer-term concern is that dry-season flows in the rivers where glacial meltwater has been the primary source of stream flow will be reduced to a trickle. As much as 15 percent of this region's rice harvest depends on the dry-season flows of the major rivers. The loss of these harvests would reduce many farmers' incomes and create food shortages for them and the cities they supply. Coming on top of a global rise in food prices in recent years, this would place further strain on the well-being of poor people throughout Southeast Asia.

The higher temperatures associated with current trends in global climate change mean evaporation rates will rise, resulting in drier conditions in fields, lower lake levels, and lower fish catches because of changing habitats for aquatic animals (see Figure 10.16B). Evaporation that results in reduced river and groundwater levels can also cause saltwater intrusions into estuaries and freshwater aquifers, further threatening fisheries.

Coral Reef Bleaching

Global climate change is expected to increase sea temperatures in Southeast Asia, threatening the coral reefs that sustain much of the region's fishing and tourism industries. A coral reef is an intricate structure that is composed of the calcium-rich skeletons of millions of tiny living creatures called coral polyps. The polyps are subject to **coral bleaching** (see Figure 10.16A), or color loss, which results when photosynthetic algae that live in the corals are expelled by a variety of human-instigated or naturally occurring changes. Both rising water temperature related to climate change and oceanic acidification resulting from the absorption of excess CO_2 in the atmosphere can cause bleaching. Ocean acidification is thought to have increased by about 30 percent over the last 340 years, a rate of change that may exceed the ability of ocean creatures to adapt. Bleaching also occurs in response to pollution from urban sources, such as sewage discharges, storm-water runoff, and industrial water pollution.

Fish catches can be affected by bleaching as well. Many of the fish caught in Southeast Asia's seas depend on healthy coral reefs for their reproduction and survival. Under normal conditions, coral that is assaulted with just one of the bleaching situations recovers within weeks or months. However, severe or repeated bleaching can cause corals to die. Unprecedented global coral bleaching events affecting roughly half of the world's coral reefs took place repeatedly from 1998 to 2017, and regional events from the Western Indian Ocean to the Great Barrier Reef off Australia have killed from one-third to one-half of these reefs. The thousands of rural communities throughout coastal Southeast Asia that rely on fish for food are thus also threatened by coral bleaching. So far, however, the greatest observable impacts on people have been in the tourism industry. In the Philippines, the coral bleaching of 1998 brought a dramatic decline in tourists who normally came to dive the country's usually spectacular reefs, resulting in a loss of about U.S.$30 million to the economy.

coral bleaching color loss that results when photosynthetic algae that live in corals are expelled due to warming sea water or acidification as excess CO_2 is absorbed

Storm Surges and Flooding

Although the relationship is not yet entirely understood, violent tropical storms seem to be increasing in both number and intensity as the climate warms. When Typhoon Haiyan hit in 2013, it became the strongest typhoon to make its way onto land in recorded history. Haiyan struck the Philippines, killing at least 5600 people and leaving millions homeless. Since then typhoons have intensified by at least 12 percent at the same time that coastal ocean waters, which are known to fuel the energy of tropical storms, have warmed. Climatologists studying data on tropical storms predict that the entire Southeast Asian region will have a somewhat higher number of typhoons over the next few years and that the duration of peak winds along coastal zones will increase. This is significant because many poor, urban migrants have crowded into precarious dwellings in low-lying coastal cities such as Manila, Bangkok, Rangoon, and Jakarta, where coping with high winds, flooding, and the aftermath of storms will likely be a frequent experience (see Figure 10.16C).

CHECK YOUR UNDERSTANDING

1. Typhoons are particularly significant for people living on which landforms?
2. How is wet rice cultivation threatened by glacial melting in the Himalayan Mountains?
3. The text describes several causes of coral bleaching. Explain the link between coral bleaching and human activity in the case of rising ocean temperatures.
4. Why would glacial melting in distant lands concern people in Southeast Asia?
5. Describe four ways in which climate change appears to be affecting life in Southeast Asia.

RESPONSES TO CLIMATE CHANGE IN THE MEKONG RIVER SYSTEM

Many of the drivers of climate change lie beyond the policy control of Southeast Asians. Glacial melting is taking place beyond their borders. Temperature increases in oceans and lakes and increased evaporation rates must be addressed at the global scale. But Southeast Asians can contribute to reducing CO_2 emissions by limiting the removal of forests for lumber exports and commercial agriculture.

Another way that the people of this region can moderate climate change is to reduce their own fossil fuel consumption by turning to alternative sources of energy. Regional demand for energy has doubled and doubled again just during the last decade as modernization increases. Both the Philippines and Indonesia have significant potential for generating electricity from *geothermal energy* (heat stored in Earth's crust). This energy is particularly accessible near active volcanoes, which both countries have in abundance. The Philippines already generates 27 percent of its electricity from geothermal energy (2015) and has 29 more projects under development. In Indonesia, by some estimates, geothermal energy could eventually provide a majority of energy needs; 62 projects are under development there. Solar energy is another attractive source, given that the entire region lies near the equator, the part of the planet that receives the most solar energy. Wind energy may be cost-effective in Lao and Vietnam, where many population centers are in high-wind areas. Despite all these options, the energy source that is most popular and expanding the fastest is hydroelectric power.

The Effects of Hydroelectric Dams

The Mekong is one of Earth's longest and most complex rivers. Beginning high on the Tibetan plateau as a tiny glacier-fed trickle, the Mekong (called the *Lancang* in China) flows through deep gorges, gaining speed and volume and picking up sediment for a thousand miles or more through Yunnan, China. It skirts the borders of far-eastern India and Myanmar (Burma) before entering Lao, the halfway point on its journey to the sea. In this first half of its path, the Mekong, with its hundreds of tributaries, drains a huge area of China, bits of India, and northern Myanmar (Burma). The river system supports a remarkable amount of biodiversity, rivaled only by that of the Congo River and the Amazon: 20,000 species of plants and 2500 species of animals live here, including freshwater dolphins, giant catfish, and spiders the size of grapefruits. The ethnic diversity is also astonishing, with hundreds of cultures intricately attuned to the variable physical environment of the river basin. In its entirety, the Mekong sustains around 60 million people, mostly rural and poor.

As the Mekong flows southeast through Lao, its silt-laden waters descend from rugged mountain territory into rolling hills and then floodplains, where the silt is deposited as floodwaters slow down and spread out. Periodic floods pick up and redeposit the nutrient-rich silt over and over, providing farmers with natural fertilizer and fisheries with food. By the time the Mekong reaches Cambodia, it is a wide, slow, meandering river in a vast floodplain. In Vietnam it divides into **distributaries** that fan out to form the Mekong delta.

Modernity is arriving late to the Mekong region, but bringing quick change. Most everyone now has access to cell phones, which must be charged with electricity. Electricity is essential to nearly every other aspect of development, hence the emphasis on hydroelectric power, which is especially appealing in China, where carbon emissions cause rampant pollution. At least 86,000 hydroelectric dams have been built across China since the 1960s. By the time the Mekong leaves China at the Laotian border, it has already passed through 20 sites where dams are operational, under construction, or planned, with more on the river's many tributaries. With the aid of Chinese funding, Lao plans nine Mekong dams (Cambodia, two more). Lao, whose main resources are its fertile soil and the waters of the Mekong, plans to profit from the sales of the power generated from these dams. Thailand is already a major purchaser of Laotian hydropower, but Laotian planners are now pausing as they realize that dams will inevitably change soil fertility and river water quality, and not for the better.

Dams have numerous consequences for the environment. Although dams can be used to control damage from flooding, floods serve a purpose. The natural seasonal ebb and flow of river water distributes moisture and nutrients downstream. It is now apparent that seasonal variations in flow are essential for the health of many plant and animal communities, and by extension, human communities. Dams restrict the movement and hence the reproduction of freshwater fish, and create vast reservoirs that flood villages and valuable cropland, rendering them unusable and prompting migration to cities. Furthermore, the standing water produces hospitable habitat for disease-bearing mosquitoes.

As scientists have gained a better understanding of ecological nuances, they have warned that even expert planners tend to underestimate the long-term damage done by the construction of dams and their operation over time. Planners, meanwhile, often overestimate the benefits that dams can produce. As the demand for energy increases, it is tempting to ignore the negative effects of creating that energy with hydroelectric dams.

Researchers connected to the Borneo Project, discussed at the beginning of the chapter, have recently suggested that localized approaches, based on small-scale appropriate green technologies, could supply needed electricity to even remote communities via mini-grid systems (employing small thermal, wind, and solar methods and even very small hydropower dams) without disrupting the environment or traditional ways of life. And such systems could create local jobs that would preclude the need to migrate to the city, so often the result of mega-hydroelectric dam construction. In this way remote places could have access to electricity to charge phones and computers without suffering radical environmental and social impacts.

CHECK YOUR UNDERSTANDING

1. Describe the possible sources of energy available in Southeast Asia that as yet are not being widely used.
2. As you consider the total route of the Mekong River, explain why it is such an important world watercourse.
3. What are some of the useful functions of natural cyclical floods?
4. Explain the ecological and social problems created by large-scale dam projects.
5. Describe how low-impact green technologies could change the present situation of rural development leading to mass migration to urban areas.

HUMAN PATTERNS OVER TIME

The human history of Southeast Asia is extremely complex. Very recent DNA and archaeological research has shown that archaic humans (Denisovans) were in the vicinity of present-day Indonesia as

distributaries branches of a river in its delta that fan out and carry water away from the main channel long as 80,000 years ago. They probably interbred in prehistory with other migrants from the Eurasian continent, as discussed below. More recently, before the time of Christ, Southeast Asia was influenced by traders from China, India, and Arabia. Later still, it was colonized by Europe (from the 1500s to the early 1900s). The Philippines were acquired by the United States at the end of the Spanish-American War and were occupied by the United States from 1898 to 1946. During World War II, much of Southeast Asia was occupied by Japan. By the late twentieth century, occupation and domination by outsiders had ended and the region was profiting from selling manufactured goods to its former colonizers. More so than Middle and South America, Africa, and South Asia, Southeast Asia has had some success in achieving widespread, if modest, prosperity from its new links to the global economy. It has done so largely by following the example of some of East Asia's most successful countries, such as Japan, Korea, Taiwan, and more recently, China.

The Peopling of Southeast Asia

The modern indigenous populations of Southeast Asia arose from the earliest humans to come into the region and from two later migrations widely separated in time. In the first migration, about 40,000 to 60,000 years ago, **Australo-Melanesians**, a group of hunters and gatherers from what are now northern India and Myanmar (Burma), moved to the exposed landmass of Sundaland (see Figure 10.9) and into present-day Australia. Their descendants still live in Indonesia's easternmost islands and in small, usually remote pockets on other islands and the Malay Peninsula, and in Australia, New Guinea, and other parts of Oceania (Figure 11.1).

In the second migration (from 6000 to 10,000 years ago, after the last ice age), people from southern China began moving into Southeast Asia. Their migration gained momentum about 5000 years ago, when a culture of skilled farmers and seafarers from southern China, the **Austronesians**, migrated first to what is now Taiwan, then to the Philippines, and then into the islands of Southeast Asia and the Malay Peninsula (see Figure 10.3). Some of these adventuresome sea travelers eventually moved westward to southern India and to Madagascar (off the east coast of Africa; see Chapter 7), and eastward to the far reaches of the Pacific islands (see Figure 11.18).

DIVERSE CULTURAL INFLUENCES

Over the last several thousand years, Southeast Asia has been—and continues to be—shaped by a steady circulation of cultural influences, both internal and external. Merchants, religious teachers, and sometimes even invading armies from the Indian subcontinent, Southwest Asia, China, Japan, and finally Europe came to the region by overland trade routes and the surrounding seas. These newcomers, whether peaceful or aggressive, introduced religions (Buddhism, Hinduism, Islam, Daoism, Christianity), trade goods (such as cotton textiles), and food plants (such as mangoes and tamarinds from India, and later numerous plants from the Americas), changing lifeways in the Indonesian and Philippine archipelagos and throughout the mainland. The monsoon winds, which blow from the west in the spring and summer, facilitated access for merchant ships from South Asia and the Persian Gulf. These ships sailed home on winds blowing from the east in the autumn and winter. These easterly winds carried ships with spices, bananas, sugarcane, silks, and other East and Southeast Asian items, as well as people, to the wider world, including eventually the Americas.

Australo-Melanesians a group of hunters and gatherers who moved 40,000 to 60,000 years ago from what are now northern India and Myanmar (Burma) to the exposed landmass of Sundaland and present-day Australia

Austronesians a group of skilled farmers and seafarers from southern China who migrated south to various parts of Southeast Asia between 6000 and 10,000 years ago

Religious Legacies

Spatial patterns of religion in Southeast Asia reveal an island–mainland division that reflects the history of influences from India, China, Southwest Asia, and Europe (see map and photos in Figure 10.32 later in this chapter). Today, Buddhism dominates mainland Southeast Asia, while Hinduism remains dominant only on the Indonesian islands of Bali and Lombok. Both Hinduism and Buddhism arrived thousands of years ago with monks and traders who traveled by sea and along overland trade routes that connected India and China through Myanmar (Burma). Early Southeast Asian kingdoms and empires switched back and forth between Hinduism and Buddhism as their official religion. Spectacular ruins of these Hindu–Buddhist empires are scattered across the region; the most famous is the city of Angkor in what is now Cambodia. At its zenith in the 1200s, Angkor had grown quickly; it rivaled the size of Los Angeles and was home to 750,000 people. The latest research shows that Angkor failed because it could not adjust fast enough to repeated climate extremes: alternating super-heavy rains and long dry periods that eventually overwhelmed the urban infrastructure and food production systems, forcing people to flee.

In Vietnam, people practice a mix of Buddhist, Confucian, and Taoist customs that reflect the thousand years (ending in 938 C.E.) when Vietnam was part of various Chinese empires. China's traders and laborers also brought cultural influences to scattered coastal zones throughout Southeast Asia.

Islam is now dominant in most of the islands of Southeast Asia, spreading mainly from India after it was conquered by Mughals in the fifteenth century. Muslim spiritual leaders and traders converted many formerly Hindu–Buddhist kingdoms in Indonesia, Malaysia, and parts of the southern Philippines. Islam remains foremost in Malaysia and in Indonesia, which has the largest Muslim population on Earth, and in the southern Philippines. Protestant Christianity was introduced by the Dutch but did not catch on. Roman Catholicism is the predominant religion in Timor-Leste and in most of the Philippines, which were colonized by Portugal and Spain, respectively.

European Colonization

Between the sixteenth and the twentieth centuries, several European countries established colonies or quasi-colonies in Southeast Asia (**Figures 10.17 and 10.18**). Drawn by the region's fabled spice trade, the Portuguese established the first permanent European settlement in Southeast Asia at the port of Malacca, Malaysia, in 1511. Although better ships and weapons gave the Portuguese an

A A bas-relief showing the Dutch massacre of civilians in the Indonesian village of Rawagede in 1947, during Indonesia's war of independence. [ROMEO GACAD/AFP/Getty Images]

B A painting commemorating a Filipino attack on U.S. forces during the Philippine–American War (1899–1902). [Prisma/UIG/Getty Images]

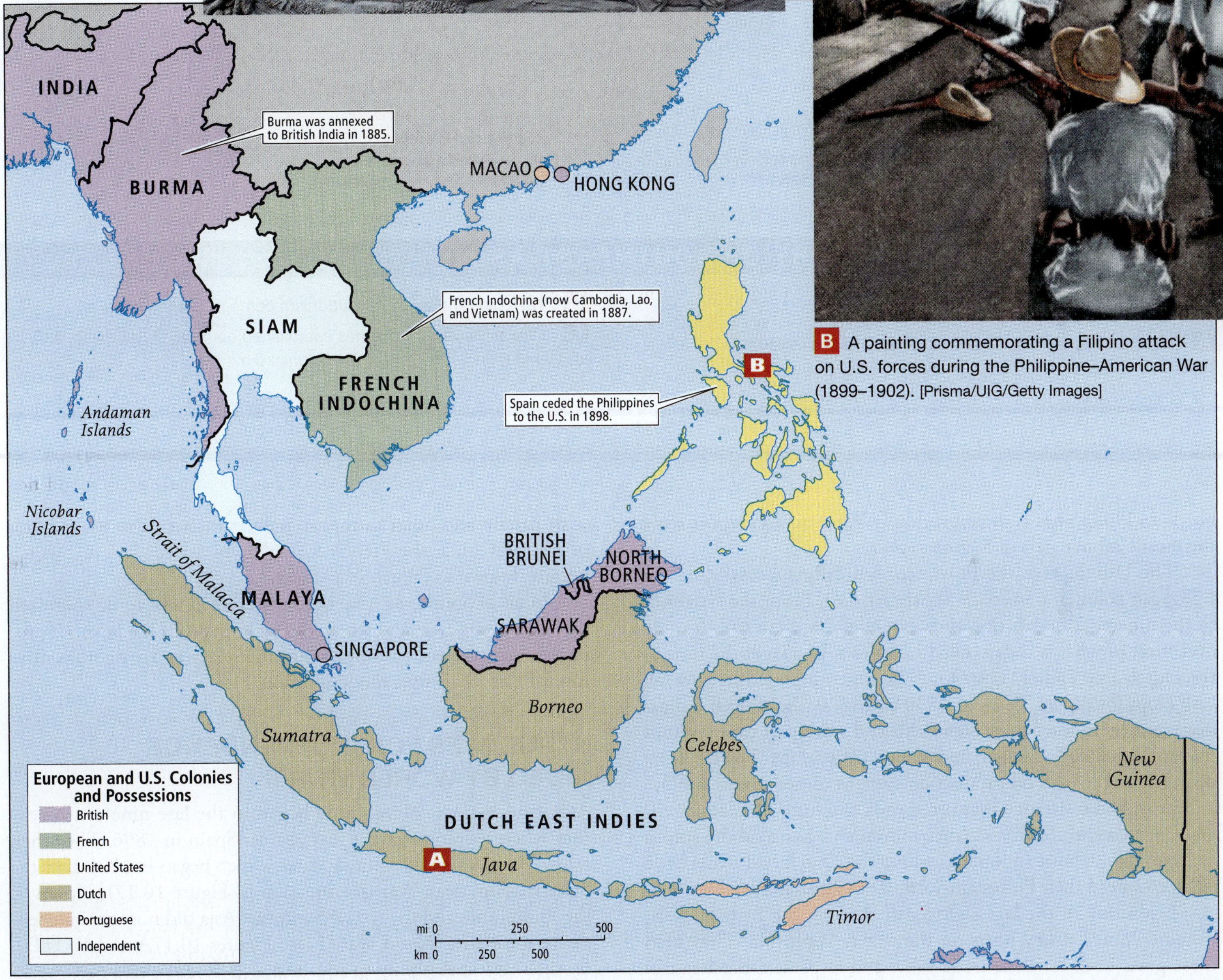

Figure 10.17 European and U.S. colonies in Southeast Asia, 1914. Of the present-day countries in Southeast Asia, only Thailand (formerly called Siam) was never colonized. [Research from: *Hammond Times Concise Atlas of World History* (Maplewood, NJ: Hammond, 1994), p. 101.]

advantage, their anti-Islamic and pro-Catholic policies provoked strong resistance in Southeast Asia. Only on the small island of Timor-Leste did the Portuguese establish Catholicism as the primary religion.

Arriving first in 1521, the Spanish had established trade links across the Pacific between the Philippines and their colonies in the Americas by 1540 (Figure 10.18C). Like the Portuguese, they practiced a style of colonial domination grounded in Catholicism, but they met less resistance because of their greater tolerance of non-Christians. The Spanish ruled the Philippines for more than 350 years; as a result, except for the southernmost islands, which retain strong indigenous and long-standing Muslim influences, the

A Young men of the Ifugao tribal people, one of several Austronesian ethnic groups in the Philippines. [John S Lander/LightRocket/Getty Images]

B The ruins of Bagan, Myanmar (Burma), home of 2200 Buddhist temples since at least 874 C.E. [Joe & Clair Carnegie/Libyan Soup/Moment Open/Getty Images]

C A monument to Ferdinand Magellan's lan in the Philippines, 1521. He died shortly therea never reaching his goal, the Spice Islands (Molu Indonesia). [John S Lander/LightRocket/Getty Images

60,000 B.C.E. | 40,000 B.C.E. | 20,000 B.C.E. | 8000 B.C.E. | 500 B.C.E. | 0 C.E. | 500 C.E. | 1000 C.E. | 1500 C.E.

60,000–40,000 B.C.E. Austro-Melanesian migrations

8000 B.C.E. Austronesian migrations

c. 200 B.C.E.–Present Hindu–Buddhist kingdoms

874–1369 C.E.

1521

Figure 10.18 VISUAL HISTORY OF SOUTHEAST ASIA

Thinking Geographically

A Describe from where those who populated this region in prehistory came.

B How did Hinduism and Buddhism come to Southeast Asia?

C Which European countries established colonies in Southeast Asia and what marked the end of the colonial era?

northern Philippines is the most deeply Westernized and certainly the most Catholic part of Southeast Asia.

The Dutch were the most economically successful of the European colonial powers in Southeast Asia. From the sixteenth to the nineteenth centuries, they extended their control of trade over most of what is today called Indonesia, known at the time as the Dutch East Indies. The Dutch became interested in growing cash crops for export. Between 1830 and 1870, they forced indigenous farmers to leave their own fields and work part time without pay on Dutch coffee, sugar, and indigo plantations. The resulting disruption of local food production systems caused severe famines and provoked resistance that often took the form of Islamic religious movements. Such resistance movements hastened the spread of Islam throughout Indonesia, where the Dutch had made little effort to spread their Protestant version of Christianity.

Beginning in the late eighteenth century, the British established colonies at key ports on the Malay Peninsula. They held these ports both for their trade value and to protect the Strait of Malacca, the shortest passage for sea trade between Britain's empire in India and China. In the nineteenth century, Britain extended its rule over the rest of modern Malaysia in order to benefit from Malaysia's tin mines and plantations. Britain also added Myanmar (then known as Burma) to its empire, which provided access to forest resources and overland trade routes into southwest China.

The French first entered Southeast Asia as Catholic missionaries in the early seventeenth century. They worked mostly in the eastern mainland area in the modern states of Vietnam, Cambodia, and Lao. In the late nineteenth century, spurred by rivalry with Britain and other European powers for access to the markets of nearby China, the French formally colonized the area, which became known as French Indochina.

In all of Southeast Asia, the only country not to be colonized by Europe was Thailand (then known as Siam). Like Japan, it protected its sovereignty through both diplomacy and a vigorous drive toward European-style modernization.

STRUGGLES FOR INDEPENDENCE AIDED BY WORLD WAR II

Agitation against colonial rule began in the late nineteenth century when Filipinos fought first against Spain in 1896. They then resisted control by the United States, which began in 1898 after the Spanish–American War (see the map in Figure 10.17). However, the Philippines and the rest of Southeast Asia did not win independence until after World War II (see Figures 10.17A and 10.18D). By then, Europe's ability to administer its colonies had been weakened by the devastation of the war, during which Japan had conquered most of the parts of Southeast Asia that had been controlled by Europe and the United States. Japan held these lands until it was defeated by the United States at the end of the war. By the mid-1950s, the colonial powers had granted self-government to most of the region, and all of Southeast Asia was independent by 1975.

The Vietnam War

The bitterest battle for independence took place in French Indochina (the territories of present-day Vietnam, Lao, and Cambodia),

D Japanese forces land in the Philippines during World War II. [Keystone/Hulton Archive/Getty Images]

E The defoliant Agent Orange was sprayed over Vietnam and Lao between 1962 and 1971 to destroy crops and expose North Vietnamese troops. [Dick Swanson/Getty Images]

F Bangkok, Thailand, is emerging as a global center for tourism, manufacturing, and trade. [Igor Prahin/Moment/Getty Images]

1800 C.E. | 1850 C.E. | 1900 C.E. | 1950 C.E. | 2000 C.E.

- **1511–1975** European colonial era
- **1830–1870** Dutch forced-labor system causes famines in Indonesia
- **1896–1975** Struggles for independence
- **1941**
- **1955–1975** Vietnam War
- **1962–1971**
- **1960–Present** Long periods of sustained economic growth with sporadic economic crises

D How did this region experience World War II and its aftermath?

E How did people in Vietnam and Lao get exposed to Agent Orange, a dangerous chemical?

acquired by France in the nineteenth century. Although all three became nominally independent in 1949, France retained political and economic power over them. Various nationalist leaders, most notably Vietnam's Ho Chi Minh, headed resistance movements against continued French domination. After failed diplomatic efforts in Europe and the United States, the resistance leaders accepted military assistance from communist China and the Soviet Union, despite ancient antipathies toward China for its previous millennia of domination. In this way, the Cold War was brought to mainland Southeast Asia.

In 1954, Ho Chi Minh defeated the French at Dien Bien Phu in northern Vietnam. The United States stepped in because it had become worried about the spread of international communism should the anticolonial resisters, now supported by communists, succeed. The **domino theory**—the idea that if one country "falls" to communism, other nearby countries will follow—was a major influence in this decision, because both North Korea and China had recently become communist. The Vietnamese resistance, which controlled the northern half of the country, attempted to wrest control of the southern half of Vietnam from the U.S.-supported and quite corrupt South Vietnamese government. The pace of the war accelerated in the mid-1960s. After many years of brutal conflict, public opinion in the United States forced a hasty U.S. withdrawal from the conflict in 1973. The civil war continued in Vietnam, finally ending in 1975, when the North defeated the South and established a new national government.

More than 4.5 million people died during the Vietnam War, including more than 58,000 U.S. soldiers. Another 4.5 million on both sides were wounded, and bombs, napalm, and chemical defoliants ruined much of the Vietnamese environment. Land mines continue to be a hazard to this day, and the effects of the highly toxic defoliant known as Agent Orange are still causing debilitating birth defects among many rural Vietnamese and Laotian people (see Figure 10.18E). The withdrawal from Vietnam in 1973 ranks as one of the most profound military defeats in U.S. history. After the war, the United States crippled Vietnam's recovery by imposing severe economic sanctions that lasted until 1993. Since then, however, the United States and Vietnam have developed a significant trade relationship.

The "Killing Fields" in Cambodia

In Cambodia, where the Vietnam War had spilled over the border, a particularly violent revolutionary faction called the Khmer Rouge seized control of the government in 1975. Inspired by a vision of a rural communist society, they attempted to destroy virtually all traces of European influence. They targeted Western-educated urbanites in particular, forcing them into labor camps where more than 2 million Cambodians—one-quarter of the population—starved or were executed in what became known as the "Killing Fields."

In 1979, Vietnam deposed the Khmer Rouge and, until 1989, ruled Cambodia through a puppet government. A 2-year civil war then ensued. Despite major UN efforts throughout the 1990s to establish a multiparty democracy in Cambodia, the country

domino theory a foreign-policy theory that suggested that if one country "fell" to communism, neighboring countries would also fall, like a line of dominoes remained plagued by political tensions between rival factions and by government corruption.

In March 2009, Kang Kek Iew, the first of the Khmer Rouge leaders to be tried for war crimes and genocide, was forced to listen to and watch lengthy accounts of the torture of men, women, and children that he supervised. He was convicted in 2010 and sentenced to 35 years in prison. Most operatives in the Killing Fields will never be prosecuted.

CHECK YOUR UNDERSTANDING

1. The modern indigenous populations of Southeast Asia arose from several strands of ancient human migrations, one as long as 80,000 years ago, ___________; others about 40,000 to 60,000 years ago, ___________; and finally others about 6000 to10,000 years ago, ___________. Who were these people?
   a. the Austronesians
   b. the Australo-Melanesians
   c. the Denisovans
2. Today, which religion dominates mainland Southeast Asia? Which remains dominant only on the Indonesian islands of Bali and Lombok?
3. Describe the human activities that brought Hinduism, Buddhism, and Islam to Southeast Asia.
4. Islam is now dominant in which countries of Southeast Asia? Which country has the largest Muslim population on Earth?
5. Between the sixteenth and the twentieth centuries, several European countries, the United States, and Japan established a series of colonies or quasi-colonies that covered almost all of Southeast Asia. What were the goals of the colonists, and what marked the end of the colonial era?
6. Describe the U.S. policy that informed the interference by the United States in the wars of independence in French Indochina.
7. If 4.5 million people died, including more than 58,000 U.S. soldiers, who won the Vietnam War?

GLOBALIZATION AND DEVELOPMENT

10.4. Describe how the formation of state-aided market economies and export-led production have been key to the economic development of this region.

10.5. Explain why regional economic cooperation is expanding, but only slowly.

state-aided market economies national governments help certain sectors of the economy with special loans

Globalization has brought both spectacular successes and occasional periodic declines to the economies of Southeast Asia. Key to the economic development of this region are strategies that were pioneered earlier in East Asia: the formation of state-aided market economies, combined with export-led economic development. Regional economic cooperation, so successful in the EU and elsewhere, has not been a central feature.

During and immediately after the era of European colonialism, trade among the countries within the region was inhibited by the fact that they all exported similar goods—primarily food and raw materials—and imposed tariffs against one another. They imported consumer products, industrial materials, machinery, and fossil fuels, mostly from the developed world.

BORROWED IDEAS FOR ECONOMIC GROWTH

Beginning in the 1960s, some national governments in Southeast Asia were able to create strong and sustained economic expansion by emulating two strategies for economic growth pioneered earlier by Japan, Taiwan, and South Korea (see Chapter 9). One was the formation of **state-aided market economies**. That is, national governments in Indonesia, Malaysia, Thailand, and to some extent the Philippines intervened strategically to force financial institutions to help certain economic sectors develop; in addition, investment by foreigners was limited so that the governments could have more control over the direction of the economy. The second strategy was to push export-led economic development, meaning investments were encouraged in industries that manufactured products for export, primarily to developed countries. These two strategies amounted to a limited and selective embrace of globalization in that global markets for the region's products were sought but foreign sources of capital were not. If money was made, it would stay in the region.

These approaches were a dramatic departure from those used in other developing areas. In Middle and South America and parts of Africa, postcolonial governments relied on import substitution industries that produced low-quality manufactured goods mainly for local use. By contrast, export-led growth allowed Southeast Asia's industries to produce high-quality goods and to earn much more money in the vastly larger markets of the developed world. Standards of living increased markedly, especially in Malaysia, Singapore, Indonesia, and Thailand. Other important results of Southeast Asia's success were a decrease in wealth disparities and improvements in vital statistics—lower infant and maternal mortality, longer life expectancy, and lower population growth—all related to higher domestic incomes. By the 1990s, Southeast Asian countries had some of the highest economic growth rates in the world. The growth was based on an economic strategy that emphasized the export of manufactured goods—initially clothing and then more sophisticated technical products.

WORKING CONDITIONS

In general, the benefits of Southeast Asia's "economic miracle" were not equally apportioned. In the region's new factories and other enterprises, it is not unusual for assembly-line employees to work 10 to 12 hours per day, 7 days per week for less than the legal minimum wage and without benefits (**Figure 10.19B** and map). Labor unions that typically would address working conditions and wage grievances are frequently repressed or banned by governments, and international consumer pressure to improve working conditions at U.S.-based companies has been only partially effective. For example, the U.S.-based shoe company Nike has tried for years to sidestep the entire issue of U.S. customer complaints by outsourcing its manufacturing to non-U.S. contractors in the region. U.S. fair-labor NGOs pursued one Nike subcontractor in Jakarta, Indonesia, where for 19 years workers were forced to work an extra hour every day without pay. The NGOs filed a lawsuit against the subcontractor, which they won, resulting in a $1 million settlement that will give the 4500 workers $222 each. However, given that there are more than 160,000 workers for Nike subcontractors in Indonesia alone, this settlement is but a drop in the bucket.

EXPORT PROCESSING ZONES

In the 1970s, some governments in the region began adopting an additional strategy for encouraging economic development. Still focused on exports, this time they sought foreign sources of capital, but the places in which those sources could invest were limited to specially designated free trade areas (Figure 10.19C). Such export processing zones (EPZs; see the discussion in Chapter 3) are places in which foreign companies can set up their facilities using inexpensive, locally available labor to produce items only for export. (Maquiladoras in Mexico are examples of EPZ companies set up by foreign firms.) Taxes are eliminated or greatly reduced as long as the products are re-exported (not sold within the country). Since the 1970s, EPZs have expanded economic development in Malaysia, Indonesia, Vietnam, and the Philippines, and are now used widely in China and Middle and South America, and to a lesser extent in sub-Saharan Africa.

THE FEMINIZATION OF LABOR

Between 80 and 90 percent of the workers in the EPZs are women, not only in Southeast Asia but in other world regions as well. The **feminization of labor** has been a distinct characteristic of globalization over the past three decades (see Chapter 9). Employers prefer to hire young, single women because they are perceived as the least expensive, least troublesome employees. Statistics do show that, generally, women across the globe will work for lower wages than men, will not complain about poor and unsafe working conditions, will accept being restricted to certain jobs on the basis of sex, and are not as likely as men to agitate for promotions; though, of course, there are many exceptions. The reasons for this widespread pattern are complex and are discussed throughout this book.

GROWTH OF THE SERVICE SECTOR

A much more powerful force that is improving working conditions and driving up pay is the growth of the **service sector** throughout the region. Service sector (often called tertiary sector) jobs are in education, health care, sales, entertainment, and financial services. They produce nontangible products and information and typically require education beyond high school. Competition for hiring these educated workers means that wages and working conditions are already better than in manufacturing and are likely to improve faster. The service sector now accounts for more than one-fifth of the value of trade between the region and the rest of the world. And services dominate the economies as a percentage of GDP and as a percentage of employment in Singapore, Malaysia, Indonesia, the Philippines, and Thailand (Figure 10.19A). In all other countries except Timor-Leste, the service sector accounts for at least 40 percent of GDP and is approaching parity with the industrial sector as a percentage of GDP and of employment. Only in Brunei is the industrial sector still dominant in GDP and employment; there it is entirely based on oil production. Agriculture still employs more than a majority of workers in Myanmar (Burma) and Lao. In Timor-Leste, where 41 percent of the people work in agriculture yet produce only 9 percent of GDP, oil is the mainstay of the economy but employs only 13 percent of the population.

ECONOMIC CRISES AND RECOVERY: THE PERILS OF GLOBALIZATION

Periodically, economic crises have swept through this region, with varying effects on society. For example, in the late 1990s as part of an effort to open up national economies to the free market, and with a general push from the International Monetary Fund (IMF), Southeast Asian governments relaxed controls on the financial sector, which had traditionally been highly regulated. Soon, Southeast Asian banks were flooded with money from investors in the rich countries of the world who hoped to profit from the region's growing economies. Flush with cash and newfound freedoms, the banks often made reckless decisions. For example, bankers made risky loans to real estate developers, often for high-rise office building construction. As a result, many Southeast Asian cities soon had far too much office space, with millions of investment dollars tied up in projects that contributed little to economic development.

One of the forces that led bankers to make such bad decisions was a kind of corruption known as **crony capitalism**. In most Southeast Asian countries, as elsewhere, corruption is related to the close personal and family relationships among high-level government officials, bankers, and wealthy business owners. In Indonesia, for example, the most lucrative government contracts and business opportunities were reserved for the children of former President Suharto, who ruled the country from 1967 to 1997. His children became some of the wealthiest people in Southeast Asia. This kind of corruption expanded considerably when the new foreign investment money was allowed in, much of which was diverted to bribery or unnecessary projects that brought prestige to political leaders.

The cumulative effect of crony capitalism and the lifting of controls on banks was that many ventures failed to produce any profits at all. In response, foreign investors panicked, withdrawing their money on a massive scale. In 1996, before the crisis, there was a net

feminization of labor the rising numbers of women in both the formal and the informal labor force

service sector the part of the economy that employs people in knowledge-based jobs, such as in education, health care, sales, entertainment, and financial services

crony capitalism a type of corruption in which government officials, bankers, and entrepreneurs have close personal as well as business relationships

PHOTO ESSAY

10.19 Globalization and Development in Southeast Asia

The countries of Southeast Asia are all engaged in development that connects them to the global economy. All have prospered as a result and one of the effects has been an improvement in human well-being: health, education, and income. But as the map indicates, stark differences in HDI remain across the region. Even within the countries labeled medium (yellow and green) there are notable differences; for example, Myanmar (Burma) only recently increased its standing from low, and Lao and Cambodia have many pockets of poverty. The photos illustrate a range of factors that account for the differences.

THINKING GEOGRAPHICALLY

- **A** Why does development so often mean a growth in service jobs?
- **B** Why do many employers prefer workers to be young, single women?
- **C** What are some reasons for using EPZs?
- **D** Why is corruption so common in crony capitalism situations?

A Service-sector growth. Singapore is a regional leader in creating service-sector jobs. Many educated young people across the region want to work in Singapore, a major shipping, tourism, high-tech, and research center. [Have a nice day Photo/Shutterstock]

B Working conditions. Many consumer goods are made in Southeast Asia. In Cambodia, the factory workers have rough commutes of an hour or more in open truck beds. A fee is charged for the ride, but the vehicles have few safety precautions. [AP Images/Heng Sinith]

C Export processing zones. Highly skilled employees of the Japanese firm Mitsubishi Motors Corporation work in an EPZ in Laem Chabang, Thailand. Autos built in such zones are usually required to be sold outside the country of origin. [AP Images/Kyodo]

D Crony capitalism. Once Malaysia's most powerful man, former Prime Minister Najib Razak arrives in court to face multiple charges of corruption, each charge carrying a 20-year jail sentence. Razak, charged with looting U.S.$4.5 billion, pled not guilty in the first of several trials. [AP images/Vincent Thian]

inflow of U.S.$94 billion to Southeast Asia's leading economies. In 1997, inflows had ceased and there was a net outflow of U.S.$12 billion. This financial debacle forced millions of people into poverty and changed the political order in some countries. There were also some geographic aspects to the crisis: the banks involved were located primarily in Singapore and the major cities of Thailand, Malaysia, and Indonesia. These were the countries to feel the immediate effects of the crisis. Urban landscapes were dotted with ghostly high-rise construction projects halted mid-build. While urban areas were hit hardest, eventually the effects filtered into the hinterland, and workers lost jobs even in remote areas.

The IMF Bailout and Its Aftermath

The IMF made a major effort to keep the region from sliding deeper into recession by instituting reforms designed to make banks more responsible in their lending practices. The IMF also required structural adjustment policies (SAPs; see Chapter 3), which required countries to cut government spending (especially on social services for poor people, such as health care and education) and abandon policies intended to protect domestic industries.

After several years of economic chaos and much debate over whether the IMF bailout and accompanying SAPs helped or hurt a majority of Southeast Asians, economies began to recover. In the largest of the region's economies, growth resumed by 1999; by 2006, the crisis had been more or less overcome. But the crisis of 1997–2006 remains an ominous reminder of the risks of globalization, and was the impetus to reform the SAP policies once so popular with the IMF (see discussion in Chapter 7).

The Impact of China's Growth

During the Southeast Asian financial crisis of 1997–2006, Singapore, Malaysia, and Thailand lost out to China in attracting new industries and foreign investors. By the early 2000s, China attracted more than twice as much foreign direct investment (FDI) as Southeast Asia. However, China's growth also became an opportunity for Southeast Asia. Singapore, Malaysia, Indonesia, Thailand, and the Philippines "piggy-backed" on China's growth by winning large contracts to upgrade China's infrastructure in areas such as wastewater treatment, gas distribution, and shopping mall development.

Southeast Asia has also used its comparative advantages over China to attract investment. In poorer countries, such as Vietnam, wages have remained lower than in China. This has attracted investments in low-skill manufacturing that had been going to China. Meanwhile some of the region's wealthier countries—Singapore, Malaysia, and Thailand—now position themselves as locations that offer more highly skilled labor than does China, and better high-tech infrastructure for the delivery of services.

Association of Southeast Asian Nations (ASEAN) an organization of Southeast Asian governments that furthers economic growth and political cooperation between member countries and with other areas of the world

Global Recessions from 2007 to the Present

The region was hit again by the effects of the global recession that began in 2007 and became full-blown by 2008. Growth slowed markedly because high oil and food prices restricted disposable cash worldwide, and consumers in wealthy countries (especially in the United States and the EU countries, which are Southeast Asia's biggest trade partners) faced crippling debt that seriously curtailed their spending. By 2010, East and Southeast Asia appeared to be recovering. In part, the recovery was based on the increased demand for goods and services within the domestic economies of China and Southeast Asia, where consumers now had some disposable cash. Southeast Asia's ability to respond to this demand was facilitated by policies that opened up trade between countries in the region and access to China.

But the positive developments of 2010 did not last. By 2012, financial troubles in Europe and slow recoveries in the United States and Japan all lowered demand for Southeast Asian products. Singapore, which has a highly sophisticated economy that is based on its banking and IT services and the transfer of goods through its giant port, is particularly susceptible to slowdowns in rich countries. Manufacturing countries like Indonesia, Malaysia, and Thailand also fare best when international demand for their products is robust.

Cambodia, Lao, and Myanmar (Burma) still rely on raw material exports to places like China. By 2015, China's economy had begun to slow, and in order to energize its economy and meet the consumer needs of its own people, China began discouraging imports in favor of Chinese-produced goods, thereby reducing demand for raw materials and manufactured goods from Southeast Asia. That trend continues as the Chinese economy waxes and wanes due to internal politics and threatened trade wars.

REGIONAL TRADE, ASEAN, AND THE TRANS-PACIFIC PARTNERSHIP

During the 1980s and 1990s, Southeast Asian countries traded more with China and the rich countries of the world than they did with each other. This issue of insufficient trade among Southeast Asian countries was the reason behind the creation of the **Association of Southeast Asian Nations (ASEAN)**, an organization of Southeast Asian nations that has grown in strength over the years. The ASEAN member countries are *Brunei, Myanmar (Burma), Cambodia, Indonesia, Lao, Malaysia, Philippines, Singapore, Thailand,* and *Vietnam*; candidate members are Papua New Guinea and Timor-Leste. As of 2018, close associates of ASEAN are Canada, China, Japan, South Korea, Australia, the European Union, India, New Zealand, Russia, and the United States.

Regional and Global Integration

ASEAN started in 1967 as an anticommunist, anti-China association, but it now focuses on agreements that strengthen regional cooperation, including agreements with China. One example is the Southeast Asian Nuclear Weapons–Free Zone Treaty signed in December 1995. Another is the ASEAN Economic Community, or AEC, a trade bloc loosely patterned after the North American Free Trade Agreement and the European Union; the AEC was formalized on December 31, 2015. ASEAN has also ratified free trade agreements with Australia, New Zealand, China, India, Japan, and South Korea, and is a major participant in the Trans-Pacific

Figure 10.20 ASEAN internal and external trade, 2017. The ASEAN countries are very active in world trade as importers and exporters. The goal of increasing intraregional trade is being reached, but it is still relatively low, especially when compared to the European Union (see Figure 4.21), where intra-EU trade amounts to twice the monetary value of extra-EU trade.

(A) ASEAN trade in goods by selected partner country/region, 2017 (US$ billion)

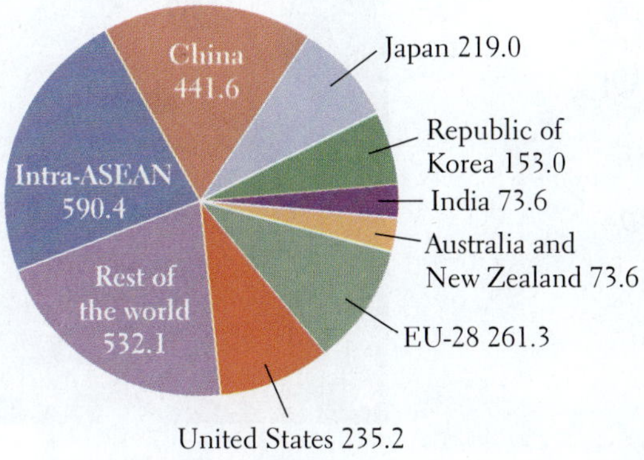

Total: 2574.8

Pie chart A indicates that almost one-quarter of the total regional trade (22.9 percent) is now between ASEAN countries, but external trade dominates.

(B) ASEAN member states' exports of goods by destination, 2017 (%)

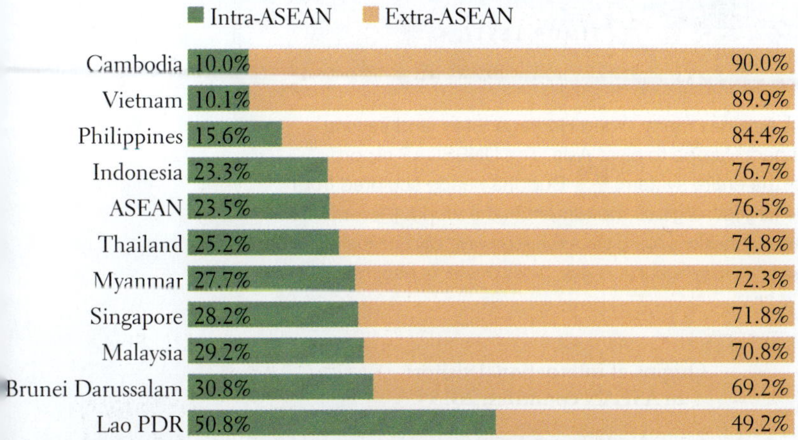

Bar graph B shows the varying dominance of external trade for each country.

(C) ASEAN net inflows by sector, 2017 (US$ million)

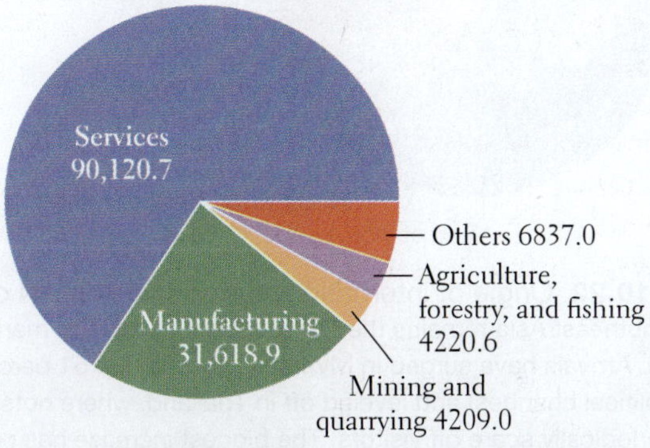

Total: 137,006.2

Pie chart C on foreign direct investment (FDI) shows that FDI is overwhelmingly engaged in the emerging service sectors of ASEAN economies. [Data from: ASEAN_Statistical_Highlights_2018.pdf.]

Partnership (TPP) negotiations, which the United States ratified in 2016 under President Obama. President Trump withdrew the United States from the TPP in 2017 three days after taking office. In 2018, when 11 countries agreed to leave the United States out of a new free trade deal, naming the new arrangement the Comprehensive and Progressive Agreement for Trans-Pacific Partnership (CPTPP), Trump began arranging to rejoin the group.

ASEAN now focuses on increasing trade within the association, long recognized as insufficient though essential to stable regional prosperity. ASEAN efforts are beginning to be successful: the total intraregional trade between ASEAN countries in 2017 (22.9 percent) was more than trade with any other outside country or region (**Figure 10.20A**). And by 2017, ASEAN was addressing a variety of other issues that were aimed at encouraging such things as the free flow of services, skilled labor, investment, and capital between countries. ASEAN's overall stated goals are to develop a sustainable regional economic community that protects consumers and intellectual property rights and levels the playing field for competition.

Regional Inequalities

For all the emphasis on regional integration, the fact is that the 11 countries in the region of Southeast Asia vary widely in levels of economic development, as alluded to in the above discussion of the uneven performance in the growth of the service sector. Perhaps most telling are figures on gross national income (GNI) per capita, adjusted for purchasing power parity (PPP) (2018). Singapore, the smallest country, has a per capita GNI (PPP) of U.S.$90,570, one of the highest such figures in the world. The poorest country in the region is Cambodia, with a per capita GNI (PPP) of just U.S.$3760, one of the lowest such figures. Statistics on other aspects of human well-being are similarly divergent, as is shown in Figure 10.30 later in this chapter. Part of the divergence in development and well-being is related to variations in the size and geographic features of countries in the region, but it is interesting to note that despite efforts at regional trade integration, there has been little region-wide effort to decrease the gross disparities in levels of development among ASEAN member states. There is no central bureaucracy similar to that in Brussels in the European Union that promotes financial transparency and unity, and no official ASEAN agency like the European Commission that works toward **social cohesion** for the entire region. Nonetheless, there are public efforts to promote social cohesion across the many ethnic groups in the region. Billboards promoting unity are common in Singapore and Malaysia, and the motto of ASEAN is "One Vision, One Identity, One Community."

Piracy on the High Seas

An example of the weakness in regional solidarity that ASEAN has not managed to address is the reemergence of piracy in the waters of Southeast Asia. The narrow Malacca strait between Sumatra (Indonesia) and Malaysia has been the site of numerous hijackings (15 in 2014, an average of 8 per year by 2018). Crews of small armed vessels take advantage of the crowded waters to sneak onto

social cohesion an expression of solidarity and the willingness of members of a society to cooperate with each other in order for all to survive and prosper

Figure 10.21 Transportation infrastructure and heritage tourism. Southeast Asia has many World Heritage Sites, and several are located along the partially completed Asian Highway, which will eventually connect Europe to all of Asia. The UNESCAP report, "Development of the Asian Highway," describes the project and includes a map of the entire system. **A** Buddha heads sculpted from stone at the Angkor Wat temple complex in Cambodia (number 1 on the map), the largest religious monument in the world. [Research from: "World Heritage List," UN World Heritage Convention, at http://whc.unesco.org/en/list/; "Tourism Attractions Along the Asian Highway," UN Economic and Social Commission for Asia and the Pacific, 2004, at http://www.unescap.org/our-work/transport/asian-highway/about; and *Asia Times Online*, at http://www.atimes.com/atimes/Asian_Economy/images/highways.html.] [joakimbkk/E+/Getty Images]

modest commercial cargo boats and tankers. More than 120,000 such small craft ply these waters every year. The pirates overpower the crew, hold them captive, and quickly move the captured ship to one of a number of secluded small ports, from which they transfer or sell the cargo with the help of accomplices. Only a few deaths have resulted, but the losses are significant and governments as well as ASEAN officials have attempted to join forces to stop the phenomenon with only partial success.

Tourism and Development

International tourism is an important and rapidly growing economic activity in most Southeast Asian countries (**Figure 10.21**). Between 1991 and 2017, the number of intra- and extra-ASEAN tourists went from about 20 million to 125 million. As in other trade matters, Southeast Asians are themselves increasingly visiting neighboring countries (**Figure 10.22**). This is a positive trend because familiarity between neighbors lays the groundwork for various forms of regional cooperation, including business ventures, cultural exchanges, law enforcement, and infrastructure improvements.

In response to its popularity with global and regional tourists, ASEAN members have been working to improve the region's transportation infrastructure. One such project is the Asian Highway,

Figure 10.22 Origin of international visitors to ASEAN countries, 2015. Southeast Asia remains the fastest-growing tourism market in the world. Arrivals have surged in Myanmar (Burma) (up 51 percent after recent political changes) and leveled off in Thailand, where riots and coups periodically scare off visitors. The biggest increase has come from intraregional travel—people in Southeast Asia visiting their neighbors. [Data from: Travel and Tourism: Economic Impact, 2018, Southeast Asia, World Travel and Tourism Council https://www.wttc.org/-/media/files/reports/economic-impact-research/regions-2018/southeastasia2018.pdf.]

a web of standardized roads that loops through the mainland and connects it with Malaysia, Singapore, and Indonesia (the latter by ferry; see the Figure 10.21 map). Eventually, the Asian Highway will facilitate ground travel through 32 Eurasian countries, from Russia to Indonesia and from Turkey to Japan.

The surge of tourism in Southeast Asia is welcomed, and it contributes about 12 percent of the region's GDP per capita and accounts for close to 5 percent of jobs. But concerns are raised that some countries are becoming too dependent on tourism, an industry that leaves economies vulnerable to events that can precipitously stop the flow of visitors (natural disasters, political upheavals), or that leave local people vulnerable to the sometimes destructive demands of tourists (see "Globalization and Gender: The Sex Industry"). Examples of disasters are the tsunami of December 2004 that killed several thousand in Thailand and Indonesia, many of whom were international tourists; the giant typhoon, Haiyan, in the Philippines in 2013; and various human-made disasters, such as the ISIS-related terrorist bombings and shootings in Indonesia (2016–2018) and in southern Thailand (2006, 2012, 2016, 2018). Perhaps more importantly, tourism can divert talent from contributing to national development, as the following vignette illustrates.

VIGNETTE Tan Phuc waits patiently for a mechanic to mount the new rear tire on his Chinese-made 125cc motorcycle. He is a motorcycle tour guide in Vietnam who likes to take his clients to places along Highway 14 in the Central Highlands, close to the historic Ho Chi Minh Trail. Tan's current client, a Dutch tourist, has spent 3 days filling his digital camera with images of vast coffee plantations covered in snowy blossoms, of silkworms frying in oil while their cocoons are unwound and spun into silk thread, and of indigenous minority children who met the gaze of his camera with casual curiosity. Tan finds his clients in bus stations and backpacker hostels and negotiates a daily rate for tours—usually between U.S.$50 and U.S.$75, not insignificant in a country that has an average per capita annual income of U.S.$2790.

Tan Phuc's life has always been affected by global forces. As a child in the Mekong delta, he sold produce to U.S. soldiers at a nearby base. During the war, he lost his brother, a soldier for South Vietnam; after the war, his father spent 10 years in a communist reeducation camp. Even with the hardships, Tan obtained an education. But when the annual inflation rate of 400 percent shrank his salary as a high school math teacher, Tan could not adequately meet his family's needs. Like other skilled Vietnamese people, Tan took advantage of Vietnam's transition to a market economy, which began in the 1980s, and established a business that caters to tourists. This means that he, along with many other educated Vietnamese people, no longer works in occupations crucial to Vietnam's future, like education. ■

CHECK YOUR UNDERSTANDING

1. Give several examples of development strategies that Southeast Asia has borrowed from other parts of the world.
2. Recessions have occurred frequently. What are some of the forces that have contributed to recessions?
3. What are the goals of the Association of Southeast Asian Nations (ASEAN)?
4. What are some of the critiques of tourism as a development strategy?

POWER AND POLITICS

10.6. Discuss the recent expansion of political freedoms in Southeast Asia, and describe how authoritarianism, corruption, and violence have at times reversed these gains.

While there has been a general shift in Southeast Asia away from authoritarianism and toward more political freedoms, there are also several cases where democracy has lost gains and dictatorship has returned. Some people argue that authoritarianism has deep cultural roots in Southeast Asia and should be given more respect. Others see a certain amount of authoritarianism as necessary to control corruption, drug addiction, and political violence. Still others argue that respect for political freedoms is more likely to expose corruption and transform militant movements into peaceful political parties.

SOUTHEAST ASIA'S AUTHORITARIAN TENDENCIES

Some Southeast Asian leaders, such as Singapore's former prime minister, the late Lee Kuan Yew, have said that Asian values are not compatible with Western ideas of democracy and political freedom. Lee asserted that Asian values are grounded in the Confucian view that individuals should be submissive to authority; therefore Asian countries should avoid the highly contentious public debate of open electoral politics. This view was recently reiterated by the Philippine president, Rodrigo Duterte, as he floated the idea that perhaps the constitution should be amended and a "revolutionary government" established. Nevertheless, despite the high respect accorded Lee (the world's longest-serving prime minister), when confronted with governments that abuse their power, people throughout Southeast Asia have not submitted but rather have rebelled (**Figure 10.23**). Even in Singapore, the Western-educated son of Lee Kuan Yew, Lee Hsien Loong, who has been the appointed, not elected, prime minister since 2004, has expressed more interest in the growth of political freedoms than did his father.

While authoritarianism seems to be giving way to more political freedom and regular elections, these processes are often complicated by corruption and violent state repression of political movements. In the Philippines, for example, after being plagued with a history of colonialism under the Spanish, then military occupation by the United States from 1898 to 1946 (the United States maintained a large military base until 1991), and a long line of dictators throughout the twentieth century, Filipinos elected governments that were somewhat successful at reforms (suppression of crime and drug abuse, health care improvements, urban infrastructure reforms), especially in the north of the island group. But most recently, the government, led by an aspiring strongman with populist inclinations, Rodrigo Duterte (elected president in 2016), reverted to autocratic solutions. For example, Duterte sought to quell illegal drug sales by murdering the dealers and users (1900 lives lost in 7 weeks). Although Duterte is from the island of Mindanao in the southern Philippines, he has continued the long-standing confrontation with Muslim militants in the southern islands. The government has been reluctant to compromise, refusing to allow greater Muslim autonomy in the south. In retaliation two sibling Islamists led an insurgency in the city of Marawi that

10.23 Power and Politics in Southeast Asia

There are significant barriers to public participation in politics in Southeast Asia, including a long history of political violence and ethnic cleansing. Also, officials have promoted the idea that authoritarian political cultures are "normal" for the region, while democracy is messy and disruptive. However, throughout the region people are pushing for more access to the political process. Political opposition leaders are increasingly vocal, citizens have mounted public protests, some countries have enacted effective political reforms, and political corruption is occasionally punished.

THINKING GEOGRAPHICALLY

A Why would a government ban a flag?

B Which pressure group may have provided these political leaders with flowers, and why?

C The equator runs just south of this battle scene. What does this picture suggest about the equipping of the troops for this battle?

D How might banning the activities of journalists outside a courtroom inhibit public participation in politics?

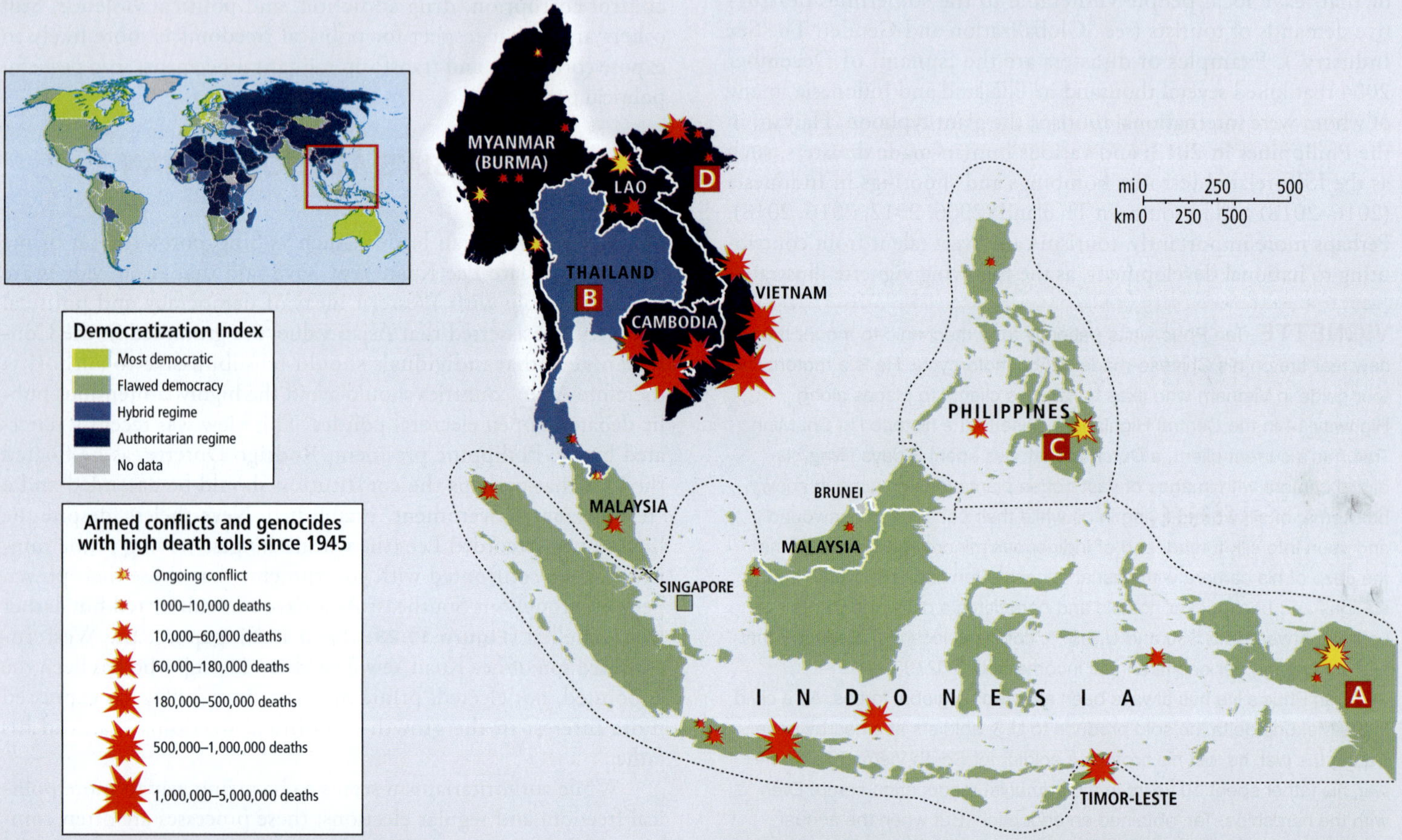

A **Indonesia.** The banned flag of the militant Free Papua Movement is carried at a demonstration in West Papua, Indonesia, shortly before police open fire. Political freedoms are growing in Indonesia, but repression of political movements has a long history. As colonial control by the Dutch ended in the 1960s, West Papua was acquired by Indonesia in a sham election rejected by West Papuans, who say that to this day they face daily surveillance and intimidation by the Indonesian military and police. The vast Freeport McMoran gold and copper mine in West Papua is an important source of tax money for Indonesia. [TJAHJONO ERANIUS/AFP/Getty Images]

B **Thailand.** Leaders of Thailand's opposition party Pheu Thai leave court after being charged with sedition by the ruling military junta for violating the junta's rule against political gatherings. Thailand, long thought to be on the road to stable democracy, has been in a political uproar since a popular populist prime minister was removed by the military in 2006. [KRIT PHROMSAKLA NA SAKOLNAKORN/AFP/Getty Images]

C **Philippines.** Philippine troops struggled to take back the city of Marawi in Mindanao from Islamic militants. Trained for jungle warfare, they were unfamiliar with urban warfare, and many hundreds of young men lost their lives. [AP Images/Bullit Marquez]

D **Vietnam.** Police and local militia members stop journalists from taking pictures outside a court in Ho Chi Minh City, Vietnam, where a pro-democracy activist was being tried. Vietnam's government remains communist and authoritarian and allows the population few political freedoms. Journalists are harassed by the police and military because they are thought to undermine the government's authority. [IAN TIMBERLAKE/AFP/Getty Images]

forced at least 120,000 to flee before the military put an uneasy and temporary stop to the violence (see Figure 10.23C).

Thailand, a constitutional monarchy, was regarded for some years as the most stable democracy in the region, despite its record of numerous coups d'état. It was always a fragile democracy because of deep divisions between those who favored authoritarianism (including the monarchists, many economic elites, and the military) and those who were part of or supported the large **populist political movement** advocated by the political party Pheu Thai. In 2006, the military took over Thailand's government after a corrupt but charismatic prime minister, Thaksin Shinawatra, was implicated in several financial scandals. In 2011, Shinawatra's sister, Yingluck, was elected prime minister, with overwhelming support from her brother's supporters. But in 2014, in the twelfth coup d'état since 1932, she was removed from power and tried on corruption charges by the military, which had once again seized control of the government. The military suspended the constitution, declared martial law, reduced the power of political parties by forbidding political meetings (see Figure 10.23B), and imposed censorship of the internet and the media. Urban prosperous elites, dominant in Bangkok and to the west and south of the country, favor the authoritarian military government, while a large and vocal rural majority north and east of Bangkok coalesces around the party Pheu Thai, which is best characterized as supporting a version of populist absolutism. Thailand remains in political transition: in 2016, a new constitution was approved that because of a proportional representation system makes it difficult for a party to gain a majority in parliament. Legislative elections long promised were finally held in 2019, and as predicted no party won a majority. Thailand, a constitutional monarchy, crowned a new king in 2019. He, who is revered as a deity, promises to calm political conflict, but with an authoritarian tinge that may further undermine democracy.

Other countries are also struggling to find a balance between meeting demands for more political freedoms and relying on authoritarianism to create stability. Indonesia has frequently used brute force to quell opposition in its far-flung island dependencies (see Figure 10.23A). Cambodia's democracy is precarious because of widespread and entrenched corruption, and violence there is common. The wealthier and usually stable countries of Malaysia and Singapore continue to use authoritarian versions of democracy that impose severe limitations on freedom of the press and freedom of speech.

Across the region, when governments aspiring to democratic policies fail to keep the peace, authoritarianism, often backed by a strong military, is seen as a justified counterforce to "too much" democracy. In virtually every country, the military has been called on to restore civil calm after periods of civil disorder. Military rank is highly regarded and many top elected officials have had military careers.

Very recently, authoritarianism has been challenged by political reform in Myanmar (Burma), where for decades, numerous regional ethnic minorities and pro-democracy movements have been repressed with an iron fist. For a time it seemed that a gradual expansion of political freedoms was taking place in Myanmar (Burma), as evidenced by the recent freeing of political prisoners such as Aung San Suu Kyi (who has led antimilitary protests and was under house arrest for 15 of the years between 1989 and 2010), a reduction in restrictions on the press, and more or less free elections in 2015. The military, calling these changes "disciplined democracy," still holds an unelected 25 percent of parliamentary seats, has the constitutional right to appoint key ministry positions (defense, home affairs, and border affairs), and must approve all changes to the constitution. Despite the fact that Aung San Suu Kyi appears to have relinquished her leadership of the democratic opposition (she has failed to speak out against ill-treatment of the Rohingya minority), political change appears to be coming in Myanmar (Burma), and the country's economy is already undergoing transformation.

Little political reform is taking place in Lao and Vietnam (see Figure 10.23D), where authoritarian socialist regimes have a firm grip on power, or in Brunei, which is governed by an authoritarian sultanate. For these and other reasons, authoritarianism is likely to remain a powerful force in the politics of this region for some time.

populist political movement a coalition of usually rural or working-class activists who claim that virtuous citizens are being exploited by a small group of elites who are presumed to be in illegitimate control of the government

soft power initiatives diplomatic overtures or offers of aid that appear to be only friendly with no ulterior geopolitical motives

CHINA'S SOFT POWER BRINGS WORRIES AND MILITARISM

While the countries of the region are unlikely to go to war against each other, they are all keeping an eye on China, who often offers friendship through **soft power initiatives**—bits of aid that appear to have no ulterior geopolitical motives. Chinese investment money and covert military moves are influencing internal affairs in virtually every part of Southeast Asia. For example, in Cambodia, dozens of Chinese-run casinos have crowded into a small coastal city, forcing hundreds of family-owned businesses to close and drastically changing local ways of life with the introduction of wealthy Chinese tourists. Service jobs provided by casinos don't pay well enough to replace the lost income from small businesses.

Chinese investment in hydroelectric dams along the Mekong River was discussed earlier in "Responses to Climate Change in the Mekong River System." China is planning to widen and deepen the Mekong to accommodate cargo boats. Regional environmentalists have already noted drastic environmental deterioration (**Figure 10.24**) and are worried that China could use the dams and its upstream position along the river to coerce the roughly 60 million people downstream in Thailand, Lao, Cambodia, and Vietnam, who depend on the river for food and income, to do China's bidding. Furthermore, the Chinese are beginning to aggressively patrol the Mekong with gunboats, raising concerns about militaristic plans.

The Mekong flows into the South China Sea where China has in recent years also been encroaching militarily. Newly discovered oil resources in the South China Sea, as well as fishing resources and vital shipping routes, have resulted in competing territorial claims by China, Taiwan, Vietnam, Malaysia, Brunei, Indonesia, and the Philippines. One contention is over the Spratly Islands and associated maritime shoals, reefs, and cays, at least ten of which are occupied by the Philippines but others of which are physically claimed by China, Taiwan, and Vietnam. It is understandable

Figure 10.24 Fisher people and farmers along the Mekong suffer losses from China-built dams. Here, a Laotian fisherman casts his nets in the Mekong in Thailand, where drastically changed water levels threaten livelihoods. [Chien-min Chung/Getty Images News/Getty Images]

that Southeast Asian nations consider military preparedness to be prudent.

Most countries have recently invested heavily in military equipment, primarily from Europe and the United States. In the midst of this buildup, even countries with no claims to the South China Sea are arming themselves. Tiny, wealthy, trade-centered Singapore recently spent one-quarter of its GDP on weapons. It has the largest military budget in the region and now produces armored troop carriers for use at home and for export.

WILL EXPANSION OF POLITICAL FREEDOMS BRING PEACE TO INDONESIA?

Indonesia is the largest country in Southeast Asia and the most fragmented—physically, culturally, and politically (see Figures 10.1 and 10.3). It comprises more than 17,000 islands (3000 of which are inhabited), stretching over 3000 miles (8000 kilometers) of ocean. It is also the most culturally diverse, with hundreds of ethnic groups and multiple religions. Although Indonesia has the largest Muslim population in the world, there are also many Christians, Buddhists, Hindus, and followers of various local religions and spiritual traditions. Indonesia has long been home to migrants from all over the world: in prehistory, during the European colonial era, during world wars in the twentieth century, and most recently by refugees seeking asylum. With all these potentially divisive forces, many wonder whether this multi-island country of 265 million people might be headed for political disintegration, or whether it might instead prove to be a model of social integration.

Until the end of World War II, Indonesia was a loose assembly of distinct island cultures that Dutch colonists managed to hold together as the Dutch East Indies. When Indonesia became an independent country in 1945, its first president, Sukarno, hoped to forge a new nation out of these many parts, founded on a strong communist ideology. To that end, he articulated a national philosophy known as ***Pancasila***, which was aimed at holding the disparate nation together, primarily through nationalism and concepts of religious tolerance.

In 1965, during the height of the Cold War, when the United States was particularly wary of communism, a staunchly anticommunist general in the Indonesian army who was supported by the United States, Haji Mohammad Suharto, ousted Sukarno in a coup and ruled the country for 31 years. It is now clear that Suharto's regime was responsible for a purge of ordinary citizens suspected of being communists, during which as many as a million people were killed.

Despite his abrupt removal from office by Suharto, Sukarno's unifying idea of Pancasila endures as a central theme of life in Indonesia (and less formally throughout the region). Pancasila embraces five precepts: *belief* in God, and the observance of *social conformity*, *corporatism* (often defined as "organic social solidarity with the state"), *consensus*, and *harmony*. These last four precepts could be interpreted as discouraging dissent or even loyal opposition, and they seem to require a perpetual stance of national boosterism. Pancasila is criticized by some as being too secular, while others see it as not going far enough to protect the freedom of people to believe in multiple deities, as Indonesia's many Hindus do, or the freedom to not believe in any deity at all. Some praise Pancasila for unifying the extremely ethnically diverse and geographically dispersed country, while others say that the precepts have had a chilling effect on participatory democracy and on criticism of the government, the president, and the army.

The first orderly democratic change of government in Indonesia did not take place until national elections in 2004. Since then, there have been several peaceful elections, and the government is stable enough to allow citizens to publicly protest some policies. But corruption remains a threat that encourages feelings of nostalgia for the Suharto model of authoritarian government. In 2019, Joko Widodo was elected to a second term, defeating an ultranationalist populist, who is now contesting election results. Widodo, who won by more than 13 percentage points, seems committed to extending democratic participation and implementing moderate policies, despite a deeply divided parliament. According to reliable reports, the elections were meticulously well run in this, the world's third-largest democracy.

Separatist movements have sprouted in four distinct areas in recent years, demonstrating Indonesia's fragility. The only rebellion to succeed was in Timor-Leste, which became an independent country in 2002. However, its case is unique in that the area was under Portuguese control until 1975, when it was forcibly integrated into Indonesia. Two other separatist movements (in the Molucca Islands and in West Papua; see Figure 10.23A) developed largely in response to Indonesia's forced **resettlement schemes**, which were begun in 1965. Also known as *transmigration schemes*, these programs have relocated approximately 20 million people from crowded islands such as Java to less densely settled islands. The policies were originally initiated under the Dutch in 1905 to relieve crowding and provide agricultural labor for plantations in thinly populated areas. After independence,

Pancasila a national philosophy that embraces five precepts: *belief* in God and the observance of *conformity*, *corporatism*, *consensus*, and *harmony*

resettlement schemes government plans to move large numbers of people from one part of a country to another to relieve urban congestion, disperse political dissidents, or accomplish other social purposes; also called *transmigration*

Indonesia used resettlement schemes both to remove troublesome people and to bring outlying areas under closer control of the central government in Jakarta. Resettlement schemes continue today, at roughly 60,000 people per year. However, now resettlement of millions from coastal zones is being forced by rising sea levels due to climate change. There is even a move to relocate the capital from Jakarta, which is sinking.

The far-western province of Aceh, in the north of Sumatra, has long been troubled by political violence and accusations of terrorism. However, the recent expansions of political freedoms in Aceh may have helped the province chart a course to peace. Conflicts there originated when revenues from extraction of Aceh's oil were going to the central government in Jakarta. The Acehnese people protested what they saw as the expropriation of oil without compensation, and Jakarta sent the military to silence them. Many who spoke out against the military presence or who supported the Free Aceh movement were accused of terrorism and arrested, jailed, forced into hiding, or even killed. The conflict seemed unresolvable. Then, in 2004, the earthquake and tsunami in Aceh, which killed more than 170,000 Acehnese, suddenly brought many outside disaster relief workers to Sumatra because the Indonesian government was unable to give sufficient aid to the victims. Global press coverage of the relief effort mentioned the recent political violence; this created a powerful incentive for separatists and the government to cooperate in order to receive outside aid. A resulting peace accord signed in Helsinki in 2005 brought many former combatants into the political process as democratically elected local leaders. Virtually all the separatists laid down their arms; but recently underdevelopment, lack of opportunities, and a rising climate of fear have instigated a revival of the Free Aceh movement and the long-delayed organization of a Truth and Reconciliation Commission for the period 2016–2021.

TERRORISM, POLITICS, AND ECONOMIC ISSUES

Like authoritarianism, terrorism has long been a counterforce to political freedoms in this region. Terrorist violence short-circuits the public debate that is at the heart of democratic processes and appears to reinforce the need for repressive authoritarian measures. However, it is important to recognize that terrorist movements often thrive in the context of both economic deprivation and political repression, two factors that are often interrelated. As was the case in Aceh and is the case in the southern Philippines (see Figure 10.23C), terrorism often draws support from people who feel shut out of opportunities for economic advancement. The peaceful solution to terrorism could be as simple as a willingness to listen to both the political and economic desires of those who might otherwise be attracted to terrorism.

CHECK YOUR UNDERSTANDING

1. What was the apparent intention of some Southeast Asian leaders, such as Singapore's former prime minister Lee Kuan Yew, who have said that Asian values are not compatible with Western ideas of democracy and political freedom?
2. Discuss why authoritarianism has been attractive across Southeast Asia as a way of dealing with diverse multicultural populations that at times come into conflict with each other.
3. Describe Pancasila and explain why forced social cohesion has been a strategy in this region. How has Pancasila inhibited democratic participation?
4. Compare authoritarianism and terrorism as counterforces to political freedoms in this region.

URBANIZATION

10.7. Describe urbanization in Southeast Asia and discuss the factors that are fueling urban growth.

10.8. Explain why the largest Southeast Asian cities receive most migrants, despite having insufficient housing, water, sanitation, and jobs.

Southeast Asia as a whole is only 47 percent urban, but the rural–urban balance is shifting steadily in response to the economic transformation of the region into an industrial hot spot.

THE FORCES OF CHANGE

The forces driving farmers into the cities are called the *push factors* in rural-to-urban migration. These factors include deforestation of subsistence lands for commercial crops like oil palm, the rising cost of farming (related to the use of new crops, technologies, and competition with agribusiness), and the loss of farm labor opportunities. *Pull factors*, in contrast, are those that attract people to the city, such as abundant manufacturing jobs, education opportunities, and the rumored excitement of urban living. In Southeast Asia, as in all other regions, these factors have come together to create steadily increasing urbanization—a few percentage points every year. Malaysia is 75 percent urban; Brunei, 78 percent; Singapore, 100 percent; Myanmar (Burma), 29 percent; Cambodia, 23 percent; Lao, 35 percent; Vietnam, 35 percent; the Philippines, 47 percent; Indonesia, 54 percent; and Thailand, 50 percent.

Employment in agriculture has been declining throughout Southeast Asia since the introduction of new production methods that increased the use of labor-saving equipment and reduced the need for human labor. Chemical pesticide and fertilizer use has also grown. While such additives can increase harvests and the supply of food to cities, they also can drive the costs of production higher than what most farmers can afford. Many family farmers have sold their land to more prosperous local farmers or to agribusiness corporations and have moved to the cities. These people, skilled at traditional farming but with little formal education, often end up in the most menial of urban jobs (**Figure 10.25C, D**) and live in circumstances that do not allow them to grow their own food (Figure 10.25A).

Labor-intensive manufacturing industries (garment and shoe making, for instance) are expanding in the cities and towns of the poorer countries, such as Cambodia, Vietnam, and parts of Indonesia and Timor-Leste. In the urban and suburban areas of the wealthier countries—Singapore, Malaysia, Thailand, and parts of Indonesia and the northern Philippines—technologically sophisticated manufacturing industries are also growing. These include automobile assembly, chemical and petroleum refining, and computer and other electronic equipment assembly. Riding on this growth in manufacturing are innumerable construction projects that often provide employment to recent migrants (see Figure 10.25C, D and the vignette on the global economy in Chapter 1).

Cities such as Jakarta, Manila, and Bangkok, among the most rapidly growing metropolitan areas in the world, are *primate cities*—cities that, with their suburbs, are vastly larger than all others in a country. Bangkok is more than 20 times larger than Thailand's next-largest metropolitan area, Udon Thani; Manila is more than 10 times larger than Davao in Mindanao, the second-largest city in the Philippines. Thanks to their strong industrial base and political power and the massive immigration they attract, primate cities can dominate whole countries; and in the case of Singapore, the city constitutes the entire country.

Rarely can primate cities provide sufficient housing, water, sanitation, or decent jobs for all the new rural-to-urban migrants. Many millions of urban residents in this region live in squalor, often on floating raft-villages on rivers and estuaries. Of all the cities in Southeast Asia, only Singapore provides well for nearly all of its citizens (see Figure 10.25B). Even there, however, a significant undocumented noncitizen population lives in poverty in hidden dormitories and on islands surrounding the city. The experience of rural-to-urban migrants who go to Bangkok or Jakarta is typical: migrants there often live in slums on the banks of polluted, trash-ridden bodies of water (see Figure 10.25A).

EMIGRATION RELATED TO URBANIZATION AND GLOBALIZATION

The same push and pull factors that convince migrants to go to the region's cities also tempt millions to leave Southeast Asia. People abandon agricultural subsistence in the countryside to seek jobs that provide a cash income they can use to feed their families and buy consumer goods. These migrants are a major force of globalization, as they supply much of the world's growing demand for low- and middle-wage workers who are willing to travel or live temporarily in foreign cities. For example, 40 percent of the foreign workers in Taiwan are Indonesian. These migrants are also a globalizing force within their home countries because their **remittances** (money sent home) boost family incomes and supply governments with badly needed **foreign exchange** (foreign currency) that countries use to purchase imports. Filipinos working abroad are their country's largest source of foreign exchange, sending home more than U.S.$6 billion annually and increasing household annual income by an average of 40 percent. And when they return home, they bring new ways of thinking and consuming, some useful, some potentially disruptive.

The Merchant Marine

Skilled male seamen from Southeast Asia make up a significant portion of the global merchant marine labor force. The **merchant marine** refers to the approximately 12,000 privately owned ships and tankers that carry cruise passengers and solid and liquid cargo across the globe. Most of these workers are unionized and hence experience considerably better conditions than do most of the world's migrant workers. Nonetheless, in the merchant marine, it is customary for workers to be paid according to their homeland's pay scales, which makes employing seamen from low-wage Southeast Asian countries attractive to ship owners. The seamen work aboard international freighters or on luxury cruise liners as deckhands, cooks, and engine mechanics; only a few become officers, because most officers are selected from European and American countries. Generally, seamen work for 6 months at a time—with only a few hours a day for breaks—saving nearly every penny. At the end of a tour of duty, they return home to their families for another 6 months, where they often contribute financially to the well-being of an extended group of kin and friends and, because of their higher than average earnings, they often send their children for advanced education.

The Maid Trade

Women constitute well over 50 percent of the more than 10 million overseas workers from Southeast Asia. Many skilled female nurses and technicians from the Philippines work in European, North American, and Southwest Asian cities. At least 5 million participate in the global "maid trade" (**Figure 10.26**). The maid trade is dominated by educated women from the Philippines and Indonesia who work under 2- to 4-year contracts in wealthy urban homes in Malaysia, Thailand, Singapore, Hong Kong, and Saudi Arabia and other Persian Gulf countries. In 2015 there were at least 80,000 housemaids living in the United Arab Emirates and more than 2 million from Indonesia and the Philippines in Saudi Arabia.

VIGNETTE Every Sunday is amah (nanny) day in Hong Kong. Gloria Cebu and her fellow Filipina maids and nannies stake out temporary geographic territory on the sidewalks and public spaces of the central business district (see Figure 10.26A). Informally arranging themselves according to the different dialects of Tagalog (the official language of the Philippines they speak at home; at work in Hong Kong, most speak English), they create room-like enclosures of cardboard boxes and straw mats, where they share food, play cards, give massages, and do each other's hair and nails. Gloria says it is the happiest time of her week, because for the other 6 days she works alone, caring for the children of two bankers.

Gloria, who is a trained law clerk, has a husband and two children back home in Manila. Because the economy of the Philippines has stagnated, she can earn more in Hong Kong as a nanny than in Manila in the legal profession. Every Sunday she sends most of her income (U.S.$125 a week) home to her family. ■

By 2000, governments in Southeast Asia were recognizing that migration constituted a loss for their countries as well as an income advantage. They began stipulating minimum wages that their citizens had to be paid while working as housemaids and required a minimum number of days off. Resistance has grown to the entire trend of sending thousands of young women to work abroad as servants

remittances money sent home by people who are working abroad, usually temporarily

foreign exchange foreign currency that countries need to purchase imports

merchant marine the approximately 12,000 ships and tankers that carry passengers and solid and liquid cargo across the globe

10.25 Urbanization in Southeast Asia

Southeast Asia is rapidly urbanizing, as manufacturing and service-sector industries pull in people from rural areas and as changes in agriculture push farmers to the cities. All cities in the region are struggling to cope with rapid growth and nearly all are unable to provide the rush of new people with adequate housing, so in most cities many live in slums lacking sanitation facilities and clean water. The map shows that many cities are close to the ocean or on rivers; hence, many families live on tiny houseboats tied together to form a floating village.

THINKING GEOGRAPHICALLY

A How does this photo demonstrate sharp class distinctions in Jakarta?

B If government-built apartment buildings in Singapore are intended for the middle class, where would you expect the poor and the upper classes to live?

C Discuss why the situation depicted in this photo might be viewed as a triumph for women's rights in Bangkok.

D What would be the consequences of these carts being replaced with one dump truck? Are the carts perhaps more efficient for some tasks?

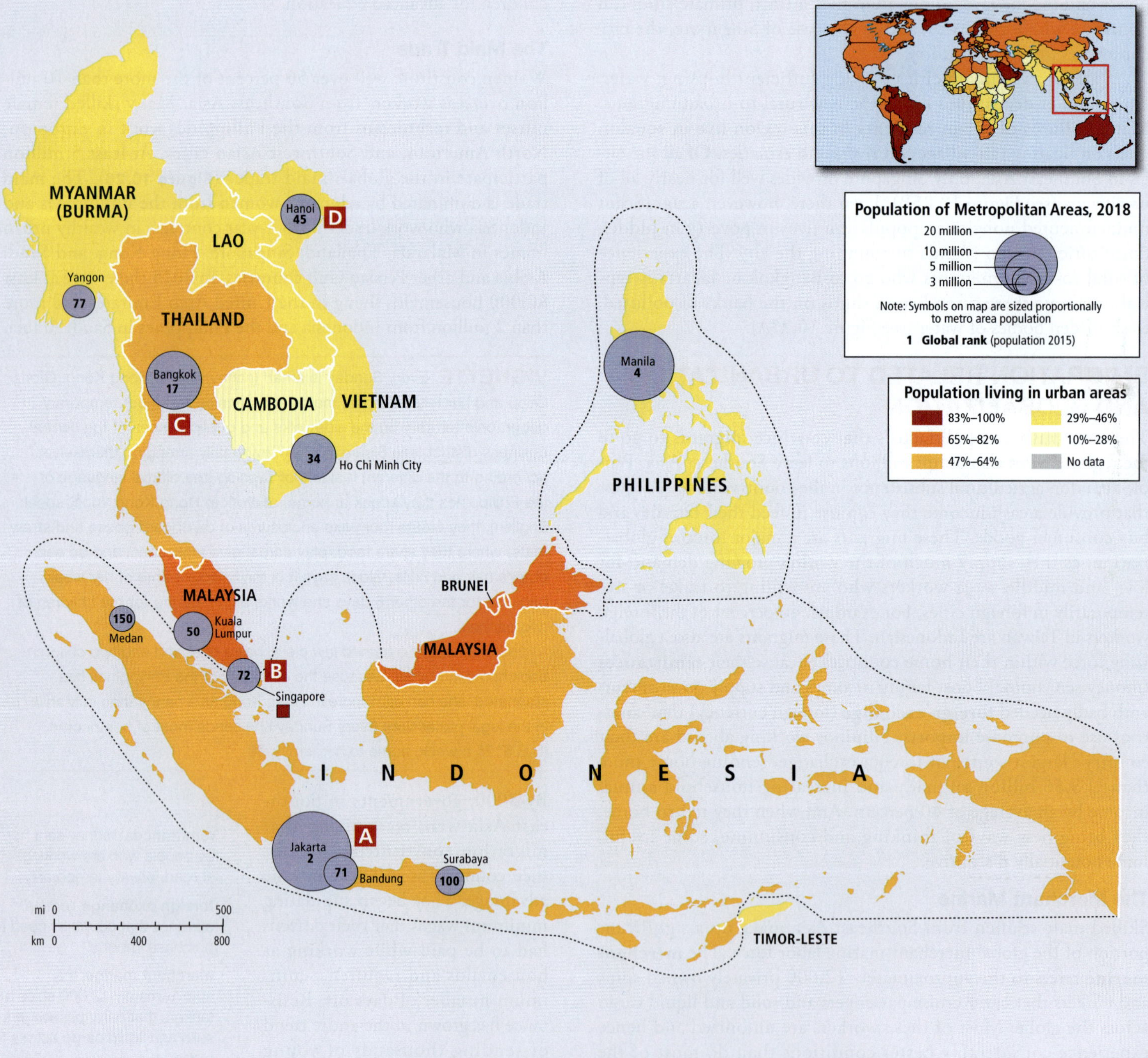

A Jakarta. In the foreground is a waterfront area in Jakarta, Indonesia, where 62 percent of the population lives in slums. A high-tech center, Jakarta has some sleek high-rise apartments for tech workers, which can be seen behind the waterfront. [BAY ISMOYO/AFP/Getty Images]

B Singapore. Of all cities in the region, Singapore has the best housing. Eighty-five percent of the population lives in apartment buildings designed, built, and managed by the government. People are proud to live in Singapore; neighborhoods put up signs to show pride in their communities or to advertise support groups for those with particular issues, such as alcoholism. [CP Cheah/Moment Open/Getty Images]

C Bangkok. A woman works at a construction site in Bangkok, Thailand. Gender roles are changing as more women move to the cities. [AFP PHOTO/PAIROJ/Getty Images]

D Hanoi. Migrant workers from rural areas pull carts to a construction site in Hanoi, Vietnam. Many migrants start out in low-wage, manual labor jobs like this one. [HOANG DINH NAM/AFP/Getty Images]

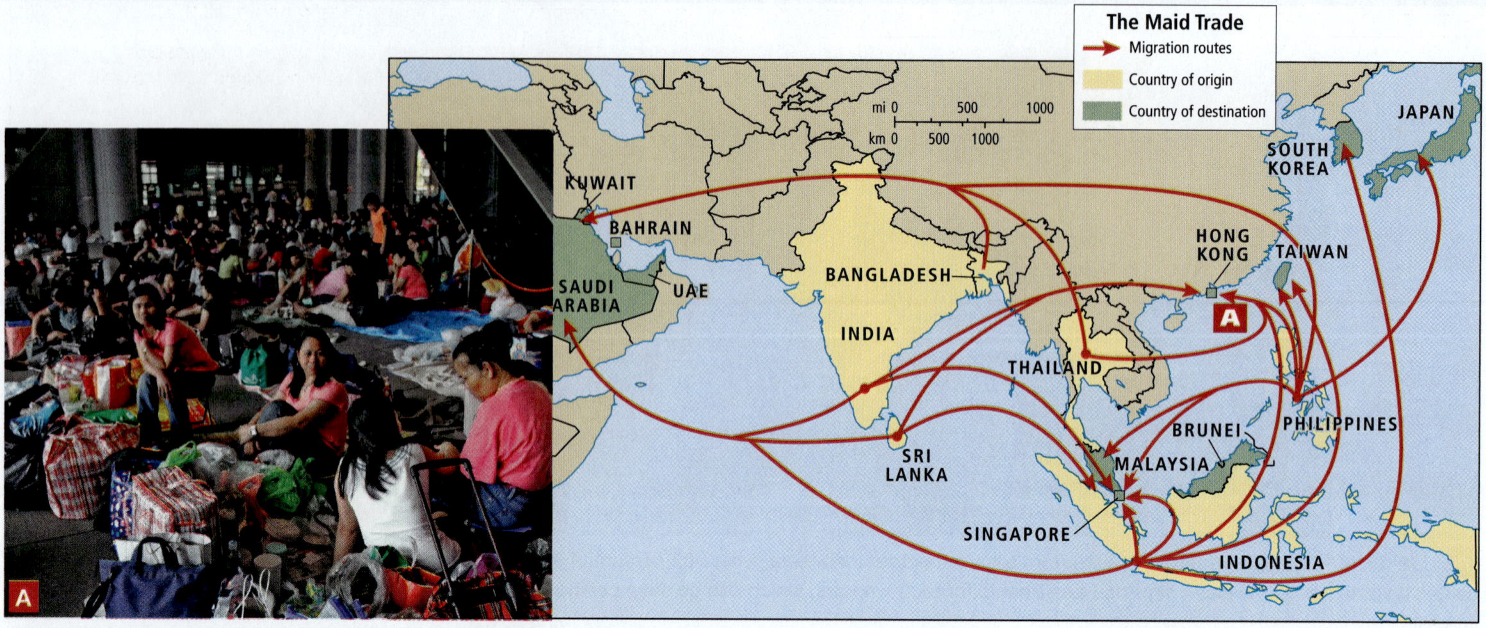

Figure 10.26 Globalization: The maid trade. The map shows the origins and destinations of maids throughout the region. By 2018 there were at least 5 million women, primarily from the Philippines, Indonesia, and Myanmar (Burma), working as domestic servants across Asia. **A** In Hong Kong it became customary for Filipina maids and nannies to gather on Sundays in central Hong Kong to relax, talk, trade goods, play games, and pack up things to send home. [Research from: Joni Seager, *The Penguin Atlas of Women in the World* (New York: Penguin Books, 2003), p. 73; and Antje Missbach and Wayne Palmer, "Indonesia: A Country Grappling with Migrant Protection at Home and Abroad," September 19, 2018, Migration Policy Institute, at https://www.migrationpolicy.org/article/indonesia-country-grappling-migrant-protection-home-and-abroad.] [TED ALJIBE/AFP/Getty Images]

and nannies. When Joko Widodo was elected president of Indonesia (2014), he directed the labor department to begin retrieving all Indonesians working as housemaids in foreign countries, saying that the custom of sending domestic workers abroad undermined the country's "self-esteem and dignity." His position was grounded in the suspicion that the wages paid were too low and the working conditions were inhumane, perhaps including sexual exploitation. Indonesia has been joined in this opposition to housemaid emigration by the Philippines. Regional leaders noted, in a 2007 ASEAN agreement for the protection of workers, that the emigration of young people revealed their own poor development policies. Emigration flows result in a loss of human capital and are a reflection of the lack of satisfactory living and employment opportunities in their home countries, as the preceding vignette illustrates. By 2018 the flow of migrant housemaids had resumed.

CHECK YOUR UNDERSTANDING

1. What are the major challenges to Southeast Asian cities that are receiving the majority of rural-to-urban migrants?
2. Identify the countries in Southeast Asia that are attracting laborers for manufacturing industries (garment and shoe making, for instance) and those that are attracting highly educated and skilled workers to their cities.
3. Which parts of the Philippines (north or south) are likely to be the most developed?
4. Identify the primate cities of the region and the special challenges they face.
5. Of the thousands of people who emigrate from the region each year to seek temporary or long-term employment abroad, who seem to have the best wage and working conditions?

POPULATION, GENDER, AND CULTURE

10.9 Explain the factors that influence population dynamics in Southeast Asia.

10.10 Discuss how the divergent array of cultural traditions existing in close proximity has the potential to breed social conflict.

Southeast Asia is home to nearly 660 million people (more than double the U.S. population), who occupy a land area that is about one-half the size of the United States. At current rates, Southeast Asia's population, which is large and growing, is projected to reach more than 780 million by 2050, by which time much of this population will live in cities (Figure 10.25 and **Figure 10.27**). However, population projections could be inaccurate, both because rates of natural increase are trending markedly lower and because many Southeast Asians are migrating to find employment outside the region.

POPULATION DYNAMICS

Population dynamics vary considerably among the countries of Southeast Asia. The variation is due to differences in economic development, government policy, prescribed gender roles, and

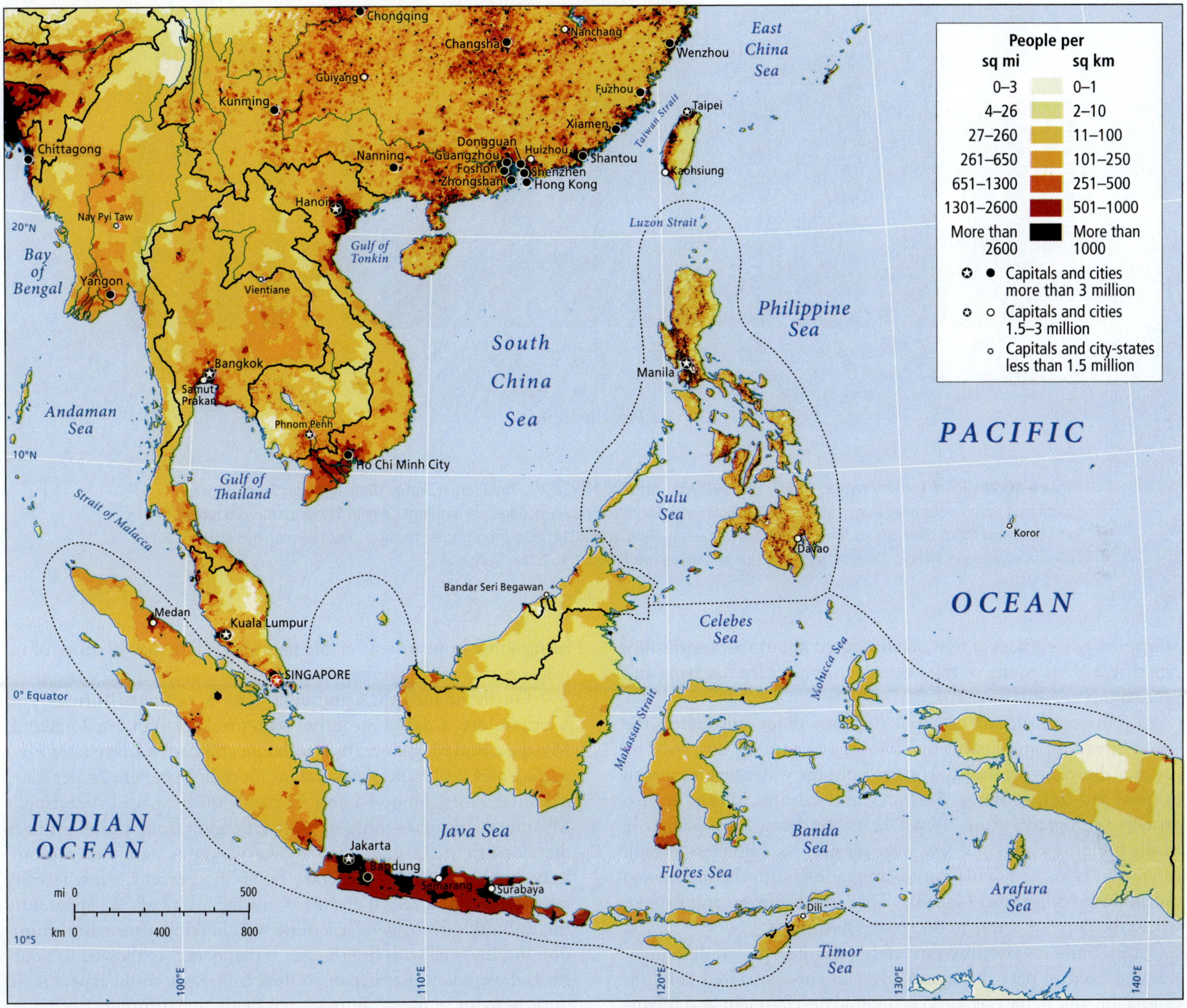

Figure 10.27 Population density in Southeast Asia. Although population density in this region is quite high, growth rates are slowing, primarily because fewer children are being born. The factors leading to declining fertility rates are many: economic development, urbanization, changing gender roles, and government policies are the most important. Only in the poorest parts of the region are fertility rates still high.

broader religious and cultural practices. Most countries are nearing the last stage of the **demographic transition**, a phenomenon where births and deaths go from being quite high to being so much lower that population growth is minuscule or slightly negative. In the last several decades, overall fertility rates in Southeast Asia have dropped rapidly (**Figure 10.28**). Whereas women formerly had 5 to 7 children and they lost many to infant mortality, they now have 1 to 3 and infant deaths are rare, with Timor-Leste (4.2, infant mortality 30/1000) being the major exception. However, because in most countries populations are still quite young (regionally 27 percent of the population is aged 15 years or younger), population is likely to grow for several decades because so many people are just coming into their reproductive years.

Brunei, Malaysia, Singapore, Thailand, and Vietnam have reduced their fertility rates so sharply—below replacement levels of 2.1 (see Figure 10.28)—that they are seeking ways to cope with aging and shrinking populations. Regionally, education levels are the highest in Singapore, where most educated women work outside the home at skilled jobs and professions. The

demographic transition a phenomenon where births and deaths go from being high to being so much lower that population growth is minuscule or slightly negative

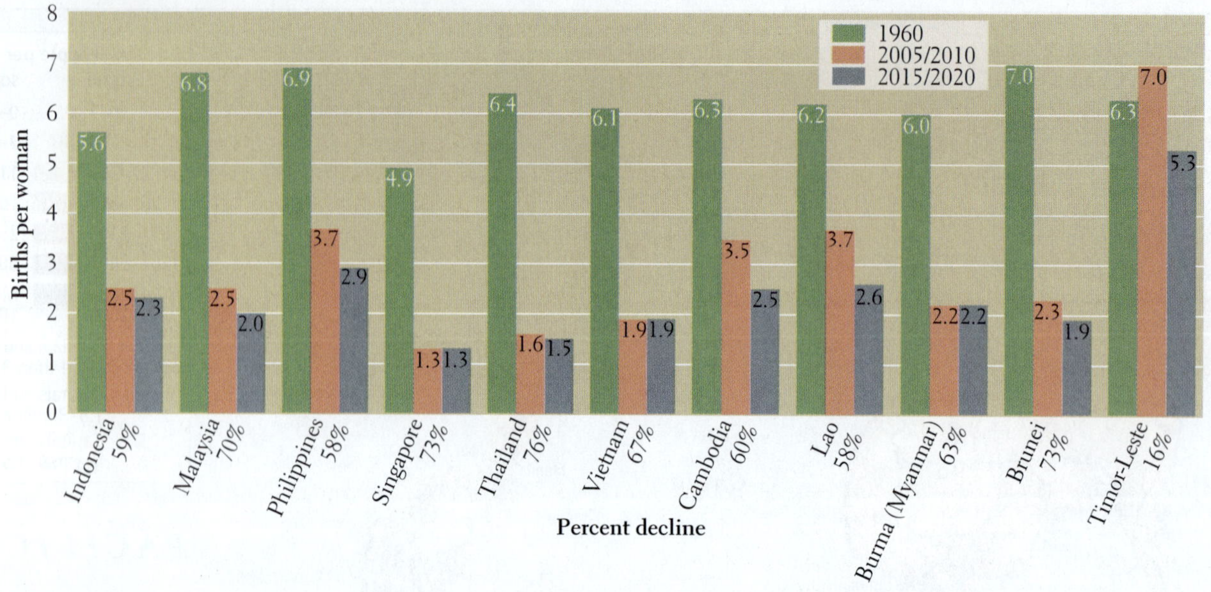

Figure 10.28 Total fertility rates, 1960, 2000–2005, and 2015–2020. Total fertility rates declined significantly for all Southeast Asian countries from 1960 to 2018. Estimates to 2020 show declines continuing. Only in Timor-Leste did rates increase, and there they did so for a few years (between 2004 and 2011) and then began to drop. [Data from: United Nations, 2015 Human Development Report, Table 18, and 2018 Human Development Report, R, Table 7, page 44.]

Singapore government is now so concerned about the low fertility rate that it offers young couples various incentives for marrying and procreating.

An important source of population growth for Singapore is the steady stream of immigrants that its vibrant economy attracts. The country draws highly skilled workers from the wider region, as well as from the United States, Australia, and Europe. Singapore also attracts immigrants from elsewhere in Southeast Asia, primarily in the building trades and low-wage services. Nonetheless, overall the region is losing population to emigration, with the mentioned exceptions of Singapore, Malaysia, Thailand, Vietnam, and Brunei, where the decrease is due to decreasing fertility.

Thailand's low fertility rate of 1.5 children per adult woman was achieved in part through a government-sponsored condom campaign, and in part by rapid economic development and urbanization, which made many couples feel that smaller families would be best. Also, as women gained more opportunities to work and study outside the home, they decided to have fewer children. High literacy rates for both men and women, along with Buddhist attitudes that accept the use of contraception, have also been credited for the decline in Thailand's fertility rate.

The poorest and most rural countries in the region show the usual correlation between poverty, high fertility, and high infant mortality. Timor-Leste, which is predominantly Roman Catholic due to the influence of its colonizer, Portugal, consists of the eastern half of the island of Timor, plus adjacent islands. Timor-Leste suffered a violent, impoverishing civil disturbance after declaring independence from Portugal in 1975. Indonesia immediately invaded and violently occupied it until 1999, when, after long international negotiations, it became independent from Indonesia. Today Timor-Leste is characterized by poverty, low use of birth control, a high fertility rate (4.2 children per adult woman), and a high infant mortality rate (30 per 1000 live births). Its oil resources bring some hope of reasonable prosperity, but the volatility of oil prices makes relying on oil risky.

On the mainland, Cambodia, Lao, Myanmar (Burma), and Vietnam are notably more poverty-stricken than are Thailand, Malaysia, and Singapore. In Cambodia and Lao, fertility rates average 2.5 and 2.6, respectively. Infant mortality rates are 24 per 1000 live births for Cambodia and 42 per 1000 for Lao. Interestingly, Myanmar (Burma)—only recently emerging from a total military dictatorship into a partially democratic regime with a strong military component—has a literacy rate of 85 percent, a low fertility rate of 2.2, but a very high infant mortality rate of 52. Apparently the military government has done well looking after basic education but less well with health care. If democracy is allowed to flourish and women to participate to their fullest, we could expect rapid improvement in overall human well-being in Myanmar (Burma).

On the other hand, in Vietnam, where people are only slightly more prosperous and urbanized, the fertility rate has already dropped to 1.9 children per adult woman and the infant mortality rate is just 15 per 1000 births, far lower than those of Myanmar (Burma), Cambodia, and Lao. These low rates for Vietnam are explained by the fact that as a socialist state, Vietnam provides basic education and health care—including birth control—to all, regardless of income. In Vietnam, literacy rates are well over 90 percent for men and women, whereas they are only 74 percent for people in Cambodia and 58 percent in Lao. In addition, Vietnam's rapidly developing economy is pulling in foreign investment, which provides more employment for women, and careers are replacing child rearing as the central focus of many women's lives.

In the 1990s, out of concern that a rapidly growing population would jeopardize Vietnam's upswing in economic development, the government of Vietnam reintroduced a two-child policy that it had previously had on and off since the 1960s. This policy brought on gender imbalance, as some couples chose abortion for female fetuses, and the birth rate dropped precipitously, below the replacement level. By 2015 all restrictions on fertility were removed.

The Philippines, which has a higher per capita income than Lao, Cambodia, Timor-Leste, or Vietnam, is an anomaly in regional fertility patterns, primarily because it is predominantly (93 percent) Roman Catholic (see the Figure 10.32 map later in the chapter), a religion that officially does not allow birth control. Fertility there is among the highest in the region (2.7 per adult female), infant mortality is in the medium range (21 per 1000 live births), and maternal deaths from childbirth are high (114 per 100,000 births), despite female literacy rates also being high (98 percent). The story of Gina Judilla offers some insights. Gina Judilla, who works outside the home, has had six children with her unemployed husband. They wanted only two, but because of the strong role of the Catholic Church and the political pressure it exerts, birth control was not available to them. Abortion is legal only to save the life of the mother, so with every succeeding pregnancy she tried folk methods of inducing an abortion. None worked. Now she can afford to send only two of her six children to school.

A recent survey showed that 48 percent of all pregnancies in the Philippines in 2010 were unintended, which often led to illegal "backstreet" abortions. In 2018, only 40 percent of Philippine women had access to modern birth control methods.

POPULATION PYRAMIDS

The youth and gender features of Southeast Asian populations are best appreciated by looking at the population pyramids for Indonesia, the largest country in the region (265 million). **Figure 10.29** shows the 2018 population pyramid and the projected population pyramid for 2050. The general pyramid shape (narrow at the top, wider toward the bottom) indicates that most people are in the younger age brackets (15 and under); then, 10 years ago a slight decline began, so the pyramid is narrower at the bottom (Figure 10.29A). The projections to 2050 (Figure 10.29B) show that eventually, with declining birth rates, those 39 and under will be outnumbered by those 40 and older. If death rates and birth rates continue to fall, Indonesia will accumulate ever-larger numbers in the upper age groups, and the pyramid will eventually be more box-shaped, as those for Europe are now. This means that Indonesia, and eventually the rest of the region, will have to contend with aging populations, just as is the case in Japan and Europe. Singapore is already concerned with this possibility.

Gender number disparities (more males than females) have existed for at least 50 years; they can be seen by carefully examining the length of the bars on the male and female sides of the pyramid for each age group 49 and younger. The differences seem slight but they are significant because the rather consistent deficit of females amounts to millions. Some were selected out before birth, and others simply did not flourish because of lack of access to food and care. Still there is more to the story of gender disparities, because for the cohorts above age 55 a deficit of males appears. This is a global phenomenon and thus far, explanations for why, on average, females live longer than males are inadequate.

GENDER ROLES IN SOUTHEAST ASIA

Gender roles are being transformed across Southeast Asia by urbanization and the changes it brings to family organization and employment. Here we look at some surprising trends affecting traditional patterns of gender roles in extended families. Moving to the city shifts people away from extended families and toward the nuclear family. The gains that women have made in political empowerment, educational achievement, and paid employment have only partially erased gender disparities within families.

Family Organization, Traditional and Modern

Throughout the region, it has been common for a newly married couple to reside with, or close to, the wife's parents. Along with this custom is a range of behavioral rules that empower the woman in a traditional marriage, despite some basic patriarchal attitudes.

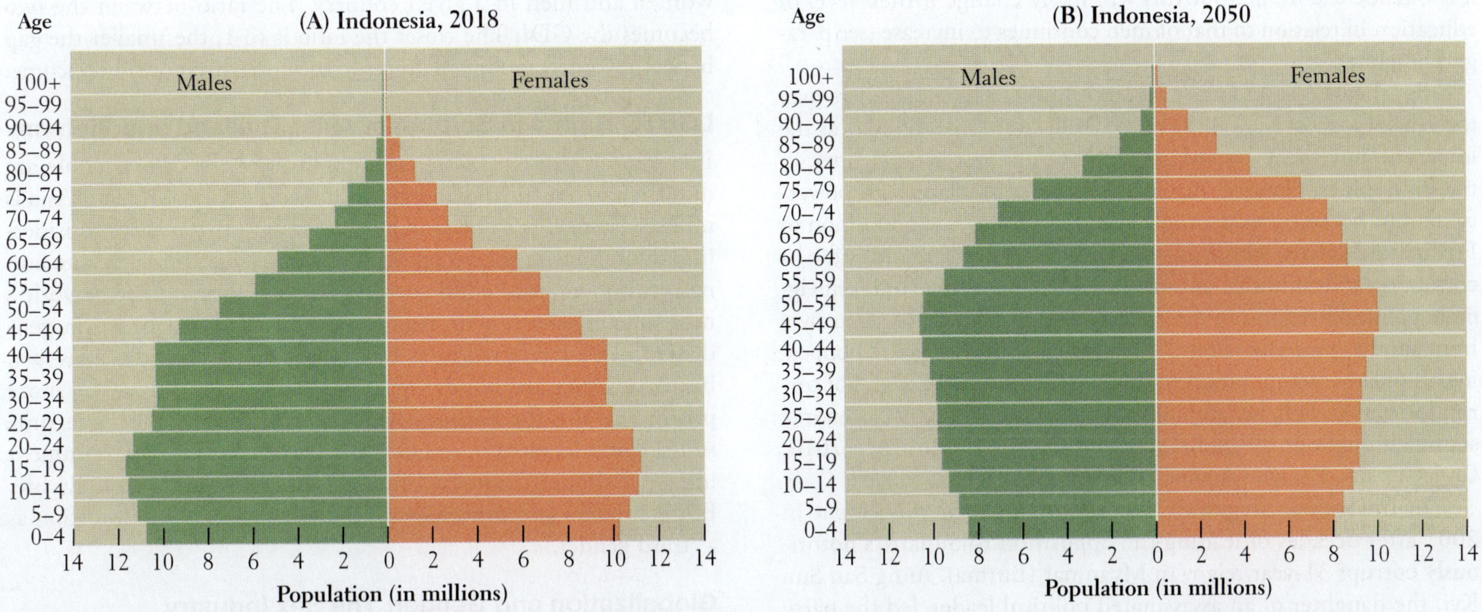

Figure 10.29 Population pyramids for Indonesia, 2018 and 2050 (projected). In 2018, Indonesia's population was 265.2 million. It is projected to be 319.7 million in 2050. [Data from: U.S. Census Bureau, *International Data Base*, 2018, at http://www.census.gov/population/international/data/idb/informationGateway.php.]

For example, a family is headed by the oldest living male, usually the wife's father. When he dies, he passes on his wealth and power to the husband of his oldest daughter, not to his own son. (A son goes to live with his wife's parents and inherits from them.) Hence, a husband may live for many years as a subordinate in his father-in-law's home. Instead of the wife being the outsider, subject to the demands of her mother-in-law (as is the case, for example, in India or Saudi Arabia), it is the husband who must show deference. The inevitable tension between the wife's father and the son-in-law is resolved by the custom of *ritual avoidance*—in daily life they simply arrange to not encounter each other much. The wife manages communication between the two men by passing messages and even money back and forth. Consequently, she has access to a wealth of information crucial to the family and has the opportunity to influence each of the two men. Another traditional custom that empowers women is that women are often the family financial managers. Husbands turn over their pay to their wives, who then apportion the money to various household and family needs.

Urbanization and the shift to the nuclear family mean that young couples now frequently live apart from the extended family, an arrangement that takes the pressure to defer to his father-in-law off the husband. Because this nuclear family unit is often dependent entirely on itself for support, wives usually work for wages outside the home. Although married women in a nuclear unit lose the power they would have if they lived among their close kin, they are empowered by the opportunity to have a career and an income. The main drawback of this compact family structure, as many young urban families have discovered in Europe and the United States, is that there is no pool of relatives available to help working parents with child care and housework. Further, no one is available to help elderly parents maintain the rural family home.

Political and Economic Empowerment of Women

Women have made some impressive gains in politics in Southeast Asia. Economically, they still earn less money than men and work less outside the home, but this will likely change if their level of education in relation to that of men continues to increase (see paragraph below).

Southeast Asia has had several prominent female leaders over the years, most of whom have risen to power in times of crisis as the leaders of movements opposing corrupt or undemocratic regimes. In the Philippines, Corazon Aquino, a member of a large and powerful family, became president in 1986 after leading the opposition to Ferdinand Marcos, whose 21-year presidency was infamous for its corruption and authoritarianism. She is credited with helping reinvigorate political freedoms in the Philippines. Gloria Macapagal-Arroyo, from another powerful family, became president in 2001 after opposing a similarly corrupt president, though she was then accused of corruption herself. Still, her administration kept the Philippine economy sound throughout the global recession starting in 2007. In 2010, Corazon's son, Benigno Aquino III, became president.

In Indonesia, Megawati Sukarnoputri became president in 2001 after decades of leading the opposition to Suharto's notoriously corrupt 31-year reign. In Myanmar (Burma), Aung San Suu Kyi, the daughter of an assassinated political leader, led the party that won elections in 1990. For more than two decades, while confined in house arrest, she led opposition to the military dictatorship and eventually succeeded in arranging for elections. In Thailand, Yingluck Shinawatra was elected prime minister in 2011 as part of a populist political movement started by her brother. She was deposed in a military coup.

All of these women leaders were wives, daughters, or siblings of powerful male political leaders, which raises some questions of nepotism. However, family favoritism cannot account for several countries where the percentage of female national legislators is well above the world average of 24 percent: Timor-Leste (33.8 percent), the Philippines (29.5 percent), Lao (27.5 percent), and Vietnam (26.7 percent).

Despite their increasing successes in politics and their acknowledged role in managing family money, women still lag well behind men in terms of economic well-being. Throughout the region, men have a higher rate of employment outside the home than women and are paid more for doing the same work. But changes may be on the way. In Brunei, Malaysia, the Philippines, and Thailand, significantly more women than men are completing training beyond secondary school. If training qualifications were the sole consideration for employment, women would have an advantage over men. While women are still disadvantaged in relation to men despite their more advanced education, the increasing power of service-sector economies—which generally require more education—should help women. The service economy already dominates in Singapore, the Philippines, Malaysia, Thailand, and Indonesia.

The Gender Development Index (GDI) measures the effects of the many life situations where females fall behind males by looking at the basic human development factors of health, education, and income by gender. These three factors are the basis for the Human Development Index (HDI) discussed earlier. In Figure 10.19 we mapped HDI in Southeast Asia and we noted that not only were there big differences in HDI across the region, but women workers in the countries with the lowest HDI rankings had special problems with working conditions. In **Figure 10.30** we focus on the Gender Development Index (GDI), which compares the HDI values for women and men in a given country. The ratio between the two becomes the GDI. The closer the ratio is to 1, the smaller the gap between women and men in access to health, education, and income.

LGBTQ Rights in Southeast Asia: Thailand and Vietnam

Lesbian, gay, bisexual, transgender, and questioning people are discriminated against throughout the world, and Southeast Asia is no exception. Both Thailand and Vietnam, however, are considering giving more legal recognition to same-sex couples in the arenas of property and child custody. Thailand is among the world's most open societies with regard to transgender people; transgender women are referred to as *kathoey*, and transgender men as *tom*, both of which are nonderogatory terms. Of the two, the more prominent are the kathoey, whose gender identity has become somewhat celebrated in recent years especially in the entertainment industry, though kathoeys often face discrimination in daily life. Both kathoeys and toms may soon benefit from legal protection as a "third gender."

Globalization and Gender: The Sex Industry

Southeast Asia has become one of several global centers for the sex industry, supported in large part by international visitors willing to pay for sex. **Sex tourism** in Southeast Asia grew out of the sexual entertainment industry that served foreign military troops stationed in Asia during World War II, the Korean War, and the Vietnam War. Now, primarily civilian men arrive from around the globe to live out their fantasies during a few weeks of vacation. The industry is found throughout the region but is most prominent in Thailand. In 2017, 35 million tourists visited Thailand alone, up from 250,000 in 1965. The largest number came from China, the fastest-growing segment of tourism arrivals. Some observers estimate that as many as 50 percent were looking for some kind of sexual experience. Even though the industry is officially illegal, some Thai government officials have publicly praised sex tourism for its role in helping the country weather economic crises, because it supports more than 2.3 million jobs (2017). Some corrupt officials also favor sex tourism because it provides them with a source of untaxed income from bribes.

One result of the popularity of sex tourism is a high demand for sex workers, which has attracted organized crime. Estimates of the numbers of sex workers vary from 30,000 to more than a million in Thailand alone. Gangs often coerce girls and women into remaining in sex work once they have been tricked or forced into the trade. Demographers estimate that 20,000 to 30,000 Burmese girls taken from Myanmar (Burma) against their will—some as young as 12—are working in Thai brothels. Their wages are too low to enable them to buy their own freedom. In the course of their work, they must service more than 10 clients per day, and they are routinely exposed to physical abuse and sexually transmitted diseases, especially HIV. One result of the rising political status of women is that there is more focus on controlling human trafficking and the abuse of women and girls in the sex industry.

VIGNETTE Watsanah K. (not her real name) attends afternoon classes in English and secretarial skills, and then goes to work at 4:00 P.M. in a bar in Patpong, Bangkok's red light district. There she will meet men from Europe, North America, Japan, Taiwan, China, Australia, and Saudi Arabia, who will pay to have sex with her. She leaves work at about 2:00 A.M., studies for a while, and then goes to sleep.

Born in northern Thailand to an ethnic minority group made up of impoverished subsistence farmers, she married at 15 and had two children shortly thereafter. Her husband developed an opium addiction, so she divorced him and left for Bangkok with her children. She found work at a factory that produced seatbelts for a nearby automobile plant. During the 2008 economic crisis, Watsanah lost her job. To feed her children, she became a sex worker.

The pay, between U.S.$400 and U.S.$800 a month, is much better than the U.S.$100 a month she earned in the factory, but the work is dangerous and demeaning. Sex work, though widely practiced and generally accepted in Thailand, is illegal, and sex workers are looked down on. As a result, Watsanah must live in constant fear of going to jail and losing her children. Moreover, she cannot always make her clients use condoms, which puts her at high risk of contracting HIV and other sexually transmitted diseases. "I don't want my children to grow up and learn that their mother is a prostitute," says Watsanah. "That's why I am studying. Maybe by the time they are old enough to know, I will have a respectable job." ■

Southeast Asia's Encounter with HIV/AIDS

As in sub-Saharan Africa (see Chapter 7), HIV/AIDS is a significant public health issue in Southeast Asia, although infection and death rates are now declining across the region. Cambodia, Thailand, and Myanmar (Burma) currently have the highest infection rates.

HIV rates are expected to increase rapidly in rural areas and in secondary cities where conservative religious leaders and faith-based international agencies restrict sex education and AIDS-prevention programs, such as the promotion of condom use. Sex education is widely viewed as promoting promiscuity. Aggressive prevention programs are even more essential because of popular customs, such as sexual experimentation (at least among men), the reluctance of women to insist that their husbands and boyfriends use condoms, intravenous drug use (primarily by men), and the high mobility of young adults. Also contributing to the spread of HIV among young people (male and female) are sex tourism and sex-related human trafficking.

Thailand is one of the few countries that has been able to drastically reduce the incidence of HIV/AIDS. It did so through a well-funded program that increased the use of condoms, decreased sexually transmitted diseases dramatically through health education, and reduced visits to sex workers by half. In Thailand in 2001, AIDS was the leading cause of death, overtaking stroke, heart disease, and cancer, but it is now just the third-leading cause of death. Estimates are that in 2018 about 500,000 Thais were infected, down from 1 million 14 years ago. Men between the ages of 20 and 40 have the highest rates of infection; most vulnerable are men who have sex with men, and transgender sex workers. HIV/AIDS rates in cities such as Bangkok, Thailand, Yangon, Myanmar (Burma), and Yogyakarta, Indonesia, are thought to be 20 to 29 percent.

SOCIOCULTURAL ISSUES

Because of the region's long and complex history, the people of Southeast Asia have a great diversity of cultures and religious traditions. Now globalization and urbanization are adding yet more diversity as new cultural influences come in from abroad and as urban women gain more independence and pursue careers less focused on children and the home.

Cultural and Religious Pluralism

Southeast Asia is a place of **cultural pluralism** in that it is inhabited by groups of people from many different backgrounds. Over the past 40,000 years, migrants have come to the region from Southwest Asia, Central Asia, Europe, India, the Tibetan Plateau, the Himalayas, China, Japan, Korea, and the Pacific. Many of these groups have remained distinct, maintaining religious and traditional practices from these diverse cultural backgrounds partly because they lived in isolated pockets separated by rugged topography or seas. Increasingly people of this

sex tourism the sexual entertainment industry that serves primarily men who travel for the purpose of living out their fantasies during a few weeks of vacation

cultural pluralism the cultural identity characteristic of a region where groups of people from many different backgrounds have lived together for a long time but have remained distinct

PHOTO ESSAY
10.30 Population, Gender, and Culture in Southeast Asia

The Gender Development Index (GDI) map shows how well countries ensure gender equality in three categories: health, education, and income. The countries with rankings closest to 1 have the most gender equality. As might be expected, Singapore ranks in group 1, or very high, as does Brunei; but Malaysia, Thailand, the Philippines, and Vietnam also rank very high. To understand this, it is important to remember that the ranks are based on how equal the access to health, education, and income for males and females is within a given country. Both genders may have very good access (Singapore) or relatively poor access to these factors (Vietnam), but if access is equal, the rank will be 1, or very high. Myanmar (Burma) ranks 2, high; Indonesia and Lao rank 3, medium; Cambodia 4, low; and Timor-Leste 5, very low.

THINKING GEOGRAPHICALLY

A What do you think accounts for the fact that women and men can have roughly equal access to health and education and yet women earn so much less in the same or similar jobs held by men?

B War and economic change often open up new opportunities for women entrepreneurs. How might these new opportunities change traditional gender roles?

C Why might the activism of Timor-Leste women, in the resistance against the Indonesian occupation of Timor-Leste, have led to their increased participation in politics postindependence?

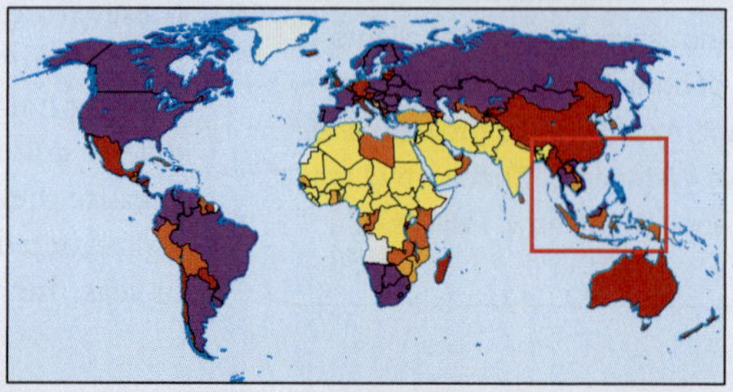

A Education versus pay. Ethnic Chinese, like this woman in Malaysia, have lived and worked in Southeast Asia for generations. Coming first as manual laborers, many are now highly educated and skilled. Males and females are near parity in Malaysia and yet, while women have on average slightly more schooling than men, female earnings are just two-thirds of male earnings. [Asia Images Group Pte Ltd/AsiaPix/Alamy]

B Entrepreneurship. A Vietnamese woman runs a small food vending business in Can Tho, Mekong Delta. Women entrepreneurs are increasingly active in Vietnam. [MyLoupe/Universal Images Group/Getty Images]

C Political activism. Women have long been politically active in Timor-Leste. They were integral to the resistance against Indonesia, and since independence have figured prominently in national politics. Lucia Lobato, a former minister for justice, was the first female presidential candidate from the Social Democrat Party and was active, though not as a candidate, in the 2019 presidential election. [LIRIO DA FONSECA/REUTERS/Newscom]

D Fishing. The Mekong River is significantly affected by environmental change due to dams and diversions, yet fishing remains an essential source of food and income. In modern times, women are often the chief fishers and fish mongers. [79Photography/Alamy]

region are demonstrating their cultural practices for their neighbors through such things as multiethnic food courts and festivals (**Figure 10.31**).

As discussed in "Religious Legacies," the major religious traditions of Southeast Asia include Hinduism, Buddhism, Confucianism, Taoism, Islam, Christianity, and animism (**Figure 10.32**). In animism, a belief system common among many indigenous peoples, natural features such as rocks, trees, rivers, crop plants, and the rains all carry spiritual meaning. These natural phenomena are the focus of festivals and rituals to give thanks for bounty and to mark the passing of the seasons, and these ideas have permeated all the imported religious traditions of the region.

The patterns of religious cultural practice are complex; the map in Figure 10.32 reveals an island–mainland division. All but animism originated outside the region and were brought primarily by traders, priests (Brahmin and Christian), and colonists. Buddhism is dominant on the mainland, especially in Myanmar (Burma), Thailand, and Cambodia. In Vietnam, people practice a mix of Buddhism, Confucianism, and Taoism that originated in China. Islam is dominant in Indonesia (the world's largest Muslim country), on the southern Malay Peninsula in far southern Thailand, and in both peninsular and island Malaysia. Islam has long been in the southern Philippines and is gaining in popularity and militancy there. Roman Catholicism is the predominant religion in Timor-Leste, the Molucca Islands, and in the middle and northern Philippines, where it was introduced by Portuguese and Spanish colonists. Hinduism first arrived with Indian traders thousands of years ago and was once much more widespread; now it is found only in small patches, chiefly on the islands of Bali and Lombok, east of Java. Recent Indian immigrants who came as laborers in the twentieth century (during the latter part of the European colonial period) have reintroduced Hinduism to Myanmar (Burma), Malaysia, and Singapore, but only as minority communities.

All of Southeast Asia's religions have changed as a result of exposure to one another. Many Muslims and Christians believe in spirits and practice rituals that have their roots in animism. Hindus and Christians in Indonesia, surrounded as they are by Muslims, have absorbed ideas from Islam, such as the seclusion of women. Muslims have absorbed ideas and customs from indigenous belief systems, especially ideas about kinship and marriage, as illustrated by marriage customs.

Figure 10.31 LOCAL LIVES: Festivals in Southeast Asia

(A) Songkran, or New Year's Day, is celebrated in Thailand. Water is thrown on friends and passersby as a sign of respect and renewal. This developed from the tradition of gathering water used to wash statues of the Buddha and gently pouring that water on the shoulders of elderly relatives. However, since the festival happens during April, a very hot time of the year, a tradition developed of throwing water on anyone. [STR/AFP/Getty Images]

(B) A calligrapher at the Temple of Literature in Hanoi, Vietnam, elegantly writes down intentions for Tet (the New Year, celebrated in January or February). Exchanges of such calligraphy are hung in people's homes; the expressed intentions surround a particular idea or a quality, such as happiness, wealth, virtue, knowledge, talent, or long life. [HOANG DINH NAM/AFP/Getty Images]

(C) An elaborately costumed dancer in the Ati-Atihan Festival in Kalibo, Aklan, in the Philippines. Originally celebrating the migration to the Philippines of a group of indigenous people from the island of Borneo, Ati-Atihan includes non-Christian traditions but is also celebrated by Christian Filipinos as a day honoring the infant Jesus. [Jeremy Jones Villasis/Getty Images]

A Catholic children in Lucban, Quezon, Philippines, prepare for a parade celebrating San Isidro Labrador, the patron saint of farmers. This is part of the Pahiyas Festival, a Catholic festival adapted from what was originally an animist harvest celebration. [JAY DIRECTO/AFP/Getty Images]

B A Muslim woman gives alms to a Buddhist monk near Borobudur, an ancient Buddhist monument on the island of Java, Indonesia. Many Islamic traditions in Java incorporate Buddhist ideas and practices. [CLARA PRIMA/AFP/GettyImages]

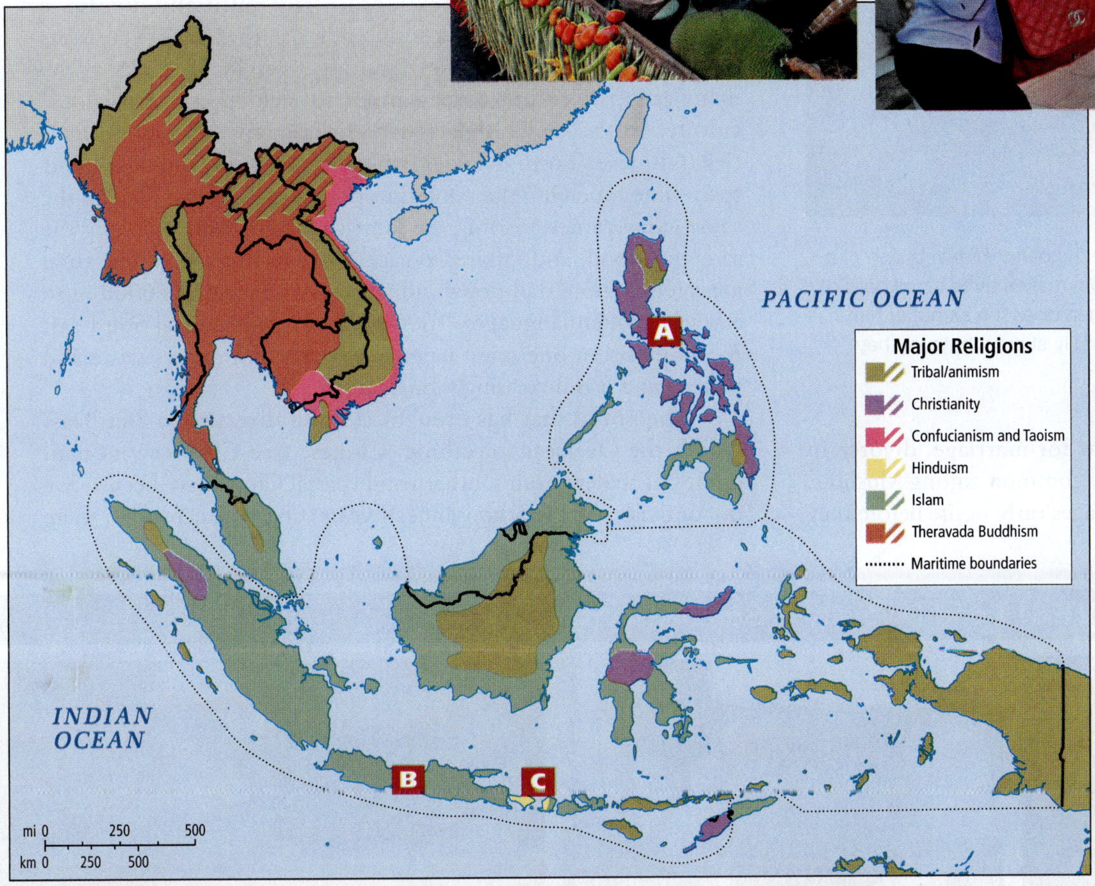

Figure 10.32 **Religions of Southeast Asia.** The religious heritage of Southeast Asia is unusually diverse: five of the world's six major religions are practiced there. Animism, the oldest belief system, is found in both island and mainland locations and has many subtle influences. [Research from: *Oxford Atlas of the World* (New York: Oxford University Press, 1996), p. 27.]

C A dancer in Bali, Indonesia, where traditional Hindu performances incorporate Buddhist and animist symbols and stories. [Paul Kennedy/Alamy]

Although arranged marriages have historically been the norm across Southeast Asia, in most urban and rural areas, most marriages are now love matches. Such is the case for Harum and Adinda, who live on the island of Java in Indonesia. They met in high school and some 10 years later, after saving a considerable sum of money, decided to formally ask both sets of parents if they could marry. Harum, an accountant, would normally be expected to pay the wedding costs, which could run to many thousands of dollars; however, Adinda was able to contribute from her salary as a teacher. Like nearly all Javanese, both are Muslim. Because Islam does not have elaborate marriage ceremonies, it is customary to enhance these festive occasions with colorful rituals from Christianity, Buddhism, and indigenous animism.

In preparation, both bride and groom participate (separately) in unique Javanese rituals that remain from the days when marriages were arranged and the bride and groom did not know each other (**Figure 10.33**). A *pemaes,* a woman who prepares a bride for her wedding and whose role is to inject mystery and romance into the marriage relationship, bathes and perfumes the bride. She also puts on the bride's makeup and dresses her, all the while making offerings to the spirits of the bride's ancestors and counseling her about how to behave as a wife and how to avoid being dominated by her husband. The groom also takes part in ceremonies meant to prepare him for marriage. Both are counseled that their relationship is bound to change over the course of the decades as they mature and as their family grows older.

Figure 10.33 A Javanese wedding. In the *dahar klimah* phase of a traditional Javanese wedding, the bridegroom makes three small balls of the food and feeds them to the bride; she then does the same for him. The ritual reminds them that they should joyfully share whatever they have. [ADEK BERRY/AFP/Getty Images]

Despite the elaborate preparations for marriage, divorce in Indonesia (and also in Malaysia) is fairly common among Muslims, who often go through one or two marriages early in life before they settle into a stable relationship. Although the prevalence of divorce is lamented by society, it is not considered outrageous or disgraceful. Apparently, ancient indigenous customs predating Islam allowed for mating flexibility early in life, and this attitude is still tacitly accepted.

Globalization Brings Cultural Diversity but Also Homogeneity

Southeast Asia's cities are extraordinarily culturally diverse, a fact that is illustrated by a multitude of ethnic food customs (**Figure 10.34**). But in some ways, this diversity decreases as the many groups continue to be exposed to each other and to global culture. For example, Malaysian teens of many different ethnicities (Chinese, Tamil, Malay, Bangladeshi) spend much of their spare time following the same European soccer teams, playing the same video games, visiting the same shopping centers, eating the same fast food, and talking to each other in English. While rural areas retain more traditional influences—of the world's 6000 or so actively spoken languages, 1000 can be found in rural Southeast Asia—in cities, one main language usually dominates trade and politics and it is increasingly English.

One group that has brought cultural diversity to Southeast Asia is the Overseas (or ethnic) Chinese (see Chapter 9). Small groups of traders from southern and coastal China have been active in Southeast Asia for thousands of years. Over the centuries, there

Figure 10.34 LOCAL LIVES: Food Diversity in Southeast Asia

(A) *Satay*, or marinated and grilled meat or fish often served with a peanut sauce, is prepared at a market in Chiang Mai, Thailand. Satay first became popular in this region in the nineteenth century on the island of Java, Indonesia, after the arrival of Muslim merchants and other immigrants from South and Southwest Asia. Some food scholars argue that satay is related to kebab, a food found throughout South and Southwest Asia. [John Elk/Getty Images]

(B) Throughout Southeast Asia, cold desserts feature shaved ice, coconut milk, evaporated or condensed milk, sweeteners, beans, noodles, corn, jelly, and other toppings. They are called *ais kacang* in Malaysia, *halo-halo* in the Philippines, *cendol* in Indonesia, and *ching bo leung* in Vietnam. Ice machines on European ships in the early twentieth century turned these desserts from beverages into ice cream–like sundaes. [SEET YING LAI PHOTOGRAPHY/Getty Images]

(C) A *banh mi* sandwich, a street food in Vietnam and Lao, is a perfect example of the union of European (in this case, French) and Southeast Asian cuisine. European ingredients include French bread (a baguette), mayonnaise, and pâté. Vietnamese and Laotian ingredients include coriander, pickled vegetables, hot peppers, cucumber, sliced pork or headcheese, and fish sauce. This sandwich now has a U.S. audience and fans. [Rebecca Skinner/Getty Images]

has been a constant trickle of immigrants from China. The forebearers of most of today's Overseas Chinese, however, began to arrive in large numbers during the nineteenth century, when the European colonizers needed labor for their plantations and mines. Later, those who fled China's Communist Revolution after 1949 joined them, and all sought permanent homes in Southeast Asian trading centers. Today, more than 26 million Overseas Chinese live and work across Southeast Asia, primarily as shopkeepers and as small business owners. A few are wealthy financiers, and a significant number are still engaged in agricultural labor.

Chinese success at small-scale commercial and business activities throughout the region has reinforced the perception that the Chinese are diligent, clever, and extremely frugal, often working very long hours. Although few are actually wealthy, with their region-wide family and friendship connections and access to start-up money, some have been well positioned to take advantage of the new growth sectors in the globalizing economies of the region. Sometimes externally funded, new Chinese-owned enterprises have put out of business older, more traditional establishments that depended on local customers and employed local people (both those of local ethnic origin and ethnic Chinese).

In recent years, when low- and middle-income Southeast Asians were hurt by the recurring financial crises, some tended to blame their problems on the Overseas Chinese. Waves of violence resulted. Chinese people were assaulted, their temples desecrated, and their homes and businesses destroyed. Conflicts involving the Overseas Chinese have taken place in Vietnam, Malaysia, and in many parts of Indonesia (Sumatra, Java, Kalimantan, and Sulawesi) as well. Some Overseas Chinese have attempted to diffuse tensions through public education about Chinese culture. Others have shown their civic awareness through philanthropy by financing economic and social aid projects to help their poorer neighbors, usually of local ethnic origins (Malay, Thai, Indonesian).

CHECK YOUR UNDERSTANDING

1. Account for the factors that affect population growth rates in this region.
2. Explain the phenomenon that accounts for population growth in Singapore, where natural increase rates are low.
3. Assess the degree to which HIV/AIDS is a threat in this region. Which nation has taken preventive measures?
4. How do you account for the many major religious traditions found in Southeast Asia?
5. Which countries are leading the way toward more acceptance of LGBTQ people?
6. Describe the political gains made by women in Southeast Asia. What factor may eventually help women to earn as much as men?
7. How has urbanization affected gender roles and family organization?
8. Describe the plight of women caught up in the sex tourism sector of Southeast Asian economies.
9. Describe some situations that illustrate cultural mixing in Southeast Asia.

SUBREGIONS OF SOUTHEAST ASIA

10.11 Distinguish between the various subregions geographically, economically, and by their sociopolitical circumstances.

The subregions of Southeast Asia share many similarities: tropical environments; ethnically diverse populations, but all with Overseas Chinese minorities that are especially active in commerce (see "Globalization Brings Cultural Diversity but Also Homogeneity"); vestiges of European colonialism (except in Thailand); difficult political trade-offs between democracy and dictatorship; and the general importance of religion. Some obvious differences among the subregions are in the pace of modernization and in the varying responses to cultural and ethnic diversity—from celebrating diversity to tolerating it, to violently rejecting it.

MAINLAND SOUTHEAST ASIA: MYANMAR (BURMA) AND THAILAND

Myanmar (Burma) and Thailand occupy the major portion of the Southeast Asian mainland and share the long, slender peninsula that reaches south to Malaysia and Singapore (**Figure 10.35**). Although Myanmar (Burma) and Thailand are adjacent and share similar physical environments, Myanmar (Burma) is poor, depends on agriculture, and is just beginning to show signs of emerging from a repressive military government, whereas Thailand has rapidly industrialized and has had a more open, less repressive society, despite occasional coups d'état and military rule (the last in 2014). Both countries trade in the global economy, but in different ways: Myanmar (Burma) supplies raw materials (including illegal drug components), while Thailand, though the world's largest exporter of rice, earns most of its income by providing low- to medium-wage labor to many multinational manufacturing firms (see Figure 10.35C), and through its thriving tourism industry that has a worldwide clientele.

The landforms of Myanmar (Burma) and northeastern Thailand consist of a series of ridges and gorges that bend out of the Plateau of Tibet and descend to the southeast, spreading across the Indochina peninsula. The Irrawaddy and Salween Rivers originate from glaciers on the Plateau of Tibet and flow south through the narrow gorges. The Irrawaddy forms a huge watery delta at the southern tip of Myanmar (Burma; see Figure 10.35B and map). The Chao Phraya flows from Thailand's northern mountains through the large plain in central Thailand and enters the sea south of Bangkok. Most farming is done in the interior lowlands in Myanmar (Burma) and in Thailand's central plain.

Ancient migrants from southern China, Tibet, and eastern India settled in the mountainous northern reaches of Myanmar (Burma) and Thailand. The rugged topography has in the past

A Myanmar's Rohingya women refugees learn to sew personal hygiene products (sanitary napkins) so they can contribute to family incomes. This project challenges discrimination on two levels: it gives job skills to long-oppressed Rohingya people and it provides products to ease the lives of women. The cloth napkins are washable and reusable. Handling the sanitary aspect of menstruation remains a problem for women in many places across the globe. [Reza Juanda/Anadolu Agency/Getty Images; MANAN VATSYAYANA/AFP/Getty Images]

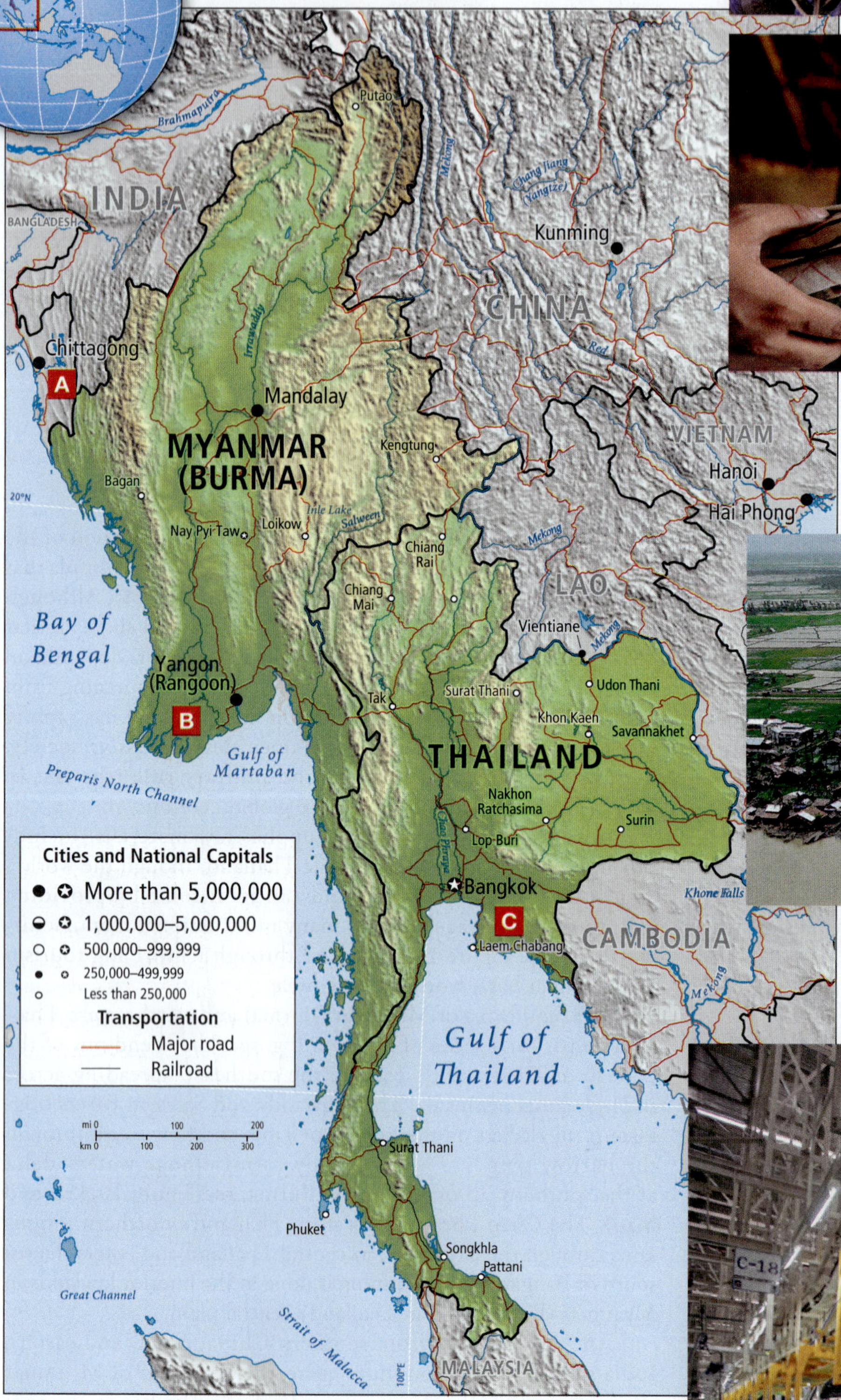

Figure 10.35 Mainland Southeast Asia: The Myanmar (Burma) and Thailand subregion.

B A village in the water-soaked Irrawaddy delta after being hit by Cyclone Nargis in 2008. The storm killed 200,000 people, many of whom would have survived if Myanmar's government had allowed in foreign aid and volunteers. [HLA HLA HTAY/AFP/Getty Images]

C A super-modernized Mitsubishi auto plant in Laem Chabang, Thailand, where workers must be educated and technically skilled. [Dario Pignatelli/Bloomberg/Getty Images]

protected these indigenous peoples from outside influences. The largest of these groups are the Shan, Karen, Mon, Chin, and Kachin, many of whom still follow traditional ways of life and practice animism. By contrast, the southern valleys and lowlands of Myanmar (Burma) and Thailand's central plain are Buddhist, modernized, and have much urban development. The original name of Burma comes from Burmans, who constitute about 70 percent of the population and live primarily in the lowlands. The name Myanmar was chosen by the military dictatorship to politically suppress the influence of Burmans. Thailand is named for the Thais, a diverse group of indigenous people who originated in southern China. Both countries are predominantly Buddhist.

Myanmar (Burma)

This country is rich in natural resources. Between 70 and 80 percent of the world's remaining teakwood still grows in its interior uplands; a single tree can be worth U.S.$200,000. Other resources include oil, natural gas (of particular interest to India, China, and the United States), tin, antimony, zinc, copper, tungsten, limestone, marble, and gemstones. However, in part because of corruption and repression by a military junta, Myanmar (Burma) ranks as one of the region's poorest countries. Per capita GNI (PPP) was estimated in 2017 to be U.S.$5830; only that of Cambodia was lower. About 70 percent of the people still live in rural villages and travel to nearby markets to sell or trade their produce for necessities (**Figure 10.36**). Their livelihoods traditionally were based on wet rice cultivation and the growing of corn, oilseeds, sugarcane, and legumes, and on logging teak and other tropical hardwoods; but for decades some have increasingly turned to opium poppies as a crop and other drug-related activities.

In 2011, Myanmar (Burma) was estimated to supply opium for more than 60 percent of the heroin market in the United States; it alternated with Afghanistan as the world's largest producer (the situation is now in flux, with production and transport under some level of control). Opium poppy cultivation, along with methamphetamine production, is the major source of income for a number of northern indigenous ethnic groups. In some villages, 90 percent of the adults are said to be drug addicted. The military government of Myanmar (Burma) has long encouraged, manipulated, and profited from the drug traffic, especially in the lightly settled uplands and mountains near the Thai border. When indigenous inhabitants protested, they were silenced by assassinations, plundering, resettlement, and the abduction and sale of their young women into the sex industries in neighboring countries.

Figure 10.36 A market at Inle Lake, Myanmar (Burma). Several indigenous groups live in villages around the lake, growing vegetables, flowers, and rice. Men produce pottery and items made of silver and brass, and women weavers have made the area Myanmar (Burma)'s second-largest producer of silk products. [Roland Neveu/LightRocket/Getty Images]

But now European coffee drinkers, who are offering farmers in the Shan state an alternative to opium production, provide a ray of hope. Since 2015, nearly 1000 farmers have worked with the United Nations Office of Drugs and Crime (UNODC) and European donors to acquire knowledge and expertise in the coffee-growing and -processing business and have formed a cooperative that produces extraordinarily high-quality coffee. The coffee is to be processed in Myanmar (Burma), not Europe, so local economies can benefit from the value chain.

In Rakhine state, other Europeans are finding a way to help. The 3 million Muslim Rohingya, originally from Bangladesh, who took refuge in Myanmar (Burma) already in the fifteenth century, have long suffered severe discrimination from the dominant Burmese ethnic group, and this ill-treatment has increased in recent years. For this and other human rights offences, the international community applied economic sanctions against Myanmar's (Burma's) military government, but even after elections in 2015 these efforts had little or no effect. Then, European NGOs began helping the Rohingya women with basic strategies to enhance their earning power, improve their health, and enhance gender equity (see Figure 10.35A).

Unlike pro-democracy movements that resulted in real change elsewhere in Southeast Asia, for more than two decades, people in Myanmar (Burma) protested futilely the rule of a corrupt and authoritarian military regime. In 1990 when, in a landslide, the people elected Aung San Suu Kyi to lead a civilian reformist government, the regime refused to step aside. Suu Kyi, who won the Nobel Peace Prize in 1991, spent 15 years under house arrest until 2010. Widespread pro-democracy protests were repeatedly and brutally repressed after 1990. Elections in 2010, still overwhelmingly controlled by the military, brought a modestly reformist former general to the presidency. Limited free elections in 2012 and 2015 gave the pro-democracy party another landslide victory, with Aung San Suu Kyi's party winning a majority in parliament; but the military retains an allotment of 25 percent of parliamentary seats. In 2012, Suu Kyi belatedly claimed her Nobel Prize in Norway and in 2016, her party's success in the election meant that she was poised to pick the president, a post she was constitutionally barred from because she had married a British citizen.

Aung San Suu Kyi's credibility has been damaged in recent years because of her failure to take a stand against government "ethnic cleansing" of the Muslim Rohingya minority in the coastal state of Rakhine. Despite being in the country for hundreds of years, the Rohingya have never been recognized as a legitimate ethnic group and are regarded as illegal immigrants. Buddhist nationalists have been particularly anti-Rohingya. According to Human Rights Watch, the Myanmar (Burma) government has institutionalized discrimination against the Rohingya.

Thailand

During the boom years of the mid-1990s, Thailand was known as one of the Asian **tiger economies**—that is, it was rapidly approaching widespread modernization and prosperity. Thailand takes pride in having avoided colonization by European nations and in having transformed itself from a traditional agricultural society into a modern nation. According to the United Nations, between 1975 and 1998, Thailand had the world's fastest-growing economy, averaging an annual growth in GNI per capita of 4.9 percent (U.S. annual growth then averaged 1.9 percent). At the same time, the country kept unemployment relatively low for a developing country—at 6 percent—and inflation in check at 5 percent or lower. The soaring economic growth resulted from the rapid industrialization that took place in Thailand when the government provided attractive conditions for large multinational corporations. Industries situated themselves in Thailand to take advantage of its literate yet low-wage workforce, its lenient laws covering environmental degradation, and its permissive regulation of manufacturing and trade (see Figure 10.35C).

Urbanization and Industrialization The Bangkok metropolitan area, which lies in the Chao Phraya delta at the north end of the Gulf of Thailand, contributes 50 percent of the country's GNI and contains 16 percent (10.5 million people) of its population (see the Figure 10.25 map). Cities elsewhere in Thailand are also growing and attracting investment, especially Chiang Mai in the north, Khon Kaen in the east, Surat Thani on the Malay Peninsula, and Laem Chabang (see Figure 10.35C and map).

Despite its urbanizing trends, about one-third of Thai working people are still farmers but, according to official statistics, they produce only 10 percent of the nation's GNI. (Remember, though, that much of what farmers, especially women, produce is not counted in the statistics.) Urbanization is proceeding rapidly; 50 percent of the population, or about 33 million people, live in cities, many in crowded, polluted, and often impoverished conditions. It is common for thousands of immigrants from the countryside to live in slums along urban riverbanks, as is shown for Indonesia in Figure 10.25A. Industrialization has had unanticipated detrimental side effects in Thailand. Many of the millions of rural people drawn to the cities seeking jobs, status, and an improved quality of life arrive only to find a difficult existence and meager earnings.

Political Instability Thailand, which is actually a monarchy where the much-revered king usually plays a low-profile role, had a reputation as a rapidly developing democracy that allowed for protest and provided mechanisms for constitutional adjustments and smooth transitions after regular elections. A widespread civilian bureaucracy provided stability and credibility for the government. Nevertheless, in 2006, Thailand's democracy was weakened when a military coup d'état, professing loyalty to the king, overthrew an apparently corrupt but popular, and populist, prime minister, Thaksin Shinawatra. Red-shirted crowds of Thaksin supporters faced off against yellow-shirted supporters of the monarchy and military. Cycles of bloody street battles and arrests ensued well into 2012. Mr. Thaksin, still highly controversial due to corruption charges and a tendency to manipulate his red-shirt supporters, was followed in office by his younger sister, Yingluck Shinawatra, who was then overthrown by another coup. These blows to Thai democracy are affecting economic development because the civil unrest discourages tourism and investment.

tiger economies Southeast Asian countries that managed to rapidly modernize and prosper after World War II

While Thailand's infrastructure is not as developed as those of other middle-income countries, hopes are high: the emerging Asian Highway network (see the Figure 10.21 map) should facilitate trade with India and China; Thailand now has three international airports, making it easier for tourists to visit; and more than one-third of the population, or 26 million people, have cell phones, which elsewhere have helped facilitate entrepreneurship and general development; but in Thailand, the military government controls the internet.

CHECK YOUR UNDERSTANDING

1. Describe how Myanmar (Burma) and Thailand differ in political structure, economy, and sociocultural variables.
2. Explain why, despite rich natural resources, Myanmar (Burma) is one of the region's poorest countries troubled by serious ethnic violence and sketchy dependence on the global drug trade.
3. Discuss how Thailand's reputation as a successful democracy on the fast track to development has been clouded by a series of coups d'état and civil unrest.

MAINLAND SOUTHEAST ASIA: LAO, CAMBODIA, AND VIETNAM

On the map in **Figure 10.37**, the countries of Lao, Cambodia, and Vietnam on the Southeast Asian mainland might appear to be ideally suited for peaceful cooperation, sharing as they do a common heritage of Buddhism (see Figure 10.32) and the Mekong River and its delta (see "Responses to Climate Change in the Mekong River System"). Nonetheless, these three countries went through a disruptive half-century of war that pitted them against each other. Until the end of World War II, all three were colonies, known collectively as *French Indochina*, and all three fought long struggles to transform themselves into independent nations. And then came the Vietnam War and its aftermath (see "The Vietnam War"). As of 2018, they remain essentially communist states with measured amounts of free market capitalism. Visitors and investors began arriving in the 1990s, and by 2000, once-somber city streets were abuzz with people and enterprises, especially in Vietnam (see Figure 10.37D).

A long, curved spine of mountains runs through Lao and into Vietnam. In the southern part of the subregion, these mountains are flanked on the west by the broad floodplain of the Mekong that is occupied by Thailand and Cambodia (see the Figure 10.37 map), and on the east by the 1000-mile-long (1600-kilometer-long) coastline and fertile river deltas of Vietnam (see Figure 10.37A). Vietnam, with 94.7 million people, is by far the most populous of these three countries. Cambodia has 16 million people; Lao, almost entirely mountainous, has only 7 million. The rugged mountainous territory of northern Vietnam and Lao is the least densely occupied area in mainland Southeast Asia. Most of the subregion's population

Figure 10.37 Mainland Southeast Asia: The Vietnam, Lao, and Cambodia subregion.

A A woman rows a boat near Chau Doc, in the Mekong delta region of south Vietnam. [Ron Watts/Getty Images]

B Farmers plant rice in the flat wetlands of Cambodia where the soil is rich due to frequent depositions by the flooding Mekong. [Pascal Deloche/Getty Images]

C A reclining Buddha statue touts the joys of relaxation in Vientiane, Lao. [Christopher Groenhout/Getty Images]

D A family of four on one motorbike in Ho Chi Minh City. [Neil Massey/Getty Images]

lives along the coastal zones of Vietnam and in the Mekong delta, where people accommodate the seasonal floods by building their homes on stilts, as in the Ganga–Brahmaputra delta in India. Farmers take advantage of the wet tropical climate and flat terrain to cultivate rice (see Figure 10.37B). The Red River delta in northern Vietnam is also an important rice-growing area. Throughout the subregion, about 65 percent of the people support themselves as subsistence farmers; in Cambodia, only 23 percent of the people live in cities. There, people farm small plots of land as they have for many generations. Development experts concerned about sustainable livelihoods for Cambodians say that for the time being this is the wisest way to provide for the population, but to provide truly adequate nutrition and surplus food for cities, managed egalitarian agricultural reform is needed.

The main trading partners for Lao and Cambodia are Thailand and China (as sources of imports) and the United States and Thailand (as destinations for exports of timber, vegetables, and inexpensive manufactured goods). Thai investors wield considerable power in all of these countries because they have interests in the production of rice, vegetables, coffee, sugarcane, and cotton for export, in addition to such resources as timber, gypsum, tin, gold, gemstones, and hydroelectric power. As discussed in "Responses to Climate Change in the Mekong River System," a major environmental issue is the planned damming of the highly biodiverse Mekong River to create hydropower that would be sold primarily to Thailand, the primary investor in the dams. Lao, which needs the money generated by the dams, has been silent about the protests of Cambodia and Vietnam—both downstream of the dams—that the dams' environmental impact would be overwhelming (see Figure 10.24).

Vietnam is the most developed of the eastern Southeast Asian mainland countries (see Figure 10.37D). After the United States withdrew its military forces in 1973, the communist government, assisted by the Soviet Union, began to invest aggressively in health care, basic nutrition, and basic education. By 2012, 87 percent of children reached the fifth grade, and adult literacy was above 94 percent. Forty percent of the Vietnamese people are still cultivators (down from 56 percent just 9 years ago), and they have extensive knowledge of such practical matters as useful plants (including medicinal herbs) and animals, home building and maintenance, and fishing. The Vietnamese have also developed a national flair for excellent cuisine; when times are good, the dinner table is filled with artfully prepared fish, vegetables, herbs, rice, and fruits.

Vietnam has mineral resources (phosphates, coal, manganese, offshore oil) and at one time had a lush forest cover, much of which was destroyed by defoliants in the latter years of the Vietnam War (See Figure 10.18E). Despite its resources, Vietnam's economy languished after the war, partly because of economic sanctions imposed by the United States and partly because of the inefficiencies of communism. In the mid-1980s, the Hanoi leadership began to introduce elements of a market economy in a program of economic and bureaucratic restructuring called ***doi moi*** (very similar to perestroika in the Soviet Union and the structural adjustment programs imposed on many countries by international lending agencies). With the lifting of the U.S. sanctions in 1994, firms from all over the world began to open branches in Vietnam. By the late 1990s, about 90 percent of the country's industrial labor force worked in the private sector, producing such exportable products as textiles and clothing, cement, fertilizers, and processed food. Vietnam has entered preferential trade agreements with the European Union and with the United States via the Trans-Pacific Partnership.

The factors that brought growth during the 1990s and 2000s also caused considerable dislocation in some parts of society. One focus of *doi moi* has been the privatization of land and other publicly held assets. The farmers who own newly privatized land have more security and so pay more attention to conservation and efficient production. But land reform left out the poorest people, who have had to resort to the informal economy to make a living. And in the late 1990s, farmers felt the squeeze when some of their lands were confiscated for use in foreign-funded enterprises such as tourist hotels, golf courses, and oil refineries.

doi moi a program of economic and bureaucratic restructuring in Vietnam, similar to perestroika in the Soviet Union and structural adjustment programs of international lending agencies

CHECK YOUR UNDERSTANDING

1. Discuss why contentions over how to use the Mekong River and its delta sustainably have been rising in Lao, Cambodia, and Vietnam.
2. How would you account for the fact that Lao and Cambodia are the poorest countries in the subregion?
3. Explain how Vietnam differs from Lao and Cambodia in population, and matters of health, education, and economic development.

ISLAND AND PENINSULAR SOUTHEAST ASIA: MALAYSIA, BRUNEI, AND SINGAPORE

Malaysia and neighboring Singapore and Brunei are the most economically successful countries in Southeast Asia. Malaysia was created in 1963 when the previously independent Federation of Malaya (the southern portion of the Malay Peninsula) was combined with Singapore (at the peninsula's southernmost tip) and the territories of Sarawak and Sabah (on the northern coast of the large island of Borneo to the east) (**Figure 10.38**). All had been British colonies since the nineteenth century. Singapore became independent from Malaysia in 1965. The tiny and wealthy sultanate of Brunei, also on the northern coast of Borneo, refused to join Malaysia and remained a British colony until its independence in 1984. Singapore and Brunei are two of the wealthiest countries in the world, both with annual per capita incomes of over $80,000 (PPP). Virtually all citizens have a high standard of living, with Brunei's wealth coming from its oil and natural gas and Singapore's from a diversified economy grounded in global trade. Malaysia also has a diversified economy, though it is much less wealthy than these two neighbors, with an annual GNI per capita (PPP) of about $28,000. All three rank in the Very High category of the Human Development Index (HDI; see Figure 10.19 map).

Malaysia

Malaysia is home to 32.5 million ethnically diverse people. Slightly over 50 percent of Malaysia's people are ethnic Malays, and nearly all Malays are Muslims. Ethnic Chinese, most of whom are Buddhist, make up nearly 24 percent of the population. Just over 7 percent of people in Malaysia are Tamil- and English-speaking Indian Hindus, and 11 percent are indigenous peoples of Austronesian ancestry, who live primarily in Sarawak and Sabah. Until the 1970s, conflicts among these many groups divided the country socially and economically. Malaysians have worked hard to improve relationships among their diverse cultural groups, and they have had noteworthy success.

Today, most Malaysians (86 percent) live on the Malay Peninsula, which has 40 percent of the country's land area. Throughout most of its history, indigenous Malays and a small percentage of Tamil-speaking Indians inhabited peninsular Malaysia. The Indians were traders who plied the Strait of Malacca, the ancient route from India to the South China Sea. Even before the eastward spread of Islam in the thirteenth century, Arab traders began using the straits in the ninth and tenth centuries, stopping at small fishing villages along the way to replenish their ships. By 1400,

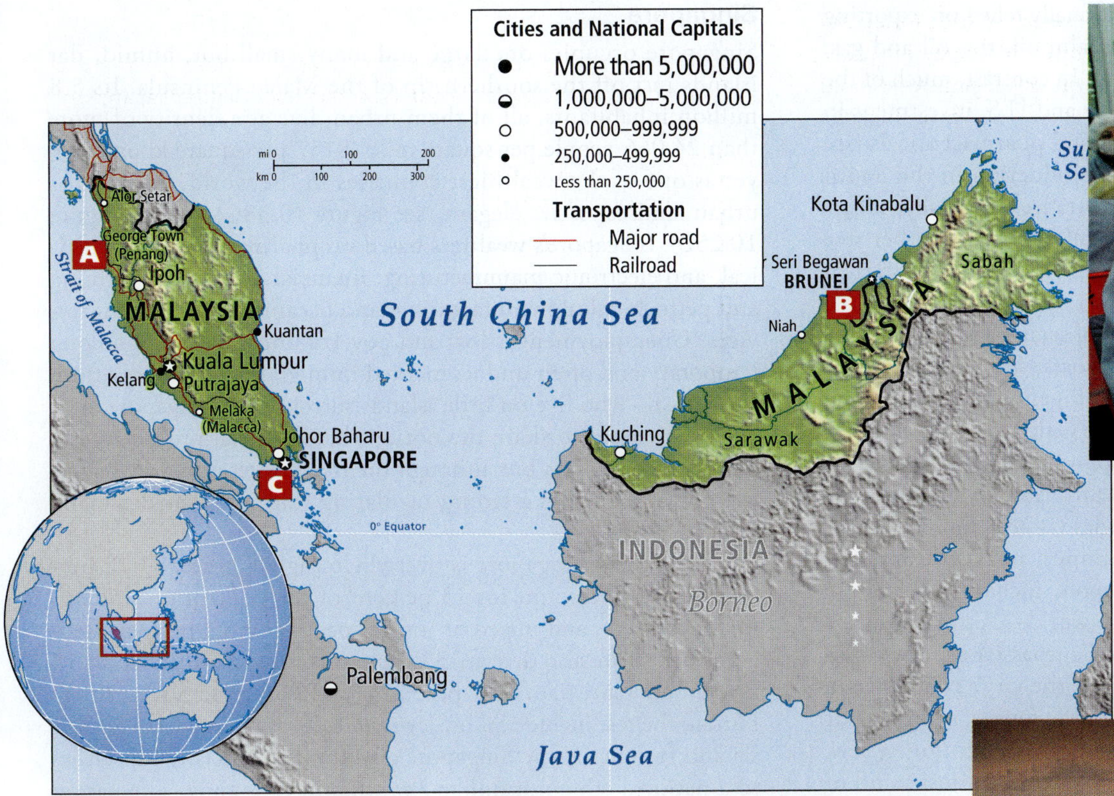

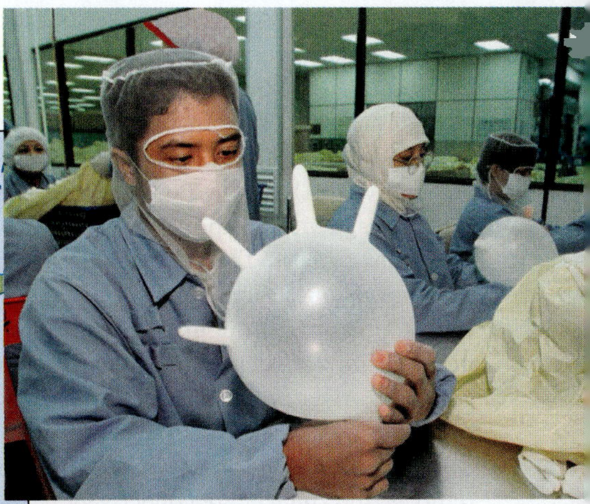

A Three workers check for defects in latex gloves being manufactured at a health-care products plant in Penang, Malaysia. [Mike Fiala/Getty Images]

B Men work on a fossil fuel (oil or gas) drilling rig in Brunei. [Dean Conger/Getty Images]

C Singapore's skyline, as seen from the top of a high-rise luxury hotel. [Cheoh Wee Keat/Getty Images]

Figure 10.38 Island and peninsular Southeast Asia: The Malaysia, Singapore, and Brunei subregion.

virtually all of the ethnic Malay inhabitants of peninsular Malaysia had converted to Islam.

During the colonial era, the British brought in Chinese Buddhists and Hindu Indians to work as laborers on peninsular Malaysian plantations. The Overseas Chinese eventually became merchants and financiers, while Indians achieved success in the professions and in small businesses. The far more numerous Malays remained poor village farmers and plantation laborers. As economic disparities widened, antagonism between the groups increased. After independence, animosities exploded into widespread rioting by the poor in 1969. Political rights were suspended, and it took 2 years for the situation to calm down. The violence so shocked and frightened Malaysians that they agreed to address some of the fundamental social and economic causes.

After a decade of discussions among all groups, in the early 1980s, Malaysia launched a long-term affirmative action program, called ***Bumiputra***, designed to help the Malays and indigenous peoples advance economically. The policy required Chinese business owners to have Malay partners. It set quotas that increased Malay access to schools and universities and to government jobs. As the program evolved in the 1980s, the goal became to bring Malaysia to the status of a fully developed nation by the year 2020. The program has succeeded in narrowing some social inequalities fairly rapidly. In 2018, the United Nations ranked Malaysia fifty-seventh, in the Very High human development (HDI) category. As incomes in Malaysia have increased, wealth disparity (the gap between rich and poor), though still considerable, has narrowed rather than widened. This narrowing is an unusual pattern and reflects the success of the *Bumiputra* affirmative action program.

Bumiputra an affirmative action program in Malaysia to address economic and social inequalities between elite Overseas Chinese businesspeople and indigenous (and other) disadvantaged ethnic groups

The economy of Malaysia has traditionally relied on exporting raw materials, such as timber, rubber, palm oil, tin, oil and gas, and iron ore; and on light manufacturing. In contrast, much of the growth of the 1990s came from Japanese and U.S. investment in manufacturing, especially of electronics and pharmaceuticals (see Figure 10.38A), and from offshore oil production in the South China Sea. Timber exports grew in importance (especially in Sarawak and Sabah). The development of oil palm and rubber tree plantations on cleared forest land (see the vignette at the start of the chapter) has brought global criticism because of its multiple negative impacts: deforestation, burning of scrap vegetation, GHG emissions, river pollution, the demise of forest species (such as the orangutan), and the displacement of indigenous people. Many countries in this region feverishly compete with each other in forest removal and, hence, are collectively responsible for massive deforestation and related air pollution. Currently, ASEAN countries are jointly reassessing their economic strategies vis-à-vis forests.

The Malaysian capital, Kuala Lumpur, boasts beautifully designed skyscrapers like those in Singapore, including what were in 2000 the two tallest buildings in the world—Petronas Towers I and II (buildings elsewhere have since surpassed the towers in height). Nonetheless, at least one-fifth of the city's residents are squatters, some on land in the shadows of the towers, some on raft houses on rivers or bays. Malaysia is hoping to benefit from a new, multibillion-dollar high-tech manufacturing corridor just south of Kuala Lumpur. Called the MSC Malaysia, this project is now in limbo due to massive diversions of investment moneys into the private pockets of politicians.

Like many on the Asian development fast track, Malaysia has volatile economic and political patterns. Currency values and stock market prices fluctuate widely. Lavish building projects may be halted for years, with skyscrapers standing vacant as they did in Kuala Lumpur from 1998 through 2001. Then there are the perpetual political scandals, most recently the jailing of the prime minister for thefts from the Malaysian development fund (see Figure 10.19D). Overall, however, Malaysia seems likely to continue to prosper and foster cultural diversity while discouraging extremist factions, religious or ethnic.

Brunei

Brunei is a country, a little smaller than Delaware, on the north coast of Borneo, where it is situated between the Malaysian states of Sarawak and Sabah. It is a sultanate, ruled by the same family for more than 600 years. The sultan is both the head of government and the head of state. Brunei gained independence from Britain in 1984. It has a small economy, with over half of its GNI from crude oil and natural gas production that make up 90 percent of the country's exports (see Figure 10.38B). Significant income from overseas investments, primarily in the petrochemical industries, helps to give Brunei the world's eighth-highest per capita GNI income ($83,750 in 2017). Medical care is free, as is education through the university level. Ethnically, Malays comprise two-thirds of the population, Chinese people about 11 percent, and a variety of others make up the rest. Sixty-seven percent of the people are Muslim, 13 percent are Buddhist, 10 percent Christian, and 10 percent practice other beliefs. More than 95 percent of the population is literate and the life expectancy is over 75 years.

Singapore

Singapore occupies one large and many small hot, humid, flat islands just off the southern tip of the Malay Peninsula. Its 5.8 million inhabitants, all of them urban, live at a density of more than 24,066 people per square mile (8157 per square kilometer), yet as one of the wealthiest countries in the world, Singapore's urban landscapes are elegant (see Figure 10.38C; see also Figure 10.25B). Singapore's wealth is based on pharmaceutical, biomedical, and electronic manufacturing; financial services; oil refining and petrochemical manufacturing; and oceanic transshipment services. Unemployment is low and poverty unusual, except among temporary and often undocumented immigrants—primarily from Indonesia—who live on little islands surrounding Singapore. Singaporeans seem to share the notion that financial prosperity is a worthy first priority, but although the free market economy reigns, the government has a strong regulating role and provides many services.

Ethnically, Singapore is overwhelmingly Chinese (74.2 percent). Malays account for 13 percent of the population, Indians for 9.2 percent, and mixed or "other" people for 3.3 percent. Religious beliefs are also diverse: 33 percent are Buddhist or Taoist, 18 percent are Christian, 14.7 percent are Muslim, and 5 percent are Hindu; the rest include Sikhs, Jews, and Zoroastrians. As in Malaysia and Indonesia, the Singapore government officially subscribes to a national ethic not unlike *Pancasila* in Indonesia. The nation commands ultimate allegiance; loyalty is to be given next to the community and then to the family, which is recognized as the basic unit of society. Individual rights are respected but not championed. The emphasis is on shared values, racial and religious harmony, and community consensus rather than majority rule, which is thought to lead to conflict.

Singapore has a meticulously planned cityscape with safe, clean streets and little congestion because of an elaborate light rail system. Eighty percent of the people live in government-built housing estates (see Figure 10.25B), and workers are required to contribute up to 25 percent of their wages to a government-run pension fund. However, home care of the elderly is the responsibility of the family and is enforced by law. Each child, on average, receives 10 years of education and can continue further if his or her grades and exam scores are high enough. Literacy is close to 100 percent. The government strictly controls virtually all aspects of society. Permits are required for almost any activity that could have a public effect: having a car radio, owning a copier, working as a journalist, having a satellite dish, being a sex worker, or performing as a street artisan or entertainer. Law and order are strictly enforced. Some years ago, a visiting U.S. teenager was sentenced to a caning for spray-painting graffiti. Drug users are severely punished, and drug dealers are sentenced to life imprisonment or death. But virtually all citizens of Singapore seem to have accepted this control and strictness in return for a safe city and one of the highest per capita incomes ($90,570) in the world.

CHECK YOUR UNDERSTANDING

1. Discuss why people in Singapore and Brunei—very small, relatively stable, and very wealthy countries—would accept the idea that democracy would be too chaotic.

2. Describe the position of ethnic Chinese in Malaysia and how public policy was used to diffuse ethnic and political tensions against them.
3. In which parts of this subregion are democratic traditions fairly strong, but authoritarian policies and corruption of government still common?

INDONESIA AND TIMOR-LESTE

Indonesia, with the world's largest Muslim population, is a recent amalgamation of island groups, inhabited by people who, before the colonial era, never thought of themselves as a national unit (**Figure 10.39**). The island of Java is the seat of government and the locus of political and economic power. Rivalry between island peoples remains strong, and many resent the dominance of the Javanese in government and business. This resentment and other burgeoning issues of identity and allegiance are serious threats to Indonesia's continued unity (see the discussion of Indonesia that begins with "Will Expansion of Political Freedoms Bring Peace to Indonesia?"). Emblematic of the fragile state of this subregion is the fact that Timor-Leste, the eastern half of the small island of Timor, is now independent. In 1999, after more than 20 years of armed conflict with Indonesia, in which the infrastructure was destroyed and as many as 250,000 people died (one-quarter of the population), the people of eastern Timor voted in a UN plebiscite to split from Indonesia and become known as Timor-Leste (East Timor). For 2 more years, the Indonesian military tried to enforce cohesion, in part because of Timor-Leste's oil and gas deposits, but in 2002, after 1500 more deaths, Timor-Leste became an independent country.

The archipelago of Indonesia consists of the large islands of Sumatra, Java, and Sulawesi; the Lesser Sunda Islands east of Java (including Bali); Kalimantan, which shares the island of Borneo with Malaysia and Brunei; the Moluccas; and West Papua on the western half of New Guinea. In all, Indonesia contains some 17,000 islands, but some are small, uninhabited bits of coral reef. The term *Indonesia* was coined in 1850 by James Logan, a Singapore-residing Englishman, from two Greek words: *indos* (Indian) and *nesoi* (islands). Indonesians themselves now refer to the archipelago as *Tanah Air Kita*, meaning "Our Land and Water." This name conveys a sense of the archipelago environment, but overstates the idea that there is a unified feeling of togetherness and environmental concern.

The island of Java is home to 57 percent of the country's population (151 million people in 2015) but Java has only 7 percent of

A A typical traditional wood dwelling in north Sumatra. [Don Mammoser/Shutterstock]

Figure 10.39 The Indonesia and Timor-Leste subregion.

Figure 10.40 Men from the Dani ethnic group take part in a mock battle in Baliem Valley, West Papua. Logging and mining interests, mostly owned by corporations from outside West Papua, as well as immigration from Java and other densely populated islands, have driven many Papuans off their land. Meanwhile, a large population of indigenous people still lives in the remote Baliem Valley of West Papua, where they practice subsistence agriculture. [Peter Ptschelinzew/Getty Images]

the country's land. However, thousands of years of ash from Java's 17 volcanoes have made the soil rich and productive, capable of supporting large numbers of people. Another 21 percent of the population lives on the adjacent volcanic island of Sumatra. Indonesia frequently resettles large numbers of people to other islands to relieve population pressure elsewhere or to isolate dissidents.

This resettlement of Javanese people to other islands has been a contentious issue because indigenous people resent being inundated with Javanese culture and concepts of economic development. They see resettlement as a Javanese effort to gain access to, and profit from, indigenously held natural resources, such as oil, precious metals, and forestlands. Indigenous people, some of whom continue to live via subsistence agriculture as they have for millennia, have been repeatedly devastated by the rapid changes brought to everyday life by the newcomers. The government's argument that resettlement provides jobs is true, strictly speaking. But often the habitat is degraded beyond repair. The central government tends to ignore these hardships, instead invoking the principles of *Pancasila*, arguing that resettlement strengthens Indonesian identity by spreading modernization and eliminating ways of life that differ from the government's vision of the norm (**Figure 10.40**).

CHECK YOUR UNDERSTANDING

1. Why is Indonesia considered a precarious political unit? What are its population size and cultural attributes?
2. How did Timor-Leste come to be an independent country? What are its population characteristics and resources?
3. How has Indonesia's strategy of using resettlement to resolve cultural strife and unemployment problems affected indigenous people in the country?

THE PHILIPPINES

The Philippines, lying at the northeastern reach of the Southeast Asian archipelago, comprises more than 7000 islands spread over about 500,000 square miles (1.3 million square kilometers) of ocean (**Figure 10.41**). The two largest islands are Luzon in the north and Mindanao in the south. Together, they make up about two-thirds of the country's total land area, which is about the size of Arizona. The Philippine Islands, part of the Pacific Ring of Fire (see Figure 1.26), are volcanic. The violent eruption of Mount Pinatubo in June 1991 devastated 154 square miles (400 square kilometers) and blanketed most of Southeast Asia with ash (see Figure 10.41A). Volcanologists had predicted the eruption, and precautions were taken, so although about 3 million people were threatened, the death toll was less than 0.028 percent—still, 847 people died. Over time, volcanic eruptions have given the Philippines fertile soil and rich deposits of minerals (gold, copper, iron, chromate, and several other elements).

Population, Urbanization, and Disaffection

When the Philippines became a virtual U.S. colony in 1898, after the Spanish-American War, it had 7 million people. Just over 120 years later, the population is 15 times larger, at about 107 million. Today, close to 50 percent of Filipinos live in cities, about the same proportion for Southeast Asia as a whole (48 percent). The capital Manila, whose metropolitan area comprises 16 cities with a total population of more than 24 million (see Figure 10.25 map), is one of the largest and most densely settled urban agglomerations in the world (107,561 people per square mile or 43,000 per square kilometer). Some urban residents were displaced to the city by the eruption of Mount Pinatubo (see Figure 10.41A), others by rapid deforestation, dam projects, and mechanized commercial agriculture.

Because of the forced nature of rural-to-urban migration, many people are underemployed urban squatters living in shelters they have built out of scraps. To the Philippine central government, the masses of urban poor represent a particular political and civil threat. Even though the excesses of the dictator Ferdinand Marcos ended in 1986, the presidencies of Corazon Aquino, Gloria

A The eruption of Mount Pinatubo in 1991, which killed 847 people, affected about 3 million others, and impacted climate on a global scale for at least 3 years. [InterNetwork Media/Photodisc/Getty Images]

B An indigenous woman fishes for squid near the Philippines' Northern Sierra Madre National Park. [UniversalImagesGroup/Getty Images]

Figure 10.41 The Philippines subregion.

Macapagal-Aroyo, Benigno Aquino III, and now the flamboyant authoritarian Rodrigo Dutertre, have filled the decades since with ineffective efforts to improve life. People's faith in their government has not been restored. Marcos' brutal regime began in 1965; since his ouster in 1986, economic progress has been slow, and violent social unrest has become very much a part of life in the Philippines.

Cultural and Economic Disparities

The cultural complexity of Philippine society is illustrated by its ethnic profile. The majority of the population (96 percent) belongs to one or more of 60 distinct indigenous ethnic groups. The Chinese make up another 1.5 percent of the population, and the remaining 2.5 percent includes Europeans, Americans, other Asians, and people native to other Southeast Asian islands.

Most Filipinos are Roman Catholic (83 percent), but the southern and central portions of the island of Mindanao and the Sulu Archipelago have been predominantly Muslim for centuries. Over the last few decades, the central government has been resettling many thousands of Catholics in Mindanao, explicitly to dilute the Muslim population. Rival family clans, with armed militias who do their bidding, dominate local politics in Mindanao. In retaliation against those clans that are allied with Catholic elected national officials (President Gloria Arroyo, and now Rodrigo Dutertre), Muslim fighters, some possibly linked to international terrorist movements, began operating in the southern Philippines, carrying out bombings that have killed scores of people and in 2018 destroyed the entire city of Marawi (see Figure 10.23C and the figure map).

Wealth and poverty are also often associated with certain ethnic groups, as is the case elsewhere in Southeast Asia. The vast majority of the wealthy are descendants of Spanish and Spanish-Filipino plantation landowners or of Chinese financiers and businesspeople. It is estimated that 30 percent of the top 500 corporations in the islands are controlled today by ethnic Chinese Filipinos, who make up less than 1 percent of the population. Wealthy families control 78 percent of all corporate wealth in the Philippines. Meanwhile, the poor are overwhelmingly indigenous people.

In contrast to Malaysia, where ethnic disparities in wealth became an openly discussed national social issue and then were addressed with overt policies, the Philippine government under Marcos did not embark on an economic policy to help disadvantaged ethnic groups, or improve the country's infrastructure, or find strategies for national reconciliation. President Marcos's reaction to protests that erupted in 1972 was to declare martial law. He hung on to power for another 14 years, during which there was little progress in infrastructure development, education advancement, job creation, population control, or wealth redistribution. Successors to Ferdinand Marcos have been hampered in their efforts to settle social unrest and attract investment by the triple economic disasters of the Mount Pinatubo eruption, the closing of the U.S. military bases, both in 1991, and a drastic cut in U.S. foreign aid once the Americans left. By 2015, there was talk (unlikely to be fulfilled) that the United States might negotiate the reopening of the Subic Bay base as part of its new interest in playing a stronger role in the greater Asian region, a development that would have major economic impacts on the Philippines.

The case of the Philippines under Marcos and since shows how poor leadership and corrupt government can hold back a country. Despite the rampant sale abroad of its once-rich timber resources, the country's economy has been so lethargic that actual unemployment rates are thought to be as high as 30 percent, with 40 percent of the population living below the poverty line. While the older rural poor must scavenge for survival (Figure 10.41B), many of the best-educated young Filipinos have chosen to temporarily migrate to more vibrant economic zones, where they often work at jobs that are far below their qualifications (see "The Maid Trade"). Living frugally on only a portion of their earnings, they send billions of dollars in remittances to their families each year.

Filipinos Envision a Better Future

Ordinary Filipinos (many once worked abroad) are keenly aware that a better social and political system is possible. In cooperation with the United Nations, the government of New Zealand, and several NGOs, a group of concerned Philippine citizens have produced 11 Human Development Reports over the last two decades. The report for 2012/13 is a model of frank talk about civic problems and the varying effects of geography on the possibilities for local change. Noting that there has been rising inequality and disparity between urban and rural areas, the report states that "human development should be the core means and the ultimate goal of development efforts" (*Philippines Human Development Report 2012–2013*, UNDP, p. 2).

The 2015 report advocated that "work" be defined broadly not only as making a living, but as a way to participate in the holistic project of human development—with social cohesion as a family and community goal. The report went on to tutor the public on the multiple ways to measure human development and on just where the Philippines fits into the global picture. The careful marshaling of evidence by these citizens and their brave dedication to rational long-term change bodes well for the future of the Philippines.

CHECK YOUR UNDERSTANDING

1. What are some of the features of life in the Philippines in the twentieth century that contributed to the problems that Filipinos must deal with today?
2. How have the firms most involved in development in the Philippines paid attention to human well-being of the majority? Which ethnicities are the strongest economically?
3. Explain the role of ethnicity and clans in influencing local politics in the Philippines.
4. Why might it be argued that Muslim leaders in the south of the Philippines have valid reasons to rebel against national leaders and policies?
5. Why is the production of local Human Development Reports by Philippine citizens especially crucial to progress in a country long plagued with authoritarian leaders?

CRITICAL THINKING QUESTIONS

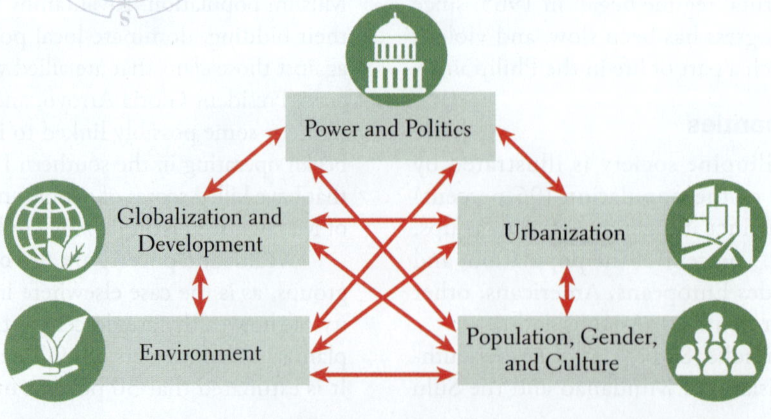

The diagram represents connections among the five geographic themes that structure this book. Listed below are some important questions that have been addressed in this chapter. Answer each question, and indicate where in the diagram you think the topics in each question belong.

1. Many of Southeast Asia's most critical environmental issues relate in some way to climate change. Explain how rising ocean temperatures and storm surges resulting from typhoons are thought to be climate-change related.
2. Explain why deforestation for commercial agriculture is viewed as particularly damaging to indigenous people in terms of flooding, drought, and food security.
3. What are two ways in which deforestation leads to increases in greenhouse gasses?
4. Explain why the formation of state-aided market economies and export-led production have both been key to the economic development of this region.
5. What are some factors that led to regional economic cooperation expanding only very slowly?
6. How has crony capitalism both fueled economic growth in this region and been a drag on growth?
7. Compare ASEAN's overall stated goal to develop a sustainable regional economic community with what has been achieved by the European Union.
8. How does the history of colonialism in this region influence the current dynamics of political power?
9. What arguments have some Southeast Asian leaders made about Asian values not being compatible with Western ideas of democracy and political freedom?
10. Why are there growing concerns about China's interest in the Mekong River and South China Sea?
11. Describe the factors that account for present levels of urbanization in Southeast Asia.
12. Why do Southeast Asian cities provide so poorly for their citizens in terms of housing, water, and sanitation?
13. Has the development of mechanized agriculture and agribusiness increased food self-sufficiency in this region?
14. Explain why development and urbanization in the region have led to the decision by many to migrate abroad for employment.
15. Where have population growth rates slowed the most in this region, and why?
16. Describe the circumstances that have led to better job opportunities for women.
17. Discuss some of the ways in which gender roles vary from those typically found in South Asia or East Asia.
18. How has the role of the Overseas Chinese evolved over time? Why have some people in this region resented them?

Key Terms

archipelago 555
Association of Southeast Asian Nations (ASEAN) 576
Australo-Melanesians 568
Austronesians 568
Bumiputra 605
coral bleaching 566
crony capitalism 573
cultural pluralism 593
demographic transition 589
detritus 557
distributaries 567
doi moi 604
domino theory 571
feminization of labor 573
foreign exchange 585
merchant marine 585
Pancasila 583
populist political movement 582
remittances 585
resettlement schemes 583
service sector 573
sex tourism 593
social cohesion 577
soft power initiatives 582
state-aided market economies 572
Sundaland 556
tiger economies 602

More Practice at SaplingPlus

Read the interactive e-text, review key concepts, and check your understanding.

Children play during high tide in a South Tarawa village. [Josh Haner/ The New York Times/Redux]

11

Oceania: Australia, New Zealand, and the Pacific

On South Tarawa island in the Kiribati chain, children wade through rising tides that have isolated their homes, eroded the beaches, and killed most shoreline coconut trees. Kiribati is a Pacific nation of 33 tiny islands barely 2 to 3 meters above sea level. Rising seas are swamping the islands, which straddle the equator and the international date line. Scientists predict that Kiribati may be uninhabitable in a few decades because of rising seas and climate change.

Aurora is a nursing student who left Kiribati and now lives in Brisbane, Australia. While she studies, Aurora pines for her island's blue lagoons and the daily close company of her raucous extended family. Aurora is studying nursing in a government program called AusAID, one of several strategies developed by former Kiribati President Anote Tong to encourage his people gradually to "migrate with dignity." Tong sees climate change as a huge challenge, and he hopes that other South Pacific islands, as well as Australia and New Zealand, will allow Kiribati people to resettle there. Through the AusAID program, Aurora and other young people receive an education that will help them support large extended families wherever they find a new home.

Casting about for other solutions, in 2014 the government of Kiribati purchased a large estate on the multi-island nation of Fiji, 1300 miles (2092 kilometers) to the south. Tong's plans were unclear: Did he plan to produce food to export to Kiribati, or transfer Fijian soil to Kiribati, or relocate Kiribati's 100,000 people to Fiji? President Tong's purchase triggered controversy at home and in Fiji. In 2016 his party lost the Kiribati presidency to a climate change skeptic who promised to focus instead on economic development, especially tourism and infrastructure improvement, while effectively ignoring the rising seas.

Learning Objectives

Environment: Physical and Human

11.1. Identify Oceania's physical features: its landforms, climate, and flora and fauna.

11.2. Describe the environmental issues faced by Oceania due to climate change.

11.3. Explain how the region's unique ecology is threatened by human activity.

11.4. Discuss the settling of Oceania, European colonization, and postcolonial developments.

Globalization and Development

11.5. Explain how Oceania's new focus on neighboring Asia is transforming patterns of trade and economic development across the region.

11.6. Explain how recent economic changes have affected traditional ways of life and human/environment relations throughout Oceania.

Power and Politics

11.7. Compare the system of democracy that prevails in New Zealand, Australia, and Hawaii with the Pacific Way, the political and cultural philosophy of the islands.

11.8. Describe how the global refugee crisis is testing this region's ability to maintain its reputation as a humanitarian refuge and zone of opportunity.

Urbanization

11.9. Explain patterns of urbanization in this lightly populated region.

Population, Gender, and Culture

11.10. Describe the population trends in different areas of Oceania.

11.11. Contrast gender roles in Australia, New Zealand, and Hawaii with those of the Pacific islands and Papua New Guinea.

11.12. Describe the indigenous cultures of the region and efforts to create unity across the region.

Zhengzhou
Fukuoka
Osaka
JAPAN
Nanjing
CHINA
Shanghai
Chang Jiang
Hangzhou
Changsha
Nanchang
Taipei
Taichung
Guangzhou (Canton)
TAIWAN
Kaohsiung
Hong Kong
Northern Mariana Islands (U.S.)
PHILIPPINES
Saipan
Tinian
South China Sea
Manila
Philippine Sea
Agana
Guam (U.S.)
Caroline Islands
FEDERATED STATES OF MICRONESIA
MARSHALL ISLANDS
Yap
Sulu Sea
PALAU
Ngerulmud
Pohnpei
Truk
Palikir
Majuro
International Date Line
BRUNEI DARUSSALAM
Bandar Seri Begawan
Micronesia
Kosrae
MALAYSIA
Celebes Sea
Tarawa
Borneo
Melanesia
NAURU
KIRIBATI (GILBERT ISLANDS)
Equator
Sulawesi (Celebes)
Admiralty Islands
Yaren
Jayapura
Kavieng
West Papua
Wewak
New Ireland
Bismarck Archipelago
Rabaul
New Guinea
Buka
Bougainville
Banda Sea
PAPUA NEW GUINEA
Surabaya
INDONESIA
Salamaua
New Britain
SOLOMON ISLANDS
TUVALU (ELLICE ISLANDS)
Tokelau (New Zealand)
Java
Bali
Dili
Arafura Sea
Malaita
TIMOR-LESTE
Daru
Funafuti
Port Moresby
Honiara
Thursday Island
Torres Strait
Timor Sea
Wallis & Futuna (France)
Darwin
Great Barrier Reef
Mata Utu
Apia
INDIAN OCEAN
Jabiru
Gulf of Carpentaria
Coral Sea Islands (Australia)
SAMOA (WESTERN SAMOA)
Katherine
Cooktown
VANUATU (NEW HEBRIDES)
FIJI
King Leopold Ranges
Cape York Peninsula
Cairns
Vanua Levu
Derby
Daly Waters
Coral Sea
Port-Vila
Viti Levu
Suva
Tennant Creek
Selwyn Range
Great Dividing Range
Mackay
Nuku'alofa
TONGA
NORTHERN TERRITORY
Great Sandy Desert
Hamersley Range
QUEENSLAND
Noumea
New Caledonia (France)
Macdonnell Ranges
Alice Springs
Gibson Desert
Great Artesian Basin
Carnarvon
Uluru (Ayers Rock)
Musgrave Ranges
WESTERN AUSTRALIA
AUSTRALIA
Brisbane
Norfolk (Australia)
Great Victoria Desert
Geraldton
Kingston
SOUTH AUSTRALIA
Darling
Kermadec Islands (New Zealand)
Broken Hill
NEW SOUTH WALES
Nullarbor Plain
Flinders Range
Perth
Murray
Esperance
Elliston
Sydney
Great Dividing Range
Great Australian Bight
Adelaide
Canberra
Albany
Australian Alps
North Island
Tasman Sea
Auckland
INDIAN OCEAN
VICTORIA
Melbourne
NEW ZEALAND
Gisborne
120°E
130°E
140°E
International Date Line
Launceston
Westport
Wellington
TASMANIA
Chatham Islands (New Zealand)
Hobart
Christchurch
South Island
Southern Alps
Dunedin
Invercargill
Bounty Islands (New Zealand)
Antipode Islands (New Zealand)
Auckland Islands (New Zealand)

Oceania, shown in **Figure 11.1**, is a vast and often idealized region that faces a number of difficult realities, among them climate change. The challenges facing Aurora and the people of the Kiribati archipelago affect both the region as a whole and individual lives.

What Makes Oceania a Region?

Oceania is made up of Australia, New Zealand, Papua New Guinea, Hawaii, and the many small islands scattered across the Pacific Ocean (**Figure 11.2**). It is a unique region in that it is composed primarily of ocean and covers the largest area of Earth's surface of any region, yet it is home to only 42.5 million people, the vast majority of whom live in Australia (24.1 million), Papua New Guinea (8.5 million), and New Zealand (4.9 million). In this book, we also include the U.S. state of Hawaii (1.4 million) because its historic and present roles are very much part of Oceania as well as the United States. Altogether, the thousands of other Pacific islands are home to only 3.6 million people. The Pacific Ocean is both a uniting feature and a barrier: the vast ocean serves as a link that unites Oceania as a region, profoundly influencing life even on dry land, but across the region, the ocean also acts as a biological and cultural barrier, inhibiting interaction and diffusion.

Perhaps nowhere else on Earth are the effects of climate change as measurably real as they are in the low-lying islands of the Pacific, where changes in rainfall patterns and sea level are strikingly evident. Because many people in the islands have fairly simple lifestyles, most Pacific islands have relatively low emissions of greenhouse gases on a per capita basis. Their emissions are also low as a whole because the islands' total population is small. But seemingly remote places in the Pacific are anything but remote from general global influences. Beyond climate change, people in this region are dealing with changes brought on by tourism, environmental degradation caused by many modern practices, freshwater scarcity, ocean pollution, and rapidly changing cultural and trading patterns.

Terms in This Chapter

Some maps in this chapter show different place-names for the same locations. This reflects the political evolution of the region, where some islands previously were grouped under one name that remains in everyday use but are now grouped into country units with another name. For example, the Caroline Islands, located north of New Guinea, still go by that name on maps and charts, but they have been divided into two countries: Palau (a small group of islands at the western end of the Caroline Islands) and the Federated States of Micronesia (an official political unit), which extends over 2000 miles west to east, from Yap to Kosrae. In addition, there are three names commonly used to refer to large cultural groupings of islands: Micronesia, Melanesia, and Polynesia. These groupings are not political units; rather, they are based on ancient ethnic and cultural links and can be seen later in the chapter in Figure 11.18.

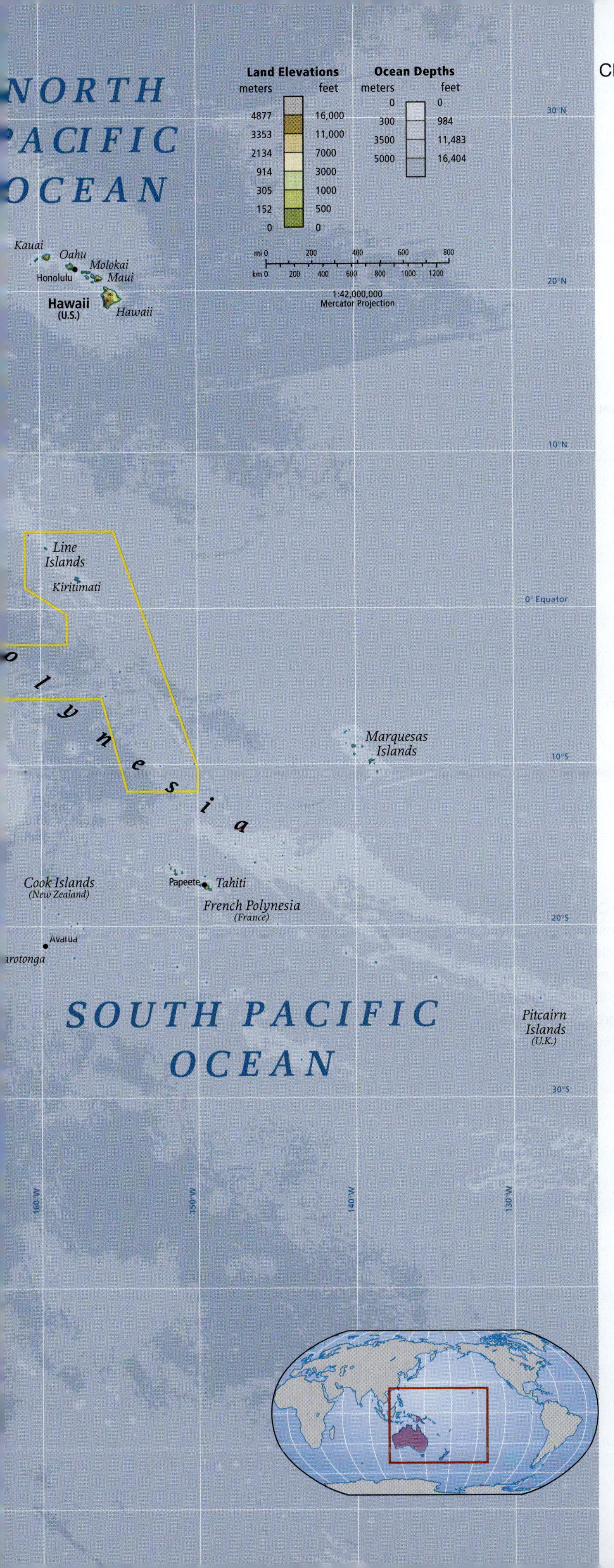

◀ **Figure 11.1** Physical map of Oceania.

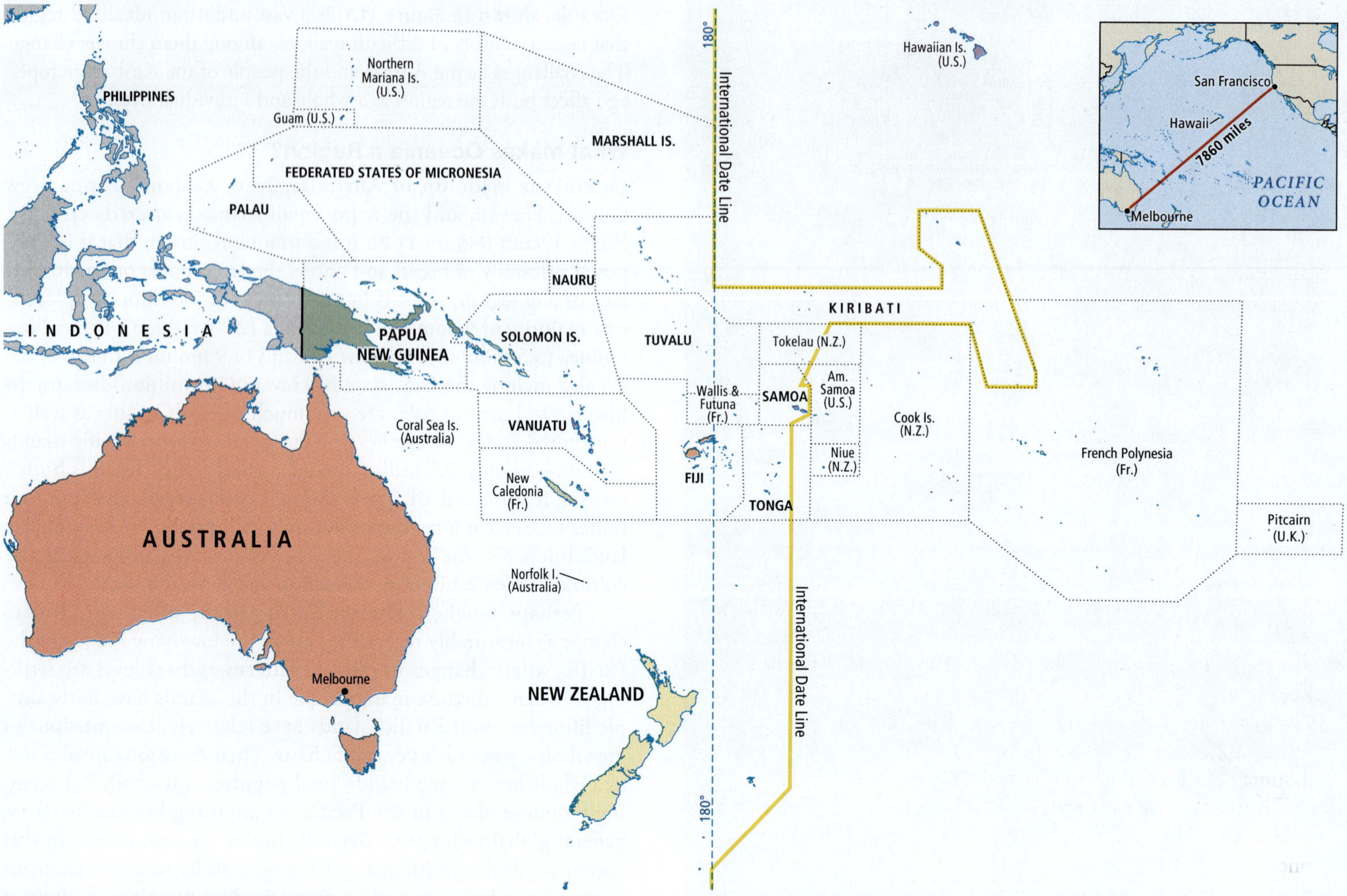

Figure 11.2 Political map of Oceania. Fourteen of the political entities in Oceania are independent countries (identified on the main map). The rest are territories of or are otherwise affiliated with other nations. Not depicted on the map are the Pacific Island Wildlife Refuges, a widely scattered, essentially uninhabited group of northern Pacific islands that constitute a U.S. territory. Also not shown is Easter Island (Rapa Nui), a Polynesian island (owned by Chile), which lies 2283 miles (3512 kilometers) off the west coast of Chile. To give a sense of the size of the Pacific, the inset map shows the distance from San Francisco to Melbourne, a flight that takes 16 hours and crosses the international date line (IDL), an imaginary line at about 180 degrees east (or west) of the Greenwich meridian. This line marks the change from one day to the next. If you leave San Francisco on a Tuesday, when you cross the IDL it will become Wednesday. Deviations in the IDL are made to avoid dividing countries, such as Kiribati.

ENVIRONMENT: PHYSICAL AND HUMAN

11.1 Identify Oceania's physical features: its landforms, climate, and flora and fauna.

11.2 Describe the environmental issues faced by Oceania due to climate change.

11.3 Explain how the region's unique ecology is threatened by human activity.

11.4 Discuss the settling of Oceania, European colonization, and postcolonial developments.

In Oceania, the Pacific Ocean has long served as a highway on which plants and animals found their way from island to island by floating on the water, swimming, or flying, and humans have used the ocean as a way to make contact with other peoples for visiting, trading, and raiding. But the wide expanses of water are also a profound barrier to the natural diffusion of plant and animal species and keep Pacific Islanders spatially isolated from one another. The vast ocean has imposed solitude and fostered unique evolutionary trends for plants and animals. The solitude has also fostered self-sufficiency and subsistence economies among its human occupants well into the modern era. Unfortunately, as we shall see, the ocean also serves as a conveyer of pollution, especially discarded plastics (see Figure 11.17 later in the chapter).

LANDFORMS: CONTINENT FORMATION

The largest landmass in Oceania is the ancient continent of Australia at the southwestern edge of the region (see Figure 11.1). The Australian

Figure 11.3 Uluru (Ayers Rock). This land formation, near the center of Australia, is a smooth remnant of ancient mountains. Geologists call such an isolated formation that rises abruptly from a virtually level plain a monadnock or inselberg. The site is held sacred by central Aboriginal Australians. It is also one of Australia's most popular tourist destinations. [Michael Dunning/Getty Images]

continent is partially composed of some of the oldest rock on Earth and has been relatively stable for more than 200 million years, with very little volcanic activity and only an occasional mild earthquake. Australia was once a part of the great landmass called **Gondwana** that formed the southern part of the ancient supercontinent Pangaea (see Figure 1.8). What became present-day Australia broke free from Gondwana and drifted until it eventually collided with the part of the Eurasian Plate on which Southeast Asia sits. That impact created the mountainous island of New Guinea to the north of Australia.

Australia is shaped roughly like a dinner plate with two bites taken out of it: one in the north (the Gulf of Carpentaria) and one in the south (the Great Australian Bight). The center of the plate is the great lowland Australian desert (often referred to as the **Outback**), with only two hilly zones and rocky outcroppings. Over millennia, the forces of erosion—both wind and water—have worn most of Australia's landforms into low, rounded formations. Some of these, like Uluru (Ayers Rock), are quite spectacular (**Figure 11.3**). The eastern rim of Australia is composed of uplands; the highest and most complex of these are the long, curving Eastern Highlands (labeled "Great Dividing Range" in Figure 11.1).

Off the northeastern coast of the continent lies the **Great Barrier Reef**, the largest coral reef in the world and a World Heritage Site since 1981 (**Figure 11.4**). It stretches along the coast of Queensland in an irregular arc for more than 1250 miles (2000 kilometers), covering 135,000 square miles (350,000 square kilometers). The Great Barrier Reef is so large that it influences Australia's climate by interrupting the westward-flowing ocean currents in the mid–South Pacific circulation pattern. Warm water is shunted to the south, where it warms the southeastern coast of Australia. Threats to the health of the Great Barrier Reef are discussed in "Coral Bleaching."

LANDFORMS: ISLAND FORMATION

The islands of the Pacific were (and are still being) created by a variety of processes related to the movement of tectonic plates. The islands found in the western reaches of Oceania—including New Guinea, New Caledonia, and the main islands of Fiji—are remnants of the Gondwana landmass; they are large, mountainous, and geologically complex. Other islands in the region are volcanic in origin and form part of the Ring of Fire (see Figure 1.9). Many in this latter group are situated in boundary zones where tectonic plates are either colliding or pulling apart. For example, the Northern Mariana Islands east of the Philippines (see Figures 11.1 and 11.2) are volcanoes that were formed when the Pacific Plate plunged beneath the Philippine Plate. The two islands of New Zealand were created when the eastern edge of the Indian–Australian Plate was thrust upward by its convergence with the Pacific Plate.

The Hawaiian Islands were produced through another form of volcanic activity associated with **hot spots**, places where particularly hot magma moving upward from Earth's

Gondwana the great landmass that formed the southern part of the ancient supercontinent Pangaea

Outback any remote, lightly populated area, usually in the dry central regions of Australia

Great Barrier Reef the longest coral reef in the world, located off the northeastern coast of Australia

hot spots individual sites of upwelling material (magma) that originate deep in Earth's mantle and surface in a tall plume; hot spots tend to remain fixed relative to migrating tectonic plates

Figure 11.4 Great Barrier Reef, Queensland, Australia. [Gonzalo Azumendi/Getty Images]

Figure 11.5 Coral Atoll of Tupai in French Polynesia. The land surface is the rim of a volcano; the central pond is the crater of the submerged volcano. [Jean-Pierre Pieuchot/Getty Images]

core breaches the crust in tall plumes. Over the past 80 million years, the Pacific Plate has moved across these hot spots, creating a string of volcanic formations 3600 miles (5800 kilometers) long. The youngest volcanoes, only a few of which are active, are on or near the islands known as Hawaii.

Volcanic islands exist in three forms: volcanic high islands, low coral atolls, and **makatea** (sometimes called seamounts), which are submerged coral platforms uplifted nearly to, or to, the surface by volcanism. High islands are usually volcanoes that rise above the sea into mountainous rocky formations, which, because of their varying height and rugged landscapes, contain a rich variety of environments. New Zealand, the Hawaiian Islands, Mo'orea (one of the Society Islands in French Polynesia), and Easter Island are all examples of high islands. An **atoll** is a low-lying island or chain of islets, formed of coral reefs that have built up on the rim of a submerged volcanic crater (**Figure 11.5**). These reefs are arranged around a central lagoon that was once the volcano's crater. Because of their low elevation, atoll islands tend to have only a small range of environments, limited rainfall, and therefore limited supplies of fresh water.

CLIMATE

Although the Pacific Ocean stretches almost from pole to pole, most of the land of Oceania is situated within the Pacific's tropical and subtropical latitudes. The tepid water temperatures of the central Pacific bring mild climates year-round to nearly all the inhabited parts of the region. The southernmost reaches of Australia and New Zealand have the widest seasonal variations in temperature (**Figure 11.6**).

Moisture, Air Movement, and Rainfall

With the exception of the Outback, the vast arid interior of Australia, much of Oceania is warm and humid nearly all the time. New Zealand and the high islands of the Pacific receive copious rainfall; before human settlement, they supported dense forest vegetation (see Figure 11.6B). Now, after 1000 years of human impact, much of that forest is gone, and is replaced by crops, sheep pastures, and cities.

Travelers approaching New Zealand, either by air or by sea, often notice a distinctive long white cloud that stretches above the north island. Seven hundred years ago, early **Maori** settlers (members of the Polynesian group) also noticed this phenomenon and they named that place **Aotearoa**, "land of the long white cloud," a name that is now applied to all of New Zealand. The cloud is the result of particularly high winds, complex landforms, and moist conditions.

The legendary **Roaring Forties** (named for the 40th parallel south) are powerful air and ocean currents that speed around the Southern Hemisphere (usually between the latitudes of 40 and 50 degrees) virtually unimpeded by landmasses. These westerly winds (winds that blow west to east), which are responsible for Aotearoa's distinctive long white cloud, deposit a drenching 130 inches (330 centimeters) of rain per year in the New Zealand highlands and more than 30 inches (76 centimeters) per year on the coastal lowlands (see Figure 11.6B). At the southern tip of New Zealand's North Island, the wind averages more than 40 miles per hour (64 kilometers per hour) nearly 120 days per year. Farmers in the area stake their cabbages to the ground so the plants will not blow away.

By contrast, two-thirds of Australia is overwhelmingly dry (see Figure 11.6A). The dominant winds affecting Australia are the north and south easterlies (winds that blow east to west) that converge east of the continent (see map in Figure 11.6). Australia's Great Dividing Range blocks the movement of moist, westward-moving air so that rain does not reach the interior (an orographic pattern; see Figure 1.10 in Chapter 1). This blockage means that a large portion of Australia receives less than 20 inches (50 centimeters) of rain per year, and humans have found rather limited uses for this interior territory. But the eastern (windward) highland slopes of the Great Dividing Range receive more abundant moisture. This relatively moist eastern rim of Australia was favored as a habitat by both indigenous people and the Europeans who displaced them after 1800. During the southern summer, the fringes of the monsoon that passes over Southeast Asia and Eurasia bring moisture across Australia's northern coast. There, annual rainfall varies from 20 to 80 inches (50 to 200 centimeters).

Overall, Australia is so arid that it has only one major river system, which is in the temperate southeast where most Australians live. There, the Darling and Murray rivers drain one-seventh of the continent, flowing west and south into the Indian Ocean near Adelaide (see Figure 11.1). One measure of Australia's overall dryness is that the entire average annual flow of the Murray–Darling river system is equal to just one day's average flow of the Amazon in Brazil.

In the island Pacific, mountainous high islands also exhibit orographic rainfall patterns, meaning the windward side is wet and the leeward side is dry (see the diagram in Figure 1.10). Rainfall amounts on the low-lying islands vary considerably across the region. Some of the islands lie directly in the path of trade winds, which usually deliver between 60 and 120 inches (152 to 305 centimeters) of rain per year. These islands support a remarkable variety

makatea coral platforms uplifted by volcanism, sometimes called seamounts

atoll a low-lying island or chain of islets, formed of coral reefs that have built up on the rim of a submerged volcano

Maori Polynesian people indigenous to New Zealand

Aotearoa Polynesian name for New Zealand meaning "land of the long white cloud"

Roaring Forties powerful air and ocean currents at about 40° S latitude that speed around the far Southern Hemisphere virtually unimpeded by landmasses

11.6 Climates of Oceania

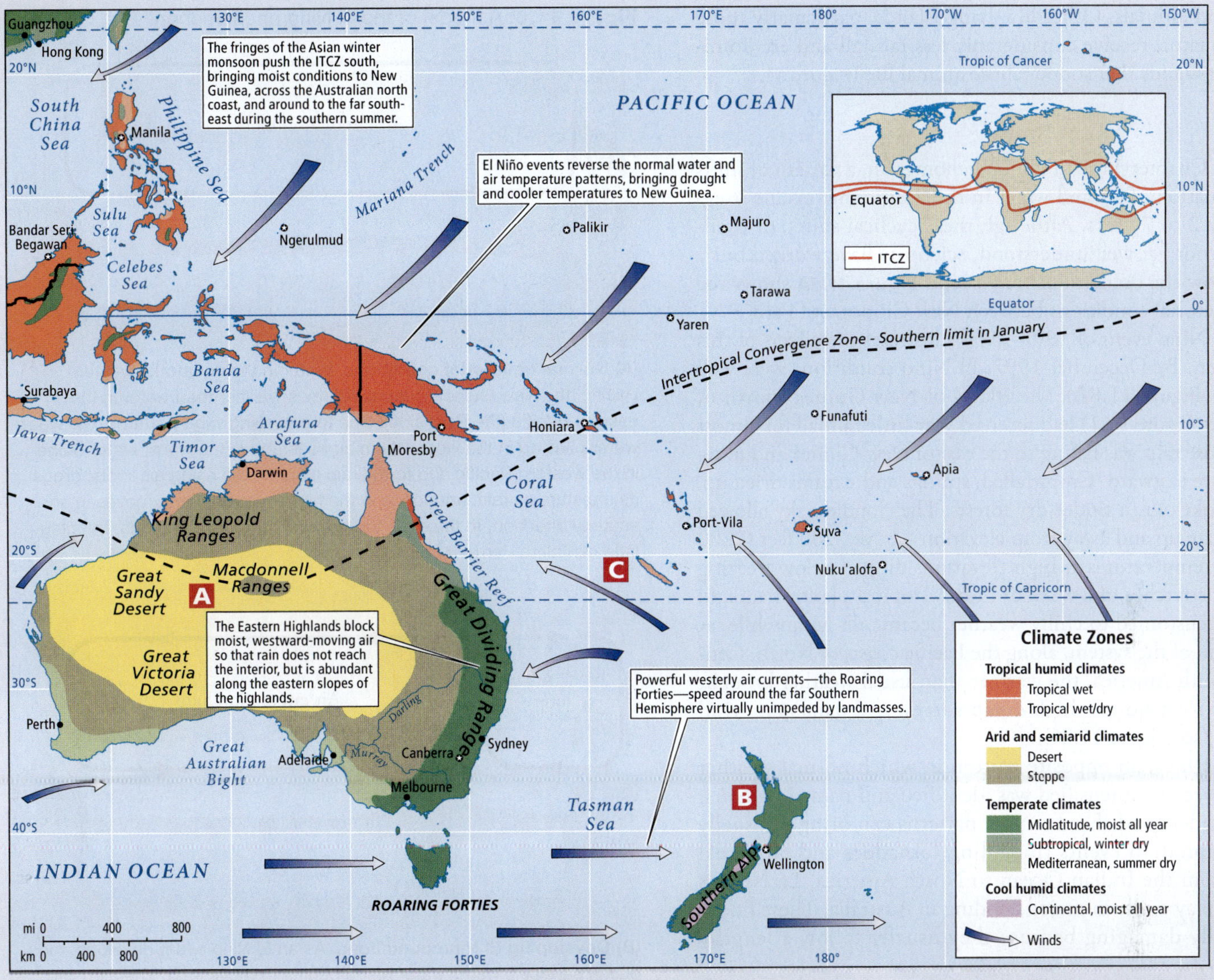

A Desert, Alice Springs, Outback of Australia [John White Photos/Getty Images]

B Midlatitude, moist all year, giant tree fern forest, west coast of New Zealand [PatrikStedrak/Getty Images]

C Tropical wet/dry, coast of Isle of Pines, New Caledonia, just east of Australia [GUIZIOU Franck/hemis.fr/Getty Images]

of plants and animals. Other low-lying islands, particularly those near the equator, receive considerably less rainfall and are dominated by grasslands that support little animal life (**Figure 11.7**).

El Niño

Recall from Chapter 3 the El Niño phenomenon, a pattern of shifts in the circulation of air and water in the Pacific that occurs irregularly every 2 to 7 years. Although these cyclical shifts, or oscillations, are not yet well understood, scientists have worked out a model of how the oscillations may occur. **Figure 11.8A** shows the normal equatorial conditions between New Guinea and Peru.

The El Niño event of 1997–1998 illustrates the effects of this phenomenon. By December 1997, El Nino conditions were well under way (Figure 11.8B).The island of New Guinea (north of Australia; see Figure 11.1) had received very little rainfall for almost a year, because rain was falling to the east of New Guinea and moving ever more eastward. Crops failed, springs and streams dried up, and fires broke out in tinder-dry forests. The cloudless sky allowed heat to radiate up and away from elevations above 7200 feet (2200 meters), so temperatures at high elevations dipped below freezing at night for stretches of a week or more. Tropical plants died, and people unaccustomed to chilly weather became ill. Meanwhile, at the other end of the system, along the Pacific coasts of North, Central, and South America, the warmer-than-usual weather brought unusually strong storms, high ocean surges, and damaging wind and rainfall (see Figure 11.8C).

In the 1980s, an opposite pattern in which normal weather conditions become intensified was identified and named La Niña. It is now understood that La Niña patterns can bring unusually severe precipitation events (including tornadoes and blizzards) to places from the Indian Ocean to North America. La Niña is thought to play a role in major flooding in Australia. These floods are especially damaging because they usually follow a lengthy

Figure 11.7 Dry-adapted vegetation on islands in the Kiribati chain near the equator. These islands lie outside of the path of the trade winds and are too low to collect orographic rainfall, and high temperatures mean high evaporation rates. [Natalia Harper/Alamy]

Figure 11.8 A model of the El Niño phenomenon.

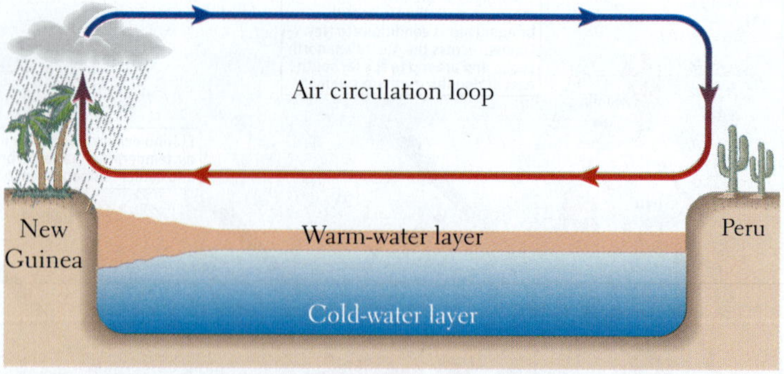

(A) Normal equatorial conditions. Water in the equatorial western Pacific (the New Guinea/Australia side) is warmer than water in the eastern Pacific (the Peru side). Due to prevailing wind patterns, the warm water piles up in the west. Warm air rises above the warm-water bulge in the western Pacific and forms rain clouds. The rising air cools, drops its moisture as rain, and, once in the higher atmosphere, moves in an easterly direction. In the east, the now dry, cool air descends, bringing little rainfall to Peru.

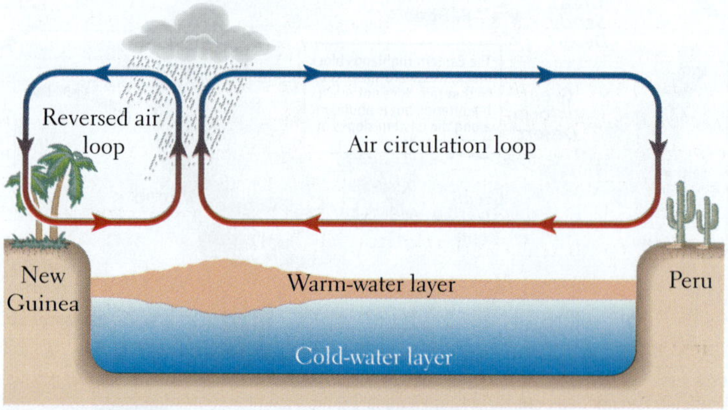

(B) Developing El Niño conditions. As an El Niño event develops, the ocean surface's warm-water bulge begins to move east. The air rising above it splits into two formations, one circling east to west in the upper atmosphere and one west to east.

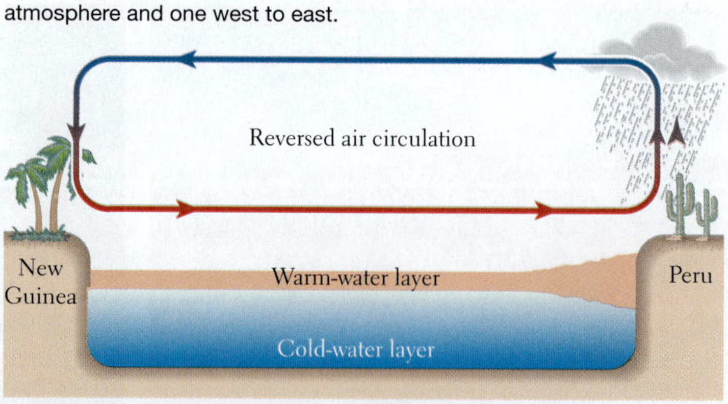

(C) Fully developed El Niño. Slowly, as the bulge of warm water at the surface of the ocean moves east, it forces the whole system into the fully developed El Niño, with air at the surface and in the upper atmosphere flowing in reverse of normal (A). Instead of warm, wet air rising over the mountains of New Guinea and condensing as rainfall, cool, dry, cloudless air descends to sit at Earth's surface in the west. Meanwhile, in the east, the normally dry, clear coast of Peru has clouds and rainfall. [Research from: El Niño diagram: Ivan Cheung, George Washington University, Geography 137, Lecture 16, October 29, 2001; confirmed by numerous other diagrams at https://www.abc.net.au/news/2017-10-19/la-nina-enso/9061284.]

El Niño–connected drought, which leaves the bare Earth surface vulnerable to erosion. Australia experienced the hottest and driest year on record in 2018, and an El Niño event is predicted by the World Meteorological Organization for 2019.

FAUNA AND FLORA

The fact that Oceania is made up of an isolated continent and numerous islands has affected its animal life (fauna) and plant life (flora). Many of its species are **endemic**, meaning that they exist in a particular place and nowhere else on Earth. This is especially true in Australia (of the 750 species of birds known in Australia, more than 325 species are endemic), but many Pacific islands also have endemic species.

Animal and Plant Life in Australia

The uniqueness of Australia's animal and plant life is the result of the continent's long physical isolation, large size, relatively homogeneous landforms, and arid climate. Since Australia broke away from Gondwana more than 65 million years ago, its animal and plant species have evolved in isolation. One spectacular result of this isolation is the presence of more than 144 living species of endemic marsupial animals. **Marsupials** are mammals whose babies at birth are still at a very immature stage; the marsupial then nurtures them in a pouch equipped with nipples. The best-known marsupials are kangaroos; other marsupials include wombats, koalas, and bandicoots. The **monotremes**, egg-laying mammals that include the duck-billed platypus and the spiny anteater, are endemic to Australia and New Guinea.

Most of Australia's endemic plant species are adapted to dry conditions. Many of the plants have deep taproots to draw moisture from groundwater, and small, hard, pale green, or shiny leaves to reflect heat and to hold moisture. Much of the continent is grassland and scrubland with bits of open woodland; there are only a few true forests, found in pockets along the Eastern Highlands and the southwestern tip and in Tasmania (**Figure 11.9**). Two plant genera account for nearly all the forest and woodland plants: Eucalyptus (450 species, often called gum trees) and Acacia (900 species, often called wattles).

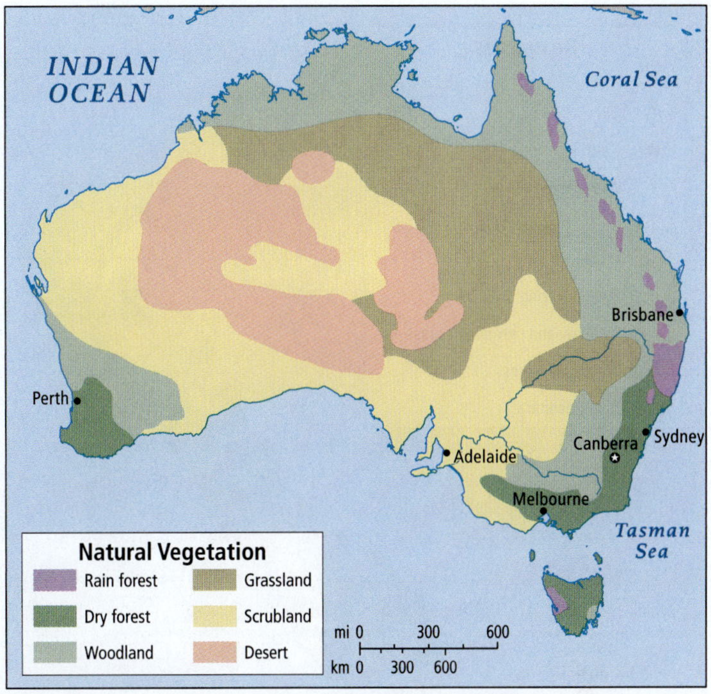

Figure 11.9 Map of Australia's natural vegetation. Much of Australia is grassland and scrubland. A few forests can be found in the Eastern Highlands, the far southwest, and Tasmania. [Research from: Tom L. McKnight, *Oceania* (Englewood Cliffs, NJ: Prentice Hall, 1995), p. 28.]

Animal and Plant Life in New Zealand and the Pacific Islands

Naturalists and evolutionary biologists have had a great interest in the species that inhabited the Pacific islands before humans arrived. Charles Darwin formulated many of his ideas about evolution after visiting the Galápagos Islands of the eastern Pacific (see the Figure 3.1 map) and the islands of Oceania.

Islands gain plant and animal populations from the sea and air around them as organisms are carried from larger islands and continents by birds, storms, or ocean currents. Once these organisms "colonize" their new home, they may evolve over time into new species that are unique to one island. High, wet islands generally contain more varied species because their more complex environments provide niches for a wider range of wayfarers and thus more opportunities for evolutionary change.

Once they arrive, human inhabitants modify the flora and fauna of islands. In prehistoric times, Asian explorers in oceangoing canoes brought plants such as bananas and breadfruit and animals such as pigs, chickens, and dogs to Oceania. European settlers later introduced grains, vegetables, fruits, invasive grasses, cattle, sheep, goats, rabbits, horses, housecats, and rats. Today, human activities from tourism to military exercises to urbanization to commercial agriculture continue to change the flora and fauna of Oceania.

Generally, the diversity of land animals and plants is richest in the western Pacific, near the larger landmasses. It thins out to the east, where the islands are smaller and farther apart. The natural rain forest flora is rich and abundant in New Zealand and New Guinea, and also on the high islands of the Pacific. However, the natural fauna is much more limited on these islands (**Figure 11.10**). While New Guinea has fauna comparable to Australia, to which it was once connected via Sundaland (see Figure 10.9), New Zealand and the Pacific islands have no indigenous land mammals, almost no indigenous reptiles, and only a few indigenous species of frogs, as they were never connected to Australia and New Guinea by a land bridge that land animals could cross. Two indigenous birds in New Zealand, the kiwi and the huge moa (a bird that grew up to 12 feet [3.7 meters] tall), were a major source of food for the Maori people. The moa was hunted to extinction before the Europeans arrived. Today, New Zealand may well be the country with the most nonnative species of mammals, fish, and fowl, nearly all brought there by European settlers (see "Invasive Species and Food Production").

endemic belonging or restricted to a particular place

marsupials mammals whose babies at birth are still at a very immature stage; the marsupial then nurtures them in a pouch equipped with nipples

monotremes egg-laying mammals, such as the duck-billed platypus and the spiny anteater

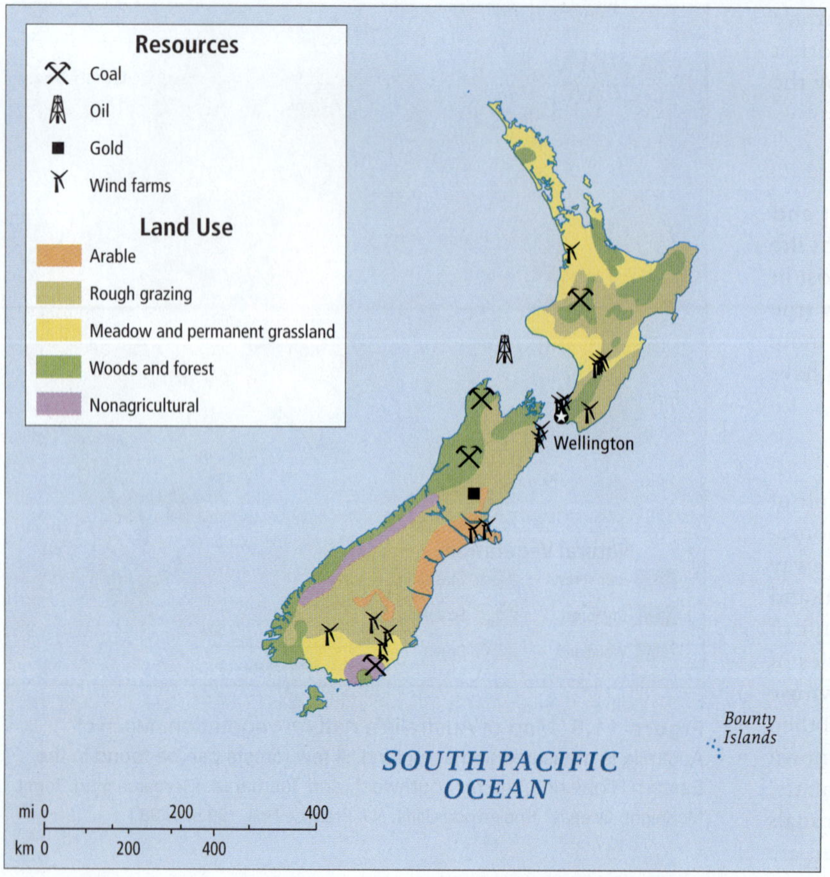

Figure 11.10 Land uses and natural resources of New Zealand. As a result of European settlement and the clearing of land for farming, only 23 percent of New Zealand remains forested. [Research from: Richard Nile and Christian Clark, *Cultural Atlas of Australia, New Zealand and the South Pacific* (New York: Facts on File, 1996), p. 194.]

CHECK YOUR UNDERSTANDING

1. Describe the basic size, population characteristics, and natural features of Oceania.
2. Describe some of the vegetation of Australia.
3. What are some of the distinguishing features of the animals of Oceania?
4. Briefly describe how the islands of the Pacific were created, referring to both tectonic activity and volcanic activity.
5. Describe the climate of the section of the Pacific where most of the people of Oceania live.
6. Due to their isolation from the life forms found on other major landmasses, the fauna and flora of the region are unique in many distinctive ways. What modern body of knowledge have they contributed to?

CLIMATE CHANGE IN OCEANIA

Oceania, on the whole, is a minor contributor of greenhouse gases; and yet, Australia has some of the world's highest greenhouse gas emissions on a per capita basis. Much of Australia's emissions, like those of the United States, result from the use of automobile-based transportation systems to connect a widely dispersed network of cities and towns. Additionally, Australia's heavy dependence on coal to generate electricity has added to high emissions, much as coal dependence has in the United States. Also, Australia is the world's largest coal exporter (the United States, China, and India use their coal internally), selling most of it to Japan, China, and India. And coal is the largest commodity export for Australia, so it is central to the national economy just at a time when coal is being criticized for the greenhouse gases it produces.

Australia has a relatively small population (24.1 million people in 2018), so it accounts for only slightly more than 1 percent of global emissions; and in fact, Australia is using less and less coal in order to reduce emissions yet further. However, despite the negligible contributions of greenhouse gases made by the islands of the Pacific, they, as well as Australia, are quite vulnerable to the environmental effects of global climate change (**Figure 11.11**).

Sea Level Rise and Storm Surges

The best scientific research indicates that global warming is raising sea levels primarily through thermal expansion as rising temperatures cause the water in the ocean to expand in size, but seas are also rising because glaciers and polar ice caps are melting. Obviously, rising seas are of great concern to residents of islands, such as the atoll Tuvalu that already is barely above the waves (see Figure 11.11B). If sea levels rise 4 inches (10 centimeters) per decade, as predicted by the International Panel on Climate Change, many of the lowest-lying Pacific atolls will disappear under water within 50 years. Other islands, some with already very endangered coastal zones, such as Kiribati (see the chapter-opening photo), will see these zones shrink further and become more vulnerable to storm surges and cyclones. In late February 2016, the worst cyclone to hit the southern Pacific struck the Fiji island chain, killing 42 and leaving at least 60,000 without shelter, water, and electricity. Wind speeds exceeded 200 miles (325 kilometers) per hour and waves were 40 feet (13 meters) high.

Wildfires and Other Water-Related Vulnerabilities

Much of Oceania is particularly vulnerable to the droughts and floods that can result from global climate change. Parts of Australia, New Zealand, and some low, dry Pacific islands are already undergoing prolonged droughts and freshwater shortages requiring drastic adjustments of daily life and livelihoods. The fear is that the severe droughts are not just the usual periodic dry spells but may represent permanent alterations in rainfall patterns that could result in worse wildfires. Such fires emerged as a major issue in Australia in February 2009, when 173 people died in a rural firestorm near Melbourne. In 2015 through 2018, repeated wildfires struck Western Australia, South Australia, Queensland, New South Wales, and Victoria—the last three of which lie in the most humid, and hence most fire-resistant, part of the country (see Figures 11.1 and 11.11C).

As fresh water becomes more scarce across the region, *virtual water* (see Chapter 1) becomes an issue. All export-related activities that permanently consume or degrade fresh water in their extraction or production processes (such as coal mining; oil and gas extraction; meat, wool, and wheat production; and tourism-related construction and maintenance) are essentially extracting virtual water from places that are already under water stress. If the true costs of this freshwater depletion were counted and added to the price of the products, these exports might no longer be competitive on the world market—at least not until all global producers understood virtual water accountability to be in their best interests and raised their prices accordingly.

Coral Bleaching

Another concern is that the warming of the oceans will not only result in stronger tropical storms, but also threaten coral reefs and the fisheries that depend on them by causing **coral bleaching** (see Figure 11.11A). When ocean waters warm just a few degrees, the living coral organisms expel the algae that live inside and give the coral its color. Coral bleaching is a phenomenon that has affected all reefs in this region in recent years, with scientists announcing, in 2016, Australia's "biggest ever environmental disaster": The Great Barrier Reef had lost 50 percent of its coral cover and was vulnerable to losing another 25 percent. Coral bleaching is so concerning because between a quarter and a third of marine species have part of their life cycle on reefs, so coral bleaching threatens many aquatic communities and ultimately the global food supply. This is especially true on some of the Pacific islands that have few other local food resources. By 2018, scientists began to see a few signs that the bleaching was slowing. In the future, barring further environmental deterioration, just possibly the coral might regenerate.

Responses to Potential Climate Change Crises

In order to slow climate change and associated global warming, Oceania is pursuing a number of alternative energy strategies to reduce greenhouse gas emissions. Although Australia remains dependent on fossil fuel sales (primarily the sale of coal, and also crude oil and natural gas) to Asia, it is pursuing renewable energy strategies for domestic use. These include increasing power production from geothermal, solar, and biomass sources, and to address water shortages, Australia is decreasing the emphasis on hydropower. New Zealand has set a goal of obtaining 95 percent of its energy from renewable sources by 2025. Much of this energy will come from wind power, which has a great deal of potential in this region, especially in areas near the Roaring Forties. On low Pacific islands, solar energy is now the most widely used alternative to costly and polluting imported fuel.

Oceania is also a world leader in implementing alternative water technologies. Some of these are simple but effective age-old methods, such as harvesting rainwater from roofs and the ground for household use. Most buildings in rural Australia, New Zealand, and many Pacific islands get at least part of their water this way, relieving surface and groundwater resources. Australia and New Zealand are now stretching water resources further by using very efficient drip irrigation technologies extensively in agriculture (see Figure 11.11D) and new low-cost water-filtration techniques that facilitate recycling of water.

INVASIVE SPECIES AND FOOD PRODUCTION

People who settled Oceania in prehistory and into the modern era introduced plants, animals, and ways of raising and processing them. The changes to the environments of Oceania wrought by these introductions were significant, and yet today, to the casual observer, the many alien features appear native. Take, for example, the case of wine grapes in Australia, or taro on Pacific islands, or the Maori *hangi* oven in New Zealand (**Figure 11.12**). In the process of human settlement, many of the unique endemic plants and animals of Oceania were displaced by **invasive species**, organisms that, lacking natural predators, spread aggressively into regions beyond where they were first introduced, adversely affecting economies or environments. Europeans brought many exotic, or alien, plants and animals to Oceania to support their food production systems. Ironically, some of these same species—for example, rabbits—are now major threats to food production.

Australia

When Europeans first settled the continent, they brought many new animals and plants with them, sometimes intentionally, sometimes unintentionally. Early British settlers who enjoyed eating rabbits brought them to Australia, where they proved to be among the most destructive of introduced species. Many rabbits were released for hunting, but with no natural predators, the rabbits multiplied quickly, consuming so much of the native vegetation that many indigenous animal species disappeared. Moreover, rabbits became a major cause of agricultural crop loss and reduced the capacity of grasslands to support herds of introduced sheep and cattle. Attempts to control the rabbit population by introducing European foxes and cats backfired as these animals became major invasive nuisances themselves. Foxes and cats have driven several native Australian predator species to extinction without having much effect on the rabbit population. Intentionally introduced diseases have proven more effective at controlling the rabbit population, though rabbits have repeatedly developed resistance to them.

Herding has also had a huge impact on Australian ecosystems. Because the climate is arid and soils in many areas are relatively infertile, the dominant land use in Australia is the grazing of introduced domesticated animals—primarily sheep, but also cattle. More than 15 percent of the land has been allocated for grazing, and Australia leads the world in exports of sheep and cattle products.

coral bleaching occurs when the living coral organisms expel the algae that live inside and give the coral its color

invasive species nonnative organisms that, lacking natural predators, spread into regions outside their native range, adversely affecting economies or environments

PHOTO ESSAY

11.11 Vulnerability to Climate Change in Oceania

Oceania is vulnerable to a wide variety of hazards related to climate change, including sea level rise, stronger tropical storm intensity, the warming of ocean waters, and decreasing precipitation. Fortunately, several countries in this region are already implementing practices that are increasing resilience to climate hazards. Countries that are larger, higher, and wealthier (Australia, New Zealand, Guam, and the U.S. state of Hawaii) have the best opportunities to find solutions, and these parts of Oceania are being asked to help the smaller and poorer places.

THINKING GEOGRAPHICALLY

A Why is it difficult to find local solutions to coral bleaching?

B Other than sea level rise, how are low-lying islands susceptible to climate change effects?

C Check the location of Sydney on the climate map (Figure 11.6) and natural vegetation map (Figure 11.9) and then suggest a reason why wildfires there are especially alarming.

D If New Zealand has a wet climate, why would irrigation be needed?

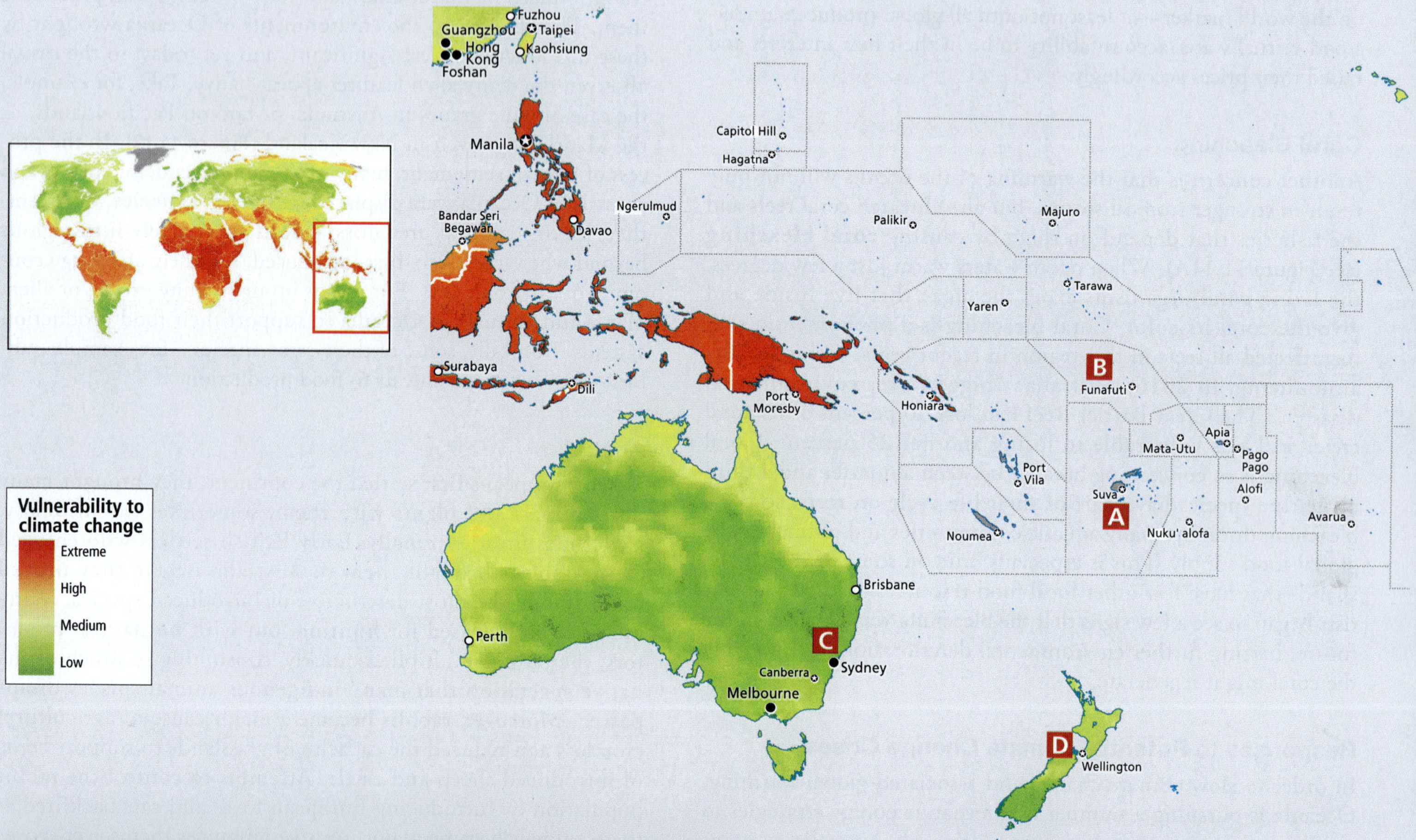

A Coral bleaching near Fiji. A marine biologist monitors the reefs off the coast of Fiji for signs of coral bleaching, a response to stress brought on by overly warm water. When the water is too warm, the coral expels microalgae that live symbiotically with the coral, giving it color and nourishing the many aquatic species living on the reefs. Despite Fiji's attempts to protect the coral, bleaching threatens many aquatic species, including fish that sustain Pacific Islanders. [Mark Conlin/Getty Images]

B A low-lying atoll in Tuvalu. With the highest point only 14.7 feet (4.5 meters) above sea level, the nine islands of Tuvalu are quite vulnerable to sea level rise. Combined with increased flooding during tropical storms, higher sea levels could make many low islands across Oceania uninhabitable. Tuvalu's government is already negotiating the future resettlement of parts of its population to nearby nations such as New Zealand. [TORSTEN BLACKWOOD/AFP/Getty Images]

C A wildfire outside Sydney, Australia. Higher temperatures and years of drought are bringing stronger and more extensive wildfires to this region. Australian and U.S. firefighters used to cooperate to fight each other's fires, but now in both places, fires are occurring year-round so coordinated firefighting is no longer possible. [AAP/Dan Himbrechts/STRINGER/REUTERS/Newscom]

D Irrigation in Marlborough, New Zealand. A vineyard is fitted with a drip irrigation system, visible at the base of the vines. These systems provide resilience in the face of drought and use substantially less water than other methods. Both New Zealand and Australia have vibrant wine-producing districts that supply the global market, but are now facing drought. [David Wall Photo/Getty Images]

Figure 11.12 LOCAL LIVES: Foodways in Oceania

A A winemaker in Australia's Hunter Valley assesses the shiraz grapes at her winery. Grapes were introduced from Europe, and Australia is now the fourth-largest exporter of wine in the world, after Italy, France, and Spain. [Peter Stoop/Fairfax Media via Getty Images]

B Taro is grown in flooded fields, and was originally brought to Oceania from Southeast Asia. It is now a major part of diets throughout the Pacific islands. While the leaves are also eaten, the cooked tuber is particularly prized as a source of calories when processed into *poi*, a paste of variable thickness that is eaten with the fingers and accompanied by various seasoned meats, fish, vegetables, and sauces. [Robert Madden/National Geographic/Getty Images]

C Roasted food is being removed from a Maori earth oven, or *hangi*, in New Zealand. First a pit is dug, and then a fire is made to heat stones placed in the pit. Baskets of food are placed over the hot stones (which are covered with cloth and then earth) for several hours until the food is cooked. [Chris Jackson/Getty Images]

Dingoes, the indigenous wild dogs of Australia (**Figure 11.13**), prey on sheep and young cattle. To separate the wild dogs from the herds, the Dingo Fence—the world's longest fence—was built. It extends 3488 miles (5614 kilometers) and is unfortunately a major ecological barrier to other wild species. Meanwhile, kangaroos (the natural prey of dingoes) have learned to live on the sheep side of the fence, where their population has boomed beyond sustainable levels.

New Zealand

New Zealand's environment has been transformed by introduced species and food production systems even more extensively than Australia's environment. No humans lived in New Zealand until about 700 years ago, when the Polynesian Maori people settled there. When they arrived, dense midlatitude rain forest covered 85 percent of the land. The Maori were cultivators who brought in yams and taro as well as other nonnative plants, rats, and birds. By the time of significant European settlement (after 1825), forest clearing and overhunting by the Maori had already degraded many environments and driven several bird species to extinction.

European settlement in New Zealand dramatically intensified environmental degradation. Attempts to re-create European farming and herding systems in New Zealand resulted in environments that today are actually hostile to many native species, a growing number of which are becoming extinct. Only 23 percent of the country remains forested (see Figure 11.10), with ranches, farms, roads, and urban areas claiming more than 90 percent of the lowland area.

Figure 11.13 The Australian dingo and the fence meant to contain it.

(A) The dingo, a native wild dog, lives mainly in the Australian Outback. Dingoes were likely brought to Australia by Aboriginal Australians, who used them as guard dogs and possibly as a food source. [Robin Smith/Getty Images]

(B) The sheep brought by Europeans were easy prey, so shepherds considered dingoes to be pests, and in the 1880s, they built a 3488-mile (5614-kilometer) Dingo Fence to keep dingoes out of southeastern Australia. It was only partially effective. [Claver Carroll/age fotostock/Alamy Stock Photo]

The ordinary house cat (*Felis catus*), brought from Europe to control rabbits, mice, and rats, is an example of an interloper whose impact has been astonishing. In New Zealand, it is estimated that feral cats kill up to 100 million birds each year. Many of the victims are endemic birds such as tuis and kukupa, with little inborn wariness for predators. For some, the answer has been to eliminate all feral cats and sterilize all housecats—a project that gained steam in New Zealand in 2013. But this solution has not been popular with New Zealand's pet lovers; eliminating or reducing the number of cats may also give rise to a burgeoning rodent population. Despite the attempts to reduce the cat population, cats remain popular: in 2016, 45 percent of New Zealanders still owned one.

Most of New Zealand's cleared land is used for export-oriented farming and ranching. Grazing has become so widespread that today there are 15 times as many sheep as people (**Figure 11.14**), and three times as many cattle. Both farming and ranching have severely degraded the environment. Soils exposed by the clearing of forests proved infertile, forcing farmers and ranchers to augment them with agricultural chemicals. The chemicals, along with feces from sheep and cattle, have seriously polluted many waterways, causing the extinction of some aquatic species.

ENVIRONMENTAL AWARENESS IN NEW ZEALAND AND THE PACIFIC ISLANDS

Perhaps because it has witnessed so many dire environmental consequences of human settlement, New Zealand is a global leader in environmental awareness. It formed the world's first environmentally focused national political party in 1973—the Green Party of Aotearoa—now the country's third largest, and spearheaded the world's first nuclear-free zone in 1984. New Zealand's government promotes a "clean and green" image internationally, and most New Zealanders acknowledge the severity of existing environmental problems and the need for further action.

Environmental awareness has been slower to develop in the Pacific islands, but as these islands became more connected to the global economy over the years, local environmental awareness increased. Then, the prospect of global climate change and its

Figure 11.14 The dominant animal in New Zealand. A flock of sheep in New Zealand, where ranches, farms, roads, and urban areas cover 90 percent of the lowlands. Forests once covered 85 percent of New Zealand, but after two centuries of export-oriented agriculture and forestry, only 23 percent of the country remains forested. In recent decades, there have been increased efforts to conserve the remaining forests. [Raimund Linke/Getty Images]

potential negative impacts on the region belatedly raised awareness. As the islands were deforested and mined or converted to commercial agriculture, unique species of plants and animals were driven to extinction. Hawaiian Islanders were among the first to notice. The extensive conversion of tropical Hawaiian forests to export crops, such as sugar cane and pineapples, has caused the extinction of numerous plant, bird, and land species. Now, Hawaii is home to more threatened or endangered species than any other U.S. state, despite having less than 1 percent of the U.S. landmass.

GLOBALIZATION AND ENVIRONMENTAL DAMAGE

Over the last century, across the Pacific, flows of resources and pollutants have increased dramatically. Mining, nuclear pollution, commercial agriculture and fishing, and tourism are all examples of how globalized demands have transformed environments in the Pacific islands. In most cases these developments have not benefited Pacific people, who, instead, have had to deal with vast negative environmental effects while the profits have gone to investors in America, Europe, and, more recently, Asia.

Mining in Papua New Guinea and Nauru

Mining has rendered the islanders of Oceania losers in multiple ways: foreign-owned mining companies that took advantage of poorly enforced or nonexistent environmental laws are responsible for major environmental damage. In the Ok Tedi Mine on Papua New Guinea, 80 million tons of mine waste devastated river systems (**Figure 11.15A, B**). The environmental degradation forced tens of thousands of indigenous subsistence cultivators into new mining market towns, where their horticulture skills were of little use and where they needed cash to buy food and pay rent. Also, most of the profits of mining go to foreign-owned mining companies, while the best-paying mining jobs go to outsiders.

The story of mining disasters across Oceania is extensive. The most extreme case of environmental damage caused by mining took place on the once densely forested Melanesian island of Nauru, which is one-third the size of Manhattan and located northeast of the Solomon Islands (see Figure 11.1). Fifty years ago, Nauru's wealth became legendary, based on proceeds from the strip-mining of high-grade phosphates derived from eons of bird droppings (guano) that are used to manufacture munitions and fertilizer. The phosphate mining companies were owned first by Germany, then Japan, and finally Australia. For a time in the early 1970s, Nauru had the highest per capita income in the world (although not at all distributed equitably). Today, the phosphate reserves are nearly depleted, the proceeds have been ill spent, and the environment destroyed. Junked mining equipment sits on miles of bleached white sand where forests once stood. Instead, Nauru now serves as a detention camp for more than 500 asylum-seekers from Iraq, Iran, Afghanistan, Somalia, Cambodia, and Myanmar (Rohingya). Nauru has become an increasingly fraught part of Australia's controversial offshoring policy for those seeking asylum, called the *Pacific Solution*, discussed in "The Challenge of Migrants and Refugees").

Nuclear Pollution

The geopolitical aspects of globalization have hit Oceania's environment especially hard. Nuclear weapons testing by France and the United States from the 1940s to the 1960s (during the Cold War), as well as the dumping of nuclear waste by various nuclear powers, have long been major environmental issues for the Pacific islands (**Figure 11.16**). In New Zealand in July 1985, the French secret service blew up the ship *Rainbow Warrior* owned by the antinuclear environmental group Greenpeace. In response, the 1986 Treaty of Rarotonga established the South Pacific Nuclear Free Zone, which was an expansion of a similar zone set up in New Zealand in 1984. Most independent countries in Oceania signed

Figure 11.15 The Ok Tedi mine and river, western Papua New Guinea.

(A) The Ok Tedi open-pit copper and gold mine. A large hole now exists where Mount Fubilan once stood. The products of the Ok Tedi Mine (gold and copper) are not used by the indigenous subsistence cultivators of Papua New Guinea, yet they were forced off their land and into new mining market towns as a result of European settlement, the clearing of forests, and chemical pollution. What are some possible ways to alert consumers of the gold and copper to the negative impacts of mining on indigenous peoples? [Andrew Peacock/Getty Images]

(B) The Ok Tedi River as it flows downstream of the mine. Each year since its opening in 1984, the mine has discharged millions of tons of contaminated mine tailings and eroded sediments, resulting in a once-deep river becoming clogged with poisonous runoff that kills many trees along its banks. Sediment from the mine has killed innumerable fish in the river, contaminated 500 square miles (1300 square kilometers) of farmland, and adversely affected 50,000 people in 120 villages. Litigation against the mine owners is ongoing. [The Asahi Shimbun Premium/Getty Images]

Figure 11.16 Nuclear bomb test blast on Bikini Atoll. In 1946, on Bikini Atoll in the Marshall Islands (then a U.S. territory), the United States conducted one of the first underwater tests of a nuclear weapon and its effects on naval vessels. Since then, the United States and France have conducted over 300 nuclear tests in Oceania, some without sufficient attention to nuclear contamination. [Scott Camazine/Getty Images]

this treaty, which bans nuclear weapons testing and nuclear waste dumping on their lands. Because of political pressure from France and the United States, however, French Polynesia and U.S. territories such as the Marshall Islands have not signed the treaty.

The Great Pacific Garbage Island

When oceanographer Charles Moore found himself surrounded by a massive floating island of degrading plastic garbage in the north Pacific in 1997, he thought it was an anomaly. But his investigations revealed that the disposable, throwaway aspect of modern living was responsible. The "island" included plastic beverage bottles and caps along with Lego blocks, trash bags, toothbrushes, footballs, and kayaks—indeed, virtually every consumer product made (**Figure 11.17**). By 2008, several such garbage masses were floating in the Pacific and Atlantic oceans. Modern plastics exposed to sunlight degrade into tiny bits that are then ingested by sea birds and animals or into microscopic fragments ingested by filter-feeding organisms. The plastics also release carcinogenic chemicals as they break down. Millions of seabirds, sea mammals, and fish die each year after exposure to the residue of this trash. Humans are affected because, as marine scientists say, whatever goes into the ocean is likely to go into the food chain and eventually end up on someone's dinner plate.

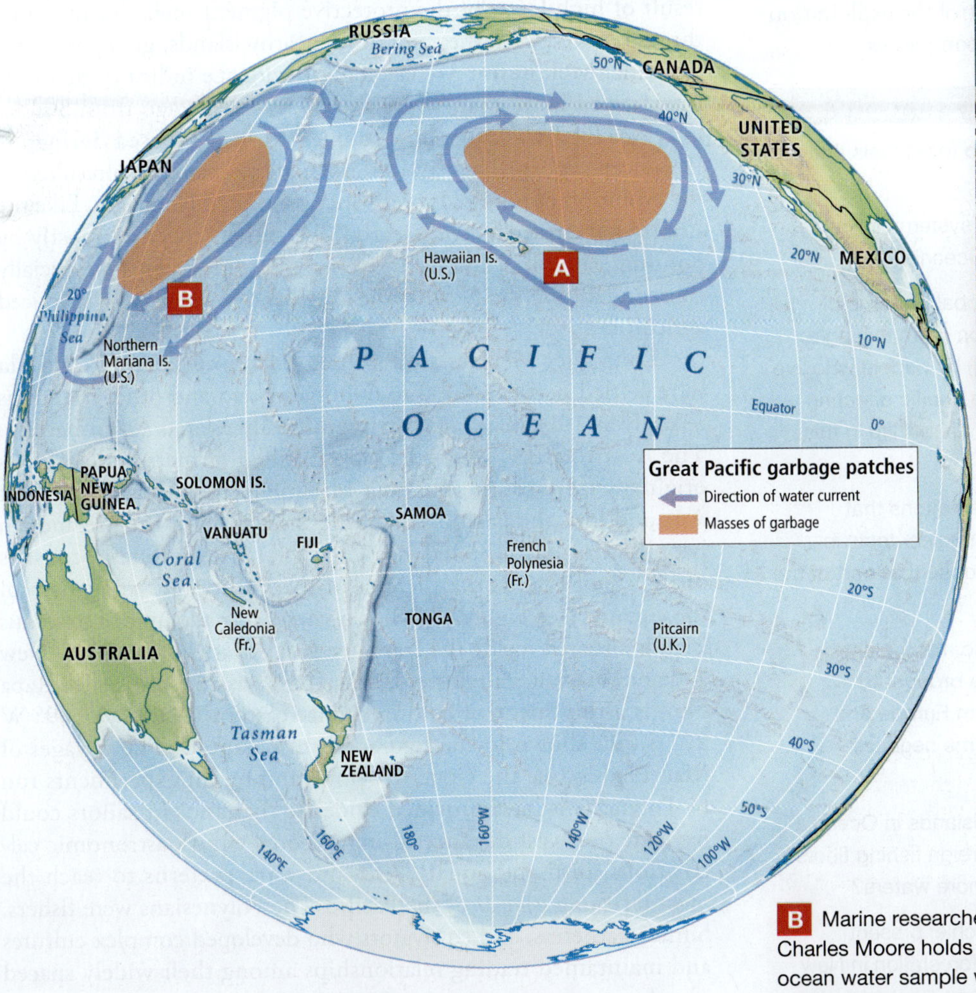

Figure 11.17 Pacific Ocean floating garbage patches.

A A floating containment barrier at the Pacific Ocean end of Ala Moana Canal in Honolulu holds garbage. [PhilAugustavo/Getty Images]

B Marine researcher Charles Moore holds an ocean water sample with debris from the Great Pacific Garbage Patches. [Jonathan Alcorn/Bloomberg via Getty Images]

The science of marine debris is young, and a wide range of experts must do the research needed. In 2014, scientists in Europe and North America discovered a significant accounting error: from the vast amount of plastic manufactured, the five huge circulating masses of trash in the world ocean should have contained 10 to 100 times more plastic than they did. Where is the missing plastic, and what is the impact of all this plastic on marine animals and those who eat them? Many ideas are currently under study on how to clean up the floating islands of garbage, but they are only beginning to be implemented.

The UN Convention on the Law of the Sea

Based on the idea that all the problems of the world's oceans are interrelated and need to be addressed as a whole, the United Nations Convention on the Law of the Sea (UNCLOS) established rules governing all uses of the world's oceans and seas; it has been ratified by 157 countries (but not the United States). The treaty allows islands to claim rights to ocean resources 200 miles (320 kilometers) out from their shores. Island countries can now make money by licensing privately owned fleets from Japan, South Korea, Russia, the United States, and elsewhere to fish within these offshore limits. As of yet, however, there is no overarching enforcement agency, and protecting the fisheries from overfishing by these rich and powerful licensees has turned out to be an enforcement nightmare for tiny island governments with few resources. Similarly, it has proven difficult to monitor and control the exploitation of seafloor mineral deposits by foreign mining companies.

CHECK YOUR UNDERSTANDING

1. Describe the renewable energy alternatives to fossil fuels that are being pursued across the region of Oceania.
2. How has the introduction of food production systems from elsewhere resulted in ecological damage in Oceania?
3. How are global patterns of consumption (from mining to vacationing) by people who live far from the Pacific affecting human and animal life in this region?
4. Why did the nations that conducted nuclear tests in Oceania choose that part of the world?
5. Why were so many plants and animals brought to Oceania from Europe and what are some negative results?
6. Why might islands in Oceania object to foreign fishing fleets in their offshore waters?
7. What is the chief present cause of deforestation in New Zealand?

Aboriginal Australians the longest-surviving inhabitants of Oceania, whose ancestors, the Australoids, migrated from Southeast Asia 50,000 to 70,000 years ago over the Sundaland landmass that was exposed during the ice ages

Melanesians a group of Australoids who settled throughout New Guinea and other nearby islands

Melanesia New Guinea and the islands south of the equator and west of Tonga (the Solomon Islands, New Caledonia, Fiji, and Vanuatu)

Micronesia the small islands that lie east of the Philippines and north of the equator

Polynesia the numerous islands situated inside an irregular triangle formed by New Zealand, Hawaii, and Easter Island

HUMAN PATTERNS OVER TIME

Oceania's past has been shaped by its ancient settlement from the Asian mainland and by the more recent arrival of Europeans in Australia, New Zealand, and the islands. Oceania's present is increasingly being influenced by economic and geographic considerations, particularly its physical proximity to Asia.

The Peopling of Oceania

The longest-surviving inhabitants of Oceania are **Aboriginal Australians**, whose ancestors (the Australoids) migrated from Southeast Asia 50,000 to 70,000 years ago (**Figure 11.18**), at a time when sea level was somewhat lower. It is possible that some memory of this ancient journey may be preserved in Aboriginal oral traditions, which recall mountains and other geographic features that are now submerged under water. At about the same time that the Aboriginal Australians were settling Australia and Tasmania, related groups were settling nearby areas. The distribution of these groups, the sequence of settlement, and the navigation skills necessary to explore and occupy this huge oceanic environment are complex. Figure 11.18 and the vignette about Mau Piailug, the modern-day navigator, are attempts to clarify this story.

Melanesians, so named for their relatively dark skin tones, a result of high levels of the protective pigment melanin, migrated throughout New Guinea and other nearby islands, giving this area its name, **Melanesia**. Archaeological evidence indicates that they first arrived more than 50,000 to 60,000 years ago from Sundaland (see Figure 10.9), a now-submerged shelf exposed during the Pleistocene epoch. They lived in isolated pockets, which resulted in the evolution of hundreds of distinct yet related languages. Like the Aboriginal Australians, the early Melanesians survived mostly by hunting, gathering, and fishing, although some groups—especially those inhabiting the New Guinea highlands—eventually practiced agriculture.

Long after Melanesia was settled, **Micronesia** and **Polynesia** were settled between 5000 to 6000 years ago and other islands as recently as 1000 years ago, by linguistically related Austronesians. The Austronesians were a group of skilled farmers and seafarers originally from southern China who migrated through Southeast Asia and into the Pacific, sometimes mixing with the Melanesian peoples they encountered (see Chapter 10). Micronesia consists of the small islands that lie east of the Philippines and north of the equator (see Figure 11.18). Polynesia is made up of numerous islands situated inside the large, irregular triangle formed by New Zealand, Hawaii, and Easter Island. (Easter Island, also called Rapa Nui, is a tiny speck of land in the far eastern Pacific, at 109° W 27° S, not shown in the figures in this chapter.) The voyages of Mau Piailug (see the vignette below) and recent experiments run by Polynesians have provided evidence that ancient sailors could navigate over vast distances using seasonal winds, astronomic calculations, bird and aquatic life, and wave patterns to reach the most far-flung islands of the Pacific. The Polynesians were fishers, hunter-gatherers, and cultivators who developed complex cultures and maintained trading relationships among their widely spaced islands.

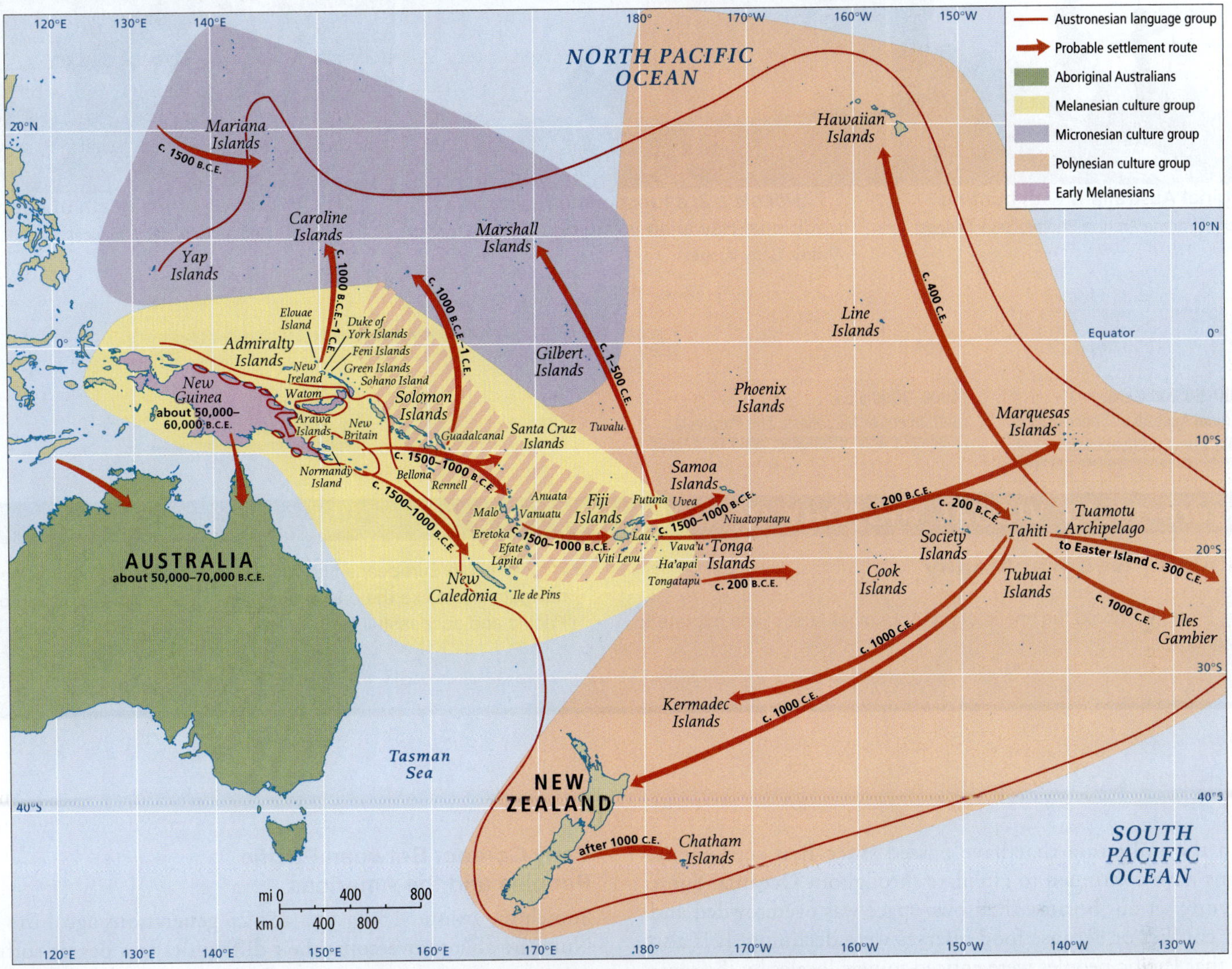

Figure 11.18 Primary indigenous culture groups of Oceania. By 50,000 to 70,000 years ago, humans had come to New Guinea and Australia. About 25,000 years ago, people began moving across the ocean to nearby Pacific islands. Movement into the more distant islands commenced with the arrival of Austronesians (5000–6000 B.P.), who went on to inhabit the farthest reaches of Oceania. The largest culture area is Polynesia (peach color), which stretches from Hawaii to Easter Island (Rapa Nui, which is so far east that it is not on this map) to New Zealand. [Research from: Richard Nile and Christian Clerk, *Cultural Atlas of Australia, New Zealand, and the South Pacific* (New York: Facts on File, 1996), pp. 58–59.]

VIGNETTE *With courage, you can travel anywhere in the world and never be lost. Because I have faith in the words of my ancestors, I'm a navigator.*

—Mau Piailug (1932–2010)

In 1976, Mau Piailug made history by sailing a reconstruction of a traditional double-hulled Pacific island voyaging canoe, the *Hōkūle'a,* across the 2400 miles (3860 kilometers) of deep ocean between Hawaii and Tahiti. He did so without a compass, charts, or other modern instruments, using only methods passed down through his family. To find his way, he relied mainly on observations of the stars, the Sun, and the Moon. When clouds covered the sky, he used the patterns of ocean waves and swells as well as the presence of seabirds to tell him of distant islands over the horizon.

Piailug reached Tahiti 33 days after leaving Hawaii and made the return trip in 22 days. His voyage resolved a major scholarly debate over how people settled the many remote islands of the Pacific without navigational instruments thousands of years before the arrival of Europeans. Some thought that navigation without instruments was impossible and argued that would-be settlers simply drifted about on their canoes at the mercy of the winds, most of them starving to death on the seas, with a few happening upon new islands by chance. It was hard to refute this argument because local navigational methods had died out almost everywhere. However, in isolated Micronesia, where Piailug lived, indigenous navigational traditions still survive.

After the successful 1976 voyage, Piailug trained several students in traditional navigational techniques. His efforts have become a symbol of cultural rebirth and a source of pride throughout the Pacific. In 2007, the protégés of Mau Piailug sailed from Hawaii through the Marshall Islands to Yokohama, Japan, to celebrate peace and the human need to stay connected with nature. In 2016, the *Hōkūle'a* arrived in the Potomac River in Washington, DC, as part of a global traverse using Polynesian *wayfinding* navigation techniques. ■

A Aboriginal Australians with hunting tools. [Pictorial Parade/Archive Photos/Getty Images]

B *Hōkūle'a 2*, a functioning replica of a traditional Hawaiian voyaging canoe. [Stephen Alvarez/Getty Images]

C An etching of the death of Captain Cook in 1783. [Hulton Archive/Getty Images]

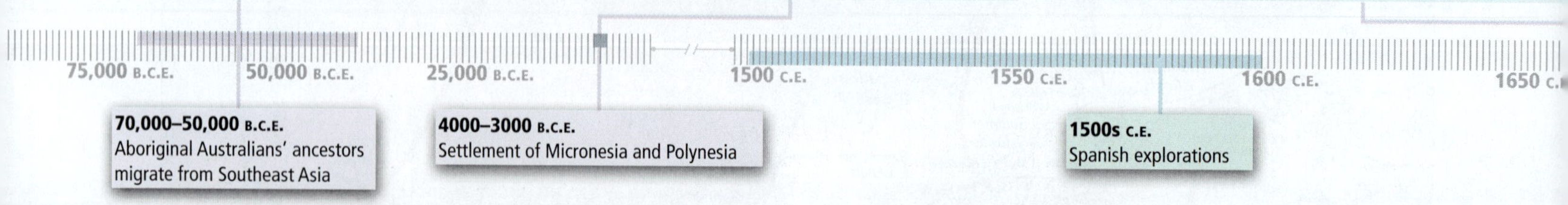

Figure 11.19 VISUAL HISTORY OF OCEANIA

Thinking Geographically

A From where did the ancestors of Aboriginal Australians migrate to Oceania?

B Why did Europeans find it hard to believe that Polynesians could navigate boats like the one in this painting over the wide span of the Pacific and successfully find their way back and forth?

In the millennia that have passed since first settlement, humans have continued to circulate throughout Oceania. Some apparently set out because their own space was overcrowded and full of conflict or because food reserves were declining. It is also likely that Pacific peoples were enticed to new locales by the same lures that later attracted some of the more romantic explorers from Europe and elsewhere: sparkling beaches, magnificent blue skies, aromatic breezes, and lovely landscapes.

The Saga of Rapa Nui (Easter Island)

One of the most fascinating mysteries about the exploration and settlement of the Pacific by indigenous people is that regarding Easter Island (Rapa Nui), a 63-square-mile (163-square-kilometer) volcanic formation 2283 miles (3512 kilometers) off the west coast of Chile. Archeological evidence suggests that the island was settled by 100 or fewer Polynesian people about 1200 C.E They arrived in double-hulled canoes. Over 900 monolithic full-bodied human figures (most are half buried) carved from local volcanic rock attest to a vibrant society that could support massive artworks, yet by the time Europeans arrived in 1722, decline was well under way, sparking a debate. Was environmental collapse the explanation for the decline, or political revolution, or diseases of some sort? Collaborative research by universities in California, Virginia, Spain, New Zealand, and Denmark have contributed to the developing understanding that Rapa Nui people experienced a slow decline because of a range of environmental problems that prohibited continual production of sufficient food; this decline preceded the arrival of Europeans.

Early Contact Between Pacific Peoples and the Americas

Somehow, around 1300 C.E. (19–23 generations ago), the Rapa Nui were able to overcome their difficulties to a degree sufficient enough to continue exploration further to the east. Genomic data show that the Rapa Nui traveled to coastal South America long before Europeans appeared in the Americas or the Pacific, and there, they mixed with Native Americans. This evidence coincides with other studies that recently have found genomic proof of ancient Polynesian ancestry among indigenous Brazilians. This may help to explain the long-puzzling fact that sweet potatoes, genetically derived from sweet potatoes domesticated in South America, were being cultivated in the Pacific long before European contact. The sweet potato transfer may have been the result of very early contact between Polynesian people and Native Americans.

Arrival of the Europeans in the Pacific

The earliest recorded contact between Pacific peoples and Europeans took place in 1521, when the Portuguese navigator Ferdinand Magellan (exploring for Spain) rounded the tip of South America, crossed an unusually calm Pacific, and landed on the island of Guam in Micronesia (**Figure 11.19**). He then moved on to the Philippines. That encounter ended badly. The islanders, intrigued by European vessels, tried to take a small skiff. For this crime, Magellan had his men kill the offenders and burn their village to the ground. A few months later, Magellan was himself killed by islanders in what later became the Philippines, which he had claimed

D An etching of British convicts being ransported aboard ships to Australia in he mid-1800s. [Hulton Archive/Getty Images]

E Silver mining in Australia, 1900. [Popperfoto/Getty Images]

F Chinese Australians participate in a parade celebrating the Chinese New Year. [Lisa Maree Williams/Getty Images]

1700 C.E. | 1800 C.E. | 1850 C.E. | 1900 C.E. | 1950 C.E. | 2000 C.E.

1700s
Early French and British explorations

1788–1868
European population of Australia goes from 0 to 1.7 million

1788–1945
Era of European orientation

1945–1970s
Era of North American orientation

1970–Present
Era of increasing orientation to Asia

C What were the chief interests of early Europeans when they came to Oceania in the sixteenth, seventeenth, and eighteenth centuries?

D Summarize the sequence of immigrants to Australia from early years of European settlement to the present.

E Until World War II (approximately), on what did the economy in most parts of Oceania depend?

F What is the present role of Asians in the makeup of Australia's population?

for Spain. Nevertheless, by the 1560s, the Spanish had set up a lucrative Pacific trade route between Manila in the Philippines and Acapulco on the west coast of Mexico. Explorers from other European states followed, first taking an interest mainly in the region's valuable spices. The British and French explored Oceania extensively in the eighteenth century (see Figure 11.19C).

The Pacific was not formally divided among the colonial powers until the nineteenth century. By that time, the United States, Germany, and Japan had joined France and Britain in taking control of various island groups. As in other regions, the European colonizers of Oceania emphasized extractive agriculture and mining. Because native people were often displaced from their lands or exposed to exotic diseases to which they had no immunity, their populations declined sharply.

The Colonization of Australia and New Zealand

Although all of Oceania has been under European or U.S. rule at some point, the most Westernized parts of the region are Australia and New Zealand. The colonization of these two countries by the British has resulted in Australia and New Zealand having many parallels with North America. In fact, the American Revolution was a major impetus for "settling" Australia because once the North American colonies became independent, the British needed another location where they could send their convicts and other outcasts. In early-nineteenth-century Britain, a relatively minor theft—for example, of a piglet—might be punished with 7 years of hard labor in Australia (see Figure 11.19D).

A steady flow of English and Irish convicts arrived in Australia until 1868. Most of the convicts chose to stay in the colony after their sentences were served, and they are given credit for Australia's rustic self-image and egalitarian spirit. They were joined by a much larger group of voluntary immigrants from the British Isles who were attracted by the availability of inexpensive farmland. Waves of these immigrants continued to arrive until World War II. New Zealand was settled in the mid-1800s, somewhat later than Australia. Although its population also derives primarily from British immigrants, New Zealand was never a penal colony.

Another similarity among Australia, New Zealand, and North America was the treatment of indigenous peoples by European settlers. In both Australia and New Zealand, native peoples were killed outright, annihilated by infectious diseases, or shifted to the margins of society. The few who lived on territory the Europeans deemed undesirable were able to maintain their traditional ways of life. However, the vast majority of the survivors lived and worked in grinding poverty, either in urban slums or on cattle and sheep ranches. Today, native peoples still suffer from discrimination and maladies such as alcoholism and malnutrition. Even so, some progress is being made toward improving their lives (see "Aboriginal Land Claims") and some reparations have been paid (see Figure 11.23C later in the chapter). In 2008, the then newly elected prime minister of Australia, Kevin Rudd, officially apologized to Aboriginal people for the treatment they have received since the land was first colonized.

Closely related to attitudes toward indigenous people were attitudes toward immigrants of any color other than white. By 1901, a whites-only policy (called the "White Australia policy") governed Australian immigration, with favored migrants coming from the British Isles and (after World War II) from southern Europe. This discrimination persisted until the mid-1970s, when the White Australia policy on immigration was ended. In New Zealand, where similar racist attitudes prevailed, there was never an official whites-only policy, and by the 1970s, students and immigrants were arriving from Asia and the Pacific islands. But controversy over immigration in Australia and New Zealand continues. Recently, debate has centered on the arrival of refugees by boat from various parts of Asia, including Iraq, Iran, Syria, Myanmar, and Cambodia (see Figure 11.23 and "The Challenge of Migrants and Refugees").

Oceania's Shifting Global Relationships

During the twentieth century, Oceania's relationship with the rest of the world went through three phases: from a predominantly European focus, to identification with the United States and Canada, and finally to the currently emerging linkage with Asia.

Until roughly World War II, the colonial system gave the region a European orientation. In most places, the economy depended largely on the export of raw materials to Europe (see Figure 11.19E). Thus, even when a colony gained independence from Britain, as did Australia in 1901 and New Zealand in 1907, people remained strongly tied to their mother countries. Even today, the queen of England remains the titular head of state in both countries. During World War II, however, the European powers provided only token resistance to Japan's invasion of much of the Pacific and its bombing of northern Australia. This impotence on the part of Europe spurred a change in the region's political and economic orientation.

After World War II, the United States, which already had a strong foothold in the Philippines, became the dominant power in the Pacific, and U.S. investment grew more important to the economies of Oceania. Australia and New Zealand joined the United States in a Cold War military alliance, and both fought alongside the United States in Korea and Vietnam, suffering considerable casualties and experiencing significant antiwar activity at home. U.S. cultural influences became strong, too, as North American products, technologies, movies, and pop music penetrated much of Oceania.

By the 1970s, another shift took place as many of the island groups were granted self-rule by their European colonizers, with Oceania steadily drawn into the growing economies of Asia. Since the 1960s, Australia's thriving mineral (coal, oil, and gas) export sector has become increasingly geared toward supplying raw materials to Asian manufacturing industries (primarily Japan in the 1960s and China since the 1990s). Similarly, New Zealand's wool and dairy exports have gone mostly to Asian markets since the 1970s. Despite occasional backlashes against "Asianization," Australia, New Zealand, and the rest of Oceania are being transformed by Asian influences. Many Pacific islands have significant Chinese, Japanese, Filipino, and Indian minorities, and the small Asian minorities of Australia and New Zealand are growing in size (see Figure 11.19F). On some Pacific islands, Asians now constitute the largest portion of the population (42 percent in Hawaii, for example).

CHECK YOUR UNDERSTANDING

1. From where did the ancestors of Aboriginal Australians migrate to Oceania?
2. Describe the four distinct indigenous cultural regions of Oceania.
3. Why did Europeans find it hard to believe that Polynesians could navigate boats like the *Hōkūle'a* over the wide span of the Pacific and reliably find their way back and forth?
4. Compare Asia's influence in Oceania before World War II to that since World War II.
5. Give several examples of how the region's environments are threatened by modern global trends.

GLOBALIZATION AND DEVELOPMENT

11.5 Explain how Oceania's new focus on neighboring Asia is transforming patterns of trade and economic development across the region.

11.6 Explain how recent economic changes have affected traditional ways of life and human/environment relations throughout Oceania.

Globalization and Oceania's increasingly strong focus on Asia—rather than longtime connections with Europe and North America—have transformed patterns of trade and economic development across the region. These changes are driven largely by Asia's growing prosperity, which means increasing demands for resources needed for its massive production of manufactured goods.

OCEANIA'S NEW ASIAN ORIENTATION

One could say that globalization in Oceania began when the first European explorers arrived in the region, beginning the trend of influence by outsiders (primarily Europeans) on settlement, culture, and economics. More recently, the United States has exerted a powerful influence on trade and politics in the region. For the past several decades, however, globalization has reoriented this region toward Asia, which buys more than 70 percent of Australia's exports (mainly coal, iron ore, and other minerals). In 2011, China and India each purchased not only the output of mines, but also major shares of Australia's particularly high-quality coal deposits. Both countries use coal to generate energy and are trying to secure future access to high-quality coal like that in Australia, because it produces relatively lower harmful emissions. Asia also buys nearly 40 percent of New Zealand's exports (primarily meat, wood, wool, wine, and dairy products). Other products (mostly agricultural) and services, such as tourism, are obtained from islands across Oceania (**Figure 11.20**).

In addition, Asia is a major source of the region's imports. Because there is relatively little manufacturing in Oceania, most manufactured goods used there are imported from China, Japan,

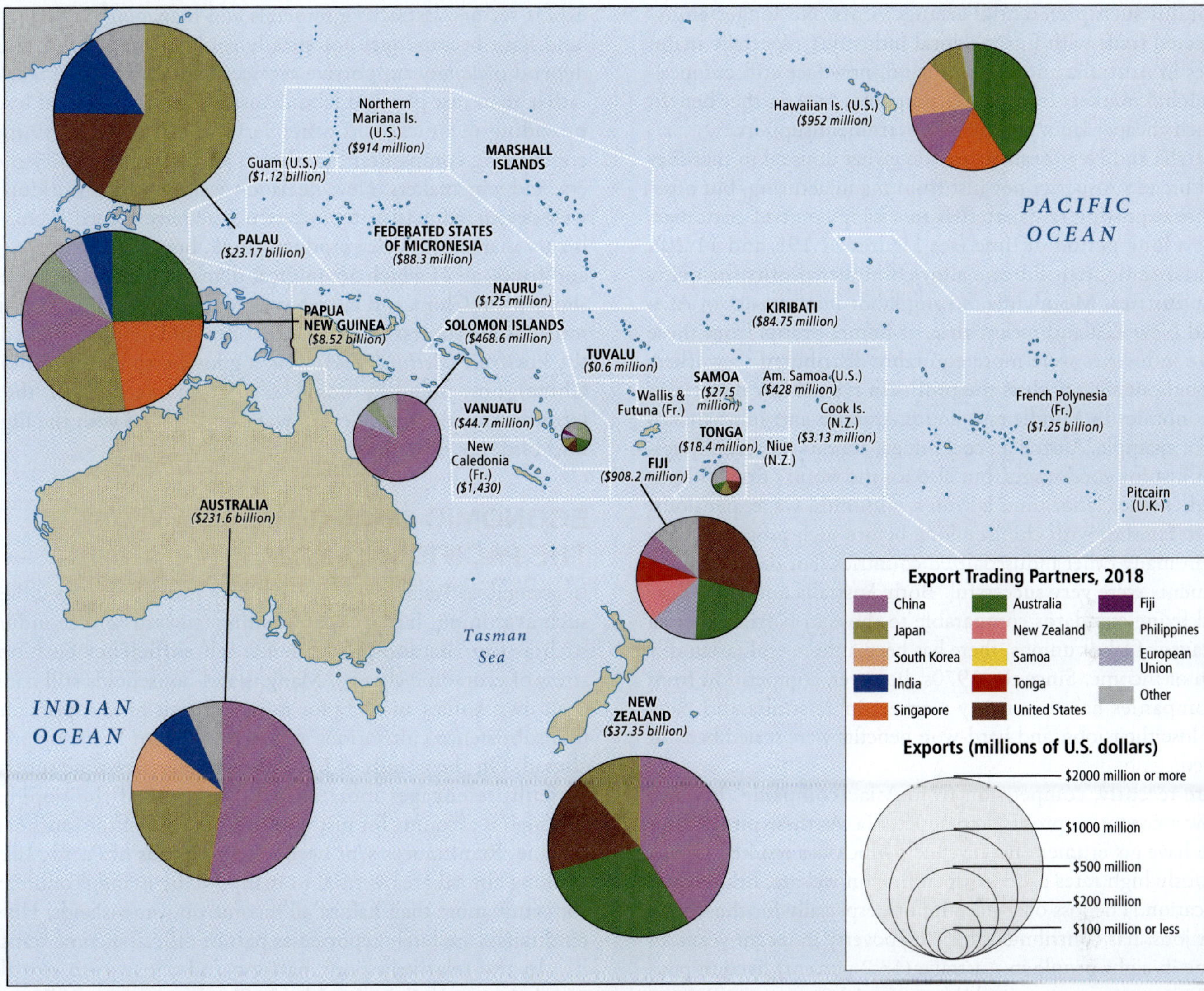

Figure 11.20 Oceania's export trading partners. The colors of each pie chart indicate a country's export trading partners. The "other" sections can include trade with Canada, Mexico, the Caribbean, non-EU Europe, sub-Saharan Africa, and other locales, some of them new trading partners. (Figures for Hawaii do not include exports to other parts of the United States.) [Data from: Central Intelligence Agency, *The World Factbook*, at https://www.cia.gov/library/publications/the-world-factbook/geos/xx.html; and U.S. Department of State, "East Asian and Pacific Affairs: Countries and Other Areas, 2018," at http://www.state.gov/p/eap/ci/index.htm.]

South Korea, Singapore, and Thailand—Oceania's leading trading partners. Both Australia and New Zealand have free trade agreements either completed or in continuous negotiation with Asia's two largest economies, China and Japan.

The Pacific islands are even further along in their reorientation toward Asia than are Australia and New Zealand. Asia's economic dominance in the islands is apparent on many fronts: not only are coconut, forest, and fish products from the Pacific islands sold to Asian markets, but Asian companies actually own more and more of these industries on the various islands. Furthermore, fleets from Asia regularly fish the offshore waters of Pacific island nations, displacing local fishers. Asians also dominate the Pacific island tourist trade, both as tourists and as investors in the tourism infrastructure. And growing numbers of Asians are taking up residence in the Pacific islands, exerting widespread economic and social influence.

ASIA'S ECONOMIC DEVELOPMENT "MIRACLE" STRESSES AUSTRALIA AND NEW ZEALAND

For Australia and New Zealand, Asia's global economic rise has generated both more trade and more competition with Asian economies in foreign markets. Throughout Oceania, local industries formerly enjoyed protected or preferential trade with Europe. They have lost that advantage because EU regulations stemming from the EU's membership in the World Trade Organization (WTO) now prohibit such preferential arrangements. No longer enjoying protected trade with Europe, local industries, especially major industries in Australia and New Zealand, now face stiff competition in global markets from large companies in Asia that benefit from much cheaper labor and major government support.

Australia and New Zealand are somewhat unusual in that they achieved broad prosperity not just from manufacturing, but especially from exporting raw materials to a wide range of customers and over a long period of time (see Figures 11.19E and 11.20). Preferential trade with Europe allowed higher profits for many export industries. Meanwhile, strong labor movements in Australia and New Zealand meant that, at home, profits from these extractive industries were more equitably distributed to workers and throughout society than the profits in typical raw materials–based economies in Middle and South America and sub-Saharan Africa. For example, Australian coal miners' unions successfully agitated not just for good wages, but also for the world's first 35-hour workweek. Other labor unions won a minimum wage, pensions, and aid to families with children long before such programs were enacted in many other industrialized countries. For decades, these arrangements were very successful. Both Australia and New Zealand had living standards comparable to those in North America and, because of labor unions, there has been a more egalitarian distribution of income. Since the 1970s, however, competition from Asian companies has seen many workers in Australia and New Zealand lose their jobs, and hard-won benefits were scaled back or eliminated.

More recently, competition from Asian companies has also led to lower corporate profits across Oceania. As these profits have fallen, so have government tax revenues, which has resulted in cuts to previously high rates of social spending on welfare, health care, and education. The loss of social support, especially for those who have lost jobs, has contributed to rising poverty in recent years. In 2018, one in eight people in Australia (13.2 percent) lived in poverty. The poverty rate among children was 17.3 percent. (Poverty rates in the United States were very similar: 12.3 percent overall and 17.5 percent for children.)

MAINTAINING RAW MATERIALS EXPORTS AS SERVICE ECONOMIES DEVELOP

Although the monetary contribution of raw materials export industries to the national economies of Australia and New Zealand remains high, these same industries are decreasing in proportion to other sectors of both countries' economies in that they now employ fewer people because of mechanization (see **Figure 11.21A**). This shift to replacing human labor with machinery has been essential for these industries to stay globally competitive with countries where living standards are lower and workers are paid far less.

MIRAB economy an economy based on migration, remittances, and an aid-supported bureaucracy

subsistence affluence a lifestyle whereby people are self-sufficient with regard to most necessities, yet have sources of cash income that they can save for travel and occasional purchases of manufactured goods

Meanwhile, the economies of both Australia and New Zealand are increasingly dominated by diverse and growing service sectors, which are also linked to the region's export sectors. Extracting minerals and managing herds and cropland have become technologically sophisticated enterprises that depend on many supportive services and an educated workforce rather than just physical labor. Australia is now a world leader in providing technical and other services not only to mining and engineering companies, but also to sheep farmers, dairy producers, and winemakers. New Zealand's well-educated workforce and well-developed marketing infrastructure have helped it break into luxury markets for dairy products (milk, butter, and cheese), meats, and fruits, all of which are in great demand by newly middle-class shoppers in China and elsewhere (see Figure 11.21B). Perhaps the most visible success has been New Zealand's global marketing of the kiwifruit (*Actinidia deliciosa*, a gooseberry native to southern China), now found in most U.S. food markets. (*Kiwi*, the slang term for anyone from New Zealand, originated with the flightless kiwi bird, not the fruit.)

ECONOMIC CHANGE IN THE PACIFIC ISLANDS

In general, as Pacific islands shift away from extractive industries, such as mining, farming, and fishing, toward service industries, such as tourism and government, self-sufficiency cushions the stress of economic change. Many island households still construct their own homes and rely for much of their food supply on fishing, subsistence cultivation, and resources sent home by migrants abroad. On the islands of Fiji, for example, part-time subsistence agriculture engages more than 60 percent of the population, although it accounts for just under 17 percent of the gross national income. Remittances sent home by thousands of Pacific Islanders working abroad are essential to many Pacific island economies and constitute more than half of all income on some islands. However, remittances are rarely reported as part of official income statistics.

In the relatively poor nations and those with less-skilled populations—the Solomon Islands, Tuvalu, and parts of Papua New Guinea, for example—conditions typify what has been termed a **MIRAB economy**—one based on migration, remittances, aid, and bureaucracy. This formula misses some useful facts: Food and shelter are often self-provided and represent a large investment of time and effort, migration does not always result in remittances, and foreign aid from former or present colonial powers supports government bureaucracies that provide crucial employment for the educated and semiskilled. Although a MIRAB economy has little potential for long-term growth because it is dependent on outside factors, these diversified sources of income mean that Islanders can be self-sufficient in terms of food and shelter while saving extra cash for travel, future education, and occasional purchases of manufactured goods. The lifestyle supported by the MIRAB system is sometimes referred to as **subsistence affluence**, meaning people are living reasonably well, due to their own production of food and other necessities.

Where there is poverty in the Pacific, it is often related to geographic isolation, which means a lack of access to information and economic opportunity. Although computers and the new global communication networks (internet and cell-phone service) are just becoming widely available in the Pacific islands, they have the potential to significantly alleviate this isolation (see Figure 11.21C).

11.21 Globalization and Development in Oceania

The map shows how people in Oceania rank on the Human Development Index (HDI). Although there are some pockets of poverty and underdevelopment in the Pacific, and there are some anomalies, such as islands crammed with refugees (Nauru, Manus), overall, the story is one of widespread realistic adjustment to a changing world—adjustment that has placed a premium on sustainability and the preservation of indigenous ways. In Australia, New Zealand, Hawaii, and Guam, there has been a shift away from hard labor on farms and in industries to high-skilled jobs in service economies. Meanwhile, Pacific Islanders emphasize sustainability and fight poverty with practical self-reliance.

THINKING GEOGRAPHICALLY

A Considering the location of the main markets for Australia's industrial products, what advantages and disadvantages of Australia's location must be considered by Australian labor unions in their negotiations with management?

B Considering its location and population size, if you owned a dairy in New Zealand, why might you consider hiring a skilled high-tech marketing expert?

C What are the likely difficulties for Pacific Islanders seeking remote tech work?

A This disused factory on Cockatoo Island was once part of Australia's largest shipyard. It is now a UNESCO World Heritage site where visitors learn about Australia's once-vibrant industrial sector, which has been shrunk by competition from Asia and Europe. [Christine Wehrmeier/Alamy]

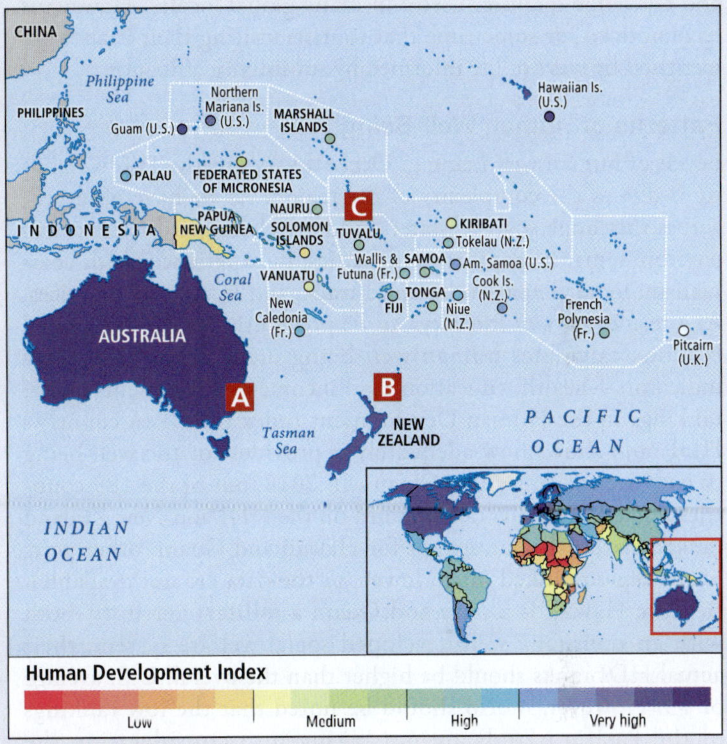

B The rapidly growing service sector in New Zealand has been particularly successful in helping upscale producers of dairy (butter, milk, cheese), meat (lamb), and fruit (wine and kiwifruit) market their products locally as well as in China, India, Europe, and the Americas. [Karel Lorier/Alamy Stock Photo]

C In the island Pacific, rapid economic change is cushioned by a strong tradition of self-sufficiency. People live lives of subsistence affluence. As a result, a low HDI rank does not mean inevitable poverty and hunger. As access to technology increases, even remote locations like Tuvalu may provide opportunities to flourish. [Dmitry Malov/Alamy Stock Photo]

Subsistence Affluence Practices Could Go Global

Those concerned with global sustainability have noted that many of the qualities of *subsistence affluence* practiced by Pacific Islanders have the potential to be expanded upon and adapted to other places. For example, local self-sufficiency based on home gardening and resource conservation strategies has been diffused to North America by Pacific Islander emigrants, and this way of life, with its low carbon footprint, matches quite closely the philosophies of Native American and other environmentally conscious groups. The low-key optimism and accepting attitude toward life that are characteristic of Oceania have also been recognized as admirable and teachable qualities useful in many global locales. Hawaiians have noticed for some time that tourists visiting their islands are intrigued by ways of life informed by subsistence affluence.

Patterns of Human Well-Being

Levels of human well-being in Oceania vary greatly. This variability is due to the complexity of the region. As we have observed, across this huge span of water and land there are differences in environments, in population densities, in experiences with colonialism, in participation in world trade, and in cultural practices. As a result, ways of life vary. As we previously noted, the United Nations calculates human well-being using three customary indicators—health, education, and income—and presents global rankings in the Human Development Index (HDI). A country's HDI rank shows how adequately it provides for the well-being of its citizens (Figure 11.21 map). In 2018, out of the 189 countries ranked, Australia ranked third on the HDI and New Zealand ranked sixteenth, but, except for Hawaii and Guam, other parts of the region ranked much lower (or the data are not available). Because Hawaii is a state and Guam a military territory, both with an unusually well-developed social welfare system, their actual HDI ranks should be higher than the overall U.S. average of 13. However, it also should be noted that the low rankings of the Pacific islands are not taking into consideration the difficult-to-measure features of subsistence affluence and the strong communitarian values of the Pacific Way (see "The Pacific Way in the Islands"). Combined, these two supportive customs can result in higher-than-expected actual well-being.

THE ADVANTAGES AND STRESSES OF TOURISM

Tourism is not generally recognized as an export sector, but it is if the tourists are coming from abroad and bringing money with them to spend on food, lodging, transport, shopping, and entertainment. Across Oceania, tourism is a growing part of economies and it is recognized as deserving special attention and regulation if it is to be a sustainable source of jobs and income. The total contribution of tourism to the gross domestic product (GDP) of the whole of Oceania in 2017 was U.S.$200 billion. This was about 12.3 percent of the total GDP for the region, and the employment supported by tourism was about 2.5 million jobs or 12.7 percent of total employment. Tourists come largely from China, Japan, Korea, Taiwan, Southeast Asia, the Americas, and Europe. In 2017 (the latest year for which complete figures are available), 26 million tourists arrived in Oceania (including Hawaii). Of those tourists, 23 percent came from all of Asia; just 11 percent came from Europe (down from 17 percent in recent years); 33 percent came from North America; 19 percent were from within Oceania; and 12 percent traveled from other locations. Australia and Hawaii are the most popular destinations, garnering about 9 million visitors each; New Zealand attracted 3.5 million in 2017. Guam is also popular, attracting 1.5 million visitors from Japan, South Korea, and China (**Figure 11.22**).

Large numbers of visitors expecting to be entertained and graciously accommodated can place a special stress on local inhabitants. In some Pacific island groups, the number of tourists far exceeds the island population. Guam, for example, receives tourists annually in numbers equivalent to seven times its population. Palau and the Northern Mariana Islands annually receive more than four times their populations. And although these visitors bring money to the islands' economies, they create problems for island ecologies, place extra burdens on scarce resources (land, water, fuel, food, and waste management systems), and expect a standard of living that may be far out of reach for local people.

Conflict over Tourism in Hawaii

Perhaps nowhere else in the region are the issues raised by tourism as clear and as acknowledged by local people as they are in Hawaii. Since the 1950s, travel and tourism have been the largest industries in Hawaii, accounting for around 22 percent of the state GDP in 2017, and tourism is related in one way or another to nearly 75 percent of all jobs in the state. (By comparison, travel and tourism account for 9 percent of GDP worldwide.) In 2017, tourism employed one out of every three Hawaiians and accounted for more than 20 percent of state tax revenues.

Nearly everywhere, the tourism sector is plagued with unpredictability: tourism is a luxury that is easily jettisoned when funds get tight. And even a tiny hint of political problems or a natural

Figure 11.22 Tourism in Oceania. The origins of the tourists reflect changing trade patterns in the region, with more and more journeying from Asia. Here, you see hotels on Honolulu's Waikiki Beach, which provide lodging for a majority of Hawaii's 9 million or more yearly visitors. Nearby is the giant Ala Moana shopping center geared to meet the tourists' shopping needs, not those of local shoppers. [Gerard SIOEN/Gamma-Rapho via Getty Images]

disaster in a host country will cause trip cancellations. Dramatic fluctuations in tourist visits, often driven by forces far removed from Oceania, can wreak havoc on local economies, affecting not just tourist facilities but supporting industries as well. For example, the construction industry thrives by building condominiums, hotels, resorts, and retirement facilities. The Asian recession of the late 1990s, the terrorist attacks of September 11, 2001, and the global recession of 2008–2009 all affected Hawaii's economy by creating dramatic slumps in tourist visits. By 2017, however, Hawaii's tourism industry rebounded: more than 9.4 million visitors arrived that year, up 25 percent over 2009.

To ordinary Hawaiians, mass tourism can sometimes feel like an invasion. For example, in 2017, Hawaii hosted seven tourists for every one Hawaiian. But to local people what was the real net benefit of all these visitors, who had to be housed, fed, entertained, and cleaned up after? This is a question Hawaiians continue to ask themselves. Furthermore, an important segment of the Honolulu tourist infrastructure—hotels, golf courses, specialty shopping centers, import shops, and nightclubs—is geared exclusively to foreign visitors, and many such facilities are owned by foreign investors. Hawaiian citizens and even some vacationers can feel out of place in facilities focused on high-end foreign consumers.

Just the demand for golf courses imposed by mass tourism has resulted in what Native (indigenous) Hawaiians view as the desecration of sacred sites. Land that in precolonial times was communally owned, cultivated, and used for sacred rituals was first confiscated by the colonial government and more recently sold to Asian golf course developers. Now the only people with access to the sacred sites are fee-paying tourist golfers.

Nevertheless, the golf industry cannot be ignored. As of 2018, there were more than 90 golf courses in Hawaii, and the golf industry alone contributed $2.5 billion to the state's economy—more than twice the amount derived from agriculture. Golf's total annual economic impact represents about 13 percent of Hawaii's tourism income, a fact that confuses the picture for those who oppose golf on environmental or cultural grounds.

The pressure to make land available for tourism of all kinds has decreased access to land by local citizens. For example, retired mainland Americans who relocate to Hawaii in search of a sunny spot—in what is often called *residential tourism*—have caused property values to rise steeply and thereby increased the costs of housing for local people.

Sustainable Tourism

Some Pacific islands have attempted to deal with the pressures of tourism by adopting the principle of **sustainable tourism**, which aims to decrease tourism's footprint and minimize disparities between hosts and visitors. Samoa, with financial aid from New Zealand, has developed sustainable tourism components (beaches, wetlands, forested island environments, and cultural attractions) and its islanders provide *knowledge-based tourism experiences* for the thoughtful traveler. These experiences are information-rich explanations of Samoa's political, social, and environmental issues. Globally there is an increasing market for knowledge-based tourism, and in parts of the European Union instruction-rich family visits to working farms has revived some rural economies (see "The Growth of Corporate Agriculture and Food Marketing" in Chapter 4).

Recognizing that unplanned tourism could leave Samoa's ecosystems vulnerable to climate change and haphazard development and overwhelm the traditional Samoan way of life, the sustainable tourism project emphasizes **human resource capacity building**—connecting Samoans with the education and information necessary to take charge of tourism's role in their communities. Visitors and tourism agencies are asked to support the project financially and collaboratively. Local people, on the one hand, must provide environmental and cultural education to visitors; on the other hand they must, themselves, receive career development and environmental management skills. For example, because islands are closed ecosystems, tourism has an enormous environmental impact through the consumption of resources, the generation of waste, and the use of scarce water. Samoans now educate their visitors on how waste can be minimized and recycled and the generation of greenhouse gases limited—knowledge the tourists can apply at home.

While Samoan tourism has the capacity to provide long-term income for the island, it must satisfy four criteria to remain sustainable: it must be financially viable by providing worthwhile jobs for local people; it must provide visitor satisfaction; it must have a strong genuine cultural component that is not perverted into a glitzy, inauthentic product; and it must be supported by the host community. Samoans and their colleagues in New Zealand hope to create a sustainability model that can serve elsewhere in the Pacific region.

THE FUTURE: DIVERSE GLOBAL ORIENTATIONS

Despite the powerful forces pushing Oceania toward Asia, important factors still favor strong ties with Europe and North America. Even with the new trade links and China's recent efforts to expand diplomatic and cultural relations with Australia, both Australia and New Zealand remain staunch military allies of the United States. Over the years, both have participated in U.S.-led wars in Korea, Vietnam, Afghanistan, and Iraq. In 2012, in an apparent effort to check the growing influence of the Chinese military in the South China Sea and the Indian Ocean, the Australian government gave the U.S. Marine Corps access to a large tract of land near Darwin (located in Australia's Northern Territory). The United States and Australia also opened discussions regarding the use of the Cocos Islands (Australian possessions in the Indian Ocean) for reconnaissance purposes.

In some of the Pacific islands, strong links to Europe and North America are also upheld by the continuing administrative control of Europe and the United States. In Micronesia, the United States governs Guam and the Northern Mariana Islands; in Polynesia, American Samoa is a U.S. territory. Just as the Hawaiian Islands are a U.S. state, the 120 islands of French Polynesia—including Tahiti and the rest of the Society Islands, the Marquesas Islands, and the Tuamotu Archipelago—are Overseas Territories of France. Any desire for independence in these possessions has, as yet, not been sufficient to override the financial benefits of aid, subsidies, and investment money provided by France and the United States.

sustainable tourism aims to decrease tourism's footprint and minimize disparities between hosts and visitors

human resource capacity building efforts to increase the understanding and skills of people so they can manage development in their own communities

The Asia Pacific Economic Cooperative (APEC), a coalition of 21 Pacific countries, was formed in 1989. Today, the member states of APEC account for approximately 40 percent of the world's population, just over 50 percent of global production, and more than 40 percent of global trade. APEC members include Oceania's most populous countries (Australia, Papua New Guinea, and New Zealand), as well as Brunei, Canada, Chile, China, Hong Kong, Indonesia, Japan, South Korea, Malaysia, Mexico, Peru, the Philippines, Russia, Singapore, Taipei, Thailand, the United States, and Vietnam. APEC was organized to enhance economic prosperity and strengthen the Asia–Pacific community, and while much is made of its potential, its inability to compel its membership to act in any significant way has led some to dismiss the group as a pointless "talk shop."

President Obama sought to design a "strategic pivot toward Asia" for the United States, in part to check the growing assertive efforts by China to be the primary leader of the countries in the Pacific region and beyond. Obama hoped to get APEC to evolve into a more potent force, perhaps to place it on a par with the European Union (after which it is partially patterned). He saw the **Trans-Pacific Partnership (TPP)** agreement, brought forward in its final negotiated form in October 2015, as a way to smooth relations between Asia, Oceania, and the United States in matters of trade, food security, climate change, product regulations, and energy-efficient transportation. But others (some among the public in the United States, Japan, and Mexico) viewed the TPP as mostly a give-away to international business that would limit competition, move jobs to low-wage Asian countries, and thus encourage higher prices for consumers. In particular, the provision that would allow multinational corporations to challenge regulations before special tribunals was intensely opposed by critics in the United States.

If approved, the TPP would have set new terms for trade between the United States and 11 other Pacific Rim countries; but when President Trump came into office he reaffirmed his "America first" stance and withdrew the United States from the TPP. After Trump withdrew, the remaining 11 signatories agreed to a modified version, the Comprehensive and Progressive Agreement for Trans-Pacific Partnership (CPTPP), which went into force December 30, 2018.

CHECK YOUR UNDERSTANDING

1. Why is Asia now playing a larger role in the affairs of Australia, New Zealand, and the Pacific islands?
2. What are some of the ways in which Asia's role is stressful for the region?
3. Describe how service industries are becoming the dominant source of income for most of Oceania's citizens.
4. What is a MIRAB economy? Does it further sustainability and contribute to subsistence affluence?
5. What is the Comprehensive and Progressive Agreement for Trans-Pacific Partnership (CPTPP) agreement? To what extent is the United States involved?

Asia Pacific Economic Cooperative (APEC) a group of 21 Pacific Rim countries organized in 1989 to increase trade and cooperation

Trans-Pacific Partnership (TPP) the largest regional trade accord in history

Pacific Way a system of political thought that emphasizes regional identity, decision making via consensus, and respect for traditional patriarchal authority

POWER AND POLITICS

11.7 Compare the system of democracy that prevails in New Zealand, Australia, and Hawaii with the Pacific Way, the political and cultural philosophy of the islands.

11.8 Describe how the global refugee crisis is testing this region's ability to maintain its reputation as a humanitarian refuge and zone of opportunity.

One of the current bones of contention between Australia and New Zealand, on the one hand, and the Pacific islands, on the other, is just what kind of government and civic participation best serve the people. This is in many ways a postcolonial debate: although both sides have had the experience of being colonies, on this particular issue, the Pacific islands are challenging the European versions of governance still largely accepted as valid by Australia and New Zealand.

PARLIAMENTARY SYSTEM IN AUSTRALIA AND NEW ZEALAND

In Australia and New Zealand, government is based on European-style parliamentary systems grounded on universal voting rights for adults, free speech, debate, and majority rule. Democratic principles influence both debate and conflict resolution from the community level to the national level, and the internal and external political issues faced by Australia and New Zealand bear a strong similarity. Both have majority populations with a European heritage and large minority populations of indigenous people—Aboriginal people in Australia and Maori (Polynesian) in New Zealand. And both are accepting immigrants from many locations. Political issues that drive internal debate in both countries include how to manage migration between the two countries and immigration from outside; how to maintain social cohesion in the face of increasing diversity; how to adjust their resource export–based economies to a greater focus on technical innovation and services; and how to reformulate international relationships as the region pivots from strong ethnic and economic connections to Europe to closer associations with Asia.

THE PACIFIC WAY IN THE ISLANDS

By contrast, the Pacific islands prefer the **Pacific Way**, which is based on traditional notions of power and problem solving and refers to a way of settling issues familiar to many Pacific Islanders. It was developed in small communities where decisions are reached in face-to-face meetings, and consensus and mutual understanding are favored rather than open confrontation and majority rule. High value is placed on respect for traditional leadership (especially the usual patriarchal leadership in families and villages) rather than on political freedoms such as free speech and individuality. As such, the Pacific Way can embody very different, but not invalid, definitions of fairness and corruption than do parliamentary systems.

The Pacific Way in Fiji

As a political and cultural philosophy, the Pacific Way was first articulated as a formal concept in Fiji around the time of Fiji's independence from the United Kingdom in 1970. It subsequently became popular in many other Pacific islands, most of which gained independence in the 1970s and 1980s.

The Pacific Way carries a flavor of postcolonial resistance to Europeanization and has often been invoked to uphold the notion of a regional identity shared by Pacific islands that grows out of their unique history and social experience. It was particularly influential among educators given the task of writing new textbooks to replace those used by the former colonial masters. The new texts focused students' attention away from Britain, France, and the United States and toward their own cultures. Appeals to the Pacific Way have also been used to uphold attempts by Pacific island governments to control their own economic development and solve their own political and social problems.

The Value of the Pacific Way to Grassroots Sustainability

Faulted by outsiders as undemocratic, closer inspection may render this judgment unduly harsh. The Pacific Way is likely to endure, especially as a concept that upholds Pacific regional identity and traditional culture. Furthermore, some organizations now use the Pacific Way as the basis for an integrated approach to economic development and the resolution of environmental issues. For example, the Secretariat of the Pacific Regional Environmental Programme (SPREP) builds on traditional Pacific island economic activities—such as fishing and local traditions that require knowledge and awareness of the environment—to promote grassroots economic development and environmental sustainability. The strategic focuses of SPREP are climate change, biodiversity, and environmental policy design. Implementation of Pacific Way procedures by SPREP in policy development has the reputation for involving the public more effectively than so-called democratic approaches. Because a high value is placed on **consensus**, all parties must adjust their points of view to reach an agreement.

In managing political conflict, however, the Pacific Way has proven to have a downside, at least from the perspective of those favoring Western-style democracy. The Pacific Way has occasionally been invoked as a philosophical basis for overriding democratic elections if the election results challenge the power of indigenous Pacific Islanders. In 1987, 2000, and 2006, indigenous Fijians used the Pacific Way to justify coups d'état against legally elected governments (**Figure 11.23C**). All three of the overthrown governments were dominated by Indian Fijians, the descendants of people from India whom the British brought to Fiji more than a century ago to work on sugar plantations.

Fiji's population is now about 58 percent indigenous Fijian and 37 percent Indian Fijian. Indigenous Fijians are generally less prosperous and tend to live in rural areas and villages where community affairs are still governed by traditional chiefs. Indian Fijians, by contrast, now hold significant economic and political power, especially in the urban centers and in areas of tourism and sugar cultivation. In response to the coups, many Indian Fijians left Fiji, resulting in a loss of badly needed skilled workers that has slowed economic development.

Across Oceania, political responses to the Fiji coups were divided. Australia, New Zealand, and the United States (via APEC and the state government of Hawaii) saw the coups as illegal and demanded that the election results stand and the Indian Fijians be returned to office. But much of Oceania has cited the tenets of the Pacific Way in arguments supporting the coup leaders. As in Fiji, those who govern many of the Pacific islands are leaders of indigenous descent who have not always had the strongest respect for political freedoms, especially when their hold on power is threatened. Their decisions have at times upheld traditional Pacific values such as stability, respect for authority, and certain kinds of environmental awareness, even though these decisions have also contributed to corruption and civil disorder (see Figure 11.23C).

The coups resulted in Fiji being suspended from the Commonwealth of Nations (a union of former British colonies) for subverting majority-rule democracy. As a result, it is ineligible for Commonwealth aid and was not allowed to participate in Commonwealth sports events until 2014. Because sports play such a central role in Pacific identity (see "Sports"), this latter sanction carried significant weight.

THE CHALLENGE OF MIGRANTS AND REFUGEES

Legal migrants with demonstrable skills are welcomed every year to Australia and New Zealand in rather large numbers. In 2015, Australia's net immigration was about 184,000 and New Zealand's net immigration was about 50,000 (around 15 percent of these migrants are simply moving between Australia and New Zealand). Both countries have a high proportion of foreign-born residents. In Australia, about 30 percent of its 24 million people are foreign-born, and in New Zealand, approximately 20 percent of its 4.9 million were born abroad (in the United States, only 13.7 percent were foreign-born in 2017, perhaps 16.7 percent, if illegal immigrants are included). Both Australia and New Zealand are just now revising their immigration rules, and both actively encourage immigration by professional workers from any country. Of those from outside Oceania, most are emigrating from India and China, with smaller percentages from the United Kingdom and Southeast Asia. Australia firmly contends that it has the right to decide who legally immigrates to its soil.

Unfortunately, ethnic and religious hostilities in places far from Oceania have unexpectedly exposed fissures in the region's reputation for generous and humane migration policies. Once the "white only" policies of the earlier twentieth century were dropped by the 1970s, far more humane policies were adopted. However, by the 2000s, ruthless human traffickers, sensing that Australia and New Zealand could be counted on to accept people fleeing Earth's many conflict zones—Afghanistan, Pakistan, Syria, Iraq, Iran, Myanmar, Cambodia—offered to take people in rickety boats to the coasts of these two countries for outrageously high fees.

In the case of **refugees** and **asylum-seekers** (both defined as different from ordinary immigrants) the two countries had compassionate policies. In 2015, New

consensus a strategy for achieving broad group unanimity, not based on majority rule, but on all parties adjusting their desires to fit those of the entire group

refugees people fleeing armed conflicts or persecution who are protected by international law (1951 Refugee Convention)

asylum-seekers refugees whose claims have not yet been evaluated—usually they are seeking protection from persecution on account of race, religion, nationality, or politics

Power and Politics in Oceania

The map shows the Democratization Index for the region. Australia, New Zealand, Hawaii, and most island groups are judged to be full democracies; Fiji is a hybrid regime; and New Guinea is a flawed democracy. Political culture in this region varies significantly between the more Europeanized areas of Australia, New Zealand, and Hawaii, and the more indigenous and traditional Pacific islands. Recently, the reputation of Australia and New Zealand, especially their support for humanitarian concerns, has made these countries the destination of refugees fleeing poverty and persecution in Earth's trouble spots.

THINKING GEOGRAPHICALLY

A **What was the colonial rationale, in place until 1993, that held the Aboriginal Australians had no prior claim to land in Australia?**

B **When did Australia's whites-only immigration policy formally end?**

C **In Fiji, three coups d'état carried out by indigenous Fijians have removed legally elected governments headed by which ethnicity?**

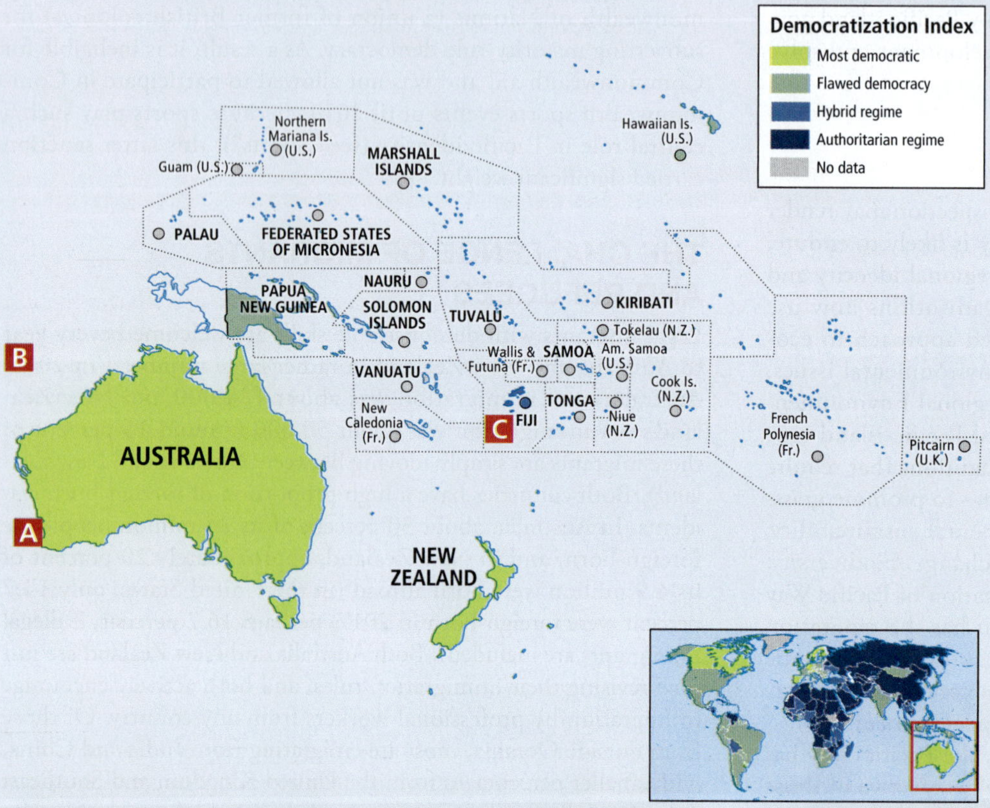

A In the 1990s Corrie Bodney, an elder of the Ballaruk Aboriginal tribe, staged a sit-in at Perth International Airport, which is located on land the Ballaruk occupied for thousands of years prior to colonization by the British. In 2013, the government of Western Australia offered several Aboriginal tribes more than U.S.$1 billion to settle a larger claim, which included the entire city of Perth. [THE WEST AUSTRALIAN/AFP/Getty Images]

B A funeral at an immigration detention facility on Christmas Island, Australia, for asylum-seekers from Iraq who drowned trying to reach Australia by boat. Despite Australia's immigrant origins, immigration is a hot-button issue. Particularly controversial is the *Pacific Solution*—a policy of paying islands, such as Nauru and Papua New Guinea, to take in asylum-seekers brought by smugglers' boats. They may wait for years in prison-like facilities to be judged worthy of asylum. Some Australians believe that refugees should be accepted more quickly; others see them as a security risk and economic drain. [Mark Kolbe/Getty Images]

C A Fijian soldier during the military coup of 2006, which overturned the fair election of officials who were Indian Fijian. Military takeovers by indigenous Fijians took place in 1987, 2000, and 2006. [WILLIAM WEST/AFP/Getty Images]

Figure 11.24 Camp for refugees. Undocumented asylum-seekers on Nauru await processing of their asylum claims. The wait is 4 years or more for some. [EPA/EOIN BLACKWELL/european pressphoto agency b.v./Alamy]

Zealand was prepared to resettle 750 refugees, Australia (a much larger country) would take 13,750, and both were prepared to help such refugees adjust to life in a new place (especially those Pacific Islanders fleeing the effects of climate change). Nonetheless, the huge and unexpected influx of more than 30,000 impoverished and desperate people from distant regions far exceeded what the public in the two countries was willing to accept. Both countries were already dealing with right-wing political objections to cultural diversity, as shown by the massacre in two New Zealand mosques in 2019, when, like the EU, they were faced with thousands more refugees than anticipated from Syria, Iraq, Iran, Myanmar, and elsewhere.

As Australia and New Zealand struggled to find a way to address the overflow and simultaneously maintain their own social cohesion, they devised the controversial **Pacific Solution** to handle undocumented asylum-seekers—a solution that, even when it was first proposed in 2001, was regarded by the international community as racist and inappropriate. Similarly viewed in the present, the Pacific Solution mandates that undocumented asylum-seekers—those who have exceeded both countries' limited quotas—be held in detention centers in Nauru and on Manus, an outlying small island off Papua New Guinea (**Figure 11.24**). Australia and New Zealand claim that basic food and shelter are supplied and basic education is provided for the children, but independent journalists documented filthy conditions and brutal treatment. For the adults, who range from physicians and teachers and technologists to unschooled farmhands, there is little to do and no future to look forward to. In April 2016, the Supreme Court of Papua New Guinea declared the Manus detention center (then holding more than 1300 people) illegal and subject to closure. Australia remains intransigent and the Manus camp was still open in 2019; but public outcry after rumors of increasing refugee suicides has resulted in many of the imprisoned, especially children, being brought to Australia for medical and mental health treatments. This effort to alleviate the situation became a major political football in Australia by 2019.

CHECK YOUR UNDERSTANDING

1. Explain the contrasting political philosophies of parliamentary democracy and the Pacific Way.
2. How might the consensus feature of the Pacific Way facilitate public participation in policy formation even better than democratic procedures do?
3. What is the role of the Pacific Way in Fiji's recent political history?
4. Why is Australia's Pacific Solution to address the overflow of impoverished asylum-seeking refugees so highly controversial?

URBANIZATION

11.9 Explain patterns of urbanization in this lightly populated region.

Oceania is following the global trend of migration from the countryside to cities; overall, 67 percent of the population now lives in urban areas. Australia and New Zealand have among the highest percentages of city dwellers outside Europe. More than 86 percent of Australians live in a string of cities along the country's relatively well-watered and fertile eastern and southeastern coasts. Similarly, 86 percent of New Zealanders live in urban areas. The vast majority in these two countries dwell in modern comfort, work in a range of occupations typical of highly industrialized societies, and have access to tax-supported health-care, education, and leisure facilities (**Figure 11.25A**). Vibrant, urban-based service economies employ about three-quarters of the population in both countries. Declining employment in mining and agriculture, where mechanization has dramatically reduced the number of workers needed, has also contributed to urbanization.

Throughout the Pacific, urban centers have transformed natural landscapes. In some small, densely populated countries, such as Guam, Palau, Nauru, and the Marshall Islands, urban landscapes dominate. The concentration of people can mean that such cities are places of opportunity, but, when poverty is widespread, they also can be sites of social conflict and environmental hazards (Figure 11.25C).

Most Pacific island towns and all the capital cities are located in ecologically fragile coastal settings like those shown in Figures 11.25C and D. Many of these waterfront towns were established during the colonial era as ports or docking facilities and were situated in places suitable for only limited numbers of people. Consequently, little land is available for development and access to housing is inadequate. Squatter settlements have been a visible feature of island urban areas for decades. The coastal locations for towns and cities in Oceania make these settlements particularly vulnerable to rising sea levels and violent storms associated with climate change.

Urbanization has enhanced the complexity of multiculturalism across the region. Many new urban residents are letting go of the rural ways of their childhoods, as well as their ethnic identity and cultural commitments. As time goes on, more urban people are

Pacific Solution Australia and New Zealand's controversial policy to pay nearby islands to indefinitely hold the excess number of undocumented immigrants in tent communities

PHOTO ESSAY

11.25 Urbanization in Oceania

There are two patterns of urbanization in Oceania. Australia, New Zealand, and Hawaii are very urbanized places and have high standards of living; Papua New Guinea and many Pacific islands have high rural densities, but their cities tend to have low standards of living and be characterized by coastal shantytowns. [Data from: Population Reference Bureau, *2018 World Population Data Sheet*, at http://www.prb.org /pdf11/2018population-data-sheet_eng.pdf.]

THINKING GEOGRAPHICALLY

A Describe the stereotypes about Australia that leave some surprised to learn that nearly 90 percent of its population lives in cities.

C and **D** What best explains why Pacific island cities have far lower standards of living than Australia, New Zealand, and Hawaii?

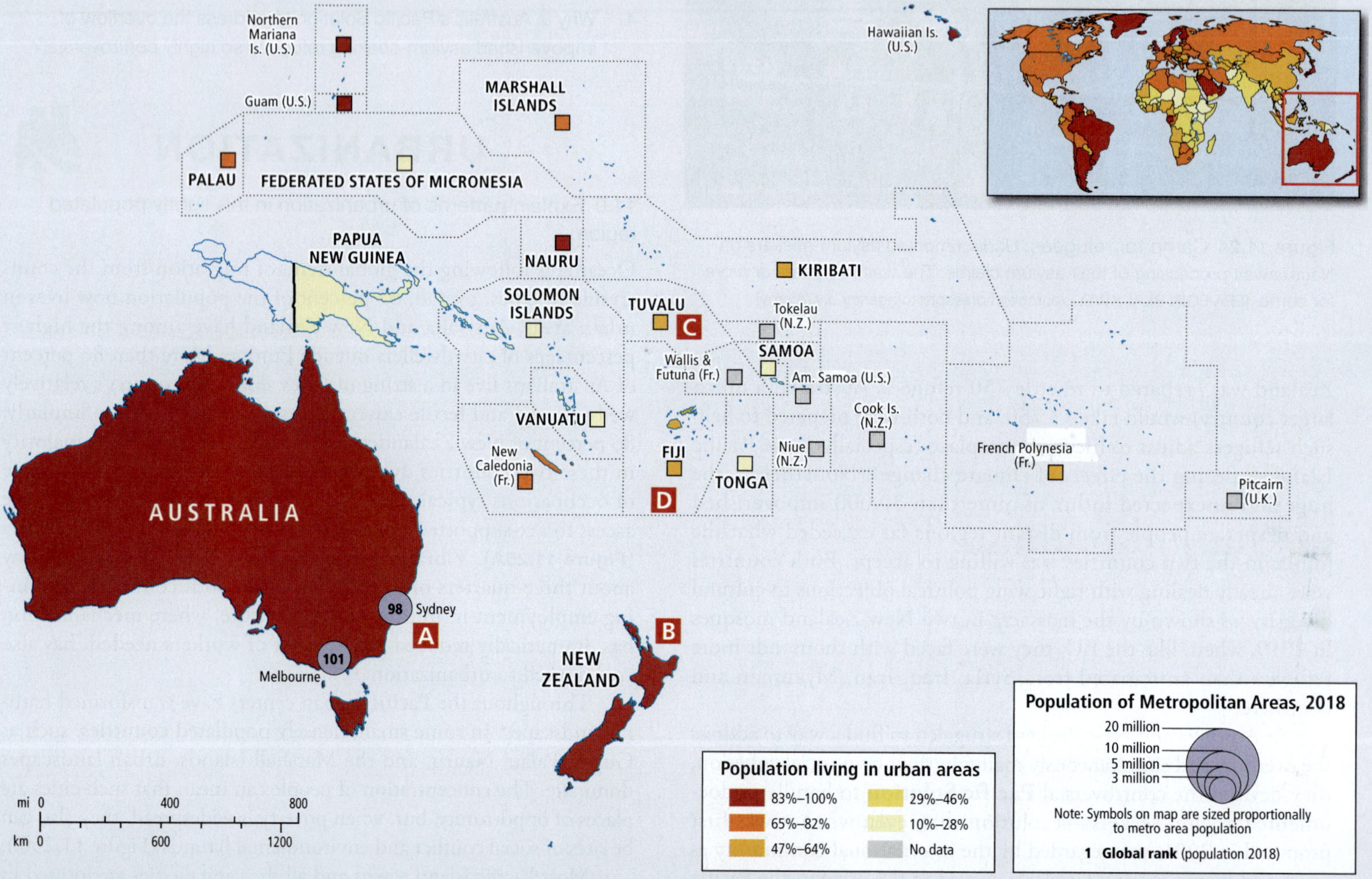

A Tourists climb the Harbor Bridge in Sydney, Australia, the largest city in Oceania. Sydney is consistently ranked among the most livable cities in the world, along with Melbourne and Perth, Australia, and Auckland, New Zealand. [TORSTEN BLACKWOOD/AFP/Getty Images]

B A New Zealand child sings and dances to music during a free concert on Christmas Day in a public park in Auckland. [Sandra Mu/Getty Images]

C Children wade through garbage during high tide at Funafuti Atoll, capital of Tuvalu. Tuvalu is very densely populated, with 4847 people per square mile (1871 per square kilometer), but the dense settlements don't really function as cities. It is one of the poorest nations in Oceania, with a GNI (PPP) per capita of about U.S.$5780. [TORSTEN BLACKWOOD/AFP/Getty Images]

D A 14-year-old indigenous Fijian girl sits in a rural squatter settlement outside Suva, the capital. GNI (PPP) in Fiji is U.S.$9000, about one-fifth of that in Australia, and access to health care and education is also low. [Alex Ellinghausen/The Sydney Morning Herald/Fairfax Media via Getty Images]

marrying across ethnic divisions, having only one or two children, and creating new patterns of social alliances and networks. Such cultural blending can ultimately result in enhanced social cohesion, but the result can also be new social tensions, as the very nature of traditional Pacific social life changes to relationships that are less intimate, less grounded in family and clan. There is some evidence that urban unemployment and unrest are on the rise, and low rates of economic growth restrict the revenue available to governments to manage urban development. Furthermore, children like those pictured in Figures 11.25C and D may have been born into a newly urbanized family living in a shantytown, and they may have never known the advantages of life on a rural Pacific island such as enjoyed by the children in the photo at the beginning of this chapter.

CHECK YOUR UNDERSTANDING

1. Describe the process of urbanization in Oceania. When did it start? How much of the population lives in cities?
2. Where is the urbanization trend the weakest?
3. Name some of the most urbanized places in Oceania. In what ecological zone are most cities located in Oceania? Why are these settings fragile?
4. Why have ethnic identity and cultural commitments decreased among urban residents in Oceania?

POPULATION, GENDER, AND CULTURE

11.10 Describe the population trends in different areas of Oceania.

11.11 Contrast gender roles in Australia, New Zealand, and Hawaii with those of the Pacific islands and Papua New Guinea.

11.12 Describe the indigenous cultures of the region and efforts to create unity across the region.

The themes of population, gender, and culture intertwine in unique ways in this region. Though few people live in Oceania, they are distributed very unevenly. Population growth is slowing and this is in large part due to changing gender roles, with women having more life choices than in the past. Women's choices are expanding because of changes in culture, that broad catchall that encompasses how people live their lives, and culture is changing for a wide range of reasons that stretch from local to global circumstances. Where populations are noticeably aging, immigration is needed, but often feared for cultural reasons.

POPULATION NUMBERS AND DENSITIES

Although Oceania occupies a huge portion of the planet, its total population is only 41 million people, close to that of the state of California (40 million). The people of Oceania live on a total land area slightly larger than the contiguous United States but spread out in bits and pieces across an ocean larger than the Eurasian landmass (see **Figure 11.26**). The Pacific islands have nearly 4.73 million people (including Hawaii's 1.43 million); of the bigger population centers, Australia has 24.1 million, Papua New Guinea has 8.5 million, and New Zealand has 4.9 million.

arable land suitable for growing crops

Population densities in Oceania vary widely. In Australia there are just 7.8 people per square mile (3 per square kilometer) for the country as a whole; and if only **arable land** (suitable for growing crops) is counted, there are about 130 per square mile (50 per square kilometer). For comparison, the United States has 380 per square mile (215 per square kilometer) of arable land. New Zealand's arable land density is quite a bit higher, at 2064.4 people per square mile (830 per square kilometer). In the Pacific islands, some are uninhabited or only sparsely settled, while others are extremely densely populated. The most densely populated islands are some of the smallest, poorest, and lowest in elevation, which means their people are dangerously exposed to rising sea levels (see Figure 11.25C). For example, the very low Solomon Islands, in the eastern Pacific, and Palau, in Micronesia, have 8866 and 4625 people, respectively, per arable square mile (3421 and 1782, respectively, per square kilometer).

CHECK YOUR UNDERSTANDING

1. What about its population features sets Oceania off from other world regions?
2. Characterize the population densities of Australia, New Zealand, and Hawaii and compare them with the rest of Oceania.

CONTRASTING POPULATION AND GENDER PATTERNS

As is the case in many regions of the world, population growth is slowing in Oceania, and in Oceania as elsewhere, important factors influencing this slower growth are related to gender. Everywhere on Earth, fertility rates tend to mirror education and employment opportunities for women. In Australia, New Zealand, and Hawaii, where opportunities for women have improved and they experience a widening range of life choices, fertility rates are low at 1.7–1.9, well below replacement rates. If this continues, populations will shrink. In the islands, where opportunities for women are restricted to traditional roles, child-rearing is the main option for women and fertility rates range between 2.0 and 4.1; these rates will result in population growth and are found in places that are already densely populated and poor. Nonetheless, even these rates are lower than they were in 2010. Change is coming.

While there is a trend toward more gender equality throughout Oceania, gender inequality persists in some places more than others. Where women are gaining political and economic power, as is the case in Australia, New Zealand, Guam, and Hawaii, gender inequality in pay and opportunity is decreasing. On the other hand, in Papua New Guinea and the Pacific islands, where women are generally not politically active and where they tend to take non-leadership roles in community affairs and hence have little influence on public policy, gender inequality is more apparent, but, nonetheless, change is on the horizon.

A closely related factor affecting population patterns is the age of the populations. On some of the poorer Pacific islands, close to 40 percent of the population is under age 15. So even if people decide to limit fertility, populations are likely to grow because a large

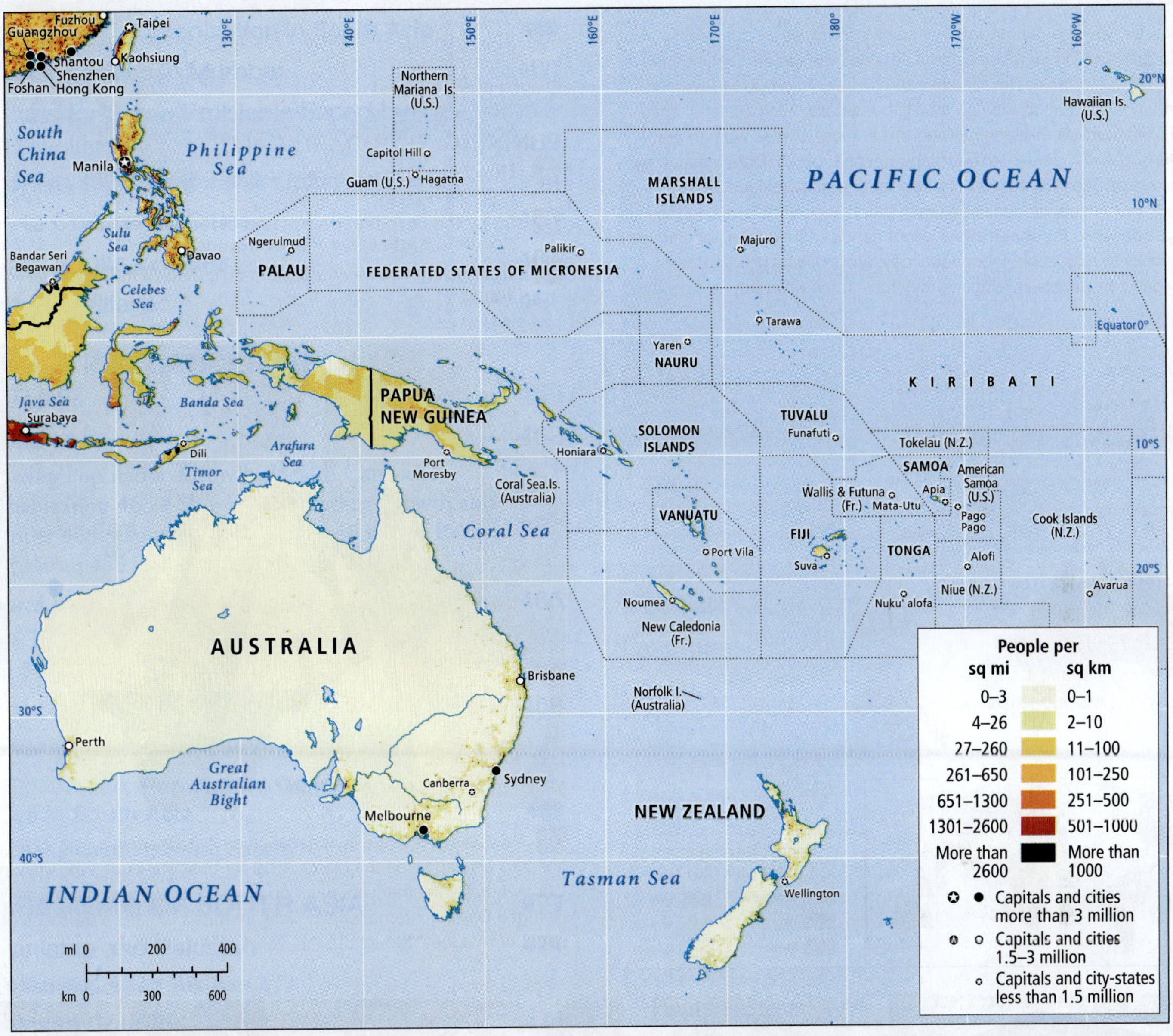

Figure 11.26 Population density map of Oceania. Population density on this map is calculated for total area, much of which (especially in Australia, but also on many islands) is not usable for agriculture (arable) or for dwelling. Population growth is slowing throughout Oceania as more people move to the cities, where health care is better (hence infant mortality lower) and women are more likely to pursue careers—factors that make large families less likely. Australia, New Zealand, Guam, and Hawaii are furthest along in the urbanization process, with smaller families and increasingly older populations. Most Pacific islands still have high growth rates, are relatively densely populated, and are relatively poor. The size of the region and the smallness of most islands make showing density difficult on printed maps.

proportion is just reaching reproductive age. This is not the case in Australia, New Zealand, and Hawaii, where just 19 percent is under age 15.

In Australia, New Zealand, and Hawaii, women's access to jobs and policy-making positions in government has improved, particularly over the last few decades. Even though New Zealand and the Australian province of South Australia were among the first places in the world to grant European women full voting rights (in 1893 and 1895, respectively), women were not very active politically; recently, however, New Zealand elected a succession of three women prime ministers, and in 2010, Australia also elected a woman prime minister. Moreover, according to the 2018 Inter-Parliamentary Union report on women in lower houses of parliament, in both countries, the proportion of women in national legislatures (38 percent in New Zealand, 28.7 percent in Australia) is well above the global average of 22 percent.

In Papua New Guinea and the Pacific islands, women generally have little political and economic power. No woman has been elected to a top-level national office, and women are a tiny minority in national legislatures when they are present at all. The one exception is Fiji, where in 2018, 10 women were elected to Parliament (women now constitute 20 percent of the delegates), perhaps indicating that, after years of military dominance and political disruption, real change is underway in Fiji (see **Figure 11.27**). Lenora Qereqeretabua, elected

PHOTO ESSAY

11.27 Gender, Population, and Culture in Oceania

The Gender Development Index (GDI) ranks countries by comparing the HDI rate for women (access to health care, education, and income) to the HDI rate for men. Hawaii ranks the highest (in group 1)—women there are nearly equal to men on the HDI. Australia, New Zealand, and Tonga all rank medium-high (in group 2)—in those places women are between 2.5 and 5 percent lower than men on the HDI. Data are lacking for the rest of Oceania, so we must rely on anecdotal evidence. In Oceania, discrimination against women has taken many forms: in the colonial era (when European ideas about gender dominated), women had little access to education, jobs, equal pay, and political power. Before Europeans came, gender relations may have been different, as discussed in "Gender Myths and Realities." This photo essay shows how women's roles are expanding as they gain access to political power and as men change their attitudes.

THINKING GEOGRAPHICALLY

A How has women's access to political power in New Zealand changed in recent years?

B Make the case that basket and cloth weaving are not solely decorative arts in the Pacific islands.

C How would you evaluate the man's comment on gender in Tonga?

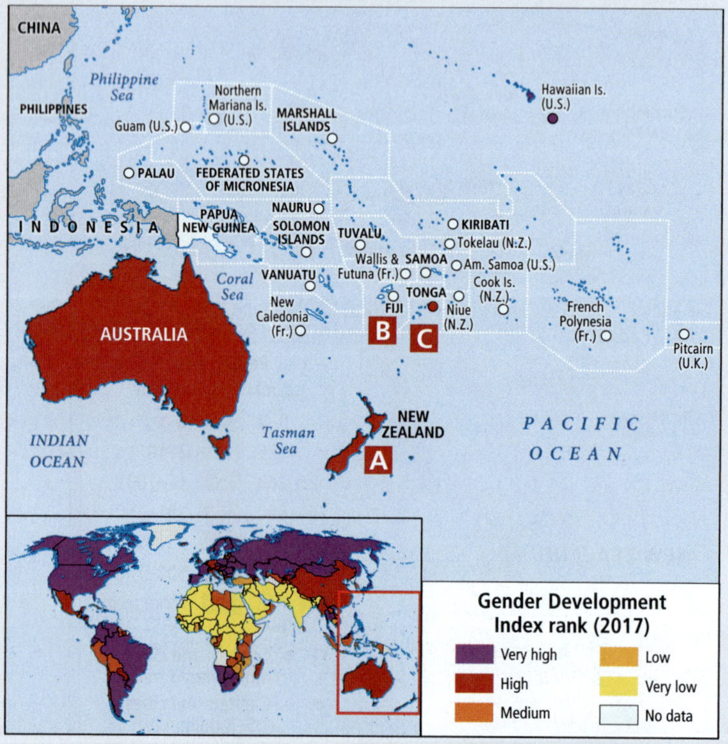

A New Zealand Prime Minister Jacinda Ardern speaks with children who survived the March 2019 terrorist attack at a mosque in Christchurch. She met with them on June 20, 2019, which is World Refugee Day, an international holiday intended to raise awareness of the plight of refugees throughout the world. [Phil Walter/Getty Images]

B Women in the Pacific islands have long had important, but underappreciated, roles as the makers of essential or decorative materials for Pacific households. Among the dominant houseware crafts are basket weaving and the making of textiles. Women first gather and prepare the materials from nature and then weave. Here two women make tapa cloth, a traditional decorative fabric. [Bhaskar Krishnamurthy/Superstock]

C Open expressions of masculinity are important in Tonga, but attitudes about male domination are changing. Here traditional male dancers perform an aggressive dance for King Tupou VI at the Royal Coronation Luncheon. But recently a group of Tonga men met to discuss how to end forms of violence against women. One man queried, "Why do we test women's virginity ('api') before marriage and we don't care about the man's sexual history? He could have several sexual relationships and children before marriage but no one cares about that. . . Our expectations and testing of a woman's virginity [while] not having any expectations on the man is unfair on the woman—is this aspect of culture still required?" [Edwina Pickles/Fairfax Media/Getty Images]

to the Fiji Parliament in 2018, is now being mentioned as a possible future prime minister. As an indigenous Fijian who ran under the Pacific Way banner, she expresses camaraderie with women Indian Fijian colleagues, which bodes well for future ethnic relations in this troubled country.

In both Australia and New Zealand, young women are pursuing higher education and professional careers and postponing marriage and childbearing until their thirties. (This is also a trend in Hawaii, Guam, and the islands with a French affiliation.) Nonetheless, in most of island Oceania it is assumed that women will be housewives. For example, the expectation is that women, not men, will interrupt their careers to stay home to care for preschool or elderly family members.

Gender Pay Gaps

Pay and employment opportunities are also factors in the choices women make. In Australia, women receive on average 15 percent less pay than men for equivalent work; and in New Zealand, about 8 percent less. These gender pay gaps, however, are smaller than in many other developed countries (in the United States women receive 18 percent less). For years women were only found in the lower levels of civil service in both Australia and New Zealand, but now opportunities for supervisory positions for women are opening up. In December 2018 New Zealand announced a major increase in these positions: now 52 percent of the executives in public service jobs are women (**Figure 11.28**). New Zealand has gone from being a laggard on women's rights to being a leader.

In the island Pacific, data on gender pay gaps are less available in part because economies are informal, with few records kept and actual payment often made in trades (barter) of garden produce or craft items or neighborly help. In the Pacific, gender roles and relationships are grounded in traditional cultural values, and roles are expected to change over the course of a lifetime. Because of the emphasis on community rather than the individual, male and female Pacific Islanders form mating unions and raise children and contribute informally to family finances and other assets. Cash or careful accounting is used only for transactions such as paying formal bills or children's school fees. Traditionally, men are the boat builders, navigators, deepwater fishers, and house builders. On some islands, men are also the usual preparers of food, while women often supply many of the ingredients through their gathering, fishing, shell-fishing, and cultivating efforts (**Figure 11.29**).

Figure 11.28 Naomi Ferguson is one of the new executives in public service in New Zealand. She is commissioner and chief executive of inland revenue. [RNZ/Philippa Tolley]

Figure 11.29 A woman fishing in Fakarava, an atoll in the Tuamotu archipelago of French Polynesia. [AP Images/Sergi Reboredo]

Most traders in marketplaces are women, and, aside from fish, the items they sell are also caught or made, and transported by women. Many young women today first fulfill traditional roles as mates and mothers and practice a wide range of domestic crafts, such as weaving and basketry (see Figure 11.27B). In middle age, however, these same women may return to school and then take up careers in government service or private enterprise. With the aid of government scholarships and the active support of women friends and relatives, some Pacific Island women pursue higher education or job training that takes them far from the villages where they raised their children. Thus, the expectation that Aurora (in this chapter's opening vignette) will study far from home and then support her family and elders is in line with evolving gender roles in Pacific ways of life. Aurora may decide to have a family that will divert her into domesticity for a few years, but she can expect that as her credentials and job experiences accumulate, she will enjoy a position of considerable power in her community, an honor typically accorded to only elderly women in the past.

CHECK YOUR UNDERSTANDING

1. Why do you think it is significant that New Zealand and the Australian province of South Australia were among the first places in the world to grant European women full voting rights (in 1893 and 1895, respectively)?
2. Why would opportunities for education and professional careers for women result in lower birth rates?

3. Assess the present situation for young women in various parts of the Pacific.
4. Describe the probable life course of a modern Pacific island young woman. Is she likely or unlikely to go on to higher education or training?

SOCIOCULTURAL ISSUES

The cultural sea change in Oceania, away from Europe and toward Asia and the Pacific, has been accompanied by new respect for indigenous Pacific peoples. Also, increasing economic interdependence with Asia has diminished historic discrimination against Asians in this region.

Ethnic Roots Reexamined

Until very recently, most people of European descent in Australia and New Zealand thought of themselves as Europeans in exile. Many considered their lives incomplete until they had made a pilgrimage to the British Isles or the European continent. In her book *An Australian Girl in London* (1902), Louise Mack wrote: "[We] Australians [are] packed away there at the other end of the world, shut off from all that is great in art and music, but born with a passionate craving to see, and hear and come close to these [European] great things. . . ."

These longings for Europe were accompanied by racist attitudes toward both indigenous peoples and Asians. Most histories of Australia written in the early twentieth century failed to even mention the Aboriginal people, and later writings described them as amoral and troublesome. At midcentury, there were numerous projects to take Aboriginal children from their parents and acculturate them to European ways in boarding schools known for abuse and brutality. From the 1920s to the 1960s, whites-only immigration policies barred Asians, Africans, and Pacific Islanders from migrating to Australia and discouraged them from entering New Zealand. As we have seen, trading patterns in that era further reinforced connections to Europe.

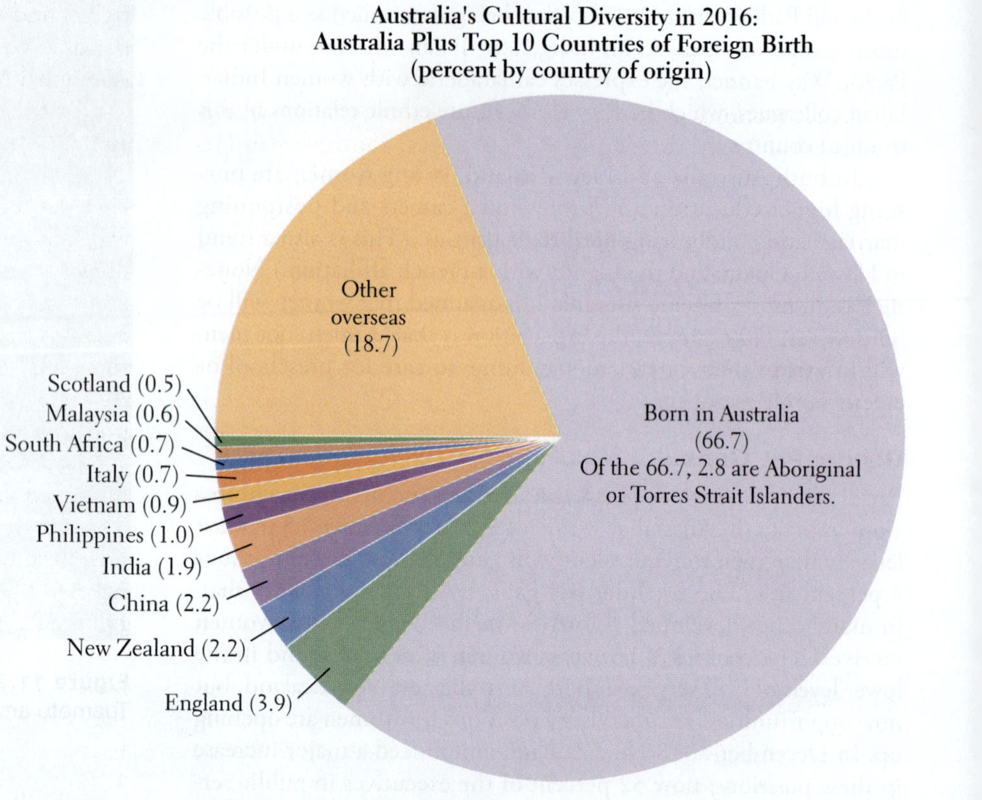

Figure 11.30 Australia's cultural diversity in 2016. Australia and New Zealand are both among the most ethnically diverse countries on Earth. Here is a pie chart of Australia showing the diversity of its population by country of birth as revealed by the 2016 census. [Data from: Australian Bureau of Statistics, 2016 census.]

Weakening of the European Connection in Australia When migration from the British Isles slowed after World War II, both Australia and New Zealand began to lure immigrants from southern and eastern Europe, many of whom had been displaced by the war. Hundreds of thousands came from Greece, Italy, and what was then Yugoslavia. The arrival of these non-English-speaking people initiated a shift toward a more multicultural society. Eventually, the whites-only immigration policy was abandoned and people began to arrive from many places. There was an influx of Vietnamese refugees in the early 1970s during the frantic exodus that followed the U.S. withdrawal from Vietnam. More recently, skilled workers from India, China, and elsewhere in Asia have been helping to meet the growing demand for information technology (IT) specialists throughout the service sector.

Both countries pride themselves on being "multicultural societies," yet a gap between this proud self-image and reality was revealed by the Australian Human Rights Commission in 2018, when it noted that 95 percent of Australian senior leaders in all walks of life were still European in background and 94 percent of Parliament is of European heritage.

As of 2016, 28.6 percent of the Australian population was foreign-born. The fastest-growing group was from India. Nevertheless, while new immigration policies are increasing the numbers of immigrants from China, Vietnam, and India, people of all Asian ancestries combined (including those born in Australia) remain a small percentage of the total population in both Australia and New Zealand. In 2016, 70 percent of Australians were of European ancestry, 2.8 percent were of Aboriginal or Torres Island descent, and 8.4 percent of Asian descent (**Figure 11.30**). In New Zealand in 2016, 74 percent were European, 15 percent were of Maori descent, and 11.8 percent were Asian. Although Europeans are slowly decreasing as a percentage of the population in both Australia and New Zealand, because of low birth rates and high immigration rates, Europeans are still projected to constitute two-thirds or more of both countries' populations by 2021.

The Social Repositioning of Indigenous Peoples in Australia and New Zealand Perhaps the most interesting population feature in Australia and New Zealand is that the number of people in both countries who claim indigenous origins is increasing. Between 1991 and 1996, the number of Australians claiming Aboriginal origins rose by 33 percent. By 2009, the Aboriginal population was estimated at 528,600. In New Zealand, between 1991 and 2009, the number claiming Maori background rose by 20 percent (to 652,900).

These increases in indigenous claims are mostly due to changing identities, not to a population boom. More positive attitudes toward indigenous peoples have encouraged the open acknowledgment of Aboriginal or Maori ancestry. Also, marriages between European and indigenous peoples are now more common. As a result, the number of people with a recognized mixed heritage is increasing.

As society has acknowledged that discrimination has been the main reason for the low social standing and impoverished state of indigenous peoples, respect for Aboriginal and Maori culture has increased. The Aboriginal Australians base their way of life on the idea that the spiritual and physical worlds are intricately related (**Figure 11.31**). The dead are present everywhere in spirit, and they guide the living in how to relate to the physical environment. Much Aboriginal spirituality refers to the Dreamtime, the time of creation when the human spiritual connections to rocks, rivers, deserts, plants, and animals were made clear. However, very few Aboriginal people continue to practice their own cultural traditions or live close to ancient homelands. Instead, many live in impoverished urban conditions where the guidelines for living a respectful Aboriginal life are breaking down. In New Zealand, where the Maori constitute about 15 percent of the country's population and Auckland has the largest Polynesian population (including Native Maori) of any city in the world, there are now many efforts to bring Maori culture more into the mainstream of national life.

Aboriginal Land Claims In 1988, during a bicentennial celebration of the founding of white Australia, a contingent of some 15,000 Aboriginal people protested that they had little reason to celebrate. During the same 200 years, they were assumed to have no prior claim to any land in Australia, they had lost basic civil rights, and they had effectively been erased from the Australian national consciousness. Into the 1960s, it was even illegal for Aboriginal Australians to drink alcohol.

British documents indicate that during colonial settlement, all Australian lands were deemed to be available for British use. The Aboriginal Australians were thought to be too primitive to have concepts of land ownership because their nomadic cultures had "no fixed abodes, fields or flocks, nor any internal hierarchical differentiation." Pressure from Aboriginal activists increased and eventually, in 1993, the Australian High Court declared this old British position void. After that, Aboriginal groups began to win some land claims as was mentioned for the Ballaruk in Western Australia (see Figure 11.23A), mostly for land in the arid interior previously controlled by the Australian government. **Figure 11.32** shows the Aboriginal Tent Embassy, versions of which have stood on the grounds of Parliament in Canberra for more than 40 years. The Aboriginal Tent Embassy was instrumental in raising public awareness of injustices and remains a national symbol of Aboriginal civil rights. Court cases and other efforts to restore Aboriginal rights and lands continue into the present.

Maori Land Claims In New Zealand, relations between the majority European-derived population and the indigenous Maori have proceeded only somewhat more amicably than in Australia. In 1840, the Maori signed the Waitangi Treaty with the British, assuming they were granting only rights of land usage, not ownership

Figure 11.31 Aboriginal rock art. A Mimi spirit painted on the ceiling of a rock cave at Kakadu National Park, Australia. To the Aboriginal Australians, Mimi spirits are teachers who pass between this world and another dimension via crevices in rocks. They are responsible for many teachings on hunting, food preparation, use of fire, dance, and sexuality. [Michael S. Nolan/AGE Fotostock]

Figure 11.32 A performer at the Aboriginal Tent Embassy in Canberra, Australia. Intermittently since 1972, and continuously since 1992, Aboriginal activists have camped out on the grounds of Australia's House of Parliament in Canberra. The first tent embassy was in response to the government's denial of land ownership and other land rights to Aboriginal Australians in territories they had continuously occupied for thousands of years. As Aboriginal land rights have gained recognition, the tent embassy has championed other causes, including opposition to mining that threatens Aboriginal communities and cultural sites, and advocacy for the Aboriginal urban poor, such as residents of the Redfern community in Sydney. The tent embassy has been targeted by arsonists; and the Australian government plans a more permanent Aboriginal structure, but will then ban camping at the embassy. [ANOEK DE GROOT/AFP/Getty Images]

(see Figure 11.35B later in the chapter). The Maori did not regard land as a tradable commodity, but rather as an asset of the people, used by families and larger kin groups to fulfill their needs. The geographer Eric Pawson writes: "To the Maori the land was sacred . . . [and] the features of land and water bodies were woven through with spiritual meaning and the Maori creation myth." The British chose to assume that the treaty had given them exclusive rights to settle the land with British migrants and to extract wealth through farming, mining, and forestry.

By 1950, the Maori had lost all but 6.6 percent of their former lands to European settlers and the government. Maori numbers had shrunk from a probable 120,000 in the early 1800s to 42,000 in 1900, and the Maori came to occupy the lowest and most impoverished rung of New Zealand society. In the 1990s, however, the Maori began to reclaim their culture, and they established a tribunal that forcefully advances Maori interests and land claims through the courts. Since then, nearly half a million acres of land and several major fisheries have been transferred back to Maori control. As of 2018, Maori in New Zealand number about 733,000 or 15 percent of the population, but, according to Australian censuses, another 100,000 Maori live in Australia. Altogether, those who identify as Maori outnumber those who are officially ethnically Maori, which seems to indicate a shift toward popular acceptance of Maori identity. Nonetheless, the Maori still have notably higher unemployment, lower education levels, and poorer health than the New Zealand population as a whole.

Gender Myths and Realities

Perceptions of Oceania are colored by many myths about how men and women are and should be. As always, the realities are more complex than the myths. Because of Oceania's cultural diversity, there are many different acceptable roles for men and women. Men in the Pacific islands, as mentioned previously, traditionally were cultivators, deepwater fishers, boat builders, and masters of seafaring. In Polynesia, men also were responsible for many aspects of food preparation, including cooking. In the modern world, men fill many positions, but idealized male images continue to be associated with power and vigorous activities.

In Australia and New Zealand, the self-assured, hypermasculine, white, working-class settler has long had prominence in the national mythologies. In New Zealand, he was a farmer and herdsman. In Australia, he was more often a many-skilled laborer—a stockman, sheep shearer, cane cutter, or digger (miner)—who possessed a laconic, laid-back sense of humor. Labeled a "**swagman**" for the pack he carried, he went from **station** (large farm) to station in the Outback, or mine to mine, working hard but sporadically, gambling and drinking, and then working again until he had enough money or experience to make it in the city (**Figure 11.33**).

In cities, the swagman often felt ill at ease and chafed to return to the wilds. Now immortalized in songs ("Waltzing Matilda," for example), novels, and films, these men are portrayed as a rough and nomadic tribe whose social life is dominated by male camaraderie and frequent brawls. No small part of this characterization of males derives from the fact that many of Australia's first immigrants came as convicts from the British Isles.

swagman an itinerant Australian male laborer who worked on large farms or in mines

station a large Australian farm or ranch, often where cattle or sheep were raised

Figure 11.33 Gender and national mythology. Australian wild horse hunter George Girdler epitomizes the hypermasculine, white, working-class settler who is central to the national mythologies of Australia and New Zealand and often serves as a role model for young men. [George Silk/The LIFE Picture Collection/Getty Images]

Today, as part of larger efforts to recognize the diversity of Australian society and the harm done by gender stereotyping, new ways of life for men are emerging and are breaking down the national image of the tough male loner. Nonetheless, the old model persists in the public images of Australian businessmen, politicians, journalists, and movie stars (Hugh Jackman, Russell Crowe, Heath Ledger).

Images of Pacific island women as gentle, simple, compliant love objects were perhaps the most enduring myths Europeans created regarding Oceania. (Tourist brochures still promote this perception.) There is ample evidence to suggest that Pacific Islanders did have more sexual partners in a lifetime than Europeans did. However, the reports of unrestrained sexuality related by European sailors were no doubt influenced by the exaggerated fantasies one might expect from all-male crews living at sea for months at a time. The notes of Captain James Cook are typical: "No women I ever met were less reserved. Indeed, it appeared to me, that they visited us with no other view, than to make a surrender of their persons." Over the years, such notions about Pacific island women have been encouraged by the paintings and prints of Paul Gauguin (**Figure 11.34**), the writings of novelist Herman Melville (*Typee*), and the studies of anthropologist Margaret Mead (*Coming of Age in Samoa*), as well as by movies and musicals such as *Mutiny on the Bounty* and *South Pacific*.

In actuality, women's roles in the island Pacific varied considerably from those in Europe, but not in the ways early European explorers imagined. Women often exercised a good bit of power in family and clan, and their power increased with motherhood and advancing age. In Polynesia, a woman could achieve the rank of ruling chief in her own right, not just as the consort of a male chief. Women were primarily craftspeople (see Figure 11.27B), but they also contributed to subsistence by gathering fruits and nuts and by fishing (see Figure 11.29). And in some places—Micronesia, for example—lineage was established through women, not men (a custom that makes sense when a woman is likely to have more than one sexual partner).

Today, there are some trends toward equality across gender lines throughout Oceania, but the persistence of inequality is tenacious. As mentioned earlier, a striking disparity is emerging: in Australia, New Zealand, Hawaii, and a very few other islands (see Figure 11.27), women are gaining political and economic empowerment; in Papua New Guinea and most of the Pacific islands, change is much slower.

Forging Unity in Oceania

Although wide ocean spaces and the great diversity of languages in the region sometimes make communication difficult, travel, sports, and festivals (**Figure 11.35**) are three forces that help bring the people of Oceania closer together.

Languages The linguist David Crystal tells us that each language through its vocabulary and structure offers a unique vision of the world. The Pacific islands—most notably Melanesia—have a rich variety of languages, each presenting a slightly different perspective. Some islands in a single chain can have several different languages. A case in point is Vanuatu, a chain of 80 mostly high volcanic islands to the east of northern Australia (see Figure 11.1). At least 108 languages are spoken by a population of just 180,000—an average of 1 language for every 1600 people. It is easy to see that such remote languages spoken by so few are endangered in a globalizing world.

While language can be an important part of a community's cultural identity, it can also be a hindrance to cross-cultural understanding. In Melanesia and elsewhere in the Pacific, the need for communication with the wider world is served by a number of **pidgin** languages that are similar enough to be mutually understood. Pidgins are made up of words borrowed from several languages by people involved in trading relationships. Over time, pidgins can grow into fairly complete languages, capable of fine nuances of expression. When a particular pidgin is in such common use that mothers talk to their children in it, then it can literally be called a "mother tongue." In Papua New Guinea, a version of pidgin English is the official language. Increasingly, English is the lingua franca (preferred language of communication) for everyone in Oceania and, as such, it threatens the 1300 indigenous languages of the region.

Interisland Travel One way in which unity is manifested in Oceania is interisland travel. Today, people travel in small planes from the outlying islands to hubs such as Fiji, where jumbo jets can be boarded for Auckland, Melbourne, and Honolulu. Cook Islanders call these little planes "the canoes of the modern age," and people travel for many reasons. Dancers from across the region attend the annual folk festival in Brisbane; businesspeople from Kiribati, Micronesia, can fly to Fiji to take a short course at the University of the South Pacific; a Cook Islands teacher can take graduate training in Hawaii; and sports fans can visit multiple locations over time. This penchant for travel is part of *subsistence affluence* (see "Economic Change in the Pacific Islands"). People who have little cash income will diligently save for a desired trip.

Sports Sports and games are a major feature of daily life throughout Oceania. The region has shared sports traditions with, and borrowed them from, cultures around the world. Surfing evolved in Hawaii and, like outrigger sailing and canoeing, derives from ancient navigational customs that matched human wits against the power of the ocean. On hundreds of Pacific islands and in Australia and New Zealand, rugby, volleyball, soccer, and cricket are important community-building activities. Baseball is a favorite in the parts of Micronesia that were U.S. trust territories. Women compete in

Figure 11.34 *Arearea* ("Amusement") by Paul Gauguin. In this 1892 painting, Tahitian women are rendered in a European Romantic pastoral style that emphasizes their gentle, compliant demeanor. [Imagno/Getty Images]

pidgin a language used for trading; one made up of words borrowed from the several languages of people involved in trading relationships

Figure 11.35 LOCAL LIVES: Festivals in Oceania

A

B

C

A A young Aboriginal dancer at the Garma Festival, which is held to encourage the practice of the traditional dance, singing, visual art, and ceremonies of the Yolngu people. The festival is held every year in Arnhem Land, which overlooks the Gulf of Carpentaria in Australia's Northern Territory. [Glenn Campbell/The Sydney Morning Herald/Fairfax Media via Getty Images]

B Waitangi Day in New Zealand, a national holiday that commemorates the signing of a treaty between the indigenous Maori of New Zealand and the British. The long boats shown here are Maori canoes, known as *waka*, and are part of a reenactment of the treaty's signing. [Kenny Rodger/Getty Images]

C Dislocate, an astonishing street theater group, presenting a mix of slapstick circus and story-telling skills, is one of Australia's leading comedy ensembles. Here it performs at the Sydney Festival, a 3-week international arts event held every January. [Wendell Teodoro/WireImage/Getty Images]

the popular sport of netball (similar to basketball but without a backboard).

Pan-Oceania sports competitions are the single most common and resilient link among the countries of the region. Attendance at regional sports events is so desirable that low-income islanders will hold yard sales and raffles to amass the cash necessary to make the trip. The centrality of such competitions in daily life encourages regional identity and provides opportunities for ordinary citizens to travel extensively around the region and to other parts of the world (**Figure 11.36**).

The haka is an example of how, in the postcolonial modern era, indigenous culture in Oceania is being revived, celebrated, and appropriated in new places by those who wish to project a multicultural image. The haka is a highly emotional and physical traditional dance performed by the Maori to motivate fellow participants and to intimidate opponents before a confrontation or major event, even an event such as a wedding. Dances like this have historically been a part of many cultures in the islands of Oceania, but the haka has now become an integral part of rugby, the region's most popular sport (**Figure 11.37**). Before almost every international match for the past century, the All Blacks (the New Zealand men's rugby team) have performed the haka: chanting, screaming, jumping, stomping their feet, poking out their tongues, widening their eyes to show the whites, and beating their thighs, arms, and chests.

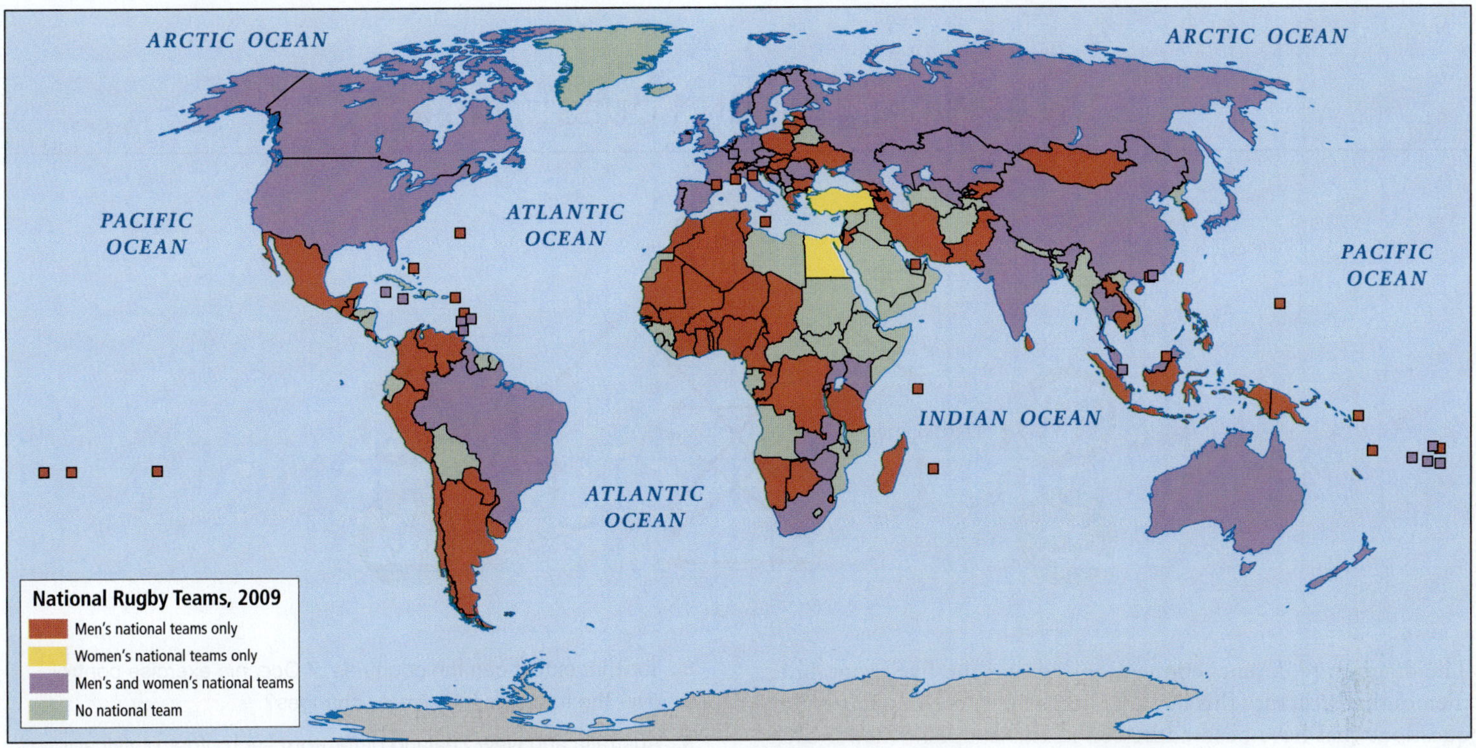

Figure 11.36 Rugby around the world. In more than 136 countries, women, men, boys, and girls play rugby. All of these countries have men's national rugby teams, and 58 have women's national teams. The men's World Cup rugby competition began in 1987, the women's in 1991. In April 2010, New Zealand boasted the top team for both men and women, but the ranking of the men's teams can change weekly. [Research from: http://www.irb.com/aboutirb/organisation/index.html.]

Outside Oceania, those who perform the haka include the rugby teams at Jefferson High in Portland, Oregon, and Middlebury College in Vermont, and the football teams at Brigham Young University and the University of Hawaii. All these teams have players who are of Polynesian heritage. Most practitioners speak of the haka as filling them with the necessary exuberance, aggression, and spirituality to play a vigorous and successful game. To see Maori-created videos of the haka, go to YouTube.

Figure 11.37 The haka, a Maori tradition. A haka performed by the New Zealand men's rugby team, the All Blacks, before a match against Australia. [Cameron Spencer/Getty Images]

CHECK YOUR UNDERSTANDING

1. What is challenging Oceania's long-standing cultural and economic links to Europe?
2. What are some recent immigration patterns in Australia and New Zealand?
3. How are indigenous people throughout the region asserting their rights?
4. Explain how mythic gender roles have characterized perceptions of the region and give two examples.
5. How do sports and festivals function as unifying forces for the region?

CRITICAL THINKING QUESTIONS

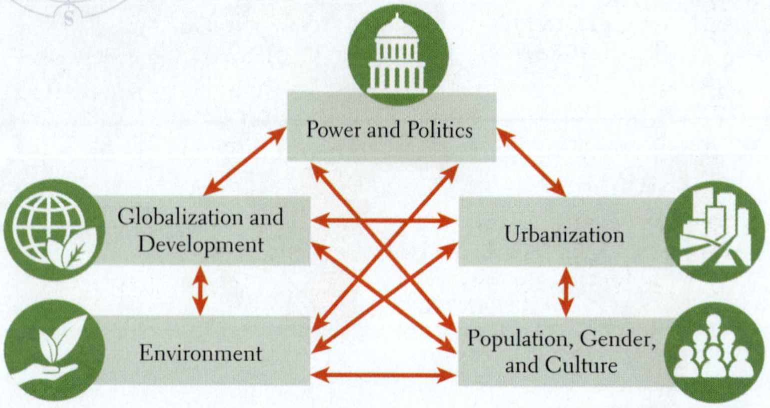

The diagram represents connections among the five geographic themes that structure this book. Listed below are some important questions that have been addressed in this chapter. Answer each question, and indicate where in the diagram you think the topics in each question belong.

1. What has been the impact of European colonialism on the demography and settlement patterns of Aboriginal Australians and Maori peoples?
2. What do you find most remarkable about the navigation skills of Pacific Islanders? How might recognition of these skills contribute to general multicultural understanding?
3. Describe and explain the changing orientation of Oceania to Europe, the Americas, and Asia over the years since the 1500s.
4. As Australia and New Zealand have moved away from their intense cultural and economic involvement with Europe, what new policies and attitudes have evolved to facilitate their deeper involvement with Asia?
5. If you were a college student in Australia or New Zealand, how might you experience these changes away from home? Think about fellow students, career choices, language learning, and travel choices.
6. Discuss the emerging cultural identity of the Pacific islands, taking note of the extent to which Australia and New Zealand share or do not share in this identity. What factors are helping to forge a sense of unity across Polynesia and beyond? (First, review the spatial extent of Polynesia.)
7. Discuss the many ways in which Asia has historic, and now increasingly economic, ties to Oceania. In your discussion, include patterns of population distribution, mineral exports and imports, technological interactions, and tourism.
8. To what extent can the countries of Oceania exercise control over the future as the climate changes?
9. Australia and New Zealand differ from each other physically. Compare and contrast the two countries in relation to water, vegetation, and prehistoric and modern animal populations.
10. Indigenous peoples worldwide are taking action to safeguard their cultures, rights, and access to land and resources. Discuss how the indigenous peoples of Australia, New Zealand, and the Pacific islands are serving as leaders in this movement and what measures they are taking to reconstitute a sense of cultural heritage.
11. How is tourism both boosting economies and straining environments and societies throughout the Pacific islands? Describe the solutions that are being proposed to reduce the negative impacts of tourism.
12. Compare and contrast how women do and do not have political and economic power in Australia, in New Zealand, in Papua New Guinea, and in the Pacific islands.
13. What are some ways that gender myths for males and females in Oceania are debilitating for each of the sexes?
14. Compared with other regions, Australia and New Zealand are somewhat unusual in having become broadly prosperous on the basis of raw materials exports. How would you explain this achievement?
15. Multiculturalism is a deeply held value in Australia and New Zealand and a point of national pride, yet there are fissures in this identity; what are they and what seems to have caused them?

Key Terms

Aboriginal Australians 630
Aotearoa 618
arable land 646
Asia Pacific Economic Cooperative (APEC) 640
asylum-seekers 641
atoll 618
consensus 641
coral bleaching 623
endemic 621
Gondwana 617
Great Barrier Reef 617
hot spots 617
human resource capacity building 639
invasive species 623
makatea 618
Maori 618
marsupials 621
Melanesia 630
Melanesians 630
Micronesia 630
MIRAB economy 636
monotremes 621
Outback 617
Pacific Solution 643
Pacific Way 640
pidgin 653
Polynesia 630
refugees 641
Roaring Forties 618
station 652
subsistence affluence 636
sustainable tourism 639
swagman 652
Trans-Pacific Partnership (TPP) 640

More Practice at SaplingPlus

Read the interactive e-text, review key concepts, and check your understanding.

Sami reindeer herders in northern Norway are concerned that their way of life is threatened.
[Scott Wallace/Getty Images]

1

Epilogue: Polar Regions

In the far northern parts of Norway, Sweden, Finland, and western Russia, the indigenous Sami people make a living by fishing, fur trapping, and sheep and reindeer herding. Mikkel, like the herder shown here, makes his living by herding reindeer. He scans the hills of Finnmark County, Norway, for signs of his 600 reindeer. Some of the young have not been getting enough to eat, and with temperatures dropping to 5°F (–15°C) in the mornings, he may have to bring them home to nurse them back to health. But the cold is actually a blessing after years of unusually warm winters that have decreased the amount of lichen and moss on which reindeer graze. Spying some reindeer about to enter the woods, Mikkel jumps on his snowmobile, hoping to head them off before they encounter the lynx and wolverines who have killed 20 of his herd just this year.

Lately, Mikkel worries less about climate change and predators than about the new development in Finnmark County. The government recently approved a controversial open-pit copper mine that will dump waste into a nearby fjord, threatening a salmon fishery. Herding areas are threatened by new roads and resorts to serve tourists eager to experience the Arctic, and wild berries that Mikkel's family depends on are now also gathered by retirees from southern Norway.

Mikkel takes a dim view of the tourist industry's appropriation of the culture of his people, the indigenous Sami, who spread across northern Norway, Sweden, Finland, and Russia. While Mikkel and his kin struggle through the cold, barely getting by in some years, tours advertise a happy, carefree image of Sami life, with trips on sleds drawn by reindeer driven by people pretending to be Sami. Christmas brings further annoyance, as the town of Rovaniemi, Finland, advertises itself as the "Official Hometown of Santa Claus," where Santa's elves are dressed as Sami, and playing cards depicting Sami as drunken fools are sold in gift shops.

Learning Objectives

Environment: Physical and Human

E1.1 Assess the evidence that climate change is transforming polar environments more than others, and that this has major consequences for the rest of Earth.

Globalization and Development

E1.2 Evaluate how the economic base of the poles is changing in response to warmer temperatures and greater demand for its resources.

Power and Politics

E1.3 Analyze the factors that are making the polar regions a growing source of conflict among nations.

Urbanization

E1.4 Weigh the factors leading to depopulation of urban centers in some polar areas, and urban growth in others.

Population, Gender, and Culture

E1.5 Describe how population and gender patterns are changing due to cultural and economic transitions as well as environmental problems faced by indigenous groups.

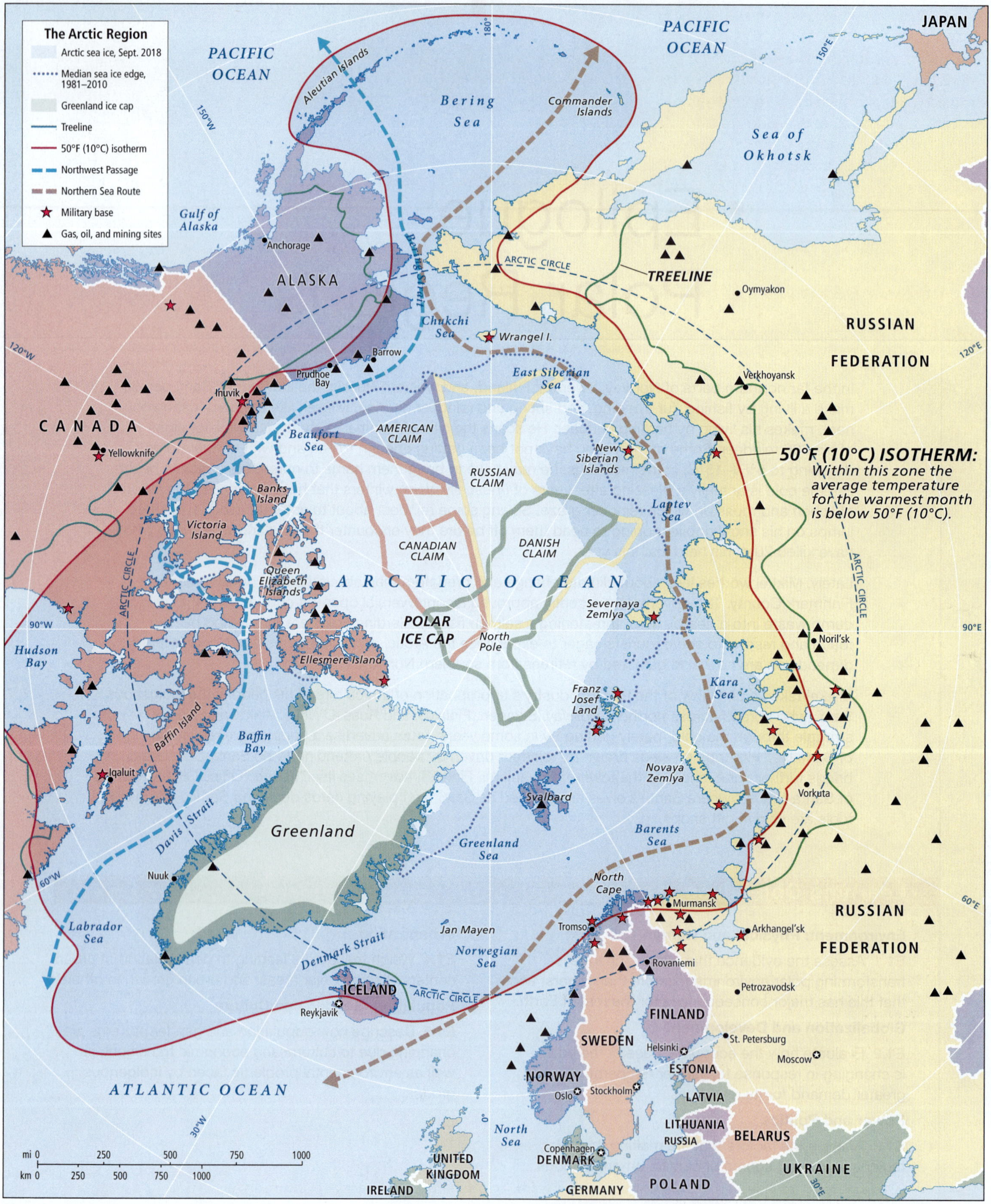

Figure E1.1 Map of the Arctic.

Despite their remoteness and small human populations, the polar regions (including the Arctic and Antarctica) are changing rapidly (**Figure E1.1**). Climate change is melting polar ice caps, bringing interest in trade across potentially more navigable seas. Newly exposed land is attracting more mining, oil, and other resource extraction industries, and politics are shifting as nations compete for control of newly accessible resources. Arctic indigenous populations, who have been here for more than 40,000 years, are often wary of the growing presence of outsiders who have so often acted in ways that harm the environment, livelihoods, and culture of longtime residents.

What Makes the Arctic and Antarctic Regions?

The Arctic is most often defined as the areas north of the Arctic Circle, where the Sun never sets for part of the summer and never rises for part of the winter. Some consider the Arctic to be the area where the average temperature for the warmest month is below 50°F (10°C), as indicated by the red line on the map (see Figure E1.1). The Antarctic, the Southern Hemisphere's counterpart to the Arctic, is usually defined by the continent of Antarctica. There is no indigenous human population, and its current inhabitants are a few thousand scientists who live at scattered research outposts for a few months or years at a time.

Figure E1.2 A fjord in Norway during the winter. [Paul Nicklen/ National Geographic Image Collection/Getty Images]

ENVIRONMENT: PHYSICAL AND HUMAN

E1.1 Assess the evidence that climate change is transforming polar environments more than others, and that this has major consequences for the rest of Earth.

Climate change is transforming the polar regions more profoundly than any other part of the planet in ways that have enormous implications for the rest of us. Among the global impacts coming from this region are sea level rise and greenhouse gas (GHG) emissions due to thawing permafrost.

PHYSICAL GEOGRAPHY

The intense cold of the polar regions results in ice caps that cover much of the Arctic Ocean, Greenland, and Antarctica. Arctic landforms are highly varied, centered on an ocean surrounded by coastal lowlands and a few mountain ranges. The tallest of the mountains are in Greenland, where they reach as high as 12,119 feet (3694 meters). There are 12 major river systems that empty into the Arctic Ocean, some of which drain vast areas south of the Arctic Circle. The largest arctic rivers are found in Russia, where the Yenisei River has a larger discharge than the Mississippi River. Many coastal areas of the polar regions have **fjords**, long, narrow inlets formed by glacial erosion (**Figure E1.2**). Fjords are most common along more mountainous coastlines in Norway, Greenland, and northern Canada, with a few in Antarctica as well.

Antarctica has a similarly wide range of landforms. Underneath a sheet of ice that averages about 1 mile (1.6 kilometers) thick, Antarctica is a complex archipelago of islands. There are a few ice-free areas near the coast that together make up about 2 percent of Antarctica. The Antarctic ice cap holds roughly 90 percent of the world's ice and 70 percent of its fresh water. With so much water locked in the form of ice, there are few rivers in Antarctica, and the longest runs for just under 20 miles (32 kilometers). Antarctica's tallest mountain, Mount Vinson, reaches 16,050 feet (4892 meters).

Polar climates are characterized by extremely low solar radiation and relatively little precipitation. Because the waters of the Arctic Ocean never get below 28°F (−2°C) and rarely above 46°F (8°C), they give the Arctic a more moderate climate, with warmer winters relative to some places just to the south. In fact, the lowest temperatures in the Northern Hemisphere are in the area between Verkhoyansk and Oymyakon in northern Russia (well south of the Arctic Circle), where the average daily low temperatures are below −53°F (−47°C) from December through February. In Antarctica, the continent's landmass and ice cap cool down more severely in the winter, resulting in the lowest temperatures ever recorded on Earth (−128°F, −89°C).

CLIMATE AND VEGETATION

Bitter cold, snow and ice, lack of sunlight and moisture, and soil that is permanently frozen a few feet below the surface (permafrost) severely limit vegetation in the polar regions. Arctic vegetation varies between coniferous forests called taiga (see Chapter 5) in areas without permafrost, and elsewhere tundra, which consists of lichen and mosses that stay close to the ground to conserve warmth and moisture. Antarctica has tundra only on the 2 percent of its area not covered by its thick ice cap. Despite holding so much of Earth's water, Antarctica is a desert; most of its interior areas receive only 2 inches (5 centimeters) of precipitation each year, less than the Sahara.

ENVIRONMENTAL ISSUES

Climate change is transforming environments more profoundly at the poles than at any other place on

fjord a long, narrow inlet formed by glacial erosion

Earth. Temperatures far above normal are melting polar glaciers and sea ice rapidly, resulting in habitat loss for many species, and contributing to sea level rise. The disproportionate increase in temperature at the poles is called **polar amplification**, which scientists think is related mainly to reductions in sea ice and snow cover, which tend to reflect heat from the sun back into space, and the expansion of open water and land, which tend to absorb heat from the sun.

Polar Melting and Sea Level Rise

While the melting of sea ice has no net effect on sea level, the melting of land ice contributes about half to two-thirds of the annual increase in global sea level (the other half to a third comes from thermal expansion of the oceans). Around half of melting land ice is in the ice sheets of Greenland and Antarctica, with the other half in melting glaciers in nonpolar regions. Warming is happening faster in the Arctic than in the Antarctic, which means that Greenland's ice sheet is melting faster and contributing much more to sea level rise than Antarctica, even though Greenland contains only one-tenth the volume of ice. The Arctic is warming faster because more of it is water, which tends to absorb more heat than land or ice. Meanwhile Antarctica's large ice sheets are reflecting more of the sun's energy and are warming more slowly.

The Arctic has lost roughly 8 percent of its land and sea ice over the past 30 years, with recent research suggesting that the Arctic Ocean may be ice-free during summers by as early as 2030. These changes are already challenging some arctic species that hunt or give birth on the ice, such as polar bears, walruses, certain kinds of seals, and some seabirds. Meanwhile the foundations of arctic food chains, algae, may be changing due to invasions of algae from warmer waters to the south.

The situation is more complex in Antarctica, where recent studies suggest that while overall temperatures are increasing on land and in the seas, temperatures are actually declining and sea ice is increasing in eastern Antarctica. Recent research by NASA suggests that in the continental interior of Antarctica temperatures are rising, increasing the potential for the air to hold moisture, leading to higher snowfall. However, this is most likely not enough to offset melting along the coasts, in west Antarctica, and on the Antarctic peninsula, resulting in an overall loss of ice on the continent (**Figure E1.3**). So far Antarctica's contribution to sea level rise has been low relative to other areas, but scientists keep a close watch on it because sea levels would rise 200 feet (60 meters) if its ice sheet melted entirely.

As a result of these changes, Antarctic species that breed and hunt on sea ice, such as several types of penguin, will decline in some areas and possibly increase in others (**Figure E1.4**). In the seas surrounding Antarctica, rising temperatures have more uniform impacts, such as reductions in **krill**, the tiny crustaceans on which whales, seals, penguins, and fish feed. This is already resulting in declining seal populations in some areas.

Figure E1.3 Scientists measure the sea level at Antarctica's Ross Sea. While Antarctica has contributed relatively little to sea level rise so far, if its massive ice sheet were to melt entirely, sea levels would rise 200 feet (60 meters). [Paul Nicklen/National Geographic/Getty Images]

Figure E1.4 Emperor penguins dive beneath sea ice in Antarctica. Emperor penguins live only in Antarctica and are proving extremely sensitive to climate change. In warmer years, declines in sea ice—the penguins' ideal hunting habitat—have resulted in widespread starvation among the penguins. In colder-than-normal years, few penguin chicks hatch. [Paul Nicklen/National Geographic/Getty Images]

polar amplification the increase in the intensity of climate change at the poles

krill tiny crustaceans that whales, seals, penguins, and fish feed on

Methane and Permafrost

Climate change at the poles may also result in increased GHG emissions because methane, a powerful GHG, is released when permafrost thaws and polar oceans warm. Some studies estimate that the amount of methane trapped in permafrost under the polar oceans and frozen in the oceans (as methane hydrates) could be large enough to trigger much more rapid warming of Earth's climate if released. Dramatic increases in methane release from the Arctic have been documented in recent years, though some studies suggest that certain bacteria in permafrost soils on land can consume some of this methane.

New Protected Areas

Conservation efforts at the poles are strengthening. In the Arctic, the United States, Canada, Greenland, Norway, and Russia created a conservation plan for polar bears in 2013 that addresses threats from shipping, oil and gas exploration, and military conflict. In 2015, Russia took steps to protect wild salmon fishing areas in its far east. In 2013, the U.S. government protected large areas of bird habitats in northern Alaska from oil and gas development. In Nunavut, Canada, a new marine reserve for whales was created in 2011. Additions to the Antarctic Treaty that have been in effect since 1998 are designed to protect the seas around Antarctica from overfishing and pollution.

CHECK YOUR UNDERSTANDING

1. Why does the Arctic have a more moderate climate than some places just to the south?
2. What factors limit the growth of vegetation in the poles?
3. What drives polar amplification?
4. How are polar wildlife populations being impacted by climate change?
5. How much sea level rise results from the melting of polar land ice?
6. Why are scientists concerned about the release of polar methane?

GLOBALIZATION AND DEVELOPMENT

E1.2 Evaluate how the economic base of the poles is changing in response to warmer temperatures and greater demand for its resources.

In the Arctic, military spending, mining, fishing, and hunting have made up the economic base for a long time, but new expansions in energy and shipping industries are appearing in response to reductions in land and sea ice. Meanwhile international fishing and a booming tourism industry have already begun to exert pressure on Antarctica, and the human population of Antarctica may increase.

The U.S. Geological Survey estimates that almost a quarter of the world's oil and gas is located beneath the Arctic, with roughly half in Russia, which has so far been the country most interested in developing its arctic resources. Recent years have seen dramatic increases in shipping in the Russian Arctic, mostly due to new development of oil, gas, and coal reserves that are now more accessible due to climate change.

Ice-free shipping routes through the Arctic Ocean have long been sought after because they offer shorter travel distances between the Atlantic Ocean and Pacific Ocean, requiring as much as 40 percent less fuel than current sea routes involving the Suez or Panama canals. However, the persistence of summer sea ice in the Arctic until at least 2030, along with the infrastructure needed to support Arctic Ocean shipping routes, means that the routes are not likely to be in large-scale service until 2040 (**Figure E1.5**).

Antarctica is also experiencing more pressure for resource extraction. The international fishing industry is rapidly increasing its catches in the Southern Ocean, though this is mostly due to a growing global demand for fish, not due to changes related to the rise in temperatures (**Figure E1.6**). Meanwhile, larger human populations in Antarctica could develop with the renegotiation of the Antarctic Treaty of 1959, which currently bans all resource extraction. The treaty won't be formally reviewed until 2048, but more than 30 countries are already building up research presences on the continent, hoping to secure claims to mineral and energy resources once they become available. These resources include oil, natural gas, and a wide variety of metallic ore deposits located along the Antarctic coastline.

Figure E1.5 A nuclear-powered ice breaker cuts a path through the Kara Sea for a convoy of ships. [Sovfoto/UIG via Getty Images]

Figure E1.6 One of the thousands of fishing vessels active in the Southern Ocean. Species like the Patagonian toothfish (marketed as Chilean sea bass) are severely overfished. Tiny shrimp known as krill, which are used in fish farms and health food supplements, are also being overfished. [Paul Sutherland/National Geographic/Getty Images]

Tourism is also growing rapidly in the Antarctic, with tens of thousands of visitors coming each summer, mostly via cruise ships leaving from Chile, South Africa, and New Zealand. Some wildlife populations, especially penguins, are getting sick from bacteria that humans are bringing. However, these impacts are relatively small compared to those related to climate change.

CHECK YOUR UNDERSTANDING

1. How is the economic base of the Arctic changing in response to climate change?
2. What economic activities are being negatively impacted by climate change?
3. What activities does the Antarctic Treaty of 1959 ban?

POWER AND POLITICS

E1.3 Analyze the factors that are making the polar regions a growing source of conflict among nations.

The polar regions are a growing source of conflict. Multiple nations with competing territorial claims are interested in controlling the rapidly changing and increasingly valuable lands, seas, and resources of both the Arctic and Antarctica (Figure E1.1 and **Figure E1.7**).

In the Arctic, eight countries have territorial claims, and all are members of an intergovernmental forum known as the *Arctic Council*. Established in 1996 to promote cooperation, coordination, and interaction, the Arctic Council specifically avoids geopolitical and security issues, which may limit its utility in the future. One source of contention is the United Nations Convention on the Law of the Sea (UNCLOS, see Chapter 11), which allows for countries to claim exclusive economic development rights for 200 miles (322 kilometers) out from their coastlines. Currently there are disputes between Canada, Denmark, Russia, and the United States over several islands and the surrounding seas to which UNCLOS would grant owner's rights. In 2007, Russia made a provocative claim to an underwater landform, the Lomonosov Ridge, that includes the North Pole, on which a submarine planted a Russian flag. Since then both Canada and Denmark, via its territory of Greenland, have made a claim to the Lomonosov ridge. Meanwhile the United States, having never ratified the UNCLOS, may in the future challenge any and all claims to arctic territory, though its ability to do so would rely almost entirely on its military and economic strength.

The Arctic has a long history of strategic and war-related activity, and all arctic nations are expanding their military presence here. The largest cities all have a significant military component to their economy, and even the most remote parts of the Arctic had some

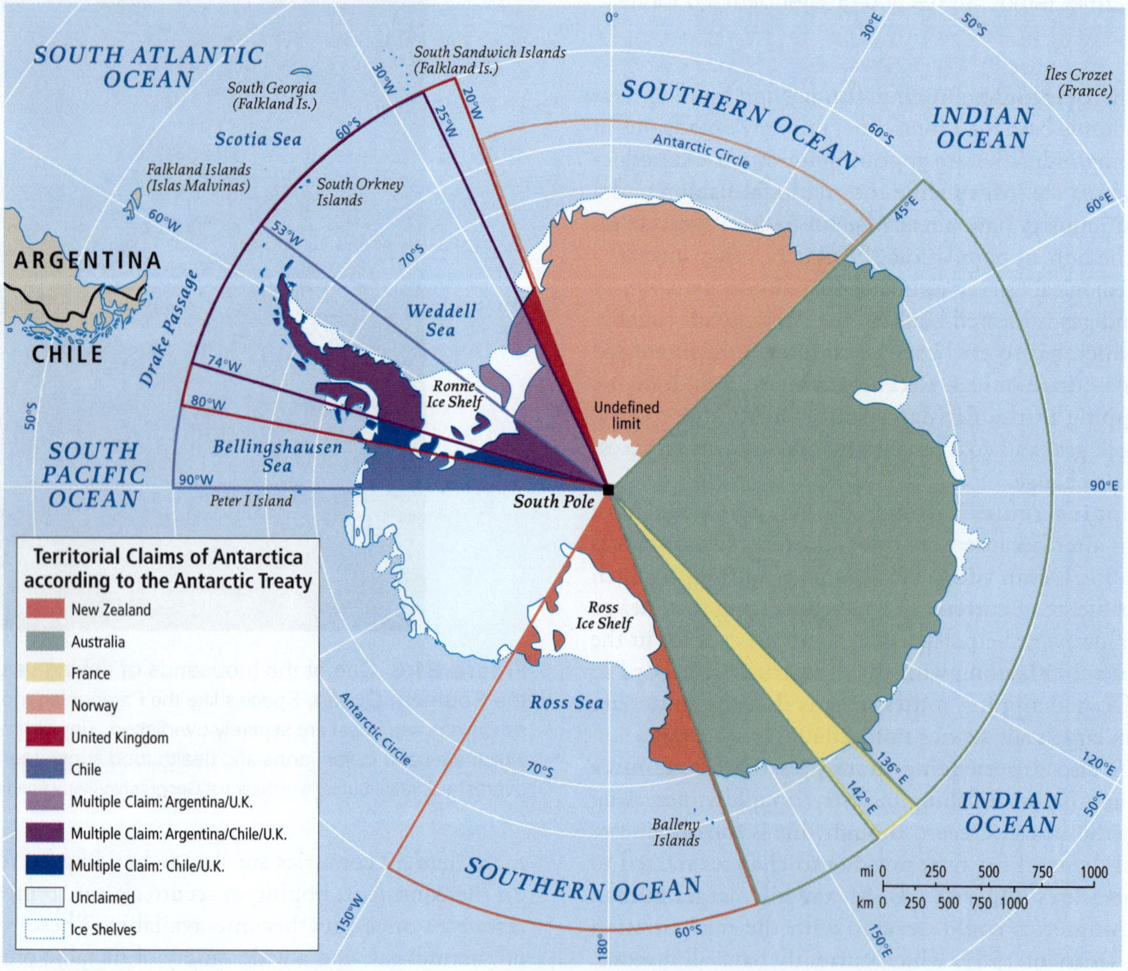

Figure E1.7 Territorial claims on Antarctica.

level of military activity during the Cold War between the United States and the Soviet Union. This included testing of nuclear weapons, constant patrols by nuclear-armed jets and submarines, and extensive military exercises.

Recent years have seen a huge increase in arctic military activity, with Russia holding massive annual exercises that have grown into the largest in its history, involving hundreds of thousands of troops, with limited participation from China and Mongolia. These are widely interpreted as a signal of Russia's resolve in defending its claims to arctic territory, and in response NATO has held its own large military exercises involving troops from over 30 countries, including all arctic countries.

Antarctica has a whole different set of nations with territorial claims, several of which overlap (see Figure E1.7). While formal military operations are not allowed, according to the Antarctic Treaty of 1959, more than 30 national militaries help maintain their home countries' research stations.

CHECK YOUR UNDERSTANDING

1. Why are the polar regions a growing source of conflict?
2. What territorial claims does the UNCLOS provide for?
3. What form has the recent increase in arctic military activity taken?

URBANIZATION

E1.4 Weigh the factors leading to depopulation of urban centers in some polar areas, and urban growth in others.

Urban centers are few and far between in the Arctic and completely absent in the Antarctic. Rural areas are generally depopulating in the Arctic, as people move to the cities or leave the region. Many indigenous people are having to abandon their rural livelihoods, based on hunting, fishing, or herding, due to difficulties related to climate change. Meanwhile, the largest arctic cities are also in decline due to changes in Russia's economy; however, several smaller Scandinavian cities are growing because of renewed interest in the area.

The largest arctic cities are in Russia, and were founded at the direction of the government in Moscow in the twentieth century to fulfill both strategic and resource-extraction needs. The biggest is Murmansk (population 300,000), which was created by the Russian Empire in 1915 as an ice-free port that would enable Russia's allies in World War I to supply it with military equipment and ammunition (**Figure E1.8**). Still a major military outpost, Murmansk's population has plummeted since the fall of the Soviet Union, as the government has been less able to support the city's industries. The next-largest arctic city is Norilsk (population 175,000), founded in the late 1920s as a state-run mining **gulag camp** that depended on prisoners supplied by the government for much of its labor. After declining since the fall of the USSR, Norilsk's population is stabilizing, despite being home to some of the worst pollution on Earth (see "Resource Use and the Environment" in Chapter 5). Vorkuta (population 58,000) was started in the 1930s as a coal-mining gulag camp and is also shrinking. Russia also has the largest cities in the near-Arctic, including Arkhangelsk (325,000), Petrozavodsk (278,000), and St Petersburg (5.2 million), all of which are growing.

Figure E1.8 Reindeer racing at a winter festival in Murmansk, Russia. [Sovfoto/UIG via Getty Images]

The largest arctic city outside of Russia is Tromsø, Norway (population 72,000), which was a fortress town at least 1000 years ago, populated by Vikings and Sami. In contrast to the arctic cities in Russia, Tromsø and other arctic cities in Scandinavia are growing moderately due to increased government and private investment in arctic resource extraction. The Finnish city of Rovaniemi (population 61,000), which lies just 6 miles (10 kilometers) south of the Arctic Circle, is an administrative and commercial center for Finland's Arctic. Both Tromsø and Rovaniemi have major universities and government agencies. Outside of Russia and Scandinavia, the next largest arctic town is Barrow, Alaska (population 4000), which serves the surrounding oil and gas industry on the north slope of Alaska.

CHECK YOUR UNDERSTANDING

1. Where are the largest urban centers in the Arctic and near-Arctic?
2. What was the driving force behind the establishment of the largest arctic cities?
3. Where are arctic cities growing and shrinking?

POPULATION, GENDER, AND CULTURE

E1.5 Describe how population and gender patterns are changing due to cultural and economic transitions as well as environmental problems faced by indigenous groups.

Polar populations are stable on the whole, with population growth in Scandinavia and North America balancing population decline in Russia. Human

gulag camp a settlement set up by the central government of the USSR, in which prisoners were sent to state-run enterprises to provide labor development is generally lower in the arctic regions of countries, relative to national averages. There is also a complex gender imbalance at the poles.

The Arctic is home to roughly 2.5 million people, most of whom live in Russia. Here populations are declining due to emigration of urbanites who came to the Arctic in the twentieth century (mainly from western Russia). Indigenous populations, though, are growing throughout the Arctic because of their relatively high childbirth rate. For example, women in Nunavut, Canada, give birth to three children, on average.

GENDER IMBALANCE

The gender imbalance in the polar regions is complex. Arctic North America, Greenland, and Scandinavia generally have more men than women, while the Russian Arctic has more women than men. Antarctica has many more men than women, due to male dominance in the scientific disciplines practiced at the research stations there.

In almost every case, arctic places have an above-average ratio of men to women with respect to national averages. For example, the gender ratio is 94:100 (94 males for every 100 females) in Murmansk, Russia, while in Russia as a whole the gender ratio is 86:100. The tendency for arctic areas to have above-average male populations is usually explained by their economic base in resource-extraction industries, which tend to employ more men, as well as by more traditional patriarchal power structures, which extend male dominance into higher-paying government and service sector jobs. These factors tend to encourage women seeking more opportunities for education and career advancement to migrate (Russia's lower ratio of men to women, relative to other parts of the Arctic, is related to higher mortality of men due to cardiovascular disease, work-related hazards, and higher rates of alcoholism and suicide).

Among indigenous populations that consume large amounts of arctic fish, environmental pollution may also be introducing a gender imbalance, with twice as many girls as boys being born in some villages. Studies suggest that chemicals used all over the world in pesticides, electrical transformers, and flame retardants act as *endocrine disruptors* that, when they reach high enough concentrations, mimic human hormones that switch the gender of a fetus from male to female during pregnancy. These chemicals become concentrated in arctic fish through several linked processes. First, they make their way to the Arctic through wind and ocean currents. Then, a process of **bioaccumulation** takes place as organisms absorb the toxins faster than their bodies can excrete them. On an ecosystem scale, a process called **biomagnification** occurs as the concentration of toxins increases at successive levels of the food chain. For example, PCBs, a toxic class of chemicals used in electrical equipment, exist in fairly low concentrations in the bodies of arctic plankton, but the smaller fish that eat the plankton have much higher concentrations, the larger fish that eat the smaller fish have even higher concentrations, and the humans who eat the larger fish have the highest concentrations of PCBs. The gender imbalance that endocrine disruptors are introducing to some indigenous arctic communities is prompting renewed efforts to make governments regulate the use of chemicals more carefully.

bioaccumulation a process in which organisms absorb toxins faster than their bodies can excrete them

biomagnification a process in which the concentration of toxins increases at successive levels of the food chain

INDIGENOUS PEOPLES AND CULTURAL SURVIVAL

The vignette at the beginning of this epilogue hints at some of the issues facing indigenous people in this vast, sparsely populated region. Climate change is making survival more difficult for the many people who hunt, fish, or herd in the Arctic. Most are struggling to adapt as species move to new areas and as new techniques to exploit them are required by warmer temperatures. For example, some hunting grounds once accessible year-round by snowmobiles now require expensive paved roads for part of the year, especially during the spring thaw that turns unpaved roads into impassable mud pits (see "Climate and Vegetation" in Chapter 5 for a discussion of the *rasputitsa*, or "quagmire season").

Climate change is just one of many threats to indigenous cultures, which for more than 500 years have endured many unwelcome influences of national governments, corporations, tourists, and other outsiders. Visitors often come to the Arctic with grand designs that overlook or are openly hostile to the people who have lived there for tens of thousands of years. For example, Norway had a policy of *Norwegianization* from 1850 to the 1980s that was aimed at destroying Sami culture. The Sami languages were made illegal, Sami lands were confiscated, and knowledge of the Norwegian language was required to buy or lease land. Norwegianization was gradually abandoned in the 1980s, and a Sami parliament was established in 1989 with authority to distribute funds designated for the preservation of Sami culture. Even so, discrimination against Sami persists.

For 500 years Russia's indigenous populations have suffered brutal repression, displacement, and forced acculturation by governments that considered them "uncivilized." During Imperial Russia's colonization of Siberia (1580 to the late 1700s), several czars waged genocidal wars against Arctic peoples, wiping out entire cultures. Survivors were severely reduced in number and lost much of their land to Russian colonists from the west. Only the most remote and warlike, such as the Chukchi of the far eastern Arctic, maintained some independence. During the Soviet era, indigenous people were further displaced by mining and other extractive industries and forced onto collective farms and reindeer ranches. Their children were sent to boarding schools in Moscow and St. Petersburg, where many of them died of illness. In the post-Soviet era, many collective farms and ranches in the Arctic have closed as the oil and gas industry expands further into reindeer herding lands.

Arctic indigenous peoples in North America have a similar history. Imperial Russia claimed possession of Alaska from the 1780s until 1867, during which time many native peoples were forced into slavery by Russian fur traders. After the United States purchased Alaska from Russia in 1867, whaling in the North American Arctic increased, reducing access to an important source of food for many arctic indigenous peoples, who faced periodic starvation. Influxes of foreign whalers, miners, fur traders, and missionaries also introduced diseases such as measles and influenza, which were deadly to the indigenous populations in Alaska and the Canadian Arctic and could kill more than 30 percent of a community in a single epidemic.

As in Russia, indigenous cultures in the North American Arctic were considered "uncivilized," and hundreds of thousands of children were forcibly removed from their families and sent to boarding schools, where they were encouraged to abandon all aspects of their native culture. These schools, established in the late nineteenth and early twentieth centuries by missionaries with government funding, were renowned for physical abuse as well as deadly epidemics that killed thousands of children. Recent decades have seen greater autonomy granted to indigenous peoples in the Canadian Arctic in the form of the Nunavut province.

The long history of oppression and dehumanization of indigenous people suggests that many will remain on a semi-impoverished periphery in the era of heightened interest in arctic resources, but fairer treatment by governments and higher human development are also coming to arctic indigenous peoples. In addition to the reforms in Norway that led to the Sami parliament, Canada's arctic peoples won greater control over their lands in the form of the Nunavut territory, created in 1999. Holding the vast majority of Canada's arctic lands, Nunavut has more autonomy and receives considerably more aid from the federal government than other parts of Canada. While human development remains the lowest in Canada, it has increased considerably in recent decades.

CHECK YOUR UNDERSTANDING

1. Where are polar populations growing and declining?
2. How are bioaccumulation and biomagnification contributing to a complex gender imbalance at the poles?
3. How is the era of heightened interest in arctic resources likely to impact indigenous people?

CRITICAL THINKING QUESTIONS

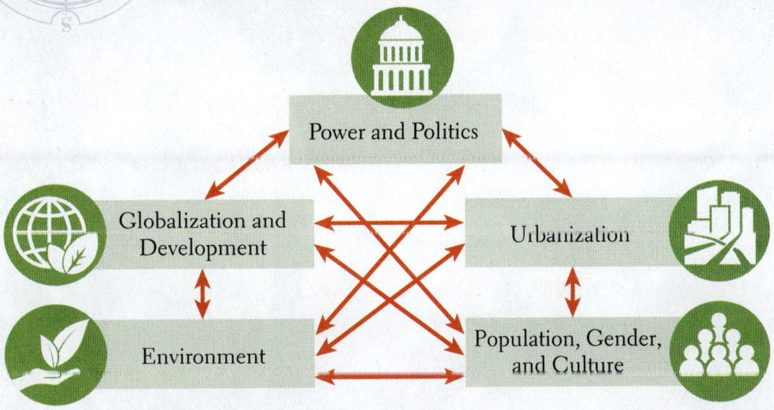

The diagram represents connections among the five geographic themes that structure this book. Listed below are some important questions that have been addressed in this epilogue. Answer each question, and indicate where in the diagram you think the topics in each question belong.

1. How could climate change at the poles influence your life, and how could your life influence climate change at the poles?
2. What new industries are increasing their presence in the Arctic and why?
3. Why are conflicts over the Arctic growing?
4. What are the differences between large arctic cities in Russia and Scandinavia?
5. What factors contribute to the complex polar gender imbalance?
6. What common experiences have arctic indigenous people endured over the past several centuries?

Key Terms

bioaccumulation 666
biomagnification 666
fjord 661
gulag camp 665
krill 662
polar amplification 662

More Practice at SaplingPlus

Read the interactive e-text, review key concepts, and check your understanding.

Cosmonauts aboard the *Salyut 7* spacecraft in 1982.
[SVF2/Sovfoto/Universal Images Group via Getty Images]

2 Epilogue: Space

On June 8, 1985, Russian cosmonauts Vladimir Dzhanibekov and Viktor Savinikh floated into the dark and frozen *Salyut 7* space station and found the crackers and salt tablets left by the previous crew. A traditional Russian greeting, guests are welcomed with bread, symbolizing wealth, and salt, symbolizing protection. The salt was particularly relevant as *Salyut 7* had lost all power and communications systems 4 months previously and was drifting out of orbit. Dzhanibekov and Savinikh had to find out what was wrong, fix it, and prepare the station for its next crew. While working, the cosmonauts had to take frequent breaks due to the absence of ventilation in the frozen station, which created a risk of carbon dioxide poisoning from their own exhaled breath. Eventually, the cause of the power loss was found—a faulty battery charger for the solar power system—and easily mended.

Salyut 7 was the most advanced Soviet space station, though at roughly half the size of a small mobile home, living quarters were cramped relative to the current International Space Station (ISS), which is the size of a six-bedroom house. With no laundry, cosmonauts wore the same underwear for several days in a row, after which it was jettisoned out the waste hatch and descended toward Earth, burning up on reentry.

Salyut 7 was celebrated for its many photographic observations of the Earth, which contributed to oil and gas exploration, agriculture, Earth science, and mapping, and were estimated to have had a substantial economic impact. The crew also kept a log of their observations, which later became the basis for a major Russian film about *Salyut 7* in 2017.

After its last manned mission in 1986, *Salyut 7* was placed into a high "storage" orbit, from which it would be retrieved for later use. However, disruptions after the fall of the USSR delayed its retrieval, and higher-than-expected solar wind drag started *Salyut 7* on a descent toward Earth. In 1991 it burned up reentering the Earth's atmosphere.

The story of *Salyut 7* captures the difficulties of life in space and how events on Earth can wreak havoc with even the best-laid plans for space exploration.

Learning Objectives

Environment: Physical and Human

E2.1 Investigate the effects on Earth of solar weather, asteroids, and comets and the usefulness of human-made satellites in better understanding processes on Earth.

Globalization and Development

E2.2 Evaluate how commercial activity in space is expanding and propelling future space exploration.

Power and Politics

E2.3 Analyze current disagreements about the militarization of space and the rights of individuals and corporations to own property in space.

Urbanization

E2.4 Assess the usefulness of space-based remote sensing technologies to better understand urban processes on Earth and to help cities adapt to change.

Population, Gender, and Culture

E2.5 Describe how gender issues have shaped space exploration, and how space exploration might change cultures across Earth.

Given the increasing environmental and social stresses on Earth, and growing evidence of vast extraterrestrial resources, it's no surprise that space exploration has captured the imagination of many people. But is this just an effort to escape serious problems here on Earth? Don't we need to focus on solving problems here before we turn our attention elsewhere? Far from an escape from Earth, most human activity in space is directed at solving problems on Earth, and it is doing so at an accelerating pace. While most space-based enterprises are commercial or military in nature, many of the largest problems on Earth, such as climate change, cannot be fully understood in ignorance of the technologies already in use in space.

low Earth orbit (LEO) an orbit that lies between 99 miles (160 kilometers) and 1243 miles (2000 kilometers) above Earth

geospace the region of outer space closest to Earth, including the upper atmosphere and extending out 40,000 miles (65,000 km), to the farthest reaches of Earth's magnetic field

interplanetary space the region of space between the Sun and planets of our solar system

interstellar space the region of space between our solar system and others within the Milky Way galaxy

intergalactic space the region of space between the Milky Way galaxy and other galaxies

solar wind tiny charged particles emitted by the Sun that travel at millions of miles per hour and form a gaslike plasma that can conduct electricity and respond to magnetic fields

Low Earth orbit (LEO), the region that lies between 99 miles (160 kilometers) and 1243 miles (2000 kilometers) above Earth, is already highly utilized, with thousands of satellites having been launched into orbit over the past 70 years and tens of thousands more planned in the next decade. These provide services central to the management of global communications, transportation, commerce, and the conduct of war **(Figure E2.1)**. If LEO were a country its economy would rank it among the top 30 in the world, ahead of countries like the United Arab Emirates and South Africa. It would also be one of the world's fastest growing economies, with sustained growth of 7 percent a year over the past decade.

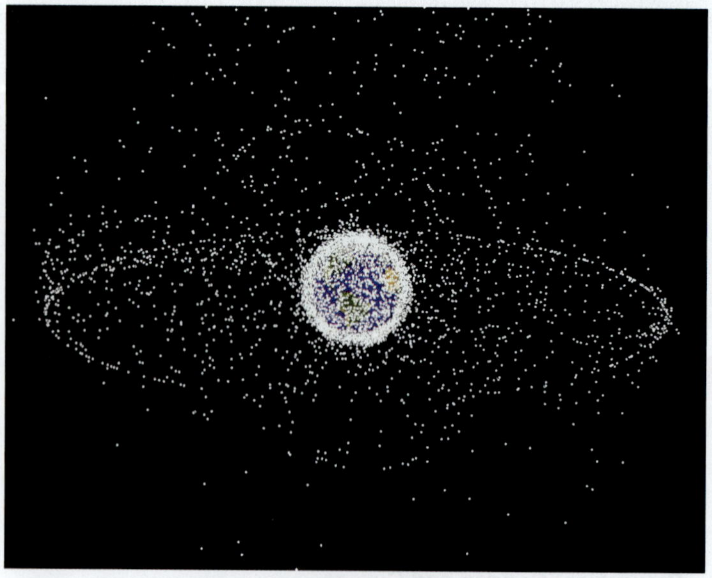

Figure E2.1 Objects orbiting Earth. Ninety-five percent of the objects orbiting Earth that are currently being tracked by NASA are space junk, or nonfunctional satellites. In this computer-generated image, the white dots represent the objects. (The dots are not to scale and are much smaller relative to Earth in real life.) Most objects are in the crowded region closest to Earth, called low Earth orbit. [NASA]

Beyond LEO, major plans are also afoot. While it's easy to write off proposals by billionaires to colonize the moon and Mars as science fiction, it's less easy to do so when they are launching satellites for less than one-tenth the cost of a decade ago, and creating habitable structures that now orbit the Earth, as some have recently done. Most involved with efforts to mine the Moon and asteroids argue that it will increase living standards while reducing the need for environmentally damaging resource extraction here on Earth.

What Makes Outer Space a Region?

Outer space is the area between Earth's atmosphere and all other objects in the universe. Strictly speaking, asteroids, planets, and stars are not part of outer space, but rather are contained within it. However, these and other celestial objects are treated as part of outer space in this epilogue.

Outer space is usually broken down into several regions. **Geospace** is the region of outer space closest to Earth. Geospace includes the upper atmosphere and extends out 40,000 miles (65,000 km), to the farthest reaches of Earth's magnetic field. Almost all human activity in outer space is taking place in geospace. **Interplanetary space** exists between the Sun and planets of our solar system; **interstellar space** exists between our solar system and others within our galaxy (the Milky Way); and **intergalactic space** exists between our galaxy and other galaxies.

ENVIRONMENT: PHYSICAL AND HUMAN

E2.1 Investigate the effects on Earth of solar weather, asteroids, and comets and the usefulness of human-made satellites in better understanding processes on Earth.

A less hospitable environment than outer space is hard to imagine. Intense cold of −454°F (−270°C) is combined with extremely low pressure, approaching a perfect vacuum, that would suck all the air from a human being's lungs, causing them to pass out within 15 seconds and most likely die within 60 seconds. And yet direct sunlight can create temperatures of 250°F (121°C) on the surface of objects such as the Moon or a space vehicle, and bare skin would burn immediately in space, likely resulting in skin cancer. Tiny bits of rock, metal, and other solid material, called *meteoroids*, are another hazard in outer space. Often created by collisions between larger objects, tiny meteoroids can travel at thousands of miles per hour in space, and any space vehicle is covered with tiny impact craters from them. There is also very little gravity in space, resulting in a bewildering array of complications for life that has evolved under gravity, even when one is shielded from other space hazards.

SPACE WEATHER

The Sun and other stars are constantly ejecting charged particles that form a gas-like plasma that can conduct electricity and respond to magnetic fields. The Sun's emission of these particles is known as **solar wind**, which is thrown out at millions of miles per hour, reaching far beyond the orbits of the most distant planets in the solar system. Earth is largely shielded from the solar wind by its magnetic field, also called the *magnetosphere* **(Figure E2.2)**. Some particles get through the magnetosphere and light up the sky near the poles, creating the *northern* and *southern lights* and occasionally disrupting electric power-delivery systems. Solar wind is lethal to terrestrial life, and all celestial objects without strong magnetic fields are bombarded by this form of intense radiation. All spacecraft are designed to provide shielding, but the most intense solar winds still penetrate, increasing the risk of radiation sickness and cancer among space travelers.

ENVIRONMENTAL ISSUES

There are significant threats to Earth's ecosystems from space, such as impacts from large asteroids, or from the proliferation of human-made satellites orbiting Earth. A **satellite** is any object in space that orbits another object of greater mass. Over 8000 artificial satellites have been launched into orbit, providing information about Earth, and to a lesser degree about outer space.

Environmental Hazards from Space

The most certain threat to Earth from space is an asteroid or comet impact, such as the asteroid between 200 and 650 feet wide (60–200 meters) that flattened 770 square miles (2000 square kilometers) of forest in Tunguska, Siberia, in 1908; or the 6-mile-wide asteroid that landed just off the coast of the Yucatán Peninsula 66 million years ago, possibly causing the extinction of the dinosaurs and numerous other life-forms. In 2013 a meteor 100 feet (30 meters) wide exploded over Chelyabinsk, Russia, creating a shockwave over a 200 square-mile area and injuring 1600 people, mostly due to broken glass from the shattered windows of the 7200 buildings near the meteor's path. The meteor entered the Earth's atmosphere without being detected, which spurred the formation of the International Asteroid Warning Network by the United Nations that year and NASA's Planetary Defense Coordination Office in 2016, which aims to detect meteors larger than 100 feet (30 meters) and if necessary divert them.

Relative to most human activities, rocket launches currently have relatively low environmental impacts, but this could change. With current technology, the number of rocket launches would have to multiply by over a thousand before they contributed even one-thousandth of a percent of global GHG emissions. Damage to the *ozone layer*, a part of the atmosphere that absorbs most of the ultraviolet radiation from the Sun that reaches Earth, is more likely but still uncertain. Some research suggests that significant losses of ozone due to rocket launches could be seen at around 100 times current launch rates, which could be reached in the next two decades.

So far the greatest human-created environmental hazard in space is the increasing amount of debris or **space junk** that is in orbit **(Figure E2.3)**. Some space junk is less problematic, as it originates from satellites located in **geostationary orbit (GEO)** above a fixed point 22,236 miles (35,786 kilometers) above Earth, widely spaced and not in motion relative to each other. However, as can be seen in Figure E2.1, many satellites are also located in low Earth orbit (LEO), the region that lies between 99 miles (160 kilometers) and 1243 miles (2000 kilometers) above Earth, in which satellites travel at speeds of around 17,500 miles per hour, roughly ten times faster than a speeding bullet. While hundreds of millions of pieces of space junk have accumulated in LEO, most of them tiny pieces broken off of nonfunctional satellites or rockets, only a few collisions involving space junk have been observed. In fact most collisions that damage functioning satellites come from natural meteoroids, many of which are smaller than a grain of sand, and even these occur at a rate of only about one a year. Nevertheless, without adequate precautions space junk could accumulate to the point where travel through LEO would be impossible, a situation now

Figure E2.2 Artist's rendition of Earth's magnetic field. Satellite is shown in orbit and is not to scale. [Science & Society Picture Library/Getty Images]

Figure E2.3 A rocket booster that fell to Earth without burning up and landed in Vietnam. Most pieces of space junk are much smaller. [STR/AFP/Getty Images]

satellite any object in space that orbits another object of greater mass

space junk human-created space debris

geostationary orbit (GEO) an orbit located at least 22,236 miles (35,786 kilometers) above a fixed point on Earth, in which satellites do not move relative to each other known as the **Kessler syndrome**, named after the NASA scientist who first proposed it. All spacefaring nations try to minimize the accumulation of space junk, with most requiring that satellites relocate themselves after their useful life is complete, either to an out-of-the-way "graveyard orbit" in GEO or back toward LEO, where they will burn up on reentering the atmosphere.

Environmental Benefits of Satellites

Many satellites provide *remote sensing* services, gathering a wide array of visual, photographic, and electronic signal data that help geographers and other scientists to monitor human impacts on the biosphere. For example, these data are used to estimate rates and locations of deforestation, to contain oil spills and forest fires, and to monitor sea level rise, the growth and melting of glaciers, and changes in temperature at the Earth's surface. Weather satellites assist forecasting by monitoring cloud formations, temperature, and precipitation, while navigation satellites, such as the Global Positioning System, provide accurate latitude and longitude as well as elevation data that are now a central part of most efforts at environmental management.

Kessler syndrome a situation where so much space junk accumulates around Earth that travel through low Earth orbit becomes impossible

CHECK YOUR UNDERSTANDING

1. How can charged particles from the Sun influence human infrastructure?
2. What hazards and benefits do human-made satellites create?
3. How likely is it that an asteroid large enough to threaten human life will pass close to the Earth in your lifetime?

GLOBALIZATION AND DEVELOPMENT

E2.2 Evaluate how commercial activity in space is expanding and propelling future space exploration.

The range of commercial activity in space is expanding from satellite-based enterprises to include space launching, space-based manufacturing, and asteroid mining. The space economy will likely reach a size of $3 trillion in the next 30 years, equivalent to the world's fifth-largest economy today, the UK, with satellite-based activities financing most of this growth.

THE SATELLITE ECONOMY

Over three-quarters of economic activity in space is financed by services provided by communications satellites. Space-based commerce began in 1962 with the launch of the first telecommunications satellites **(Figure E2.4)**, but commerce didn't dominate space until the late 1990s when for-profit satellite launches overtook those for military and civilian government purposes. Commercial satellite launches now far outnumber all other types of launches **(Figure E2.5)**.

Roughly a quarter of all satellites ever launched have provided *remote sensing* services to national militaries, corporations, and the general public, and they are a major part of the space economy. In the early years most were "spy satellites" gathering a wide array of visual, photographic, and electronic signal data, and almost all belonged to the United States and Russia (see Figure E2.5). The extent of secret surveillance that spy satellites are capable of is not publicly available information, but information from these satellites has been leaked to the internet by people and organizations that object to what they see as a breach of their right to privacy (see Chapter 2). Most remote sensing satellites are now owned by corporations, but military spy satellites are some of the most advanced and expensive satellites in orbit, and their creation accounts for a large amount of the money earned by satellite manufacturers. Remote sensing satellites still form the majority of satellites launched into orbit each year, though in the near future they will be vastly outnumbered by communications satellites.

After remote sensing satellites, communications satellites are the next-largest group, comprising about 16 percent of all satellites ever launched and the majority of satellites currently being launched or planned for the next decade. Most are used to extend the range of television, radio, and internet communication signals that must traverse long distances. Because most broadcasted signals travel in a straight line, they are limited in range by the

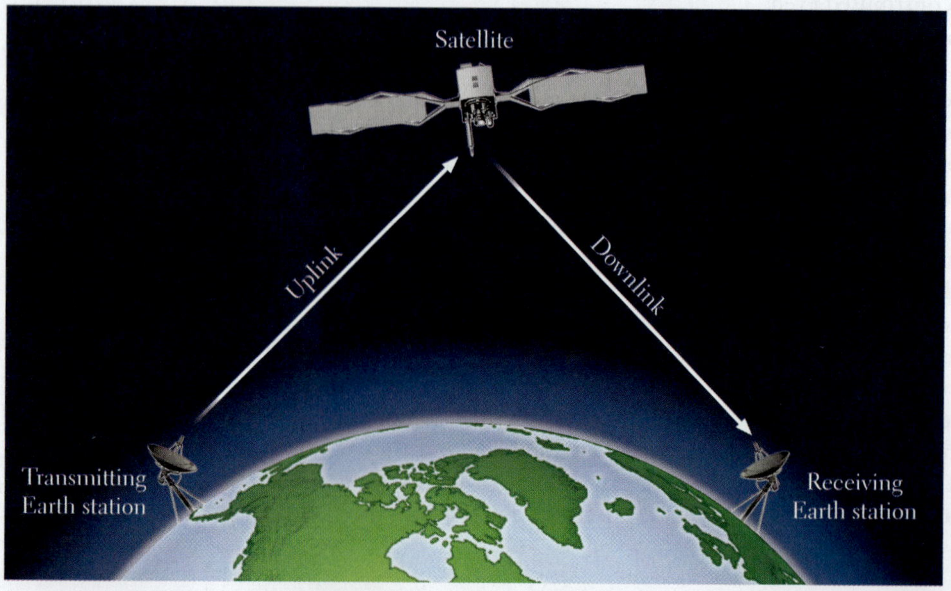

Figure E2.4 Satellites and signals. Because of their high altitude, satellites can extend the range of communications signals on Earth that would otherwise be blocked by topography and the curvature of Earth. [Research from: http://www.radio-electronics.com/info/satellite/communications_satellite/communications-satellite-technology.php.]

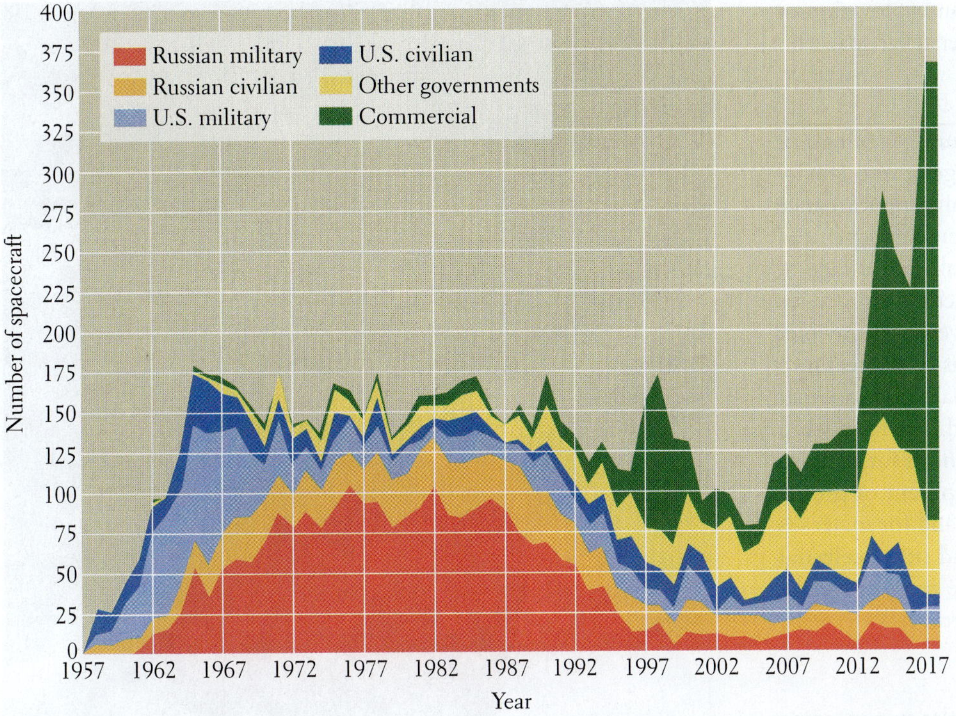

Figure E2.5 Space launches from 1957 to 2017. Notice the dominance of the Russian and U.S. military space launches until the late 1980s, and the recent growth of other governmental and commercial space launches since then. [Data from: http://claudelafleur.qc.ca/images/YR-5714.jpg.]

curvature of Earth as well as by hills and mountain ranges. Communications satellites receive signals from the surface of the planet and bounce them back, thus allowing them to avoid these obstacles (see Figure E2.4). Over 20,000 communications satellites in orbit are planned for launch over the next decade, most to provide global wireless high-speed internet access to even the most remote areas on Earth. This is more than three times the number of satellites ever launched, and more than ten times the number of currently functioning satellites.

THE IMPACTS OF SATELLITES IN RURAL AREAS

Satellites have already had transformative influences on society, especially in rural areas. Most of the money driving the current growth in satellite services comes from rural people, primarily in the developed world, who pay over $130 billion a year for access to satellite television and internet. This sector is now shifting over to internet-based television services given the massive investments being made in satellite networks that provide global high-speed internet access. Currently just over half the world's population has no access to high-speed internet, with over 700 million people in China, and more than 1 billion people in Africa lacking access. Economists estimate that for every 10 percent of the population that gains access to high-speed internet, the rate of economic growth increases by 1–1.5 percent.

Satellites are already providing important services to rural areas in the developing world. Rural schools often have satellite-based internet links to give access to online lessons, and rural health-care centers can access online medical records and databases to improve the quality of care. Politics as well is changing as satellite-connected rural polling stations are better able to monitor and avoid election fraud, a service that benefited Burkina Faso in its elections of 2012 and 2015.

Disaster response is also improving, with satellite communications providing better continuity of business and governance during hurricanes as well as rapid and precise monitoring of floods and other events that damage infrastructure. This has occurred during and after every major flood or hurricane across the globe since the mid-1960s, though much of this information was closely guarded by national militaries until the 1970s.

Satellites have long provided information relevant to agriculture, and since 1985, 34 of the world's poorest countries have benefited from the prediction of potentially low crop yields as part of the Famine Early Warning Systems Network created by USAID and NASA, which is based largely on analysis of satellite imagery. Meanwhile precision agriculture systems, such as those that have helped make the tiny country of the Netherlands the world's second-largest food exporter by value after the United States, use satellite data to give farmers a detailed picture of the needs of their crops, enabling them to get the most benefit from the application of irrigation water and agricultural chemicals.

SPACE LAUNCHES

Commercial activity has expanded into the building and launching of rockets and other spacecraft, with private companies now providing these services at significantly lower costs than the government-sponsored space programs that once dominated this sector. SpaceX, a U.S. company founded in 2002 by former PayPal founder and Tesla Motors CEO Elon Musk (of South Africa), has been largely responsible for driving costs down with reusable rocket technology. From the 1970s to 2000 it cost $8400 to launch a pound into LEO. In 2006 SpaceX cut that to $4500 per pound, and then to $640 per pound in 2019, with costs for future launches potentially as low as $23 per pound. This has opened space access to countries like Bangladesh and Indonesia, spurred satellite manufacturers to increase their production, and forced Russia to refocus on satellite manufacturing and NASA to focus on exploration of the more distant regions of interplanetary and interstellar space.

MANUFACTURING IN SPACE

With the cost of accessing space plummeting, a number of manufacturing processes are developing to take advantage of zero gravity and the extreme environments of space. Already high-performance fiber-optic cables, solar panels, and three-dimensional printed mammalian organs are being made in space to take advantage of the zero-gravity environment. Some toxic manufacturing processes may be relocated to space to make use of high levels of ultraviolet radiation, which can rapidly break down dangerous substances into more inert forms, reducing their pollution potential. Processes that require cooling may also take place in space, with the microchip manufacturing industry potentially avoiding the need for trillions of gallons of cooling water each year. The development of profitable manufacturing in space will increase the demand for materials in space, spurring on the search for asteroids and other celestial bodies that can supply resources, thus avoiding the costly and risky process of launching materials into space from Earth.

MINING IN SPACE

There are more than 100 million asteroids in the solar system, some of which contain precious metals in quantities valued in the trillions of dollars. The enormous profit potential of asteroid mining is a growing force of interplanetary space exploration **(Figure E2.6)**, with Japan's space program already having retrieved samples from an asteroid and brought them to Earth in 2010 and again in 2019. A growing number of companies are currently planning to mine asteroids, primarily with robotic spacecraft operated from Earth's surface.

The effects asteroid mining could have on Earth are hard to predict. A single asteroid could hold more gold, platinum, or even iron than has ever been mined in human history. While most materials mined from such an asteroid would likely be used in space, if even a relatively small percentage were brought back to Earth, this could collapse the prices of many metals and jeopardize millions of mining jobs. However, a larger supply of these metals could support new industries that would employ millions more people, for example in the manufacture of fuel cells that use platinum or electronic circuits that use gold. Even without the trillion-dollar profits that are predicted, asteroid mining will probably take place in order to supply deep space exploration missions, given that some asteroids hold large amounts of water, which can be split into hydrogen for fuel and oxygen for respiration.

While tests and plans for asteroid mining are now well advanced, many technical barriers remain. Nevertheless, both the United States and Luxembourg are developing laws to regulate space, and policies that encourage asteroid mining.

CHECK YOUR UNDERSTANDING

1. What is currently financing most activities in space?
2. Where do the consumers financing the growth in satellite services tend to live?
3. What technology is driving down the cost of space launches?
4. What type of manufacturing is asteroid mining likely to supply?

Figure E2.6 Mining outpost. An artist's rendition of a mining outpost on an asteroid near Earth that is about 2000 feet (610 meters) wide and 1000 feet (305 meters) long. [VICTOR HABBICK VISIONS/Getty Images]

POWER AND POLITICS

E2.3 Analyze current disagreements about the militarization of space and the rights of individuals and corporations to own property in space.

While all spacefaring nations have agreed that no parts of outer space can be claimed as national territory, there is disagreement over the rights of individuals or corporations to own property in space. Meanwhile agreements to limit the use of weapons of mass destruction (including nuclear weapons) in space have not limited the militarization of space, with national militaries continuing to play a central role in space exploration.

SPACE AND THE LAW

Following intense lobbying by several asteroid-mining companies, the Space Act of 2015 was passed by the U.S. Congress and signed into law by President Obama, giving companies the rights to exploit any minerals or other nonliving resources they find in space. The law also states that the U.S. government will not claim any part of space as national territory, which is an attempt to avoid a clear break with the **Outer Space Treaty of 1967**. This treaty, which the United States and all spacefaring nations have signed, forbids nations from claiming territory in space and declares that the exploration of all celestial objects "shall be carried out for the benefit and the interests of all countries and shall be the province of all mankind." The U.S. government seemed to uphold this principle when it cited the treaty during its successful defense in court in 2001 against a parking ticket issued it by beef jerky entrepreneur and space enthusiast Gregory Nemitz, who had registered a claim to the asteroid Eros, which NASA had just landed a probe on. Nevertheless, both Russia and Brazil have criticized the Space Act of 2015 for promoting ownership of space resources in a way that violates the Outer Space Treaty of 1967.

The U.S. government has so far made no move to withdraw from the Outer Space Treaty of 1967, possibly due to other provisions of the treaty that ban the use or deployment of weapons of mass destruction in space. While the treaty doesn't ban weapons from space (and the United States, Russia, and China have already deployed weapons there), many argue that it reduces the potential for armed conflict in space by preventing nations and corporations from making any broad territorial claims.

THE MILITARIZATION OF SPACE

National militaries have played a pivotal role in space exploration, and the first object to enter space was a German V2 rocket during World War II. After the war, former Nazi military scientists played key roles in both the U.S. and Soviet space programs, which were contained within military agencies or evolved out of them. Much space exploration has been shaped by competing military interests, such as the Cold War between the United States and the Soviet Union, and the more recent U.S. military rivalry with China. Much of the hardware that makes space exploration possible was developed by national militaries, including the intercontinental ballistic missiles (originally designed to carry nuclear warheads) that have propelled most launches into space, as well as the many space-based systems for remote sensing, communication, and navigation. Every major military power on Earth now considers these space-based technologies to be an essential part of warfare, and all the major spacefaring nations now have militarized "space forces" that constantly train for war in space to protect their satellites and, if deemed necessary, disable those of other countries. Of Earth's military "space powers," the United States is the largest, with a recently declared "space force" that is in fact a continuation of activities by the U.S. Air Force since the 1950s. The U.S. military has roughly double the number of functioning satellites in orbit (around 200) as the next-largest "space powers," Russia and China, combined.

CHECK YOUR UNDERSTANDING

1. What are the main provisions of the most important international legal agreement currently regulating the use of space?
2. What rights does the Space Act of 2015 give?
3. What countries currently have the largest "space forces" and what is their purpose?

URBANIZATION

E2.4 Assess the usefulness of space-based remote sensing technologies to better understand urban processes on Earth and to help cities adapt to change.

The challenges of life in the harsh environment of space currently limit human settlements to tiny space stations orbiting the Earth, and in the near future the moon. Urban settlements on the lunar or Martian surfaces are more distant prospects. However, the use of space-based remote sensing technologies to better understand urban processes on Earth and to help cities adapt to change is a well-developed and growing field within geography and related disciplines.

THE INTERNATIONAL SPACE STATION

At a cost of more than U.S.$150 billion, the International Space Station (ISS) is the largest peacetime international collaboration in history. Its two largest participants in terms of funding and personnel, the United States and Russia, were once adversaries in the Cold War and are still occasional rivals. Since its first components were lifted into orbit in 1998, the ISS has served as a platform for several thousand experiments on microgravity, radiation, space-based manufacturing, and a wide variety of other subjects. As long as a football field from end to end, the ISS is the largest human-made object in orbit and is visible with the naked eye. Crews often do not share a common language, are isolated from family and peers, and adapt to a noisy and sometimes hazardous indoor environment plagued by mold and fungi. The ISS's record of unbroken collaboration in these difficult circumstances has earned it multiple peace-related awards, including a nomination for the Nobel Peace Prize, and provides an important basis for future international cooperation in space.

But is a space station a city? The three to seven astronauts who live on board the ISS at any one time, or even the over 230 people who have visited the ISS over its more than two decades of operation, certainly don't qualify. However, the amount of money invested in the ISS is as large as the entire residential real estate market of a large U.S. city, such as Nashville, Tennessee, or the entire GNI of Hungary. As the ISS comes to the end of its useful life in the next decade, planned Earth-orbiting space stations are cheaper and more numerous, and at least some are being funded by private space entrepreneurs focusing on low-gravity manufacturing and tourism. Only one country, China, currently maintains its own space station, but several private space stations are under development. Two inflatable unmanned space habitats created by Bigelow Aerospace, funded by U.S. billionaire Robert Bigelow, are currently orbiting Earth. The planned successor to the ISS will orbit the moon, acting as a base for future exploration of the lunar and Martian surfaces.

HUMAN SETTLEMENTS BEYOND EARTH'S ORBIT

While the government space programs of the United States, Russia, China, Japan, the EU, Israel, and India have explored asteroids, the moon, and Mars with probes, private space efforts are actively

Outer Space Treaty of 1967 treaty that the United States and all spacefaring nations have signed that forbids nations from claiming territory in space pursuing colonization of these extremely hazardous places. While Elon Musk's SpaceX envisions colonization of Mars, Jeff Bezos of Amazon and the space company Blue Origin envisions human colonization of the moon and space itself. In any of these places, the physical challenges posed to humans are formidable, involving exposure to radiation, meteoroids, the effects of microgravity, and the need to minimize the use of materials from Earth. Shielding against radiation and space debris has long been included in spacecraft and space stations, but long-term settlements would need more extensive protection. Human settlements on asteroids, moons, and Mars will likely take advantage of underground architecture, which provides shielding of all kinds as well as insulation from the huge temperature swings found in interplanetary space. But these protections will do little to mitigate the effects of long-term exposure to microgravity, which results in loss of bone density, declining muscle mass, and permanently impaired vision after even short visits to space. In addition to these biological obstacles to urbanization in space, the high costs of launching material into space will require extensive capabilities in materials recycling and space-based resource acquisition. Potentially large earnings from space-based manufacturing and space mining could fund further advances in technological capability that may overcome these obstacles to the development of long-term space-based settlements.

REMOTE SENSING OF URBAN AREAS

While actual urban settlements in space may be decades or more away, observations of Earth from satellites in space have been contributing to our understanding of urbanization since the 1970s. Remotely sensed data are often made into spatially precise maps that are particularly useful for understanding urban growth over time, especially when combined with computer models that help us understand why cities grow the way they do. Satellite data are also used in urban disaster management because of the large amount of detailed spatial information that can be rapidly supplied. For example, during the aftermath of Hurricane Katrina, residents and emergency response personnel in New Orleans, Louisiana, needed maps of flooding in the city. Remote sensing scientists at the U.S. Geological Survey's Center for Earth Resources Observations and Science (EROS) were able to provide detailed and comprehensive maps showing the exact location and depth of the flooding within days. This information proved invaluable to emergency responders as they struggled to assist residents trapped in their homes and on rooftops. Over the course of the next weeks and months, scientists at EROS were able to provide detailed maps and digital images of the disaster **(Figure E2.7)**, which became essential tools for city planners and disaster management personnel as they tried to understand why the city's flood-protection systems had failed. Looking forward, geographers and urban planners will be using remotely sensed satellite data on a daily basis to help cities across the world adapt to new technologies such as driverless vehicles that depend on satellite data to safely navigate in urban areas, as well as to climate change and growing urban populations.

Figure E2.7 Satellite image of New Orleans after Hurricane Katrina. A satellite image of the 800-foot-wide (240-meter) break in East New Orleans' Industrial Canal in Louisiana and surrounding flooding on August 31, 2005, two days after Hurricane Katrina hit the city. The image was collected from QuickBird, a commercial satellite capable of capturing objects larger than 5.2 feet (1.6 meters) from an orbit 280 miles (450 kilometers) above the surface of Earth. [DigitalGlobe/Getty Images]

CHECK YOUR UNDERSTANDING

1. Why is the International Space Station widely celebrated?
2. What are the challenges of life in the harsh environment of space?
3. What are some factors that could encourage long-term human habitation in space?
4. How have space-based remote sensing technologies helped us better understand urban processes on Earth?
5. How can satellite imagery be used in urban disaster management?

POPULATION, GENDER, AND CULTURE

E2.5 Describe how gender issues have shaped space exploration, and how space exploration might change cultures across Earth.

While it is not yet clear that a long-term, sustained human population is possible in space, observation of Earth from space has been providing insights into the study of human populations for decades. Meanwhile, space exploration and research have been profoundly shaped by culturally defined notions of gender. These and other aspects of human culture are on a course for transformation in light of looming advances in space exploration.

FERTILITY IN SPACE

At present, human abilities to produce healthy populations in space appear uncertain. Studies with animals show that while reproduction is possible in space, exposure to radiation and zero gravity significantly reduces sperm and egg counts and may result in infertile offspring. Studies of animals born in space reveal many birth defects, including poor equilibrium due to delayed development of cells that sense gravity. NASA offers astronauts the option of having samples of their sperm and eggs frozen for later use before they go into space, though many have conceived and given birth to healthy children after they have returned.

USING SATELLITES TO STUDY POPULATION

Remote sensing satellites have been providing important data for estimating population size, location, levels of economic development, and potential health hazards. The World Bank is now testing the use of satellite images to help direct its economic development efforts in rural Sri Lanka, by analyzing the size and height of buildings, the use of electric lights at night, and the presence of automobiles as indicators of population density, distribution, living standards, and the position along the demographic transition. Conditions that contribute to disease can also be detected by using satellites. For example, remotely sensed images of algal blooms in the waters of coastal Bangladesh have helped predict the locations of outbreaks of cholera, and satellite data on vegetation cover and precipitation patterns have been used to guide mosquito-eradication efforts aimed at containing outbreaks of malaria in Thailand.

GENDER IN SPACE

There is a significant gender gap in space-related research and exploration. In 1963, two years after the Soviet Union sent the first man (Yuri Gagarin) into space, they sent the first woman (Valentina Tereshkova) into space. She was the twelfth human in space, and it was another 19 years and 76 crewed space missions before the Soviet Union sent the second woman into space. The United States had a 20-year ban on women in the space program, despite early studies that showed that women had particular advantages for space travel, such as smaller size, a lower metabolic rate, and excellent performance under stress. The perception by NASA engineers that women were unsuited to the stresses of spaceflight delayed women's participation in the U.S. space program until the late 1970s, and the first U.S. woman wasn't launched into space until 1983. Women still represent less than 14 percent of human spaceflights since 1983, but that share has increased to 31 percent of spaceflights since 2013. The gender gap in space exploration continues, in part because of the requirement of most space programs that astronauts have at least a bachelor's degree in a STEM (science, technology, engineering, or mathematics) field. Women are historically underrepresented in these fields due to a variety of cultural influences, including teachers, friends, and families who discourage girls and young women from taking advanced math and science courses.

To date there have been no openly LGBT people in space, although U.S. astronaut Sally Ride was revealed to have been the first lesbian in space after her death in 2012.

CULTURAL ISSUES

All of the risks and uncertainties of space travel and exploration are dwarfed by the implications of finding even just a single new planet hospitable to humans, or a single form of intelligent life. What will human occupation of other worlds mean for life on Earth? If the past history of European colonialism is any example, the effects will be complex and not completely understood, even hundreds of years after the first space colonies are established. Given the risks and the drain on resources, human and otherwise, many question the wisdom of exploring deep space at all.

All major space programs have a focus on Mars, which is the next most habitable planet in the solar system after Earth. However, settling Mars presents such enormous technical challenges that astronomers have long looked beyond the solar system into deep space for more habitable planets. NASA's Kepler space observatory is a Sun-orbiting space telescope launched in 2009 and designed to find Earth-sized planets in the Milky Way galaxy that have similar temperature ranges as those on Earth, given their location in their solar systems. By 2019, scientists using Kepler had confirmed the existence of more than 46 such planets, and based on this and other data, astronomers estimate that there are around 40 billion habitable Earth-sized planets in the Milky Way galaxy. The closest to Earth is Proxima Centauri b, which orbits the star nearest the sun at 4 light years away. NASA estimates that humans will need another 50 years of technological development to create a spacecraft capable of traveling one-tenth of the speed of light, which would take 40 years to reach Proxima Centauri b.

What Will Space Exploration Do to Life on Earth?

The extension of human civilization into outer space has already brought significant changes to Earth, and future space exploration is on course to parallel or even exceed the transformations brought about by European colonialism in the Americas. Already the focus on asteroid mining mirrors the objectives of early Spanish colonialists, who came to the Americas seeking gold and other precious metals, often leaving chaos and disorder in their wake. The exploration of space could bring new biological resources, similar to the exchange of food crops between Europe and the Americas that eventually transformed the entire world. The fate of indigenous American cultures, who were nearly wiped out by diseases brought by European explorers, will hopefully be avoided thanks to extensive contamination control efforts currently under development by astrobiologists and bioengineers.

The potential for huge profits to be reaped from space exploration offers many interesting scenarios of cultural transformation as well. With a single asteroid capable of delivering almost unimaginable riches, disparities in wealth could increase as mines on Earth close while asteroid-mining operations expand, bringing in enormous profits. On the other hand, resources from space could also provide the basis for a new age of technological advancement and higher living standards, much like the resources of the Americas fueled Europe's Industrial Revolution of the 1800s.

Perhaps the biggest implications of human space exploration involve potential encounters with intelligent extraterrestrial life, a subject that has permeated popular culture for more than a century. Humans have been searching for signals of extraterrestrial life using radio antennae since the 1890s, and this ongoing work has been funded by the National Science Foundation, NASA, and private sources since the 1950s. The U.S. military has investigated the topic on multiple occasions, most recently from 2007 to 2012 through the Advanced Aerospace Threat Identification Program (AATIP). In 2017 the program released video taken by cameras on board U.S. Navy jets of seemingly intelligently controlled flying objects demonstrating capabilities beyond any known aircraft. Most of the program's $22 million supported research into these flight characteristics and analysis of materials recovered from vehicles that may or may not be of terrestrial origin. AATIP funds were administered by Bigelow Aerospace, known for its billionaire founder's enduring obsession with extraterrestrial life and its inflatable space habitats currently in use on the ISS. Since its formal end, much of AATIP's research has been taken up by private investors and corporations.

Is Exploration Beyond Geospace Wise?

While geospace is now a central part of global communications infrastructure, moving beyond this area will require major efforts at a time when humans are already straining Earth's resources. With asteroid mining in its infancy, Earth's resources will have to be relied on for the next decade at least. Nevertheless, some billionaires such as Jeff Bezos of Amazon and the space launch company Blue Origin envision that space-based manufacturing will accelerate development of space mining, and eventually allow the most polluting and energy-intensive processes to be relocated to geospace or the Moon, thus preserving the Earth's biosphere. But even without access to the resources of space, many justify further exploration with the technological "spin-offs" that this activity has already produced, such as more efficient solar panels, satellite monitoring of deforestation and land use change, better water filtration systems . . . the list goes on. Some claim technological spin-offs have yielded as much as $7 for every $1 spent, but the fact is that the impacts of space research and exploration, especially on ecosystems, are hard to quantify. Regardless, even in the present era when only governments and billionaires can afford to participate, objections to further space exploration on purely environmental grounds are largely being disregarded. Meanwhile as the cost of accessing space continues to fall, more people are doing so, with potentially profound implications for life on Earth.

CHECK YOUR UNDERSTANDING

1. What are the challenges to fertility in space?
2. How are satellite data being used to study human populations?
3. How have gender issues shaped space exploration?
4. How can the history of European colonialism in the Americas provide insights into how contact with other planets might transform life on Earth?
5. What are some arguments for and against exploration beyond geospace?

CRITICAL THINKING QUESTIONS

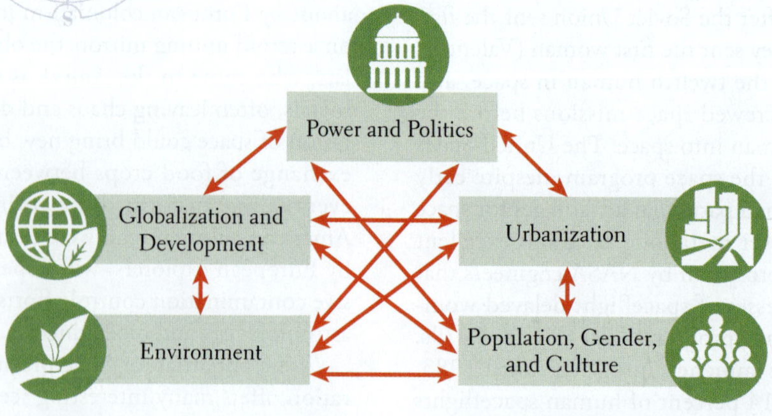

The diagram represents connections among the five geographic themes that structure this book. Listed below are some important questions that have been addressed in this epilogue. Answer each question, and indicate where in the diagram you think the topics in each question belong.

1. If the Kessler syndrome occurred, how would it impact your life?
2. How has your life been influenced by the use of data from satellites (remote sensing)?
3. If asteroid mining became a major industry, what places described earlier in this book would be the most affected?
4. What factors make war in space more likely or unlikely?
5. Given what you have learned about gender from earlier chapters of this book, do you think that the gender gap in space exploration and research will persist, or will it shrink until men and women are equally represented in these fields?
6. What problems and opportunities for Earth could human habitation of other worlds bring?

Key Terms

geospace 670
geostationary orbit (GEO) 671
intergalactic space 670
interplanetary space 670
interstellar space 670
Kessler syndrome 672
low Earth orbit (LEO) 670
Outer Space Treaty of 1967 675
satellite 671
solar wind 670
space junk 671

More Practice at SaplingPlus

Read the interactive e-text, review key concepts, and check your understanding.

Glossary

Note: These definitions are only partial and are to be understood in the context in which they appear in the text.

1884 Berlin Conference (p. 382) a meeting that divided Africa among European trading companies and colonial powers, establishing many boundaries that persist to this day

Aboriginal Australians (p. 630) the longest-surviving inhabitants of Oceania, whose ancestors, the Australoids, migrated from Southeast Asia 50,000 to 70,000 years ago over the Sundaland landmass that was exposed during the ice ages

acculturation (p. 169) adaptation of a minority culture to the host culture enough to function effectively and be self-supporting; cultural borrowing

acid rain (p. 71) precipitation that has formed through the interaction of rainwater or moisture in the air with sulfur dioxide and nitrogen oxides emitted during the burning of fossil fuels, making it acidic

Active Aging (p. 226) a three-pronged EU policy aimed at changing society's attitudes toward the abilities and needs of people as they age

agribusiness (p. 86) the business of farming conducted by large-scale operations that purchase, produce, finance, package, and distribute agricultural products

agroforestry (p. 372) growing economically useful crops of trees on farms, in conjunction with the usual plants and crops

Ainu (p. 529) an indigenous minority group in Japan characterized by their light skin and thick, wavy hair, who are thought to have migrated thousands of years ago from the northern Asian steppes

alluvium (p. 530) river-borne sediment

Altiplano (p. 181) an area of high plains in the central Andes of South America

animism (p. 409) a belief system in which spirits, including those of the deceased, are thought to exist everywhere and to offer protection to those who pay their respects

Aotearoa (p. 618) Polynesian name for New Zealand meaning "land of the long white cloud"

apartheid (p. 385) a system of laws mandating racial segregation in South Africa from 1948 until 1994

aquifers (p. 69) ancient natural underground reservoirs of water

arable land (p. 646) suitable for growing crops

archipelago (p. 555) a group, often a chain, of islands

Asia–Pacific region (p. 547) a huge trading area that includes East Asia, South Asia, Southeast Asia, and Oceania

Asia Pacific Economic Cooperative (APEC) (p. 640) a group of 21 Pacific Rim countries organized in 1989 to increase trade and cooperation

assimilation (pp. 169, 229) the loss of old ways of life and the adoption of the lifestyle of another culture

Association of Southeast Asian Nations (ASEAN) (p. 576) an organization of Southeast Asian governments that furthers economic growth and political cooperation between member countries and with other areas of the world

asylum-seekers (pp. 229, 641) refugees who left their homes because of violent persecution and seek new legal status in an adopted country; refugees whose claims have not yet been evaluated—usually they are seeking protection from persecution on account of race, religion, nationality, or politics

atoll (p. 618) a low-lying island or chain of islets, formed of coral reefs that have built up on the rim of a submerged volcano

Australo-Melanesians (p. 568) a group of hunters and gatherers who moved approximately 40,000 to 60,000 years ago from what are now northern Indian and Burma (Myanmar) to the exposed landmass of Sundaland and present-day Australia

austerity (p. 149) reductions in government deficits through cuts in spending and/or higher taxes

Austronesians (p. 568) a group of skilled farmers and seafarers from southern China who migrated south to various parts of Southeast Asia between 6000 and 10,000 years ago

authoritarianism (p. 39) a political system that subordinates individual freedom to the power of the state or of elite regional and local leaders

Aztecs (p. 143) indigenous people of high-central Mexico noted for their advanced civilization before the Spanish conquest

banlieues (p. 220) the troubled suburbs of Paris where a high proportion of recent immigrants live, excluded from decent employment, higher education, and many other advantages of urban life

Berber (p. 350) an indigenous group of people in North Africa that are most numerous in Morocco and Algeria

bioaccumulation (p. 666) a process in which organisms absorb toxins faster than their bodies can excrete them

biodiversity (p. 134) the variety of life forms to be found in a given area

biomagnification (p. 666) a process in which the concentration of toxins increases at successive levels of the food chain

biosphere (p. 18) the entirety of Earth's integrated physical spheres, with humans and other impacts included as part of nature

birth rate (p. 47) the number of births per 1000 people in a given population, per unit of time (usually per year)

Bolsheviks (p. 270) a faction of communists who came to power during the Russian Revolution

brain drain (p. 161) the migration of educated and ambitious young adults to cities or foreign countries, depriving the communities from which the young people come of talented youth

Brexit (p. 193) the 2016 decision by UK voters to leave the European Union; details of how this exit will proceed remain unresolved as of late 2019

brownfields (p. 100) old industrial sites whose degraded conditions pose obstacles to redevelopment

Buddhism (p. 444) a religion of Asia that originated in India in the sixth century B.C.E. as an effort to reform and reinterpret Hinduism

buffer state (p. 298) a country that is situated between two rival and more powerful political entities and serves as a neutral territory, limiting direct contact and potential conflict between the two powers

Bumiputra (p. 605) an affirmative action program in Malaysia to address economic and social inequalities between elite Overseas Chinese business people and indigenous (and other) disadvantaged ethnic groups

capitalism (pp. 42, 208) an economic system based on the private ownership of the means of production and distribution of goods, driven by the profit motive, and characterized by a competitive marketplace; an economic system characterized by privately owned businesses and industrial firms that adjust prices and output to match the demands of the market

capitalists (p. 270) usually a wealthy minority that owns the majority of factories, farms, businesses, and other means of production

carbon sequestration (p. 372) the removal and storage of carbon taken from the atmosphere

cartel (p. 325) a group of producers strong enough to control production and set prices for products

caste system (p. 443) a complex, ancient Hindu system for dividing society into hereditary hierarchical classes

Caucasia (p. 260) the mountainous region between the Black Sea and the Caspian Sea

central planning (p. 209) a communist economic model in which a central bureaucracy dictates prices and output with the stated aim of allocating goods equitably across society according to need

centrally planned economy (p. 270) an economic system in which the state owns all land and means of production, while government officials direct all economic activity, including the locating of factories, residences, and transportation infrastructure

chaebol (p. 543) huge corporations in South Korea that are involved in a variety of economic sectors

Christianity (p. 318) a monotheistic religion based on the belief in the teachings of Jesus of Nazareth, a Jew who described God's relationship to humans as primarily one of love and support, as exemplified by the Ten Commandments

civil disobedience (p. 446) protesting of laws or policies by peaceful direct action

civil society (p. 42) the social groups and traditions that function independently of the state to foster a sense of unity and common purpose among the general population

clear-cutting (p. 74) a method of logging that involves cutting down all trees on a given plot of land regardless of age, health, or species

climate (p. 13) the long-term balance of temperature and precipitation that characteristically prevails in a particular region

climate change (p. 18) a slow shifting of climate patterns due to the general cooling or warming of the atmosphere

Cohesion Policy (p. 211) (often referred to as *social cohesion*) EU guidelines for programs aimed at removing social and economic disparities across Europe

Cold War (pp. 208, 271) a period of conflict, tension, and competition between the United States and the Soviet Union that lasted from 1945 to 1991; the geopolitical rivalry that pitted the United States and western Europe, who were espousing free market capitalism and democracy, against the USSR and its allies, who were promoting a centrally planned economy and a communist state

colonialism (p. 33) a period starting in about 1500 C.E., when European countries began conquering distant parts of the world and extracting their resources

commodities (p. 386) raw materials that are traded, usually to other countries, for processing or manufacturing into more valuable goods

commodity dependence (p. 387) economic dependence on exports of agricultural and mineral raw materials

Common Agricultural Program (CAP) (p. 214) an EU program meant to guarantee secure and safe food at affordable prices, produced sustainably by farmers who earn a fair income

communism (pp. 42, 208, 270) an economic system based on a state-controlled economy, a socialized system of public services, and a centralized government in which citizens participate only indirectly through the Communist Party; a political ideology and economic system, based largely on the writings of the German revolutionary Karl Marx, in which the state owns all farms, industry, land, and buildings and provides for the needs of the people (a version of socialism); an ideology that calls on workers to overthrow capitalism and establish an egalitarian society where workers share what they produce

Communist Party (p. 270) the political organization that ruled the USSR from 1917 to 1991; other communist countries, such as China, Mongolia, North Korea, and Cuba, also have ruling communist parties

Confucianism (p. 500) a Chinese philosophy that teaches the importance of stability and social order based on traditional institutions such as the patriarchal family, community, and the state

consensus (p. 641) a strategy for achieving broad group unanimity, not based on majority rule, but on all parties adjusting their desires to fit those of the entire group

continental climate (p. 196) a climate pattern in which summers are fairly hot and moist and winters become longer and colder the deeper into the interior of the continent one goes

contested space (p. 157) any area that two or more groups claim or want to use in different and often conflicting ways, such as the Amazon or Palestine

conurbation (pp. 97, 516) an area formed when several cities expand so that their edges meet and coalesce

coral bleaching (pp. 566, 623) color loss that results when photosynthetic algae that live in the corals are expelled due to warming sea water or acidification as excess CO2 is absorbed; occurs when the living coral organisms expel the algae that live inside and give the coral its color

coup d'état (p. 157) a military- or civilian-led forceful takeover of a government

Creoles (p. 144) people mostly of European descent born in the Americas

crony capitalism (p. 573) a type of corruption in which politicians, bankers, and entrepreneurs, sometimes members of the same family, have close personal as well as business relationships

cultural homogenization (p. 229) the tendency toward uniformity of ideas, values, technologies, and institutions among associated culture groups; the loss of ethnic distinctiveness

cultural landscape (p. 4) the combination of human and natural features in a landscape, including aspects designed intentionally by people and parts that evolved unintentionally out of human interactions with nature

cultural pluralism (p. 593) the cultural identity characteristic of a region where groups of people from many different backgrounds have lived together for a long time but have remained distinct

Cultural Revolution (p. 503) a political movement launched in 1966 to force the entire population of China to support the continuing revolution

culture (p. 54) all the ideas, materials, and institutions that people have invented that are not directly part of our biological inheritance

czar (p. 269) the ruler of the Russian empire

decolonization (p. 209) the process of dissolving or ending colonial relationships, institutions, and mind-sets

death rate (p. 47) the ratio of total deaths to total population in a specified community, usually expressed in numbers per 1000 or in percentages

demilitarized zone (p. 504) a border area between rival states where military activities are prohibited; usually refers to the border between North and South Korea

democratization (p. 39) the transition toward political systems guided by competitive elections

demographic transition (p. 47, 589) a phenomenon where births and deaths go from being high to being so much lower that population growth is miniscule or slightly negative; it is usually accompanied by a cluster of other changes, such as change from a subsistence to a cash economy, increased education rates, and urbanization

desalination (p. 315) the removal of salt from seawater—accomplished through the use of expensive and energy-intensive technologies—to make the water suitable for drinking or irrigating

desertification (p. 313) a set of ecological changes where a dry area with some vegetation is turned into desert

detritus (p. 557) dead organic material (such as plants and insects) that collects on the ground and in the soil

development (p. 31) a complex transition in how people make a living, characterized in part by a shift from economies based on extractive resources to those based on human resources

diaspora (p. 318) the dispersion of Jews around the globe after they were expelled from the Eastern Mediterranean by the Roman Empire, beginning in 73 C.E.; the term can now refer to other geographically dispersed culture groups

dictator (p. 157) a ruler who claims absolute authority, governing with little respect for the law or the rights of citizens

diffusion (p. 320) the process by which ideas, things, and people move across space and time

digital divide (p. 81) the discrepancy in access to information technology between small, rural, and poor areas and large, wealthy cities that contain major government research laboratories and universities

distributaries (p. 567) branches of a river in its delta that fan out and carry water away from the main channel

divide and rule (p. 383) the deliberate intensification of divisions and conflicts among "native" groups by European colonial powers

doi moi (p. 604) a program of economic and bureaucratic restructuring in Vietnam, very similar to perestroika in the Soviet Union and the structural adjustment programs of international lending agencies

domino theory (p. 571) a foreign-policy theory that suggested that if one country "fell" to communism, others in the neighboring region would also fall, like a line of dominoes

double day (p. 231) the longer workday of women with jobs outside the home who also work as caretakers, housekeepers, and/or cooks for their families

dowry (p. 467) a price paid by the family of the bride to the groom (the opposite of bride price); formerly, a custom practiced only by the rich

economic core (p. 78) the dominant economic region within a larger region

economic diversification (p. 328) the expansion of an economy to include a wider array of activities

economic migrants (p. 229) those who seek employment opportunities better than what they have at home

economies of scale (p. 212) reductions in the unit cost of production that occur when goods or services are efficiently mass-produced, resulting in increased profits per unit

ecotourism (p. 141) tourism built around natural environments

El Niño (p. 135) periodic climate-altering changes in the circulation of the Pacific Ocean and associated airmasses, now understood to operate on a global scale

endemic (p. 621) belonging or restricted to a particular place

Enlightenment (p. 205) the European intellectual movement that emphasized the power of the individual to use reason and logic (science) to understand the world

ethnic group (p. 55) a group of people who share a common ancestry and sense of common history, a set of beliefs, a way of life, a technology, and usually a common geographic location of origin

euro (€) (p. 214) the official (but not required) currency of the European Union

European Central Bank (ECB) (p. 214) directs monetary policy and ensures the soundness of the banking system among eurozone countries

eurozone (p. 214) all the countries using the euro currency

European Union (EU) (p. 193) a supranational organization that unites most of the countries of Europe

evangelical Protestantism (p. 171) a Christian movement that focuses on personal salvation and empowerment of the individual through miraculous healing and transformation

exclave (p. 249) a portion of a country that is separated from the main part

export processing zones (EPZs) (p. 153) specially created legal spaces or industrial parks within a country where, to attract foreign-owned factories, duties and taxes are not charged

export-led growth (p. 505) an economic development strategy that relies heavily on the production of manufactured goods destined for sale abroad

failed state (p. 339) a country where the government has lost control and is no longer able to fulfill its basic responsibilities as a defender of a sovereign state

fallow period (p. 373) the time (ideally, a decade or more) during which an agricultural plot is allowed to rest and regrow a natural vegetation cover

favelas (p. 161) Brazilian urban slums and shantytowns built by the poor; called colonias, barrios, or barriadas in other countries

female genital mutilation (FGM) (p. 406) removing the labia and clitoris and sometimes stitching the vulva nearly shut

female seclusion (p. 344) the requirement that women stay out of public view

feminization of labor (p. 573) the rising numbers of women in both the formal and the informal labor force

Fertile Crescent (p. 316) an arc of relatively lush, fertile land, where nomadic peoples began the earliest known agricultural communities and where the world's first cities emerged

fjord (p. 661) a long, narrow inlet formed by glacial erosion

floating population (p. 517) the Chinese term for people who live and work in a place other than their household registration location; many are people who have left economically depressed rural areas for the cities

floodplain (p. 12) the flat land around a river where sediment is deposited during flooding

food security (p. 373) the condition of stable access to sufficient, safe, and nutritious food

foreign exchange (p. 585) foreign currency that countries need to purchase imports

formal region (p. 9) an area defined by a single specific trait that has been mapped and described, such as the area in which Spanish is the primary language of government

forward capital (p. 187) a capital city built in the hinterland to draw migrants and investment for economic development and sometimes for political/strategic reasons

fossil fuel (p. 308) a source of energy formed from the remains of dead plants and animals; fossil fuel includes oil, coal, and natural gas

fracking (p. 74) a drilling technology that uses high pressure injection of water to fracture rock formations, releasing natural gas and oil that they contain

free trade (p. 37) the unrestricted international exchange of goods, services, and capital

functional region (p. 9) an area defined by an activity organized around a single place, such as the area served by a single post office

gender (p. 50) the ways a particular social group defines the differences between the sexes

genocide (p. 398) the deliberate destruction of an ethnic, racial, religious, or political group

gentrification (p. 100) the renovation of old urban districts by affluent investors, a process that often displaces poorer residents

geographic information science (GISc) (p. 8) the body of science that supports multiple spatial analysis technologies and keeps them at the cutting edge

geopolitics (p. 42) the study of how geography and politics influence each other

geospace (p. 670) the region of outer space closest to Earth, including the upper atmosphere and extending out 40,000 miles (65,000 km) to the furthest reaches of Earth's magnetic field

geostationary orbit (p. 671) an orbit located at least 22,236 miles (35,786 kilometers) above a fixed point on Earth, in which satellites do not move relative to each other

gerrymandering (p. 108) the practice of manipulating the boundaries of political districts to benefit candidates of a particular group or political party

glasnost (p. 271) literally, "openness"; the policies instituted in the late 1980s under Mikhail Gorbachev that encouraged more transparency and openness in Soviet government and society

globalization (p. 33) the growth of worldwide linkages and the changes these linkages are bringing about in daily life, especially in economies, but also in culture and biology

global warming (p. 18) the warming of Earth's climate as atmospheric levels of greenhouse gases increase

Gondwana (p. 617) the great landmass that formed the southern part of the ancient supercontinent Pangaea

grassroots economic development (p. 392) economic development projects designed to use local skills to create products or services for local consumption with simple technology

Great Barrier Reef (p. 617) the longest coral reef in the world, located off the northeastern coast of Australia

Great Leap Forward (p. 503) a failed economic reform program under Mao Zedong intended to quickly raise China's industrial level

green (p. 198) an adjective indicating a person or group that is environmentally conscious

greenhouse gases (GHGs) (p. 18) gases, such as carbon dioxide and methane, released into the atmosphere by human activities, which become harmful when released in excessive amounts

Green political parties (p. 198) groups that consistently advocate for emission controls, community recycling, grassroots work on local environments, and general humanitarian policies

green revolution (p. 30) increases in food production brought about through the use of new seeds, fertilizers, mechanized equipment, irrigation, pesticides, and herbicides

gross domestic product (GDP) per capita (p. 31) the total market value of all goods and services produced within a particular country's borders and within a given year, divided by the number of people in the country

gross national income (GNI) per capita (p. 31) the sum of a country's gross domestic product plus all net income received from overseas, divided by the midyear population

groundwater (p. 378) water naturally stored in aquifers as many as 5000 years ago during wetter climate conditions

growth poles (p. 517) zones of development whose success draws more investment and migration to a region

guest workers (p. 229) legal workers from outside a country who help fulfill the need for temporary workers but who are expected to return home when they are no longer needed

gulag camp (p. 665) a settlement set up by the central government of the USSR, in which prisoners were sent to state-run enterprises to provide labor

Gulf states (p. 308) Arab countries that border the Persian Gulf; Saudi Arabia, Kuwait, Bahrain, Oman, Qatar, and the United Arab Emirates

hacienda (p. 147) a large agricultural estate in Middle or South America, more common in the past; usually not specialized by crop and not focused on market production

hajj (p. 319) the pilgrimage to the city of Makkah (Mecca) that all Muslims are encouraged to undertake at least once in a lifetime

hijra (p. 470) a trans female identity numbering 6–7 million people, found mostly in north India, Nepal, Pakistan, and Bangladesh

Hinduism (p. 442) a major world religion practiced by approximately 900 million people, 800 million of whom live in India

housing segregation (p. 100) a common phenomenon in cities, in which ethnic or racial or economic classes live in separate neighborhoods and the segregation is maintained by discriminatory customs, not law

Holocaust (p. 208) during World War II, a massive execution by the Nazis of 6 million Jews and 5 million gentiles (non-Jews)

Horn of Africa (p. 372) the triangular Somali peninsula that juts out from northeastern Africa below the Red Sea and wraps around the Arabian Peninsula

hot spots (p. 617) individual sites of upwelling material (magma) that originate deep in Earth's mantle and surface in a tall plume; hot spots tend to remain fixed relative to migrating tectonic plates

***hukou* system** (p. 517) the system in China where each person's permanent residence is registered and any person who wants to migrate must obtain permission from authorities to do so

human geography (p. 3) the study of patterns and processes that have shaped human understanding, use, and alteration of Earth's surface

humanism (p. 205) a philosophy and value system that emphasizes the dignity and worth of the individual, regardless of wealth or social status

human resource capacity building (p. 639) efforts to increase the understanding and skills of people so they can manage development in their own communities

import substitution industrialization (ISI) (p. 149) policies that encourage local production of machinery and other items that previously had been imported at great expense from abroad

Incas (p. 143) indigenous people who ruled the largest pre-Columbian state in the Americas, with a domain stretching from what is now southern Colombia to northern Chile and Argentina

income disparity (p. 152) the gap in income between rich and poor

indigenous (p. 131) native to a particular place or region

Indus Valley civilization (p. 440) the first substantial settled agricultural communities, which appeared about 4500 years ago along the Indus River in modern-day Pakistan and northwest India

Industrial Revolution (p. 29) a series of innovations and ideas that occurred broadly between 1750 and 1850, which changed the way goods were manufactured

informal economy (p. 32) all aspects of the economy that take place outside official channels or "off the books"

infrastructure (p. 77) road, rail, and communication networks and other facilities necessary for economic activity

inland delta (p. 370) a place where a river reaches a flat inland area and splits into multiple channels

intergalactic space (p. 670) the region of space between the Milky Way galaxy and other galaxies

interplanetary space (p. 670) the region of space between the Sun and planets of our solar system

interstellar space (p. 670) the region of space between our solar system and others within the Milky Way galaxy

intertropical convergence zone (ITCZ) (p. 370) a band of warm winds that converge from both the north and the south converge at the equator, pushing air upward and causing copious rainfall

intifada (p. 337) a prolonged Palestinian uprising against Israel

invasive species (p. 623) nonnative organisms that, lacking natural predators, spread into regions outside of their native range, adversely affecting economies or environments

inversion (p. 498) the unusual condition when warm air overlies cool air, impeding normal circulation and trapping pollution

Iron Curtain (p. 208) a fortified border zone that separated western Europe from eastern Europe during the Cold War

Islam (p. 307) a monotheistic religion that emerged in the seventh century C.E.; its tenets are based on the writings of the Prophet Muhammad

Islamic jihadism (p. 220) a personal or group struggle to promote Islamic revivalism, often with the threat of or actual use of force

Islamism (p. 307) a grassroots religious revival in Islam that seeks political power to curb secular influences in society and to replace secular governments and civil laws with governments and laws guided by Islamic principles

isthmus (p. 132) a narrow strip of land that joins two larger land areas

Jainism (p. 444) a religion of Asia that originated as a reformist movement within Hinduism more than 2000 years ago

jati (p. 443) in Hindu India, the subcaste into which a person is born, which traditionally defined the individual's experience for a lifetime

jihad (p. 330) in Arabic, struggle; either understood as a personal struggle to be a good Muslim, or against enemies of Islam

Judaism (p. 318) a monotheistic religion characterized by the belief in one god, a strong ethical code summarized in the Ten Commandments, and an enduring ethnic identity

just-in-time system (p. 508) the system pioneered in Japanese manufacturing that clusters companies that are part of the same production system close together so that they can deliver parts to each other precisely when they are needed

Kessler syndrome (p. 672) a situation where so much space junk accumulates around Earth that travel through low Earth orbit becomes impossible

kleptocracy (p. 386) raw materials that are traded, usually to other countries, for processing or manufacturing into more valuable goods

krill (p. 662) tiny crustaceans that whales, seals, penguins, and fish feed on

landforms (p. 12) physical features of Earth's surface, such as mountain ranges, river valleys, basins, and cliffs

Latino (p. 65) a term used to refer to all Spanish-speaking people from Middle and South America, although their ancestors may have been European, African, Asian, or Native American

latitude (p. 5) the distance in degrees north or south of the equator; lines of latitude run parallel to the equator, and are also called parallels

less developed countries (LDCs) (p. 31) countries that have labor-intensive and low-wage, often agricultural, economies

liberation theology (p. 170) a movement within the Roman Catholic Church that uses the teachings of Jesus to encourage the poor to organize to change their own lives and to encourage the rich to promote social and economic equity

lingua franca (p. 411) a language of trade

loess (p. 530) windblown material that forms deep soils in China, North America, and central Europe

longitude (p. 5) the distance in degrees east and west of Greenwich, England; lines of longitude, also called meridians, run from pole to pole (the line of longitude at Greenwich is 0° and is known as the prime meridian)

low Earth orbit (LEO) (p. 670) an orbit that lies between 99 miles (160 kilometers) and 1243 miles (2000 kilometers) above Earth

Maastricht Treaty (p. 209) the 1992 treaty that established the European Union, signed by 15 countries of the European Community in the city of Maastricht, Netherlands

machismo (p. 167) a set of values that defines manliness in Middle and South America

makatea (p. 618) coral platforms uplifted by volcanism, sometimes called seamounts

Maori (p. 618) Polynesian people indigenous to New Zealand

map projections (p. 5) the various ways of showing the spherical Earth on a flat surface

maquiladoras (p. 138) foreign-owned, tax-exempt factories, often located in Mexican towns near the U.S. border, that hire workers at low wages to assemble manufactured goods, which are then exported mainly to the United States and Canada

marianismo (p. 167) a set of values based on the life of the Virgin Mary, the mother of Jesus, that defines the proper social roles for women in Middle and South America

marketization (p. 149) the development of a free market economy in support of free trade

marsupials (p. 621) mammals whose babies at birth are still at a very immature stage; the marsupial then nurtures them in a pouch equipped with nipples

Mediterranean climate (p. 196) a climate pattern of warm, dry summers and mild, rainy winters

Melanesia (p. 630) New Guinea and the islands south of the equator and west of Tonga (the Solomon Islands, New Caledonia, Fiji, and Vanuatu)

Melanesians (p. 630) a group of Australoids who settled throughout New Guinea and other nearby islands

mercantilism (p. 206) a strategy for increasing a country's power and wealth by acquiring colonies and managing all aspects of their production, transport, and trade for the colonizer's benefit

merchant marine (p. 585) the approximately 12,000 ships and tankers that carry passengers and solid and liquid cargo across the globe

mestizos (p. 144) people of mixed European, African, and indigenous descent

metropolitan areas (p. 97) cities of 50,000 or more and their surrounding suburbs and towns

microcredit (p. 452) a program based on peer support that makes very small loans available to very-low-income entrepreneurs

Micronesia (p. 630) the small islands that lie east of the Philippines and north of the equator

Middle America (p. 132) Mexico, Central America, and the islands of the Caribbean

middle income countries (MICs) (p. 31) countries that have shifted to more employment in the secondary sector (industry) and higher-wage tertiary sector (services)

midlatitude temperate climate (p. 196) as in south-central North America, China, and Europe, a climate that is moist all year with relatively mild winters and long, mild-to-hot summers

MIRAB economy (p. 636) an economy based on migration, remittances, and aid-supported bureaucracy

mixed agriculture (p. 383) farming that involves raising a variety of crops and animals on a single farm, often to take advantage of several environmental riches

Mongols (p. 269) nomadic pastoral people centered in East and Central Asia, who by the thirteenth century had established by conquest an empire that stretched from Europe to the Pacific

mono-crop agriculture (p. 374) farming in which a single crop species is grown in a large field

monotheism (p. 317) the belief system based on the idea that there is only one god

monotremes (p. 621) egg-laying mammals, such as the duck-billed platypus and the spiny anteater

monsoon (p. 430) a seasonal period of precipitation or dryness: the wet summer monsoon begins with the arrival of rain in June; the dry weather monsoon begins by November

more developed countries (MDCs) (p. 31) countries that have economies often generated by tertiary and quaternary sectors and that generally provide adequate education, health care, and other social services to help their people contribute to economic development

Mughals (p. 441) a dynasty of Central Asian origin that ruled India from the sixteenth century to the nineteenth century

Muslims (p. 318) followers of Islam

nation (p. 38) a community of people usually having a common culture, origin, and language, occupying a specific territory; states often are composed of multiple nations

nation-state (p. 38) a state that coincides with a single nation, occupying roughly the same territory

nationalism (p. 217) strong devotion and loyalty to the interests or culture of a particular country, nation, or cultural group

nationalize (p. 150) to seize private property and place under government ownership, with some compensation

neoliberalism (p. 148) a capitalist economic and political system that favors the fulfillment of human needs by free and open markets strongly linked to global markets through free trade, with governments having a minimal role

nomadic pastoralists (p. 266) people whose way of life and economy are centered on tending grazing animals that are moved seasonally to gain access to the best pasture

nongovernmental organizations (NGOs) (p. 42) associations outside the formal institutions of government that promote activism on political, social, economic, or environmental issues

nonpoint sources of pollution (p. 263) diffuse sources of environmental contamination, such as untreated automobile exhaust, raw sewage, and agricultural chemicals

North Atlantic Drift (p. 196) the easternmost end of the Gulf Stream, a broad, warm-water current that brings large amounts of warm water to the coasts of Europe

North Atlantic Treaty Organization (NATO) (p. 221) a military alliance of European (including Turkey) and North American countries, created during the Cold War to counter the influence of the Soviet Union and to provide international security and cooperation

nuclear family (p. 111) a family consisting of a married father and mother and their children

occupied Palestinian Territories (oPT) (p. 308) Palestinian lands occupied by Israel since 1967

offshore outsourcing (p. 450) the contracting of business or production functions to foreign companies where labor and other costs are lower

oligarchs (p. 272) in Russia, those who acquired great wealth during the privatization of Russia's resources and who use that wealth to exercise power

OPEC (Organization of the Petroleum Exporting Countries) (p. 325) a cartel of oil-producing countries—currently, Algeria, Angola, Iran, Iraq, Kuwait, Libya, Nigeria, Qatar, Saudi Arabia, the United Arab Emirates, Indonesia, Ecuador, and Venezuela—that was established to regulate the production and price of oil

organically grown (p. 87) products produced without chemical fertilizers and pesticides

orographic precipitation (p. 15) precipitation produced when a moving moist air mass encounters a mountain range, rises, cools, and releases condensed moisture that falls as rain

Ottoman Empire (p. 320) influential Islamic empire centered in today's Turkey that lasted from the 1200s to the early 1900s

Outback (p. 617) any remote, lightly populated area, usually in the central regions of Australia

Outer Space Treaty of 1967 (p. 675) treaty that the United States and all spacefaring nations have signed that forbids nations from claiming territory in space

Overseas Chinese (p. 529) Chinese emigrants and their descendants, especially those in Southeast Asia

Pacific Rim (p. 103) a term that refers to all the countries that border the Pacific Ocean

Pacific Solution (p. 643) Australia and New Zealand's controversial policy to pay nearby islands to indefinitely hold the excess number of undocumented immigrants in tent communities

Pacific Way (p. 640) a system of political though that emphasizes regional identity, decision making via consensus, and respect for traditional patriarchal authority

Pancasila (p. 583) a national philosophy that embraces five precepts: *belief* in God and the observance of *conformity, corporatism, consensus,* and *harmony*

Paris Accord, 2017 (p. 198) a UN-sponsored agreement by 181 countries to lower greenhouse gas emissions

Partition (p. 446) the breakup following Indian independence that resulted in the establishment of Hindu India and Muslim Pakistan

pastoralism (p. 375) a way of life based on herding; practiced primarily on savannas, on desert margins, or in the mixture of grass and shrubs called open bush

patriarchal (p. 344) relating to a social organization in which the father is supreme in the clan or family

perestroika (p. 271) literally, "restructuring"; the restructuring of the Soviet economic system that was done in the late 1980s in an attempt to revitalize the economy

permafrost (p. 261) permanently frozen soil that lies just beneath the surface

physical geography (p. 3) the study of Earth's physical processes: how they work and interact, how they affect humans, and how they are affected by humans

pidgin (p. 653) a language used for trading; one made up of words borrowed from the several languages of people involved in trading relationships

plantation (p. 147) a large factory farm that grows and partially processes a single cash crop

plate tectonics (p. 12) the scientific theory that Earth's surface is composed of large plates that float on top of an underlying layer of molten rock; the movement and interaction of the plates create many of the large features of Earth's surface, particularly mountains

polar amplification (p. 662) the increase in the intensity of climate change at the poles

political ecology (p. 32) the study of the interactions among development, politics, human well-being, and the environment

political freedoms (p. 39) the rights and capacities that support individual and collective liberty and public participation in political decision making

polygyny (p. 406) the practice of having multiple wives

Polynesia (p. 630) the numerous islands situated inside an irregular triangle formed by New Zealand, Hawaii, and Easter Island

population pyramid (p. 50) a graph that depicts the age and sex structures of a political unit, usually a country

populist (p. 208) movement appealing to the interests and prejudices of ordinary people as distinct from the rich and powerful

populist political movement (p. 582) a coalition of usually rural or working class activists who claim that virtuous citizens are being exploited by a small group of elites who are presumed to be in illegitimate control of the government

postcolonial (p. 220) the conditions and attitudes that persist in a society after colonization is over

primary sector (p. 31) extractive economic activity such as mining, forestry, and agriculture

primate city (p. 161) a city, plus its suburbs, that is vastly larger than all others in a country and in which economic and political activity is centered

privatization (pp. 149, 272) the selling of formerly government-owned industries and firms to private companies or individuals; the sale of industries that were formerly owned and operated by the government to private companies or individuals

public space (p. 334) a place, such as a square or a park, that is open for all citizens to join together for personal enjoyment or political participation

pull factors (p. 105) positive factors that draw in migrants

purchasing power parity (PPP) (p. 31) the amount that the local currency equivalent of U.S.$1 will purchase in a given country

purdah (p. 467) the practice of concealing women from the eyes of nonfamily men

push factors (p. 105) negative factors that cause people to leave home, family, and friends

qanats (p. 538) underground tunnels, built by ancient cultures and still used today, that carry groundwater for irrigation in dry regions

quaternary sector (p. 31) knowledge-based economic activities such as information technology and research and development

Qur'an (or Koran) (p. 311) the holy book of Islam, believed by Muslims to contain the words Allah revealed to Muhammad

race (p. 59) a social or political construct that is based on apparent characteristics such as skin color, hair texture, and face and body shape, but that is of no biological significance

rain shadow (p. 15) an area of low rainfall on the drier side of a mountain range affected by orographic rainfall

rate of natural increase (RNI) (p. 47) the rate of population growth measured as the excess of births over deaths per 1000 individuals per year, without regard for the effects of migration

redlining (p. 100) The practice used by banks of drawing red lines on urban maps to show where they will not approve loans for home buyers regardless of qualifications

refugees (pp. 229, 641) those who have fled from their homes because of war or violent social unrest or discrimination; people fleeing armed conflicts or persecution who are protected by international law (1951 Refugee Convention)

region (p. 9) a unit of Earth's surface that contains distinct patterns of physical features and/or distinct patterns of human development

regional self-sufficiency (p. 509) an economic policy in communist China that encouraged each region to develop its own industrial and agricultural resources

regional specialization (p. 509) specialization (rather than self-sufficiency) that takes advantage of regional variations in climate, natural resources, and location

religious nationalism (p. 454) the association of a country with a particular religion, with the intent of excluding or disadvantaging other religions

remittances (p. 585) money sent home by people who are working abroad, usually temporarily

resettlement schemes (p. 583) government plans to move large numbers of people from one part of a country to another to relieve urban congestion, disperse political dissidents, or accomplish other social purposes; also called *transmigration*

Roaring Forties (p. 618) powerful air and ocean currents at about 40° S latitude that speed around the far Southern Hemisphere virtually unimpeded by landmasses

Roma (p. 208) the now-preferred term in Europe for Gypsies; some prefer to be called Gypsies as a point of ethnic identity

Russian Federation (p. 261) Russia and its political subunits, which include 21 internal republics

Russification (p. 281) the czarist and Soviet policy of encouraging ethnic Russians to settle in non-Russian areas as a way to assert political control

Sahel (p. 370) a band of arid grassland that runs east–west along the southern edge of the Sahara

Salafism (p. 330) a religious movement that started in Egypt and that has influenced the ideology of radical Islamism

salinization (p. 312) a process where evaporation of water leaves salty minerals behind in the soil; may occur naturally or as a result of irrigation

satellite (p. 671) any object in space that orbits another object of greater mass

scale (of a map) (p. 4) the proportion that relates the dimensions of the map to the dimensions of the area it represents; also, variable-sized units of geographical analysis from the local to the regional to the global

Schengen Agreement (p. 211) an agreement first signed in 1985 and implemented in 1995 that allows for free movement across common borders in the European Union

secondary sector (p. 31) industrial economic activity such as processing, manufacturing, and construction

secular states (p. 331) countries that have no state religion and in which religion has no direct influence on affairs of state or civil law

service sector (p. 573) the part of the economy that employs people in knowledge-based jobs, such as in education, health care, sales, entertainment, and financial services

sex (p. 50) the biological category of male, female, intersex; does not indicate how people may behave or identify themselves

sex ratio (p. 464) the ratio of females to males in a population

sex tourism (p. 593) the sexual entertainment industry that serves primarily men who travel for the purpose of living out their fantasies during a few weeks of vacation

sex work (p. 461) the provision of sexual acts for a fee

shari'a (p. 319) Islamic religious law that guides daily life according to the interpretations of the Qur'an

sheikhs (p. 354) traditionally, patriarchal leaders over ancestral tribal groups; today, the title is still used to denote someone in a leadership position

Shi'ite (or Shi'a) (p. 320) the smaller of two major groups of Muslims; Shi'ites are found primarily in Iran and southern Iraq

shifting cultivation (pp. 143, 373) a system of agriculture in which small plots of forest are cleared, the dried brush is burned to release nutrients, and the clearings are planted; each plot is used for 2 or 3 years and then abandoned for many years of fallowing

Siberia (p. 261) the territories of Russia that are located east of the Ural Mountains

Sikhism (p. 444) a religion of South Asia that combines beliefs of Islam and Hinduism

Silk Road (p. 266) historic trading route between China and the Mediterranean; named after the importance of silk as a desired commodity of trade

Slavs (p. 268) a group of people who originated in what are now Poland, Ukraine, and Belarus

slum (p. 43) densely populated area characterized by crowding, run-down housing, and inadequate access to food, clean water, education, and social services

smog (p. 71) a combination of industrial emissions, car exhaust, and water vapor that frequently hovers as a yellow-brown haze over many cities, causing a variety of health problems

social cohesion (pp. 234, 577) an expressed goal of the EU-28 involving balanced and sustainable development, reduced disparities between regions and countries, and equal opportunities for all people; an expression of solidarity and the willingness of members of a society to cooperate with each other in order for all to survive and prosper

social exclusion (p. 229) the condition of being systematically left out of important opportunities for participation in society

social protection (p. 234) the tax-supported systems that provide citizens with benefits such as health care, long-term care, pensions, and relief from poverty and social exclusion

social safety net (p. 94) the services provided by the government—such as welfare, unemployment benefits, and health care—that prevent people from falling into extreme poverty

socialism (p. 148, 209) an economic system in which all people have access to basic goods and services and large-scale industries may be collectively owned, with any profits benefitting society as a whole

soft power initiatives (p. 582) diplomatic overtures or offers of aid that appear to be only friendly with no ulterior geopolitical motives

solar wind (p. 670) tiny charged particles emitted by the Sun that travel at millions of miles per hour and form a gaslike plasma that can conduct electricity and respond to magnetic fields

South America (p. 132) the continent south of Central America

Soviet Union (p. 259) see *Union of Soviet Socialist Republics*

space junk (p. 671) human-created space debris

special economic zones (SEZs) (p. 517) free trade zones within China, which are commonly called export processing zones (EPZs) elsewhere

state (p. 38) an organized political community in a defined territory and under one government that has *sovereignty*, the ability to conduct its internal affairs as it sees fit without interference from outside

state-aided market economy (pp. 505, 572) an economic system based on market principles but with strong government guidance; in contrast to the limited government, free market economic system of the United States and, to a lesser degree, Europe; national governments help certain sectors of the economy with special loans

station (p. 652) a large Australian farm or ranch, often where cattle or sheep were raised

steppes (p. 261) semiarid grass-covered plains

structural adjustment programs (SAPs) (p. 149) economic reorganization toward less government involvement in industry, agriculture, and social services; sometimes imposed by the World Bank and the International Monetary Fund as conditions for receiving loans

subduction zone (p. 133) a zone where one tectonic plate slides under another

subsidies (p. 214) monetary assistance granted by a government to an individual or group in support of an activity, such as farming, that is viewed as being in the public interest

subsistence affluence (p. 636) a lifestyle whereby people are self-sufficient with regard to most necessities, yet have sources of cash income that they can save for travel and occasional purchases of manufactured goods

subsistence agriculture (p. 373) farming that provides food for only the farmer's family and is usually done on small farms

suburbs (p. 97) populated areas along the peripheries of cities

Sundaland (p. 556) part of the submerged Eurasian continental shelf that, when previously above sea level, served as a bridge between the mainland and islands

Sunni (p. 320) the larger of two major groups of Muslims

sustainable development (p. 32) the improvement of current standards of living in ways that will not jeopardize those of future generations

sustainable tourism (p. 639) aims to decrease tourism's footprint and minimize disparities between hosts and visitors

swagman (p. 652) an itinerant Australian male laborer who worked on large farms or in mines

taiga (p. 263) subarctic coniferous forests

Taliban (p. 457) an archconservative Islamist movement that gained control of the government of Afghanistan in the mid-1990s

temperature-altitude zones (p. 135) regions of the same latitude that vary in climate according to altitude

tertiary sector (p. 31) service-based economic activity such as transportation, education, health care, tourism, and financial services

theocratic states (p. 331) countries that require all government leaders to subscribe to a state religion and all citizens to follow rules decreed by that religion

tiger economies (p. 602) Southeast Asian countries that managed to rapidly modernize and prosper after World War II

total fertility rate (TFR) (p. 48) the average number of children that women in a country are likely to have at the present rate of natural increase

trade deficit (p. 82) the extent to which the money earned by exports is exceeded by the money spent on imports

trade winds (p. 135) winds that blow from the northeast and the southeast toward the equator

trafficking (p. 289) the recruiting, transporting, and harboring of people through coercion for the purpose of exploiting them

Trans-Pacific Partnership (TPP) (p. 640) the largest regional trade accord in history

transparency (p. 395) in politics, the state of being open to observation and participation by the public

tsunami (p. 488) a large sea wave, usually caused by an earthquake with an epicenter underneath the ocean

tundra (p. 122, 261) a region of winters so long and cold that the ground is permanently frozen several feet below the surface; a cold, treeless area, between the ice cap and the continental climate forest, where the subsoil is permanently frozen

typhoon (p. 490) a tropical storm in the western Pacific Ocean; called a cyclone or hurricane elsewhere

underemployment (p. 275) the condition in which people are working too few hours to make a decent living or are highly trained but working at menial jobs

Union of Soviet Socialist Republics (USSR) (p. 259) the multinational union formed from the Russian empire in 1922 and dissolved in 1991; commonly known as the *Soviet Union*

United Nations (UN) (p. 42) an assembly of 193 member states that focuses on economic development, general health and well-being, democratization, peacekeeping assistance, humanitarian aid, and scientific research

United Nations Human Development Index (HDI) (p. 32) calculates a country's level of development, with a combination of statistical indicators for per capita income (PPP), life expectancy, and education

United States–Mexico–Canada Agreement (USMCA) (p. 81) a revised version of the North American Free Trade Agreement made in 1994 that added Mexico to the 1989 economic arrangement between the United States and Canada

urban growth poles (p. 164) locations within cities that are attractive to investment, innovative immigrants, and trade, and thus attract economic development like a magnet

urban sprawl (p. 73) the encroachment of suburbs on agricultural land

urbanization (p. 43) the process whereby cities, towns, and suburbs grow as populations shift from rural to urban livelihoods

value chains (p. 391) links between various aspects of a production line to maximize efficiency and profits

varna (p. 444) the four hierarchically ordered divisions of society in Hindu India underlying the caste system: Brahmins (priests), Kshatriyas (warriors/kings), Vaishyas (merchants/landowners), and Sudras (laborers/artisans)

veil (p. 346) a piece of clothing that covers a woman's hair, face, or much of her body

vernacular region (p. 9) an area defined by perceptions of shared characteristics

virtual water (p. 216) the volume of water required to produce, process, and deliver a good or service that a person consumes

Wahhabism (p. 330) a religious movement that started in Saudi Arabia and that has influenced the ideology of radical Islamism

weather (p. 13) the short-term and spatially limited expression of climate that can change in a matter of minutes

welfare states (p. 232) government accepts responsibility for the well-being of people, guaranteeing basic rights to education, affordable housing and food, unemployment insurance, and health care for all

West Bank barrier (p. 337) an Israeli-built concrete wall or fence that now surrounds much of the West Bank and encompasses many of the Jewish settlements there

wet rice cultivation (p. 494) a prolific type of rice production that requires the plant roots to be submerged in water for part of the growing season

world city (p. 516) one of the most important cities in the global system in terms of economic, political, and cultural dominance

yurt (or *ger*) (p. 537) round, heavy, felt tent stretched over collapsible willow lattice frames used by nomadic herders in northwestern China, Mongolia, and Central Asia

Zionism (p. 335) a movement that started in nineteenth-century Europe to create a Jewish homeland in Palestine

Index

Page numbers in *italic* indicate figures; page numbers in **bold** indicate key terms.

Aadhaar project, 457
Abkhazia, *279*, 282
Abolitionism, 78
Aboriginal Australians, **630**, 632–633, *632*
Aboriginal land claims, 651
Aboriginal peoples, 65
  in Continental Interior, 122
  impact of European settlers on, 76–77, 79–80
  in Québec, *114*
Aboriginal Tent Embassy (Australia), 651
Abortion, 110, 117, 468, 524, 590–591
Abraham, 318
Accra, Ghana, *401*
Acculturation, **169**
Acid rain, **71**, 202, 265, 498
Aconcagua, 135
Active Aging, **226**
Addresses, of slum dwellers, 460
Adriatic Sea, 195
Affordable Care Act (ACA), 65, 93, 94
Afghanistan, 47, 48, 89, 272, 292, *435*, 437, 468, 470–472, *471*, 473–474
  language and ethnicity in, *473*
  population pyramid for, *50*
  U.S. war in, 89, *90*
  war and reconstruction in, 457–458
Africa, 12
  See also North Africa and Southwest Asia; Sub-Saharan Africa
  Asian investment in, 390–391
  colonization of, 236
  religions in, *409*
African Americans
  civil rights for, 93–94
  Great Migration of, 117
  incarceration of, 96
  law enforcement and, 109
  middle class, 109
  New Great Migration of, 117–118
  population by state, *77*
  slavery and, 77–78
African National Congress (ANC), 385
African Plate, 12, *13*, 66, 309
African professionals, 391
African religions, 170, *187*, 409
Africans
  See also African Americans
  slavery and, 144, *144*, 167, 207
African slave trade, 380, *381*, 382
African Tectonic Plate, 195
African Union, 353
Afterlife, 56
Aga Khan, *471*
Agent Orange, 571
Age structure, 50
  See also Population pyramid
Aging
  Active Aging, **226**
  dilemmas of, 101
Aging population
  in East Asia, 522–524, *525*
  in Europe, 224–226, *227*
  in North America, 101
Agribusiness, **86**, 106
Agriculture, 28–29
  See also Food production
  commercial, 373–375, 563
  corporate, 215–216
  deforestation and, 138, *139*
  in East Asia, 494, *495*, 496
    in China, *495*, *507*, 508–509
  in Europe, 214–216, 238–239, 242–243, 251–252
  green revolution in, 30, 86, 154, 159, 450
  mechanization, 63, 86, 120
  in Middle and South America, 143, 147, 154–155, *156*
    in Brazil, 185–186
    in Caribbean, 171
    in Central America, 176, *177*
    in Central Andes, 182
    in Mexico, 106, 174
    in Southern Cone, 183
  modernized, 29–30
  mono-crop, **374**
  in North Africa and Southwest Asia, 309, 311–313, 316–317
    in Eastern Mediterranean, 357
    in Egypt, 353
    in Maghreb, 350
    in Saudi Arabia, 355
  in North America, 75–76, 86–87, *87*
    in American South, 118
    in California, 79
    in Great Plains, 118
    in Southern California and Southwest, 124
  in Oceania, 623, 626–627
  organic, 30, 87, *88*
  plantation system of, 77–78
  in Russia and post-Soviet states, *267*, 277–278
  shifting cultivation, *21*, 143, 373, *374*, 562–563
  in South Asia, 433, 450, *452*
  in Southeast Asia, 561–563, *562*
  in sub-Saharan Africa, 373–375, *377*, 379–382
    early agriculture, 379–380
    East Africa, 415, 417–418
      in Kenya, 417–418
      in Malawi, 421
      in Tanzania, 417–418
  subsistence, 373
  sustainable, 30, 182
  water demands of, 27
Agroforestry, **372**, *373*
Aguarico River, 129
Ahraura, India, 465
AIDS. *See* HIV/AIDS
Ainu, **529**, *529*
Air pollution, 69, 437–438
  in East Asia, 497–499
  in Europe, 202
  indoor, 373
  in North America, 71–72, *71*
  in Russia and post-Soviet states, 265
  in sub-Saharan Africa, 373
Airports, 88
Air pressure, 14–15
Airtel, 391
Air transportation, in North America, 88
Alabama, 72
Alan Ross Texas Freedom parade, *98*
Al Aqsa Mosque (Jerusalem), 337–338
Alaska, 121, 122, 665, 666–667
  wildlife protection in, 663
Alawites, 338
Albania, 193, 194, 249, 251
  climate change vulnerability in, *199*
Al-Bashir, Omar, 353
Albers projection, 5, *7*
Aleppo, Syria, 357
Alexander the Great, 441
Alexandria, Egypt, *314*, 353
Algeria, 209, *219*, 220, *310*, 321, 324, 349, 350, 350–351
Alhambra, Spain, *204*

Alice Springs, Australia, *619*
Al Jazeera, 331
Allende, Salvadore, 157
Allies, 208
Alluvium, **530**
Almonds, 120, *120*
Alps, 195, *195*, 198
Al Qaeda, 90, 338, 355, 473
Al Shabaab, 410, 417
Alternative energy sources, 69, 201–202, 438
Altiplano, **181**
Altruism, 60
Amazon Basin, 134–135, *134*, 138, *139*, *141*, 185, 186
Amazon River, 134–135
American family, 111
American Indians, 65
  See also Native Americans
American South (Southeast), 116–118, *117*
Amsterdam, 236
Amur River, 297, 533
Anatolian Plate, *13*
ANC (African National Congress), 385
Andean Community of Nations, 153
Anderson, Benedict, 55
Andes Mountains, *133*, 135
Angkor, Cambodia, 568
Angkor Wat (Cambodia), *578*
Anglican Church, 409–410
Angola, *219*, *388*, 420
Animal raising, 118–119
Animals
  Columbian exchange of, 144
  in East Asia, 496
  in Oceania, 621
  of polar regions, *662*
  in South Asia, 439
  in sub-Saharan Africa, 378–379, 405, 418
Anime, *545*
Animist, **409**
Antarctica, *662*, *664*
  See also Polar regions
Antarctic Plate, *13*
Antarctic Treaty, 664
Anthropocene, 18, 20
Anticorruption movements, in South Asia, *455*, 456–457
Antigua, 171
Antigua, Guatemala, *169*
Apartheid, **385**, *385*, 420
APEC (Asia Pacific Economic Cooperative), **640**
Apocalyptica (music group), *248*
Appalachian Mountains, 66, *66*
Apple, 37
Aquaculture, 436
Aquifers, **69**, 72
Aquino, Benigno, III, 592
Aquino, Corazon, 592
Arab culture, 308
Arabian peninsula (subregion), 354–356, *355*
Arabian Plate, 12, *13*, 309
Arabic language, 308
Arab–Israeli conflict. *See* Israeli–Palestinian conflict
Arable land, **646**
Arab nationalism, 351
Arab Spring movements, 39, 220, 328–329, *330*, 331, *332*, 338, 339, 350, 351, 356
Aral Sea, 265–266, *265*, *267*
Archipelago, **555**
Arctic, *660*
  See also Polar regions
Arctic climates, 16, *17*, *68*, *262*
Arctic Ocean, 247–248, 261, 293
Ardern, Jacinda, *648*
Argentina, 34, 150, *156*, 157, 167, 169, 170–171, 183, 184, *185*
Arid climates, 16
Arizona, 125
Armani, Giorgio, 245
Armed conflict. *See* Conflicts
Armenia, 220, 259, 299–300, *299*
Arnhem Land (Australia), *654*
Arunachal Pradesh, 474–475
Asado, *156*
ASEAN (Association of Southeast Asian Nations), 37, **576**–577
Asia
  See also East Asia; South Asia; Southeast Asia
  economic development "miracle" in, 635–636
  globalization and, 83
  Oceania and, 634–636
Asian brown cloud, 437
Asian carp, 73, *73*
Asian Highway, 578–579
Asians, in Caribbean, 169
Asia Pacific Economic Cooperative (APEC), **640**
Assad regime, 338–339
Assam, India, 478–479
Assembly, freedom of, 39
Assimilation, **169**, 229–230, **229**
Association of Southeast Asian Nations (ASEAN), 37, **576**–577
Assyrian Empire, *317*
Astana, Kazakhstan, *284*
Asteroids, 670, 671, 672
Aswan dams (Egypt), 315
Asylum seekers, *228*, **229**, 248, **641**, *642*
Asymmetry, 93
Atacama, Chile, *136*
Atacama Desert, 181
Atatürk, Mustafa Kemal, 359, 360
Ati-Atihan Festival, *596*
Atlanta, Georgia, 46, 72, *118*
Atlantic Ocean, 202, 247
Atlantic Plate, 133
Atlantic Provinces, 112–114, *113*
Atlas Mountains, *311*
Atoll, **618**
Aung San Suu Kyi, 554, 582, 592, 601
Austerity, **149**
Austin, Texas, 125
Australia
  Asia and, 635–636
  CO_2 emissions, *23*
  colonization of, 633–634
  cultural diversity in, *650*
  Europe and, 650
  food production in, 623
  indigenous people in, 650–651
  invasive species in, 623, 626
  vegetation in, *621*
Australian desert, 617
Australo-Melanesians, **568**
Austria, *195*, 208, 209, 222
Austronesians, **568**
Authoritarianism, **39**, 278
  in Southeast Asia, 579, 582
  in sub-Saharan Africa, 394–395
Automation, 80–81, *84*
Automobile industry, 83, *84*
Automobiles, 88
Autonomous regions (China), 527
Avci, Elif, 358
Avci, Veli-Çetin, 358
Average population density, 46
Axis powers, 208
Ayers Rock (Australia), 617
Azerbaijan, 220, 259, 283, 299–300, *299*
Aztecs, **143**, 144

Babylonian Empire, *317*, 361
Bachelet, Michelle, 183
Baghdad, 361
Bahamas, 171
Bahrain, 355
Baidu, 514
Baku, Azerbaijan, 283
Bali, 568
Balkans, 194
Balkan Stability Pact, 251
Baltic Sea, 202, 247
Baltic states, 247, 248–249
Bamboo, 372–373
Bananas, 171
Bangalore, India, 481–482
Bangkok, Thailand, *565*, *587*
Bangladesh, 58, 433, *435*, 436, 447, 456, 479
Banh mi, *598*
Bank failures, 83
Banking and finance, 252
  in China, 509, 533
  European Central Bank, 214
  microcredit, 392, *451*, 452-54
  mobile phones and, 367, 454
  recession of 2007–2009 and, 217, 240
  in Singapore, 576
  women in, 85
Bank of America, 95
Banlieues, **220**
Banquiao Dam, 498
Baptists, 110
Barbados, 171, 173, *173*, 174
Barbary Coast, 351
Barbecue, *106*, *156*
Barbuda, 171
Barcelona, Spain, *223*, 244
Barrios, 43
Barrow, Alaska, 665
Basel, Switzerland, *196*
Al-Bashir, Omar, 353
Basque country, 244
Baxter, Darrick, 58
Bazaar, *341*
Beans, 75
Beber, **350**
Bechtel, 28, 141
Bedouins, 309, 320, 321
Beef, 118, 119
Bees, 120, *120*
Beijing, China, *484*, *519*, 533
Beijing–Tibet Railway, *528*
Belarus, 201, 259, 274, *287*, 293, 297–298, *297*
Belém, Brazil, *136*
Belgium, 206, 209, 220, 232, 235, 236
Belief systems, 56
Belize, 167
Belt and Road Initiative, *507*, 511
Benares, India, 427
Benelux, 236
Bengal, India, 446
Benguela Current, 372
Benin, 379
Bennett, Louise, 242
Berber–Arab culture, 350
Berbers, *351*
Bergolio, Jorge, 170–171
Bering land bridge, 75
Bering Strait, 66
Berlin, Germany, *223*, *240*
Berlin Conference, **382**
Berlin Wall, 239, *240*
Bharatiya Janata Parishad (BJP), 454, 456
Bhutan, 443, 444, 445, 446, 448, 463, 470, 474, 475
Bible Belt, 110
Big Mac, 31
Bikini Atoll, *629*
Bilingualism, 126
Bingham Canyon, *121*
Bin Laden, Osama, 89, 355
Bioaccumulation, **666**
Biodiversity, **134**
Biomagnification, **666**
Biosphere, **18**
  human impact on, *20–21*
Birth control pill, 101
Birth rate, 47, **47**
  See also Fertility rates
  in Europe, 224–225
  in North Africa and Southwest Asia, 342
  in North America, 101
  in Russia and post-Soviet states, 286
  in South Asia, 463
  in sub-Saharan Africa, 399
Bismarck, Otto von, 382
Bison, 120
Black Hawk Down, 417
Black Sea, 195, 201
Blue Mosque, *322*
Blue Nile, 352
Bodney, Corrie, *642*
Boers, 384–385, 419
Boko Haram, 398, 410
Bolivia, 138, 144, 150, 157, 181, 183
Bollywood, 460
Bolsheviks, **270**
Bombay, India, 427
Borders, in Europe, 226–230
Borneo, *550*
Borneo Project, 551, 567
Borscht, *290*
Bosnia and Herzegovina, 194, *218*, 249
Bosporus Straits, 359
Boston, Massachusetts, 46, 113
Boston Common, 112, *113*
Botswana, *370*, 420
Brahmaputra River, 429, 431, *433*
Brahmins, 444, *468*

Brain drain, **161**
Brasília, Brazil, 187
Bratusek, Alenka, 231
Brazil, 31, 42, 89, 138, 144, 152, 157, 167, 169, 185–188, *186*
activism in, 186
agriculture in, 185–186
BRICS and, 42, 89, 152–153, 274
economy of, 185
gender-based violence in, *168*
internet use in, 154
ISI and, 149
landless movement in, 159
mega events in, 187–188
physical geography of, 185
population pyramid for, *165*
race in, 170
shantytowns in, 161, *163*
social stratification in, 185
urbanization in, 186–187
urban migrants in, *162*
Brazilian Highlands, 185
Breadbasket, 78, 118–120
Brexit, **193**, 194, 214, 221, 236, 241–242
BRICS (Brazil, Russia, India, China, South Africa), 42, 89, 152–153, 274
Brides, shortage, in China, 524
Britain. *See* United Kingdom
British colonies, 76
in South Africa, 385
in South Asia, 427, 445, *445*, 456
in sub-Saharan Africa, 383–384, 420
British East India Company, 445–446, 456
British Empire, 59, 207–208, 241, 321, 445
British India, 33
British Isles, 240–242
British Overseas Territories, 241
British Petroleum (BP), 73
Brothels, 461
Brownfields, **100**
Bruegel, Pieter (the elder), *204*
Brunei, 604, *605*, 606
Brussels, Belgium, 222, 236
Budapest, Hungary, 222
Buddhism, 56, *57*, **444**
Buenos Aires, Argentina, 161, 184, *185*
Buffalo Commons, 120
Buffer state, **298**
Bukhara, Uzbekistán, *268*
Bulgaria, 194, 249, 252
Bull riding, *108*
Bumiputra, **605**
Burj al Arab hotel, *327*
Burj Khalifa, 340
Burka, 346
Burma. *See* Myanmar (Burma)
Burning Man, *108*
Burundi, 399, 415
Bush, George W., 89, 180
Bush administration, 88
Bushmeat, 379
Business process management, 450
Bus rapid transport (BRT), 164
Byaslag, *495*

Cacao, 144, 236
CAFTA (Central American Free Trade Agreement), 108, 178
Cahokia, 76
Cairo, Egypt, 43, 340, 353
Calcutta, India, 427
Calgary, 122
Calgary Stampede, *108*
California
agriculture in, 79
almond industry in, 120
settlement of, 78–79
Southern, 124–126, *124*
water depletion in, 72
Caliphate, 398
Calligraphy, *596*
Cambodia, *565*, 602–604, *603*
"killing fields" in, 571–572
Cambridge, Massachusetts, 113
Camels, *318*
Cameroon, 413
Canada, 65
See also North America
Aboriginal peoples in, 79–80
agriculture in, *87*
climate change and, 69
CO_2 emissions, *23*
cultural issues in, 104–107
immigration and diversity, 104–105
race and ethnicity, 109
religion, 109–111
democratic system of government in, 93–94
economic interdependence between U.S. and, 83, *85*
family size in, *102*
HDI of, 34
health-care system in, 94, *95*
influence abroad of, 88–89
oil reserves of, 74, 89
political freedoms in, 89, 92
population of, 101, *103*
relationships between U.S. and, 93–95
taxes in, 235
vulnerability to climate change in, 69–71, *70*
Canada–U.S. Free Trade Agreement, 81
Canadian Shield, 121–122
Candomblé, *187*
Cantonese, 536
CAP (Common Agriculture Program), 214–215
Cape of Good Hope, 384
Cape Town, South Africa, 46, *369*
Capitalism, **42**, **208**
crony, 573, *575*
Capitalists, **270**
Caputova, Zuzana, 232
Caracas, Venezuela, *179*, 180
Carbon dioxide, 19, *23*, 138, 266, 372
See also Greenhouse gases (GHG)
Carbon sequestration, **372**
Caribbean, 171–174, *172*
Barbados, 173, *173*, 174
colonialism in, 33
colonization of, 236
Cuba, 172–173
culture of, 169
Haiti, 173–174, *173*
political, social, and economic change in, 171–172
Puerto Rico, 172, 173
Caribbean Plate, *13*, 133
Carnival, 169, *169*
Carolinas, 77
Carp, *496*
Carpathian Mountains, 195
Carpet weaving, 461
Carrying capacity, 30
Cartagena, Colombia, *181*
Cartel, **325**
Carthage, 349, 350
Casablanca, Morocco, *341*
Caspian Sea, 201, 261, 264
Cassava root, *374*
Caste system, **443**–444, 456, *468*
Castles, *206*
Castro, Fidel, 159
Castro, Raúl, 159
Catalonia, 244
Catholic Charities, 191
Catholics, 110
See also Roman Catholic Church
Cattle, 118, 119, *119*, 147
Caucasia, **260**, 283, 299–300, *299*
Caucasus Mountains, 263
Caviar, *290*
Cell phones, 331, 367, 391, *453*
Cenderawasih Bay, *565*
Central Africa (subregion), 413–415, *414*
Central Africa Republic, 413
Central America, 65, 169, 176–179, *177*
agriculture in, 176
civil conflicts in, 178
Costa Rica, 176, 178
immigrants from, 107
migration to U.S. from, 178
Panama, 176, 178
physical geography of, 176
social and economic conditions in, 176
Central American Free Trade Agreement (CAFTA), 107, 178
Central Andes (subregion), 181–183, *182*
Central Asia, 266, *267*, 274, 277, 278, 283, 300–302, *300*
Central banks, 214
Central China (subregion), 533–535, *534*
Central Europe (subregion), 9, 193, 194, 222, 229, 235, 249–253, *250*, 259
agriculture in, 215, 251–252
air pollution in, 202
democracy in, 217
industry in, 251–252
market reforms in, 252
migrants from, 228
Soviet Union and, 208, 209, 271, 272
Central India (subregion), 479–480, *479*
Central lowlands, 67, *67*
Centrally planned economy, **270**
Central planning, **209**, 448
Central Siberia, 295–296, *295*
Central Siberian Plateau, 261
Ceylon. *See* Sri Lanka
Chad, 353
Chador, 346, 413
Chaebol, **543**–544
Chang Jiang River, 487
Chapatis, *466*
Charcoal, *372*, 373
Chávez, Hugo, 150, 180
Chechnya, 281–282, 292
Chennai, India, 427, 481–482
Chernobyl disaster, 264, *264*
Chevron, 131, 180
Chiang Mai, Thailand, *598*
Chiapas, Mexico, 159
Chicago, Illinois, 78
Childbirth, to teens, *53*
Childcare, 226
Child labor, *207*, 382, 461
Child marriage, *53*
Children
illegal immigrant, 107
in North Africa and Southwest Asia, 348
one-child policy (China), 524, *525*
in South Asia, 460–461
Chile, 34, 167, 169, 183
climate in, 135
coup in, 157
earthquakes in, 133
health care in, *151*
independence for, *143*
U.S. involvement in, 157
China, 31, 38, 42, 89
See also East Asia
aging population in, 523–524
agriculture in, *495*, *507*, 508–509
air pollution in, 497–498, *498*
BRICS and, 42, 89, 152–153, 274
Central, 533–535
Central and Coastal Plains, 534–535
climate, *489*
climate change and, *492–493*
CO_2 emissions, *23*, *26*
commune system in, 508–509
communist revolution in, 502–503
Cultural Revolution in, *501*, 503
economic reforms in, 509–510, 517, 520–521
economy of, 65, 505, 508–510, 517, 520–521, 532–533, 535
environmental issues in, 490–491, 497–498
ethnicity in, 526–527
exports, 239
Far North and West, 537–539, *538*
Far Northeast, 532–533
flooding in, 491, *493*
foreign influence of, 89
globalization and, 83
Hong Kong and, 521
human development in, 34
human patterns over time in, 499–505
Imperial, 499–500, *501*, 502
indigenous people in, 529
industry in, 509
influence of, in Middle and South America, 152–153
languages in, 526–527
Muslims in, 527–528
Northeast, 530–533, *531*
one-child policy, 524, *525*
Overseas Chinese, **529**
politics in, 514–515
population and gender in, 522–530
soft power of, 582–583
Southeast Asian financial crisis and, 576
Southern, 535–537, *535*
sub-Saharan Africa and, 420–421
Tibet and, 528
U.S. relations with, 89
U.S. trade war with, 83
Venezuela and, 180
in Zambia, 421

Chinatowns, *525*
Chinese workers, in Angola, *388*
Chipko movement, *439*
Chiquita, 178
Chocolate, 236
Cholera, 28
Chongqing, China, 533
Christianity, 56, *57*, 109–110, 170–171, 290, 317–318, **318**
  evangelical, 410
  Protestant Reformation, 206
  Roman Catholic Church, 144, 165, 167, 170–171, 225, 245
  in sub-Saharan Africa, 409–410
Christian missionaries, 384
Christmas Island, Australia, *642*
Church of the Holy Spirit, *208*
Circular migration, 402, 406
Cities
  See also Urbanization; *specific cities*
  in East Asia, *518–519*
  environment in, 46
  in Europe, 207–208, 222
  growth of, 43–46
  infrastructure of, *45*
  in Middle and South America, 161–164
  in North Africa and Southwest Asia, 339–340
  in North America, 97–100
  in Russia and post-Soviet states, 283, 293
  in South Asia, 458–461, 460
  in sub-Saharan Africa, 399, 402, 413
  world, 208, 516
Citigroup, 95
City-state, 38
Civil disobedience, **446**
Civilization
  in Africa, 379
  development of, 316–317
Civil rights, 205
  See also Political freedoms
Civil society, *41*, **42**
Civil war
  See also Conflicts
  in Afghanistan, 457
  in Angola, *219*, 421
  in Bosnia, *218*
  in Cambodia, 571
  in Colombia, 160, 180
  in Congo, 395
  in Liberia, *397*, 413
  in Libya, 39
  in Middle and South America, 178
  in Nepal, 458, 470
  in Pakistan, 57, 447
  in Somalia, 417
  in South Africa, 385
  in Spain, 243
  in Sri Lanka, 456, 482–483
  in Sudan, 352, 353, 415, 416
  in Syria, *41*, 329, 338, 357, 360
  in Tajikistan, 292, 301
  in Vietnam, 571
  in Zambia, 420
Civil War, U.S., 77, 78
Clam chowder, *106*
Clean water, 402
Clean Water Act (CWA), 73
Clear-cutting, **74**–75, *74*, 123
Cleveland, Ohio, 78, 115
Climate, **13**–14
  See also Climate change
  of East Asia, 488–490, *489*
  of Europe, 195–196, *197*
  of Middle and South America, 135–137, *136*
  of North Africa and Southwest Asia, 309, *310*
  of North America, 67, *68*, 69
  of Oceania, 618–621, *619*
  of polar regions, 661
  of Russia and the post-Soviet states, 261, *262*, 263
  of South Asia, 430–431, *432*
  of Southeast Asia, 557, *558*
  of sub-Saharan Africa, 370, *371*, 372
Climate change, 18–**19**
  air pollution and, 69
  Central Andes and, 182
  contributions to, 437–438
  deforestation and, 372, 559–561
  desertification and, 313
  drivers of, 19
  European response to, 198
  food production and, 561–563
  impacts, 19, 22
  in Oceania, 622–623
  polar regions and, 661–663
  responding to, 26
  responses to
    in Meking River system, 566–567
    in Oceania, 623
  in Southeast Asia, 559–567
  tropical forests and, 138
  vulnerability to, 22–26, *24–25*
    in East Asia, *492–493*, 494
    in Europe, 198, 199, *199*
    in Middle and South America, *140*
    in North Africa and Southwest Asia, 313, *314*, 315
    in North America, 69–71, *70*
    in Oceania, *624–625*
    in Russia and post-Soviet states, 266, *267*
    in South Asia, 431, 433, *434–435*, 436–440
    in Southeast Asia, *564–565*
    in sub-Saharan Africa, *376–377*
  water and, 375, 378, 563, 566
  wildlife and, 378–379
Climate region, 14, 16–17
Clinton, Hillary, 92
Coal, 26, 72
Coal mining, 75, *75*, 78
Coastal erosion, 69
Coastal lowlands, 67, *67*, *430*
Coca, *146*
Cocaine, 96, 159–160, *160*, 183
Cochabamba, Bolivia, 28, 138, 141
Cockatoo Island, *637*
Cocos Plate, *13*, 133
Coffee, 145, 171
Cohesion Policy, **211**
Cohousing, 101, *102*
Cold War, 39, 42, 89, 157, 159, 208–209, **208**, 259, 271–272, **271**, 293
  in 1980, *271*
  Europe and, 221
  in North Africa and Southwest Asia, 323
  Pakistan and, 474
Colombia, *41*, 157, 160, 179, 180–181
Colonialism, **33**, *36*, 56, 59
  decolonization and, 209, 420
  European, 33, *36*, 207–208, *207*, 236, 239, 282–285, 321, *383*
    in Middle and South America, 143–144, 146–147, *147*
    in North Africa and Southwest Asia, 321, 321–323, *322*
    in North America, 73, 76–80
    in Oceania, 633–634
    in Southeast Asia, 568–570, *569*
    in Southern Africa, 419–420
    in sub-Saharan Africa, 367, 382–385, 386–387, 414, 419–420
  Japan and, 503
  Russia and, 269
Colorado River, 72, 125, *125*
Colorado River Aqueduct, *124*
Color revolutions, 278
Columbia, *158*
Columbian exchange, 144, 205
Columbus, Christopher, 131, 144
Commercial agriculture, **373**–375, 563
Committee of the Regions, 221
Commodities, **386**
Commodity dependence, **387**, 420–421
Common Agricultural Program (CAP), **214**–215
Common Market, 209
Commons, 112, *113*
Commonwealth of Nations, 641
Commune system, 508–509
Communism, **42**, **270**
  in China, 508–509
  defined, **208**–209
  in East Asia, 505, 511, 541
  in Eastern Europe, 194, 207–208
  impact of, 207–208
  in Russia and post-Soviet states, 270–271, 272
  in Southeast Asia, 604
    Vietnam War and, 571
  Soviet promotion of, 272
Communist Manifesto, The (Marx), 207–208
Communist Party, **270**
Communist Revolution, in Russia, 269–270
Comoros, 415, 418
Confit, *215*
Conflicts
  See also Civil war
  in Afghanistan and Kashmir, 473–474
  in Central America, 178
  in Colombia, 180–181
  in East Africa, 415–417
  in Horn of Africa, 415
  in Russia and post-Soviet states, 278, 280–282
  since 1990, *40*
  in South Asia, 436, 454, *455*, 457–458
  in Sri Lanka, 482–483
  in sub-Saharan Africa, 375, 398, 413, 415–417
  in West Africa, 413
  in Western Sahara, 351
Confucianism, **500**
Confucian temples, *500*
Congo (Brazzaville), 413
Congo (Kinshasa), *41*, *388*, 395, *397*, 398, *401*, 413, 414, 415
Congo Free State, 414
Congo River, 369–370
Conrad, Joseph, 414
Consensus, **641**
Constantinople, Turkey, 204
Container ships, 200–201
Contested space, **157**
Continental climates, 16, *17*, *136*, **196**, *197*, *262*, *310*, *489*
Continental crust, 12
Continental Interior (subregion), 121–122, *121*
Continents
  See also Plate tectonics
  Pangaea hypothesis, 12
Contraception, 404
Contras (Nicaragua), 178–179
Conurbation, **97**, **516**
Cook, James, 653
Cool humid climates, 16
Copenhagen, Denmark, *45*, 222
Copper mines, *148*
Coral, *565*
Coral bleaching, **566**, **623**, *624*
Corn, 29, 75, 106, 120, 144, *156*
Corn tortillas, *156*
Corporate agriculture, 215–216
Corporations, multinational, 33, 37
Correa, Rafael, 183
Corruption, 157
  in Brazil, 188
  in Catholic Church, 170
  in Central Europe, 249, 252
  in China, 514, 535
  in Greece, 245
  in Haiti, 174
  in Italy, 245
  in Mexico, 174
  in Middle and South America, 138, *140*, 144, 149, 157, 159, 161
  in post-Soviet states, 259
  in Russia, 274, 282
  in South Asia, *455*, 456–457
  in Southeast Asia, 572, 573
  in South Korea, 544
  in sub-Saharan Africa, 385–386, 387, *388*, 394–395
  in Syria, 338
  in Venezuela, 180
  in Yemen, 357
Corsica, *197*
Cosmonauts, *668*, 669
Costa Rica, 176, 178
Côte d'Ivoire, 412, 413
Cotton, 437
Council of the European Union, 220
Council on Hemispheric Affairs, 181
Coup d'état, **157**
Court of Justice (ECJ), 221
Court of Justice of the European Union (CJEU), 221
Creoles, **144**
Crimea, 280

Critchfield, Richard, 465
Croatia, 194
Crony capitalism, **573**, *575*
Crops
  See also Agriculture; Food production
  Columbian exchange of, 144
  GMO, 87
  North America, 75–76
Cross-border relations, in North Africa and Southwest Asia, 315–316
Crowley, William, *163*
Crowley's model of urban land use, *163*
Crude oil imports, *92*
Cruise ships, 172
Crystal, David, 653
Cuba, *158*, 159, 171, 172–173
Cuisines. *See* Foodways
Cultural diversity. *See* Diversity
Cultural groups, 55
Cultural hearth, 112
Cultural homogenization, **229**
Cultural landscape, **4**
Cultural pluralism, **593**, 596
Cultural Revolution, *501*, **503**
Culture, **54**–60
  See also Sociocultural issues
  Arab, 308
  Berber–Arab, 350
  East Asian, 500
  ethnicity and, 55
  European, 203–204
  French, 238
  globalization and, 598–599
  Japanese, 546
  language and, 56, 58–59
  Latino, 125–126
  Middle and South America, 167, *168*, 169–171
  North Africa and Southwest Asia, 350
  North America, *102*, 104–107
  Oceania, *648*, 650–655
  race and, 59–60
  religion and, 56
  South Asia, 465–468
  Southeast Asia, 568–570, 593, *594–595*, 596–599
  in space, 677–678
  sub-Saharan Africa, *408*, 409–411
  values and, 55
Culture of poverty, 109
Curitaba, Brazil, 164
Cusco, Peru, 181, *182*
Cyprus, 214, 217, 242
Cyrillic alphabet, 268, *268*
Czar, **269**
Czechia, 194, *209*
Czech Republic, 195, *213*, 249, 252

Daimler Trucks, 239–240
Dalai lama, 528, *539*
Dalits, 444, 456
Dallas, Texas, *98*, 99
Damascus, 317
Dams, 69, 315, 315–316, *316*, 387, *428*, 436, 567
  in China, 498
  Three Gorges Dam, 496–497
Dani people, *608*
Danube River, 195
Darfur, 353
Dark Ages, 205
Day of the Dead, *169*
Da Yunhe (Grand Canal), 532
Dead Sea, 357
Dead zones, 73
Death rate, 47, **47**, 286
Debt
  in Europe, 214, 245–246
  in Middle and South America, *150*
  in sub-Saharan Africa, 387, 390
  in U.S., 92–93
Deccan Plateau, 429, *429*
Decolonization, 209, **209**, 420
Decriminalization, of drugs, 96–97
Deepwater Horizon oil spill, 73
Deforestation, 19, *20*, *21*, 67, 138
  alternatives to, 372–373
  of Amazon, *139*
  resistance to, 439
  in South Asia, 438–439
  in Southeast Asia, 559–561
  in sub-Saharan Africa, 372–373, 394
Deindustrialization, 211, 445–446
Delhi, India, 12, 13, 437, 438, 458, 476–477
Del Monte, 154
Demilitarized zone (DMZ), **504**
Democracy, 39, 217
  See also Political freedoms
  in Cambodia, 571, 582
  in Indonesia, 583
  in Mongolia, 540
  in North Africa and Southwest Asia, 329, 334
  in Oceania, 641, *642*
  in Russia and post-Soviet states, 259
  in Southeast Asia, *580*, 590, 601
  in sub-Saharan Africa, 398
  in Thailand, 582, 601, 602
Democracy index, *40*, *91*
Democratic Party, 93, 108
Democratic Republic of Congp. *See* Congo (Kinshasa)
Democratization, **39**
Democratization Index, *218*, *333*, *396*, *642*
Demographic dividend, 464
Demographic transition, **47**–48, *49*, **589**
  in Middle and South America, 165
  in North America, 101
  in Southeast Asia, 589
  in sub-Saharan Africa, 403–404
Denmark, *213*, 246, 247
Desalination, **315**
Desertification, **313**, 375
Deserts, 16, *17*, *68*, *136*, *310*, *371*, *432*, *619*
De Soto, Hernando, 76
Detritus, **557**
Detroit, Michigan, 46, 78, 100, *100*, 115, *116*, 145
Development, 30, **31**
  See also Economy; Globalization
  of civilization, 316–317
  classifications of, 31
  in East Asia, 505–511, *506–507*
  in Europe, 210–216, *213*
  gender equality and, 54
  globalization and, 33–38, *34–36*
  government and, 38
  measuring, 31–32
  in Middle and South America, 148–156
  in North Africa and Southwest Asia, 324–328, *327*
  in North America, 80–88, *84*
  in Oceania, 634–640, *637*
  in polar regions, 663–664
  in Russia and post-Soviet states, 272–278, *276*
  in South Asia, 447–454, *451*
  in Southeast Asia, 572–579, *574–575*
  in space, 672–674
  in sub-Saharan Africa, *388*, 391–394
  sustainable, 32–33
  transitions of, 31
Development institutions, 37, 38
Dhaka, Bangladesh, *45*
Dharavi (Mumbai), 460
Dia de los Muertos, *169*
Diaspora, **318**
Dictators, **157**
Diet. *See* Foodways
Diffusion, **320**
Digital divide, **81**
Dingo, *627*
"Dirty War" (Argentina), 157
Disease
  See also Health care
  European, 76, 144
  HIV/AIDS, 404, 405–406, *405*, *408*, 421, 593
  poverty-related, 30
  in sub-Saharan Africa, 405
  waterborne, 28, 141, 161
Dislocate (theater troupe), *654*
Distributaries, **567**
Diversity
  backlash against, 107–109
  in New England, 113–114
  in North America, 104–105, *105*
  Russian domination and, 280–281
Divide and rule, **383**
Djenné, 380
Djibouti, 415
DMZ (demilitarized zone), **504**
Doctors Without Borders, 42
Doi moi, **604**
Dolpo-pa people, 475
Dome of the Rock, 337
Domesticated plants, *146*
Dominican Republic, 171
Domino theory, **571**
Dongguan, Guangdong Province, *507*
Double day, 51, **231**
Dowry, **467**–468
Dragon Con, *118*
Dravidian language, 442
Drought, 22, *140*, 198, 431, 622–623
Drug cartels, 95–96
Drug courts, 96
Drugs
  cocaine, 96, 159–160, *160*, 183
  heroin, 96, 159–160
  illegal, *158*
  legalization of, 96–97
  marijuana, 96–97, 159–160
  in North America, 95–97
  opium, 502, 593, 601
Drug trade, 159–160, *160*, 180–181
Druze, 357
Dubai, *327*, 340, 355–356
Dubrovnik, 222
Durban, South Africa, *421*
Dust Bowl, 78
Dutch colonies, 76
Dutch East India Company, 384–385
Dutch East Indies, 209
Duvalier, Jean-Claude "Baby Doc," 174
Duvalier, Jean-Claude "Papa Doc," 174
Dzhanibekov, Vladimir, 669

Earth, 670
  magnetic field of, *671*
  object orbiting, *670*
Earthquakes, 12, 26, 74, 133, 309, 342, 429–430, 472, 490, 497, 536, 556
East Africa (subregion), 415–419, *416*
East Asia (world region), 38, 485–549, *486–487*
  globalization and development in, 505–511, *506–507*
  human patterns over time in, 499–505
  imperialism in, 502
  physical geography of, 488–499
  political map of, *488*
  population, gender, and culture in, 522–530, *525*
  power and politics in, 511–516, *512–513*
  subregions of, *530*
    Central China, 533–535, *534*
    China's Northeast, 530–533, *531*
    Far North and West China, 537–539, *538*
    Japan, 544–547, *545*
    Korea, North and South, 541–544, *542*
    Mongolia, 539–541, *540*
    Southern China, 535–537, *535*
    Taiwan, 547–548, *547*
  urbanization in, 516–522, *518–519*
  U.S. relations with, 89
  visual history of, *500–501*
East China Sea, 488
Easter Island, 632
Eastern Europe, 194, 208
Eastern Hemisphere, 5
Eastern Highlands (Australia), 617
Eastern Mediterranean, 357–358, *357*
East Germany, 208, 239
East India Company, 384–385
Ecological footprint, 18, *19*
Economic and technology development zones (ETDZs), *520*
Economic Community of West African States (ECOWAS), *393*
Economic core, **78**, 80
Economic crises
  in Southeast Asia, 573, 576
  in sub-Saharan Africa, 387
Economic development
  in Asia, 635–636
  in Europe, 210–216, *213*
  measures of, 31–32
  in Middle and South America, 144–147
Economic diversification, **328**
Economic elites, 92
Economic inequality, *53*, 54

Economic integration, 211–212, 214
Economic migrants, **229**
Economic reforms
in China, 509–510, 517, 520–521
in East Asia, 517
in South Asia, 448
Economies of scale, **212**
Economist magazine, 31, 40
Economy
See also Agriculture; Banking and finance; Debt; Development; Employment; Energy resources; Globalization; Human well-being
centrally planned, **270**
in East Asia, 505–511
in China, 65, 505, 508–510, 517, 520–521, 532–533, 535
foreign direct investment in, *520*
in Hong Kong, 537
in Japan, 508, 521, 523, 546–547
Korea, North and South, 543–544
in Mongolia, 539–540
in Taiwan, 548
in Tibet, 539
of Europe, 210–216, 223
in Europe
in British Isles, 240, 241, 242
in Central Europe, 249, 252
in Estonia, 249
in France, 238–239
in Germany, 240
in Greece, 245–246
in Italy, 244–245
in Scandinavia, 247
in South Europe, 243
in Spain and Portugal, 244
formal, 32
global, 33
informal, 32, 154, 275, 402
knowledge, 81
in Middle and South America
in Brazil, 185
in Caribbean, 171–172, 172–173
in Mexico, 174–175
in Southern Cone, 183
in North Africa and Southwest Asia, 328
in Egypt, 353–354
in Turkey, 359–360
in North America, 80–88
in Canada, 83, *85*
in United States, 65, 80–81
in Oceania, 634–640
in Russia and post-Soviet states, 263, 272–275
in Belarus, 298
in Central Asia, 301
in Russia, 293, 295, 297
satellite, 672–673
in South Asia, 448–450, 452–454
in Afghanistan and Pakistan, 473
in Bangladesh, 479
in Himalayan subregion, 475
in India, 448–450, 475–476, 480, 481–482
in Sri Lanka, 481
in Southeast Asia, 572–579
in Brunei, 606
in Cambodia, 603–604
in Malaysia, 606
in Southeast Asia
in Myanmar (Burma), 599, 601
in Philippines, 608, 610–611
in Singapore, 606
in Thailand, 599, 602
in Vietnam, 603–604
in sub-Saharan Africa, 389–390, *389*, 391–394
in East Africa, 415
in Madagascar, 418
in Southern Africa, 420–421
women in, 83, 85
Ecotourism, **141**–142, *141*, 172
Ecuador, 129, 131, 138, 157, 181, 183
Edmonton, Alberta, 122
Education, 39
female access to, 51, *53*, 54
HIV/AIDS, 406
in North Africa and Southwest Asia, *329*
women and, 85, 347
Egypt, 39, 89, 220, 309, 321, 329, *332*, 340, *345*, 348, 352–354, *352*
agriculture in, 353
economic development in, 353–354
Electricity, 19, 26, 437, 458, 479, 482, 494, 496–497, 498, 534, 566, 567, 622
See also Hydroelectric dams
Electric vehicles, 72, 438, *438*, 448, 490
Elephants, 379, 418
Elites
economic, 92, 148
in Middle and South America, 164
El Niño, **135**, 620–621, *620*
El Salvador, *168*, 170, 178
Emigration. *See* Immigration; Migration
Emperor penguins, *662*
Employment
See also Labor force; Unemployment
of children, *207*, 382, 461
in health care sector, *102*
immigrants and, 106–107
in India, 450
in Japan, 546–547
manufacturing, 80–81
Maquila, *176*
in Middle and South America, 154, 162, 165, 167
in North Africa and Southwest Asia, 340, 342, 350
in North America, 81, 122
offshore outsourcing, 450
in Russia and post-Soviet states, 274–275
in South Africa, 420
in South Asia, 448, 450, 452, 461, 477
in Southeast Asia, 573
in sub-Saharan Africa, 379, 389, 390, 399, 402, 415, 420
of women, 85, 101, 288–289, *347*, 404, 573
Endangered languages, 58
Endangered species, 379
Endemic, **621**
Energy efficiency, in cities, 46
Energy issues, Europe, 201–202
Energy resources
See also Fossil fuels; Hydroelectric dams
alternative, 69, 201–202, 438
Brazil, 185
in sub-Saharan Africa, 393–394
England, 207, 241
See also United Kingdom
English language, 59, 126, 241, 442, 473
Enlightenment, **206**
Entrepreneurship, 85, *595*
Environment
See also Acid rain; Climate change; Deforestation; Water resources
in East Asia, 490–494
in Europe, 196, 198–202
human impact on, 15, 18–26, *20–21*, 28–29
in Middle and South America, 137–142
in North Africa and Southwest Asia, 309, 311–316, *311*
in North America, 69–75
in polar regions, 661–663
in Russia and the post-Soviet states, 263–266
in South Asia, 431–440
in space, 671–672
in sub-Saharan Africa, 372–379
urban planning and, 46
Environmentalism, *45*
Environmental knowledge, 58
Environmental protection
in Costa Rica, 178
ecotourism and, 141–142
laws, 37
EPZs (export processing zones), **153**, 573, *575*
Equator, 5, *7*
Equatorial conditions, *620*
Equatorial Guinea, 413
Erdoğan, Recep Tayyip, 220
Eritrea, 415
Eritrean cuisine, *374*
Erosion, 12, 13
Espanola, New Mexico, 76
Estonia, 194, *223*, 247, 248–249, 259, 293
ETDZs (economic and technology development zones), *520*
Ethanol fuels, 26, 29, *120*, 185
Ethiopia, 315, 415
Ethiopian cuisine, *374*
Ethnic groups, **55**, 249–251, *251*
Ethnicity
in Afghanistan, *473*
in China, 526–527
culture and, 55
in East Asia, 529
median household income by, in U.S., *109*
in Nigeria, *384*
in North America, 107–109
in Oceania, 650
in Russia and post-Soviet states, *281*
in South Asia, 442
in sub-Saharan Africa, 410–411
Ethnic violence, in Central Europe, 249–250
Ethnocentrism, 55
EU. *See* European Union
EU-27, 194
EU-28, 194, 201
Euphrates River, 309, *316*, 361
Eurasian Plate, *13*, 195
Euro, **214**, 241
Europe, *192–193*
See also European Union; *specific countries*
Australia and, 650
culture of, 203–204
debt crises in, 214
economy of, 210–216
environment of, 198–202, 216
feudalism in, 205
globalization and development in, 210–216, *213*
human patterns over time in, 202–210
immigration to, 214, 226–230
Industrial Revolution in, 33
Maghreb and, 350
Marshall Plan for, 209
physical geography of, 194–196
political map of, *194*
population, gender, and culture in, 224–236, *233*
postcolonial, 220
power and politics in, 217–221, *218–219*
pre-1989 alignment of countries in, *10*
refugees to, 193
subregions of
Central Europe, 249–253, *250*
North Europe, 246–249, *246*
South Europe, 242–247, *243*
West Europe, 236–242, *237*
taxes in, 235
urbanization in, 205–206, 222–224, *223*
visual history of, *204–205*
European Central Bank (ECB), **214**
European colonies, 33, *36*, 207–208, 236, 239, 321
in North America, 73, 76–78
in Oceania, 633–634
in Southeast Asia, 568–570, *569*
in sub-Saharan Africa, 367, 382–385, *383*, 419–420
European Commission, 220
European conquest, of Middle and South America, 143–144
European Council, 220
European Economic and Social Committee (EESC), 221
European Economic Area (EEA), 217
European Economic Community (EEC), 209
European Environment Agency, 198
European imperialism, 502
European Parliament, 220
European Russia, 261, 293, *294*
European settlement
of North America, 76–78, 79–80
in Oceania, 632–633
in Southern Africa, 419–420
European Union (EU), 37, **193**
in 2016, *11*
birth of, 209–210
Brexit and, 193, 194, 221, 236, 241–242

budget, 210, *210*
challenges facing, 193
economic comparison of U.S. and, 211
energy independence in, 201
expansion, politics of, 217, 220
future of, 221
as global peacemaker, 221
governing institutions, 220–221, *221*
growth of, 210
population pyramid for, *225*
Russian oil and gas imports, 273, *273*
as superpower, 42
trade agreements, *212*
Turkey and, 358–359
Eurozone, **214**
Evangelical Christianity, 410
Evangelical Protestantism, **171**
Evaporation, 566
Exclave, **249**
Exploration, by Europeans, 207–208
Exponential population growth, *48*
Export-led economic development, **572**
Export-led growth, **505**
Export processing zones, **153**, 573, *575*
Exports and imports, *212*
See also International trade
food, 315
oil and gas, 325–326
raw materials, 636
in South Asia, 437
in sub-Saharan Africa, 420–421
Export subsidies, 37
Exposure, 22
Extended family, 167
External landscape processes, 12–13
Extractive economic activities, 31
Extractive resources, 31

Facebook, 232, 361
Factory farms, 86–87, 118, 147
Failed state, **339**
Fair trade, 38, 236
Falafel, *313*
Fallow period, **373**
Families
American, 111
extended, 167
in North Africa and Southwest Asia, 343–344
nuclear, 111
in Southeast Asia, 591–592
Family farms, 86, 101, 215
Family leave, 232
Famine, 29, 446, 450
Farakka Barrage, 436
FARC (Revolutionary Armed Forces of Colombia), 180
Far Eastern India, 478–479
Farms
See also Agriculture; Food production
factory, 86–87, 118, 147
family, 215
family-run, 86, 101
in North America, 99
in South Asia, 450
subsidies for, 81
Far North and West China, 537–539, *538*
Faroe Islands, 247
Fast-food industry, 147

Favelas, 43, **161**, 163–164, *163*
Federal Conservation program, 120
Federal government
Canada, 93
U.S., 93–94
Federal lands, 122
Federated States of Micronesia, 615
Female genital mutilation (FGM), 348, **406**–407, *407*, *408*
Female literacy, in South Asia, 464, *464*
Female seclusion, **344**
Feminization of labor, **573**
Ferguson, Naomi, *649*
Fernández de Kirchner, Cristina, 183
Fertile Crescent, **316**–317, *317*, 357
Fertility rates, 48
in Europe, 224–225, 226
in North Africa and Southwest Asia, 342
in North America, 101
in Oceania, 646
in South Asia, 463–464, *463*
in Southeast Asia, 589–590, *589*, *590*
in space, 677
in sub-Saharan Africa, 403
Fertilizers, 86, 374–375
Fes, 350
Festivals
in East Asia, *499*
in Europe, *224*
in Middle and South America, 169, *169*
in North Africa and Southwest Asia, *319*
in North America, *108*
in Oceania, *654*
in Russia and post-Soviet states, *289*
in South Asia, *466*
in Southeast Asia, *596*
in sub-Saharan Africa, *410*
Feudalism, 205
FIFA World Cup, 187
Fiji, 617, 640–641, *642*, *645*
Finance, 510–511
Finland, *35*, 246, 247
Fires, 622–623, *625*
First Nation, 65
First Opium War, *501*, 522
Firzik, Marina, 191
Fisheries/fishing, 113, 123, 182, 247–248, *378*, 417, 496, *595*, *663*
Fish farming, 436
Five Pillars of Islam, 318–319
Fjord, **661**, *661*
Floating population, **517**
Flooding, 13, 67, *140*, 431, 491, *493*, *565*, **566**, 622–623
Floodplain, **12**
Florida, 72
impact of climate change on, *70*
Spanish settlers in, 76
Food, 28–30
deforestation and, *21*
GMO, 30
insecurity, 28, 421
Food and Agricultural Organization (FAO), 28, *29*
Food imports, 315
Food marketing, 215–216

Food prices, 86
Food processing, *184*
Food production, 28–30
See also Agriculture
in East Asia, 494, 496
in Europe, 214–216
in Middle and South America, 154–155
in North Africa and Southwest Asia, 311–313
in North America, 86–87
in Oceania, 623, 626–627
in Russia and post-Soviet states, *276*, 277–278, *277*
in Southeast Asia, 561–563, *562*
in sub-Saharan Africa, 402
sustainability and, 86–87
Food security, 29–30, 214, **373**
in East Asia, 494
in sub-Saharan Africa, 387, 389, 402
Foodways
in East Asia, *495*
in Europe, *215*
in Middle and South America, *156*
in North Africa and Southwest Asia, *313*
in North America, *106*
in Oceania, *626*
in Russia and post-Soviet states, *290*
in South Asia, *466*
in Southeast Asia, *598*
in sub-Saharan Africa, *374*
Forced labor, 54
Foreign-born population, in U.S., *105*
Foreign direct investment (FDI)
in East Asia, *520*
in Southeast Asia, *577*, 603–604
in sub-Saharan Africa, 390–391, 415, 421
Foreign exchange, **585**
Foreign involvement, in Middle and South America, 157
Foreign policy, U.S., 88–89
Foreign workers
in Japan, 547
from Turkey, 358
Forestry/forests, 113, 138, 195–196, 265, 372, 413, 439
Forests, 479
Formal economy, 32, 174
Formal region, **9**
Forward capital, **187**
Fossil fuels, 19, 26, 69, 72, 201, 202, 272–274, **308**, 313
See also Coal mining; Natural gas; Oil resources
in North Africa and Southwest Asia, 324–326, *324*, 328
Fossil water, 311
Fracking, 26, **74**
France, 214, 236, 238–239
agriculture in, 238–239
Algeria and, *219*, 220, 321, 351
colonies of, 239
fertility rate in, 225, 226
Francis, Pope, 170–171
Franco, Marielle, *168*
Free enterprise system, 80
Free market, 235
Free Papua Movement, *580*

Free trade, **37**
agreements, 80, 81–**82**, 106, 153
backlash against, 83
debate over, 37
globalization and, 81–83
reforms, 38
Free Trade Area of the Americas (FTAA), 153
Free trade zones, 153
French colonies, 76
French Guiana, 180
French Revolution, 207
French settlement, of Québec, 114–115
Freshwater, pollution of, 202
Frontal precipitation, 15
Fronts, 15
Frozen conflict, 282
FTAA (Free Trade Area of the Americas), 153
Fukushima disaster, 201
Fulani people, 375
Funafuti Atoll, *645*
Functional region, **9**
Futnani Chambers building, *479*

G8 (Group of Eight), 274
Gabon, 413
Gandhi, *441*
Gandhi, Mohandas, 446, 456
Ganga–Brahmaputra Delta, 477
Ganga River, 429, 436, 436–437, *437*, *466*
Ganges River, 427
Garma Festival, *654*
Garreau, Joel, 112
Gaza Strip, 334
Gazprom, 274, *276*
GDI. *See* Gender Development Index (GDI)
GDP (gross domestic product), **31**
informal economy as percentage of, *390*
GDP per capita, **31**
in India, *449*
Gender, 46, **50**–54
See also Gender imbalance; Gender roles; Men; Women
demographic transition and, 47–48
in East Asia, *525*
equality, 54, 102
in Europe, *233*
fertility and, 101
globalization and, 592–593
HIV and, 406
informal economy and, 32
in Middle and South America, 167, *168*
in North Africa and Southwest Asia, 334, 342–343, 343–348, *345*
in North America, 94–95, 101, *102*
in Oceania, 646–647, *648*, 649, 652–653
policies, in EU, 231–232
rural–urban migration and, 161
in Russia and post-Soviet states, 288–290, *291*
in Scandinavia, 248
in South Asia, 463, 468–471, *469*
in Southeast Asia, 591–593, *594–595*
in space, 677
in sub-Saharan Africa, 399, 406–409, *408*

Gender-based violence, *52*, 54
Gender Development Index (GDI), 52, *52*, 54, *102*, *168*, *232*, *233*, *291*, 343, *345*, *408*, *469*, *525*, *594*, *648*
Gendered spaces, 344, 346
Gender imbalance
in Afghanistan, 50
in China, 524
in polar regions, 666
in Qatar and UAE, 342–343
in Vietnam, 590–591
Gender pay gap, 54, 85, *86*, 231, *231*, 346–347, *595*, 649
Gender roles, 51, *52–53*, 54
Confucian, 500
in Europe, 230–232
in Middle and South America, 167
in Mongolia, 541
in North Africa and Southwest Asia, 344, 346
in Southeast Asia, 591–593
General Court (GC), 221
General Motors (GM), *84*
Genetically modified (GMO) crops, 87
Genetic modification (GMO), 30
Genocide, **398**, 399
in Europe, 218, 335
in Rwanda, 399
in Southeast Asia, 572
in sub-Saharan Africa, 411
in Turkey, 300
Gentrification, **100**
Geographers, 4
Geographic information science (GISc), **8**
Geographic information systems (GIS), 8
Geography, 2–3
See also Physical geography
study of, 3–12
visual tools in, 4–9
Geopolitics, **42**, 323, 324, 395
Canadian role in, 89
in North Africa and Southwest Asia, 334–339
of the sea, 516
in sub-Saharan Africa, 395, 398
U.S. role in, 88–89
Georgia, colony of, 77
Georgia (country), 72, 220, 259, 278, 282, 299–300, *299*
Geospace, **670**, 678
Geostationary orbit (GEO), **671**
Geothermal energy, 567
Ger, *541*
Germany, 214, 231, 239–240
division of, 208, 239
Nazi, 208, 217, 271
population pyramid for, *225*
postwar, 217
renewable energy in, 202
Uplands, *195*
Gerrymandering, **108**, *108*
Ghana, 380, *401*, 412
Ghettos, 43
GHG. *See* Greenhouse gases (GHG)
Gift of the Givers, 42, *43*
Gini index, 152, *152*
Girls
education of, 51, *53*, 54
in North Africa and Southwest Asia, 348
GIS. *See* Geographic information systems (GIS)
GISc. *See* Geographic information science (GISc)
Giza pyramids, *354*
Glacial melting, 22, *140*, 433, 566
Glaciers, 66–67, 266, *267*
Glasnost, 217, **271**
Glastonbury Festival, *224*
Global economy, 33, 36–37
See also Economy; Globalization
Globalization, 33–38, **33**, 65, 149
See also Development; Economy; Free trade
Asia and, 83
backlash against, 83
climate change and, 138
culture and, 598–599
debate over, 37
development and, 34–37, *34–36*
in East Asia, 505–511, *506–507*
environmental damage and, 628–630
in Europe, 210–214, *213*, 216
fisheries and, 496
free trade and, 81–83
gender and, 592–593
government and, 38
impact on Great Plains, 119–120
in Middle and South America, 148–156, *151*
in North Africa and Southwest Asia, 324–328, *327*
in North America, 80–88, *84*
in Oceania, 628–630, 634–640, *637*
in polar regions, 663–664
in Russia and post-Soviet states, 272–278, *276*
in South Asia, 447–454, *451*
in Southeast Asia, 572–579, *574–575*
in space, 672–674
in sub-Saharan Africa, 386–391, *388*
Global recessions, 576
Global relationships, of Oceania, 634
Global scale, 10
Global Slavery Index, 382
Global warming, 18–**19**
See also Climate change
GNI. *See* Gross national income (GNI) per capita
Golden Temple, *482*
Gold rush, 78
Gondwana, **617**
Good Friday Agreement, 241
Gorakhpur, India, *53*
Gorbachev, Mikhail, 271, 272, 288
Gospel of success, 171, 410
Gospels, 318
Government
democratic systems of, in North America, 93–94
globalization, development, and, 38
Government subsidies, for farmers, 81
Grabar-Kitarovic, Kolinda, 231
Grameen Bank, *451*, 453–454
Grand Marais, Minnesota, *67*
Grandmother hypothesis, 51, *52*
Grassroots economic development, **392**
Grassroots sustainability, 641
Gravlax, *215*
Great Barrier Reef, **617**, *617*
Great Basin, 69
Great Britain. *See* United Kingdom
Great Depression, 270
Great Dividing Range (Australia), 617
Great Escarpment, 369
Great Lakes, 66
Great Leap Forward, **503**
Great Man-Made River (North Africa), 315
Great Migration, 117
Great Pacific Garbage Island, 629–630, *629*
Great Plains (subregion), 78, 79, 86, 118–120, *119*
Great Recession, 83
Great Rift Valley (Africa), 369
Great Wall, *500*
Great Zimbabwe Empire, 380
Greece, 242, 245–246
ancient, 203, 242
debt crisis in, 214, 245–246
Turkey and, 359
Green, **198**
Green Belt Movement, 372
Greenhouse effect, *22*
Greenhouse gases (GHG), 18–**19**, *23*, 26, 138, 198, 201
See also Carbon dioxide; Climate change
from North America, 69
from Russia and post-Soviet states, 266, 295
in South Asia, 437
Greenland, 247
Greenpeace, 628
Green political parties, **198**
Green revolution, **30**, 86, 154, 159, 450
Gross domestic product (GDP), **31**
Gross domestic product (GDP) per capita, **31**
"Gross national happiness," *474*, 475
Gross national income (GNI) per capita, **31**, 32, *32*, *326*
EU, 211
in North Africa and Southwest Asia, *326*
in Russia and post-Soviet states, 275
in sub-Saharan Africa, *387*
U.S., 211
Groundwater, **378**
Group of Eight (G8), 274
Growth poles, **517**
Growth rate, 47
Guangdong, China, *34*
Guatemala, *165*, 170, 178
Guerilla gardeners, *200*
Guest workers, **229**, 358
Guevara, Che, *158*
Guianas, 169, 180
Guilin, Guangxi Zhuan, *489*
Guinea Highlands, 375
Gulags, 271, 296, **665**
Gulf of Aden, 417
Gulf of Carpentaria, 617
Gulf of Mexico, 67, 69
Gulf states, *23*, **308**, 340, 346, 354–356
See also specific countries
Gulf Stream, 22, 69, 73
Gulf War, 328, 338, 361
Guyana, 167, 169, 180
Gypsies, 208
Gyula sausage, *250*

Habitat loss, 69, 73, 99, 662
Haciendas, **147**, 154, 159
Hainan, China, 488, *489*
Haiti, 89, *90*, 138, *140*, 171, 173–174, *173*
Hajj, **319**
Haka, *655*
Han, 526–527
Hangi, *626*
Hanoi, *587*
Hapsburg Empire, 249
Harappa culture, 440
Harbin, China, *499*
Haryana, 475–476
Hausa-Fulani, 384
Hawaii, 65
tourism in, 638–639
HDI. *See* Human Development Index (HDI)
Health care
access to, 48, 65, *84*, 101
in Canada, 94, *95*
in Chile, *151*
employment in, *102*
expenditures on, 235
in North Africa and Southwest Asia, *329*
in South Asia, 463
in sub-Saharan Africa, 405–406
in U.S., 94, *95*
Heavy metal music, 248, *248*
Hemispheres, 5
Herbicides, 30
Herding, 375
Hernandez, Carlos, 63, 65
Hernandez, Orbalin, 153
Heroin, 96, 159–160
Hezbollah, 338
High-altitude climates, 16, *17*, *432*
Higher education
in Gulf States, 356
institutions of, 113
women in, 85, 347
Highlands, of Middle and South America, 133–134
High-speed rail, 200, *200*, *501*
Highways, 88, 97, 118, 138, 200, *507*
Hijab, 346
Hijra, **470**
Himalaya Mountains, 12, 428, 429, *429*, 431, 433
Himalayan Country (subregion), 474–475, *474*
Hinduism, 56, *57*, **442**–443, *455*, 470–471
Hindutva, 454, 470
Hispaniola, 144
HIV/AIDS, 404, 405, 405–406, *405*, *408*, 421, 593

Ho Chi Minh, 571
Ho Chi Minh City (Vietnam), *556*, *581*
Ho Chi Minh Trail, 579
Hog farms, *253*
Holi, 1
Holocaust, **208**, 328
Homeless Workers Movement, 159, 163
Homo erectus, 379
Homo sapiens, 379
Homo sapiens sapiens, 59
Honduras, *25*, 178
Hong Kong, 488, 502, 521, 537
Horn of Africa, **372**, 415
Horses, 76
Hot spots, **617**
Household duties, 51
Housewife contract, 235
Housing industry, in U.S., 83
Housing segregation, **100**
Huang He River, 487, *532*
Hub-and-spoke network, 88
Hukou system, **517**
Human development, 34
  in Central Andes, 182
  in Europe, *213*
  in India, 449–450
  measures of, 32
  population growth and, **47**–48
Human Development Index (HDI), **32**, 34, *35*, *84*, *151*, *168*, 211, *213*, 235, 275, *276*, *327*, *388*, *574*, *637*
Human geography, **3**
Human impact, on environment, 15, 18–26, *20–21*, 28–29
Humanism, **206**
Human origins, 367, 379
Human patterns over time
  in East Asia, 499–505
  in Europe, 202–210
  in Middle and South America, 142–148
  in North Africa and Southwest Asia, 316–324
  in North America, 75–80
  in Oceania, 630–634
  in Russia and post-Soviet states, 266, 268–272
  in South Asia, 440–447
  in Southeast Asia, 567–568
  in sub-Saharan Africa, 379–386
Human resource capacity building, **639**
Human resources, 31, 113–114
Human rights abuses, in Russia and post-Soviet states, 292
Human settlements, in space, 675–676
Human trafficking, *52*, 54, 382
Human well-being, 32, 33
  in East Asia, 540
  in Europe, 193, 235, 247
  in North Africa and Southwest Asia, 327, 334, 360
  in Oceania, 638
  in Russia and post-Soviet states, 275, 277, 302
  in South Asia, 475
  in Southeast Asia, 574, 578, 590
Hungary, 9–10, 249, *250*
Hunger, 28, *29*, 450
Hunting, 379
Hunza Valley, *471*
Hurricane Katrina, *70*
Hurricane Maria, 22
Hurricanes, 22, 69, *70*, 135, 137
Hurricane Sandy, 26, *140*
Hussein, Saddam, 338
Hybrid regimes, 278
Hydraulic fracturing, 26
Hydroelectric dams, 123, 387, 496–497, 567, 582
Hydroelectric energy, 26
Hyperinflation, 180

Ibadism, 356
Ice ages, 66
Iceland, *53*, 231, 246
Iftar, *319*
Igbo people, 383
Illegal drugs, 95–97, 126, *158*, 159–160, *160*, 180
Illegal immigration, 106, 107
Il Palio, *224*
IMF. *See* International Monetary Fund (IMF)
Immigrants
  See also Migrants
  assimilation of, 229–230
  burden of taxpayers of, 106
  children, 107
  in Europe, 226–230, 235
  job losses and, 106–107
  Latino, 117
  legal, 106
  in Oceania, 641, 643
  statistics on, 107
  undocumented, 106, 107
  wages of, 107, 120
Immigration
  See also Migration
  in Europe, 214, 226, *228*, 244, 245
  illegal, 106
  in Middle and South America, 169
  in North America, 104–105, 104–107
  reasons for, 105–106
  in Southeast Asia, 585
  in UK, 242
Imperial China, 499–500, *501*, 502
Imperialism, 502
  See also Colonialism
Import quotas, 37
Imports. *See* Exports and imports
Import substitution industrialization (ISI), **149**, 328
Inca Empire, 181
Incarceration rates, in United States, 96
Incas, **143**
Income, 449–450
Income inequality, **152**
  in North America, *84*, 109
  in South Asia, 450
  in U.S., 65, 83
Independence movements
  in South Asia, 446
  in Southeast Asia, 570–572
  in sub-Saharan Africa, 420
Independent living initiative, 226
India, 42, 89, 152, 209
  Africa and, 391
  BRICS and, 42, 89, 152–153, 274
  British, 33
  caste system in, 443–444, 456
  climate change vulnerability in, *24*
  CO_2 emissions, *23*, 26
  economic growth in, 448–450
  economy of, 475–476, 480, 481–482
  human development in, 34, *35*
  middle class in, 450
  Partition of, 446–447, *447*
  sex ratio imbalance in, 464
  sexual harassment in, 1
Indian–Australian Plate, 12, *13*
Indian National Congress, 446
Indian Ocean, 15, 369, 428
Indian subcontinent, 427
  See also South Asia
Indigenous, **131**
Indigenous belief systems, 409
Indigenous peoples, 132
  See also Aboriginal peoples; Native Americans; *specific groups*
  in China, 529
  in Middle and South America, 142–143, 144, 169
  in Oceania, 630–632, *631*, 650–651
  in polar regions, 666–667
  in Southeast Asia, *550*, 551, *554*
Indigenous religion, *57*
Indo-Gangetic Plain, 429, *429*
Indonesia, *565*, *580*, 607–608, *607*
Indulgences, 206
Indus River, 429
Indus River Valley, *429*
Industrial economic activities, 31
Industrial economy, 47, 185
Industrialization, in North America, 80
Industrial Revolution, **29**, 33, 208, 242, 380
Industry
  See also Manufacturing
  in Central Europe, 251–252
  in East Asia, 509
  in India, 437, 480
  in Italy, 245
  in Japan, 508
  in Mexico, 174–175
  in Russia and post-Soviet states, 272–275, *285*
  in South Asia, 447
  in South Europe, 244
  in Soviet Union, 270–271
  in Thailand, 602
Indus Valley civilization, **440**
Indus Waters Treaty, 436
Inequality
  in Brazil, 185
  in China, *510*
  of feudalism, 205
  gender, 231
  in Middle and South America, 152
  in Philippines, 609–610
  in Russia and post-Soviet states, 272, 275, 277
  in South Asia, 460
  in Southeast Asia, 577
  in Southern Africa, 420
  in sub-Saharan Africa, 385–386, 420
Infant mortality rate, 48
  in Caribbean, 171
  in Middle and South America, 164
  in South Asia, 463
  in sub-Saharan Africa, 404
Infectious diseases, 405
Informal economy, **32**, 174
  in Middle and South America, 154
  regional, as percentage of GDP, *390*
  in Russia and post-Soviet states, 275
  in sub-Saharan Africa, 389–390, 402
Information technology (IT), 81, 123–125, 154, 450, *451*, 514
Infrastructure, *45*, *77*, 146–147, 164, 387
Injera, *374*
Inland delta, **370**
Inner cities, 100
Inner-city decay, 99–100
Inner Mongolia, 539
Intergalactic space, **670**
Intergovernmental Panel on Climate Change (IPCC), 378–379
Interisland travel, in Oceania, 653
Internally displaced people, 342, 398
Internal processes, 12
Internal republics, 280–281
International cooperation, 42
International Date Line, 5, *7*
International immigrants, 228
International Labor Organization, 54
International Monetary Fund (IMF), 37, 38, 149, 152, 153, 182, 387, 395, 418, **576**
International Space Station, *269*, 675
International trade, 33
  See also Globalization; Trade
  in East Asia, 517
  in Europe, 251–252
  Europe and, 236
  sub-Saharan Africa and, 386
  Suez Canal and, 354
Internet, 81, 154, *155*
  in East Asia, *515*
  in North Africa and Southwest Asia, 331, 334
  in South Asia, *453*
  in sub-Saharan Africa, 391
Interplanetary space, **670**
Interstellar space, **670**
Intertropical convergence zone (ITCZ), 15, **370**, *371*, 430–431, *430*, 431, 490, *558*
Intifada, **337**
Invaders, in South Asia, 441–442
Invasive species, 73, *73*, **623**, 626–627
Investment restrictions, 37
IPCC (Intergovernmental Panel on Climate Change), 378–379
IRA (Irish Republican Army), 241
Iran, 89, 339, *343*, 358, 360–361
Iran–Contra affair, 178–179
Iranian Revolution, 360
Iranian theocracy, 331, 361
Iran–Iraq War, 361
Iraq, 89, 338, 358, 361
Iraq War, 88, 89, 360, 361
Ireland, 225, 240–241
Irish Republican Army (IRA), 241
Iron, 78

Iron Curtain, **208**, *209*
Irrawaddy River, *555*
Irrigation, 24, 46
advances in, 72
in Central Asia, *265*
in China, 491, *493*, 532, 533, 537, 538
in East Asia, 494, 509
of Great Plains, *119*
in Japan, 544
in Middle and South America, 143, 163, 174, 181
in New Zealand, *625*
in North Africa and Southwest Asia, 309, 311, 313, 315–316, 355, 359
in North America, 69, 71, 72, 79, *87*, 122, 124
in Oceania, 623, *625*
in Russia and post-Soviet states, 266, *267*
in South Asia, 433, 436, 437, 450, 473, 475, 477
in Southeast Asia, 563
in Southern California and Southwest, 124
in sub-Saharan Africa, 370, 375, 378, 402
ISIL. *See* Islamic State
ISIS. *See* Islamic State
Islam, 56, *57*, **307**, 317–318
in China, 527–528
diffusion of, 320, *321*
in Europe, 205
pillars of, 318–319
religious law, 319–320
in Russia and post-Soviet states, 290, 292
in South Asia, 456
in sub-Saharan Africa, 409–410, 417
Sunni–Shi'ite division, 319–320
women and, 348
Islamic jihadism, **220**
Islamic Revolution, 360
Islamic State, 338, 339, 398
Islamism, **307**, 330–331, 355, 398
Island and Peninsular Northeast Asia: Malaysia, Brunei, and Singapore, 604–606, *604*
Israel, 89, 313, 324, *327*, 334, *334*, 357–358
creation of state of, 335
economy of, 328
Palestine and, 334–338, *335*, *336*
population pyramid for, *343*
territorial disputes, 337–338
two-state solution and, 335, 337
Israeli–Palestinian conflict, 334–338, *335*, *336*
ISS (International Space Station), *269*, 675
Istanbul, Turkey, 204
Isthmus, **132**
Itaipu Dam, *184*
Italy, 214, 217, 242, 244–245
ITCZ. *See* Intertropical convergence zone (ITCZ)

Jainism, **444**
Jakarta, *587*
Jakobsdottir, Katrin, *53*
Jamaica, 169
James Bay, 115
Japan, 544–547, *545*
aging population in, 522–523, *525*
Ainu in, 529, *529*
CO_2 emissions, *23*
culture of, 546
economic development in, 508
economy of, 521, 523, 546–547
expansions, *504*
human development in, *34*
imperialism of, 502
limited resources of, 544
physical geography of, 544
politics in, 511, 514
working in, 546–547
as world leader, 503–504
Jati, **443**, 444, *468*
J curve, 47, *48*
Jericho, 317
Jerusalem, *335*, 337–338
Jesus of Nazareth, 318
Jews, 318
Jiaozuo, *493*
Jihad, **330**–331
Jing, Chai, 485
Jin Mao Tower, *534*
Job losses
See also Unemployment
to automation, *84*
free trade and, 81–82
globalization and, 83
immigrants and, 106–107
Johnson-Sirleaf, Ellen, 399, 413
Jordan, *318*, 324, 329, 342, 357
JPMorgan Chase, 95
Juan de Fuca Plate, *13*
Judaism, 56, 317–318, **318**
Just-in-time system, **508**, *508*

Kabaddi, *466*
Kabila, Joseph, *41*
Kabila, Laurent, 395
Kabul, Afghanistan, *21*, 472
Kaliningrad, 293
Karachi, Pakistan, 46, 458
Kara Sea, *663*
Kasank National Park, 379
Kashi, *527*
Kashmir, 457, 473–474
Kayford Mountain, West Virginia, *75*
Kazakhstan, 259, 261, *262*, 263, 264, 274, 280–281, *284*, 286, *287*, 300–302, *300*
Kenya, *364*, 365, *397*, *401*, 415, 417–418
Kerala, 482
Kessler syndrome, **672**
Keystone Pipeline, 74
Khartoum, 352
Khat, 356
Khmer Rouge, 571–572
Khomeini, Ayatollah, 331
Khubz, *313*
Kiev, Ukraine, *284*
"Killing Fields," 571–572
Kimchi, *495*
Kim Il Sung, 542
Kim Jong Il, 542
Kim Jong Un, 542
Kinshasa, Congo, *41*, 46, 395, *401*, *415*
Kirchner, Néstor, 183
Kiribati, *620*
Kleptocracies, **386**
Knights, 205
Knowledge economy, 81
Knowledge-intensive services, 222
Koli, 461, *461*
Kolkata, 427, 436, 458, *466*, 477–478
Köppen classification system, 14
Koran. *See* Qur'an (Koran)
Korea, North and South, 504–505, 541–544, *542*
Korean War, 504–505
Kosovo, 251
Krill, **662**
Krishna, 1
Kshatriyas, 444
Kuala Lumpur, Malaysia, 606
Kumbh Mela, *466*
Kurdistan, 360
Kurds, 55, 220, *332*, 338, 359, 360, 361
Kuwait, 326, 361
Kyoto, *546*
Kyoto Protocol, 26
Kyrgyzstan, 259, 278, 286–287, *287*, 292, 300–302, *300*

Labor
cheap overseas, 83
child, *207*, 382, 461
feminization of, 573
forced, 54
slave, 33
Labor force
in Europe, 225
in Japan, 546–547
older adults in, 226
women in, 85, 101, 288–289, *347*, 404, 573
Lagos, Nigeria, *401*
Lake Chad, 375
Lake Tanganyika, 370
Lake Victoria, 369, 379
Landforms, 8, **12**
of East Asia, 488
of Europe, 194–195, *195*
of Middle and South America, 133–135
of North Africa and Southwest Asia, 309
of North America, 66–67
of Oceania, 616–618
of Russia and the post-Soviet states, 261
of South Asia, 429–430, *429*, *430*
of Southeast Asia, 555–556
of sub-Saharan Africa, 369–370, *369*
Land-for-peace formula, 337
Landless movement, 159
Language, 56, 58–59
in Afghanistan, *473*
Arabic, 308
in China, 526–527, 536
English, 59, 126, 241, 442, 473
lingua franca, 411
major language families, *58*
Mandarin, 526–527
in Oceania, 653
pidgin, **653**
in South Asia, 442, *443*
in sub-Saharan Africa, 410–411, *411*
La Niña, 620
Laos, 602–604, *603*
La Paz, Bolivia, *140*
Laramie, Wyoming, 122
Larose, Louisiana, *67*
Latin America. *See* Middle and South America (world region)
Latino culture, 125–126
Latinos, **65**, 117
Latitude, **5**, *7*
La Tomatina, *224*
Latvia, 194, 247, 248–249, 259, 293
Law enforcement, race and, 109
Lebanon, 309, 324, 339, 342, 357
Lee Hsien Loong, 579
Lee Kuan Yew, 579
Legalization, of drugs, 159–160
Legend, *6*
Legend (of map), 4, 6, 270
Leopold II, 395, 414
Lesbian, gay, bisexual, transgender, and questioning (LGBTQ) people
gender-based violence against, 54
in Middle and South America, 167
politics and, 95
in Russia and post-Soviet states, 290
in South Asia, 470–471, 592
in sub-Saharan Africa, 407, 409, 410
Less developed countries (LDCs), **31**, 37, 43
See also specific countries
Lewis, Arthur, 171
LGBTQ. *See* Lesbian, gay, bisexual, transgender, and questioning (LGBTQ) people
Lhasa, Tibet, *539*
Liberation theology, **170**
Liberia, *397*
Libya, 39, 324, 326, *333*, 349, 350, 351, 413
Life expectancy, 47
in Europe, 224
in Middle and South America, 171
in Russia and post-Soviet states, 286–287
Lima, Peru, *52*, 161, *162*
Lingua franca, **411**
Lions, *416*
Lisbon, Portugal, *243*
Literacy rates
in Caribbean, 171
female, 464, *464*
in South Asia, 464, *464*
Lithium, 183
Lithuania, 194, 247, 248–249, 259, 293
Livability, 99–100
Livestock, 118–119
Living wages, 37
Ljubljana, 222
Lomo saltado, *156*
Loango, 379, 413
Local scale, 10
Loess, **530**
Loess Plateau, 531, *532*
Logging, 74–75, *74*, 78, 123, 138, *560*
Lombok, 568
London, 208, 222, *237*, 241
Longitude, **5**, *7*

Long-lot system, 114
Los Angeles, 46, 125
Lousiana, *67*
Lo Van, Tom, 120
Low Countries, 236
Low Earth orbit (LEO), **670**
Lowlands, of Middle and South America, 134–135
Lumumba, Patrice, 395
Luther, Martin, 206
Luxembourg, 236

Maasai Mara game reserve, *377*, 379
Maasai people, 407
Maastricht Treaty, **209**, 214
Maathai, Wangari, 372
Macao, 537
Macedonia, 194
Machismo, **167**, *168*
Machu Picchu, *142*, 181
Macron, Emmanuel, *219*
Madagascar, *372*, 415, 418, *418*
Madras, 427
Maghreb (subregion), 349–351, *349*
Magma, 12
Maid trade, 585, 588, *588*
Mainland Southeast Asia: Lao, Cambodia, and Vietnam, 602–604, *603*
Mainland Southeast Asia: Myanmar (Burma) and Thailand, 599–602, *600*
Makatea, **618**
Makkah (Mecca), 308, 319, 320
Malaria, 28, 405
Malawi, *388*, 394, 420, 421
Malaysia, 32 33, 36, 604 606, *605*
Male-on-female violence, 1
Mali, 375
Mali Empire, 380
Malnutrition, 30, 450
Malta, 242
Managua, Nicaragua, 161
Mandarin Chinese, 59, 526–527
Mandela, Nelson, 385
Manufactured goods, 145–146
Manufacturing
See also Industry
in India, 437
job losses in, *84*, 115
in Mexico, *151*
in North America, 80–81, 115–116, 145
in South Africa, 420
in South Asia, *451*
in space, 674
Maori, **618**
Maori land claims, 651–652
Mao Zedong, *501*, 502–503
Map projections, **5**, 8
Maps, 3, 4–8
Map scale, 5
Maquiladoras, 126, **138**, 174–175, *176*
Mardi Gras, *108*, 169
Mardrid, Spain, 244
Marianismo, **167**, *168*
Marijuana, 96–97, 159–160
Marketization, **149**
Marrakech, 350
Marriage
age at, 51
bride shortage in China, 524
child, *53*
forced, 54
polygyny, 348, 406, *407*, *408*
same-sex, 95, 167
in South Asia, 467–468, *468*
in sub-Saharan Africa, 406
Marseille, France, 239
Marshall Plan, 209, 240
Marsupials, **621**
Marx, Karl, 42, 207–208, 270
Maryland, colony of, 78
Masculinity, *648*
Maslenitsa, *289*
Massachusetts, colony of, 76
Mato Grosso, 185
Mauritius, 415, 418
Maya civilization, 15
McDonald's, 31, *84*
Means of production, 209
Measles, 76
Meat consumption, 26
Meatpacking industry, 119–120
Meat production, 118
Mecca. *See* Makkah (Mecca)
Mechanization, 207
in agriculture, 63, 86, 120
Media landscape
in North Africa and Southwest Asia, 331, 334
in Russia and post-Soviet states, 282
Median household income, by race and ethnicity, in U.S., *109*
Medical tourism, 173
Medicine. *See* Health care
Medieval era, 205
Medina, Jorge, 167, 339
Mediterranean climates, 16, **196**, *197*, *310*
Mediterranean Sea, 202, 353
Megalopolis, 97
Mekong River, *556*, 566–567, *583*, *595*
Melanesians, **630**
Mella, Juan Antonio, *158*
Men
gender-based violence against, 54
gender expectations for, 51
income of, 51
masculinity, *648*
Menchú, Rigoberta, 178
Menopause, 51
Mercantilism, **207**
Mercator, Gerhardus, 5
Mercator projection, 5, *7*
Mercedes-Benz, 239
Merchant marine, **585**
Mercosur, 153
Meridians, 5
Merkel, Angela, 231
Mesopotamia, 440
Mestizo cultures, 132
Mestizos, 144, 169
Methamphetamine (meth), 96, 159–160
Methane, 26, 662
Métis, 65
Metropolitan areas, **97**
See also Cities; Urbanization
Mexicali, Mexico, 153
Mexico, 31, 65, 174–176, *175*
agriculture in, 106
American Southwest and, 125–126
drug cartels in, 95–96, *158*, 160
economy of, 174–175
free trade and, 81–82
internet use in, 154
ISI and, 149
machismo in, *168*
manufacturing in, *151*
migration and, 175–176
Spanish colonization of, 79
water resources in, 138, *140*, 141
Zapatista rebellion, 159
Mexico City, *45*, 141, 144, *156*, 161
Michigan Central Railroad Station, *116*
Microcredit, 392, *451*, **452**–454
Micronesia, **630**
Mid-Atlantic colonies, 78
Middle America, **132**
Middle and South America (world region), 129–189, *130–131*
globalization and development in, 148–156, *151*
human patterns over time in, 142–148
physical geography of, 133–137
climate, 135–137, *136*
environment, 137–142
landforms, 133–135, *133*, *134*
population, gender, and culture in, 164–171, *165*, *168*
power and politics in, 157–161, *158*
subregions of
Brazil, 185–188, *186*
Caribbean, 171–174
Central America, 176–179, *177*
Central Andes, 181–183, *182*
Mexico, 174–176, *175*
northern Andes and Caribbean coast of South America, 179–181, *179*
Southern Cone, 183–185, *184*
urbanization in, 161–164, *162*
visual history of, *142–143*
Middle class, 46, 109
India, **450**, *459*
Middle East, 220, 307, 308
See also North Africa and Southwest Asia
fossil fuels from, 201
Middle income countries (MICs), **31**, 37
Middle West, 101
Midlatitude temperate climates, 16, *17*, **196**, *197*, *619*
Midwest, 101, 103
Migrants, *45*, *62*, 63, 86, 124
See also Immigrants
economic, **229**
remittances by, 153–154
urban, 161, *162*
Migration
See also Immigration
in China, *521*
circular, 402, 406
to Europe, 214
Mexican, 175–176
in Middle and South America, 153–154, 178
to North America, 75–76
rural-to-urban, 161
from Southeast Asia, 585
in sub-Saharan Africa, 404–405
to United States, 178
urban, 43–46, 340, 342
Milan, Italy, 245
Militarism, in Southeast Asia, 582–583
Military bases, U.S., 88–89, *90*, *91*
Military expenditures, *329*
Milosevic, Slobodan, 251
Mineral resources
See also Mining
extraction of, 148
in Middle and South America, *156*
in Russia and post-Soviet states, 264–265
Ming dynasty, 502
Minifundios, 176, *177*
Mining, *21*, *121*, 122, 148, *148*, 202, 264–265, 628, *628*
in South Africa, 385
in space, 674, *674*
MIRAB economy, **636**
Mississippi River Delta, 13, 67, *67*
Mobile phone users, 331
in South Asia, *453*
in sub-Saharan Africa, 367, 391, *392*
Modi, Narenda, 427, 454
Mogadishu, 417
Mohenjo-daro, *440*
Moldova, 220, 259, 277, 293, 297–298, *297*
Money, in politics, 92
Mongolia, *489*, 504–505, 539–541, *540*, *541*
communist era in, 540
postcommunist era, 540–541
Mongols, **269**, 320, 502
Mono-crop agriculture, **374**
Monotheism, **317**–318
Monotremes, **621**
Monroe Doctrine, 157
Monsoons, 15, **430**–431, *433*, 490
Montenegro, 194, 217, 249
Montreal, Canada, 76
Moors, 244
Morales, Evo, 183
More developed countries (MDCs), **31**, 37
Mormons, 110
Morocco, *24*, 220, 309, *310*, 324, 329–330, *341*, 351
Moscow, Russia, 269, *269*, *275*, *276*, 283, *284*, 293
Motherhood, in South Asia, 467
Mothers, working, 226
Mountaintop removal, *75*
Mount Ararat, *311*
Mount Kenya, 369
Mount Kilimanjaro, 369, *369*
Movement, freedom of, 39
Movement of Landless Rural Workers (MST), 159
Mozambique, 399, 420
Mt. Fuji, *501*
Mugabe, Robert, 398
Mughals, **441**–442
Muhammad (prophet), 307, 318, 319
Mulley, Phillip, 439
Multicultural, 55
Multimodal transport, 200

Multinational corporations, 33, 37, 154, 222
Mumbai, India, 22, 26, 43, 46, 427, *435*, 458, 460, 470
Murmansk, 293
Muslim Brotherhood, 331, 337, 339
Muslim fundamentalists, in Afghanistan, 272
Muslim immigrants, 193, 228–229
  in Europe, 229, *230*, 239
  terrorism and, 229–230
Muslims, **318**
  See also Islam
  postcolonialism and, 220
  in Western China, 527–528
  women, 465, 467
Myanmar (Burma), *21*, 209, *558*, 599–602, *600*

Naadam, *499*
NAFTA. *See* North American Free Trade Agreement (NAFTA)
Nairobi, Kenya, 43, *364*, *401*
Nakhodka, 296
Namibia, 420
Napoleon, 269
Narmada River, 425, 429, 479–480
NASA, 19
Nation, **38**
National Aeronautics and Space Administration (NASA), 19
National Climate Change Vulnerability and Risk Assessments in Europe, 2018, 198
National debt. *See* Debt
National Front, 239
National Institutes of Health (NIH), 106
Nationalism, **217**
  Arab, 351
  religious, 454, 456
  in Russia and post-Soviet states, 290
Nationalization, 395
Nationalize, **150**
Nation-state, **38**, 217
Native Americans, 65, 73
  in Continental Interior, 122
  in Great Plains, 120
  impact of European settlers on, 76–77, 79–80
  relocation of, 79
NATO (North Atlantic Treaty Organization), 89, **221**, 245, 358
Natural gas, 26, 264, 272–274, *273*, 300, 302, *324*
Natural habitats. *See* Habitat loss
Natural resources
  Central America, 176
  Iran, 360
  of Middle and South America, 144–146
  Middle and South America
    Brazil, 185
    Caribbean, 171
    Central Andes, 183
  North Africa and Southwest Asia, 324–326, *324*
  North America, 69
  Oceania, *622*, 636
  Russia and the post-Soviet States, 264–265
Nauru, 628
Nauryz, *289*, *319*
Nazca Plate, *13*, 133
Nazi Germany, 208, 217, 271
Neo-colonialism, 395
Neoliberalism, **148**–150, 172, 182–183
Nepal, 458, 470, 475
Netherlands, 207, 236
  climate change vulnerability in, *199*
  land reclamation in, 236, *238*
Nevis, 171
New Amsterdam, 76
New Caledonia, 617
New Delhi, India, *441*, 476–477
New England and the Atlantic Provinces (subregion), 78, 112–114, *113*
New England clam chowder, *106*
New Great Migration, 117–118
New Guinea, 617
New Jersey, colony of, 78
New Orleans, Louisiana, *108*
New York, colony of, 78
New York City, 26, 46, 97, *98*, 99, 241
New Zealand, *645*
  animal and plant life in, 621
  Asia and, 635–636
  colonization of, 633–634
  economy of, *637*
  environmental awareness in, 627–628
  food production in, 626–627
  indigenous people in, 650–651
  invasive species in, 626–627
  land uses and natural resources, *622*
NGOs (nongovernmental organizations), **42**, *43*, 391, 393, 395
Nicaragua, 150, 178–179
Nicholas II, 269
Niger, *35*, 38, 375
Nigeria, 384, *401*, 412
  colonization in, 383–384
  conflicts in, 383–384
  ethnic geography, *384*
  population growth in, 404
Niger River, 370, 375, *378*, 412
Niger River Delta, *370*
Nike, 37
Nile Delta, 315, 353
Nile region, 352–354
Nile River, 309, 353
Nile Waters Agreement, 315
Nine Nations of North America, The (Garreau), 112
Ningxia Huiza Autonomous Region, 491
Niqab, 346
Nobility, 205
Nomadic pastoralists, **266**
No Mas movement, 181
Non, *290*
Nongovernmental organizations (NGOs), **42**, *43*, 391, 393, 395
Nonpoint sources of pollution, **263**
Nonviolence, as political strategy, 446
Norilsk, Russia, *264*
Norilsk Nickel, *264*, 265
North Africa and Southwest Asia (world region), 305–363, *306–307*
  geopolitical situations in, 334–339
  globalization and development in, 324–328, *327*
  human patterns over time in, 316–324
  physical geography of, 309–316
  political map of, *308*
  population, gender, and culture in, 342–348, *345*
  power and politics in, 328–339, *332–333*
  refugees from, 193
  Spain and, 244
  state formation in, 321–323
  subregions of
    Arabian peninsula, 354–356, *355*
    Eastern Mediterranean, 357–358, *357*
    Maghreb, 349–351, *349*
    Nile: Sudan and Egypt, 352–354, *352*
    Northeast, 358–361, *358*
  urbanization in, 339–342, *341*
  visual history of, *322–323*
  Western domination in, 321–323
North America (world region), 63–127, *64–65*
  See also Canada; United States
  aging in, 101
  climate of, 67, *68*, 69
  colonization of, 76–78
  cultural issues in, 104–107
    American family, 111
    immigration and diversity, 104–107
    race and ethnicity, 107–109
    religion, 109–111
  environment of, 69–75
  food production in, 86–87, *87*
  foodways in, *106*
  globalization and development in, 80–88, *84*
  human patterns over time in, 75–80
  landforms of, 66–67
  physical geography of, 66–75
  political map of, *66*
  population, gender, and culture in, 100–111
  power and politics in, 88–97, *90–91*
  regional composition of, 80
  subregions of, 111–126, *112*
    American South (Southeast), 116–118, *117*
    Continental Interior, 121–122, *121*
    Great Plains, *119*
    Great Plains Breadbasket, 118–120
    New England and the Atlantic Provinces, 112–114
    Old Economic Core, 115–116, *116*
    Pacific Northwest, 122–124, *123*
    Québec, 114–115
    Southern California and the Southwest, 124–126
  transportation networks in, 88, 97
  urbanization in, 97–100, *98–99*
  visual history of, *76–77*
North American Free Trade Agreement (NAFTA), 37, 38, 81, 106, 153, 159, 173
North American Plate, *13*, 66, 133
North Atlantic Drift, **196**, 241
North Atlantic Treaty Organization (NATO), 89, **221**, 245, 358
North China Plain, 532
Northeastern South Asia (subregion), 477–479, *478*
Northern Andes and Caribbean coast of South America (subregion), 179–181, *179*
Northern Hemisphere, 5, 8
Northern Ireland, 217, *219*, 241
North Europe, 194, 202, 235, 246–249, *246*
North European Plain, 195, *195*, 261
North Korea, *513*, 541–544
North Macedonia, 217
North Sea, 202, 247
Northwest India (subregion), 475–477, *476*
Northwest Territory, 80, 122
Norway, *34*, 217, 232, 246, 247, *658*, *661*
Novosibirsk, Russia, 296
Nuclear family, **111**
Nuclear pollution, 628–629
Nuclear power, 264, 361
Nuclear power plants, 201
Nuclear testing, *629*
Nuclear weapons, 264
Nunavut, 65
Nunavut Territory, 80

Obama, Barack, 94, 159, 173
Obamacare. *See* Affordable Care Act
Occidental Petroleum, 129
Occupied Palestinian Territories (oPT), **308**, 357
Ocean currents, 22, 620–621
Oceania (world region), 65, 613–657, *614–615*
  fauna and flora of, 621
  globalization and development in, 628–630, 634–640, *637*
  human patterns over time in, 630–634
  physical geography of, 616–630
  political map of, *616*
  population, gender, and culture in, 646–655, *648*
  power and politics in, 640–643, *642*
  urbanization in, 643–646, *644–645*
  visual history of, *632–633*
Oceanic crust, 12
Oceans
  See also Sea level rise; *specific oceans*
  floating garbage islands in, 629–630, *629*
  United Nations Convention on the Law of the Sea (UNCLOS), 630
Offshore outsourcing, **450**
Ogaden War, 417
Ogallala aquifer, 72, *72*
Oil resources
  in Brazil, 185
  in Central America, 176
  crude oil imports, *92*
  in Cuba, 173
  in Europe, 201
  extraction of, 73–74, 89, 122
  in North Africa and Southwest Asia, 324–326, *324*, 328
    Arabian peninsula, 354–356
    Iran, 360
    Iraq, 361

prices, 325–326
production and consumption, by country, *325*
in Russia and post-Soviet states, 264, 272–274, *273*, 300, 302
in Scandinavia, 247
in South America, 129, 131
U.S. imports of, 89
in Venezuela, 150, 180
Oil spills, 73, *128*, 129
Okavango Delta, *370*
Okolloh, Ory, 365
Ok Tedi mine, *628*
Ok Tedi River, *628*
Old Economic Core (subregion), 80, 81, 115–116, *116*
Older adults
See also Aging population
in work force, 226
Oligarchs, **272**, *276*
Olympic National Park, *74*
Olympic Peninsula, *74*
Olympics, 187, 188
One-child policy, 524, *525*
OPEC (Organization of the Petroleum Exporting Countries), 89, 180, **325**–326
Operation Bootstrap, 173
Operation Condor, 157
Opium, 502, 593, 601
Opium Wars, 502, 522
oPT. *See* Occupied Palestinian Territories (oPT)
Orange Revolution, 278
Orangi Pilot Project, 460
Orangutans, *561*
Oregon Territory, 78
Organic agriculture, 30
Organically grown, **87**, *88*
Organization of the Petroleum Exporting States (OPEC), 89, 180, **325**–326
Orographic precipitation, **15**
Ortega, Daniel, 179
Orthodox Jews, *345*
Other backward class (OBC), 456
Ottoman Empire, **320**, 322, *322*, *323*, 359
Outback, **617**
Outer space. *See* Space
Outer Space Treaty of 1967, **675**
Outsourcing, 450
Overseas Chinese, **529**
Owens Valley, California, *18*

Pacific Islands
animal and plant life in, 621
economic changes in, 636, 638
environmental awareness in, 627–628
Pacific Mountain Zone, 261
Pacific Northwest (subregion), 79, 122–124, *123*
Pacific Plate, *13*, 66
Pacific Rim, **103**
Pacific Solution, *642*, **643**
Pacific Way, 640–641, **640**
"Pajama traders," 521
Pakistan, 58, 89, 436, 437, 456, 464, 471–474, *471*, *473*
Palestine, 334–338, *334*, *335*, *336*, 342
Palestinian Territories, 308, 353, 357
See also Occupied Palestinian Territories (oPT)
Palm Jumeirah, *341*
Pampas, *134*
Panama, 176, 178
Panama Canal, 146, 178
Pancasila, **583**
Pangaea, 66, 369
Pangaea hypothesis, 12, *13*
Papua New Guinea, 628
Paraguay, 157, 183
Parallels, 5
Parental leave, 232
Paris, France, 208, 236, 238
Paris Accord (2017), **198**
Paris Agreement, 26, 198
Parliamentary system, in Australia and New Zealand, 640
Partition, **446**, *447*
Passover, *319*
Pastoralism, **375**
Patkar, Medha, *424*, 425
Patriarchal, **344**
Peace, 39
Peanuts, 144
Pennsylvania, colony of, 78
Peopling. *See* Human patterns over time
Peppers, 144
Per capita emissions, 46
Perestroika, **271**
Permafrost, **261**, 662
Permeable national borders, 126
Persian Gulf, 89, 315, 324, *332*
Peru, 135, 138, 144, *156*, 181, 182
Peru Current, 135
Pesticides, 86
Peter the Great, *268*
Petrobras, 188
Pharmaceuticals, water pollution and, 73
Philippine Plate, *13*
Philippines, *555*, *558*, *581*, 608–610, *609*
Phoenix, Arizona, 46, *99*
Photo interpretation, 8–9
Physical geography, **3**, 12–15
See also Climate; Landforms; Plate tectonics; Vegetation
of East Asia, 488–499
of Europe, 194–196
of Middle and South America, 133–137
of North Africa and Southwest Asia, 309–316
of North America, 66–75
of Oceania, 616–630
of polar regions, 661–663
of Russia and the post-Soviet states, 261–263
of South Asia, 428–440
of Southeast Asia, 555–567
of sub-Saharan Africa, 368–379
Piailug, Mau, 631
Pidgin, **653**
Pilgrimage, 319
Pineapples, 144, *146*
Pink Tide, 150, 159, 186
Pinochet, Augusto, 157
Pinyin, 487
Piracy, 417, 577–578
Pizarro, Francisco, *143*
Plantations, 77–78, **147**, 171
Plateau of Tibet, 539
Plate tectonics, **12**, *13*, 133–134
Poaching, 379, 418
Poland, 9–10, 202, 217, *233*, 249
Polar amplification, **662**
Polar melting, 662
Polar regions (world region), 659–667, *660*
globalization and development in, 663–664
indigenous people in, 666–667
physical geography of, 661–663
population, gender, and culture in, 665–667
power and politics in, 664–665
urbanization in, 665
Police shootings, 109
Political activism, *595*
Political conflicts, *21*
Political ecology, **32**–33
Political elections, in North America, 92
Political freedoms, **39**, *41*
See also Democracy
in China, 514–515
in East Asia, *512*
in Europe, 217
expansion of, factors encouraging, 39, 42
in Middle and South America, 157
in North America, 89, 92
in Southeast Asia, 583–584
Political reform, in North Africa and Southwest Asia, 329–330
Political violence, *41*
Politicians, female, 94–95, *95*, 231–232, **232**, *292*, 334, 399, 468
Politics. *See* Power and politics
Pollution
air, 69, 71–72, *71*, 202, 265, 373, 437–438, 497–499
nonpoint sources of, **263**
nuclear, 628–629
in Russia and post-Soviet states, 263, 264
water, *71*, 72–73, *140*, 202, 436–437
Polo, Marco, 244
Polygyny, 348, **406**, *407*, *408*
Polynesia, **630**
Poor, 43
Pope, 170–171
Popular democracy, 217
Population
See also Aging population
in East Asia, 522–526, *525*
in Europe, 224–230
in Great Plains, 120
in Middle and South America, 164–167, *168*
in North Africa and Southwest Asia, 342–343, *345*
in North America, 100–103, *102*, *103*, *104*
in Oceania, 646–647, *648*
in polar regions, 665–666
in Russia and post-Soviet states, 286–287, *288*, *291*
in South Asia, 462–465, *462*
in Southeast Asia, 588–591, *589*, *594–595*
in space, 676–677
in sub-Saharan Africa, 402–406, *408*
world density, *48–49*
Population boom, green revolution and, 30
Population density, 46
Population distribution
in East Asia, 524, 526, *526*
in Middle and South America, 165–166, *166*
in North Africa and Southwest Asia, 343, *343*, *344*
in North America, 101, 103, *103*
in Oceania, *647*
in Russia and post-Soviet states, 287, *288*
in sub-Saharan Africa, 403–404, *403*
Population growth
Aboriginal peoples, 80
exponential, *48*
global patterns of, 46–47
global spatial variations in, 46
human development and, **47**–48
in Middle and South America, *165*
Native American, 80
negative, 46
Population pyramid, **50**, 286, 463–464
Belarus, *287*
Canada, *103*
China, *523*
EU, *225*
Germany, *225*
Indonesia, 591
Iran, *343*
Israel, *343*
Kazakhstan, *287*
Kyrgyzstan, *287*
Nigeria, *404*
Pakistan, *465*
Qatar, *343*
Russia, *287*
South Africa, *404*
Sri Lanka, *465*
Sweden, *225*
United States, *103*
Populations, 46
Populist movements, 157, **208**, **582**
Portland, Oregon, 99
Ports, Europe, 200–201
Portugal, 144, 207, 214, *219*, 242, 243–244
Portuguese colonies, 144, *145*
Portuguese traders, in Africa, 380
Postcolonial, **220**
Postcolonial exploitation, in Central Africa, 414–415
Postcolonialism, 220
Postindustrial economy, 47
Potatoes, 144, *146*
Potjiekos, *374*
Pound, 241
Poutine, *106*
Poverty, 30, 37
See also Inequality
in American South, 117–118
in Brazil, 187
culture of, 109
in Europe, 211
in South Asia, 475
in sub-Saharan Africa, 387, 420

Poverty Reduction Strategy Papers (PRSPs), 38, 152, 390
Power and politics, 38–39, *40–41*
See also Geopolitics; Political freedoms
in East Asia, 511–516, *512–513*
in Europe, 207, 217–221, *218–219*
language and, 58
in Middle and South America, 157–161, *158*, 171
in North Africa and Southwest Asia, 328–339, *332–333*, 361
in North America, 88–97, *90–91*
in Oceania, 640–643, *642*
in polar regions, 664–665
power and, *40–41*
religion and, 110–111
role of money in, 92
in Russia and post-Soviet states, 278–283, *279*
in South Asia, 454–458, *455*
in Southeast Asia, 579–584, *580–581*, 602
in space, 674–675
in sub-Saharan Africa, 394–399, *396–397*
Precipitation, 15, 135, 431, 619
Predjama Castle (Slovenia), *206*
Pregnancy, teen, *53*
Prescription drugs, 96
Press, freedom of, 39
Prícope, 413
Primary elections, 92
Primary sector, **31**
Primate city, **161**, *162*
Prime meridian, 5, *7*
Printing press, 206
Prison population, in U.S., 96
Private space, 344, *346*
Privatization, 27–28, **149**, **272**
Prosperity, 39, 42
Prostitution. *See* Sex workers
Protected areas, in South Asia, 439–440
Protected geographical indication (PGI), *250*
Protectionism, 37
Protestant Reformation, 206
Protestants, 110
evangelical, **171**
Protests, in China, 514–515
Public health, in sub-Saharan Africa, 405–406
Public space, **334**, 344, *346*
Public transportation, 46, *98*, 99, 164
Puerto Misahualli, 142
Puerto Rico, 22, 65, *143*, 172, 173
Pull factors, **105**
Punjab, 475–476
Purchasing power parity (PPP), **31**, 32
Purdah, **467**
Push factors, **105**–106
Pussy Riot, *256*, 257, 259
Putin, Vladimir, 257, 278, 282, 290, *291*
Pyroclassic flows, 134
Al-Qaddafi, Muammar, 351

Qanats, **538**
Qatar, *34*, 326, *327*, 331, 339, *343*, 346, 355
Qing empire, 502
Quarternary sector, **31**
Québec (subregion), 93, 114–115
Québec City, 76, *114*
Quiet Revolution (Canada), 115
Quinoa, *146*
Quotas, trade, 81
Qur'an (Koran), 308, **311**, 318, 330

Race, **59**–60
law enforcement and, 109
median household income by, in U.S., *109*
in Middle and South America, 169–170
in North America, 107–109
"Race to the bottom," 37
Racism, 59–60, 96
Radha, 1
Radiation, solar, *22*
Radical Islamism, 398
Railroads, 78–79, *79*, 88
Rain. *See* Acid rain; Precipitation
Rain belt, 15
Rainbow Warrior, 628
Rain forests, 134, 138, 185–186, 412
Rain shadow, **15**, *18*
Rajasthan, 475
Ramadan, *335*, *410*
Ranching, 147
Rantchev, Nikolai, *52*
Rapa Nui, 632
Rape, 54
Rate of natural increase (RNI), **47**
Ravenna, 222
Raw materials, 386
Ray, Satyajit, 477
Razak, Najib, *575*
Recession of 2007–2009, 83
Redlining, **100**
Redneck Fishing Tournament, *73*
Red Sea, 369
Red Square, Moscow, *269*
Refugees, *41*, 106, *190*, 191, **229**, *229*, 342, **641**, *648*
to Europe, 193, 226–230, *228*, 245
in Oceania, 641, 643, *643*
Somali, *377*
in sub-Saharan Africa, 398
Sudan, *314*
Syrian, 338–339
Region, 9–11, **9**
Regional conflicts. *See* Conflicts
Regional economic development, in sub-Saharan Africa, 391–394
Regional self-sufficiency, **509**
Regional specialization, **509**
Regional trade
in Southeast Asia, 576–577, *577*
in sub-Saharan Africa, 391–392
Regional trade agreements, 153
Regional trade blocs, 37
Regions
climate, 14, 16–17
traits of, 11
of world, *2–3*
Reign of Terror, 207
Reindeer racing, *665*
Religion, 56, *57*
See also specific religions
in Africa, *409*
freedom of, 39
indigenous people in, 409
in Middle and South America, 170–171
monotheism, 317–318
in North America, 109–111, *110*
politics and, 110–111
Protestant Reformation, 206
in Russia and post-Soviet states, 290
in South Asia, 436, 442–444, *443*, 454, 456
in Southeast Asia, 568, 593, 596, *597*
in sub-Saharan Africa, 398, 409–410
Religious nationalism, **454**, *455*, 456
Remittances, 153–154, **585**
Remote sensing, of urban areas, 676
Renewable energy, 26, 201–202, 438
Republican Party, 93, 108
Republic of Ireland, 240–241
Reservoirs, 69, 315
Resettlement schemes, **583**
Resilience, 22
Resource extraction, 216
Central Andes, 183
in Middle and South America, 144–146, 146–147, 148
in Russia and post-Soviet states, 264–265
Resources
See also Natural resources
extractive, 31
human, 31
Retirement, 101
Réunion, 415
Revolutionary Armed Forces of Colombia (FARC), 180
Revolutionary Guard, 360
Revolutionary movements, in Middle and South America, 157, *158*, 159
Rhine River, 195
Rhodesia, 420
See also Zimbabwe
Rice, 437
Rice cultivation, 494, 563, *563*
Ring of Fire, 12, *14*, 133
Rio de Janeiro, Brazil, 43, *134*, 141, *163*, 164, *169*, 185, 187–188
Riparian areas, 72–73
Rivers
See also specific rivers
conflicts over, 436
in Europe, 195, 200
in South Asia, 429, 436
in sub-Saharan Africa, 369–370
Road system
See also Transportation networks
in Amazon, 138
in East Asia, *507*
in Europe, 200
in North America, 88, 97
Roaring Forties, **618**
Robinson projection, 5, *7*
Robots, *84*, *523*, *545*
Rocky Mountains, 78, 121, 122
Rocky Mountain zone, 66, *66*, 69
Rodeos, *108*
Roma, **208**
Roman Catholic Church, 144, 165, 167, 170–171, 225
Protestant Reformation and, 206
Vatican City, 245
Roman Catholicism, 110
Romance languages, 203
Roman Empire, 203, 242, 318, 379
Romania, 194, 195, *213*, *233*, 249, 252, 252–253, *253*, 349
Rosalino, Aguilar Busto, 154
Rose Revolution, 278
Ross Sea, *662*
Rotterdam, Netherlands, 236
Rouhani, Hassan, 361
Roundup, 30
Royal Niger Company, 383–384
Rub'al Khali, 309, *311*
Rubber plantations, 413, 482
Rudd, Kevin, 633
Rugby, *655*
RugMark, 461
Rural economic development
in South Asia, 450, 452
in sub-Saharan Africa, 399
Rural-to-urban migration, 161
Rushdie, Salman, 242
Russia, 9, 42, 89, 152, 260, 293–297
See also Russia and the post-Soviet States (world region)
agriculture in, 277–278, *277*
BRICS and, 274
Central Siberia, 295–296, *295*
CO_2 emissions, *23*
Communist Revolution in, 269–270
corruption, 274
European, 293, *294*
Far East, 296–297, *296*
forestry in, 265
fossil fuels from, 201, 272–274, *273*
global economy and, 274
influence in Europe, 2016, *11*
politics in, 278–283
urbanization in, 293
Russia and the post-Soviet States (world region), 257–303, *258–259*, *260*
environment of, 263–266
globalization and development in, 272–278, *276*
human patterns over time in, 266, 268–272
physical geography of, 261–263
political map of, *260*
population, gender, and culture in, 286–292
power and politics in, 278–283, *279*
subregions of
Belarus, Ukraine, and Moldova, 297–298, *297*
Caucasia: Georgia, Armenia, and Azerbaijan, 299–300, *299*
Central Asian states, 300–302, *300*
Russia, 293–297
urbanization in, 283–285, *284*
visual history of, *268–269*
Russian Empire, 266, 268–269, *270*
Russian Far East, 296–297, *296*
Russian Federation, **261**
Russian Mafia, 274
Russification, **281**

Rust Belt, 115
Rwanda, 399, 415

Sachet water, *401*, 402
Sahara Desert, 309, 315
Sahel, **370**, 375, 398, 407
Salafism, **330**
Salinization, **312**
Salt March (India), *441*, 446
Salyut 7 spacecraft, *668*
Same-sex marriage, 95, 167
Samurai, *499*
San'a, Yemen, *332*
Sand dunes, *311*
Sanders, Bernie, 92
Sandinista revolution, 178–179
San Francisco, California, 46
Sanitation, 141, 164
Santiago, Chile, 161, *162*
Santos, Juan Manuel, *41*
São Paulo, Brazil, 46, 164, 167, 185
São Tomé, 413
SAPs. *See* Structural adjustment programs (SAPs)
Sarajevo, *218*
Sardar Sarova Dam, *428*
Sardinia, 245
Satanic Verses, The (Rushdie), 242
Satay, *598*
Satellite economy, 672–673
Satellite imagery, 8
Satellites, **671**, 672, *672*
impact in rural areas, 673
to study population, 677
Saud family, 354–355
Saudi Arabia, *310*, 315, 321, 326, 331, 339, *345*, 347–348, 354–356
Savannas, 413
Savinikh, Viktor, 669
Scale, **4**–5, 6, *6*
Scale of analysis, 5, 10
Scales of imagery, 5
Scandinavia, 232, 235, 246–248
Schengen Agreement, **211**, 212, 214, 226, 227
Schengen Area, *228*
Scotland, 241
Sea level rise, 19, 22, 69, 433, *435*, 436, 622, 662, *662*
Seasons, 14, *15*
Seawater pollution, 202
Secondary sector, **31**
Secretariat of the Pacific Regional Environmental Programme (SPREP), 641
Secular rational values, 55
Secular states, **331**
Security Council, 42
Seko, Mobutu Sese, 387, 395
Self-expression values, 55
Self-sufficiency, 448, *637*
Semana Santa, *169*
Semiarid climates, 16
Sen, Amartya, 477
Sendero Luminoso (Shining Path), 182
Senegal, 399
Sensitivity, 22
Seoul, South Korea, *519*
September 11, 2001, 42, 89, 126
Serbia, 194, 217, 249, 250–251
Serfs, 269
Service and Food Workers Union, *41*
Service-oriented economic activities, 31
Service sector, **573**, *575*
employment in, 82
in Europe, 222, 224
in Mexico, 175
in North America, 81, 82
in Oceania, *637*
in South Asia, 460
wages in, *84*
Seven Years War, 114
Sex, **50**–51
Sex imbalance, 50
Sex industry, 592–593
Sex ratio, **464**
Sex-ratio imbalance, 1
Sex structure, 50
Sex tourism, **593**
Sex trafficking, 54, 289
Sexual harassment, in India, 1
Sexual orientation. *See* Lesbian, gay, bisexual, transgender, and questioning (LGBTQ) people
Sex workers, *291*, **461**
Seychelles, 415, 418
SEZ (special economic zones), **517**
Shanghai, China, *45*, 46, *506*, *519*, 521–522, *534*
Shanghai Tower, *534*
Shanghai World Financial Center, *534*
Shantytowns, 43, 44, 161, *162*, 163–164, *163*
Shari'a, **319**, 417
Sheep, *627*
Sheikhs, **354**
Shifting cultivation, *21*, **143**, **373**, *374*, 562–563
Shi'ite Muslims, 319–**320**, 338, 339, 361
Shintoism, 57
Shipping lanes, 200–201
Shishmaref, Alaska, *70*
Shock therapy, 272
Siam. *See* Thailand
Siberia, **261**, *262*, 263, 269, 283, 295–296, *295*
Sichuan Province (China), 533–534
Sicily, Italy, 245
Sierra Leone, 413
Sierra Madre, *133*
Sikhism, **444**
Silicon Vale, 242
Silicon Valley, 123
Silicon Wadi, *327*
Silk Road, **266**, 268, *268*, 357, *527*
Singapore, *525*, *587*, 604, *605*, 606
Single-parent households, 109
Skin color, 59–60, *59*, 169–170
Slash and burn agriculture, 143
Slave labor, 33
Slave rebellions, 77
Slavery, 207
in Middle and South America, 144, 167
modern, 54, 382
modern, in Niger, *382*
in North America, 76, 77–78
in sub-Saharan Africa, 380, 382
trade routes for, *144*
Slave trade, 380, *381*, 382
Slavs, **268**
Sleeping sickness, 405, 412
Slovakia, 195, 249
Slovenia, 191, 194, *206*, 215–216, 231–232, 249, 250–251
Slums, **43**, 161, 402, *459*, 460
Smallpox, 76, 144
Smart growth, 99–100
Smog, **71**
Social cohesion, 211, 229, **234**, **577**
Social contract, 144
Social exclusion, **229**
Social forestry movement, 439
Socialism, **148**–149, 149–150, 182–183, **209**
Social media, 331, 361
Social movements, in South Asia, 456–457
Social participation initiative, 226
Social protections, **234**
in Europe, 224, 232, 234–235, *234*
in Scandinavia, 248
Social safety net, **94**, *94*
Social Security Trust Fund, 92
Social welfare, 234
Sociocultural issues
See also Culture
in East Asia, 526–530
in Europe, 203–204
in Middle and South America, 167, 169–171
in North Africa and Southwest Asia, 343–348, 350
in North America, 104–107
in Oceania, 650–655
in Russia and post-Soviet states, 288–292
in South Asia, 465–468
in Southeast Asia, 593, 596–599
in space, 677–678
in sub-Saharan Africa, 409–411
SodaStream, *335*
Soft power initiatives, 582–583, **582**
Soil erosion, 30, 119, 120
Solar energy, 13–14, 26, 69, 125, 201, *201*, 438, *438*, *493*
Solar radiation, *22*
Solar wind, **670**–671
Solidarity, 217
Solzhenitsyn, Aleksandr, 296
Somalia, 415
civil war in, 417
piracy in, 417
Somali refugees, *377*
Somali Subplate, 12, *13*
Songkran, *596*
Sony, *507*
Soufrière Hills Volcano, *134*
South. *See* American South (Southeast)
South Africa, 42, 89, 152, 398, *421*
apartheid in, 385, *385*, 420
Boers in, 419–420
BRICS and, 274
colonization of, 384–385
economy of, 420
population pyramid for, *404*
South America, **132**
See also Middle and South America (world region)
South American Plate, *13*, 133
South Asia (world region), 425–483, *426–427*
British colonial rule of, 445, *445*
deindustrialization of, 445–446
globalization and development in, 447–454, *451*
human patterns over time in, 440–447
monsoon season in, 15
physical geography of, 428–440
political map of, *428*
population, gender, and culture in, 462–471, *462*, *469*
post-independence, 447
power and politics in, 454–458, *455*
precolonial, 441–442, *442*
refugees from, 193
subregions of
Afghanistan and Pakistan, 471–474, *471*
Central India, 479–480, *479*
Himalayan Country, 474–475, *474*
Northeastern South Asia, 477–479, *478*
Northwest India, 475–477, *476*
South India and Sri Lanka, 481–483, *481*
urbanization in, 458–461, *459*, 463
visual history of, *440–441*
South China Sea, 488
Southeast. *See* American South (Southeast)
Southeast Asia (world region), 551–611, *552–553*
colonization of, 236, 568–570, *569*
financial crisis, 573, 576
globalization and development in, 572–579, *574–575*
human patterns over time in, 567–568
independence movements in, 570–572
militarism in, 582–583
monsoon season in, 15
physical geography of, 555–567
population, gender, and culture in, 588–599, *589*, *594–595*
power and politics in, 579–584, *580–581*
subregions of
Indonesia and Timor-Leste, 607–608, *607*
Island and Peninsular Northeast Asia: Malaysia, Brunei, and Singapore, 604–606, *604*
Mainland Southeast Asia: Lao, Cambodia, and Vietnam, 602–604, *603*
Mainland Southeast Asia: Myanmar (Burma) and Thailand), 599–602, *600*
Philippines, 608–610, *609*
urbanization in, 584–588, *586–587*
visual history of, *570–571*
Southeastern Anatolia Project, 316
Southeastern coast, of China, 536–537
Southeastern Europe, 194
Southern Africa (subregion), 419–422, *419*
Southern Baptists, 110
Southern California, 124–126, *124*
Southern China, 535–537, *535*

Southern Common Market (Mercosur), 37
Southern Cone, 183–185, *184*
Southern Hemisphere, 5, 8
Southern Ocean, *663*
South Europe (subregion), 194, 229, 242–247, *243*
South India, 481–483, *481*
South Korea, 541–544
South Ossetia, 282
South Sudan, *24*, 353, *397*, 415, 415–417
South Tarawa, *612*, 613
Southwest, 79, 124–126, *124*
Southwest Asia, refugees from, 193
Sovereignty, 38–39, 42
Sovereign wealth funds, 355
Soviet bloc, 208
Soviet Union, 9–10, **259**, 283
  See also Russia; Russia and the post-Soviet States
  breakup of, 249
  Cold War and, 208–209, 271–272
  Cold War in, 42
  collapse of, 272
  creation of, 208
  Cuba and, 159
  Cub and, 172
  fall of, 39, 42, 89, 215, 217
  pre-1989 alignment of countries in, *10*
  regional economic development, 283
  under Stalin, 270–271
  World War II and, 271
Soybeans, 186
Space (world region), 669–680
  cultural issues in, 677–678
  environment of, 670–672
  exploration, 677–678
  fertility in, 677
  gender in, 677
  geospace, 670
  globalization and development in, 672–674
  intergallactic, 670
  interplanetary, 670
  interstellar, 670
  law and, 674–675
  Low Earth Orbit (LEO), **670**
  manufacturing in, 674
  militarization of, 675
  mining in, 674
  population, gender, and culture in, 676–678
  power and politics in, 674–675
  urbanization in, 675–676
  weather, 670–671
Space junk, 671–672, **671**
Space launches, 673–674, *673*
Spain, *24*, *199*, 207, 214, *223*, *224*, 242, 243–244
Spanish colonies, 76, 144, *145*, 147
Spanish conquest, 181
Spanish empire, 183
Spanish explorers, 76
Spatial distribution, 4
Spatial interactions, 4
Special economic zones (SEZs), **517**
Speech, freedom of, 39
Sphere of influence, 220
Sphinx, *354*
Sports, 653–655
Sport tourism, 172
SPREP (Secretariat of the Pacific Regional Environmental Programme), 641
Squash, 75
Sri Lanka, 440, 456, 481–483
Stalin, Joseph, 263, 270–271
State, **38**
State-aided market economies, **505**, **572**
Stateless nation, 38
State-owned industries, 274
Station, **652**
St. Augustine, Florida, 76
Steel industry, 78, 80–81
Steppes, 16, *17*, **261**, *262*, *310*, *489*
St. John's, Newfoundland, 76
St. Kitts, 171
St. Lucia, 36, *36*, 172
St. Martin, 171
Stockholm, Sweden, *233*
Stone Forest, *536*
Storms, 22
Storm surges, **566**, 622
Storm-water runoff, 73
St. Petersburg, 283, 293
Structural adjustment programs (SAPs), 37, 38, **149**, 152, 154, 387, 576
Subduction, 12
Subduction zone, **133**
Subprime mortgage crisis, 83
Sub-Saharan Africa (world region), 31, 365–423, *366–367*
  Asian investment in, 390–391
  colonization of, 367, 382–385, 386–387, 414, 419–420
  democracy in, *91*
  demographic transition in, 47
  globalization and development in, 386–391, *388*
  human development in, 34
  human patterns over time in, 379–386
  physical geography of, 368–379
  political map of, *368*
  population, gender, and culture in, 402–411, *408*
  post-independence, 367, 385–386, 414–415
  power and politics in, 394–399, *396–397*
  refugees from, 193
  subregions of
    Central Africa, 413–415, *414*
    East Africa, 415–419, *416*
    Southern Africa, 419–422, *419*
    West Africa, 412–413, *412*
  urbanization in, 43, 399–402, *400–401*
  U.S. relations with, 89, *91*
  visual history of, *380–381*
Subsidence, 67
Subsidies, **214**
Subsistence affluence, **636**, *637*, 638
Subsistence agriculture, 28, **373**
Subsistence economies, 48
Subtropical temperate climates, 16, *371*, *558*
Suburbs, 73, 93, **97**, 98
Sudan, *314*, 315–316, 321, 348, 352–354, *352*
Sudras, 444
Suez Canal, 354
Sugar, 172, 182, 208
Sugar cane, 33, *143*
Sumatra, Indonesia, *556*
Sumerian Empire, *317*
Sundaland, **556**, *557*
Sunni Muslims, 319–**320**, 361
Sunni–Shi'ite division, 328, 338, 339
Supercontinents, 12
Suriname, 180
Survival values, 55
Sushi, *495*
Sustainability, food production and, 86–87, 494
Sustainable agriculture, 30, 182
Sustainable development, 32–33, **32**
Sustainable tourism, **639**
Svalbard, 247
Swagman, **652**
Sweden, 54, *233*, 246, 247, 248
  fertility rate in, 226
  population pyramid for, *225*
Switzerland, 217
Sydney, Australia, *644*
Syria, 39, 220, 324, 338–339, 357
Syrian refugees, *41*, 193, 245, 338–339
Tagine stew, *313*
Tagore, Rabindranath, 477
Taiga, **263**
Tailings, 75
Taiwan, 488, 504–505, *513*, 529, 547–548, *547*
Tajikistan, 259, *267*, 300–302, *300*
Taj Majal, *440*, 442
Taliban, 292, **457**–458, 468, 470, 472
Tallinn, Estonia, *223*
Tanjung Puting National Park, *561*
Tanzania, 415, 417–418
Tariffs, 37, 81, 83, 149, 215
Taro, *626*
Taxes
  in Canada, 235
  in Europe, 235
  in U.S., 235
Taylor, Charles, 399, 413
Tea, *377*
Technology
  See also Information technology (IT)
  European, 76–77
  manufacturing, 80–81
Technology transfer, 83
Teen pregnancy, *53*
Television, 348
Temperate climates, 16, *68*
Temperature, 14–15
Temperature-altitude zones, 16, **135**, *137*
Tennessee, 72
Tenochtitlán, 144
Tent villages, 43
Teotihucan, Mexico, *142*
Terra cotta army, *500*
Terrorism
  Muslim immigrants and, 193, 229–230
  in Southeast Asia, 584
  War on Terror, 89
Terrorist attacks, 42, 89, 126, 220, 227, 229–230
Tertiary sector, **31**, 81
  See also Service sector
Tet, *596*
Texaco, 131
Texas, drug courts in, 96
Textile industry, 446, 479
Textile weaving, *441*
Thailand, *558*, *565*, *581*, 599–602, *600*
Thali, *466*
Theocracy, Iranian, 331, 361
Theocratic states, **331**
Three Gorges Dam, 496–497, *497*
Tiananmen Square, *513*
Tianjin, 533
Tibet, 487, 528–529, *528*, 539, *539*
Tierra del Fuego, Argentina, *133*, *136*
Tigris River, 309, *316*, 361
Timber, *559*, *560*
Timkat, *410*
Timor-Leste, 607–608, *607*
Tito, Josep Broz, 249–250
Tokyo, Japan, *519*, *545*
Tomatoes, 144, *146*
Tonga, *648*
Tornados, 22
Tortellini, *215*
Total fertility rate (TFR), **48**
Tourism
  in Caribbean, 171–172, 173
  in Dubai, 355–356
  in Egypt, 354, *354*
  in Europe, 222, 238, 239, 243
  medical, 173
  in Mexico, 175
  in Oceania, 638–639, *638*
  sex, 593
  in Southeast Asia, 578–579
  in sub-Saharan Africa, 379, 418
  sustainable, 639
Tower of Belem, *243*
Town charters, 205
Town square, 112
TPP (Trans-Pacific Partnership), **640**
Trade
  See also Exports and imports; International trade
  barriers, 81, 83
  China, 152–153
  EU and, 211, *212*
  fair, 38
  free, 37–38
  illegal drug, 95–96
  regional
    in sub-Saharan Africa, 391–392
    trade agreements, 153
  religion and, 56
  Silk Road, 266, 268, *268*
  in South Asia, 441–442
  tropical timber, *559*, *560*
  between U.S. and Canada, *85*
Trade deficit, **82**
Trade protections, 37
Trade routes, *145*
Trade winds, **135**
Traditional agriculture, 373
Traditional values, 55
Trafficking, **289**
Trail of Tears, 79
Trans-Alaska Pipeline, 73–74, *74*
Trans-Amazon Highway, 138

Trans-Canada Highway System, 88
Transcaucasia, 299
Transcultural, 55
Transgender people, 51, *53*, 54
in Middle and South America, 167
in South Asia, *469*, 470
Transnistria, 298
Trans-Pacific Partnership (TPP), **640**
Transparency, **395**
Transportation
nineteenth-century, *79*
urban, 164
Transportation networks, *393*
in China, 517
in East Asia, *507*
in Europe, 198
in Middle and South America, 164
in North America, 88, 97
in Russia and post-Soviet states, 263, *285*
in Southeast Asia, *578*
in sub-Saharan Africa, 392–393, 402
Trieste, 222
Trinidad and Tobago, *23*, 169, 171
Tripoli, Libya, *333*, 350
Tropical humid climates, 16
Tropical rain forest, 134, 138, 185–186, 412
Tropical savanna, 16
Tropical timber trade, *559*, *560*
Tropical wet climates, 16
Tropical wet/dry climates, 16, *136*, *371*, *432*, *489*, *558*, *619*
Trudeau, Pierre, 93
Trump, Donald, 108, 211
Trump administration, 26, 83
Trypanosomiasis, 405
Tsawwassen, 99
Tsetse fly, 405
Tsunamis, 430
Tuberculosis, 405
Tulip Revolution, 278
Tundra, 16, *17*, **122**, **261**
Tunisia, 324, 349
Turkey, 217, 220, 309, *310*, 316, 321, 324, 334, 358–360
Turkmenistan, 259, 274, 300–302, *300*
Tuvalu, *625*, *645*
Twain, Mark, 414
Twitter, 361
Typhoons, *565*, **566**

Uganda, *24*, *377*, 415
UK. *See* United Kingdom
Ukraine, 201, 220, 259, 274, 277–281, *279*, *280*, *284*, 293, 297–298, *297*
Ulan Bator, Mongolia, 498, *541*
Uluru (Ayers Rock), *617*
Umhangla, *410*
UN. *See* United Nations
UNASUR, 153
Underdeveloped countries. *See* Less developed countries (LDCs)
Underemployment, **275**
Undernourishment, 28, *29*
Undocumented workers, 63, 106, 107, 124
Unemployment
in Middle and South America, 153
in North Africa and Southwest Asia, 329, 353–354
in North America, 94, 107
in Russia and post-Soviet states, 275, 288
in sub-Saharan Africa, 387, 420
Union of Soviet Socialist Republics (USSR), **259**
See also Russia and the post-Soviet States; Soviet Union
Unions, 208
United Arab Emirates (UAE), 326, 328, 340, 355
United Fruit, 178
United Kingdom, 241–242
Brexit, 193, 194, 214, 221, 236, 241–242
immigrants in, 242
Iraq and, 361
United Nations (UN), **42**
United Nations Convention on the Law of the Sea (UNCLOS), 630
United Nations Human Development Index (HDI). *See* Human Development Index (HDI)
United States
agriculture in, 86–87, *87*
alternative energy production in, 69
challenges to global power of, 89
cities in, 97
climate change and, *25*, 69
CO_2 emissions, *23*
Cold War and, 42, 208–209
colonial, 76–78
Cuba and, 172, 173
cultural issues in, 104–107
American family, 111
immigration and diversity, 104–105, *105*
race and ethnicity, 107–109
religion, 109–111
democratic system of government in, 93–94
economic comparison of EU and, 211
economic interdependence between Canada and, 83, *85*
economy of, 65, 80–81
foreign involvement of, in Middle and South America, 157
HDI of, 34
health care system in, 94, *95*
influence abroad of, 88–89
Iran and, 360
Iraq and, 361
migration to, 175–176
Pakistan and, 474
Panama and, 178
political elections in, 92
political freedoms in, 89, 92
population of, 101, *103*
population pyramid for, *50*
power and politics in, 88–97
debt and politics, 92–93
drugs and politics, 95–97
gender issues, 94–95
political freedoms, 89, 92
relationships between Canada and, 93–95
slavery in, 77–78
Somalia and, 417
southern, 77–78
as superpower, 42, 88
taxes in, 235
trade deficit of, 82
trade war with China, 83
urbanization in, 97–100, *98–99*
Venezuela and, 180
vulnerability to climate change in, 69–71, *70*
water pollution in, 72–73
westward expansion, 78–79
United States Agency for International Development (USAID), 387, 470
United States–Mexico–Canada Free Trade Agreement (USMCA), 37, 81–82, 106, 153
Unreasonable searches and seizures, freedom from, 39
Uplands, in Germany, *195*
Ural Mountains, 261
Uranium, 201
Urban environmentalism, *45*
Urban gardening, 402
Urban growth poles, **164**
Urbanization, 43–46, *44–45*
agriculture and, 29
in East Asia, 516–522, *518–519*
in Europe, 205–206, 222–224, *223*
farmland and urban sprawl, 99
fertility and, 101
in Middle and South America, 161–164, *162*
Brazil, 186–187
in North Africa and Southwest Asia, 339–342, *341*
in North America, 69, 73, 97–100, *98–99*
in Oceania, 643–646, *644–645*
in polar regions, 665
in Russia and post-Soviet states, 283–285, *284*
smart growth and livability, 99–100
in South Asia, 458–461, *459*, 463
in Southeast Asia, 584–588, *586–587*
Philippines, 608–609
Thailand, 602
of space, 675–676
in sub-Saharan Africa, 399, 399–402, *400–401*
water and, 28
Urban migration, 43, 46
Urban planning, 46, 161, 163, *163*
Urban sprawl, 46, **73**, 99, *99*
Urban transportation, 164
Urban villages, 461
Uruguay, 34, 157, 167, 183
U.S.–Canada border, 93
U.S. Congress, 95
Ushahidi platform, 365, *368*
U.S. households by type, *111*
U.S. Interstate Highway System, 88, 97, 118
U.S.–Mexico border, 126
U.S. military, 88–89, *90*, *91*, *332*
USSR. *See* Soviet Union
Uttar Pradesh, India, 477
Uygurs, 512, 527–528, 537
Uzbekistan, 259, *267*, *268*, 292, 300–302, *300*
Vaishyas, 444
Value chains, **391**
Values, 55
Vancouver, Canada, *98*, 99
Vanilla, 144
Varanasi, 427
Varna, **444**
Vatican City, 245
Vedas, 442
Vegetation, 8
of Europe, 195–196
of North Africa and Southwest Asia, 309
of Oceania, 621
of polar regions, 661
of Russia and the post-Soviet states, 261, 263
of South Asia, 430–431
of Southeast Asia, 557
of sub-Saharan Africa, 370, 372
Veil, **346**
Venezuela, 150, *151*, 157, 179, *179*, 180
Venice, Italy, 222, *233*, 244
Vernacular region, **9**
Victoria Falls, 369, *369*
Victory Day, *289*
Vienna, 222
Vietnam, *581*, 602–604, *603*
Vietnam War, 97, 570–571
Village life, in South Asia, 465
Violence
See also Civil war; Conflicts
in Algeria, 350–351
in Congo, 395
gender-based, *52*, 54, *168*
male-on-female, 1
political, *41*
in sub-Saharan Africa, 414
against women, *469*
Virginia, colony of, 76, 77
Virgin Mary, 167
Virgin of Guadalupe, *168*
Virtual water, 27, **216**, 315, 437
Vitamin D, 59
Vladivostok, 296
Volcanoes, 12, 133–134, *556*
Volga River, 261, *269*
Vologda region, *262*
Voter turnout, 92
Voting rights, in North America, 92

Wadi Mur, Yemen, *314*
Wadis, 309
Wages
gender gap in, 54, 85, *86*, 231, *231*, 346–347, *595*, 649
immigrants and, 107, 120
living, 37
service sector, 81, *84*
in South Europe, 244
in U.S., 211
Wahhabism, **330**, 355
Waitangi Day, *654*
Wales, 241
Walmart, 37, 81, 82, *82*, 175
War on drugs, 96, 160
War on Terror, 89
Wars, *21*
See also Civil war; Conflicts; *specific wars*

Washington, DC, 46
Wastage, 125
Wastewater, 28, 436
Waterborne diseases, 28, 141, 161
Water conservation, 311
Water depletion, 72
Water footprint, 27
Water pollution, 26, *140*
  in Europe, 202, *203*
  in North America, *71*, 72–73
  in South Asia, 436–437
  urbanization and, 28
Water resources, 26–28
  access, 27–28, *27*
  agricultural demands for, 118, 120
  in Central Andes, 182, 183
  clean, 402
  desalination, 315
  in Middle and South America, 138, 141
  in North Africa and Southwest Asia, 311–313, *312*
    Turkey, 359
  in North America, *70*, 72–73
  prices, 26
  privatization of, 27–28
  quality, 28
  renewable, *27*
  sachet, 402
  scarcity, 125
  in South Asia, 436–437
  in Southeast Asia, 563, 566
  in sub-Saharan Africa, 375, 378, *378*, 402
  urbanization and, 28
  use per capita calculation, 27
  virtual, **216**, 437
  wastage, 125
Water scarcity, 22, 72, 138
Water shortages, 490–491
Wealth, of Gulf States, 354–356
Weather
  El Niño and, 135
  space, 670–671
Weathering, 12, 13, **13**
Wegener, Alfred, 12
Welfare programs
  in Canada, 94
  in U.S., 94
Welfare states, **232**
Wells Fargo, 95
West Africa (subregion), 236, 412–413, *412*
West Bank, 334, 337, 342
West Bank barrier, **337**
West Bengal, 477
West Coast, of U.S., 80
Western Europe, 194
Western Ghats, 429, *430*
Western Hemisphere, 5
Western Sahara, 351
West Europe (subregion), 194, 235, 236–242, *237*
West Germany, 208, 209, 239, 240
West Papua, Indonesia, *580*
West Siberian Plain, 261, 263
Westward expansion, 78–79
Wet rice cultivation, **494**, 563
Wheat, 118
White minority rule, 420
White Nile, 352, 369
White supremacy, 108
WHO (World Health Organization), 437
Widowhood, in South Asia, 467
Wildebeest, *377*
Wildfires, 198, 622–623, *625*
Wildlife
  See also Animals
  East Africa, 418, *418*
  in South Asia, 439
  in Southeast Asia, *561*
  in sub-Saharan Africa, 378–379, 405, *416*
Wind, 15
Windhoek, Namibia, 28
Wind power, 26, 69, 201, 202, 438
Wine, 252
Wine industry, *625*, *626*
Wollard, Gary, *119*
Women
  See also Gender
  double day for, 288
  in economy, 83, 85
  education of, 85, 347
  in elected office, 231–232, *232*, *292*
  European, 230–232
  gender expectations for, 51
  as heads of state, 231–232, 334
  income of, 51
  in labor force, 85, 101, 404, 573
  in Middle and South America, in maquiladoras, 175
  Muslim, 465, 467
  in North Africa and Southwest Asia, 334, 342, 343–348, *345*, *347*
  in Oceania, 646–647, *648*, 649
  political status of, 289–290
  in politics, 94–95, *95*
  rights of, 346–348
  rural–urban migration and, 161
  in Russia and post-Soviet states, 288–290
  in South Asia, 465, 467–471, 468–471, *469*
  in Southeast Asia, 591–593
  in sub-Saharan Africa, 375, 399, 406–409
  Taliban and, 468, 470
  trafficking of, 289
  unpaid work by, 32
  violence against, *469*
  wages of, 85, *86*
  in work force, 226
Women-owned businesses, 85
Workers
  foreign, in Japan, 547
  migrant, *62*, 63, 86
  in Southeast Asia, *595*
  undocumented, 63, 124
Working
  in Gulf States, 356
  in Southeast Asia, 572–573, *575*
Working class, in Europe, 208
Working mothers, 226
Work patterns, in Russia and post-Soviet states, 274–275
World Agroforestry Centre, *373*
World Bank, 37, 38, 153, 387, 418
World cities, 208, **516**
World Cup, 187
World Health Organization (WHO), 437
World ocean, 114
World population density, *48–49*
World regional scale, 10
World regions, *2–3*
World Trade Organization (WTO), 37, 274, 391
World Values Survey, 55, *56*
World War I, 208, *208*, 321, 359
World War II, 80, 89, 208, 214, 217, 271, 321, 328, *512*, 570
WTO (World Trade Organization), 37, 274, 391

Xinjiang Uygur Autonomous Region, *527*, 537–538, *538*
Xizang. *See* Tibet

Yangtze River, 487
Yasukini Shrine, *512*
Yellow River, 487, *532*
Yellow Sea, 488
Yemen, *314*, *332*, 339, 340, 356
Yom Kippur War, 325
Yoruba people, 383
Yosakoi, *545*
Yucatan Lowlands, *134*
Yucatan Peninsula, 15
Yugoslavia, 193, 194, 208, 217, 249–250
Yunnan–Guizhou Plateau, 488, 536, *536*
Yunna Province (China), 529, 536
Yunus, Muhammad, 453, 454
Yurts, **537**, *541*

Zahran, Bayan Mahmoud, 348
Zambezi River, 369, *369*
Zambia, *369*, 420, 421
Zapata, Emiliano, 159
Zapatista rebellion, 159
Zharkeshov, Yernar, 285
Zheng He, 502
Zimbabwe, 380, 398, 420, 421
Zionism, **335**
Zoroastrians, 320
Zuckerberg, Mark, 232

PHYSICAL GEOGRAPHY MAP